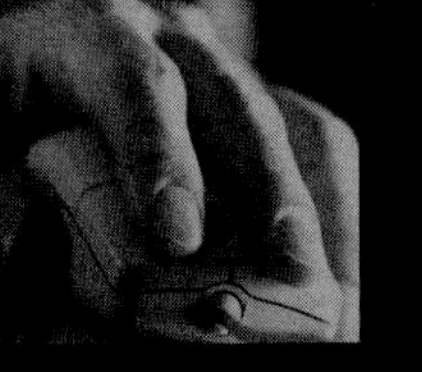

Website for *Essential Biology*, Third Edition and *Essential Biology with Physiology*, Second Edition

Log in.

Explore.

Succeed.

To help you succeed in introductory biology, your professor has arranged for you to enjoy access to great media resources, on an interactive CD-ROM and on the *Essential Biology* website. You'll find that these resources that accompany your textbook will enhance your course materials.

Minimum system requirements:

WINDOWS
Windows 98, 2000, XP
64 MB RAM
1024 x 768 screen resolution
Internet Explorer 5.5 and 6.0;
Netscape 7.0
Plug-Ins: Latest versions of Flash, Shockwave, and Quicktime
Internet Connection: 56k modem or faster

MACINTOSH
Mac OS 10.2.8, 10.3, and 10.4
64 MB RAM
1024 x 768 screen resolution
Safari 1.2 (only OS 10.3) and 2.0 (only OS 10.4);
Netscape 7.0
Plug-Ins: Latest versions of Flash, Shockwave, and Quicktime
Internet Connection: 56k modem or faster

Got technical questions?

For technical support, please visit http://247.aw.com.
Email technical support is 24/7.

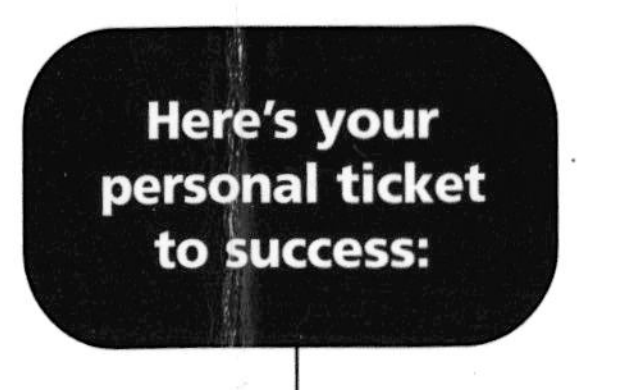

How to log in to the Essential Biology website:

1. Go to www.campbellbiology.com
2. Click on the appropriate book cover. Cover must match the textbook edition being used for your class.
3. Click "Register."
4. Scratch off the silver foil coating below to reveal your pre-assigned access code.
5. Enter your pre-assigned access code exactly as it appears below.
6. Complete the online registration form to create your own personal user Login Name and Password.
7. Once your personal Login Name and Password are confirmed by email, go back to www.campbellbiology.com, click on the appropriate book cover, type in your new Login Name and Password, and click "Log In."

Your Access Code is:

If there is no silver foil covering the access code above, the code may no longer be valid. In that case, you need to either:
• Purchase a new student access kit at your campus bookstore.
• Purchase access online using a major credit card.
Go to www.campbellbiology.com, click on the appropriate book cover and click Buy Now.

Important: Please read the Subscription and End-User License Agreement located on the "Log In" screen before using the *Essential Biology* website or CD-ROM. By using the website or CD-ROM, you indicate that you have read, understood, and accepted the terms of this agreement.

Features of the Student website and CD-ROM for *Essential Biology*, Third Edition and *Essential Biology with Physiology*, Second Edition

Chapter Guides

Each Chapter Guide provides students with a road map to help them identify study needs, and then directs them to the media activities, resources, and quizzes that help students master the content.

eTutors

eTutor animations and student tutorials bring difficult concepts to life, including cell structure and function, cellular respiration, photosynthesis, mitosis, meiosis, protein synthesis, and how neurons work. Each step-by-step interactive student tutorial is reinforced by a review sheet and a gradable quiz. Access the eTutor animations and student tutorials from the student website.

MP3 Tutors

MP3 Tutor sessions are narrated by *Essential Biology* co-author and award-winning teacher Eric Simon, who walks students through an "audio tour" of 22 of the most difficult—yet essential—biological concepts for the non-majors biology course. Access MP3 Tutor sessions from the student website.

Discovery Channel Video Clips

These brief 2-5 minute video clips from Discovery Channel cover topics from fighting cancer to antibiotic resistance to introduced species. Access Discovery Channel Video Clips from the student website.

You Decide

These activities help students learn how to interpret research data and make informed decisions by examining topics such as low fat versus low carb diets and global warming.

Graph It

These interactive tutorials provide students with hands-on experience interpreting graphs and data.

Activities

Explore approximately 200 activities, including videos, animations, and interactive review exercises that reinforce the concepts in each chapter.

Case Studies in the Process of Science

Perform 56 virtual investigations that develop scientific thinking skills such as data collection and analysis.

Quizzes

Students can assess their understanding with over 200 multiple-choice questions. Each chapter includes a Pre-Test to diagnose current knowledge, an Activities Quiz that tests understanding of the Activities in each chapter, and a comprehensive Chapter Quiz.

Cumulative Quizzes

Students can create customized tests on multiple chapters at once by choosing the chapters and the number of questions desired.

Additional Features of the Website and CD-ROM

- E-Book
- All art from the textbook
- Word study tools, including flash cards, key terms, and a glossary with audio pronunciations
- Web links and references
- Research Navigator—access to research databases of source material from EBSCO Academic Journal and Abstract Database, New York Times Search by Subject Archive, Financial Times Article Archive, and "Best of the Web" Link Library
- Lab Bench—Lab activities that connect laboratory procedures to biological concepts.

How to Start the Student CD-ROM

WINDOWS:

1. Insert the Student CD-ROM into the CD-ROM drive. If you have autorun turned on, the installer should launch automatically. If it does not, follow these steps:
   a) Navigate through My Computer to your default CD-ROM drive.
   b) Double-click on the CD-ROM icon.
   c) Double-click on the Windows_Installer.exe icon and follow the instructions in the dialog boxes.
2. Click the "Check Browser" button to launch your default browser and see if it meets the minimum system requirements.
3. If you need to install a new browser, click the "Install Browser" button and select a browser to install.
4. After you have a browser that meets the minimum requirements, click on the "Install Plug-ins" button and launch each of the installers for the required plug-ins.
5. Once all plug-ins have been installed, return to the Main Screen and click the "Launch" button to begin using the CD-ROM.

Once you have a valid browser and have installed all the required plug-ins you can skip the above steps and click on the "Launch" button directly to begin using the CD-ROM. Or, if you have the valid browser and plug-ins installed and you don't have autorun installed, you can click on essential_biology.html to skip the installation process.

MACINTOSH:

1. Insert the Student CD-ROM into the CD-ROM drive. Double-click the Macintosh_Installer.
2. Click the "Check Browser" button to launch your default browser and see if it meets the minimum system requirements.
3. If you need to install a new browser, click the "Install Browser" button and select a browser to install.
4. After you have a browser that meets the minimum requirements, click on the "Install Plug-ins" button and launch each of the installers for the required plug-ins.
5. Once all plug-ins have been installed, return to the Main Screen and click the "Launch" button to begin using the CD-ROM.

Once you have a valid browser and have installed all the required plug-ins you can skip these steps and click on the "Launch" button to begin using the CD-ROM. Or, if you have the valid browser and plug-ins installed, you can click on essential_biology.html to skip the installation process.

ESSENTIAL BIOLOGY

WITH PHYSIOLOGY

Second Edition

Neil A. Campbell
Jane B. Reece
Berkeley, California
Eric J. Simon
New England College
Henniker, New Hampshire

San Francisco • Boston • New York
Cape Town • Hong Kong • London • Madrid
Mexico City • Montreal • Munich • Paris
Singapore • Sydney • Tokyo • Toronto

In loving memory of my brother, Michael Long Reece (1951-2005) J.B.R.

For Amanda, Reed, and Forest, who make it all worthwhile E.J.S.

Editor-In-Chief: Beth Wilbur
Senior Acquisitions Editor: Chalon Bridges
Director of Development: Deborah Gale
Senior Project Manager: Ginnie Simione Jutson
Managing Editor: Michael Early
Development Editors: Evelyn Dahlgren, Susan Weisberg
Developmental Artist: Russell Chun
Photo Editor: Donna Kalal
Photo Researcher: Maureen Spuhler
Copyeditor: Janet Greenblatt
Marketing Managers: Christy Lawrence, Lauren Harp, Jeff Hester
Supplements Project Editor: Susan Berge
Supplements Production Supervisor: Jane Brundage
Media Supplements Production Supervisor: Jennifer Mattson
Publishing Assistant: Benjamin Lau
Permissions Editor: Sue Ewing
Media Developmental Manager: Pat Burner
Senior Media Producer: Jonathan Ballard
Media Project Manager: Brienn Buchanan
Web Development: Linda Young
Production Management and Composition: Carlisle Publishing Services
Illustrations: Precision Graphics
Text Design: Mark Ong, Jana Anderson
Cover Design: Yvo Riezebos
Manufacturing Buyer: Stacy Wong
Director, Image Resource Center: Melinda Patelli
Image Rights and Permissions Manager: Zina Arabia
Cover Printer: Phoenix Color
Printer: Courier, Kendallville

On the cover: Photograph of a "devil's flower mantis" (*Blepharopsis medica*). Courtesy Getty Images, Inc./Image Bank.

Credits continue in Appendix C.

Library of Congress Cataloging-in-Publication Data

Campbell, Neil A., 1946-2004
Essential biology and physiology / Neil A. Campbell, Jane B. Reece, Eric J. Simon.—2nd ed.
p. cm.
ISBN 0-8053-6841-8
1. Biology. 2. Physiology I. Reece, Jane B. II. Simon, Eric J. (Eric Jeffrey), 1967- . III. Title.
QH307.2.C365 2007
570—dc22

2006023388

PEARSON
Benjamin Cummings

Benjamin Cummings
1301 Sansome Street
San Francisco, CA 94111
www.aw-bc.com

ISBN 0-8053-6841-8 (Student edition)
ISBN 0-8053-0635-8 (Professional copy)
ISBN 0-8053-9422-2 (a la Carte edition)
ISBN 0-13-238024-2 (NASTA edition)

1 2 3 4 5 6 7 8 9 10—CRK—10 09 08 07 06

See how biology is relevant to your world

BIOLOGY AND SOCIETY

Cells That Cure

A middle-aged man staggers, gasps for air, and falls to the ground clutching his chest. A passerby quickly puts out the call to 911—heart attack! As rescuers rush to the scene, the real medical drama plays out within the cells of the victim's heart. Starved for oxygen, his heart muscle cells begin to die. If too many cells perish, the heart will never again be able to function. Even if resuscitation efforts are successful and the man survives this acute medical emergency, he will most likely suffer from chronic problems. Because heart muscle cells—unlike many cells of the human body—do not regenerate over time, his heart may be permanently scarred. Constantly undersupplied by blood, the damaged heart muscle will gradually lose strength, leaving the patient at high risk for more heart failure. In severe cases, a heart transplant is the only possibility for a long-term cure.

In recent years, a new type of treatment has emerged that may offer hope for the rebuilding of dead heart muscle. In this procedure, called "cell therapy," cells are taken from elsewhere in the patient's body and delivered to the ailing heart. Typically, immature muscle cells are harvested from the patient's thigh and grown in a lab to increase their numbers. Tens or hundreds of millions of the new muscle cells are then injected directly into heart scar tissue during bypass or other heart surgery (**Figure 4.1**).

Although new and exciting, the technique of cell therapy remains in the testing phase. Worldwide, more than a dozen clinical trials are currently under way—including ones in Boston, Florida, and Houston. Even though many patients seem to respond to the treatment, scientists do not yet understand *how* cell therapy works. The intense effort that scientists are now directing toward the study of heart cells and how cell therapy affects them illustrates the main point of this chapter: To understand how life on Earth works—including your own body—you first need to learn about cells. Cells are the building blocks of all life, which makes them as fundamental to biology as atoms are to chemistry. Moreover, the cell is the smallest entity that exhibits all the characteristics of life. In this chapter, we'll take a tour of cells and explore their structure and function. ■

SEM 100×

Myoblast

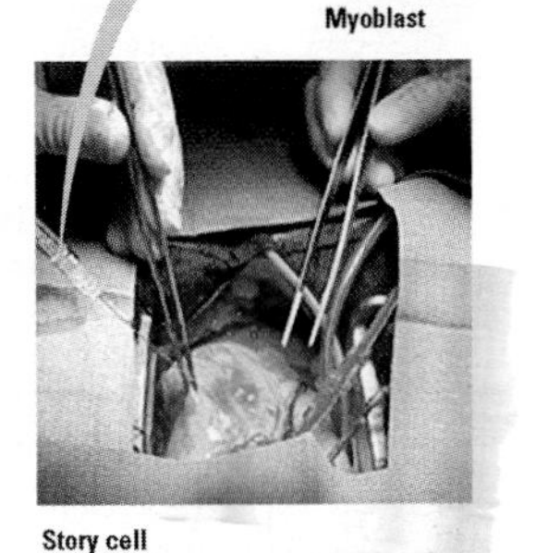

Story cell

Figure 4.1 A cellular cure? Injecting immature muscle cells (inset) into the heart may help regenerate cells damaged by a heart attack.

The **Biology and Society** opening essay shows you how content in the chapter can be applied to aspects of everyday life. Topics range from fad diets to blood doping to cell therapy to stone-washed jeans.

Essential Biology **integrates human applications** throughout the narrative. Here the text discusses lactose intolerance.

Disaccharides A **disaccharide**, or double sugar, is constructed from two monosaccharides through a dehydration reaction. An example of a disaccharide is maltose, also called malt sugar, which consists of two glucose monomers (Figure 3.10). Maltose, naturally found in germinating seeds, is used in making beer, malt whiskey and liquor, malted milk shakes, and malted milk ball candy.

Lactose, another disaccharide, is made from the monosaccharides glucose and galactose. Lactose is sometimes called "milk sugar" because it is primarily found in dairy products. You have probably heard of lactose intolerance, the inability to properly digest lactose. Cells of the small intestine produce a molecule called lactase that breaks down lactose into glucose and galactose monomers through a hydrolysis reaction. In lactose-intolerant people, these cells produce insufficient amounts of lactase. As a result, lactose is not properly broken down and absorbed, causing uncomfortable symptoms, such as bloating, gas, and diarrhea. People with lactose intolerance must avoid dairy foods, use dairy substitutes (such as soy milk), or supplement their diet with lactase in pill or liquid form (Figure 3.11).

The most common disaccharide is sucrose, common table sugar, which consists of a glucose linked to a fructose. Sucrose is the main carbohydrate in plant sap, and it nourishes all the parts of the plant. Sucrose is extracted from the stems of sugarcane or, more commonly in the United States, the roots of sugar beets. However, sucrose is rarely used as a sweetener in processed foods. Much more common is high-fructose corn syrup (HFCS), made through a commercial process that converts natural glucose in corn syrup to the much sweeter fructose. If you read the label on any soft drink can or bottle, you're likely to find that HFCS is the first or second ingredient listed.

The United States is one of the world's leading markets for sweeteners, with the average American consuming about 64 kilograms (Kg)—that's a whopping 140 pounds!—per year, mainly as sucrose and HFCS (Figure 3.12). This national "sweet tooth" persists in spite of our growing awareness about how sugar can negatively affect our health. Sugar is a major cause of tooth decay. High sugar consumption also tends to replace eating more varied and nutritious foods. The description of sugars as "empty calories" is accurate in the sense that most sweeteners contain only negligible amounts of nutrients other than carbohydrates. For good health, we also require proteins, fats, vitamins, and minerals. And we need to include substantial amounts of complex carbohydrates—that is, polysaccharides—in our diet.

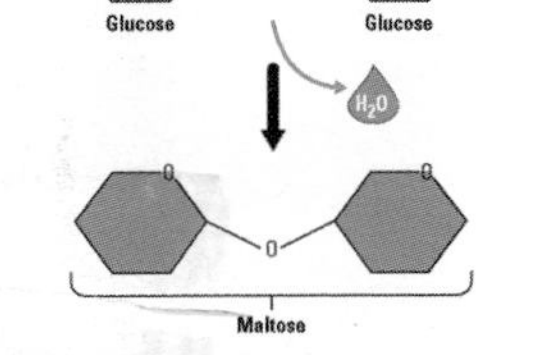

Figure 3.10 Disaccharide (double sugar) formation. To form a disaccharide, two simple sugars are joined by a dehydration reaction, in this case forming a bond between two glucose monomers to make the double sugar maltose.

Figure 3.11 Lactose intolerance. People who are lactose intolerant can eat foods that contain lactose (such as ice cream) by taking pills that contain the enzyme lactase (inset).

Figure 3.12 A year's supply of sugar. Americans consume an average of 64 kg (140 lb) of sweetener per person per year—over a third of a pound per day. The 64 kg of sucrose (table sugar) in this photograph will give you some idea of that level of consumption. Reading food labels will make you aware of the amount of sugar in processed foods.

40 UNIT 1 Cells

TO THE STUDENT:
HOW TO USE THIS BOOK (continued)

See biology come alive with integrated text and art

Selected text and figure captions include art to help you follow the figure easily.

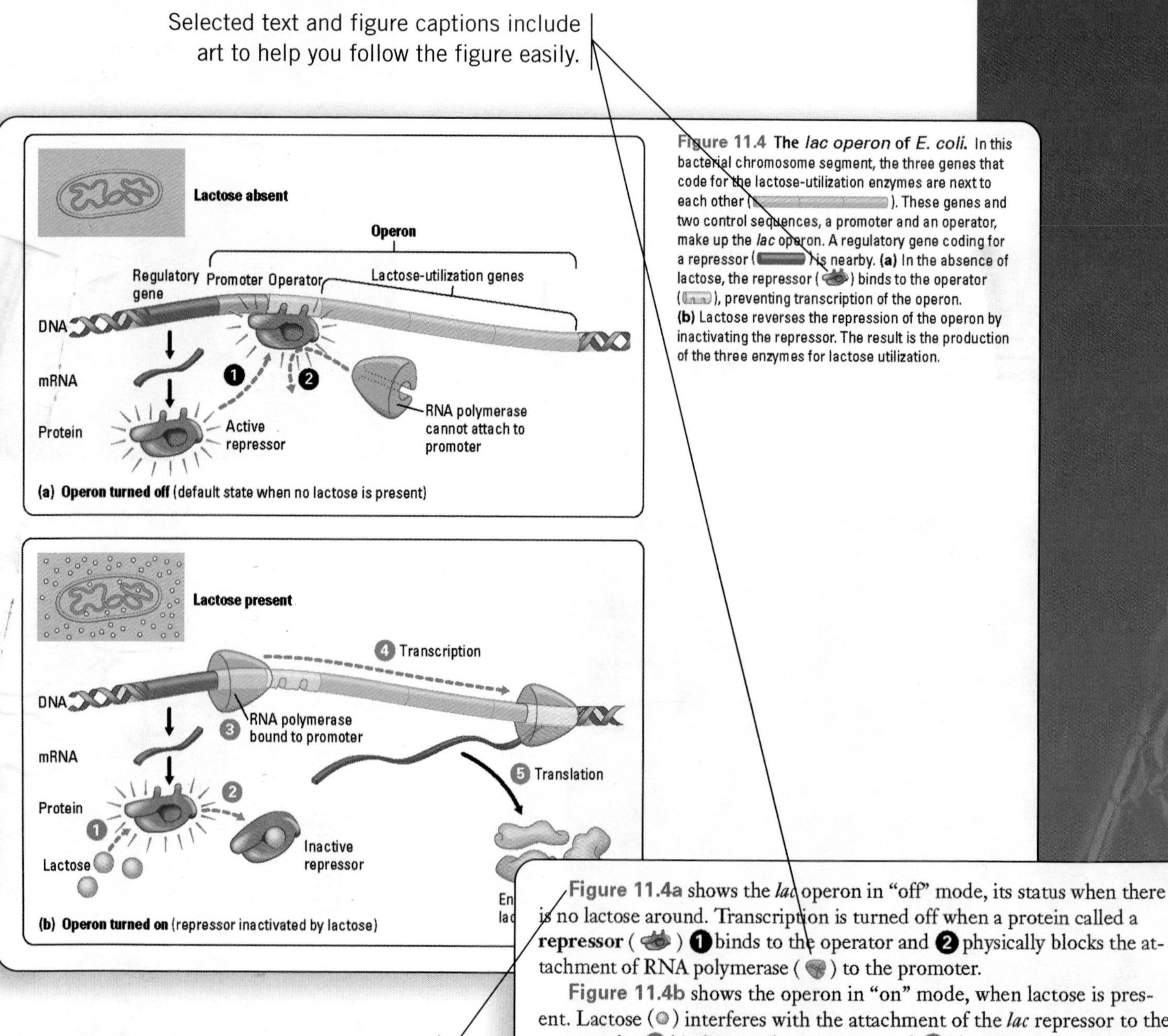

Figure 11.4 The *lac operon* of *E. coli.* In this bacterial chromosome segment, the three genes that code for the lactose-utilization enzymes are next to each other ( ). These genes and two control sequences, a promoter and an operator, make up the *lac* operon. A regulatory gene coding for a repressor ( ) is nearby. **(a)** In the absence of lactose, the repressor ( ) binds to the operator ( ), preventing transcription of the operon. **(b)** Lactose reverses the repression of the operon by inactivating the repressor. The result is the production of the three enzymes for lactose utilization.

Figure 11.4a shows the *lac* operon in "off" mode, its status when there is no lactose around. Transcription is turned off when a protein called a **repressor** ( ) ❶ binds to the operator and ❷ physically blocks the attachment of RNA polymerase ( ) to the promoter.

Figure 11.4b shows the operon in "on" mode, when lactose is present. Lactose ( ) interferes with the attachment of the *lac* repressor to the operator by ❶ binding to the repressor and ❷ changing its shape. In its new shape, the repressor is inactive; it cannot bind to the operator, and the operator switch remains on. ❸ RNA polymerase can now bind to the promoter and from there ❹ transcribe the genes of the operon into mRNA. ❺ Translation produces all three enzymes needed for lactose utilization.

Green figure numbers in the text and legends help you move between text and images with ease.

Discover how scientists pursue questions

Can Alzheimer's Disease Be Detected Early?

Alzheimer's disease (AD) is a devastating illness that gradually destroys a person's memory and ability to think. As the disease progresses, the brain becomes riddled with deposits of a protein molecule called beta-amyloid, the accumulation of which is believed to cause brain damage. One of the major hurdles in the fight against AD is the lack of a diagnostic test; doctors cannot be entirely sure that a person has AD until autopsy. Development of an early test for Alzheimer's remains an important but so far elusive goal.

Progress toward a diagnostic test began with the **observation** by researchers at the University of Pittsburgh and Uppsala University in Sweden that a protein molecule they dubbed PIB binds to beta-amyloid. The PIB molecule includes an atom of carbon-11, a rare radioactive isotope of carbon. When PIB is injected into the brain, the radiation emitted by the carbon-11 isotope can be detected and visualized on a computer screen using a PET scanner (a medical diagnostic device that detects and displays radioactive emissions). The researchers then asked the **question**, Can PIB be used to detect Alzheimer's disease in living patients? Their **hypothesis** was that following a PIB injection, PET scans would reveal higher concentrations of PIB in the brains of Alzheimer's patients than in the brains of nonaffected individuals.

The **experiment** that researchers used to test this hypothesis involved 16 patients diagnosed with mild Alzheimer's disease and 9 healthy control subjects. The PET scans of the AD patients did indeed show higher concentrations of PIB in regions of the brain known to be affected by Alzheimer's (**Figure 2.6**). These **results**, published in 2004, suggest that the radioactive PIB molecule may hold the key to early detection of this crippling disease. ■

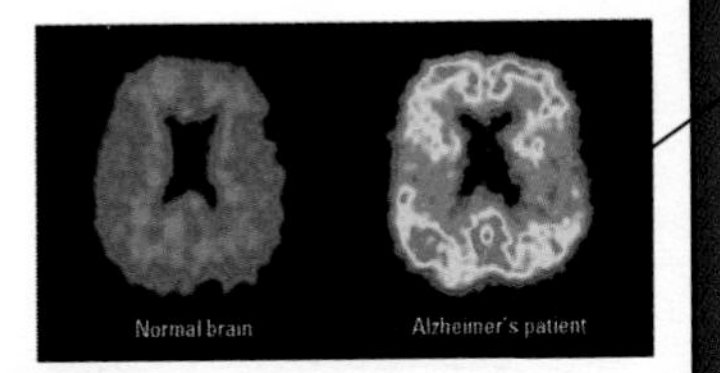

Figure 2.6 A diagnosis for Alzheimer's disease? These images are PET scans of the brains of a healthy person (left) and a person in the early stages of Alzheimer's disease (right). Both patients were injected with PIB, a protein molecule that binds to certain other molecules associated with AD. PIB includes a radioactive isotope that the PET scanner can detect. Notice that the brain of the AD patient has high levels of PIB (red and yellow areas), while the unaffected patient's brain has lower levels (blue).

NEW! **The Process of Science** sections give you real-world examples of how the scientific method is applied. Both classic and contemporary experiments show how scientists pursue questions such as, "Can Alzheimer's Disease be detected early?"

Relate biological concepts to evolution

As the capstone of each chapter, the **Evolution Connection** demonstrates how the theme of evolution runs throughout all of biology. As a result, you can see biology as a coherent study of changing life on a changing planet rather than a collection of facts.

Evolving Enzymes

Organisms use a great variety of enzymes. How could such diversity arise? Recall that most enzymes are proteins, and proteins are encoded by genes. Data gathered from the analysis of genetic sequences suggest that many of our genes arose through molecular evolution: One ancestral gene randomly duplicated, and the two copies of that gene diverged over time via genetic mutation, eventually becoming two distinct genes that produce two distinct enzymes.

Enzyme evolution can be greatly accelerated in the laboratory. For example, one group of researchers working with the bacteria *Escherichia coli* (*E. coli*) sought to modify lactase. Recall from Chapter 3 that lactase is an enzyme that can split the disaccharide lactose into its two component monosaccharides (**Figure 5.21**). In a lab procedure called directed evolution, many copies of the gene for the original enzyme were mutated at random. The mutated enzymes that could best perform a new function (in this case, to split a bond in a different substrate molecule) were artificially selected using a screening test. Those enzymes were then subjected to another round of duplication, mutation, and selection. After seven rounds, this process produced a novel enzyme that recognized a new substrate.

The processes of natural selection and directed evolution both result in the production of new enzymes with new functions. There are, however, several important differences between the two schemes. The most obvious is the amount of time involved: many, many years for the former, just weeks for the latter. But the most significant difference between the two processes is mentioned in the name: The laboratory experiment was directed; the researchers engineered the evolutionary outcome to suit their chosen purpose. Natural selection, on the other hand, selects enzyme variants that work best in the organism's natural environment. ■

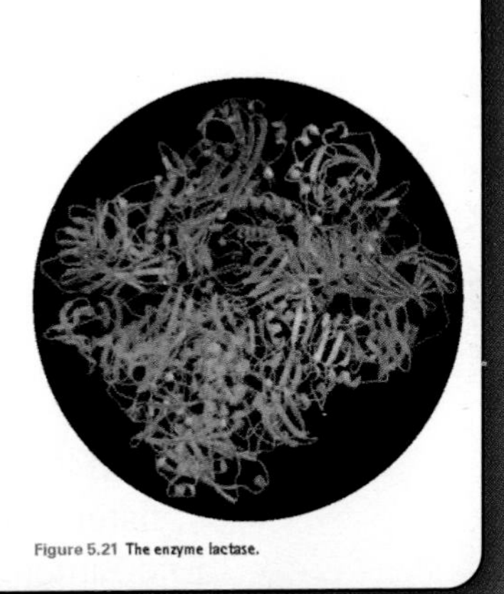

Figure 5.21 The enzyme lactase.

Chapter Review

SUMMARY OF KEY CONCEPTS

For study help and activities, go to campbellbiology.com or the student CD-ROM.

Biology and Society: Feeling the Burn

- Aerobic metabolism occurs when cells receive enough oxygen to support their energy needs. If a body exceeds its aerobic capacity, the demand for oxygen is greater than the body's ability to deliver it, and metabolism becomes anaerobic.

Energy Flow and Chemical Cycling in the Biosphere

- **Producers and Consumers** Autotrophs (producers) make organic molecules from inorganic nutrients via photosynthesis. Heterotrophs (consumers) must consume organic material and obtain energy via cellular respiration.
- **Chemical Cycling between Photosynthesis and Cellular Respiration**

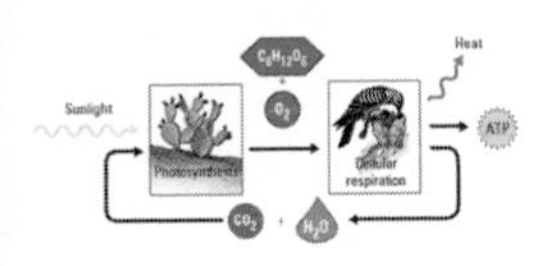

Activity *Build a Chemical Cycling System*

Cellular Respiration: Aerobic Harvest of Food Energy

eTutor *Cellular Respiration*

MP3 Tutor *Cellular Respiration Part 1*

MP3 Tutor *Cellular Respiration Part 2*

- **The Relationship between Cellular Respiration and Breathing** The bloodstream distributes O_2 from the lungs to all the cells of the body and transports CO_2 waste from the cells to the lungs for disposal.
- **The Overall Equation for Cellular Respiration**

- **The Role of Oxygen in Cellular Respiration**

Oxidation: Glucose loses electrons (and hydrogens)
$C_6H_{12}O_6$ → CO_2
Electrons (and hydrogens) → ATP
O_2 → H_2O
Reduction: Oxygen gains electrons (and hydrogens)

Redox reactions transfer electrons from food molecules to an electron acceptor called NAD^+, forming NADH. The NADH then passes the high-energy electrons to an electron transport chain that eventually "drops" them to O_2. The energy released during this electron transport is used to regenerate ATP from ADP. The affinity of oxygen for electrons keeps the redox reactions of cellular respiration working.

- **The Metabolic Pathway of Cellular Respiration** You can follow the flow of molecules through the process of cellular respiration in the following diagram:

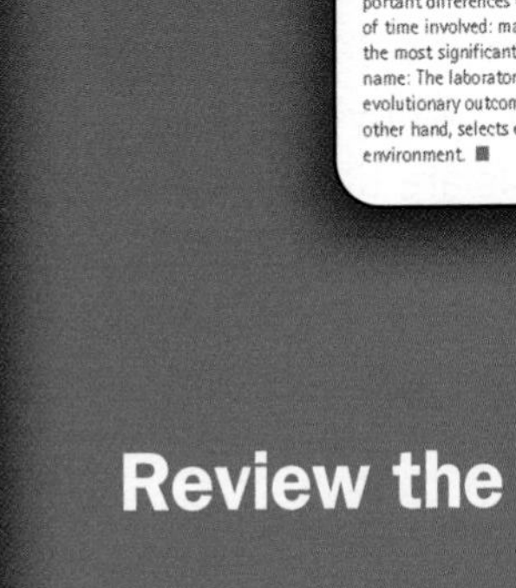

The electron transport chains pump H^+ across the membrane as electrons flow stepwise from NADH to oxygen. Backflow of H^+ across the membrane powers the ATP synthases, which attach phosphate to ADP to make ATP.

Activity *Overview of Cellular Respiration*

CHAPTER 6 Cellular Respiration: Obtaining Energy from Food 101

Review the material covered in each chapter

Visual summaries in the Chapter Review help you visualize key concepts. The chapter summary also refers you to media activities and the student website.

TO THE STUDENT:
MAKE THE BEST USE OF YOUR STUDY TIME

Get interactive tutoring on the toughest topics in biology

NEW! eTutor icons direct you to 3-D, interactive, self-paced tutorials that will help you master tough topics like cellular respiration, mitosis, and meiosis. Prepare for exams by completing the tutorial, printing a personal review sheet, and taking a quiz.

A tutor that goes anywhere you go!

NEW! Whenever you see this icon, you can listen to **MP3 Tutor Sessions** narrated by coauthor Eric Simon. These portable tutorials are like private office hours with the author and they allow you to study anytime, anywhere. MP3 Tutor Sessions are located on the companion website and CD-ROM.

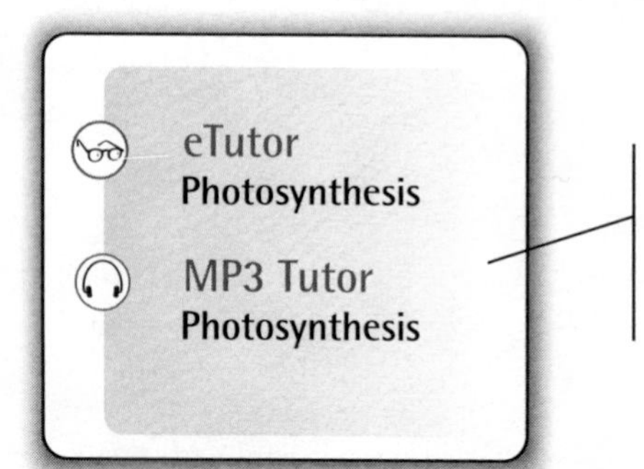

Media boxes in the text guide you to tutorials that will help you master the most challenging topics.

Study efficiently using the companion website

HOME | FAQs | SITE REQS | TECH SUPPORT | FEEDBACK | CREDITS | AW-BC.COM | TUTOR CENTER

ESSENTIAL BIOLOGY

CAMPBELL REECE SIMON

ESSENTIAL BIOLOGY WITH PHYSIOLOGY

Chapter 7: Photosynthesis: Using Light to Make Food GO

Home ➤ 7: Photosynthesis: Using Light to Make Food ➤ Chapter Guide

- Chapter Guide
- Art
- Word Study Tools
- Web Links & References
- Instructor Media
- eTutors
- MP3 Tutors
- Discovery Videos
- You Decide
- Graph It!
- E-Book
- Cumulative Test
- Glossary
- Research Navigator
- LabBench
- About the Book

① Focus Your Effort

Take the Pre-Test to get your own personalized Study Plan.

② Direct Your Learning

- Biology and Society: Plant Power for Power Plants
- The Basics of Photosynthesis
  - eTutor: Photosynthesis
  - MP3 Tutor: Photosynthesis
  - Activity: The Plants in Our Lives
  - Activity: The Sites of Photosynthesis
  - Activity: Overview of Photosynthesis
- The Light Reactions: Converting Solar Energy to Chemical Energy
  - Activity: Light Energy and Pigments
  - Activity: The Light Reactions
- Evolution Connection: The Oxygen Revolution
- Chapter Review

③ Test Yourself

The Activities Quiz will test your knowledge of the content in the Activities above.

The Chapter Quiz will test your knowledge on the content in the textbook chapter.

④ Extend Your Knowledge

- Case Studies in the Process of Science: How Does Paper Chromatography Separate Plant Pigments?
- Case Studies in the Process of Science: How Is the Rate of Photosynthesis Measured?
- Discovery Education Channel Video: Space Plants

Use the pre-test as your road map to identify your study needs.

Utilize the resources that support your learning style, from MP3 Tutors for auditory learners to eTutors for visual learners.

Use the chapter quiz to check your understanding and prepare for exams.

See science in action with Discovery Channel Video Clips

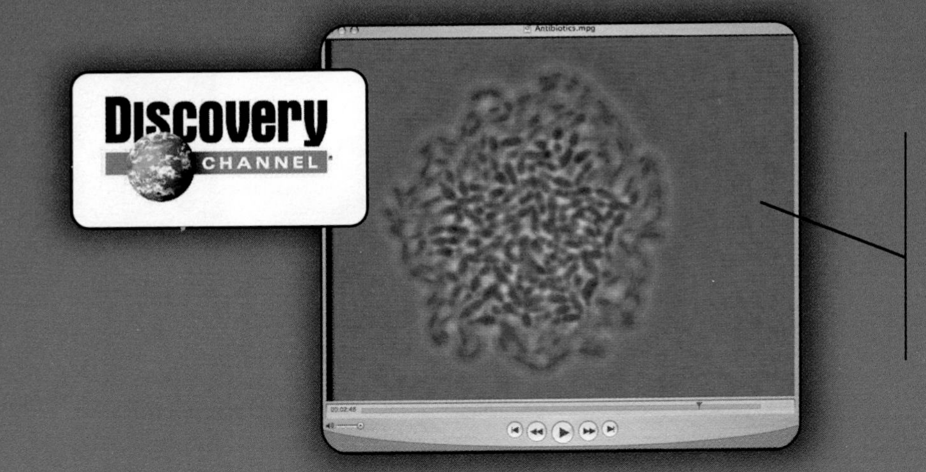

These brief 3–5 minute Discovery Channel video clips cover topics from fighting cancer to antibiotic resistance and introduced species. The video clips are located on the companion website and are included with each new copy of the text.

TO THE INSTRUCTOR:

CREATE DYNAMIC LECTURES IN HALF THE TIME!

NEW! *Essential Biology* Media Manager 2.0 Instructor CD-ROMs/DVD (0-8053-0435-5)

All instructor and student media resources are organized in one convenient, easy-to-use package. Inside, you'll find:

- All of the art, tables, and photos from the book with customizable labels.
- A new "Quick Start" CD-ROM containing all of the textbook figures embedded within PowerPoint® for fast lecture preparation.
- An intuitive "shopping cart" tool allowing you to quickly select and download any figure, photo, animation, or video from the collection.

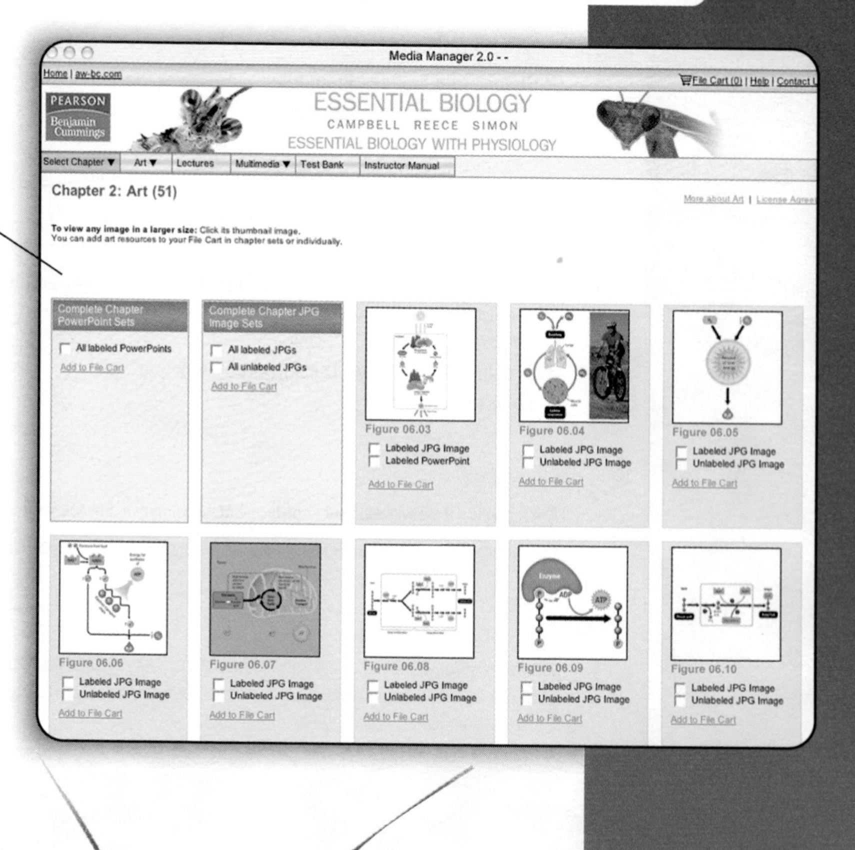

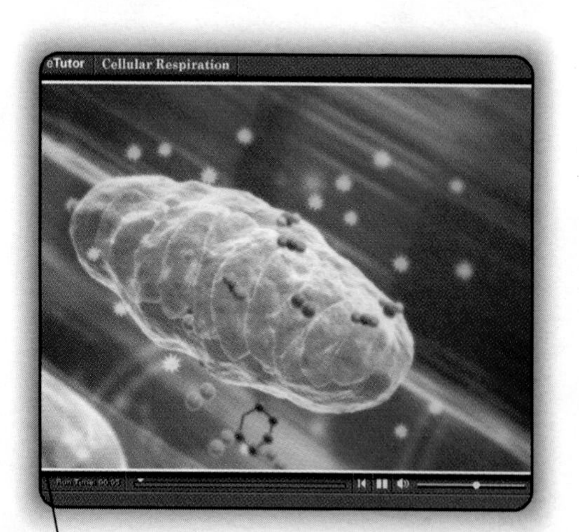

NEW! eTutor Animations invigorate classroom lectures with effective, 3-minute "movie quality" 3D graphics that capture students' attention. Using the eTutor animations, you can bring essential, difficult-to-teach biological concepts to life during your lectures. Topics include: A Tour of the Cell, Cellular Respiration, Photosynthesis, Mitosis, Meiosis, Protein Synthesis, and How Neurons Work.

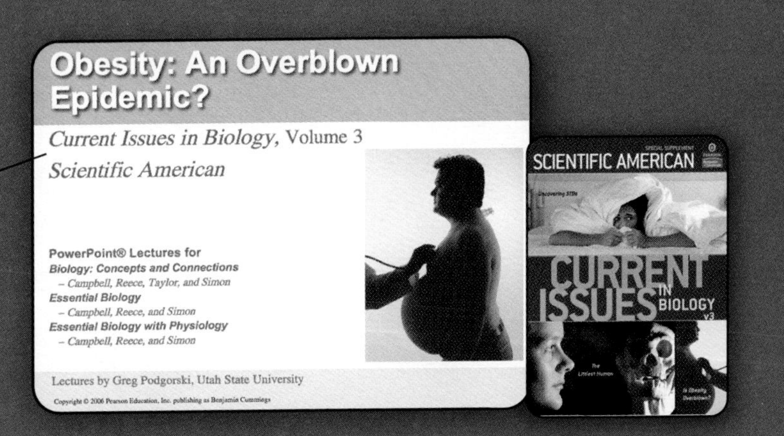

Timesaving, Customizable PowerPoint® Lectures are provided for each textbook chapter, along with additional slides that accompany the engaging articles in *Current Issues in Biology* Volumes 1–3 from *Scientific American.*

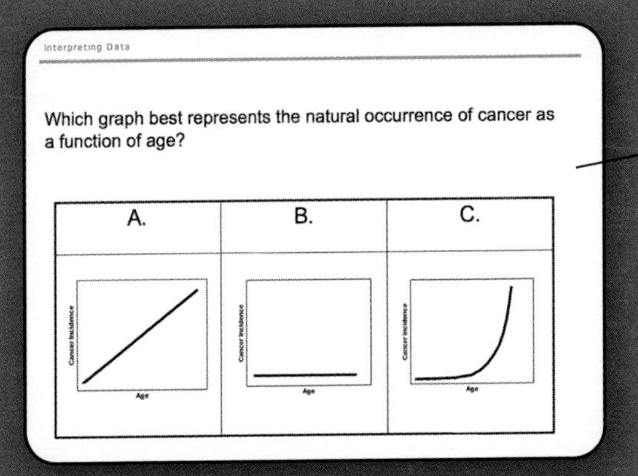

Active Lecture Questions stimulate effective classroom discussions and are conveniently embedded into PowerPoint® for use with or without clickers.

Introduced Species

Competition, Predation, and Extinction

Biology Lecture Launchers
from *Benjamin Cummings* and *Discovery Channel School*

Presentation prepared by Andrew Stull

Discovery Channel Video Clips help you create exciting lectures that will engage your students. Short 3–5 minute videos cover high-interest topics such as cancer treatment and antibiotic resistance. Each video is supported by PowerPoint® slides with teaching tips and questions that help you facilitate discussions and connect the video footage to your lecture topic.

New kinds of test questions for effective assessment

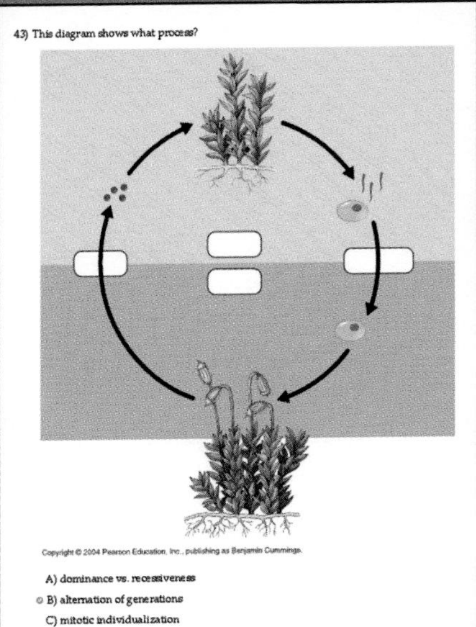

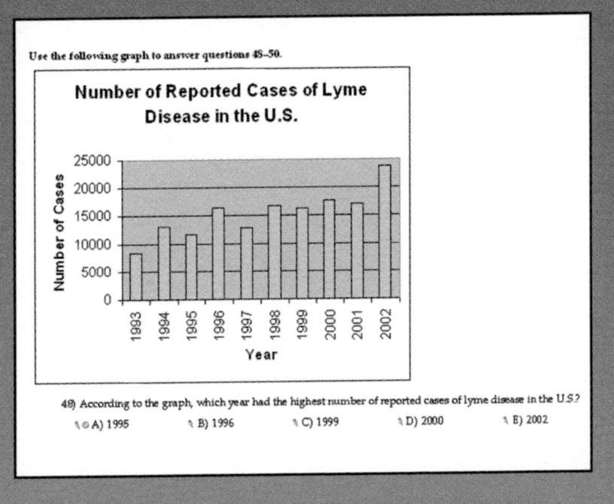

Do you want to move beyond the usual multiple-choice test questions? Three new categories of test questions are included in the accompanying Test Bank:

- **Interpreting art questions** foster visual learning
- **Interpreting graphs and data questions** focus on quantitative skills
- **Scenario questions** encourage critical thinking

The Test Bank, which is included in the Media Manager, also provides hundreds of standard short-answer, multiple-choice, and true/false questions as well as an optional section with questions that test students on the CD-ROM/website activities. The Test Bank is also available in print and in the instructor section of CourseCompass™, Blackboard, and WebCT™.

Preface

Welcome to the golden age of biology! There has never been a more exciting time to teach and learn about life. Reading the newspaper or watching the news reminds us daily that the subject of biology is woven into the fabric of our society as never before. As the pace of biological discovery accelerates, so does the number of ways that it touches our lives. Medicine, agriculture, forensics, ecology, psychology, history—these are just a few of the subjects to which biology has made significant contributions in recent years. Through *Essential Biology with Physiology*, we are privileged to help instructors educate the next generation of "citizen scientists."

While the present golden age is rich with learning opportunities, it also creates a teaching and learning challenge: how to keep up with the discovery explosion that could suffocate students under an avalanche of information. Neil Campbell conceived of *Essential Biology with Physiology* as a tool for helping instructors and students focus on the most important areas of biology within a single term by organizing the material within six core areas: cells, genes, evolution, ecology, animals, and plants. Neil's vision, which we carry on and extend in this edition, has enabled us to keep *Essential Biology with Physiology* manageable in size without being superficial in developing the concepts that are most fundamental to understanding life. As Neil did, we take the "less is more" mantra in education today to mean fewer topics, not more diluted explanations.

In continuing conversations with instructors around the nation, we have noticed some important trends in *how* these vital topics are taught. In particular, many instructors articulated a desire to achieve three goals in their course: to relate the core content to students' lives, to clarify the process of science, and to demonstrate how the theme of evolution integrates all of biology. In this edition of *Essential Biology with Physiology*, we sought to help instructors tackle each of these vital issues head-on:

- Each chapter begins with a "Biology and Society" opening essay that relates the core content of that chapter to an issue that concerns students. Our goal is to help students make biologically informed decisions throughout their lives—to evaluate various health and environmental issues, for example.
- A new feature for this edition is "The Process of Science" sections. Throughout the book, classic and modern experiments related to the chapter content are presented in a way that highlights the scientific method. We hope that such repetition will help inform students how the scientific method is actually applied by working scientists.
- An integrated view of life depends on a theme that cuts across all topics, and that theme is evolution. As a unifying theme of this textbook, evolution elevates biology from a collection of facts to a coherent study of changing life on a changing planet. Thus, each chapter of *Essential Biology with Physiology* ends with an "Evolution Connection" that relates the content of the chapter to the central theme of biology.

This edition continues to make use of innovative features from earlier editions, such as text/art integration that teaches important topics through both words and pictures, frequent references within the text to important applications of the material to students' lives, and the visual summary diagrams that close each chapter. In addition to "The Process of Science" sections, here are just a few more improvements to this edition that you'll find in the supplements package for *Essential Biology with Physiology*, Second Edition:

- A new Media Manager helps instructors gather and organize the wide array of electronic supplements that complement the book.
- A new set of eTutor Instructor Animations and Student Tutorials brings interactive multimedia tools to bear on biology's toughest topics, those that vex students and teachers year after year (such as photosynthesis, cellular respiration, and mitosis).
- MP3 Tutors, narrated by coauthor Eric Simon, help explain important concepts to students in a medium that matches their on-the-go lifestyle.

We see our responsibility as science educators to be especially important in communicating with students who are not biology majors, because their attitudes about science and scientists are likely to be shaped by a single required science course—this course. Long after students have forgotten most of the specific content of their college courses, they will be left with general impressions that will influence their interests, opinions, values, and actions. We hope that this textbook and its supplements will help students fold biological perspectives into their personal worldviews. Please let us know how we are doing and how we can improve the next edition of *Essential Biology with Physiology*.

Jane Reece
Benjamin Cummings
1301 Sansome Street
San Francisco, CA 94111

Eric Simon
Dept. of Biology
New England College
Henniker, NH 03242

Acknowledgments

As authors, we are incredibly privileged to have worked with the very best publishing professionals and biology colleagues. Though the responsibility for any shortcomings of this book lies solely with us, its merits reflect the contributions of many associates who helped us with the textbook, supplements, and integral media activities that complement the lessons.

First and foremost, we must acknowledge our huge debt to Neil Campbell, the founding author of this book and a source of inspiration for both of us. Although this edition of the book has been carefully and thoroughly revised—to improve its science, its connections to students' lives, and its pedagogy—it remains infused with Neil's vision. Furthermore, most of the prose and pictures he created for earlier editions remain. Neil remains the true lead author of the book.

This book would never have come into existence without the efforts of the *Essential Biology with Physiology* team at Benjamin Cummings. Beth Wilbur oversaw the project as editor-in-chief. Beth's talents as a consensus builder and team leader helped to keep the project always moving forward. Senior acquisitions editor Chalon Bridges brought a clear vision, unending enthusiasm, and positive energy to the book; her bright nature and dedication to the nonmajors biology course inspired us, especially during the most difficult stages of writing. Of course, publishing excellence flows from the top, and we are grateful to Linda Davis, president of Addison Wesley and Benjamin Cummings, for the example she provides to her entire company. Further guidance was provided by Deborah Gale, director of development, and Lauren Fogel, director of media development. Benjamin Cummings authors are very fortunate to have such supportive leadership.

The editorial handprints of developmental editor Evelyn Dahlgren can be seen on every page of this book—and the final book is much the better for it. As the intensity of the project ramped up, so did Evelyn's output and thoughtfulness. Senior project manager Ginnie Simione Jutson ably guided the many facets of the project, both text and media, with the perfect combination of firmness and compassion. Without her, we would be lost.

If you find yourself admiring a wonderful piece of art in this book, you can thank the gifted Russell Chun, developmental artist. Senior photo editor Donna Kalal, aided by photo researcher Maureen Spuhler, used her keen eye and discerning taste to provide the wealth of photographs used in the book and its supplements. Freelance developmental editor Susan Weisberg maintained her high standards, even when the schedule was at its most demanding. Editorial assistant Benjamin Lau smoothed out the day-to-day operations of the whole team and answered every challenge set to him. The dedication to perfection shown by the entire editorial team inspired the authors to do our very best.

After our words and drawings were committed to paper, the production team transformed them into the book you hold in your hands. Michael Early, managing editor, rose to the challenge of our very tight schedule with aplomb. For the production and composition of the book, we thank Lori Dalberg of Carlisle Publishing Services, whose professionalism and commitment to the quality of the finished product eased our authorial burdens tremendously. The authors owe much to the copyeditor, Janet Greenblatt, for polishing our words, and to permissions editor Sue Ewing. If you like the look of the book, it is thanks to design director Mark Ong, text designer Jana Anderson, and cover designer Yvo Riezebos. We thank Precision Graphics for creating the illustrations. In the final stages of production, the talents of manufacturing buyer Stacy Wong shone.

More and more, the success of a textbook depends on the quality of its supplements. Luckily, the *Essential Biology with Physiology* supplements team is as committed to the core goals of accuracy and readability as the authors and editorial team. Supplements editor Susan Berge expertly coordinated the supplements, a difficult task given their number and variety, and production supervisors Jane Brundage and Jennifer Mattson handled the production of these materials. We owe particular gratitude to the supplements authors, especially Brad Williamson, who worked on the eTutors and MP3 Tutors; Jay Comeaux of Louisiana State University for his thoughtful work on the Discovery Channel Videos transcripts and questions; and Ed Zalisko of Blackburn College, who wrote the Student Study Guide and Instructor Guide. We thank the team of Test Bank and media quiz question contributors: Eugene Fenster of Longview Community College, James Bray of Blackburn College, Jon Hoekstra of Gainesville State College, Dawn Keller of Hawkeye College, Mark Manteuffel of St. Louis Community College, Richard Myers of Missouri State University, Cara Shillington of Eastern Michigan College, and the aforementioned Ed Zalisko. As teachers, we know the value of well-crafted PowerPoint lectures, so we thank Chris Romero of Front Range Community College and John Hammett for writing and editing those that accompany this book. We are grateful to Greg Podgorski of Utah State University for his work on the Study Card and Scientific American Current Issues in Biology Volumes 3 and 4. We would also like to acknowledge copyeditor Jan McDearmon for her work on many of the supplements.

Playing key roles in of the development and production of the media were contributor Brad Williamson, developmental manager Pat Burner, developmental artist Russell Chun, senior media producer Jon Ballard, project manager Brienn Buchanan, audio engineer Aaron Gass, and web developer Linda Young. We also thank Cedric Buckley of Jackson State University for his work on the student CD-ROM and website. We acknowledge and greatly appreciate the talents and hard work of Animated Biomedical Productions and Groove XI for their amazing work on the eTutor instructor animations and student tutorials.

As educators and writers, we rarely think about marketing. But for what we try to do as authors, "market" translates as "the students and instructors we are trying to serve." Christy Lawrence, director of marketing, Lauren Harp, executive marketing manager, and Jeff Hester, marketing manager, helped us achieve our authorial goals by keeping us constantly focused on the needs of students and instructors. For their amazing efforts in marketing, we also thank creative director Lillian Carr, marketing specialist Jane Campbell, and marketing communications specialist Kristi Hlaing. We also send a hearty thank you to the Benjamin Cummings field staff for representing *Essential Biology with Physiology* on campuses. These representatives are our lifeline to the greater educational community, telling us what you like (and don't like) about this book and media. Their enthusiasm for sharing our words buoyed us during the writing process, and the from-the-classroom feedback they provided was instrumental in shaping the vision of the book.

Also contributing substantial experience and wisdom to the manuscript were biology educators Martha Taylor, Brad Williamson, Jay Withgott, James Newcomb, and Kevin Padian. Eric Simon would like to thank the following colleagues for specific content suggestions and other support: T. Ryan Gregory (University of Guelph, Ontario, Canada), Mark J. Daly (Center for Human Genetic Research, Harvard Medical School), Kevin McMahon (New Hampshire State Police Forensics Laboratory), Kendra Hill and Lan Xu (San Diego State University), Marshall Simon, Nikos Kyrpides (Genomes OnLine Database), Robert M. Martore (South Carolina Office of Fisheries Management), Dr. Michael C. Ain (Johns Hopkins Children's Center), Jamey Barone, and Barbara Gaskell at Premiere Printing. Eric Simon would also like to thank his colleagues at New England College for their support and for providing a model of excellence in education: Amanda and Giulia Bussone, Natalia Plotnikova, Debra Dunlop, Sachie Howard, Mark Mitch, James Newcomb, Maria Colby, Ed Cooper, and Stephen Fritz. In particular, Eric Simon thanks Amanda Marsh for her constant support, compassion, and wisdom.

Furthermore, at the end of these acknowledgments you'll find a list of the many instructors who provided valuable information about their courses, reviewed chapters, and/or conducted class tests of *Essential Biology with Physiology* with their students. We thank them for their efforts and support.

Most of all, we thank our families, friends, and colleagues who continue to tolerate our obsession with doing our best for science education.

Jane Reece and Eric Simon

Reviewers of This Edition

Andrea Bixler,
Clarke College

Judy Bluemer,
Morton College

Carol A Britson,
University of Mississippi

Steve Brumbaugh,
Green River Community College

Bane Cheek,
Polk Community College

Thomas F. Chubb,
Villanova University

Jay L. Comeaux,
Louisiana State University

Elizabeth Desy,
Southwest State University

Richard Driskill,
Delaware State University

Lianne Drysdale,
Ozarks Technical Community College

Tamar Liberman Goulet,
University of Mississippi

Blanche C. Haning,
Vance-Granville Community College

Reba Harrell,
Hinds Community College

Juliana Hinton,
McNeese State University

Mary K. Kananen,
Pennsylvania State University, Altoona

Dawn Keller,
Hawkeye College

Peter King,
Francis Marion University

Ruhul H. Kuddus,
Utah Valley State College

James V. Landrum,
Washburn University

Mark Manteuffel,
St. Louis Community College

Lance D. McBrayer,
Georgia Southern University

Timothy D. Metz,
Campbell University

James Newcomb,
New England College

Jon R. Nickles,
University of Alaska, Anchorage

Sandra M. Pace,
Rappahannock Community College

Kathleen E. Pelkki,
Saginaw Valley State University

Paula A. Piehl,
Potomac State College of West Virginia University

Gregory Podgorski,
Utah State University

Elena Pravosudova,
Sierra College

Hallie Ray,
Rappahannock Community College

Nathan S. Reyna,
Howard Payne University

Todd Rimkus,
Marymount University

Tyson Sacco,
Cornell University

Michael Scott,
Lincoln University

Eric Scully,
Towson State University

Lois Sealy,
Valencia Community College

Sandra S. Seidel,
Elon University

Cara Shillington,
Eastern Michigan University

Brian Shmaefsky,
Kingwood College

Bethany Stone,
University of Missouri, Columbia

Rainy Inman Shorey,
Ferris State University

Linda Tichenor,
University of Arkansas, Fort Smith

Michael Twaddle,
University of Toledo

Lisa A. Werner,
Pima Community College

Rick Wiedenmann,
New Mexico State University at Carlsbad

Peter J. Wilkin,
Purdue University North Central

Judy A. Williams,
Southeastern Oklahoma State University

Shirley Zajdel,
Housatonic Community College

Reviewers of Previous Editions

Marilyn Abbott,
Lindenwood College

Tammy Adair,
Baylor University

Felix O. Akojie,
Paducah Community College

William Sylvester Allred, Jr.,
Northern Arizona University

Estrella Z. Ang,
University of Pittsburgh

David Arieti,
Oakton Community College

C. Warren Arnold,
Allan Hancock Community College

Mohammad Ashraf,
Olive-Harvey College

Bert Atsma,
Union County College

Yael Avissar,
Rhode Island College

Barbara J. Backley,
Elgin Community College

Gail F. Baker,
LaGuardia Community College

Kristel K. Bakker,
Dakota State University

Linda Barham,
Meridian Community College

Charlotte Barker,
Angelina College

S. Rose Bast,
Mount Mary College

Sam Beattie,
California State University, Chico

Rudi Berkelhamer,
University of California, Irvine

Penny Bernstein,
Kent State University, Stark Campus

Suchi Bhardwaj,
Winthrop University

Karyn Bledsoe,
Western Oregon University

Donna H. Bivans,
East Carolina University

Andrea Bixler,
Clarke College

Brian Black,
Bay de Noc Community College

Allan Blake,
Seton Hall University

Sonal Blumenthal,
University of Texas at Austin

Lisa Boggs,
Southwestern Oklahoma State University

Dennis Bogyo,
Valdosta State University

Virginia M. Borden,
University of Minnesota, Duluth

James Botsford,
New Mexico State University

Cynthia Bottrell,
Scott Community College

Richard Bounds,
Mount Olive College

Cynthia Boyd,
Hawkeye Community College
Robert Boyd,
Auburn University
B. J. Boyer,
Suffolk County Community College
Mimi Bres,
Prince George's Community College
Patricia Brewer,
University of Texas at San Antonio
Jerald S. Bricker,
Cameron University
George M. Brooks,
Ohio University, Zanesville
Janie Sue Brooks,
Brevard College
Steve Browder,
Franklin College
Evert Brown,
Casper College
Mary H. Brown,
Lansing Community College
Richard D. Brown,
Brunswick Community College
Joseph C. Bundy,
University of North Carolina at Greensboro
Carol T. Burton,
Bellevue Community College
Rebecca Burton,
Alverno College
Warren R. Buss,
University of Northern Colorado
Miguel Cervantes-Cervantes,
Lehman College, City University of New York
Bane Cheek,
Polk Community College
Thomas F. Chubb,
Villanova University
Reggie Cobb,
Nash Community College
Pamela Cole,
Shelton State Community College
William H. Coleman,
University of Hartford
James Conkey,
Truckee Meadows Community College
Karen A. Conzelman,
Glendale Community College
Ann Coopersmith,
Maui Community College
James T. Costa,
Western Carolina University
Pat Cox,
University of Tennessee, Knoxville
Pradeep M. Dass,
Appalachian State University
Paul Decelles,
Johnson County Community College
Galen DeHay,
Tri County Technical College
Cynthia L. Delaney,
University of South Alabama
Jean DeSaix,
University of North Carolina at Chapel Hill
Elizabeth Desy,
Southwest State University
Edward Devine,
Moraine Valley Community College
Dwight Dimaculangan,
Winthrop University
Deborah Dodson,
Vincennes Community College
Diane Doidge,
Grand View College
Don Dorfman,
Monmouth University
Terese Dudek,
Kishawaukee College
Shannon Dullea,
North Dakota State College of Science
David A. Eakin,
Eastern Kentucky University
Brian Earle,
Cedar Valley College
Ade Ejire,
Johnston Community College
Dennis G. Emery,
Iowa State University
Virginia Erickson,
Highline Community College
Carl Estrella,
Merced College
Marirose T. Ethington,
Genesee Community College
Paul R. Evans,
Brigham Young University
Zenephia E. Evans,
Purdue University
Jean Everett,
College of Charleston
Dianne M. Fair,
Florida Community College at Jacksonville
Joseph Faryniarz,
Naugatuck Valley Community College
Phillip Fawley,
Westminster College
Lynn Fireston,
Ricks College
Jennifer Floyd,
Leeward Community College
Dennis M. Forsythe,
The Citadel
Carl F. Friese,
University of Dayton
Suzanne S. Frucht,
Northwest Missouri State University
Edward G. Gabriel,
Lycoming College
Anne M. Galbraith,
University of Wisconsin, La Crosse
Kathleen Gallucci,
Elon University
Gregory R. Garman,
Centralia College
Gail Gasparich,
Towson University
Kathy Gifford,
Butler County Community College
Sharon L. Gilman,
Coastal Carolina University
Mac Given,
Neumann College
Patricia Glas,
The Citadel
Ralph C. Goff,
Mansfield University
Marian R. Goldsmith,
University of Rhode Island
Andrew Goliszek,
North Carolina Agricultural and Technical State University
Curt Gravis,
Western State College of Colorado
Larry Gray,
Utah Valley State College
Tom Green,
West Valley College
Robert S. Greene,
Niagara University
Ken Griffin,
Tarrant County Junior College
Denise Guerin,
Santa Fe Community College
Paul Gurn,
Naugatuck Valley Community College
Peggy J. Guthrie,
University of Central Oklahoma
Henry H. Hagedorn,
University of Arizona
Blanche C. Haning,
North Carolina State University
Laszlo Hanzely,
Northern Illinois University
Sherry Harrel,
Eastern Kentucky University
Frankie Harriss,
Independence Community College
Lysa Marie Hartley,
Methodist College
Janet Haynes,
Long Island University
Michael Held,
St. Peter's College
Consetta Helmick,
University of Idaho
Michael Henry,
Contra Costa College
Linda Hensel,
Mercer University
Jana Henson,
Georgetown College
James Hewlett,
Finger Lakes Community College
Richard Hilton,
Towson University
Phyllis C. Hirsch,
East Los Angeles College
A. Scott Holaday,
Texas Tech University
R. Dwain Horrocks,
Brigham Young University
Howard L. Hosick,
Washington State University
Carl Huether,
University of Cincinnati
Celene Jackson,
Western Michigan University
John Jahoda,
Bridgewater State College
Richard J. Jensen,
Saint Mary's College
Tari Johnson,
Normandale Community College
Tia Johnson,
Mitchell Community College
Greg Jones,
Santa Fe Community College
John Jorstad,
Kirkwood Community College
Tracy L. Kahn,
University of California, Riverside
Robert Kalbach,
Finger Lakes Community College
Mary K. Kananen,
Pennsylvania State University, Altoona
Thomas C. Kane,
University of Cincinnati
Arnold J. Karpoff,
University of Louisville
John M. Kasmer,
Northeastern Illinois University
Valentine Kefeli,
Slippery Rock University
John Kelly,
Northeastern University
Cheryl Kerfeld,
University of California, Los Angeles
Henrik Kibak,
California State University, Monterey Bay
Kerry Kilburn,
Old Dominion University
Joyce Kille-Marino,
College of Charleston
Peter Kish,
Oklahoma School of Science and Mathematics
Robert Kitchin,
University of Wyoming
Richard Koblin,
Oakland Community College
H. Roberta Koepfer,
Queens College
Michael E. Kovach,
Baldwin-Wallace College
Jocelyn E. Krebs,
University of Alaska, Anchorage
Nuran Kumbaraci,
Stevens Institute of Technology

Roya Lahijani,
Palomar College
Vic Landrum,
Washburn University
Gary Kwiecinski,
The University of Scranton
Lynn Larsen,
Portland Community College
Siu-Lam Lee,
University of Massachusetts, Lowell
Thomas P. Lehman,
Morgan Community College
William Leonard,
Central Alabama Community College
Shawn Lester,
Montgomery College
Leslie Lichtenstein,
Massasoit Community College
Barbara Liedl,
Central College
Harvey Liftin,
Broward Community College
David Loring,
Johnson County Community College
Lewis M. Lutton,
Mercyhurst College
Maria P. MacWilliams,
Seton Hall University
Michael Howard Marcovitz,
Midland Lutheran College
Angela M. Mason,
Beaufort County Community College
Roy B. Mason,
Mt. San Jacinto College
John Mathwig,
College of Lake County
Bonnie McCormick,
University of the Incarnate Word
Tonya McKinley,
Concord College
Mary Anne McMurray,
Henderson Community College
Katrina McCrae,
Abraham Baldwin Agricultural College
Ed Mercurio,
Hartnell College
David Mirman,
Mt. San Antonio College
Nancy Garnett Morris,
Volunteer State Community College
Angela C. Morrow,
University of Northern Colorado
Patricia S. Muir,
Oregon State University
Jon R. Nickles,
University of Alaska, Anchorage
Jane Noble-Harvey,
University of Delaware
Jeanette C. Oliver,
Flathead Valley Community College
David O'Neill,
Community College of Baltimore County
Lois H. Peck,
University of the Sciences, Philadelphia
Kathleen E. Pelkki,
Saginaw Valley State University
Jennifer Penrod,
Lincoln University
Rhoda E. Perozzi,
Virginia Commonwealth University
John S. Peters,
College of Charleston
Pamela Petrequin,
Mount Mary College
Bill Pietraface,
State University of New York, Oneonta
Rosamond V. Potter,
University of Chicago
mailto:karen.powell
Western Kentucky University
Elena Pravosudova,
Sierra College
Hallie Ray,
Rappahannock Community College
Jill Raymond,
Rock Valley College
Dorothy Read,
University of Massachusetts, Dartmouth
Philip Ricker,
South Plains College
Todd Rimkus,
Marymount University
Lynn Rivers,
Henry Ford Community College
Jennifer Roberts,
Lewis University
Laurel Roberts,
University of Pittsburgh
April Rottman,
Rock Valley College
Maxine Losoff Rusche,
Northern Arizona University
Michael L. Rutledge,
Middle Tennessee State University
Mike Runyan,
Lander University
Travis Ryan,
Furman University
Sarmad Saman,
Quinsigamond Community College
Leba Sarkis,
Aims Community College
Walter Saviuk,
Daytona Beach Community College
Neil Schanker,
College of the Siskiyous
Robert Schoch,
Boston University
John Richard Schrock,
Emporia State University
Julie Schroer,
Bismarck State College
Karen Schuster,
Florida Community College at Jacksonville
Brian W. Schwartz,
Columbus State University
Sandra Seidel,
Elon University
Wayne Seifert,
Brookhaven College
Patty Shields,
George Mason University
Cara Shillington,
Eastern Michigan University
Cahleen Shrier,
Azusa Pacific University
Jed Shumsky,
Drexel University
Greg Sievert,
Emporia State University
Jeffrey Simmons,
West Virginia Wesleyan College
Frederick D. Singer,
Radford University
Anu Singh-Cundy,
Western Washington University
Sandra Slivka,
Miramar College
Margaret W. Smith,
Butler University
Thomas Smith,
Armstrong Atlantic State University
Deena K. Spielman,
Rock Valley College
Minou D. Spradley,
San Diego City College
Eric Stavney,
Highline Community College
Robert Stamatis,
Daytona Beach Community College
Mark T. Sugalski,
New England College
Marshall D. Sundberg,
Emporia State University
Adelaide Svoboda,
Nazareth College
Sharon Thoma,
Edgewood College
Kenneth Thomas,
Hillsborough Community College
Sumesh Thomas,
Baltimore City Community College
Betty Thompson,
Baptist University
Paula Thompson,
Florida Community College
John Tjepkema,
University of Maine at Orono
Bruce L. Tomlinson,
State University of New York, Fredonia
Leslie R. Towill,
Arizona State University
Bert Tribbey,
California State University, Fresno
Robert Turner,
Western Oregon University
Virginia Vandergon,
California State University, Northridge
William A. Velhagen, Jr.,
Longwood College
Jonathan Visick,
North Central College
Michael Vitale,
Daytona Beach Community College
Lisa Volk,
Fayetteville Technical Community College
Stephen M. Wagener,
Western Connecticut State University
James A. Wallis,
St. Petersburg Community College
Jennifer Warner,
University of North Carolina at Charlotte
Dave Webb,
St. Clair County Community College
Harold Webster,
Pennsylvania State University, DuBois
Ted Weinheimer,
California State University, Bakersfield
Lisa A. Werner,
Pima Community College
Joanne Westin,
Case Western Reserve University
Wayne Whaley,
Utah Valley State College
Joseph D. White,
Baylor University
Quinton White,
Jacksonville University
Leslie Y. Whiteman,
Virginia Union University
Rick Wiedenmann,
New Mexico State University at Carlsbad
Judy A. Williams,
Southeastern Oklahoma State University
Dwina Willis,
Freed Hardeman University
David Wilson,
University of Miami
Mala S. Wingerd,
San Diego State University
E. William Wischusen,
Louisiana State University
Darla J. Wise,
Concord College
Michael Womack,
Macon State College
Bonnie Wood,
University of Maine at Presque Isle
Mark L. Wygoda,
McNeese State University
Samuel J. Zeakes,
Radford University
Uko Zylstra,
Calvin College

Detailed contents

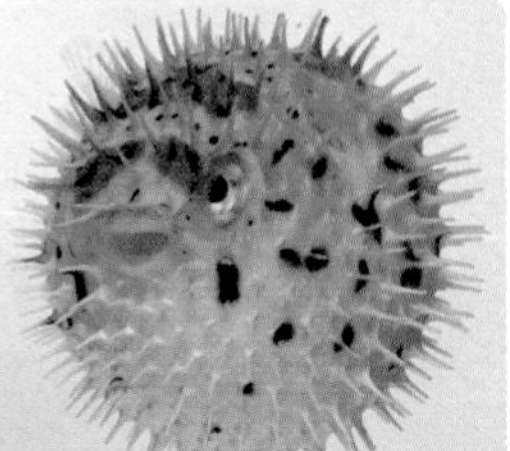

UNIT 1 Cells

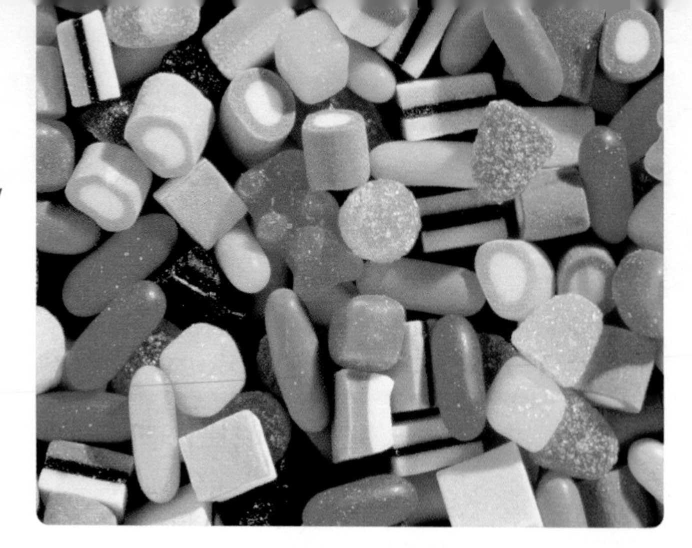

4 A Tour of the Cell 54

5 The Working Cell 72

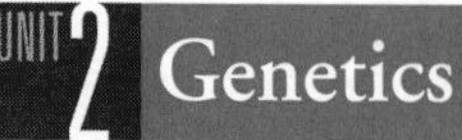

Genetics

8 Cellular Reproduction: Cells from Cells 120

9 Patterns of Inheritance 143

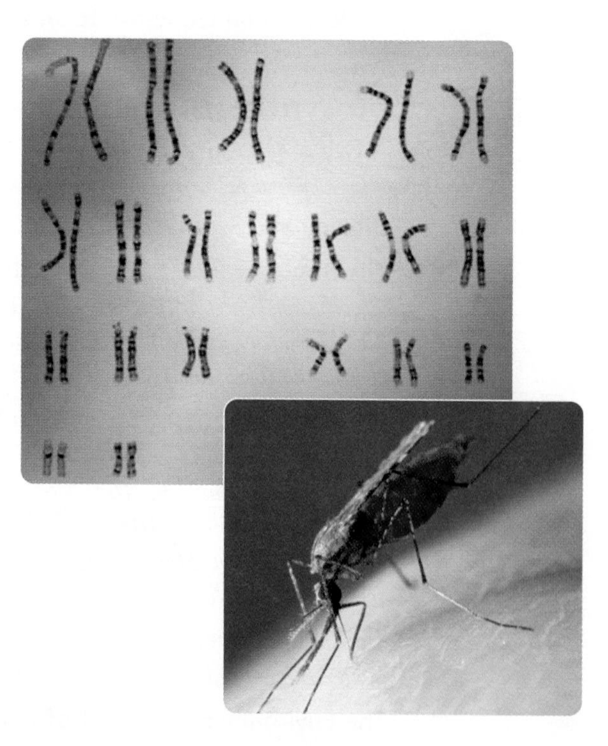

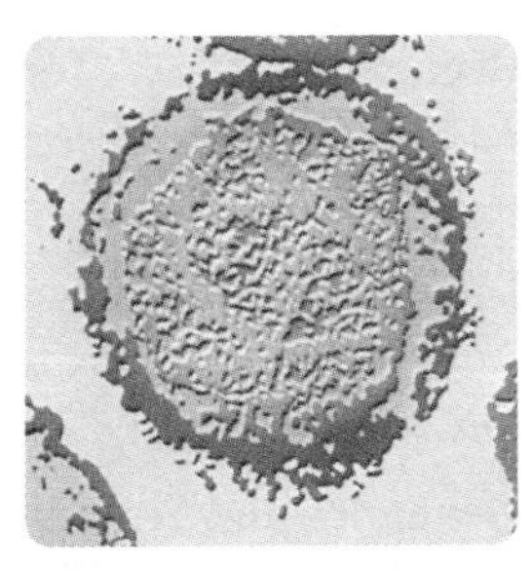

11 How Genes Are Controlled 199

12 DNA Technology 219

UNIT 3 Evolution and Diversity

13 How Populations Evolve 243

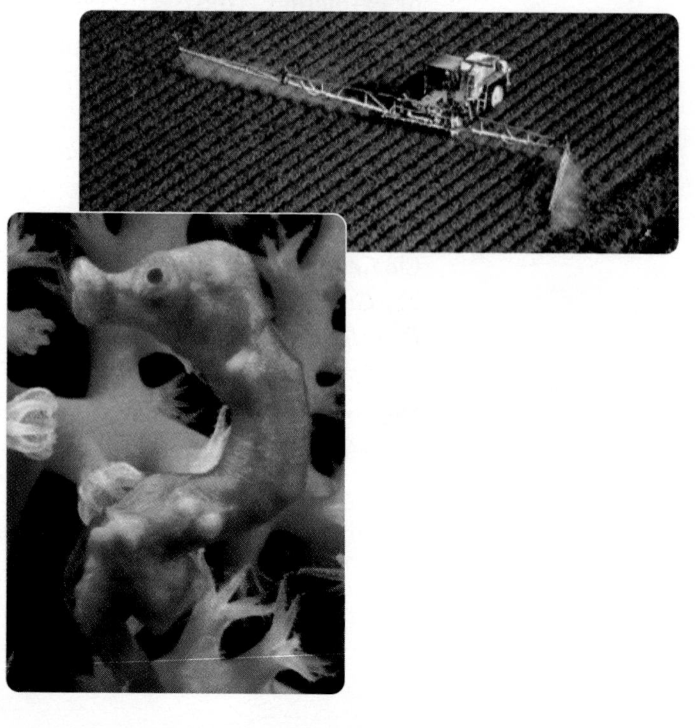

14 How Biological Diversity Evolves 269

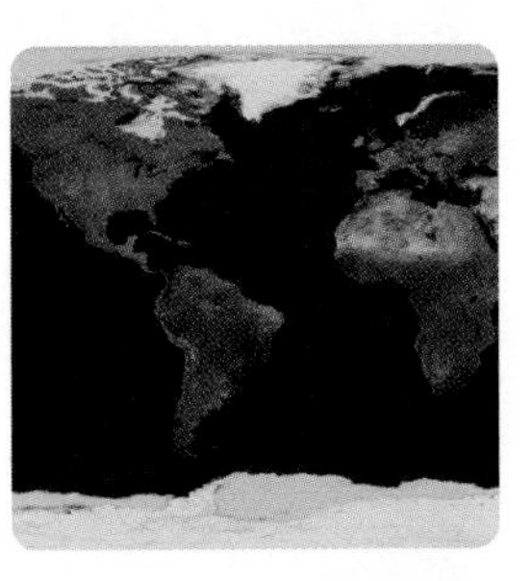

15 The Evolution of Microbial Life 295

16 Plants, Fungi, and the Move onto Land 319

17 The Evolution of Animals 342

UNIT 4 Ecology

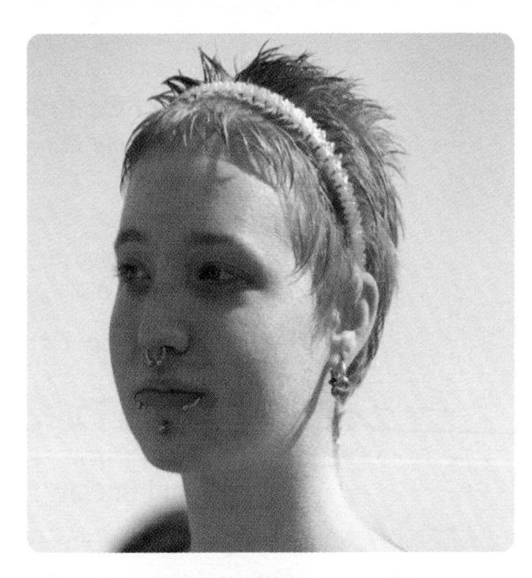

22 Nutrition and Digestion 485

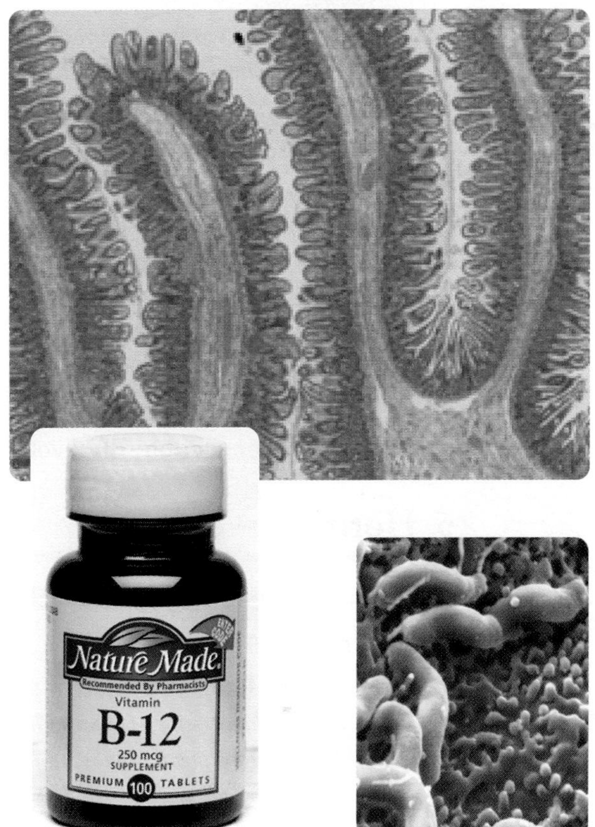

23 Circulation and Respiration 504

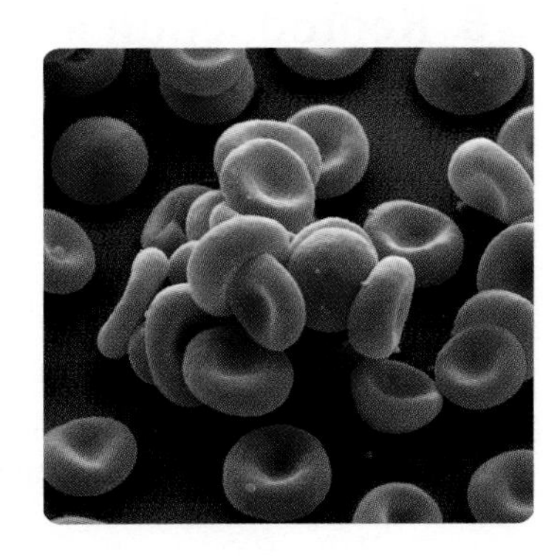

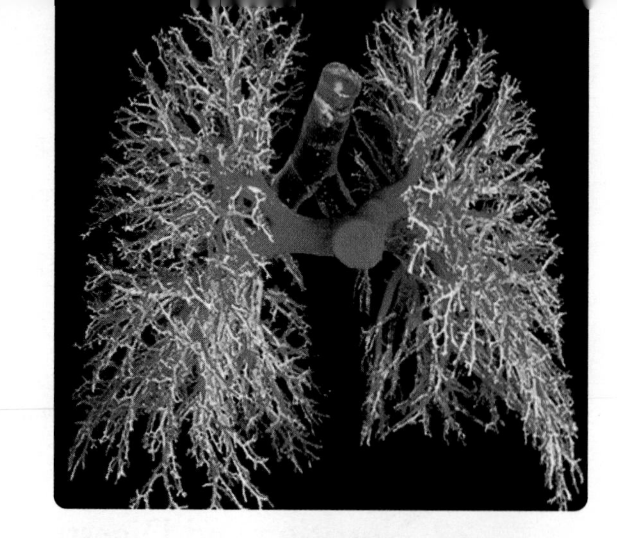

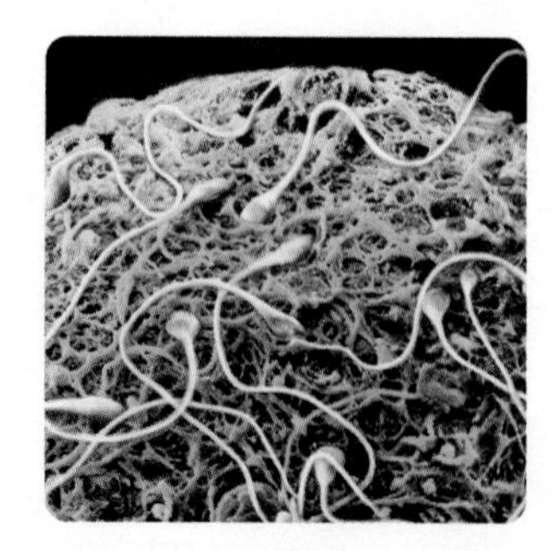

27 Nervous, Sensory, and Motor Systems 584

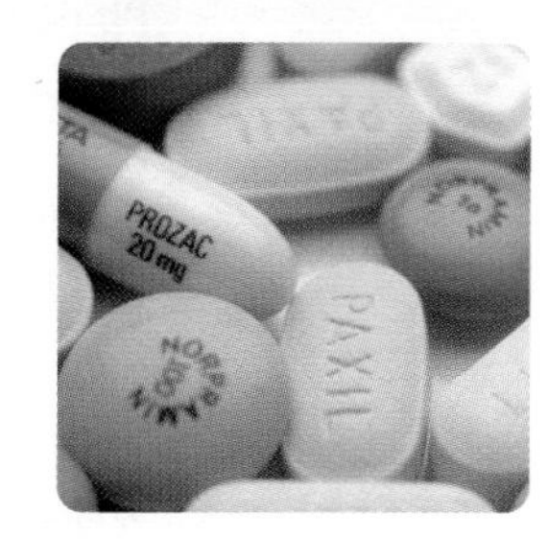

UNIT 6 Plant Structure and Function

28 The Life of a Flowering Plant 618

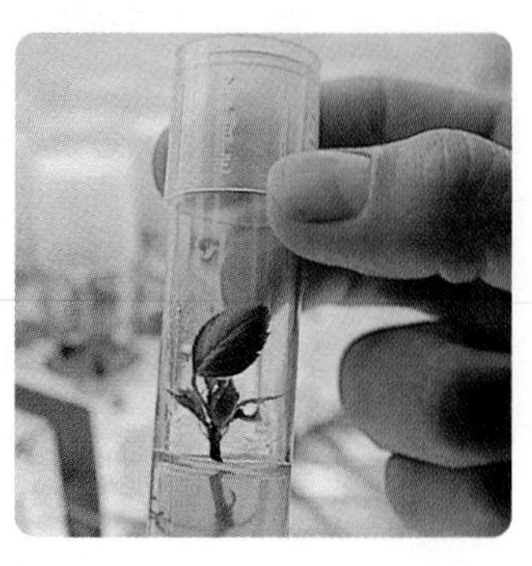

1 Introduction: Biology Today

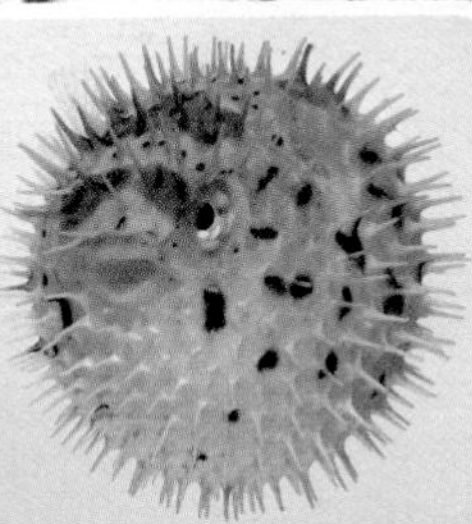

Scientists have determined **DNA sequences** from humans, puffer fish, mosquitoes, and rice.

Amoebas, molds, trees, and people are all made from **similar cells.**

Biologists have identified about **1.8 million species** of living organisms.

All organisms share a common chemical language for their genetic material, DNA.

BIOLOGY AND SOCIETY

Living in a Golden Age of Biology

We are living in a golden age of biology. The largest and best-equipped community of scientists in history is beginning to solve biological puzzles that once seemed unsolvable. We are moving ever closer to understanding how a single cell becomes a plant or animal; how plants trap solar energy and store that energy in food; how organisms form networks in biological communities such as forests and coral reefs; and how the great diversity of life on Earth evolved from the first microbes. Exploring life has never been more exhilarating. Welcome to the big adventure of the twenty-first century!

Modern biology is as important as it is inspiring, with exciting breakthroughs changing our very culture. Genetics and cell biology are revolutionizing medicine and agriculture. Molecular biology is providing new tools for investigating ancestry and solving crimes. Ecology is helping us evaluate environmental issues, such as the causes and consequences of global warming. Neuroscience and evolutionary biology are reshaping psychology and sociology. These are just a few examples of how biology is woven into the fabric of society as never before. It is no wonder that biology is part of our daily lives **(Figure 1.1)**.

We wrote this book to help students who are not biology majors develop an appreciation for the science of life and apply that understanding as they evaluate social issues. We believe that such a biological perspective is essential for any educated person, which is why we named our book *Essential Biology*. So, whatever your reasons for taking this course, even if only to meet your college's science requirement, you'll soon discover that this is the best time ever to study biology. To help you get started, this first chapter of *Essential Biology* defines biology and then expands on important concepts within this definition. First, we'll survey the properties of life and the scope of life. Next, we'll introduce evolution as the theme that unifies all of biology. Finally, we'll set the study of life in the broader context of science as a process of inquiry and illustrate some of the connections between biology and society. ■

Figure 1.1 **A small sample of biology in our everyday lives.**

The Scope of Life

Biology is the scientific study of life. It's a huge subject that gets bigger every year. We can think of biology's enormous scope as having two major dimensions. First, life is structured on a size scale ranging from the molecular to the global. The second dimension of biology's scope stretches across the enormous diversity of life on Earth, now and throughout life's history. But before we discuss the great number of differences between living creatures, let's discuss the properties that unify all living things.

The Unity of Life

The phenomenon we call **life** largely defies a simple, one-sentence definition. Yet almost any child perceives that a dog or a bug or a plant is alive, while a rock is not. We recognize life largely by what living things do. All life is unified by a common set of characteristics.

Figure 1.2 highlights some of the properties and processes we associate with life: (a) *Order*. All living things exhibit complex but ordered organization, as seen in the highly ordered structure of a sunflower. (b) *Regulation*. The environment outside an organism (a living thing) frequently changes, but mechanisms regulate the organism's internal environment, keeping it within limits that sustain life. For example, a jackrabbit can adjust its body temperature by regulating the amount of blood flowing through its ears. When the rabbit's body temperature rises, more blood flows through the vessels in its ears, allowing excess heat to be released to the air. (c) *Growth and development*. Information carried by genes—the units of inheritance that transmit information from parents to offspring—controls the pattern of growth and development in all organisms, including the Nile crocodile. (d) *Energy utilization*. Organisms take in energy and transform it in performing all of life's activities. For example, a hummingbird obtains energy in the form of plant nectar and uses it to power flight and other work. (e) *Response to the environment*. All organisms respond to environmental stimuli. For example, a Venus flytrap closes its trap rapidly in response to the environmental stimulus of an insect landing on it. (f) *Reproduction*. Organisms reproduce their own kind. Thus, pandas reproduce only pandas—never crocodiles or hummingbirds. (g) *Evolution*. Reproduction underlies the capacity of populations to change (evolve) over time. For example, the appearance of the pygmy seahorse has evolved in a way that camouflages the animal in its environment. Evolutionary change has been a central, unifying feature of life since life arose nearly 4 billion years ago.

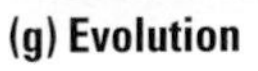

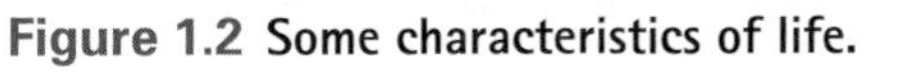

Figure 1.2 Some characteristics of life.

Life at Its Many Levels

In *Essential Biology*, we will probe life all the way down to the submicroscopic scale of molecules such as DNA, the chemical responsible for inheritance. At the other extreme of biological size and complexity, our exploration will take us up to the global scale of the entire biosphere, which consists of all the environments on Earth that support life—including soil; oceans, lakes, and other bodies of water; and the inner atmosphere. To illustrate, let's start with the biosphere and work our way down to smaller and smaller levels of biological organization.

Figure 1.3 begins with a partial view of the biosphere and then zooms to a scene of an African savanna. The savanna is an example of an ecosystem. An ecosystem consists of all organisms living in a particular area, as well as the nonliving, physical components of the environment that affect the organisms, such as water, air, soil, and sunlight. All the organisms in the savanna (zebras, insects, grasses, bacteria, and so on) are collectively called a community. Within communities are various populations, groups of interacting individuals of one species, such as a herd of zebras. Below population in the hierarchy is the organism, an individual living thing, such as a zebra.

Life's hierarchy continues to unfold within an individual organism. The zebra's body consists of several organ systems, such as the digestive and circulatory systems. Each organ system consists of organs, such as the heart and blood vessels of the circulatory system. As we continue downward through life's hierarchy, each organ is made up of several different tissues, each of which consists of a group of similar cells performing a specific function. The cell is the basic unit of life.

Finally, we reach the chemical level in the hierarchy. Each cell consists of an enormous number of chemicals that function together to give the cell the properties we recognize as life. At the bottom of Figure 1.3 is a computer graphic of DNA, the chemical of inheritance and the substance of genes. DNA is an example of a molecule, a cluster of even smaller chemical units called atoms. In the tiny segment of the DNA molecule shown, each sphere represents an atom.

From the interactions within the biosphere to the molecular machinery within cells, biologists are investigating life at its many levels. Let's take a closer look here at just two biological levels near opposite ends of the size scale: ecosystems and cells.

Ecosystems Life does not exist in a vacuum. Each organism interacts continuously with its environment, which includes other organisms as well as nonliving factors. The roots of a tree, for example, absorb water and minerals from the soil. Leaves take in carbon dioxide gas from the air. Chlorophyll, the green pigment of the leaves, absorbs sunlight, which drives the plant's production of sugar from carbon dioxide and water. This food production is called photosynthesis. The tree also releases oxygen to the air, and its roots help form soil by breaking up rocks. Both organism and environment are affected by the interactions between them. The tree also interacts with other living things, including microorganisms (microscopic organisms) in the soil that are associated with the plant's roots and animals that eat its leaves and fruit. We are, of course, among those animals.

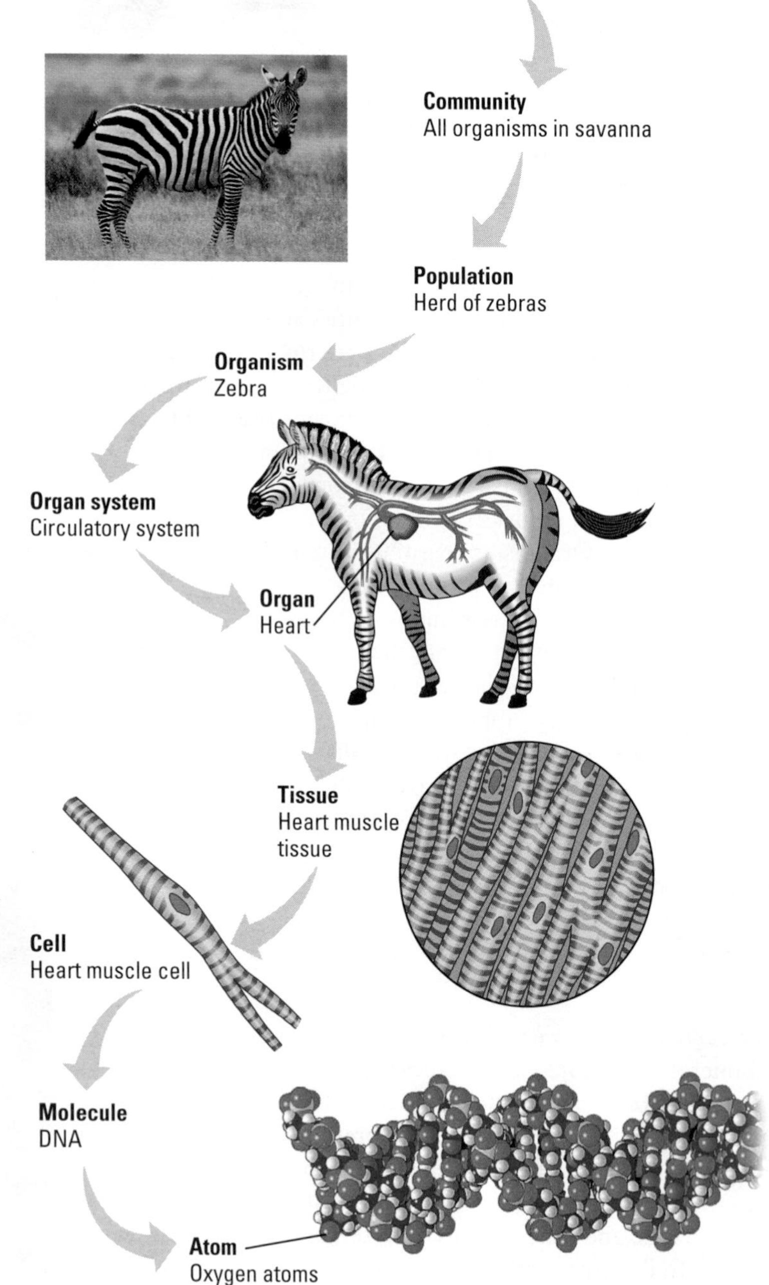

Figure 1.3 Zooming in on life. Biologists explore life at levels ranging from the biosphere to the atoms and molecules that make up cells.

The dynamics of any ecosystem depend on two main processes **(Figure 1.4)**. The first major process is the cycling of nutrients. For example, minerals that plants take up from the soil will eventually be recycled to the soil by microorganisms that decompose leaf litter and other organic refuse. The second major process in an ecosystem is the flow of energy from sunlight to producers and then on to consumers and decomposers. Producers are photosynthetic organisms, such as plants; consumers are the organisms, such as animals, that feed on plants, either directly (by eating plants) or indirectly (by eating animals that eat plants); decomposers, such as fungi, recycle the remains of deceased organisms, changing complex dead material into simple mineral nutrients. Thus, energy flows through an ecosystem (entering as sunlight and exiting as heat), whereas nutrients are recycled.

The biosphere is enriched by a great variety of ecosystems. A tropical rain forest in South America is an ecosystem. Very different from tropical forests are the ecosystems of deserts, such as those of the southwestern United States. A coral reef, such as the Great Barrier Reef off the eastern coast of Australia, is an ecosystem, and so is any small pond that may exist on your campus or in your city. Even a woodland patch in New York's Central Park qualifies as an ecosystem, small and artificial as it is by forest standards and disrupted as it is by human visitors. The fact is, humans are organisms that now have some presence, often disruptive, in all ecosystems. And the collective clout of 6 billion humans and their machines has an impact on the entire biosphere. For example, our fuel-burning, forest-chopping actions are changing the atmosphere and the planet's climate in ways that we do not yet fully understand, though we already know that this global vandalism jeopardizes the diversity of life on Earth.

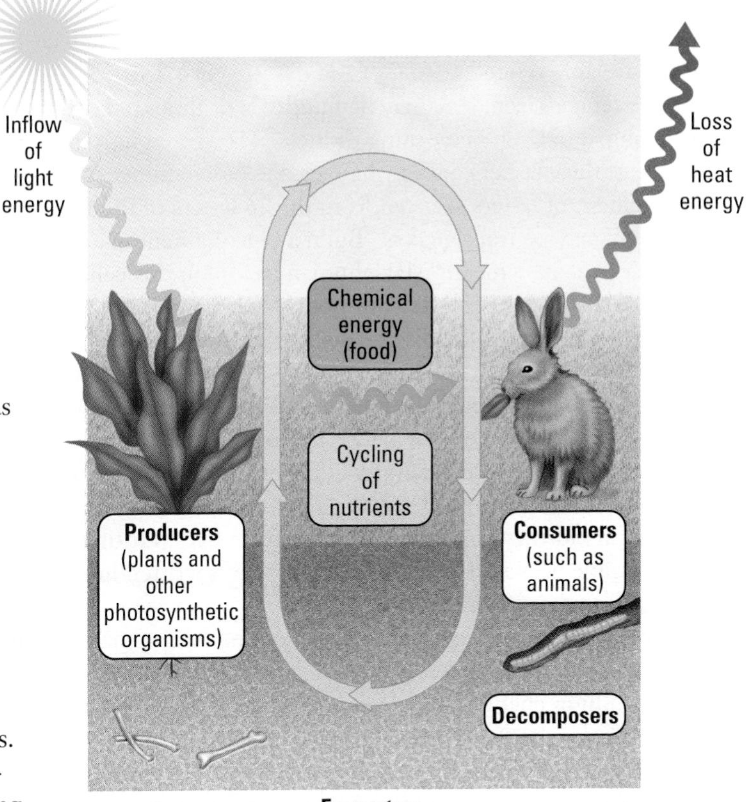

Figure 1.4 Energy and nutrient flow in an ecosystem. Living is work, and work requires that organisms obtain and use energy. Most ecosystems are solar powered. The energy that enters an ecosystem as sunlight exits as heat, which all organisms dissipate to their surroundings whenever they perform work. In contrast, the nutrients within an ecosystem are recycled.

Cells and Their DNA Let's downsize now from ecosystems to cells. The cell has a special place in the hierarchy of biological organization: It is the lowest level of structure that can perform all activities required for life, including reproduction.

All organisms are composed of cells. They occur singly as a great variety of unicellular organisms, mostly microscopic. And cells are also the subunits that make up the tissues and organs of plants, animals, and other multicellular organisms. In either case, the cell is the organism's basic unit of structure and function. The ability of cells to divide to form new cells is the basis for all reproduction and for the growth and repair of multicellular organisms, including humans.

We can distinguish two major kinds of cells: prokaryotic and eukaryotic **(Figure 1.5)**. The prokaryotic cell is much simpler and usually much smaller than the eukaryotic cell. The cells of bacteria are prokaryotic. Most other forms of life, including plants, animals, and fungi, are composed of eukaryotic cells. In contrast to the prokaryotic cell, the eukaryotic cell is subdivided by internal membranes into many different functional compartments, or organelles. For example, the nucleus, the largest organelle in most eukaryotic cells, houses DNA, the heritable material that directs the cell's many activities. Prokaryotic cells also have DNA, but it is not packaged within a nucleus.

Though very different in structural complexity, prokaryotic and eukaryotic cells have much in common at the molecular level. Most importantly, all cells use DNA as the chemical material of genes, the discrete units of hereditary information. Of course, bacteria and humans inherit different genes, but that information is encoded in a chemical language common to all organisms. In fact, the language of life has an alphabet of

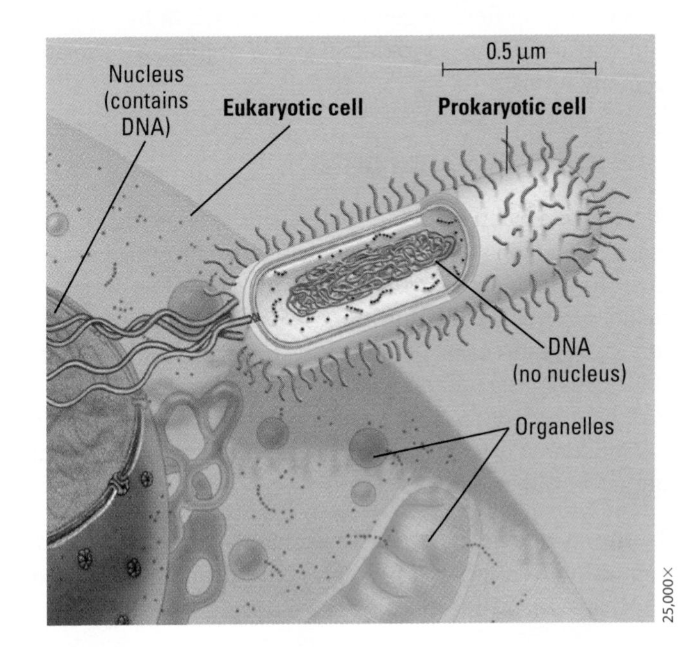

Figure 1.5 Two main kinds of cells: prokaryotic and eukaryotic. The scale bar represents 0.5 micrometers (µm). There are 1,000 µm in a millimeter (mm). (You can review metric measurements in Appendix A.) The number along the bottom right edge represents the magnification of the image. In this case, the image shown is approximately 25,000 times as big as the actual cells.

just four letters. The chemical names of DNA's four molecular building blocks are abbreviated as A, G, C, and T **(Figure 1.6)**.

An average-sized gene may be hundreds or thousands of chemical "letters" long. That gene's meaning to the cell is written in its specific sequence of these letters, just as the message of a sentence is encoded in its arrangement of letters selected from the 26 letters of the English alphabet. One gene may be translated as "Build a blue pigment in a bacterial cell." Another gene may mean "Make human insulin in this cell." Insulin is a chemical that helps regulate your body's use of sugar as a fuel. People who have certain forms of the disease diabetes produce an inadequate supply of insulin. To overcome this shortage, diabetics can inject themselves with insulin produced by genetically engineered bacteria. These bacteria produce insulin because they contain a gene for insulin production transplanted from a human cell. This example of genetic engineering was one of the earliest successes of biotechnology, a field that has transformed the pharmaceutical industry and extended millions of lives **(Figure 1.7)**. And it is only possible because biological information is written in the universal chemical language of DNA.

The entire "book" of genetic instructions that an organism inherits is called its genome. The nucleus of each human cell packs a genome that is about 3 billion chemical letters long. In recent years, scientists have tabulated virtually the entire sequence of these letters, and the press and world leaders have acclaimed this international achievement as the greatest scientific triumph ever. But unlike past cultural zeniths, such as the landing of Apollo astronauts on the moon, the sequencing of the human genome is more a commencement than a climax. As the quest continues, biologists will learn the functions of thousands of genes and how their activities are coordinated in the development of an organism. Additionally, the genomes of other organisms (such as *E. coli* bacteria, fruit flies, and dogs) have been sequenced, allowing scientists to compare the genomes of different species. This emerging field of genomics—a branch of biology that studies whole genomes—is a striking example of human curiosity about life at its many levels.

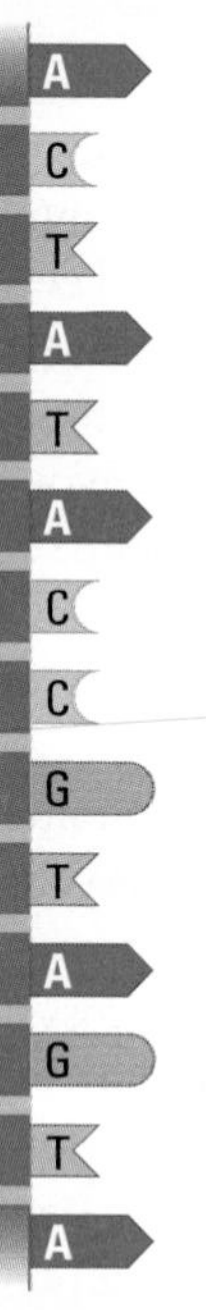

Figure 1.6 The language of DNA. These simple shapes and letters symbolize the four kinds of chemical building blocks that are chained together in DNA. A gene is a segment of DNA composed of hundreds or thousands of these building blocks, of which we see only a short stretch here. Each gene encodes information in its specific sequence of the four chemical letters, which are universal among all life on Earth.

Figure 1.7 DNA technology in the drug industry. In modern biotechnology manufacturing facilities, genetically modified microorganisms produce useful pharmaceuticals. In this facility in India, yeast produce large quantities of human insulin.

Life in Its Diverse Forms

Diversity is a hallmark of life. The zebra shown in Figure 1.3 is just one of about 1.8 million species that biologists have identified and named (the scientific name for this zebra is *Equus burchelli*). The diversity of known life includes over 290,000 plants, almost 52,000 vertebrates (animals with backbones), and more than 1 million insects (more than half of all known forms of life). Biologists add thousands of newly identified species to the list each year. Estimates of the total number of species range from 10 million to over 200 million. Whatever the actual number, the vast diversity of life gives biology a very wide scope.

Grouping Species: The Basic Concept Biological diversity can be something to relish and preserve, but it can also be a bit overwhelming **(Figure 1.8)**. Confronted with complexity, people are inclined to

Figure 1.8 A small sample of biological diversity. Shown here are just some of the tens of thousands of species in the butterfly and moth collection at the National Museum of Natural History in Washington, D.C. As diverse as the species are, they are all variations on a common anatomical theme. One of biology's major goals is to explain how such diversity arises while also accounting for characteristics common to different species.

categorize diverse items into a smaller number of groups. Grouping species that are similar is natural for us. We may speak of "squirrels" and "butterflies," even though we recognize that each group actually includes many different species. We may even sort groups into broader categories, such as rodents (which include squirrels) and insects (which include butterflies). Taxonomy, the branch of biology that names and classifies species, formalizes this hierarchical ordering according to a scheme you will learn about in Chapter 14. Here we consider only the broadest units of classification.

The Three Domains of Life Biologists divide the diversity of life into three main groups. The three groups, called domains, are Bacteria, Archaea, and Eukarya **(Figure 1.9)**. The first two domains, Bacteria and Archaea, identify two very different groups of organisms that have prokaryotic cells. All the eukaryotes (organisms with eukaryotic cells) are placed within the domain Eukarya. Depending on the classification scheme used, domain Eukarya is divided into at least four smaller categories called kingdoms. Most members of three of the kingdoms—Plantae, Fungi, and Animalia—are multicellular. These three kingdoms are distinguished partly by how the organisms obtain food. Plants produce their own sugars and other foods by photosynthesis. Fungi are mostly decomposers, obtaining food by digesting dead organisms. Animals obtain food by ingesting (eating) and digesting other organisms. (This is, of course, the kingdom to which we belong.) Those eukaryotes that do not fit into the other three kingdoms are referred to as the protists. Protists are generally single-celled; they include microscopic protozoans, such as

Domain Bacteria

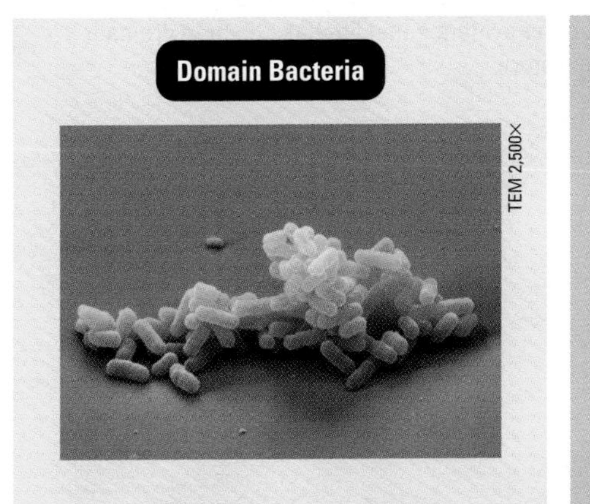

Domain Archaea

Domain Eukarya

Kingdom Plantae

Kingdom Fungi

Kingdom Animalia

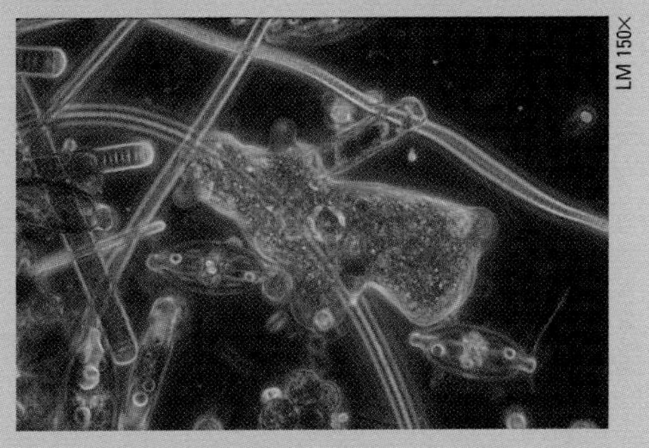

Protists (multiple kingdoms)

Figure 1.9 The three domains of life.

amoebas. But protists also include certain multicellular forms, such as seaweeds. The protists will almost surely be reorganized into separate, multiple kingdoms in coming years.

Unity in the Diversity of Life If life is so diverse, how can biology have any unifying themes? What, for instance, can a tree, a mushroom, and a human possibly have in common? As it turns out, a great deal! Underlying the diversity of life is a striking unity, especially at the lower levels of biological organization. We have already seen one example: the universal genetic language of DNA. That fundamental language connects all kingdoms of life, even uniting prokaryotes such as bacteria with eukaryotes such as humans. What can account for this combination of unity and diversity in life? The scientific explanation is the biological process called evolution.

Evolution: Biology's Unifying Theme

The history of life, as documented by fossils and other evidence, is a saga of a restless Earth billions of years old, inhabited by a changing cast of living forms **(Figure 1.10)**. Life evolves. Just as each individual has a family history, each species is one twig of a branching tree of life extending back in time through ancestral species more and more remote. Species that are very similar, such as the brown bear and the polar bear, share a common ancestor that represents a relatively recent branch point on the tree of life **(Figure 1.11)**. But through an ancestor that lived much farther back in time, all bears are also related to squirrels, humans, and all other mammals. Hair and milk-producing mammary glands are just two of a long list of uniquely mammalian traits. It

Figure 1.10 Digging into the past. In the desert of Niger, Africa, paleontologist (fossil specialist) Paul Sereno excavates the leg bones of a giant plant-eating dinosaur called *Jobaria.* The fossil record supports other evidence that life has changed dramatically over Earth's long history.

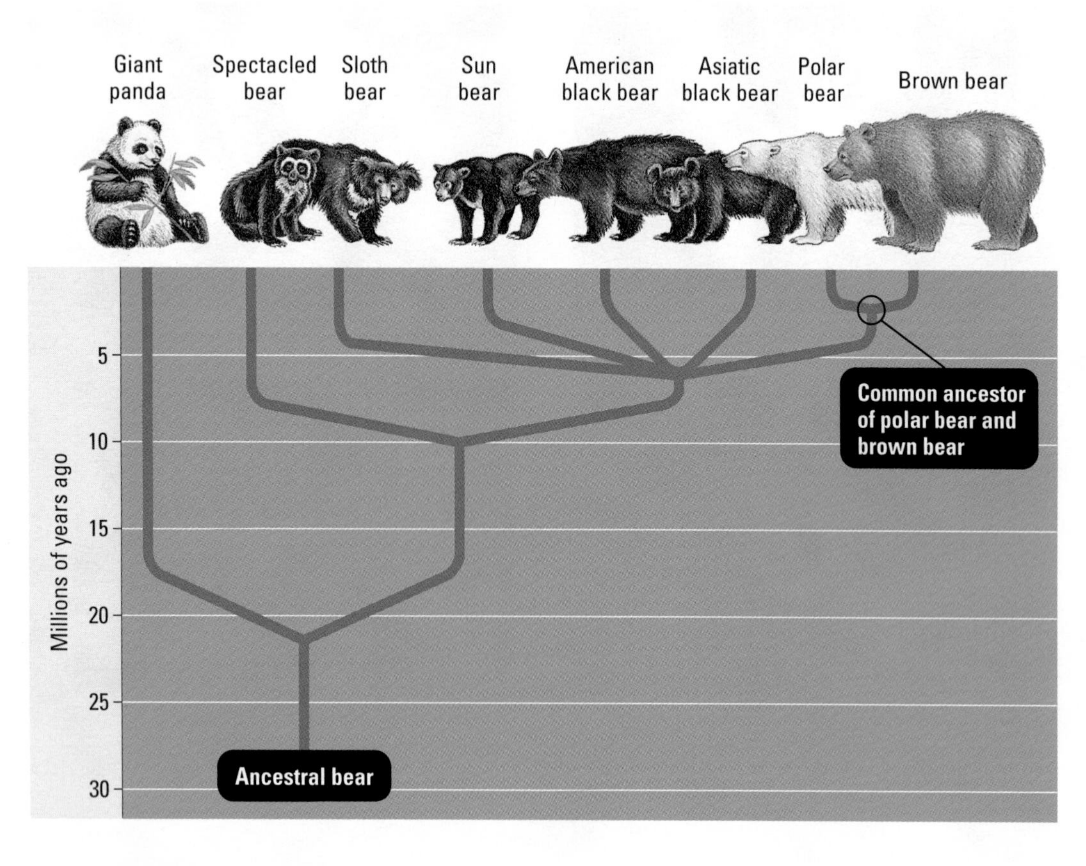

Figure 1.11 An evolutionary tree of bears. This tree is based on both the fossil record and the evaluation of relationships among living bears by comparing genetic information (DNA sequences).

is what we would expect if all mammals descended from a common ancestor, a prototypical mammal. And mammals, birds, reptiles, and all other vertebrates share a common ancestor even more ancient. Evidence of a still broader relationship can be found in similarities that are seen within all eukaryotic cells. Trace life back far enough, and there are only fossils of the primeval prokaryotes that inhabited Earth over 3 billion years ago. All of life is connected. And the basis for this kinship is evolution, the process that has transformed life on Earth from its earliest beginnings to the extensive diversity we see today. Evolution is the theme that unifies all of biology.

The Darwinian View of Life

The evolutionary view of life came into focus in 1859 when British biologist Charles Darwin published *The Origin of Species* **(Figure 1.12)**. His book developed two main points. First, Darwin marshaled the available evidence in support of the evolutionary view that species living today descended from ancestral species. (We'll examine some of the evidence for evolution in Chapter 13.) Darwin called this process "descent with modification." It is an insightful phrase, as it captures the duality of life's unity (descent) and diversity (modification). In the Darwinian view, for example, the diversity of bears is based on different modifications of a common ancestor from which all bears descended. As the second main point in *The Origin of Species*, Darwin proposed a mechanism for descent with modification. He called this process natural selection.

ON

THE ORIGIN OF SPECIES

BY MEANS OF NATURAL SELECTION,

OR THE

PRESERVATION OF FAVOURED RACES IN THE STRUGGLE FOR LIFE.

BY CHARLES DARWIN, M.A.,

LONDON:
JOHN MURRAY, ALBEMARLE STREET.
1859.

Figure 1.12 A portrait of the young Charles Darwin (1809–1882). Darwin's publication of *The Origin of Species* guaranteed his immortality as the most influential scientist in the development of modern biology. He is buried next to Isaac Newton in London's Westminster Abbey.

Natural Selection

As you'll learn in Chapter 13, Charles Darwin gathered important evidence for his theories during an around-the-world voyage. He was particularly struck by the diversity of animals on the Galápagos Islands, off the coast of Ecuador. Darwin regarded adaptation to the environment and the origin of new species as closely related processes. If some geographic barrier—an ocean separating islands, for instance—isolated two populations of a single species, the populations could diverge more and more in appearance as each

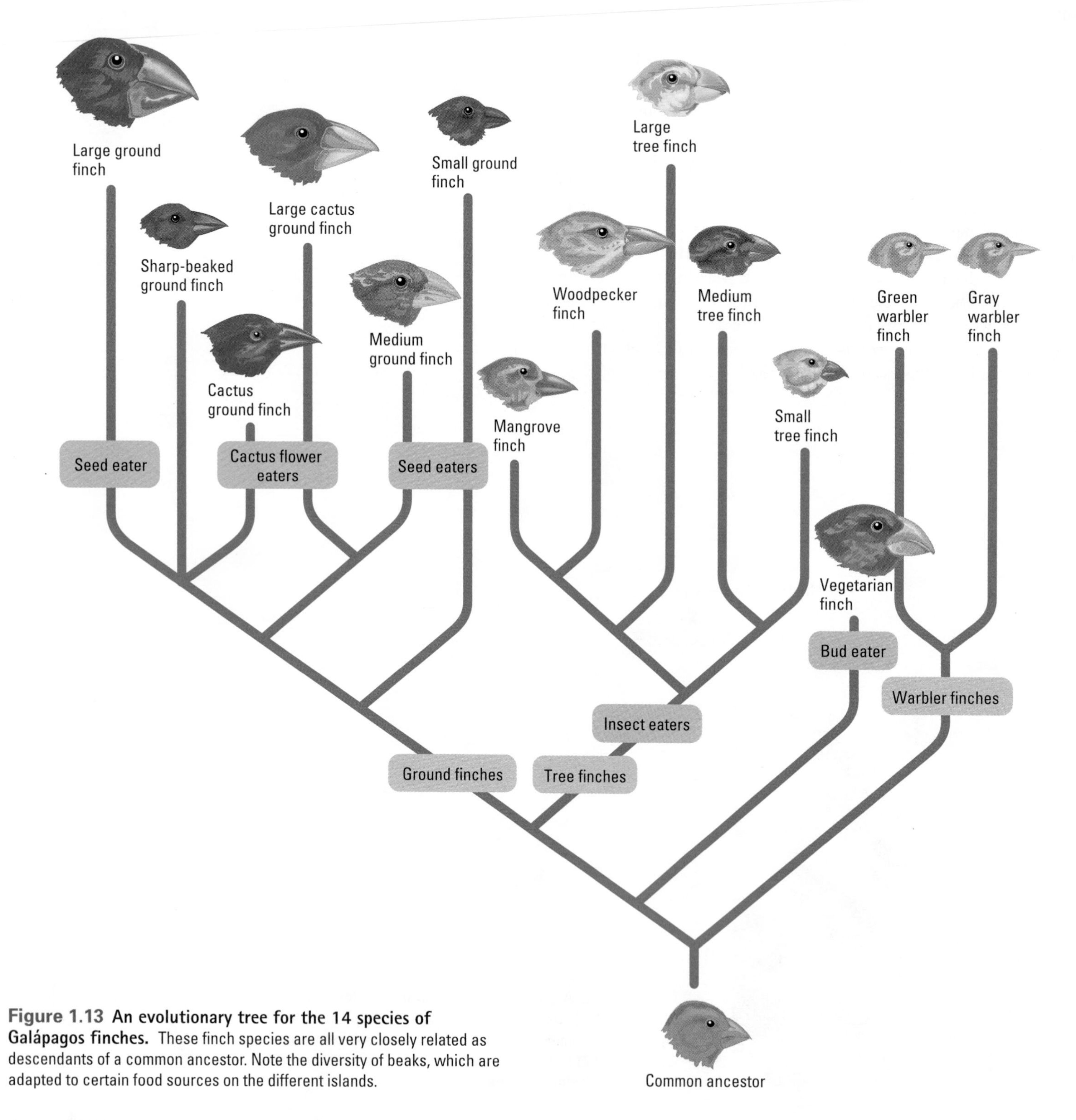

Figure 1.13 An evolutionary tree for the 14 species of Galápagos finches. These finch species are all very closely related as descendants of a common ancestor. Note the diversity of beaks, which are adapted to certain food sources on the different islands.

adapted to local environmental conditions. Over many generations, the two populations could become dissimilar enough to be designated separate species—14 species in the case of the birds called Galápagos finches, which have beak shapes and other refinements that are adapted to their particular environments **(Figure 1.13)**. Darwin realized that explaining such adaptation was the key to understanding descent with modification, or evolution. This focus on adaptation helped Darwin envision his concept of natural selection as the mechanism of evolution.

Darwin's Inescapable Conclusion Darwin synthesized the theory of natural selection from two observations that by themselves were neither profound nor original. Others had the pieces of the puzzle, but Darwin saw how they fit together. As the late evolutionary biologist Stephen Jay Gould put it, Darwin based his mechanism of natural selection on "two undeniable facts and an inescapable conclusion." Let's look at his logic:

FACT 1: **Overproduction and competition.** Any population of a species has the potential to produce far more offspring than the environment can possibly support with resources such as food and shelter. This overproduction leads to competition among the varying individuals of a population for these limited resources.

FACT 2: **Individual variation.** Individuals in a population of any species vary in many inherited traits. No two individuals in a population are exactly alike. You know this variation to be true of human populations; careful observers find variation in populations of all species.

THE INESCAPABLE CONCLUSION: **Unequal reproductive success.** In the struggle for existence, those individuals with traits best suited to the local environment will, on average, have the greatest reproductive success: They will leave the greatest number of surviving, fertile offspring. Therefore, the very traits that enhance survival and reproductive success will be disproportionately represented in succeeding generations of a population.

It is this unequal reproductive success that Darwin called **natural selection.** And the product of natural selection is adaptation, the accumulation of favorable variations in a population over time. Examples include finch beaks well equipped for available food sources and—as we'll see later in this chapter—markings that reduce predation.

As the process leading to adaptation, natural selection is also the mechanism of evolution. **Figure 1.14** presents a hypothetical example of a beetle population that colonizes a location were the soil has been blackened by a recent brush fire. ❶ Initially, the population varies extensively in the coloration of individuals, from very light gray to charcoal. ❷ For hungry birds that prey on the beetles, it is easiest to spot the beetles that are lightest in color. ❸ The selective predation favors survival and reproductive success of the darker beetles. Thus, genes for dark color are passed along to the next generation in greater frequency than genes for light color. ❹ Generation after generation, the beetle population adapts to its environment through natural selection.

Observing Artificial Selection Darwin found convincing evidence for the power of unequal reproduction in examples of artificial selection, the selective breeding of domesticated plants and animals by humans. We humans have been modifying other species for millennia by selecting breeding stock with certain traits. The plants and animals we grow for food

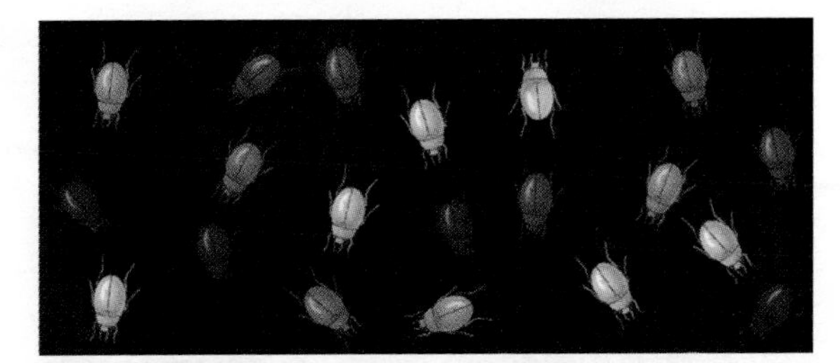

❶ **Population with varied inherited traits**

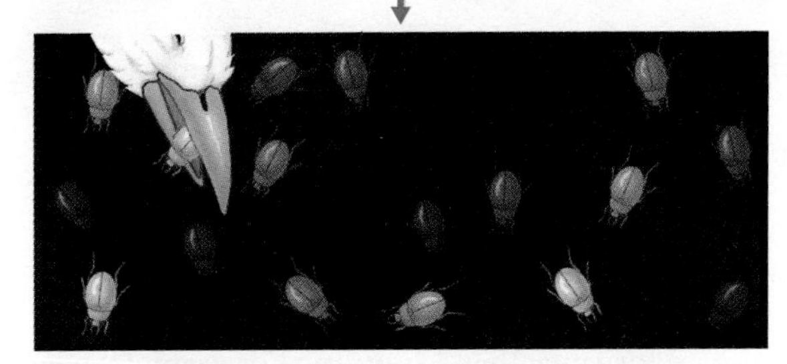

❷ **Elimination of individuals with certain traits**

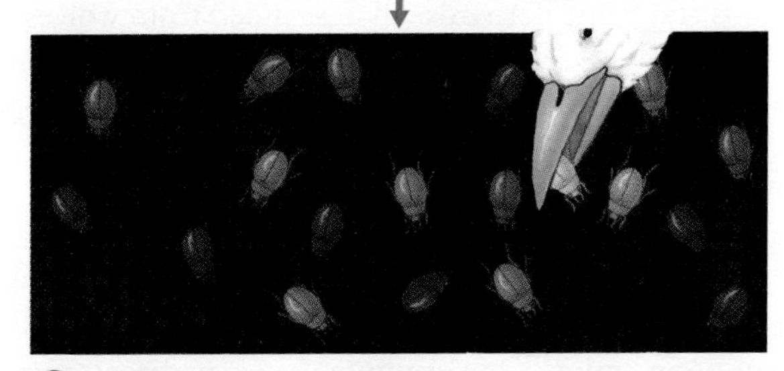

❸ **Reproduction of survivors**

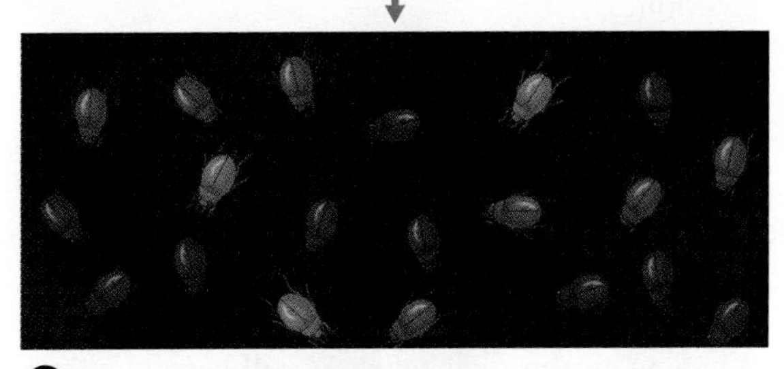

❹ **Increasing frequency of traits that enhance survival and reproductive success**

Figure 1.14 Natural selection.

Figure 1.15 Examples of artificial selection. **(a)** These vegetables (and more) all have a common ancestor in one species of wild mustard (inset). Humans have customized the plant by selecting different parts of the plant to accentuate as food. **(b)** The ancestry of all modern breeds of domesticated dogs can be traced back to wolves (inset). The tremendous variety of modern dogs reflects thousands of years of artificial selection by humans.

bear little resemblance to their wild ancestors **(Figure 1.15a)**. The power of selective breeding is especially apparent in our pets, which have been bred more for fancy than for utility. For example, people in different cultures have customized hundreds of dog breeds as different as basset hounds and Saint Bernards, all descended from wolves **(Figure 1.15b)**. Darwin could see that in artificial selection, humans were substituting for the environment in screening the heritable traits of populations.

Observing Natural Selection If artificial selection could achieve so much change so rapidly, Darwin reasoned, then natural selection should be capable of considerable adaptation of species over hundreds or thousands of generations. We now recognize many examples of natural selection in action. One example that may directly affect your life is how resistance to antibiotics by disease-causing bacteria evolves by natural selection.

Antibiotics are drugs that help cure certain infections by impairing bacteria. Antibiotics have saved millions of humans, including one of the authors of this book, who was rescued at age 13 from bacterial meningitis by injections of the antibiotic penicillin. But there's a dark side to the widespread use of antibiotics: It has driven the evolution of antibiotic-resistant populations of the very bacteria the drugs are meant to kill.

When an antibiotic is taken, it will usually kill most, but not all, of the infecting bacteria. The environment, which now contains the antibiotic, selects among the varying bacteria of a population for those mutant individuals that can survive the drug. In some cases, the mechanism of resistance depends on the ability of the bacteria to destroy the antibiotic—or even use the drug as food! While the drug kills most of the bacteria, those few bacteria that are resistant soon multiply and become the norm in the population rather than the exceptions.

The evolution of antibiotic-resistant bacteria is a huge problem in public health. For example, there are now some strains of tuberculosis-causing bacteria that are resistant to all three of the antibiotics currently used to treat the disease **(Figure 1.16)**. Unfortunate people infected with one of these resistant strains have no better chance of surviving than did tuberculosis patients a century ago.

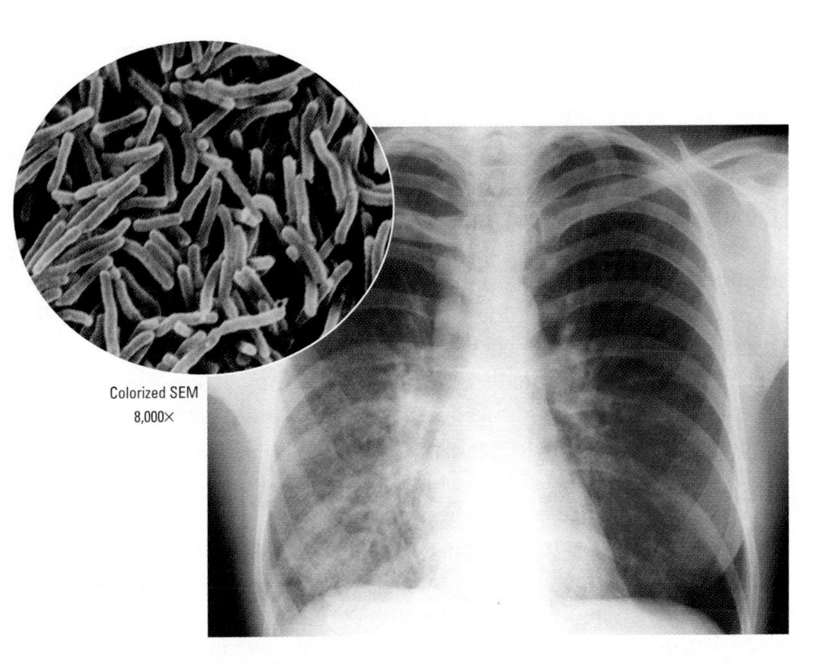

Figure 1.16 Natural selection in action. Antibiotic-resistant forms of tuberculosis-causing bacteria (inset) have made the disease a threat again in the United States. This colorized X-ray shows the lungs of a tuberculosis patient. The infection is shown in red.

The problem of antibiotic-resistant bacteria is influencing the way many physicians prescribe antibiotics. Doctors who understand that the abuse of these drugs is speeding the evolution of resistant bacteria are less likely to prescribe antibiotics needlessly—for instance, for a patient complaining of a common cold or flu, diseases caused by viruses (not bacteria), against which antibacterial drugs are powerless.

It is important to note that adaptation of antibiotic-resistant bacteria does not mean that the drugs *created* the favorable characteristics. Instead, the environment screened the heritable variations that already existed among individuals of a population and favored the ones best suited to the present conditions. In Chapter 13, you will learn more about how natural selection works and see several other examples of how natural selection affects your life.

Darwin's publication of *The Origin of Species* fueled an explosion in biological research and knowledge that continues today. Over the past 145 years, a tremendous amount of evidence has accumulated in support of Darwin's theory of evolution by natural selection, making it one of biology's best demonstrated, most comprehensive, and longest-lasting theories. In every chapter of *Essential Biology*, we will demonstrate connections to evolution—the unifying theme of biology.

The Process of Science

Recall the first definition you read in this chapter: Biology is the scientific study of life. Now that we have explored the question, What is life? we can turn our attention to the next obvious question: What is science? And how do we tell the difference between science and other ways of trying to make sense of nature?

The word *science* is derived from a Latin verb meaning "to know." **Science** is a way of knowing, one that is based on inquiry. It developed from people's curiosity about themselves and the world around them. This basic human drive to understand is manifest in two main scientific approaches: discovery science and hypothesis-driven science. Most scientists practice a combination of these two forms of inquiry.

Discovery Science

Science seeks natural causes for natural phenomena. This limits the scope of science to the study of structures and processes that we can observe and measure, either directly or indirectly with the help of tools, such as microscopes, that extend our senses. This dependence on observations that other people can confirm demystifies nature and distinguishes science from belief in the supernatural. Science can neither prove nor disprove that angels, ghosts, deities, or spirits, whether benevolent or evil, cause storms, rainbows, illnesses, or cures, for such explanations are outside the bounds of science.

Verifiable observations and measurements are the data of **discovery science (Figure 1.17)**. In our quest to describe nature accurately, we discover its structure. In biology, discovery science enables us to describe life at its many levels, from ecosystems down to cells and molecules. Darwin's careful description of the diverse plants and animals he collected in South America is an example of discovery science. A more recent example is the sequencing of the human genome, a detailed dissection and description of the genetic material.

Discovery science can lead to important conclusions based on a type of logic called inductive reasoning. An inductive conclusion is a generalization that summarizes a large number of observations. "All organisms are made of cells" is an example. That induction was based on two centuries of biologists discovering cells in every biological specimen they observed with microscopes. The careful observations of discovery science and the inductive conclusions they sometimes produce are fundamental to our understanding of nature.

Figure 1.17 Careful observation and measurement: the raw data for discovery science. Dr. Jane Goodall spent decades recording her observations of chimpanzee behavior during field research in the jungles of Gambia.

Hypothesis-Driven Science

The observations of discovery science engage inquiring minds to ask questions and seek explanations. Ideally, such investigation consists of what is called the scientific method. As a formal process of inquiry, the **scientific method** consists of a series of steps **(Figure 1.18)**. These steps guide scientific investigations, but working scientists typically do not follow them rigidly; different scientists proceed through the scientific method in different ways.

Most modern scientific investigations can be described as **hypothesis-driven science.** A **hypothesis** is a tentative answer to some question—an explanation on trial. We all use hypotheses in solving everyday problems. Let's say, for example, that your flashlight fails during a campout. That's an observation. The question is obvious: Why doesn't the flashlight work? A reasonable hypothesis based on past experience is that the batteries in the flashlight are dead.

Once a hypothesis is formed, an investigator can use deductive logic to test it. Deduction contrasts with induction, which, remember, is reasoning from a set of specific observations to reach a general conclusion. In deduction, the reasoning flows in the opposite direction, from the general to the specific. From general premises, we extrapolate to the specific results we should expect if the premises are true. For example, if all organisms are made of cells (premise 1), and humans are organisms (premise 2), then humans are composed of cells (deductive prediction about a specific case).

In the process of science, the deduction usually takes the form of predictions about what experimental results or observations we should expect if a particular hypothesis (premise) is correct. We then test the hypothesis by performing an experiment to see whether or not the

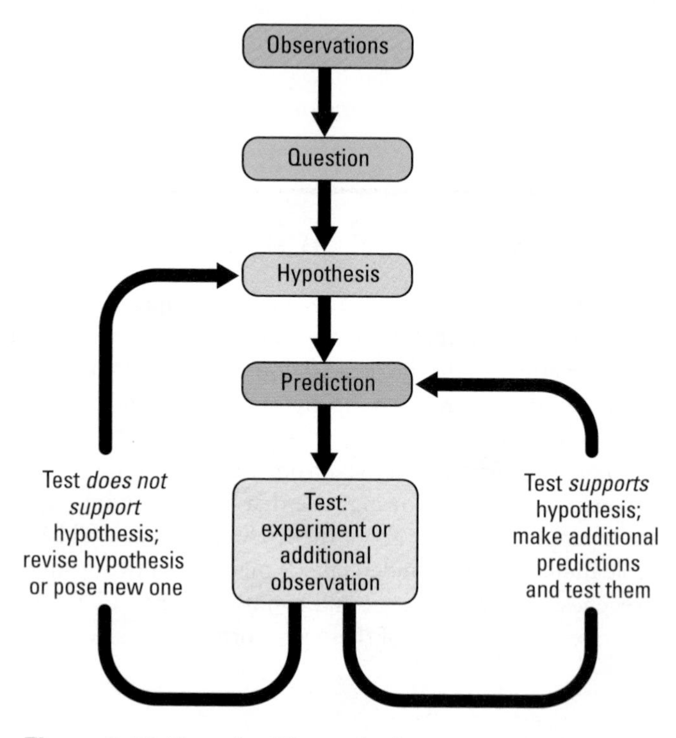

Figure 1.18 The scientific method.

results are as predicted. This deductive testing takes the form of "If . . . then" logic:

Observation: My flashlight doesn't work.

Question: What's wrong with my flashlight?

Hypothesis: The flashlight's batteries are dead.

Prediction: If I replace the batteries, the flashlight will work.

Experiment: I replace the batteries with new ones.

Predicted result: The flashlight should work.

Let's say the flashlight still doesn't work. We can test an alternative hypothesis if new flashlight bulbs are available **(Figure 1.19)**. We could also blame the dead flashlight on campground ghosts playing tricks, but that hypothesis is untestable and therefore outside the realm of science.

THE PROCESS OF SCIENCE

Can Colors Protect a Snake? One way to learn more about how hypothesis-based science works is to examine a **case study**, an in-depth examination of an actual investigation. The rest of this section is a case study about what harmless snakes might gain by imitating poisonous ones. Later chapters will include other case studies in the process of science, which illustrate how hypothesis-driven science has been used in experiments both modern and classic. To emphasize how the scientific method has been employed in each case, the key steps will always be highlighted in blue.

The case study we will examine here was inspired by some general observations. Many poisonous animals are brightly colored, with distinctive patterns in some species. This appearance is called warning coloration because it marks the animal as dangerous to potential predators. But there are also mimics. These imposters look like a poisonous species but are really harmless to predators. The question that follows from these observations is: What is the function of such mimicry? It is a reasonable hypothesis that such deception is an evolutionary adaptation that reduces the harmless animal's risk of being eaten.

In 2001, a team of biologists designed a simple set of experiments to test the hypothesis that mimics benefit because predators confuse them with the actual harmful species. Researchers David and Karin Pfennig, along with one of their college students, tested this hypothesis by studying mimicry in snakes that live in North and South Carolina. They began

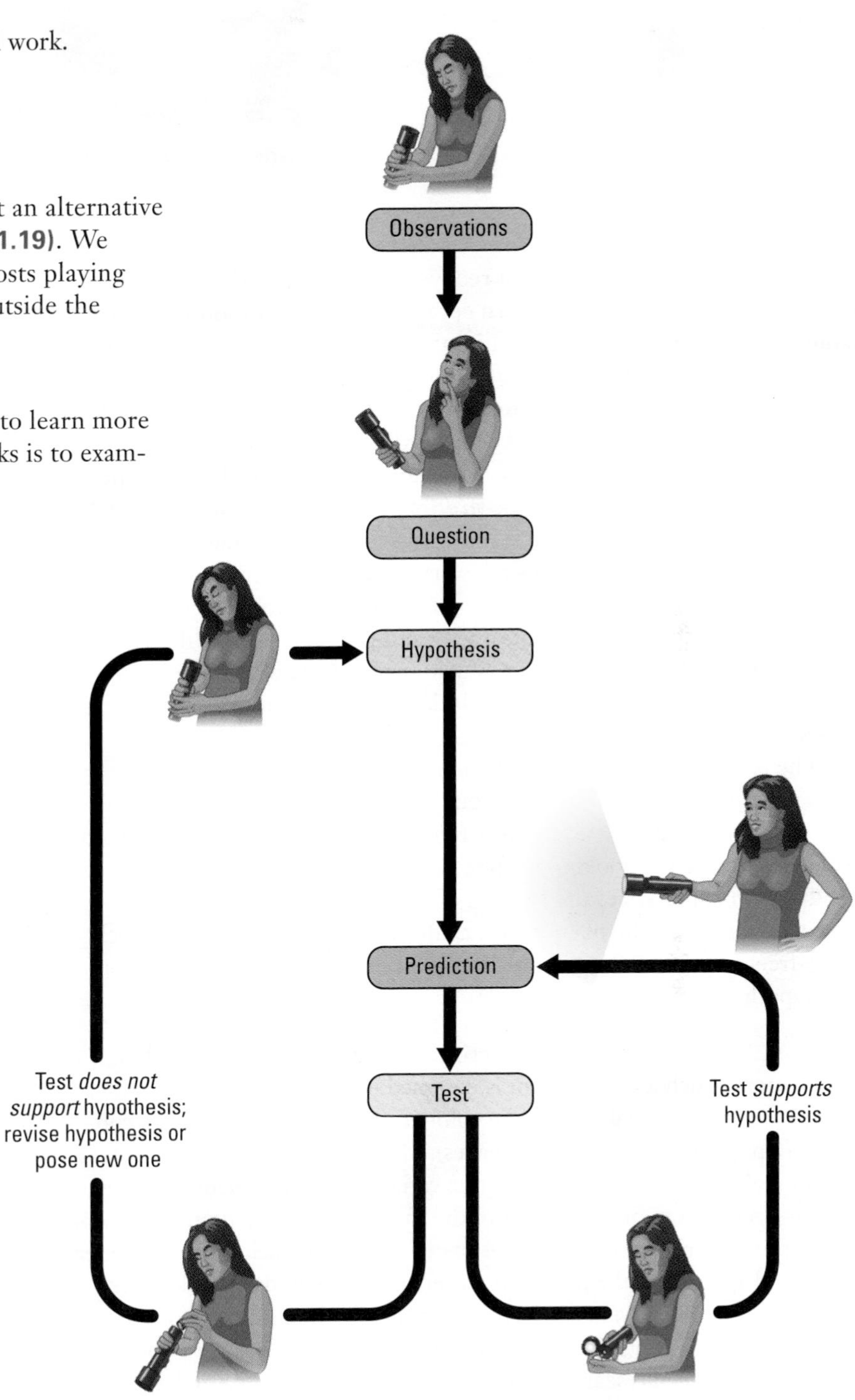

Figure 1.19 Applying the scientific method to a campground problem.

(a) Eastern coral snake (poisonous)

(b) Scarlet king snake (nonpoisonous)

Figure 1.20 Mimicry in snakes. **(a)** The dangerous eastern coral snake. **(b)** The scarlet king snake, a nonpoisonous species that mimics the eastern coral snake.

with the following **observations**: A poisonous snake called the eastern coral snake is marked by rings of red, yellow, and black **(Figure 1.20a)**. Predators rarely attack the eastern coral snake. A nonpoisonous snake named the scarlet king snake mimics the ringed coloration of the coral snake **(Figure 1.20b)**.

These observations raise a **question**: What is the function of the king snake's mimicry of the coral snake? According to the mimicry **hypothesis**, the king snake's resemblance to the coral snake repels predators. The hypothesis leads to the **prediction** that predators will attack snakes with bright rings of red, yellow, and black less frequently than they will attack snakes lacking such warning coloration. To test this prediction, the researchers performed a clever **experiment**. They made hundreds of artificial snakes out of wire and plasticine, a claylike substance. The artificial snakes were of two types: those with plain brown coloration and those with the red, black, and yellow ring pattern of scarlet king snakes.

The researchers placed equal numbers of the two types of artificial snakes in various field sites throughout North and South Carolina. After four weeks, the team retrieved the artificial snakes and counted how many had been attacked by looking for bite or claw marks. The bar graph in **Figure 1.21** summarizes the **results** of the artificial snake experiment. The data show that the plain brown artificial snakes were attacked much more frequently than the ones with colored rings. The results thus support the hypothesis. ■

The case study we have just described is an example of a **controlled experiment**. Such an experiment is designed to compare an experimental group (the artificial snakes with the colored rings, in this case) with a control group (plain brown artificial snakes). Ideally, a control group and an experimental group differ only in the one variable the experiment is designed to test—in our example, the effect of the snakes' coloration on the behavior of their predators. During the experiment, conditions such as light, temperature, and appetite of the predators varied, but both kinds of snakes were subject to the same variations. Thus, the control group canceled out the effects of all variables other than the one being tested. The use of a controlled experiment enabled researchers to compare the two groups and draw conclusions about the effect of the colored rings on predator behavior.

The snake study reinforces the important point that scientists must test their hypotheses. Without such testing, ideas about nature, such as

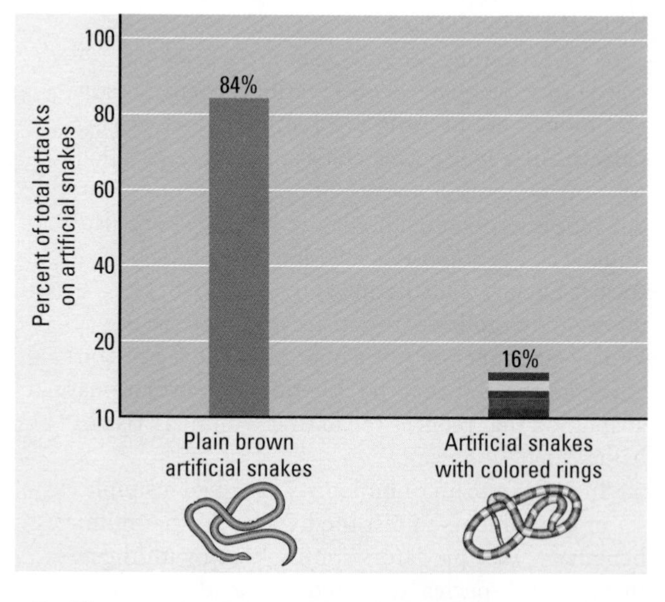

Figure 1.21 Effect of artificial snake coloration on attacks by predators. Results of mimicry experiments using artificial snakes show a dramatic difference in the frequency of attacks on plain brown snakes versus snakes with colored rings.

speculations on the function of mimicry, are "just so" stories. And explaining that something is true just because "it's so" is not very convincing.

The Culture of Science

It is not unusual for several scientists to ask the same questions. Such convergence contributes to the progressive and self-correcting qualities of science. Scientists build on what has been learned from earlier research, and they pay close attention to contemporary scientists working on the same problem. Scientists share information through publications, seminars, meetings, and personal communication. The Internet has added a new medium for this exchange of ideas and data.

Both cooperation and competition characterize the scientific culture **(Figure 1.22)**. Scientists working in the same research field subject one another's work to careful scrutiny. It is common for scientists to check the conclusions of others by attempting to repeat experiments. This obsession with evidence and confirmation helps characterize the scientific style of inquiry. Scientists are generally skeptics.

We have seen that science has two key features that distinguish it from other styles of inquiry: (1) a dependence on observations and measurements that others can verify and (2) the requirement that ideas (hypotheses) are testable by experiments that others can repeat.

Figure 1.22 Science as a social process. Here, New York University plant biologist Gloria Coruzzi (left) mentors one of her students in the methods of molecular biology.

Science, Technology, and Society

Science and technology are interdependent. New technologies, such as more powerful microscopes and computers, advance science. And scientific discoveries can lead to new technologies. In most cases, technology applies scientific discoveries to the development of new goods and services. For example, just over 50 years ago two scientists, James Watson and Francis Crick, discovered the structure of DNA through the process of science. Their discovery eventually led to a variety of DNA technologies, including the genetic engineering of microorganisms to mass-produce human insulin and the use of DNA fingerprinting for investigating crimes **(Figure 1.23)**. Perhaps Watson and Crick envisioned that their discovery would someday inform new technologies, but that probably did not motivate their research, nor could they have predicted exactly what the applications would be. The direction technology takes depends less on the curiosity that drives basic science than it does on the current needs of humans and the changing climate of culture.

Technology has improved our standard of living in many ways, but it is a double-edged sword. Technology that keeps people healthier has enabled the population to grow more than tenfold in the past three centuries, to double to 6 billion in just the past 40 years. The environmental consequences are sometimes devastating. Acid rain, deforestation, global warming, nuclear accidents, toxic wastes, and extinction of species are just a few of the repercussions of more and more people wielding more and more technology. Science can help us identify such problems and provide insight about what course of action may prevent further damage. But solutions to these problems have as much to do with politics, economics, culture, and the values of societies as with science and technology. Now that science and technology have become such powerful functions of society, each of us has the responsibility to become a "citizen scientist" by developing a reasonable amount of scientific and technological literacy.

Figure 1.23 DNA technology and the law. Forensic technicians can use traces of DNA extracted from a blood sample or other body tissue collected at a crime scene to produce a molecular "fingerprint." The stained bands visible in this photograph represent fragments of DNA, and the pattern of bands varies from person to person. The legal applications of DNA technology have become widespread in the past decade. It is just one example of biology's prominent role in society today.

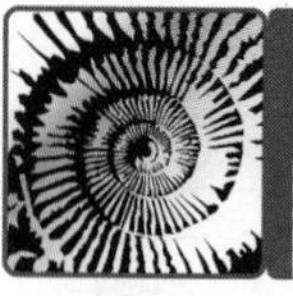

EVOLUTION CONNECTION

Theories in Science

Many people associate facts with science, but accumulating facts is not what science is primarily about. A telephone book is an impressive catalog of factual information, but it has little to do with science. It is true that facts, in the form of verifiable observations and repeatable experimental results, are the prerequisites of science. What really advances science, however, is some new theory that ties together a number of observations that previously seemed unrelated. The cornerstones of science are the explanations that apply to the greatest variety of phenomena. People like Newton, Darwin, and Einstein stand out in the history of science not because they discovered a great many facts, but because their theories had such broad explanatory power.

What is a scientific theory, and how is it different from a hypothesis? A **theory** is much broader in scope than a hypothesis. This is a hypothesis: "Mimicking coral snakes is an adaptation that helps protect some nonpoisonous snakes from predators." But this is a theory: "Adaptations evolve by natural selection."

Because theories are so comprehensive, they only become widely accepted in science if they are supported by an accumulation of extensive and varied evidence. The use of the term *theory* in science for a comprehensive explanation supported by abundant evidence contrasts with our everyday usage, which equates theories more with speculations or hypotheses. Natural selection qualifies as a scientific theory because of its broad application and because it has been validated by a continuum of observations and experiments.

Scientific theories are not the only way of "knowing nature," of course. A comparative religion course would be a good place to learn about the diverse legends that tell of a supernatural creation of Earth and its life. Science and religion are two very different ways of trying to make sense of nature. Art is still another way. A broad education should include exposure to these different ways of viewing the world. Each of us synthesizes our worldview by integrating our life experiences and multidisciplinary education. As a science textbook and part of that broad education, *Essential Biology* showcases life in the scientific context of evolution, the one theme that continues to hold all of biology together, no matter how big and complex the subject becomes. ■

UNIT 1

Cells

2 Essential Chemistry for Biology

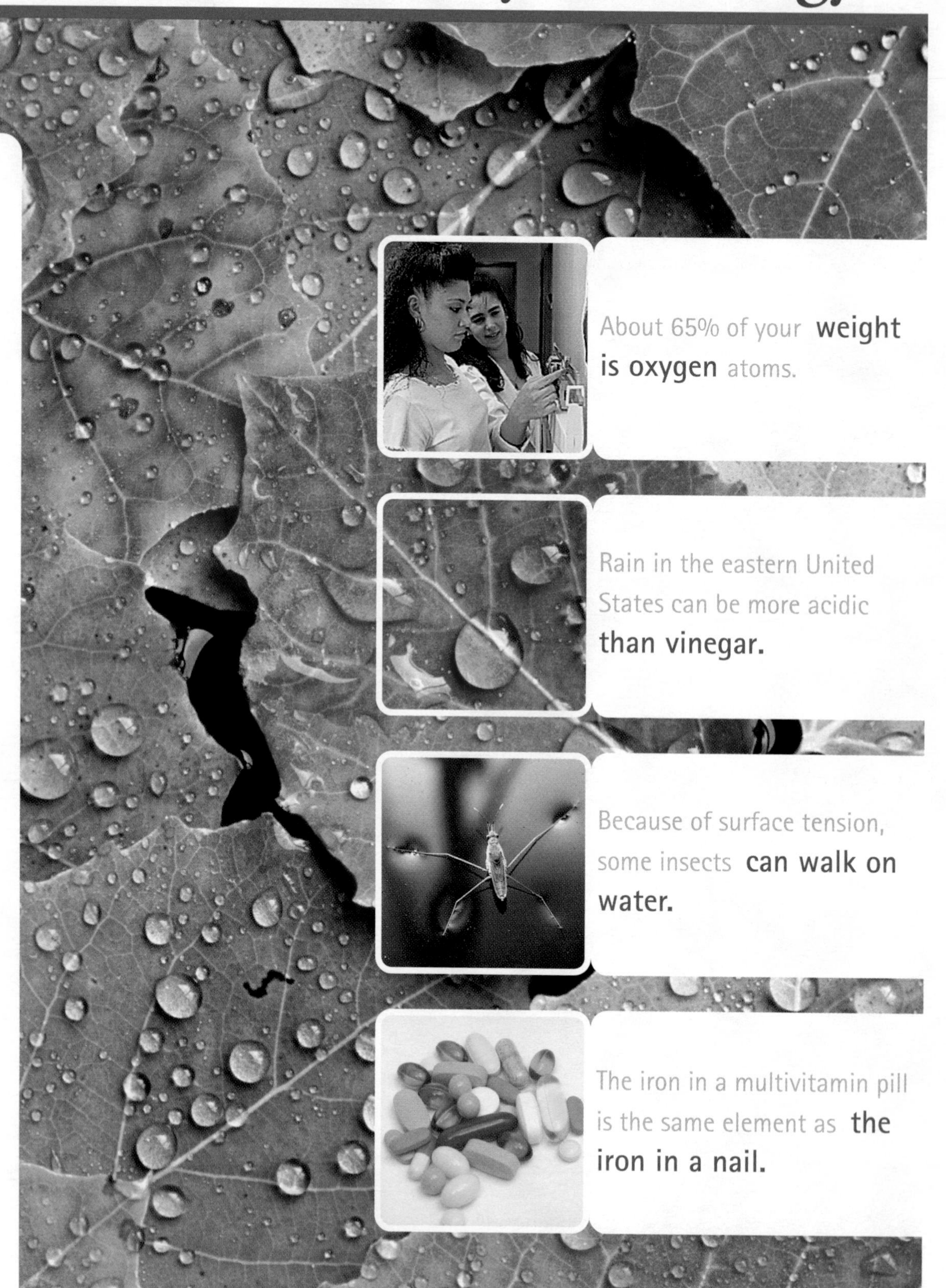

About 65% of your weight is oxygen atoms.
Rain in the eastern United States can be more acidic than vinegar.
Because of surface tension, some insects can walk on water.
The iron in a multivitamin pill is the same element as the iron in a nail.

BIOLOGY AND SOCIETY

AIDS: A Matter of Chemistry

In 1981, the Centers for Disease Control (CDC) began tracking high numbers of otherwise healthy young men who were coming down with extremely rare diseases. What really caught the attention of the CDC was the age and health status of the men involved. In the past, the rare infections the researchers were tracking had only affected people whose immune systems had been severely compromised, such as the elderly or people on chemotherapy. But none of the newly affected patients fit this profile, leaving the CDC baffled.

When doctors examined these patients, they discovered that every man affected had a failing immune system that left him susceptible to infections that would normally be defeated in a healthy person. Researchers soon discovered the cellular basis of the new illness: The patients' helper T cells—one of the key players in the body's immune system—were being wiped out. In 1982, the disorder was named acquired immunodeficiency syndrome (AIDS). In 1984, scientists determined that AIDS is caused by the human immunodeficiency virus (HIV).

But why does HIV target only helper T cells? It turns out that protein molecules on the outside of HIV bind to certain molecules found exclusively on the outside of the helper T cells **(Figure 2.1)**. Only when this "molecular handshake" is completed can HIV enter and destroy helper T cells. This chemical interaction is the key event in the development of AIDS. Understanding how these molecules interact may lead to the creation of drugs that interfere with this interaction, potentially stopping the disease.

The discovery of the cause of AIDS illustrates an important point: Biology includes the study of life at many levels. A health epidemic affecting an entire population (AIDS) was studied by working with individual organisms (patients) and then isolated to one organ system (the immune system). The problem was then traced to a single type of cell (helper T cells) and eventually to individual molecules on those cells. The story of AIDS thus shows that many important biological questions can be reduced to questions of chemistry.

Living organisms are, at their most basic level, chemical systems, making a knowledge of chemistry essential to understanding biology. In this chapter, you will learn some essential chemistry that you'll be able to apply throughout your study of life. ■

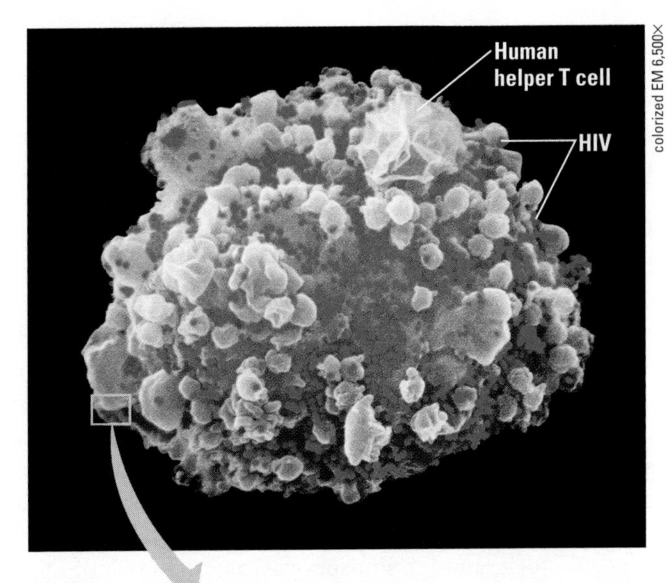

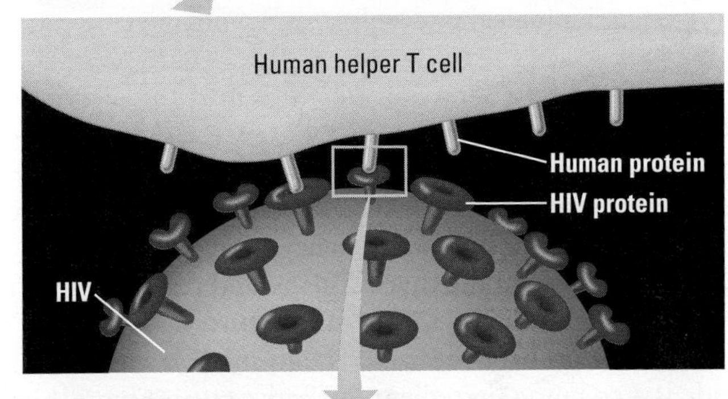

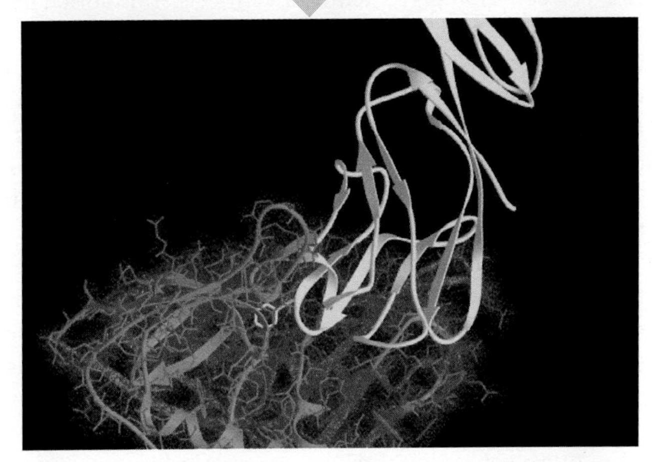

Figure 2.1 The chemical interaction at the heart of AIDS. The electron micrograph at the top shows a human helper T cell under attack by HIV (small red spheres). The middle section shows the interaction between a protein molecule on the outside of HIV and a protein molecule on the outside of the helper T cell. The zoom-up at the bottom is a computer model of the "molecular handshake," the interaction between chemicals that underlies HIV's ability to devastate the human immune system.

Some Basic Chemistry

Take any biological system apart and you eventually end up at the chemical level. In fact, you could think of your body as a giant bag of chemical reactions. Beginning at this basic biological level, let's explore the chemistry of life.

Matter: Elements and Compounds

Humans and other organisms and everything around them are all made of matter, the physical "stuff" of the universe. Defined more formally, **matter** is anything that occupies space and has mass. Matter is found on Earth in three physical states: solid, liquid, and gas.

Matter is composed of chemical **elements,** substances that cannot be broken down into other substances. There are 92 naturally occurring elements on Earth; examples are carbon, oxygen, gold, and fluorine. Each element has a symbol, the first letter or two of its English, Latin, or German name. For instance, the symbol for gold, Au, is from the Latin word *aurum*. All the elements are listed in the periodic table of the elements, a familiar fixture in any chemistry or biology lab (**Figure 2.2**; see Appendix B for a full version).

Of the 92 naturally occurring elements, 25 are essential to life. Four of these elements—oxygen (O), carbon (C), hydrogen (H), and nitrogen (N)—make up about 96% of the weight of the human body, as well as most other living matter (**Figure 2.3**). Much of the remaining 4% is accounted for by 7 elements, most of which are probably familiar to you, such as calcium (Ca) and phosphorus (P). Calcium, important for building strong bones and teeth, is found abundantly in milk and dairy products as well as sardines and green, leafy vegetables (collards, kale, and broccoli, for example). Phosphorus, a component of DNA and other important biological molecules, can be obtained by eating eggs, beans, and nuts.

Less than 0.01% of your weight is made up of 14 trace elements, listed in the legend for Figure 2.3. **Trace elements** are required in only very small amounts, but you cannot live without them. The average human, for example, needs only a tiny bit of iodine, about 0.15 milligram (mg) each day. An iodine deficiency in the diet, however, prevents normal functioning of the thyroid gland and results in goiter, an abnormal enlargement of the thyroid gland (**Figure 2.4**). Another trace element that you've probably heard of is fluorine (F), which in a form called fluoride is a common ingredient in Earth's crust. Fluoride helps to prevent dental cavities by affecting the metabolism of oral bacteria and by promoting the replacement of lost minerals on the surface of the teeth. For more than 50 years, the American Dental Association has supported fluoridation of community drinking water at a concentration that can reduce tooth decay. The widespread addition of fluoride to drinking water and toothpaste has been a major factor in the decline in tooth decay in the United States and other industrialized countries.

Elements can combine to form **compounds,** substances that contain two or more elements in a fixed ratio. In everyday life, compounds are much more common than pure elements. Familiar examples are relatively simple compounds such as table salt and water. Table salt is sodium chloride, NaCl, consisting of equal parts of the elements sodium (Na) and chlorine (Cl). A molecule of water, H_2O, has two atoms of hydrogen and one atom of oxygen. Most of the compounds in living organisms contain several different elements. DNA, for example, contains carbon, nitrogen, oxygen, hydrogen, and phosphorus.

Atoms

Each element consists of one kind of atom, which is different from the atoms of other elements. An **atom,** named from a Greek word meaning "indivisible," is the smallest unit of matter that still retains the properties of an element. In other words, the smallest amount of the element carbon is one carbon atom. Just how small is this "piece" of carbon? It would take about a million carbon atoms to stretch across the period at the end of this sentence.

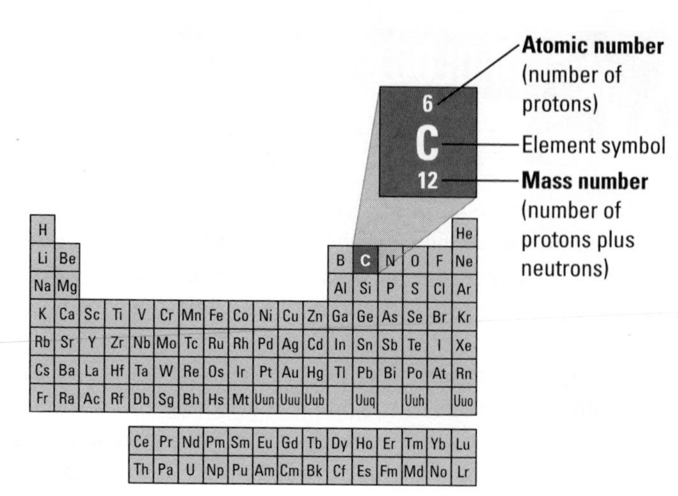

Figure 2.2 Abbreviated periodic table of the elements. The periodic table lists all of the chemical elements, both natural and human-made. In the full periodic table (see Appendix B), each entry contains the element symbol in the center, with the atomic number above and the mass number below (both are discussed later in this chapter). The element highlighted here is carbon (C).

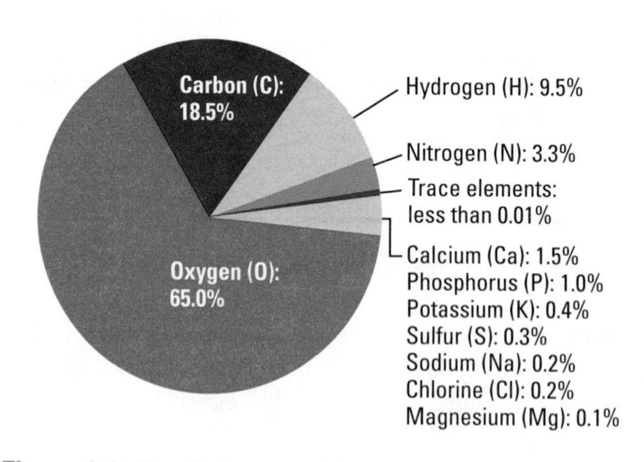

Figure 2.3 Chemical composition of the human body by weight. The four elements oxygen, carbon, hydrogen, and nitrogen make up 96.3% of human body weight. The trace elements, which make up less than 0.01%, are boron (B), chromium (Cr), cobalt (Co), copper (Cu), fluorine (F), iodine (I), iron (Fe), manganese (Mn), molybdenum (Mo), selenium (Se), silicon (Si), tin (Sn), vanadium (V), and zinc (Zn).

The Structure of Atoms Atoms are composed of subatomic particles, of which the three most important are protons, electrons, and neutrons. A **proton** is a subatomic particle with a single unit of positive electrical charge (+). An **electron** is a subatomic particle with a single unit of negative electrical charge (−). A **neutron** is electrically neutral (has no electrical charge).

Let's look at the structure of an atom of the element helium (He), the "lighter-than-air" gas used to make party balloons rise **(Figure 2.5)**. Each atom of helium has 2 neutrons (●) and 2 protons (⊕) tightly packed into the **nucleus,** the atom's central core. Two electrons (⊖) orbit the nucleus at nearly the speed of light. The attraction between the negatively charged electrons and the positively charged protons keeps the electrons in orbit. When an atom has an equal number of protons and electrons (as helium does), its net electrical charge is zero and so the atom is neutral.

Elements differ in the number of subatomic particles in their atoms. The number of protons in an atom, called the **atomic number,** determines which element it is. An atom's **mass number** is the sum of the number of protons and neutrons in its nucleus. Both the atomic number and the mass number can be read from the periodic table (see Figure 2.2 and Appendix B). For helium, the atomic number is 2 and the mass number is 4 (2 protons + 2 neutrons). **Mass** is a measure of the amount of material in an object. A proton and a neutron have nearly identical mass, called 1 atomic mass unit (amu) for convenience. An electron has very little mass—only about 1/2,000 the mass of a proton.

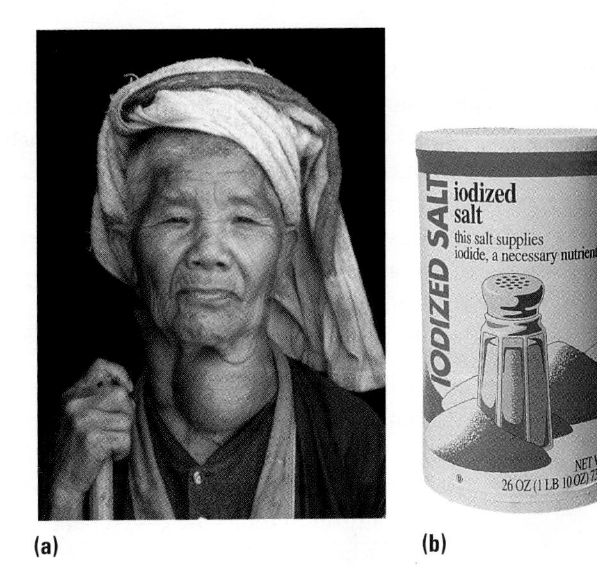

Figure 2.4 Why is salt "iodized"? (a) Goiter, an enlargement of the thyroid gland, shown here in a Burmese woman, can occur when a person's diet does not include enough iodine, a trace element. **(b)** Goiter has been nearly eliminated in many nations by adding iodine to table salt (iodized salt). Unfortunately, goiter still affects many thousands of people in developing countries.

Isotopes Some elements can exist in different forms called isotopes. The different **isotopes** of an element have the same numbers of protons and electrons but different numbers of neutrons; in other words, isotopes are forms of an element that differ in mass. **Table 2.1** shows the numbers of subatomic particles in the three isotopes of carbon. Carbon-12 (named for its mass number 12), with 6 neutrons and 6 protons, makes up about 99% of all naturally occurring carbon. Most of the other 1% consists of carbon-13, with 7 neutrons and 6 protons. A third isotope, carbon-14, with 8 neutrons and 6 protons, occurs in minute quantities. Notice that all three isotopes have 6 protons—otherwise, they would not be carbon. Both carbon-12 and carbon-13 are stable isotopes, meaning their nuclei remain intact more or less forever. The isotope carbon-14, on the other hand, is unstable, or radioactive. A **radioactive isotope** is one in which the nucleus decays, giving off particles and energy.

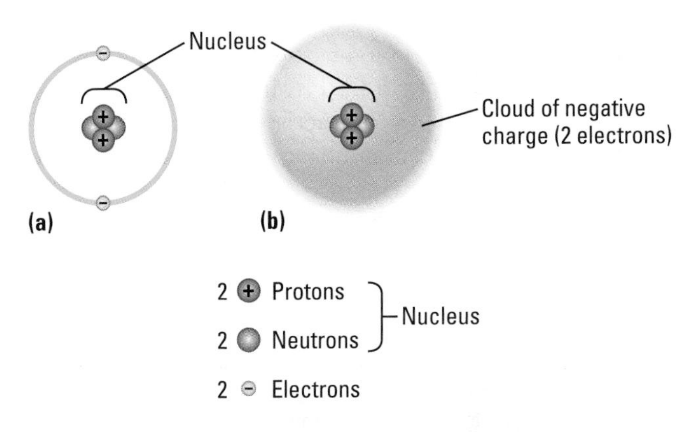

Figure 2.5 Two simplified models of a helium atom. **(a)** This model shows the subatomic particles in an atom of helium. **(b)** This model, slightly more realistic, shows the electrons as a spherical cloud of negative charge surrounding the nucleus. Neither model is drawn to scale. In real atoms, the electron cloud is much bigger compared to the nucleus. If the electron cloud were the size of a football stadium, the nucleus would only be the size of a fly on the field.

Radioactive isotopes have many uses in biological research and medicine. Living cells cannot distinguish radioactive isotopes from nonradioactive isotopes of the same element. Consequently, organisms take up and use compounds containing radioactive isotopes in the usual way. Once taken up, the location and concentration of radioactive isotopes can be detected because of the radiation they emit. This makes radioactive isotopes useful as tracers—biological spies, in effect—for monitoring the fate of atoms in living organisms.

Although radioactive isotopes have many beneficial uses, uncontrolled exposure to them can harm living organisms by damaging cellular molecules, especially DNA. In 1986, the explosion of a nuclear reactor at Chernobyl, Ukraine, released large amounts of radioactive isotopes, killing 30 people within a few weeks and exposing thousands to an increased risk of developing cancer. In fact, the incidence of thyroid cancer among Ukrainian

Table 2.1 Isotopes of Carbon

	Carbon-12	Carbon-13	Carbon-14
Protons	6 } mass number 12	6 } mass number 13	6 } mass number 14
Neutrons	6	7	8
Electrons	6	6	6

children increased tenfold in the ten years following the accident. A 2005 United Nations study predicted that 4,000 people will eventually die as a result of radiation exposure from the Chernobyl disaster (although some experts think the number of casualties might actually be much higher).

Natural sources of radiation can also pose a threat. Radon, a radioactive gas, may cause lung cancer. Radon can contaminate buildings in regions where underlying rocks naturally contain the radioactive element uranium. Homeowners can buy a radon detector or hire a company to test their home to ensure that radon levels are safe.

Can Alzheimer's Disease Be Detected Early?

Alzheimer's disease (AD) is a devastating illness that gradually destroys a person's memory and ability to think. As the disease progresses, the brain becomes riddled with deposits of a protein molecule called beta-amyloid, the accumulation of which is believed to cause brain damage. One of the major hurdles in the fight against AD is the lack of a diagnostic test; doctors cannot be entirely sure that a person has AD until autopsy. Development of an early test for Alzheimer's remains an important but so far elusive goal.

Progress toward a diagnostic test began with the **observation** by researchers at the University of Pittsburgh and Uppsala University in Sweden that a protein molecule they dubbed PIB binds to beta-amyloid. The PIB molecule includes an atom of carbon-11, a rare radioactive isotope of carbon. When PIB is injected into the brain, the radiation emitted by the carbon-11 isotope can be detected and visualized on a computer screen using a PET scanner (a medical diagnostic device that detects and displays radioactive emissions). The researchers then asked the **question**, Can radioactive isotopes be used to detect Alzheimer's disease in living patients? Their **hypothesis** was that PIB could be used this way. They made the **prediction** that following a PIB injection, PET scans would reveal higher concentrations of PIB in the brains of Alzheimer's patients than in the brains of nonaffected individuals.

The **experiment** that researchers used to test this hypothesis involved 16 patients diagnosed with mild Alzheimer's disease and 9 healthy control subjects. The PET scans of the AD patients did indeed show higher concentrations of PIB in regions of the brain known to be affected by Alzheimer's (**Figure 2.6**). These **results**, published in 2004, suggest that the radioactive PIB molecule may hold the key to early detection of this crippling disease. ■

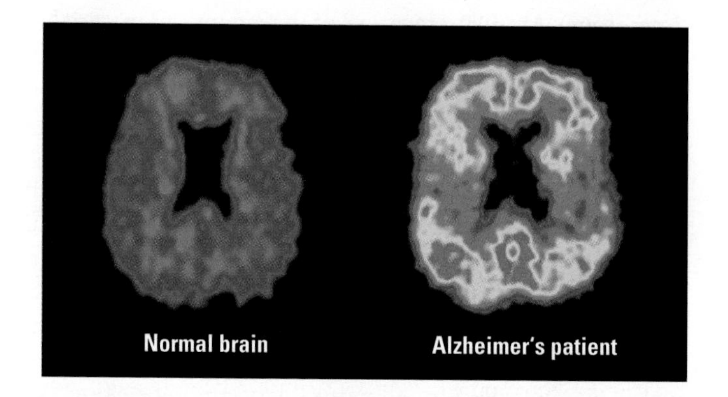

Figure 2.6 A diagnosis for Alzheimer's disease? These images are PET scans of the brains of a healthy person (left) and a person in the early stages of Alzheimer's disease (right). Both patients were injected with PIB, a protein molecule that binds to certain other molecules associated with AD. PIB includes a radioactive isotope that the PET scanner can detect. Notice that the brain of the AD patient has high levels of PIB (red and yellow areas), while the unaffected patient's brain has lower levels (blue).

Electron Arrangement and the Chemical Properties of Atoms Of the three subatomic particles we've discussed, electrons are the ones that primarily determine how an atom behaves when it encounters other atoms. Electrons vary in the amount of energy they possess. The farther an electron is from the nucleus, the greater its energy. Electrons do not orbit an atom at just any energy level, but only at specific levels called electron shells. Depending on the number of electrons, atoms may have one, two, or more electron shells, with electrons in the outermost shell having the highest energy. Each shell can accommodate up to a specific number of electrons. The innermost shell is full with only 2 electrons, while the second and third shells can each hold up to 8 electrons.

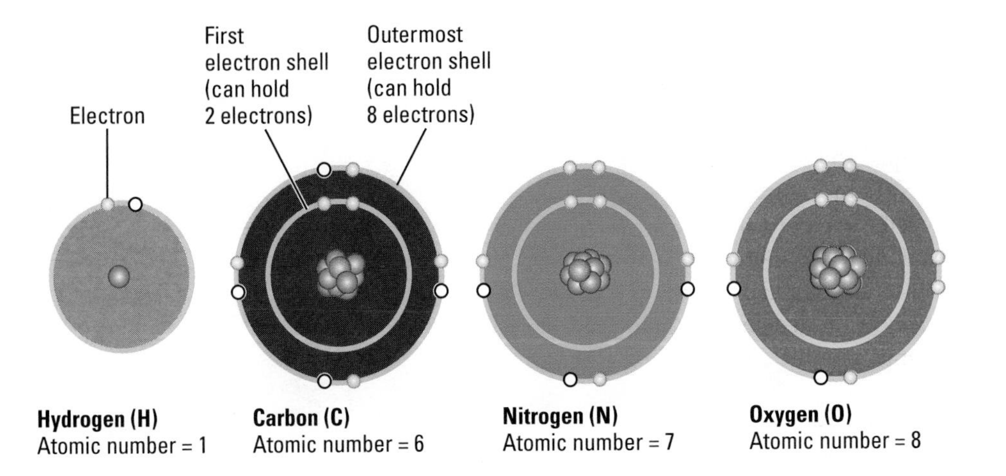

Figure 2.7 Atoms of the four elements most abundant in living matter. All four atoms are chemically reactive because their outermost electron shells are not filled. The small empty circles (O) in these diagrams represent unfilled "spaces" in the outer electron shells.

The number of electrons in the outermost shell determines the chemical properties of an atom. Atoms whose outer shells are not full tend to interact with other atoms—that is, to participate in chemical reactions. **Figure 2.7** shows the electron shells of four biologically important elements. Because the outer shells of all four atoms are not filled, these atoms react readily with other atoms. The hydrogen atom is highly reactive because it has only 1 electron in its single electron shell, which can accommodate 2 electrons. Atoms of carbon, nitrogen, and oxygen are also highly reactive because their outer shells, which can hold 8 electrons, are not filled. In contrast, the helium atom in Figure 2.5 has a single, first-level shell that is full with 2 electrons. As a result, helium is chemically unreactive.

Chemical Bonding and Molecules

Chemical reactions enable atoms to give up or acquire electrons, thereby completing their outer shells. Atoms do this by either transferring or sharing outer electrons. These interactions usually result in atoms staying close together, held by attractions called **chemical bonds.**

Ionic Bonds Table salt is an example of how the transfer of electrons can bond atoms together. The two ingredients of table salt are the elements sodium (Na) and chlorine (Cl). When a sodium atom donates an electron to a chlorine atom, the electron transfer results in both atoms having full outer shells of electrons **(Figure 2.8)**. Before the electron transfer, each of these atoms is electrically neutral. Because electrons are negatively charged particles, the electron transfer moves one unit of negative charge from sodium to chlorine. The atoms are now **ions,** the term for atoms that are electrically charged as a result of gaining or losing electrons. The loss of an electron gives the sodium ion a charge of +1, while chlorine's gain of an electron gives it a charge of −1. The sodium ion (Na^+) and chloride ion (Cl^-) are held together by an **ionic bond,** the attraction between oppositely charged ions. Compounds, such as table salt, that are held together by ionic bonds are called ionic compounds. Fluorine in Earth's crust is often found in the form of ionic compounds such as calcium fluoride (CaF_2), the result of bonds between calcium ions (Ca^{2+}) and fluoride ions (F^-). (Note that negatively charged ions often have names ending in "ide," like "chloride" or "fluoride.")

Covalent Bonds In contrast to the complete *transfer* of electrons that leads to ionic bonds, a **covalent bond** forms when two atoms *share* one or more pairs of outer-shell electrons. Atoms held together by covalent bonds form a **molecule.** For example, a covalent bond connects each

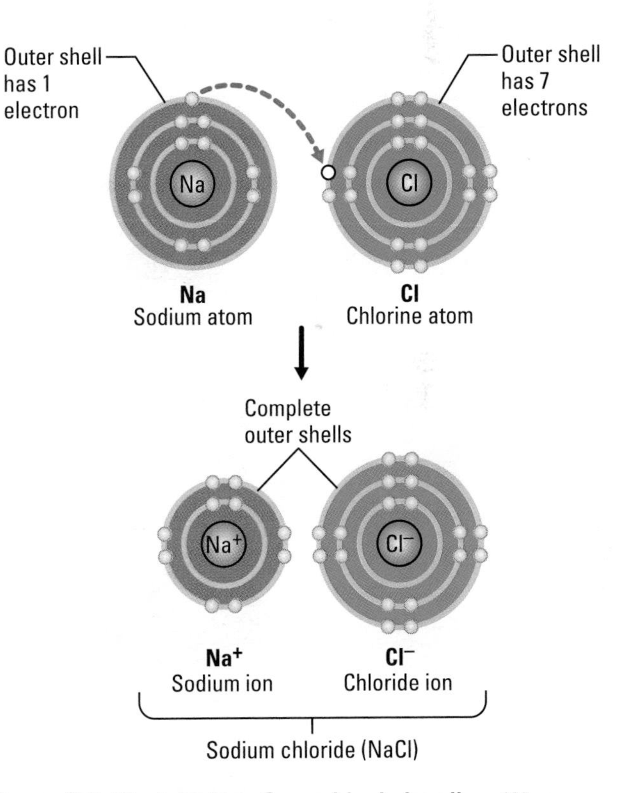

Figure 2.8 Electron transfer and ionic bonding. When a sodium atom and a chlorine atom meet, an outer electron is stripped from sodium, filling the chlorine atom's outer shell with 8 electrons. Sodium, having lost 1 electron, now has only two shells. The electron transfer between the two atoms results in two ions with opposite charges. The attraction between the two ions holds them together in an ionic bond.

hydrogen atom to the carbon in the molecule CH_4, a common gas called methane **(Figure 2.9)**. You can see that each of the four hydrogen atoms in a molecule of methane shares one pair of electrons with the single carbon atom.

The number of covalent bonds an atom can form is equal to the number of additional electrons needed to fill its outer shell. Note in Figure 2.9 that hydrogen (H) can form one covalent bond, oxygen (O) can form two, and carbon (C) can form four. The single covalent bond in H_2 completes the outer shells of both hydrogen atoms. In contrast, an oxygen atom needs two electrons to complete its outer shell. In an O_2 molecule, the two oxygen atoms share two pairs of electrons, forming a double covalent bond.

Though H_2 and O_2 are molecules, neither qualifies as a compound because these molecules are each composed of only one element. An example of a molecule that is a compound is ammonia (NH_3).

Hydrogen Bonds Water (H_2O) is also a compound. Its structure consists of two hydrogen atoms joined to one oxygen atom by single covalent bonds (the "sticks" represent bonds between the atoms, which are shown as "balls"):

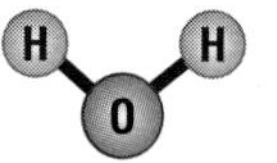

However, the electrons of the covalent bonds are not shared equally between the oxygen and hydrogen. The two yellow arrows in the following

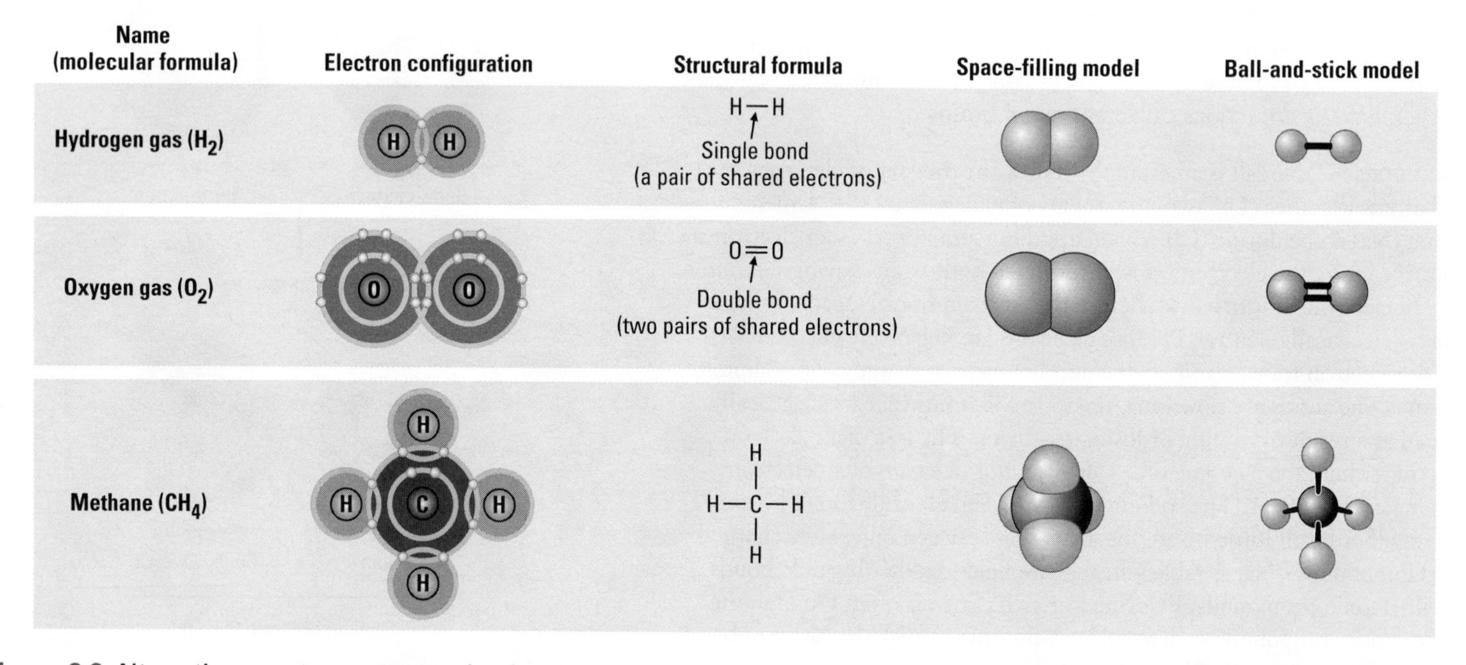

Figure 2.9 Alternative ways to represent molecules. A molecular formula, such as CH_4 or O_2, tells you the number of each kind of atom in a molecule but not how they are attached together. You can see from the electron configuration that each atom completes its outer shell by sharing electrons with its covalent partner. In a structural formula, each line represents a covalent bond, a pair of shared electrons. A space-filling model, in which the color-coded balls symbolize atoms, shows the shape of a molecule. In a ball-and-stick model, the "balls" represent atoms and the "sticks" represent the bonds between the atoms.

diagram indicate the stronger pull on the shared electrons that oxygen has compared with its hydrogen partners:

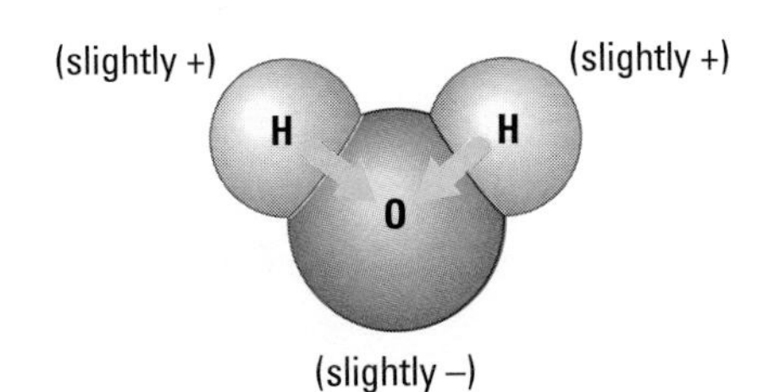

The unequal sharing of negatively charged electrons, combined with its V shape, makes a water molecule polar. A **polar molecule** has opposite charges on opposite ends. In the case of water, the oxygen end of the molecule has a slight negative charge, while the region around the two hydrogen atoms is slightly positive.

The polarity of water results in weak electrical attractions between neighboring water molecules. The molecules tend to orient such that the hydrogen atom of one molecule is near the oxygen atom of an adjacent water molecule. These weak attractions are called **hydrogen bonds (Figure 2.10)**. As you will see later in this chapter, the ability of water to form hydrogen bonds has many implications for our study of life.

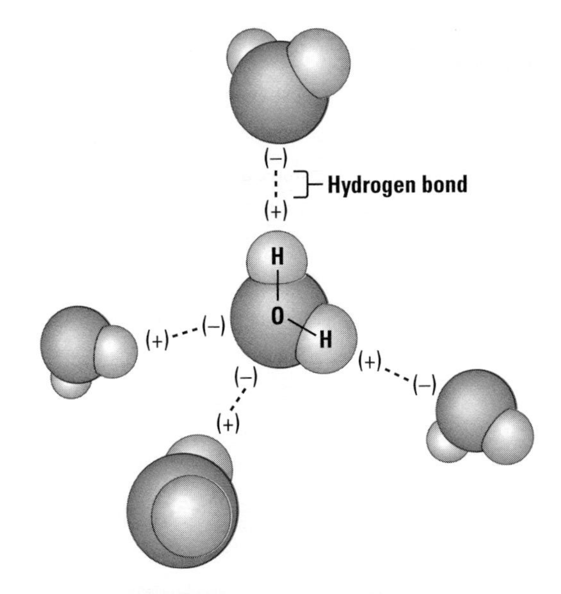

Figure 2.10 Hydrogen bonding in water. The charged regions of the polar water molecules are attracted to oppositely charged areas of neighboring molecules. Each molecule can hydrogen-bond to a maximum of four partners. (The dashed lines represent hydrogen bonds.)

Chemical Reactions

The chemistry of life is dynamic. Your cells are constantly rearranging molecules by breaking existing chemical bonds and forming new ones. Such changes in the chemical composition of matter are called **chemical reactions.** A simple example is the reaction between oxygen gas and hydrogen gas that forms water (this is an explosive reaction, which, fortunately, does not occur in your cells):

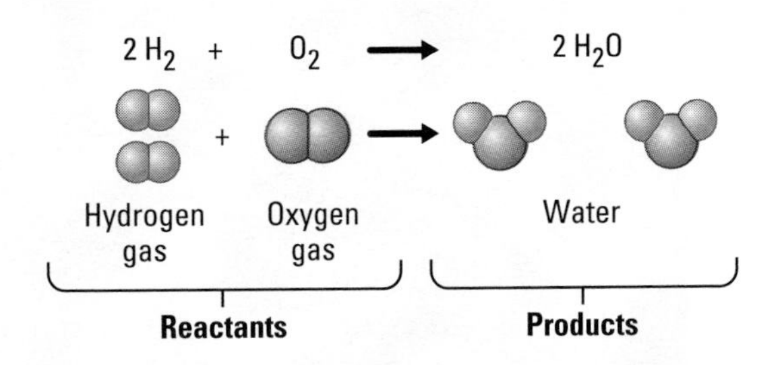

Let's translate the chemical shorthand: Two molecules of hydrogen gas ($2\,H_2$) react with one molecule of oxygen gas (O_2) to form two molecules of water ($2\,H_2O$). The arrows indicate the conversion of the starting materials, the **reactants** ($2\,H_2$ and O_2), to the **products** ($2\,H_2O$).

Notice that the same numbers of hydrogen and oxygen atoms are present in reactants and products, although they are grouped differently. Chemical reactions cannot create or destroy matter, but only rearrange it. These rearrangements usually involve the breaking of chemical bonds in reactants and the forming of new bonds in products.

The water molecules we have built here are a good conclusion to this section on basic chemistry. Water is a substance so important in biology that we'll take a closer look at its life-supporting properties in the next section.

CHECKPOINT

1. What four elements are most abundant in living matter?
2. Why is water (H_2O) classified as a compound, but oxygen gas (O_2) is not?
3. A nitrogen atom has 7 protons, and the most common isotope of nitrogen has 7 neutrons. A radioactive isotope of nitrogen has 9 neutrons. What are the atomic numbers and mass numbers of the stable and radioactive forms of nitrogen?
4. Why are radioactive isotopes useful as tracers in research on the chemistry of life?
5. Explain what holds together the ions in a crystal of table salt (NaCl).
6. What is chemically nonsensical about this structure?

   H–C=C–H

7. Why is it unlikely that two neighboring water molecules would be arranged like this?

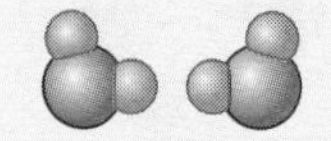

Answers: 1. Oxygen, carbon, hydrogen, and nitrogen **2.** Because water consists of two different elements, while oxygen gas consists of only one **3.** Nitrogen-14 has an atomic number of 7 and a mass number of 14. Nitrogen-16 has an atomic number of 7 and a mass number of 16. **4.** Organisms incorporate radioactive isotopes of an element into their molecules just as they do the nonradioactive isotopes, and researchers can detect the presence of the radioactive isotopes. **5.** Ionic bonds, the electrical attraction between the sodium ions (Na^+) and the chloride ions (Cl^-), hold the ions together. **6.** Each carbon atom has only three covalent bonds instead of the required four. **7.** The positively charged hydrogen regions would repel each other.

Water and Life

Life on Earth began in water and evolved there for 3 billion years before spreading onto land. Modern life, even land-dwelling life, is still tied to water. You've had personal experience with this dependence on water every time you seek liquids to quench your thirst and replenish your body's water content. Inside your body, your cells are surrounded by a fluid that's composed mostly of water, and your cells themselves range from 70% to 95% in water content.

The abundance of water is a major reason that Earth is habitable **(Figure 2.11)**. In fact, it is because life as we know it is so dependent on water that the search for water on other planets (particularly Mars) is such a priority for the U.S. space program. Water is so common in our environment that it's easy to overlook its extraordinary behavior. We can trace water's unique life-supporting properties to the structure and interactions of its molecules.

Figure 2.11 A watery world. Three-quarters of Earth's surface is submerged in water. Although most of this water is in liquid form, water is also present on Earth as ice and vapor (including clouds). Water is the only common substance that exists in the natural environment in all three physical states of matter: solid, liquid, and gas. You can see each of these three states of water in this view of Earth from space.

Water's Life-Supporting Properties

MP3 Tutor The Properties of Water

The polarity of water molecules and the hydrogen bonding that results (see Figure 2.10) explain most of water's life-supporting properties. We'll explore four of those properties here: the cohesive nature of water, the ability of water to moderate temperature, the biological significance of ice floating, and the versatility of water as a solvent.

The Cohesion of Water Water molecules stick together as a result of hydrogen bonding. Hydrogen bonds between molecules of liquid water last for only a few trillionths of a second, yet at any instant, many of the molecules are hydrogen-bonded to others. This tendency of molecules to stick together, called **cohesion,** is much stronger for water than for most other liquids. The cohesion of water is important in the living world. Trees, for example, depend on cohesion to help transport water from their roots to their leaves **(Figure 2.12)**.

Related to cohesion is surface tension, a measure of how difficult it is to stretch or break the surface of a liquid. Hydrogen bonds give water unusually high surface tension, making it behave as though it were coated with an invisible film **(Figure 2.13)**.

Figure 2.12 Cohesion and water transport in plants. The evaporation of water from leaves pulls water upward from the roots through microscopic tubes in the trunk of the tree. Because of cohesion, the pulling force is relayed through the tubes all the way down to the roots. As a result, water rises against the force of gravity.

How Water Moderates Temperature If you've ever burned your finger on a metal pot while waiting for the water in it to boil, you know that water heats up much more slowly than metal. In fact, because of hydrogen bonding, water has a stronger resistance to temperature change than most other substances.

Temperature and heat are related, but different. A swimmer crossing San Francisco Bay has a higher temperature than the water, but the bay contains far more heat because of its immense volume. **Heat** is the amount of energy associated with the movement of the atoms and molecules in a body of matter. **Temperature** measures the intensity of heat—that is, the average speed of molecules rather than the total amount of heat energy in a body of matter.

When water is heated, the heat energy first disrupts hydrogen bonds and then makes water molecules jostle around faster. The temperature of the water doesn't go up until the water molecules start to speed up. Because heat is first used to break hydrogen bonds rather than raise the temperature, water absorbs and stores a large amount of heat while warming up only a few degrees in temperature. Conversely, when water cools, hydrogen bonds form, a process that releases heat. Thus, water can release a relatively large amount of heat to the surroundings while the water temperature drops only slightly.

Earth's giant water supply—the oceans, seas, lakes, and rivers—allows temperatures to stay within limits that permit life by storing a huge amount of heat from the sun during warm periods and giving off heat to warm the air during cold conditions. That's why coastal areas generally have milder climates than inland regions. Water's resistance to temperature change also stabilizes ocean temperatures, creating a favorable environment for marine life.

Another way that water moderates temperature is by **evaporative cooling.** When a substance evaporates (changes physical state from a liquid to a gas), the surface of the liquid remaining behind cools down. This occurs because the molecules with the greatest energy (the "hottest" ones) tend to vaporize first. It's as if the five fastest runners on your track team quit school, lowering the average speed of the remaining team. On a global scale, surface evaporation cools tropical oceans. On the scale of individual organisms, evaporative cooling prevents some land-dwelling creatures from overheating. It's why sweating helps you maintain a constant body temperature, even

Figure 2.13 A water strider walking on water. The cumulative strength of hydrogen bonds between water molecules allows this insect to walk on pond water without breaking the surface.

when exercising on a hot day **(Figure 2.14)**. And the old expression "It's not the heat, it's the humidity" has its basis in the difficulty of sweating water into air that is already saturated with water vapor.

Figure 2.14 Sweating as a mechanism of evaporative cooling.

The Biological Significance of Ice Floating When most liquids get cold, their molecules move closer together. If the temperature is cold enough, the liquid freezes and becomes a solid. Water, however, behaves differently. When water molecules get cold enough, they move apart, forming ice. A chunk of ice has fewer molecules than an equal volume of liquid water; it floats because it is less dense than the liquid water around it. Like water's other life-supporting properties, floating ice is a consequence of hydrogen bonding. In contrast to the short-lived hydrogen bonds in liquid water, those in solid ice last longer, with each molecule bonded to four neighbors. As a result, ice is a spacious crystal **(Figure 2.15)**.

How does the fact that ice floats help support life on Earth? Imagine what would happen if ice sank. All ponds, lakes, and even the oceans would eventually freeze solid. During summer, only the upper few inches of the oceans would thaw. Instead, when a deep body of water cools, the floating ice insulates the liquid water below, allowing life to persist under the frozen surface.

Water as the Solvent of Life If you've ever enjoyed a glass of sweetened ice tea or mixed a sore throat gargle, you know that you can dissolve sugar or salt in water. This results in a mixture known as a **solution,** a liquid consisting of a homogeneous mixture of two or more substances. The dissolving agent is called the **solvent,** and a substance that is dissolved is a **solute.** When water is the solvent, the resulting solution is called an **aqueous solution.**

The fluids of organisms are aqueous solutions. Water can dissolve an enormous variety of solutes necessary for life and is the solvent inside all cells, in blood, and in plant sap. As a solvent, it is a medium for chemical reactions. Water can dissolve ionic salts and many polar molecules, such as

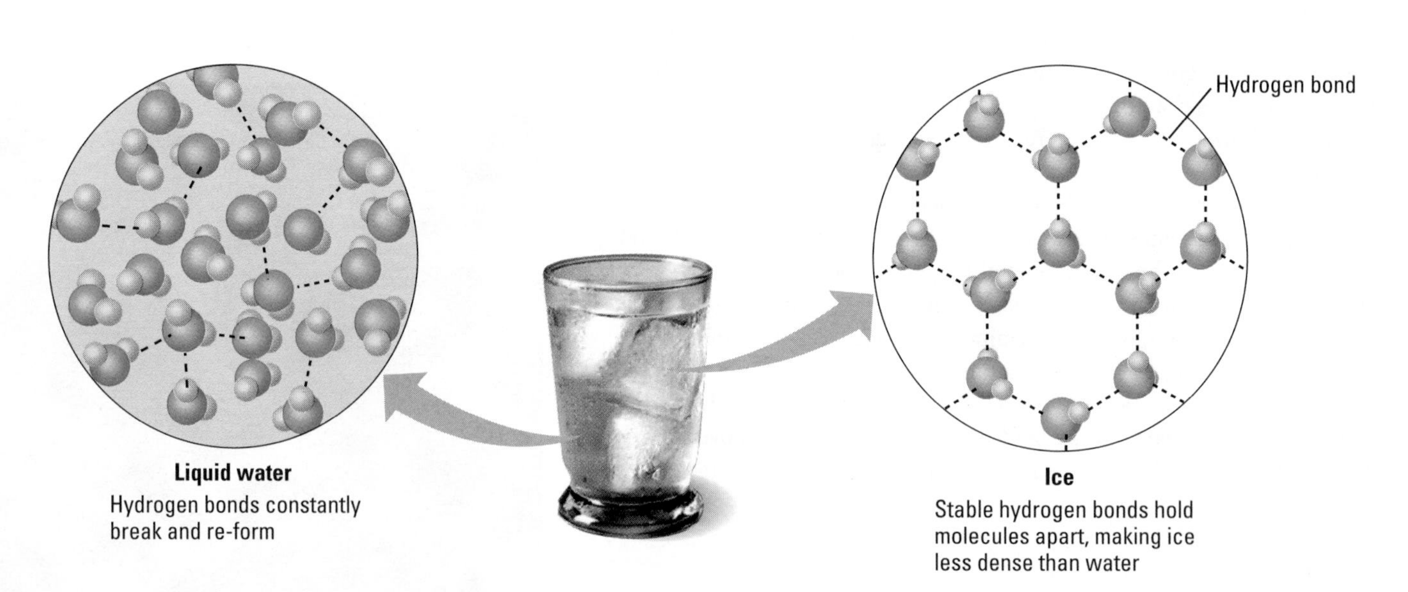
Hydrogen bond
Liquid water
Hydrogen bonds constantly break and re-form
Ice
Stable hydrogen bonds hold molecules apart, making ice less dense than water

Figure 2.15 Why ice floats. Compare the tightly packed molecules in liquid water with the spaciously arranged molecules in the ice crystal. The more stable hydrogen bonding in ice holds the molecules apart, resulting in ice being less dense than liquid water. The expansion of water as it freezes can crack boulders when water is in crevices, and it can also break the water pipes of an unheated house in winter.

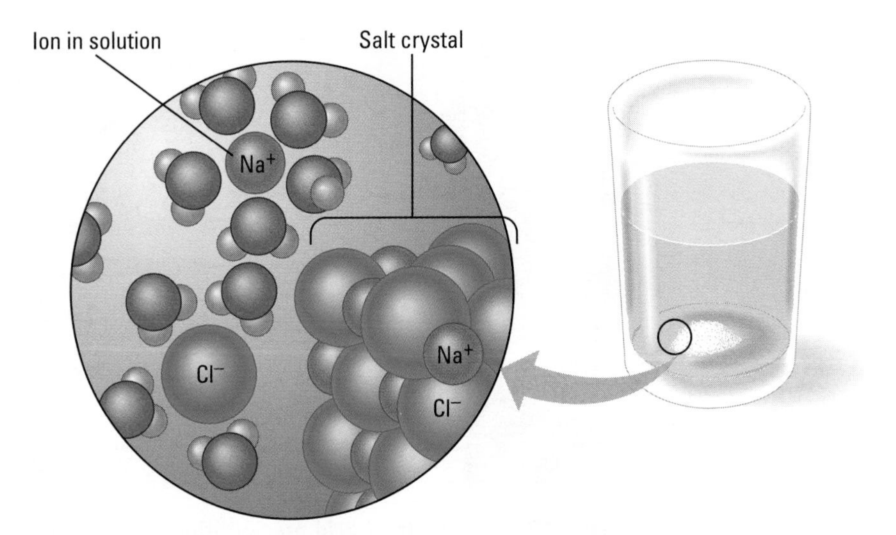

Figure 2.16 A crystal of table salt (NaCl) dissolving in water. The sodium and chloride ions at the surface of the crystal have affinities for different parts of the water molecules. The positive sodium ions (Na^+) attract the electrically negative oxygen regions (red) of the water molecules. The negative chloride ions (Cl^-) attract the positively charged hydrogen regions (gray) of water. As a result, H_2O molecules surround the ions, dissolving the crystal in the process.

sugars, by orienting local regions of positive and negative charge toward the charged regions of polar molecules **(Figure 2.16)**.

Acids, Bases, and pH

In the aqueous solutions within organisms, most of the water molecules are intact. However, some of the water molecules actually break apart into hydrogen ions (H^+) and hydroxide ions (OH^-). A balance of these two ions is critical for the proper functioning of chemical processes within organisms.

A chemical compound that releases H^+ to a solution is called an **acid.** One example of a strong acid is hydrochloric acid (HCl), the acid in your stomach. In solution, HCl breaks apart into the ions H^+ and Cl^-. A **base** (or alkali) is a compound that accepts H^+ and removes it from solution. Some bases, such as sodium hydroxide (NaOH), do this by releasing OH^-, which combines with H^+ to form H_2O.

To describe the acidity of a solution, we use the **pH scale,** a measure of the hydrogen ion (H^+) concentration in a solution. The scale ranges from 0 (most acidic) to 14 (most basic). Each pH unit represents a tenfold change in the concentration of H^+ **(Figure 2.17)**. For example, lemon juice at pH 2 has 100 times more H^+ than an equal amount of tomato juice at pH 4. Pure water and aqueous solutions that are neither acidic nor basic are said to be neutral; they have a pH of 7. They do contain H^+ and OH^-, but the concentrations of the two ions are equal. The pH of the solution inside most living cells is close to 7.

Even a slight change in pH can be harmful to an organism because the molecules in cells are extremely sensitive to H^+ and OH^- concentrations. Biological fluids contain **buffers,** substances that prevent harmful changes in pH by accepting H^+ when that ion is in excess and donating H^+ when it is depleted. This buffering process, however, is not foolproof. The biological damage from an unfavorable pH is apparent in the toll that acid

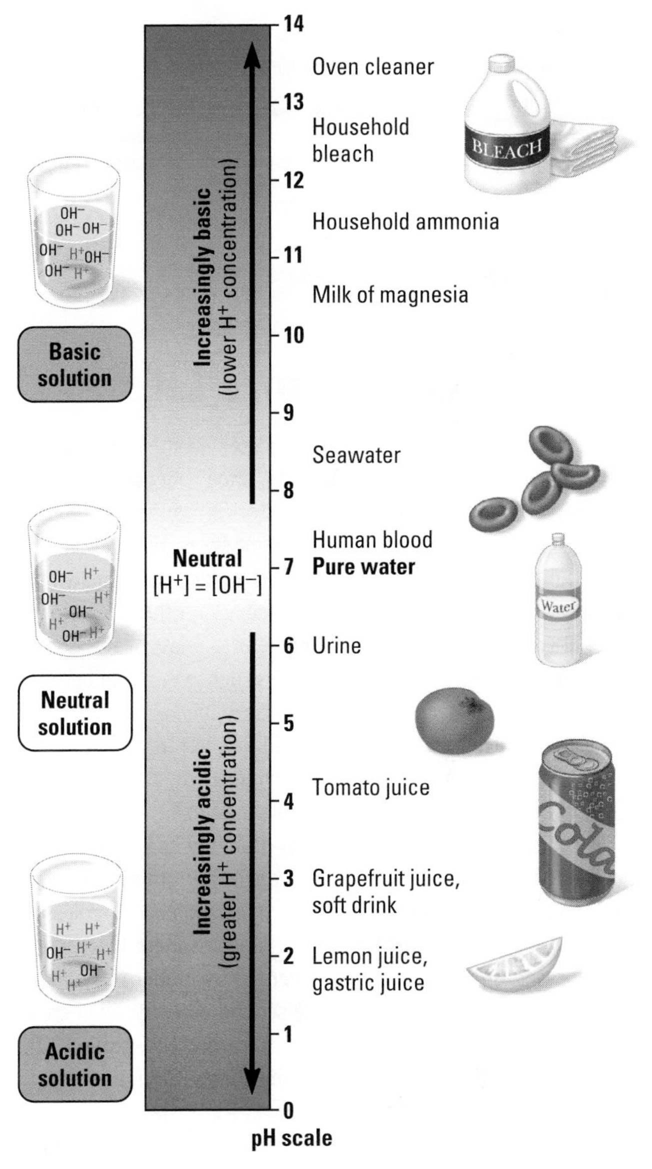

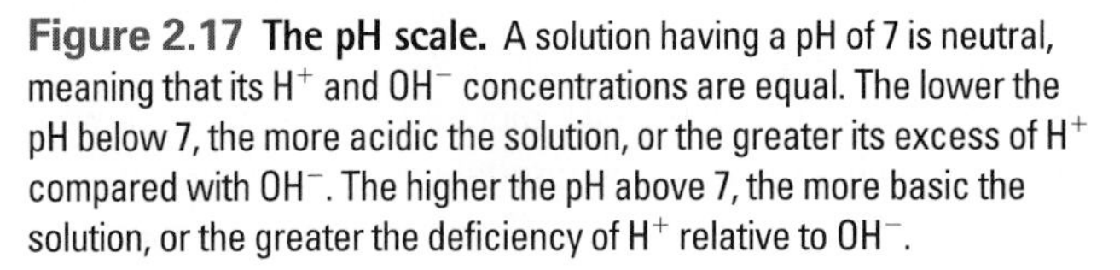
Figure 2.17 The pH scale. A solution having a pH of 7 is neutral, meaning that its H^+ and OH^- concentrations are equal. The lower the pH below 7, the more acidic the solution, or the greater its excess of H^+ compared with OH^-. The higher the pH above 7, the more basic the solution, or the greater the deficiency of H^+ relative to OH^-.

precipitation can take on an ecosystem such as a pond or forest (**Figure 2.18**). It is a daunting reminder that the chemistry of life is linked to the chemistry of the environment. It reminds us, too, that chemistry happens on a global scale, since industrial processes in one region of the world often cause acid precipitation to fall in another part of the world.

Figure 2.18 The effects of acid precipitation on a forest. Acid fog and acid rain contributed to the death of many of the fir trees in this forest in the Czech Republic. Acid precipitation results from water in the atmosphere reacting with certain pollutants spewed as exhaust from automobiles, factories, and power plants. The acid rain, snow, or fog can descend on land or lakes hundreds of miles downwind from the sources of pollution. Acid precipitation is a problem in the United States as well; rain with a pH between 2 and 3—more acidic than vinegar—has been recorded in the eastern United States.

CHECKPOINT

1. Explain why, if you pour very carefully, you can actually "stack" water slightly above the rim of a cup.
2. Why is it more dangerous to stay neck-deep in a 105°F (41°C) hot tub for an hour than it is to sit for an hour outside when the air temperature is 105°F?
3. Explain why ice floats.
4. Why are blood and most other biological fluids classified as aqueous solutions?
5. Compared with a basic solution of pH 8, the same volume of an acidic solution at pH 5 has __________ times more hydrogen ions (H^+).

Answers: 1. Surface tension due to water's cohesion will hold the water together. **2.** Evaporative cooling in the hot tub is limited to the skin of the head and neck. **3.** Ice is less dense than liquid water because the more stable hydrogen bonds "lock" the molecules into a spacious crystal. **4.** The solvent is water. **5.** 1,000

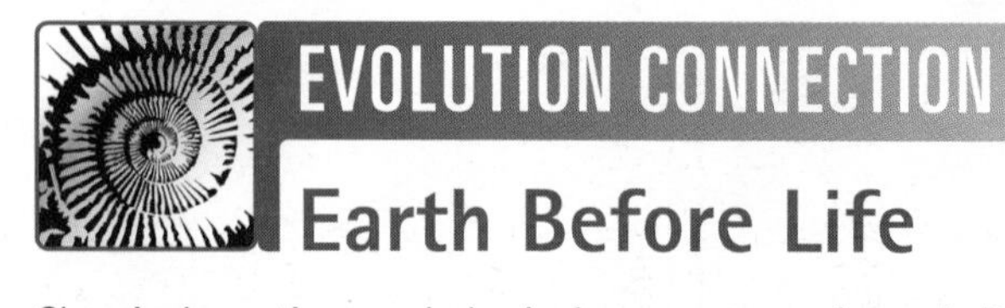

EVOLUTION CONNECTION

Earth Before Life

Chemical reactions and physical processes on the early Earth created an environment that made life possible. And life, once it began, transformed the planet's chemistry. Biological and geologic histories are inseparable.

Earth began as a cold world when gravity drew together dust and ice orbiting a young sun about 4.5 billion years ago. The planet eventually melted from the heat produced by compaction, radioactive decay, and the impact of meteorites. Molten material sorted into layers of varying density. Most of the iron and nickel sank to the center and formed a dense core. Less dense material became concentrated in a layer called the mantle, which surrounds the core. And the least dense material solidified to form a thin crust. The present continents, including North America, are attached to plates of crust that float on the flexible mantle.

The first atmosphere, which was probably composed mostly of hot hydrogen gas (H_2), escaped. The gravity of Earth was not strong enough to hold such small molecules. Volcanoes belched gases that formed a new atmosphere (**Figure 2.19**). Based on analysis of gases vented by modern volcanoes, scientists have speculated that the second early atmosphere consisted mostly of water vapor (H_2O), carbon monoxide (CO), carbon dioxide (CO_2), nitrogen (N_2), methane (CH_4), and ammonia (NH_3). The first seas formed from torrential rains that began when Earth had cooled enough for water in the atmosphere to condense. In addition to an atmosphere very different from the one we know, lightning, volcanic activity, and ultraviolet radiation were much more intense when Earth was young. In such a seemingly inhospitable world, life began about 3.5–4.0 billion years ago. In Chapter 15, we'll examine some of the hypotheses and experiments of scientists who investigate the origins of the first life on Earth. ■

Figure 2.19 The gaseous exhaust of a volcano. An artist's conception of the Earth 3 billion years ago.

Chapter Review

SUMMARY OF KEY CONCEPTS

For study help and activities, go to campbellbiology.com or the student CD-ROM.

Some Basic Chemistry

• **Matter: Elements and Compounds** Matter consists of elements and compounds, which are combinations of two or more elements. Of the 25 elements essential for life, oxygen, carbon, hydrogen, and nitrogen are the most abundant in living matter.

Case Study in the Process of Science *How Are Space Rocks Analyzed for Signs of Life?*

• **Atoms**

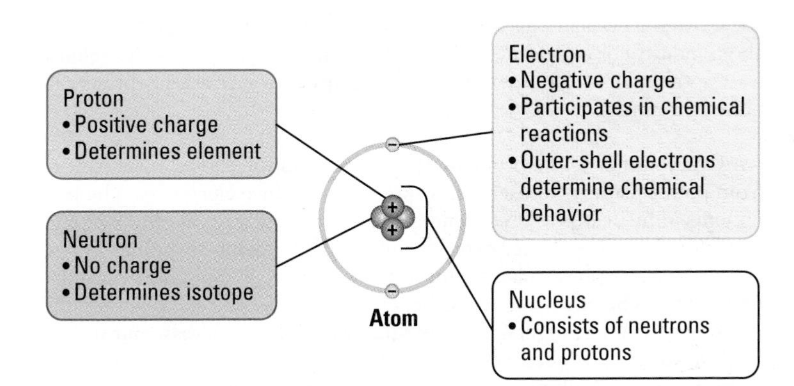

Activity *The Structure of Atoms*

Activity *Electron Arrangement*

Activity *Build an Atom*

• **Chemical Bonding and Molecules** Electron transfers that complete outer electron shells result in charged atoms, or ions. Oppositely charged ions are held together by ionic bonds. Salts are ionic compounds. In covalent bonds, atoms complete their outer electron shells by sharing electrons. A molecule consists of two or more atoms connected by covalent bonds. Water is a polar molecule: The slightly positively-charged H atoms in one water molecule may be attracted to the partial negative charge of O atoms in neighboring water molecules, forming weak but important hydrogen bonds:

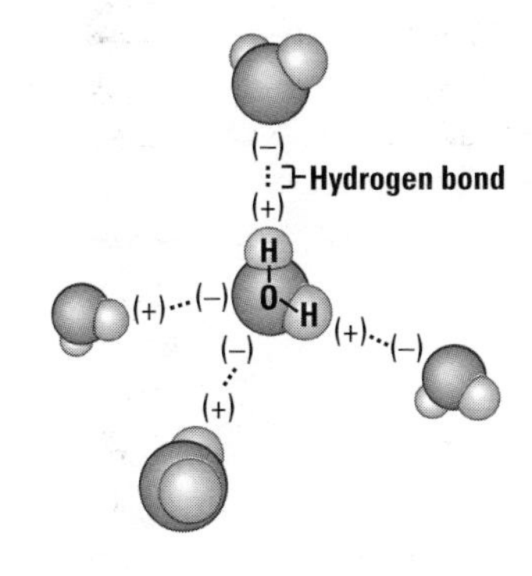

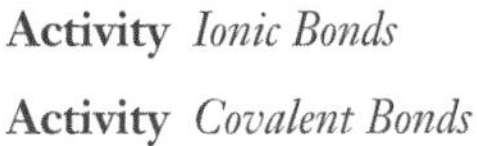

Activity *Ionic Bonds*

Activity *Covalent Bonds*

• **Chemical Reactions** By breaking bonds in reactants and forming new bonds in products, chemical reactions rearrange matter.

Water and Life

• **Water's Life-Supporting Properties** The ability of leaves to pull water up microscopic tubes is an example of how water's cohesion supports life. Water moderates temperature by absorbing heat in warm environments and releasing heat in cold environments. Evaporative cooling also helps stabilize the temperatures of oceans and organisms. The fact that ice floats because it is less dense than liquid water prevents the oceans from freezing solid. Blood and other biological fluids are aqueous solutions with a diversity of solutes dissolved in water, a versatile solvent.

MP3 Tutor *The Properties of Water*

Activity *The Structure of Water*

Activity *The Cohesion of Water in Trees*

• **Acids, Bases, and pH**

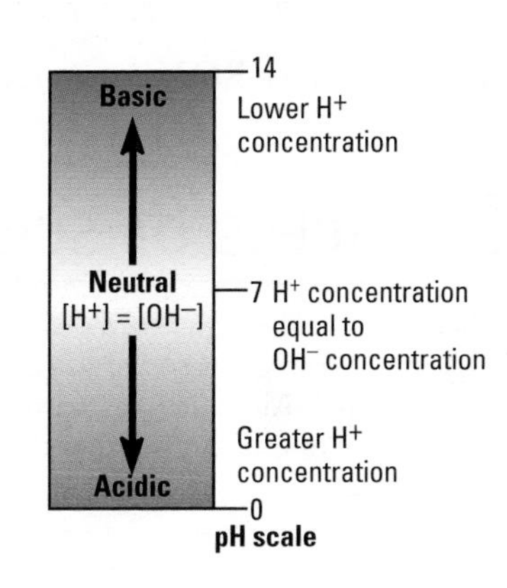

Activity *Acids, Bases, and pH*

Case Study in the Process of Science *How Does Acid Precipitation Affect Trees?*

SELF-QUIZ

1. Which of the following elements is least abundant in your body?
   a. carbon
   b. oxygen
   c. phosphorus
   d. iron

2. Which of the following are compounds? $MgCl_2$, H_2, Fe, C_2H_6.

3. A chemical compound is to a(an) __________ as a body organ is to a tissue.

4. An atom can be changed into an ion by adding or removing __________. An atom can be changed into a different isotope by adding or removing __________. But if you change the number of __________, the atom becomes a different element.

5. A sulfur atom has 6 electrons in its third (outermost) shell, which can hold 8 electrons. As a result, it forms __________ covalent bonds with other atoms.

6. Explain the difference between an ionic bond and a covalent bond in terms of what happens to the electrons in the outer shell of the participating atoms.

7. Which of the following is not a chemical reaction?
   a. Sugar ($C_6H_{12}O_6$) and oxygen gas (O_2) combine to form carbon dioxide (CO_2) and water (H_2O).
   b. Sodium metal and chlorine gas unite to form sodium chloride.
   c. Hydrogen gas combines with oxygen gas to form water.
   d. Ice melts to form liquid water.

8. Some people in your study group say they don't understand what a polar molecule is. You explain that a polar molecule
   a. is slightly negative at one end and slightly positive at the other end.
   b. has an extra electron, giving it a positive charge.
   c. has an extra electron, giving it a negative charge.
   d. has covalent bonds.

9. Explain how the unique properties of water result from the fact that water is a polar molecule.

10. A can of cola consists mostly of sugar dissolved in water, with some carbon dioxide gas that makes it fizzy and makes the pH less than 7. Describe the cola using the following terms: solute, solvent, acidic, aqueous solution.

Answers to the Self-Quiz questions can be found in Appendix D.

Go to the website or CD-ROM for more Self-Quiz questions.

THE PROCESS OF SCIENCE

11. Animals obtain energy through a series of chemical reactions in which sugar ($C_6H_{12}O_6$) and oxygen gas (O_2) are reactants. This process produces water (H_2O) and carbon dioxide (CO_2) as waste products. How might you use a radioactive isotope to find out whether the oxygen in CO_2 comes from sugar or oxygen gas?

12. The following diagram shows the arrangement of electrons around the nucleus of a fluorine atom (left) and a potassium atom (right). Predict what would happen if a fluorine atom and a potassium atom came into contact. What kind of bond do you think they would form? What kind of compound?

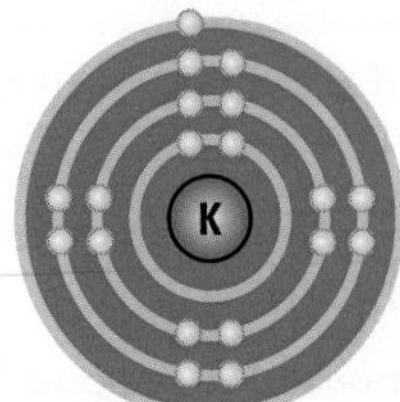

Fluorine atom Potassium atom

BIOLOGY AND SOCIETY

13. Critically evaluate this statement: "It's paranoid and ignorant to worry about industry or agriculture contaminating the environment with chemical wastes; this stuff is just made of the same atoms that were already present in our environment."

14. One solution to the problem of acid precipitation caused by emissions from power plants is to use nuclear power to produce electricity. The proponents of nuclear power contend that it is the only way that the United States can increase its energy production while reducing air pollution, because nuclear power plants emit little or no acid-precipitation-causing pollutants. What are some of the benefits of nuclear power? What are the possible costs and dangers? Do you think we ought to increase our use of nuclear power to generate electricity? Why or why not? If a new power plant were to be built near your home, would you prefer it to be a coal-burning plant or a nuclear plant? Why?

3 The Molecules of Life

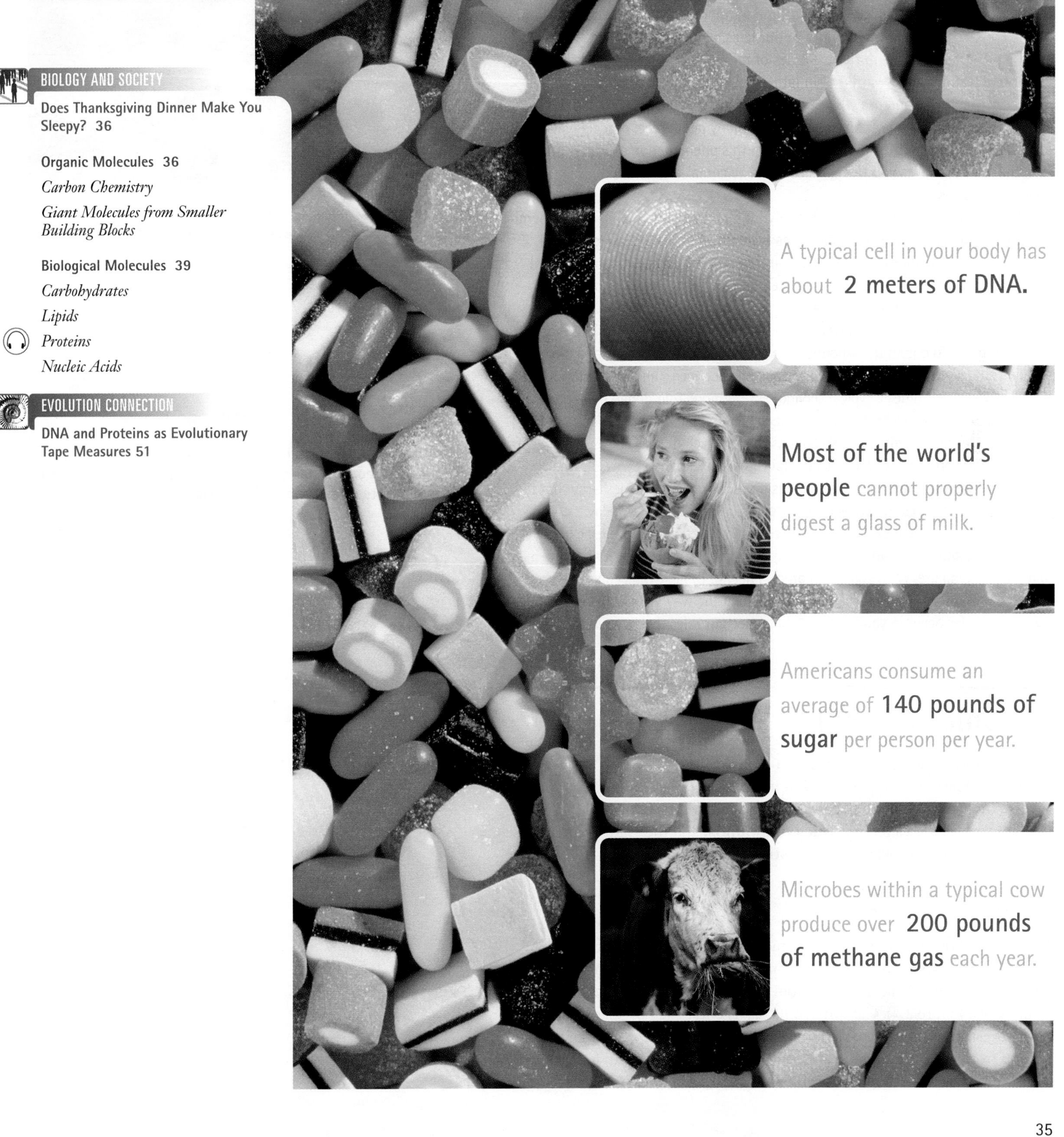

A typical cell in your body has about **2 meters of DNA.**
Most of the world's people cannot properly digest a glass of milk.
Americans consume an average of **140 pounds of sugar** per person per year.
Microbes within a typical cow produce over **200 pounds of methane gas** each year.

BIOLOGY AND SOCIETY

Does Thanksgiving Dinner Make You Sleepy?

For many Americans, Thanksgiving is a time to appreciate the company of family and friends, to reflect on the bounty of our lives, and to eat. A lot! A typical Thanksgiving meal is rich in three classes of molecules that play important roles in our body: carbohydrates (found in large amounts in mashed potatoes and rolls), fats (in butter and gravy), and proteins (in turkey meat) **(Figure 3.1)**.

After finishing a huge Thanksgiving meal, many people feel especially lethargic and a few even doze off. You may have heard this phenomenon attributed to the idea that turkey makes you sleepy. Is the "turkey coma" an urban myth? Or is there a biological basis to this claim?

It is true that turkey meat is relatively high in tryptophan, one of the chemical building blocks of proteins (shown as a chemical structure in Figure 3.1). Once tryptophan is digested, the body can convert it to serotonin, a chemical that can act on the brain to promote sleep. In fact, tryptophan is sold as a dietary supplement to treat insomnia.

There is little evidence, however, that a turkey dinner encourages sleep more than any other meal. To obtain the amount of tryptophan found in one pill of a typical sleeping supplement, you would have to consume between $\frac{3}{4}$ and $1\frac{1}{2}$ pounds of turkey—two to four times a typical serving! Furthermore, tryptophan must be taken on an empty stomach to be an effective sleep aid. And while turkey does contain relatively high levels of tryptophan, so do most protein-rich foods, such as other meats, nuts, and dairy products. In fact, fish contains more tryptophan by weight than turkey does. Thanksgiving sleepiness is thus more likely due to the lethargy that follows any large meal, especially one that occurs during a busy, active day.

In this chapter, we'll explore the structure and function of proteins, carbohydrates, lipids, and other carbon-containing molecules that are essential to any meal and indeed to all life. We'll start with an overview of organic molecules and then examine each of several classes individually. Along the way, we'll emphasize where these molecules occur in your diet and the important roles they play in your body. ■

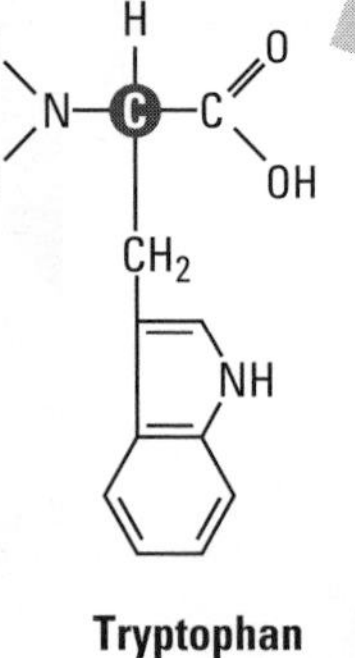

Figure 3.1 A sleep-inducing meal? Turkey meat, like other high-protein foods, contains tryptophan. Nonetheless, tryptophan is probably not responsible for Thanksgiving sleepiness.

Organic Molecules

A cell is mostly water, but the rest of it consists mostly of carbon-based molecules. Carbon is unparalleled in its ability to form the large, complex, diverse molecules that are necessary for life functions. Compounds that contain carbon are called **organic compounds**, and the study of organic compounds is called **organic chemistry**.

Carbon Chemistry

Why are carbon atoms so versatile as molecular ingredients? Recall that an atom's bonding ability is related to the number of electrons it must share to complete its outer shell. A carbon atom has 4 electrons in an outer shell that holds 8 (see Figure 2.7). Carbon completes its outer shell by sharing electrons with other atoms in four covalent bonds. Each carbon thus acts as an intersection from which an organic molecule can branch off in up to four directions. And because carbon can use one or more of its bonds to attach to other

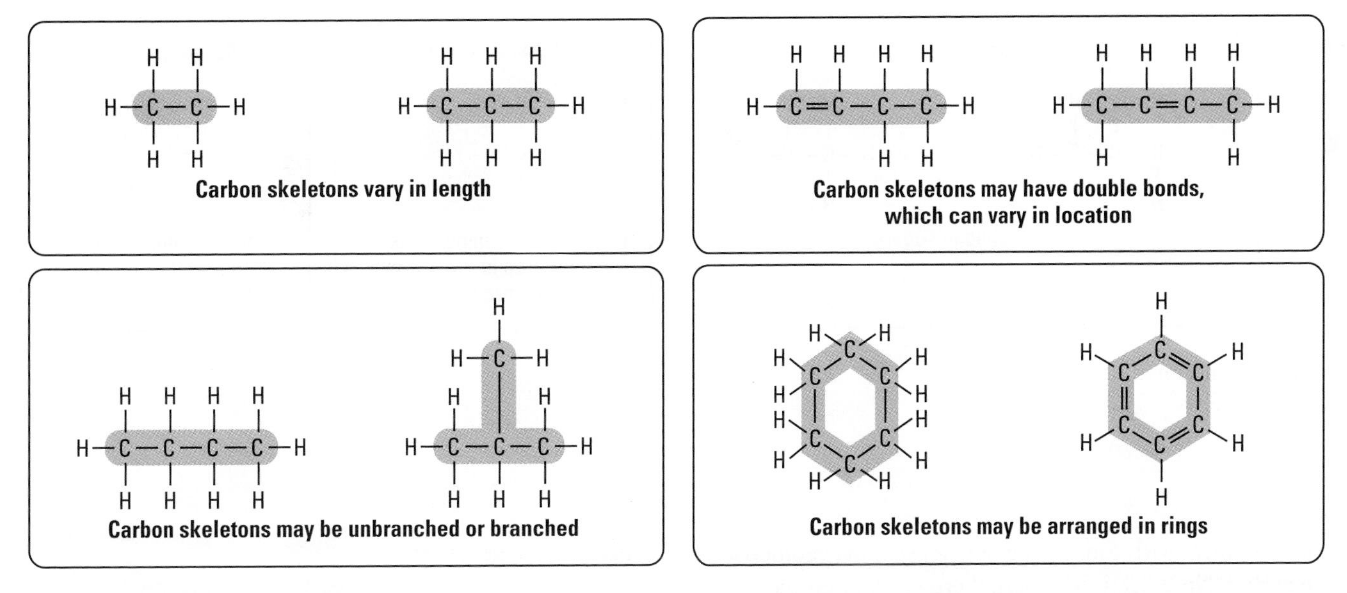
Carbon skeletons vary in length
Carbon skeletons may have double bonds, which can vary in location
Carbon skeletons may be unbranched or branched
Carbon skeletons may be arranged in rings

Figure 3.2 Variations in carbon skeletons. All of these examples are hydrocarbons, organic molecules consisting only of carbon and hydrogen. Notice that each carbon atom forms four bonds, and each hydrogen atom forms one bond. Remember that one line represents a single bond (sharing one pair of electrons) and two lines represent a double bond (sharing two pairs of electrons).

carbon atoms, it is possible to construct an endless diversity of carbon skeletons varying in size and branching pattern **(Figure 3.2)**. The carbon atoms of organic molecules can also use one or more of their bonds to partner with other elements, most commonly hydrogen, oxygen, and nitrogen.

In terms of chemical composition, the simplest organic compounds are **hydrocarbons,** organic molecules containing only carbon and hydrogen atoms. And the simplest hydrocarbon is methane, a single carbon atom bonded to four hydrogen atoms **(Figure 3.3)**. Methane is one of the most abundant hydrocarbons in natural gas and is also produced by prokaryotes that live in swamps and in the digestive tracts of grazing animals, such as cows. Larger hydrocarbons (such as octane, with eight carbons) are the main molecules in the gasoline we burn in cars and other machines **(Figure 3.4)**. Hydrocarbons are also important fuels in your body; the energy-rich parts of fat molecules have a hydrocarbon structure.

Each type of organic molecule has a unique three-dimensional shape. Notice in Figure 3.3 that carbon's four bonds point to the corners of an

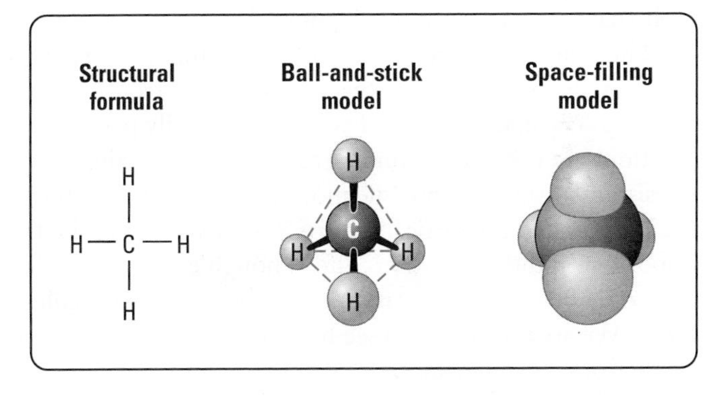
Structural formula
Ball-and-stick model
Space-filling model

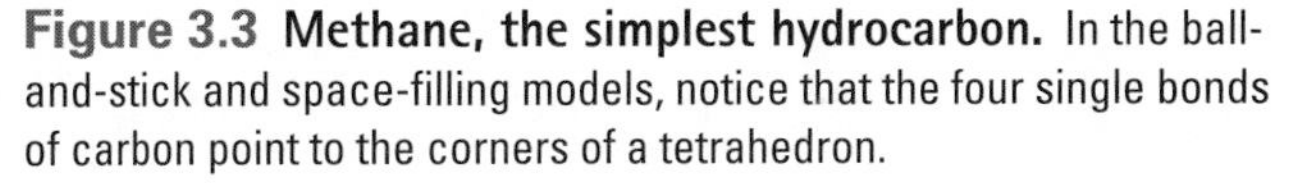

Figure 3.3 Methane, the simplest hydrocarbon. In the ball-and-stick and space-filling models, notice that the four single bonds of carbon point to the corners of a tetrahedron.

Octane
in gasoline
Dietary fat
in food

Figure 3.4 Hydrocarbons as fuel. Energy-rich hydrocarbons provide fuel for machines and, in the form of the hydrocarbon content of fats, the body's cells.

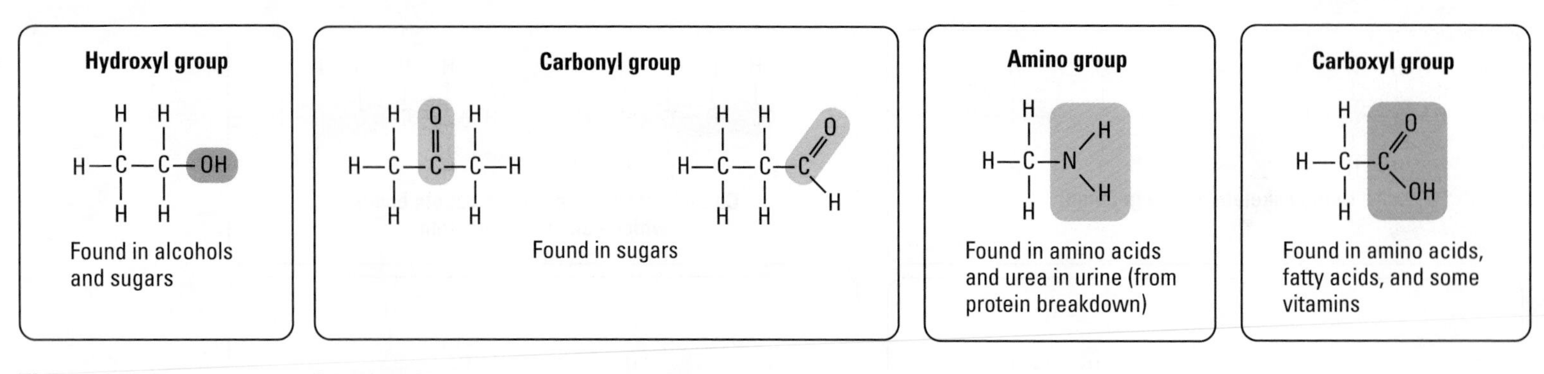

Figure 3.5 Some common functional groups.

imaginary tetrahedron (an object with four triangular sides). This geometric pattern occurs at each carbon "intersection" where there are four covalent bonds, and thus large organic molecules can have very elaborate shapes. As you will see in subsequent chapters, a recurring theme in biology is the importance of the shape of molecules. Many vital processes within living organisms rely on the ability of molecules to recognize one another based on their shape (see, for example, Figure 2.1).

The unique properties of an organic compound depend not only on its carbon skeleton but also on the atoms attached to the skeleton. In an organic molecule, the groups of atoms that usually participate in chemical reactions are called **functional groups.** Each functional group behaves consistently from one organic molecule to another, helping give each molecule its unique properties. **Figure 3.5** shows four of the functional groups important in the chemistry of life. Though each example in the figure contains only one functional group, many biological molecules have two or more. We are now ready to see how your cells make large molecules out of smaller organic molecules.

Giant Molecules from Smaller Building Blocks

On a molecular scale, many of life's molecules are gigantic; in fact, biologists call them **macromolecules** (*macro* means "big"). DNA is a macromolecule, as are the carbohydrates in starchy foods and the proteins that compose your hair. Even though they are quite large, the structure of most macromolecules can be easily understood because they are **polymers,** large molecules made by stringing together many smaller molecules called **monomers.** A polymer is like a pearl necklace made by joining together many pearl monomers.

Cells link monomers together through a **dehydration reaction,** a chemical reaction that removes a molecule of water **(Figure 3.6a)**. For each monomer added to a chain, a water molecule (H_2O) is formed by the release of two hydrogen atoms and one oxygen atom from the monomers. This same dehydration reaction occurs regardless of the specific monomers and the type of polymer the cell is producing.

Organisms not only make macromolecules, but also have to break them down. For example, many of the molecules in your food are macromolecules. You must digest these giant molecules to make their monomers available to your cells, which can then rebuild the monomers into your own brand of macromolecules. This digestion occurs by a process called **hydrolysis (Figure 3.6b)**. Hydrolysis means to break

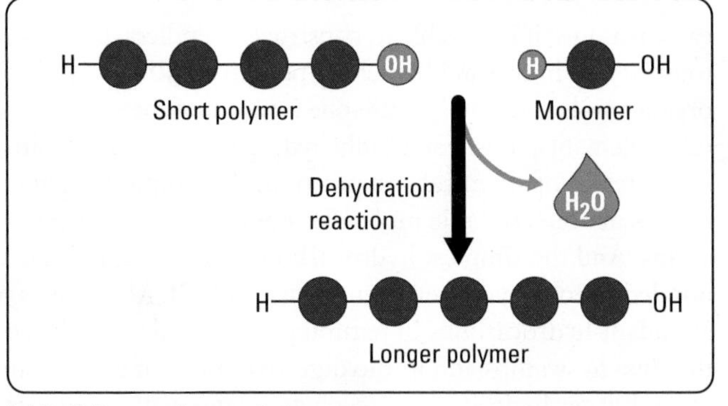

(a) Building a polymer chain

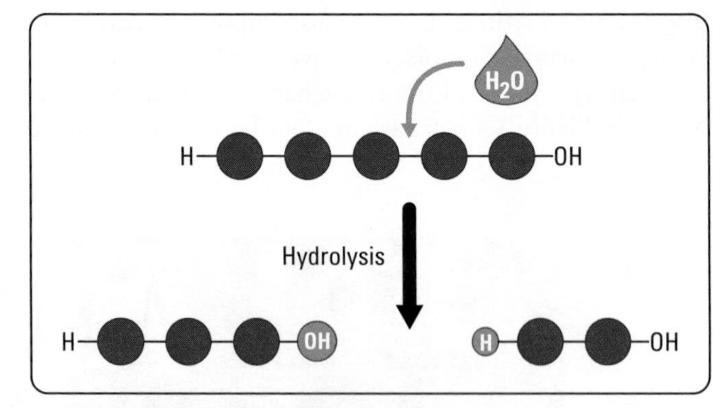

(b) Breaking a polymer chain

Figure 3.6 Synthesis and digestion of polymers. (a) The only atoms shown in these diagrams are hydrogens and hydroxyl groups (—OH) in strategic locations on the monomers. A polymer grows in length when an incoming monomer and the monomer at the end of the existing chain each contribute to the formation of a water molecule. The monomers replace those lost covalent bonds with a bond to each other. **(b)** Hydrolysis reverses the process by breaking down the polymer with the addition of water molecules, which break the bonds between monomers.

(lyse) with water *(hydro)*. Cells break bonds between monomers by adding water to them, a process essentially the reverse of a dehydration reaction.

CHECKPOINT

1. Draw a structural formula for C_2H_4. Remember that each carbon has four bonds; each hydrogen has one.
2. When two molecules of glucose ($C_6H_{12}O_6$) are joined together in a dehydration reaction, what are the formulas of the two products? (Remember that chemical reactions involve the rearrangement of atoms and that no atoms are gained or lost.)

Answers: 1. $H_2C{=}CH_2$ (each C bonded to two H, C=C double bond) **2.** $C_6H_{12}O_6 + C_6H_{12}O_6 \rightarrow C_{12}H_{22}O_{11} + H_2O$

Figure 3.7 Honey, a mixture of two simple sugars. The sweet taste of honey comes from its main ingredients: the monosaccharides glucose and fructose.

Biological Molecules

In the remainder of the chapter, we'll explore the four categories of large molecules in cells: carbohydrates, lipids, proteins, and nucleic acids, the category that includes DNA. For each category, you'll learn about the structure and function of the molecules by first learning about the monomers used to build them.

Carbohydrates

Carbohydrates, commonly known as "carbs," include the small sugar molecules dissolved in soft drinks as well as the long starch molecules we consume in pasta and potatoes. In animals, carbohydrates serve as a primary source of dietary energy; in plants, carbohydrates are used as a building material to form much of the plant body.

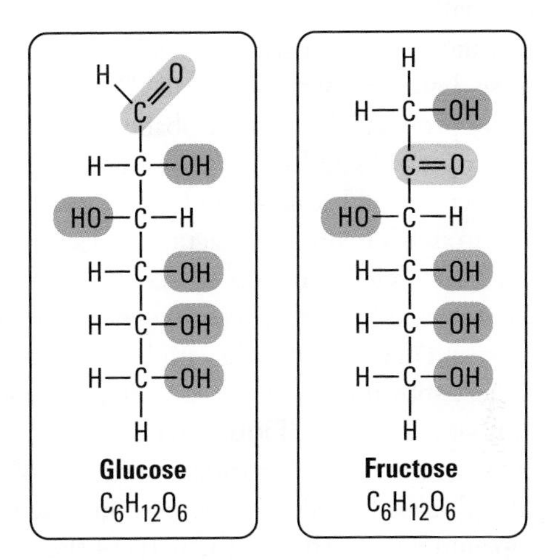

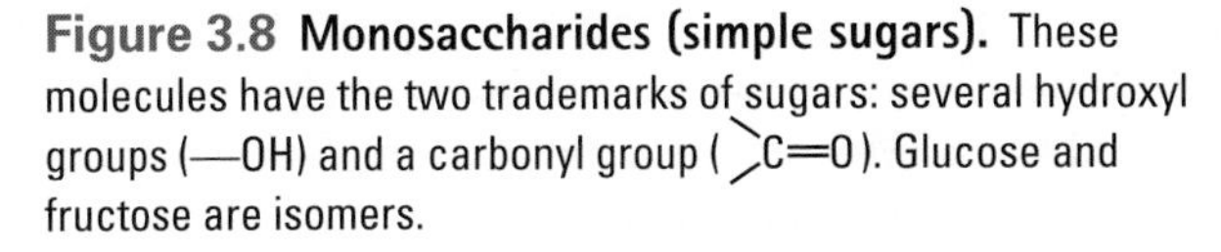

Figure 3.8 Monosaccharides (simple sugars). These molecules have the two trademarks of sugars: several hydroxyl groups (—OH) and a carbonyl group (>C=O). Glucose and fructose are isomers.

Monosaccharides Simple sugars, or **monosaccharides** (from the Greek *mono*, single, and *sacchar*, sugar), include glucose, found in sports drinks, and fructose, found in fruit. Both of these simple sugars are found in honey **(Figure 3.7)**. The molecular formula for glucose is $C_6H_{12}O_6$. Fructose has the same formula, but its atoms are arranged differently **(Figure 3.8)**. Glucose and fructose are examples of **isomers,** molecules that have the same molecular formula but different structures. (Isomers are like anagrams—*heart* and *earth*, for example—words that contain the same letters in a different order.) Because shape is so important, seemingly minor differences in the arrangement of atoms give isomers different properties. In this case, the rearrangement of chemical groups makes fructose taste considerably sweeter than glucose.

It is convenient to draw sugars as if their carbon skeletons were linear. However, in aqueous solutions, many monosaccharides form rings, as shown for glucose in **Figure 3.9**.

Monosaccharides, particularly glucose, are the main fuel molecules for cellular work. Analogous to an automobile engine consuming gasoline,

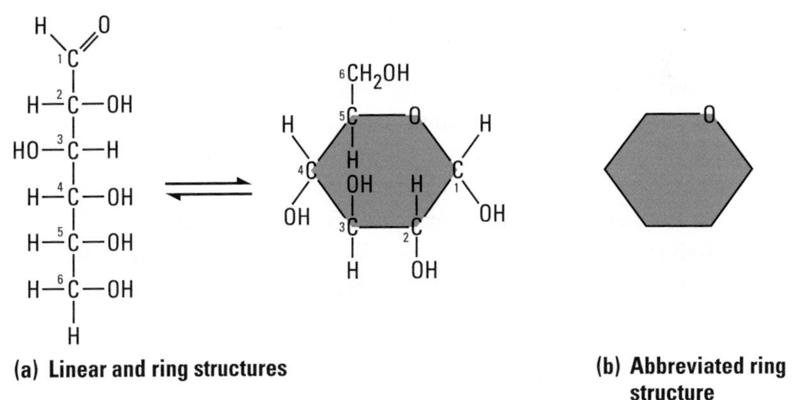

Figure 3.9 The ring structure of glucose. (a) Dissolved in water, one part of a glucose molecule can bond to another part to form a ring. The carbon atoms are numbered here so you can relate the linear and ring versions of the molecule. As the double arrows indicate, ring formation is a reversible process, but at any instant in an aqueous solution, most glucose molecules are rings. **(b)** From now on, we'll use this abbreviated ring symbol for glucose. Each unmarked corner represents a carbon and its attached atoms.

your cells break down glucose molecules and extract their stored energy, giving off carbon dioxide as "exhaust." The rapid conversion of glucose to cellular energy is why intravenous dextrose (a form of glucose) is administered to sick or injured patients in an ambulance or emergency room; an aqueous solution of glucose is injected into the bloodstream to provide an immediate energy source to tissues in need of repair. In addition to their use as an energy source, monosaccharides also provide cells with carbon skeletons that can be used as raw material for manufacturing other kinds of organic molecules.

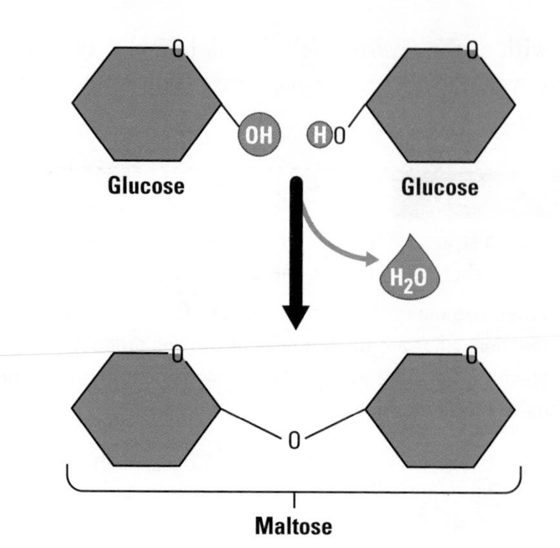

Figure 3.10 **Disaccharide (double sugar) formation.** To form a disaccharide, two simple sugars are joined by a dehydration reaction, in this case forming a bond between two glucose monomers to make the double sugar maltose.

Disaccharides A **disaccharide,** or double sugar, is constructed from two monosaccharides through a dehydration reaction. An example of a disaccharide is maltose, also called malt sugar, which consists of two glucose monomers **(Figure 3.10)**. Maltose, naturally found in germinating seeds, is used in making beer, malt whiskey and liquor, malted milk shakes, and malted milk ball candy.

Lactose, another disaccharide, is made from the monosaccharides glucose and galactose. Lactose is sometimes called "milk sugar" because it is primarily found in dairy products. You have probably heard of lactose intolerance, the inability to properly digest lactose. Cells of the small intestine produce a molecule called lactase that breaks down lactose into glucose and galactose monomers through a hydrolysis reaction. In lactose-intolerant people, these cells produce insufficient amounts of lactase. As a result, lactose is not properly broken down and absorbed, causing uncomfortable symptoms, such as bloating, gas, and diarrhea. People with lactose intolerance must avoid dairy foods, use dairy substitutes (such as soy milk), or supplement their diet with lactase in pill or liquid form **(Figure 3.11)**.

Figure 3.11 **Lactose intolerance.** People who are lactose intolerant can digest foods that contain lactose (such as ice cream) by taking pills that contain the enzyme lactase (inset).

The most common disaccharide is sucrose, common table sugar, which consists of a glucose linked to a fructose. Sucrose is the main carbohydrate in plant sap, and it nourishes all the parts of the plant. Sucrose is extracted from the stems of sugarcane or, more commonly in the United States, the roots of sugar beets. However, sucrose is rarely used as a sweetener in processed foods. Much more common is high-fructose corn syrup (HFCS), made through a commercial process that converts natural glucose in corn syrup to the much sweeter fructose. If you read the label on any soft drink can or bottle, you're likely to find that HFCS is the first or second ingredient listed.

The United States is one of the world's leading markets for sweeteners, with the average American consuming about 64 kilograms (Kg)—that's a whopping 140 pounds!—per year, mainly as sucrose and HFCS **(Figure 3.12)**. This national "sweet tooth" persists in spite of our growing awareness about how sugar can negatively affect our health. Sugar is a major cause of tooth decay. High sugar consumption also tends to replace eating more varied and nutritious foods. The description of sugars as "empty calories" is accurate in

Figure 3.12 **A year's supply of sugar.** Americans consume an average of 64 kg (140 lb) of sweetener per person per year—over a third of a pound per day. The 64 kg of sucrose (table sugar) in this photograph will give you some idea of that level of consumption. Reading food labels will make you aware of the amount of sugar in processed foods.

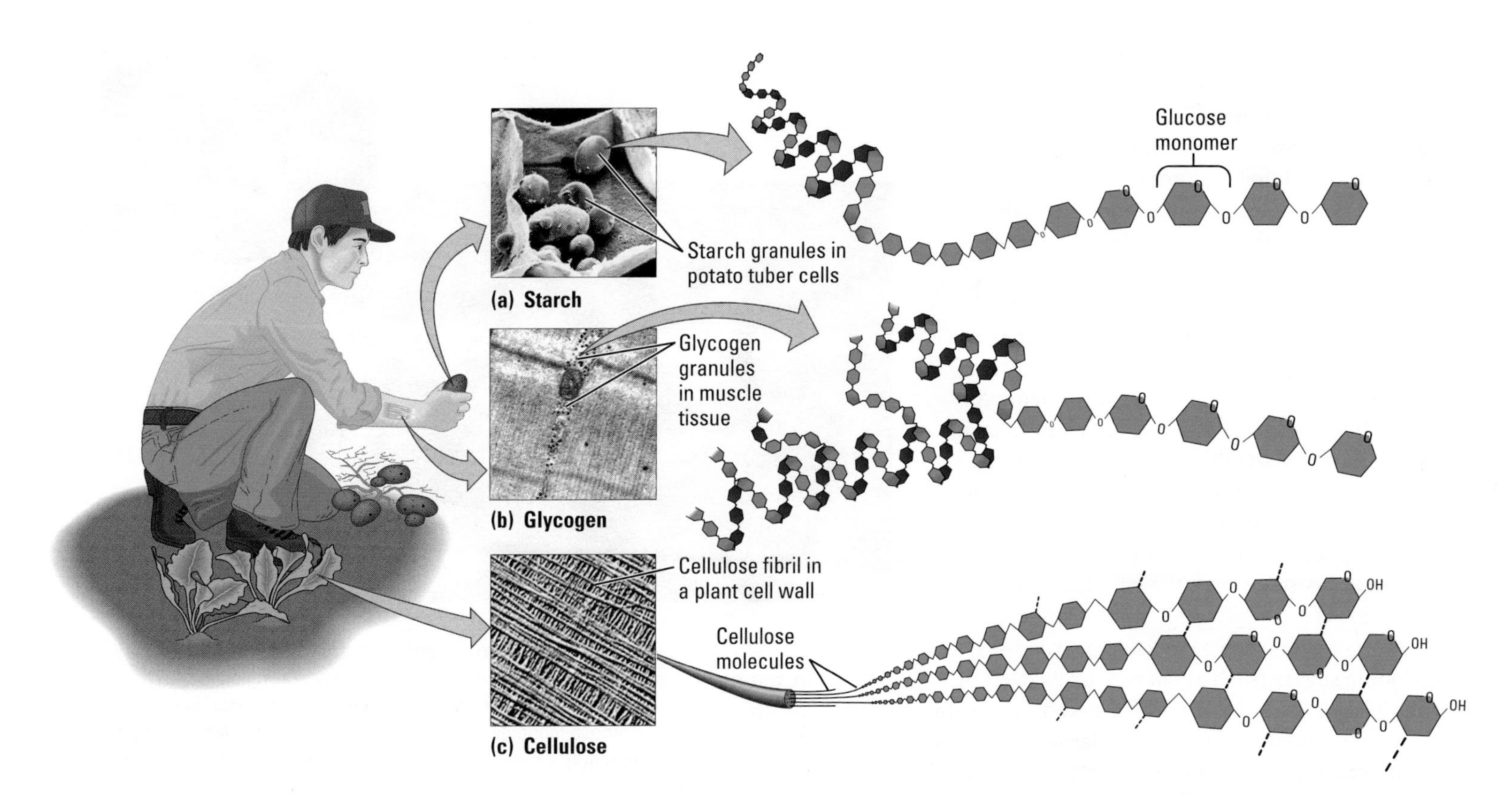

Figure 3.13 Polysaccharides. (a) Plants store glucose by polymerizing it in the form of starch. **(b)** Animals also store glucose, but in the form of glycogen, a polysaccharide more extensively branched than starch. **(c)** The cellulose of plant cell walls is an example of a structural polysaccharide. The cellulose molecules are assembled into fibrils that make up the main fabric of the walls. Wood, a cell wall material consisting of such cellulose fibrils along with other polymers, is strong enough to support trees hundreds of feet high. We take advantage of that structural strength in our use of lumber as a building material.

the sense that most sweeteners contain only negligible amounts of nutrients other than carbohydrates. For good health, we also require proteins, fats, vitamins, and minerals. And we need to include substantial amounts of complex carbohydrates—that is, polysaccharides—in our diet.

Polysaccharides Complex carbohydrates, or **polysaccharides,** are long chains of sugar units—polymers of monosaccharides. One familiar example is starch, found in roots and other plant organs. **Starch** consists of many glucose monomers strung together **(Figure 3.13a)**. Plant cells store starch in granules, where it is available as a sugar stockpile that can be broken down as needed to provide energy and raw material for building other molecules. Potatoes and grains, such as wheat, corn, and rice, are the major sources of starch in the human diet. Humans and most other animals are able to use plant starch as food by hydrolyzing the bonds between glucose monomers within their digestive systems.

Animals store excess sugar in the form of a polysaccharide called **glycogen.** Glycogen is similar in structure to starch in that it is also a polymer of glucose monomers, but glycogen is more extensively branched **(Figure 3.13b)**. Most of our glycogen is stored as granules in our liver and muscle cells, which break down the glycogen to release glucose when it is needed for energy. This is the basis for "carbo loading," the consumption of large amounts of starchy foods the night before an athletic event. The starch is converted to glycogen, which is then available for rapid use during physical activity the next day.

In addition to playing an important role in nutrition, certain polysaccharides serve as structural components. **Cellulose,** the most abundant organic compound on Earth, forms cable-like fibrils in the tough walls that enclose plant cells and is a major component of wood **(Figure 3.13c)**.

Cellulose resembles starch and glycogen in being a polymer of glucose, but its glucose monomers are linked together in a different orientation. Unlike the glucose linkages in starch and glycogen, those in cellulose cannot be broken by most animals. The cellulose in plant foods, which passes unchanged through our digestive tract, is commonly known as dietary "fiber" or "roughage." Because it remains undigested, fiber does not serve as a nutrient, although it does appear to help keep our digestive system healthy. Most Americans do not get the recommended levels of fiber in their diet. Foods rich in fiber include fruits and vegetables, whole grains, bran, and beans. Grazing animals and wood-eating insects such as termites, which do derive nutrition from cellulose, have prokaryotes inhabiting their digestive tracts that break the cellulose down **(Figure 3.14)**.

Simple sugars (such as glucose or fructose) and double sugars (such as sucrose or lactose) dissolve readily in water, forming sugary solutions, such as in soft drinks. In contrast, cellulose and some forms of starch are such large molecules that they do not dissolve in water. In spite of this difference, almost all carbohydrates are **hydrophilic,** which literally means "water-loving." Hydrophilic molecules adhere water to their surface. It is the hydrophilic quality of cellulose that makes a fluffy bath towel so water absorbent.

Figure 3.14 Prokaryotes in grazing animals. Prokaryotes live within the digestive tract of many grazing animals. The prokaryotes break down the cellulose a cow eats, converting it to glucose monomers, which the cow can digest. A by-product of this reaction is the production of large amounts of methane gas.

Low-Carb Diets In recent years, carbohydrates have gained considerable attention as the target of weight loss programs. You or someone you know has probably tried a "low-carb diet" to lose weight. For many people, cutting back on carbohydrates is a valid way to lose weight for the simple reason that the majority of calories in a typical American diet come from carbohydrates: Cutting carbs means cutting calories. Consumers need to be wary, however, of new products that boast that they are "low carb" or low in "net carbs." These terms are not regulated by the government and may be placed on *any* product, even extremely unhealthy ones. To lose weight, the U.S. Department of Agriculture (USDA) recommends cutting back on the calories you take in (by reducing consumption of carbohydrates or other calorie-rich foods) while simultaneously increasing calories burned (through exercise).

CHECKPOINT

1. If the monosaccharides glucose and fructose both have the same molecular formula ($C_6H_{12}O_6$), how can they have different properties?
2. How do manufacturers produce the high-fructose corn syrup listed as an ingredient on a soft drink bottle? Why is this profitable?

Answers: 1. The atoms are arranged differently, forming molecules with different shapes and different chemical properties. **2.** A commercial process converts glucose in the syrup to the much sweeter fructose. Less syrup will then have to be used, saving money for the manufacturer.

Lipids

In contrast to carbohydrates and most other biological molecules, **lipids** are **hydrophobic,** which means that they do not mix with water (from the Greek *hydro*, water, and *phobos*, fearing). You have probably observed this chemical behavior in an unshaken bottle of salad dressing: The oil, which is a type of lipid, separates from the vinegar, which is mostly water. If you shake the bottle, you can force a temporary mixture long enough to douse your salad with dressing, but what remains in the bottle will quickly separate again once you stop shaking it. Lipids are a diverse set of molecules; two examples are fats and steroids.

Fats Dietary **fat** consists largely of molecules of triglyceride. Each **triglyceride** is made of a glycerol molecule joined with three fatty acid molecules via dehydration reactions **(Figure 3.15).** The major portion of a fatty acid is a long hydrocarbon, which, like the hydrocarbons of gasoline, stores a lot of energy. In fact, a pound of fat packs more than twice as much energy as a pound of carbohydrate such as starch. The downside to this energy efficiency is that it is very difficult for a person trying to lose weight to "burn off" excess body fat. It is important to understand that a reasonable amount of body fat is both normal and healthy as a fuel reserve. We stock these long-term food stores in specialized reservoirs called adipose cells, which swell and shrink when we deposit and withdraw fat from them. In addition to storing energy, adipose tissue cushions vital organs and insulates us, helping maintain a warm body temperature even when the outside air is cold.

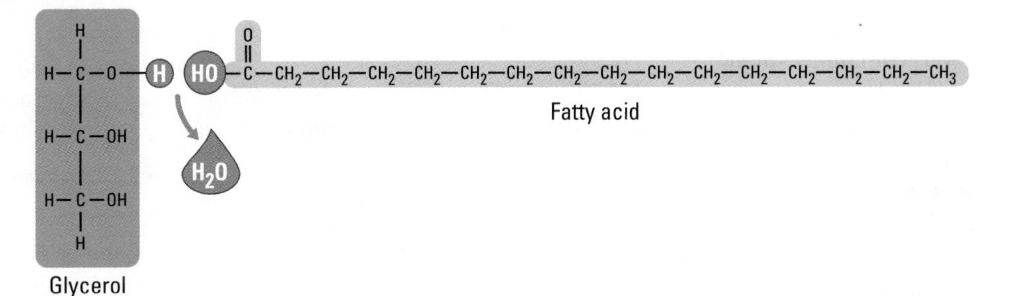

(a) A dehydration reaction linking a fatty acid to glycerol

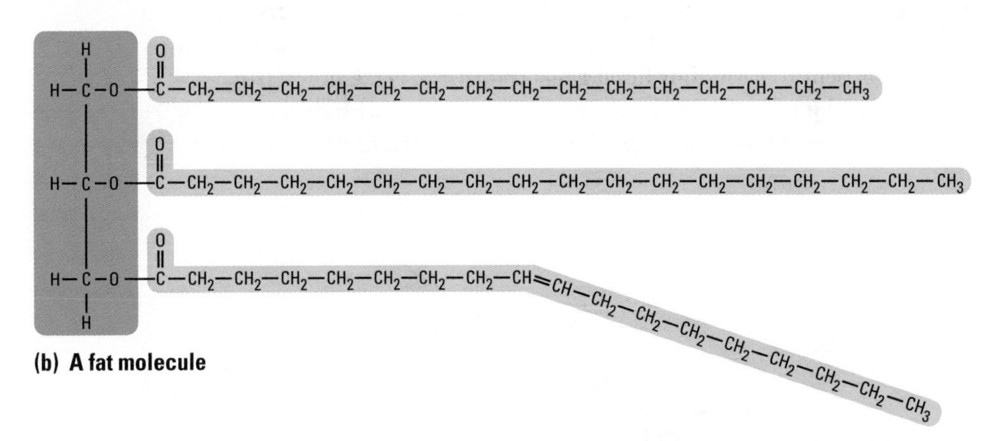
(b) A fat molecule

Figure 3.15 The synthesis and structure of a fat, or triglyceride. **(a)** This diagram shows the first of three fatty acids that will attach to glycerol through a dehydration reaction. **(b)** The finished fat has a glycerol "head" and three fatty acid "tails." The fatty acids consist mainly of energy-rich hydrocarbons.

Notice in Figure 3.15b that one of the fatty acids bends where there is a double bond in the carbon skeleton. That fatty acid is said to be **unsaturated** because it has fewer than the maximum number of hydrogens at the location of the double bond. The other two fatty acids in the fat molecule lack double bonds in their hydrocarbon portions. Those fatty acids are **saturated,** meaning that they contain the maximum number of hydrogen atoms. A saturated fat is one with all three of its fatty acids saturated. If one or more of the fatty acids is unsaturated, then it's an unsaturated fat, such as the one in Figure 3.15b. A polyunsaturated fat has several double bonds within its fatty acids.

Most animal fats, such as lard and butter, have a relatively high proportion of saturated fatty acids. The linear shape of saturated fatty acids allows them to stack easily, so saturated fats tend to be solid at room temperature. Diets rich in saturated fats may contribute to cardiovascular disease by promoting **atherosclerosis.** In this condition, lipid-containing deposits called plaque build up within the walls of blood vessels, reducing blood flow and increasing risk of heart attacks and strokes. In contrast, plant and fish fats are relatively high in unsaturated fatty acids. The bent shape of unsaturated fatty acids makes them less likely to form solids, so unsaturated fats are usually liquid at room temperature. Vegetable oils (such as corn and canola oil) and fish oils (such as cod liver oil) are examples.

While plant oils tend to be low in saturated fat, tropical plant oils are an exception. Cocoa butter, a main ingredient in chocolate, contains a mix of saturated and unsaturated fat that gives it a melting point near body temperature. Thus, chocolate stays solid at room temperature but melts in the mouth. This pleasing "mouth feel" is one of the reasons chocolate is so appealing.

Sometimes, such as when producing margarine and peanut butter, a food manufacturer wishes to use a vegetable oil but needs the food product to be solid. To achieve this, the manufacturer can convert unsaturated fats to saturated fats by adding hydrogen, a process called **hydrogenation.** Unfortunately, hydrogenation also creates **trans fat,** a type of unsaturated fat that is even more unhealthy than saturated fat. Starting in 2006, the FDA will require trans fats to be specifically listed in the nutrition label of all foods containing them.

While saturated and trans fats should generally be avoided, it is not true that *all* fats are unhealthy. In fact, some fats perform important functions within the body and are beneficial and even essential to a healthy diet. For example, omega-3 fatty acids, found in foods such as nuts and oily fish such as salmon **(Figure 3.16)**, have been shown to reduce the risk of coronary heart disease and relieve the symptoms of arthritis and inflammatory bowel disease.

Figure 3.16 Beneficial fats. The foods pictured here are all rich in omega-3 fatty acids, which have been shown to reduce the risk of heart disease.

Steroids Classified as lipids because they are hydrophobic, **steroids** are very different from fats in structure and function. The carbon skeleton of a steroid is bent to form four fused rings **(Figure 3.17)**. Cholesterol, which gets a lot of bad press because of its association with cardiovascular disease, is a steroid. But cholesterol is also an essential molecule in your

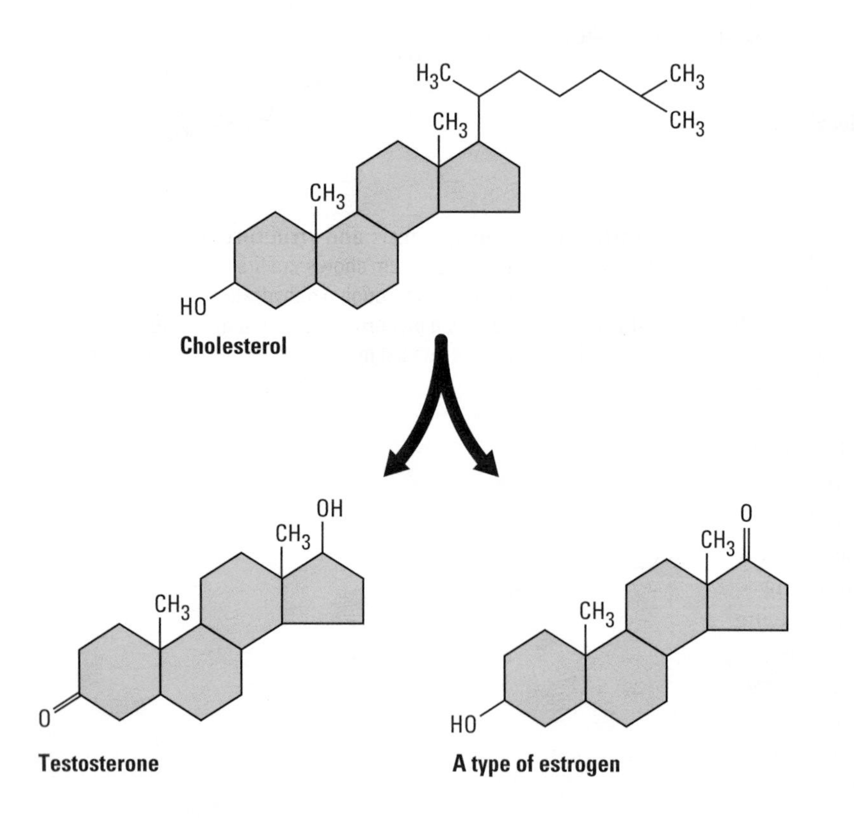

Figure 3.17 Examples of steroids. All steroids have a carbon skeleton consisting of four fused rings, abbreviated here with all the atoms of the rings omitted. Different steroids vary in the functional groups attached to this core set of rings, and these variations affect their function. For example, the subtle difference between testosterone and estrogen influences the development of the anatomical and physiological differences between male and female mammals, including humans.

Figure 3.18 Steroids and the modern athlete. Baseball player Jose Canseco, an admitted user of performance-enhancing steroids, testifies before Congress during a 2005 inquiry into the abuse of drugs by professional athletes.

body. As we will see in Chapter 4, cholesterol and phospholipids, another type of fat, are key components of the membranes that surround your cells. Cholesterol is also the "base steroid" from which your body produces other steroids, including estrogen and testosterone, the steroids that function as sex hormones.

The controversial drugs called **anabolic steroids** are synthetic variants of testosterone, the male sex hormone. Testosterone causes a general buildup in muscle and bone mass during puberty in males and maintains masculine traits throughout life. Because anabolic steroids structurally resemble testosterone, they also mimic some of its effects. Some athletes use anabolic steroids to build up their muscles quickly and enhance their performance.

In 2003, the discovery that some athletes were using a new anabolic steroid called THG rocked the sports world. THG is a chemically modified ("designer") steroid intended to avoid detection by drug tests. New tests revealed widespread use of THG and other steroids among athletes from many sports **(Figure 3.18)**.

Using anabolic steroids is indeed a fast way to increase body size beyond what hard work alone can produce. But at what cost? Steroid abuse can cause serious physical and mental problems, including violent mood swings ("'roid rage"), depression, liver damage, high cholesterol, shrunken testicles, a reduced sex drive, and infertility. These last symptoms occur because anabolic steroids often cause the body to reduce its normal output of sex hormones. Most athletic organizations now ban the use of anabolic steroids because of their many potential health hazards coupled with the unfairness of an artificial advantage.

CHECKPOINT

1. On a food package label, what is the meaning of "unsaturated fats"?
2. In classifying the molecules of life, what do dietary fats and human sex hormones have in common?

Answers: 1. Unsaturated fats have fewer than the maximum number of hydrogens. Double bonds occur between some of the carbons in the fatty acids instead of bonds between carbon and hydrogen. **2.** Both fats and sex hormones, which are steroids, are classified as lipids because they are hydrophobic.

(a) Structural proteins provide support. Examples are the proteins found in hair, horns, feathers, spider webs, and connective tissues such as tendons and ligaments.

(b) Storage proteins, found in seeds and eggs, provide a source of amino acids for developing plants and animals.

(c) Contractile proteins are found primarily in muscles.

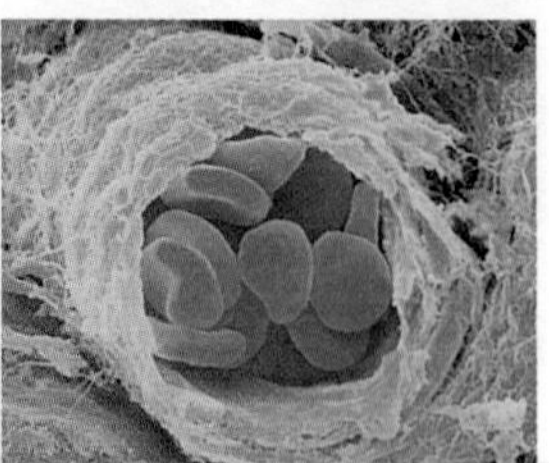

(d) Transport proteins include hemoglobin, the iron-containing protein in blood that conveys oxygen from your lungs to other parts of the body. The red blood cells in this photograph contain hemoglobin.

Figure 3.19 Some functions of proteins.

Proteins

MP3 Tutor
Protein Structure and Function

Proteins are the most elaborate of life's molecules. A **protein** is a polymer constructed from amino acid monomers. Your body has tens of thousands of different kinds of proteins, and each kind of protein has a unique three-dimensional shape that corresponds to a specific function. Proteins perform most of the tasks the body needs to function. **Figure 3.19** surveys the functions of four types of proteins: structural proteins, storage proteins, contractile proteins, and transport proteins. Other types of proteins include defensive proteins, such as antibodies of the immune system, and signal proteins, which convey messages from one cell to another. Enzymes (such as lactase), another important type of protein, change the rate of a chemical reaction without being changed in the process (as you will see in Chapter 5). Now let's take a look at the architecture of proteins.

The Monomers: Amino Acids All proteins are constructed from a common set of 20 kinds of amino acids. Each **amino acid** consists of a central carbon atom bonded to four covalent partners (carbon, remember, always forms four covalent bonds). Three of those attachments are common to all 20 amino acids: a carboxyl group (—COOH), an amino group (—NH_2), and a hydrogen atom. The variable component of amino acids, the side group (also called the radical group), is attached to the fourth bond of the central carbon. Each type of amino acid has a unique side group, giving that amino acid its special chemical properties **(Figure 3.20)**. You read about one amino acid, tryptophan, in the opening essay of this chapter.

Proteins as Polymers Cells link amino acids together by—you guessed it—dehydration reactions. The resulting bond between adjacent amino acids is called a **peptide bond (Figure 3.21)**. Proteins usually consist of 100 or more amino acids, forming a chain called a **polypeptide.**

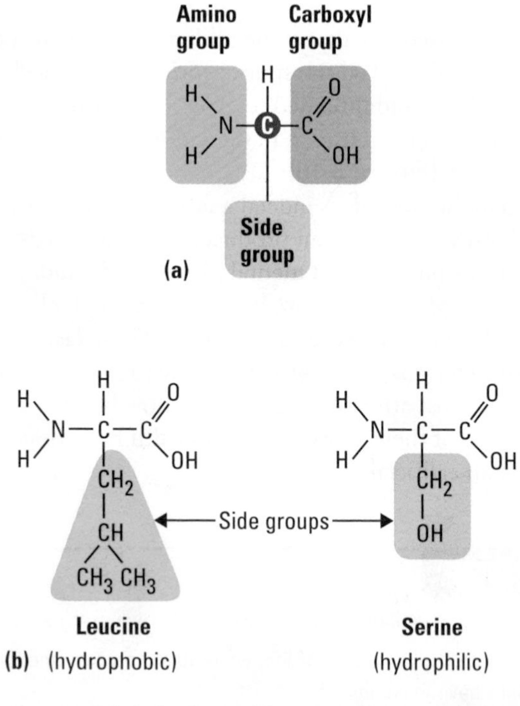

Figure 3.20 Amino acids. (a) The general structure of an amino acid. **(b)** The 20 amino acids vary only in their side groups, which give these monomers their unique properties. For example, the side group of the amino acid leucine is pure hydrocarbon. That region of leucine is hydrophobic, because hydrocarbons don't mix with water. In contrast, the side group of the amino acid serine has a hydroxyl (—OH) group, which is hydrophilic.

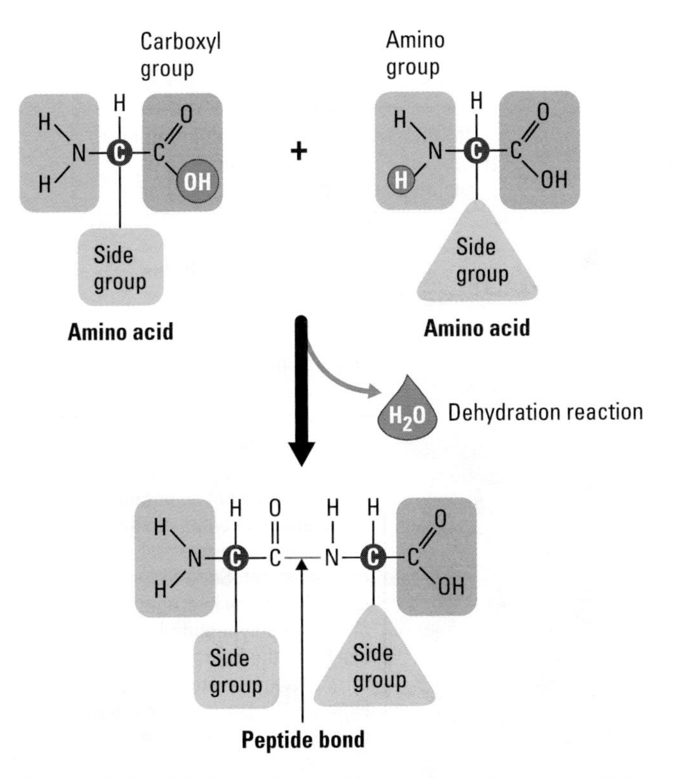

Figure 3.21 Joining amino acids. A dehydration reaction links adjacent amino acids by a peptide bond.

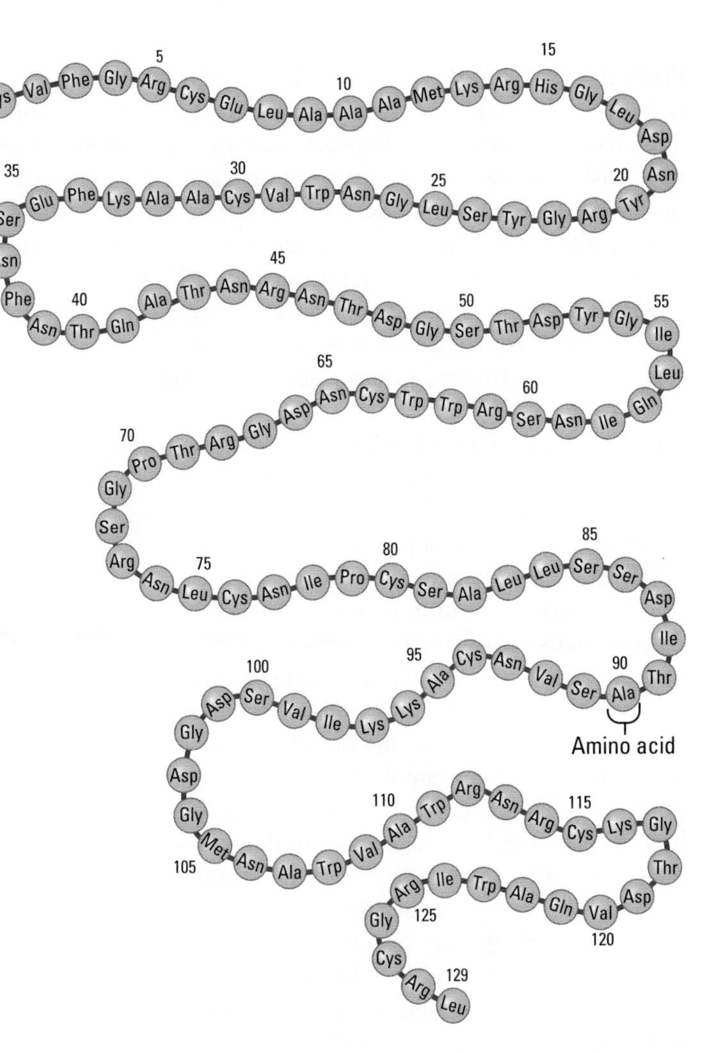

Figure 3.22 The primary structure of a protein. This is the unique amino acid sequence, or primary structure, of a protein called lysozyme. (The chain was drawn in serpentine fashion so that it would fit on the page. The actual shape of lysozyme is much more complex.) The names of the amino acids are given as their three-letter abbreviations, with the positions of lysozyme's 129 amino acids numbered along the chain.

Your body has tens of thousands of different kinds of protein. How is it possible to make such a huge variety of proteins from just 20 kinds of amino acids? The answer is arrangement. You know that you can make many different words by varying the sequence of just 26 letters. Though the protein alphabet is slightly smaller (just 20 "letters"), the "words" are much longer, with a typical polypeptide being at least 100 amino acids in length. Just as each word is constructed from a unique succession of letters, each protein has a unique linear sequence of amino acids. This specific amino acid sequence is called the protein's **primary structure** (**Figure 3.22**).

Changing a single letter can drastically affect the meaning of a word—"tasty" versus "nasty," for instance. Similarly, even a slight change in primary structure can affect a protein's ability to function. Consider, for example, the substitution of one amino acid for another at a particular position in hemoglobin, the blood protein that carries oxygen. Such an amino acid swap is the cause of sickle-cell disease, an inherited blood disorder (**Figure 3.23**).

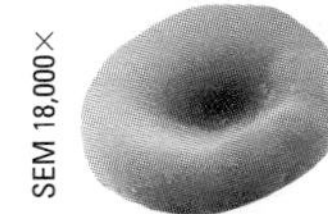

Normal red blood cell

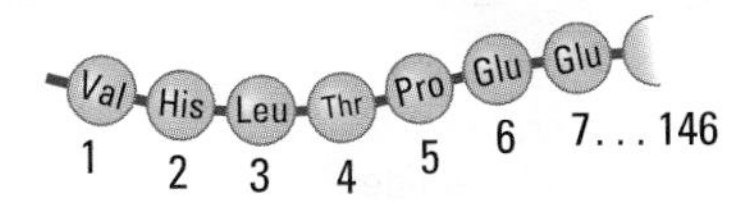
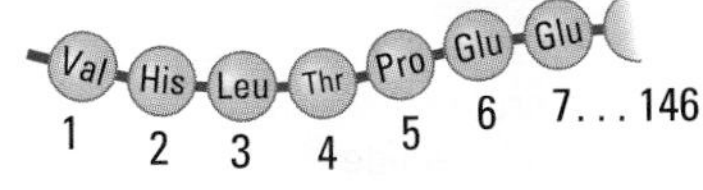

Normal hemoglobin

(a) Normal hemoglobin. Red blood cells of humans are normally disk-shaped. Each cell contains millions of molecules of the protein hemoglobin, which transports oxygen from the lungs to other organs of the body. Next to the photograph, you can see the first 7 of the 146 amino acids in a polypeptide chain of hemoglobin.

Sickled red blood cell

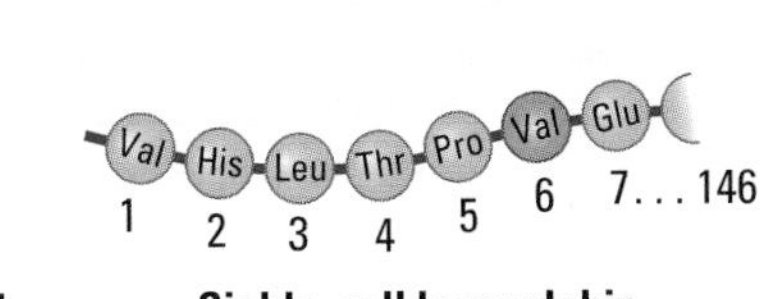

Sickle-cell hemoglobin

(b) Sickle-cell hemoglobin. A slight change in the primary structure of hemoglobin causes sickle-cell disease. The inherited substitution of one amino acid—valine in place of the amino acid glutamic acid—occurs in the number 6 position of the polymer. The abnormal hemoglobin molecules tend to crystallize, deforming some of the cells into a sickle shape. The life of someone with the disease is characterized by dangerous episodes when the angular cells clog tiny blood vessels, impeding blood flow.

Figure 3.23 A single amino acid substitution in a protein causes sickle-cell disease.

Protein Shape At this point, you might be thinking that a polypeptide chain is the same thing as a protein, but that's not quite true. The distinction between the two is analogous to the relationship between a long strand of yarn and a sweater of particular size and shape that you could knit from the yarn. A functional protein is not just a polypeptide chain, but one or more polypeptides precisely twisted, folded, and coiled into a molecule of unique shape. If we dissect the overall shape of a protein, we can recognize at least three levels of structure: primary, secondary, and tertiary. Proteins with more than one polypeptide chain have a fourth level: quaternary structure. You can examine these levels of protein structure in **Figure 3.24**.

When a cell makes a polypeptide, the chain usually folds spontaneously to form the functional shape for that protein. It is a protein's three-dimensional shape that enables the molecule to carry out its specific function in a cell. In almost every case, a protein's function depends on its ability to recognize and bind to some other molecule. For example, the receptors on brain cells that recognize serotonin—the sleep-inducing chemical you read about in the opening essay—are actually proteins. If the protein receptor's shape were to be altered, then it would not be able to perform this recognition function. With proteins, *function follows form*—that is, what a protein does is a consequence of its shape.

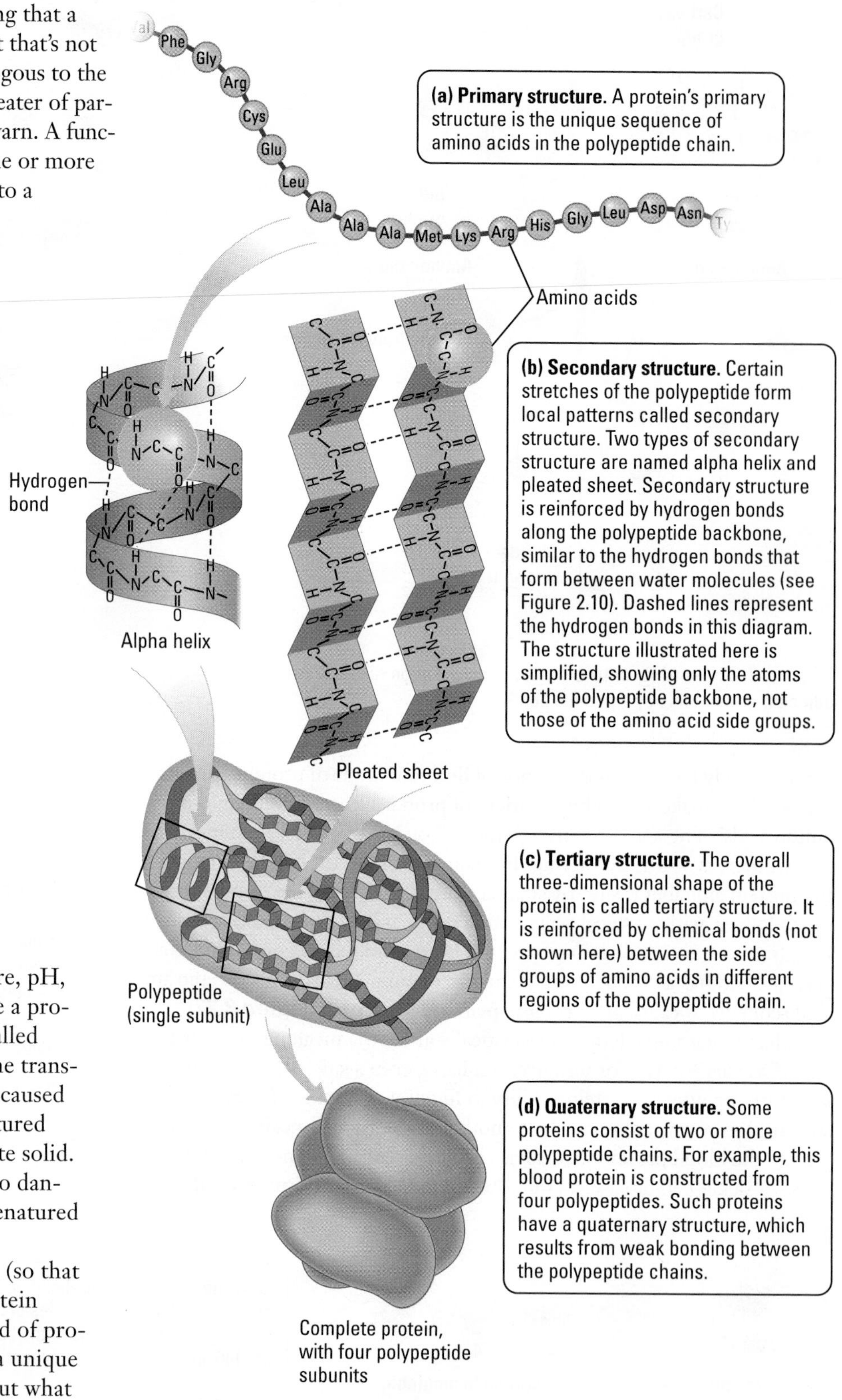

What Determines Protein Structure? A protein's shape is sensitive to the surrounding environment. An unfavorable change in temperature, pH, or some other quality of the environment can cause a protein to unravel and lose its normal shape. This is called **denaturation** of the protein. If you cook an egg, the transformation of the egg white from clear to opaque is caused by proteins in the egg white denaturing. The denatured proteins become insoluble in water and form a white solid. One of the reasons why extremely high fevers are so dangerous is that some proteins in the body become denatured above about 104°F.

Given an environment suitable for that protein (so that it doesn't denature), the primary structure of a protein causes it to fold into its functional shape. Each kind of protein has a unique primary structure and therefore a unique shape that enables it to do a certain job in a cell. But what determines primary structure, a protein's specific amino acid sequence? Each polypeptide chain has a sequence specified by an inherited gene. And that relationship between genes and proteins brings us to this chapter's last category of molecules.

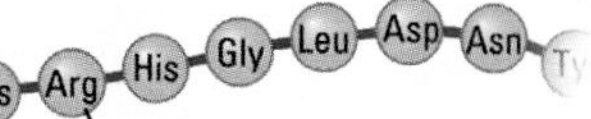

(a) Primary structure. A protein's primary structure is the unique sequence of amino acids in the polypeptide chain.

Amino acids

(b) Secondary structure. Certain stretches of the polypeptide form local patterns called secondary structure. Two types of secondary structure are named alpha helix and pleated sheet. Secondary structure is reinforced by hydrogen bonds along the polypeptide backbone, similar to the hydrogen bonds that form between water molecules (see Figure 2.10). Dashed lines represent the hydrogen bonds in this diagram. The structure illustrated here is simplified, showing only the atoms of the polypeptide backbone, not those of the amino acid side groups.

Hydrogen bond

Alpha helix

Pleated sheet

(c) Tertiary structure. The overall three-dimensional shape of the protein is called tertiary structure. It is reinforced by chemical bonds (not shown here) between the side groups of amino acids in different regions of the polypeptide chain.

Polypeptide (single subunit)

(d) Quaternary structure. Some proteins consist of two or more polypeptide chains. For example, this blood protein is constructed from four polypeptides. Such proteins have a quaternary structure, which results from weak bonding between the polypeptide chains.

Complete protein, with four polypeptide subunits

Figure 3.24 The four levels of protein structure. The example shown in this figure is a protein that transports certain hormones and vitamins in the bloodstream.

CHECKPOINT

1. __________ is to polysaccharide as amino acid is to __________.
2. Which of the following is *not* composed of proteins: hair, muscle, cellulose, enzymes, antibodies?
3. In what way is the production of a polypeptide similar to the production of a polysaccharide?
4. Why does a denatured protein no longer function?

Answers: 1. Monosaccharide (or simple sugar); protein (or polypeptide) **2.** Cellulose **3.** Both of these polymers consist of monomers linked by dehydration reactions. **4.** Protein function depends on shape, which is altered by denaturation.

Nucleic Acids

Nucleic acids are information storage molecules that provide the directions for building proteins. The name *nucleic* comes from their location in the nuclei of eukaryotic cells. There are actually two types of nucleic acids: **DNA** (those most famous of chemical initials, which stand for deoxyribonucleic acid) and **RNA** (for ribonucleic acid). The genetic material that humans and other organisms inherit from their parents consists of giant molecules of DNA. Within the DNA are genes, specific stretches of DNA that program the amino acid sequences (primary structure) of proteins. Those programmed instructions, however, are written in a kind of chemical code that must be translated from "nucleic acid language" to "protein language." A cell's RNA molecules help make this translation **(Figure 3.25)**. You'll learn more about how DNA and RNA work in Chapter 10.

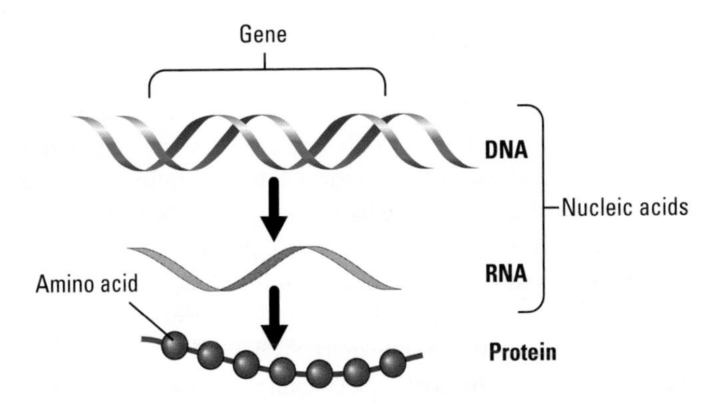

Figure 3.25 Building a protein. Within the cell, a gene (a segment of DNA) provides the directions to build a molecule of RNA, which can then be translated into a protein.

Nucleic acids are polymers made from monomers called **nucleotides (Figure 3.26)**. Each nucleotide contains three parts. At the center of each nucleotide is a five-carbon sugar, deoxyribose in DNA and ribose in RNA. Attached to the sugar is a negatively charged phosphate group containing a phosphorus atom bonded to oxygen atoms (PO_4^-). Also attached to the sugar is a nitrogen-containing base **(nitrogenous base)** made of one or two rings. (It is called a base because it behaves like a base, accepting H^+ in aqueous solutions.) The sugar and phosphate are the same in all nucleotides; only the base varies. Each DNA nucleotide has one of the following four bases: adenine (abbreviated A), guanine (G), cytosine (C), or thymine (T). Thus, all genetic information is written in a four-letter alphabet—A, G, C, T—the bases that distinguish the four nucleotides that make up DNA **(Figure 3.27)**.

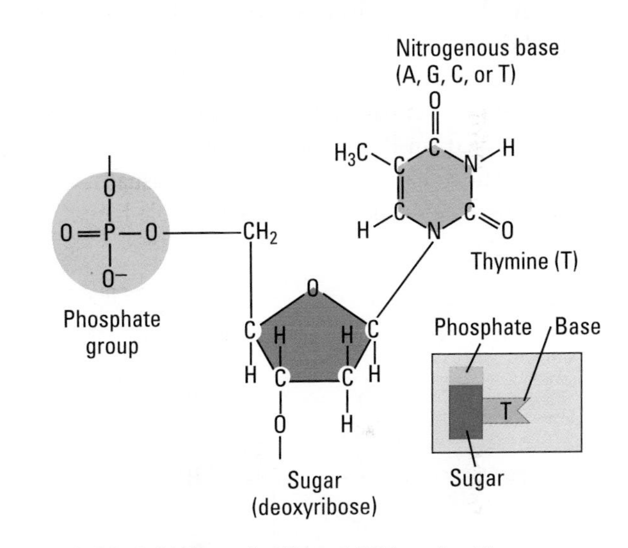

Figure 3.26 A DNA nucleotide. A DNA nucleotide monomer consists of three parts: a sugar (deoxyribose); a phosphate; and a nitrogenous base. (At the bottom of this figure, the three parts of a nucleotide are symbolized with shapes and colors rather than actual chemical structures.)

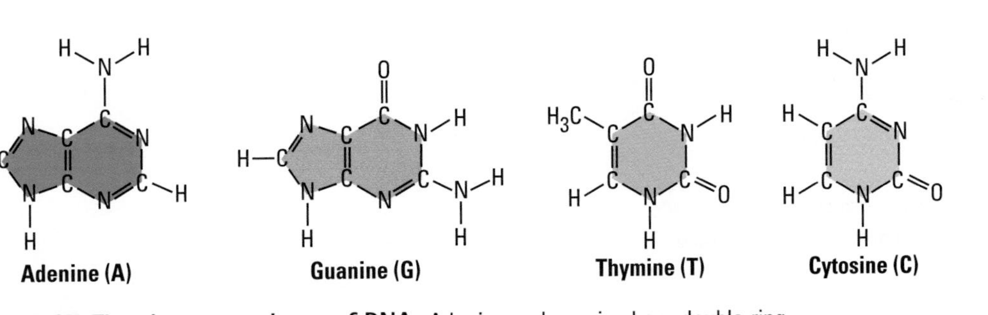
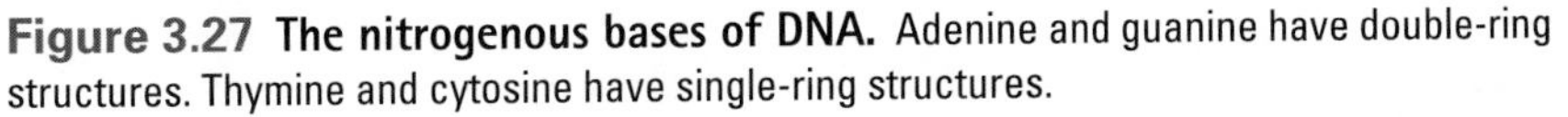

Figure 3.27 The nitrogenous bases of DNA. Adenine and guanine have double-ring structures. Thymine and cytosine have single-ring structures.

Nucleotide monomers are linked into long chains called polynucleotides, or DNA strands **(Figure 3.28a)**. Nucleotides are joined together by covalent bonds between the sugar of one nucleotide and the phosphate of the next. This results in a **sugar-phosphate backbone,** a repeating pattern of sugar-phosphate-sugar-phosphate, with the bases hanging off the backbone like appendages. Polynucleotides vary in length from long to very long, so the number of possible polynucleotide sequences is very great. One long polynucleotide may contain many genes, each a specific series of hundreds or thousands of nucleotides. And each of these genes stores information in its unique sequence of nucleotide bases. In fact, it is this information that cells translate into an amino acid sequence to make a specific protein.

A molecule of DNA is double-stranded, with two polynucleotides wrapped around each other to form a **double helix (Figure 3.28b)**. In the central core of the helix, the bases along one DNA strand hydrogen-bond to bases along the other strand. This base pairing is specific: The base A can pair only with T, and G can pair only with C. Thus, if you know the sequence of bases along one DNA strand, you also know the sequence along the complementary strand in the double helix. As we will see in Chapter 10, this unique base pairing is the basis of DNA's ability to act as the molecule of inheritance.

What about RNA? As its name ribonucleic acid implies, its sugar is ribose rather than deoxyribose. By comparing the RNA nucleotide in **Figure 3.29** with the DNA nucleotide in Figure 3.26, you can see that the RNA ribose sugar has an extra —OH group compared with the DNA deoxyribose sugar (*deoxy* means "without an oxygen"). Another difference between RNA and DNA is that instead of the base thymine, RNA has a similar but distinct base called uracil (U). Except for the presence of ribose and uracil, an RNA polynucleotide chain is identical to a DNA polynucleotide chain. However, RNA is usually found in single-stranded form, while DNA usually exists as a double helix.

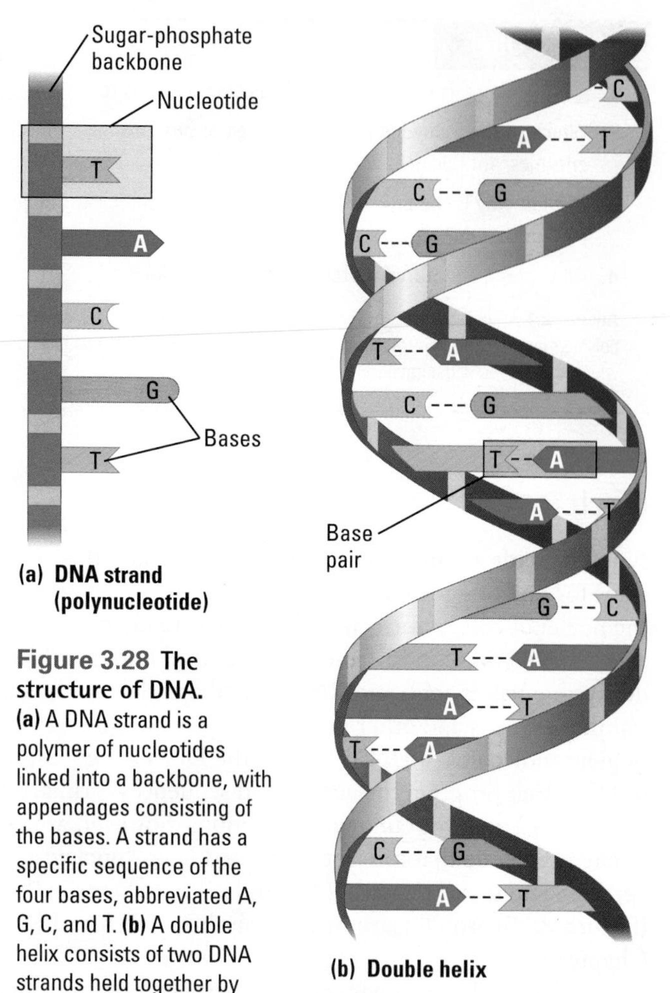

Figure 3.28 The structure of DNA. **(a)** A DNA strand is a polymer of nucleotides linked into a backbone, with appendages consisting of the bases. A strand has a specific sequence of the four bases, abbreviated A, G, C, and T. **(b)** A double helix consists of two DNA strands held together by bonds between bases. The bonds are individually weak—they are hydrogen bonds, likes those between water molecules—but they zip the two strands together with a cumulative strength that gives the double helix its stability. The base pairing is specific: A always pairs with T; G always pairs with C.

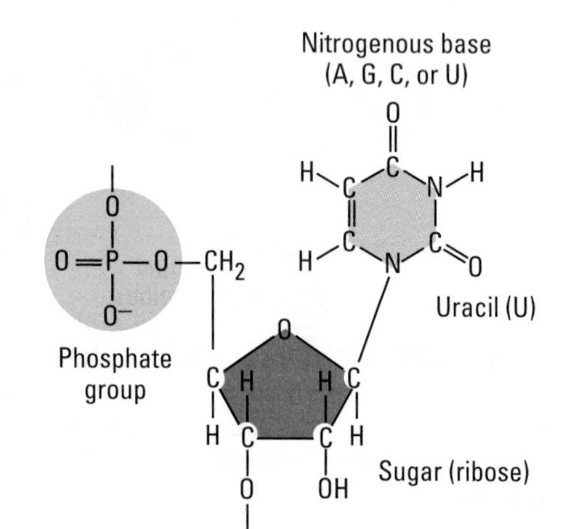

Figure 3.29 An RNA nucleotide. Notice that this RNA nucleotide differs from the DNA nucleotide in Figure 3.26 in two ways. The RNA sugar is ribose rather than deoxyribose, and the base is uracil (U) instead of thymine (T). The other three kinds of RNA nucleotides have the bases A, C, and G, as in DNA.

CHECKPOINT

1. One molecule of DNA contains __________ polynucleotide strands, each of which is composed of __________ different kinds of nucleotides. (Provide two numbers.)
2. In a double helix, a region along one DNA strand has the sequence GAATGC. What is the base sequence along the complementary region of the other strand of the double helix?

Answers: 1. two; four **2.** CTTACG

EVOLUTION CONNECTION

DNA and Proteins as Evolutionary Tape Measures

Genes (DNA) and their products (proteins) are historical documents. These information-rich molecules are the records of an organism's hereditary background. The linear sequences of nucleotides in DNA molecules are passed from parents to offspring, and these DNA sequences determine the amino acid sequences of proteins in the offspring (see Figure 3.25). The DNA and proteins of siblings are more similar than the DNA and proteins of unrelated individuals of the same species. This concept of molecular genealogy also extends to relationships between species.

Testable hypotheses are at the heart of science, and analysis of DNA and protein sequences adds a new tool for testing evolutionary hypotheses. For example, fossil evidence and anatomical similarity support the hypothesis that humans and monkeys are closely related animals. This hypothesis is testable: We can use it to make predictions about what to expect if the hypothesis is correct. For example, if humans and monkeys are closely related, then they should share a greater proportion of their inherited DNA and protein sequences than they do with more distantly related species. If molecular analysis does not confirm this prediction, then the test casts doubt on the hypothesis of a close evolutionary relationship. In fact, however, molecular analysis of DNA and protein sequences in humans and monkeys *does* support the hypothesis that these two species are very closely related **(Figure 3.30)**. Molecular biology has added a new tape measure to the toolkit biologists use to assess evolutionary relationships. ■

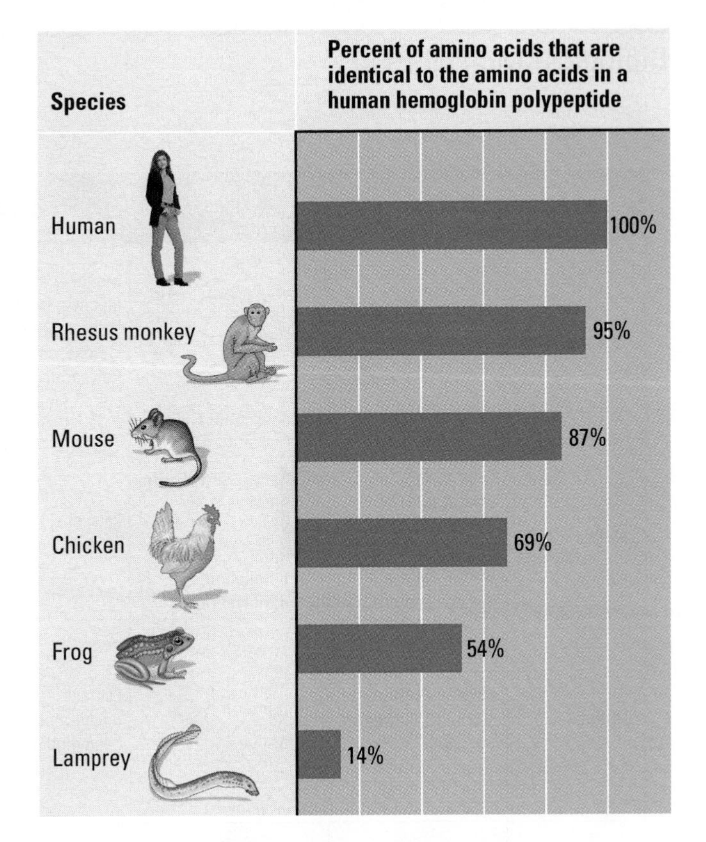

Figure 3.30 Comparing the amino acid sequences of a protein among six vertebrates. Inherited DNA determines the amino acid sequences of proteins. In this case, the sequences are for one of the polypeptides making up hemoglobin, the oxygen-transporting protein in the blood of vertebrates (animals with backbones). You can see that 95% of the amino acids of the rhesus monkey hemoglobin polypeptide are identical to the human polypeptide, compared with only 14% of the lamprey amino acids.

Chapter Review

SUMMARY OF KEY CONCEPTS

For study help and activities, go to campbellbiology.com or the student CD-ROM.

Organic Molecules

- **Carbon Chemistry** Carbon atoms can form large, complex, diverse molecules by bonding to four partners, including other carbon atoms. In addition to variations in the size and shape of carbon skeletons, organic molecules also vary in the presence and locations of different functional groups.

Activity *Diversity of Carbon-Based Molecules*

Case Study in the Process of Science *What Factors Determine the Effectiveness of Drugs?*

Activity *Functional Groups*

- **Giant Molecules from Smaller Building Blocks**

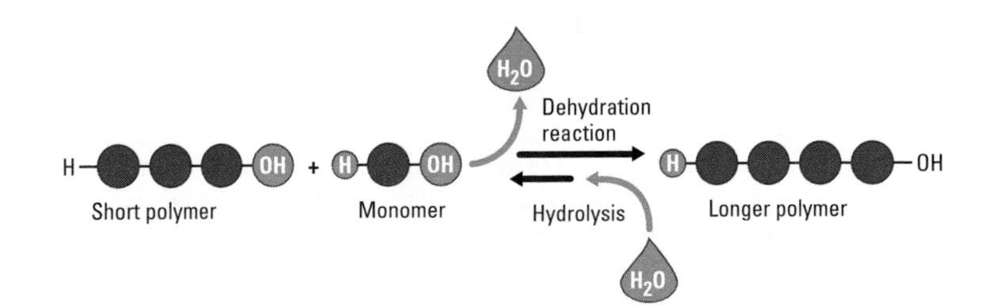

Activity *Making and Breaking Polymers*

Biological Molecules

Biological macromolecule	Function	Monomer	Examples
Carbohydrates	Dietary energy; storage; plant structure	Monosaccharide	Monosaccharides: glucose, fructose. Disaccharides: lactose, sucrose. Polysaccharides: starch, cellulose.
Lipids	Long-term energy storage (for fats); hormones (for steroids)	Fatty acid; Glycerol; Components of a triglyceride	Fats, oils, steroids
Proteins	Enzymes, structure, storage, contraction, transport, etc.	Amino group; Carboxyl group; Side group; Amino acid	Lactase (an enzyme), hemoglobin
Nucleic acids	Information storage	Phosphate; Base; Sugar; Nucleotide	DNA, RNA

• **Carbohydrates** Simple sugars (monosaccharides) provide cells with energy and carbon skeletons for building other organic molecules. Double sugars (disaccharides), such as sucrose (table sugar), consist of two monosaccharides joined by a dehydration reaction. Polysaccharides are macromolecules, long polymers of sugar monomers. Starch and glycogen are storage polysaccharides in plants and animals, respectively. The cellulose of plant cell walls is an example of a structural polysaccharide.

Activity *Models of Glucose*

Activity *Carbohydrates*

• **Lipids** Along with other kinds of lipids, fats are hydrophobic. Fats are the major form of long-term energy storage in animals. A fat, or triglyceride, consists of three fatty acids joined to a glycerol. Most animal fats are saturated, meaning that their fatty acids have the maximum number of hydrogens. Plant oils contain mostly unsaturated fats, having fewer hydrogens in the fatty acids because of double bonding in the carbon skeletons. Steroids, including cholesterol and the sex hormones, are also lipids.

Activity *Lipids*

• **Proteins** There are 20 types of amino acids, the monomers found in proteins. They are linked by dehydration reactions to form polymers called polypeptides. A protein consists of one or more polypeptides folded into a specific three-dimensional shape. Contributing to this shape are four levels of structure: the protein's amino acid sequence (primary structure), localized folding in certain regions of the protein (secondary structure), the overall shape of the folded polypeptide (tertiary structure), and an association of subunits in proteins with more than one polypeptide (quaternary structure). Shape is sensitive to environment, and if a protein loses its shape because of an unfavorable environment, the protein's function is also disrupted.

MP3 Tutor *Protein Structure and Function*

Activity *Protein Functions*

Activity *Protein Structure*

• **Nucleic Acids** Nucleic acids include RNA and DNA. DNA takes the form of a double helix, two DNA strands (polymers of nucleotides) held together by bonds between nucleotide components called bases. There are four kinds of DNA bases: adenine (A), guanine (G), thymine (T), and cytosine (C). A always pairs with T, and G always pairs with C. These base-pairing rules enable DNA to act as the molecule of inheritance. RNA has U (uracil) instead of T.

Activity *Nucleic Acid Functions*

Activity *Nucleic Acid Structure*

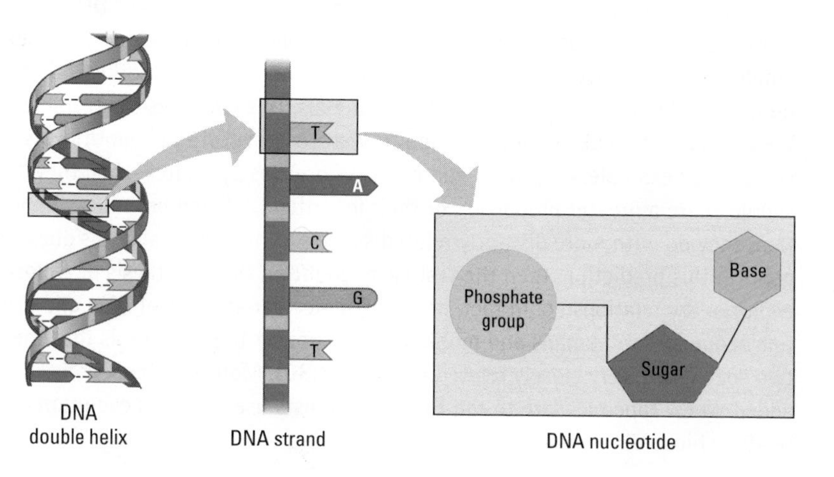

SELF-QUIZ

1. Monomers are joined together to form larger polymers through __________ reactions. Polymers are broken down into the monomers that make them up through the chemical reaction called __________.

2. Which of the following terms includes all the others in the list?
   a. polysaccharide b. carbohydrate
   c. monosaccharide d. disaccharide

3. One molecule of fat is made by joining three molecules of __________ to one molecule of __________.

4. Which of the following statements about saturated fats is true?
   a. Saturated fats contain one or more double bonds along the hydrocarbon tails.
   b. Saturated fats contain the maximum number of hydrogens along the hydrocarbon tails.
   c. Saturated fats make up the majority of most plant oils.
   d. Saturated fats are healthier for you than unsaturated fats.

5. Humans and other animals cannot digest wood because they
   a. cannot digest any carbohydrates.
   b. cannot chew it fine enough.
   c. lack the enzyme needed to break down cellulose.
   d. get no nutrients from it.

6. Changing one amino acid within a protein could change what about a protein?
 a. the primary structure
 b. the overall shape of the protein
 c. the function of the protein
 d. all of the above

7. Most proteins can easily dissolve in water. Knowing that, where within the overall three-dimensional shape of a protein would you most likely find hydrophobic amino acids?

8. A shortage of phosphorus in the soil would make it especially difficult for a plant to manufacture
 a. DNA.
 b. proteins.
 c. cellulose.
 d. fatty acids.
 e. sucrose.

9. A glucose molecule is to __________ as a __________ is to a nucleic acid.

10. Name three similarities between DNA and RNA. Name three differences.

Answers to the Self-Quiz questions can be found in Appendix D.

Go to the website or CD-ROM for more Self-Quiz questions.

THE PROCESS OF SCIENCE

11. A food manufacturer is advertising a new cake mix as fat-free. Scientists at the U.S. Food and Drug Administration (FDA) are testing the product to see if it truly lacks fat. Hydrolysis of the cake mix yields glucose, fructose, glycerol, a number of amino acids, and several kinds of molecules with long hydrocarbon chains. Further analysis shows that most of the hydrocarbon chains have a carboxyl group at one end. What would you tell the food manufacturer if you were the spokesperson for the FDA?

12. Lactase is an enzyme that breaks down the disaccharide lactose into the monosaccharides glucose and galactose. Imagine that you have produced several mutant versions of lactase, each of which differs from normal lactase by a single amino acid. Describe a test that could indirectly determine which of the mutations significantly alters the three-dimensional shape of the protein.

BIOLOGY AND SOCIETY

13. Some amateur and professional athletes take anabolic steroids to help them "bulk up" or build strength. The health risks of this practice are extensively documented. Apart from these health issues, what is your opinion about the ethics of athletes using chemicals to enhance performance? Is this a form of cheating, or is it just part of the preparation required to stay competitive in a sport where anabolic steroids are commonly used? Defend your opinion.

14. Recent court rulings have found that tobacco companies can be held liable for the health consequences of their products. While lung cancer does kill many people every year, heart disease kills many more. Imagine you're a juror sitting on a trial where a fast-food manufacturer is being sued for producing a harmful product. To what extent do you think manufacturers of unhealthy foods should be held responsible for the health consequences of their products? In what ways is this situation similar to the tobacco lawsuits? In what ways is it different? As a jury member, how would you vote?

15. Each year, industrial chemists develop and test thousands of new organic compounds for use as pesticides, such as insecticides, fungicides, and weed killers. In what ways are these chemicals useful and important to us? In what ways can they be harmful? Is your general opinion of pesticides positive or negative? What influences have shaped your feelings about these chemicals?

4 A Tour of the Cell

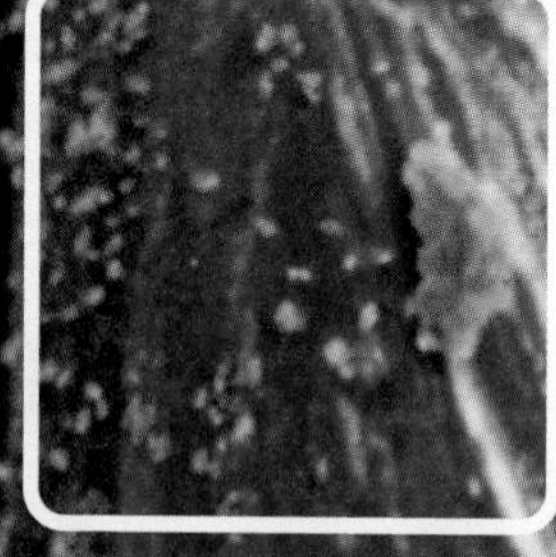

An electron microscope can visualize objects a million times smaller than the **head of a pin.**

If you stacked up **8,000 cell membranes,** they would only be as thick as a page in this book.

The cells of a whale are about **the same size** as the cells of a mouse.

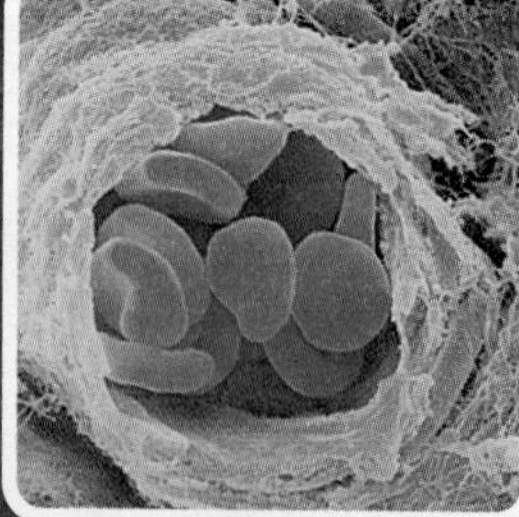

Every second, your body produces about **2 million red blood cells.**

BIOLOGY AND SOCIETY

Cells That Cure

A middle-aged man staggers, gasps for air, and falls to the ground clutching his chest. A passerby quickly puts out the call to 911—heart attack! As rescuers rush to the scene, the real medical drama plays out within the cells of the victim's heart. Starved for oxygen, his heart muscle cells begin to die. If too many cells perish, the heart will never again be able to function. Even if resuscitation efforts are successful and the man survives this acute medical emergency, he will most likely suffer from chronic problems. Because heart muscle cells—unlike many cells of the human body—do not regenerate over time, his heart may be permanently scarred. Constantly undersupplied by blood, the damaged heart muscle will gradually lose strength, leaving the patient at high risk for more heart failure. In severe cases, a heart transplant is the only possibility for a long-term cure.

In recent years, a new type of treatment has emerged that may offer hope for the rebuilding of dead heart muscle. In this procedure, called "cell therapy," cells are taken from elsewhere in the patient's body and delivered to the ailing heart. Typically, immature muscle cells are harvested from the patient's thigh and grown in a lab to increase their numbers. Tens or hundreds of millions of the new muscle cells are then injected directly into heart scar tissue during bypass or other heart surgery (**Figure 4.1**).

Although new and exciting, the technique of cell therapy remains in the testing phase. Worldwide, more than a dozen clinical trials are currently under way—including ones in Boston, Florida, and Houston. Even though many patients seem to respond to the treatment, scientists do not yet understand *how* cell therapy works. The intense effort that scientists are now directing toward the study of heart cells and cell therapy illustrates the main point of this chapter: To understand how life on Earth works—including your own body—you first need to learn about cells. Cells are the building blocks of all life, which makes them as fundamental to biology as atoms are to chemistry. Moreover, the cell is the smallest entity that exhibits all the characteristics of life. In this chapter, we'll take a tour of cells and explore their structure and function. ■

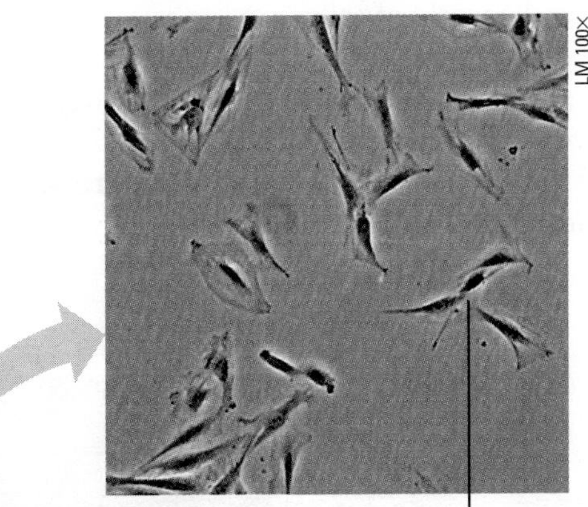

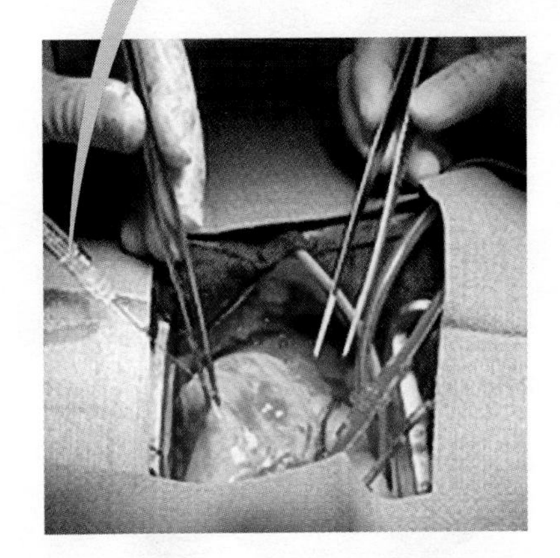

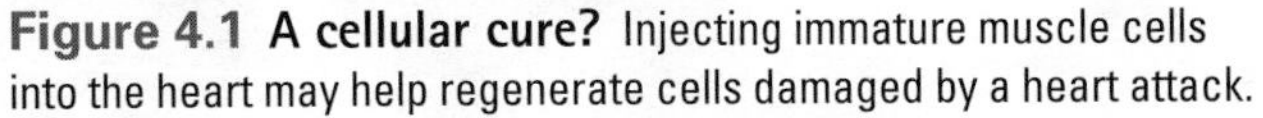

Figure 4.1 A cellular cure? Injecting immature muscle cells into the heart may help regenerate cells damaged by a heart attack.

The Microscopic World of Cells

Each cell in the human body is a miniature marvel of great complexity. If a complicated machine with millions of parts—say, a jumbo jet—were reduced to microscopic size, it would still seem simple and crude compared with the complexity of a living cell.

Organisms are either single-celled, such as most prokaryotes (bacteria and archaea) and protists, or multicelled, such as plants, animals, and most fungi. Your own body is a cooperative society of trillions of cells of many different specialized types. Three examples are the muscle cells that keep your heart beating, the nerve cells that control your muscles, and the red blood cells that carry oxygen throughout your body. Everything you do—every action and every thought—reflects processes occurring at the cellular level. For exploring this world of cells, our main tools are microscopes.

Microscopes as a Window on the World of Cells

Our understanding of nature often parallels the invention and refinement of instruments that extend human senses to new limits. The development of the microscope, for example, has provided an increasingly clear window on the world of cells.

The type of microscope used by Renaissance scientists, as well as the microscope you will use if your biology course includes a lab, is called a **light microscope (LM)**. Visible light is projected through the specimen, such as a single-celled protist **(Figure 4.2a)**. Glass lenses then enlarge the image and project it into a human eye or a camera.

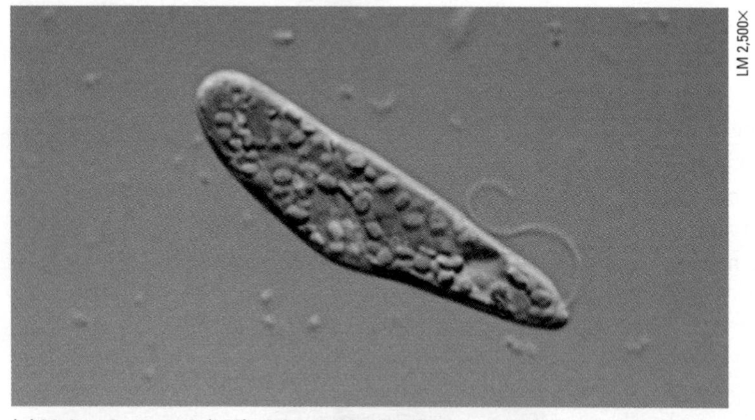

(a) Light micrograph (LM) of the protist *Euglena*

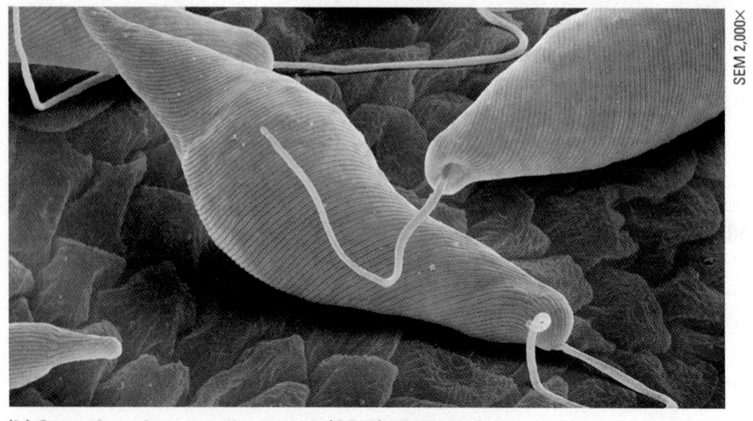

(b) Scanning electron micrograph (SEM) of *Euglena*

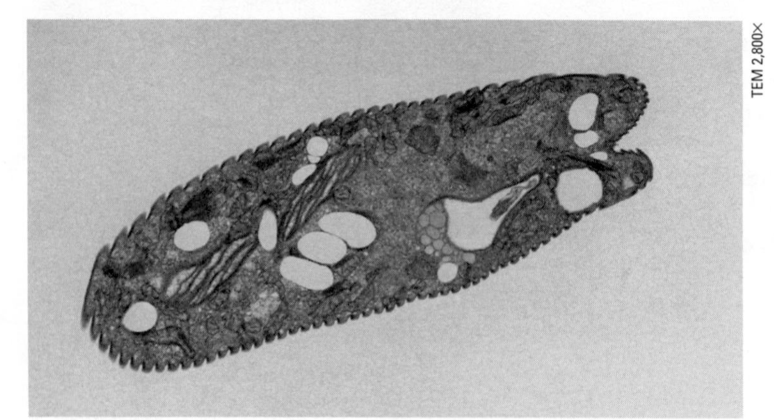

(c) Transmission electron micrograph (TEM) of *Euglena*

Figure 4.2 Different views of a protist, *Euglena*. These photographs were taken with different types of microscopes: **(a)** light microscope (LM); **(b)** scanning electron microscope (SEM); and **(c)** transmission electron microscope (TEM). Such photographs taken with microscopes are called micrographs. Throughout this textbook, each micrograph will have a notation along its side. For example, "LM 1,000×" indicates that the micrograph was taken with a light microscope and the objects are magnified to 1,000 times their original size.

Two important factors in microscopy are magnification and resolving power. **Magnification** is an increase in the object's apparent size compared with its actual size. The clarity of that magnified image depends on **resolving power,** which is the ability of an optical instrument to show two objects as separate. For example, what appears to the unaided eye as one star in the sky may be resolved as two stars with a telescope. Each optical instrument—be it an eye, a telescope, or a microscope—has a limit to its resolving power. The human eye can resolve points that are as close together as $\frac{1}{10}$ millimeter (mm), which equals 10^{-4} meter (m). For light microscopes, the resolving power is about 0.2 micrometer (μm), the size of a small bacterial cell (1 μm = $\frac{1}{1,000}$ mm, or 10^{-3} mm, which equals 10^{-6} m). This limits the useful magnification to about 1,000×. With greater magnification, the image becomes blurry.

Cells were first described in 1665 by the British scientist Robert Hooke, who used a microscope to examine a thin slice of cork from the bark of an oak tree. For the next two centuries, scientists found cells in every organism they examined with a microscope. By the mid-1800s, this accumulation of evidence led to the **cell theory,** which includes the induction that all living things are composed of cells. The cell theory was later expanded to include the notion that all cells arise from previously existing cells (a topic covered in Chapter 8).

Our knowledge of cell structure took a giant leap forward as biologists began using electron microscopes in the 1950s. Instead of using light, the **electron microscope (EM)** uses a beam of electrons to resolve objects. The electron microscope has a much better resolving power than the light microscope. In fact, the most powerful modern electron microscopes can distinguish objects as small as 0.2 nanometer (1 nm = $\frac{1}{1,000}$ μm, or 10^{-3} μm, which equals 10^{-9} m). This is a thousandfold improvement over the light microscope. The period at the end of this sentence is about a million times bigger than an object 0.2 nm in diameter. The highest-power electron micrographs you will see in this book have magnifications of about 100,000×. Such power reveals the details of diverse parts, or **organelles** ("little organs"), within a cell **(Figure 4.3)**.

Figures 4.2b and **c** show images taken with two kinds of electron microscopes. Biologists use the **scanning electron microscope (SEM)** to study the detailed architecture of the surface of a cell. The **transmission electron microscope (TEM)** is especially useful for exploring the internal structure of a cell. Preparing specimens for both types of electron microscopes requires killing and preserving cells before they can be examined. Thus, the light microscope is still very useful as a window on living cells.

Measurement Equivalents

1 meter (m) = 100 cm = 1,000 mm = about 39.4 inches

1 centimeter (cm) = 10^{-2} ($\frac{1}{100}$) meter (m) = about 0.4 inch

1 millimeter (mm) = 10^{-3} ($\frac{1}{1,000}$) m = $\frac{1}{10}$ cm

1 micrometer (μm) = 10^{-6} m = 10^{-3} mm

1 nanometer (nm) = 10^{-9} m = 10^{-3} μm

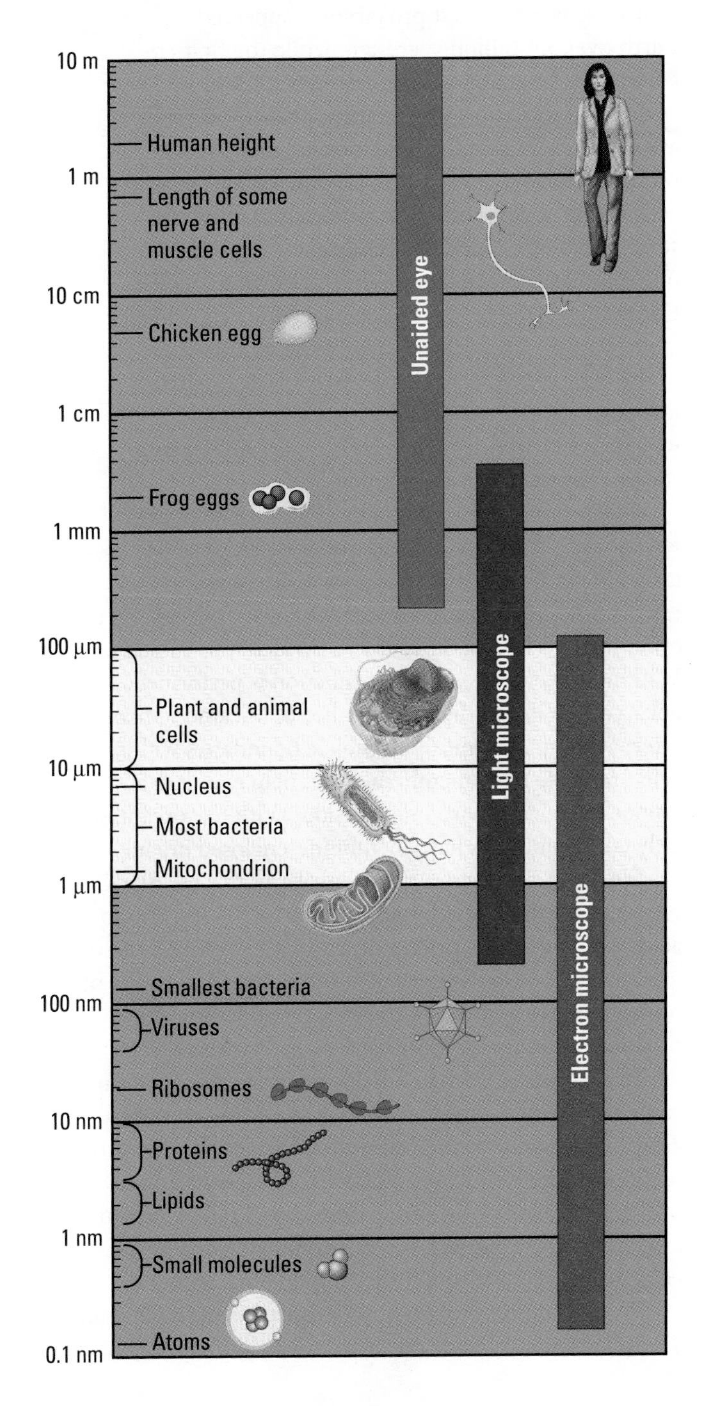

Figure 4.3 The size range of cells. Starting at the top of this scale with 10 m and going down, each reference measurement along the left side marks a tenfold decrease in size. Most cells are between 1 and 100 μm in diameter (the yellow area in the scale), a size range that can be viewed with either a light microscope or an electron microscope, but not the unaided eye. (For a complete list of metric measurements and their equivalents, see Appendix A.)

The Two Major Categories of Cells

The countless cells that exist on Earth fall into two basic categories: prokaryotic cells and eukaryotic cells. Bacteria and archaea consist of **prokaryotic cells** (see Figures 1.5 and 1.9) and so are called prokaryotes. All other organisms—protists, plants, fungi, and animals, including humans—are composed of **eukaryotic cells** and are called eukaryotes.

Prokaryotic and eukaryotic cells differ in several important respects **(Figure 4.4)**. Prokaryotic cells are usually much smaller, about one-tenth the length of a typical eukaryotic cell. As indicated in the fossil record, prokaryotes are older in an evolutionary sense: The first prokaryotes appeared on Earth over 3.5 billion years ago, while the first eukaryotes did not appear until around 2.1 billion years ago. And, most importantly, prokaryotic cells are structurally simpler. The most significant structural difference between prokaryotic and eukaryotic cells is that prokaryotic cells generally lack internal structures surrounded by membranes, while eukaryotic cells have several membrane-enclosed organelles. For example, the nucleus of a eukaryotic cell, which is surrounded by a double membrane, houses most of the cell's DNA. A prokaryotic cell lacks such a nucleus; its DNA is coiled in a nucleoid region, which, unlike a true nucleus, is not partitioned from the rest of the cell by membranes.

The interior of a prokaryotic cell is like an open warehouse. There are distinct spaces where specific tasks are performed, but these spaces are not separated from each other by barriers. A eukaryotic cell, on the other hand, is like an office that is divided into cubicles. Within each cubicle, a specific function is performed. A eukaryotic cell thus divides the labor of life among many internal compartments. The cubicle boundaries within cells are made from membranes that help maintain a unique chemical environment inside. With few exceptions, only eukaryotic cells have membrane-enclosed organelles.

Figure 4.5 depicts a typical prokaryotic cell. Structures called ribosomes build proteins by linking amino acids into sequences programmed by the DNA. A plasma membrane surrounds the cell and regulates the traffic of molecules into and out of the cell. Surrounding the plasma membrane of most bacteria is a rigid cell wall, which protects the cell and helps maintain its shape. In some prokaryotes, another layer, a sticky outer coat called a capsule, surrounds the cell wall. Capsules provide protection and help prokaryotes stick to surfaces. Some prokaryotes have short projections called pili, which may also attach to surfaces. The prokaryotic flagella of some cells propel them through their liquid environment.

We'll examine prokaryotes in more detail in Chapter 15. Eukaryotic cells are our main focus in this chapter.

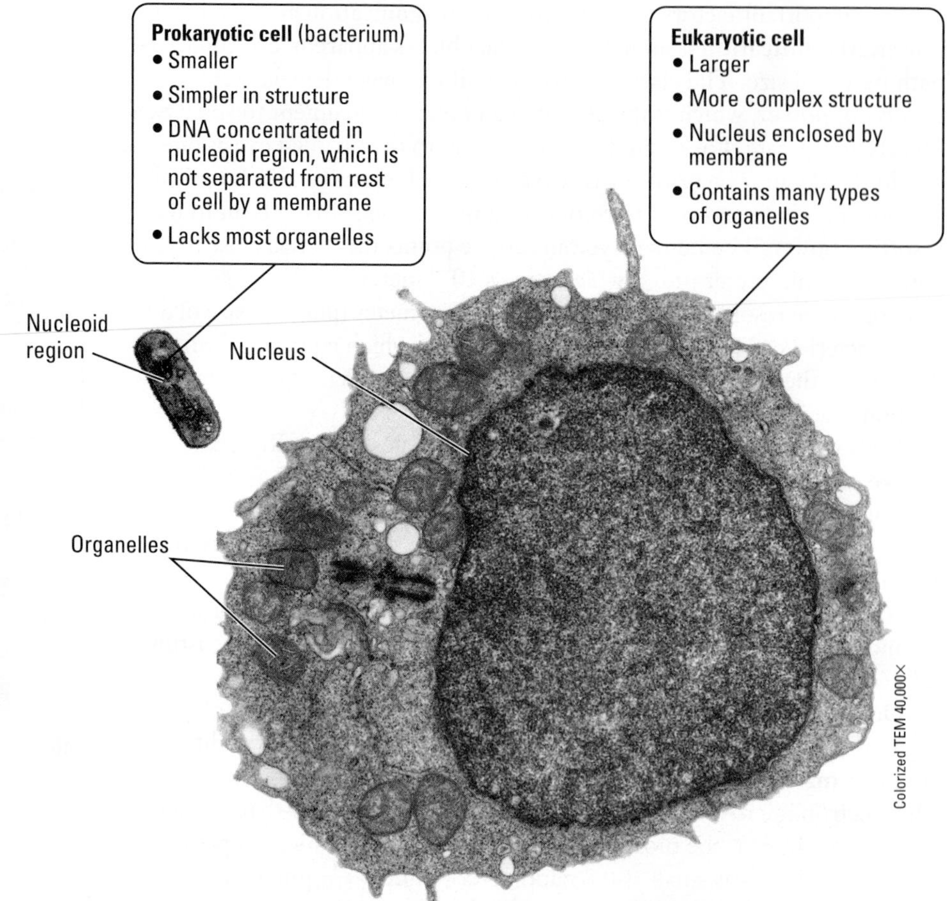

Figure 4.4 Contrasting the size and complexity of prokaryotic and eukaryotic cells. The smaller cell is a bacterium, an example of a prokaryote, and the larger cell is a eukaryotic cell. Eukaryotic cells are generally about ten times larger in diameter than prokaryotic cells (see Figure 4.3 for size ranges).

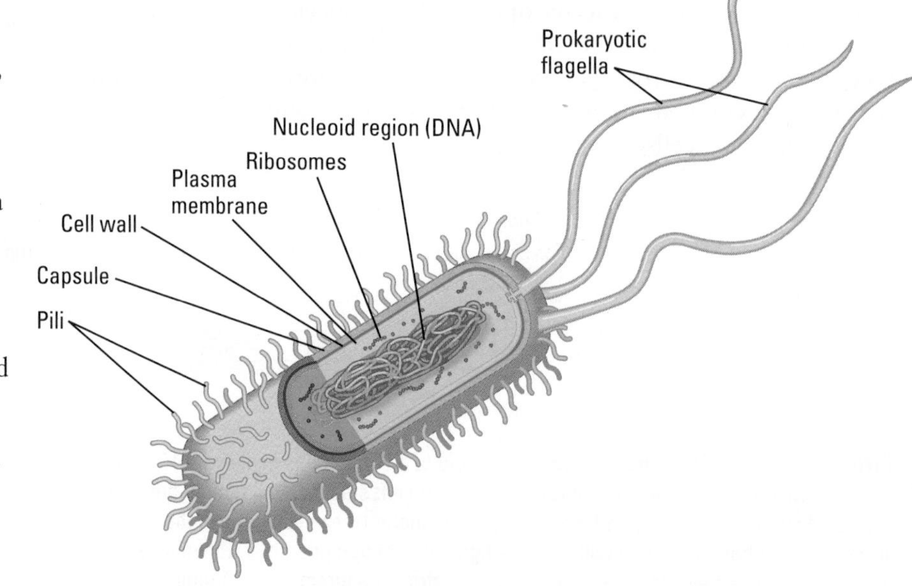

Figure 4.5 An idealized prokaryotic cell.

A Panoramic View of Eukaryotic Cells

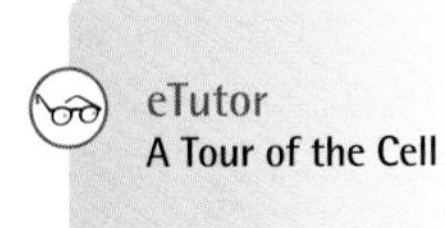

Figure 4.6 provides a panoramic view of two idealized eukaryotic cells, an animal cell and a plant cell. The similarities are more obvious than the differences. Both cells have a very thin outer membrane, the **plasma membrane,** which regulates the traffic of molecules between the cells and their surroundings. Each cell also has a prominent **nucleus,** the membrane-enclosed organelle that contains DNA.

The entire region of the cell between the nucleus and plasma membrane is called the **cytoplasm.** It consists of various organelles suspended in a fluid, the **cytosol.** The structure of each organelle has become adapted during evolution to perform specific functions. Most of the organelles are enclosed by membranes, but some are not.

As you can see in Figure 4.6, most organelles are found in both animal and plant cells. One important difference is the presence of chloroplasts in plant cells but not in animal cells. Chloroplasts are the organelles that convert light energy to the chemical energy of food. Also notice that unlike animal cells, plant cells have a protective cell wall outside of the plasma membrane. We'll see other differences and similarities between plant and animal cells as we now take a closer look at the architecture of eukaryotic cells, beginning with the plasma membrane.

CHECKPOINT

1. Which type of microscope would you use to study (a) the changes in shape of a living human white blood cell; (b) the finest details of surface texture of a human hair; (c) the detailed structure of an organelle in the cytoplasm of a human liver cell?
2. Using a light microscope to examine a thin section of a large spherical cell, you find that the cell is 0.3 mm in diameter. The nucleus is about one-fourth as wide. What is the diameter of the nucleus in micrometers?
3. How is the nucleoid region of a prokaryotic cell unlike the nucleus of a eukaryotic cell?
4. Name three structures in plant cells that animal cells lack.

Answers: 1. (a) light microscope; (b) scanning electron microscope; (c) transmission electron microscope **2.** About 0.075 mm, which equals 75 µm, **3.** There is no membrane enclosing the prokaryotic nucleoid region. **4.** Chloroplasts, a central vacuole, and a cell wall

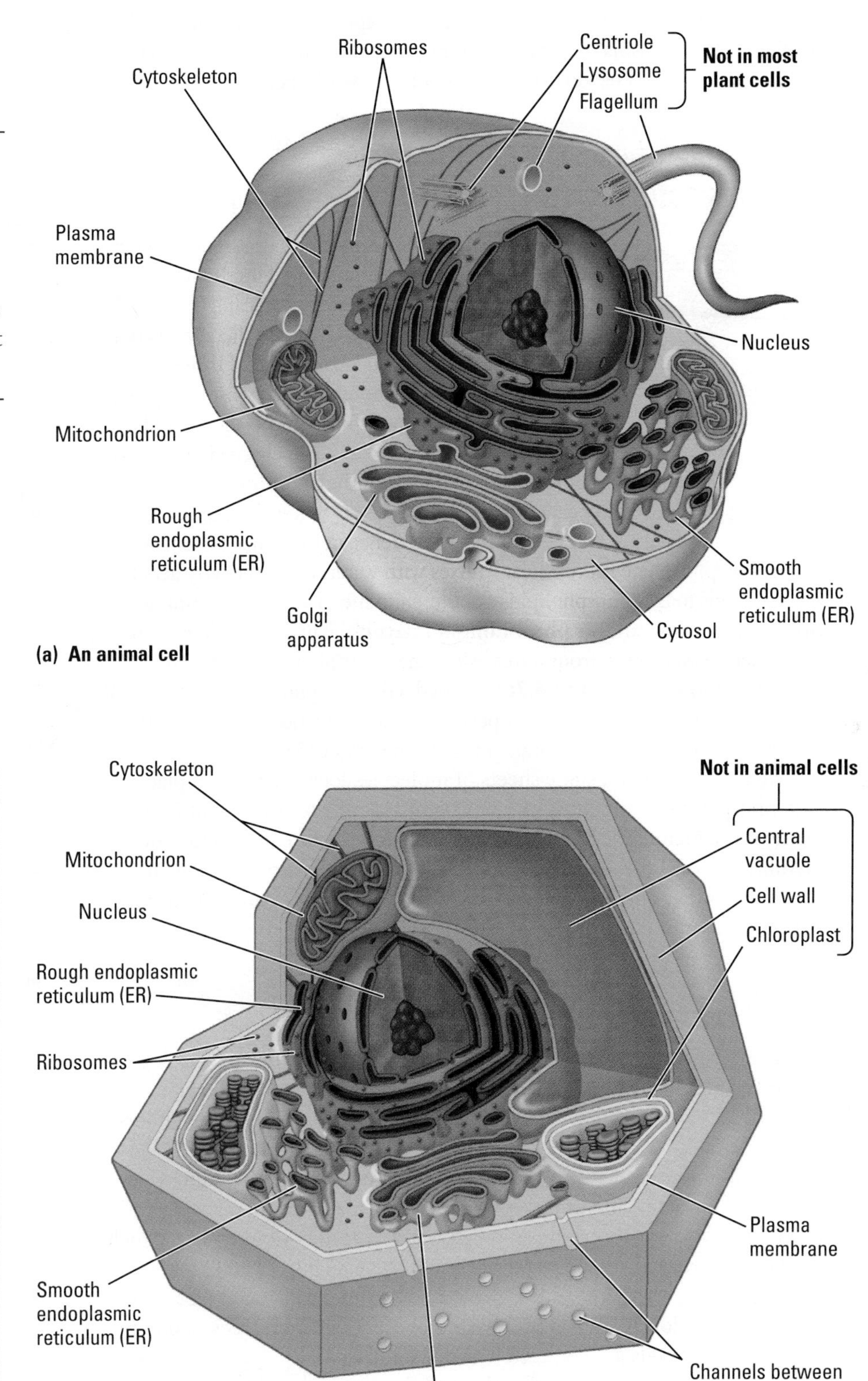

Figure 4.6 A panoramic view of an idealized animal cell and plant cell. For now, the labels on the drawings are just words, but these organelles will come to life as we take a closer look at how each part of the cell functions. To keep from getting lost on our tour of the cell, we'll carry miniature versions of these overview diagrams as our road maps, with the structure we're interested in highlighted.

Membrane Structure

Before we enter the cell to explore the various organelles, let's make a quick stop at the surface of this microscopic world. The plasma membrane is the edge of life, the boundary that separates the living cell from its nonliving surroundings. The plasma membrane is a remarkable film, so thin that you would have to stack 8,000 of them to equal the thickness of the page you're reading. Yet the plasma membrane can regulate the traffic of chemicals into and out of the cell. The key to how a membrane works is its structure.

The Plasma Membrane: A Fluid Mosaic of Lipids and Proteins

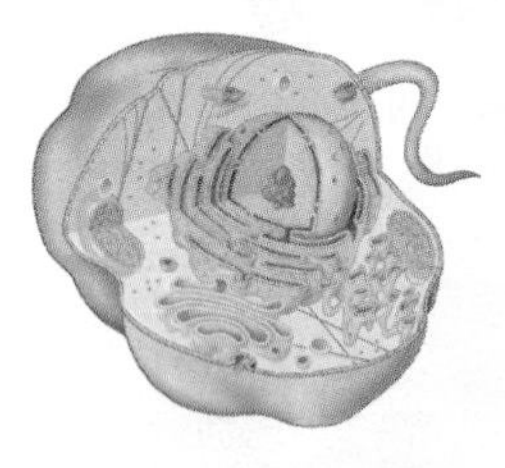

The plasma membrane as well as other membranes of the cell are composed mostly of lipids and proteins. The lipids belong to a special category called **phospholipids.** They are related to dietary fats but have only two fatty acid tails instead of three (see Figure 3.15). A phospholipid has a phosphate group (a combination of phosphorus and oxygen) in place of the third fatty acid. The phosphate group is electrically charged, which makes it hydrophilic ("water-loving"). The rest of the phospholipid, however, consisting of the two fatty acid tails, is hydrophobic ("water-fearing"). Thus, phospholipids have a kind of chemical ambivalence in their interactions with water. The phosphate group "head" mixes with water, while the fatty acid tails avoid it. This makes phospholipids good membrane material. By forming a two-layered membrane, or **phospholipid bilayer,** the hydrophobic parts of the molecules stay away from water, while the hydrophilic portions remain surrounded by water (**Figure 4.7a**). Embedded in the phospholipid bilayer of most membranes are proteins that perform various functions (**Figure 4.7b**). You'll learn more about membrane proteins in Chapter 5.

Membranes are not static sheets of molecules locked rigidly in place. The phospholipids and most of the proteins are free to drift about in the plane of the membrane. This behavior is captured in the description of a membrane as a **fluid mosaic**—fluid because the molecules can move freely past one another and mosaic because of the diversity of proteins that float like icebergs in the phospholipid sea.

Cell Surfaces

Most cells secrete materials for coats of one kind or another that are external to the plasma membrane. These extracellular coats help protect and support cells and facilitate certain interactions between cellular neighbors in tissues.

Recall, for example, that plant cells have a cell wall surrounding the plasma membrane. The walls protect the cells, maintain their shape, and keep the cells from absorbing so much water that they would burst. Much thicker and stronger than the plasma membrane, plant cell walls are made from cellulose fibrils embedded in a matrix of other molecules (see Figure 3.13c). We exploit this strength by using lumber cut from trees as building material. Plant cells are connected to one another via channels that pass through the cell walls, connecting the cytoplasm of each cell to its neighbors

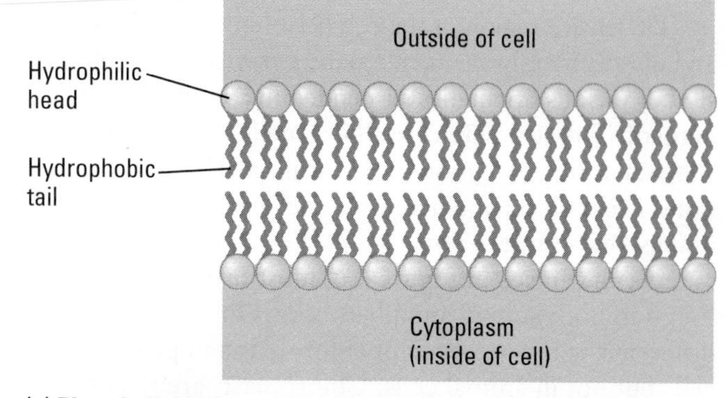

(a) Phospholipid bilayer of membrane. At the interface between two aqueous compartments, phospholipids arrange themselves into a bilayer. The symbol for phospholipids that we'll use throughout this book looks like a lollipop with two wavy sticks. The "head" of the lollipop is the end with the phosphate group, which is hydrophilic. The two sticks, the hydrocarbon tails of the phospholipid, are hydrophobic. Notice how the bilayer arrangement keeps the heads exposed to water while keeping the tails in the dry interior of the membrane.

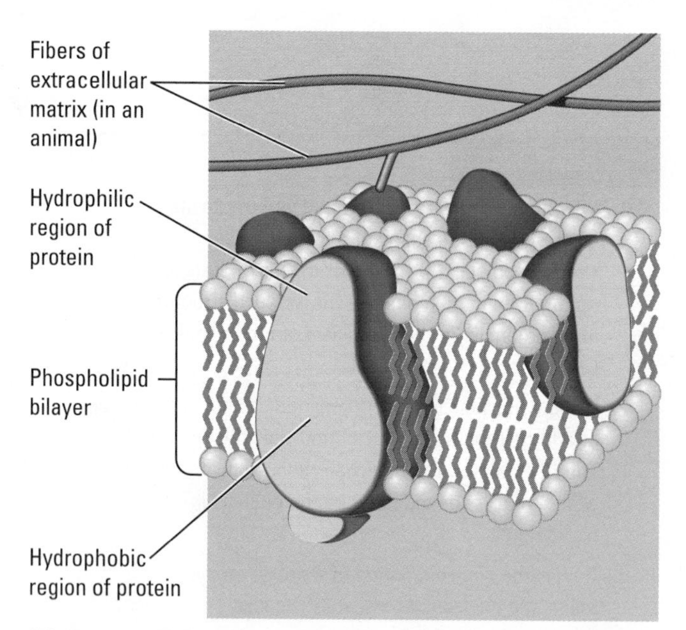

(b) Fluid mosaic model of membrane. Membrane proteins, like the phospholipids, have both hydrophilic and hydrophobic regions. The membrane behaves as a fluid mosaic because its phospholipids and many of its diverse proteins drift within the membrane. Notice that some of the proteins are bound to fibers of the extracellular matrix, which helps hold cells in place in animal tissue.

Figure 4.7 Plasma membrane structure.

(see Figure 4.6b). These channels allow water and other small molecules to move between cells, integrating the activities of a tissue.

Although they lack a cell wall, most animal cells secrete a sticky coat called the **extracellular matrix.** This layer helps hold cells together in tissues, and it can also have protective and supportive functions. Cells are often bound to the extracellular matrix by surface proteins in the plasma membrane (as shown in Figure 4.7b). In addition, the surfaces of most cells contain **cell junctions,** structures that connect them to other cells. Cell junctions allow adjacent cells to function in a coordinated way as part of a tissue.

CHECKPOINT

1. Why do phospholipids tend to organize into a bilayer in an aqueous solution?
2. Explain how each word in the term *fluid mosaic* describes the structure of a membrane.
3. What polysaccharide is the primary component of plant cell walls?

Answers: 1. The bilayer structure shields the hydrophobic tails of the phospholipids from water while exposing the hydrophilic heads to water. **2.** A membrane is fluid because its components are not locked into place. A membrane is mosaic because it contains a variety of different proteins embedded within it. **3.** Cellulose

The Nucleus and Ribosomes: Genetic Control of the Cell

If we think of the cell as a factory, then the nucleus is its executive boardroom. The top managers are the genes, the inherited DNA molecules that direct almost all the business of the cell. Genes store the information necessary to produce proteins, which then do most of the actual work of the cell. A **gene** is a stretch of DNA that contains the code for the structure of a specific protein.

Structure and Function of the Nucleus

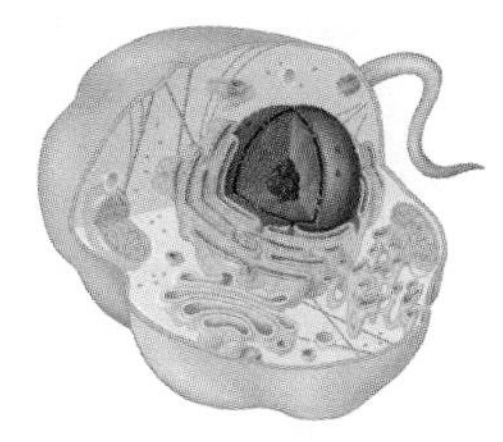

The nucleus is bordered by a double membrane called the **nuclear envelope (Figure 4.8)**. Each membrane of the nuclear envelope is similar in structure to the plasma membrane. Pores through the envelope allow the passage of material between the nucleus and the cytoplasm. Within the nucleus, long DNA molecules and associated proteins form long fibers called **chromatin.** Each long fiber constitutes one **chromosome.** The number of chromosomes in a cell depends on the species; for example, each human body cell has 46 chromosomes, while rice cells have 24 and porcupine cells have 34. Whatever their number, each individual chromosome is made of one long strand of chromatin, a combination of DNA and proteins.

In association with the chromatin, the nucleus contains a ball-like mass of fibers and granules called the **nucleolus.** It produces the component parts of ribosomes.

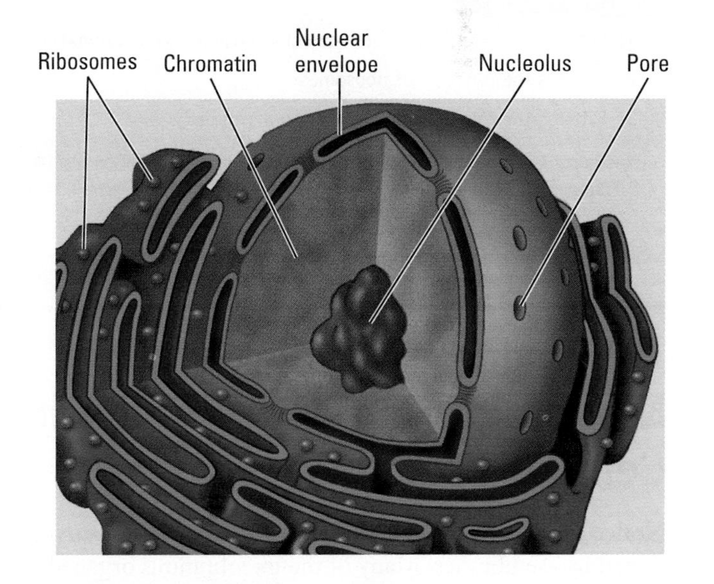

Figure 4.8 The nucleus.

Ribosomes

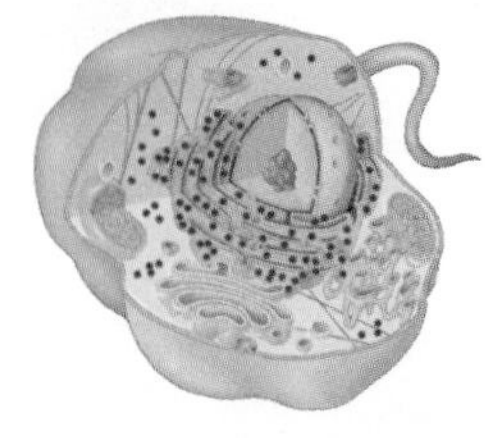

The small dots in the cells in Figure 4.6 and outside the nucleus in Figure 4.8 are **ribosomes.** Although they are assembled from components made in the nucleus, ribosomes do not begin to work until the components move through the pores of the nucleus into the cytoplasm. Ribosomes are responsible for protein synthesis. Some ribosomes are suspended in the cytosol, the fluid of the cytoplasm. They make proteins that will remain dissolved in the cytosol. Other ribosomes are attached to the outside of a membranous organelle called the endoplasmic reticulum. These ribosomes make proteins destined to be incorporated into membranes or secreted by the cell.

The ribosomes of prokaryotes and eukaryotes, while similar in overall structure, differ considerably in their makeup. These differences play an important role in human health: Certain antibiotic drugs can bind to and disrupt bacterial ribosomes while ignoring the ribosomes of human cells. Antibiotics such as erythromycin, streptomycin, tetracycline, and chloramphenicol all work this way.

How DNA Controls the Cell

How do the DNA "executives" in the nucleus direct the "workers" in the cytoplasm? Follow the sequence of events in the eukaryotic cell shown in **Figure 4.9**. ❶ DNA programs protein production in the cytoplasm by transferring its coded information to a molecule called messenger RNA (mRNA). Like a middle manager, the RNA molecule then carries the order to "build this type of protein" from the nucleus to the cytoplasm. ❷ The mRNA exits through pores in the nuclear envelope and travels to the cytoplasm, where it binds to ribosomes. ❸ As a ribosome (disproportionately large in this figure) moves along the mRNA, the genetic message is translated into a protein of specific amino acid sequence. You'll learn how the message is translated in Chapter 10.

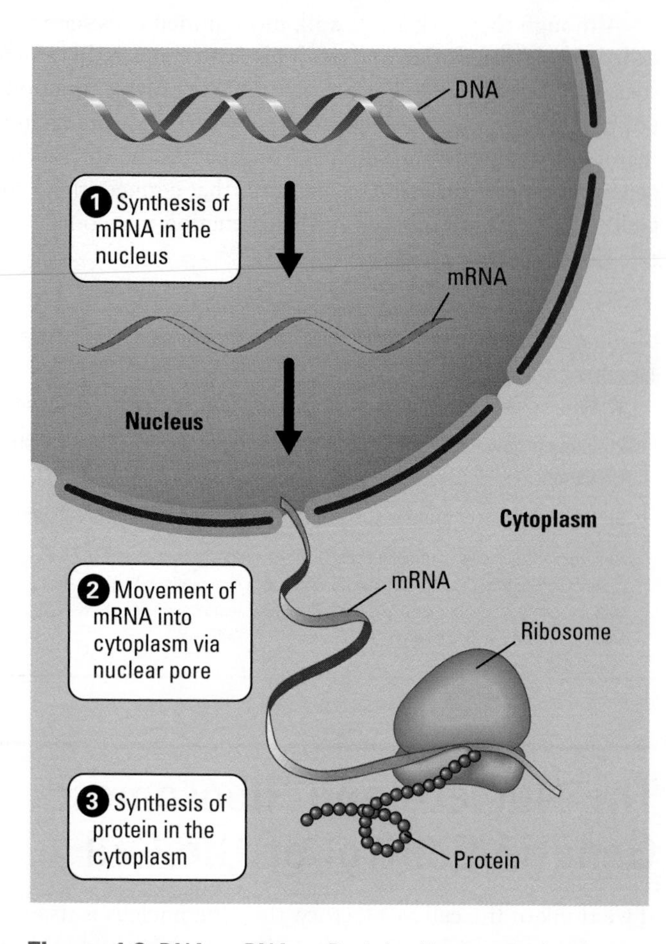

Figure 4.9 DNA → RNA → Protein: How genes in the nucleus control the cell. The structure and behavior of a cell reflect the kinds of proteins it has, but it is the inherited DNA that determines a cell's protein composition.

CHECKPOINT

1. What is the relationship between chromosomes and chromatin?
2. What is the function of ribosomes?
3. What is the role of mRNA in making a protein?

Answers: 1. Chromosomes are made of chromatin, which is a combination of DNA and proteins. **2.** Protein synthesis **3.** The mRNA carries the genetic message from a gene (DNA) to ribosomes that translate it into protein.

The Endomembrane System: Manufacturing and Distributing Cellular Products

Notice again in Figure 4.6 that the cytoplasm of a eukaryotic cell is partitioned by membranes. Many of the membranous organelles belong to the **endomembrane system.** This system includes the endoplasmic reticulum, the Golgi apparatus, lysosomes, and vacuoles.

The Endoplasmic Reticulum

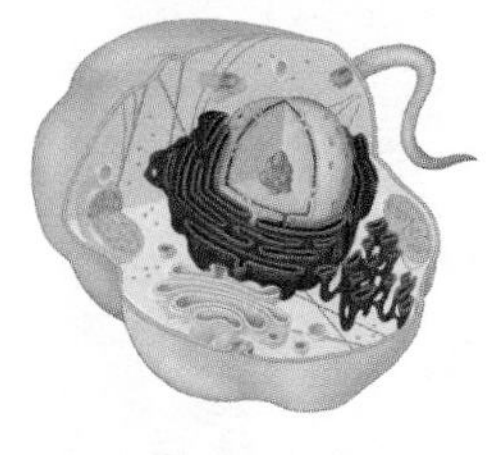

The **endoplasmic reticulum (ER)** is one of the main manufacturing facilities within a cell. It produces an enormous variety of molecules. The ER is a membranous labyrinth of tubes and sacs running throughout the cytoplasm. The ER membrane separates its internal compartment from the surrounding cytosol **(Figure 4.10)**. There are two distinct types of ER: rough ER and smooth ER. These two ER components are physically connected, but they differ in structure and function.

Rough ER The "rough" in **rough ER** refers to the appearance of this organelle in electron micrographs (see Figure 4.10). The roughness is due to ribosomes that stud the outside of the ER membrane. These ribosomes produce two main types of proteins: membrane proteins and secretory proteins. Some newly manufactured membrane proteins are embedded right in the ER membrane. Thus, one function of rough ER is the production of new membrane. Secretory proteins are those the cell will actually export (secrete) to the fluid outside the cell. Cells that secrete a lot of protein—such as the cells of your salivary glands, which secrete an enzyme into your mouth—are especially rich in rough ER. Some of the products manufactured by rough ER are dispatched to other locations in the cell via **transport vesicles,** membranous spheres that bud from the ER **(Figure 4.11)**.

Smooth ER The "smooth" in **smooth ER** refers to the fact that this organelle lacks the ribosomes that populate the surface of rough ER (see Figure 4.10). A diversity of enzymes built into the smooth ER membrane enables this organelle to perform many functions. One is the synthesis of lipids, including steroids (see Figure 3.17). For example, the cells in your ovaries or testes that produce sex hormones, which are steroids, are enriched with smooth ER. In liver cells, the functions of smooth ER include the detoxification of drugs and other poisons that might be present in the bloodstream. For example, certain ER enzymes detoxify sedatives such as barbiturates, stimulants such as amphetamines, and some antibiotics (which is why they don't persist in the bloodstream after combating an infection). As liver cells are exposed to a drug, the amounts of smooth ER and its detoxifying enzymes increase. This can strengthen the body's tolerance for the drug, meaning that higher doses will be required in the future to achieve the desired effect. The growth of smooth ER in response to one drug can also increase tolerance to other drugs, including important medicines. Barbiturate use, for example, may decrease the effectiveness of certain antibiotics by accelerating their breakdown in the liver. Furthermore, increasing tolerance to drugs is one of the hallmarks of addiction—a potentially serious consequence of the continued use of certain drugs.

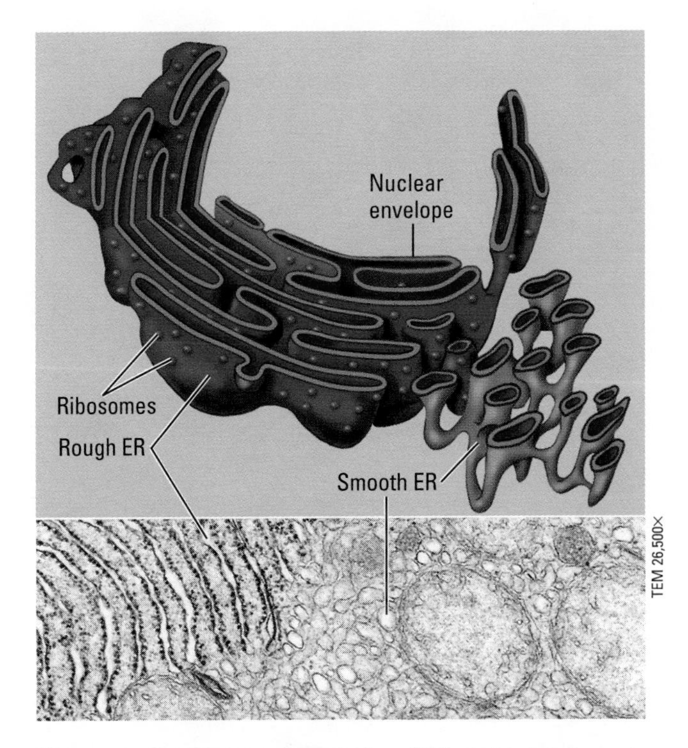

Figure 4.10 Endoplasmic reticulum (ER). In this drawing and micrograph (bottom), the flattened sacs of rough ER and the tubes of smooth ER are continuous, though different in structure and function. Notice that the ER is also continuous with the nuclear envelope, which is actually part of the endomembrane system.

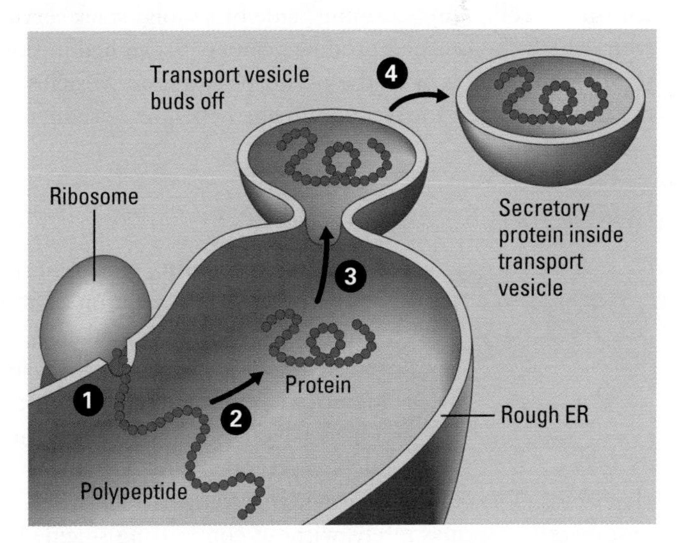

Figure 4.11 How rough ER manufactures and packages secretory proteins. ❶ A ribosome links amino acids to form a polypeptide chain with a unique amino acid sequence (see Chapter 3). The chain threads through the membrane and into the cavity of the ER. ❷ Secretory proteins are often modified in the ER. ❸ Secretory proteins depart in transport vesicles that ❹ bud off from the ER. If the proteins require no further refinement in other organelles, they are secreted from the cell when the vesicles fuse with the plasma membrane, the outer membrane of the cell.

The Golgi Apparatus

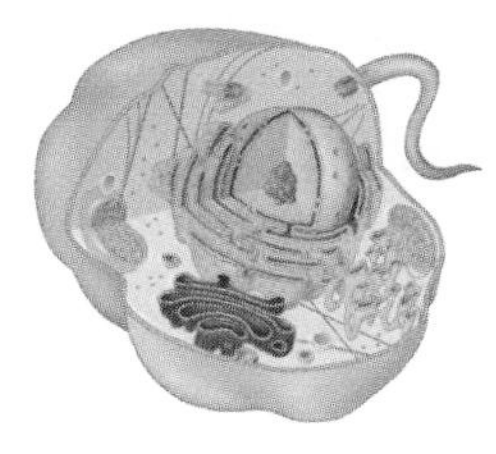

The **Golgi apparatus,** an organelle named for its discoverer, Italian scientist Camillo Golgi, is a refinery, warehouse, and shipping center. Working in close partnership with the ER, the Golgi apparatus receives, refines, stores, and distributes

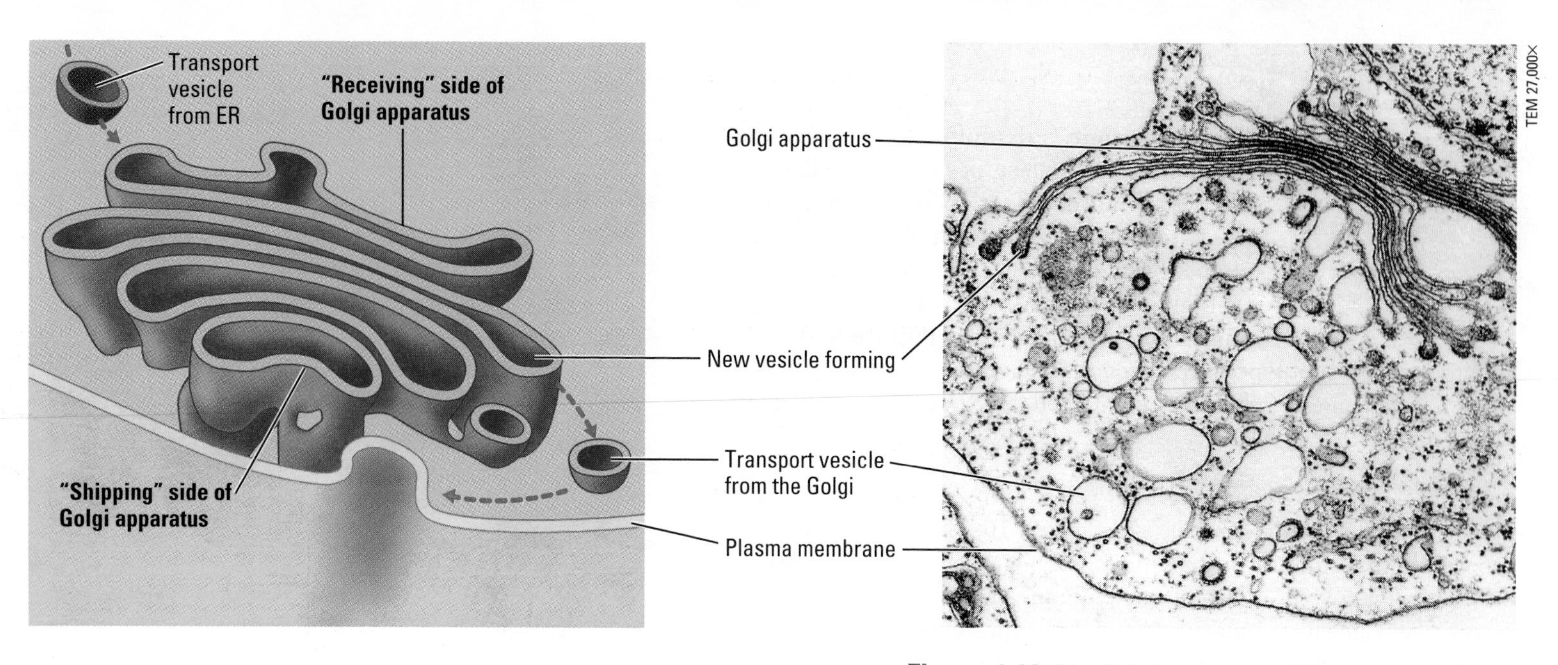

Figure 4.12 The Golgi apparatus. This component of the endomembrane system consists of flattened sacs arranged something like a stack of pita bread. A cell may contain just a few Golgi stacks or hundreds of them.

chemical products of the cell **(Figure 4.12)**. Products made in the ER reach the Golgi in transport vesicles. One side of a Golgi stack serves as a receiving dock for these vesicles. Enzymes of the Golgi modify many of the ER products during their stay in the Golgi. For example, the Golgi chemically tags protein products to mark their final destination within the cell. The "shipping" side of a Golgi stack serves as a depot from which the finished products can be dispatched in transport vesicles to other organelles or to the plasma membrane. Vesicles that bind with the plasma membrane secrete finished chemical products to the outside of the cell.

Lysosomes

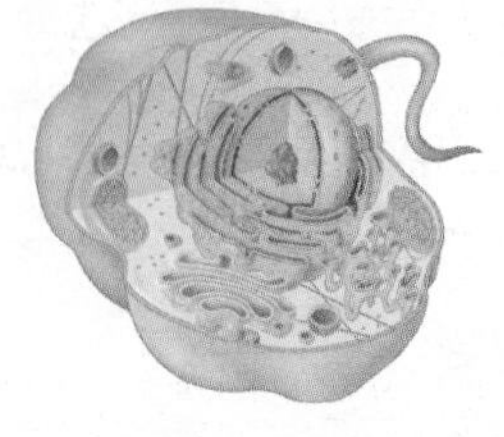

The name **lysosome,** which means "breakdown body," is a good description of how these organelles function in animal cells. (They are absent from most plant cells.) A lysosome is a membrane-enclosed sac of digestive enzymes. These enzymes can break down macromolecules such as proteins, polysaccharides, fats, and nucleic acids. The lysosome provides a compartment where the cell can digest macromolecules safely, without committing suicide by unleashing these digestive enzymes on the cell itself.

Lysosomes have several types of digestive functions. Many cells engulf nutrients into tiny cytoplasmic sacs called **food vacuoles.** Lysosomes fuse with the food vacuoles, exposing the food to enzymes that digest it **(Figure 4.13a)**. Small molecules that result from this digestion, such as amino acids, leave the lysosome and nourish the cell. Lysosomes also help destroy harmful bacteria. Our white blood cells ingest bacteria into vacuoles, and lysosomal enzymes that are emptied into these vacuoles rupture the bacterial cell walls. Lysosomes also serve as recycling centers for damaged organelles. Without harming the cell, a lysosome can engulf and digest parts of another organelle, making its molecules available for the construction of new organelles **(Figure 4.13b)**. Lysosomes also have

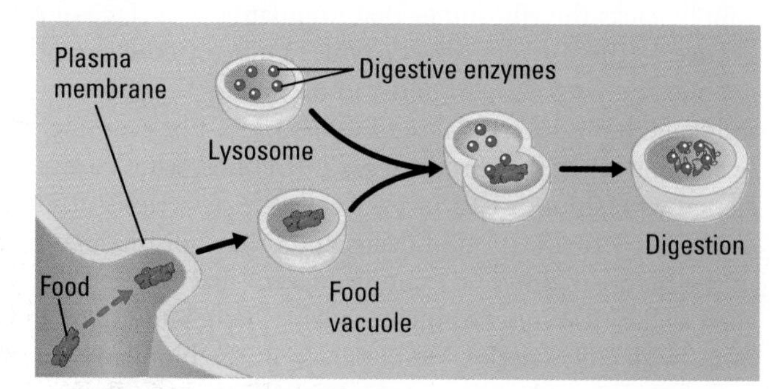

(a) Lysosome digesting food

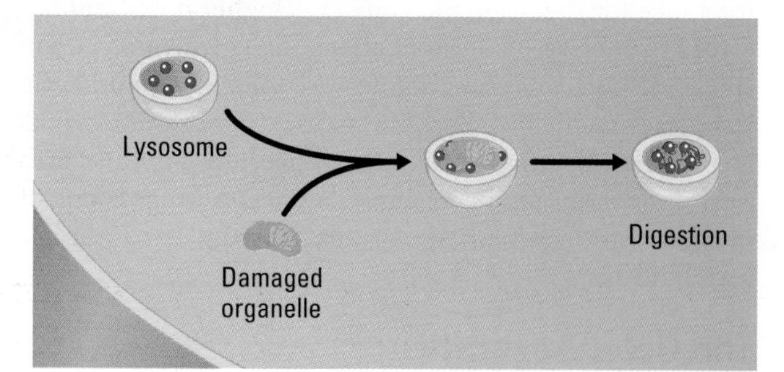

(b) Lysosome breaking down damaged organelle

Figure 4.13 Two functions of lysosomes. As sacs of digestive enzymes, lysosomes **(a)** digest food and **(b)** help recycle the molecules of the cell itself by breaking down damaged organelles.

sculpturing functions in embryonic development. For example, lysosomal enzymes destroy cells of the webbing that joins the fingers of early human embryos. In this case, the lysosomes act as "suicide packs," breaking open and causing the programmed death of whole cells.

The importance of lysosomes to cell function and human health is made strikingly clear by the serious hereditary disorders called lysosomal storage diseases. A person with such a disease is missing one or more of the digestive enzymes normally found within lysosomes. The abnormal lysosomes become engorged with indigestible substances, and this eventually interferes with other cellular functions. Most of these diseases are fatal in early childhood. In Pompe's disease, the lack of an enzyme results in the accumulation of harmful amounts of the polysaccharide glycogen in muscle cells; this causes weakening of the muscles, particularly in the heart, and often leads to early heart failure. Another example is Tay-Sachs disease, which ravages the nervous system. In this disorder, lysosomes lack a lipid-digesting enzyme, and nerve cells in the brain are damaged as they accumulate excess lipids. Fortunately, storage diseases are rare in the general population. For Tay-Sachs disease, people who carry the abnormal gene that causes the disease can be identified through genetic testing.

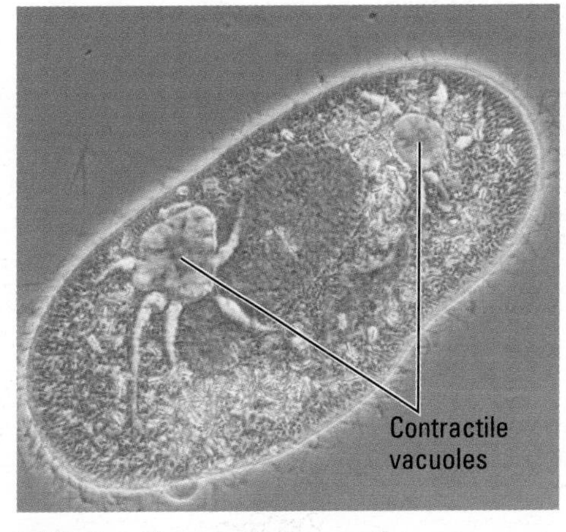

(a) Contractile vacuoles in a protist

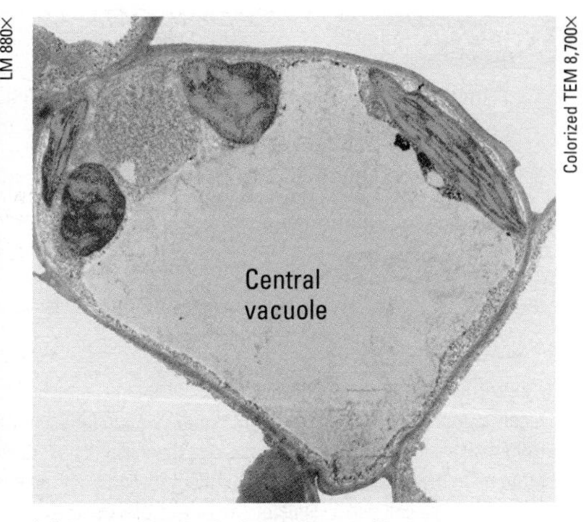

(b) Central vacuole in a plant cell

Figure 4.14 Two types of vacuoles. **(a)** This single-celled organism, a protist named *Paramecium*, has two contractile vacuoles. **(b)** The central vacuole (the large light green area in this colorized micrograph) is often the largest organelle in a mature plant cell.

Vacuoles

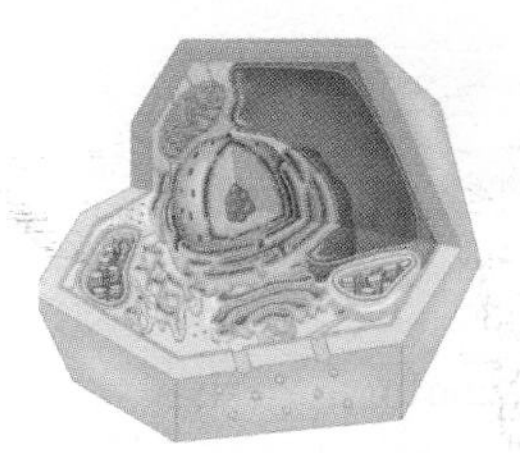

Vacuoles are membranous sacs that bud from the ER, Golgi, or plasma membrane. Vacuoles come in different sizes and have a variety of functions. For example, Figure 4.13a shows a food vacuole budding from the plasma membrane. Certain freshwater protists have contractile vacuoles that function as pumps to expel excess water that flows into the cell from the outside environment **(Figure 4.14a)**. Another type of vacuole is a plant cell's **central vacuole,** which can account for more than half the volume of a mature cell **(Figure 4.14b)**.

The plant cell vacuole is a versatile compartment. It is the place where the plant stores organic nutrients. For example, proteins are stockpiled in the vacuoles of cells in seeds, such as beans and peas. Central vacuoles also contribute to plant growth by absorbing water and causing cells to expand. Central vacuoles in flower petals may contain pigments that attract pollinating insects. Central vacuoles may also contain poisons that protect against plant-eating animals.

Figure 4.15 will help you review how all the organelles of the endomembrane system are related. Note that it is possible for a product made in one part of the endomembrane system to eventually exit the cell or become part of another organelle without ever crossing a membrane. Also note that membrane originally fabricated by the ER can eventually turn up as part of the plasma membrane through the fusion of secretory vesicles. In this way, even the plasma membrane is related to the endomembrane system.

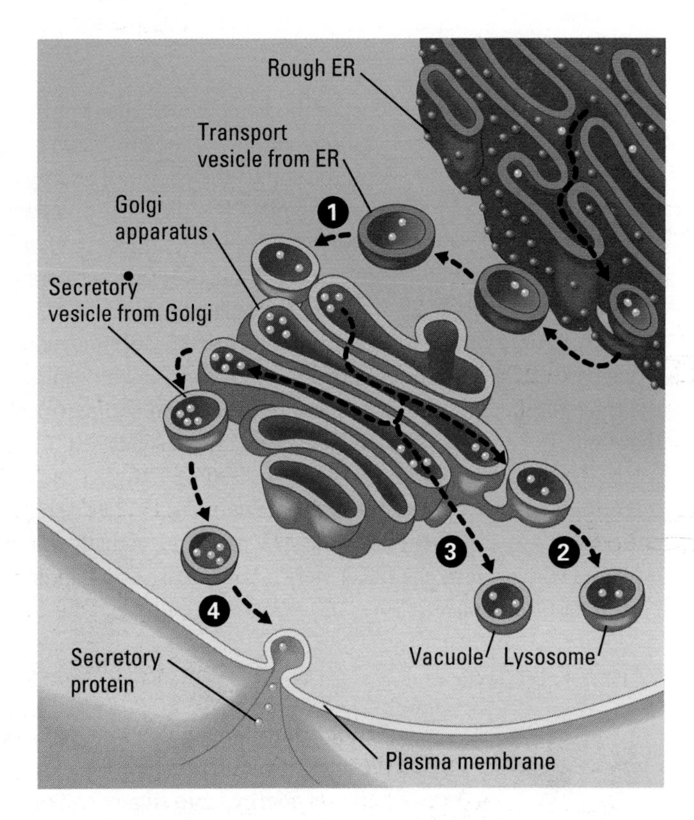

Figure 4.15 Review of the endomembrane system. The arrows show some of the pathways of cell product distribution and membrane migration via transport vesicles. For example, ❶ digestive enzymes made by the rough ER are transported via vesicles to the Golgi for processing. ❷ Lysosomes containing the processed digestive enzymes bud off from the Golgi. ❸ Other cell products end up in vacuoles for storage. ❹ Still other cell products are secreted from the cell.

CHECKPOINT

1. Which structure includes all the others in the list: rough ER, smooth ER, endomembrane system, the Golgi apparatus?
2. What makes rough ER rough?
3. What is the relationship between the Golgi apparatus and the ER in a protein-secreting cell?
4. How can defective lysosomes result in excess accumulation of a particular chemical compound in a cell?

Answers: **1.** Endomembrane system **2.** Ribosomes attached to the membranes **3.** The Golgi receives transport vesicles that bud from the ER and that contain proteins synthesized in the ER. The Golgi finishes processing the proteins and then dispatches transport vesicles that secrete the proteins to the outside of the cell. **4.** If the lysosomes lack an enzyme needed to break down the compound, the cell will accumulate an excess of that compound.

Chloroplasts and Mitochondria: Energy Conversion

A cell requires a continuous energy supply to do the work of life. The two types of cellular power stations are the organelles called chloroplasts and mitochondria.

Chloroplasts

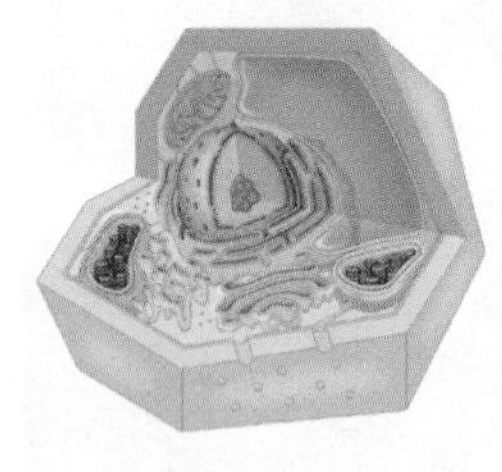

Most of the living world runs on the energy provided by photosynthesis, the conversion of light energy from the sun to the chemical energy of sugar and other organic molecules. **Chloroplasts,** which are unique to the photosynthetic cells of plants and protists, are the organelles that perform photosynthesis.

The chloroplast is partitioned into three major compartments by internal membranes **(Figure 4.16)**. One compartment is the space between the two membranes that envelop the chloroplast. The **stroma,** the thick fluid within the chloroplast, is the second compartment. Suspended in that fluid, the interior of a network of membrane-enclosed tubes and disks forms the third compartment. Notice in Figure 4.16 that the disks occur in interconnected stacks called **grana** (singular, *granum*). The grana are the chloroplast's solar power packs, the structures that actually trap light energy and convert it to chemical energy. You'll learn in Chapter 7 how the chloroplast functions.

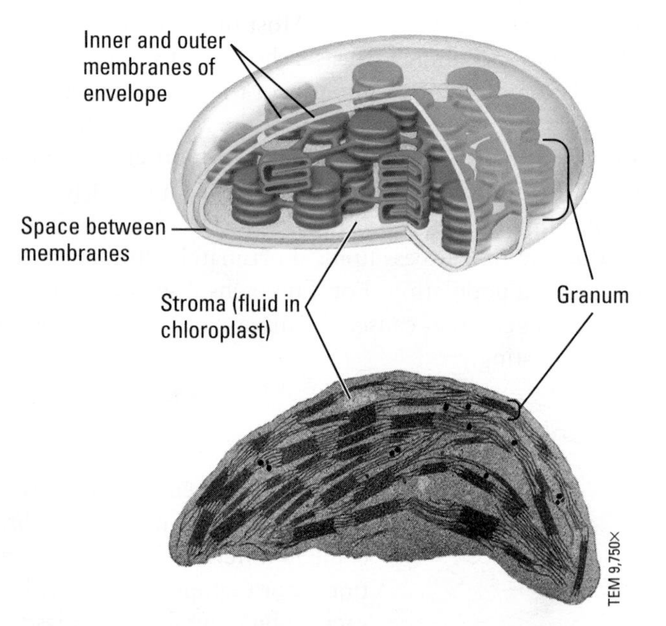

Figure 4.16 The chloroplast: site of photosynthesis.

Mitochondria

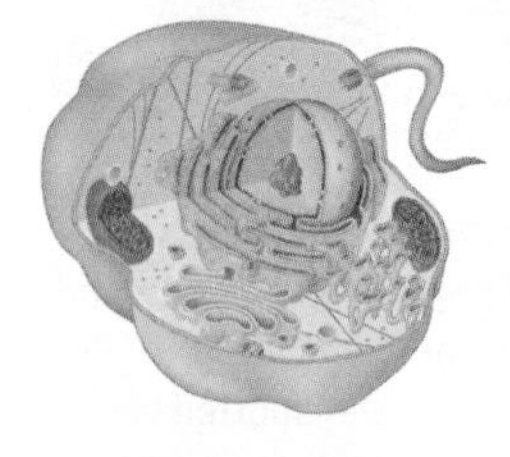

Mitochondria (singular, *mitochondrion*) are the sites of cellular respiration. This process harvests energy from sugars and other food molecules and converts it to another form of chemical energy called ATP. Cells use molecules of ATP as the direct energy source for most of their work. In contrast to chloroplasts, mitochondria are found in almost all eukaryotic cells, including your own. An envelope of two membranes encloses the mitochondrion, which contains a thick fluid called the **matrix (Figure 4.17)**. The inner

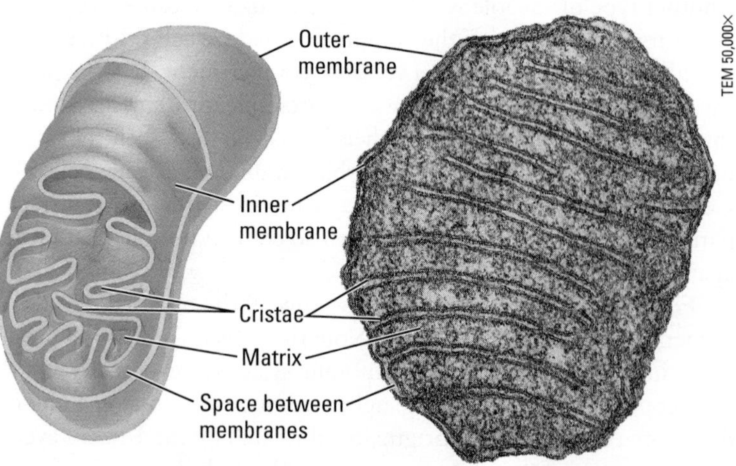

Figure 4.17 The mitochondrion: site of cellular respiration.

membrane of the envelope has numerous infoldings called **cristae.** Many of the enzymes and other molecules that function in cellular respiration are built into the inner membrane. By increasing the surface area of this membrane, the cristae maximize ATP output. In Chapter 6, you'll learn more about how mitochondria convert food energy to ATP energy.

Besides their ability to provide cellular energy, mitochondria and chloroplasts share another feature unique among eukaryotic organelles: They contain DNA that encodes some of their proteins. This DNA is evidence that mitochondria and chloroplasts evolved from free-living prokaryotes in the distant past (see Chapter 15 for a discussion of this hypothesis). The small genomes of mitochondria can affect human health. In 2004, researchers discovered that a mutation within one mitochondrial gene can cause a condition called metabolic syndrome. Metabolic syndrome can lead to significant health problems, including hypertension and diabetes. The role of the mitochondrial genome in human health and disease remains a topic of active investigation among cell biologists.

CHECKPOINT

1. What does photosynthesis accomplish?
2. What is cellular respiration?

Answers: **1.** The conversion of light energy to chemical energy stored in food molecules **2.** A process that converts the chemical energy of sugars and other food molecules to chemical energy in the form of ATP

The Cytoskeleton: Cell Shape and Movement

If someone asked you to describe a house, you would most likely mention the number of rooms and their location. You probably would not think to mention the beams and floor joists that support the house. Yet these structures perform an extremely important function. Similarly, cells have an infrastructure called the **cytoskeleton,** a network of fibers extending throughout the cytoplasm. The cytoskeleton serves as both skeleton and "muscles" for the cell, functioning in both support and movement.

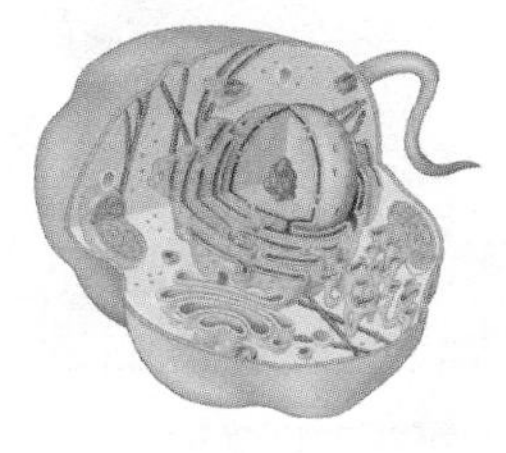

Maintaining Cell Shape

One function of the cytoskeleton is to give mechanical support to the cell and maintain its shape. This is especially important for animal cells, which lack rigid cell walls. The cytoskeleton contains several types of fibers made from different types of protein. One of the most important types of fibers is **microtubules** **(Figure 4.18a)**. Microtubules are straight, hollow tubes composed of

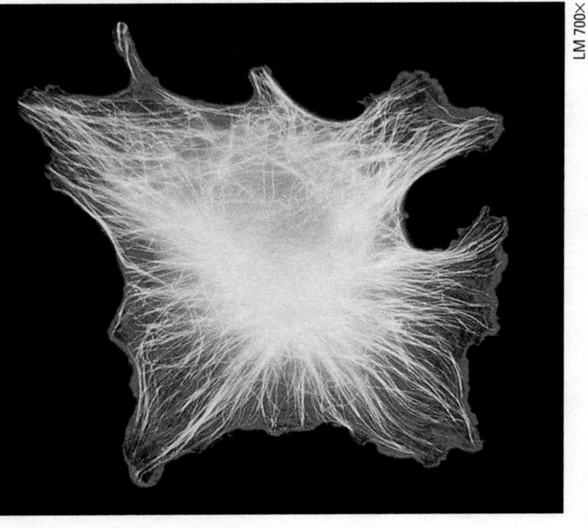

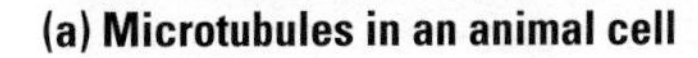
(a) Microtubules in an animal cell

LM 85×

(b) Amoeboid movement

Figure 4.18 The cytoskeleton. (a) In this micrograph, the microtubules of the cytoskeleton have been labeled with a yellow fluorescent dye. **(b)** Rapid degradation and rebuilding of microtubules is responsible for the crawling movement of organisms like the protist *Amoeba.*

globular proteins called tubulins. The other kinds of cytoskeletal fibers, called filiments, are thinner and solid.

Just as the bony skeleton of your body helps fix the positions of your organs, the cytoskeleton provides anchorage and reinforcement for many organelles in a cell. For instance, the nucleus is often held in place by a cytoskeletal cage of filaments. Other organelles move along tracks made from microtubules. For example, a lysosome might reach a food vacuole by moving along a microtubule. Microtubules also guide the movement of chromosomes when cells divide.

Although providing support like an animal's skeleton, the cell's cytoskeleton is more dynamic. It can quickly dismantle in one part of the cell by removing protein subunits and re-form in a new location by reattaching the subunits. Such rearrangement can provide rigidity in a new location, change the shape of the cell, or even cause the whole cell or some of its parts to move. This process contributes to the amoeboid (crawling) motions of the protist *Amoeba* and some of our white blood cells **(Figure 4.18b)**.

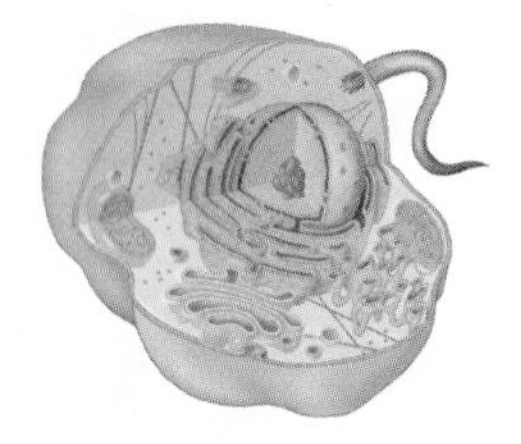

Cilia and Flagella

In some eukaryotic cells, a specialized arrangement of microtubules functions in the beating of flagella and cilia. Cilia and flagella are motile appendages—extensions from a cell that aid in locomotion. Eukaryotic **flagella** (singular, *flagellum*) propel the cell by an undulating whiplike motion. They often occur singly, such as in the sperm cells of humans and other animals **(Figure 4.19a)**. **Cilia** (singular, *cilium*) are generally shorter and more numerous than flagella and promote movement by a coordinated back-and-forth motion, like

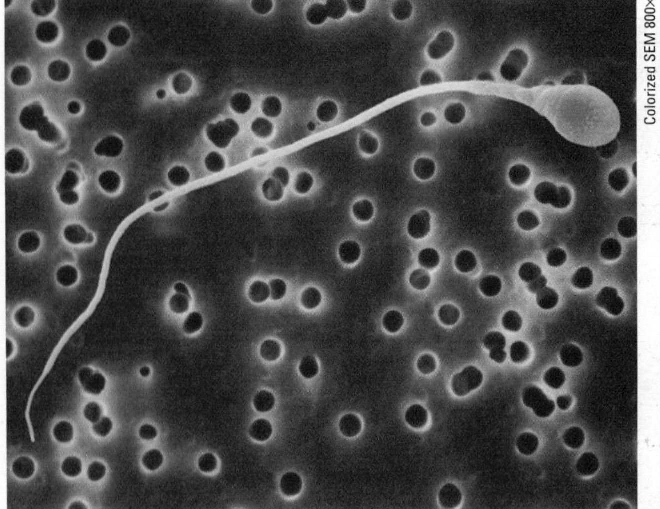

(a) Flagellum of a human sperm cell. A flagellum usually undulates, its snakelike motion driving a cell such as this sperm cell through its fluid environment.

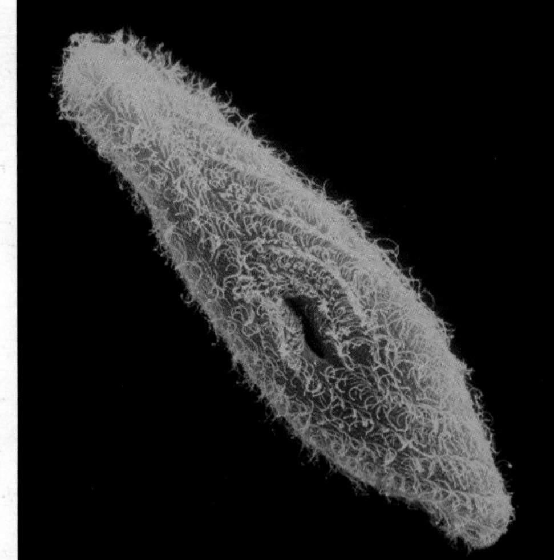

(b) Cilia on a protist. Cilia are shorter and more numerous than flagella and move with a back-and-forth motion. A dense nap of beating cilia covers this *Paramecium*, a freshwater protist that can dart rapidly through its watery home.

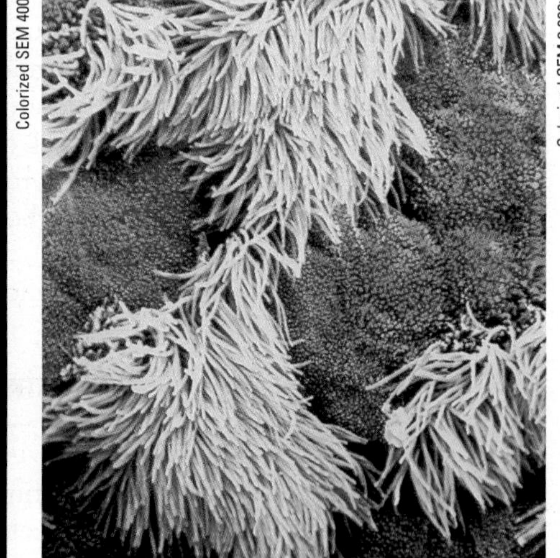

(c) Cilia lining the respiratory tract. The cilia lining your respiratory tract sweep mucus with trapped debris out of your lungs.

Figure 4.19 Flagella and cilia.

the rhythmic oars of a galley ship. Both cilia and flagella propel various protists through water **(Figure 4.19b)**.

Some cilia extend from nonmoving cells that are part of a tissue layer. There they function to move fluid over the surface of the tissue. For example, the ciliated lining of your windpipe helps cleanse your respiratory system by sweeping mucus with trapped debris out of your lungs **(Figure 4.19c)**. Tobacco smoke—whether inhaled first- or secondhand—irritates these ciliated cells, inhibiting or destroying the cilia. This interferes with the normal cleansing mechanisms and allows more toxin-laden smoke particles to reach the lungs. Frequent coughing—common in heavy smokers—then becomes the body's attempt to cleanse the respiratory system.

Because human sperm rely on flagella for movement, it's easy to understand why problems with flagella can lead to male infertility. Sperm with malfunctioning flagella are unable to travel up the female reproductive tract to fertilize an egg. Interestingly, some men with a certain type of hereditary sterility also suffer from respiratory problems. The explanation for this lies in the similarities between flagella (found in sperm) and cilia (found lining the respiratory tract). Because of a defect in the structure of their flagella and cilia, their sperm do not swim and their cilia do not sweep mucus out of their lungs.

CHECKPOINT

1. Name two functions that the cytoskeleton performs within the cell.
2. Compare and contrast cilia and flagella.

Answers: 1. Among other functions, the cytoskeleton serves as an anchor onto which organelles can attach and provides a track along which organelles can move. **2.** Cilia are short and numerous and move back and forth. Flagella are longer, often occurring singly, and they undulate. Cilia and flagella have the same basic structure and help move cells or move fluid over cells.

EVOLUTION CONNECTION

The Origin of Membranes

The plasma membrane is the boundary of all living cells. It is therefore logical to suppose that membranes first formed early in the evolution of life on Earth. Membranes allow cells to regulate their chemical exchanges with the environment—a basic requirement for life. A membrane that can regulate the flow of materials across it can enclose a solution that is different in composition from the surrounding solution, while still permitting the uptake of nutrients and the elimination of waste products. Phospholipids, the key ingredients of biological membranes, were probably among the organic molecules that formed from chemical reactions on early Earth before the emergence of life. These phospholipids could have spontaneously self-assembled into simple membranes, as we can demonstrate in a test tube **(Figure 4.20)**. This assembly requires neither genes nor other information beyond the intrinsic properties of the phospholipids themselves. The origin of membranes on a primordial Earth was likely an early step in the evolution of the first cells. We will consider this hypothesis further in Chapter 15. ■

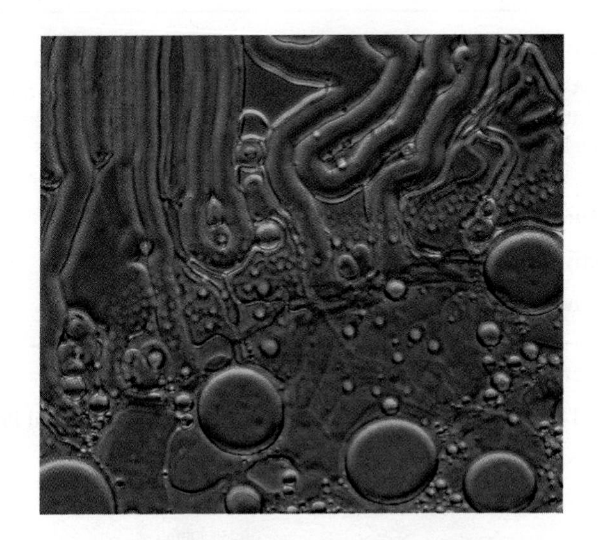

Figure 4.20 The spontaneous formation of membranes: a key step in the origin of life. When phospholipid molecules are mixed with water in a test tube, they spontaneously congregate and form water-filled bubbles (several are visible near the bottom of the photo). Thus the phospholipids have formed primitive membranes. As simple as they are, these membranes have some ability to control the traffic of substances into and out of the sphere.

Chapter Review

SUMMARY OF KEY CONCEPTS

For study help and activities, go to campbellbiology.com or the student CD-ROM.

The Microscopic World of Cells

- **Microscopes as a Window on the World of Cells** Using early microscopes, biologists discovered that all organisms are made of cells. Resolving power limits the useful magnification of microscopes. A light microscope (LM) has useful magnifications up to about 1,000×. Electron microscopes, both scanning (SEM) and transmission (TEM), are much more powerful.

Activity *Metric System Review*

Case Study in the Process of Science *What Is the Size and Scale of Our World?*

- **The Two Major Categories of Cells**

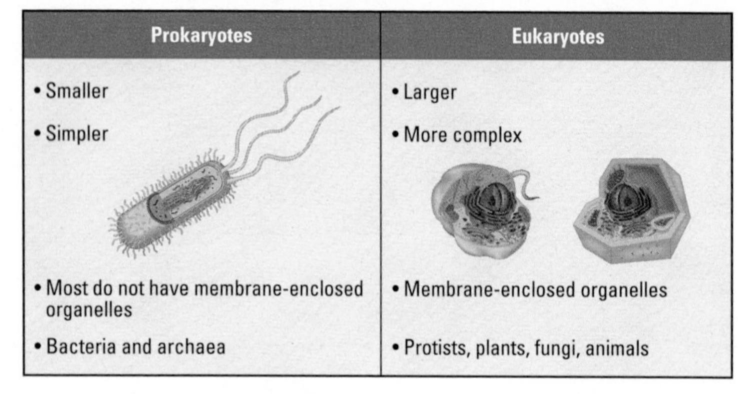

Prokaryotes	Eukaryotes
• Smaller • Simpler	• Larger • More complex
• Most do not have membrane-enclosed organelles	• Membrane-enclosed organelles
• Bacteria and archaea	• Protists, plants, fungi, animals

Activity *Prokaryotic Cell Structure and Function*

- **A Panoramic View of Eukaryotic Cells** Many cellular functions are partitioned by membranes in the complex organization of eukaryotic cells. The largest organelle is usually the nucleus. Other organelles are located in the cytoplasm, the region between the nucleus and the plasma membrane.

eTutor *A Tour of the Cell*

Activity *Comparing Cells*

Activity *Build an Animal Cell and a Plant Cell*

Membrane Structure

- **The Plasma Membrane: A Fluid Mosaic of Lipids and Proteins**

Activity *Membrane Structure*

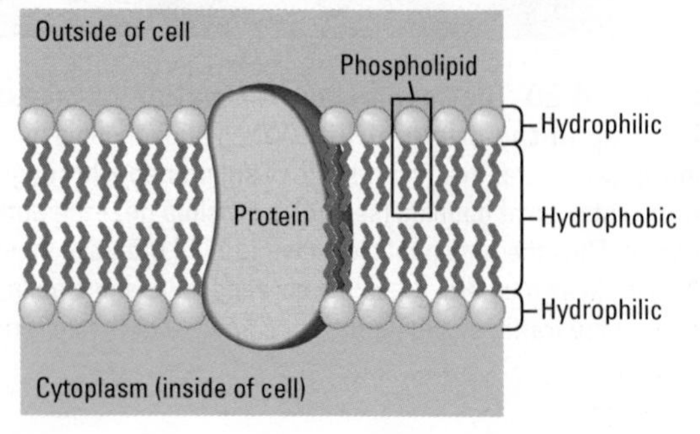

- **Cell Surfaces** Most cells secrete an extracellular coat that helps protect and support the cell. The walls that encase plant cells support plants against the pull of gravity and also prevent cells from absorbing too much water. Animal cells are coated by a sticky extracellular matrix.

The Nucleus and Ribosomes: Genetic Control of the Cell

- **Structure and Function of the Nucleus** An envelope consisting of two membranes encloses the nucleus. Within the nucleus, DNA and proteins make up chromatin fibers; each very long fiber is a single chromosome. The nucleus also contains the nucleolus, which produces components of ribosomes.
- **Ribosomes** Ribosomes produce proteins in the cytoplasm.
- **How DNA Controls the Cell** Genetic messages are transmitted to the ribosomes via messenger RNA, which travels from the nucleus to the cytoplasm.

Activity *Overview of Protein Synthesis*

The Endomembrane System: Manufacturing and Distributing Cellular Products

- **The Endoplasmic Reticulum** The ER consists of membrane-enclosed tubes and sacs within the cytoplasm. Rough ER, named for the ribosomes attached to its surface, makes membrane and secretory proteins. The functions of smooth ER include lipid synthesis and detoxification.
- **The Golgi Apparatus** The Golgi refines certain ER products and packages them in transport vesicles targeted for other organelles or export from the cell.
- **Lysosomes** Lysosomes, sacs containing digestive enzymes, function in digestion within the cell and chemical recycling.
- **Vacuoles** These membrane-enclosed organelles include the contractile vacuoles that expel water from certain freshwater protists and the large, multifunctional central vacuoles of plant cells.

Activity *The Endomembrane System*

Chloroplasts and Mitochondria: Energy Conversion

- **Chloroplasts** The sites of photosynthesis in plant cells, chloroplasts convert light energy to the chemical energy of food. Grana, stacks of membranous sacs within the chloroplasts, trap the light energy.
- **Mitochondria** These are the sites of cellular respiration, which converts food energy to ATP energy. ATP drives most cellular work. Both chloroplasts and mitochondria contain small amounts of DNA.

Activity *Build a Chloroplast and a Mitochondrion*

The Cytoskeleton: Cell Shape and Movement

- **Maintaining Cell Shape** Straight, hollow microtubules are an important component of the cytoskeleton, an organelle that gives support to and maintains the shape of cells.
- **Cilia and Flagella** Cilia and eukaryotic flagella are both motile appendages made primarily of microtubules. Cilia are short, numerous,

and move by coordinated beating. Flagella are long, often occur singly, and propel a cell through whiplike movements.

Activity *Cilia and Flagella*

Activity *Review: Animal Cell Structure and Function*

Activity *Review: Plant Cell Structure and Function*

SELF-QUIZ

1. Emily would like to film the movement of chromosomes during cell division. Her best choice for a microscope would be a
   a. light microscope, because of its magnifying power.
   b. transmission electron microscope, because of its resolving power.
   c. scanning electron microscope, because the chromosomes are on the cell surface.
   d. light microscope, because the specimen must be kept alive.

2. You look into a light microscope and view an unknown cell. What might you see that would tell you whether the cell is prokaryotic or eukaryotic?
   a. a rigid cell wall
   b. a nucleus
   c. a plasma membrane
   d. ribosomes

3. Prokaryotic cells are characteristic of the domains __________ and __________.

4. Which best describes the structure of the plasma membrane?
   a. proteins sandwiched between two layers of phospholipid
   b. proteins embedded in two layers of phospholipid
   c. a layer of protein coating a layer of phospholipid
   d. phospholipids embedded in two layers of protein

5. The ER has two distinct regions that differ in structure and function. Lipids are synthesized within the __________, and proteins are synthesized within the __________.

6. A type of cell called a lymphocyte makes proteins that are exported from the cell. You can track the path of these proteins within the cell from production through export by labeling them with radioactive isotopes. Which of the following organelles do you expect would be radioactively labeled in your experiment: chloroplasts, Golgi, plasma membrane, smooth ER, rough ER, nucleus, mitochondria? In what order would you expect the label to appear within your chosen organelles?

7. Name two similarities in the structure or function of chloroplasts and mitochondria. Name two differences.

8. Match the following organelles with their functions:
   a. ribosomes — 1. movement
   b. microtubules — 2. photosynthesis
   c. mitochondria — 3. protein synthesis
   d. chloroplasts — 4. digestion
   e. lysosomes — 5. cellular respiration

9. DNA controls the cell by transmitting genetic messages that result in protein production. Place the following organelles in the order that represents the flow of genetic information from the DNA through the cell: nuclear pores, ribosomes, nucleus, rough ER, Golgi.

Answers to the Self-Quiz questions can be found in Appendix D.

Go to the website or CD-ROM for more Self-Quiz questions.

THE PROCESS OF SCIENCE

10. The cells of plant seeds store oils in the form of droplets enclosed by membranes. Unlike the membranes you learned about in this chapter, the oil droplet membrane consists of a single layer of phospholipids rather than a bilayer. Draw a model for a membrane around an oil droplet. Explain why this arrangement is more stable than a bilayer.

11. Imagine that you are a pediatrician and one of your patients is a newborn who may have a lysosomal storage disease. You remove some cells from the patient and examine them under the microscope. What would you expect to see? Design a series of tests that could reveal whether the patient is indeed suffering from a lysosomal storage disease.

BIOLOGY AND SOCIETY

12. Doctors at a university medical center removed John Moore's spleen, which is a standard treatment for his type of leukemia. The disease did not recur. Researchers kept the spleen cells alive in a nutrient medium. They found that some of the cells produced a blood protein that showed promise as a treatment for cancer and AIDS. The researchers patented the cells. Moore sued, claiming a share in profits from any products derived from his cells. The U.S. Supreme Court ruled against Moore, stating that his lawsuit "threatens to destroy the economic incentive to conduct important medical research." Moore argued that the ruling left patients "vulnerable to exploitation at the hands of the state." Do you think Moore was treated fairly? Is there anything else you would like to know about this case that might help you decide?

5 The Working Cell

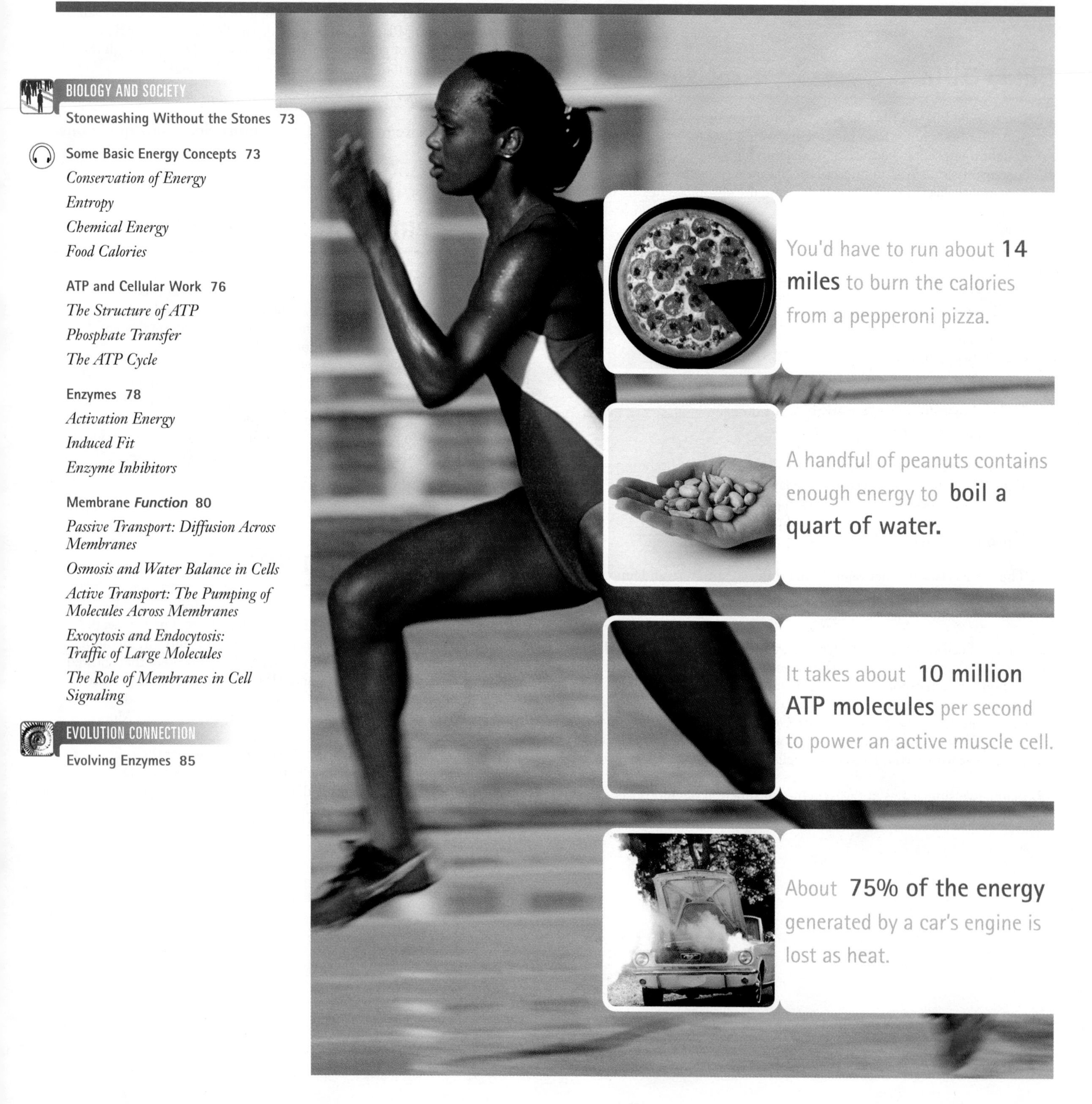

You'd have to run about **14 miles** to burn the calories from a pepperoni pizza.
A handful of peanuts contains enough energy to **boil a quart of water.**
It takes about **10 million ATP molecules** per second to power an active muscle cell.
About **75% of the energy** generated by a car's engine is lost as heat.

BIOLOGY AND SOCIETY

Stonewashing Without the Stones

When Levi Strauss began selling blue jeans in the 1870s, they were designed as work wear that could withstand the rugged demands of miners and other physical laborers who flocked to California during the Gold Rush. "Levi's" were constructed from denim, a sturdy cotton fabric, held together with rivets and dyed a distinctive indigo color. Today, blue jeans are still worn by people in physically demanding jobs, but they are also a popular fashion choice. Their original purpose—toughness—has taken a back seat to comfort and style.

Accordingly, the clothing industry has introduced new types of denim fabric that place a greater emphasis on look and feel. One style is known as stonewashed **(Figure 5.1)**. In the process of stonewashing, rolls of fabric are mixed with pumice stones inside huge tumblers. Although this process produces good-looking and comfortable jeans, it has several drawbacks: It can damage the fabric, it requires the wasteful mining of stones, and it is punishing on the equipment.

Recently, better results have been achieved using cellulase, a protein that breaks down the polysaccharide cellulose, the main component of cotton and other natural fibers (see Figure 3.13c). Cellulase is an enzyme—a catalyst that speeds up chemical reactions—used by bacteria and fungi to break down plant material. Washing denim in cellulase breaks down some of the cellulose fibers in the fabric. This softens the denim and releases indigo dye, resulting in a lighter color. Once the desired softness and color are reached, the cellulase is removed by rinsing. The textile industry calls this process "biostoning." Unlike pumice stones, the enzyme is friendly to the environment, the equipment, and the fabric, and biostoning allows manufacturers to closely control the final color.

Enzymes such as cellulase help living organisms control their internal chemical environments. In this chapter, we'll explore how cells are able to maintain this vital control—over energy production, over chemical reactions, and over the movement of materials. ■

Figure 5.1 Stonewashed versus regular jeans.

Some Basic Energy Concepts

MP3 Tutor
Basic Energy Concepts

Energy makes the world go round—both the cellular world and the larger world outside. But what exactly is energy? Our first step in understanding the working cell is to learn a few basic concepts about energy.

Conservation of Energy

Energy is defined as the capacity to perform work. Work is performed whenever an object is moved against an opposing force. In other words, work moves things in ways they would not move if left alone.

As an example, imagine a diver climbing to the top of a platform and diving off **(Figure 5.2)**. To get the diver to the top of the platform, work must be performed to overcome the opposing force of gravity. In the young man climbing up the steps to the diving platform, chemical energy from the food he ate for lunch is being converted to **kinetic energy,** the energy of motion.

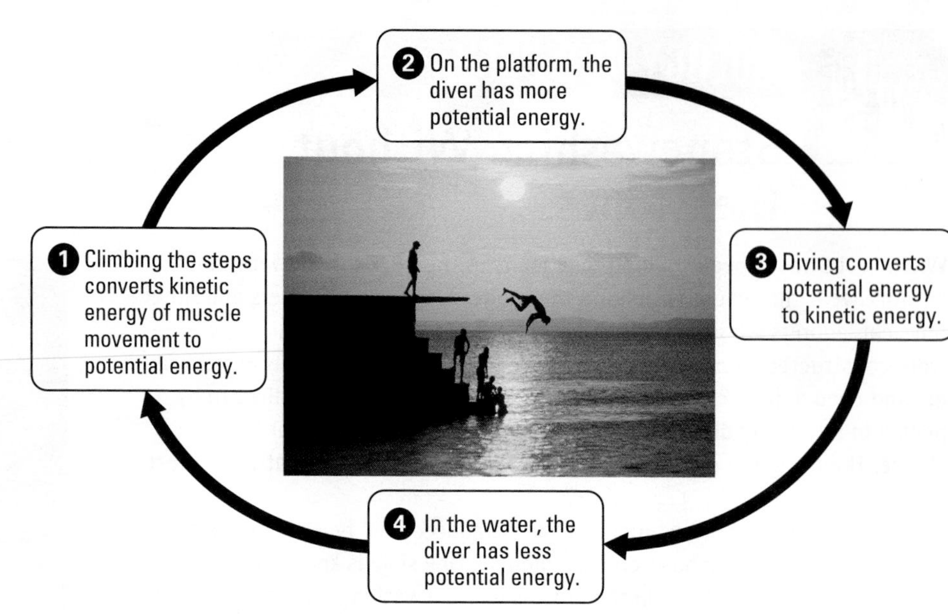

1 Climbing the steps converts kinetic energy of muscle movement to potential energy.
2 On the platform, the diver has more potential energy.
3 Diving converts potential energy to kinetic energy.
4 In the water, the diver has less potential energy.

Figure 5.2 Energy conversions during a dive.

What happens to the kinetic energy once the diver reaches the top of the platform? Has the energy disappeared? The answer is no. You may be familiar with the principle of conservation of matter, which states that matter cannot be created nor destroyed but can only be converted from one form to another. A similar principle is known as **conservation of energy.** It states that it is not possible to destroy or create energy; energy, too, can only be converted from one form to another. A power plant, for example, does not make energy; it merely converts it from one form (such as the energy stored in coal) to a more convenient form (such as electricity). That's exactly what happens in the diver's climb up the steps. The kinetic energy of his muscle movement is now stored in a form called potential energy. **Potential energy** is energy that an object has because of its location or arrangement, such as the energy contained by water behind a dam or by a compressed spring. In our example, the diver on the platform has potential energy because of his elevated location. The young man diving is converting his potential energy to kinetic energy. Life depends on the conversion of energy from one form to another.

Entropy

If energy cannot be destroyed, where has it gone once the diver hits the water? It has been converted to **heat,** a type of kinetic energy contained in the random motion of atoms and molecules. The friction between the falling body and its surroundings generated heat in the air and then in the water.

All energy conversions generate some heat. While heat production does not destroy energy, it does make it less useful. Heat, of all energy forms, is the most difficult to "tame"—the most difficult to harness for useful work. It is energy in its most chaotic form, the energy of aimless molecular movement.

Scientists use the term **entropy** as a measure of disorder, or randomness. Every time energy is converted from one form to another, entropy increases. The energy conversions during the climb up and dive off the platform increased entropy because all of the diver's potential energy was lost to his surroundings as heat. To climb up the steps for another dive, the diver must use additional stored food energy.

Chemical Energy

How can molecules derived from the food we eat provide energy for our working cells? The molecules of food, gasoline, and other fuels have a special form of potential energy called **chemical energy,** which arises from the arrangement of atoms. Carbohydrates, fats, and gasoline have structures that make them especially rich in chemical energy.

Living cells and automobile engines use the same basic process to make the chemical energy stored in their fuels available for work. In both cases, this process breaks the organic fuel into smaller waste molecules that have

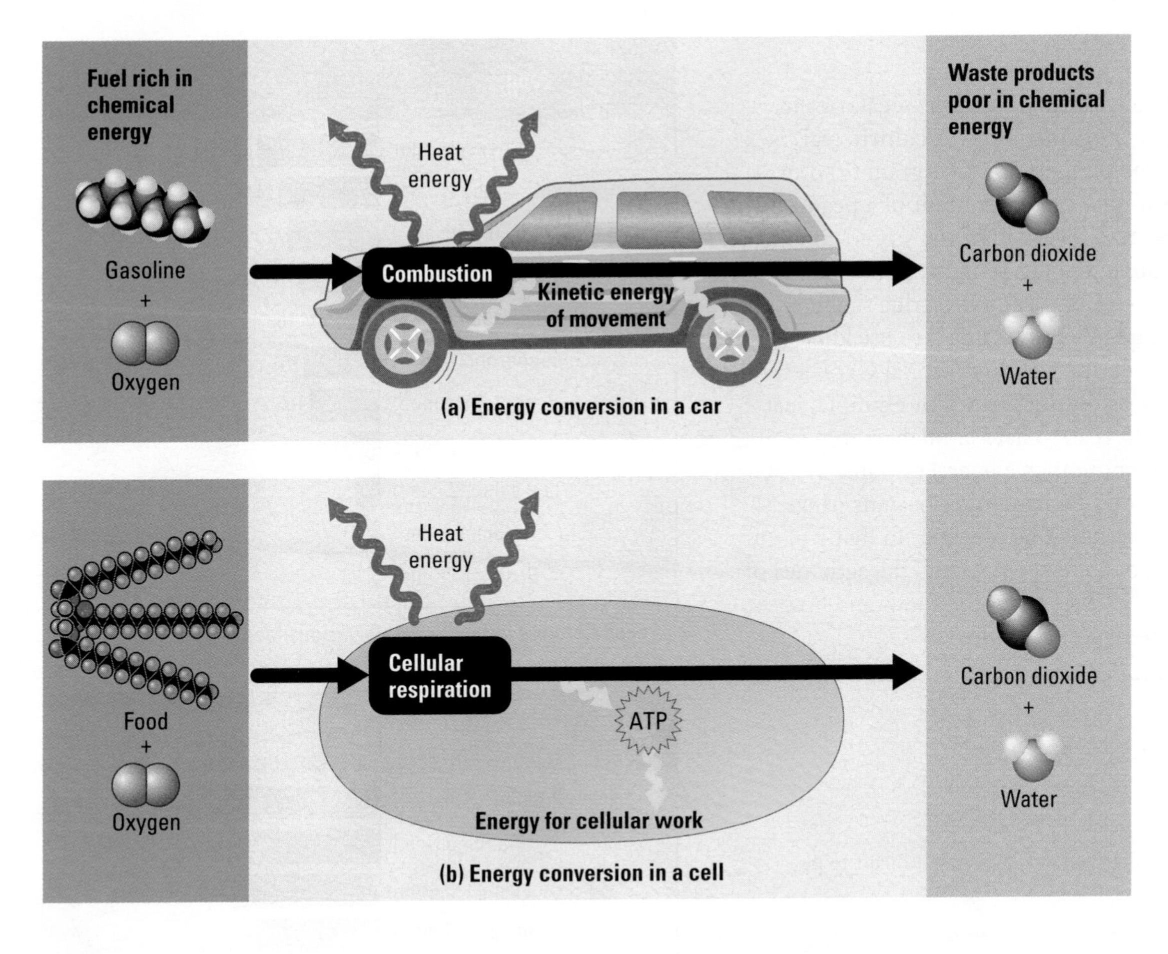

Figure 5.3 Energy conversions in (a) a car and (b) a cell. Like an automobile engine, cellular respiration uses oxygen to harvest the chemical energy of organic fuel molecules. Cellular respiration breaks sugars and other organic molecules down to smaller waste products. This chemical breakdown releases energy that was stored in the fuel molecules. Cells use this energy for their work.

much less chemical energy than the fuel molecules did, thereby releasing energy that can be used to perform work.

For example, the engine of an automobile mixes oxygen with gasoline in an explosive chemical reaction that breaks down the fuel molecules and pushes the pistons that eventually move the wheels. The waste products emitted from the exhaust pipe of the car are mostly carbon dioxide and water **(Figure 5.3a)**. About 25% of the energy an automobile engine extracts from its fuel is converted to the kinetic energy of the car's movement. The rest is converted to heat—so much, in fact, that the engine would melt if it weren't that the car's radiator and fan dispense excess heat into the atmosphere.

Your cells also use oxygen to help harvest chemical energy **(Figure 5.3b)**. And just as in a car engine, the "exhaust" is mostly carbon dioxide and water. The "combustion" of fuel in cells is called cellular respiration, a slow, gradual, more efficient "burning" of fuel compared with the explosive combustion in an automobile engine. Cellular respiration is the energy-releasing chemical breakdown of fuel molecules and the storage of that energy in a form the cell can use to perform work. We will discuss the details of cellular respiration, performed within the mitochondria of cells (see Figure 4.17), in the next chapter. You convert about 40% of your food energy to useful work, such as the contraction of your muscles. The other 60% of the energy released by the breakdown of fuel molecules generates body heat. Humans and many other animals can use this heat to keep the body at an almost constant temperature (37°C, or 98.6°F, in the case of humans), even when the surrounding air is much colder. The liberation of heat energy also explains why you feel so hot after vigorous exercise. Sweating and other cooling mechanisms enable your body to lose the excess heat, much as a car's radiator keeps the engine from overheating.

Food Calories

Read any packaged food label and you'll find the number of calories in each serving of that food. Calories are units of energy. A **calorie** (cal) is the amount of energy that can raise the temperature of 1 gram (g) of water by 1°C. You could actually measure the caloric content of a peanut by burning it under a container of water to convert all of the stored chemical energy to heat and then measuring the temperature increase of the water.

Calories are tiny units of energy, so using them to describe the fuel content of foods is not practical. Instead, it's conventional to use kilocalories (kcal), units of 1,000 calories. In fact, the Calories (capital C) on a food package are actually kilocalories. That's a lot of energy. For example, just one peanut has about 5 food Calories (5 kcal). That's enough energy to increase the temperature of 1 kg (a little more than a quart) of water by 5°C in our peanut-burning experiment. And just a handful of peanuts packs enough Calories, if converted to heat, to boil 1 kg of water. In living organisms, food isn't used to boil water, of course, but to fuel the activities of life. **Figure 5.4** shows the number of Calories in several foods and how many Calories are burned off by some typical activities.

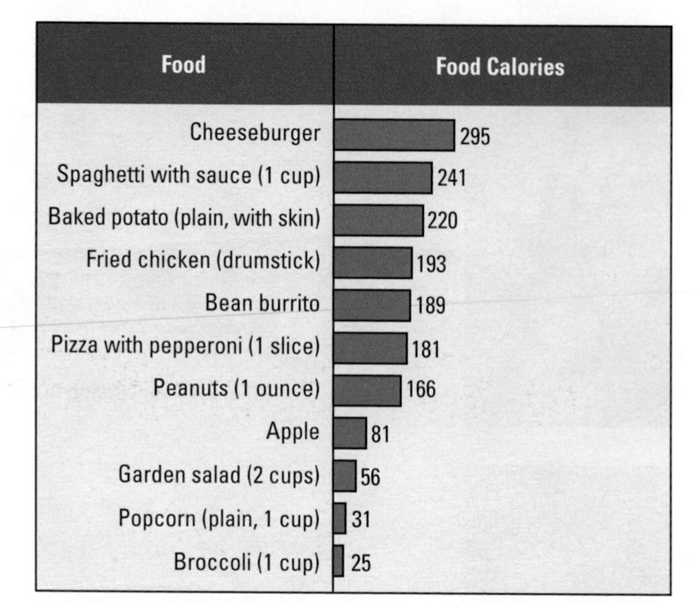

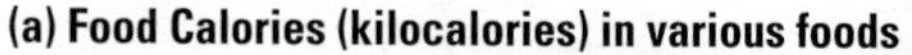

(a) Food Calories (kilocalories) in various foods

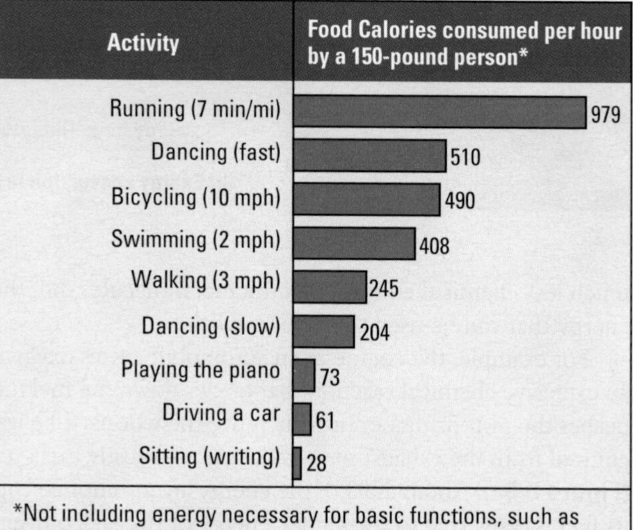

Activity
Food Calories consumed per hour by a 150-pound person*
Running (7 min/mi)
979
Dancing (fast)
510
Bicycling (10 mph)
490
Swimming (2 mph)
408
Walking (3 mph)
245
Dancing (slow)
204
Playing the piano
73
Driving a car
61
Sitting (writing)
28
*Not including energy necessary for basic functions, such as breathing and heartbeat

(b) Food Calories (kilocalories) we burn in various activities

Figure 5.4 Some caloric accounting.

CHECKPOINT

1. How can an object at rest have energy?
2. Describe the energy transformations that occur when you climb to the top of a stairway.
3. Which form of energy is the most randomized and the most difficult to put to useful work?

Answers: 1. It can have potential energy because of its location or the arrangement of its atoms. **2.** You convert the chemical energy of food to the kinetic energy of your upward climb. At the top of the stairs, some of the energy has been stored as potential energy because of your higher elevation. The rest has been converted to heat. **3.** Heat energy

ATP and Cellular Work

The carbohydrates, fats, and other fuel molecules we obtain from food do not drive the work in our cells directly. Instead, the chemical energy released by the breakdown of organic molecules during cellular respiration is used to generate molecules of ATP. These molecules of ATP then power cellular work. ATP acts like an energy shuttle, storing energy obtained from food and then releasing it as needed at a later time.

The Structure of ATP

The abbreviation ATP stands for adenosine triphosphate. **ATP** consists of an organic molecule called adenosine plus a tail of three phosphate groups (Ⓟ) **(Figure 5.5)**. You may

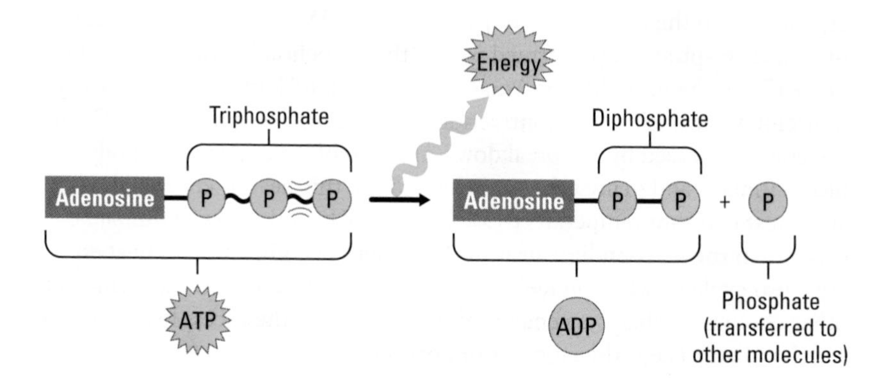

Figure 5.5 ATP power. Each Ⓟ in the triphosphate tail of ATP represents a phosphate group, a phosphorus bonded to oxygen atoms. The triphosphate tail is unstable, partly because of repulsion between the phosphate groups, which are negatively charged. There is a tendency for the phosphate group at the end of the triphosphate tail to break away from ATP and bond to other molecules instead. This phosphate transfer, catalyzed by enzymes, provides energy for cellular work. The leftover molecule is ADP.

have noticed that ATP is closely related in structure to a nucleotide. The three phosphate groups in ATP are chemically identical to the ones in the sugar-phosphate backbone of DNA or RNA.

The triphosphate tail is the "business" end of ATP, the part that provides energy for cellular work. Each phosphate group is negatively charged. Negative charges repel each other. The crowding of negative charges in the triphosphate tail contributes to the potential energy of ATP. It's analogous to storing energy by compressing a spring; if you release the spring, it will "relax," and you can use that springiness to do some useful work. For ATP power, it is release of the phosphate at the tip of the triphosphate tail that makes energy available to working cells. What is left behind is now called **ADP**, adenosine diphosphate (two phosphate groups instead of three; see Figure 5.5).

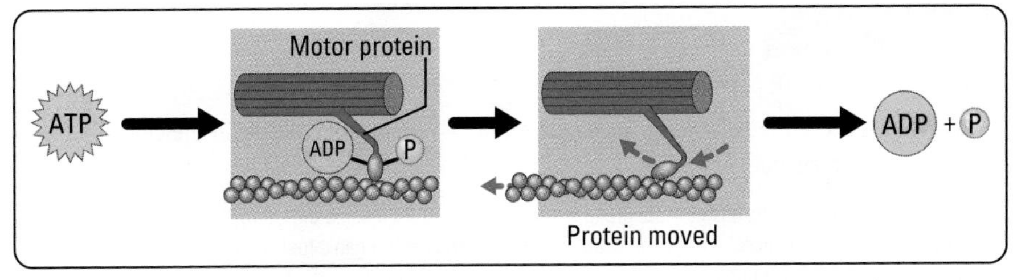

(a) Mechanical work

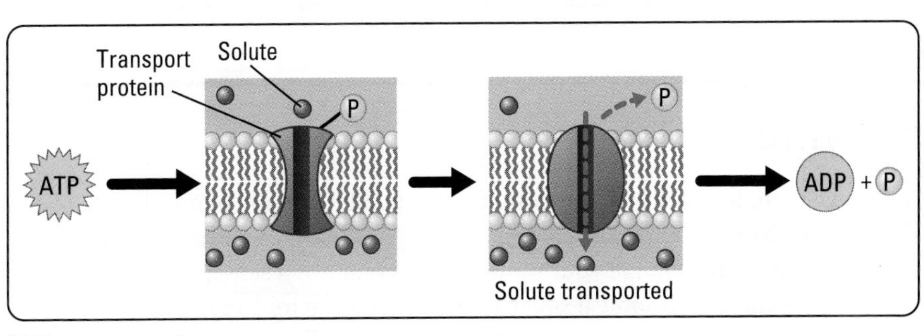

(b) Transport work

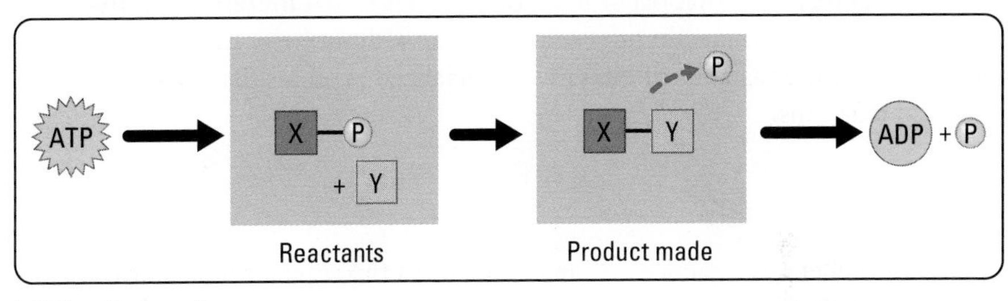

(c) Chemical work

Figure 5.6 How ATP drives cellular work. Each type of work is powered when enzymes transfer phosphate from ATP to a recipient molecule: **(a)** a motor protein (mechanical work), **(b)** a transport protein (transport work), or **(c)** a chemical reactant (chemical work).

Phosphate Transfer

When ATP drives work in cells, phosphate groups don't just fly off into space. ATP energizes other molecules in cells by transferring phosphate groups to those molecules. This helps cells perform three main kinds of work: mechanical work, transport work, and chemical work.

Imagine a bicyclist pedaling up a hill. In the muscle cells of the rider's legs, ATP transfers phosphate groups to special motor proteins. The proteins change their shape, causing the muscle cells to contract and perform mechanical work **(Figure 5.6a)**. As an example of transport work, ATP enables brain cells to pump ions across their membranes **(Figure 5.6b)**. This prepares the brain cells to transmit signals. And ATP drives the chemical work of making a cell's giant molecules. An example is the linking of amino acids to make a protein **(Figure 5.6c)**. In Chapter 3, we saw that dehydration reactions link amino acids; it is ATP that provides the energy for this process. Notice again in Figure 5.6 that all these types of work occur when target molecules accept phosphate from ATP.

The ATP Cycle

Your cells spend ATP continuously. Fortunately, it is a renewable resource. ATP can be restored by adding a phosphate group back to ADP. That takes energy, like recompressing a spring. And that's where food reenters the story. The chemical energy that cellular respiration harvests from sugars and other organic fuels is put to work regenerating a cell's supply of ATP. Cells thus have an ATP cycle: Cellular work spends ATP, which is restored when ADP and phosphate are recombined using energy released by cellular respiration **(Figure 5.7)**. Thus, ATP functions in what is called **energy coupling:** the transfer of energy from processes that yield energy, such as the breakdown of organic fuels, to processes that consume energy, such as muscle contraction and other types of cellular work. The third phosphate group acts as an energy shuttle within the ATP cycle.

The ATP cycle runs at an astonishing pace. A working muscle cell recycles all of its ATP about once each minute. That amounts to about 10 million ATP molecules spent and regenerated per second per cell.

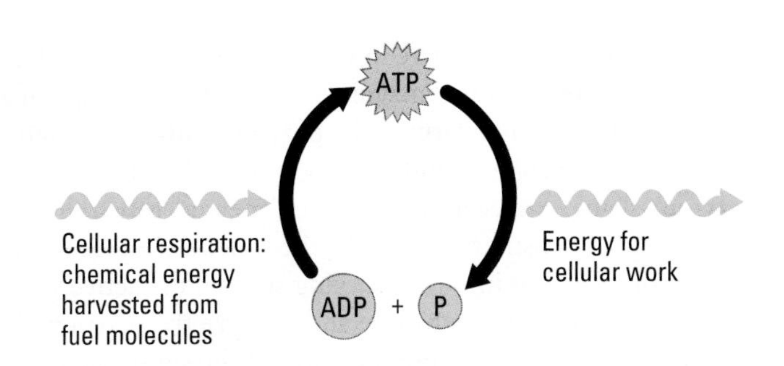

Figure 5.7 The ATP cycle.

CHECKPOINT

1. Explain how ATP powers cellular work.
2. What is the source of energy for regenerating ATP from ADP? Where in the cell does this take place?

Answers: 1. ATP transfers a phosphate group to other molecules, increasing the energy content of those molecules. **2.** Chemical energy harvested from sugars and other organic fuels via cellular respiration in the mitochondria

Enzymes

As you've seen, living organisms contain a vast collection of chemicals. Within the "bag of chemicals" that makes up your body, countless chemical reactions constantly change your molecular makeup. In a sense, a living organism is a complex "chemical square dance," with the molecular "dancers" constantly changing partners via chemical reactions. The sum total of all the chemical reactions that occur in an organism is called **metabolism.** Interestingly, almost no metabolic reactions occur without help. Instead, most require the assistance of **enzymes,** specialized proteins that speed up chemical reactions.

Activation Energy

To initiate a chemical reaction, chemical bonds in the reactant molecules must be broken. (The first step in swapping partners during a square dance is to let go of your current partner's hand.) This process requires that the molecules absorb energy from their surroundings. This energy is called **activation energy** because it activates the reactants and triggers a chemical reaction.

One way to speed up a chemical reaction is to increase the supply of energy by heating the mixture; that's why Bunsen burners are used in chemistry labs. Boiling your cells, however, is not an option for speeding up your metabolism. Enzymes enable metabolism to occur at cooler temperatures by reducing the amount of activation energy required to break the bonds of reactant molecules. If you think of the requirement for activation energy as a barrier to a chemical reaction, then the function of an enzyme is to lower that barrier **(Figure 5.8)**. An enzyme does this by binding to reactant molecules and putting them under some kind of physical or chemical stress, making it easier to break their bonds and start a reaction.

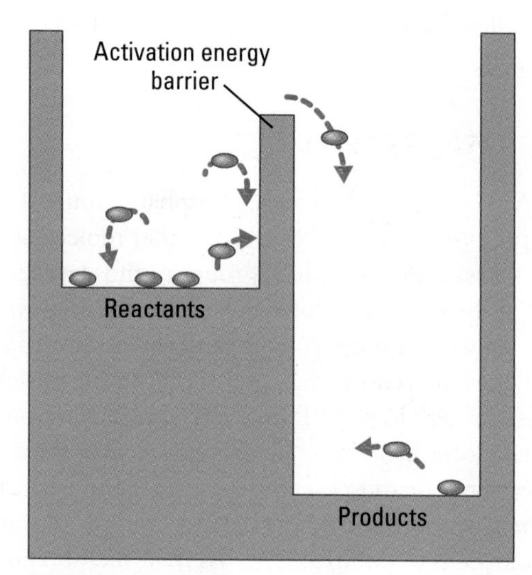

(a) Without enzyme

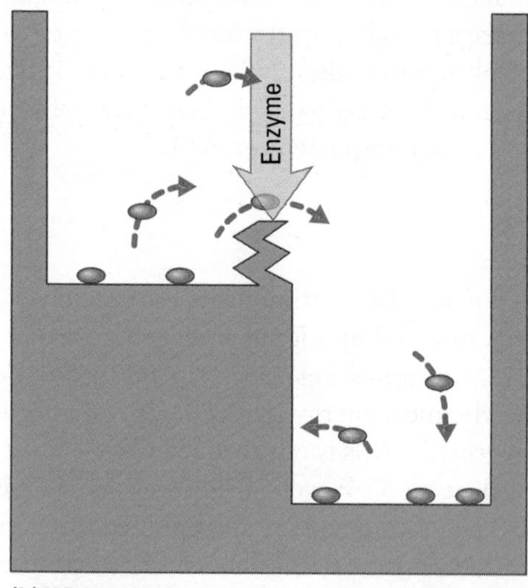

(b) With enzyme

Figure 5.8 Activation energy: a jumping-bean analogy. **(a)** A chemical reaction requires activation energy to break the bonds of the reactant molecules. The jumping beans in this analogy represent reactant molecules that must overcome the barrier of activation energy before they can reach the "product" side. **(b)** An enzyme speeds the process by lowering the barrier of activation energy.

Induced Fit

Each enzyme is very selective in the reaction it catalyzes. This specificity is based on the ability of the enzyme to recognize the shape of a certain reactant molecule, which is termed the enzyme's **substrate.** And the ability of the enzyme to recognize and bind to its specific substrate depends on the enzyme's shape. A special region of the enzyme, called the **active site,** has a shape and chemistry that fits it to the substrate molecule. When a substrate molecule slips into this docking station, the active site changes shape slightly to embrace the substrate and catalyze the reaction. This interaction is called **induced fit,** because the entry of the substrate induces the enzyme to change its shape slightly and make the fit between substrate and active site even snugger. Think of a handshake: As your hand makes contact with your friend's, it changes shape slightly to make a better fit.

After the products are released from the active site, the enzyme is once again available to accept another molecule of its specific substrate. In fact, the ability to function over and over again is a key characteristic of enzymes. **Figure 5.9** follows the action of a specific enzyme called sucrase, which hydrolyzes the disaccharide sucrose (the substrate). Like sucrase, many enzymes are named for their substrates, but with an *-ase* ending.

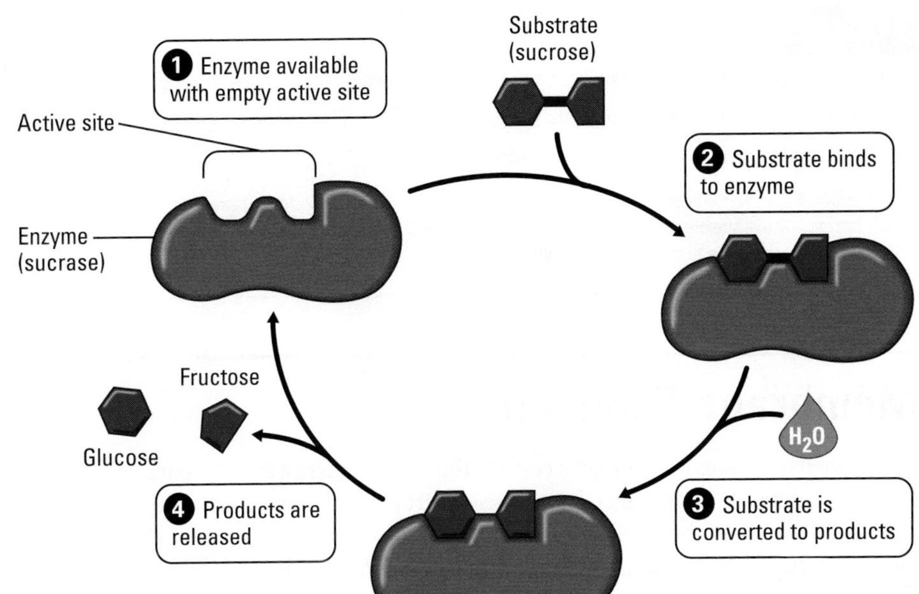

Figure 5.9 How an enzyme works. Our example is the enzyme sucrase, named for its substrate, sucrose. 1 With its active site empty, sucrase is receptive to a molecule of its substrate. 2 The substrate has a shape that fits the shape of the active site. Sucrose enters the active site, and 3 the enzyme catalyzes the chemical reaction—in this case, the hydrolysis of sucrose. 4 The products—glucose and fructose, in this example—exit the active site, and the sucrase is available to receive another molecule of its substrate.

Enzyme Inhibitors

Certain molecules can inhibit a metabolic reaction by binding to an enzyme and disrupting its function **(Figure 5.10)**. Some of these **enzyme inhibitors** are actually substrate imposters that plug up the active site. (You could not shake a friend's hand if someone else puts a banana in your hand first!) Other inhibitors bind to the enzyme at some site remote to the active site, but the binding changes the shape of the enzyme so that its active site is no longer receptive to the substrate. (Imagine being unable to shake your friend's hand because someone else is tickling your ribs.) In some cases, the binding is reversible, enabling certain inhibitors to regulate metabolism. For example, if metabolism is producing more of a certain product than a cell needs, that product may inhibit an enzyme required for its production. This **feedback regulation** keeps the cell from wasting resources that could be put to better use.

Some enzyme inhibitors act as poisons that block metabolic processes essential to the survival of an organism. For example, an insecticide called malathion inhibits an enzyme required for normal functioning of the insect nervous system. Many antibiotics that kill disease-causing bacteria are also enzyme inhibitors. For instance, penicillin inhibits an enzyme that many bacteria use to make their cell walls (see Chapter 4).

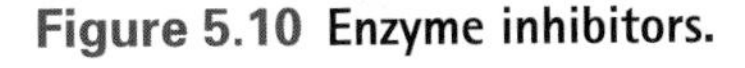

Figure 5.10 Enzyme inhibitors.

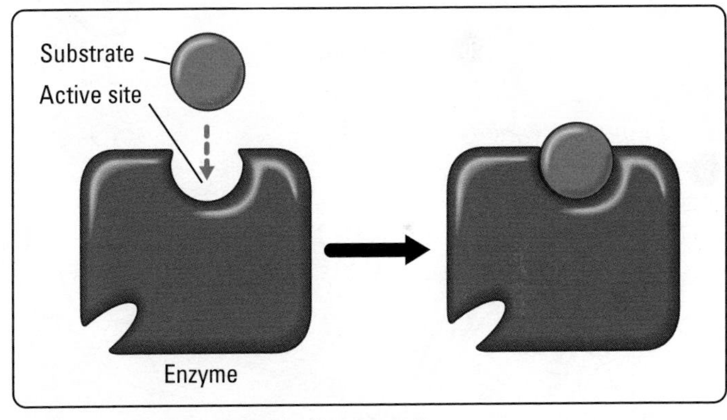

(a) Enzyme and substrate binding normally

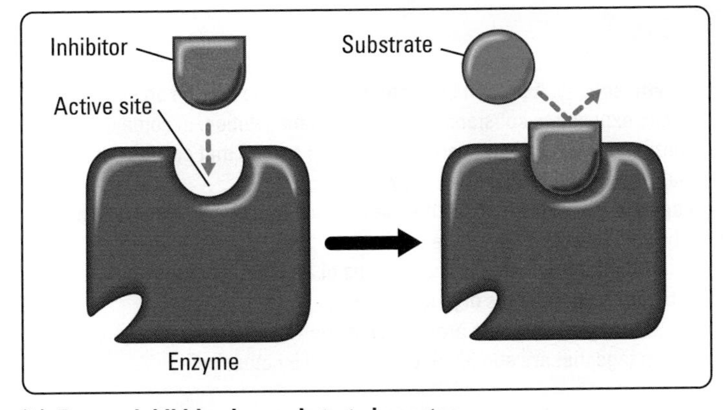

(b) Enzyme inhibition by a substrate imposter

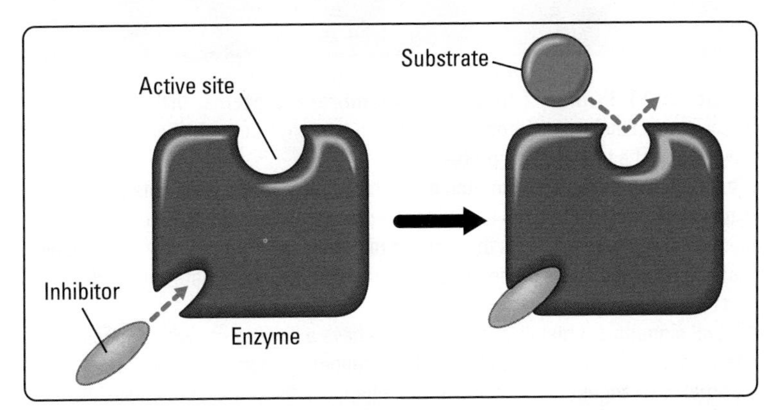

(c) Enzyme inhibition by a molecule that causes the active site to change shape

CHECKPOINT

1. What effect does an enzyme have on the activation energy of a chemical reaction?
2. How does an enzyme recognize its substrate?
3. How does the antibiotic penicillin work?

Answers: 1. An enzyme lowers the activation energy of a chemical reaction. **2.** The substrate and the enzyme's active site are complementary in shape and chemical nature. **3.** It inhibits an enzyme that certain bacteria use to make their cell walls.

Membrane Function

So far, we have discussed how cells control the flow of energy and the pace of chemical reactions. Working cells must also exert control of another kind: Cells must be able to regulate the flow of materials to and from the environment. In Chapter 4, you learned about the structure of the plasma membrane (see Figure 4.7). The many proteins embedded within the phospholipid bilayer of the plasma membrane perform a variety of functions **(Figure 5.11)**.

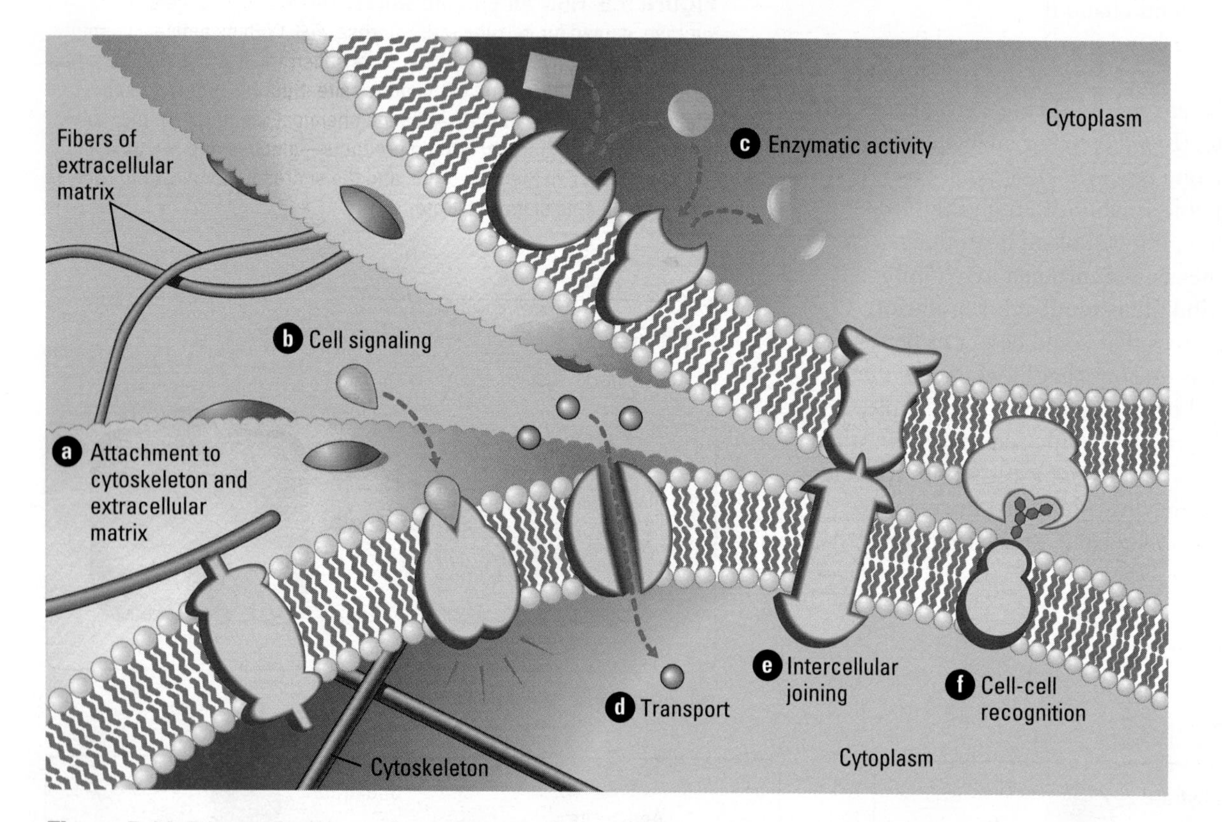

Figure 5.11 Primary functions of membrane proteins. In this diagram, all six types of membrane proteins are shown for convenience. An actual cell may have just a few of these proteins.
a Attachment to the cytoskeleton and extracellular matrix. Elements of the cytoskeleton may be bonded to membrane proteins, a function that helps maintain cell shape and fixes the location of certain membrane proteins. Proteins that adhere to the fibers of the extracellular matrix can coordinate extracellular and intracellular changes.
b Cell signaling. A membrane protein may have a binding site with a specific shape that fits the shape of a chemical messenger, such as a hormone. The external messenger (signal) may cause a change in the protein that relays the message to the inside of the cell.
c Enzymatic activity. A protein built into the membrane may be an enzyme with its active site exposed to substances in the adjacent solution. In some cases, several enzymes in a membrane are organized as a team that carries out the sequential steps of a metabolic pathway.
d Transport. A protein that spans the membrane may provide a channel across the membrane that is selective for a particular solute.
e Intercellular joining. Membrane proteins of adjacent cells may be hooked together to form various kinds of junctions.
f Cell-cell recognition. Some proteins with short chains of sugars serve as identification tags that are specifically recognized by other cells.

One of the most important jobs of the plasma membrane is to regulate the passage of materials into and out of the cell. **Transport proteins** (shown in Figure 5.11d) are critical to this task. In this section, you'll learn about the most important mechanisms of transport across membranes.

Passive Transport: Diffusion Across Membranes

Molecules are restless. The heat energy they contain makes them vibrate and wander randomly. One result of this motion is **diffusion**, the tendency for molecules of any substance to spread out into the available space. Each molecule moves randomly, and yet the overall diffusion of a population of molecules may be directional. For example, imagine many molecules of perfume isolated inside a bottle. If you remove the bottle top, every molecule of perfume will move randomly about, but the net direction of the perfume molecules will be out of the bottle to fill the room. You could, with great effort, return the perfume to its bottle, but this would never occur spontaneously.

For an example closer to a living cell, imagine a membrane separating pure water from a solution of a dye dissolved in water **(Figure 5.12)**. Assume that this membrane is permeable to the dye molecules—meaning that the dye molecules can pass through the membrane. Although each dye molecule moves randomly, there will be a net migration across the membrane to the side that began as pure water. The spreading of the dye across the membrane will continue until both solutions have equal concentrations of the dye. Once that point is reached, there will be a dynamic equilibrium, with as many dye molecules moving per second across the membrane in one direction as the other.

These examples illustrate a simple rule of diffusion: A substance will diffuse from where it is more concentrated to where it is less concentrated. Put another way, a substance tends to diffuse down its concentration gradient.

Diffusion across a membrane is an example of **passive transport**—passive because the cell does not expend any energy for it to happen. The membrane does, however, play a regulatory role by being selectively permeable. For example, small molecules such as carbon dioxide (CO_2) and oxygen (O_2) generally pass through more readily than large molecules such as proteins (otherwise, the cell would lose its macromolecules). However, the membrane is relatively impermeable to even some very small substances, such as hydrogen ions (H^+) and other inorganic ions, which are too hydrophilic to pass through the phospholipid bilayer of the membrane. Such substances can be transported via **facilitated diffusion** by specific transport proteins that act as selective corridors (see Figure 5.11d).

Passive transport is extremely important to all cells. In our lungs, for example, passive transport along concentration gradients is the sole means by which O_2, essential for metabolism, enters red blood cells and CO_2, a metabolic waste, passes out of them. One of the most important substances that crosses membranes by passive transport is water.

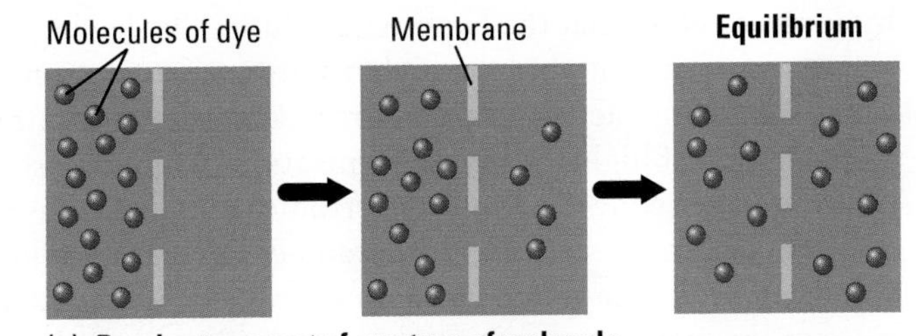

(a) Passive transport of one type of molecule

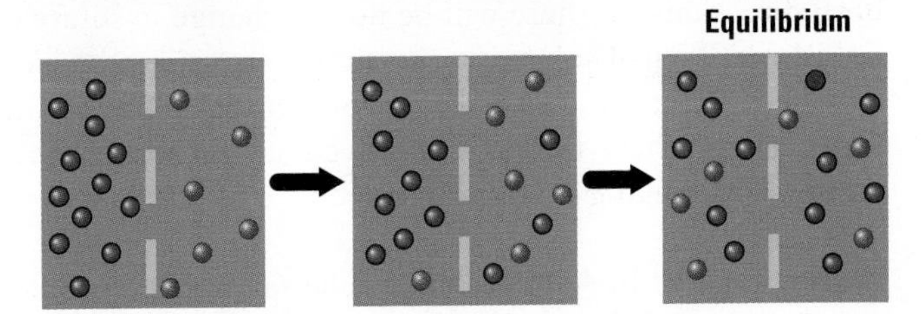

(b) Passive transport of two types of molecules

Figure 5.12 Passive transport: diffusion across a membrane. **(a)** The membrane is permeable to these dye molecules, which diffuse down their concentration gradient. At equilibrium, the molecules are still restless, but the rate of transport is equal in both directions. **(b)** If solutions have two or more solutes, each will diffuse down its own concentration gradient.

Osmosis and Water Balance in Cells

The passive transport of water across a selectively permeable membrane is called **osmosis (Figure 5.13)**. Consider the case of a membrane separating two solutions with different concentrations of a solute—say, the sugar sucrose. The membrane is permeable to water but not to the solute. The solution with a higher concentration of solute is said to be **hypertonic.** The solution with the lower solute concentration is

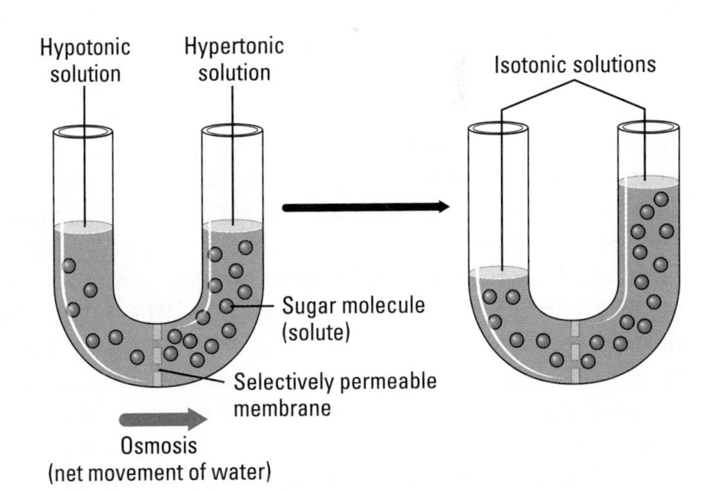

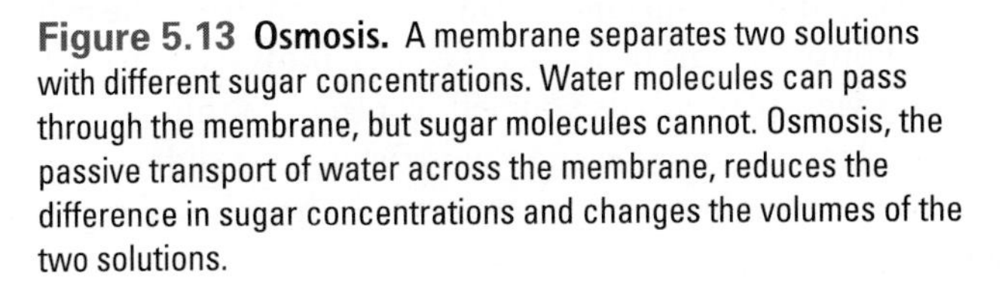

Figure 5.13 Osmosis. A membrane separates two solutions with different sugar concentrations. Water molecules can pass through the membrane, but sugar molecules cannot. Osmosis, the passive transport of water across the membrane, reduces the difference in sugar concentrations and changes the volumes of the two solutions.

hypotonic. Note that the hypotonic solution, by having the lower solute concentration, has the higher water concentration. Therefore, water will diffuse across the membrane along its concentration gradient from an area of higher water concentration (hypotonic solution) to one of lower water concentration (hypertonic solution). This reduces the difference in solute concentrations and changes the volumes of the two solutions. Once the solute concentrations become the same on both sides of the membrane, water molecules will move at the same rate in both directions, so there will be no net change in solute concentration. Solutions of equal solute concentration are said to be **isotonic**. You may have seen this term on bottles of contact lens saline solution, which is formulated to have the same solute concentration as the surface of the human eye, making it nonirritating.

Water Balance in Animal Cells The survival of a cell depends on its ability to balance water uptake and loss. When an animal cell, such as a red blood cell, is immersed in an isotonic solution, the cell's volume remains constant because the cell gains water at the same rate that it loses water (**Figure 5.14a**, top). By definition, the cell is isotonic to its surroundings because the two solutions have the same total concentration of solutes. Many marine animals, such as sea stars and crabs, are isotonic to seawater. What happens if an animal cell finds itself in a hypotonic solution, which has a lower solute concentration than the cell? The cell gains water, swells, and may burst (lyse) like an overfilled water balloon (**Figure 5.14b**, top). A hypertonic environment is also harsh on an animal cell; the cell shrivels and can die from water loss (**Figure 5.14c**, top).

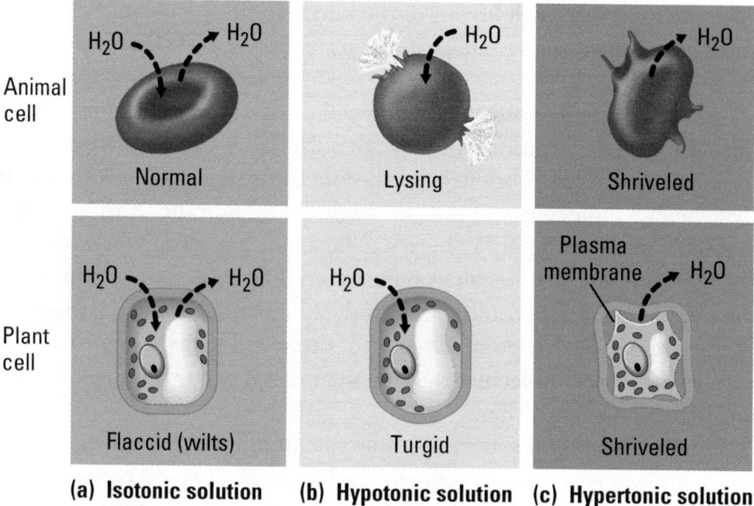

Figure 5.14 The behavior of animal and plant cells in different osmotic environments.

For an animal to survive if its cells are exposed to a hypotonic or hypertonic environment, the animal must have a way to balance an excessive uptake or excessive loss of water. The control of water balance is called **osmoregulation.** For example, a freshwater fish, whose environment is hypotonic to its body, has kidneys and gills that work constantly to prevent an excessive buildup of water in the body. And if you look at Figure 4.14a, you'll see *Paramecium's* contractile vacuole, which bails out the excess water that continuously enters the cell from the hypotonic pond water.

Water Balance in Plant Cells Problems of water balance are somewhat different for plant cells because of their rigid cell walls. A plant cell immersed in an isotonic solution is flaccid (floppy), and a plant wilts in this situation (Figure 5.14a, bottom). In contrast, a plant cell is turgid (firm) and healthiest in a hypotonic environment, with a net inflow of water (Figure 5.14b, bottom). Although the elastic cell wall expands a bit, the back pressure it exerts prevents the cell from taking in too much water and bursting, as an animal cell would in this environment. Turgor is necessary for plants to retain their upright posture and the extended state of their leaves **(Figure 5.15)**. However, in a hypertonic environment, a plant cell is no better off than an animal cell. As a plant cell loses water, it shrivels, and its plasma membrane pulls away from the cell wall (Figure 5.14c, bottom). This process, called **plasmolysis,** usually kills the cell. You can view plasmolysis yourself by placing a thin layer of red onion cells under a microscope and then washing the slide with salt water. The red plasma membrane will draw inward, shriveling away from the cell wall.

Figure 5.15 Plant turgor. An underwatered plant wilts as a result of a drop in turgor.

Active Transport: The Pumping of Molecules Across Membranes

In contrast to passive transport, **active transport** requires that a cell expend energy to move molecules across a membrane. In active transport, cellular energy is used to drive a transport protein that actively pumps a solute across a membrane *against* the solute's concentration gradient—that is, away from the side where it is less concentrated and toward the side where it is more concentrated **(Figure 5.16)**. Membrane proteins usually use ATP as their energy source for active transport.

Active transport enables cells to maintain internal concentrations of small solutes that differ from environmental concentrations. For example, compared with its surroundings, an animal nerve cell has a much higher concentration of potassium ions and a much lower concentration of sodium ions. The plasma membrane helps maintain these differences by pumping sodium out of the cell and potassium into the cell. This particular case of active transport (called the sodium-potassium pump) is vital in the propagation of nerve signals.

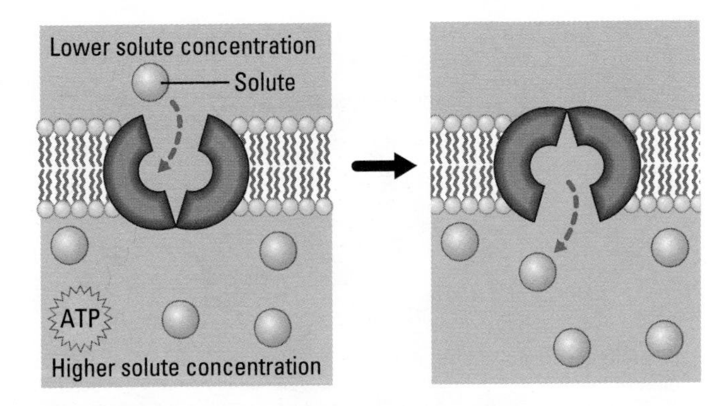

Figure 5.16 Active transport. Like enzymes, transport proteins are specific in their recognition of atoms or molecules. This transport protein (purple) has a binding site that accepts only a certain solute. Using energy from ATP, the protein pumps the solute against its concentration gradient.

Exocytosis and Endocytosis: Traffic of Large Molecules

So far, we've focused on how water and small solutes enter and leave cells by moving through the plasma membrane. The story is different for large molecules such as proteins, which are much too big to fit through the membrane itself. Their traffic into and out of the cell depends on the ability of the membrane to form sacs, thereby packaging larger molecules into vesicles. You have already seen an example in the packaging and secretion of proteins. During protein production by the cell, secretory proteins exit the cell from transport vesicles that fuse with the plasma membrane, spilling the contents outside the cell (see Figures 4.11 and 4.15). That process is called **exocytosis (Figure 5.17a)**. When you cry, for example, cells in your tear glands use exocytosis to export the salty tears. The reverse process, **endocytosis,** takes material into the cell within vesicles that bud inward from the plasma membrane **(Figure 5.17b)**.

There are three types of endocytosis. In **phagocytosis** ("cellular eating"), a cell engulfs a particle and packages it within a food vacuole **(Figure 5.18)**. In **pinocytosis** ("cellular drinking"), the cell "gulps" droplets of fluid by forming tiny vesicles. Because any and all solutes

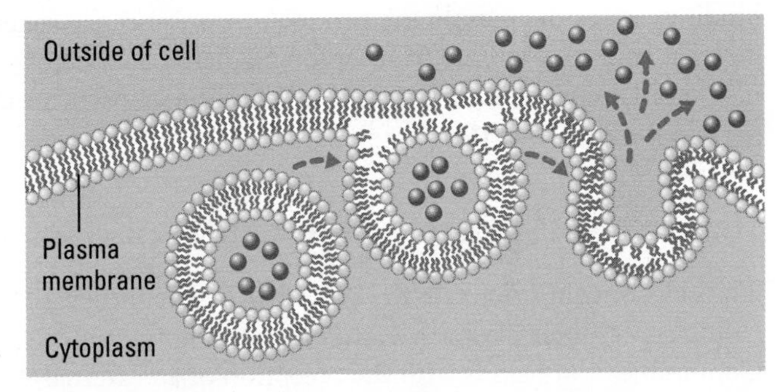

(a) Exocytosis

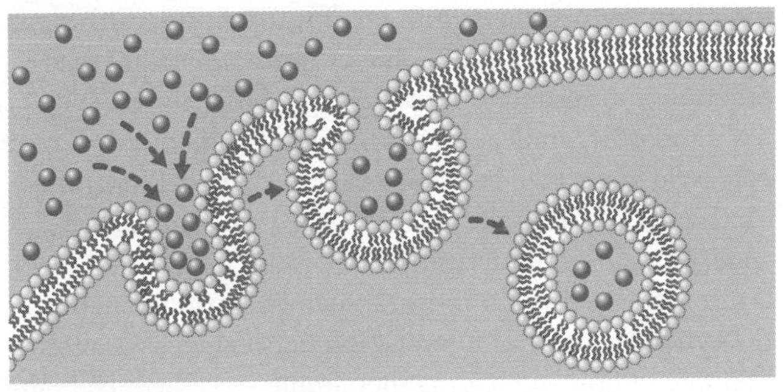

(b) Endocytosis

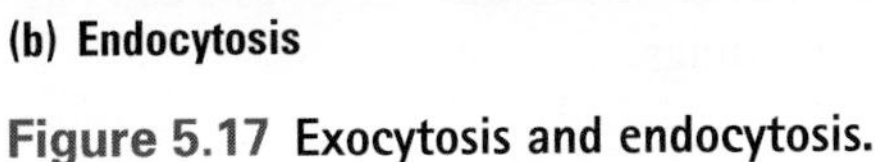

Figure 5.17 Exocytosis and endocytosis.

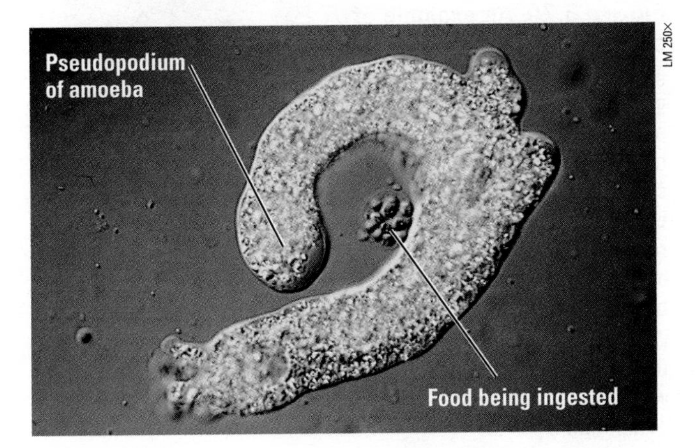

Figure 5.18 Phagocytosis. An amoeba uses a cellular extension called a pseudopodium to engulf food and package it in a food vacuole. A similar process is used by the white blood cells of your immune system to destroy invaders.

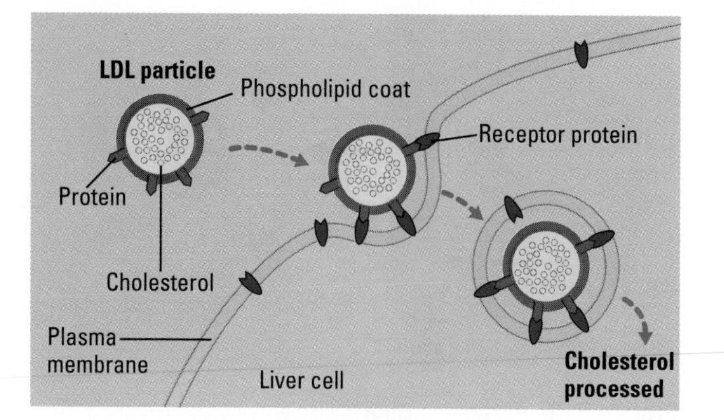

Figure 5.19 Cholesterol uptake by human cells. Normal liver cells remove excess cholesterol from the blood by receptor-mediated endocytosis. This mechanism prevents the excessive cholesterol levels in the blood that contribute to cardiovascular disease. Cholesterol circulates in the blood mainly in particles called low-density lipoproteins, or LDLs. Proteins embedded in LDL attach to receptor proteins built into liver cell membranes. This enables liver cells to take up LDLs and process the cholesterol. In one type of hereditary disorder, LDL receptors on liver cell membranes are missing or reduced in number. This defect results in a heavy load of blood cholesterol, which accumulates on artery walls and causes early onset of cardiovascular disease.

dissolved in the droplet are taken into the cell, pinocytosis is nonspecific in the substances it transports. In contrast, **receptor-mediated endocytosis** is very specific: It is triggered by the binding of certain external molecules to specific receptor proteins built into the plasma membrane. This binding causes the membrane protein to transport the specific substance into the cell. An example of receptor-mediated endocytosis is the mechanism human liver cells use to take up cholesterol particles that circulate in the blood **(Figure 5.19)**.

The Role of Membranes in Cell Signaling

The cells of your body talk to each other by chemical signaling across their plasma membranes (see Figure 5.11b). Communication begins with the reception of an external signal, such as a hormone, by a specific receptor protein built into the plasma membrane **(Figure 5.20)**. This signal triggers a chain reaction in one or more molecules that function in transduction (passing the signal along). The proteins and other molecules of the **signal transduction pathway** relay the signal and convert it to chemical forms that can function within the cell. This leads to chemical responses, such as the activation of certain metabolic functions, and structural responses, such as rearrangements of the cytoskeleton. Communication between cells is another example of how the plasma membrane serves as a cell's interface with its surroundings.

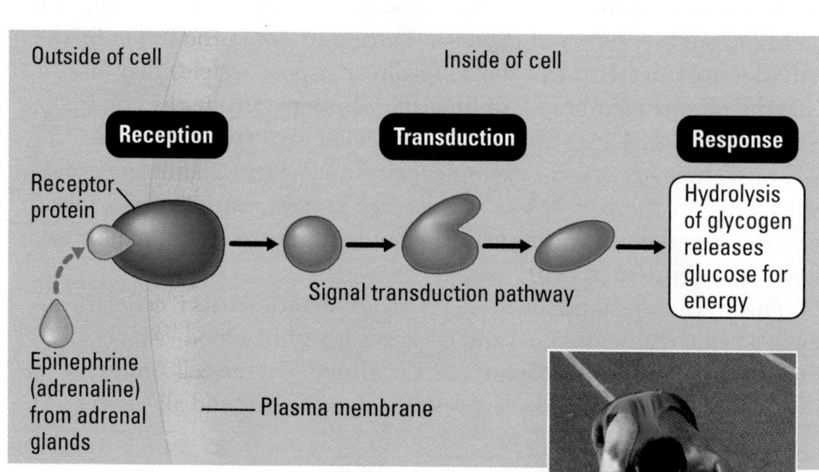

Figure 5.20 An example of cell signaling. When a person gets "psyched up" for an athletic contest, certain cells in the adrenal glands secrete a hormone called epinephrine (also called adrenaline) into the bloodstream. When that signal reaches muscle cells, it is recognized by receptor proteins built into the plasma membrane. This triggers responses (such as the breakdown of glycogen into glucose) within the muscle cells without the hormone even entering. This chain of events is part of the "fight-or-flight" response that enables you to attack or run when you're in danger—or keep your edge during an intense competition.

CHECKPOINT

1. Explain why it is not enough just to say that a solution is hypertonic.
2. What is the usual energy source for active transport?
3. What is the primary difference between passive and active transport in terms of concentration gradients?

Answers: 1. Hypertonic and hypotonic are relative terms. A solution that is hypertonic to tap water could be hypotonic to seawater. In using these terms, you must provide a comparison, as in "The solution is hypertonic to the cell." **2.** ATP **3.** Passive transport moves atoms or molecules along the concentration gradient (from higher to lower concentration), while active transport moves them against the concentration gradient.

EVOLUTION CONNECTION

Evolving Enzymes

Organisms use a great variety of enzymes. How could such diversity arise? Recall that most enzymes are proteins, and proteins are encoded by genes. Data gathered from the analysis of genetic sequences suggest that many of our genes arose through molecular evolution: One ancestral gene randomly duplicated, and the two copies of that gene diverged over time via genetic mutation, eventually becoming two distinct genes that produce two distinct enzymes.

Enzyme evolution can be greatly accelerated in the laboratory. For example, one group of researchers working with the bacteria *Escherichia coli* (*E. coli*) sought to modify lactase. Recall from Chapter 3 that lactase is an enzyme that can split the disaccharide lactose into its two component monosaccharides **(Figure 5.21)**. In a lab procedure called directed evolution, many copies of the gene for the original enzyme were mutated at random. The mutated enzymes that could best perform a new function (in this case, to split a bond in a different substrate molecule) were artificially selected using a screening test. Those enzymes were then subjected to another round of duplication, mutation, and selection. After seven rounds, this process produced a novel enzyme that recognized a new substrate.

The processes of natural selection and directed evolution both result in the production of new enzymes with new functions. There are, however, several important differences between the two schemes. The most obvious is the amount of time involved: many, many years for the former, just weeks for the latter. But the most significant difference between the two processes is mentioned in the name: The laboratory experiment was directed; the researchers engineered the evolutionary outcome to suit their chosen purpose. Natural selection, on the other hand, selects enzyme variants that work best in the organism's natural environment. ■

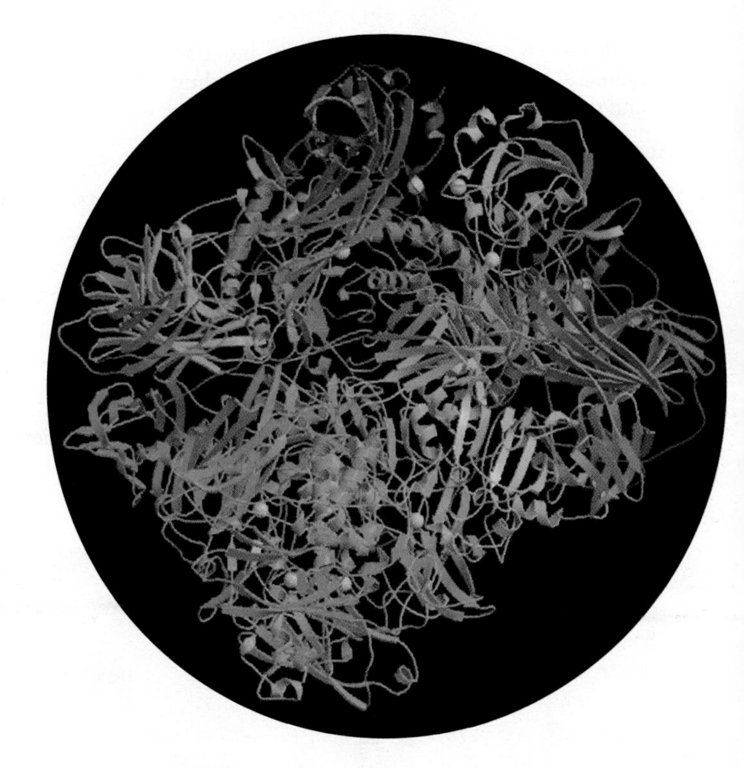

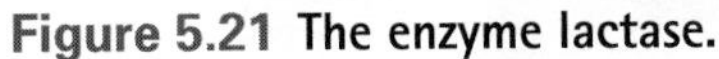

Figure 5.21 The enzyme lactase.

Chapter Review

SUMMARY OF KEY CONCEPTS

For study help and activities, go to campbellbiology.com or the student CD-ROM.

Some Basic Energy Concepts

MP3 Tutor *Basic Energy Concepts*

- **Conservation of Energy** Machines and organisms can transform kinetic energy (energy of motion) to potential energy (stored energy) and vice versa. In all such energy transformations, total energy is conserved. Energy cannot be created or destroyed.

- **Entropy** Every energy conversion releases some randomized energy in the form of heat. This is an example of the tendency for the entropy, or disorder, of the universe to increase.

- **Chemical Energy** Molecules store varying amounts of potential energy in the arrangement of their atoms. Organic molecules are relatively rich in such chemical energy.

Activity *Energy Concepts*

- **Food Calories** Actually kilocalories, food Calories are units used for the amount of energy in our foods and also for the amount of energy we expend in various activities.

ATP and Cellular Work

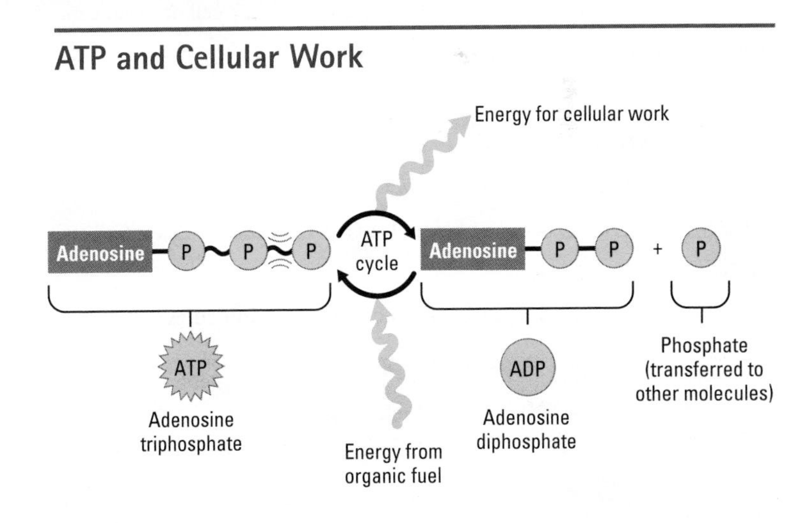

Activity *The Structure of ATP*

Enzymes

- **Activation Energy** Enzymes are biological catalysts that speed up metabolic reactions by lowering the activation energy required to break the bonds of reactant molecules.

• **Induced Fit** The entry of a substrate into the active site of an enzyme causes the enzyme to change shape slightly, allowing for a better fit and thereby promoting the interaction of enzyme with substrate.

Activity *How Enzymes Work*

Case Study in the Process of Science *How Is the Rate of Enzyme Catalysis Measured?*

• **Enzyme Inhibitors** Enzyme inhibitors are molecules that can disrupt metabolic reactions by binding to enzymes, either at the active site or elsewhere.

Membrane Function

Proteins embedded in the plasma membrane perform a wide variety of functions, including regulating transport.

• **Passive Transport, Osmosis, and Active Transport**

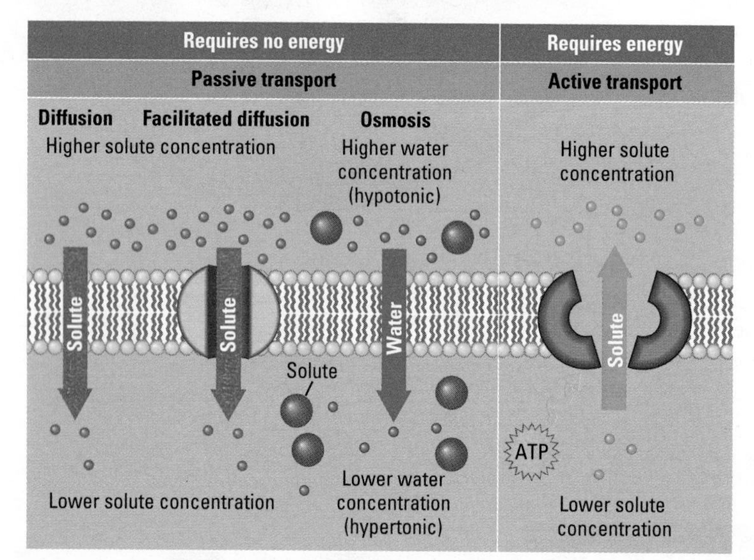

Most animal cells require an isotonic environment. Plant cells need a hypotonic environment, which keeps the walled cells turgid. Osmoregulation is the control of water balance within a cell or organism.

Activity *Membrane Structure*

Activity *Diffusion*

Activity *Facilitated Diffusion*

Activity *Osmosis and Water Balance in Cells*

Case Study in the Process of Science *How Does Osmosis Affect Cells?*

Activity *Active Transport*

• **Exocytosis and Endocytosis: Traffic of Large Molecules**

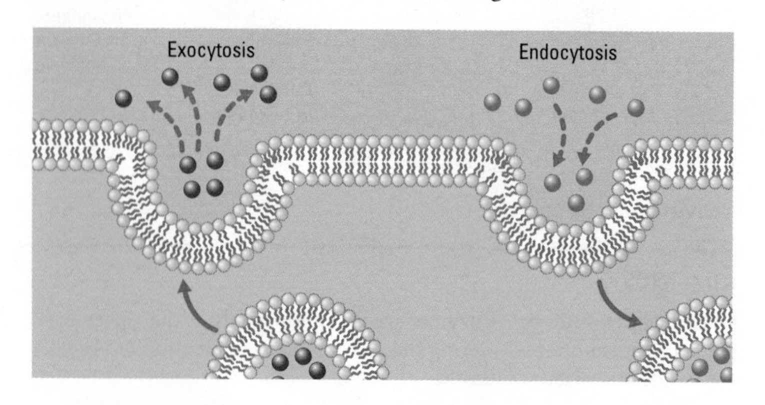

The three kinds of endocytosis are phagocytosis ("cellular eating"), pinocytosis ("cellular drinking"), and receptor-mediated endocytosis, which enables the cell to take in specific large molecules.

Activity *Exocytosis and Endocytosis*

• **The Role of Membranes in Cell Signaling** Receptors on the cell surface trigger signal transduction pathways that control processes within the cell.

Activity *Cell Signaling*

Case Study in the Process of Science *How Do Cells Communicate with Each Other?*

SELF-QUIZ

1. You have to expend considerable energy to carry a heavy sofa up a flight of stairs. Since energy cannot be destroyed, what happens to all the energy you expend?

2. __________ is the capacity to perform work, while __________ is a measure of randomness.

3. The label on a candy bar says that it contains 150 Calories. If you were able to convert all of that energy to heat, you could raise the temperature of how much water by 15°C?

4. Why does removing a phosphate group from the triphosphate tail in a molecule of ATP release energy?

5. Your digestive system is equipped with a diversity of enzymes that break the polymers in your food down to monomers that your cells can assimilate. A generic name for a digestive enzyme is hydrolase. What is the chemical basis for that name?

6. Explain how an inhibitor can disrupt the action of an enzyme even though it does not bind to the enzyme's active site.

7. If someone sitting at the other end of a restaurant smokes a cigarette, you may still breathe in some of that smoke. The movement of smoke through the air of the restaurant is an example of what type of transport?
   a. osmosis
   b. diffusion
   c. facilitated diffusion
   d. active transport

8. The total solute concentration in a red blood cell is about 2%. Sucrose cannot pass through a red blood cell's plasma membrane, but water and urea can. Osmosis will cause such a cell to shrink the most when the cell is immersed in which of the following?
   a. a hypertonic sucrose solution
   b. a hypotonic sucrose solution
   c. a hypertonic urea solution
   d. a hypotonic urea solution

9. Which of the following types of cellular transport require(s) an expenditure of energy?
   a. facilitated diffusion
   b. active transport
   c. osmosis
   d. a and b

10. A __________ is a process that links the reception of a cell signal to a response within the cell.

Answers to the Self-Quiz questions can be found in Appendix D.

Go to the website or CD-ROM for more Self-Quiz questions.

THE PROCESS OF SCIENCE

11. HIV, the virus that causes AIDS, depends on an enzyme called reverse transcriptase in order to multiply. Reverse transcriptase reads a molecule of RNA and creates a molecule of DNA from it. A molecule of AZT, the first drug approved to treat AIDS, has a shape very similar to that of the DNA base thymine. Propose a model for how AZT is able to inhibit HIV.

12. Gaining and losing weight are matters of caloric accounting: Calories in the food you eat minus Calories that you spend in activity. One pound of human body fat contains approximately 3,500 Calories. Using Figure 5.4, compare various ways you could burn off that many Calories. How far would you have to run, swim, or walk to burn the equivalent of 1 pound of fat? For how much time would you have to do each activity? Which method of burning Calories appeals the most to you? The least?

BIOLOGY AND SOCIETY

13. Obesity is a serious health problem for many Americans. It seems that there are as many fad diets as there are people trying to lose weight. The best-selling diet book to date is Dr. Atkins's *New Diet Revolution*, which proposes a low-carbohydrate diet. Most people who follow the Atkins diet compensate for the reduced carbohydrates in their diet by increasing their intake of protein and fat. What advantages are there to such a diet? What disadvantages? Do you think that the government should regulate the claims of diet books? How would you propose that the claims be tested? Do you think that diet proponents should be required to obtain and publish data before they can make their claims? Have you ever tried a fad diet?

14. Lead acts as an enzyme inhibitor, and it can interfere with the development of the nervous system. One manufacturer of lead-acid batteries instituted a "fetal protection policy" that banned female employees of childbearing age from working in areas where they might be exposed to high levels of lead. Under this policy, women were involuntarily transferred to lower-paying jobs in lower-risk areas. A group of employees challenged the policy in court, claiming that it deprived women of job opportunities available to men. The U.S. Supreme Court ruled the policy illegal. Nonetheless, many people are uncomfortable about the "right" to work in an unsafe environment. What rights and responsibilities of employers, employees, and government agencies are in conflict in this situation? Whose responsibility should it be to determine what makes a safe environment and who should or should not work there? What criteria should be used to decide?

6 Cellular Respiration: Obtaining Energy from Food

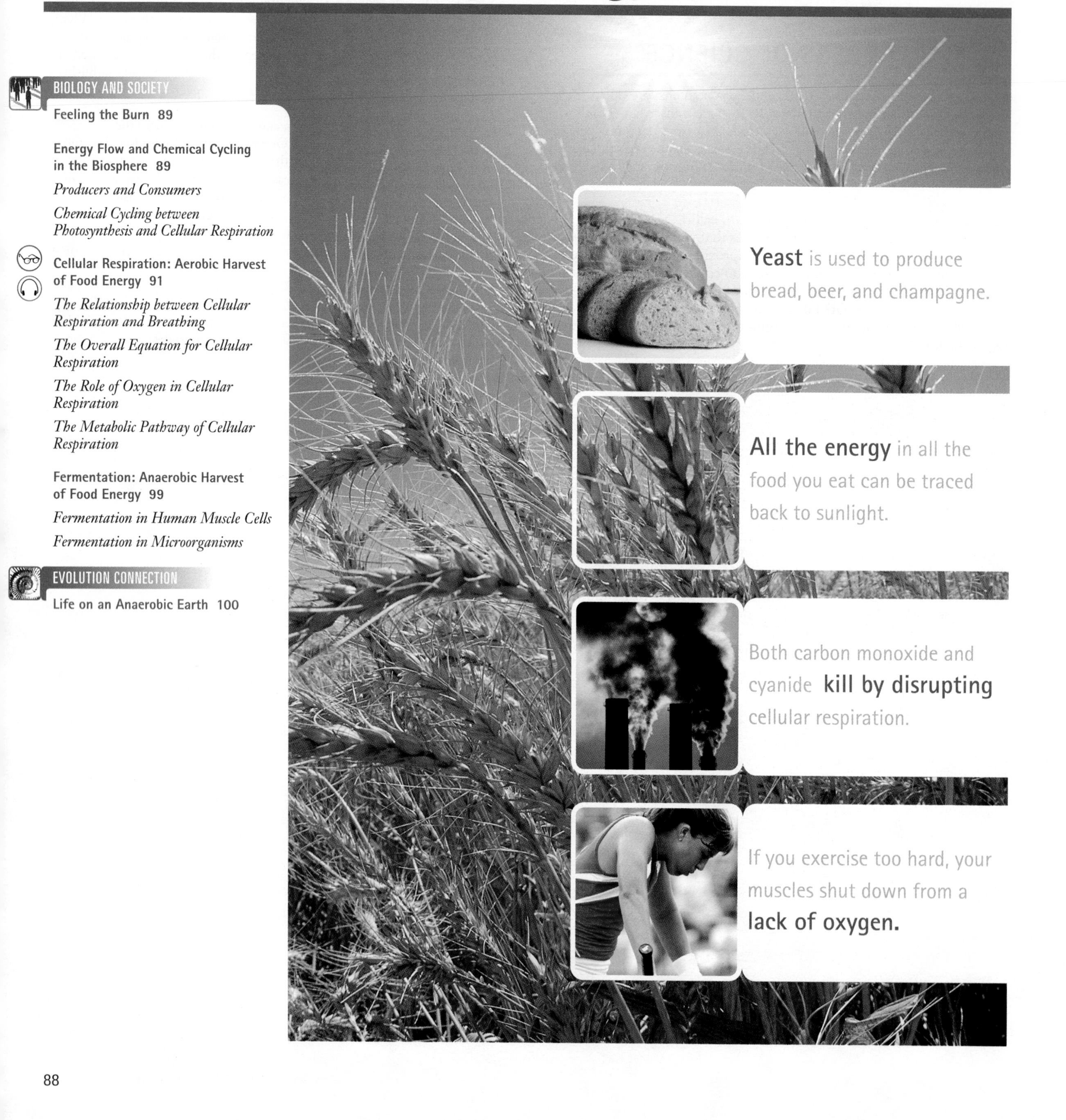

Yeast is used to produce bread, beer, and champagne.
All the energy in all the food you eat can be traced back to sunlight.
Both carbon monoxide and cyanide kill by disrupting cellular respiration.
If you exercise too hard, your muscles shut down from a lack of oxygen.

BIOLOGY AND SOCIETY

Feeling the Burn

A marathon runner, drenched in sweat and panting heavily, staggers across the finish line and collapses **(Figure 6.1)**. Even though he had the stamina to run over 26 miles, now he can barely stand. What's wrong with him?

To understand the runner's distress, we must consider what goes on within muscle cells. When you exercise, your muscles need energy to perform work. Muscle cells obtain this energy from the sugar glucose through a series of chemical reactions that require oxygen (O_2). Therefore, to keep moving, your muscles need a steady supply of O_2.

When there is enough oxygen reaching your cells to support their energy needs, metabolism is said to be **aerobic.** As your muscles work harder, you breathe faster and deeper to inhale even more O_2. If you continue to pick up the pace, you will approach your **aerobic capacity,** the maximum rate at which O_2 can be taken in and used by your muscle cells and therefore the most strenuous exercise that your body can maintain aerobically.

If you exceed your aerobic capacity, the demand for oxygen in your muscles outstrips the body's ability to deliver it; metabolism now becomes **anaerobic**. With insufficient O_2, your muscle cells switch to an "emergency mode" in which they break down glucose very inefficiently and produce lactic acid as a by-product. As lactic acid accumulates, it impairs muscle activity and causes the "burn" associated with heavy exercise ("Feel the burn! It's a *good* burn!"). The problem is that your muscles can work under these conditions for only a few minutes. If too much lactic acid builds up, your muscles give out.

Short-distance sprinters don't have to worry about building up too much lactic acid before the end of the race. However, endurance athletes have to be sure to stay within their aerobic capacity until the final sprint, at which time they can crank it up to anaerobic levels. Among well-trained athletes, collapsing at the end of the race is a sign that the switch to an anaerobic pace was well timed.

The metabolic pathways that provide energy to an athlete are used by nonathletes as well. In fact, we need energy to walk, talk, and think—in short, to stay alive. The human body has trillions of cells, all hard at work, all demanding fuel continuously. In this chapter, you'll learn how cells harvest food energy and put it to work. ■

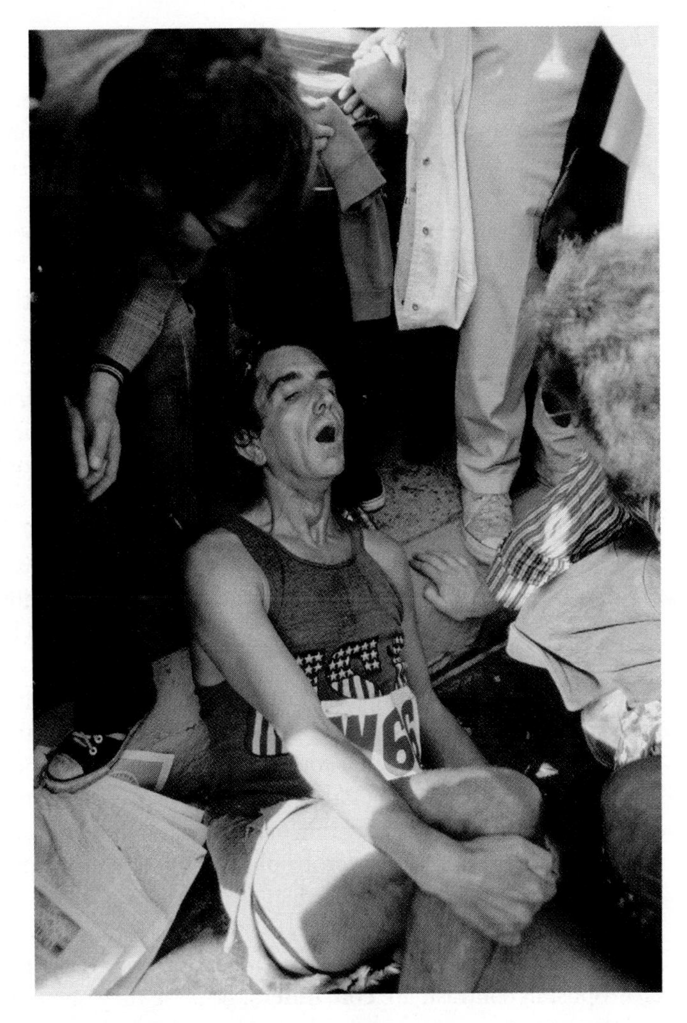

Figure 6.1 Feeling the burn. A buildup of lactic acid can cause muscles to give out and "burn."

Energy Flow and Chemical Cycling in the Biosphere

In an indirect way, the fuel molecules in food represent solar energy. We can trace the energy stored in all our food to the sun. Humans and other animals depend on plants to convert the energy of sunlight to the chemical energy of sugars and other organic molecules. And we depend on plants for more than our food. You're probably wearing clothing made of another product of photosynthesis—cotton. Most of our homes are framed with lumber, which is wood produced by photosynthetic trees. Even the text you are now reading is printed on paper, still another material that can be traced to photosynthesis in plants. But for animals, photosynthesis is primarily about feeding the biosphere.

Producers and Consumers

The star we call the sun is at the center of our solar system, almost 100 million miles away. A giant thermonuclear reactor, the sun converts matter to energy, radiating some of it as visible light. A miniscule portion of that light reaches planet Earth. And a tiny fraction of the light that illuminates Earth powers life. The process that makes this possible is called photosynthesis. *Photo* means "light," and *synthesis* means "to put together." In fact, **photosynthesis** uses light energy from the sun to power a chemical process that puts together organic molecules.

Here on land, photosynthesis occurs mainly in green cells within the leaves of plants. Leaves owe their greenness to chlorophyll, a pigment molecule contained in chloroplasts. Chloroplasts are the organelles that house the equipment for photosynthesis (see Figure 4.16). Chloroplasts trap light energy and use it to produce sugars and other energy-rich organic molecules. (You'll learn more about photosynthesis in Chapter 7.)

Plants and other **autotrophs** ("self-feeders") are organisms that make all their own organic matter—including carbohydrates, lipids, proteins, and nucleic acids—from inorganic nutrients. The term is a bit misleading in its implication that plants do not require nutrients; they do. But those nutrients are entirely inorganic: carbon dioxide from the air, and water and minerals from the soil. (Even though it contains carbon, carbon dioxide is considered to be an inorganic compound.) In contrast, humans and other animals are **heterotrophs** ("other-feeders"), organisms that cannot make organic molecules from inorganic ones. That's why we must eat—to get our nutrients from food. Heterotrophs depend on autotrophs for their organic fuel and material for growth and repair.

Most ecosystems depend entirely on photosynthesis for food. For this reason, biologists refer to plants and other autotrophs as **producers.** Heterotrophs, in contrast, are **consumers,** because they obtain their food by eating plants or by eating animals that have eaten plants **(Figure 6.2)**. We animals depend on food not only for fuel, but also for the raw organic materials we need to build our cells and tissues.

Figure 6.2 Producer and consumer. Plants and other photosynthetic organisms use light energy to drive the production of their own organic material. Animals and other consumers depend on this photosynthetic product for energy and building material. In this photo, a porcupine (consumer) eats a fruit produced by a photosynthetic plant (producer).

Chemical Cycling between Photosynthesis and Cellular Respiration

The ingredients for photosynthesis are carbon dioxide (CO_2) and water (H_2O). Carbon dioxide, a gas found in air, enters plants through tiny pores in the surface of leaves. Water is absorbed from the damp soil by the plant's roots, and it moves up the plant's veins to the leaves (see Figure 2.12). Chloroplasts rearrange the atoms of these ingredients to produce sugars, most importantly glucose ($C_6H_{12}O_6$), and other organic molecules. A by-product of photosynthesis is oxygen gas (O_2) **(Figure 6.3)**.

Plants have chloroplasts and so are capable of producing fuel from sunlight. Animals and plants use these organic fuel compounds to obtain energy. A chemical process called cellular respiration harvests energy that is stored in sugars and other organic molecules. Cellular respiration uses O_2 to help convert energy extracted from organic fuel to another form of chemical energy called ATP. Cells spend ATP for almost all their work. In both plants and animals, the production of ATP during cellular respiration occurs mainly in the organelles called mitochondria (see Figure 4.17).

Figure 6.3 Energy flow and chemical cycling in ecosystems. Energy enters a forest or other ecosystem as sunlight and exits in the form of heat. Organisms temporarily trap the energy for their work. Photosynthesis in the chloroplasts of plants converts light energy to chemical energy. Cellular respiration in the mitochondria of eukaryotes (including plants and animals) harvests the food energy to generate ATP. These molecules of ATP directly drive most cellular work. Chemical elements essential for life recycle between cellular respiration and photosynthesis.

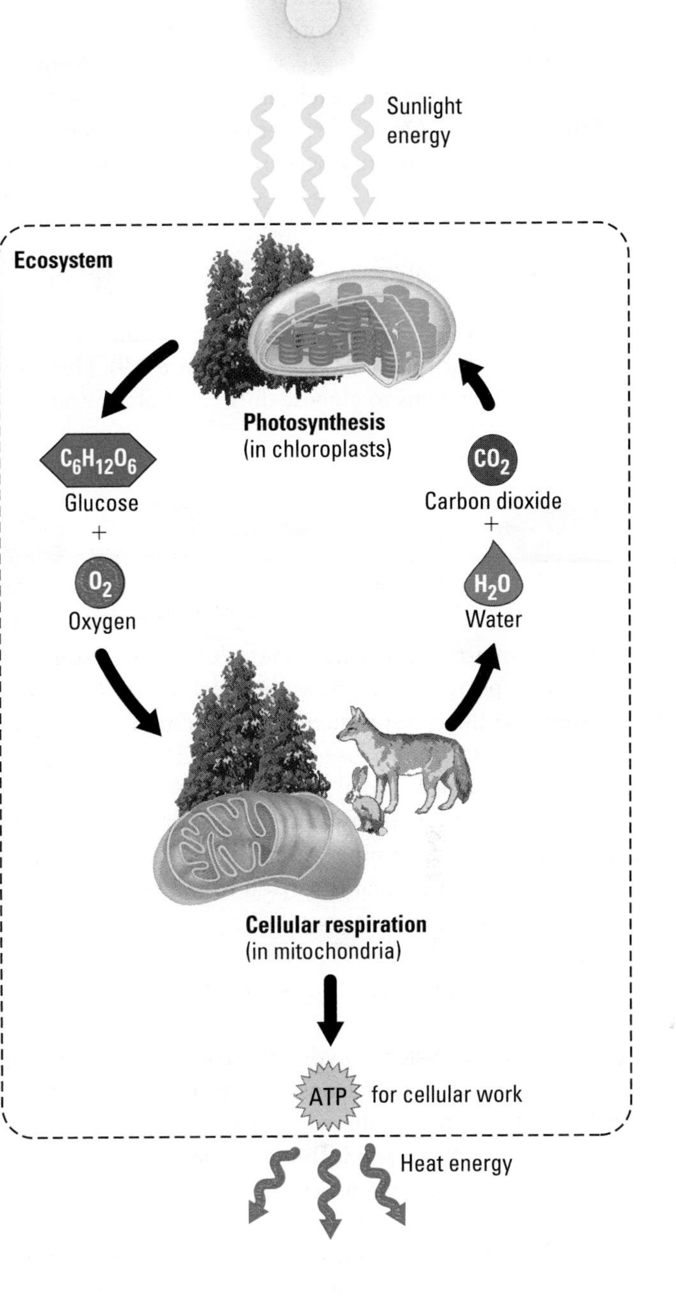

Notice in Figure 6.3 that the waste products of cellular respiration are CO_2 and H_2O—the very same chemical ingredients used for photosynthesis. Plants store chemical energy via photosynthesis and then harvest this energy via cellular respiration. However, plants usually make more organic molecules than they need for fuel. This photosynthetic surplus provides the organic material for the plant to grow. It is also the source of food for humans and other consumers. Analyze nearly any food chain, and you can trace the energy and raw materials for growth back to solar-powered photosynthesis.

CHECKPOINT

1. Although they are "self-feeders," photosynthetic autotrophs are not totally self-sufficient. What chemical ingredients do they require from the environment in order to synthesize sugar?
2. Why are plants called producers? Why are animals called consumers?
3. What is misleading about the following statement? "Plants perform photosynthesis, whereas animals perform cellular respiration."

Answers: 1. CO_2 and H_2O. (Plants also require soil minerals.) **2.** Plants produce organic molecules by photosynthesis. Consumers must acquire organic material by consuming it rather than making it. **3.** It implies that cellular respiration does not occur in plants. It does.

Cellular Respiration: Aerobic Harvest of Food Energy

eTutor
Cellular Respiration

MP3 Tutor
Cellular Respiration Part 1

MP3 Tutor
Cellular Respiration Part 2

Internal combustion engines, like the ones found in cars, use O_2 (via the air intakes) to break down gasoline. A cell also requires O_2 to break down its fuel. Cellular respiration—a living version of internal combustion—is the main way that chemical energy is harvested from food and converted to ATP energy. Cellular respiration is an aerobic process, which is just another way of saying that it requires oxygen. Putting all this together, we can now define **cellular respiration** as the aerobic harvesting of chemical energy from organic fuel molecules.

The Relationship between Cellular Respiration and Breathing

We sometimes use the word *respiration* to mean breathing. While respiration on the organismal level should not be confused with cellular respiration, the two processes are closely related **(Figure 6.4)**. Cellular respiration requires a cell to exchange two gases with its surroundings. The cell takes in oxygen in the form of the gas O_2. It gets rid of waste in the form of the gas carbon dioxide, or CO_2. Breathing exchanges these same gases between your blood and the outside air. Oxygen present in the air you inhale diffuses across the lining of your lungs and into your bloodstream. And the CO_2 in your bloodstream diffuses across the lining of your lungs and exits when you exhale.

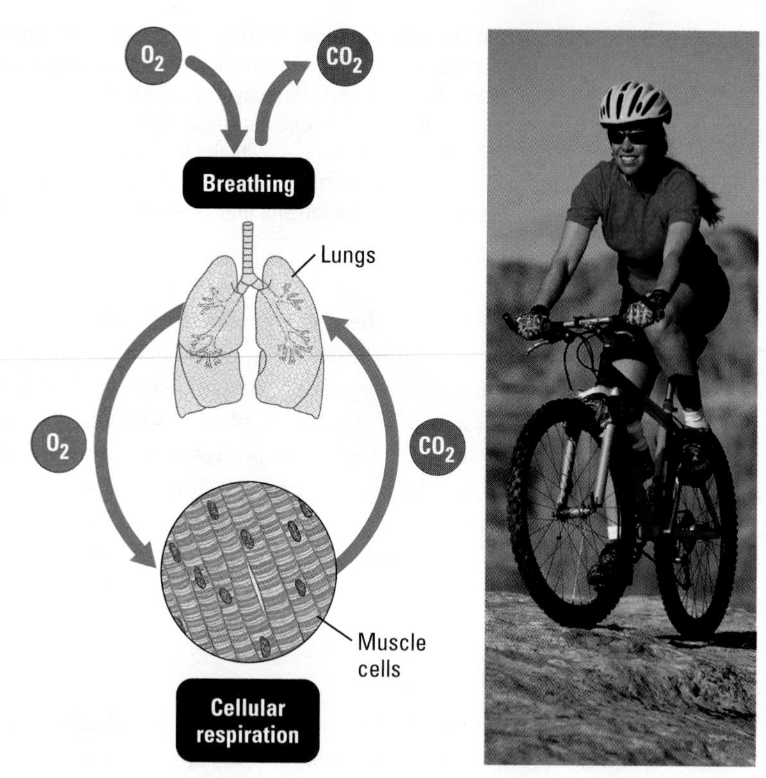

Figure 6.4 How breathing is related to cellular respiration. When you inhale, you breathe in O_2. The O_2 is delivered to your cells, where it is used in cellular respiration. Carbon dioxide, a waste product of cellular respiration, diffuses from your cells to your blood and travels to your lungs, where it is exhaled.

The Overall Equation for Cellular Respiration

A common fuel molecule for cellular respiration is glucose, a six-carbon sugar with the formula $C_6H_{12}O_6$ (see Figure 3.9). Here is the overall equation for what happens to glucose during cellular respiration:

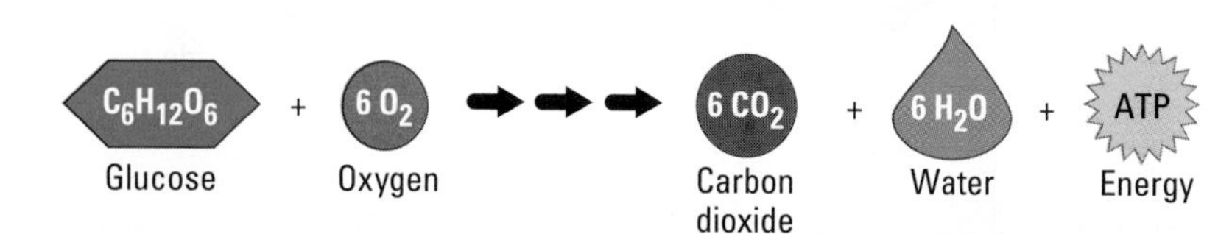

The series of arrows indicates that cellular respiration consists of many chemical steps, not just a single chemical reaction. Remember, the main function of cellular respiration is to generate ATP for cellular work. In fact, the process can produce up to 38 ATP molecules for each glucose molecule consumed.

Notice that cellular respiration also transfers hydrogen atoms from glucose to oxygen, forming water. That hydrogen transfer turns out to be the key to why oxygen is so vital to the harvest of energy during cellular respiration.

The Role of Oxygen in Cellular Respiration

In tracking the transfer of hydrogen from sugar to oxygen, we are also following the transfer of electrons. The atoms of sugar and other molecules are bonded together by shared electrons (see Figure 2.9). During cellular respiration, hydrogen and its bonding electrons change partners, from sugar to oxygen, forming water as a product.

Redox Reactions Chemical reactions that transfer electrons from one substance to another substance are called oxidation-reduction reactions, or **redox reactions** for short. The loss of electrons during a redox reaction is called **oxidation.** Glucose is oxidized during cellular respiration, losing electrons to oxygen. The acceptance of electrons during a redox reaction is called **reduction.** Oxygen is reduced during cellular respiration, accepting electrons (and hydrogen) lost from glucose:

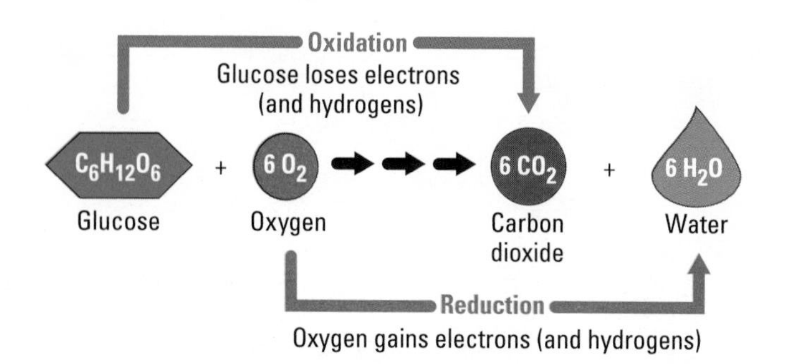

When hydrogen and its bonding electrons change partners, from sugar to oxygen, energy is released.

Why does electron transfer to oxygen release energy? In redox reactions, oxygen is an "electron grabber." An oxygen atom attracts electrons more strongly than almost any other type of atom. When electrons move (along with hydrogen) from glucose to oxygen, it is as though they were falling. They are not really falling in the sense of an apple dropping from a tree. However, in both cases, potential energy is unlocked. Instead of gravity, it is the attraction of electrons to oxygen that causes the "fall" and energy release during cellular respiration.

A very rapid electron "fall" generates an explosive release of energy in the form of heat and light. For example, a spark will trigger a reaction between hydrogen gas (H_2) and oxygen gas (O_2) that produces water (H_2O). The reaction also releases a large amount of energy as the electrons of the hydrogen "fall" into their new bonds with oxygen **(Figure 6.5)**. As another example, some fancy drinks and desserts are topped with a flaming cube of sugar. During this dramatic display, the hydrogen and electrons from the sugar "fall" to the oxygen in the air. The flame represents energy released as heat and light.

It would be difficult to capture the burst of energy released from such an explosive reaction and put it to useful work. Cellular respiration is a more controlled "fall" of electrons—more like a stepwise cascade of electrons down an energy staircase. Instead of liberating food energy in a burst of flame, cellular respiration unlocks chemical energy in smaller amounts that cells can put to productive use.

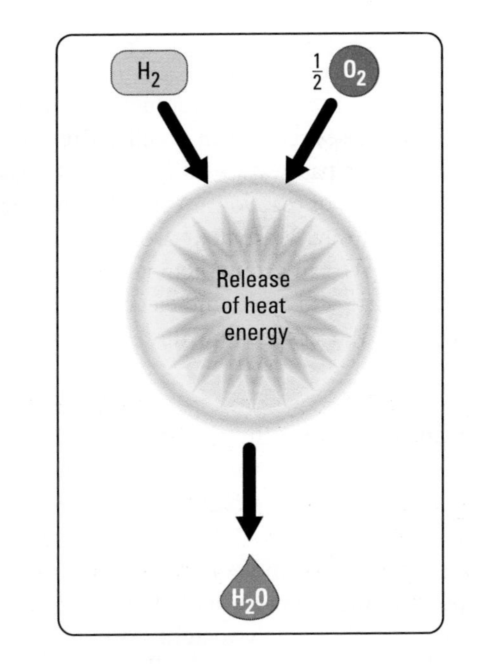

Figure 6.5 A rapid electron "fall." An all-at-once redox reaction, such as the reaction of hydrogen and oxygen to form water, releases a burst of energy. It is difficult to use such an explosion for productive work.

NADH and Electron Transport Chains Let's take a closer look at the path that electrons take on their way down from glucose to oxygen **(Figure 6.6)**. The first stop is a positively charged electron acceptor called NAD^+ (nicotinamide adenine dinucleotide). The transfer of electrons from organic fuel to NAD^+ reduces the NAD^+ to **NADH** (the H represents the transfer of hydrogen along with the electrons). In our staircase analogy, the electrons have now taken one baby step down in their trip from glucose to oxygen. The rest of the staircase consists of an **electron transport chain.** Each link in an electron transport chain is actually a molecule, usually a protein. In a series of redox reactions, each member of the chain can first accept and then donate electrons. With each transfer, the electrons give up a small amount of energy that can then be used to generate ATP from ADP. At the "uphill" end, the first molecule of the chain accepts electrons from NADH. Thus, NADH carries electrons from glucose and other fuel molecules and deposits them at the top of an electron transport chain. The electrons then cascade down the chain, from molecule to molecule. The molecule at the bottom of the chain finally "drops" the electrons to oxygen. The oxygen also picks up hydrogen, forming water.

The overall effect of all this electron traffic during cellular respiration is a downhill trip for electrons from glucose to oxygen via NADH and electron transport chains. During the stepwise freeing of chemical energy during electron transport, our cells make most of their ATP from ADP. It is actually oxygen, the "electron grabber," that makes it all possible. By pulling electrons down the transport chain from fuel molecules, oxygen functions somewhat like gravity pulling objects downhill. This is how the oxygen we breathe functions in our cells and why we cannot survive more than a few minutes without it. Viewed this way, drowning is deadly because it deprives cells of their final "electron grabbers" (oxygen) needed to drive cellular respiration.

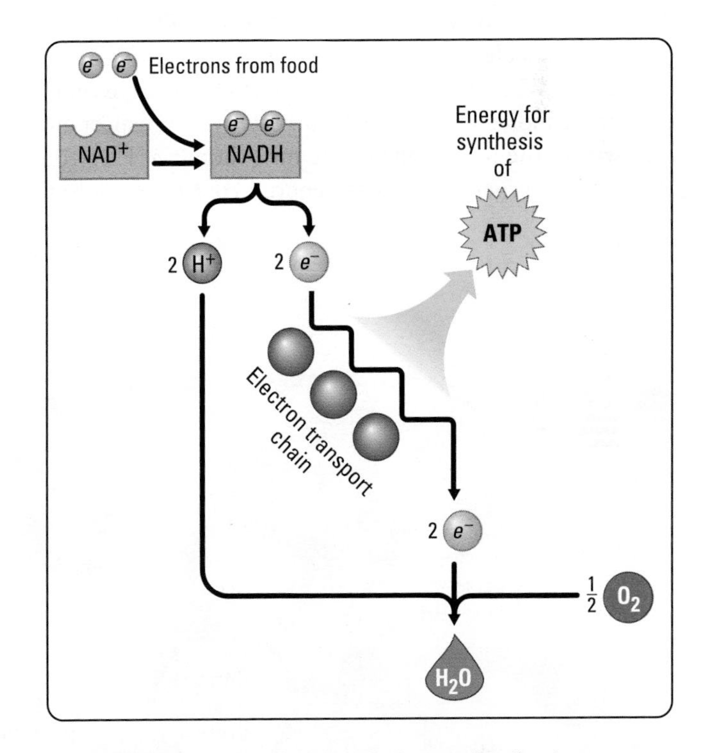

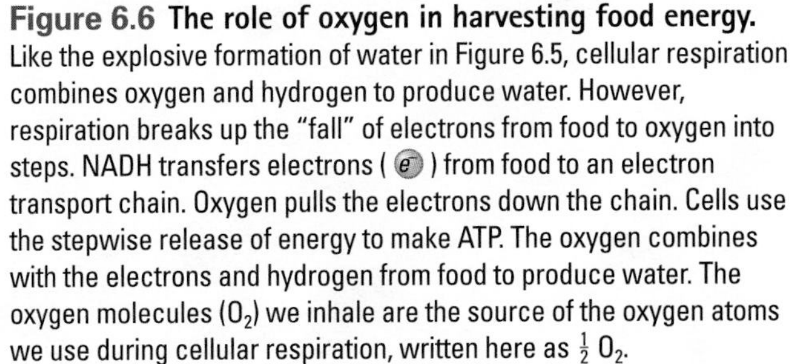

Figure 6.6 The role of oxygen in harvesting food energy. Like the explosive formation of water in Figure 6.5, cellular respiration combines oxygen and hydrogen to produce water. However, respiration breaks up the "fall" of electrons from food to oxygen into steps. NADH transfers electrons (e^-) from food to an electron transport chain. Oxygen pulls the electrons down the chain. Cells use the stepwise release of energy to make ATP. The oxygen combines with the electrons and hydrogen from food to produce water. The oxygen molecules (O_2) we inhale are the source of the oxygen atoms we use during cellular respiration, written here as $\frac{1}{2}$ O_2.

The Metabolic Pathway of Cellular Respiration

Cellular respiration is an example of metabolism, the general term for all the chemical processes that occur in cells. More specifically, cellular respiration is a metabolic pathway. That means that it is not a single chemical reaction, but a series of reactions. A specific enzyme catalyzes each reaction in a metabolic pathway. More than two dozen reactions are involved in cellular respiration. We can group them into three main metabolic stages: glycolysis, the citric acid cycle, and electron transport (which you've already encountered). Let's see how these stages cooperate to harvest food energy.

A Road Map for Cellular Respiration **Figure 6.7** is a map that will help you follow glucose through the metabolic pathway of cellular respiration. The map also shows you where the three stages of respiration occur in your cells.

During **glycolysis,** a molecule of glucose is split into two molecules of a compound called pyruvic acid. The enzymes for glycolysis are located in the cytosol. The **citric acid cycle** (also called the **Krebs cycle**) completes the breakdown of sugar all the way to CO_2, the waste product of cellular respiration. The enzymes for the citric acid cycle are dissolved in the fluid within mitochondria. Glycolysis and the citric acid cycle generate a small amount of ATP directly. They generate much more ATP indirectly, via redox reactions that transfer electrons from fuel molecules to NAD^+, forming NADH. The third stage of cellular respiration is electron transport. Electrons captured from food by NADH "fall" down electron transport chains to oxygen. The proteins and other molecules that make up electron transport chains are embedded within the membrane of the mitochondria. Electron transport from NADH to oxygen releases the energy your cells use to make most of their ATP.

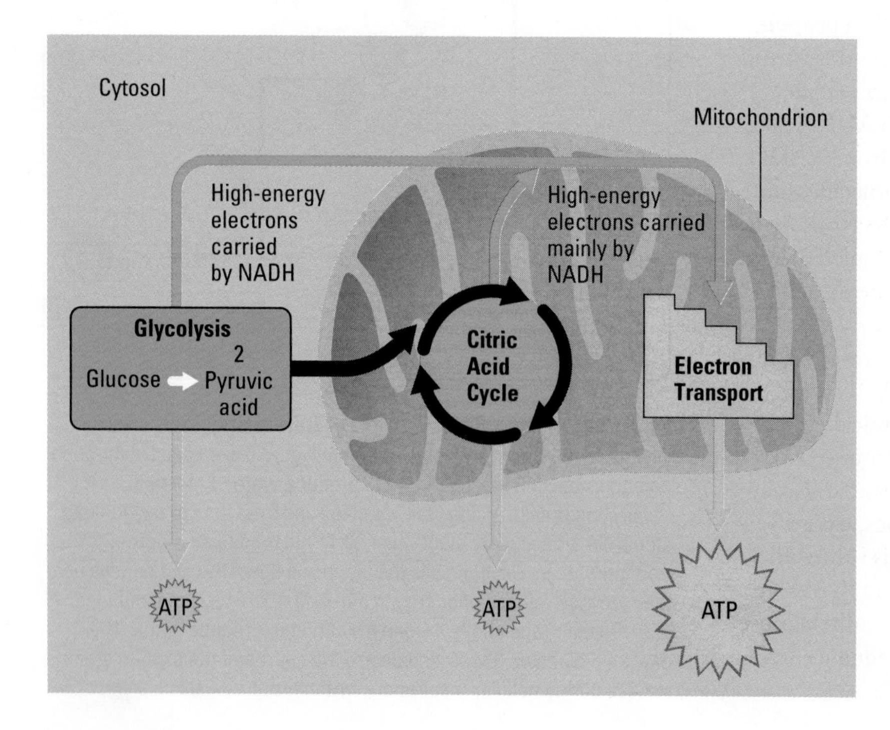

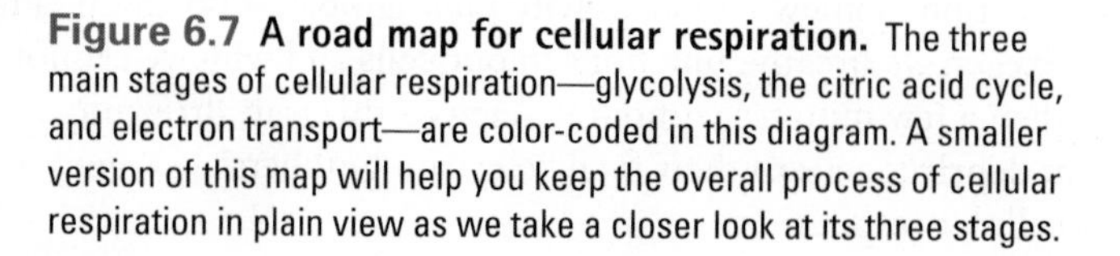
Figure 6.7 A road map for cellular respiration. The three main stages of cellular respiration—glycolysis, the citric acid cycle, and electron transport—are color-coded in this diagram. A smaller version of this map will help you keep the overall process of cellular respiration in plain view as we take a closer look at its three stages.

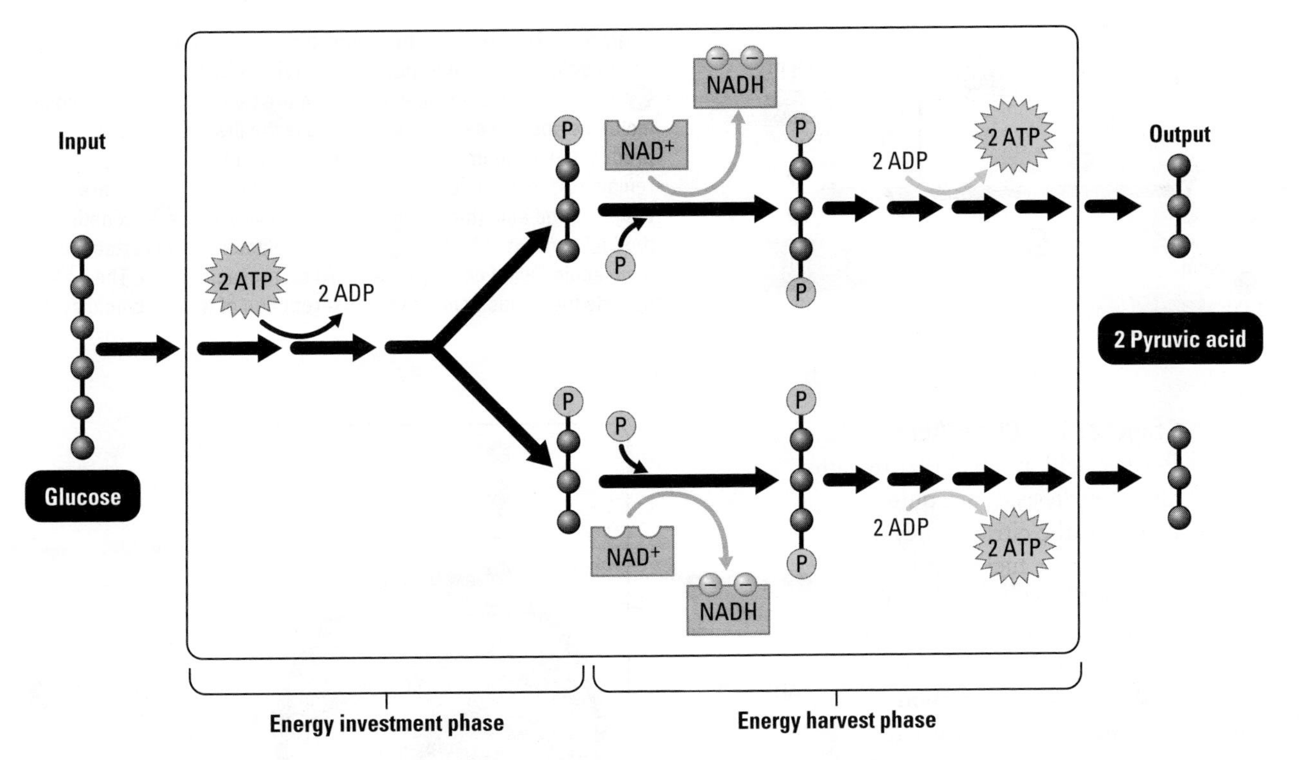

Figure 6.8 Glycolysis. Each ● in this diagram represents a carbon atom. A team of enzymes splits glucose, eventually forming two molecules of pyruvic acid. Along the way, energy is stored as ATP and NADH. Note that the cell actually invests some ATP to get glycolysis started. Enzymes attach phosphate groups (Ⓟ) to the fuel molecules during this energy investment phase. That investment is paid back with dividends during the energy harvest phase. Glycolysis generates some ATP directly, but it also donates high-energy electrons (⊖) to NAD^+, forming NADH.

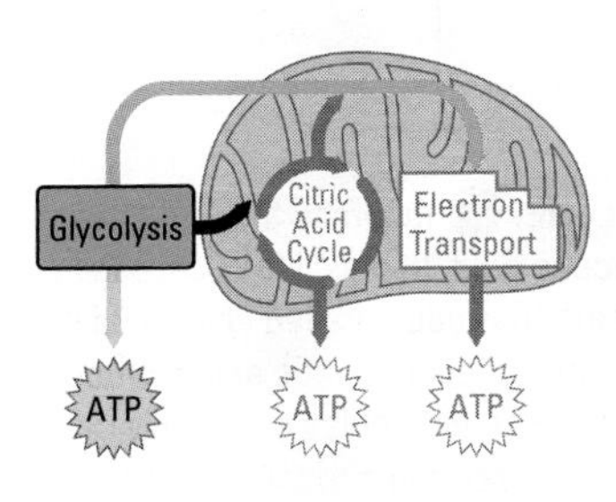

Stage 1: Glycolysis The word *glycolysis* means "splitting of sugar." That is exactly what happens **(Figure 6.8)**. During glycolysis, a six-carbon glucose molecule is broken in half, forming two three-carbon molecules. Notice in Figure 6.8 that the initial split requires an energy investment of two ATP molecules per glucose. The three-carbon molecules then donate high-energy electrons to NAD^+, the electron carrier. Glycolysis also makes four ATP molecules directly when enzymes transfer phosphate groups from fuel molecules to ADP **(Figure 6.9)**. Glycolysis thus produces a net of two molecules of ATP per molecule of glucose. (This fact will become important later during our discussion of fermentation.) What remains of the fractured glucose at the end of glycolysis are two molecules of pyruvic acid. The pyruvic acid still holds most of the energy of glucose, and that energy is harvested in the citric acid cycle.

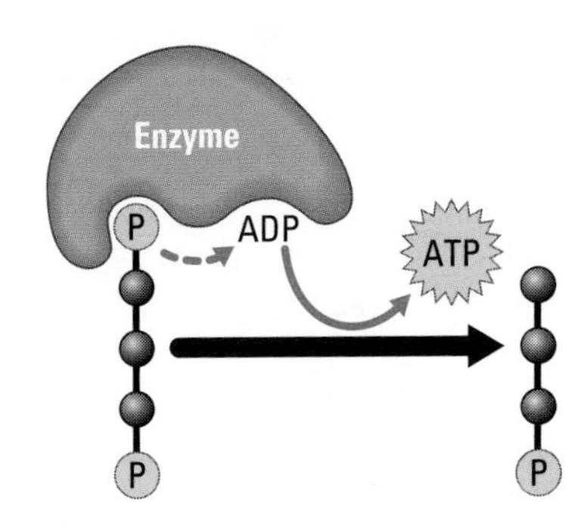

Figure 6.9 ATP synthesis by direct phosphate transfer. Glycolysis generates ATP when enzymes transfer phosphate groups directly from fuel molecules to ADP.

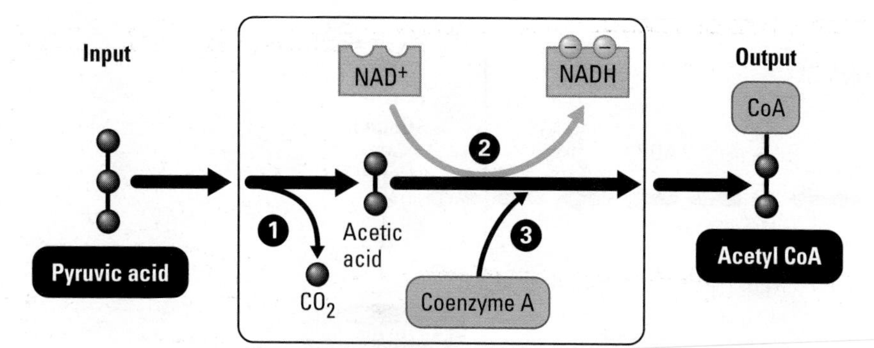

Figure 6.10 The link between glycolysis and the citric acid cycle: the conversion of pyruvic acid to acetyl CoA. ❶ Each pyruvic acid molecule (there are two per starting glucose molecule) loses a carbon as CO_2. This is the first of this waste product we've seen so far in the breakdown of glucose. The remaining fuel molecules, each with only two carbons left, are called acetic acid (the same acid that's in vinegar). ❷ Oxidation of the fuel generates NADH. ❸ Finally, the acetic acid is attached to a molecule called coenzyme A (CoA) to form acetyl CoA. The CoA escorts the acetic acid into the first reaction of the citric acid cycle.

Stage 2: The Citric Acid Cycle

Pyruvic acid, the fuel that remains after glycolysis, is not quite ready for the citric acid cycle. First, the fuel must be "prepped"—converted to a form the citric acid cycle can use **(Figure 6.10)**. The actual fuel consumed by the citric acid cycle is a two-carbon compound called acetic acid. Before entering the citric acid cycle, acetic acid is bonded to a carrier molecule called coenzyme A (CoA) to form acetyl CoA.

The citric acid cycle finishes extracting the energy of sugar by breaking the acetic acid molecules (two per glucose) all the way down to CO_2 **(Figure 6.11)**. The cycle uses some of this energy to make two ATP molecules (one per acetic acid molecule) by the direct method. However, the citric acid cycle captures much more energy in the form of six molecules of NADH and two molecules of a second electron carrier, $FADH_2$. Electron transport then converts NADH and $FADH_2$ energy to ATP energy.

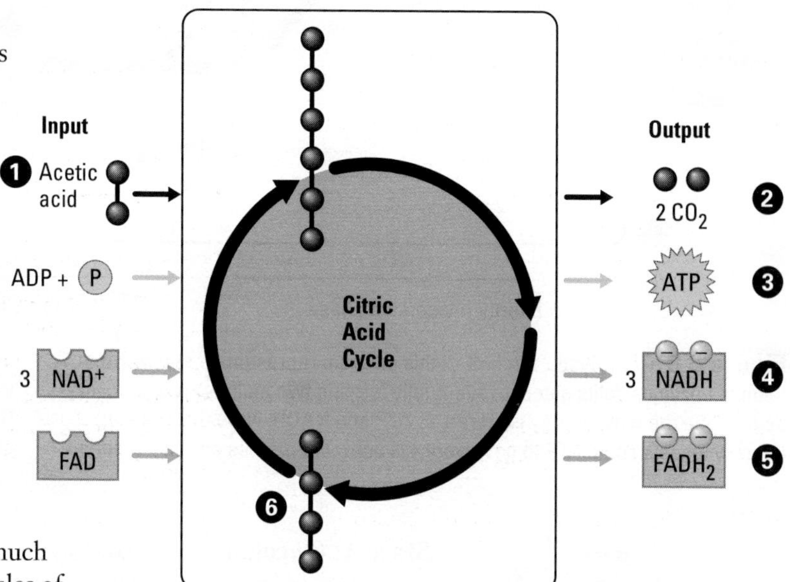

Figure 6.11 The citric acid cycle. ❶ Inputs to the citric acid cycle include fuel in the form of acetic acid. It joins a four-carbon acceptor molecule to form a six-carbon product called citric acid (for which the citric acid cycle is named). For every acetic acid molecule that enters the cycle as fuel, ❷ two CO_2 molecules eventually exit as exhaust (waste product). Along the way, the citric acid cycle harvests energy from the fuel. ❸ Some of the energy is used to produce ATP directly. ❹ Most of the energy is trapped by NADH. ❺ Some energy is captured by $FADH_2$, another electron carrier that works something like NADH. ❻ All the carbon atoms that entered the cycle as fuel are accounted for as CO_2 exhaust, and the four-carbon acceptor molecule is recycled. We have tracked only one acetic acid molecule through the citric acid cycle here. But recall that glycolysis splits glucose in two (see Figure 6.8). So the citric acid cycle actually turns twice for each glucose molecule that fuels a cell.

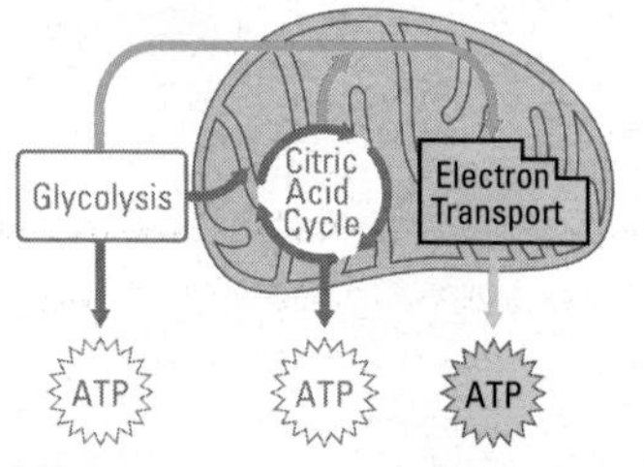

Stage 3: Electron Transport

The molecules of electron transport chains are built into the inner membranes of mitochondria. An electron transport chain functions as a chemical machine that uses the energy released by the "fall" of electrons to pump hydrogen ions (H^+) across the inner mitochondrial membrane. Remember from Chapter 2 that H^+ is dissolved as a solute in varying amounts in biological fluids. When electron transport chains pump H^+ across the membrane, the ions become more concentrated on one side of the membrane than on the other. Such a concentration gradient stores potential energy.

The energy stored by electron transport behaves something like the elevated reservoir of water behind a dam. There is a tendency for H^+ to gush back to where they are less concentrated, just as there is a tendency

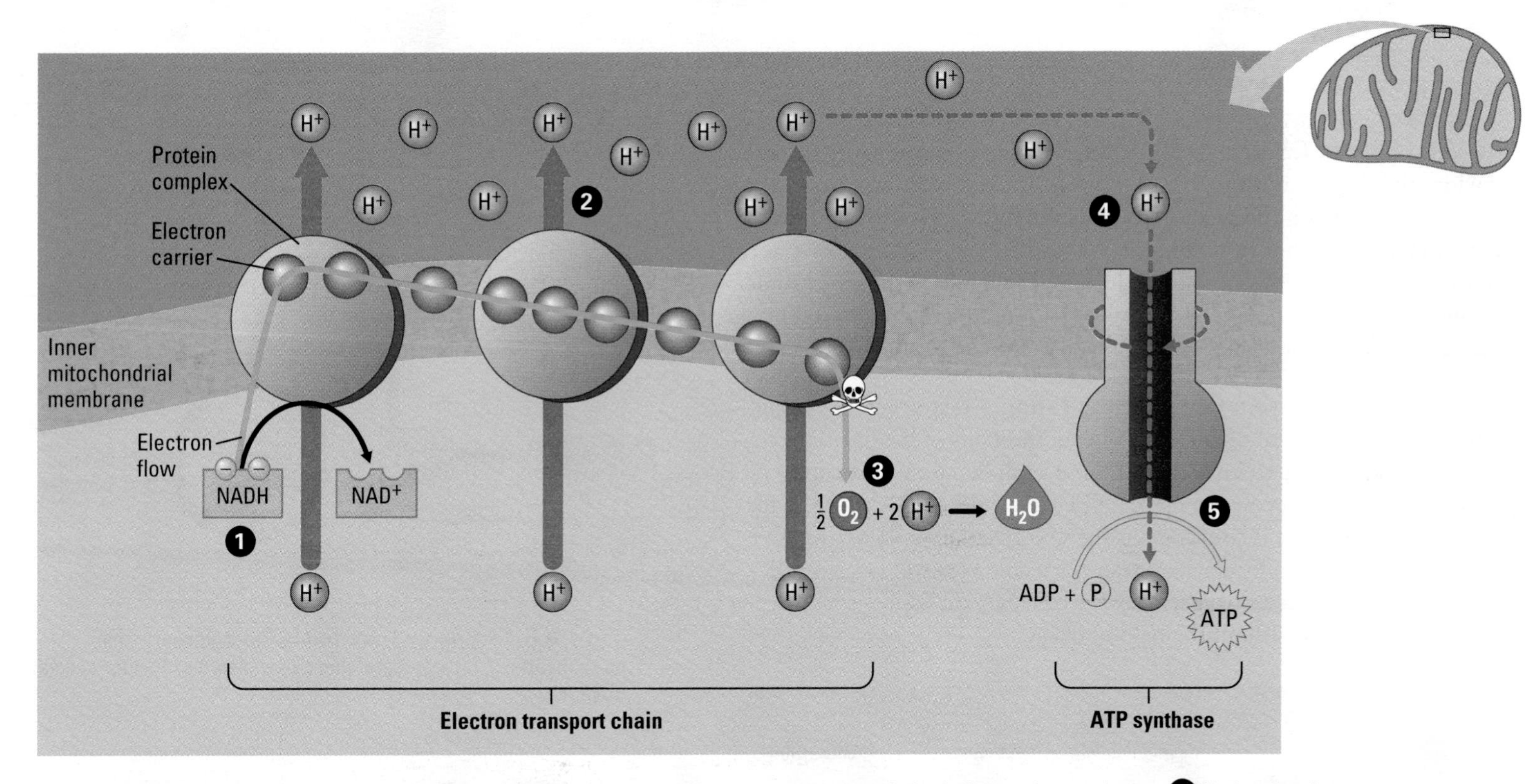

Figure 6.12 How electron transport drives ATP synthase machines. This figure shows a simplified view of how the energy stored in NADH can be used to generate molecules of ATP. ❶ NADH transfers electrons from food to electron transport chains. ❷ Electron transport chains use this energy supply to pump H^+ across the inner membrane of the mitochondrion. The infoldings of the inner membrane increase surface area, maximizing the number of electron transport chains built into the membrane. ❸ Notice that the oxygen you breathe pulls electrons down the transport chain. The deadly poisons cyanide and carbon monoxide can interrupt this step. ❹ H^+ flows back through an ATP synthase, spinning part of the synthase much the way water turns a turbine when it flows through the gates in a dam. ❺ The ATP synthase uses the energy of the H^+ gradient to regenerate ATP from ADP.

for water to flow downhill. The membrane, analogous to the dam, temporarily restrains hydrogen ions.

The energy of dammed water can be harnessed to do work. Gates in the dam allow the water to rush downhill, turning giant turbines as it goes. The spinning turbines perform work and generate electricity. Your mitochondria have structures that act like turbines. Each of these miniature machines, called an **ATP synthase,** is constructed from several proteins built into the inner mitochondrial membrane, the same membrane where electron transport chains are located (**Figure 6.12**). Electron transport provides energy for operating the ATP synthase machines, but only indirectly. Hydrogen ions pumped by electron transport rush back "downhill" through an ATP synthase. This action spins a component of the ATP synthase, just as water turns the turbines in a dam. The rotation activates catalytic sites in the synthase that attach phosphate groups to ADP molecules to generate ATP.

Some of the deadliest poisons do their damage by disrupting electron transport in mitochondria. For example, both carbon monoxide and cyanide kill by blocking the transfer of electrons from electron transport chains to oxygen (marked by ☠ in Figure 6.12). With its energy-harvesting mechanism shut down, the mitochondrial membrane can no longer convert food energy to ATP energy. Cells stop working, and the organism dies.

The Versatility of Cellular Respiration We have seen so far that food provides the energy to make the ATP our cells use for all their work. We have concentrated on the sugar glucose as the fuel that is broken down in cellular respiration, but respiration is a versatile metabolic furnace that can "burn" many other kinds of food molecules. **Figure 6.13** diagrams some metabolic routes for the use of diverse carbohydrates, fats, and proteins as fuel for cellular respiration.

Adding Up the ATP from Cellular Respiration Taking cellular respiration apart to see how all the molecular nuts and bolts of its metabolic machinery work, it's easy to lose sight of the overall function: generating ATP. **Figure 6.14** will help you add up the ATP molecules a cell can make for each glucose molecule it consumes as fuel. Notice that most of that ATP production is powered by electron transport. And electron transport depends on the presence of oxygen. Next, we'll see what happens when cells harvest food energy without the help of oxygen.

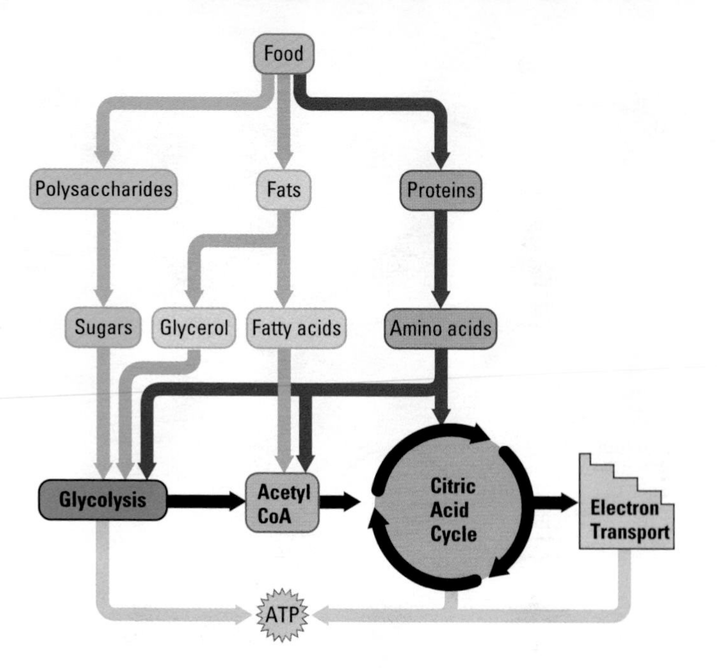

Figure 6.13 Energy from food. The monomers from carbohydrates, fats, and proteins can all serve as fuel for cellular respiration.

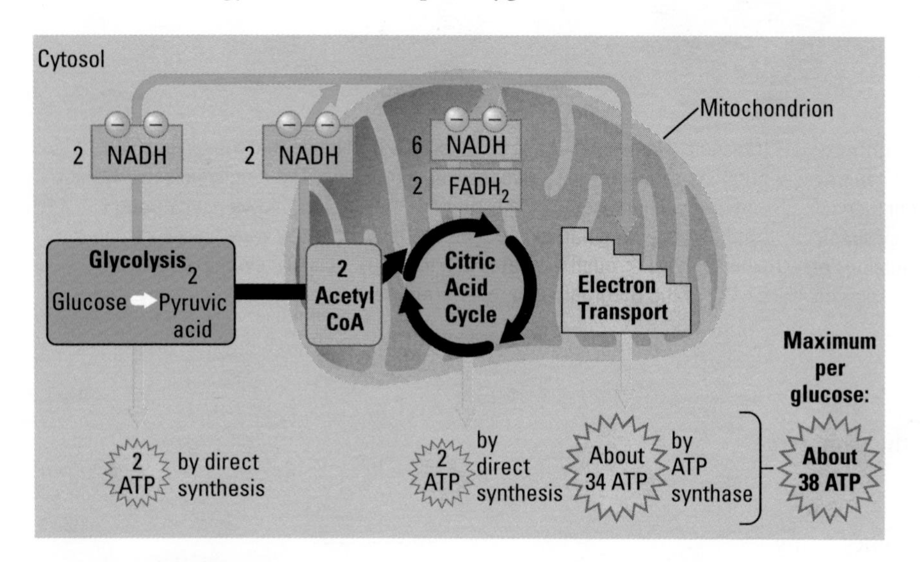

Figure 6.14 A summary of ATP yield during cellular respiration. A cell can convert the energy of each glucose molecule to as many as 38 ATP molecules (the actual number can vary by a few ATP molecules). Glycolysis and the citric acid cycle each contribute 2 ATP by direct synthesis (see Figure 6.9). The other 34 ATP molecules are produced by the ATP synthase machines. The "fall" of electrons from food to oxygen powers ATP synthase. The electrons are carried from the organic fuel to electron transport chains by NADH and $FADH_2$. Each electron pair "dropped" down a transport chain from NADH can power the synthesis of up to 3 ATP. Each electron pair transferred to an electron transport chain from $FADH_2$ is worth up to 2 ATP.

CHECKPOINT

1. How is your breathing related to your cellular respiration?
2. At the "downhill" end of the electron transport chain, when electrons from NADH are finally passed to oxygen, what waste product of cellular respiration is produced? (*Hint:* Review Figure 6.6.)
3. What is the potential energy source that drives ATP production by ATP synthase?
4. Of the three main stages of cellular respiration represented in Figure 6.7, which one uses oxygen directly to extract chemical energy from organic compounds?
5. The oxidation of acetic acid by NAD^+ extracts some chemical energy from the acetic acid. How can the cell harness that energy to make ATP?
6. Of the three stages of cellular respiration, which occurs in the cytosol, outside mitochondria?

Answers: 1. In breathing, your lungs exchange CO_2 and O_2 between your body and the atmosphere. In cellular respiration, your cells consume the O_2 in extracting energy from food and release CO_2 as a waste product. **2.** Water (H_2O) **3.** A concentration gradient of H^+ across the inner membrane of a mitochondrion **4.** Electron transport **5.** The NADH can supply electrons to the electron transport chain, which generates an H^+ gradient that drives ATP synthesis. **6.** Glycolysis

Fermentation: Anaerobic Harvest of Food Energy

Although you must breathe to stay alive, some of your cells can work for short periods without O_2. This anaerobic ("without oxygen") harvest of food energy is called **fermentation.**

Fermentation in Human Muscle Cells

When you walk between classes, your leg muscles require a constant supply of ATP, which is generated by cellular respiration. To keep this process going, blood provides your muscle cells with enough O_2 to keep electrons "falling" down transport chains in your mitochondria. But if you start to run because you're late for class, your muscles are forced to work under anaerobic conditions. That's because they are spending ATP at a rate that outpaces your bloodstream's delivery of O_2 from your lungs to your muscles.

After functioning anaerobically for about 15 seconds, muscle cells will begin to generate ATP by the process of fermentation. Fermentation relies on glycolysis, the same metabolic pathway that functions as the first stage of cellular respiration. Glycolysis does not require O_2 (see Figure 6.8); and glycolysis, remember, produces a small amount of ATP directly: 2 ATP molecules for each glucose molecule broken down to pyruvic acid. That isn't very efficient compared with the 38 or so ATP molecules each glucose generates during cellular respiration, but it can energize your leg muscles fast enough and long enough for you to make it to class. However, your cells will have to consume more glucose fuel per second, since so much less ATP per glucose molecule is generated under anaerobic conditions.

There is more to fermentation than just glycolysis. To harvest food energy during glycolysis, NAD^+ must be present as an electron acceptor (see Figure 6.8). This is no problem under aerobic conditions, because the cell regenerates NAD^+ when NADH drops its electron cargo down electron transport chains to O_2 (see Figure 6.6). However, this recycling of NAD^+ cannot occur under anaerobic conditions because there is no O_2 to accept the electrons. Instead, NADH disposes of electrons by adding them to the pyruvic acid produced by glycolysis **(Figure 6.15)**. This restores NAD^+ and keeps glycolysis working.

The reduction of pyruvic acid (addition of electrons) produces a waste product called lactic acid. A temporary accumulation of lactic acid in muscle cells may contribute to the soreness or burning you feel immediately after an exhausting run or other anaerobic spurt of activity. The lactic acid is eventually transported in the blood from muscles to the liver, where liver cells convert it back to pyruvic acid. That metabolic process requires O_2, which is why you breathe hard for some time even after you've stopped vigorous activity. If lactic acid accumulates to a critical level, extreme fatigue may cause the muscles to temporarily shut down.

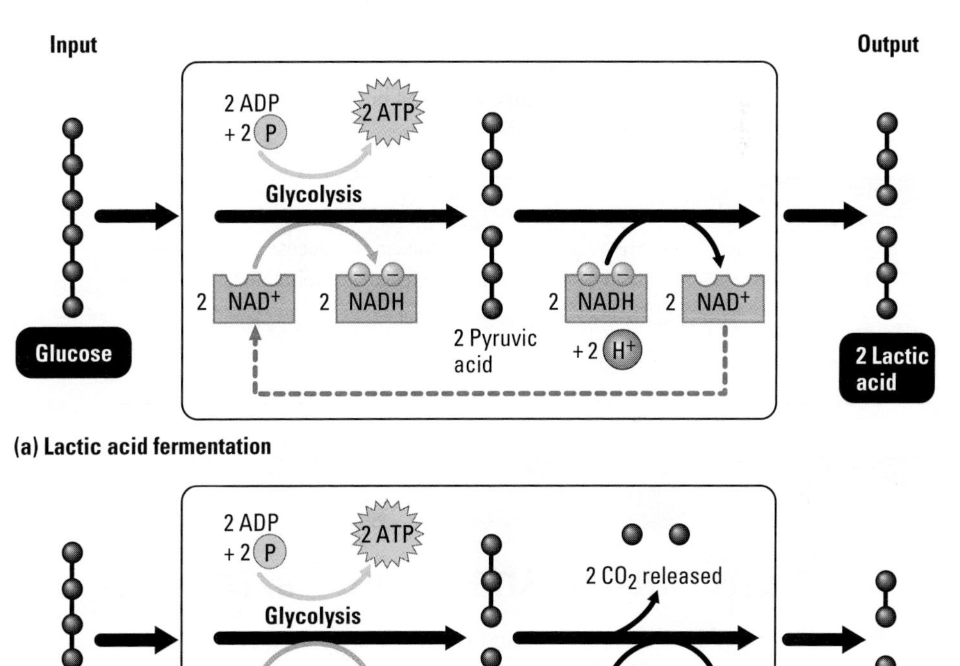

Figure 6.15 Fermentation. Glycolysis produces ATP without the help of O_2. This process requires a continuous supply of NAD^+ to accept electrons from glucose. The NAD^+ is regenerated when NADH transfers the electrons it removed from food to pyruvic acid, thereby producing **(a)** lactic acid, **(b)** ethyl alcohol, or other waste products, depending on the species of organism.

Fermentation in Microorganisms

Like human muscle cells, certain fungi and bacteria produce lactic acid as their waste product during fermentation. We have domesticated such microbes to transform milk into cheese, sour cream, and yogurt. These foods owe their sharp or sour flavor mainly to lactic acid. The food industry also uses fermentation to produce soy sauce from soybeans, to pickle cucumbers, olives, and cabbage, and to produce meat products like sausage, pepperoni, and salami.

Yeast, a microscopic fungus, is capable of both cellular respiration and fermentation. If you keep yeast cells in an anaerobic environment, they are forced to ferment sugars and other foods to stay alive. In contrast to muscle cells, the yeast style of fermentation produces ethyl alcohol as a waste product instead of lactic acid. This alcoholic fermentation also releases CO_2 (see **Figure 6.15b**). For thousands of years, humans have put yeast to work producing alcoholic beverages such as beer and wine **(Figure 6.16a)**. And as every baker knows, the CO_2 bubbles from fermenting yeast also cause bread dough to rise **(Figure 6.16b)**. (The alcohol produced in fermenting bread is released during baking.)

Yeast is an example of what is called a **facultative anaerobe,** an organism with the metabolic versatility to harvest food energy by either cellular respiration or fermentation. In contrast are **obligate anaerobes,** organisms that are actually poisoned by oxygen, such as certain bacteria that live in stagnant ponds or deep in the soil. At the cellular level, our muscle cells, like yeast cells, behave as facultative anaerobes. But as whole organisms, we are best described, in metabolic terms, as **obligate aerobes** because we depend on oxygen and cellular respiration to stay alive.

(a) A fermentation tank at a brewery

(b) Bread showing air bubbles produced by fermenting yeast

Figure 6.16 Using fermentation to make food.

CHECKPOINT

1. How many molecules of ATP are generated per molecule of glucose during fermentation? How many can be generated during cellular respiration?
2. __________ acid is to human muscle cells as ethyl __________ is to yeast.

Answers: 1. 2; up to 38 **2.** Lactic; alcohol

EVOLUTION CONNECTION

Life on an Anaerobic Earth

The citric acid cycle and the electron transport chain operate only under aerobic conditions, in the presence of O_2. Glycolysis, on the other hand, occurs under both aerobic and anaerobic conditions. The role of glycolysis in both cellular respiration and fermentation has an evolutionary basis. Ancient bacteria probably used glycolysis to make ATP before O_2 was present in high concentrations in Earth's atmosphere. The oldest known fossils of bacteria date back over 3.5 billion years, but significant levels of O_2 did not accumulate in the atmosphere until about 2.5 billion years ago. For a billion years, bacteria must have generated ATP exclusively from glycolysis, which does not require O_2. Glycolysis is the most widespread metabolic pathway, which also suggests that it evolved very early in ancestors common to all kingdoms of life. And the fact that glycolysis occurs in the cytosol, and not in mitochondria, implies great antiquity, too; mitochondria evolved long after the origin of prokaryotic life. Glycolysis is a metabolic heirloom from the earliest cells that continues to function today in the harvest of food energy. ■

Chapter Review

SUMMARY OF KEY CONCEPTS

For study help and activities, go to campbellbiology.com or the student CD-ROM.

Biology and Society: Feeling the Burn

- Aerobic metabolism occurs when cells receive enough oxygen to support their energy needs. If a body exceeds its aerobic capacity, the demand for oxygen is greater than the body's ability to deliver it, and metabolism becomes anaerobic.

Energy Flow and Chemical Cycling in the Biosphere

- **Producers and Consumers** Autotrophs (producers) make organic molecules from inorganic nutrients via photosynthesis. Heterotrophs (consumers) must consume organic material and obtain energy via cellular respiration.
- **Chemical Cycling between Photosynthesis and Cellular Respiration**

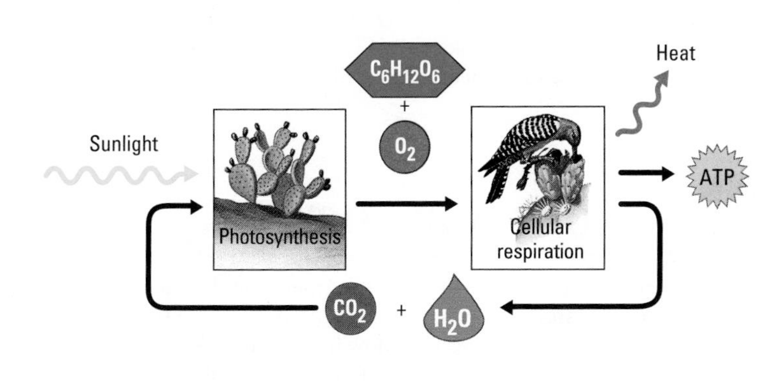

Activity *Build a Chemical Cycling System*

Cellular Respiration: Aerobic Harvest of Food Energy

eTutor *Cellular Respiration*

MP3 Tutor *Cellular Respiration Part 1: Glycolysis*

MP3 Tutor *Cellular Respiration Part 2: Citric Acid Cycle and Electron Transport Chain*

- **The Relationship between Cellular Respiration and Breathing** The bloodstream distributes O_2 from the lungs to all the cells of the body and transports CO_2 waste from the cells to the lungs for disposal.
- **The Overall Equation for Cellular Respiration**

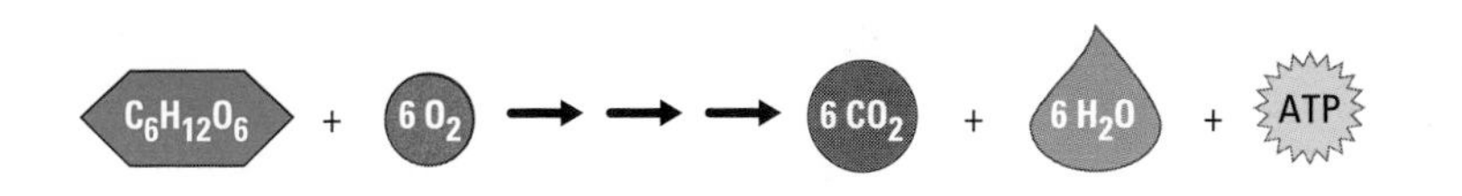

- **The Role of Oxygen in Cellular Respiration**

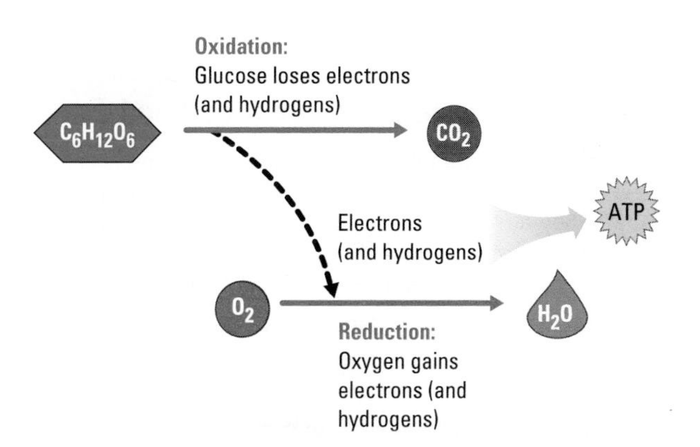

Redox reactions transfer electrons from food molecules to an electron acceptor called NAD^+, forming NADH. The NADH then passes the high-energy electrons to an electron transport chain that eventually "drops" them to O_2. The energy released during this electron transport is used to regenerate ATP from ADP. The affinity of oxygen for electrons keeps the redox reactions of cellular respiration working.

- **The Metabolic Pathway of Cellular Respiration** You can follow the flow of molecules through the process of cellular respiration in the following diagram:

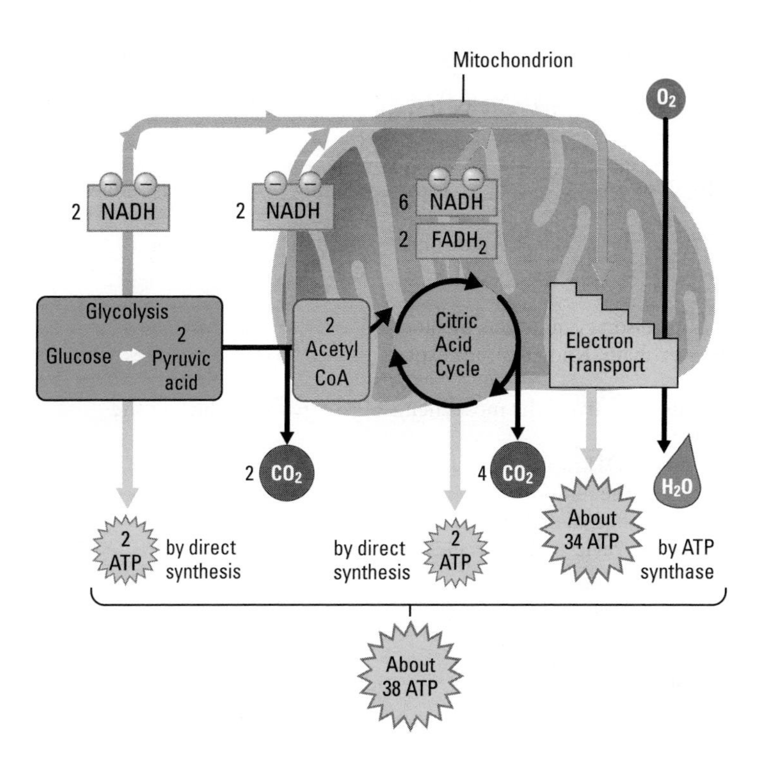

The electron transport chains pump H^+ across the membrane as electrons flow stepwise from NADH to oxygen. Backflow of H^+ across the membrane powers the ATP synthases, which attach phosphate to ADP to make ATP.

Activity *Overview of Cellular Respiration*

Activity *Glycolysis*

Activity *The Citric Acid Cycle*

Activity *Electron Transport*

Case Study in the Process of Science *How Is the Rate of Cellular Respiration Measured?*

Fermentation: Anaerobic Harvest of Food Energy

- **Fermentation in Human Muscle Cells** When muscle cells consume ATP faster than O_2 can be supplied for cellular respiration, they regenerate ATP by fermentation. The waste product under these anaerobic conditions is lactic acid. The ATP yield per glucose is much lower during fermentation than during cellular respiration.
- **Fermentation in Microorganisms** Yeast and other facultative anaerobes can survive with or without O_2. Wastes from fermentation can be ethyl alcohol, lactic acid, or other compounds, depending on the species. Some microorganisms are obligate anaerobes, which are poisoned by O_2.

Activity *Fermentation*

SELF-QUIZ

1. Which of the following statements is a correct distinction between autotrophs and heterotrophs?
   a. Only heterotrophs require chemical compounds from the environment.
   b. Cellular respiration is unique to heterotrophs.
   c. Only heterotrophs have mitochondria.
   d. Only autotrophs can live on nutrients that are entirely inorganic.

2. Of the three stages of cellular respiration, which produces the most ATP molecules per glucose?

3. In glycolysis, __________ is oxidized and __________ is reduced.

4. The final electron acceptor of electron transport chains in mitochondria is __________.

5. The poison cyanide acts by blocking a key step in the electron transport chain. Knowing this, explain why cyanide kills so quickly.

6. Cells can harvest the most chemical energy from which of the following?
   a. an NADH molecule
   b. a glucose molecule
   c. six carbon dioxide molecules
   d. two pyruvic acid molecules

7. Cellular respiration, which consumes oxygen, occurs in all of the following except
   a. aerobic consumers.
   b. plants.
   c. human muscle cells.
   d. obligate anaerobes.

8. __________ is a metabolic pathway common to both fermentation and cellular respiration.

9. Sports physiologists at an Olympic training center wanted to monitor athletes to determine at what point their muscles were functioning anaerobically. They could do this by checking for a buildup of
   a. ADP.
   b. lactic acid.
   c. carbon dioxide.
   d. oxygen.

10. A glucose-fed yeast cell is moved from an aerobic environment to an anaerobic one. For the cell to continue to generate ATP at the same rate, approximately how much glucose must it consume in the anaerobic environment compared with the aerobic environment?

Answers to the Self-Quiz questions can be found in Appendix D.

Go to the website or CD-ROM for more Self-Quiz questions.

THE PROCESS OF SCIENCE

11. Your body makes NAD^+ from two B vitamins, niacin and riboflavin. You need only tiny amounts of these vitamins. The U.S. Food and Drug Administration's recommended dietary allowances are 20 mg daily for niacin and 1.7 mg daily for riboflavin. These amounts are thousands of times less than the amount of glucose your body needs each day to fuel its energy requirements. How many NAD^+ molecules are needed for the breakdown of each glucose molecule? Why do you think your daily requirement for these substances is so small?

BIOLOGY AND SOCIETY

12. Nearly all human societies use fermentation to produce alcoholic drinks such as beer and wine. The technology dates back to the earliest civilizations. Suggest a hypothesis for how humans first discovered fermentation. In preindustrial cultures, why do you think wine was a more practical beverage than the grape juice from which it was made?

13. The consumption of alcohol by a pregnant woman can cause a series of birth defects called fetal alcohol syndrome (FAS). Symptoms of FAS include head and facial irregularities, heart defects, mental retardation, and behavioral problems. The U.S. Surgeon General's Office recommends that pregnant women abstain from drinking alcohol, and the government has mandated that a warning label be placed on liquor bottles. Imagine you are a server in a restaurant. An obviously pregnant woman orders a strawberry daiquiri. How would you respond? Is it the woman's right to make those decisions about her unborn child's health? Do you bear any responsibility in the matter? Is a restaurant responsible for monitoring the dietary habits of its customers?

7 Photosynthesis: Using Light to Make Food

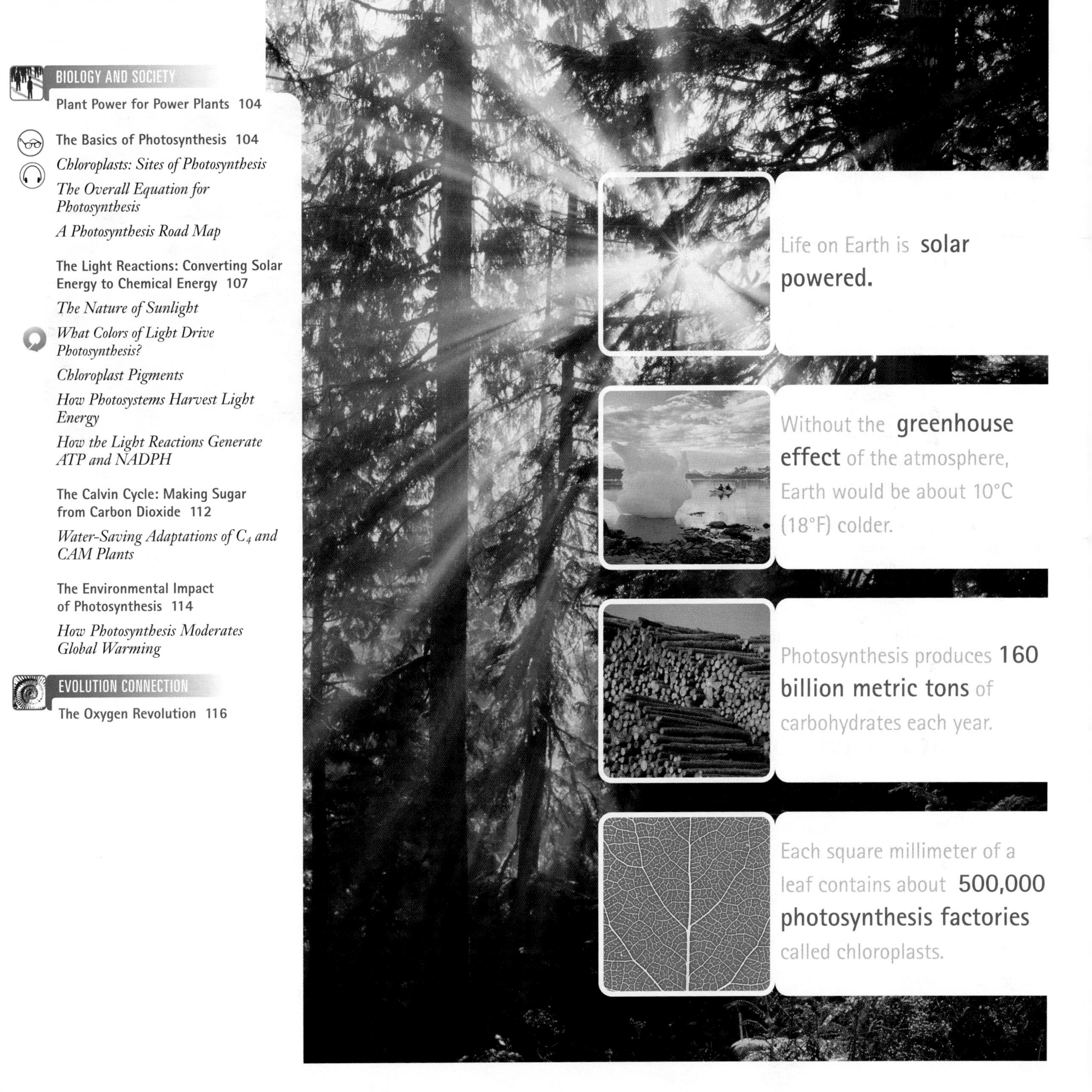

Life on Earth is **solar powered.**

Without the **greenhouse effect** of the atmosphere, Earth would be about 10°C (18°F) colder.

Photosynthesis produces **160 billion metric tons** of carbohydrates each year.

Each square millimeter of a leaf contains about **500,000 photosynthesis factories** called chloroplasts.

BIOLOGY AND SOCIETY

Plant Power for Power Plants

On a global scale, the productivity of photosynthesis is astounding. By converting the energy of sunlight to chemical energy, Earth's plants and other photosynthetic organisms make about 160 billion metric tons of organic material per year (a metric ton is 1,000 kg, about 1.1 tons). That much organic material is equivalent in weight to about 25 stacks of this book reaching from Earth to the sun!

All of the food consumed by humans can be traced to photosynthetic plants. But plants provide much more than just nourishment. They also supply many of the raw materials we need to survive. For example, during the vast majority of human history, burning plant material was the only nonsolar source of heat, light, and fuel for cooking. Over the last century, wood has been largely displaced as an energy source by fossil fuels such as coal, gas, and oil (which can also be traced to photosynthetic sources, albeit ancient ones). But recently, the benefits of plant matter as an energy source have regained attention.

Figure 7.1 shows an "energy plantation" in upstate New York. There, willow trees are being tested as a source of energy. The trees are cut once every three years, and the harvested wood is sent to power plants to generate electricity. Because they grow exceptionally fast and resprout after cutting, willows are a renewable resource. Other tree species, including sycamore, eucalyptus, and black locust, are being tested as fuel sources elsewhere.

Figure 7.1 One-year-old willow trees growing in an "energy plantation" in upstate New York.

Burning wood for energy has several advantages over burning fossil fuels. Wood has very little of the sulfur impurities that cause acid rain. Energy plantations also provide habitat for wildlife, reduce erosion, and help farmers diversify. Perhaps most significantly, vigorously growing young plants remove a lot of carbon dioxide from the air, potentially reducing the levels of gases in the atmosphere that cause global warming. Today, so-called biomass energy accounts for only about 4% of all energy consumed in the United States. But this renewable, relatively low-emissions resource may play a significant role in satisfying our future energy needs.

In this chapter, you'll learn how plants use light to make their own food. Because photosynthesis can seem like a complex process, we'll begin by examining some basic concepts that will orient the discussion. Then we'll look at the specific mechanisms involved in photosynthesis, breaking the larger process into two parts. Finally, we'll see how photosynthesis affects our global environment. ■

The Basics of Photosynthesis

eTutor
Photosynthesis

MP3 Tutor
Photosynthesis

Almost all plants are photosynthetic autotrophs—meaning that they generate their own organic matter from inorganic ingredients via photosynthesis; certain groups of protists and bacteria fall into this category, too (**Figure 7.2**). This section presents an overview of the process of photosynthesis. In later sections, we'll take a closer look.

Chloroplasts: Sites of Photosynthesis

In Chapter 4, you learned that photosynthesis occurs within **chloroplasts,** organelles present in certain plant cells. All green parts of a plant have chloroplasts and can carry out photosynthesis. In most plants, however,

(a) Mosses, ferns, and flowering plants

(b) Kelp

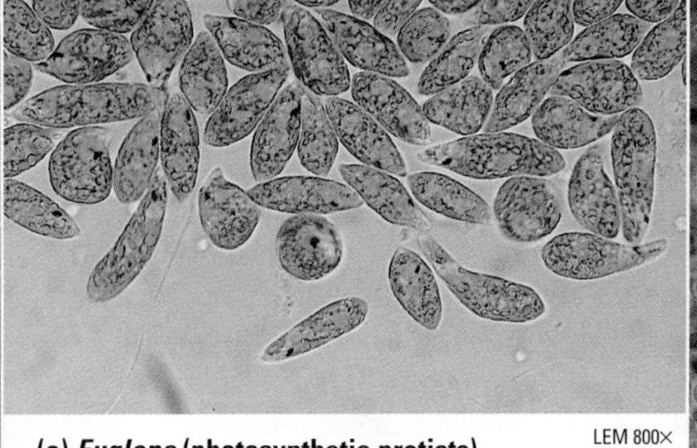

(c) *Euglena* (photosynthetic protists)

(d) Cyanobacteria (photosynthetic bacteria) LEM 2

Figure 7.2 Photosynthetic autotrophs: producers for most ecosystems. **(a)** On land, plants are the predominant producers of food. Three major groups of plants—mosses, ferns, and flowering plants—are represented in this scene. In oceans, lakes, ponds, streams, and other aquatic habitats, photosynthetic organisms include **(b)** large algae, such as this kelp, **(c)** certain microscopic protists, such as *Euglena*, and **(d)** bacteria called cyanobacteria.

leaves have the most chloroplasts and are the major sites of photosynthesis. The green color in plants comes from pigment molecules in the chloroplasts called chlorophyll. Chlorophyll absorbs the light energy that the chloroplasts put to work in making food.

Chloroplasts are concentrated in the interior cells of leaves **(Figure 7.3)**. Carbon dioxide (CO_2) enters, and oxygen (O_2) exits, by way of tiny pores called **stomata** (singular, *stoma*, meaning "mouth"). Most stomata are found on the undersurface of leaves. In addition to carbon dioxide, photosynthesis

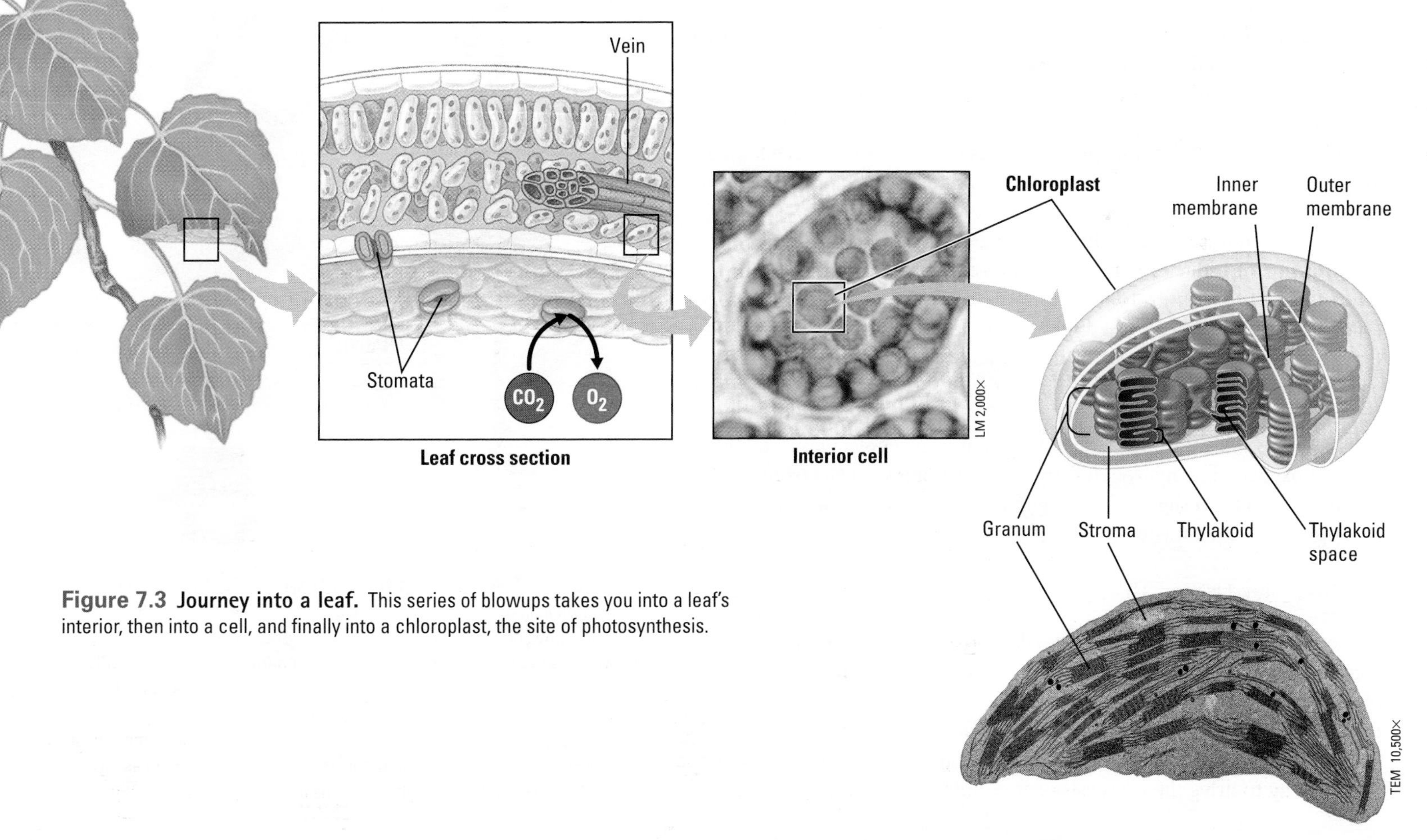

Figure 7.3 Journey into a leaf. This series of blowups takes you into a leaf's interior, then into a cell, and finally into a chloroplast, the site of photosynthesis.

also requires water as an inorganic ingredient. This needed water is mainly absorbed by the plant's roots and travels via veins to the leaves.

Membranes within the chloroplast form the apparatus where many of the reactions of photosynthesis occur. Like the mitochondrion, the chloroplast has a double-membrane envelope (see Figures 4.16 and 4.17). The chloroplast's inner membrane encloses a compartment filled with **stroma,** a thick fluid. Suspended in the stroma is an elaborate system of interconnected membranous sacs called **thylakoids.** The thylakoids are concentrated in stacks called **grana** (singular, *granum*). The chlorophyll molecules that capture light energy are built into the thylakoid membranes. The structure of the chloroplast—with its stacks of disks—aids its function by providing a large surface area for the reactions of photosynthesis.

The Overall Equation for Photosynthesis

The following chemical equation, simplified to highlight the relationship between photosynthesis and cellular respiration, provides a summary of the reactants and products of photosynthesis:

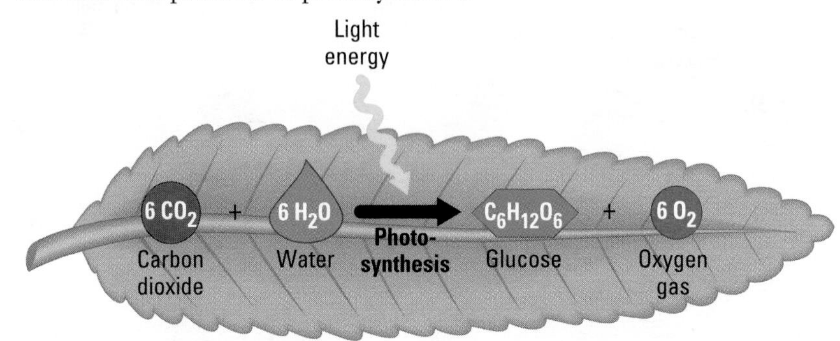

Notice that the reactants of photosynthesis, carbon dioxide (CO_2) and water (H_2O), are also the waste products of cellular respiration (see Figure 6.3). And photosynthesis produces what respiration uses, namely glucose ($C_6H_{12}O_6$) and oxygen (O_2). In other words, photosynthesis takes the "exhaust" of cellular respiration and rearranges its atoms to produce food and oxygen. It's a chemical transformation that requires much energy, and sunlight absorbed by chlorophyll provides that energy.

You learned in Chapter 6 that cellular respiration is a process of electron transfer, or reduction and oxidation (redox). A "fall" of electrons from food molecules to oxygen to form water releases the energy that mitochondria can use to make ATP (see Figure 6.5). The opposite occurs in photosynthesis: Electrons are boosted "uphill" and added to carbon dioxide to produce sugar. Hydrogen is moved along with the electrons, so the redox process takes the form of hydrogen transfer from water to carbon dioxide. This requires the chloroplast to actually split water molecules into hydrogen and oxygen. The hydrogen is transferred along with electrons to carbon dioxide to form sugar. The oxygen escapes through stomata into the atmosphere as O_2, a waste product of photosynthesis.

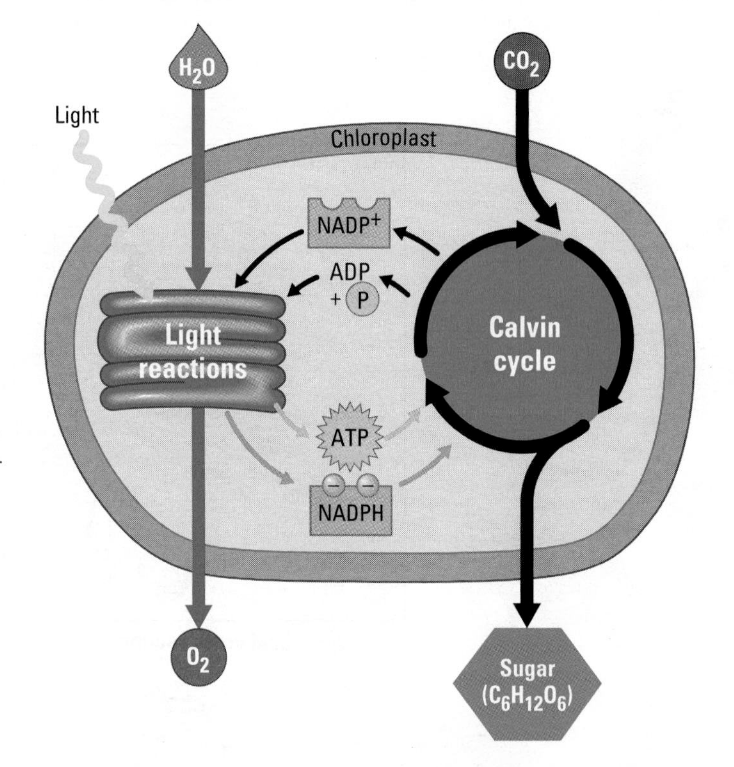

Figure 7.4 A road map for photosynthesis. The thylakoids, located inside the chloroplast, are the sites of the light reactions. Chlorophyll built into the thylakoid membranes absorbs light energy, which is then converted to the chemical energy of ATP and NADPH. The Calvin cycle uses these two products of the light reactions to power the production of sugar from CO_2. The enzymes for the Calvin cycle are dissolved in the stroma, the thick fluid within the chloroplast. We'll carry a smaller version of this road map for orientation as we take a closer look at the light reactions and the Calvin cycle.

A Photosynthesis Road Map

The equation for photosynthesis is a deceptively simple summary of a complex process. Actually, photosynthesis is not a single process, but two processes, each with many steps. These two stages of photosynthesis are called the light reactions and the Calvin cycle **(Figure 7.4)**.

The **light reactions** convert solar energy to chemical energy. They use light energy to drive the synthesis of two molecules: ATP and NADPH.

We have already met ATP as the molecule that drives most cellular work. **NADPH,** chemical cousin of the NADH that appeared in Chapter 6, is an electron carrier. In cellular respiration, remember, NADH carries electrons from food molecules. In photosynthesis, light drives electrons from water to $NADP^+$ (the oxidized form of the carrier) to form NADPH (the reduced form of the carrier). Although the light reactions convert light energy to the chemical energy of ATP and NADPH, this stage of photosynthesis does not produce sugar.

It is the **Calvin cycle** that actually makes sugar from carbon dioxide. The ATP generated by the light reactions provides the energy for sugar synthesis. And the NADPH produced by the light reactions provides the high-energy electrons for the reduction of carbon dioxide to glucose. Thus, the Calvin cycle depends on light, but only indirectly in that it requires the supply of ATP and NADPH produced by the light reactions.

The road map in Figure 7.4 will help you keep oriented as we now take a closer look at how these two stages of photosynthesis work. The following two sections focus on these two stages: the light reactions first and then the Calvin cycle.

CHECKPOINT

1. For chloroplasts to produce sugar from carbon dioxide in the dark, they would require an artificial supply of the molecules __________ and __________.
2. What are the primary inputs and outputs of the Calvin cycle?

Answers: 1. ATP; NADPH **2.** Inputs: CO_2, ATP, NADPH; output: glucose

The Light Reactions: Converting Solar Energy to Chemical Energy

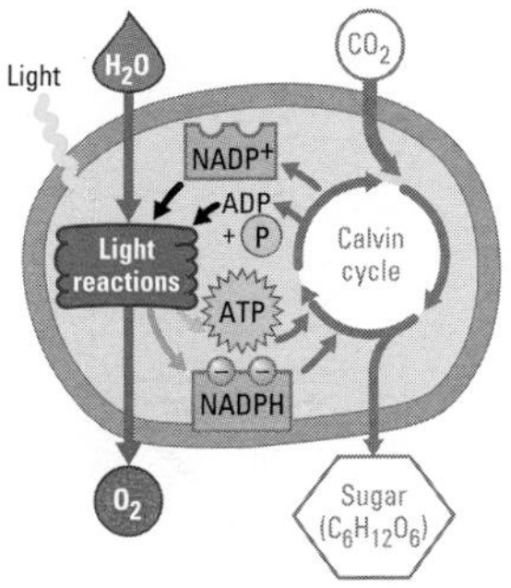

Chloroplasts are chemical factories powered by the sun, an energy source nearly 100 million miles from Earth. In this section, we'll track sunlight into a chloroplast to see how it is converted to the chemical energy of ATP and NADPH.

The Nature of Sunlight

Sunlight is a type of energy called radiation, or electromagnetic energy. Electromagnetic energy travels through space as rhythmic waves analogous to those made by a pebble dropped in a pond. The distance between the crests of two adjacent waves is called a **wavelength.** The full range of radiation, from the very short wavelengths of gamma rays to the very long wavelengths of radio signals, is called the **electromagnetic spectrum.** Visible light composes only a small fraction of the spectrum. It consists of those wavelengths that our eyes see as different colors **(Figure 7.5)**.

When sunlight shines on a pigmented material, certain wavelengths (colors) of the visible light are absorbed and disappear from the light that is reflected by the material. For example, we see a pair of jeans as blue because pigments in the fabric absorb the other colors, leaving only light in the blue part of the spectrum to be reflected from the fabric to our eyes.

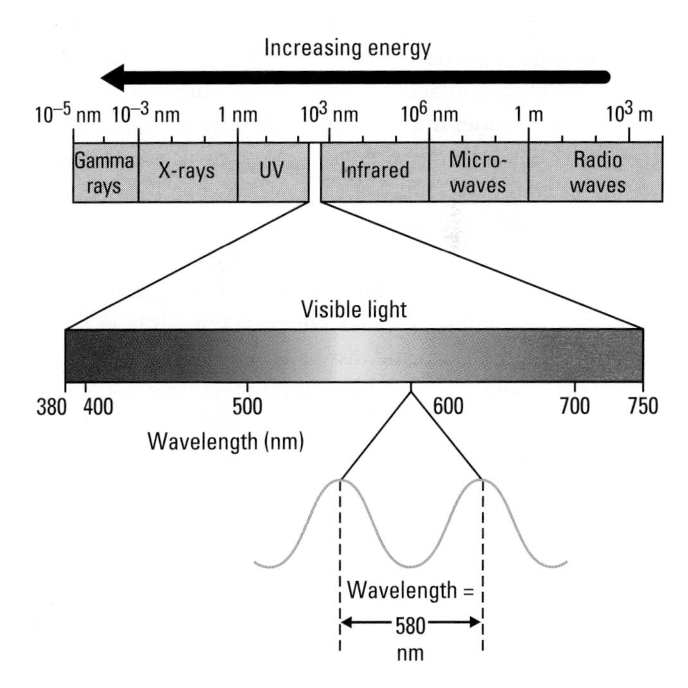

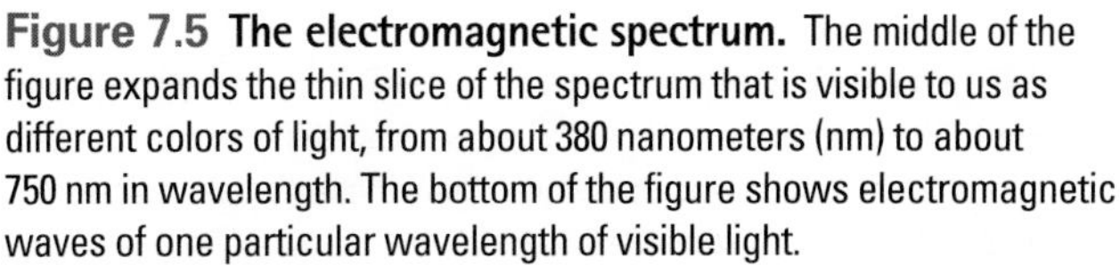

Figure 7.5 The electromagnetic spectrum. The middle of the figure expands the thin slice of the spectrum that is visible to us as different colors of light, from about 380 nanometers (nm) to about 750 nm in wavelength. The bottom of the figure shows electromagnetic waves of one particular wavelength of visible light.

What Colors of Light Drive Photosynthesis? In 1883, German biologist Theodor Engelmann made the **observation** that certain bacteria tend to cluster in areas with higher oxygen concentrations. He already knew that light passed through a prism would separate into different wavelengths (colors). Engelmann soon began to **question** whether he could use this information to determine which wavelengths of light are best for promoting photosynthesis.

Engelmann's **hypothesis** was that oxygen-seeking bacteria would congregate near regions of algae undergoing the most photosynthesis (and hence producing the most oxygen). Engelmann began his **experiment** by laying a string of freshwater algal cells within a drop of water on a microscope slide. He then added oxygen-sensitive bacteria to the drop. Next, using a prism, he created a spectrum of light and shined it on the slide. His **results**, summarized in **Figure 7.6**, showed that most bacteria congregated around algae exposed to red-orange and blue-violet light, with very few bacteria moving to the area of green light.

Other experiments have since verified that chloroplasts absorb only the select wavelengths of light that drive photosynthesis. This selective absorption explains why leaves appear green to us; light of that color is poorly absorbed by chloroplasts and is thus reflected toward the observer **(Figure 7.7)**. Of course, energy cannot be destroyed. If light is absorbed, that energy must be converted to other forms. Chloroplasts contain pigments that are able to convert some of the solar energy they absorb to chemical energy. ■

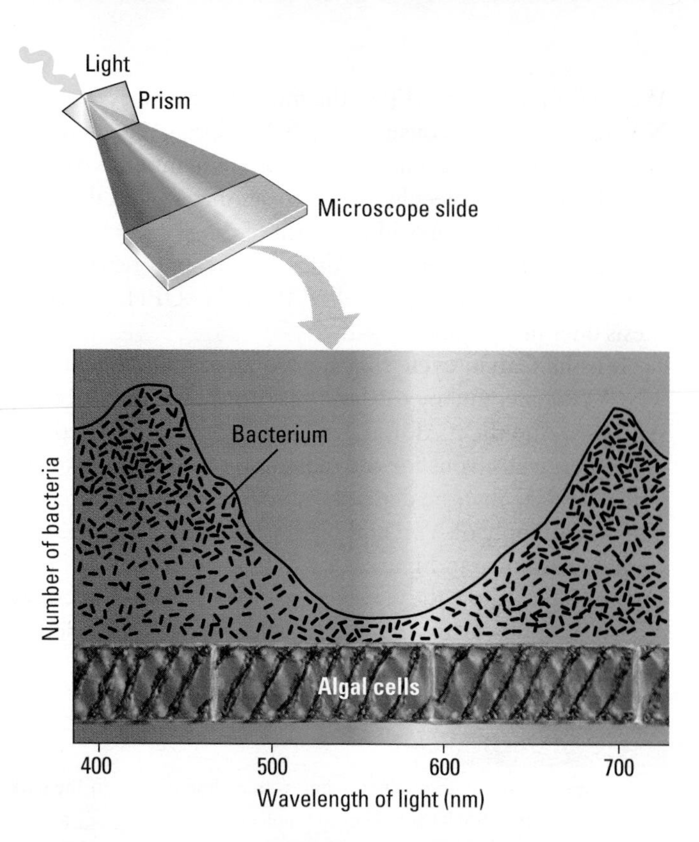

Figure 7.6 Investigating how light wavelength affects photosynthesis. Using a string of algal cells suspended in water on a microscope slide, Engelmann observed that oxygen-seeking bacteria migrate toward algae exposed to certain colors of light. Engelmann's results suggested that blue-violet and orange-red wavelengths best drive photosynthesis, while green wavelengths hardly do so.

Chloroplast Pigments

Different pigments absorb light of different wavelengths, and chloroplasts contain several kinds of pigments. One, **chlorophyll *a***, absorbs mainly blue-violet and red light. Chlorophyll *a* is the pigment that participates directly in the light reactions. A very similar molecule, chlorophyll *b*, absorbs mainly blue and orange light. Chlorophyll *b* does not participate directly in the light reactions, but it broadens the range of light that a plant can use by conveying absorbed energy to chlorophyll *a*, which then puts the energy to work in the light reactions.

Chloroplasts also contain a family of yellow-orange pigments called carotenoids, which absorb mainly blue-green light. Some pass energy to chlorophyll *a*. Other carotenoids have a protective function: They absorb and dissipate excessive light energy that would otherwise damage chlorophyll. (Similar carotenoids, which we obtain from carrots and certain other plants, may help protect our eyes from very bright light.) The spectacular colors of fall

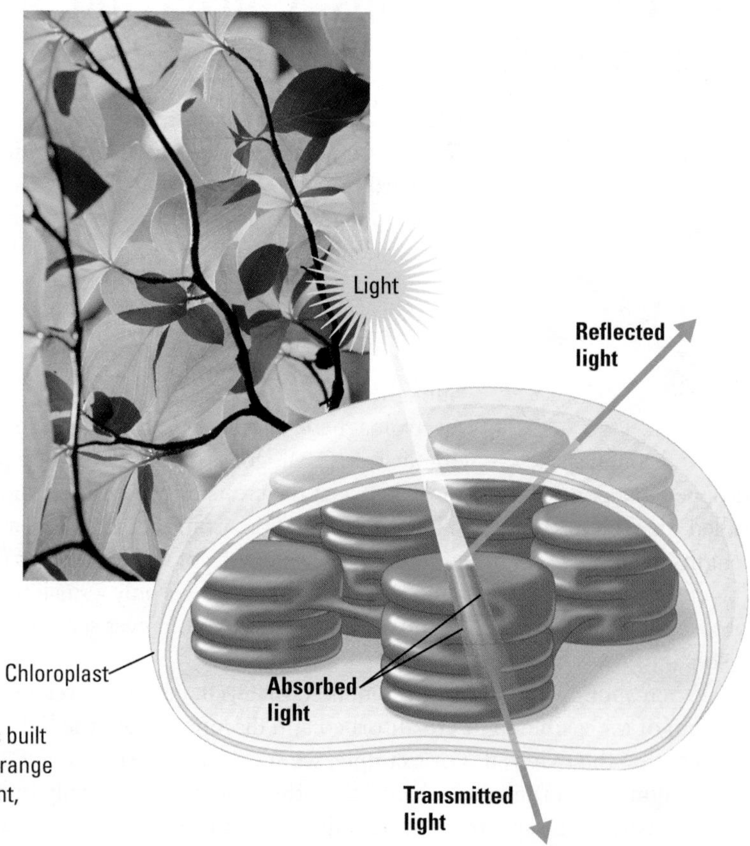

Figure 7.7 Why are leaves green? Chlorophyll and the other pigments built into the membranes of grana mainly absorb light in the blue-violet and red-orange part of the electromagnetic spectrum. The pigments do not absorb much green light, which is reflected or transmitted to our eyes.

foliage in certain parts of the world are due partly to decreases in green chlorophyll, allowing the yellow-orange hues of longer-lasting carotenoids to show through **(Figure 7.8)**.

All of these chloroplast pigments are built into the thylakoid membranes (see Figure 7.3). There the pigments are organized into light-harvesting complexes called photosystems.

Figure 7.8 Photosynthetic pigments. Falling autumn temperatures cause a decrease in the levels of green chlorophyll within the leaves of deciduous trees. This allows the colors of the carotenoids to become prominent.

How Photosystems Harvest Light Energy

The theory of light as waves explains most of light's properties. However, light also behaves as discrete packets of energy called photons. A **photon** is a fixed quantity of light energy. The shorter the wavelength of light, the greater the energy of a photon. For example, a photon of violet light packs nearly twice as much energy as a photon of red light (see Figure 7.5).

When a pigment molecule absorbs a photon, one of the pigment's electrons gains energy, and we say that the electron has been raised from a ground state to an excited state. The excited state is very unstable, so an excited electron usually loses its excess energy and falls back to its ground state almost immediately **(Figure 7.9a)**. Most pigments merely release heat energy as their light-excited electrons fall back to their ground state. (That's why a dark surface, such as a black automobile hood, gets so hot on a sunny day.) Some pigments emit light as well as heat after absorbing photons. The fluorescent light emitted by a glow stick is caused by a chemical reaction that excites electrons within a fluorescent dye **(Figure 7.9b)**. The excited electrons quickly fall back down to their ground state, releasing energy in the form of fluorescent light.

Within the thylakoid membrane, chlorophyll is organized with other molecules into photosystems. Each **photosystem** has a cluster of a few hundred pigment molecules, including chlorophylls *a* and *b* and carotenoids

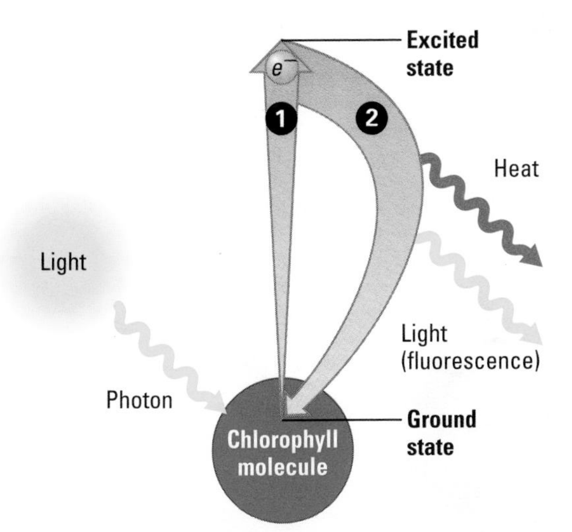

(a) Absorption of a photon. ❶ The absorption of a photon drives an electron (e^-) from its ground state to an excited state. ❷ In a billionth of a second, the excited electron falls back to its ground state, releasing heat and light.

(b) Fluorescence of a glow stick. When you break the glass vial within a glow stick, two chemicals combine. This starts a chemical reaction that excites electrons within a fluorescent dye. As the electrons fall from their excited state to the ground state, the excess energy is emitted as light.

Figure 7.9 Excited electrons within pigments.

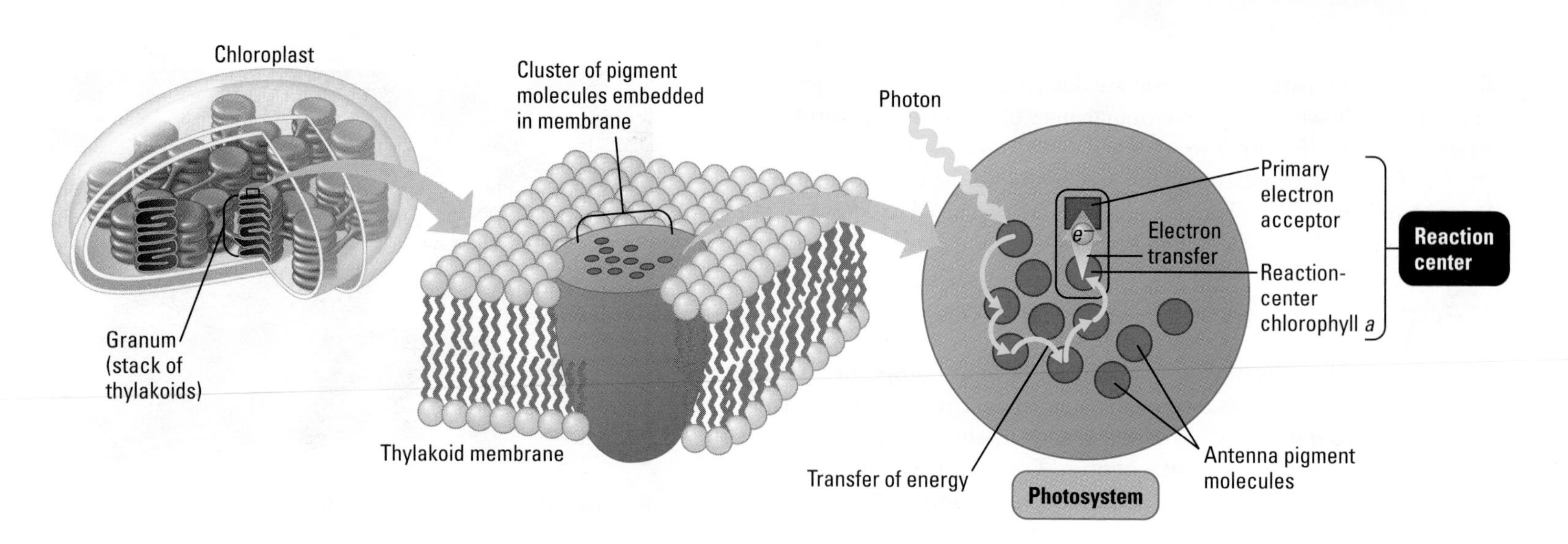

(Figure 7.10). This cluster of pigment molecules functions as a light-gathering antenna. When a photon strikes one pigment molecule, the energy jumps from molecule to molecule until it arrives at the **reaction center** of the photosystem. The reaction center consists of a chlorophyll *a* molecule that sits next to another molecule called a **primary electron acceptor.** This acceptor traps the light-excited electron from the reaction-center chlorophyll. Another team of molecules built into the thylakoid membrane then uses that trapped energy to make ATP and NADPH.

Figure 7.10 A photosystem. The pigment cluster of the photosystem functions as a light-gathering antenna that focuses energy onto the reaction center. At the reaction center of the photosystem, a chlorophyll *a* molecule transfers its light-excited electron (e⁻) to a primary electron acceptor.

How the Light Reactions Generate ATP and NADPH

Two types of photosystems cooperate in the light reactions **(Figure 7.11).** ❶ Photons excite electrons in the chlorophyll of the water-splitting photosystem, which are then trapped by the primary electron acceptor. The water-splitting photosystem replaces its light-excited electrons by extracting electrons from water. This is the step that releases O_2 during photosynthesis. ❷ Energized electrons from the water-splitting photosystem pass down an electron transport chain to the NADPH-producing photosystem.

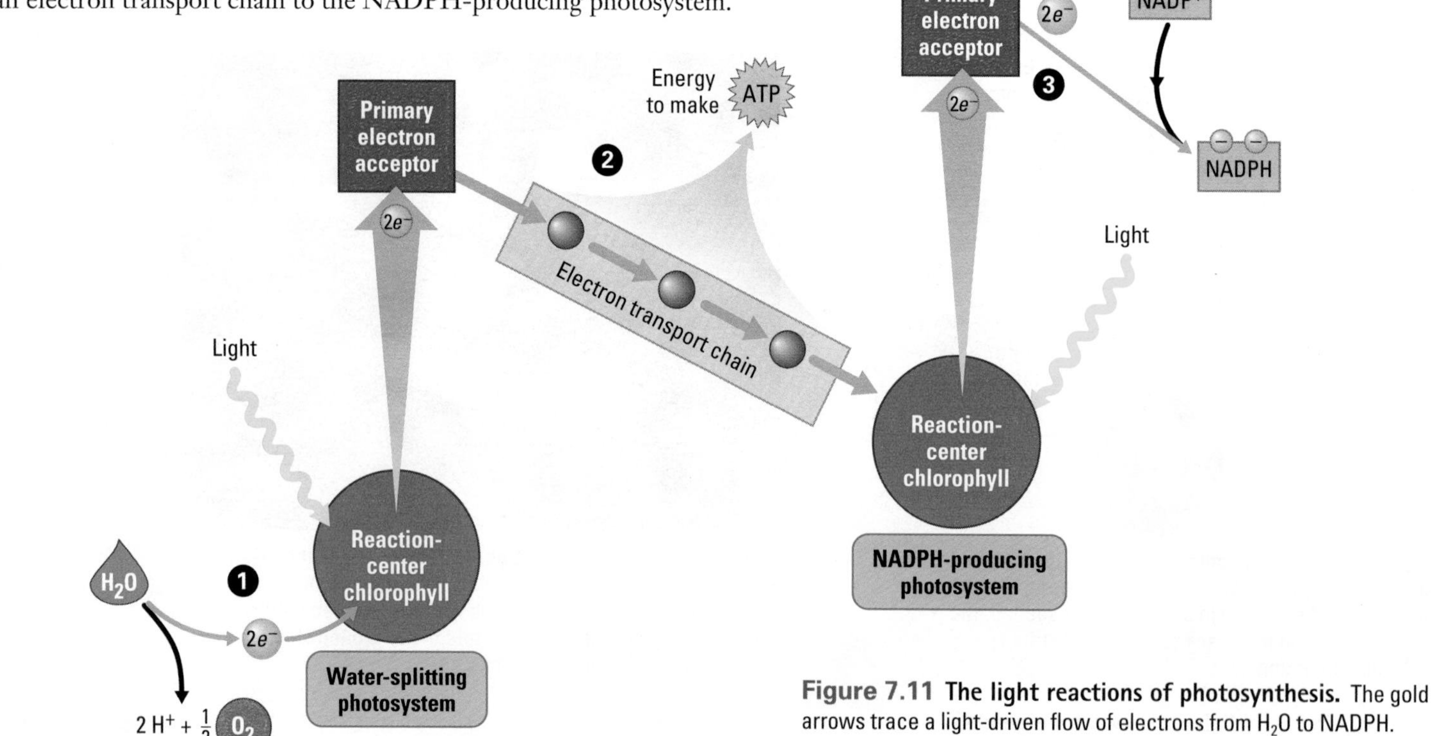

Figure 7.11 The light reactions of photosynthesis. The gold arrows trace a light-driven flow of electrons from H_2O to NADPH.

The chloroplast uses the energy released by this electron "fall" to make ATP. ❸ The NADPH-producing photosystem transfers its light-excited electrons to $NADP^+$, reducing it to NADPH. The traffic of electrons through the two photosystems is analogous to the cartoon in **Figure 7.12**.

Figure 7.13 places the light reactions in the thylakoid membrane. Notice that the mechanism of ATP production during the light reactions is very similar to the ATP production we saw in cellular respiration (see Figure 6.12). In both cases, an electron transport chain pumps hydrogen ions (H^+) across a membrane—the inner mitochondrial membrane in the case of respiration and the thylakoid membrane in photosynthesis. And in both cases, ATP synthases use the energy stored by the H^+ gradient to make ATP. The main difference is that food provides the high-energy electrons in cellular respiration, while it is light-excited electrons that flow down the transport chain during photosynthesis.

We have seen how the light reactions convert solar energy to the chemical energy of ATP and NADPH. Notice again, however, that the light reactions produce no sugar. That's the job of the Calvin cycle, as we'll see next.

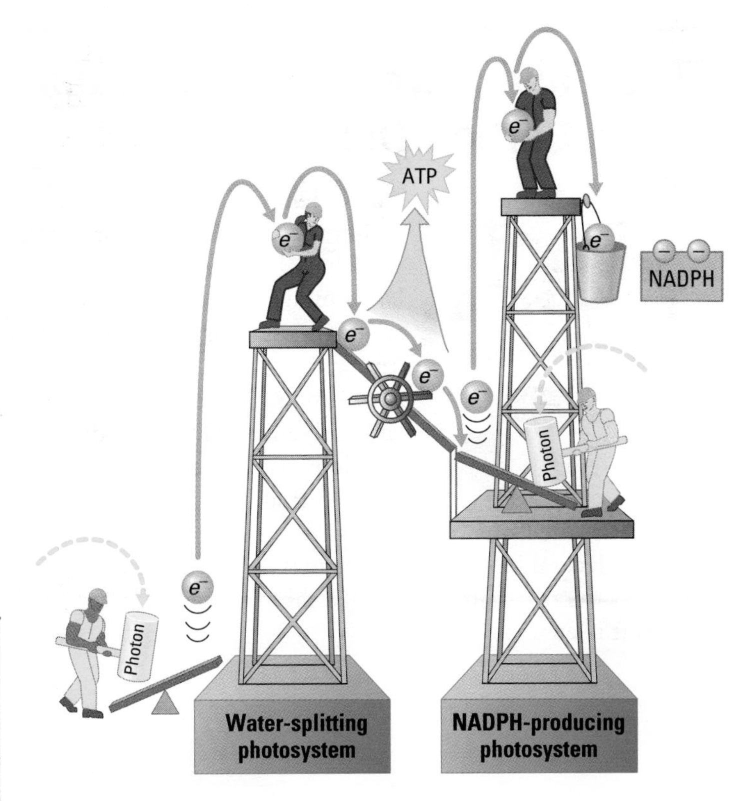

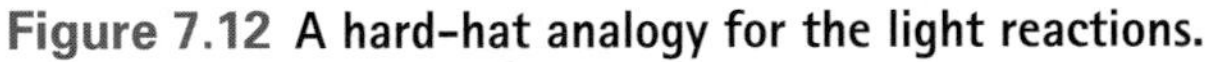
Figure 7.12 A hard-hat analogy for the light reactions.

CHECKPOINT

1. Why are leaves green?
2. Why is water required as a reactant in photosynthesis?
3. In addition to conveying electrons from the water-splitting photosystem to the NADPH-producing photosystem, the electron transport chains of chloroplasts also provide the energy for the synthesis of ___________.

Answers: 1. Chloroplast pigments in leaves selectively absorb most wavelengths of light, but green light is reflected off leaves. **2.** It is the splitting of water that provides electrons for converting CO_2 to sugar (via electron transfer by NADPH). **3.** ATP

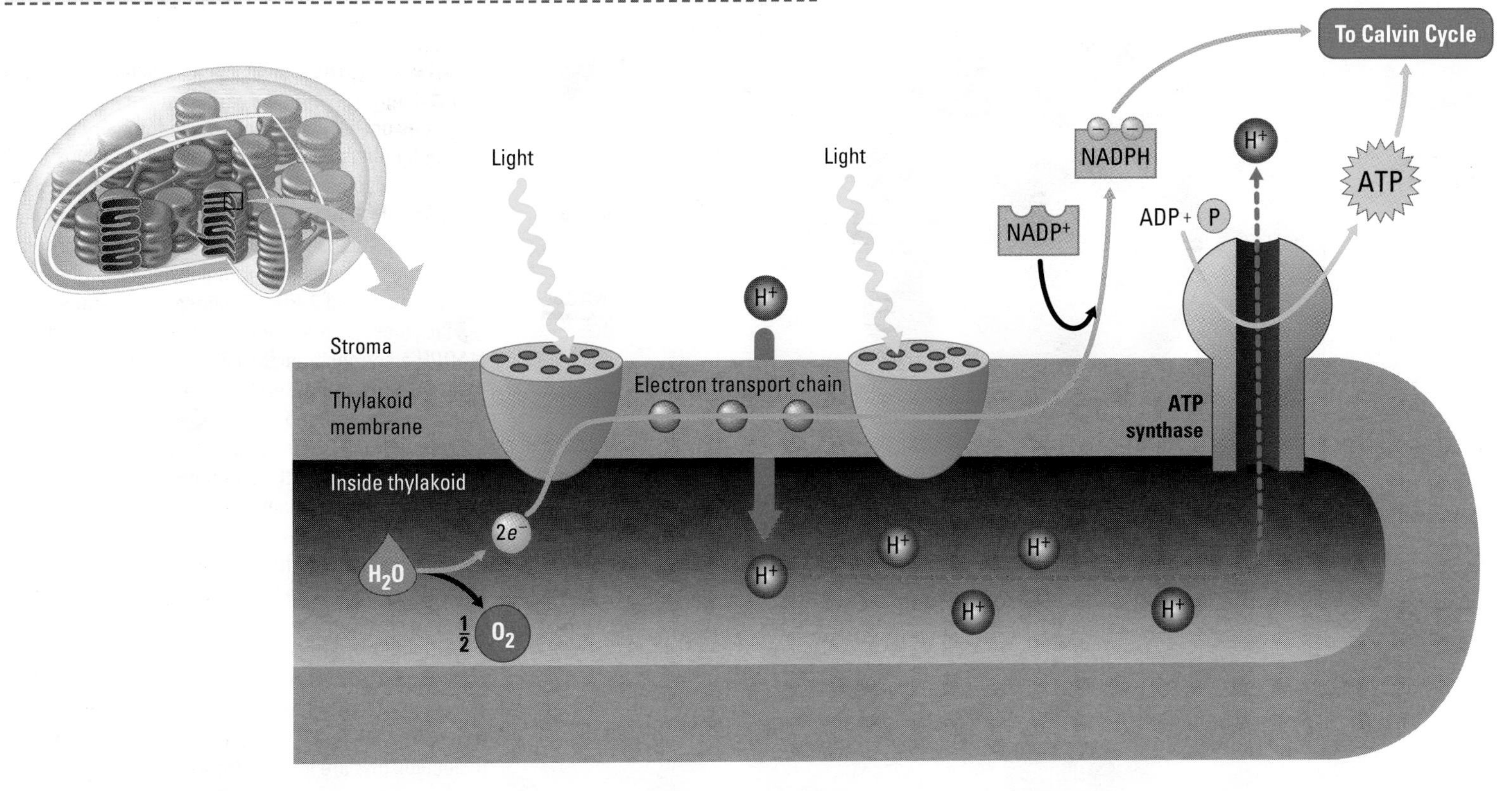

Figure 7.13 How the thylakoid membrane converts light energy to the chemical energy of NADPH and ATP. The two photosystems and the electron transport chain that connects them transfer electrons from H_2O to $NADP^+$, reducing it to NADPH. The electron transport chain also functions as a hydrogen ion (H^+) pump. ATP synthase molecules, much like the ones in mitochondria, use the energy of the H^+ gradient to make ATP.

The Calvin Cycle: Making Sugar from Carbon Dioxide

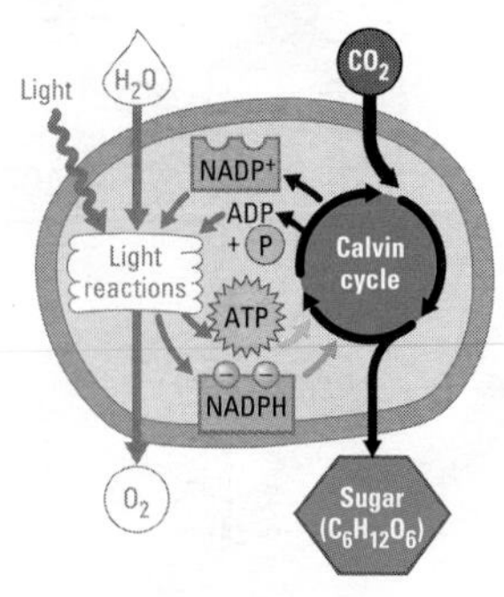

The Calvin cycle functions like a sugar factory within the stroma of a chloroplast. It is called a cycle because, like the citric acid cycle in cellular respiration, the starting material is regenerated with each turn of the cycle (**Figure 7.14**). And with each turn, there are chemical inputs and outputs. The inputs are CO_2 from the air and ATP and NADPH produced by the light reactions. Using carbon from CO_2, energy from ATP, and high-energy electrons from NADPH, the Calvin cycle constructs an energy-rich sugar molecule called glyceraldehyde 3-phosphate (G3P). The plant cell can then use G3P as the raw material to make the glucose and other organic molecules it needs.

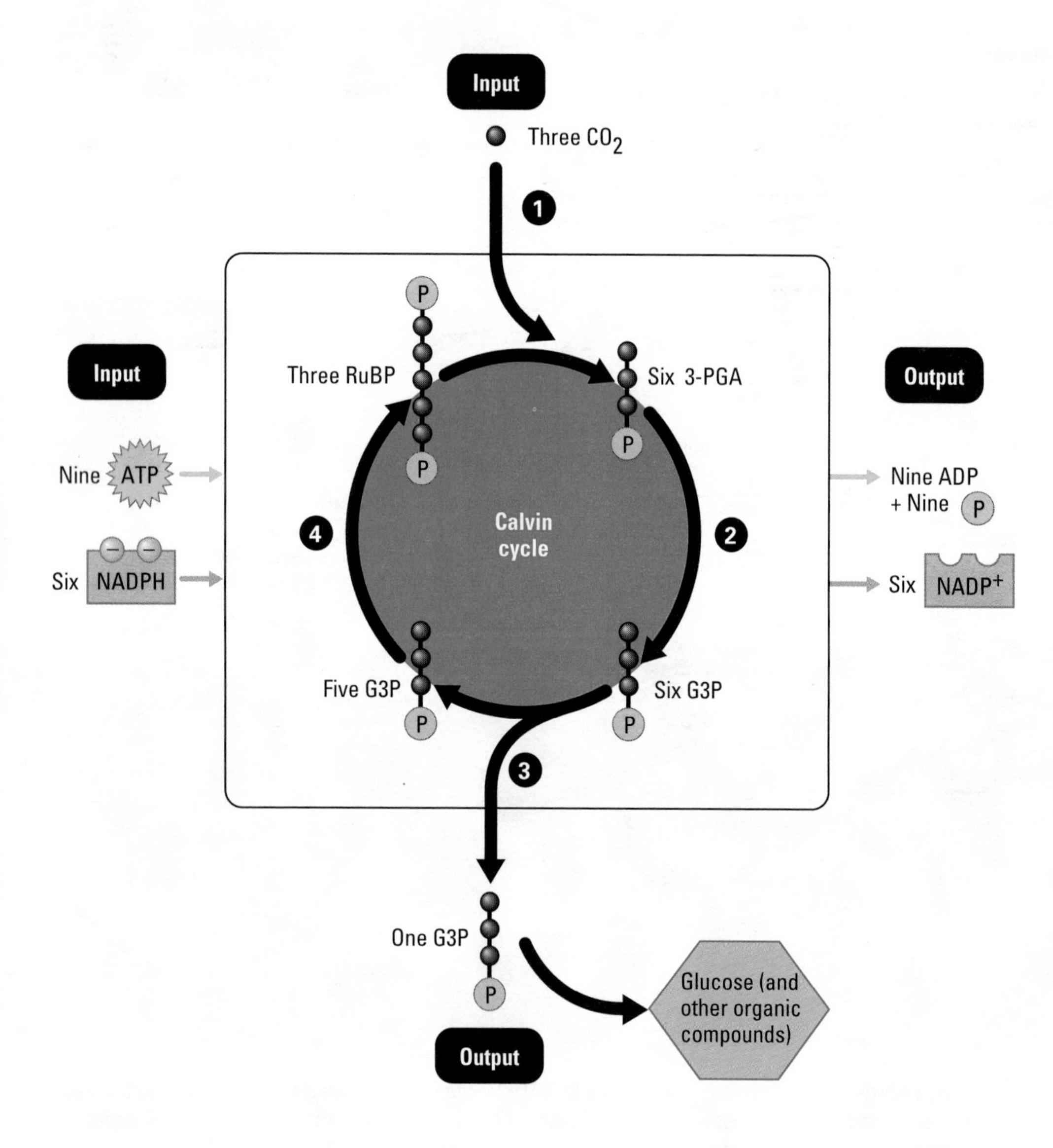

Figure 7.14 The Calvin cycle. The gray balls (●) in this diagram symbolize carbon atoms. The cycle produces one G3P sugar molecule for every three CO_2 molecules that enter the cycle. ❶ Carbon enters the cycle as CO_2. An enzyme adds the CO_2 to RuBP (ribulose bisphosphate), a five-carbon sugar. (In the diagram, Ⓟ represents a phosphate group.) The product then breaks into a three-carbon compound called 3-PGA (3-phosphoglyceric acid). ❷ Enzymes use the ATP energy and high-energy NADPH electrons from the light reactions to convert the 3-PGA to a three-carbon sugar, G3P (glyceraldehyde 3-phosphate). ❸ The cycle has converted three CO_2 molecules to one molecule of the sugar G3P. This is the direct product of photosynthesis, but plant cells can use the G3P to make glucose and other organic compounds for growth and fuel. ❹ The cycle regenerates its starting material. Note that of the six G3P molecules produced in step 3, only one of them represents net sugar output. That's because we started with a total of 15 sugar carbons in the three RuBP molecules that accepted CO_2 back in step 1. Enzymes now regenerate the RuBP by rearranging the five G3P molecules that are left after one of those sugars exits the cycle.

Water-Saving Adaptations of C_4 and CAM Plants

During the long evolution of plants in diverse environments, natural selection has refined photosynthetic adaptations that enable certain plants to continue producing food even in arid conditions.

Plants in which the Calvin cycle uses CO_2 directly from the air are called **C_3 plants** because the first organic compound produced is the three-carbon compound 3-PGA (see Figure 7.14). C_3 plants are common and widely distributed; they include soybeans, oats, wheat, and rice. One of the problems that farmers face in growing C_3 plants, however, is that dry weather can reduce the rate of photosynthesis and decrease crop productivity. On a hot, dry day, plants close their stomata, the pores in the undersurface of a leaf. Closing stomata is an adaptation that reduces water loss, but it also prevents CO_2 from entering the leaf. As a result, CO_2 levels get very low in the leaf, and sugar production ceases.

In contrast to C_3 plants, so-called **C_4 plants** have special adaptations that save water without shutting down photosynthesis. When the weather is hot and dry, a C_4 plant keeps its stomata closed most of the time, thus conserving water. At the same time, it continues making sugars by photosynthesis, using the route shown in **Figure 7.15a**. A C_4 plant has an enzyme that incorporates carbon from CO_2 into a four-carbon (4-C) compound instead of into 3-PGA. This enzyme has an intense affinity for CO_2 and can continue to mine it from the air spaces of the leaf even when the stomata are closed. The four-carbon compound the enzyme produces acts as a carbon shuttle; it donates the CO_2 to the Calvin cycle in a nearby cell, which therefore keeps on making sugars even though the plant's stomata are closed. Corn, sorghum, and sugarcane are examples of agriculturally important C_4 plants. All three evolved in hot regions of the tropics where there are frequent dry seasons.

Another photosynthetic adaptation that conserves water evolved in pineapples, many cacti, and most of the so-called succulent plants (those with very juicy tissues), such as aloe and jade plants. Collectively called **CAM plants,** most such species are adapted to very dry climates. A CAM plant conserves water by opening its stomata and admitting CO_2 mainly at night **(Figure 7.15b)**. When CO_2 enters the leaves, it is incorporated into a four-carbon compound, as in C_4 plants. The four-carbon compound in a CAM plant banks CO_2 at night and releases it to the Calvin cycle during the day. This keeps photosynthesis operating during the day, even though the leaf admits no more CO_2 because the stomata are closed. Note that in all plants—C_3, C_4, and CAM types—it is the Calvin cycle that is ultimately responsible for the synthesis of sugar.

CHECKPOINT

1. In terms of the spatial organization of photosynthesis within the chloroplast, what is the advantage of the light reactions producing NADPH and ATP on the stroma side of the thylakoid membrane?
2. What is the function of NADPH in the Calvin cycle?
3. How do special enzymes enable C_4 and CAM plants to conserve water during photosynthesis?

Answers: 1. The Calvin cycle, which consumes the NADPH and ATP, occurs in the stroma. **2.** It provides the high-energy electrons that are added to CO_2 to form sugar. **3.** By allowing photosynthesis to continue even when stomata are closed during dry conditions

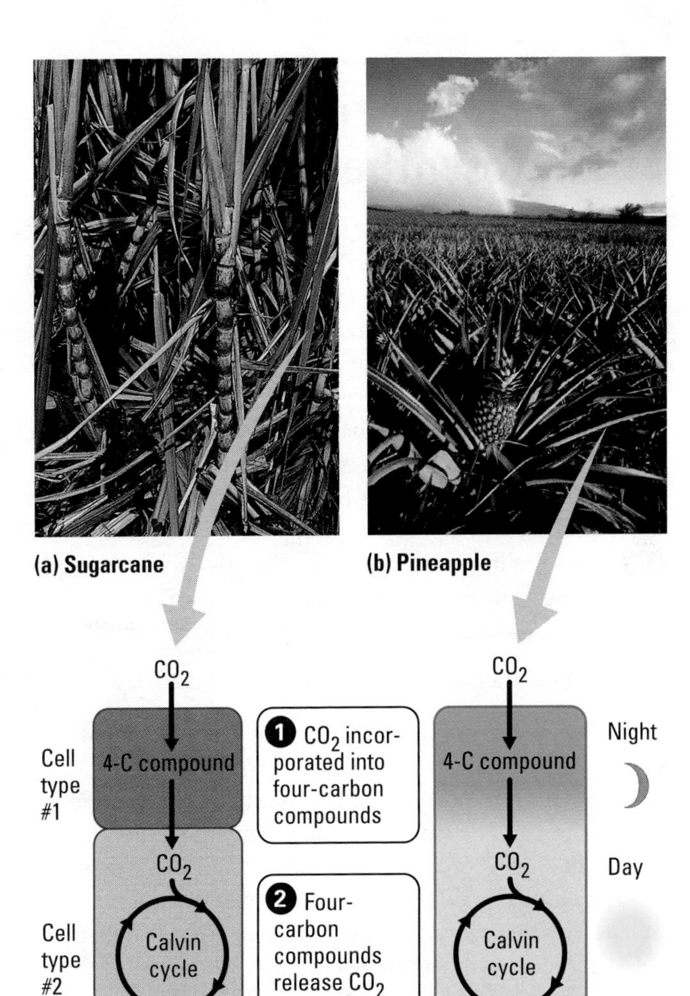

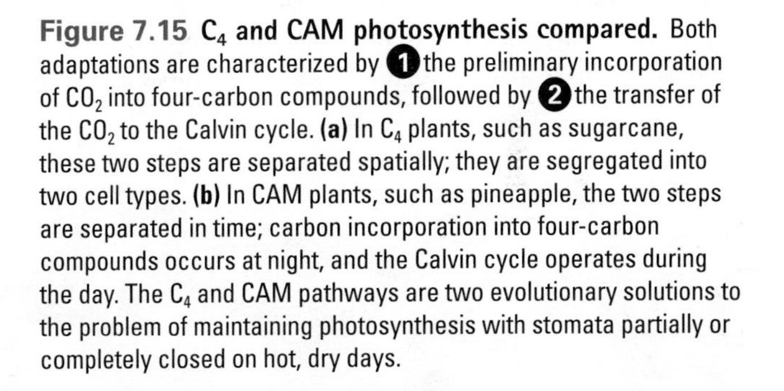

Figure 7.15 C_4 and CAM photosynthesis compared. Both adaptations are characterized by ❶ the preliminary incorporation of CO_2 into four-carbon compounds, followed by ❷ the transfer of the CO_2 to the Calvin cycle. **(a)** In C_4 plants, such as sugarcane, these two steps are separated spatially; they are segregated into two cell types. **(b)** In CAM plants, such as pineapple, the two steps are separated in time; carbon incorporation into four-carbon compounds occurs at night, and the Calvin cycle operates during the day. The C_4 and CAM pathways are two evolutionary solutions to the problem of maintaining photosynthesis with stomata partially or completely closed on hot, dry days.

The Environmental Impact of Photosynthesis

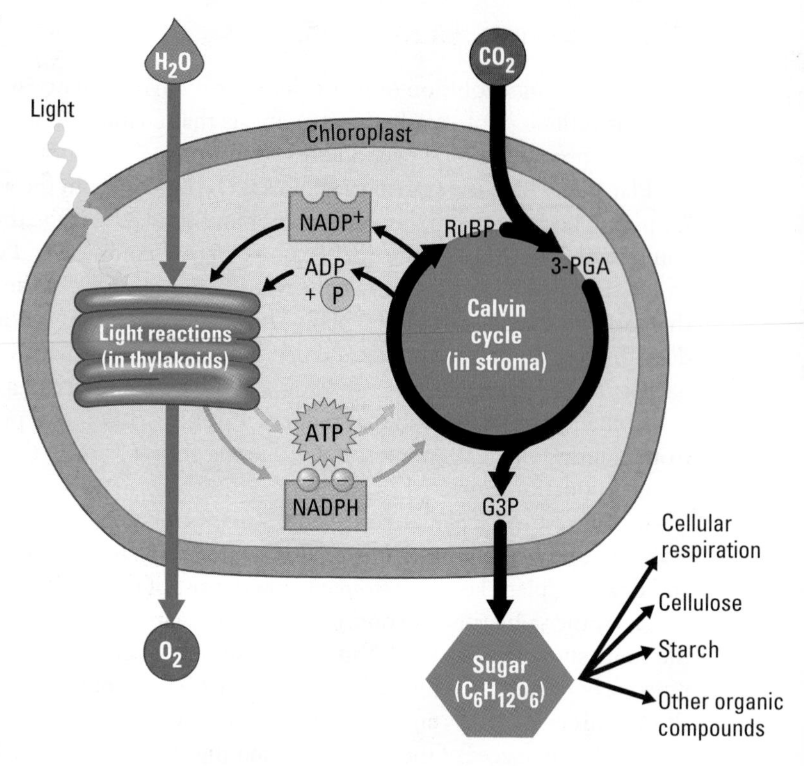

Figure 7.16 A review of photosynthesis.

Figure 7.16 reviews how the light reactions and the Calvin cycle cooperate in converting light energy to the chemical energy of food. What the diagram doesn't show is the transfer of this organic material from the producers (plants and other photosynthetic organisms) to consumers (such as the animals that eat plants). Even the energy we acquire when we eat meat was originally captured by photosynthesis. The energy in a hamburger, for instance, came from sunlight that was originally converted to chemical energy in the chloroplasts of grasses eaten by cattle.

In addition to producing food, photosynthesis also has an enormous impact on the atmosphere by swapping O_2 for CO_2. This O_2 sustains cellular respiration in many organisms. But gas exchange by plants also helps moderate temperatures on Earth, as you'll see next.

How Photosynthesis Moderates Global Warming

The greenhouse in **Figure 7.17** is used to grow plants when the temperature outside is too cold. The glass or plastic walls of a greenhouse allow solar radiation to pass through. The sunlight heats the soil, which in turn warms the air. The walls trap the warm air, raising the temperature inside.

An analogous process, commonly called the **greenhouse effect,** operates on a global scale **(Figure 7.18)**. Solar radiation warms Earth. Heat is radiated by the warmed planet, and some of this radiation is absorbed by gases in the atmosphere, which then radiate some of the heat back to Earth. This natural heating effect is highly beneficial. Without it, Earth would be much colder and less hospitable to life.

The gases in the atmosphere that absorb heat radiation are called **greenhouse gases.** Some occur naturally, such as water vapor, carbon dioxide (CO_2), and methane (CH_4), while others are synthetic, such as chlorofluorocarbons (CFCs, found in some aerosol sprays and refrigerants such as Freon). Carbon dioxide is one of the most important greenhouse

Figure 7.17 Orchids growing in a greenhouse.

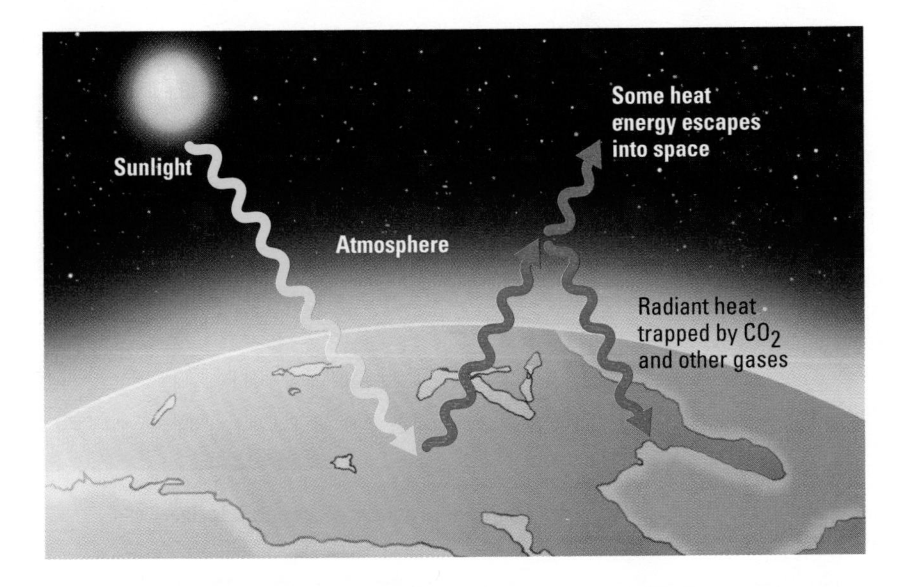

Figure 7.18 **Atmosphere CO_2 and global warming.**

gases. Photosynthetic organisms absorb billions of tons of CO_2 each year. Most of that carbon returns to the atmosphere via cellular respiration, the action of decomposers, and fires. But a substantial amount of it remains locked in large tracts of forests and undecomposed organisms. And large amounts of carbon are in long-term storage in fossil fuels buried deep under Earth's surface.

Before 1850, carbon dioxide was estimated to make up less than 0.03% of the air we breathe. Since the start of the Industrial Revolution, the atmospheric concentration of CO_2 has increased over 35%—from 280 parts per million (ppm) in the late 1700s to 380 ppm in 2006—mostly from the combustion of carbon-based fossil fuels, such as coal, oil, and gasoline. Increasing concentrations of greenhouse gases have been linked to **global warming,** a slow but steady rise in Earth's surface temperature. Predicted changes of just a few degrees over the next 50 years may have dramatic and wide-ranging consequences, including melting of polar ice, rising sea levels, extreme weather patterns, droughts, and the spread of tropical diseases.

Unfortunately, the rise in atmospheric CO_2 levels during the last century has coincided with widespread deforestation. This aggravates the global warming problem by reducing the removal of CO_2 from the atmosphere by forests. As forests are cleared for lumber or agriculture and population growth increases the demand for fossil fuels, CO_2 levels will continue to rise. We will discuss the causes and consequences further in Chapter 20.

What can be done to slow this increase in atmospheric CO_2? Potential solutions include promoting "energy plantations" (discussed in the opening essay), slowing the destruction of forests, and exploring technologies that utilize solar energy. Almost all life on Earth depends on the ability of plants and other photosynthetic organisms to convert light energy to the chemical energy of food molecules. Their contribution to life on Earth may also come to include increased removal of CO_2 from the atmosphere.

CHECKPOINT

How might the combustion of fossil fuels and wood be contributing to global warming?

Answer: 1. By raising concentrations of atmospheric CO_2 and increasing the greenhouse effect

EVOLUTION CONNECTION

The Oxygen Revolution

The atmospheric oxygen we breathe and use for cellular respiration is a by-product of the water-splitting step of photosynthesis. The first photosynthetic organisms with the metabolic equipment to split water were prokaryotes called cyanobacteria **(Figure 7.19)**. They changed Earth forever by adding O_2 to the atmosphere.

Cyanobacteria evolved between 2.7 and 3.5 billion years ago. As their numbers grew, the gradual accumulation of O_2 in the atmosphere created a crisis for other ancient forms of life, because oxygen attacks the bonds of organic molecules. The corrosive oxygen-containing atmosphere probably caused the extinction of many prokaryotic forms unable to cope. Other species survived in habitats that remained anaerobic (such as deep in the soil), where we find their descendants living today. These organisms die if they are exposed to oxygen. Still other species adapted to the changing environment, actually putting the oxygen to use in extracting energy from food, the key process of cellular respiration. The "oxygen revolution" was a major episode in the history of life on Earth. ■

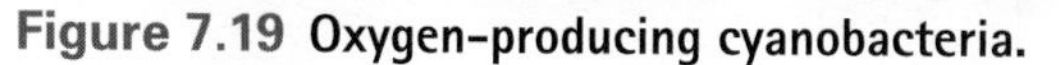
Figure 7.19 **Oxygen-producing cyanobacteria.**

Chapter Review

SUMMARY OF KEY CONCEPTS

For study help and activities, go to campbellbiology.com or the student CD-ROM.

The Basics of Photosynthesis

eTutor *Photosynthesis*

MP3 Tutor *Photosynthesis*

- **Chloroplasts: Sites of Photosynthesis** Chloroplasts contain a thick fluid called stroma surrounding a network of membranes called thylakoids.

Activity *Plants in Our Lives*

Activity *The Sites of Photosynthesis*

- **The Overall Equation for Photosynthesis**

$$6\,CO_2 + 6\,H_2O \xrightarrow[\text{Photosynthesis}]{\text{Light energy}} C_6H_{12}O_6 + 6\,O_2$$

Carbon dioxide, Water, Glucose, Oxygen gas

- **A Photosynthesis Road Map**

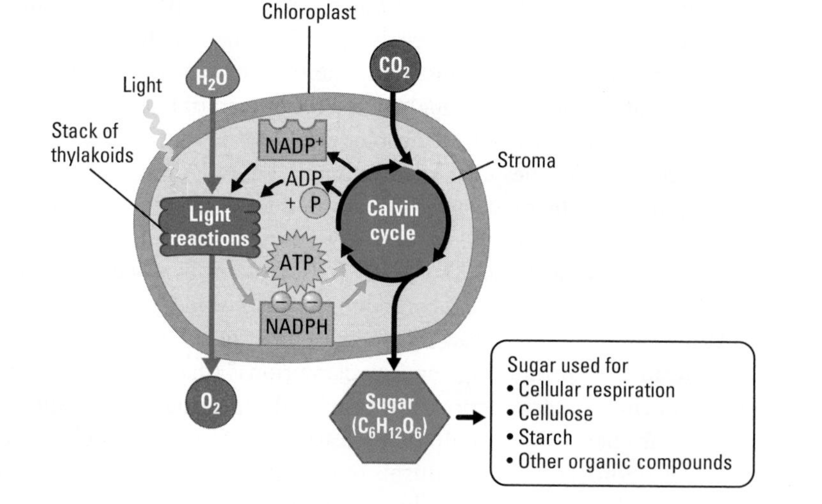

Activity *Overview of Photosynthesis*

The Light Reactions: Converting Solar Energy to Chemical Energy

- **The Nature of Sunlight** Visible light is part of the spectrum of electromagnetic energy (radiation). It travels through space as waves.

• **Chloroplast Pigments** Pigment molecules absorb light energy of certain wavelengths and reflect other wavelengths. We see the reflected wavelengths as the color of the pigment. Several chloroplast pigments absorb light of various wavelengths, but it is the green pigment chlorophyll *a* that participates directly in the light reactions.

Activity *Light Energy and Pigments*

Case Study in the Process of Science *How Does Paper Chromatography Separate Plant Pigments?*

• **How Photosystems Harvest Light Energy; How the Light Reactions Generate ATP and NADPH**

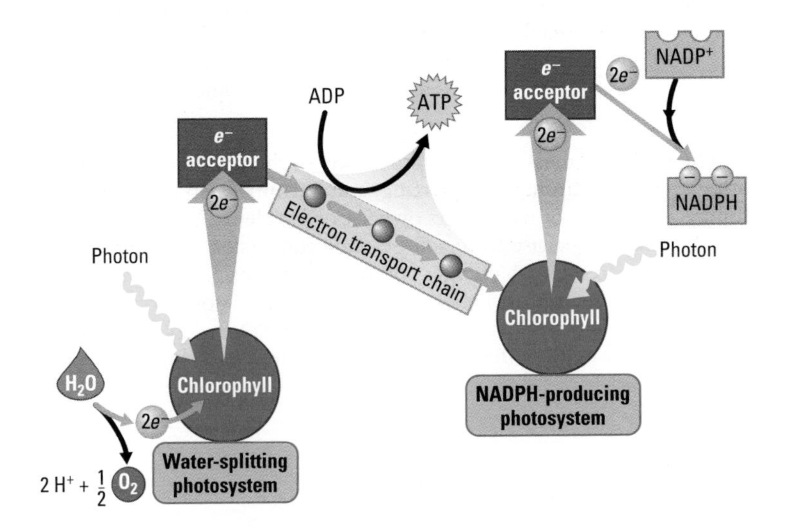

Case Study in the Process of Science *How Is the Rate of Photosynthesis Measured?*

Activity *The Light Reactions*

The Calvin Cycle: Making Sugar from Carbon Dioxide

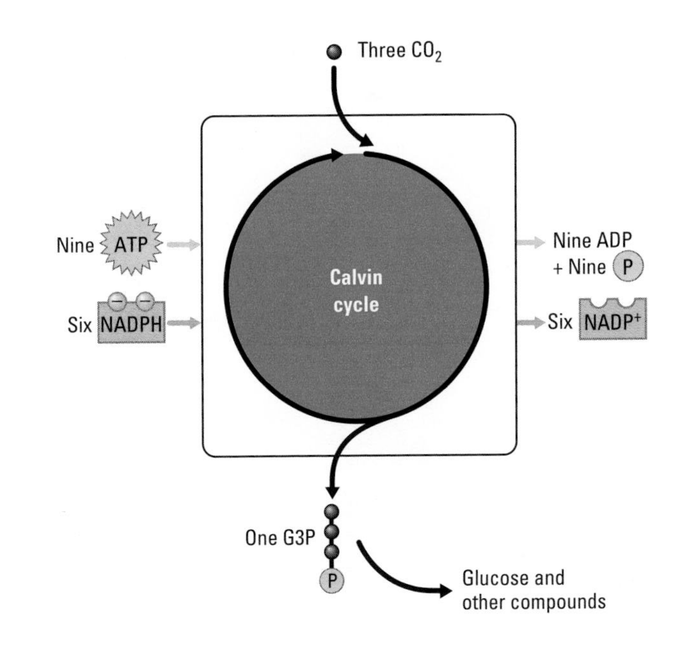

Activity *The Calvin Cycle*

• **Water-Saving Adaptations of C_4 and CAM Plants** Photosynthetic adaptations of C_4 and CAM plants enable sugar production to continue even when stomata are closed, thereby reducing water loss in arid environments.

Activity *Photosynthesis in Dry Climates*

The Environmental Impact of Photosynthesis

• The process of photosynthesis provides organic material and chemical energy for life on Earth. Photosynthesis also swaps O_2 for CO_2 in the atmosphere.

• **How Photosynthesis Moderates Global Warming** Atmospheric CO_2 traps heat and raises our planet's temperature. Photosynthesis makes use of some of this CO_2. Levels of CO_2 in the atmosphere are increasing, raising concerns about global warming. Deforestation and the burning of fossil fuels appear to be contributing to these global changes in atmosphere and climate.

Activity *The Greenhouse Effect*

SELF-QUIZ

1. The light reactions take place in the region of the chloroplast called the __________, while the Calvin cycle takes place in the __________.

2. Which of the following are inputs to photosynthesis? Which are outputs?
   a. CO_2
   b. O_2
   c. sugar
   d. H_2O
   e. light

3. What color of light is the least effective in driving photosynthesis? Why?

4. When light strikes chlorophyll molecules, they lose electrons, which are ultimately replaced by splitting a molecule of __________.

5. Which of the following are produced by reactions that take place in the thylakoids and are consumed by reactions in the stroma?
   a. CO_2 and H_2O
   b. $NADP^+$ and ADP
   c. ATP and NADPH
   d. glucose and O_2

6. The reactions of the Calvin cycle are not directly dependent on light, and yet they usually do not occur at night. Why?

7. Why is it difficult for most plants to carry out photosynthesis in very hot, dry environments, such as deserts?

8. What is the primary advantage offered by the C_4 and CAM pathways?

9. Of the following metabolic processes, which one is common to photosynthesis and cellular respiration?
   a. reactions that convert light energy to chemical energy
   b. reactions that split H_2O molecules and release O_2
   c. reactions that store energy by pumping H^+ across membranes
   d. reactions that convert CO_2 to sugar

10. The combustion of fossil fuels may be contributing to global warming mainly by raising atmospheric concentrations of __________.

Answers to the Self-Quiz questions can be found in Appendix D.

Go to the website or CD-ROM for more Self-Quiz questions.

THE PROCESS OF SCIENCE

11. Tropical rain forests cover only about 3% of Earth's surface, but they are estimated to be responsible for more than 20% of global photosynthesis. For this reason, rain forests are often referred to as the "lungs" of the planet, providing O_2 for life all over Earth. However, most experts believe that rain forests make little or no net contribution to global O_2 production. From your knowledge of photosynthesis and cellular respiration, can you explain why they might think this? (*Hint*: What happens to the food produced by a rain forest tree when it dies or is eaten by animals?)

12. Suppose you wanted to discover whether the oxygen atoms in the glucose produced by photosynthesis come from H_2O or CO_2. Explain how you could use a radioactive isotope to find out.

BIOLOGY AND SOCIETY

13. There is growing evidence that Earth is getting warmer owing to an intensified greenhouse effect resulting from increased CO_2 emissions from industry, vehicles, and the burning of forests. Global warming could influence agriculture and perhaps even melt polar ice and flood coastal regions. In response, 178 countries accepted the Kyoto agreement, which calls for mandatory reductions of greenhouse gas emissions in 30 developed nations by 2012. But in 2002, the Bush administration rejected the Kyoto agreement, instead proposing a more modest set of voluntary goals, allowing businesses to decide whether they wish to participate or not and providing a series of tax incentives to encourage them to do so. President Bush stated that the primary reasons for rejecting the agreement were that it would hurt the American economy and that some developing countries (such as India) were exempted from it, even though they produce a lot of pollution. Do you agree with the administration's decision? In what ways might efforts to reduce greenhouse gases hurt the economy? How can those costs be weighed against the costs of global warming? Do you think that poorer, less developed nations should carry an equal burden to reduce their emissions?

14. The use of biomass energy avoids many of the problems associated with gathering, refining, transporting, and burning fossil fuels. Yet biomass energy is not without its own set of problems. What challenges do you think would arise from a large-scale conversion to biomass energy? How do these challenges compare with those encountered with fossil fuels? Which set of challenges do you think is more likely to be eventually overcome? Do you think any one type of energy has more benefits and fewer costs than the others? Which one, and why?

UNIT 2 Genetics

8 Cellular Reproduction: Cells from Cells

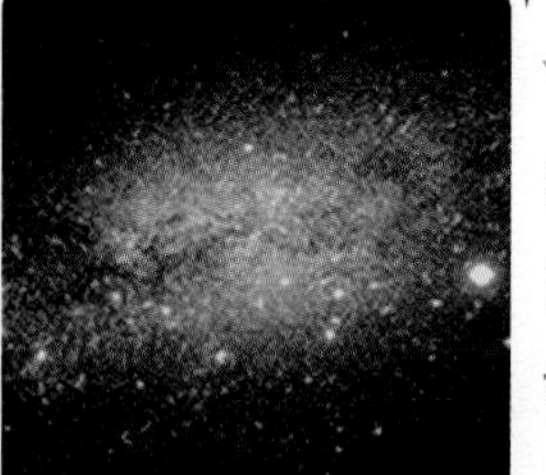

You began life as a single cell, but there are now more cells in your body than **stars in the Milky Way.**

Half of all Americans will **develop cancer** during their lifetime.

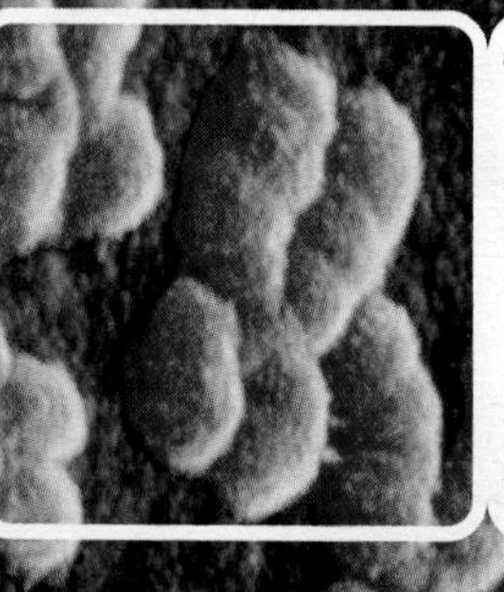

The **dance of the chromosomes** in a dividing cell is so precise that only one error occurs in 100,000 cell divisions.

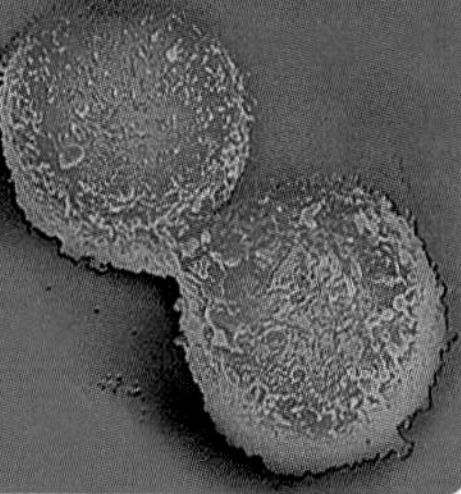

Just in the past second, **millions of your cells** have divided in two.

BIOLOGY AND SOCIETY

A $50,000 Egg!

A few years ago, an advertisement appeared in several Ivy League college newspapers with this heading: "Egg Donor Needed—Large Financial Incentive." The ad sought a woman between the ages of 21 and 32 who was tall, athletic, healthy, and had scored high on her SATs. If such a woman were willing to "donate" some of her eggs, the couple who placed the ad would pay her $50,000. The ad caught the attention of the national media because of the large sum of money involved. "Sterile Couple Seeks Ivy Dream Girl's Eggs" read one tabloid headline. While the amount of money offered is unusual, this type of transaction is actually fairly common in our society.

Infertility, the inability to produce children after one full year of trying, affects one in ten American couples. Many infertile couples turn to in vitro fertilization (IVF), a laboratory procedure that involves joining sperm and egg in a petri dish, allowing the fertilized egg to grow into an eight-cell embryo, and then implanting the embryo into the woman's uterus (**Figure 8.1**). A woman may provide her own eggs for an IVF procedure or she may obtain them from an egg donor. When seeking a donor, some women have a friend or relative willing to provide eggs, but most don't; those that don't have to obtain them from a stranger. Because the medical procedure for removing eggs involves some pain and risk for the donor, the demand for donated eggs far outstrips the supply. Since the 1990s, increasingly large sums of money have been offered to potential donors in an attempt to increase the supply of available eggs.

The economics of egg "donation" raise ethical and legal questions that our society must face. Are we comfortable with the idea of buying and selling human eggs? How much money is enough—or too much? Because IVF is an expensive procedure, couples who seek eggs are usually financially secure. The women paid to donate eggs, on the other hand, are usually not. Do we wish to encourage a system in which poor women provide eggs to rich women?

Whether a result of IVF or natural reproduction, all multicellular organisms start out as a single cell that divides into two cells, which divide to make four cells, then eight, and so on. The perpetuation of life depends on the production of new cells. In this chapter, we'll look at how individual cells reproduce and then see how cell reproduction underlies the process of sexual reproduction. ■

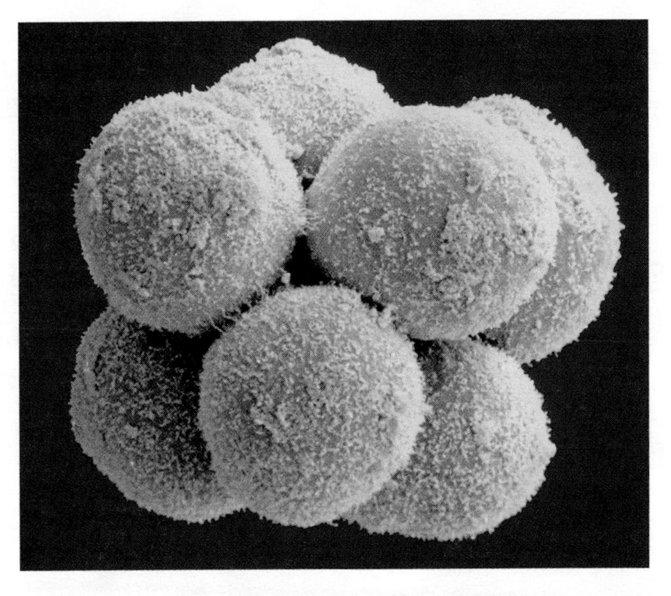

Figure 8.1 An eight-cell human embryo. This human embryo was created through in vitro fertilization. The original cell of the embryo has divided three times to form eight cells. These cells will continue to divide and then specialize, developing into a baby, which eventually grows to be an adult.

What Cell Reproduction Accomplishes

When you hear the word *reproduction*, you probably think of the birth of new organisms. But reproduction actually occurs much more often at the cellular level. Consider the skin on your arm. The surface is a protective layer of dead cells, but underneath are layers of living cells busy carrying out the chemical reactions you studied in Unit One. The living cells of your skin are also engaged in another vital activity: They are reproducing themselves. The new cells are moving outward toward the skin's surface, replacing dead cells that have rubbed off. This renewal of your skin goes on throughout your life. And when your skin is injured, additional cell reproduction helps heal the wound.

Within your body, millions of cells must divide every second to maintain a total number of about 60 trillion cells. The replacement of lost or damaged cells is just one of the important roles that cell reproduction, or

cell division, plays in your life. Another function of cell division is growth. All of the trillions of cells in your body result from repeated cell divisions that began in your mother's body with a single fertilized egg cell.

Passing On Genes from Cell to Cell

When a cell divides, the two "daughter" cells that result are genetically identical to each other and to the original "parent" cell. (By convention, biologists use the word *daughter* in this context; it does not imply gender.) Before the parent cell splits into two, it duplicates its **chromosomes,** the DNA-containing structures that carry the organism's genetic legacy. Then, during the division process, one set of chromosomes is distributed to each daughter cell. As a rule, the daughter cells receive identical sets of chromosomes with identical genes.

The Reproduction of Organisms

Some organisms reproduce by simple cell division. Single-celled organisms, such as amoebas, reproduce this way, and the offspring are genetic replicas of the parent **(Figure 8.2a).** Because it does not involve fertilization of an egg by a sperm, this type of reproduction is called **asexual reproduction.** Offspring produced by asexual reproduction inherit all their chromosomes from a single parent. Many multicellular organisms can reproduce asexually as well. For example, some sea star species have the ability to grow new individuals from fragmented pieces **(Figure 8.2b).** And if you've ever grown a houseplant from a clipping, you've observed asexual reproduction in plants **(Figure 8.2c).** In asexual reproduction, there is one simple principle of inheritance: The lone parent and each of its offspring have identical genes. The type of cell division responsible for asexual reproduction and for the growth and maintenance of multicellular organisms is called mitosis.

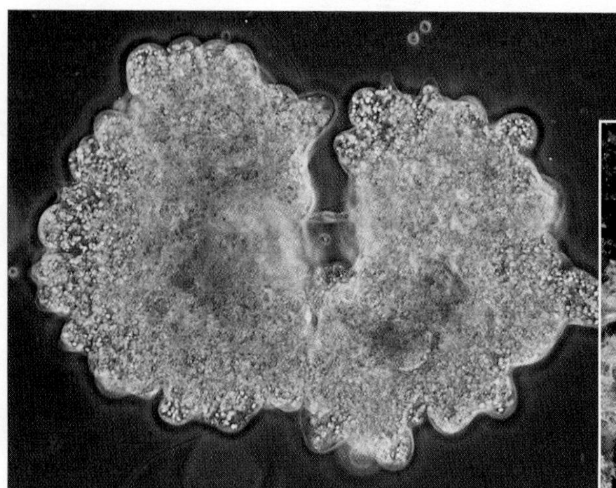

(a) Amoeba. This amoeba, a single-celled organism, is reproducing by dividing in half. Its chromosomes have been duplicated, and the two identical sets of chromosomes have been allocated to opposite sides of the parent. When division is complete, the two daughter amoebas will be genetically identical to each other.

(b) Sea star. If a sea star is split into two pieces, each piece may be able to regrow into a whole new organism. In this photo, one sea star arm is in the process of regenerating the rest of the sea star body.

(c) African violet. Some houseplants, like this African violet, can be grown asexually from a clipping (the large leaf in this photo). The resulting plant will be genetically identical to the plant from which the clipping was taken.

Figure 8.2 Asexual reproduction.

Sexual reproduction is different; it requires fertilization of an egg by a sperm. The production of egg and sperm cells involves a special type of cell division called meiosis, which occurs only in reproductive organs (such as testes and ovaries in humans). As we'll discuss later, a sperm or egg cell has only half as many chromosomes as the parent cell that gave rise to it.

Note that two kinds of cell division are involved in the lives of sexually reproducing organisms: meiosis for reproduction and mitosis for growth and maintenance. The remainder of the chapter is divided into two main sections, one for each type of cell division.

CHECKPOINT

Ordinary cell division produces two daughter cells that are genetically identical. Name three functions of this type of cell division.

Answer: Cell replacement, growth of an organism, asexual reproduction of an organism

The Cell Cycle and Mitosis

eTutor
Mitosis

MP3 Tutor
Mitosis

Almost all of the genes of a eukaryotic cell—around 25,000 genes in humans—are located on chromosomes in the cell nucleus. (The main exceptions are genes on small DNA molecules found in mitochondria and chloroplasts.) Because chromosomes are the lead players in cell division, let's focus on them before turning our attention to the cell as a whole.

Eukaryotic Chromosomes

Each eukaryotic chromosome contains one very long DNA molecule, typically bearing thousands of genes. The number of chromosomes in a eukaryotic cell depends on the species **(Figure 8.3)**. (Notice that the number of chromosomes does not correspond to the size or complexity of an organism.) Chromosomes are made up of a material called **chromatin,** a combination of DNA and protein molecules. The protein molecules help organize the chromatin and help control the activity of its genes.

Most of the time, the chromosomes exist as a diffuse mass of fibers that are much longer than the nucleus they are stored in. In fact, the total DNA in a single human cell's 46 chromosomes could stretch for over 2 meters! As a cell prepares to divide, its chromatin fibers coil up, forming compact chromosomes. When they are in this state, chromosomes are clearly visible under the light microscope, as shown in the plant cell in **Figure 8.4** (each dark purple thread is an individual chromosome). When a cell is not dividing, the chromosomes are too thin to be clearly seen in a light micrograph.

Such long molecules of DNA can fit into the tiny nucleus because within each chromosome the DNA is packed into an elaborate, multilevel system of coiling and folding. A crucial aspect of DNA packing is the association of the DNA with small proteins called **histones,** found only in eukaryotes. (Bacteria have analogous proteins, but prokaryotes lack the degree of DNA packing found in eukaryotes.)

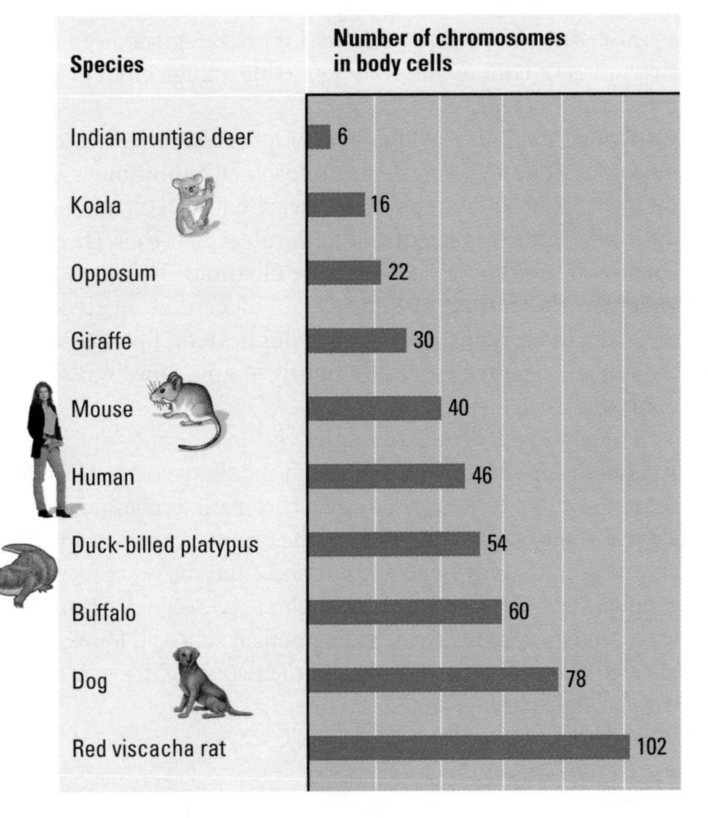

Figure 8.3 The number of chromosomes in the cells of selected mammals.

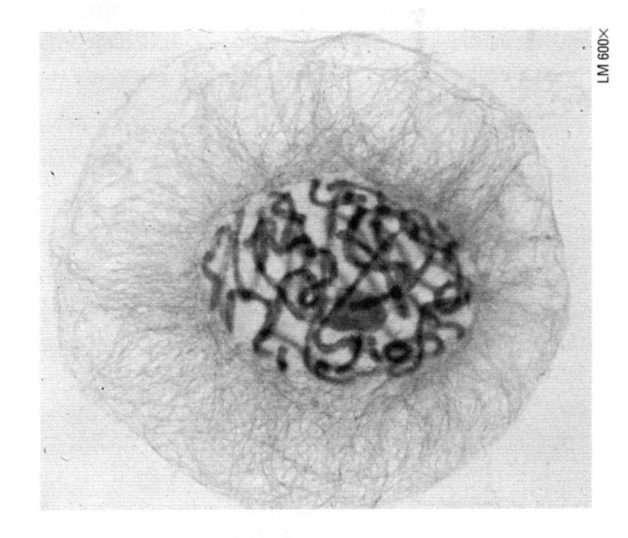

Figure 8.4 A plant cell just before division. The purple threads are the chromosomes (the colors result from staining). The thinner red threads in the surrounding cytoplasm are the cytoskeleton.

Figure 8.5 presents a simplified model for the main levels of DNA packing. At the first level of packing, shown near the top, histones attach to the DNA. In electron micrographs, the combination of DNA and histones has the appearance of beads on a string. Each "bead," called a **nucleosome,** consists of DNA wound around histone molecules. At the next level of packing, the beaded string is wrapped into a tight helical fiber. Then this fiber coils further into a thick supercoil. Looping and folding can further compact the DNA, as you can see in the chromosome at the bottom of the figure. Viewed as a whole, Figure 8.5 gives a sense of how successive levels of coiling and folding enable a huge amount of DNA to fit into a cell nucleus.

Before a cell begins the division process, it duplicates all of its chromosomes. The DNA molecule of each chromosome is copied through the process of DNA replication (see Chapter 10), and new protein molecules attach as needed. The result is that each chromosome now consists of two copies called **sister chromatids,** which contain identical genes. At the bottom of Figure 8.5, you can see an electron micrograph of a human chromosome that has duplicated. The two sister chromatids are joined together especially tightly at a narrow "waist" called the **centromere.**

When the cell divides, the sister chromatids of a duplicated chromosome separate from each other, as shown in the simple diagram in **Figure 8.6**. Once separated from its sister, each chromatid is considered a full-fledged chromosome, and it is identical to the original chromosome. One of the new chromosomes goes to one daughter cell, and the other goes to the other daughter cell. In this way, each daughter cell receives a complete and identical set of chromosomes. A dividing human skin cell, for example, has 46 duplicated chromosomes, and each of the two daughter cells that result from it has 46 single chromosomes.

The Cell Cycle

How do chromosome duplication and cell division fit into the life of a cell? The rate at which a cell divides depends on its role within the organism's body. Some cells divide once a day, others less often, and highly specialized cells, such as mature muscle cells, not at all. Eukaryotic cells that do divide undergo a **cell cycle,** an orderly sequence of events that extends from the time a cell first arises until it divides.

As **Figure 8.7** shows, most of the cell cycle is spent in **interphase.** This is a time when a cell performs its normal functions within the organism. For example, a cell in your stomach lining might make and release enzyme molecules that aid in digestion. During interphase, a cell roughly doubles everything in its cytoplasm. It increases its supply of proteins, increases the number of many of its organelles (such as mitochondria and ribosomes), and grows in size. Typically, interphase lasts for at least 90% of the cell cycle.

From the standpoint of cell reproduction, the most important event of interphase is chromosome duplication, when the DNA in the nucleus is precisely doubled. This occurs approximately in the middle of interphase, and the period when it is occurring is called the S phase (for DNA *synthesis*). The interphase periods before and after the S phase are called

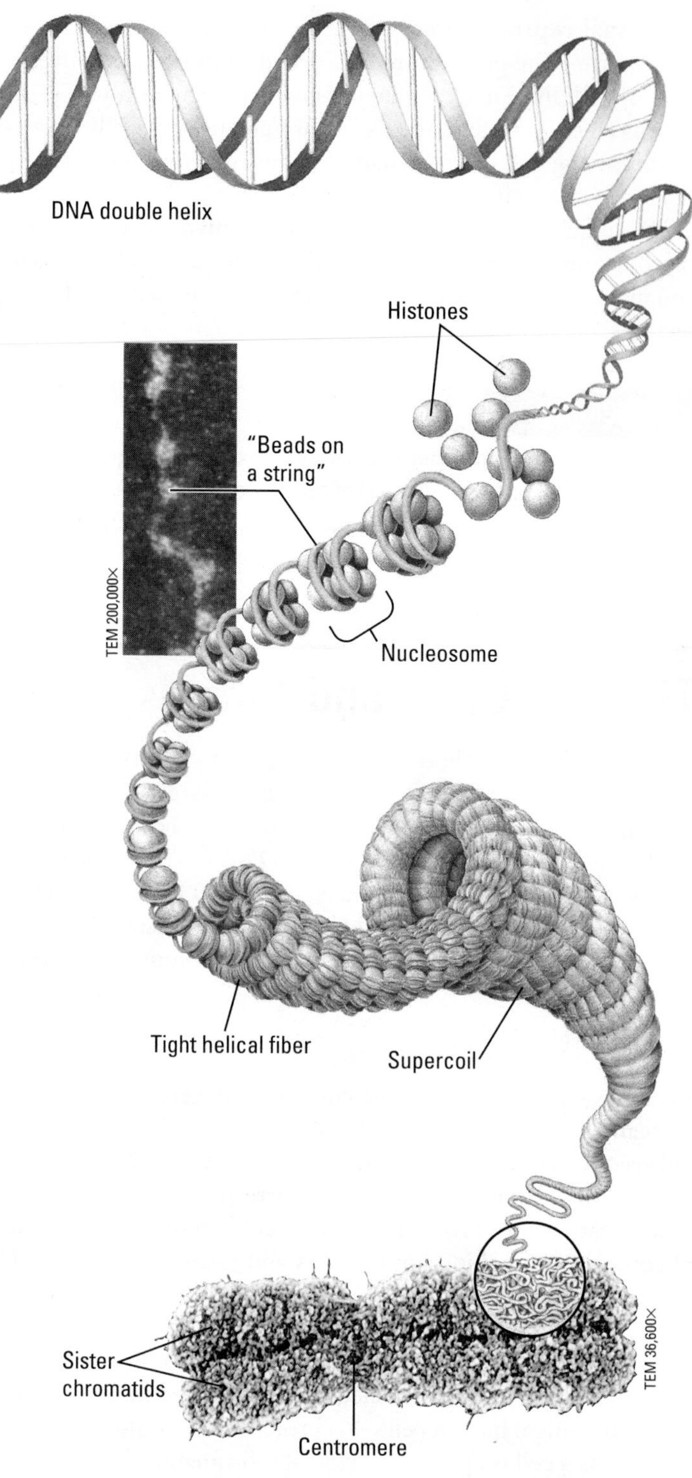

Figure 8.5 DNA packing in a eukaryotic chromosome. Successive levels of coiling of DNA and associated proteins ultimately results in highly condensed chromosomes. When not dividing, the DNA of active genes is only lightly packed, in the "beads on a string" arrangement. At the bottom of the figure, you can see a highly compacted chromosome from a cell preparing to divide. The constricted region is the centromere. The fuzzy appearance comes from the intricate twists and folds of the chromatin fibers.

the G_1 and G_2 phases, respectively (G stands for *gap*). During G_2, each chromosome in the cell consists of two identical sister chromatids, and the cell is preparing to divide.

The part of the cell cycle when the cell is actually dividing is called the **mitotic phase** (M phase). It includes two overlapping processes, mitosis and cytokinesis. In **mitosis,** the nucleus and its contents, notably the duplicated chromosomes, divide and are evenly distributed, forming two daughter nuclei. In **cytokinesis,** the cytoplasm is divided in two. Cytokinesis usually begins before mitosis is completed. The combination of mitosis and cytokinesis produces two genetically identical daughter cells, each with a single nucleus, surrounding cytoplasm with organelles, and a plasma membrane.

Mitosis is a remarkably accurate mechanism for allocating identical copies of a large amount of genetic material to two daughter cells. Experiments with yeast cells, for example, indicate that an error in chromosome distribution occurs only once in about 100,000 cell divisions. Mitosis is unique to eukaryotes. Prokaryotes have only a single small chromosome (see Chapter 10) and use a simpler mechanism for allocating DNA to daughter cells.

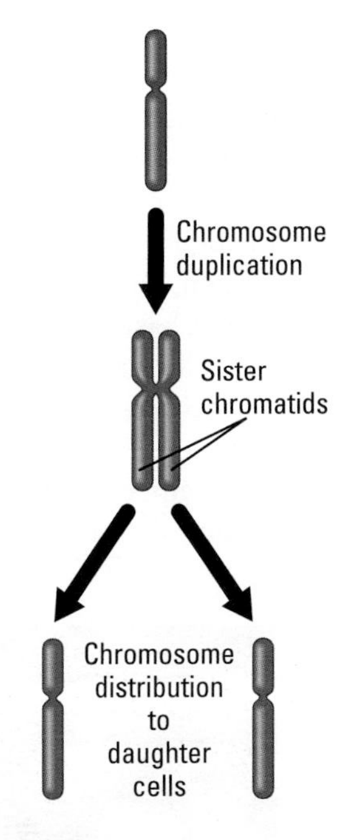

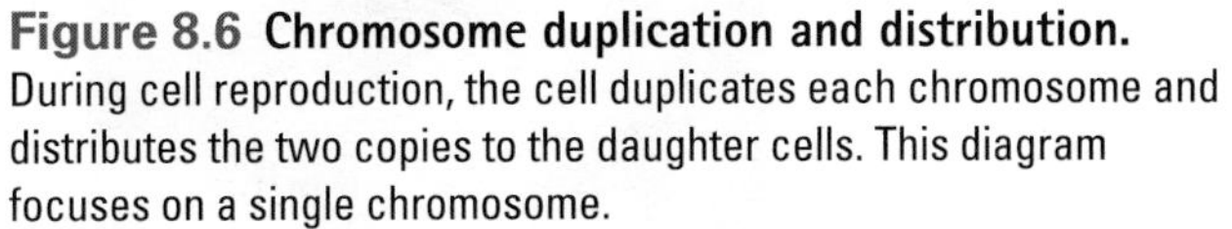

Figure 8.6 Chromosome duplication and distribution. During cell reproduction, the cell duplicates each chromosome and distributes the two copies to the daughter cells. This diagram focuses on a single chromosome.

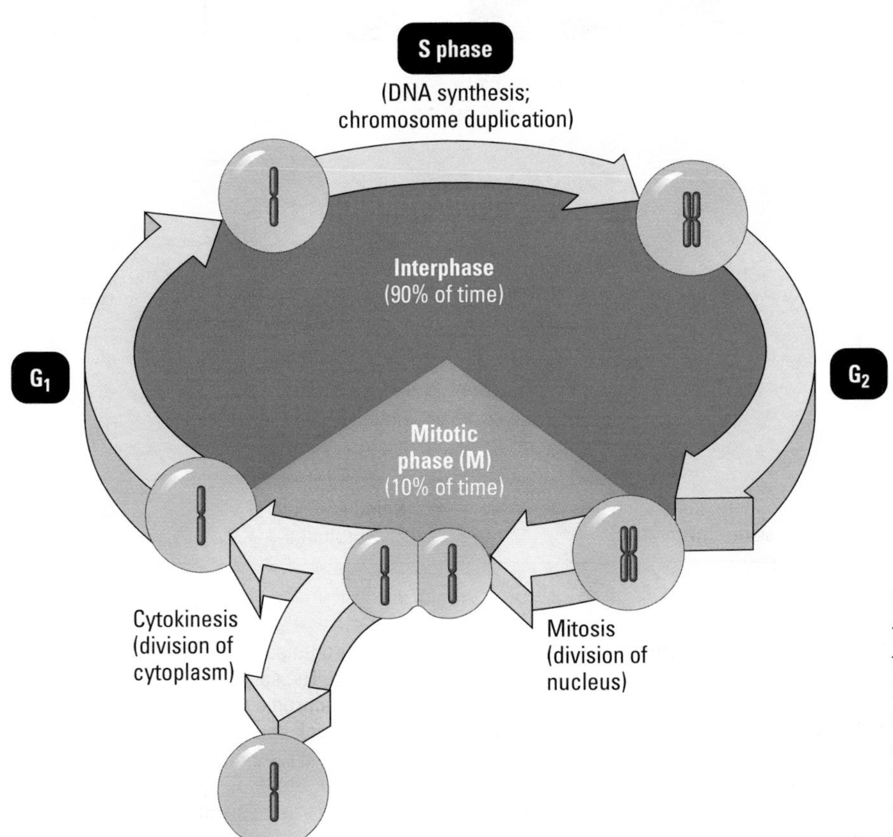

Figure 8.7 The eukaryotic cell cycle. The cell cycle extends from the "birth" of a cell, resulting from cell reproduction, to the time the cell itself divides in two. The cell spends most of the cycle in interphase. Duplication of the chromosomes occurs during the S phase. The cell metabolizes and grows throughout interphase. The actual division process occurs during the mitotic phase, which includes mitosis and cytokinesis. (During interphase, the chromosomes are diffuse masses of thin fibers; they do not actually appear in the rodlike form you see here.)

Mitosis and Cytokinesis

The light micrographs in **Figure 8.8** show the cell cycle for an animal cell, with most of the figure devoted to the mitotic phase. With the onset of mitosis, striking changes are visible in the nucleus and other cellular structures. The text under the figure describes the events occurring at each stage. Mitosis is a continuum, but biologists distinguish four main stages: **prophase, metaphase, anaphase,** and **telophase.**

The chromosomes are the stars of the mitotic drama, and their movements depend on the **mitotic spindle,** a football-shaped structure of microtubules that guides the separation of the two sets of daughter chromosomes. The spindle microtubules grow from two **centrosomes,** clouds of cytoplasmic material that in animal cells contain centrioles.

Figure 8.8 Cell reproduction: A dance of the chromosomes. After the chromatin doubles during interphase, the elaborately choreographed stages of mitosis—prophase, metaphase, anaphase, and telophase—distribute the duplicate sets of chromosomes to two separate nuclei. Cytokinesis then divides the cytoplasm, yielding two genetically identical daughter cells. The micrographs here show cells from a newt. The drawings include details not visible in the micrographs. For simplicity, only four chromosomes appear in the drawings.

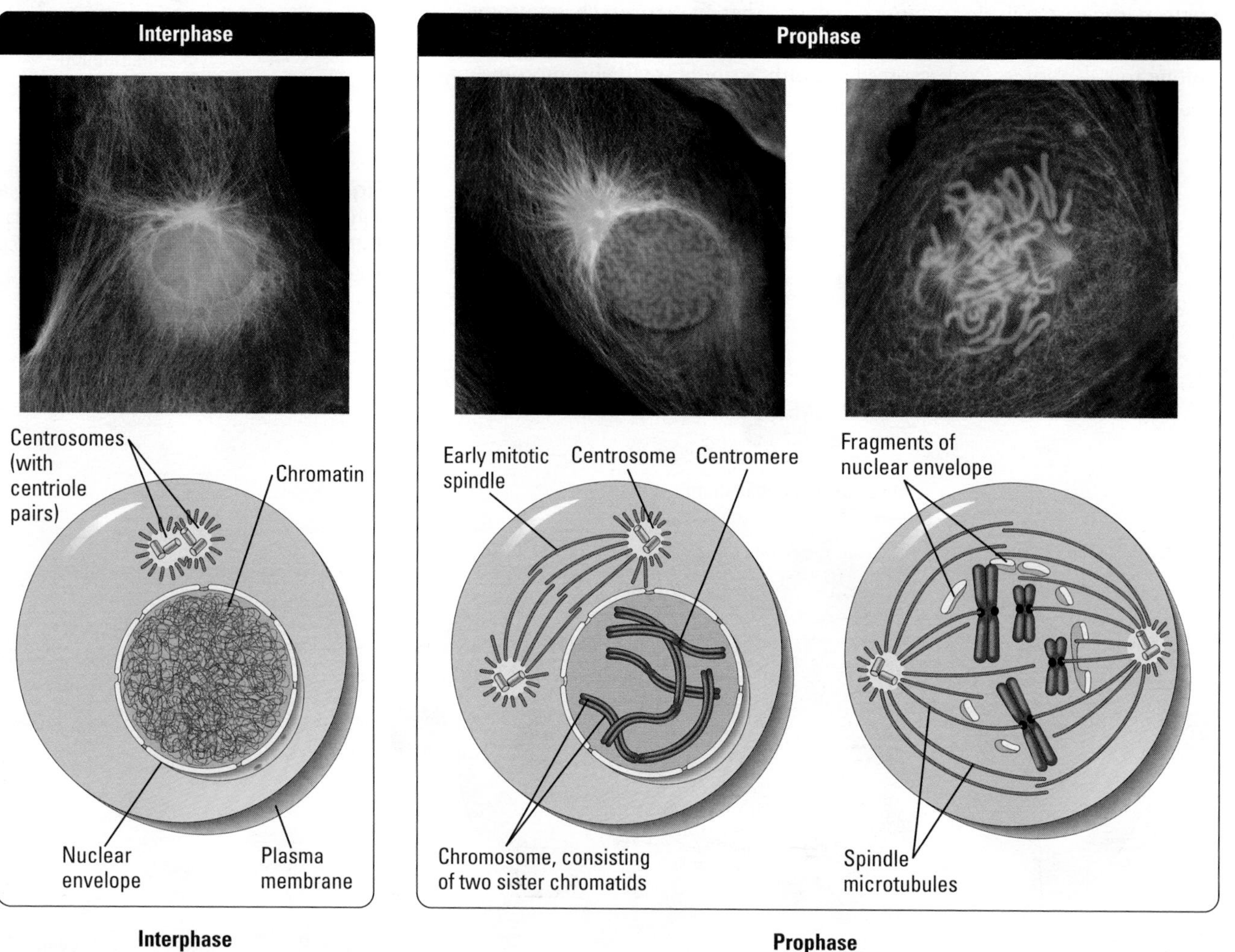

Interphase

Interphase is the period of cell growth, when the cell makes new molecules and organelles. At the point shown here, late interphase (G_2), the cytoplasm contains two centrosomes. Within the nucleus, the chromosomes are duplicated, but they cannot be distinguished individually because they are still in the form of loosely packed chromatin fibers.

Prophase

During prophase, changes occur in both nucleus and cytoplasm. In the nucleus, the chromatin fibers coil, so that the chromosomes become thick enough to be seen with the light microscope. Each chromosome appears as two identical sister chromatids joined together, with a narrow "waist" at the centromere. In the cytoplasm, the mitotic spindle begins to form as microtubules grow out from the centrosomes, which are moving away from each other.

Late in prophase, the nuclear envelope breaks up. The spindle microtubules can now reach the chromosomes, which have become thick and have a protein structure (black dot) at their centromeres. Some of the spindle microtubules capture chromosomes by attaching to these structures, throwing the chromosomes into agitated motion. Other microtubules make contact with microtubules coming from the opposite spindle pole. The spindle moves the chromosomes toward the center of the cell.

(Centrioles are can-shaped structures made of microtubules; see Figure 4.6.)

Cytokinesis, the actual division of the cytoplasm into two cells, typically occurs during telophase. In animal cells, the cytokinesis process is known as cleavage. The first sign of cleavage is the appearance of a **cleavage furrow,** an indentation at the equator of the cell. A ring of microfilaments in the cytoplasm just under the plasma membrane contracts, like the pulling of a drawstring, deepening the furrow and pinching the parent cell in two (**Figure 8.9a**, see page 128).

Cytokinesis in a plant cell occurs differently. Membrane-enclosed vesicles containing cell wall material collect at the middle of the cell. The vesicles gradually fuse, forming a membranous disk called the **cell plate.** The cell plate grows outward, accumulating more cell wall material as more

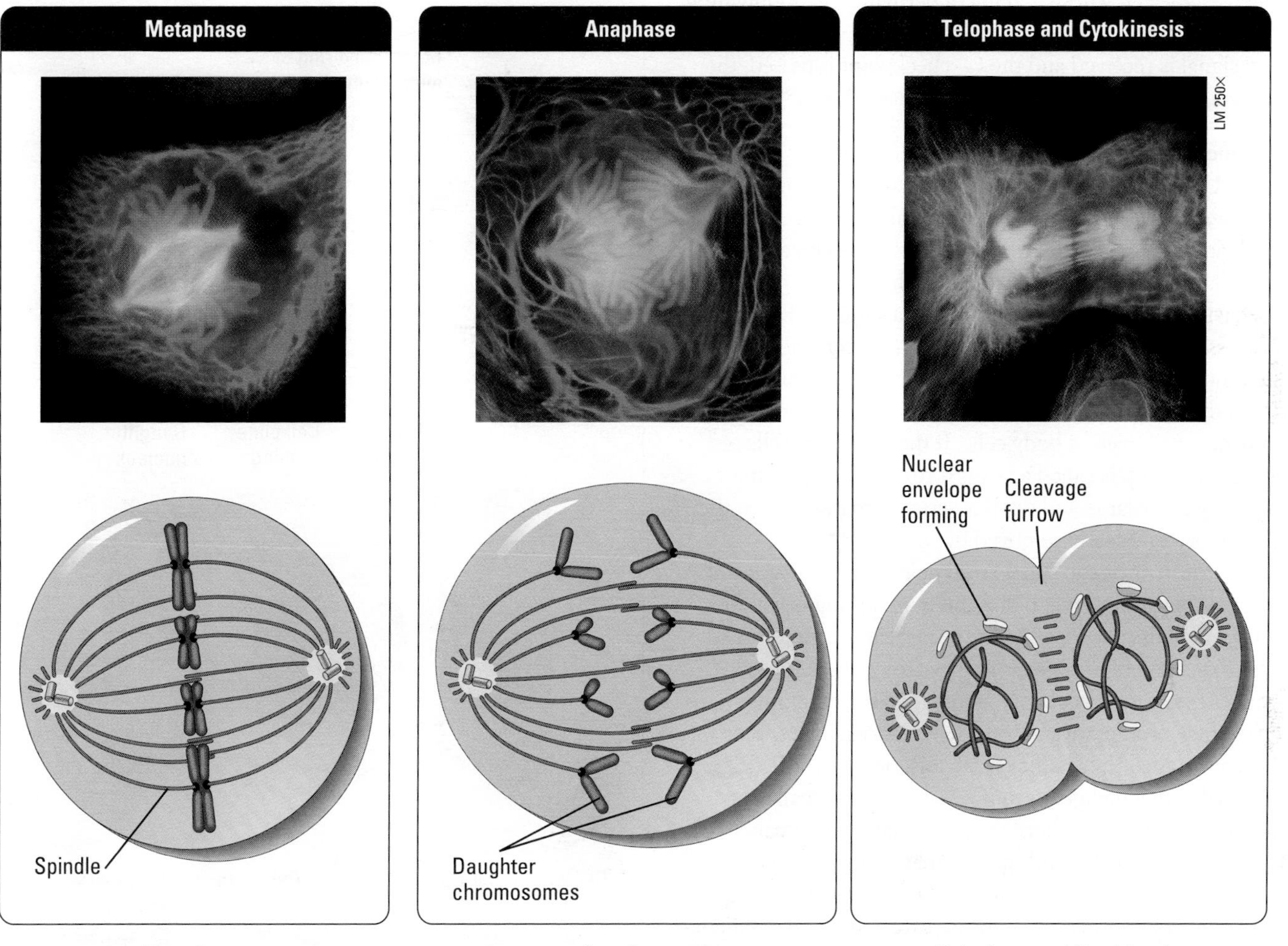

Metaphase

The mitotic spindle is now fully formed. The chromosomes convene on an imaginary plate equidistant from the two poles of the spindle. The centromeres of all the chromosomes are lined up at this plate. For each chromosome, the spindle microtubules attached to the two sister chromatids pull toward opposite poles. This tug of war keeps the chromosomes in the middle of the cell.

Anaphase

Anaphase begins suddenly, when the sister chromatids of each chromosome separate. Each is now considered a full-fledged (daughter) chromosome. Motor proteins at the centromeres "walk" the daughter chromosomes along their microtubules toward opposite poles of the cell (see motor proteins in Figure 5.6a). Meanwhile, these microtubules shorten. However, the microtubules *not* attached to chromosomes lengthen, pushing the poles farther apart and elongating the cell.

Telophase and Cytokinesis

Telophase begins when the two groups of chromosomes have reached the cell poles. Telophase is the reverse of prophase: Nuclear envelopes form, the chromosomes uncoil, and the spindle disappears. Mitosis, the division of one nucleus into two genetically identical daughter nuclei, is now finished.

Cytokinesis, the division of the cytoplasm, usually occurs with telophase. In animals, a cleavage furrow pinches the cell in two, producing two daughter cells.

vesicles join it. Eventually, the membrane of the cell plate fuses with the plasma membrane, and the cell plate's contents join the parental cell wall, resulting in two daughter cells **(Figure 8.9b).**

Cancer Cells: Growing Out of Control

For a plant or animal to grow and maintain its tissues normally, it must be able to control the timing of cell division. The sequential events of the cell cycle are directed by a **cell cycle control system** that consists of special proteins within the cell. These proteins integrate information from the environment and from other body cells and send "stop" and "go-ahead" signals at certain key points during the cell cycle using signal transduction pathways (see Figure 5.20). For example, the cell cycle normally halts within the G_1 phase of interphase unless the cell receives a go-ahead signal via certain cell cycle control system proteins. If that signal never arrives, the cell will switch into a permanently nondividing state called G_0. Some of our nerve and muscle cells, for example, are arrested at G_0. If the go-ahead signal is received and the G_1 checkpoint is passed, the cell will usually complete the rest of the cycle.

What Is Cancer? Cancer, which currently claims the lives of one out of every five people in the United States and other industrialized nations, is a disease of the cell cycle. Cancer cells do not respond normally to the cell cycle control system; they divide excessively and can invade other tissues of the body. If unchecked, cancer cells may continue to divide until they kill the organism.

The abnormal behavior of cancer cells begins when a single cell undergoes transformation, a process that converts a normal cell to a cancer cell. The body's immune system normally recognizes a transformed cell as abnormal and destroys it. However, if the cell evades destruction, it may proliferate to form a **tumor,** an abnormally growing mass of body cells. If the abnormal cells remain at the original site, the lump is called a **benign tumor.** Benign tumors can cause problems if they grow large and disrupt certain organs, such as the brain, but often they can be completely removed by surgery.

In contrast, a **malignant tumor** can spread into neighboring tissues and other parts of the body, displacing normal tissue and interrupting organ function **(Figure 8.10).** Cancer cells may separate from the original tumor or secrete signal molecules that cause blood vessels to grow toward the tumor. A few tumor cells may then enter the circulatory system and move to other parts of the body, where they may proliferate and form new tumors. The spread of cancer cells beyond their original site via the circulatory system is called **metastasis.** An individual with a malignant tumor is said to have **cancer.**

Cancers are named according to where they originate. Liver cancer, for example, always begins in liver tissue and may or may not spread from there. Cancers are grouped into four categories based on their sites of origin. **Carcinomas** are cancers that originate in the external or internal coverings of the body, such as the skin or the lining of the intestine. **Sarcomas** arise in tissues that support the body, such as bone and muscle. Cancers of blood-forming tissues, such as bone marrow and lymph nodes, are called **leukemias** and **lymphomas.**

Cancer Treatment Once a tumor starts growing in the body, how can it be treated? The three main types of cancer treatment are sometimes referred to as "slash, burn, and poison." Surgery to remove a tumor ("slash") is usually the first step. "Burn" and "poison" refer to treatments that attempt to stop cancer cells from dividing. In **radiation therapy** ("burn"), parts of the body that have cancerous tumors are exposed to high-energy radiation, which damages DNA and disrupts cell division. Because cancer cells divide more often than most

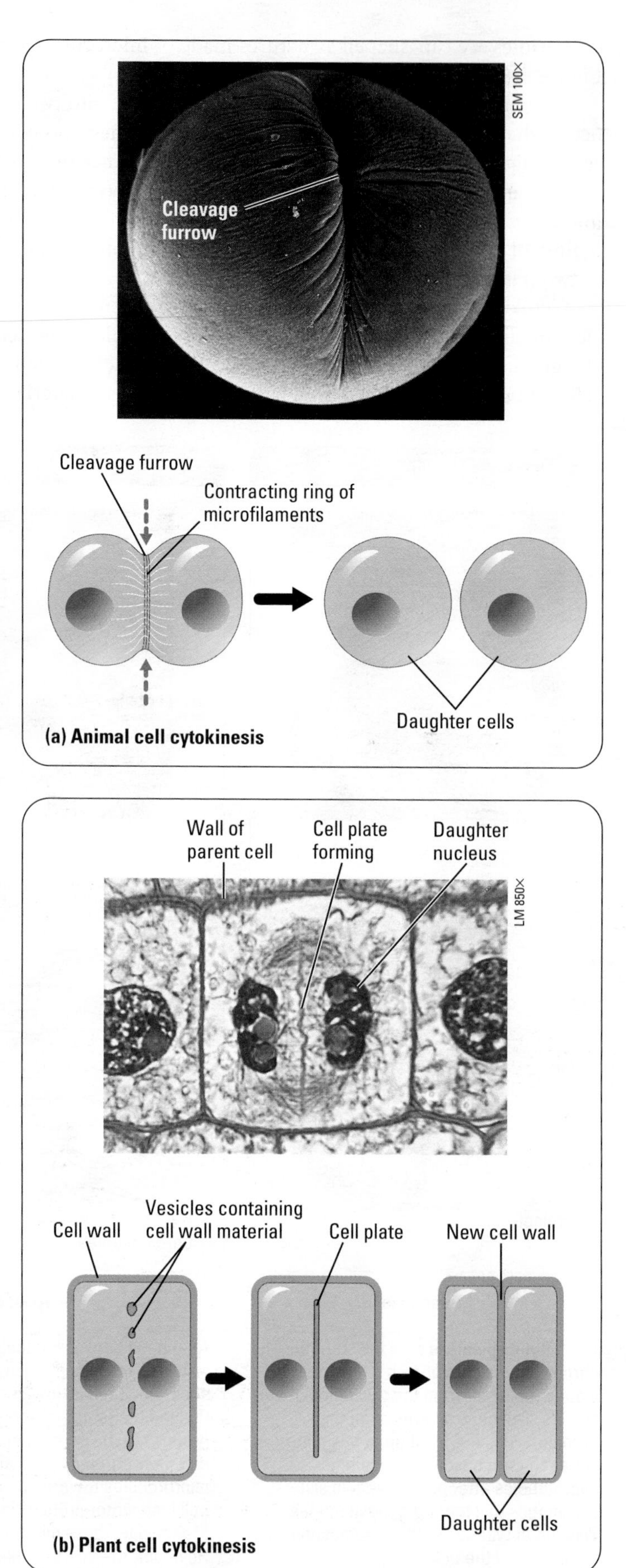

Figure 8.9 Cytokinesis in animal and plant cells.

normal cells, they are more likely to be dividing at any given time. Therefore, radiation can often destroy cancer cells without seriously injuring the normal cells of the body. However, there is sometimes enough damage to normal body cells to produce bad side effects such as nausea and hair loss. Additionally, damage to cells of the ovaries or testes can cause sterility.

Chemotherapy ("poison") uses the same basic strategy as radiation; in this case, drugs are administered that disrupt cell division. These drugs work in a variety of ways. Some prevent cell division by interfering with the mitotic spindle. Paclitaxel (trade name Taxol) freezes the spindle after it forms, keeping it from functioning. Paclitaxel is made from a chemical found in the bark of the Pacific yew, a tree found mainly in the northwestern United States. It has fewer side effects than many other anticancer drugs and seems to be effective against some hard-to-treat cancers of the ovary and breast. Another drug, vinblastine, prevents the mitotic spindle from growing in the first place. Vinblastine was first obtained from the periwinkle plant, which is native to the tropical rain forests of Madagascar.

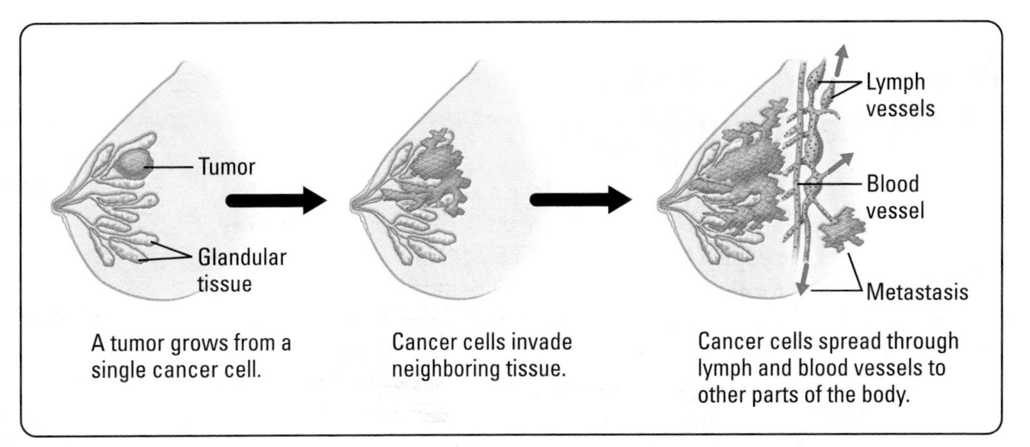
Tumor
Glandular tissue
Lymph vessels
Blood vessel
Metastasis
A tumor grows from a single cancer cell.
Cancer cells invade neighboring tissue.
Cancer cells spread through lymph and blood vessels to other parts of the body.

Figure 8.10 Growth and metastasis of a malignant tumor of the breast.

In the laboratory, researchers can grow cancer cells in culture. The cells are placed in a glass container, and nutrients are provided by an artificial liquid medium **(Figure 8.11)**. Normal mammalian cells grow in culture for only about 20 to 50 cell generations, after which they cease to divide. But cancer cells are "immortal"—they can continue to divide indefinitely, as long as they have a supply of nutrients. It is by studying cancer cells in culture that researchers are learning about the molecular changes that make a cell cancerous. We will return to the topic of cancer in Chapter 11, after exploring genes in more detail.

Cancer Prevention and Survival Although cancer can strike anyone, there are certain lifestyle changes you can make to reduce your chances of developing cancer or increase your chances of surviving it. Not smoking, exercising adequately, avoiding overexposure to the sun, and eating a high-fiber, low-fat diet can all help prevent cancer. Seven types of cancer can be easily detected: skin and oral (via physical exam), breast (via self-exams and mammograms for higher-risk women), prostate (via rectal exam), cervical (via Pap smear), testicular (via self-exam), and colon (via colonoscopy). Regular visits to the doctor can help identify tumors early, thereby significantly increasing the possibility of successful treatment.

CHECKPOINT

1. During what parts of the cell cycle does each chromosome consist of two chromatids?
2. An organism called a plasmodial slime mold is one huge cytoplasmic mass with many nuclei. Explain how a change in the cell cycle could cause this "monster cell" to arise.
3. In what sense are the two daughter cells produced by mitosis identical?
4. Based on what you have learned about cancer prevention and detection, why is lung cancer so deadly?

Answers: 1. From the end of the S phase of interphase until the beginning of anaphase in mitosis **2.** Mitosis occurs repeatedly without cytokinesis. **3.** They have identical genes (DNA). **4.** Lung cancer cannot be detected (and therefore treated) early.

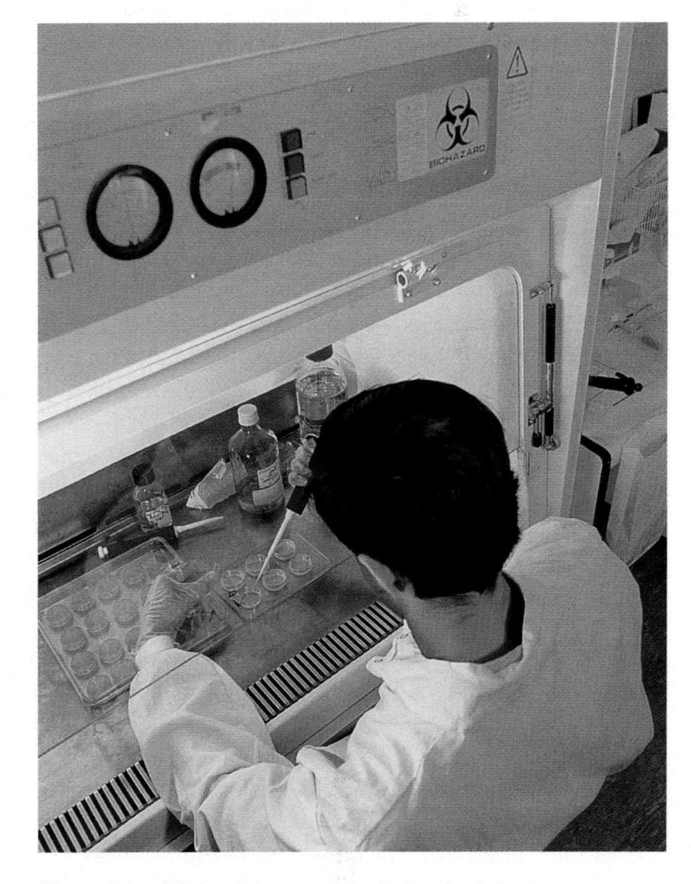

Figure 8.11 Growing cancer cells in the lab. This researcher is working under a special hood that helps maintain sterile conditions for his experiment.

Meiosis, the Basis of Sexual Reproduction

eTutor
Meiosis

MP3 Tutor
Meiosis

Only maple trees produce more maple trees, only goldfish make more goldfish, and only people make more people. These simple facts of life have been recognized for thousands of years and are reflected in the age-old saying "Like begets like." But in a strict sense, "Like begets like" applies only to asexual reproduction, such as the reproduction of the African violet in Figure 8.2c. In that case, because offspring inherit all their DNA from a single parent, they are exact genetic replicas of that one parent and of each other, and their appearances are very similar.

The family photo in **Figure 8.12** makes the point that in a sexually reproducing species, like does not exactly beget like. You probably resemble your parents more closely than you resemble a stranger, but you do not look exactly like your parents or your siblings. Each offspring of sexual reproduction inherits a unique combination of genes from its two parents, and this combined set of genes programs a unique combination of traits. As a result, sexual reproduction can produce tremendous variety among offspring.

Sexual reproduction depends on the cellular processes of meiosis and fertilization. But before discussing these processes, we need to return to chromosomes and their role in the life cycles of sexually reproducing organisms.

Figure 8.12 The varied products of sexual reproduction. A multi-ethnic family poses for a snapshot. Each child has inherited a unique combination of genes from the parents and displays a unique combination of traits.

Homologous Chromosomes

If we examine a number of cells from any individual organism, we discover that virtually all of them have the same number and types of chromosomes. Likewise, if we examine cells from different individuals of a single species—sticking to one gender, for now—we find that they have the same number and types of chromosomes. Viewed with a microscope, your chromosomes would look just like those of Britney Spears (if you're a woman) or Brad Pitt (if you're a man).

A typical human body cell, called a **somatic cell,** has 46 chromosomes. If we break open a human cell in metaphase of mitosis, take a picture of the chromosomes with a microscope, and arrange them in matching pairs, we produce a display called a **karyotype (Figure 8.13).** Notice that every (or almost every) chromosome has a twin that resembles it in size and shape. The two chromosomes of such a matching pair, called **homologous chromosomes,** carry versions of the same genes. If a gene influencing eye color is located at a particular place on one chromosome—for example, within the yellow band in the drawing in Figure 8.13—then the homologous chromosome has a version of that same gene in the same location. Altogether, we humans have 23 homologous pairs of chromosomes (see Figure 8.3). Other species have different numbers of chromosomes, but those, too, usually match in pairs.

For a human female, the 46 chromosomes fall neatly into 23 entirely homologous pairs, with the members of each pair essentially identical in appearance. For a male, however, the chromosomes in one pair do not look alike. This nonmatching pair, only partly homologous, is the male's sex chromosomes. **Sex chromosomes** determine a person's sex (male versus female). As in all mammals, human males have one X chromosome

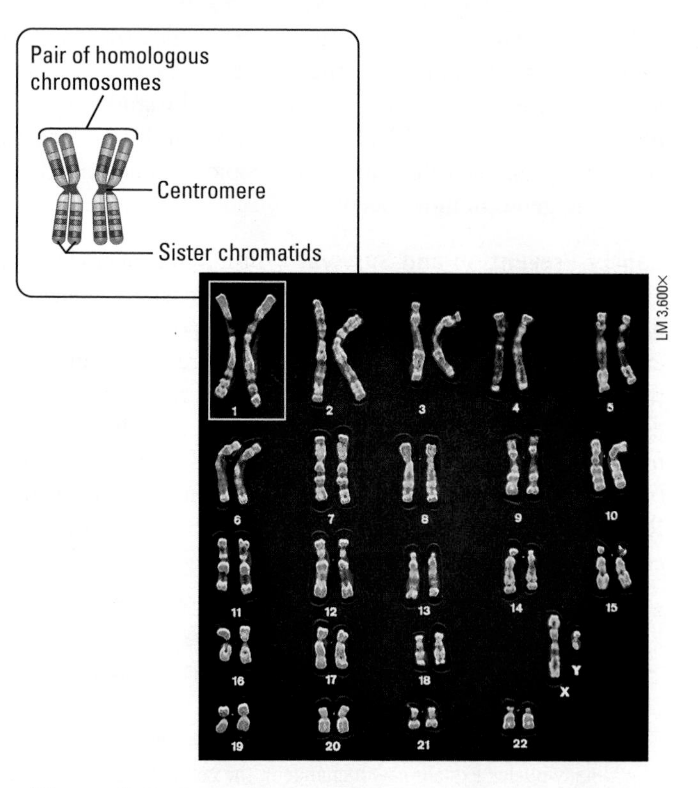

Figure 8.13 Pairs of homologous chromosomes. In this karyotype (chromosome display) of a man, the chromosomes were stained with special dyes to aid in matching them up. The result: 22 completely homologous pairs (autosomes) and a 23rd pair that consists of an X chromosome and a Y chromosome (sex chromosomes). Each chromosome consists of two sister chromatids closely attached all along their lengths. Notice that with the exception of X and Y, the homologous chromosomes of each pair match in size, centromere position, and staining pattern.

and one Y chromosome (see Figure 8.13). Females have two X chromosomes. (While X and Y sex chromosomes determine sex in mammals, other organisms have different systems; in this chapter, we focus on humans.) The remaining chromosomes, found in both males and females, are called **autosomes.** For both autosomes and sex chromosomes, we inherit one chromosome of each pair from our mother and the other from our father.

Gametes and the Life Cycle of a Sexual Organism

The **life cycle** of a multicellular organism is the sequence of stages leading from the adults of one generation to the adults of the next. Having two sets of chromosomes, one inherited from each parent, is a key factor in the life cycle of humans and all other species that reproduce sexually. Let's follow the chromosomes through the human life cycle **(Figure 8.14).**

Humans (as well as most other animals and many plants) are said to be **diploid** organisms because all body cells contain homologous pairs of chromosomes. The total number of chromosomes, 46 in humans, is the diploid number, represented as $2n$. The exceptions are the egg and sperm cells, known as **gametes.** Made by meiosis in an ovary or testis, each gamete has a single set of chromosomes: 22 autosomes plus a sex chromosome, X or Y. A cell with a single chromosome set is called a **haploid** cell; it has only one member of each homologous pair. For humans, the haploid number, n, is 23.

In the human life cycle, sexual intercourse allows a haploid sperm cell from the father to reach and fuse with a haploid egg cell of the mother in the process known as **fertilization.** The resulting fertilized egg, called a **zygote,** is diploid. It has two homologous sets of chromosomes, one set from each parent. The life cycle is completed as a sexually mature adult develops from the zygote. Mitotic cell division ensures that all somatic cells of the human body receive a copy of all of the zygote's 46 chromosomes. Thus, every one of the trillions of cells in your body can trace its ancestry back through mitotic divisions to the single zygote cell produced when your father's sperm and mother's egg fused about nine months before you were born.

All sexual life cycles involve an alternation of diploid and haploid stages. Producing haploid gametes by meiosis keeps the chromosome number from doubling in every generation **(Figure 8.15).**

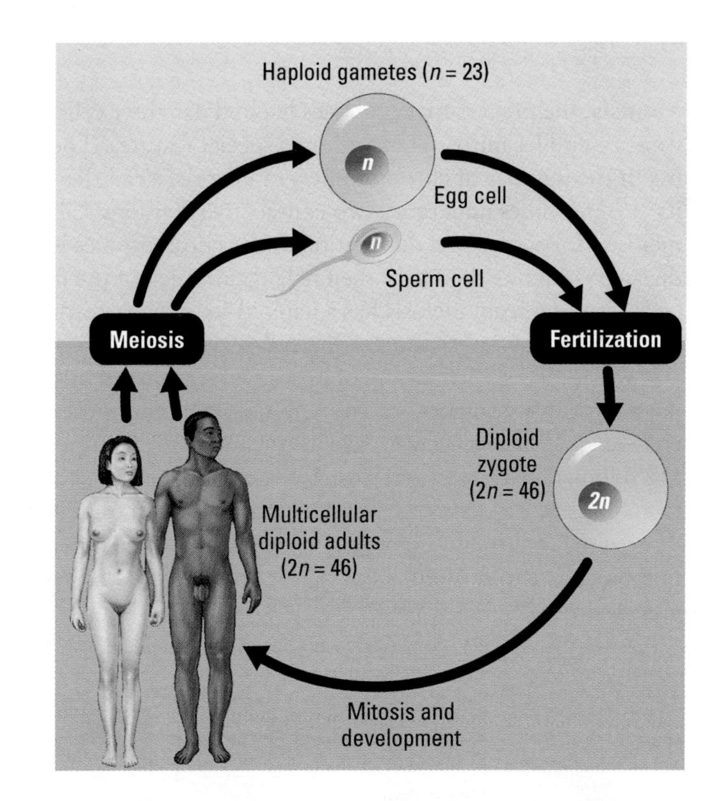

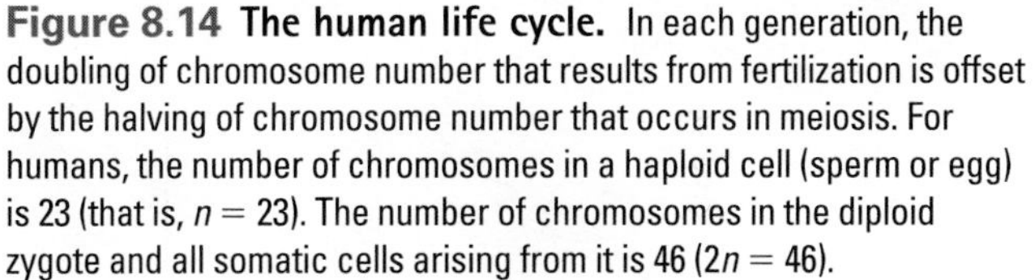
Figure 8.14 The human life cycle. In each generation, the doubling of chromosome number that results from fertilization is offset by the halving of chromosome number that occurs in meiosis. For humans, the number of chromosomes in a haploid cell (sperm or egg) is 23 (that is, $n = 23$). The number of chromosomes in the diploid zygote and all somatic cells arising from it is 46 ($2n = 46$).

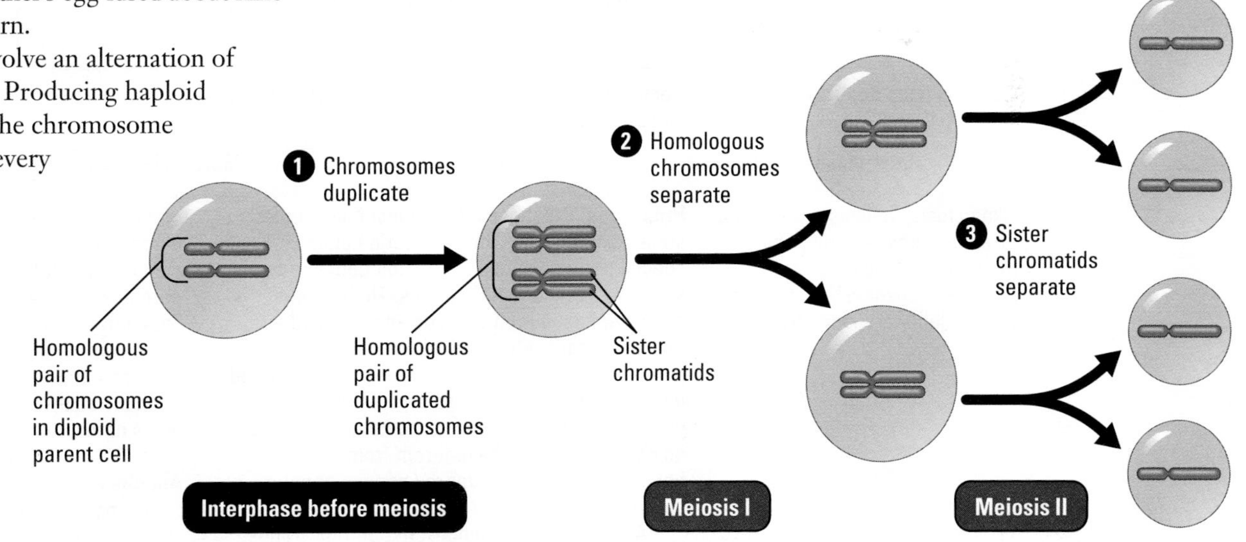

Figure 8.15 How meiosis halves chromosome number. This simplified diagram tracks just one pair of homologous chromosomes. 1 Each of the chromosomes is duplicated during the preceding interphase. 2 The first division, meiosis I, segregates the two chromosomes of the homologous pair, packaging them in separate (haploid) daughter cells. But each chromosome is still doubled. 3 Meiosis II separates the sister chromatids. Each of the four daughter cells is haploid and contains only one single chromosome from the homologous pair.

The Process of Meiosis

Meiosis, the process that produces haploid daughter cells in diploid organisms, resembles mitosis, but with two special features. The first is the halving of the number of chromosomes. In meiosis, a cell that has duplicated its chromosomes undergoes two consecutive divisions, called meiosis I and meiosis II. Because one duplication of the chromosomes is followed by two divisions (double, then half, then half again), each of the four daughter cells resulting from meiosis has a haploid set of chromosomes—only half as many chromosomes as the starting cell.

The second special feature of meiosis is an exchange of genetic material—pieces of chromosomes—between homologous chromosomes. This exchange, called crossing over, occurs during the first prophase of meiosis. We'll look more closely at crossing over later. For now, study **Figure 8.16**, including the text below it, which describes the stages of meiosis in detail.

As you go through Figure 8.16, keep in mind the difference between homologous chromosomes and sister chromatids: The two chromosomes

Figure 8.16 The stages of meiosis. The drawings here show the two cell divisions of meiosis, starting with a diploid animal cell containing four chromosomes. Each homologous pair consists of a red chromosome and a blue chromosome of the same size. The colors remind us that the members of a homologous pair have been inherited from different parents.

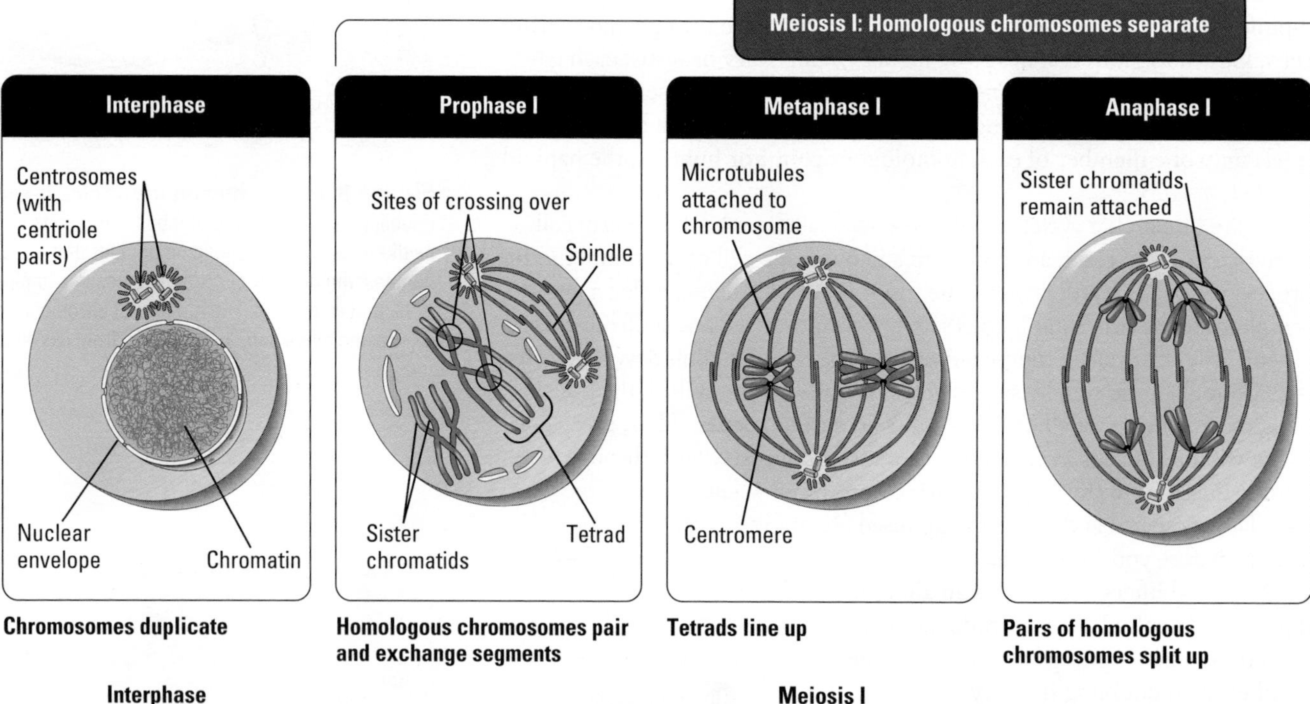

Chromosomes duplicate

Homologous chromosomes pair and exchange segments

Tetrads line up

Pairs of homologous chromosomes split up

Interphase

Like mitosis, meiosis is preceded by an interphase during which the chromosomes duplicate. Each chromosome then consists of two identical sister chromatids.

Meiosis I

Prophase I Prophase I is the most complicated stage of meiosis. As the chromatin condenses, special proteins cause the homologous chromosomes to stick together in pairs. The resulting structure has four chromatids and is called a tetrad. Within each tetrad, chromatids of the homologous chromosomes exchange corresponding segments—they "cross over." Because the versions of the genes on a chromosome (or one of its chromatids) may be different from those on its homologue, crossing over rearranges genetic information.

As prophase I continues, the chromosomes condense further, a spindle forms, and the tetrads are moved toward the center of the cell.

Metaphase I At metaphase I, the tetrads are aligned in the middle of the cell. The sister chromatids of each chromosome are still attached at their centromeres, where they are anchored to spindle microtubules. Notice that for each tetrad, the spindle microtubules attached to one homologous chromosome come from one pole of the cell, and the microtubules attached to the other chromosome come from the opposite pole. With this arrangement, the homologous chromosomes of each tetrad are poised to move toward opposite poles of the cell.

Anaphase I As in anaphase of mitosis, chromosomes now migrate toward the poles of the cell. *But in contrast to mitosis, the sister chromatids migrate as a pair instead of splitting up.* They are separated not from each other, but from their homologous partners. So in the drawing, you see two still-doubled chromosomes moving toward each pole.

of a homologous pair are individual chromosomes that were inherited from different parents, one from the mother and one from the father. The homologues in Figure 8.16 (and later figures) are colored red and blue to remind you that they differ in this way. In the interphase just before meiosis, each homologue replicates to form sister chromatids that remain together until anaphase of meiosis II. Before crossing over occurs, sister chromatids are identical and carry the same versions of all their genes.

CHECKPOINT

1. ___________ is to somatic cells as haploid is to ___________.
2. If a single diploid cell with 18 chromosomes undergoes meiosis and produces sperm, the result will be ___________ sperm, each with ___________ chromosomes. (Provide two numbers.)

Answers: 1. Diploid; gametes **2.** four; nine

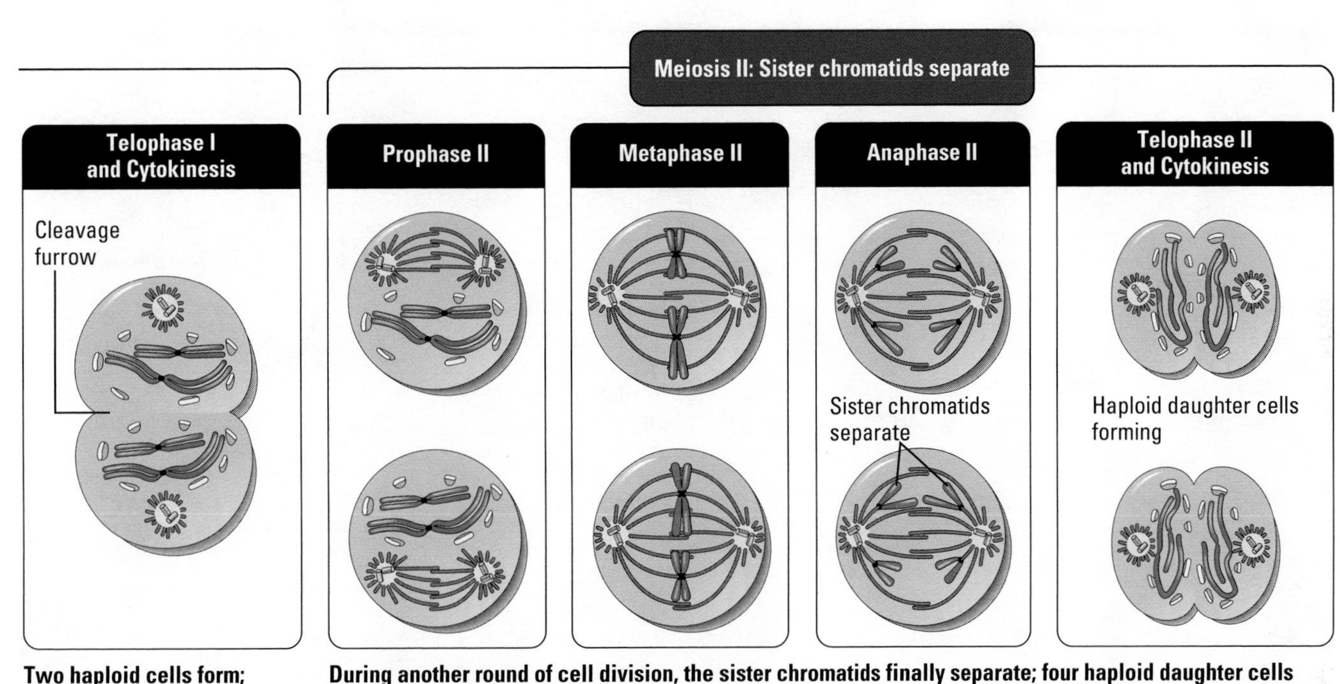

Two haploid cells form; chromosomes are still double

During another round of cell division, the sister chromatids finally separate; four haploid daughter cells result, containing single chromosomes

Meiosis II

Telophase I and Cytokinesis In telophase I, the chromosomes arrive at the poles of the cell. When they finish their journey, each pole has a haploid chromosome set, although each chromosome is still in duplicate form. Usually, cytokinesis occurs along with telophase I, and two haploid daughter cells are formed.

Depending on the species, the nuclei may or may not return to an interphase state. But in either case, there is no further chromosome duplication.

Meiosis II is essentially the same as mitosis. The important difference is that meiosis II starts with a haploid cell.

During prophase II, a spindle forms and moves the chromosomes toward the middle of the cell. During metaphase II, the chromosomes are aligned as they are in mitosis, with the microtubules attached to the sister chromatids of each chromosome coming from opposite poles. In anaphase II, the centromeres of sister chromatids finally separate, and the sister chromatids of each pair, now individual daughter chromosomes, move toward opposite poles of the cell. In telophase II, nuclei form at the cell poles, and cytokinesis occurs at the same time. There are now four daughter cells, each with the haploid number of single chromosomes.

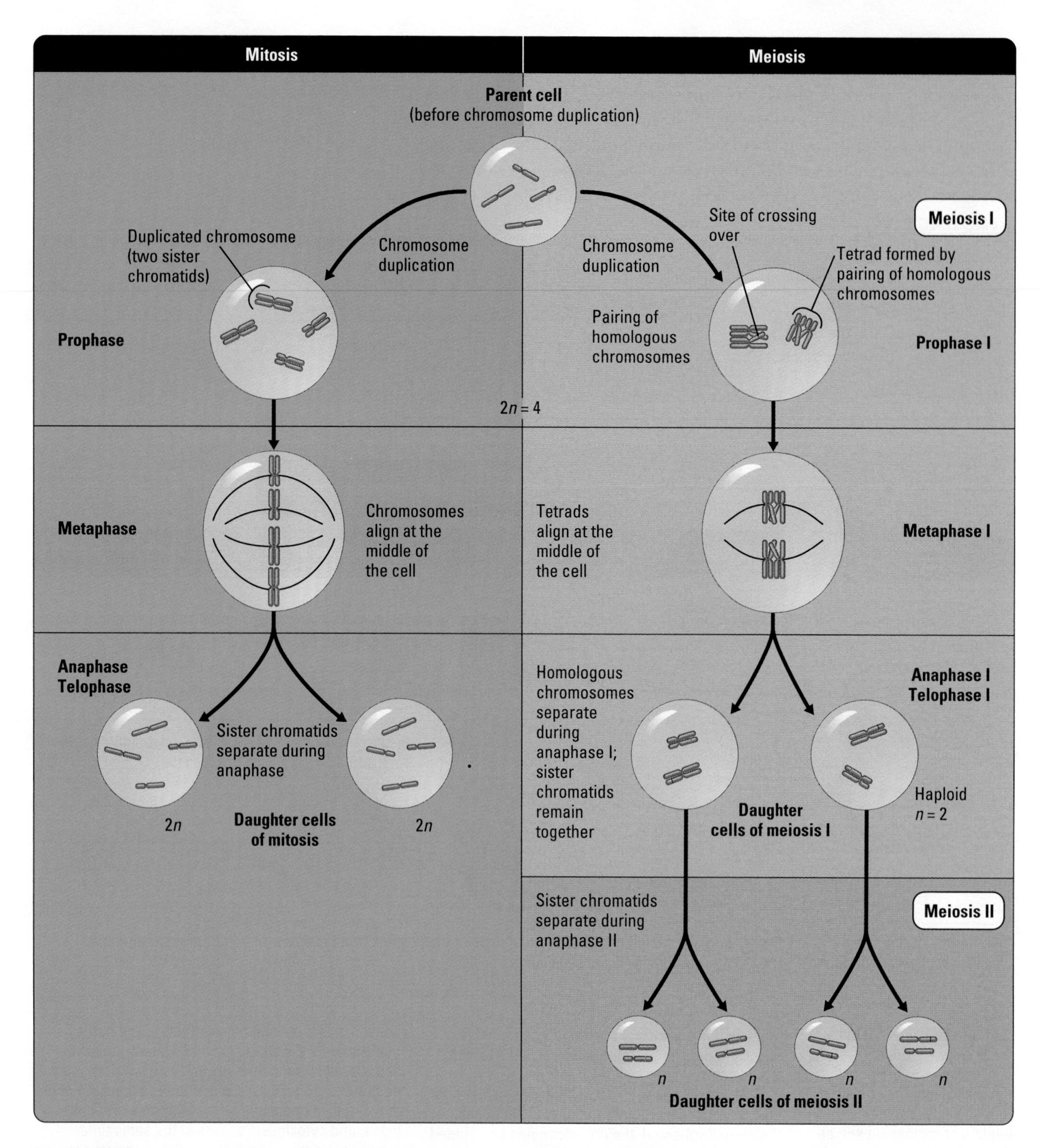

Figure 8.17 Comparing mitosis and meiosis. The events unique to meiosis occur during meiosis I: In prophase I, duplicated homologous chromosomes pair to form tetrads, and crossing over occurs between homologous (nonsister) chromatids. In metaphase I, tetrads (rather than individual chromosomes) are aligned at the center of the cell. During anaphase I, sister chromatids of each chromosome stay together and go to the same pole of the cell as homologous chromosomes separate. At the end of meiosis I, there are two haploid cells, but each chromosome still has two sister chromatids. Meiosis II is virtually identical to mitosis and separates sister chromatids. But unlike mitosis, meiosis II yields daughter cells with a haploid set of chromosomes.

Review: Comparing Mitosis and Meiosis

MP3 Tutor
Mitosis-Meiosis Comparison

We have now described the two ways that cells of eukaryotic organisms divide. Mitosis, which provides for growth, tissue repair, and asexual reproduction, produces daughter cells genetically identical to the parent cell. Meiosis, needed for sexual reproduction, yields genetically unique haploid daughter cells—cells with only one member of each homologous chromosome pair.

For both mitosis and meiosis, the chromosomes duplicate only once, in the preceding interphase. Mitosis involves one division of the nucleus and cytoplasm, producing two diploid cells. Meiosis entails two nuclear and cytoplasmic divisions, yielding four haploid cells.

Figure 8.17 compares mitosis and meiosis, tracing these two processes for a diploid parent cell with four chromosomes. As before, homologous chromosomes are those matching in size. (Imagine that the red chromosomes were inherited from your mother and the blue chromosomes from your father.) Notice that all the events unique to meiosis occur during meiosis I.

CHECKPOINT Complete the following table to compare mitosis and meiosis:

	Mitosis	Meiosis
1. Number of chromosomal duplications		
2. Number of cell divisions		
3. Number of daughter cells produced		
4. Number of chromosomes in daughter cells		
5. How chromosomes line up during metaphase		
6. Genetic relationship of daughter cells to parent cell		
7. Functions performed in the human body		

Answers: 1. 1; 1 **2.** 1; 2 **3.** 2; 4 **4.** 2*n*; *n* **5.** individually; by homologous pair **6.** identical; unique **7.** repair, growth, development; gamete formation

The Origins of Genetic Variation

As we discussed earlier, offspring that result from sexual reproduction are genetically different from their parents and from one another. When we discuss evolution in Unit Three, we'll see that this genetic variety in offspring is the raw material for natural selection. For now, let's take another look at meiosis and fertilization to see how genetic variety arises.

Independent Assortment of Chromosomes **Figure 8.18** illustrates one way in which meiosis contributes to genetic variety. The figure shows how the arrangement of homologous chromosome pairs at metaphase of meiosis I affects the resulting gametes. Once again, our example is from an organism with a diploid chromosome number of 4, with red and blue used to differentiate homologous chromosomes.

Figure 8.18 Results of alternative arrangements of chromosomes at metaphase of meiosis I. In this figure, we consider the consequences of meiosis in a diploid organism with four chromosomes (two homologous pairs). The positioning of each homologous pair of chromosomes at metaphase of meiosis I is random; the two red chromosomes can be on the same side (possibility 1) or on opposite sides (possibility 2). The arrangement of chromosomes at metaphase I determines which chromosomes will be packaged together in the haploid gametes. Because possibilities 1 and 2 are equally likely, the four possible types of gametes will be made in approximately equal numbers.

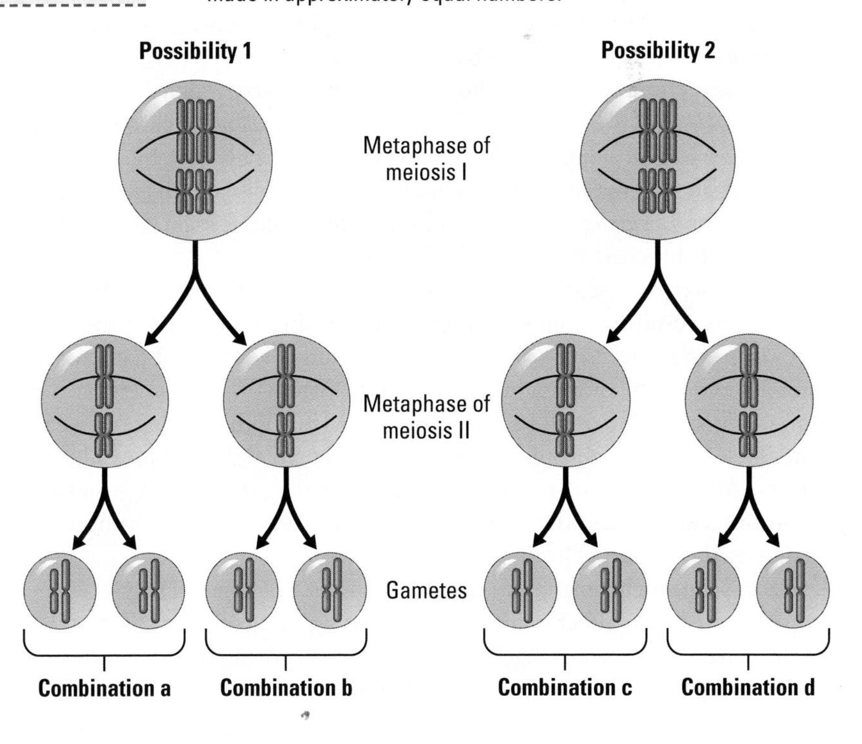

The orientation of the homologous pairs of chromosomes **(tetrads)** at metaphase I is a matter of chance, like the flip of a coin. Thus, in this example, there are two possible ways that the two tetrads can align during metaphase I. In possibility 1, the tetrads are oriented with both red chromosomes on the same side (blue/red and blue/red). In this case, each of the gametes produced at the end of meiosis II has only red or only blue chromosomes (combinations a and b). In possibility 2, the tetrads are oriented differently (blue/red and red/blue). This arrangement produces gametes with one red and one blue chromosome (combinations c and d). Thus, with the two possible arrangements shown in this example, the organism will produce gametes with four different combinations of chromosomes. For a species with more than two pairs of chromosomes, such as the human, every chromosome pair orients independently of all the others at metaphase I. (Chromosomes X and Y behave as a homologous pair in meiosis.)

For any species, the total number of chromosome combinations that can appear in gametes is 2^n, where n is the haploid number. For the organism in this figure, $n = 2$, so the number of chromosome combinations is 2^2, or 4. For a human ($n = 23$), there are 2^{23}, or about 8 million, possible chromosome combinations! This means that every gamete a human produces contains one of about 8 million possible combinations of maternal and paternal chromosomes.

Random Fertilization How many possibilities are there when a gamete from one individual unites with a gamete from another individual during fertilization? A human egg cell, representing one of about 8 million possibilities, is fertilized at random by one sperm cell, representing one of about 8 million other possibilities. By multiplying 8 million by 8 million, we find that a man and a woman can produce a diploid zygote with any of 64 trillion combinations of chromosomes! So we see that the random nature of fertilization adds a huge amount of potential variability to the offspring of sexual reproduction.

Crossing Over So far, we have focused on genetic variability in gametes and zygotes at the whole-chromosome level. We'll now take a closer look at **crossing over,** the exchange of corresponding segments between two homologous chromosomes, which occurs during prophase I of meiosis. **Figure 8.19** shows crossing over between two homologous chromosomes and the results in the gametes. At the time that crossing over begins, homologous chromosomes are closely paired all along their lengths, with a precise gene-by-gene alignment. The sites of crossing over appear as X-shaped regions; each is called a **chiasma** (plural, *chiasmata*). The homologous chromatids remain attached to each other at chiasmata until anaphase I.

The exchange of segments by homologous chromatids adds to the genetic variety resulting from sexual reproduction. In Figure 8.19, if there were no crossing over, meiosis could produce only two types of gametes. These would be the ones ending up with the "parental" types of chromosomes, either all blue or all red (as in Figure 8.18). With crossing over, gametes arise that have chromosomes that are part red and part blue. These chromosomes are called "recombinant" because they result from **genetic recombination,** the production of gene combinations different from those carried by the parental chromosomes.

Because most chromosomes contain thousands of genes, a single crossover event can affect many genes. When we also consider that multiple crossovers can occur in each tetrad, it's not surprising that gametes and the offspring that result from them can be so varied.

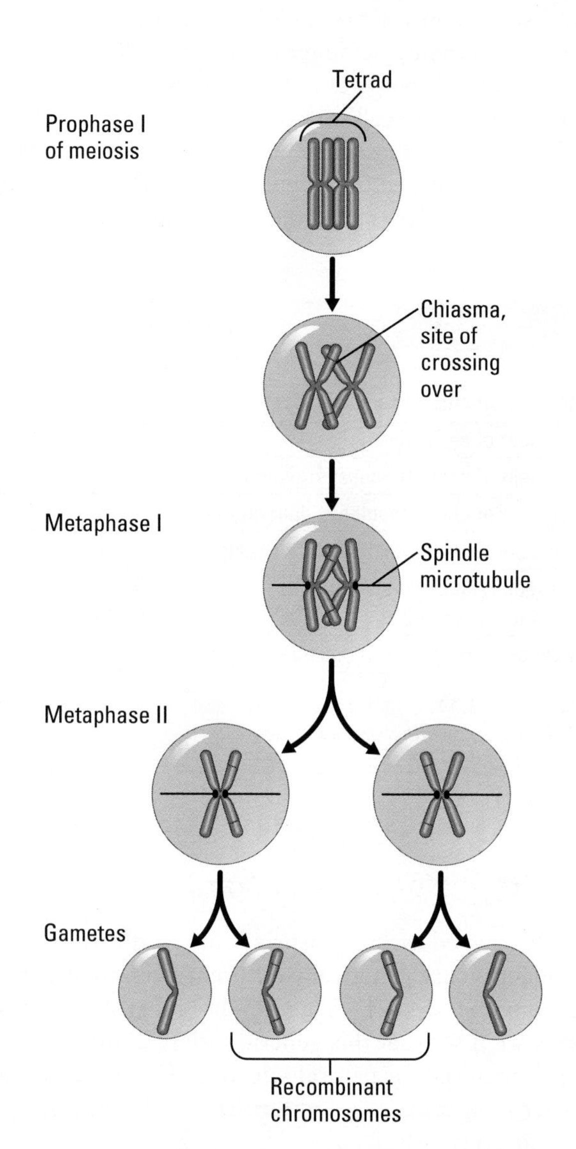

Figure 8.19 The results of crossing over during meiosis. This diagram focuses on a single pair of homologous chromosomes (a tetrad). Early in prophase I of meiosis, homologous (nonsister) chromatids exchange corresponding segments, remaining attached at the crossover points. Sister chromatids are joined at their centromeres. Following these chromosomes through the rest of meiosis, we see that crossing over gives rise to *recombinant* chromosomes—individual chromosomes that combine genetic information originally derived from different parents. With multiple pairs of homologous chromosomes, the result is a huge variety of gametes.

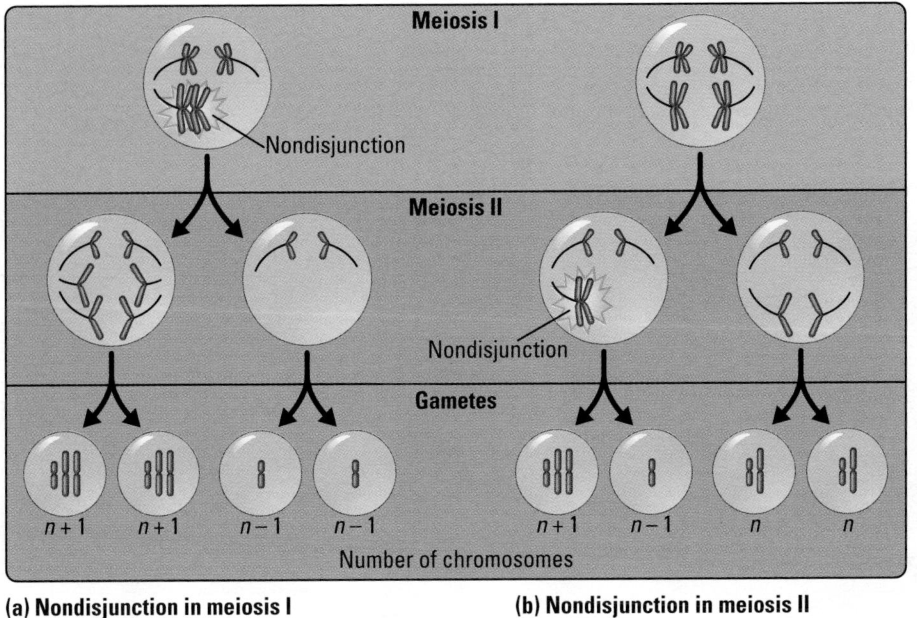

Figure 8.20 Two types of nondisjunction. In both parts of the figure, the cell at the top is diploid ($2n$), with two pairs of homologous chromosomes. **(a)** A pair of homologous chromosomes fails to separate during anaphase of meiosis I, even though the rest of meiosis occurs normally. In this case, all the resulting gametes end up with abnormal numbers of chromosomes. **(b)** Meiosis I is normal, but a pair of sister chromatids fail to move apart in one of the cells during anaphase of meiosis II. In this case, two gametes have the normal complement of two chromosomes each, but the other two gametes are abnormal.

When Meiosis Goes Awry

So far, our discussion of meiosis has focused on the process as it normally and correctly occurs. But what happens when an error occurs in the process? Such a mistake can result in genetic abnormalities that range from mild to severe to fatal.

How Accidents During Meiosis Can Alter Chromosome Number

Within the human body, meiosis occurs repeatedly as the testes or ovaries produce gametes. Almost always, the meiotic spindle distributes chromosomes to daughter cells without error. But occasionally there is an accident, called a **nondisjunction,** in which the members of a chromosome pair fail to separate at anaphase. Nondisjunction can occur during meiosis I or II **(Figure 8.20).** In either case, gametes with abnormal numbers of chromosomes are the result.

Figure 8.21 shows what can happen when an abnormal gamete produced by nondisjunction unites with a normal gamete during fertilization. When a normal sperm fertilizes an egg cell with an extra chromosome, the result is a zygote with a total of $2n + 1$ chromosomes. Mitosis then transmits the abnormality to all embryonic cells. If the organism survives, it will have an abnormal karyotype and probably a syndrome of disorders caused by the abnormal number of genes.

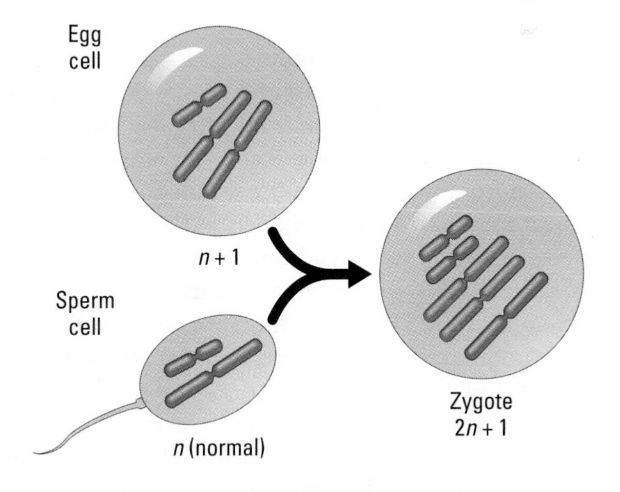

Figure 8.21 Fertilization after nondisjunction in the mother. Assuming that the organism has a diploid number of 4 ($2n = 4$), the sperm is a normal haploid cell ($n = 2$). The egg cell, however, contains an extra copy of the larger chromosome as a result of nondisjunction in meiosis; it has a total of $n + 1 = 3$ chromosomes. When the sperm and egg fuse during fertilization, the result is an abnormal zygote with an extra chromosome; it has $2n + 1 = 5$ chromosomes.

Nondisjunction explains how abnormal chromosome numbers come about, but what causes nondisjunction in the first place? We do not know the answer to that question. We do know, however, that meiosis begins in a woman's ovaries before she is born but is not completed until years later, at the time of an ovulation. Because only one egg cell usually matures each month, a cell might remain arrested in the middle of meiosis for decades. Perhaps damage to the cell during this time leads to meiotic errors. It seems that the longer the time lag, the greater the chance that there will be errors such as nondisjunction when meiosis is completed. In fact, the risk of the most common genetic condition caused by nondisjunction increases with maternal age, as we'll see next.

Down Syndrome: An Extra Chromosome 21 Figure 8.13 showed a normal human complement of 23 pairs of chromosomes. Compare it with **Figure 8.22**; besides having two X chromosomes (because it's from a female), the karyotype in Figure 8.22 has three number 21 chromosomes, making 47 chromosomes in total. This condition is called **trisomy 21**.

In most cases, a human embryo with an atypical number of chromosomes develops so abnormally that it is spontaneously aborted (miscarried) long before birth. However, some aberrations in chromosome number, including trisomy 21, seem to upset the genetic balance less drastically, and individuals carrying them survive. These people usually have a characteristic set of symptoms, called a syndrome. A person with trisomy 21, for instance, is said to have **Down syndrome** (named after John Langdon Down, who described it in 1866).

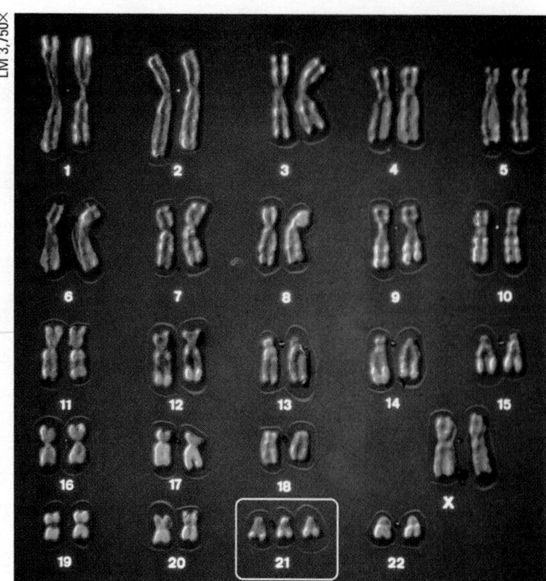

Figure 8.22 Trisomy 21 and Down syndrome. This child displays the characteristic facial features of Down syndrome. The karyotype (right) shows trisomy 21; notice the three copies of chromosome 21.

Affecting about one out of every 700 children, trisomy 21 is the most common chromosome number abnormality and the most common serious birth defect in the United States. Chromosome 21 is one of our smallest chromosomes, but an extra copy produces a number of effects. Down syndrome includes characteristic facial features—frequently a fold of skin at the inner corner of the eye, a round face, a flattened nose bridge, and small, irregular teeth—as well as short stature, heart defects, and susceptibility to respiratory infection, leukemia, and Alzheimer's disease.

People with Down syndrome usually have a life span shorter than normal. They also exhibit varying degrees of mental retardation. However, individuals with the syndrome may live to middle age or beyond, and many are socially adept and able to hold a job. A few women with Down syndrome have had children, though most people with the syndrome are sexually underdeveloped and sterile. Half the eggs produced by a woman with Down syndrome will have an extra chromosome 21, so there is a 50% chance that she will transmit the disorder to her child.

As indicated in **Figure 8.23**, the incidence of Down syndrome in the offspring of normal parents increases markedly with the age of the mother. Down syndrome affects less than 0.05% of children (fewer than one in 2,000) born to women under age 30. The risk climbs to 1.25% (one in 80) for mothers in their early 30s and is even higher for older mothers. Because of this relatively high risk, pregnant women over 35 are candidates for fetal testing for trisomy 21 and other chromosomal abnormalities (see Chapter 9).

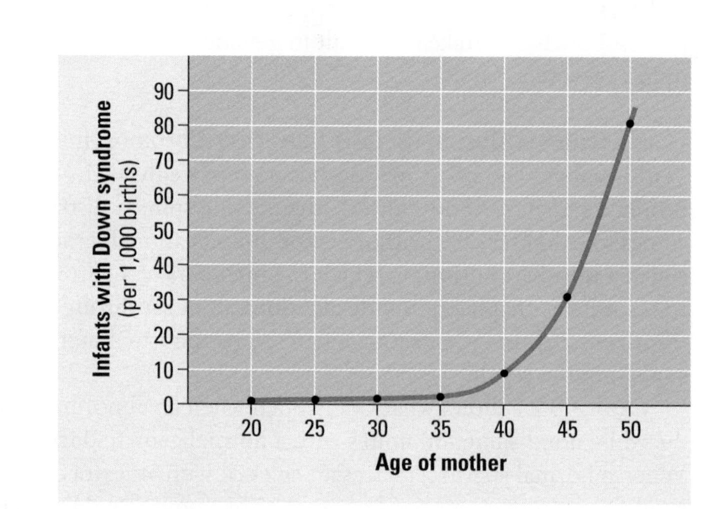

Figure 8.23 Maternal age and Down syndrome. The chance of having a baby with Down syndrome rises with the age of the mother.

Abnormal Numbers of Sex Chromosomes Nondisjunction in meiosis does not affect just autosomes, such as chromosome 21. It can also lead to abnormal numbers of sex chromosomes (X and Y). Unusual numbers of sex chromosomes seem to upset the genetic balance less than unusual numbers of autosomes. This may be because the Y chromosome is very small and carries fewer genes than other chromosomes (and most of those genes simply confer maleness). A peculiarity of X chromosomes in humans and other mammals also helps an individual tolerate unusual numbers of X chromosomes: In mammals, the cells normally operate with only one functioning X chromosome because any other copies of the chromosome become inactivated in each cell (see Chapter 11).

Table 8.1	Abnormalities of Sex Chromosome Number in Humans		
Sex Chromosomes	**Syndrome**	**Origins of Nondisjunction**	**Frequency in Population**
XXY	Klinefelter syndrome (male)	Meiosis in egg or sperm formation	$\frac{1}{2,000}$
XYY	None (normal male)	Meiosis in sperm formation	$\frac{1}{2,000}$
XXX	None (normal female)	Meiosis in egg or sperm formation	$\frac{1}{1,000}$
XO	Turner syndrome (female)	Meiosis in egg or sperm formation	$\frac{1}{5,000}$

Table 8.1 lists the most common sex chromosome abnormalities. An extra X chromosome in a male, making him XXY, produces a condition called Klinefelter syndrome. Men with this disorder have male sex organs, but the testes are abnormally small, the individual is sterile, and he often has breast enlargement and other feminine body contours. The extra X chromosome does not seem to affect intelligence. Klinefelter syndrome is also found in individuals with more than three sex chromosomes, such as XXYY, XXXY, or XXXXY. These abnormal numbers of sex chromosomes probably result from multiple nondisjunctions. Such men are more likely to have developmental disabilities than XY or XXY individuals.

Human males with a single extra Y chromosome (XYY) do not have any well-defined syndrome, although they tend to be taller than average. Females with an extra X chromosome (XXX) cannot be distinguished from XX females except by karyotype.

Females who are lacking an X chromosome are designated XO; the O indicates the absence of a second sex chromosome. These women have Turner syndrome. They have a characteristic appearance, including short stature and often a web of skin extending between the neck and shoulders. Women with Turner syndrome are sterile because their sex organs do not fully mature at adolescence, and they have poor development of breasts and other secondary sex characteristics. However, they are usually of normal intelligence. The XO condition is the sole known case where having only 45 chromosomes is not fatal in humans.

The sex chromosome abnormalities described here illustrate the crucial role of the Y chromosome in determining a person's sex. In general, a single Y chromosome is enough to produce "maleness," regardless of the number of X chromosomes. The absence of a Y chromosome results in "femaleness."

CHECKPOINT

1. Name two events during meiosis that contribute to genetic variety among gametes. During what stages of meiosis does each occur?
2. How does the karyotype of a human female differ from that of a male?
3. What is the chromosomal basis of Down syndrome?
4. Explain how nondisjunction in meiosis could result in a diploid gamete.

Answers: **1.** Crossing over between homologous chromosomes during prophase I and independent orientation of tetrads at metaphase I **2.** A female has two X chromosomes; a male has an X and a Y. **3.** Three copies of chromosome 21 (trisomy 21) **4.** A diploid gamete would result if there were nondisjunction of all the chromosomes during meiosis I or II.

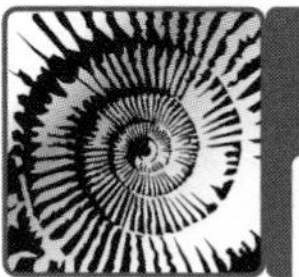

EVOLUTION CONNECTION

New Species from Errors in Cell Division

Errors in meiosis or mitosis do not always lead to problems. In fact, biologists believe that such errors have been instrumental in the evolution of many species. Numerous plant species, in particular, seem to have originated from accidents during cell division that resulted in extra sets of chromosomes. The new species is **polyploid,** meaning that it has more than two sets of homologous chromosomes in each somatic cell. At least half of all species of flowering plants are polyploid, including such useful ones as wheat, potatoes, apples, and cotton.

Let's consider one scenario by which a diploid ($2n$) plant species might generate a tetraploid ($4n$) plant. Imagine that, like many plants, our diploid plant produces both sperm and egg cells and can self-fertilize. If meiosis fails to occur in the plant's reproductive organs and gametes are instead produced by mitosis, the gametes will be diploid. The union of a diploid ($2n$) sperm with a diploid ($2n$) egg during self-fertilization will produce a tetraploid ($4n$) zygote, which may develop into a mature tetraploid plant that can itself reproduce by self-fertilization. The tetraploid plants will constitute a new species, one that has evolved in just one generation.

Although polyploid animal species are less common than polyploid plants, they are known to occur among the fishes and amphibians. Moreover, researchers in Chile have identified the first candidate for polyploidy among the mammals, a rat whose cells seem to be tetraploid (**Figure 8.24**). Tetraploid organisms are sometimes strikingly different from their recent diploid ancestors—larger, for example. Scientists don't yet understand exactly how polyploidy brings about such differences.

You'll learn more about the evolution of polyploid species in Chapter 14. In Chapter 9, we continue our study of genetic principles by looking at the rules governing the inheritance of biological traits and the connection between these traits and the organism's chromosomes. ■

Figure 8.24 Chock full of chromosomes—a tetraploid mammal? The somatic cells of this red viscacha rat from Argentina have about twice as many chromosomes as those of closely related species. (The head of its sperm is unusually large, presumably a necessity for holding all that genetic material.) Scientists think that this rat is a tetraploid species that arose when an ancestor somehow doubled its chromosome number, probably by errors in mitosis or meiosis within the animal's reproductive organs. Researchers are studying the rat's chromosomes to verify that it actually has two homologous sets.

Chapter Review

SUMMARY OF KEY CONCEPTS

For study help and activities, go to campbellbiology.com or the Student CD-ROM.

What Cell Reproduction Accomplishes

- **Passing On Genes from Cell to Cell** Cell reproduction, also called cell division, involves the duplication of all the chromosomes, followed by the distribution of the two identical sets of chromosomes to two "daughter" cells. The daughter cells are genetically identical.

- **The Reproduction of Organisms** Some organisms use mitosis (ordinary cell division) to reproduce. This is called asexual reproduction, and it results in offspring that are genetically identical to the lone parent and to each other. Mitosis also enables multicellular organisms to grow and develop and to replace damaged or lost cells. Organisms that reproduce sexually, by the union of a sperm with an egg cell, carry out meiosis, a type of cell division that yields gametes with only half as many chromosomes as body (somatic) cells.

Activity *Asexual and Sexual Reproduction*

The Cell Cycle and Mitosis

eTutor *Mitosis*

MP3 Tutor *Mitosis*

- **Eukaryotic Chromosomes** The many genes of a eukaryotic genome are grouped into multiple chromosomes in the nucleus. Each chromosome contains one very long DNA molecule, with many genes, that is tightly packed around histone proteins. Individual chromosomes are visible with a light microscope only when the cell is in the process of dividing; otherwise, they are in the form of thin, loosely packed chromatin fibers. Before a cell starts dividing, the chromosomes duplicate, producing sister chromatids (containing identical DNA) joined together at the centromere.

- **The Cell Cycle**

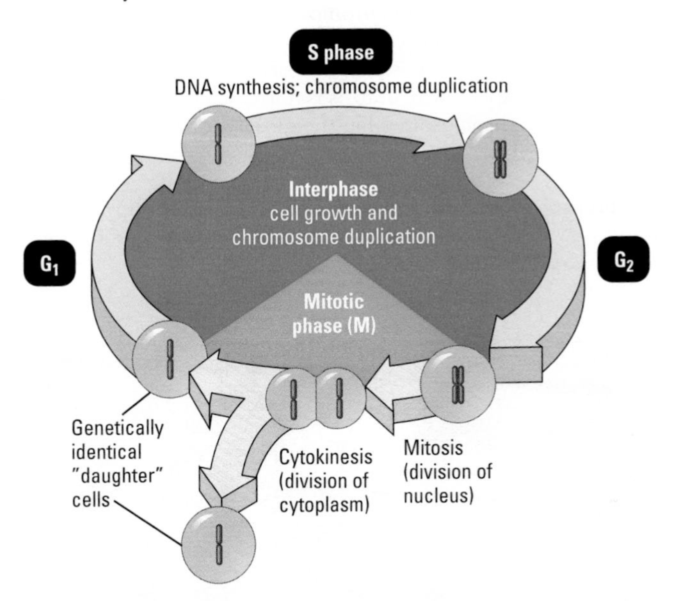

Activity *The Cell Cycle*

- **Mitosis and Cytokinesis** Mitosis is divided into four phases: prophase, metaphase, anaphase, and telophase. At the start of mitosis, the chromosomes coil up and the nuclear envelope breaks down (prophase). Then a mitotic spindle made of microtubules moves the chromosomes to the middle of the cell (metaphase). The sister chromatids then separate and are moved to opposite poles of the cell (anaphase), where two new nuclei form (telophase). Cytokinesis overlaps the end of mitosis. Mitosis and cytokinesis produce genetically identical cells. In animals, cytokinesis occurs by cleavage, which pinches the cell in two. In plants, a membranous cell plate splits the cell in two.

Activity *Mitosis and Cytokinesis Animation*

Activity *Mitosis and Cytokinesis Video*

Case Study in the Process of Science *How Much Time Do Cells Spend in Each Phase of Mitosis?*

- **Cancer Cells: Growing Out of Control** When the cell cycle control system malfunctions, a cell may divide excessively and form a tumor. Cancer cells may grow to form malignant tumors, invade other tissues (metastasize), and even kill the organism. Surgery can remove tumors, and radiation and chemotherapy are effective as treatments because they interfere with cell division. You can protect yourself against some forms of cancer through lifestyle changes and regular screenings.

Activity *Causes of Cancer*

Meiosis, the Basis of Sexual Reproduction

eTutor *Meiosis*

MP3 Tutor *Meiosis*

- **Homologous Chromosomes** The somatic cells (body cells) of each species contain a specific number of chromosomes; human cells have 46, made up of 23 pairs of homologous chromosomes. The chromosomes of a homologous pair carry genes for the same characteristics at the same places. Mammalian males have X and Y sex chromosomes (only partly homologous), while females have two X chromosomes.

- **Gametes and the Life Cycle of a Sexual Organism**

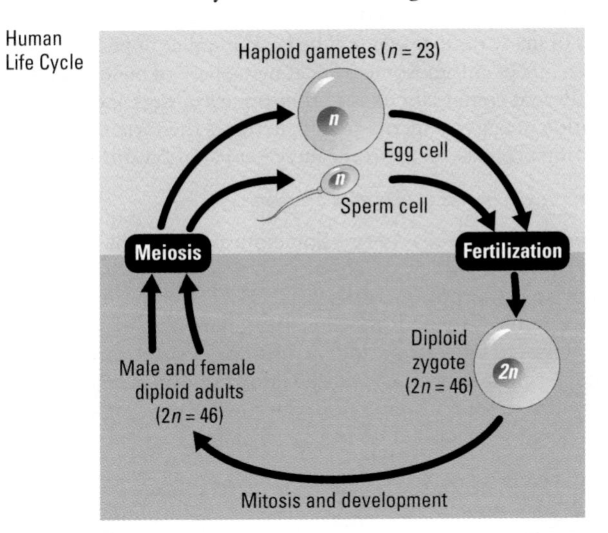

Activity *Human Life Cycle*

- **The Process of Meiosis** Meiosis, like mitosis, is preceded by chromosome duplication. But in meiosis, the cell divides twice to form four daughter cells. The first division, meiosis I, starts with the pairing of homologous chromosomes. In crossing over, homologous chromosomes exchange corresponding segments. Meiosis I separates the members of the homologous pairs and produces two daughter cells, each with one set of (duplicated) chromosomes. Meiosis II is essentially the same as mitosis; in each of the cells, the sister chromatids of each chromosome separate.

Activity *Meiosis Animation*

Case Study in the Process of Science *How Can the Frequency of Crossing Over Be Estimated?*

- **Review: Comparing Mitosis and Meiosis**

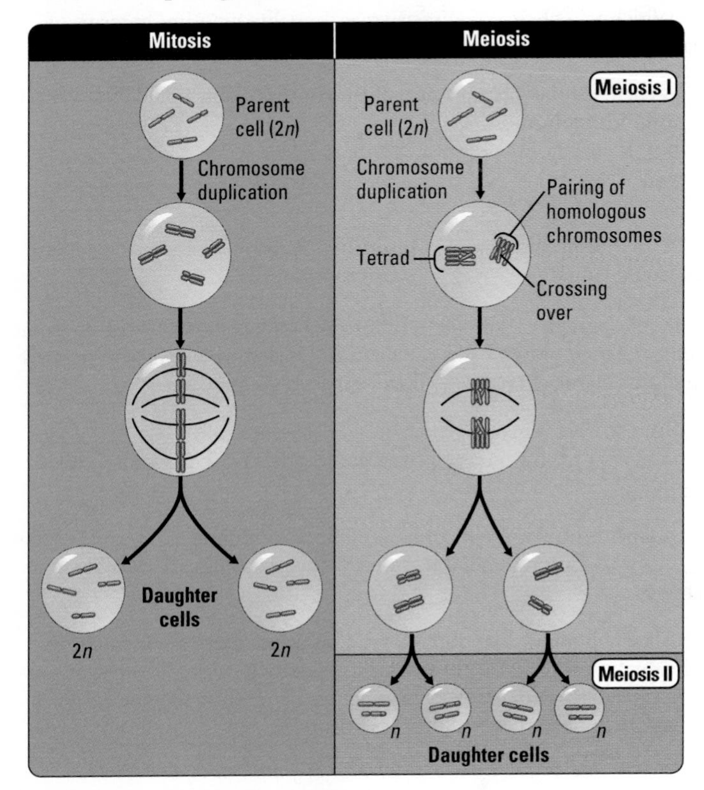

MP3 Tutor *Mitosis-Meiosis Comparison*

• **The Origins of Genetic Variation** Because the chromosomes of a homologous pair come from different parents, they carry different versions of many of their genes. The large number of possible arrangements of chromosome pairs at metaphase of meiosis I leads to many different combinations of chromosomes in eggs and sperm. Random fertilization of eggs by sperm greatly increases the variation. Crossing over during prophase of meiosis I increases variation still further.

Activity *The Origins of Genetic Variation*

• **When Meiosis Goes Awry** Sometimes a person has an abnormal number of chromosomes, which causes problems. Down syndrome is caused by an extra copy of chromosome 21. The abnormal chromosome count is a product of nondisjunction, the failure of a homologous pair of chromosomes to separate during meiosis I or of sister chromatids to separate during meiosis II. Nondisjunction can also produce gametes with extra or missing sex chromosomes, which lead to varying degrees of malfunction in humans but do not usually affect survival.

Activity *Polyploid Plants*

SELF-QUIZ

1. Which of the following is not a function of mitosis in humans?
   a. repair of wounds
   b. growth
   c. production of gametes from diploid cells
   d. replacement of lost or damaged cells

2. Why is it difficult to observe individual chromosomes during interphase?

3. A biochemist measures the amount of DNA in cells growing in the laboratory. The quantity of DNA in a cell would be found to double
   a. between prophase and anaphase of mitosis.
   b. between the G_1 and G_2 phases of the cell cycle.
   c. during the M phase of the cell cycle.
   d. between prophase I and prophase II of meiosis.

4. Which two phases of mitosis are essentially opposites in terms of changes in the nucleus?

5. A chemical that disrupts microfilament formation would interfere with
   a. DNA replication.
   b. formation of the mitotic spindle.
   c. cleavage.
   d. crossing over.

6. If an intestinal cell in a dog contains 78 chromosomes, a dog sperm cell would contain __________ chromosomes.

7. A micrograph of a dividing cell from a mouse shows 19 chromosomes, each consisting of two sister chromatids. During which stage of meiosis could this picture have been taken? (Explain your answer.)

8. Tumors that remain at their site of origin are called __________, while tumors from which cells migrate to other body tissues are called __________.

9. A fruit fly somatic cell contains eight chromosomes. This means that __________ different combinations of chromosomes are possible in its gametes.

10. Although nondisjunction is a random event, there are many more individuals with an extra chromosome 21, which causes Down syndrome, than individuals with an extra chromosome 3 or chromosome 16. Propose an explanation for this.

Answers to the Self-Quiz questions can be found in Appendix D.

Go to the website or CD-ROM for more Self-Quiz questions.

THE PROCESS OF SCIENCE

11. A mule is the offspring of a horse and a donkey. A donkey sperm contains 31 chromosomes and a horse egg 32 chromosomes, so the zygote contains a total of 63 chromosomes. The zygote develops normally. The combined set of chromosomes is not a problem in mitosis, and the mule combines some of the best characteristics of horses and donkeys. However, a mule is sterile; meiosis cannot occur normally in its testes or ovaries. Explain why mitosis is normal in cells containing both horse and donkey chromosomes but the mixed set of chromosomes interferes with meiosis.

12. You prepare a slide with a thin slice of an onion root tip. You see the following view in a light microscope. Identify the stage of mitosis for each of the outlined cells, a–d.

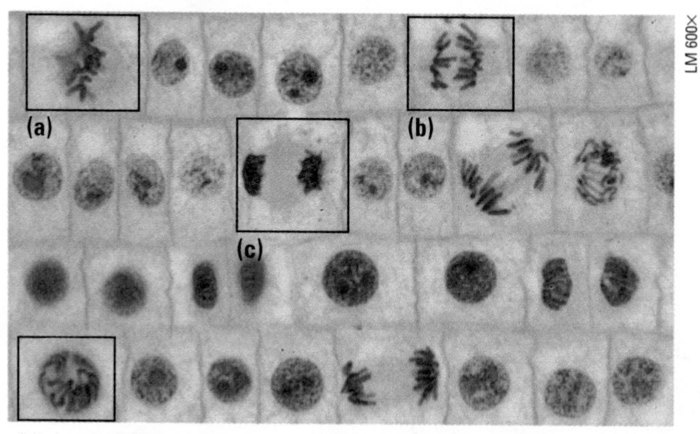

BIOLOGY AND SOCIETY

13. Every year, about a million Americans are diagnosed with cancer. This means that about 75 million Americans now living will eventually have cancer, and one in five will die of the disease. There are many kinds of cancers and many causes of the disease. For example, smoking causes most lung cancers. Overexposure to ultraviolet rays in sunlight causes most skin cancers. There is evidence that a high-fat, low-fiber diet is a factor in breast, colon, and prostate cancers. And agents in the workplace, such as asbestos and vinyl chloride, are also implicated as causes of cancer. Hundreds of millions of dollars are spent each year in the search for effective treatments for cancer, yet far less money is spent on preventing cancer. Why might this be true? What kinds of lifestyle changes could we make to help prevent cancer? What kinds of prevention programs could be initiated or strengthened to encourage these changes? What factors might impede such changes and programs? Should we devote more of our resources to treating cancer or preventing it? Defend your position.

14. The practice of buying and selling gametes, particularly eggs from fertile women, is becoming increasingly common in the United States and other industrialized countries. The story that opened this chapter described a couple who was willing to pay $50,000 for the eggs of a woman with particular physical and mental traits. Do you have any objections to this transaction? Would you be willing to sell your gametes? At any price? Whether you are willing to do so or not, do you think that other people should be restricted from doing so? Why do you think the couple sought out a woman with those specific traits? Do you agree with their reasoning?

9 Patterns of Inheritance

The same genetic defect that **causes sickle-cell disease** can also protect against malaria.

The first genetics research "lab" was a **monk's abbey garden.**

Intermarriage caused the disease **hemophilia** to be inherited by many members of Europe's royal families.

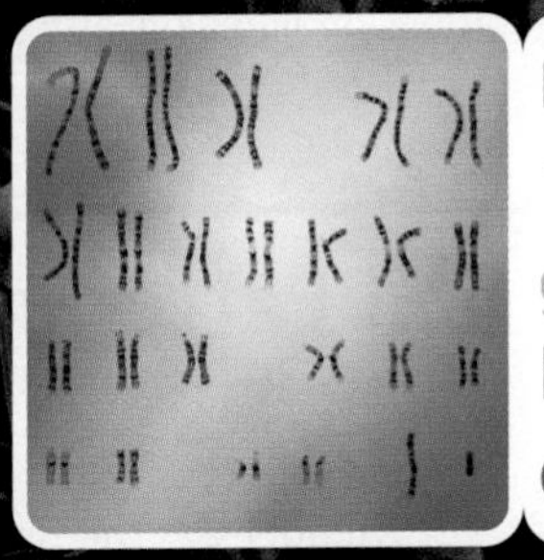

In some isolated communities, 1% of boys are born with a genetic defect that causes **Duchenne muscular dystrophy.**

BIOLOGY AND SOCIETY

Testing Before Birth

Few events in life are as exciting as the news that you're having a baby. So many decisions to make! How should you prepare the nursery? What about names? Recent advances in **genetics,** the science of heredity, have added yet another question: Should you have the fetus genetically tested? The answer may be yes for parents with an increased risk of passing on a genetic disease to their offspring. For example, women over 35 have a heightened risk of bearing children with Down syndrome, and other couples may be aware that certain genetic diseases run in their families. These prospective parents may wish to know their offspring's genetic makeup.

Genetic testing before birth usually requires the collection of fetal cells. In a procedure called amniocentesis, performed between weeks 14 and 20 of pregnancy, a physician inserts a needle through the mother's abdomen into her uterus, taking care to avoid the fetus **(Figure 9.1)**. The physician extracts 10 milliliters (mL), about 2 teaspoonfuls, of the amniotic fluid that bathes the developing fetus. Cells from the fetus (mostly shed skin cells) can then be isolated from this fluid. In an alternative procedure, chorionic villus sampling (CVS), a physician inserts a narrow, flexible tube through the mother's vagina and into her uterus. A small amount of chorionic tissue (part of the placenta) is removed by suction. CVS can be performed as early as the 8th week of pregnancy.

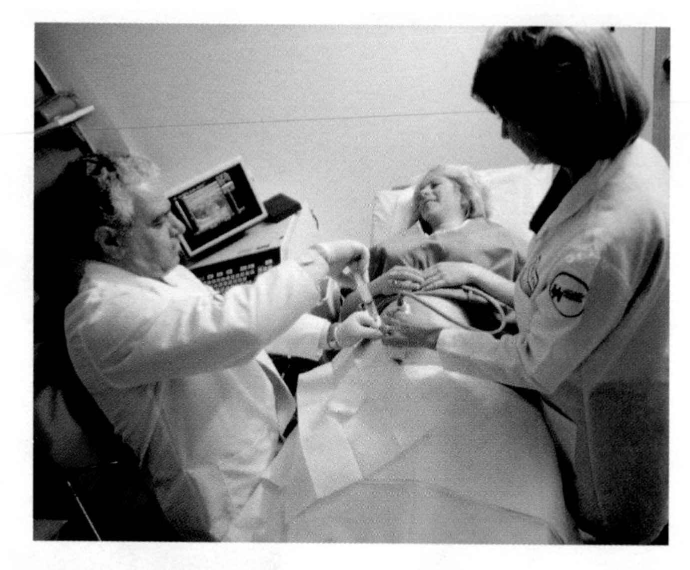

Figure 9.1 Amniocentesis. A physician inserts a needle into a mother's uterus and extracts about 10 mL (2 teaspoonfuls) of amniotic fluid. Cells collected from the fluid can be genetically tested.

Once cells are obtained, they can be screened for genetic diseases. Some genetic disorders, such as the invariably fatal Tay-Sachs disease (see Chapter 4), can be detected by the presence of certain chemicals in the amniotic fluid. To detect chromosomal abnormalities, such as Down syndrome, chromosomes can be stained and photographed (see Figure 8.13).

Unfortunately, both amniocentesis and CVS pose some risk of complications, such as maternal bleeding, miscarriage, or premature birth. Complication rates for CVS and amniocentesis are about 2% and 1%, respectively. Because of the risks, these techniques are usually reserved for situations in which the possibility of a genetic disease or other type of birth defect is significantly higher than average.

If fetal tests reveal a serious genetic disease, the parents must make a choice—either to terminate the pregnancy or to prepare themselves for a baby with serious problems. Identifying a genetic disease early can give families time to prepare—emotionally, medically, and financially.

The decision of whether or not to undergo fetal testing is just one example of how genetics affects our lives. In this chapter, you will learn the basic rules of how genetic information is passed from generation to generation and how the behavior of chromosomes accounts for these rules. Along the way, you'll see many examples of the effects of gene mutations on human health. ■

Heritable Variation and Patterns of Inheritance

Over a century ago, a monk named Gregor Mendel was the first to analyze patterns of inheritance in a systematic, scientific way **(Figure 9.2)**. During the 1860s, using the garden of his abbey in Brunn, Austria (now Brno, in

Figure 9.2 Gregor Mendel.

the Czech Republic), Mendel deduced the fundamental laws of genetics by conducting careful experiments with garden peas. He bred plants that differed in one or more inherited characteristics, such as flower color or seed shape. This work led to the discovery that the inheritance of many genetic characteristics follows a few simple rules. Mendel's work is a classic in the history of biology. Strongly influenced by his university studies in physics, mathematics, and chemistry, his research was both experimental and mathematically rigorous, and these qualities were largely responsible for his success.

In a paper published in 1866, Mendel correctly argued that parents pass on to their offspring discrete "heritable factors" that are responsible for inherited traits, such as purple flowers or round seeds in pea plants. (It is interesting to note that Mendel's publication came just seven years after Darwin's 1859 publication of *The Origin of Species*, making the 1860s a banner decade in the development of modern biology.) In his paper, Mendel stressed that the heritable factors (today called genes) retain their individual identities generation after generation, no matter how they are mixed up or temporarily masked.

In an Abbey Garden

Mendel probably chose to study garden peas because they were easy to grow and they came in many readily distinguishable varieties. And, importantly, Mendel was able to exercise strict control over the matings of his pea plants. The petals of the pea flower almost completely enclose the egg- and sperm-producing parts—the carpel and stamens, respectively (**Figure 9.3**). Consequently, in nature, pea plants usually **self-fertilize** because sperm-carrying pollen grains released from the stamens land on the tip of the egg-containing carpel of the same flower. Mendel could ensure self-fertilization by covering a flower with a small bag so that no pollen from another plant could reach the carpel. When he wanted **cross-fertilization** (fertilization of one plant by pollen from a different plant), he pollinated the plants by hand, as shown in **Figure 9.4**. Thus, whether Mendel let a pea plant self-fertilize or cross-fertilized it with a known source of pollen, he could always be sure of the parentage of his new plants.

Mendel's success was due not only to his experimental approach and choice of organism, but also to his selection of characteristics. Each of the characteristics he chose to study, such as flower color, occurred in two distinct forms. Mendel worked with his plants until he was sure he had **true-breeding** (purebred) varieties—that is, varieties for which self-fertilization produced offspring all identical to the parent. For instance, he identified a purple-flowered variety that, when self-fertilized, always produced offspring plants that all had purple flowers.

Now Mendel was ready to ask what would happen when he crossed different true-breeding varieties with each other. For example, what offspring would result if plants with purple flowers and plants with white flowers were cross-fertilized as shown in Figure 9.4? In the language of breeders and geneticists, the offspring of two different true-breeding varieties are called **hybrids,** and the cross-fertilization itself is referred to as a genetic **cross.** The parental plants are called the **P generation,** and their hybrid offspring are the **F_1 generation** (F for *filial*, from the Latin for "son or daughter"). When F_1 plants self-fertilize or fertilize each other, their offspring are the **F_2 generation.**

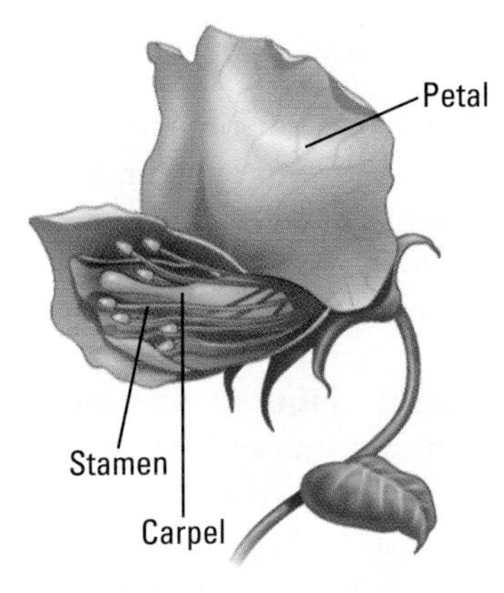

Figure 9.3 The structure of a pea flower. To reveal the reproductive organs, the stamens and carpel, one of the petals has been removed in this drawing. The carpel produces egg cells; the stamens make pollen, which carries sperm.

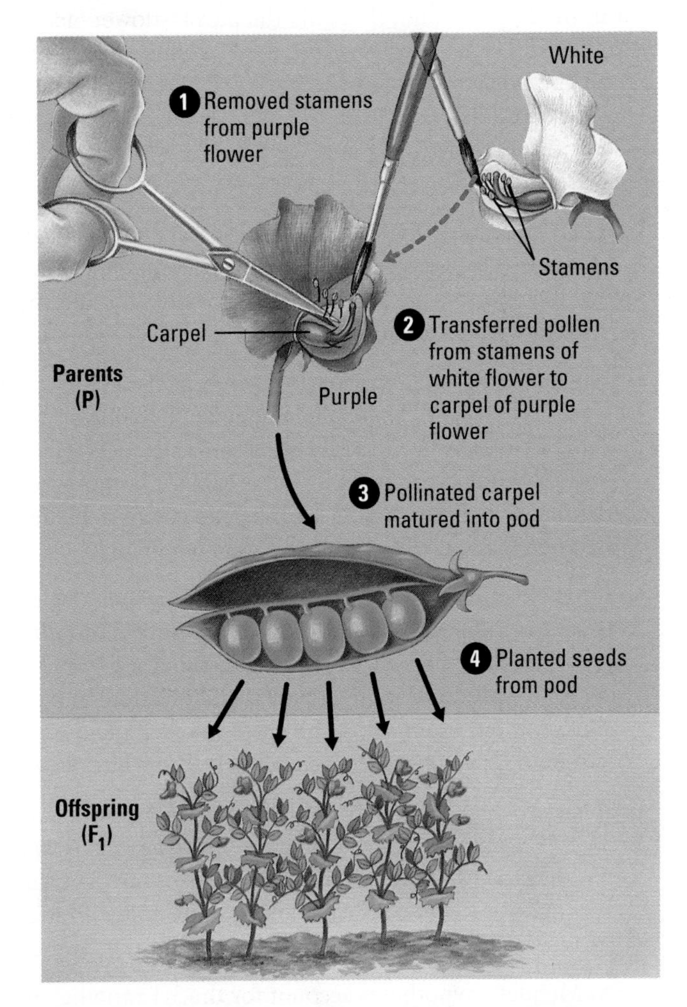

Figure 9.4 Mendel's technique for cross-fertilizing pea plants. 1 To prevent self-fertilization, Mendel cut off the stamens from an immature flower before they produced pollen. This stamenless plant would be the female parent in the experiment. 2 To cross-fertilize this female, he dusted its carpel with pollen from another plant. After pollination, 3 the carpel developed into a pod, containing seeds (peas). 4 He planted the seeds, and they grew into offspring plants.

Mendel's Law of Segregation

Mendel performed many experiments in which he tracked the inheritance of characteristics, such as flower color, that occur as two alternative traits **(Figure 9.5)**. The results led him to formulate several hypotheses about inheritance. Let's look at some of his experiments and follow the reasoning that led to his hypotheses.

Monohybrid Crosses **Figure 9.6a** starts with a cross between a pea plant with purple flowers and one with white flowers. This is called a **monohybrid cross** because the parent plants differ in only one characteristic. Mendel saw that the F_1 plants all had purple flowers. Was the heritable factor for white flowers now lost as a result of the cross? By mating the F_1 plants with each other, Mendel found the answer to be no. Of the F_2 plants, about one-fourth had white flowers (that is, there were about three purple F_2 plants for every white plant, or a 3:1 ratio of purple to white). Mendel concluded that the heritable factor for white flowers did not disappear in the F_1 plants, but that only the purple-flower factor was affecting F_1 flower color. He also deduced that the F_1 plants must have carried two factors for the flower-color characteristic, one for purple and one for white. From these results and others, Mendel developed four hypotheses. Using modern terminology (including "gene" instead of "heritable factor"), here are his hypotheses:

1. There are alternative forms of genes, the units that determine heritable traits. For example, the gene for flower color in pea plants exists in one form for purple and another for white. The alternative forms of genes are called **alleles.**
2. For each inherited characteristic, an organism inherits two alleles, one from each parent. These alleles may be the same or different. An organism that has two identical alleles for a gene is said to be **homozygous** for that gene (and is called a homozygote). An organism that has two different alleles for a gene is said to be **heterozygous** for that gene (and is called a heterozygote).
3. If the two alleles of an inherited pair differ, then one determines the organism's appearance and is called the **dominant allele;** the other has no noticeable effect on the organism's appearance and is called the **recessive allele.** We use uppercase italic letters to represent dominant alleles (in our example, P, the purple-flower allele) and lowercase italic letters to represent recessive alleles (p, the white-flower allele).
4. *A sperm or egg carries only one allele for each inherited characteristic because the two members of an allele pair segregate (separate) from each other during the production of gametes.* This hypothesis is now known as the **law of segregation.** When sperm and egg unite at fertilization, each contributes its allele, restoring the paired condition in the offspring.

Do Mendel's hypotheses account for the 3:1 ratio he observed in the F_2 generation? His hypotheses predict that when alleles segregate during gamete formation in the F_1 plants, half the gametes will receive a purple-flower allele (P) and the other half a white-flower allele (p). During pollination among the F_1 plants, the gametes unite randomly. An egg with a purple-flower allele has an equal chance of being fertilized by a sperm with a purple-flower allele or one with a white-flower allele (that is, a P egg may fuse with a P sperm or a p sperm). Because the same is true for an egg

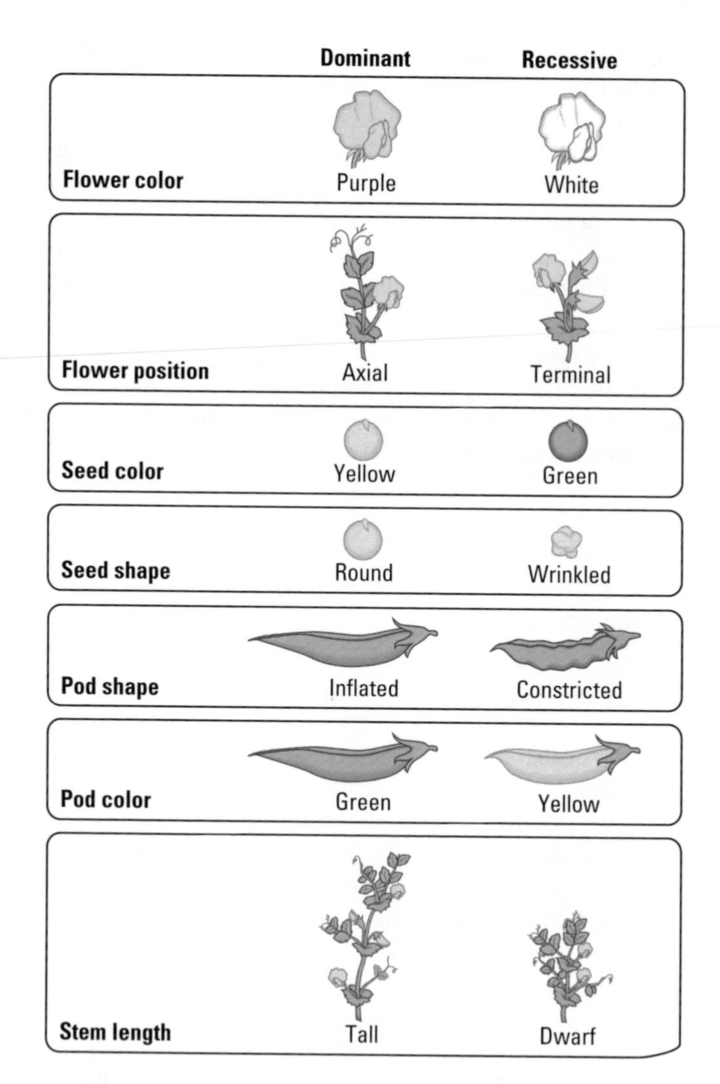

Figure 9.5 The seven characteristics of pea plants studied by Mendel. Each characteristic comes in the two alternatives shown here. The alternative on the left in each pair (such as purple flower color) is dominant; the one on the right (such as white flower color) is recessive. You will learn about dominant and recessive genes shortly.

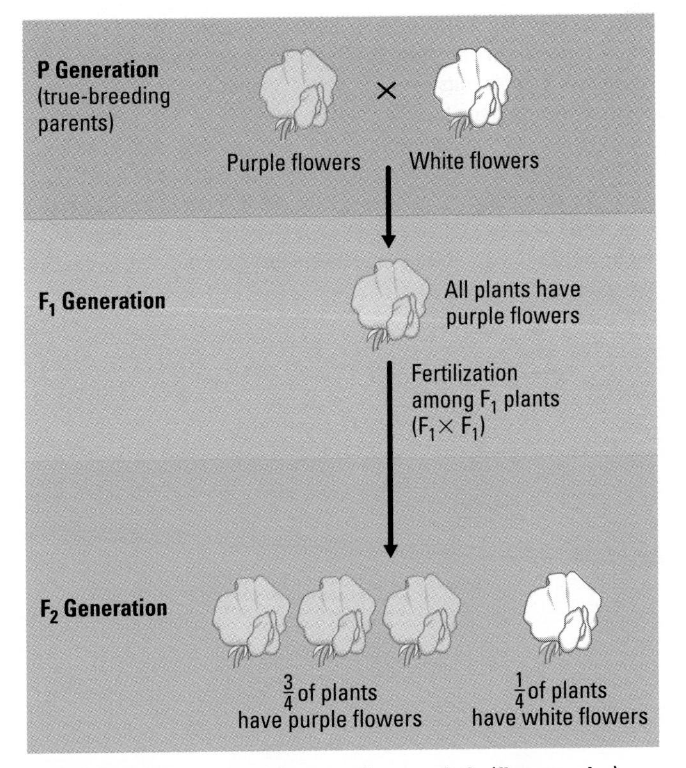

(a) Mendel's crosses tracking one characteristic (flower color)

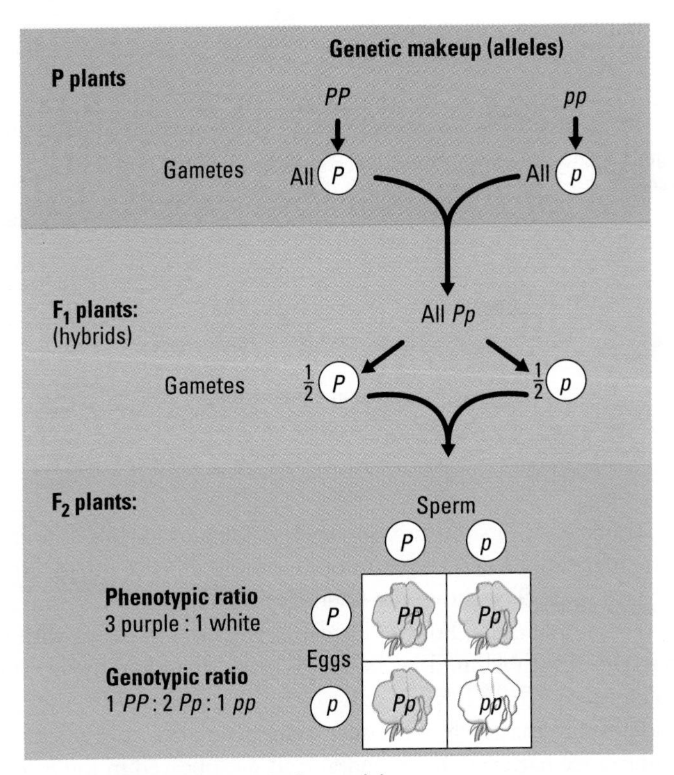

(b) Explanation of the results in part (a)

Figure 9.6 The law of segregation. **(a)** Crossing true-breeding purple-flowered plants with true-breeding white-flowered plants produced F_1 plants with purple flowers. Mendel then allowed the F_1 plants to self-pollinate. Of the 929 F_2 plants produced, 705—about three-quarters—had purple flowers, while 224—about one-quarter—had white flowers. In other words, there was a ratio of about 3:1 of the two varieties in the F_2 generation. **(b)** This diagram illustrates Mendel's explanation for the inheritance pattern shown in part (a) using modern terminology. Uppercase and lowercase letters distinguish dominant from recessive alleles (*P* for purple flowers and *p* for white flowers). The top of the diagram shows the alleles carried by the parental plants. Both plants were true-breeding, so one parental variety must have had two alleles for purple flowers (*PP*), while the other variety had two alleles for white flowers (*pp*). Gametes, symbolized with circles, each contained only one allele for the flower-color gene. Union of the parental gametes produced F_1 hybrids having a *Pp* combination (purple flowers). When the hybrid plants produced gametes, the two alleles segregated, half the gametes receiving the *P* allele and the other half receiving the *p* allele. Random combination of these gametes resulted in the 3:1 ratio that Mendel observed in the F_2 generation. The box at the bottom of the figure is a Punnett square, a useful tool for showing all possible combinations of alleles in offspring. Each square represents an equally probable product of fertilization. For example, the box in the upper right corner of the Punnett square shows the genetic combination resulting from a *p* sperm fertilizing a *P* egg.

with a white-flower allele (a *p* egg with a *P* sperm or *p* sperm), there are a total of four equally likely combinations of sperm and egg. **Figure 9.6b** illustrates these combinations using a diagram called a **Punnett square,** a handy tool for predicting the results of a genetic cross.

What will be the physical appearance of these F_2 offspring? One-fourth of the plants have two alleles specifying purple flowers (*PP*); clearly, these plants will have purple flowers. One-half (two-fourths) of the F_2 offspring have inherited one allele for purple flowers and one allele for white flowers (*Pp*); like the F_1 plants, these plants will also have purple flowers, the dominant trait. (Note that *Pp* and *pP* are equivalent and always written as the former.) Finally, one-fourth of the F_2 plants have inherited two alleles specifying white flowers (*pp*) and will express this recessive trait. Thus, Mendel's model accounts for the 3:1 ratio that he observed in the F_2 generation.

Because an organism's appearance does not always reveal its genetic composition, geneticists distinguish between an organism's expressed, or

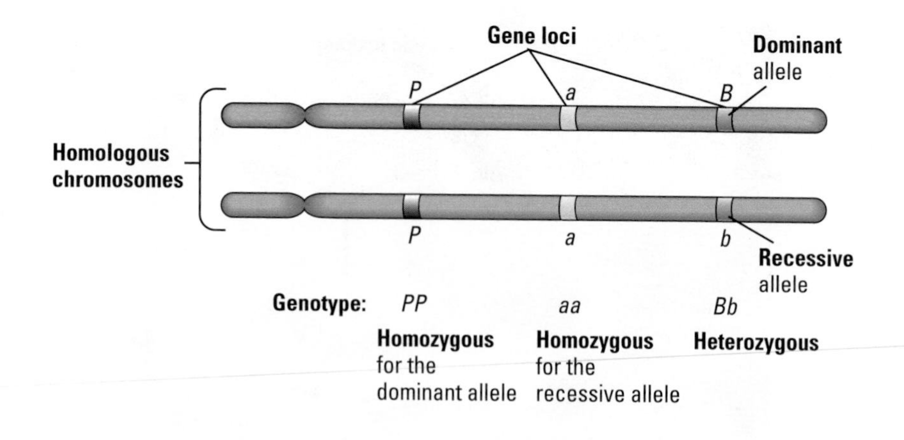

Figure 9.7 The relationship between alleles and homologous chromosomes. The two chromosomes shown here make up a homologous pair. The labeled bands on the chromosomes represent three gene loci. The matching colors of corresponding loci on the two homologues highlight the fact that homologous chromosomes carry alleles for the same genes at the same positions along their lengths. However, the two chromosomes may bear either identical alleles or different ones at any one locus. In other words, the organisms may be homozygous or heterozygous for the gene at that locus.

physical, traits, called its **phenotype** (such as purple or white flowers), and its genetic makeup, its **genotype** (in our example, *PP*, *Pp*, or *pp*). Now we can see that Figure 9.6a shows the phenotypes and Figure 9.6b the genotypes in our sample cross. For the F_2 plants, the ratio of plants with purple flowers to those with white flowers (3:1) is called the phenotypic ratio. The genotypic ratio is 1(*PP*):2(*Pp*):1(*pp*).

Mendel found that each of the seven characteristics he studied had the same inheritance pattern: One parental trait disappeared in the F_1 generation, only to reappear in one-fourth of the F_2 offspring. The underlying mechanism is stated by Mendel's law of segregation: Pairs of alleles segregate (separate) during gamete formation; the fusion of gametes at fertilization creates allele pairs again. Research since Mendel's day has established that the law of segregation applies to all sexually reproducing organisms, including humans.

Genetic Alleles and Homologous Chromosomes Before continuing with Mendel's experiments, let's see how some of the concepts we discussed in Chapter 8 fit with what we've said about genetics so far. The diagram in **Figure 9.7** shows a pair of homologous chromosomes (homologues)—chromosomes that carry alleles of the same genes. Recall from Chapter 8 that every diploid individual, whether pea plant or human, has chromosomes in homologous pairs. One member of each pair comes from the organism's female parent; the other member of each pair comes from the male parent. The labeled bands on the chromosomes in the figure represent three gene **loci** (singular, *locus*), specific locations of genes along the chromosome. You can see the connection between Mendel's law of segregation and homologous chromosomes: Alleles (alternative forms) of a gene reside at the same locus on homologous chromosomes. We will return to the chromosomal basis of Mendel's law later in the chapter.

Mendel's Law of Independent Assortment

Two other pea plant characteristics Mendel studied were seed shape and seed color. Mendel's seeds were either round or wrinkled in shape and either yellow or green in color. From tracking these characteristics one at a time in monohybrid crosses, Mendel knew that the allele for round shape (designated *R*) was dominant to the allele for wrinkled shape (*r*) and that the allele for yellow seed color (*Y*) was dominant to the allele for green seed color (*y*). What would result from a **dihybrid cross,** the mating of parental varieties differing in two characteristics? Mendel crossed homozygous

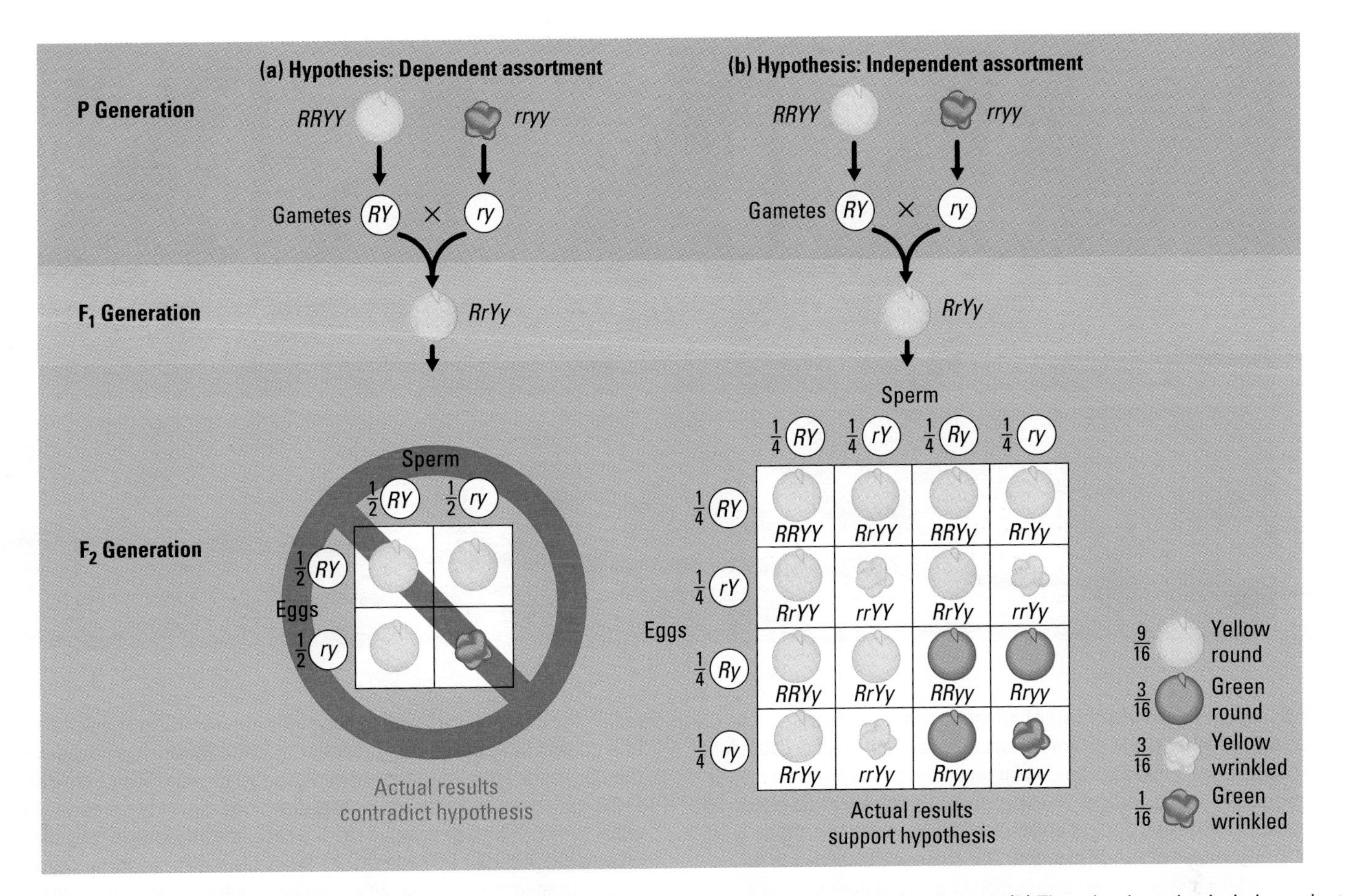

Figure 9.8 Testing alternative hypotheses for gene assortment in a dihybrid cross. **(a)** One hypothesis leads to the prediction that every plant in the F_2 generation will have seeds exactly like one of the parents, either round and yellow or wrinkled and green. **(b)** The other hypothesis, independent assortment, predicts F_2 plants with four different seed phenotypes. The results of experiments support this hypothesis.

plants having round yellow seeds (genotype *RRYY*) with plants having wrinkled green seeds (*rryy*). As shown in **Figure 9.8**, the union of *RY* and *ry* gametes from the P generation yielded hybrids heterozygous for both characteristics (*RrYy*)—that is, dihybrids. As we would expect, all of these offspring, the F_1 generation, had round yellow seeds. But were the two characteristics transmitted from parents to offspring as a package, or was each characteristic inherited independently of the other?

The question was answered when Mendel allowed fertilization to occur among the F_1 plants. If the genes for the two characteristics were inherited together (Figure 9.8a), then the F_1 hybrids would produce only the same two kinds of gametes that they received from their parents. In that case, the F_2 generation would show a 3:1 phenotypic ratio (three plants with round yellow seeds for every one with wrinkled green seeds), as in the Punnett square in Figure 9.8a. If, however, the two seed characteristics sorted independently, then the F_1 generation would produce four gamete genotypes—*RY*, *rY*, *Ry*, and *ry*—in equal quantities. The Punnett square in Figure 9.8b shows all possible combinations of alleles that can result in the F_2 generation from the union of four kinds of sperm with four kinds of eggs. If you study the Punnett square, you'll see that it predicts nine different genotypes in the F_2 generation. These nine genotypes will produce four different phenotypes in a ratio of 9:3:3:1.

The Punnett square in Figure 9.8b also reveals that a dihybrid cross is equivalent to two monohybrid crosses occurring simultaneously. From the

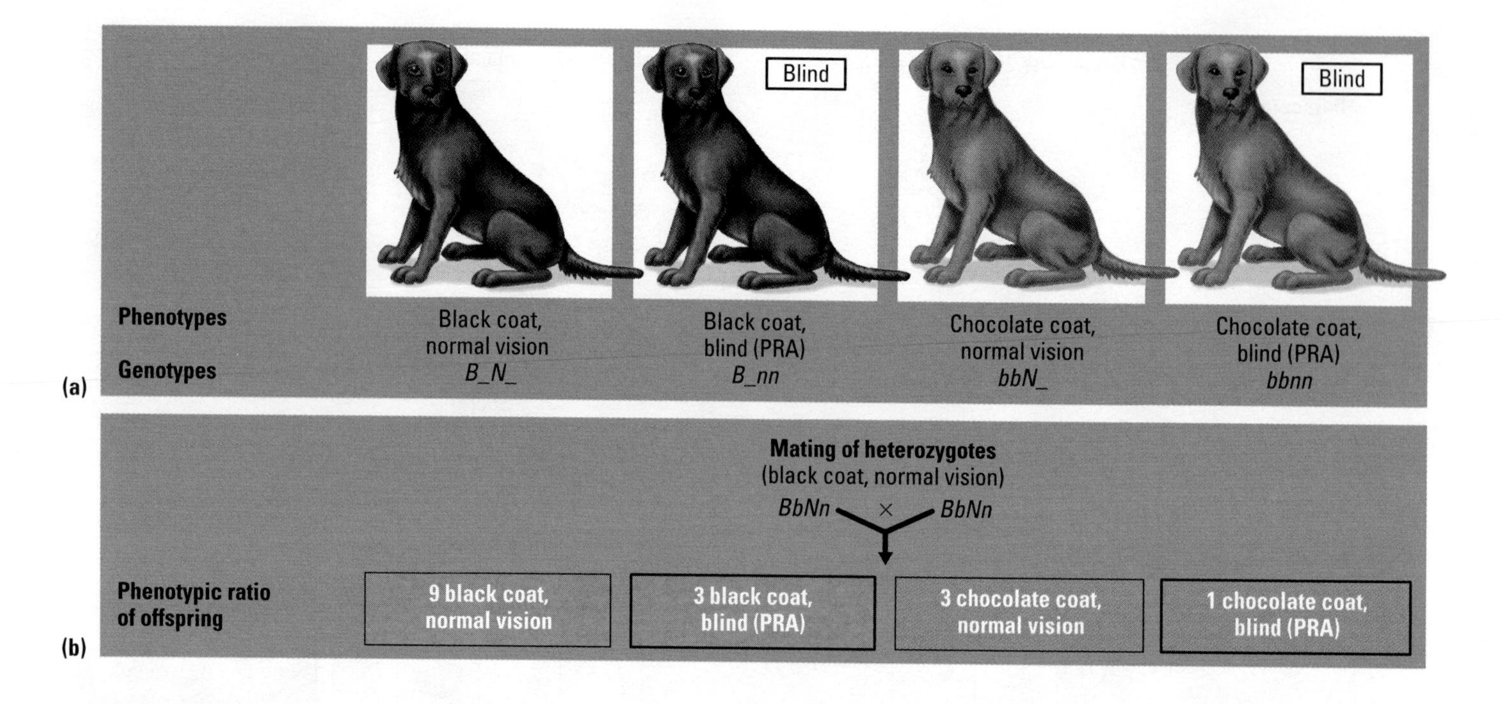

Figure 9.9 Independent assortment of genes in Labrador retrievers. **(a)** The inheritance of two hereditary characteristics in Labrador retrievers, black versus chocolate coat color and normal vision versus the eye disorder progressive retinal atrophy (PRA), is controlled by separate genes. Black Labs have at least one copy of an allele called *B*, which gives their hairs densely packed granules of a dark pigment. The *B* allele is dominant to *b*, which leads to a less tightly packed distribution of pigment granules. As a result, the coats of dogs with genotype *bb* are chocolate in color. The allele that causes PRA, called *n*, is recessive to allele *N*, which is necessary for normal vision. Thus, only dogs of genotype *nn* become blind from PRA. (Blanks in the genotypes indicate alleles that can be dominant or recessive.) **(b)** If you mate two doubly heterozygous (*BbNn*) Labs, the phenotypic ratio of the offspring (F_2) is 9:3:3:1. These results resemble the F_2 results in Figure 9.8, demonstrating that the coat color and PRA genes are inherited independently.

9:3:3:1 ratio, we can see that the ratio of plants with round seeds to those with wrinkled seeds is 12:4, as is the ratio of yellow-seeded plants to green-seeded ones. These 12:4 ratios each reduce to 3:1, which is the F_2 ratio for a monohybrid cross. Mendel tried his seven pea characteristics in various dihybrid combinations and always observed a 9:3:3:1 ratio (or two simultaneous 3:1 ratios) of phenotypes in the F_2 generation. These results supported the hypothesis that *each pair of alleles assorts independently of the other pairs of alleles during gamete formation.* In other words, the inheritance of one characteristic has no effect on the inheritance of another. This is called the **law of independent assortment.** For another application of this law, examine **Figure 9.9**, showing the independent inheritance of two characteristics in Labrador retrievers.

Using a Testcross to Determine an Unknown Genotype

Suppose you have a Labrador retriever with a chocolate coat. Consulting Figure 9.9, you can tell that its genotype must be *bb*, the only combination of alleles that produces the chocolate-coat phenotype. But what if you had a black Lab? It could have one of two possible genotypes—*BB* or *Bb*—and there is no way to tell which is correct by looking at the dog. To determine your dog's genotype, you could perform a **testcross**, a mating between an individual of dominant phenotype but unknown genotype (your black Lab) and a homozygous recessive individual—in this case, a *bb* chocolate Lab.

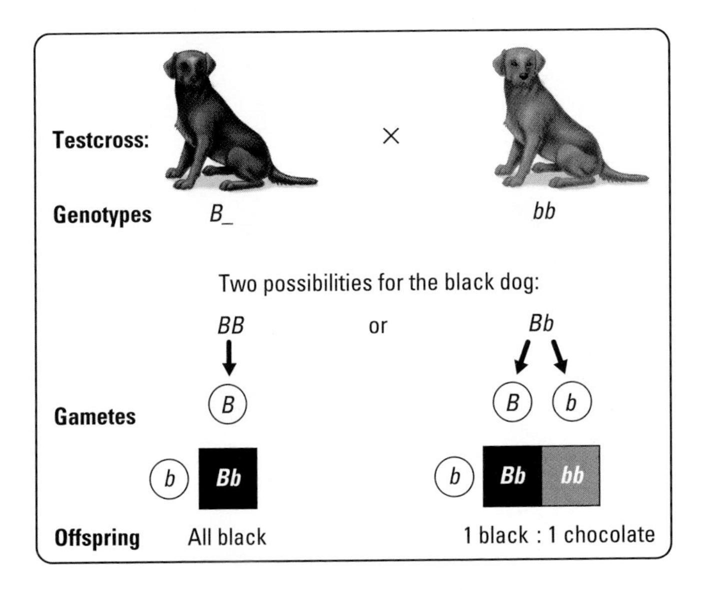

Figure 9.10 A Labrador retriever testcross. To determine the genotype of a black Lab, it can be crossed with a chocolate Lab (homozygous recessive, *bb*). If all the offspring are black, the black parent must have genotype *BB*. If half the offspring are chocolate, the black parent must be heterozygous (*Bb*).

Figure 9.10 shows the offspring that could result from such a mating. If, as shown on the left, the black parent's genotype is *BB*, we would expect all the offspring to be black, because a cross between genotypes *BB* and *bb* can produce only *Bb* offspring. On the other hand, if the black parent is *Bb*, we would expect both black (*Bb*) and chocolate (*bb*) offspring. Thus, the appearance of the offspring may reveal the original black dog's genotype. The figure also shows that we would expect the chocolate and black offspring of a *Bb* × *bb* cross to appear in a 1:1 phenotypic ratio.

Mendel used testcrosses to determine whether he had true-breeding varieties of plants. The testcross continues to be an important tool of geneticists for determining genotypes.

The Rules of Probability

Mendel's strong background in mathematics served him well in his studies of inheritance. He understood, for instance, that the segregation of allele pairs during gamete formation and the re-forming of pairs at fertilization obey the rules of probability—the same rules that apply to the tossing of coins, the rolling of dice, and the drawing of cards. Mendel also appreciated the statistical nature of inheritance. He knew that he needed to obtain large samples—count many offspring from his crosses—before he could begin to interpret inheritance patterns.

An important lesson we can learn from coin tossing is that for each and every toss of the coin, the probability of heads is $\frac{1}{2}$. In other words, the outcome of any particular toss is unaffected by what has happened on previous attempts. Each toss is an independent event.

If two coins are tossed simultaneously, the outcome for each coin is an independent event, unaffected by the other coin. What is the chance that both coins will land heads-up? The probability of such a compound event is the product of the separate probabilities of the independent events—for the coins, $\frac{1}{2} \times \frac{1}{2} = \frac{1}{4}$. This is called the **rule of multiplication,** and it holds true for independent events that occur in genetics as well as coin tosses, as shown in **Figure 9.11**. In our dihybrid cross of Labradors (see Figure 9.9), the genotype of the F_1 dogs for coat color was *Bb*. What is the probability that a particular F_2 dog will have the *bb* genotype? To produce a *bb* offspring, both egg and sperm must carry the *b* allele. The probability that an egg will have the *b* allele is $\frac{1}{2}$, and the probability that a sperm will have the *b* allele is also $\frac{1}{2}$. By the rule of multiplication, the probability that two *b* alleles will come

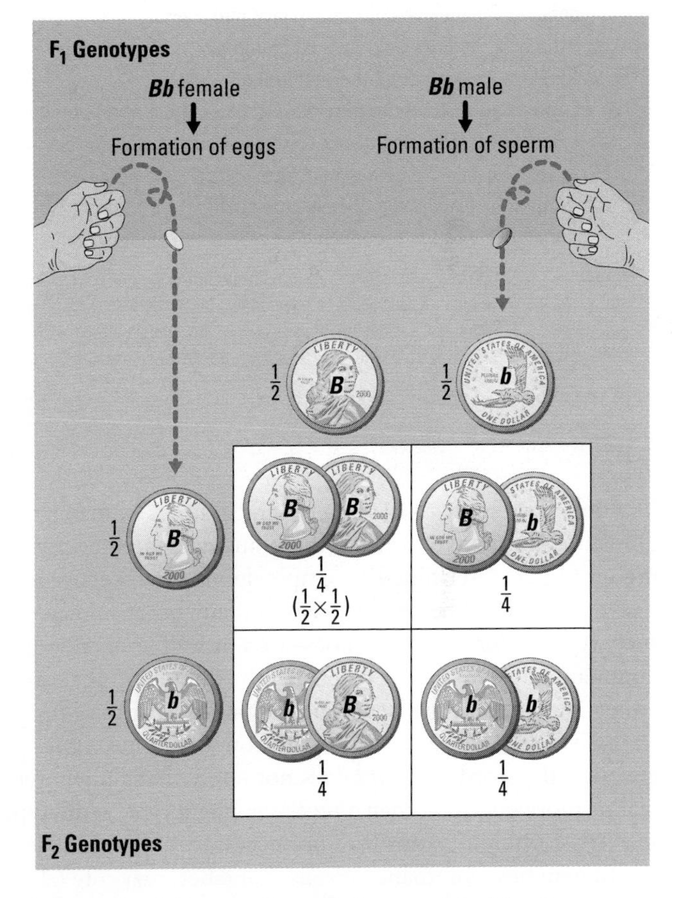

Figure 9.11 Segregation of alleles and fertilization as chance events. When a heterozygote (*Bb*) forms gametes, segregation of alleles is like the toss of a coin. An egg cell has a 50% chance of receiving the dominant allele and a 50% chance of receiving the recessive allele. The same odds apply to a sperm cell. Like two separately tossed coins, segregation during sperm and egg formation occurs as two independent events. To determine the probability that an individual offspring will inherit the dominant allele from both parents, we multiply the probabilities of each required event: $\frac{1}{2} \times \frac{1}{2} = \frac{1}{4}$.

together at fertilization is $\frac{1}{2} \times \frac{1}{2} = \frac{1}{4}$. This is exactly the answer given by the Punnett square in Figure 9.11. If we know the genotypes of the parents, we can predict the probability for any genotype among the offspring. By applying the rules of probability to segregation and independent assortment, we can solve some rather complex genetics problems.

CHECKPOINT

1. What is meant by "self-fertilization" of a plant?
2. How can two plants that have different genotypes for flower color be identical in phenotype?
3. Which of Mendel's laws is represented by each statement?
   a. Each pair of alleles segregates independently during gamete formation.
   b. Alleles segregate during gamete formation; fertilization creates pairs of alleles once again.
4. An imaginary creature called a glump has blue skin, controlled by the allele *B*, which is dominant to allele *b*. Glumps of genotype *bb* are albino (white). The independently inherited characteristic of eye color is specified by alleles *E* and *e*. Glumps of genotype *EE* have orange eyes; glumps that are homozygous recessives for the eye-color gene have pink eyes.
   a. If a homozygous blue-skinned glump mates with an albino glump, what will be the phenotypes of their offspring?
   b. You use a testcross to determine the genotype of an orange-eyed glump. Half of the offspring of the testcross have orange eyes and half have pink eyes. What is the genotype of the orange-eyed parent?
   c. A glump of genotype *BbEe* is crossed with one that is *bbEe*. What is the probability of their producing an albino baby glump with pink eyes?

Answers: 1. An egg cell in the plant's carpel is fertilized by a sperm-carrying pollen grain from a stamen of the same plant. **2.** One could be homozygous for the dominant allele, while the other is heterozygous. **3.** Statement **a** is the law of independent assortment; statement **b** is the law of segregation. **4. a.** All will have blue skin (*BB* x *bb* = *Bb*). **b.** *Ee* **c.** $\frac{1}{8}$ (that is, $\frac{1}{2} \times \frac{1}{4}$)

Family Pedigrees

Mendel's laws apply to the inheritance of many human traits. **Figure 9.12** illustrates alternative forms of three human characteristics that are each thought to be determined by simple dominant-recessive inheritance at one gene locus. (The genetic basis of many other human characteristics—such as eye color and hair color—are not well understood.) If we call the dominant allele of any such gene *A*, the dominant phenotype results from either the homozygous genotype *AA* or the heterozygous genotype *Aa*. Recessive phenotypes always result from the homozygous genotype *aa*. In genetics, the word *dominant* does not imply that a phenotype is either normal or more common than a recessive phenotype; **wild-type traits** (those seen most often in nature) are not necessarily specified by dominant alleles. In genetics, dominance means that a heterozygote (*Aa*), carrying only a single copy of a dominant allele, displays the dominant phenotype. By contrast, the phenotype of the corresponding recessive allele is seen only in a homozygote (*aa*). Recessive traits are often more common in the population than dominant ones. For example, the absence of freckles is more common than their presence.

How do we know how particular human traits are inherited? Researchers working with pea plants or Labrador retrievers can perform testcrosses. But geneticists who study humans obviously cannot control the mating of their subjects. Instead, they must analyze the results of matings

Dominant Traits | **Recessive Traits**

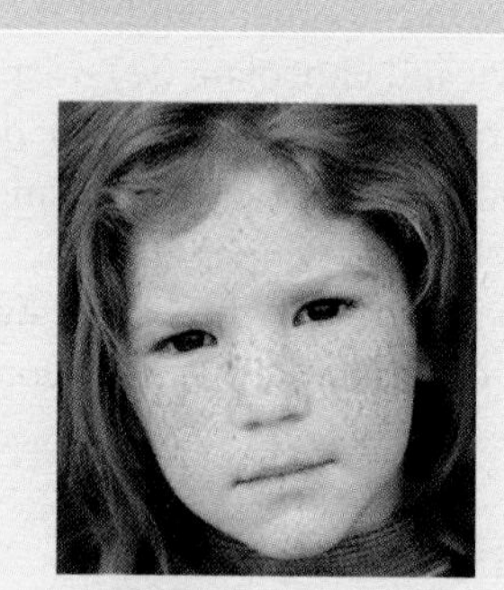
Freckles

No freckles

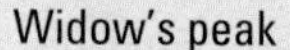
Widow's peak

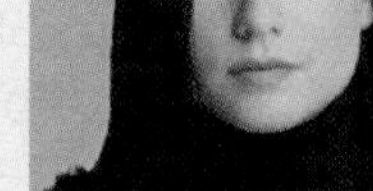
Straight hairline

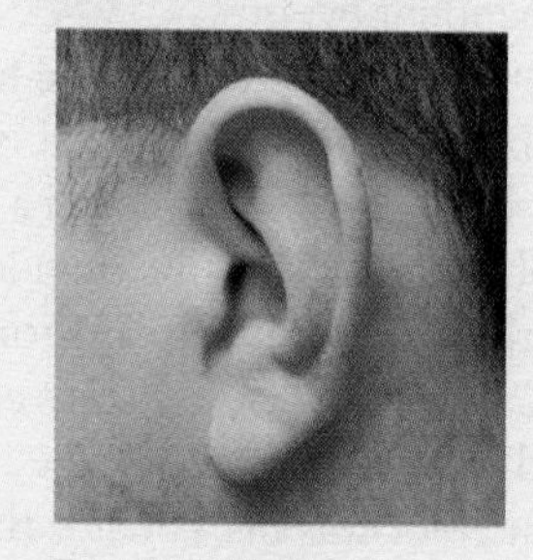
Free earlobe

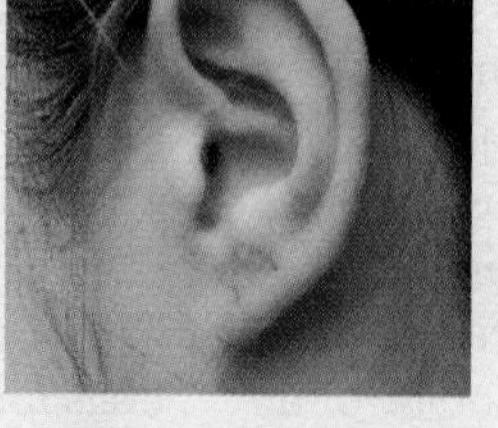

Attached earlobe

Figure 9.12 Examples of inherited traits in humans.

that have already occurred. First, the geneticist collects as much information as possible about a family's history for the trait. Then the researcher assembles this information into a family tree—the family **pedigree**. Finally, to analyze the pedigree, the geneticist uses Mendel's concept of dominant and recessive alleles and his law of segregation.

Let's apply this approach to the example in **Figure 9.13**, which shows part of a real pedigree. This pedigree is from a family that lived on Martha's Vineyard, an island off the coast of Massachusetts, where a particular kind of inherited deafness was once prevalent. In the pedigree, □ represents a male, ○ represents a female, and colored symbols (■ and ●) indicate deafness. The earliest generation studied is at the top of the pedigree. Notice that deafness did not appear in this generation and that it showed up in only two of the seven children in the third generation. By applying Mendel's law, we can deduce that the deafness allele is recessive; that's the most likely way that Jonathan Lambert could be deaf while both of his parents were not. Once we know that, we can label all the deaf individuals as homozygous recessive (*dd*). Mendel's laws also enable us to deduce the other genotypes that are shown. Jonathan and Elizabeth's hearing children, for example, must be heterozygous (*Dd*) because they all inherited one copy of the recessive (*d*) gene from their father. People who have one copy of the allele for a recessive disorder and do not exhibit symptoms are called **carriers** of the disorder.

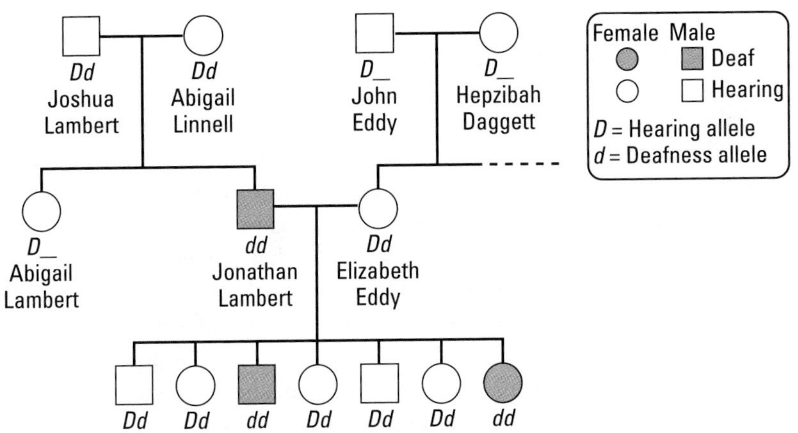

Figure 9.13 A family pedigree showing inheritance of deafness. Because only people who are homozygous for the recessive allele are deaf, Jonathan Lambert's genotype must have been *dd*. Therefore, both his parents must have carried a *d* allele along with the *D* allele that gave them normal hearing. Two of Jonathan's children were deaf (*dd*), so his wife, who had normal hearing, must also have carried a *d* allele. Likewise, all of the couple's normal children must have been heterozygous (*Dd*). What are the genotypes of Elizabeth Eddy's parents and Jonathan's sister Abigail? These three people had normal hearing, so they must have carried at least one *D* allele. And at least one of Elizabeth's parents must have had a *d* allele. But more than that we cannot say without additional information.

Human Disorders Controlled by a Single Gene

The hereditary deafness found on Martha's Vineyard is just one of thousands of human genetic disorders currently known to be inherited as dominant or recessive traits controlled by a single gene locus; ten more examples are listed in **Table 9.1**. These disorders show simple inheritance

Table 9.1 Some Autosomal Disorders in Humans

Disorder	Major Symptoms	Incidence
Recessive Disorders		
Albinism	Lack of pigment in skin, hair, and eyes	$\frac{1}{22,000}$
Cystic fibrosis	Excess mucus in lungs, digestive tract, liver; increased susceptibility to infections; death in early childhood unless treated	$\frac{1}{1,800}$ European-Americans
Galactosemia	Accumulation of galactose in tissues; mental retardation; eye and liver damage	$\frac{1}{100,000}$
Phenylketonuria (PKU)	Accumulation of phenylalanine in blood; lack of normal skin pigment; mental retardation unless treated	$\frac{1}{10,000}$ in U.S. and Europe
Sickle-cell disease (homozygous)	Sickled red blood cells; damage to many tissues	$\frac{1}{500}$ African-Americans
Tay Sachs disease	Lipid accumulation in brain cells; mental deficiency; blindness; death in childhood	$\frac{1}{3,500}$ Ashkenazi Jews
Dominant Disorders		
Achondroplasia	Dwarfism	$\frac{1}{25,000}$
Alzheimer's disease (one type)	Mental deterioration; usually strikes late in life	Not known
Huntington's disease	Mental deterioration and uncontrollable movements; strikes in middle age	$\frac{1}{25,000}$
Hypercholesterolemia	Excess cholesterol in blood; heart disease	$\frac{1}{500}$

patterns like the ones Mendel studied in pea plants. The genes involved are all located on autosomes, chromosomes other than the sex chromosomes X and Y.

Recessive Disorders Most human genetic disorders are recessive. They range in severity from relatively harmless conditions to deadly diseases. The vast majority of people afflicted with recessive disorders are born to normal parents who are both heterozygotes—that is, who are carriers of the recessive allele for the disorder but are normal in appearance.

Using Mendel's laws, we can predict the fraction of affected offspring likely to result from a marriage between two carriers. Suppose one of Jonathan Lambert's hearing sons (*Dd*) married a hearing woman whose pedigree indicated that her genotype was also *Dd*. What is the probability that they would have a deaf child? As the Punnett square in **Figure 9.14** shows, each child of two carriers has a $\frac{1}{4}$ chance of inheriting two recessive alleles. Thus, we can say that about one-fourth of the children of this marriage are likely to be deaf. We can also say that a hearing ("normal") child from such a marriage has a $\frac{2}{3}$ chance of being a carrier (that is, on average, two out of three of the offspring with the hearing phenotype will be *Dd* carriers). We can apply this same method of pedigree analysis and prediction to any genetic trait controlled by a single gene locus.

The most common lethal genetic disease in the United States is cystic fibrosis. Though the disease affects only about one in 17,000 African-Americans and about one in 90,000 Asian-Americans, it occurs in approximately one out of every 2,500 Caucasian (European ancestry) births. The cystic fibrosis allele is recessive and is carried by about one in every 25 European-Americans. A person with two copies of this allele has cystic fibrosis, which is characterized by an excessive secretion of very thick mucus from the lungs, pancreas, and other organs. This mucus can interfere with breathing, digestion, and liver function and makes the person vulnerable to recurrent bacterial infections. Untreated, most children with cystic fibrosis die by the time they are 5 years old. Although there is no cure for this fatal disease, a special diet, antibiotics to prevent infection, frequent pounding of the chest and back to clear the lungs, and other treatments can prolong life well into adulthood.

Like cystic fibrosis, most genetic disorders are not evenly distributed across all ethnic groups. Such uneven distribution is the result of prolonged geographic isolation of certain populations. For example, the isolated lives of the Martha's Vineyard inhabitants between 1700 and 1900 fostered marriage between close relatives. Consequently, the frequency of deafness remained high, and the deafness allele was rarely transmitted to outsiders.

With the increased mobility in most societies today, it is relatively unlikely that two carriers of a rare, harmful allele will meet and mate. However, the probability increases greatly if close relatives marry and have children. People with recent common ancestors are more likely to carry the same recessive alleles than are unrelated people. Therefore, a mating of close relatives, called **inbreeding,** is more likely to produce offspring homozygous for a harmful recessive trait. Wildlife biologists have observed increased incidence of harmful recessive traits among many types of inbred animals. For example, dogs that have been inbred for appearance may have serious genetic disorders, such as weak hip joints or eye problems. The detrimental effects of inbreeding are also seen in some endangered species, such as cheetahs (see Chapter 13).

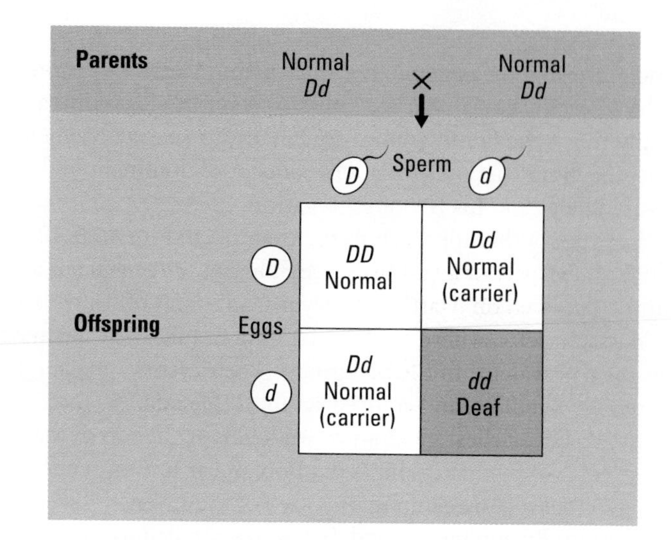

Figure 9.14 Predicted offspring when both parents are carriers for a recessive disorder. Carriers are heterozygotes who have one copy of a potentially harmful recessive allele. Because carriers also have the dominant allele, they do not display the disorder. One-fourth of the offspring of two carriers would be expected to be homozygous recessive and therefore deaf.

Dominant Disorders A number of human disorders are caused by dominant alleles. Some are nonlethal conditions, such as extra fingers and toes, or fingers and toes that are webbed. One serious disorder caused by a dominant allele is **achondroplasia,** a form of dwarfism. Among people with this disorder, the head and torso of the body develop normally, but the arms and legs are short **(Figure 9.15)**. About one out of 25,000 people have achondroplasia. The homozygous dominant genotype for this characteristic causes death of the embryo, and therefore only heterozygotes, individuals with a single copy of the defective allele, have this disorder. (This also means that a person with achondroplasia has a 50% chance of passing the condition on to any children.) Therefore, all those who do not have achondroplasia, more than 99.99% of the population, are homozygous for the recessive allele. This example makes it clear that a dominant allele is not necessarily more plentiful in a population than the corresponding recessive allele.

Dominant alleles that are lethal are, in fact, much less common than lethal recessives. One reason for this difference is that the dominant lethal allele cannot be carried by heterozygotes without affecting them. Many lethal dominant alleles result from mutations in a sperm or egg that subsequently kill the embryo. And if the afflicted individual is born but does not survive long enough to reproduce, he or she will not pass on the lethal allele. This is in contrast to lethal recessive mutations, which are perpetuated from generation to generation by the reproduction of heterozygous carriers.

A lethal dominant allele can escape elimination, however, if it does not cause death until a relatively advanced age. The allele that causes **Huntington's disease,** a degeneration of the nervous system that usually does not begin until middle age, is one example. As the disease progresses, it produces uncontrollable movements in all parts of the body. The loss of brain cells leads to memory loss and impaired judgment and contributes to depression. Diminished motor skills eventually affect the ability to swallow and speak. Death usually ensues 10 to 20 years after the onset of symptoms. Unfortunately, by the time the symptoms of Huntington's disease become evident, the afflicted individual may have had children, half of whom (on average) will have received the lethal dominant allele for the disease. This example demonstrates that a dominant allele is not necessarily "better" than the corresponding recessive allele.

Figure 9.15 Achondroplasia, a dominant trait. Dr. Michael C. Ain, a pediatric orthopedic surgeon at the Johns Hopkins Children's Center, specializes in the repair of bone defects caused by achondroplasia and related disorders.

CHECKPOINT

1. What is a wild-type trait?
2. A man and a woman who are both carriers of cystic fibrosis allele *c*, which is recessive to the normal allele *C*, have had three children without cystic fibrosis. If the couple has a fourth child, what is the probability that the child will have the disorder?
3. Peter is a 28-year-old man whose father died of Huntington's disease. Neither Peter's much older sister, who is 44, nor his 60-year-old mother shows any signs of the disease. What is the probability that Peter has inherited Huntington's disease?

Answers: 1. A trait that is the prevailing one in nature **2.** $\frac{1}{4}$ (The genotypes and phenotypes of their other children are irrelevant.) **3.** $\frac{1}{2}$

Variations on Mendel's Laws

Mendel's two laws explain inheritance in terms of discrete factors—genes—that are passed along from generation to generation according to simple rules of probability. Mendel's laws are valid for all sexually reproducing organisms, including garden peas, Labrador retrievers, and human beings. But just as the basic rules of musical harmony cannot account for all the rich sounds of a symphony, Mendel's laws stop short of explaining some patterns of genetic inheritance. In particular, they do not explain the many cases of inherited characteristics that exist in more than two clear-cut variants—such as flower color that can be red, white, or pink or human skin color in all its range of shades. In fact, for most sexually reproducing organisms, cases where Mendel's rules can strictly account for the patterns of inheritance are relatively rare. More often, the observed inheritance patterns are more complex. We will now add several extensions to Mendel's laws that account for this complexity.

Incomplete Dominance in Plants and People

The F_1 offspring of Mendel's pea crosses always looked like one of the two parent plants. In such situations, the dominant allele has the same effect on the phenotype whether present in one or two copies. But for some characteristics, the F_1 hybrids have an appearance in between the phenotypes of the two parents, an effect called **incomplete dominance.** For instance, when red snapdragons are crossed with white snapdragons, all the F_1 hybrids have pink flowers **(Figure 9.16)**. And in the F_2 generation, the genotypic ratio and the phenotypic ratio are the same: 1:2:1.

We also see examples of incomplete dominance in humans. One case involves a recessive allele (*h*) that causes **hypercholesterolemia,** a condition characterized by dangerously high levels of cholesterol in the blood. Normal individuals are *HH*. Heterozygotes (*Hh*; about one in 500 people) have blood cholesterol levels about twice normal. They are unusually prone to atherosclerosis, the blockage of arteries by cholesterol buildup in artery walls, and they may have heart attacks from blocked heart arteries by their mid-30s. Hypercholesterolemia is even more serious in homozygous individuals (*hh*; about one in a million people). These homozygotes have about five times the normal amount of blood cholesterol and may have heart attacks as early as age 2.

If we look at the molecular basis for hypercholesterolemia, we can understand the intermediate phenotype of heterozygotes **(Figure 9.17)**. The *H* allele specifies a cell-surface receptor protein that certain cells use to mop up excess cholesterol from the blood. With only half as many receptors as *HH* individuals, heterozygotes can remove much less excess cholesterol.

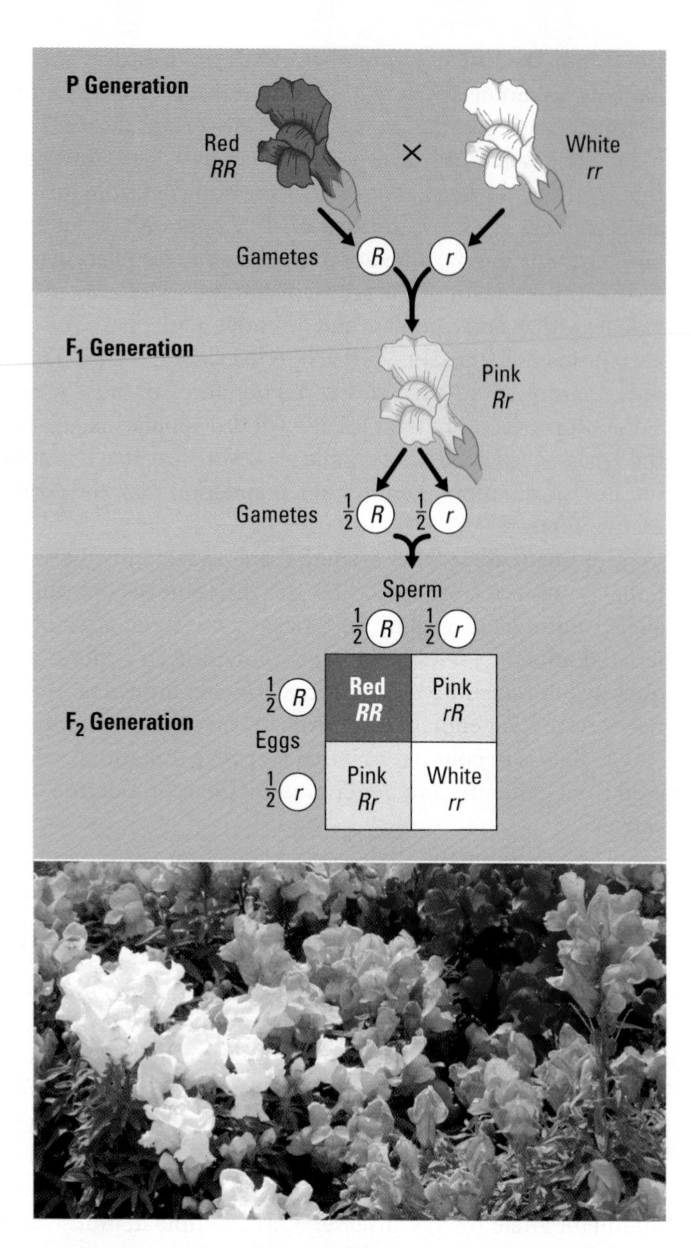

Figure 9.16 Incomplete dominance in snapdragons. In incomplete dominance, a heterozygote has a phenotype in between the phenotypes of the two kinds of homozygotes. The phenotypic ratios in the F_2 generation are the same as the genotypic ratios, 1:2:1. Compare this diagram with Figure 9.6, where one of the alleles displays complete dominance.

ABO Blood Type: An Example of Multiple Alleles and Codominance

So far, we have discussed inheritance patterns involving only two alleles per gene. But most genes occur in more than two forms, known as multiple

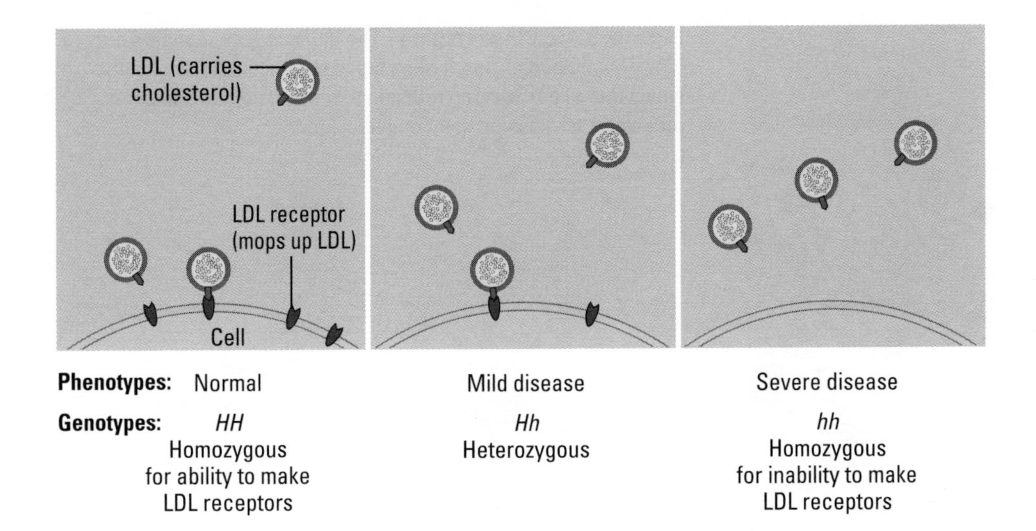

Figure 9.17 Incomplete dominance in human hypercholesterolemia. The dominant allele, which normal individuals carry in duplicate (*HH*), specifies a cell-surface protein called an LDL receptor. LDLs, or low-density lipoproteins, are cholesterol-containing particles in the blood. The LDL receptors pick up LDL particles from the blood and promote their uptake by cells that break down the cholesterol (see Figure 5.19). This process helps prevent the accumulation of cholesterol in the arteries. Heterozygotes (*Hh*) have only half the normal number of LDL receptors, and homozygotes (*hh*) have none, thereby allowing dangerous levels of LDL to build up in the blood.

alleles. Although each individual carries, at most, only two different alleles for a particular gene, in cases of multiple alleles, more than two possible alleles exist in the population.

The **ABO blood groups** in humans are an example of multiple alleles. There are three common alleles for the characteristic of ABO blood type, which in various combinations produce four phenotypes: A person's blood group may be O, A, B, or AB. These letters refer to two carbohydrates, designated A and B, that may be found on the surface of red blood cells **(Figure 9.18)**. A person's red blood cells may be coated with carbohydrate A (type A), carbohydrate B (type B), both (type AB), or neither (type O). Matching compatible blood groups is critical for safe blood transfusions. Our immune system produces blood proteins called antibodies that can bind specifically to the blood cell carbohydrates we lack. If a donor's blood cells have a carbohydrate that is foreign to the recipient, then the recipient's antibodies will cause the donated blood cells to clump together. This clumping can kill the

Genotype	Phenotype (Blood Type)	Red Blood Cells
I^AI^A or I^Ai	A	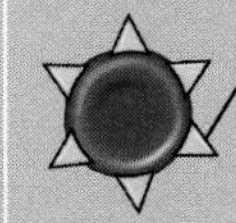 Carbohydrate A
I^BI^B or I^Bi	B	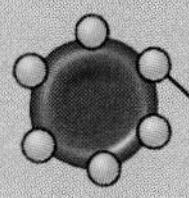 Carbohydrate B
I^AI^B	AB	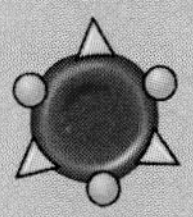
ii	O	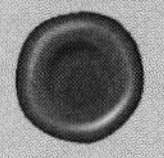

Figure 9.18 Multiple alleles for the ABO blood groups. The three versions of the gene responsible for blood type may produce carbohydrate A (allele I^A), carbohydrate B (allele I^B), or neither carbohydrate (allele i). Because each person carries two alleles, six genotypes are possible that result in four different phenotypes.

Blood Group (Phenotype)	Antibodies Present in Blood	Reaction When Blood from Groups Below Is Mixed with Antibodies from Groups at Left			
		O	A	B	AB
O	Anti-A Anti-B				
A	Anti-B				
B	Anti-A				
AB	—				

Figure 9.19 Blood-typing. The clumping reaction that occurs between antibodies and foreign blood cells is the basis of blood-typing and of the adverse reaction that occurs when someone receives a transfusion of incompatible blood.

recipient. The clumping reaction is also the basis of a blood-typing lab test **(Figure 9.19)**.

The four blood types result from various combinations of the three different alleles, symbolized as I^A (for the ability to make substance A), I^B (for B), and *i* (for neither A nor B). Each person inherits one of these alleles from each parent. Because there are three alleles, there are six possible genotypes, as listed in Figure 9.18. Both the I^A and I^B alleles are dominant to the *i* allele. Thus, I^AI^A and I^Ai people have type A blood, and I^BI^B and I^Bi people have type B. Recessive homozygotes (*ii*) have type O blood; they make neither the A nor the B carbohydrate. Finally, people of genotype I^AI^B make *both* carbohydrates. In other words, the I^A and I^B alleles exhibit **codominance,** meaning that both alleles are expressed in heterozygous individuals (I^AI^B), who have type AB blood.

Another example of codominance is the genetics of coat color among some cows and horses. The *R* allele specifies red hairs and *r* specifies white. An *RR* cow has an all-red coat, an *rr* cow has all white, but an *Rr* cow has a roan coat (both red and white hairs). Be careful to distinguish codominance (the expression of both alleles) from incomplete dominance (the expression of one intermediate trait).

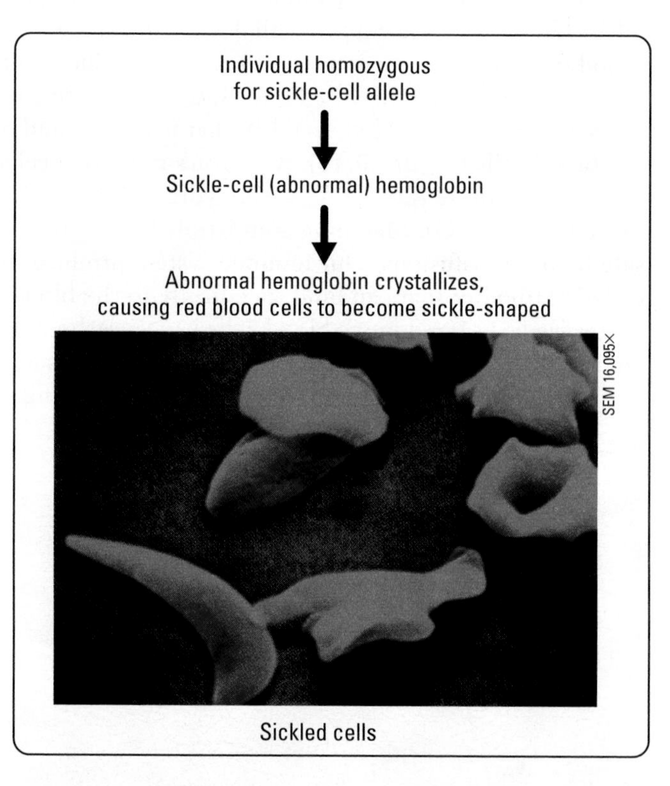

Figure 9.20 Sickle-cell disease, multiple effects of a single human gene. People who are homozygous for the sickle-cell allele have sickle-cell disease. They produce abnormal hemoglobin molecules, which cause their red blood cells, normally smooth disks, to become sickle-shaped. Sickled cells are destroyed rapidly by the body, and their destruction may cause anemia and general weakening of the body (hence the alternative name for the disorder, sickle-cell anemia). Also, because of their angular shape, sickled cells do not flow smoothly in the blood and tend to accumulate and clog tiny blood vessels. Blood flow to body parts is reduced, resulting in periodic fever, pain, and damage to various organs, including the heart, brain, and kidneys.

Pleiotropy and Sickle-Cell Disease

Most of our genetic examples to this point have been cases in which each gene specifies only one hereditary characteristic. But in many cases, one gene influences several characteristics. The impact of a single gene on more than one characteristic is called **pleiotropy.**

An example of pleiotropy in humans is **sickle-cell disease,** a disorder characterized by a diverse set of symptoms. The direct effect of the sickle-cell allele is to make red blood cells produce abnormal hemoglobin molecules (see Figure 3.23). These abnormal molecules tend to link together and crystallize. As the hemoglobin crystallizes, the normally disk-shaped red blood cells deform to a sickle shape with jagged edges **(Figure 9.20)**. Sickling of the cells, in turn, can lead to a cascade of symptoms, such as weakness, pain, organ damage,

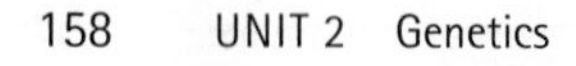

and paralysis. Blood transfusions and drugs may relieve some of the symptoms, but there is no cure, and sickle-cell disease kills about 100,000 people in the world annually.

In most cases, only people who are homozygous for the sickle-cell allele suffer from the disease. Heterozygotes, who have one sickle-cell allele and one normal allele, are usually healthy. However, the two alleles are actually codominant: Both alleles are expressed in heterozygous individuals, and their red blood cells contain both normal and abnormal hemoglobin. A simple blood test can distinguish homozygotes from heterozygotes.

Sickle-cell disease is by far the most common inherited illness among black people, striking one in 500 African-American children. About one in ten African-Americans is a heterozygote. Among Americans of other ancestries, the sickle-cell allele is very rare.

One in ten is an unusually high frequency of heterozygotes for an allele with such harmful effects in homozygotes. We might expect that the frequency of the sickle-cell allele in the population would be much lower because many homozygotes die before passing their genes to the next generation. The high frequency appears to be a vestige of the roots of African-Americans. Sickle-cell disease is most common in tropical Africa, where the deadly disease malaria is also prevalent. The microorganism that causes malaria spends part of its life cycle inside red blood cells. When it enters the red blood cells of a person with the sickle-cell allele, it triggers sickling. The body destroys most of the sickled cells and the microbes within them, and the microbe does not grow well in the cells that remain. Consequently, in many parts of Africa, sickle-cell carriers (heterozygotes) who contract malaria live longer and have more offspring than noncarriers with malaria. In this way, malaria has caused the frequency of the sickle-cell allele to remain relatively high in much of the African continent. To put it in evolutionary terms, as long as malaria is a danger, individuals with the sickle-cell allele have a selective advantage.

Polygenic Inheritance

Mendel studied genetic characteristics that could be classified on an either-or basis, such as purple or white flower color. However, many characteristics, such as human skin color and height, vary along a continuum in a population. Many such features result from **polygenic inheritance,** the additive effects of two or more genes on a single phenotypic characteristic. (This is the converse of pleiotropy, in which a single gene affects several characteristics.)

Let's consider a hypothetical model of how polygenic inheritance might work. Assume that skin pigmentation in humans is controlled by three genes that are inherited separately, like Mendel's pea genes. (Actually, genetic evidence indicates that *at least* three genes control this characteristic.) The dark-skin allele for each gene (*A*, *B*, and *C*) contributes one "unit" of darkness to the phenotype and is incompletely dominant to the other alleles (*a*, *b*, and *c*). A person who is *AABBCC* would be very dark, while an *aabbcc* individual would be very light. An *AaBbCc* person would have skin of an intermediate shade. Because the alleles have an additive effect, the genotype *AaBbCc* would produce the same skin color as any other genotype with just three dark-skin alleles, such as *AABbcc*. **Figure 9.21** shows how inheritance of these three genes could lead to a range of skin color in a population. Seven levels of pigmentation would arise at the frequencies indicated by the bars in the graph.

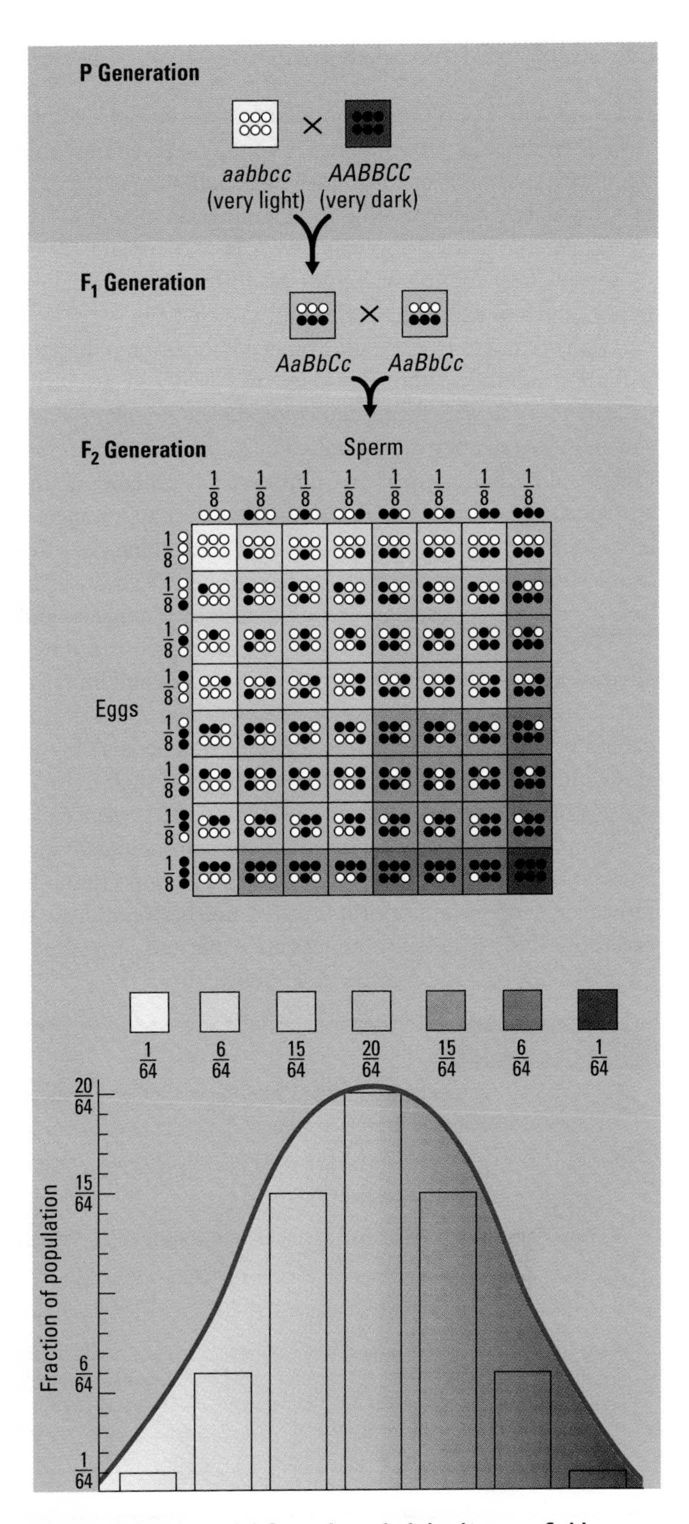

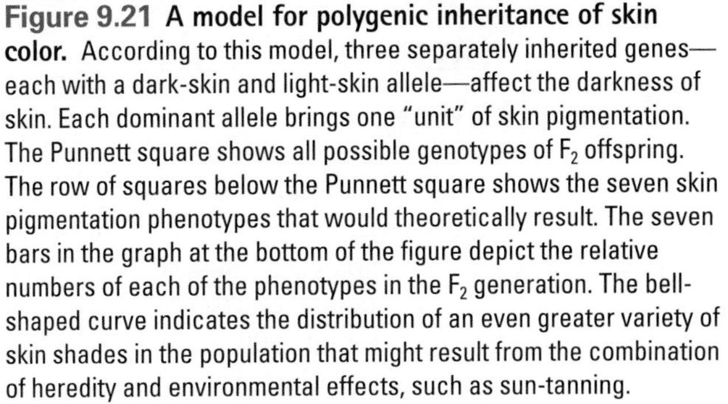

Figure 9.21 A model for polygenic inheritance of skin color. According to this model, three separately inherited genes—each with a dark-skin and light-skin allele—affect the darkness of skin. Each dominant allele brings one "unit" of skin pigmentation. The Punnett square shows all possible genotypes of F_2 offspring. The row of squares below the Punnett square shows the seven skin pigmentation phenotypes that would theoretically result. The seven bars in the graph at the bottom of the figure depict the relative numbers of each of the phenotypes in the F_2 generation. The bell-shaped curve indicates the distribution of an even greater variety of skin shades in the population that might result from the combination of heredity and environmental effects, such as sun-tanning.

The Role of Environment

If we examine a real human population for the skin-color phenotype, we would see more shades than just seven. The true range might be similar to the entire spectrum of color under the bell-shaped curve in Figure 9.21. In fact, no matter how carefully we characterize the genes for skin color, a purely genetic description will always be incomplete. This is because some intermediate shades of skin color result from the effects of environmental factors, such as exposure to the sun.

Many characteristics result from a combination of heredity and environment. For example, although a single tree is locked into its inherited genotype, its leaves vary in size, shape, and color, depending on exposure to wind and sun and the tree's nutritional state. For humans, nutrition influences height; exercise alters build; experience improves performance on intelligence tests; and social and cultural factors can greatly affect appearance. As geneticists learn more and more about our genes, it's becoming clear that many human characteristics—such as a person's susceptibility to heart disease, cancer, alcoholism, and schizophrenia—are influenced by both genes and environment.

Whether human characteristics are more influenced by genes or by the environment—nature or nurture—is a very old and hotly contested issue. For some characters, such as the ABO blood group, a given genotype mandates a very specific phenotype. In contrast, a person's blood count of red and white cells varies quite a bit, depending on such factors as the altitude, the customary level of physical activity, and the presence of infectious agents.

Simply spending time with identical twins will convince anyone that environment, and not just genes, affects a person's traits **(Figure 9.22)**. However, there is an important difference between these two sources of variation: Only genetic influences are inherited. Any effects of the environment are not passed on to the next generation.

Figure 9.22 As a result of environmental influences, even identical twins can look different.

CHECKPOINT

1. Why is a testcross unnecessary to determine whether a snapdragon with red flowers is homozygous or heterozygous?
2. Maria has type O blood, and her sister has type AB blood. What are the genotypes of the girls' parents?
3. How does sickle-cell disease exemplify the concept of pleiotropy?
4. Based on the skin-tone model in Figure 9.21, put the following individuals in order by skin tone, from lightest to darkest: *AAbbCC, aaBBcc, AabBCc, Aabbcc, AaBBCC.*

Answers: 1. Only plants homozygous for the dominant allele have red flowers; heterozygotes have pink flowers. **2.** One parent is I^Ai, and the other parent is I^Bi. **3.** Homozygotes for the sickle-cell allele have abnormal hemoglobin, and its effect on the shape of red blood cells leads to a cascade of symptoms affecting many organs of the body. **4.** *Aabbcc, aaBBcc, AabBCc, AAbbCC, AaBBCC*

The Chromosomal Basis of Inheritance

MP3 Tutor Chromosomal Basis of Inheritance

Mendel published his results in 1866, but not until long after he died did biologists understand the significance of his work. Cell biologists worked out the processes of mitosis and meiosis (see Chapter 8) in the late 1800s. Then, around 1900, researchers began to notice parallels between the behavior of chromosomes and the behavior of Mendel's heritable factors. One of biology's most important concepts—the chromosome theory of inheritance—began to emerge.

The **chromosome theory of inheritance** states that genes are located at specific positions on chromosomes and that the behavior of chromosomes during meiosis and fertilization accounts for inheritance patterns. Indeed, it is chromosomes that undergo segregation and independent assortment during meiosis and thus account for Mendel's laws. You can see the chromosomal basis of Mendel's laws by following the fates of two genes during meiosis and fertilization in pea plants **(Figure 9.23)**. Recasting Mendel's laws in terms of the interactions of chromosomes provides us with a greater understanding of the underlying principles of genetics and shows us how Mendel's work can be extended in several important ways.

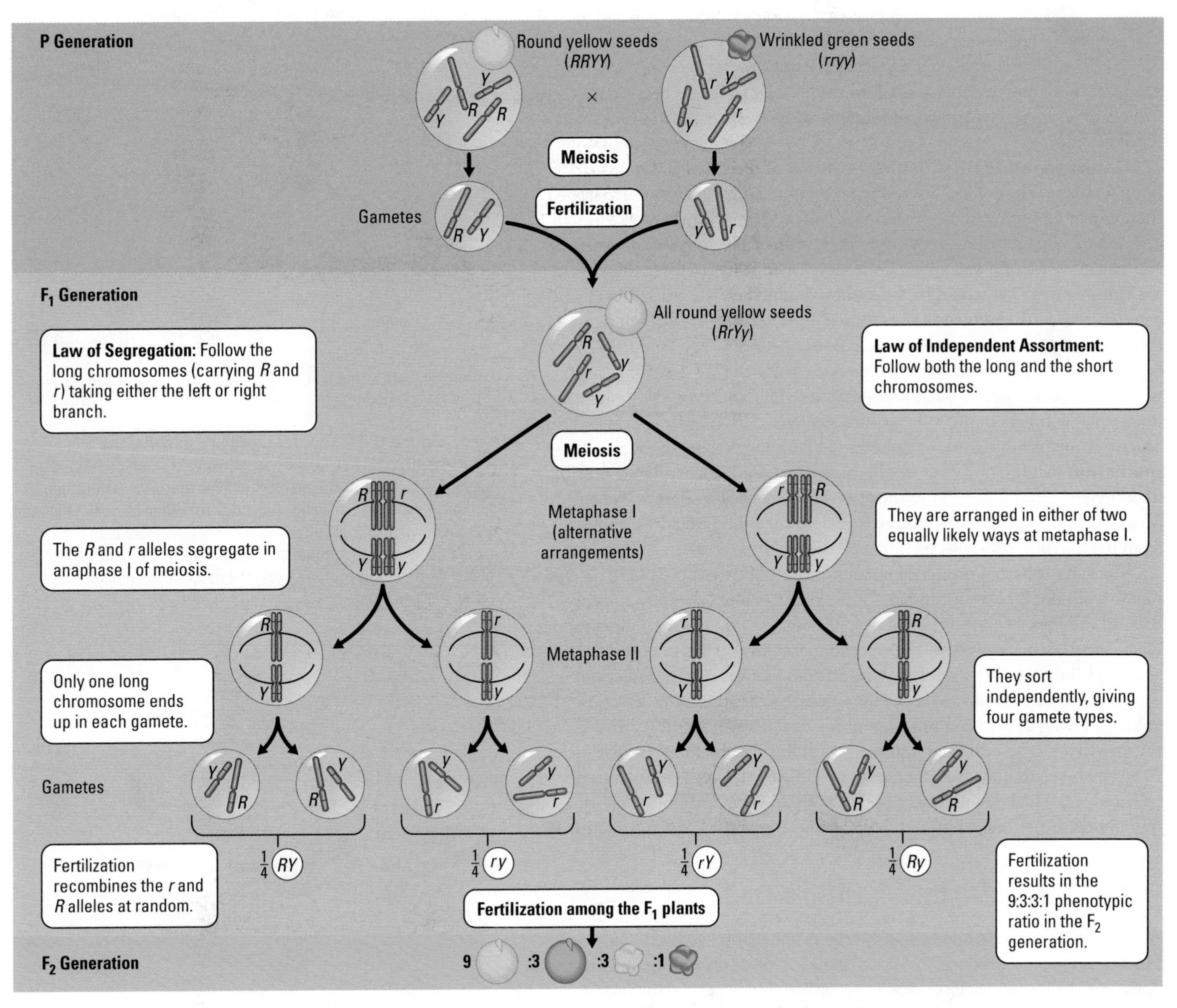

Figure 9.23 The chromosomal basis of Mendel's laws. Here we correlate the results of one of Mendel's dihybrid crosses (see Figure 9.8) with the behavior of chromosomes. Starting with two true-breeding parental plants, the diagram follows two genes through the F_1 and F_2 generations. The two genes specify seed shape (alleles *R* and *r*) and seed color (alleles *Y* and *y*) and are on different chromosomes.

Linked Genes

Realizing that genes are *on chromosomes* when they segregate leads to several important insights. For example, alleles that start out together on the same chromosome would be expected to travel together during meiosis and fertilization. In general, genes that are located close together on a chromosome, called **linked genes,** tend to be inherited together as a set. They do not follow Mendel's law of independent assortment. In fact, the unusual inheritance patterns that result from linked genes led to their discovery.

Are Some Genes Linked? The New York City laboratory of American embryologist Thomas Hunt Morgan was the site of some important early studies on how chromosome behavior can affect heredity. In the early 1900s, Morgan and his colleagues used the fruit fly *Drosophila melanogaster* in many of their experiments. Often seen flying around overripe fruit, *Drosophila* is a good research animal for studies of inheritance because it is easily and inexpensively grown and can produce several generations in a matter of months.

Morgan's work began with **observations** of fruit flies with unusual body features. These flies differed from true-breeding wild-type *Drosophila* flies, which have gray bodies (genotype *GG*) and long wings (*LL*). Morgan decided to cultivate true-breeding mutant (non-wild-type) fruit flies that had black bodies (*gg*) and short, underdeveloped wings (*ll*). He then asked the **question**: What would happen when doubly heterozygous flies (*GgLl*) were crossed with double mutants (*ggll*)? In other words, Morgan performed a dihybrid testcross (**Figure 9.24a**).

Morgan's **hypothesis** was that the body-color and wing-shape genes would act in the standard Mendelian manner: The wild-type fly would produce equal numbers of the four possible gametes (*GL*, *Gl*, *gL*, and *gl*) and the mutant would produce only *gl* gametes. This hypothesis led to the **prediction** that the offspring would show equal numbers of all four possible phenotypes ($\frac{1}{4}$ gray/long, $\frac{1}{4}$ gray/short, $\frac{1}{4}$ black/long, $\frac{1}{4}$ black/short). When Morgan conducted his **experiment** and examined the 2,300 offspring flies from the dihybrid testcross, he obtained the **results** shown in **Figure 9.24b**.

Morgan's testcross produced equal numbers of two of the four expected phenotypes but far fewer numbers of the other two phenotypes. In fact, 83% of the flies displayed the phenotypes of the parents (gray/long and black/short), while only 17% showed nonparental phenotypes (gray/short and black/long). Clearly, Morgan's results did not match his prediction.

To explain his results, Morgan hypothesized that the genes for body color and wing shape were linked—in this case, *G* with *L* and *g* with *l*—causing them to be inherited together. Such linkage would mean that meiosis in the heterozygous fruit flies would yield gametes with two genotypes (*GL* and *gl*). The large numbers of flies with gray/long and black/short traits in the experiment resulted from fertilization among these gametes. ■

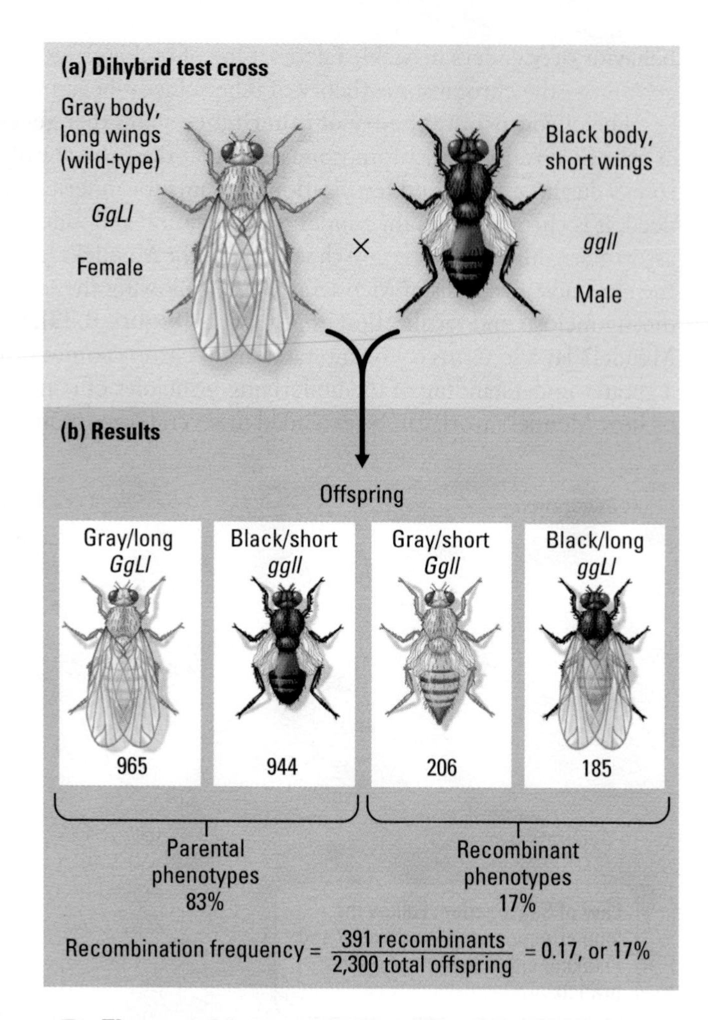

Figure 9.24 Thomas Morgan's experiment and results. **(a)** In a famous experiment, Morgan performed a dihybrid testcross on *Drosophila* fruit flies that were heterozygous for body color (*Gg*) and wing shape (*Ll*). **(b)** When Morgan tabulated the phenotypes of 2,300 offspring, he observed more than the expected number of parental phenotypes (gray/long and black/short) and fewer than expected nonparental phenotypes (gray/short and black/long).

Genetic Recombination: Crossing Over

But what of the smaller numbers of Morgan's gray/short and black/long flies? How could they be produced if the two genes involved are linked? In Chapter 8, we saw that during meiosis, crossing over between homologous chromosomes shuffles chromosome segments in the haploid daughter cells (see Figure 8.18), thereby producing new combinations of alleles. **Figure 9.25** reviews crossing over, showing how two linked genes can give rise to four different

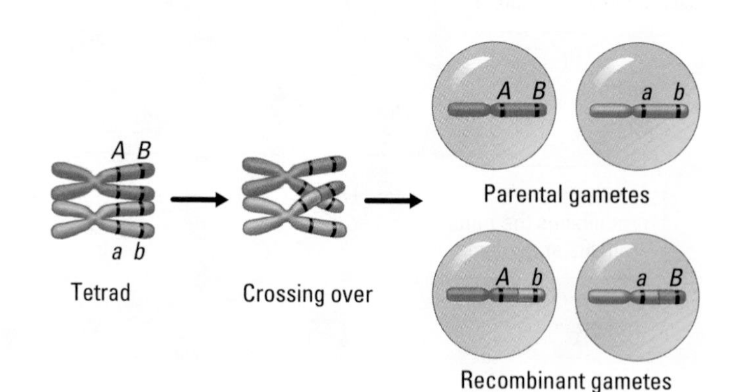

Figure 9.25 Review: Crossing over can produce recombinant gametes.

Figure 9.26 Understanding Morgan's results. This figure helps explain the unexpected results observed by Morgan in his dihybrid testcross (shown in Figure 9.24). The disproportionately large numbers of the parental phenotypes among the offspring is explained by gene linkage, the fact that the gene loci for the two characteristics are located nearby on the same chromosome and tend to remain together during meiosis and fertilization (*G* with *L* and *g* with *l*). As a result of crossing over, some of the gametes end up with recombinant chromosomes, carrying new combinations of alleles, either Gl or gL. When the recombinant gametes participate in fertilization, recombinant offspring can result.

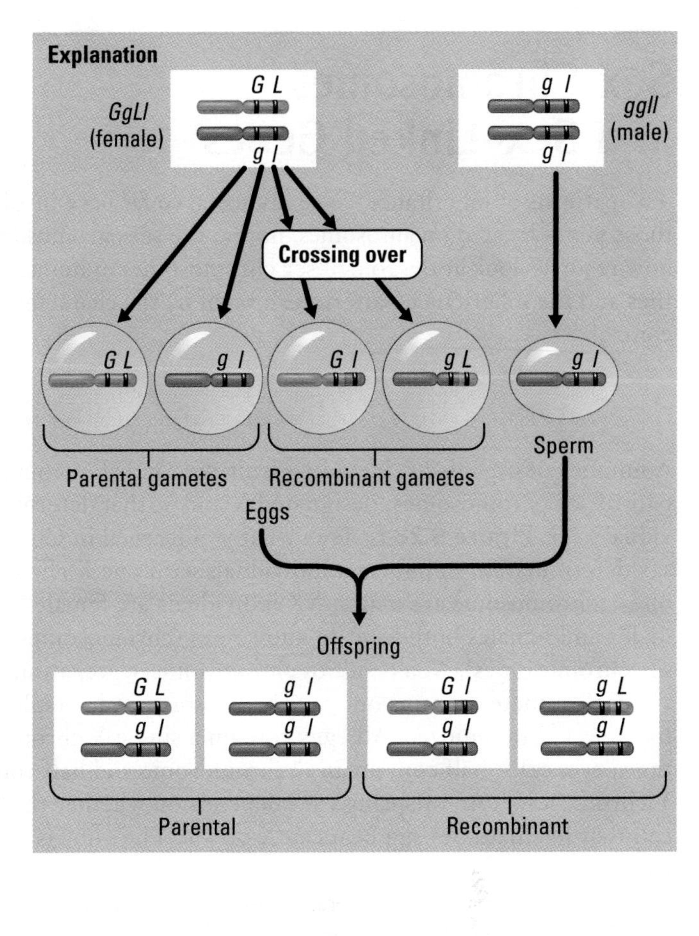

gamete genotypes. Two of the gamete genotypes reflect the presence of parental-type chromosomes, which have not been altered by crossing over. In contrast, the other two gamete genotypes are recombinant (nonparental). The chromosomes of these gametes carry new combinations of alleles that result from the exchange of chromosome segments in crossing over.

Morgan hypothesized that crossing over accounted for the observed nonparental offspring, those with the recombinant phenotypes (gray/short and black/long). Morgan reasoned that the 17% of offspring with recombinant phenotypes must have resulted from fertilization involving recombinant gametes **(Figure 9.26)**. The percentage of recombinant offspring among the total is called the **recombination frequency.** In this case, using the data from Figure 9.24, the recombination frequency is the sum of the recombinant flies, 206 + 185, divided by 2,300, which equals 17%.

Linkage Maps

While working with *Drosophila*, Alfred H. Sturtevant, one of Morgan's students, developed a way to use crossover data to map gene loci. This technique is based on the assumption that the farther apart two genes are on a chromosome, the higher the probability that a crossover will occur between them. In other words, the greater the distance between two genes, the more points there are between them where crossing over can occur. (This assumption is not entirely accurate, but it is good enough to provide useful data.) With this principle in mind, geneticists can use recombination data to assign genes to relative positions on chromosomes—that is, to map genes. **Figure 9.27** shows some of the data used to map three genes that reside on one of the *Drosophila* chromosomes. The result is called a **linkage map.**

The linkage-mapping method has proved extremely valuable in establishing the relative positions of many genes in many organisms. The real beauty of the technique is that a wealth of information about genes can be learned simply by breeding and observing the organisms; no fancy equipment is required. Today, with DNA technology, geneticists can determine the physical distances in nucleotides between linked genes. However, linkage mapping still played a critical role in the mapping of the human genome, which we'll discuss in Chapter 12.

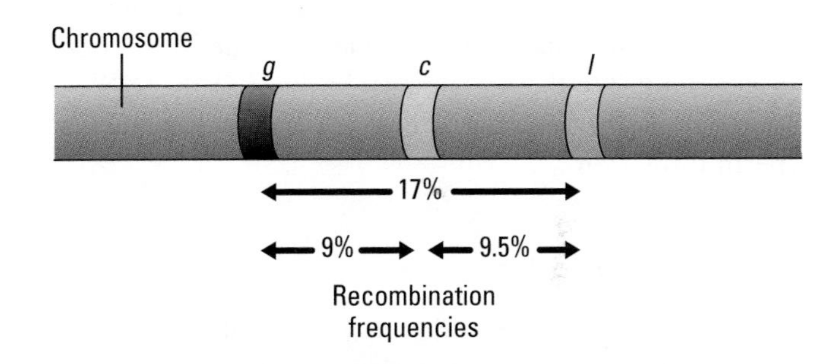

Figure 9.27 Using crossover data to map genes. This diagram represents a part of the *Drosophila* chromosome that carries three linked genes. Their recessive alleles specify black body (*g*), cinnabar eyes (*c*), and short wings (*l*). The corresponding dominant alleles specify the wild-type traits of gray body, red eyes, and long wings. Under the chromosome are the actual recombination frequencies (crossover frequencies) between these genes, taken two at a time: 17% between *g* and *l*, 9% between *g* and *c*, and 9.5% between *l* and *c*. Geneticists reasoned that these values represent the relative distances between the genes. Because the recombination frequencies between *g* and *c* and between *l* and *c* are approximately half the frequency between *g* and *l*, gene *c* must lie roughly midway between *g* and *l*. Thus, the sequence of these genes on their chromosome must be *g-c-l* (or the equivalent *l-c-g*).

CHECKPOINT

1. Which of Mendel's laws have their physical basis in the following phases of meiosis?
   a. The orientation of homologous chromosome pairs in metaphase I
   b. The separation of homologues in anaphase I
2. What are linked genes?
3. If the order of three genes on a chromosome is *A-B-C*, between which two genes will the recombination frequency be highest?

Answers: 1.a. The law of independent assortment **b.** The law of segregation **2.** Genes located near each other on the same chromosome that tend to be inherited together **3.** Between *A* and *C*

Sex Chromosomes and Sex-Linked Genes

The patterns of inheritance we've discussed so far have involved only those genes located on autosomes, not on the sex chromosomes. We're now ready to look at the role of sex chromosomes in humans and fruit flies and the inheritance patterns exhibited by the characteristics they control.

Sex Determination in Humans and Fruit Flies

A number of organisms, including fruit flies and all mammals, have a pair of sex chromosomes, designated X and Y, that determine an individual's sex. **Figure 9.28** reviews what you learned in Chapter 8 about sex determination in humans. Individuals with one X chromosome and one Y chromosome are males; XX individuals are females. Human males and females both have 44 autosomes (chromosomes other than sex chromosomes). As a result of chromosome segregation during meiosis, each gamete contains one sex chromosome and a haploid set of autosomes (22 in humans). All eggs contain a single X chromosome. Of the sperm cells, half contain an X chromosome and half contain a Y chromosome. An offspring's sex depends on whether the sperm cell that fertilizes the egg bears an X or a Y. The same is true for *Drosophila*.

The genetic basis of sex determination in humans is not yet completely understood, but one gene on the Y chromosome plays a crucial role. This gene, called *SRY*, triggers the development of the testes. In the absence of a functioning version of *SRY*, an individual develops ovaries rather than testes. Other genes on the Y chromosome are also necessary for normal sperm production. The X-Y system in other mammals is similar to that in humans. In the fruit fly's X-Y system, some genetic details are different, although a Y chromosome is still essential for sperm formation.

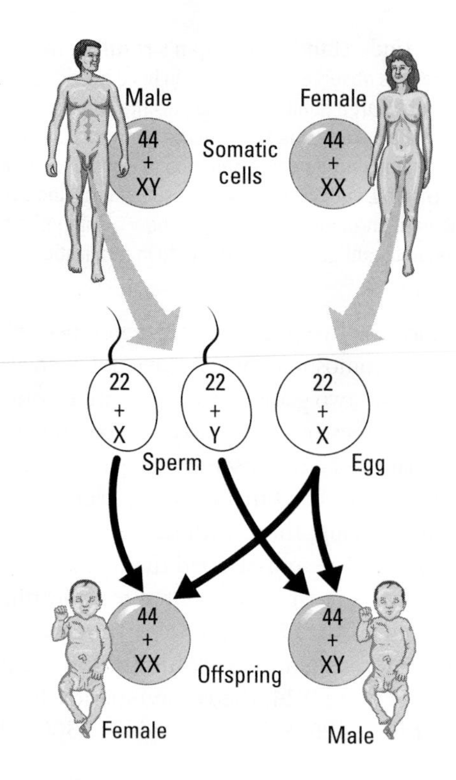

Figure 9.28 Sex determination in humans: a review.

Sex-Linked Genes

Besides bearing genes that determine sex, the so-called sex chromosomes also contain genes for characteristics unrelated to maleness or femaleness. Any gene located on a sex chromosome is called a **sex-linked gene.** (Note that the use of "linked" here is somewhat different from its use when we refer to "linked genes.") The X chromosome contains many more genes than the Y; therefore, most sex-linked genes unrelated to sex determination are found on the X chromosome.

Sex linkage was discovered by T. H. Morgan while he was studying the inheritance of white eye color in fruit flies. Wild-type fruit flies have red eyes; white eyes are very rare (**Figure 9.29**). White eye color turned out to be a recessive trait whose gene is on the fly's X chromosome; the gene and trait are thus said to be sex-linked (or X-linked).

Figure 9.30a shows what happens when a white-eyed male fly is mated with a homozygous red-eyed female. All the offspring have red eyes, suggesting that the wild type is dominant. When those offspring are bred to each other, the classic 3:1 phenotypic ratio of wild-type to

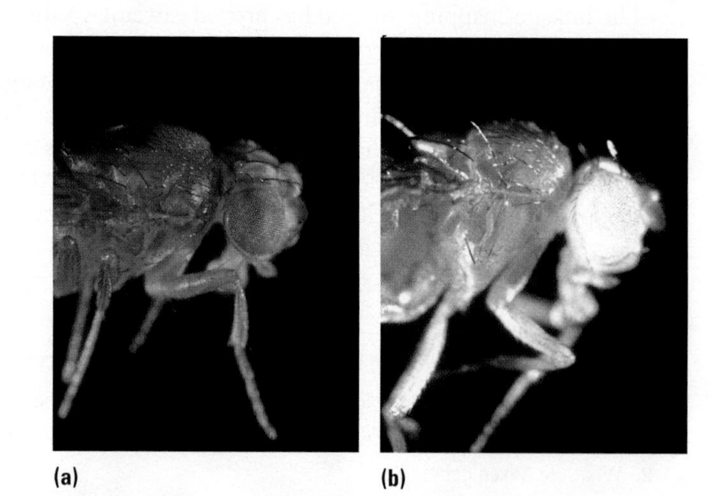

Figure 9.29 Fruit fly eye color. (a) The wild-type color for fruit fly eyes is red. **(b)** White eye color is a rare variant. Study of the inheritance patterns of the eye-color characteristic led to the conclusion that the gene controlling eye color is sex-linked.

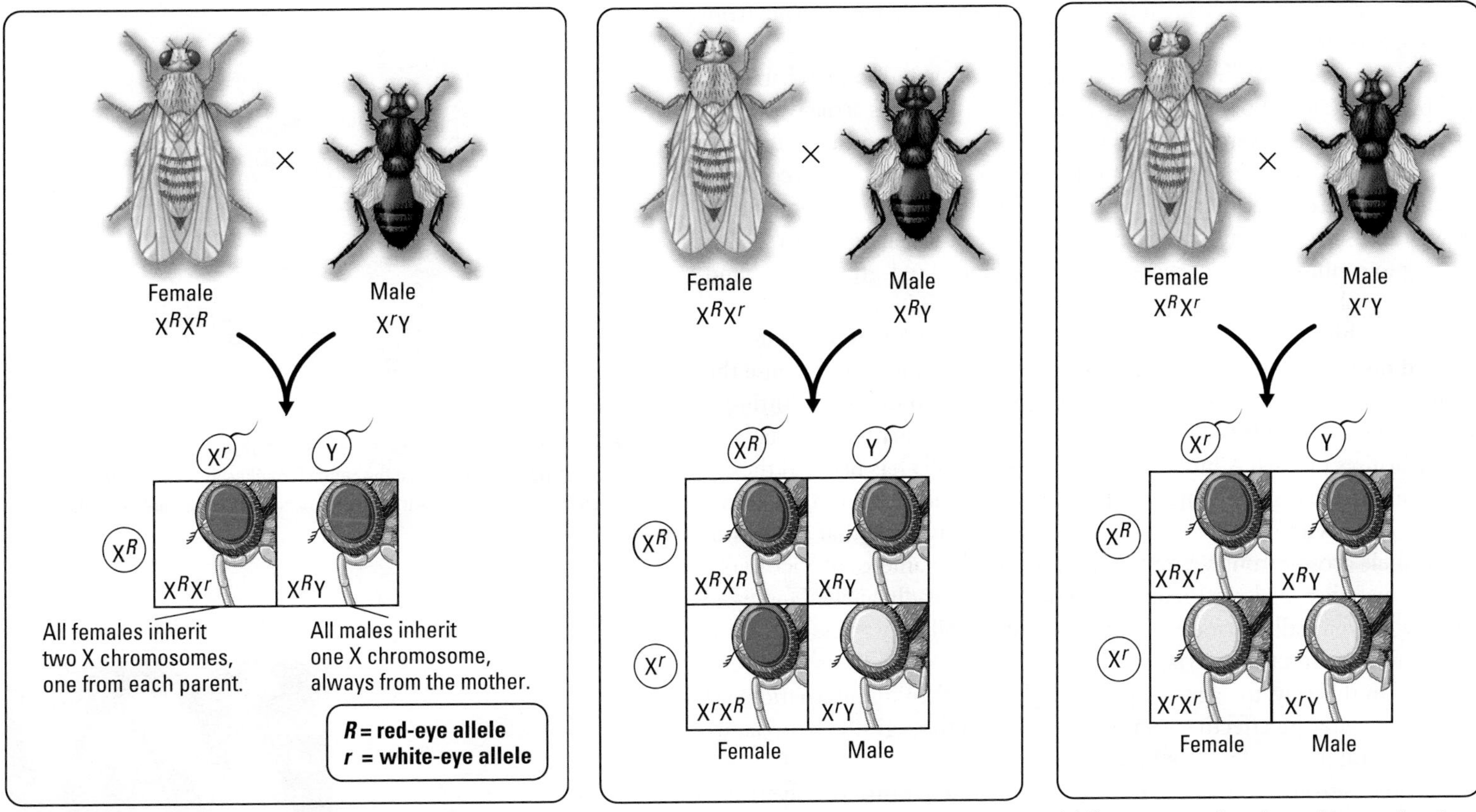

Figure 9.30 Inheritance of white eye color, a sex-linked trait. We use the uppercase letter *R* for the dominant, wild-type red-eye allele and *r* for the recessive, white-eye allele. To indicate that these alleles are on the X chromosome, we show them as superscripts to the letter X. Thus, red-eyed male fruit flies have the genotype X^RY, and white-eyed males are X^rY. The Y chromosome does not have a gene locus for eye color; therefore, the male's phenotype results entirely from his single X-linked gene. In the female, X^RX^R and X^RX^r flies have red eyes, and X^rX^r flies have white eyes.

white-eyed appears among the offspring **(Figure 9.30b)**. However, there is a surprising twist: The white-eye trait shows up only in males. All the females have red eyes, while half the males have red eyes and half have white eyes. This is because the gene involved in this inheritance pattern is located exclusively on the X chromosome; there is no corresponding eye-color locus on the Y. Thus, females (XX) carry two copies of the gene for this characteristic, while males (XY) carry only one. Because the white-eye allele is recessive, a female will have white eyes only if she receives that allele on both X chromosomes **(Figure 9.30c)**. For a male, however, a single copy of the white-eye allele confers white eyes. Since a male has only one X chromosome, there can be no wild-type allele present to offset the recessive allele.

Sex-Linked Disorders in Humans

A number of human conditions, including red-green color blindness, hemophilia, and a type of muscular dystrophy, result from sex-linked recessive alleles. Recessive sex-linked traits are expressed much more frequently in men than in women. For example, color blindness is about 20-fold more common among males than females. Like a male fruit fly, if a man inherits only one sex-linked recessive allele—from his mother—the allele will be expressed. In contrast, a woman has to inherit two such alleles—one from each parent—to exhibit the trait.

Red-green color blindness is a common sex-linked disorder characterized by a malfunction of light-sensitive cells in the eyes. It is actually a class of disorders involving several X-linked genes. A person with normal color vision can see more than 150 colors. In contrast, someone with red-green color blindness can see fewer than 25. For some affected people, red hues appear gray; others see gray instead of green; still others are green-weak or red-weak, tending to confuse shades of these colors. (If you have red-green color blindness, you probably cannot see the numeral in **Figure 9.31**.)

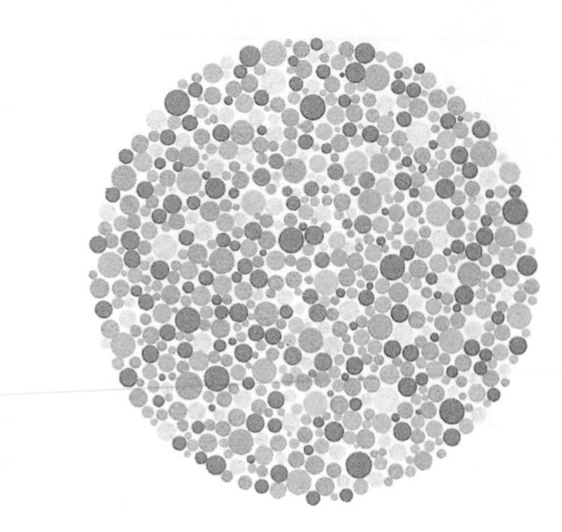

Figure 9.31 A test for red-green color blindness. Can you see a green numeral 7 against the reddish background? If not, you probably have some form of red-green color blindness, a sex-linked trait.

Hemophilia is a sex-linked recessive trait with a long, well-documented history. Hemophiliacs bleed excessively when injured because they have inherited an abnormal allele for a factor involved in blood clotting. The most seriously affected individuals may bleed to death after relatively minor bruises or cuts. A high incidence of hemophilia has plagued the royal families of Europe. The first royal hemophiliac seems to have been a son of Queen Victoria (1819–1901) of England. It is likely that the hemophilia allele arose through a mutation in one of the gametes of Victoria's mother or father, making Victoria a carrier of the deadly allele. Hemophilia was eventually introduced into the royal families of Prussia, Russia, and Spain through the marriages of two of Victoria's daughters who were carriers. In this way, the age-old practice of strengthening international alliances by marriage effectively spread hemophilia through the royal families of several nations **(Figure 9.32)**.

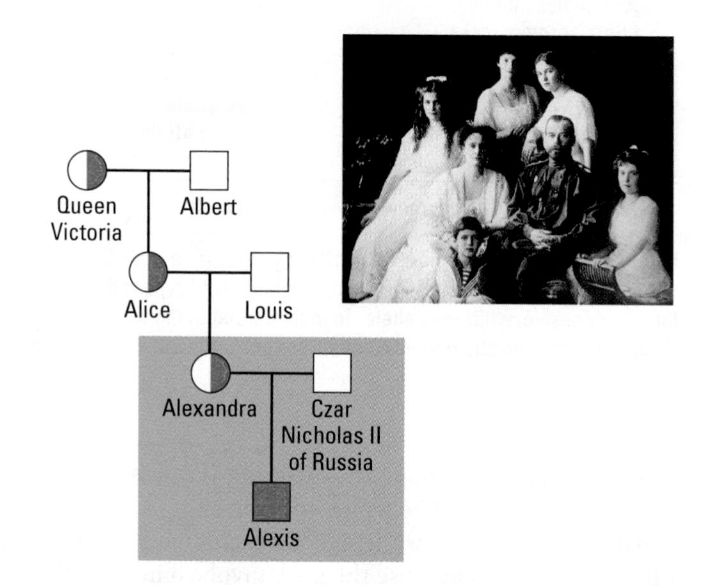

Figure 9.32 Hemophilia in the royal family of Russia. The photograph shows Queen Victoria's granddaughter Alexandra, her husband Nicholas, who was the last czar of Russia, their son Alexis, and their daughters. The pedigree uses half-colored symbols to represent heterozygous carriers of the hemophilia allele. As you can see in the pedigree, Alexandra, like her mother and grandmother, was a carrier, and Alexis had the disease.

Another sex-linked recessive disorder is **Duchenne muscular dystrophy,** a condition characterized by a progressive weakening and loss of muscle tissue. Almost all cases are males, and the first symptoms appear in early childhood, when the child begins to have difficulty standing up. He is inevitably wheelchair-bound by age 12. Eventually, he becomes severely weakened, and normal breathing becomes difficult. Death usually occurs by age 20. For such a severe disease, Duchenne muscular dystrophy is relatively common. In the general U.S. population, about one in 3,500 male babies are affected, and the disease is even more common in some inbred populations. In one Amish community in Indiana, for instance, one out of every 100 males are born with the disease. With the help of DNA technology (discussed in Chapter 12), the gene that, when mutated, causes Duchenne muscular dystrophy has been mapped at a particular point on the X chromosome. The gene's wild-type allele codes for a protein that is present in normal muscle but missing in Duchenne patients.

CHECKPOINT

1. At the moment of conception in humans, what determines the sex of the offspring?
2. What is meant by a sex-linked gene?
3. A white-eyed *Drosophila* female is mated with a red-eyed (wild-type) male. What do you predict for the numerous offspring?
4. Neither Rudy nor Carla has Duchenne muscular dystrophy, but their first son does have it. If the couple has a second child, what is the probability that he or she will also have the disease?

Answers: **1.** Whether the egg is fertilized by a sperm bearing an X chromosome (producing a female offspring) or a Y chromosome (producing a male) **2.** A gene that is located on a sex chromosome, usually the X chromosome **3.** All female offspring will be heterozygous ($X^R X^r$), with red eyes; all male offspring will be white-eyed ($X^r Y$). **4.** $\frac{1}{4}$ ($\frac{1}{2}$ chance the child will be male times $\frac{1}{2}$ chance that he will inherit the X carrying the disease allele)

EVOLUTION CONNECTION

The Telltale Y Chromosome

The Y chromosome of human males is only about one-third the size of the X chromosome and carries only $\frac{1}{100}$ as many genes. As mentioned earlier, most of the Y genes seem to determine maleness and male fertility and are not present on the X. In prophase I of meiosis, only two tiny regions of the X and Y chromosomes can cross over (recombine). Crossing over requires that the DNA in the recombining regions line up and match very closely, and for the human X and Y chromosomes, this can only happen at their tips.

Nevertheless, biologists believe that X and Y were once a fully homologous pair, having evolved from a pair of autosomes about 300 million years ago. Since that time, four major episodes of change, the most recent about 40 million years ago, have rearranged pieces of the Y chromosome in a way that prevents the matching required for recombination with the X. Over millions of years, many Y genes have disappeared, shrinking the chromosome. Meanwhile, the lack of substantial exchange between X and Y chromosomes has prevented male-determining genes from migrating to the X chromosome—which could have had the disastrous consequence of making XX individuals male.

Because most of the DNA of the Y chromosome passes more or less intact from father to son—the main changes are rare mutations—the Y chromosome provides a window to the ancestry of the male lines of humanity. Recently, researchers used comparisons of Y DNA to confirm the claim by the Lemba people of southern Africa that they are descended from ancient Jews (**Figure 9.33**). Certain Y DNA sequences had previously been shown to be distinctive of Jews, particularly of the priestly caste called Cohanim (descendants of Moses' brother Aaron, according to the Bible). These same sequences were found at equally high frequencies among the Lemba. And it was a study of Y DNA that supported the oral tradition among the descendants of the slave Sally Hemings that Thomas Jefferson was their ancestor. On a grander scale, Y chromosome studies are providing evidence bearing on the question of when and where fully modern humans first evolved. ■

Figure 9.33 A Lemba man. DNA sequences from the Y chromosomes of Lemba men suggest that the Lemba are descended from ancient Jews.

Chapter Review

SUMMARY OF KEY CONCEPTS

For study help and activities, go to campbellbiology.com or the student CD-ROM.

Heritable Variation and Patterns of Inheritance

- Gregor Mendel was the first to study genetics, the science of heredity, by analyzing patterns of inheritance. In his 1866 paper, he emphasized that heritable factors (genes) retain permanent identities.
- **In an Abbey Garden** Mendel started with true-breeding varieties of pea plants representing two alternative variants of a hereditary characteristic, such as flower color. He then crossed the different varieties and traced the inheritance of traits from generation to generation.
- **Mendel's Law of Segregation** Pairs of alleles separate during gamete formation; fertilization restores the pairs.

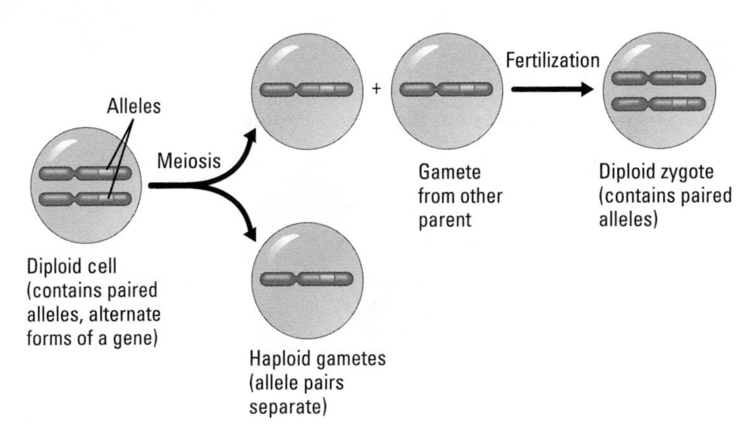

If an individual's genotype (genetic makeup) has two different alleles for a gene and only one influences the organism's phenotype (appearance), that allele is said to be dominant and the other allele recessive. Alleles of a gene reside at the same locus, or position, on homologous chromosomes. Where the allele pair match, the organism is homozygous; where they're different, the organism is heterozygous.

Activity *Monohybrid Cross*

- **Mendel's Law of Independent Assortment** By following two characteristics at once, Mendel found that the alleles of a pair segregate independently of other allele pairs during gamete formation.

Activity *Dihybrid Cross*

Activity *Gregor's Garden*

- **Using a Testcross to Determine an Unknown Genotype**

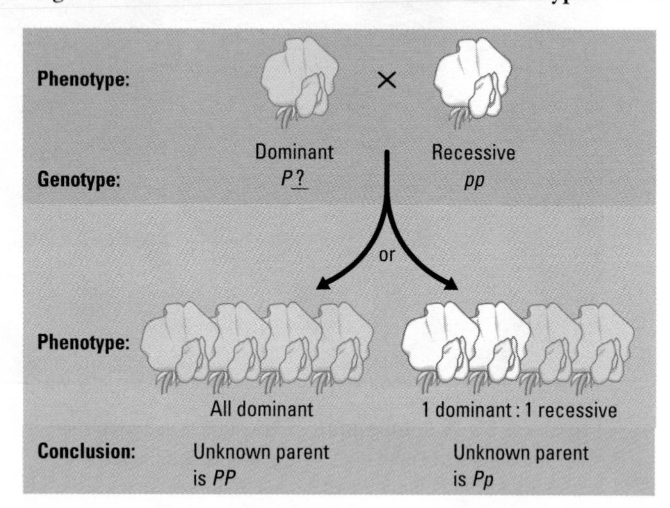

- **The Rules of Probability** Inheritance follows the rules of probability. The chance of inheriting a recessive allele from a heterozygous parent is $\frac{1}{2}$ The chance of inheriting it from both of two heterozygous parents is $\frac{1}{2} \times \frac{1}{2} = \frac{1}{4}$, illustrating the rule of multiplication for calculating the probability of two independent events.

- **Family Pedigrees** The inheritance of many human traits, from freckles to genetic diseases, follows Mendel's laws and the rules of probability. Geneticists use family pedigrees to determine patterns of inheritance and individual genotypes among humans.

- **Human Disorders Controlled by a Single Gene** For traits that vary within a population, the one most commonly found in nature is called the wild type. Many inherited disorders in humans are controlled by a single gene (represented by two alleles). Most such disorders, such as cystic fibrosis, are caused by autosomal recessive alleles. A few, such as Huntington's disease, are caused by dominant alleles.

Variations on Mendel's Laws

- **Incomplete Dominance in Plants and People**

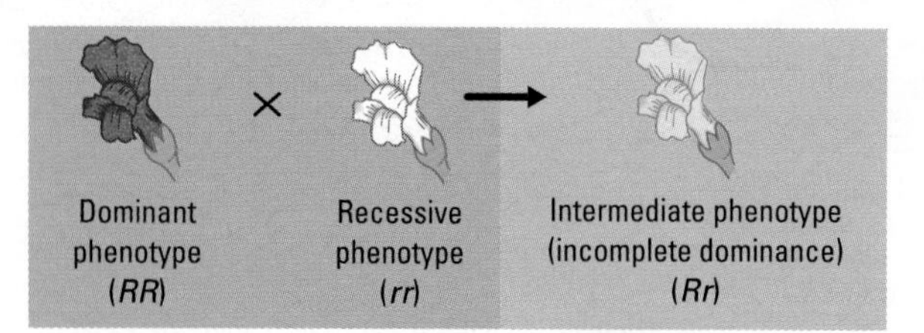

Activity *Incomplete Dominance*

- **ABO Blood Type: An Example of Multiple Alleles and Codominance** Within a population, there are often multiple kinds of alleles for a characteristic, such as the three alleles for the ABO blood groups. The alleles determining the A and B blood factors are codominant; that is, both are expressed in a heterozygote.

- **Pleiotropy and Sickle-Cell Disease**

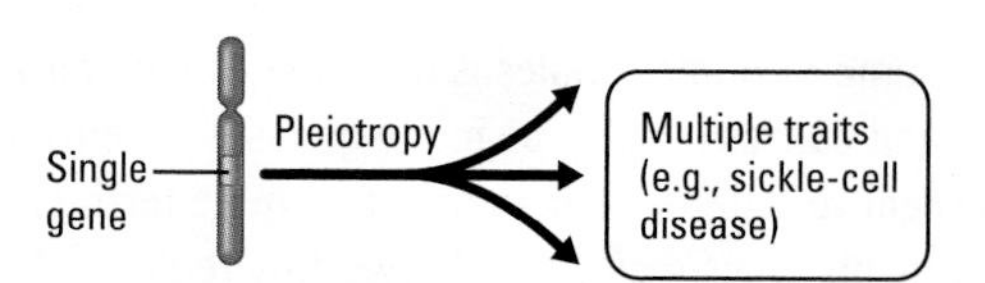

The presence of two copies of the sickle-cell allele at a single gene locus brings about the many symptoms of sickle-cell disease. But having just one copy of the sickle-cell allele may be beneficial because it provides some protection against the disease malaria.

- **Polygenic Inheritance**

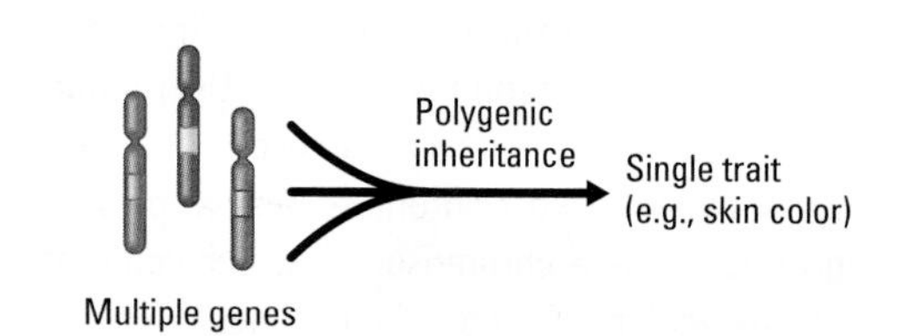

- **The Role of Environment** Many human characteristics result from a combination of genetic and environmental effects, but only genetic influences are biologically heritable.

The Chromosomal Basis of Inheritance

MP3 Tutor *Chromosomal Basis of Inheritance*

- Genes are located on chromosomes, whose behavior during meiosis and fertilization accounts for inheritance patterns (see Figure 9.23).

- **Linked Genes** Certain genes are linked: They tend to be inherited together because they lie close together on the same chromosome.

- **Genetic Recombination: Crossing Over** Crossing over can separate linked alleles, producing gametes with recombinant chromosomes and offspring with recombinant phenotypes.

- **Linkage Maps** The fact that crossing over between linked genes is more likely to occur between genes that are farther apart enables geneticists to map the relative positions of genes on chromosomes.

Activity *Linked Genes and Crossing Over*

Sex Chromosomes and Sex-Linked Genes

- **Sex Determination in Humans and Fruit Flies** In humans, the Y chromosome has a gene that triggers the development of testes; an absence of this gene triggers the development of ovaries.

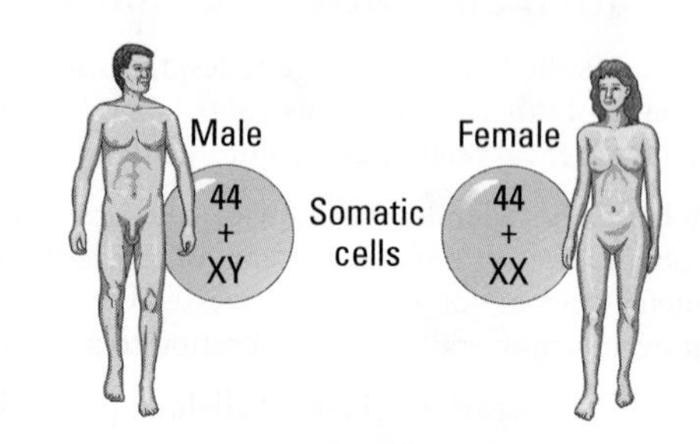

- **Sex-Linked Genes** Genes on the sex chromosomes are said to be sex-linked. In both fruit flies and humans, the X chromosome carries many genes unrelated to sex. Their inheritance pattern reflects the fact that females have two homologous X chromosomes, but males have only one.

Activity *Sex-Linked Genes*

Case Study in the Process of Science *What Can Fruit Flies Reveal about Inheritance?*

- **Sex-Linked Disorders in Humans** Most sex-linked human disorders, such as red-green color blindness and hemophilia, are due to recessive alleles and are seen mostly in males. A male receiving a single sex-linked recessive allele from his mother will have the disorder; a female has to receive the allele from both parents to be affected.

SELF-QUIZ

1. Most genes come in alternative forms called __________. A diploid individual with two identical versions of a gene is said to be __________, while an individual with two different versions of a gene is said to be __________.

2. The genetic makeup of an organism is called its __________, while the physical traits of an organism are called its __________.

3. Edward was found to be heterozygous (*Ss*) for the sickle-cell trait. The alleles represented by the letters *S* and *s* are
   a. on the X and Y chromosomes.
   b. linked.
   c. on homologous chromosomes.
   d. both present in each of Edward's sperm cells.

4. Whether an allele is dominant or recessive depends on
   a. how common the allele is, relative to other alleles.
   b. whether it is inherited from the mother or the father.
   c. whether it or another allele determines the phenotype when both are present.
   d. whether or not it is linked to other genes.

5. Two fruit flies with eyes of the usual red color are crossed, and their offspring are as follows: 77 red-eyed males, 71 ruby-eyed males, 152 red-eyed females. The gene that controls whether eyes are red or ruby is __________, and the allele for ruby eyes is __________.
   a. autosomal (carried on an autosome); dominant
   b. autosomal; recessive
   c. sex-linked; dominant
   d. sex-linked; recessive

6. All the offspring of a white hen and a black rooster are gray. The simplest explanation for this pattern of inheritance is
   a. pleiotropy.
   b. sex linkage.
   c. codominance.
   d. incomplete dominance.

7. A man who has type B blood and a woman who has type A blood could have children of which of the following phenotypes?
   a. A, B, or O
   b. AB only
   c. AB or O
   d. A, B, AB, or O

Answers to the Self-Quiz questions can be found in Appendix D.

Go to the website or CD-ROM for more Self-Quiz questions.

MORE GENETICS PROBLEMS

8. In fruit flies, the genes for wing shape and body stripes are linked. In a fly whose genotype is *WwSs*, *W* is linked to *S*, and *w* is linked to *s*. Show how this fly can produce gametes containing four different combinations of alleles. Which are parental-type gametes? Which are recombinant gametes? What process produces recombinant gametes?

9. Adult height in humans is at least partially hereditary; tall parents tend to have tall children. But humans come in a range of sizes, not just tall or short. What extension of Mendel's model could produce this variation in height?

10. A true-breeding brown mouse is repeatedly mated with a true-breeding white mouse, and all their offspring are brown. If two of these brown offspring are mated, what fraction of the F_2 mice will be brown?

11. How could you determine the genotype of one of the brown F_2 mice in problem 10? How would you know whether a brown mouse is homozygous? Heterozygous?

12. Tim and Jan both have freckles (a dominant trait), but their son Michael does not. Show with a Punnett square how this is possible. If Tim and Jan have two more children, what is the probability that *both* of them will have freckles?

13. Incomplete dominance is seen in the inheritance of hypercholesterolemia. Mack and Toni are both heterozygous for this characteristic, and both have elevated levels of cholesterol. Their daughter Katerina has a cholesterol level six times normal; she is apparently homozygous, *hh*. What fraction of Mack and Toni's children are likely to have elevated but not extreme levels of cholesterol, like their parents? If Mack and Toni have one more child, what is the probability that the child will suffer from the more serious form of hypercholesterolemia seen in Katerina?

14. A female fruit fly with forked bristles on her body is mated with a male fly with normal bristles. Their offspring are 121 females with normal bristles and 138 males with forked bristles. Explain the inheritance pattern for this trait.

15. A couple are both phenotypically normal, but their son suffers from hemophilia, a sex-linked recessive disorder. Draw a pedigree that shows the genotypes of the three individuals. What fraction of the couple's children are likely to suffer from hemophilia? What fraction are likely to be carriers?

16. Heather was surprised to discover that she suffered from red-green color blindness. She told her biology professor, who said, "Your father is color-blind too, right?" How did her professor know this? Why did her professor not say the same thing to the color-blind males in the class?

Answers to More Genetics Problems can be found in Appendix D.

THE PROCESS OF SCIENCE

17. In 1981, a stray cat with unusual curled-back ears was adopted by a family in Lakewood, California. Hundreds of descendants of this cat have since been born, and cat fanciers hope to develop the "curl" cat into a show breed. The curl allele is apparently dominant and carried on an autosome. Suppose you owned the first curl cat and wanted to

develop a true-breeding variety. Describe tests that would determine whether the curl gene is dominant or recessive and whether it is autosomal or sex-linked.

18. Imagine that you have a large collection of fruit flies divided into ten different strains. Each strain is true-breeding (homozygous) and differs from wild-type flies in just one characteristic. The only special equipment available is a magnifying glass that lets you determine the sex and traits of any fly, a large number of bottles to perform controlled matings, and an anesthetic liquid that enables you to examine and sort live flies. Using only this equipment, how much could you learn about the genetic makeup of the flies? Describe a series of experiments that would give you knowledge about fruit fly genetics.

BIOLOGY AND SOCIETY

19. Gregor Mendel never saw a gene, yet he concluded that "heritable factors" were responsible for the patterns of inheritance he observed in peas. Similarly, maps of *Drosophila* chromosomes (and the very idea that genes are carried on chromosomes) were conceived by observing the patterns of inheritance of linked genes, not by observing the genes directly. Is it legitimate for biologists to claim the existence of objects and processes they cannot actually see? How do scientists know whether an explanation is correct?

20. Many infertile couples turn to in vitro fertilization to try to have a baby. In this technique, sperm and ova are collected and used to create eight-cell embryos for implantation into a woman's uterus. At the eight-cell stage, one of the fetal cells can be removed without causing harm to the developing fetus. Once removed, the cell can be genetically tested. Some couples may know that a particular genetic disease runs in their family. They might wish to avoid implanting any embryos with the disease-causing genes. Do you think this is an acceptable use of genetic testing? What if a couple wanted to use genetic testing to select embryos for traits unrelated to disease, such as freckles? Do you think that couples undergoing in vitro fertilization should be allowed to perform whatever genetic tests they wish? Or do you think that there should be limits on what tests can be performed? How do you draw the line between genetic tests that are acceptable and those that are not?

10 The Structure and Function of DNA

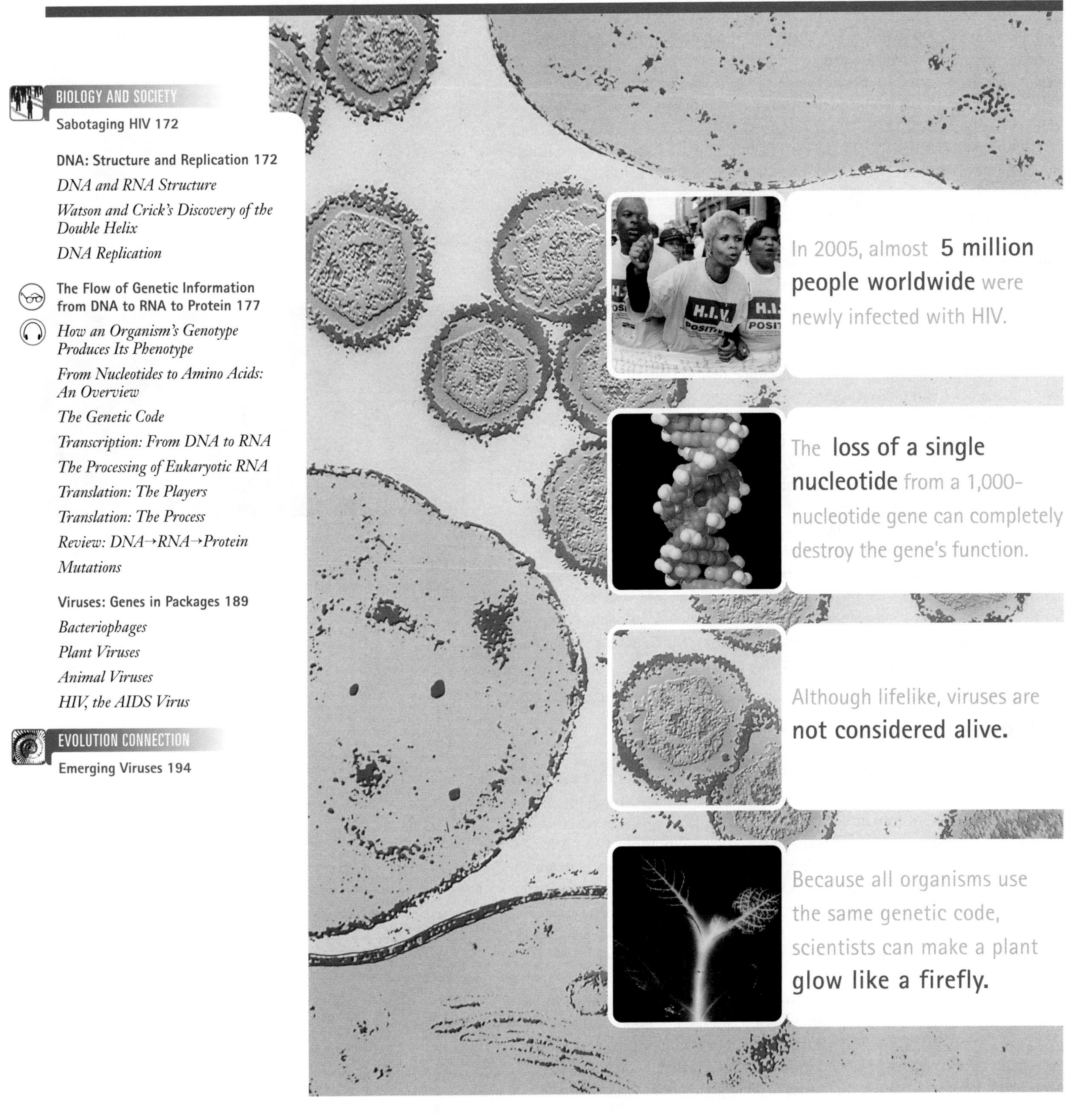

In 2005, almost **5 million people worldwide** were newly infected with HIV.

The **loss of a single nucleotide** from a 1,000-nucleotide gene can completely destroy the gene's function.

Although lifelike, viruses are **not considered alive.**

Because all organisms use the same genetic code, scientists can make a plant **glow like a firefly.**

BIOLOGY AND SOCIETY

Sabotaging HIV

AIDS, acquired immunodeficiency syndrome, is one of the most significant health challenges facing the world today. The cause of AIDS is infection by HIV, the human immunodeficiency virus. Since it was first recognized in 1981, HIV has infected an estimated 40 million people worldwide and caused 3 million deaths. While there is no cure for AIDS, its spread can be slowed by anti-HIV drugs. One such drug, AZT, has proved to be the most effective weapon yet found in the fight against AIDS.

How does AZT stop HIV? To answer this question, we have to think on the molecular level. Like all viruses, HIV must infect a host cell because it cannot reproduce on its own. After gaining entry into a human immune cell, HIV depends on a special viral enzyme to convert its RNA genome into a molecule of DNA. As you learned in Chapter 3, all DNA is made from four chemical building blocks called nucleotides: A, T, C, and G. The special viral enzyme uses nucleotides from the cytoplasm of the infected cell to build a DNA molecule.

The key to AZT's effectiveness is its shape (another example of the connection between form and function in biology). The shape of a molecule of AZT is very similar to the shape of part of the T (thymine) nucleotide (**Figure 10.1**). In fact, AZT's shape is so similar to the T nucleotide that AZT binds to the viral enzyme instead of T. But unlike thymine, AZT cannot be incorporated into a growing DNA chain. Thus, AZT "gums up the works"; it acts as a molecular saboteur that interferes with the synthesis of HIV DNA. Because this synthesis is an essential step in the reproductive cycle of HIV, AZT may block the spread of the virus.

AZT is a good example of how a detailed understanding of biological molecules can help improve human health. In this chapter, you will learn about **molecular biology,** the study of heredity at the molecular level. You will learn in detail how the DNA of genes exerts its effects on the cell and on the whole organism. You'll learn, too, how DNA replicates—the molecular basis for the similarities between parents and offspring—and how it can mutate. And because viruses have played key roles in the history of molecular biology, and continue to be important as both research organisms and disease-causing pathogens, these tiny entities are also a major topic of this chapter. ■

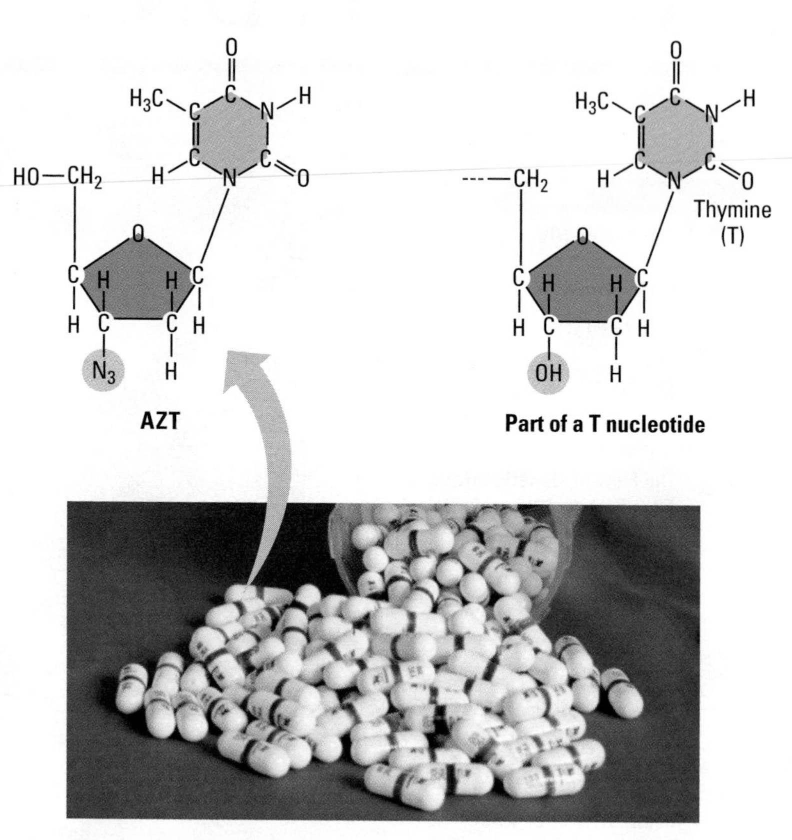

Figure 10.1 AZT and the T nucleotide. The anti-HIV drug AZT (left) has a chemical shape very similar to part of the T (thymine) nucleotide of DNA.

DNA: Structure and Replication

DNA was known to be a chemical in cells by the end of the 19th century, but Mendel and other early geneticists did all their work without any knowledge of DNA's role in heredity. By the late 1930s, experimental studies had convinced most biologists that a specific kind of molecule, rather than some complex chemical mixture, was the basis of inheritance. Attention focused on chromosomes, which were already known to carry genes. By the 1940s, scientists knew that chromosomes consisted of two types of chemicals: DNA and protein. And by the early 1950s, a series of discoveries had convinced the scientific world that DNA acts as the hereditary material.

What came next was one of the most celebrated quests in the history of science: the effort to figure out the structure of DNA. A good deal was already known about DNA. Scientists had identified all its atoms and knew how they were covalently bonded to one another. What was not understood was the specific three-dimensional arrangement of atoms that gave

DNA its unique properties—the capacity to store genetic information, copy it, and pass it from generation to generation. A race was on to discover how the structure of this molecule could account for its role in heredity. We will describe that momentous discovery shortly. First, let's review the underlying chemical structure of DNA and its chemical cousin RNA.

DNA and RNA Structure

Recall from Chapter 3 that both DNA and RNA are nucleic acids, which consist of long chains (polymers) of chemical units (monomers) called **nucleotides.** (For an in-depth refresher, see the information on nucleic acids in Chapter 3, particularly Figures 3.26–3.29.) A very simple diagram of a nucleotide polymer, or **polynucleotide,** is shown in **Figure 10.2**. This sample polynucleotide chain shows only one of many possible arrangements of the four different types of nucleotides (abbreviated A, C, T, and G) that make up DNA. Polynucleotides tend to be very long and can have any sequence of nucleotides, so a great number of polynucleotide chains are possible.

The nucleotides are joined to one another by covalent bonds between the sugar of one nucleotide and the phosphate of the next. This results in a **sugar-phosphate backbone,** a repeating pattern of sugar-phosphate-sugar-phosphate. The nitrogenous bases are arranged like

Figure 10.2 The structure of DNA. A molecule of DNA contains two polynucleotides, each a chain of nucleotides. Each nucleotide consists of a nitrogenous base, a sugar (blue), and a phosphate group (gold). The chemical structure at the right shows the details of a DNA nucleotide.

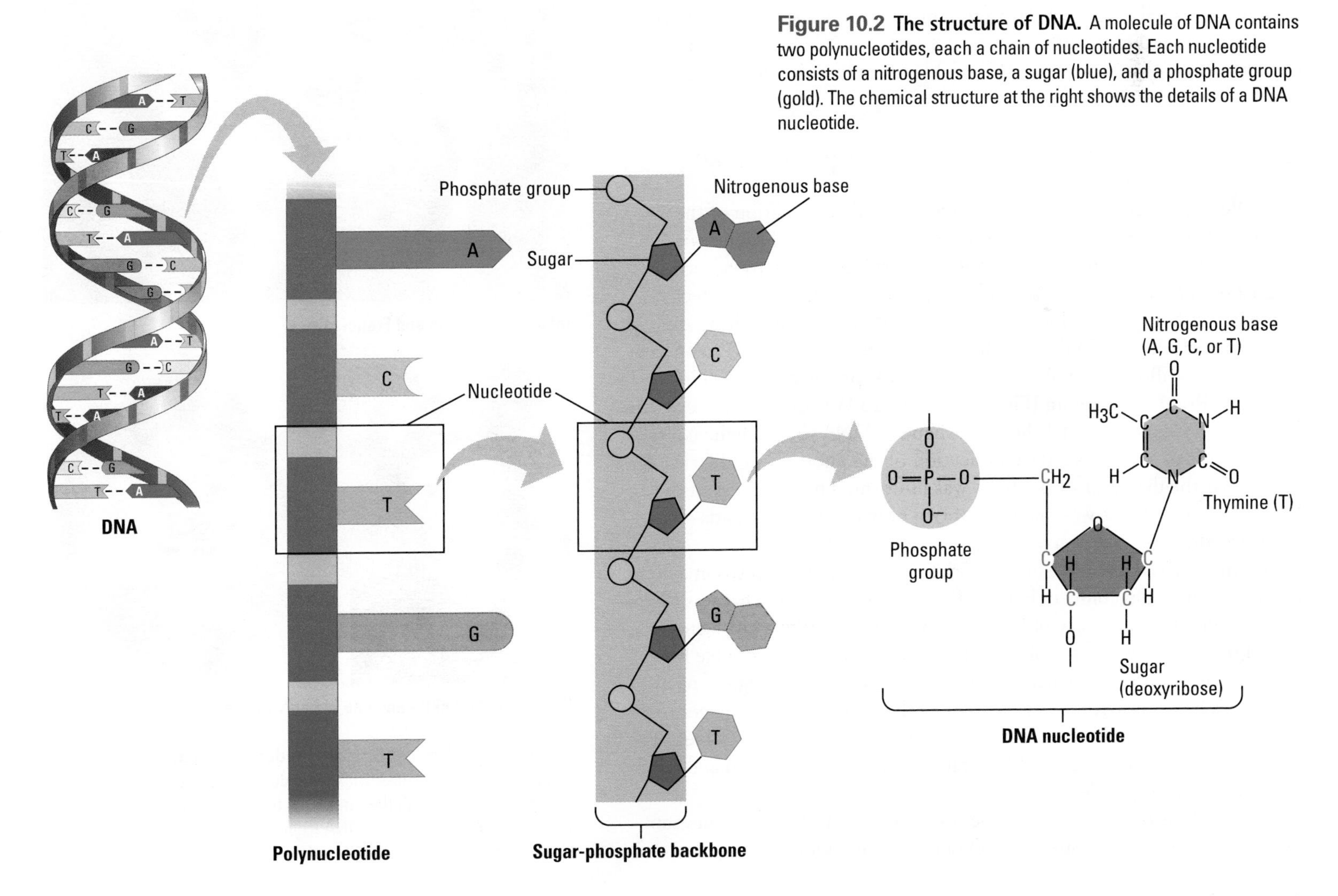

ribs that hang off this backbone. Zooming in on our polynucleotide in Figure 10.2, we see that each nucleotide consists of three components: a nitrogenous base, a sugar (blue), and a phosphate group (gold). Examining a single nucleotide even more closely (Figure 10.2, right), we see the chemical structure of its three components. The phosphate group, with a phosphorus atom (P) at its center, is the source of the *acid* in nucleic acid. (The phosphate has given up a hydrogen ion, H^+, leaving a negative charge on one of its oxygen atoms.) The sugar has five carbon atoms (shown in red): four in its ring and one extending above the ring. The ring also includes an oxygen atom. The sugar is called *deoxyribose* because, compared to the sugar ribose, it is missing an oxygen atom. The full name for **DNA** is *deoxyribonucleic acid*, with the *nucleic* part coming from DNA's location in the nuclei of eukaryotic cells. The nitrogenous base (thymine, in our example) has a ring of nitrogen and carbon atoms with various functional groups attached. In contrast to the acidic phosphate group, nitrogenous bases are basic; hence their name.

The four nucleotides found in DNA differ only in their nitrogenous bases (see Figure 3.27 for a review). At this point, the structural details are not as important as the fact that the bases are of two types. **Thymine (T)** and **cytosine (C)** are single-ring structures. **Adenine (A)** and **guanine (G)** are larger, double-ring structures. (The one-letter abbreviations can be used for either the bases alone or for the nucleotides containing them.) Recall from Chapter 3 that RNA has the nitrogenous base **uracil (U)** instead of thymine (uracil is very similar to thymine). And, as already mentioned, RNA contains a slightly different sugar than DNA (ribose instead of deoxyribose). Other than that, RNA and DNA polynucleotides have the same chemical structure.

Watson and Crick's Discovery of the Double Helix

The celebrated partnership that resulted in the determination of the physical structure of DNA began soon after a 23-year-old American named James D. Watson journeyed to Cambridge University, where Englishman Francis Crick was studying protein structure with a technique called X-ray crystallography **(Figure 10.3a)**. While visiting the laboratory of Maurice Wilkins at King's College in London, Watson saw an X-ray crystallographic photograph of DNA, produced by Wilkins's colleague Rosalind Franklin **(Figure 10.3b)**. To Watson's trained eye, the photograph clearly revealed the basic shape of DNA to be a helix (spiral). On the basis of Watson's later recollection of the photo, he and Crick deduced that the diameter of the helix was uniform. The thickness of the helix suggested that it was made up of two polynucleotide strands—in other words, a **double helix.**

Using wire models, Watson and Crick began trying to construct a double helix that would conform both to Franklin's data and to what was then known about the chemistry of DNA. After failing to make a satisfactory model that placed the sugar-phosphate backbones inside the double helix, Watson tried putting the backbones on the outside and forcing the nitrogenous bases to swivel to the interior of the molecule. It occurred to him that the four kinds of bases might pair in a specific way. This idea of *specific base pairing* was a flash of inspiration that enabled Watson and Crick to solve the DNA puzzle.

At first, Watson imagined that the bases paired like with like—for example, A with A, C with C. But that kind of pairing did not fit with the fact that

(a) James Watson and Francis Crick

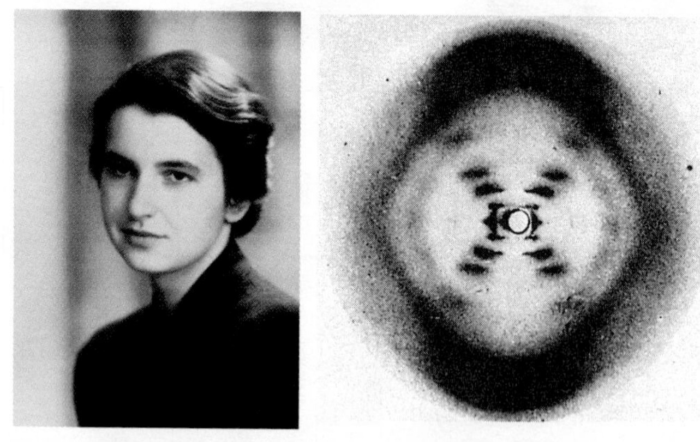

(b) Rosalind Franklin and one of her X-ray images

Figure 10.3 Discoverers of the double helix. (a) James Watson (left) and Francis Crick, who deduced the structure of DNA, are shown in 1953 with their model of the double helix. **(b)** By generating X-ray images of DNA (right), Rosalind Franklin provided Watson and Crick with some key data about the structure of DNA.

the DNA molecule has a uniform diameter. An AA pair (made of two double-ringed bases) would be almost twice as wide as a CC pair (made of two single-ringed bases), causing bulges in the molecule. It soon became apparent that a double-ringed base on one strand must always be paired with a single-ringed base on the opposite strand. Moreover, Watson and Crick realized that the individual structures of the bases dictated the pairings even more specifically. Each base has chemical side groups that can best form hydrogen bonds with one appropriate partner (to review hydrogen bonds, see Figure 2.10). Adenine can best form hydrogen bonds with thymine, and guanine with cytosine. In the biologist's shorthand, A pairs with T, and G pairs with C. A is also said to be "complementary" to T, and G to C.

You can picture the model of the DNA double helix proposed by Watson and Crick as a rope ladder having rigid, wooden rungs, with the ladder twisted into a spiral **(Figure 10.4)**. Figure 10.5 shows three more detailed representations of the double helix. The ribbonlike diagram in **Figure 10.5a** symbolizes the bases with shapes that emphasize their complementarity. **Figure 10.5b** is a more chemically precise version showing only four base pairs, with the helix untwisted and the individual hydrogen bonds specified by dashed lines; you can see that the double helix has an antiparallel arrangement—that is, the two sugar-phosphate backbones are oriented in opposite directions. **Figure 10.5c** is a computer graphic showing part of a double helix in atomic detail.

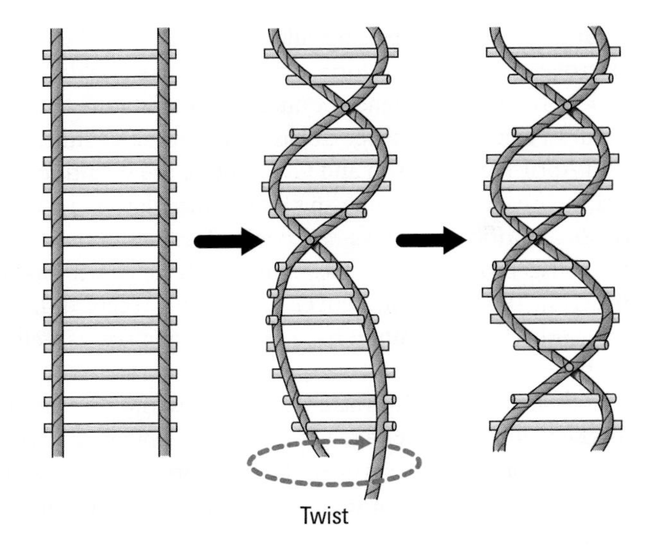

Figure 10.4 A rope-ladder model of a double helix. The ropes at the sides represent the sugar-phosphate backbones. Each wooden rung stands for a pair of bases connected by hydrogen bonds.

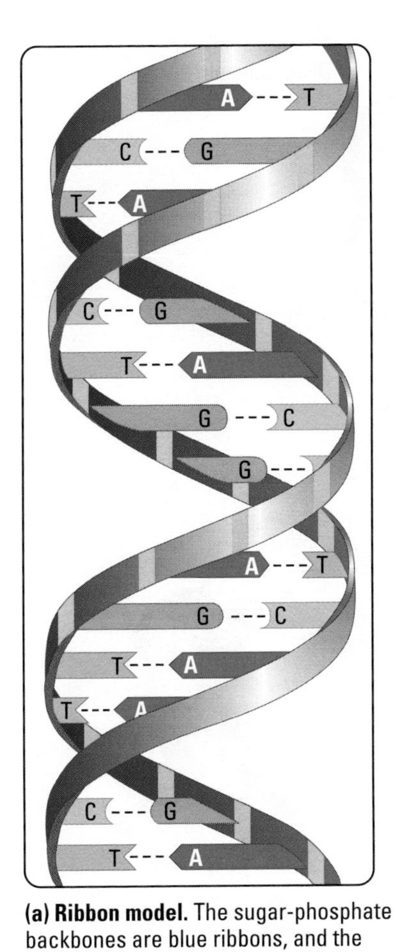

(a) Ribbon model. The sugar-phosphate backbones are blue ribbons, and the bases are complementary shapes in shades of green and orange.

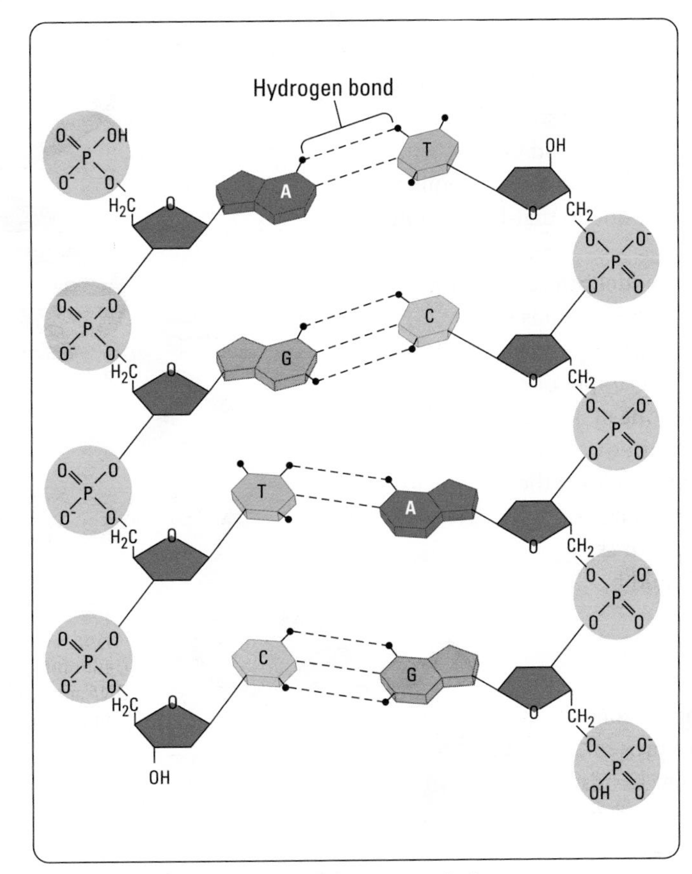

(b) Atomic model. In this more chemically detailed structure, you can see the individual hydrogen bonds (dashed lines). You can also see that the strands run in opposite directions; notice that the sugars on the two strands are upside down with respect to each other.

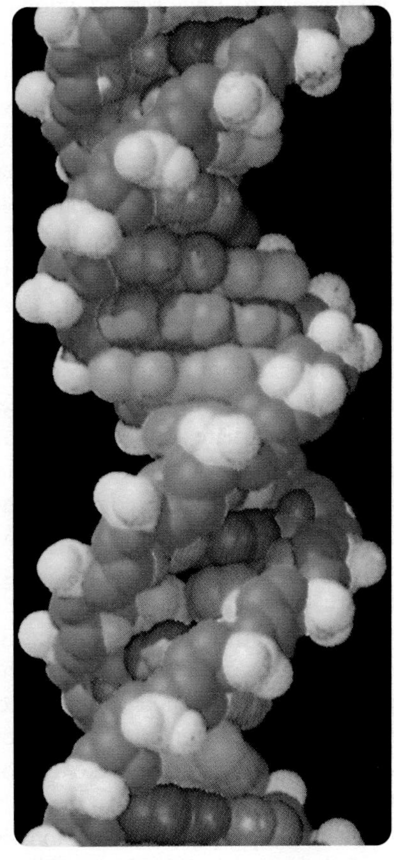

(c) Computer model. Each atom is shown as a sphere, creating a space-filling model.

Figure 10.5 Three representations of DNA.

Although the base-pairing rules dictate the side-by-side combinations of nitrogenous bases that form the rungs of the double helix, they place no restrictions on the sequence of nucleotides along the length of a DNA strand. In fact, the sequence of bases can vary in countless ways.

In April 1953, Watson and Crick shook the scientific world with a succinct, two-page paper proposing their molecular model for DNA in the British scientific journal *Nature*. Few milestones in the history of biology have had as broad an impact as their double helix, with its AT and CG base pairing. In 1962, Watson, Crick, and Wilkins received the Nobel Prize for their work. (Franklin might have received the prize as well, but she had died from cancer in 1958.)

In their 1953 paper, Watson and Crick wrote that the structure they proposed "immediately suggests a possible copying mechanism for the genetic material." In other words, the structure of DNA also points toward a molecular explanation for life's unique properties of reproduction and inheritance, as we see next.

DNA Replication

When a cell or a whole organism reproduces, a complete set of genetic instructions must pass from one generation to the next. For this to occur, there must be a means of copying the instructions. Watson and Crick's model for DNA structure immediately suggested to them that DNA replicates by a template mechanism, with each DNA strand serving as a mold, or template, to guide reproduction of the other strand. The logic behind the Watson-Crick proposal for how DNA is copied is quite simple. If you know the sequence of bases in one strand of the double helix, you can very easily determine the sequence of bases in the other strand by applying the base-pairing rules: A pairs with T (and T with A), and G pairs with C (and C with G). For example, if one polynucleotide has the sequence ATCG, then the complementary polynucleotide in that DNA molecule must have the sequence TAGC.

Figure 10.6 shows how the template model can account for the direct copying of a piece of DNA. The two strands of parental DNA separate, and each becomes a template for the assembly of a complementary strand from a supply of free nucleotides. The nucleotides are lined up one at a time along the template strand in accordance with the base-pairing rules. Enzymes link the nucleotides to form the new DNA strands. The completed new molecules, identical to the parental molecule, are known as daughter DNA molecules (no gender should be inferred from this name).

Although the general mechanism of DNA replication is conceptually simple, the actual process is complex and requires the cooperation of more than a dozen enzymes and other proteins. The enzymes that make the covalent bonds between the nucleotides of a new DNA strand are called **DNA polymerases.** As an incoming nucleotide base-pairs with its complement on the template strand, a DNA polymerase adds it to the end of the growing daughter strand (polymer). The process is both fast and amazingly accurate; typically, DNA replication proceeds at a rate of

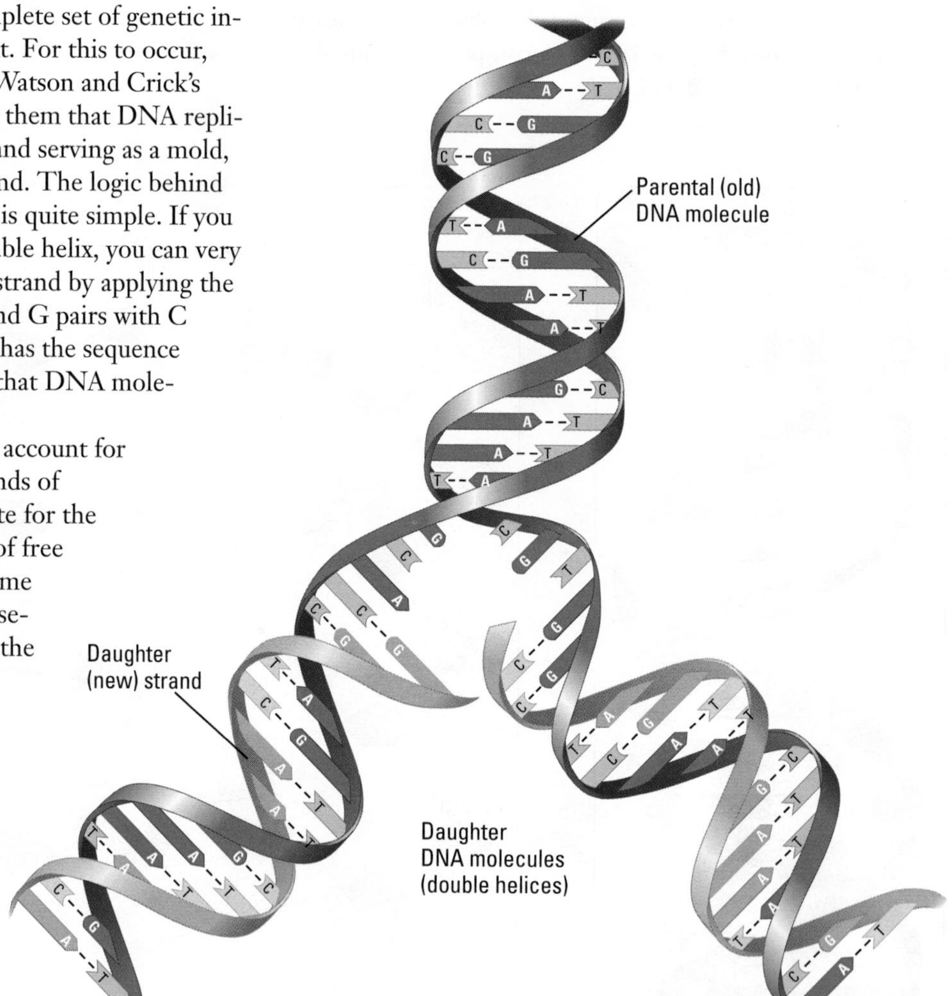

Figure 10.6 DNA replication. The two strands of the original (parental) DNA molecule (blue) serve as templates for making new (daughter) strands (orange). Replication results in two daughter DNA molecules, each consisting of one old strand and one new strand. The parental DNA untwists as its strands separate, and the daughter DNA rewinds as it forms.

50 nucleotides per second, with only about one in a billion incorrectly paired. In addition to their roles in DNA replication, DNA polymerases and some of the associated proteins are also involved in repairing damaged DNA. DNA can be harmed by toxic chemicals in the environment or by high-energy radiation, such as X-rays and ultraviolet light **(Figure 10.7)**.

DNA replication begins at specific sites on a double helix, called origins of replication. It then proceeds in both directions, creating what are called replication "bubbles" **(Figure 10.8)**. The parental DNA strands open up as daughter strands elongate on both sides of each bubble. The DNA molecule of a eukaryotic chromosome has many origins where replication can start simultaneously, shortening the total time needed for the process. Eventually, all the bubbles merge, yielding two completed double-stranded daughter DNA molecules.

DNA replication ensures that all the body cells in a multicellular organism carry the same genetic information. It is also the means by which genetic information is passed along to offspring.

Figure 10.7 Damage to DNA by ultraviolet light. The ultraviolet (UV) radiation in sunlight can damage the DNA in skin cells. Fortunately, the cells can repair some of the damage, using enzymes that include some that catalyze DNA replication. You can protect yourself from UV radiation by wearing protective clothing and sunscreen.

CHECKPOINT

1. Compare and contrast the chemical components of DNA and RNA.
2. Along one strand of a DNA double helix is the nucleotide sequence GGCATAGGT. What is the sequence for the other DNA strand?
3. How does complementary base pairing make the replication of DNA possible?
4. What is the function of DNA polymerase in DNA replication?

Answers: 1. Both are polymers of nucleotides. A nucleotide consists of a sugar + a nitrogenous base + a phosphate group. In RNA, the sugar is ribose; in DNA, it is deoxyribose. Both RNA and DNA have the bases A, G, and C; for a fourth base, DNA has T and RNA has U. **2.** CCGTATCCA **3.** When the two strands of the double helix separate, each serves as a template on which nucleotides can be arranged by specific base pairing into new complementary strands. **4.** This enzyme covalently connects nucleotides one at a time to one end of a growing daughter strand as the nucleotides line up along a template strand according to the base-pairing rules.

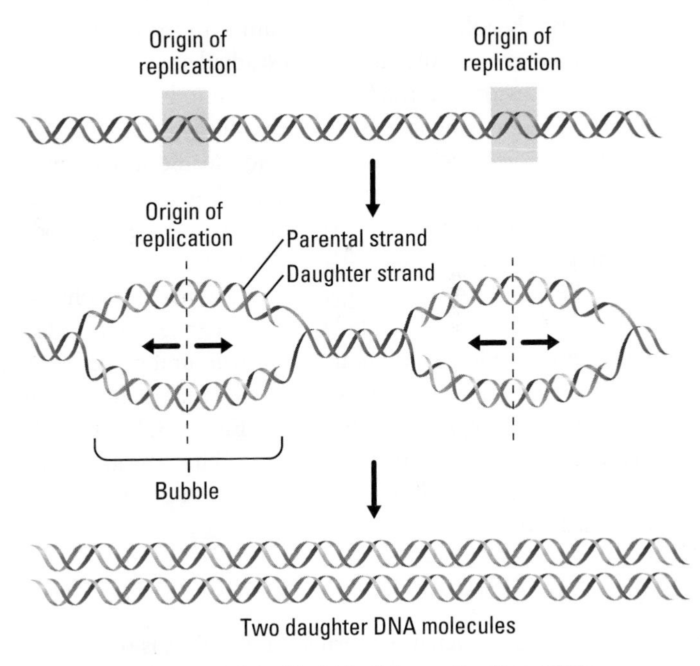

Figure 10.8 Multiple "bubbles" in replicating DNA.

The Flow of Genetic Information from DNA to RNA to Protein

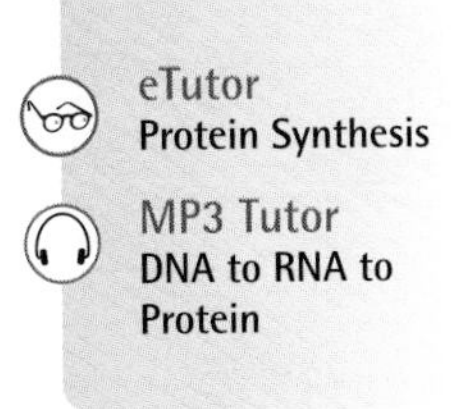

We are now ready to address the question of how DNA functions as the inherited directions for a cell and for the organism as a whole. What exactly are the instructions carried by the DNA, and how are these instructions carried out? You learned the general answers to these questions in Chapter 4, but here we will explore them in more detail.

How an Organism's Genotype Produces Its Phenotype

Knowing the structure of DNA, we can now define genotype and phenotype more precisely than we did in Chapter 9. An organism's *genotype*, its genetic makeup, is the sequence of nucleotide bases in its DNA. The *phenotype* is the organism's specific traits. An organism's phenotype arises from the actions of a wide variety of proteins. For example, structural proteins help make up the body of an organism, and enzymes catalyze its metabolic activities.

What is the connection between the genotype and the protein molecules that more directly determine the phenotype? Recall from Chapter 4 that DNA specifies the synthesis of proteins. However, a gene does not build a protein directly, but rather dispatches instructions in the form of RNA, which in turn programs protein synthesis. This central concept in biology is summarized in **Figure 10.9**. The molecular "chain of command" is from DNA in the nucleus to RNA to protein synthesis in the cytoplasm. The two main stages are **transcription,** the transfer of genetic information from DNA into an RNA molecule, and **translation,** the transfer of the information in the RNA into a protein.

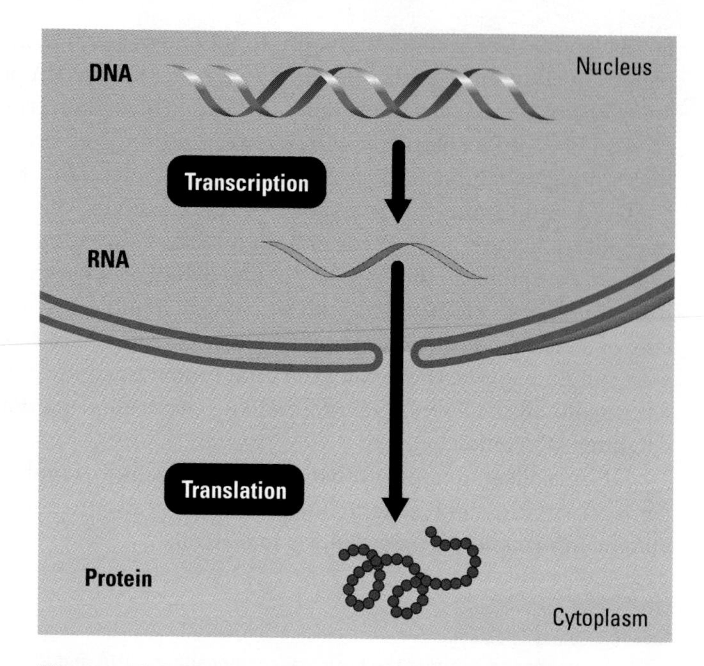

Figure 10.9 The flow of genetic information in a eukaryotic cell: a review. A sequence of nucleotides in the DNA is transcribed into a molecule of RNA in the cell's nucleus (purple area). The RNA travels to the cytoplasm (blue-green area), where it is translated into the specific amino acid sequence of a protein.

The relationship between genes and proteins was first proposed in 1909, when English physician Archibald Garrod suggested that genes dictate phenotypes through enzymes, the proteins that catalyze chemical processes. Garrod hypothesized that inherited diseases reflect a person's inability to make a particular enzyme. He gave as one example the hereditary condition called alkaptonuria, in which the urine appears dark red because it contains a chemical called alkapton. Garrod reasoned that normal individuals have an enzyme that breaks down alkapton, whereas alkaptonuric individuals lack the enzyme. Garrod's hypothesis was ahead of its time, but research conducted decades later proved him right. In the intervening years, biochemists accumulated evidence that cells make and break down biologically important molecules via metabolic pathways, as in the synthesis of an amino acid or the breakdown of a sugar. As you learned in Chapter 5, each step in a metabolic pathway is catalyzed by a specific enzyme. If a person lacks one of the enzymes, the pathway cannot be completed.

The major breakthrough in demonstrating the relationship between genes and enzymes came in the 1940s from the work of American geneticists George Beadle and Edward Tatum with the orange bread mold *Neurospora crassa*. Beadle and Tatum studied strains of the mold that were unable to grow on the usual growth medium. Each of these strains turned out to lack an enzyme in a metabolic pathway that produced some molecule the mold needed, such as an amino acid. Beadle and Tatum also showed that each mutant was defective in a single gene. Accordingly, they hypothesized that the function of an individual gene is to dictate the production of a specific enzyme.

Beadle and Tatum's "one gene–one enzyme" hypothesis has since been modified. First it was extended beyond enzymes to include all types of proteins. For example, alpha-keratin, the structural protein of your hair, is the product of a gene. Then it was discovered that many proteins have two or more different polypeptide chains (see Figure 3.24); in such cases, each polypeptide is specified by its own gene. Thus, Beadle and Tatum's hypothesis is now stated as follows: the function of a gene is to dictate the production of a *polypeptide*.

From Nucleotides to Amino Acids: An Overview

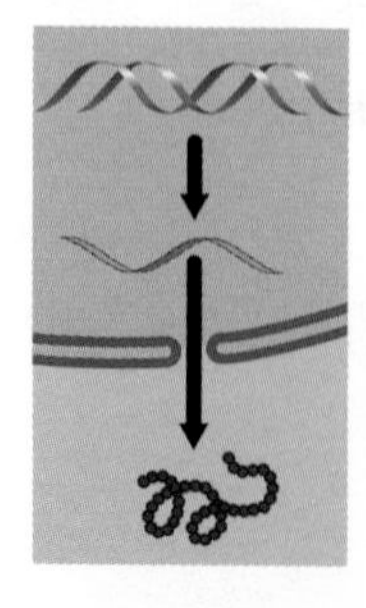

Genetic information in DNA is transcribed into RNA and then translated into polypeptides. But how do these processes occur? Transcription and translation are linguistic terms, and it is useful to think of nucleic acids and polypeptides as having languages, too. To understand how genetic information passes from genotype to phenotype, we need to see how the chemical language of DNA is translated into the different chemical language of polypeptides.

What exactly is the language of nucleic acids? Both DNA and RNA are polymers made of monomers strung together in specific sequences that convey information, much as specific sequences of letters convey information in English. In DNA, the monomers are the four types of nucleotides, which differ in their nitrogenous bases (A, T, C, and G). The same is true for RNA, although it has the base U instead of T.

The language of DNA is written as a linear sequence of nucleotide bases, a sequence such as the one you see on the enlarged DNA strand in **Figure 10.10**. Specific sequences of bases, each with a beginning and an end, make up the genes on a DNA strand. A typical gene consists of thousands of nucleotides, and a single DNA molecule may contain thousands of genes.

When a segment of DNA is transcribed, the result is an RNA molecule. The process is called transcription because the nucleic acid language of DNA has simply been rewritten (transcribed) as a sequence of bases of RNA; the language is still that of nucleic acids. The nucleotide bases of the RNA molecule are complementary to those on the DNA strand. As you will soon see, this is because the RNA was synthesized using the DNA as a template.

Translation is the conversion of the nucleic acid language to the polypeptide language. Like nucleic acids, polypeptides are polymers, but the monomers that make them up—the letters of the polypeptide alphabet—are the 20 amino acids common to all organisms. Again, the language is written in a linear sequence, and the sequence of nucleotides of the RNA molecule dictates the sequence of amino acids of the polypeptide. But remember, RNA is only a messenger; the genetic information that dictates the amino acid sequence originates in DNA.

What are the rules for translating the RNA message into a polypeptide? In other words, what is the correspondence between the nucleotides of an RNA molecule and the amino acids of a polypeptide? Keep in mind that there are only four different kinds of nucleotides in DNA (A, G, C, T) and RNA (A, G, C, U). In translation, these four must somehow specify 20 amino acids. If each nucleotide base coded for one amino acid, only 4 of the 20 amino acids could be accounted for. What if the language consisted of two-letter code words? If we read the bases of a gene two at a time, AG, for example, could specify one amino acid, while AT could designate a different amino acid. However, when the four bases are taken two by two, there are only 16 (that is, 4^2) possible arrangements—still not enough to specify all 20 amino acids.

Triplets of bases are the smallest "words" of uniform length that can specify all the amino acids. There can be 64 (that is, 4^3) possible code words of this type—more than enough to specify the 20 amino acids. Indeed, there are enough triplets to allow more than one coding for each amino acid. For example, the base triplets AAA and AAG both code for the same amino acid.

Experiments have verified that the flow of information from gene to protein is based on a triplet code. The genetic instructions for the amino acid sequence of a polypeptide chain are written in DNA and RNA as a series of three-base words called **codons.** Three-base codons in the DNA are transcribed into complementary three-base codons in the RNA, and then the RNA codons are translated into amino acids that form a polypeptide. That is, one DNA codon (three nucleotides) → one RNA codon (three nucleotides) → one amino acid. Next we turn to the codons themselves.

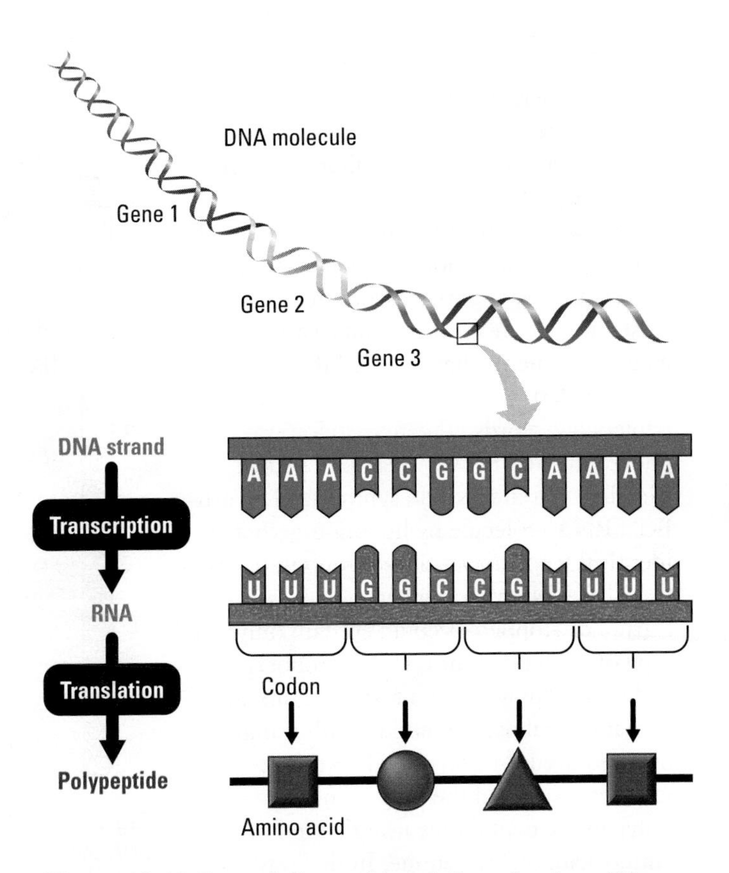

Figure 10.10 Transcription and translation of codons. This figure focuses on a small region of one of the genes carried by a DNA molecule. The enlarged segment from one strand of gene 3 shows its specific sequence of bases. The red strand underneath represents the results of transcription: an RNA molecule. Its base sequence is complementary to that of the DNA. The purple chain represents the results of translation: a polypeptide. The brackets indicate that three RNA nucleotides (a codon) code for each amino acid.

The Genetic Code

In 1799, a large stone tablet was found in Rosetta, Egypt, carrying the same lengthy inscription in three ancient scripts: Greek, Egyptian hieroglyphics, and Egyptian written in a simplified script. The Rosetta stone provided the key that enabled scholars to crack the previously indecipherable hieroglyphic code.

In cracking the **genetic code,** the set of rules relating nucleotide sequence to amino acid sequence, scientists wrote their own Rosetta stone. It was based on a series of elegant experiments that revealed the amino acid translations of each of the nucleotide-triplet code words. The first codon was deciphered in 1961 by American biochemist Marshall Nirenberg. He synthesized an artificial RNA molecule by linking together identical RNA nucleotides having uracil as their base. No matter where this message started or stopped, it could contain only one type of triplet codon: UUU. Nirenberg added this "poly U" to a test-tube mixture containing ribosomes and the other ingredients required for polypeptide synthesis. This mixture translated the poly U into a polypeptide containing a single kind of amino acid, phenylalanine. In this way, Nirenberg learned that the RNA codon UUU specifies the amino acid phenylalanine (Phe). By variations on this method, the amino acids specified by all the codons were determined.

First base	Second base: U	Second base: C	Second base: A	Second base: G	Third base
U	UUU Phenylalanine (Phe)	UCU Serine (Ser)	UAU Tyrosine (Tyr)	UGU Cysteine (Cys)	U
U	UUC Phenylalanine (Phe)	UCC Serine (Ser)	UAC Tyrosine (Tyr)	UGC Cysteine (Cys)	C
U	UUA Leucine (Leu)	UCA Serine (Ser)	UAA Stop	UGA Stop	A
U	UUG Leucine (Leu)	UCG Serine (Ser)	UAG Stop	UGG Tryptophan (Trp)	G
C	CUU Leucine (Leu)	CCU Proline (Pro)	CAU Histidine (His)	CGU Arginine (Arg)	U
C	CUC Leucine (Leu)	CCC Proline (Pro)	CAC Histidine (His)	CGC Arginine (Arg)	C
C	CUA Leucine (Leu)	CCA Proline (Pro)	CAA Glutamine (Gln)	CGA Arginine (Arg)	A
C	CUG Leucine (Leu)	CCG Proline (Pro)	CAG Glutamine (Gln)	CGG Arginine (Arg)	G
A	AUU Isoleucine (Ile)	ACU Threonine (Thr)	AAU Asparagine (Asn)	AGU Serine (Ser)	U
A	AUC Isoleucine (Ile)	ACC Threonine (Thr)	AAC Asparagine (Asn)	AGC Serine (Ser)	C
A	AUA Isoleucine (Ile)	ACA Threonine (Thr)	AAA Lysine (Lys)	AGA Arginine (Arg)	A
A	AUG Met or start	ACG Threonine (Thr)	AAG Lysine (Lys)	AGG Arginine (Arg)	G
G	GUU Valine (Val)	GCU Alanine (Ala)	GAU Aspartic acid (Asp)	GGU Glycine (Gly)	U
G	GUC Valine (Val)	GCC Alanine (Ala)	GAC Aspartic acid (Asp)	GGC Glycine (Gly)	C
G	GUA Valine (Val)	GCA Alanine (Ala)	GAA Glutamic acid (Glu)	GGA Glycine (Gly)	A
G	GUG Valine (Val)	GCG Alanine (Ala)	GAG Glutamic acid (Glu)	GGG Glycine (Gly)	G

Figure 10.11 The dictionary of the genetic code, listed by RNA codons. The three bases of an RNA codon are designated here as the first, second, and third bases. Practice using this dictionary by finding the codon UGG. This is the only codon for the amino acid tryptophan (Trp), but most amino acids are specified by two or more codons. For example, both UUU and UUC stand for the amino acid phenylalanine (Phe). Notice that the codon AUG not only stands for the amino acid methionine (Met) but also functions as a signal to "start" translating the RNA at that place. Three of the 64 codons function as "stop" signals that mark the end of a genetic message, but do not encode any amino acids.

As **Figure 10.11** shows, 61 of the 64 triplets code for amino acids. The triplet AUG has a dual function: It not only codes for the amino acid methionine (Met) but can also provide a signal for the start of a polypeptide chain. Three of the other codons do not designate amino acids. They are the stop codons that instruct the ribosomes to end the polypeptide.

Notice in Figure 10.11 that there is redundancy in the code but no ambiguity. For example, although codons UUU and UUC both specify phenylalanine (redundancy), neither of them ever represents any other amino acid (no ambiguity). The codons in the figure are the triplets found in RNA. They have a straightforward, complementary relationship to the codons in DNA. The nucleotides making up the codons occur in a linear order along the DNA and RNA, with no gaps or "punctuation" separating the codons.

Almost all of the genetic code is shared by all organisms, from the simplest bacteria to the most complex plants and animals. The universality of the genetic vocabulary suggests that it arose very early in evolution and was passed on over the eons to all the organisms living on Earth today. As you will learn in Chapter 12, such universality is extremely important to modern DNA technologies. Because the code is the same in different species, genes can be transcribed and translated after transfer from one species to another, even when the organisms are as different as a bacterium and a human or a firefly and a tobacco plant **(Figure 10.12)**. This allows scientists to mix and match genes from various species—a procedure with many useful applications.

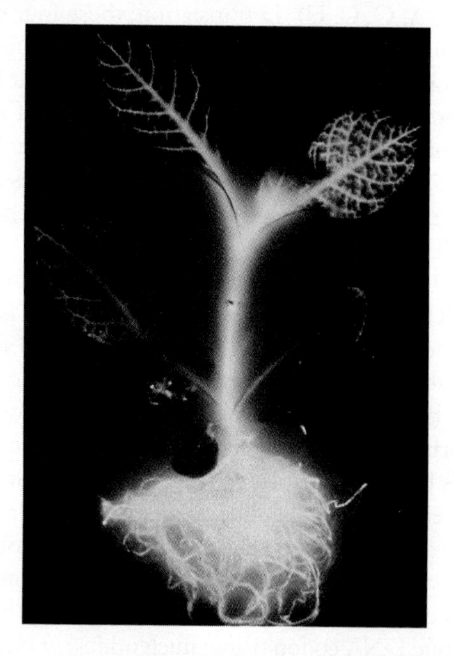

Figure 10.12 A tobacco plant expressing a firefly gene. Because diverse organisms share a common genetic code, it is possible to program one species to produce a protein characteristic of another species by transplanting DNA. This photo shows the results of an experiment in which researchers incorporated a gene from a firefly into the DNA of a tobacco plant. The gene codes for the firefly enzyme that produces a glow.

CHECKPOINT

1. What are transcription and translation?
2. How many nucleotides are necessary to code for a polypeptide that is 100 amino acids long?
3. An RNA molecule contains the nucleotide sequence CCAUUUACG. Using Figure 10.11, translate this sequence into the corresponding amino acid sequence.

Answers 1. Transcription is the transfer of genetic information from DNA to RNA—the synthesis of RNA on a DNA template. Translation is the use of the information in an RNA molecule for the synthesis of a polypeptide. **2.** 300 **3.** Pro-Phe-Thr

Transcription: From DNA to RNA

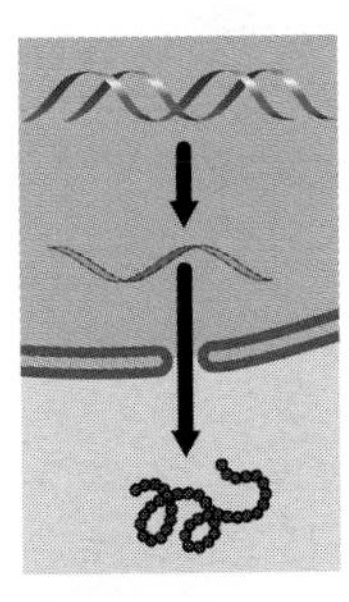

Let's look more closely at transcription, the transfer of genetic information from DNA to RNA. An RNA molecule is transcribed from a DNA template by a process that resembles the synthesis of a DNA strand during DNA replication. **Figure 10.13a** is a close-up view of this process. As with replication, the two DNA strands must first separate at the place where the process will start. In transcription, however, only one of the DNA strands serves as a template for the newly forming molecule. The nucleotides that make up the new RNA molecule take their places one at a time along the DNA template strand by forming hydrogen bonds with the nucleotide bases there. Notice that the RNA nucleotides follow the same base-pairing rules that govern DNA replication, except that U, rather than T, pairs with A. The RNA nucleotides are linked by the transcription enzyme **RNA polymerase.**

Figure 10.13b is an overview of the transcription of an entire gene. Special sequences of DNA nucleotides tell the RNA polymerase where to start and where to stop the transcribing process.

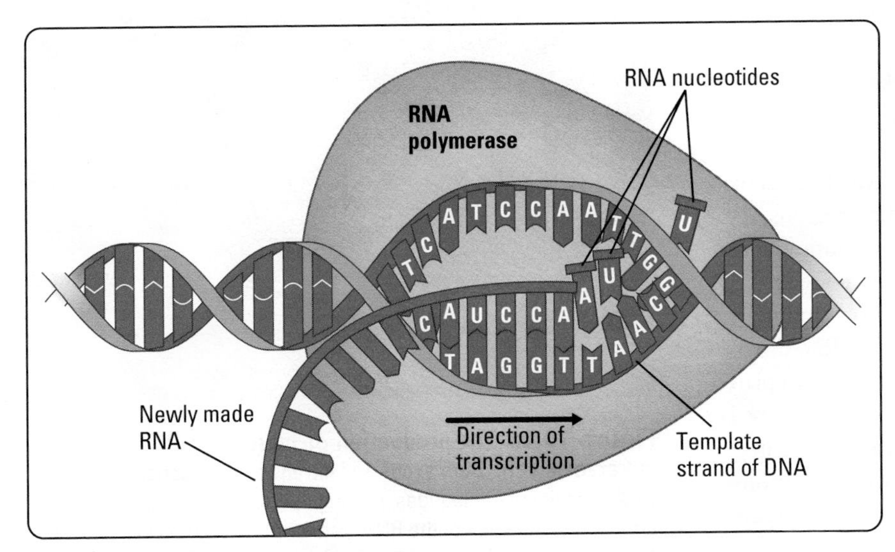

(a) A close-up view of transcription. As RNA nucleotides base-pair one by one with DNA bases on one DNA strand (called the template strand), the enzyme RNA polymerase (orange) links the RNA nucleotides into an RNA chain.

(b) Transcription of a gene. The transcription of an entire gene occurs in three phases: initiation, elongation, and termination of the RNA. The section of DNA where the RNA polymerase starts is called the promoter; the place where it stops is called the terminator.

Figure 10.13 Transcription.

Initiation of Transcription The "start transcribing" signal is a nucleotide sequence called a promoter, which is located in the DNA at the beginning of the gene. A promoter is a specific place where RNA polymerase attaches. The first phase of transcription, called initiation, is the attachment of RNA polymerase to the promoter and the start of RNA synthesis. For any gene, the promoter dictates which of the two DNA strands is to be transcribed (the particular strand varies from gene to gene).

RNA Elongation During the second phase of transcription, elongation, the RNA grows longer. As RNA synthesis continues, the RNA strand peels away from its DNA template, allowing the two separated DNA strands to come back together in the region already transcribed.

Termination of Transcription In the third phase, termination, the RNA polymerase reaches a special sequence of bases in the DNA template called a **terminator.** This sequence signals the end of the gene. At this point, the polymerase molecule detaches from the RNA molecule and the gene.

In addition to producing RNA that encodes amino acid sequences, transcription makes two other kinds of RNA that are involved in building polypeptides. We discuss these kinds of RNA a little later.

The Processing of Eukaryotic RNA

In prokaryotic cells, which lack nuclei, the RNA transcribed from a gene immediately functions as the messenger molecule that is translated, called **messenger RNA (mRNA).** But this is not the case in eukaryotic cells. The eukaryotic cell not only localizes transcription in the nucleus but also modifies, or processes, the RNA transcripts there before they move to the cytoplasm for translation by the ribosomes.

One kind of RNA processing is the addition of extra nucleotides to the ends of the RNA transcript. These additions, called the **cap** and **tail**, protect the RNA from attack by cellular enzymes and help ribosomes recognize the RNA as mRNA.

Another type of RNA processing is made necessary in eukaryotes by noncoding stretches of nucleotides that interrupt the nucleotides that actually code for amino acids. It is as if unintelligible sequences of letters were randomly interspersed in an otherwise intelligible document. Most genes of plants and animals, it turns out, include such internal noncoding regions, which are called **introns.** (Much remains unknown about the evolution and function of introns.) The coding regions—the parts of a gene that are expressed—are called **exons.** As **Figure 10.14** illustrates, both exons and introns are transcribed from DNA into RNA. However, before the RNA leaves the nucleus, the introns are removed, and the exons are joined to produce an mRNA molecule with a continuous coding sequence. This process is called **RNA splicing.** RNA splicing is believed to play a significant role in humans in allowing our approximately 25,000 genes to produce many thousands more polypeptides. This is accomplished by varying the exons that are included in the final mRNA.

With capping, tailing, and splicing completed, the "final draft" of eukaryotic mRNA is ready for translation.

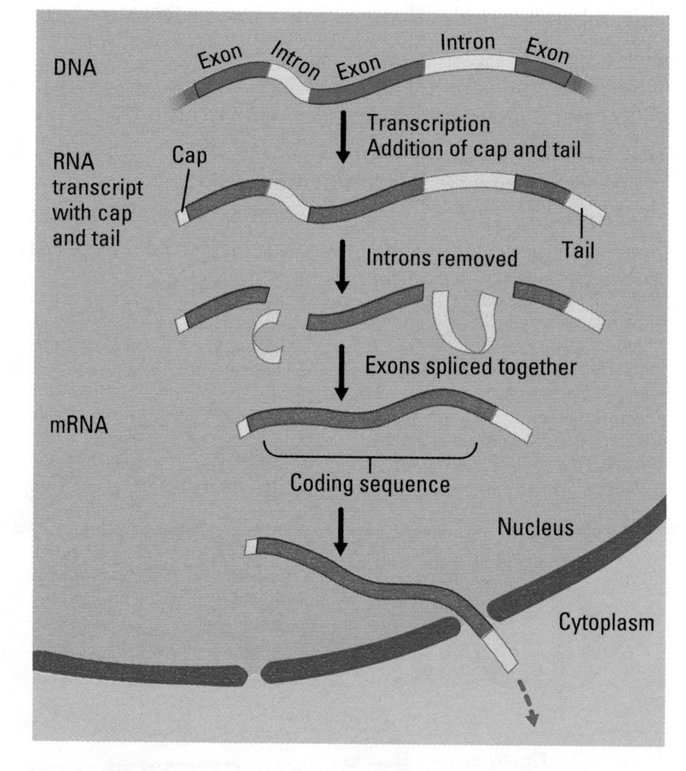

Figure 10.14 The production of messenger RNA in a eukaryotic cell. Both exons and introns are transcribed from the DNA. Additional nucleotides, making up the cap and tail, are attached at the ends of the RNA transcript. The exons are spliced together. The product, a molecule of messenger RNA (mRNA), then travels to the cytoplasm of the cell. There the coding sequence will be translated.

Translation: The Players

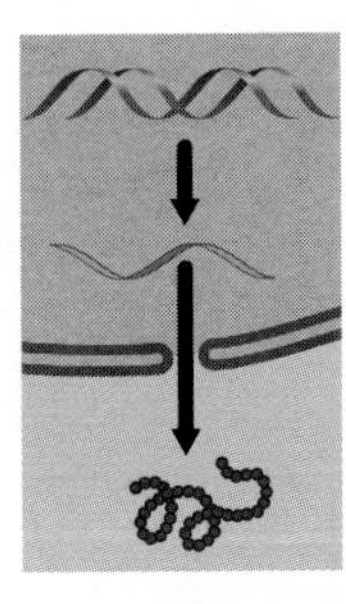

As we have already discussed, translation is a conversion between different languages—from the nucleic acid language to the protein language—and it involves more elaborate machinery than transcription.

Messenger RNA (mRNA) The first important ingredient required for translation is the mRNA produced by transcription. Once it is present, the machinery used to translate mRNA requires enzymes and sources of chemical energy, such as ATP. In addition, translation requires two heavy-duty components: ribosomes and a kind of RNA called transfer RNA.

Transfer RNA (tRNA) Translation of any language into another language requires an interpreter, someone or something that can recognize the words of one language and convert them to the other. Translation of the genetic message carried in mRNA into the amino acid language of proteins also requires an interpreter. To convert the three-letter words (codons) of nucleic acids to the one-letter, amino acid words of proteins, a cell uses a molecular interpreter, a type of RNA called **transfer RNA,** abbreviated **tRNA (Figure 10.15)**.

A cell that is ready to have some of its genetic information translated into polypeptides has in its cytoplasm a supply of amino acids, either obtained from food or made from other chemicals. The amino acids themselves cannot recognize the codons arranged in sequence along messenger RNA. It is up to the cell's molecular interpreters, tRNA molecules, to match amino acids to the appropriate codons to form the new polypeptide. To perform this task, tRNA molecules must carry out two distinct functions: (1) to pick up the appropriate amino acids and (2) to recognize the appropriate codons in the mRNA. The unique structure of tRNA molecules enables them to perform both tasks.

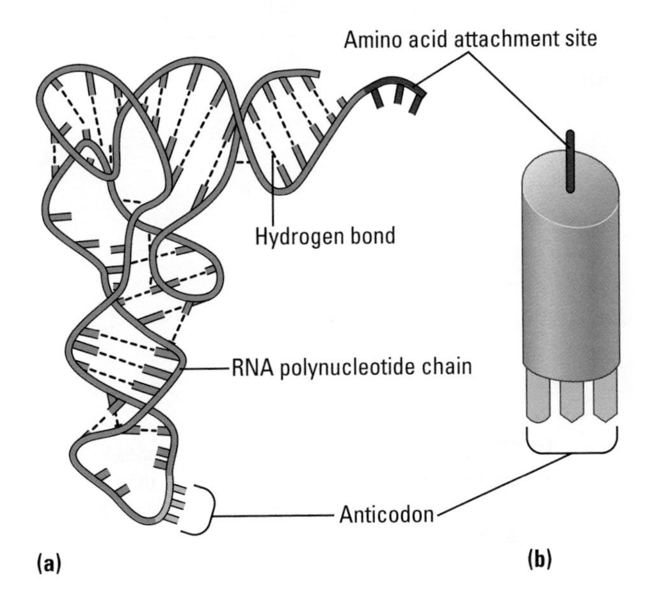

Figure 10.15 The structure of tRNA. (a) The RNA polynucleotide is a "rope" whose appendages are the nitrogenous bases. Dashed lines are hydrogen bonds, which connect some of the bases. The site where an amino acid will attach is a three-nucleotide segment at one end (purple). Note the three-base anticodon at the bottom of the molecule (light green). **(b)** This is the representation of tRNA that we use in later diagrams.

As shown in Figure 10.15a, a tRNA molecule is made of a single strand of RNA—one polynucleotide chain—consisting of about 80 nucleotides. The chain twists and folds upon itself, forming several double-stranded regions in which short stretches of RNA base-pair with other stretches. At one end of the folded molecule is a special triplet of bases called an **anticodon.** The anticodon triplet is complementary to a codon triplet on mRNA. During translation, the anticodon on the tRNA recognizes a particular codon on the mRNA by using base-pairing rules. At the other end of the tRNA molecule is a site where an amino acid can attach. Although all tRNA molecules are similar, there is a slightly different version of tRNA for each amino acid.

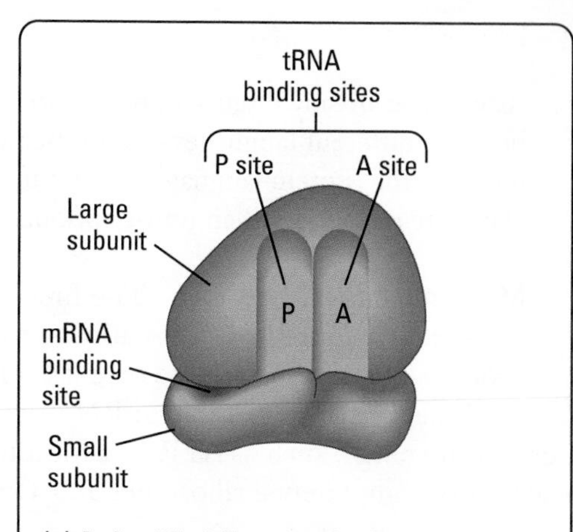

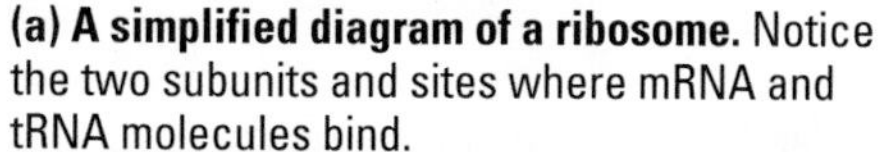
(a) A simplified diagram of a ribosome. Notice the two subunits and sites where mRNA and tRNA molecules bind.

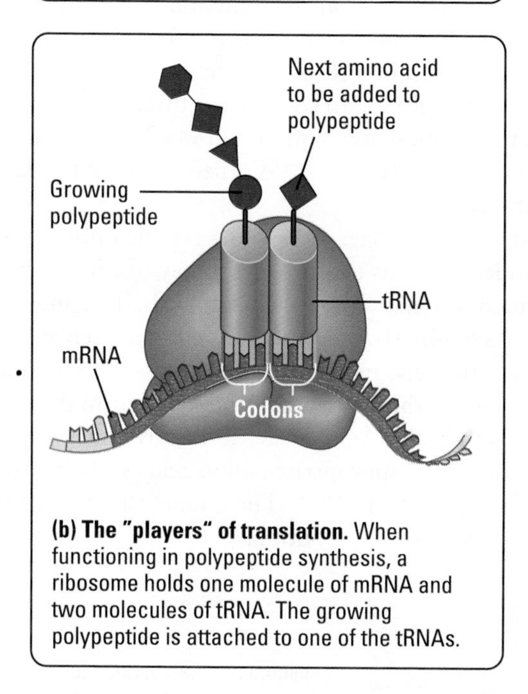

(b) The "players" of translation. When functioning in polypeptide synthesis, a ribosome holds one molecule of mRNA and two molecules of tRNA. The growing polypeptide is attached to one of the tRNAs.

Figure 10.16 The ribosome.

Ribosomes Ribosomes are the organelles that coordinate the functioning of the mRNA and tRNA and actually make polypeptides. As you can see in **Figure 10.16a**, a ribosome consists of two subunits. Each subunit is made up of proteins and a considerable amount of yet another kind of RNA, **ribosomal RNA (rRNA).** A fully assembled ribosome has a binding site for mRNA on its small subunit and binding sites for tRNA on its large subunit. **Figure 10.16b** shows how two tRNA molecules get together with an mRNA molecule on a ribosome. One of the tRNA binding sites, the P site, holds the tRNA carrying the growing polypeptide chain, while another, the A site, holds a tRNA carrying the next amino acid to be added to the chain. The anticodon on each tRNA base-pairs with a codon on mRNA. The subunits of the ribosome act like a vise, holding the tRNA and mRNA molecules close together. The ribosome can then connect the amino acid from the A site tRNA to the growing polypeptide.

Now let's examine translation in more detail, starting at the beginning.

Translation: The Process

Translation can be divided into the same three phases as transcription: initiation, elongation, and termination.

Initiation This first phase brings together the mRNA, the first amino acid with its attached tRNA, and the two subunits of a ribosome. An mRNA molecule, even after splicing, is longer than the genetic message it carries **(Figure 10.17)**. Nucleotide sequences at either end of the molecule are not part of the message but, along with the cap and tail in eukaryotes, they help the mRNA bind to the ribosome. The initiation process determines exactly where translation will begin so that the mRNA codons will

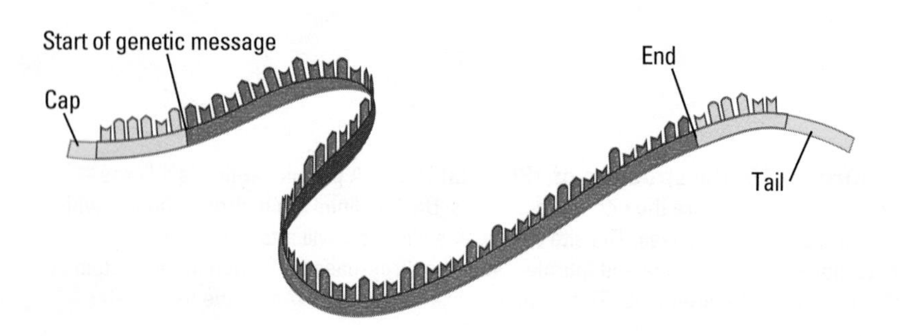

Figure 10.17 A molecule of mRNA. The pink ends are nucleotides that are not part of the message; that is, they are not translated. These nucleotides, along with the cap and tail (yellow), help the mRNA attach to the ribosome.

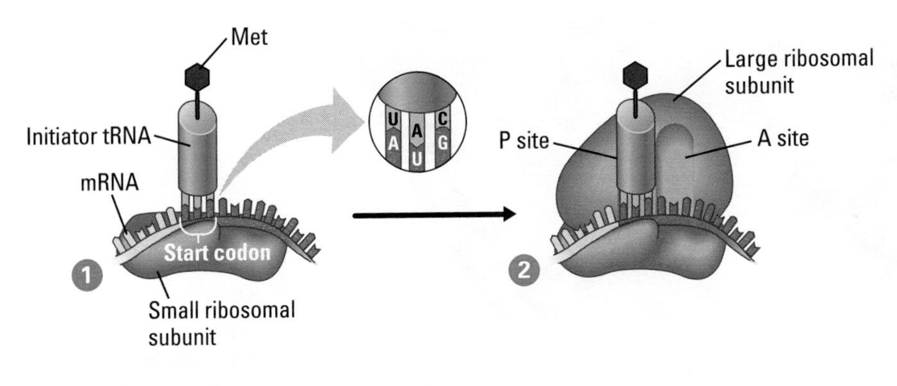

Figure 10.18 The initiation of translation. ① An mRNA molecule binds to a small ribosomal subunit. A special initiator tRNA then binds to the start codon, where translation is to begin on the mRNA. The initiator tRNA carries the amino acid methionine (Met); its anticodon, UAC, binds to the start codon, AUG. ② A large ribosomal subunit binds to the small one, creating a functional ribosome. The initiator tRNA fits into the P site on the ribosome.

be translated into the correct sequence of amino acids. Initiation occurs in two steps, as shown in **Figure 10.18**.

Elongation Once initiation is complete, amino acids are added one by one to the first amino acid. Each addition occurs in a three-step elongation process **(Figure 10.19)**.

Step ① Codon recognition. The anticodon of an incoming tRNA molecule, carrying its amino acid, pairs with the mRNA codon in the A site of the ribosome.

Step ② Peptide bond formation. The polypeptide leaves the tRNA in the P site and attaches to the amino acid on the tRNA in the A site. The ribosome catalyzes bond formation. Now the chain has one more amino acid.

Step ③ Translocation. The P site tRNA now leaves the ribosome, and the ribosome translocates (moves) the remaining tRNA, carrying the growing polypeptide, to the P site. The mRNA and tRNA move as a unit. This movement brings into the A site the next mRNA codon to be translated, and the process can start again with step 1.

Termination Elongation continues until a **stop codon** reaches the ribosome's A site. Stop codons—UAA, UAG, and UGA—do not code for amino acids but instead tell translation to stop. The completed polypeptide, typically several hundred amino acids long, is freed, and the ribosome splits into its subunits.

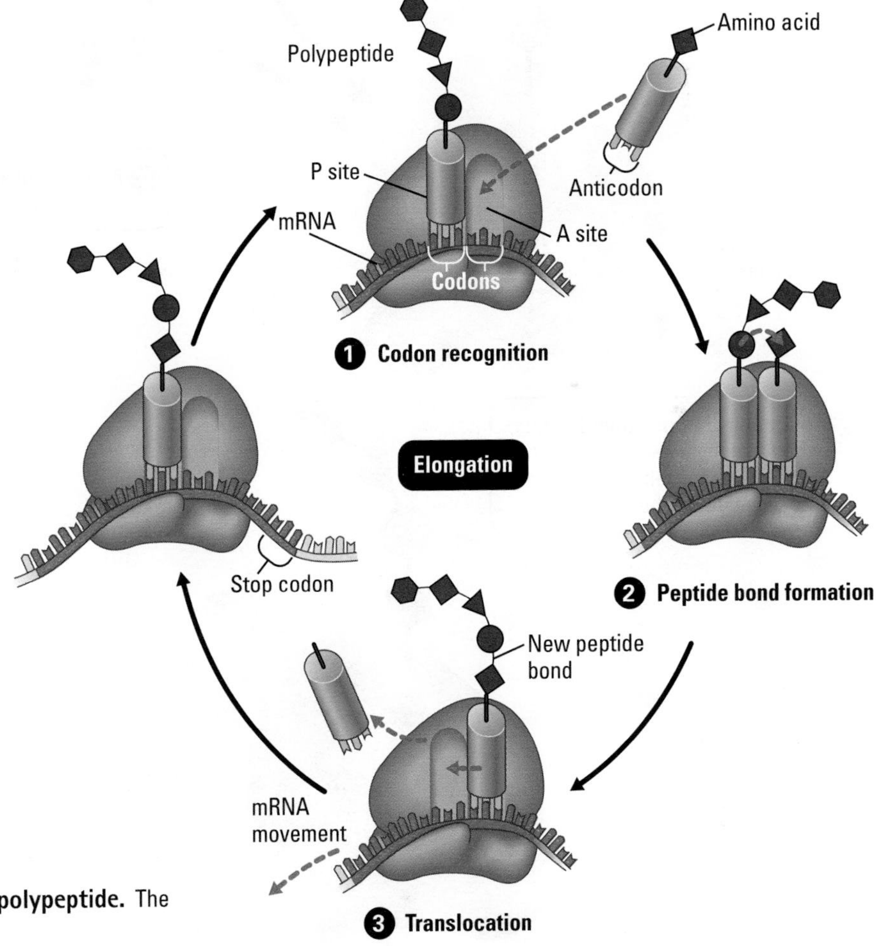

Figure 10.19 The elongation of a polypeptide. The dashed red arrows indicate movement.

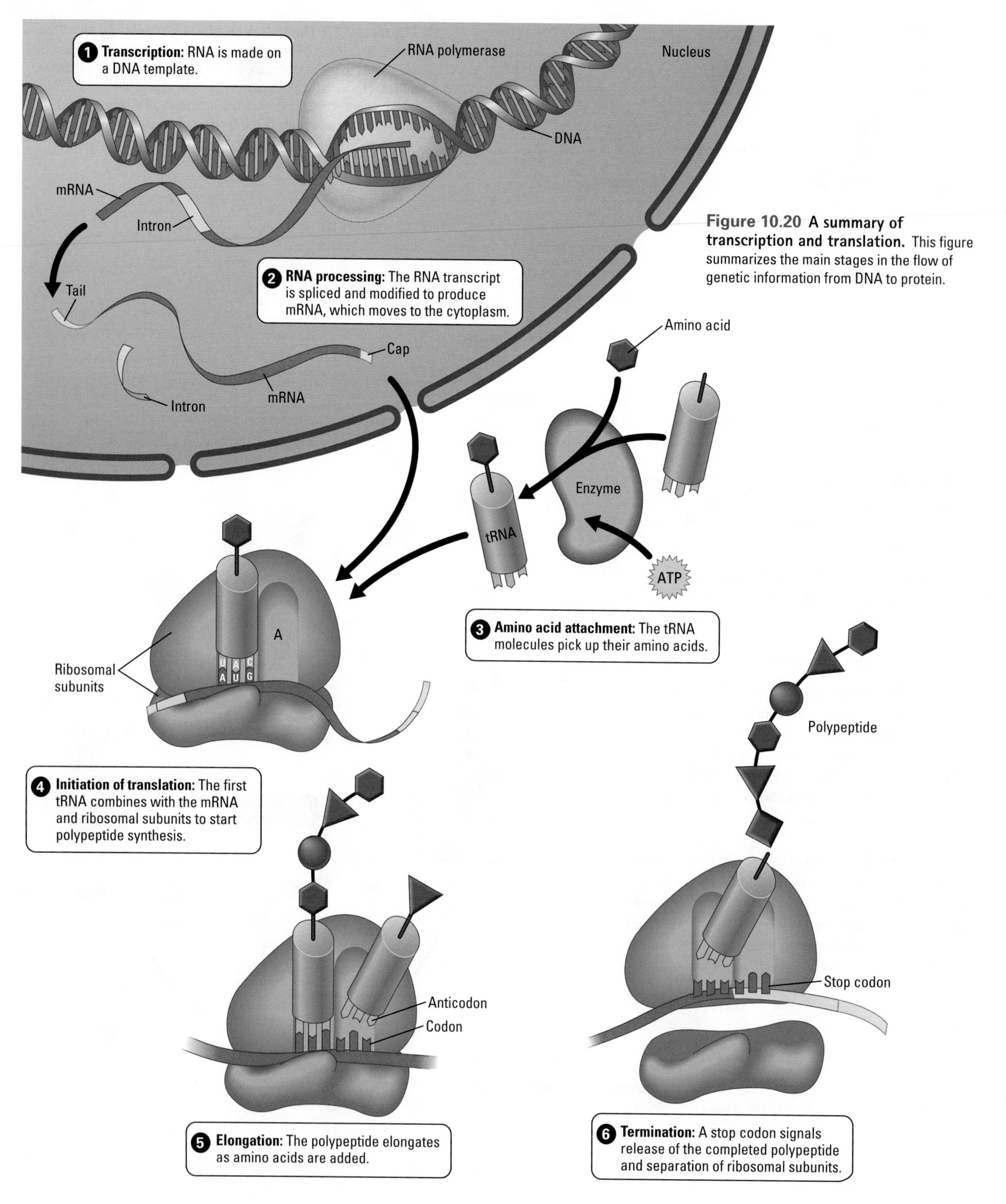

1 Transcription: RNA is made on a DNA template.
RNA polymerase
Nucleus
DNA
mRNA
Intron
Tail
2 RNA processing: The RNA transcript is spliced and modified to produce mRNA, which moves to the cytoplasm.
Cap
Intron
mRNA
Amino acid
Enzyme
tRNA
ATP
3 Amino acid attachment: The tRNA molecules pick up their amino acids.
A
U
A
C
A
U
G
Ribosomal subunits
4 Initiation of translation: The first tRNA combines with the mRNA and ribosomal subunits to start polypeptide synthesis.
Polypeptide
Anticodon
Codon
Stop codon
5 Elongation: The polypeptide elongates as amino acids are added.
6 Termination: A stop codon signals release of the completed polypeptide and separation of ribosomal subunits.

Figure 10.20 A summary of transcription and translation. This figure summarizes the main stages in the flow of genetic information from DNA to protein.

Review: DNA → RNA → Protein

Figure 10.20 reviews the flow of genetic information in the cell, from DNA to RNA to protein. In eukaryotic cells, transcription—the stage from DNA to RNA—occurs in the nucleus, and the RNA is processed before it enters the cytoplasm. Translation is rapid; a single ribosome can make an average-sized polypeptide in less than a minute. As it is made, a polypeptide coils and folds, assuming a three-dimensional shape, its tertiary structure. Several polypeptides may come together, forming a protein with quaternary structure.

What is the overall significance of transcription and translation? These are the processes whereby genes control the structures and activities of cells—or, more broadly, the way the genotype produces the phenotype. The chain of command originates with the information in a gene, a specific linear sequence of nucleotides in DNA. The gene serves as a template, dictating the transcription of a complementary sequence of nucleotides in mRNA. In turn, mRNA specifies the linear sequence in which amino acids appear in a polypeptide. Finally, the proteins that form from the polypeptides determine the appearance and capabilities of the cell and organism.

CHECKPOINT

1. How does RNA polymerase "know" where to start transcribing a gene?
2. What is an anticodon?
3. What is the function of the ribosome in protein synthesis?
4. Once polypeptide synthesis starts, what are the three main steps by which it grows (elongates)?
5. Which of the following does not participate directly in translation: ribosomes, transfer RNA, messenger RNA, DNA, enzymes, ATP?

Answers: 1. It recognizes the gene's promoter, a specific nucleotide sequence. **2.** An anticodon is the base triplet of a tRNA molecule that couples the tRNA to a complementary codon in the mRNA. The base pairing of anticodon to codon is a key step in translating mRNA to a polypeptide. **3.** The ribosome holds mRNA and tRNAs together and connects amino acids from the tRNAs to the growing polypeptide chain. **4.** Codon recognition by an incoming tRNA, peptide bond formation by the ribosome, and translocation of the tRNA and mRNA through the ribosome **5.** DNA.

Mutations

Since discovering how genes are translated into proteins, scientists have been able to describe many heritable differences in molecular terms. For instance, when a child is born with sickle-cell disease (see Figure 9.20), the condition can be traced back through a difference in a protein to one tiny change in a gene. In one of the polypeptides in the hemoglobin protein, the sickle-cell child has a single different amino acid, as you may recall from Chapter 3. This difference is caused by a single nucleotide difference in the coding strand of DNA **(Figure 10.21)**. In the double helix, one base pair is changed.

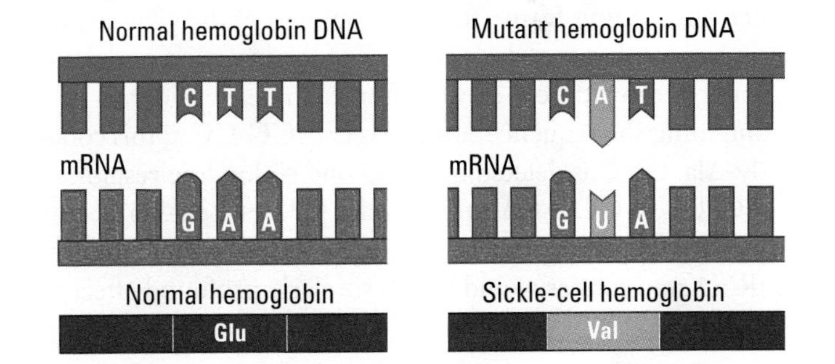

Figure 10.21 The molecular basis of sickle-cell disease. The sickle-cell allele differs from its normal counterpart, a gene for hemoglobin, by only one nucleotide. This difference changes the mRNA codon from one that codes for the amino acid glutamic acid (Glu) to one that codes for valine (Val).

The sickle-cell allele is not a unique case. We now know that the various alleles of many genes result from changes in single base pairs in DNA. Any change in the nucleotide sequence of DNA is called a **mutation.** Mutations can involve large regions of a chromosome or just a single nucleotide pair, as in the sickle-cell allele. Occasionally, a base substitution leads to an improved protein or one with new capabilities that enhance the success of the mutant organism and its descendants. Much more often, though, mutations are harmful. Let's consider how mutations involving only one or a few nucleotide pairs can affect gene translation.

Types of Mutations Mutations within a gene can be divided into two general categories: base substitutions and base insertions or deletions **(Figure 10.22)**. A base substitution is the replacement of one base, or nucleotide, by another. Depending on how a base substitution is translated, it can result in no change in the protein, in an insignificant change, or in a change that might be crucial to the life of the organism. Because of the redundancy of the genetic code, some substitution mutations have no effect. For example, if a mutation causes an mRNA codon to change from GAA to GAG, no change in the protein product would result, because GAA and GAG both code for the same amino acid (Glu). Such a change is called a silent mutation.

Other changes of a single nucleotide do change the amino acid coding. Such mutations are called missense mutations. For example, if a mutation causes an mRNA codon to change from GGC to AGC, the resulting protein will have a serine (Ser) instead of a glycine (Gly) at this position (see Figure 10.22a). Some missense mutations have little or no effect on the resulting protein, but others, as we saw in the sickle-cell case, cause changes in the protein that prevent it from performing normally.

Some base substitutions, called nonsense mutations, change an amino acid codon into a stop codon. For example, if an AGA (Arg) codon is mutated to a UGA (stop) codon, the result will be a prematurely terminated protein, which probably will not function properly.

Mutations involving the insertion or deletion of one or more nucleotides in a gene often have disastrous effects. Because mRNA is read as a series of nucleotide triplets during translation, adding or subtracting nucleotides may alter the **reading frame** (triplet grouping) of the genetic message. All the nucleotides that are "downstream" of the insertion or deletion will be regrouped into different codons. For example, consider an mRNA molecule containing the sequence AAG-UUU-GGC-GCA; this codes for Lys-Phe-Gly-Ala. If a U is deleted in the second codon, the resulting sequence will be AAG-UUG-GCG-CA, which codes for Lys-Leu-Ala (see Figure 10.22b). The altered polypeptide is likely to be nonfunctional. Inserting one or two mRNA nucleotides would have a similarly profound effect.

Mutagens What causes mutations? Mutagenesis, the creation of mutations, can occur in a number of ways. Mutations resulting from errors during DNA replication or recombination are known as spontaneous mutations, as are other mutations of unknown cause. Other sources of mutation are physical and chemical agents called **mutagens.** The most common physical mutagen is high-energy radiation, such as X-rays and ultraviolet (UV) light. Chemical mutagens are of various types. One type, for example, consists of chemicals that are similar to normal DNA bases but that base-pair incorrectly when incorporated into DNA.

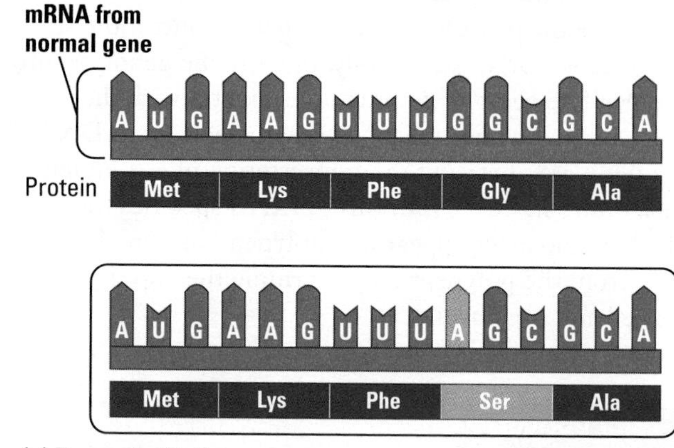

(a) Base substitution

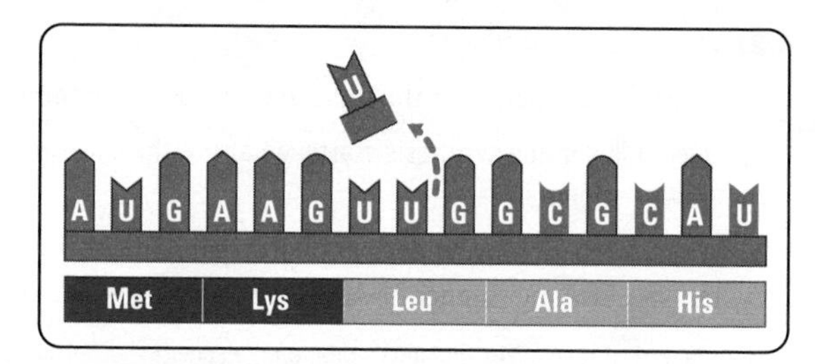

(b) Nucleotide deletion

Figure 10.22 Two types of mutations and their effects. Mutations are changes in DNA, but they are represented here as reflected in mRNA and its polypeptide product. **(a)** In the base substitution shown here, an A replaces a G in the fourth codon of the mRNA. The result in the polypeptide is a serine (Ser) instead of a glycine (Gly). This amino acid substitution may or may not affect the protein's function. **(b)** When a nucleotide is deleted (or inserted), the reading frame is altered, so that all the codons from that point on are misread. The resulting polypeptide is likely to be completely nonfunctional.

Many mutagens can act as carcinogens, agents that cause cancer. What can you do to avoid exposure to mutagens? Several lifestyle practices can help, including wearing protective clothing and sunscreen to minimize direct exposure to the sun's UV rays and not smoking. But such precautions are not foolproof; for example, you cannot entirely avoid UV radiation.

Although mutations are often harmful, they can also be extremely useful, both in nature and in the laboratory. Mutations are the source of the rich diversity of genes in the living world, a diversity that makes evolution by natural selection possible **(Figure 10.23)**. Mutations are also essential tools for geneticists. Whether naturally occurring or created in the laboratory, mutations are responsible for the different alleles needed for genetic research.

Figure 10.23 Mutations and diversity. Mutations are the ultimate source of the diversity of life on and around the coral reef in this underwater scene.

CHECKPOINT

1. What would happen if a mutation changed a start codon to some other codon?
2. What happens when one nucleotide is lost from the middle of a gene?

Answers: 1. Messenger RNA transcribed from the mutated gene would be nonfunctional because ribosomes would not initiate translation. **2.** In the mRNA, the reading frame downstream from the deletion is shifted, leading to a long string of incorrect amino acids in the polypeptide. The polypeptide will probably be nonfunctional.

Viruses: Genes in Packages

Viruses sit on the fence between life and nonlife. A virus is lifelike in having genes and a highly organized structure, but it differs from a living organism in not being made of cells and not being able to reproduce on its own. In many cases, a **virus** is nothing more than "genes in a box": a bit of nucleic acid wrapped in a protein coat **(Figure 10.24)**. A virus can survive only by infecting a living cell with genetic material that directs the cell's molecular machinery to make more viruses. In this section, we're going to take a look at the relationship between viral structure and the processes of nucleic acid replication, transcription, and translation. We'll consider viruses that infect different types of host organisms, starting with bacteria.

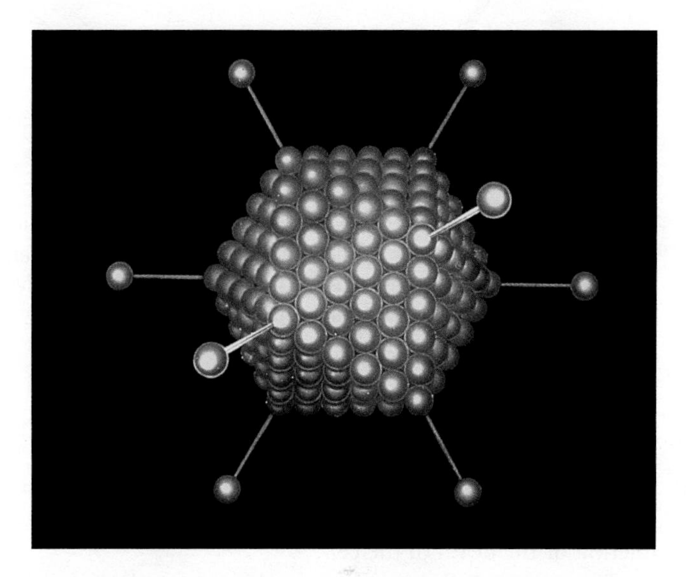

Figure 10.24 Adenovirus. A virus that infects the human respiratory system, an adenovirus consists of DNA enclosed in a protein shell shaped like a 20-sided polyhedron, shown here in a computer-generated model. At each vertex of the polyhedron is a protein spike, which helps the virus attach to a susceptible cell.

Bacteriophages

Viruses that attack bacteria are called **bacteriophages** ("bacteria-eaters"), or **phages** for short **(Figure 10.25)**. Once they infect a bacterium, most phages enter a reproductive cycle called the **lytic cycle.** The lytic cycle

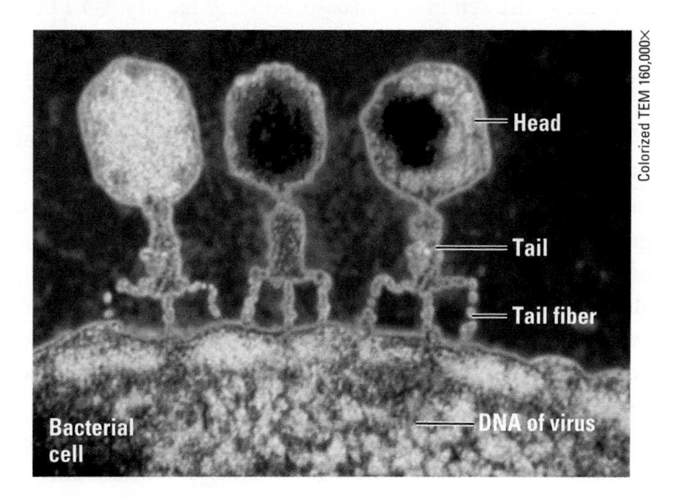

Figure 10.25 Bacteriophages infecting a bacterial cell. The "landing crafts" settling on the surface of the "planet" in this electron micrograph are actually viruses in the process of infecting a bacterium. The viruses are called bacteriophage T4, and the bacterium is *E. coli.* The phage consists of a molecule of DNA enclosed within an elaborate structure made of proteins. The "legs" of the phage (called tail fibers) bend when they touch the cell surface. The tail is a hollow rod enclosed in a springlike sheath. As the legs bend, the spring compresses, the bottom of the rod punctures the cell membrane, and the viral DNA passes from inside the head of the virus into the cell. Each phage is only about 200 nm tall (0.0002 mm).

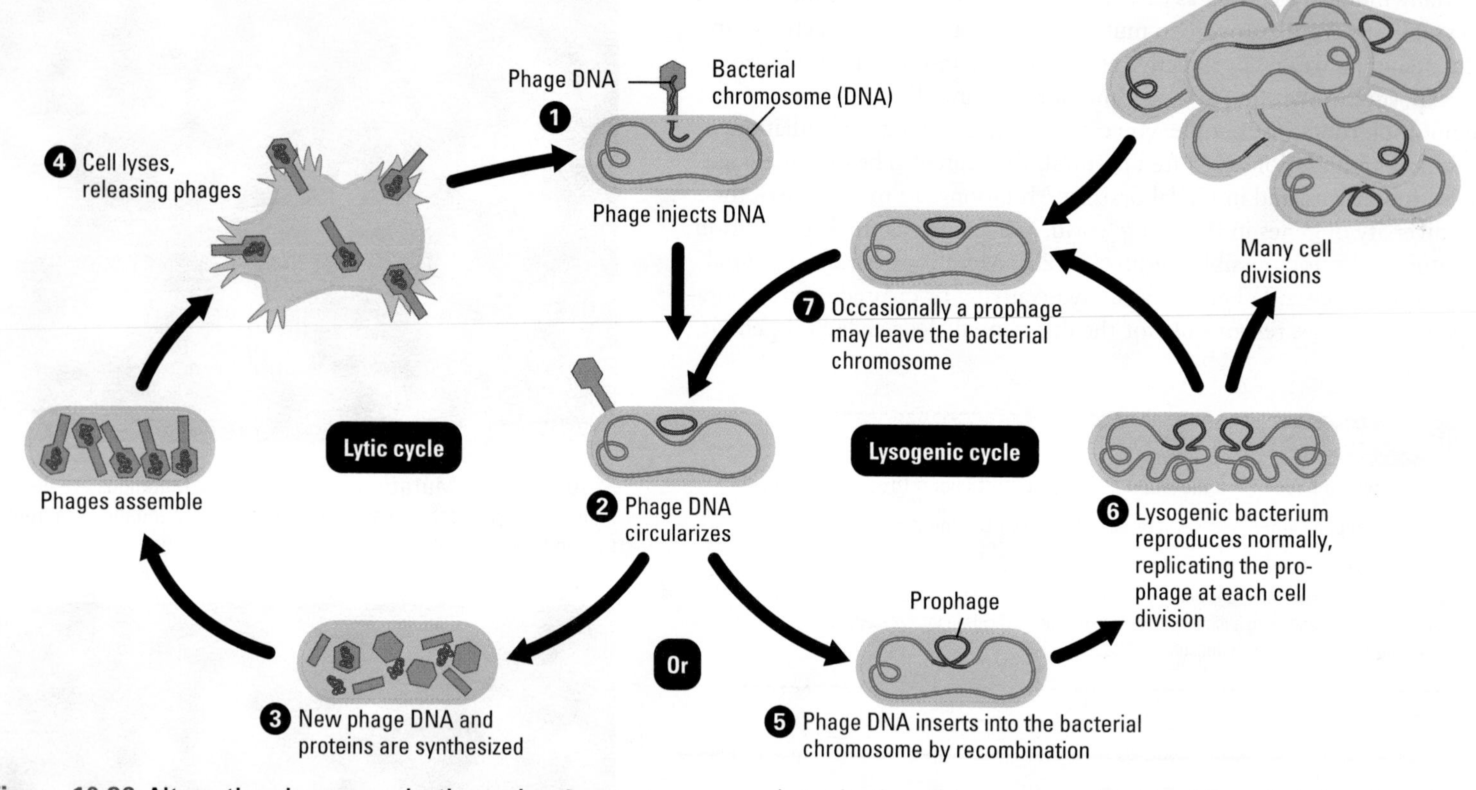

Figure 10.26 Alternative phage reproductive cycles. Certain phages can undergo alternative reproductive cycles. After entering the bacterial cell, the phage DNA can either integrate into the bacterial chromosome (lysogenic cycle) or immediately start the production of progeny phages (lytic cycle), destroying the cell. In most cases, the phage follows the lytic pathway, but once it enters a lysogenic cycle, the phage's DNA may be carried in the host cell's chromosome for many generations.

gets its name from the fact that, after many copies of the phage are produced within the bacterial cell, the bacterium lyses (breaks open). Some viruses can also reproduce by an alternative route—the **lysogenic cycle.** During a lysogenic cycle, viral DNA replication occurs without phage production or the death of the cell.

Figure 10.26 illustrates the two kinds of cycles for a phage named lambda that can infect *E. coli* bacteria. Lambda has a head (containing DNA) and a tail. Before embarking on one of the two cycles, ❶ lambda binds to the outside of a bacterium and injects its DNA inside. ❷ The injected lambda DNA forms a circle. In the lytic cycle, this DNA immediately turns the cell into a virus-producing factory. ❸ The cell's own machinery for DNA replication, transcription, and translation is hijacked by the virus and used to produce copies of the virus. ❹ The cell lyses, releasing the new phages.

In the lysogenic cycle, ❺ the viral DNA inserts by genetic recombination into the bacterial chromosome. Once inserted into the bacterial chromosome, the phage DNA is referred to as a **prophage**, and most of its genes are inactive. Survival of the prophage depends on the reproduction of the cell where it resides. ❻ The host cell replicates the prophage DNA along with its cellular DNA and then, upon dividing, passes on both the prophage and the cellular DNA to its two daughter cells. A single infected bacterium can quickly give rise to a large population of bacteria that all carry prophages. The prophages may remain in the bacterial cells indefinitely. ❼ Occasionally, however, a prophage leaves its chromosome; this event may be triggered by environmental conditions such as exposure to a

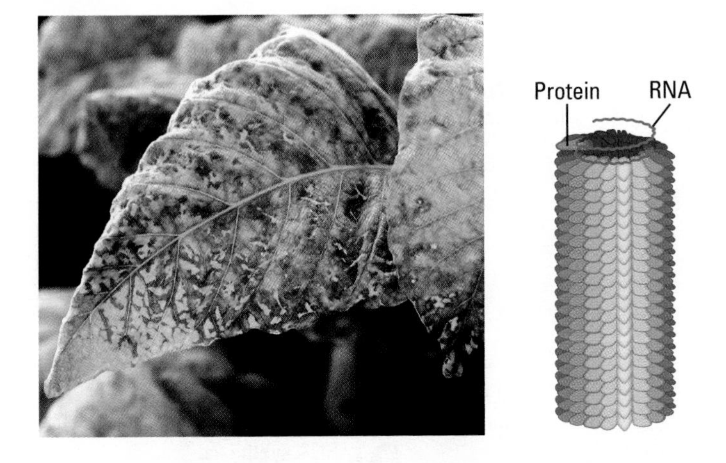

Figure 10.27 Tobacco mosaic virus. The photo shows the mottling of leaves in tobacco mosaic disease. The rod-shaped virus causing the disease has RNA as its genetic material.

mutagen. Once separate, the lambda DNA usually switches to the lytic cycle, which results in the production of many copies of the virus and bursting of the host cell.

Sometimes the few prophage genes active in a lysogenic bacterial cell can cause medical problems. For example, the bacteria that cause diphtheria, botulism, and scarlet fever would be harmless to humans if it were not for the prophage genes they carry. Certain of these genes direct the bacteria to produce toxins that make people ill.

Plant Viruses

Viruses that infect plant cells can stunt plant growth and diminish crop yields. Most plant viruses discovered to date have RNA rather than DNA as their genetic material. Many of them, like the tobacco mosaic virus shown in **(Figure 10.27)**, are rod-shaped with a spiral arrangement of proteins surrounding the nucleic acid.

To infect a plant, a virus must first get past the plant's outer protective layer of cells (the epidermis). For this reason, a plant damaged by wind, chilling, injury, or insects is more susceptible to infection than a healthy plant. Some insects also carry and transmit plant viruses. Farmers and gardeners, too, may spread plant viruses through the use of pruning shears and other tools. And infected plants may pass viruses to their offspring.

As with animal viruses, there is no cure for most viral diseases of plants, and agricultural scientists focus on reducing the number of plants that are infected and on breeding varieties of crop plants that resist viral infection. For example, strains of tomato, squash, and cantaloupe have been bred to be resistant to the tobacco mosaic virus, which can infect a wide variety of agricultural crops.

More recently, genetic engineering methods (discussed in detail in Chapter 12) have been used to create plant breeds that are resistant to some plant viruses. This has been done with the papaya, Hawaii's second largest crop. The spread of papaya ringspot potyvirus (PRSV) by aphids had wiped out the papaya in certain island regions. But since 1998, farmers have been able to plant a newly engineered PRSV-resistant strain of papaya, and papayas are now being reintroduced into their old habitats.

Animal Viruses

Viruses that infect animal cells are common causes of disease. We have all suffered from viral infections. **Figure 10.28** shows the structure of an influenza (flu) virus. Like many animal viruses, this one has an outer envelope

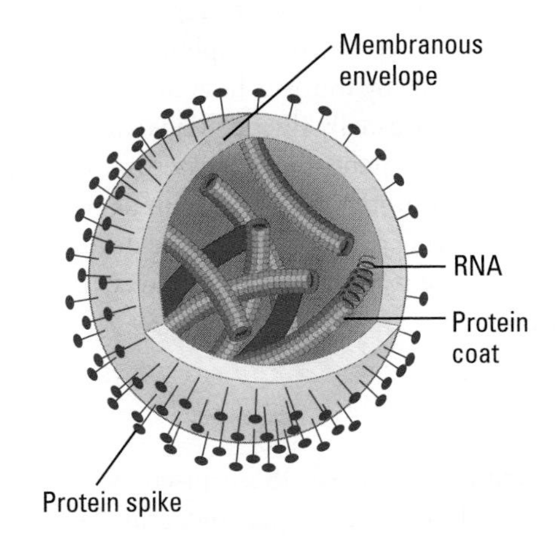

Figure 10.28 An influenza virus. The genetic material of this virus consists of eight separate molecules of RNA, each wrapped in a protein coat. Around the outside of the virus is an envelope made of membrane, studded with protein spikes.

made of phospholipid membrane, with projecting spikes of protein. The envelope helps the virus enter and leave a cell. Also, like many other animal viruses, flu viruses have RNA rather than DNA as their genetic material. Other RNA viruses include those that cause the common cold, measles, and mumps, as well as ones that cause more serious human diseases, such as AIDS and polio. Diseases caused by DNA viruses include hepatitis, chicken pox, and herpes infections.

Figure 10.29 shows the reproductive cycle of an enveloped RNA virus (the mumps virus, which causes a once-common childhood disease that has been largely eliminated by vaccines). When the virus contacts a susceptible cell, protein spikes on its outer surface attach to receptor proteins on the cell's plasma membrane. The viral envelope fuses with the cell's membrane, allowing the protein-coated RNA to ❶ enter the cytoplasm. ❷ Enzymes then remove the protein coat. ❸ An enzyme that entered the cell as part of the virus uses the virus's RNA genome as a template for making complementary strands of RNA. The new strands have two functions: ❹ They serve as mRNA for the synthesis of new viral proteins, and ❺ they serve as templates for synthesizing new viral genome RNA. ❻ The new coat proteins assemble around the new viral RNA. ❼ Finally, the viruses leave the cell by cloaking themselves in plasma membrane. In other words, the virus obtains its envelope from the cell, leaving the cell without necessarily lysing it.

Not all animal viruses reproduce in the cytoplasm. For example, the viruses called herpesviruses—which cause chicken pox, shingles, cold sores, and genital herpes—are enveloped DNA viruses that reproduce in a cell's nucleus, and they get their envelopes from the cell's nuclear membranes. Copies of the herpesvirus DNA usually remain behind as mini-chromosomes in the nuclei of certain nerve cells. There they remain latent until some sort of physical stress, such as a cold or sunburn, or emotional stress triggers the herpesvirus DNA to begin producing the virus, resulting in unpleasant symptoms. Once acquired, herpes infections may flare up repeatedly throughout a person's life. Over 75% of American adults are thought to carry herpes simplex 1 (which causes cold sores), and over 20% carry herpes simplex 2 (which causes genital herpes).

The amount of damage a virus causes the body depends partly on how quickly the immune system responds to fight the infection and partly on the ability of the infected tissue to repair itself. We usually recover completely from colds because our respiratory tract tissue can efficiently replace damaged cells by mitosis. In contrast, the poliovirus attacks nerve cells, which do not usually divide. The damage to such cells by polio, unfortunately, is permanent. In such cases, we try to prevent the disease with vaccines. The antibiotic drugs that help us recover from bacterial infections are powerless against viruses. The development of antiviral drugs has been slow because it is difficult to find ways to kill a virus without killing the host cells.

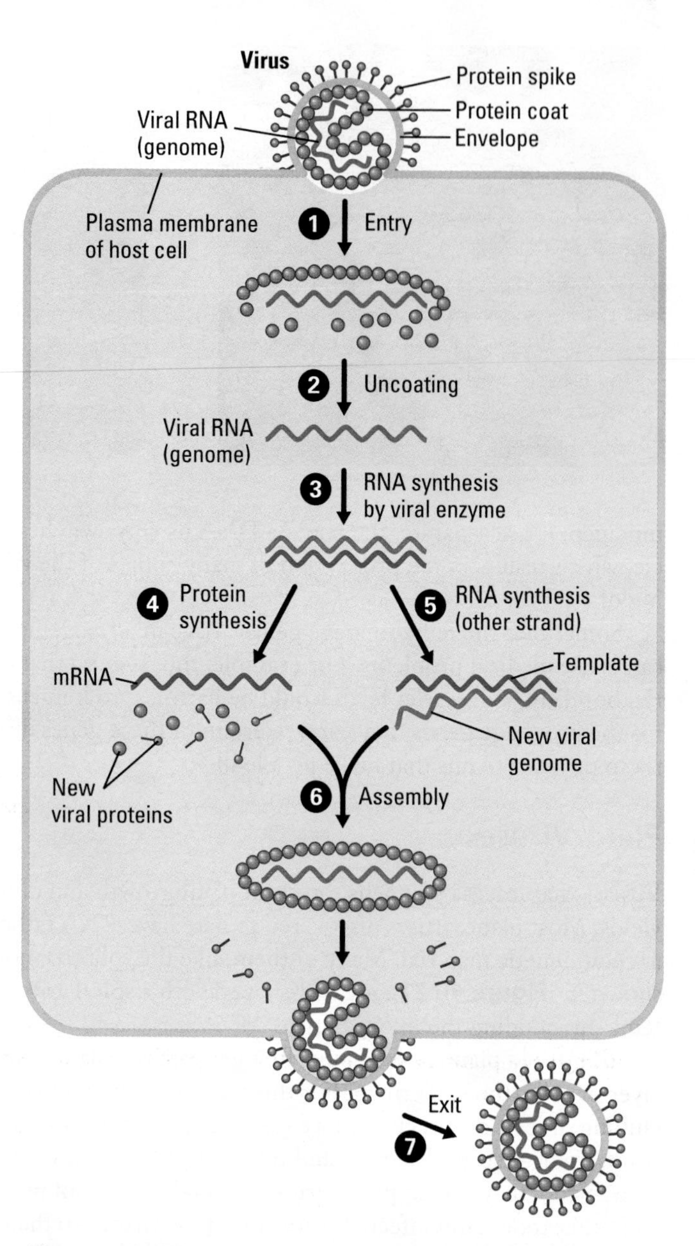

Figure 10.29 The reproductive cycle of an enveloped virus. This virus is the one that causes mumps. Like the flu virus, it has a membranous envelope with protein spikes, but its genome is a single molecule of RNA.

HIV, the AIDS Virus

The devastating disease AIDS is caused by a type of RNA virus with some special twists. In outward appearance, the AIDS virus **(Figure 10.30a)** somewhat resembles the flu or mumps virus. Its envelope enables HIV to enter and leave a cell much the way the mumps virus does. But HIV has a different mode of reproduction. It is a **retrovirus,** an RNA virus that reproduces by means of a DNA molecule. Retroviruses are so named because they reverse the usual DNA→RNA flow of genetic information. They carry molecules of an enzyme called **reverse transcriptase,** which catalyzes reverse transcription, the synthesis of DNA on an RNA template.

Figure 10.30b illustrates what happens after HIV RNA is uncoated in the cytoplasm of a cell. The reverse transcriptase (green) ❶ uses the RNA as a template to make a DNA strand and then ❷ adds a second, complementary DNA strand. ❸ The resulting double-stranded viral DNA then enters the cell nucleus and inserts itself into the chromosomal DNA, becoming a **provirus.** Occasionally, the provirus is ❹ transcribed into RNA and ❺ translated into viral proteins. ❻ New viruses assembled from these components eventually leave the cell and can then infect other cells. This is the standard reproductive cycle for retroviruses.

AIDS stands for acquired immunodeficiency syndrome, and **HIV** for human immunodeficiency virus; these terms describe the main effect of the virus on the body. HIV infects and eventually kills several kinds of white blood cells that are important in the body's immune system **(Figure 10.30c)**. The loss of such cells causes the body to become susceptible to other infections that it would normally be able to fight off. Such secondary infections cause the syndrome (a collection of symptoms) that eventually kills AIDS patients.

Two main types of anti-HIV drugs are currently used in treating AIDS. Both interfere with the reproduction of the virus. The first type inhibits the action of the HIV enzyme reverse transcriptase; AZT, mentioned at the start of this chapter, is the most widely used drug of this type. The second type inhibits the action of enzymes called proteases, which are needed for making HIV proteins. Many HIV-infected people in the United States and other industrialized countries take a "drug cocktail" that contains both reverse transcriptase inhibitors and protease inhibitors, and the combination seems to be much more effective than the individual drugs in keeping the virus at bay and extending patients' lives. However, even in combination, the drugs do not completely rid the body of the virus. Typically, HIV reproduction and the symptoms of AIDS return if a patient discontinues the medications. AIDS has no cure yet, leaving prevention (namely, the avoidance of unprotected sex and needle sharing) as the only healthy option.

CHECKPOINT

1. Describe one way some viruses can perpetuate their genes without destroying the cells they infect.
2. How do some viruses reproduce without ever having DNA?
3. What are three ways that viruses get into a plant?
4. Why is HIV called a retrovirus?

Answers: 1. Some viruses, such as phage lambda and HIV, can insert their DNA into the DNA of the cell they infect. The viral DNA is replicated along with the cell's DNA every time the cell divides. **2.** The genetic material of these viruses is RNA, which is replicated inside the infected cell by special enzymes encoded by the virus. The viral genome (or its complement) serves as mRNA for the synthesis of viral proteins. **3.** Through lesions caused by injuries, through transfer by insects that feed on the plant, and through contaminated farming or gardening tools **4.** Because it synthesizes DNA from its RNA genome. This is the reverse ("retro") of the usual DNA → RNA information flow.

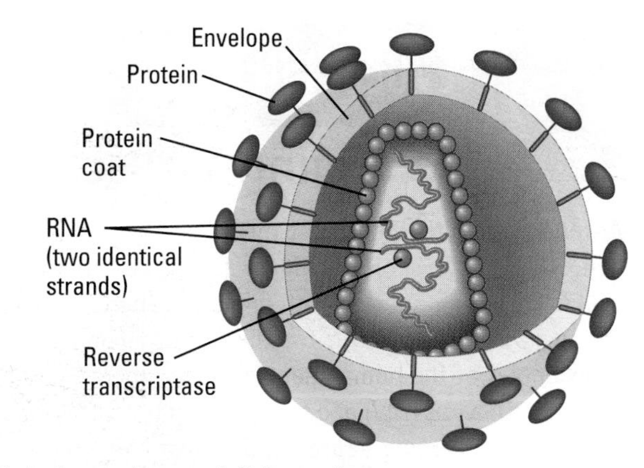

(a) HIV, the human immunodeficiency virus

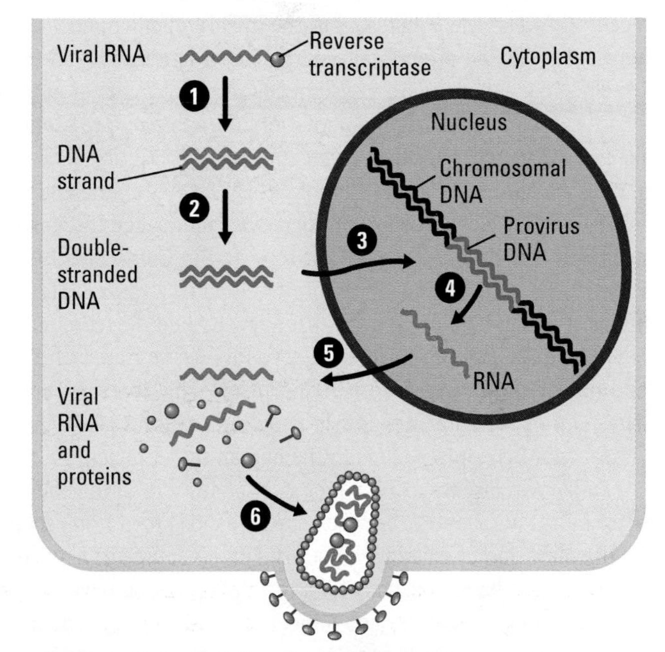

(b) The behavior of HIV nucleic acid in an infected cell

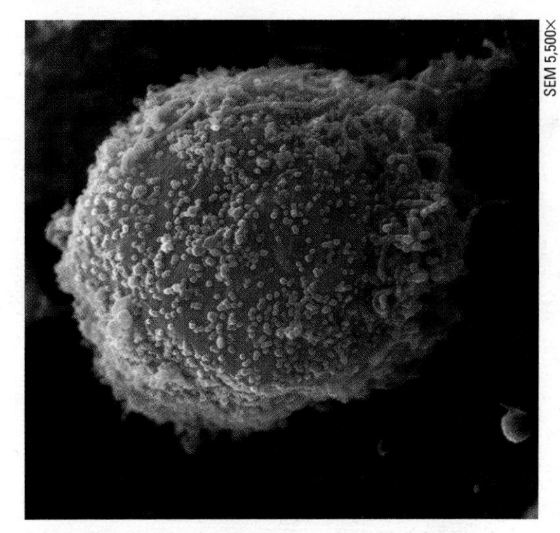

(c) HIV (blue spots) infecting a white blood cell

Figure 10.30 HIV, the AIDS virus.

EVOLUTION CONNECTION

Emerging Viruses

Acknowledging the persistent threat that viruses pose to human health, geneticist and Nobel Prize winner Joshua Lederberg once warned: "We live in evolutionary competition with microbes. There is no guarantee that we will be the survivors." Lederberg cited the AIDS epidemic and recurrent flu epidemics as examples of the human population's vulnerability to viral attacks.

The AIDS virus (HIV) and the new varieties of flu that frequently appear are examples of **emerging viruses,** viruses that have appeared suddenly or have recently come to the attention of medical scientists. Another example of an emerging virus is the deadly Ebola virus **(Figure 10.31a)**, recognized initially in 1976 in central Africa. The Ebola virus is one of several emerging viruses that cause hemorrhagic fever, an often fatal syndrome characterized by fever, vomiting, massive bleeding, and circulatory system collapse. A number of other dangerous new viruses cause encephalitis, inflammation of the brain. One such virus is the West Nile virus **(Figure 10.31b)**, which appeared for the first time in North America in 1999 and spread to all 48 contiguous U.S. states by the end of 2004.

A viral disease that has emerged even more recently is severe acute respiratory syndrome (SARS), first reported in China in February 2003 **(Figure 10.32)**. Within three months, about 8,450 people were known to be infected, some 10% of whom subsequently died. Researchers quickly identified the agent causing SARS as a coronavirus, so named for its halo-like "corona" of spikes (see Figure 10.32, inset). This virus, with a single-stranded RNA genome, was not previously known to cause disease in humans.

How do such viruses burst on the human scene, giving rise to rare or previously unknown diseases? Three processes contribute to the emergence of viral diseases: mutations, contact between species, and spread from isolated populations.

The mutation of existing viruses is a major source of new viral diseases. RNA viruses tend to have unusually high rates of mutation because errors in replicating their RNA genomes are not subject to the kind of proofreading mechanisms that help reduce errors during DNA replication. Some mutations enable existing

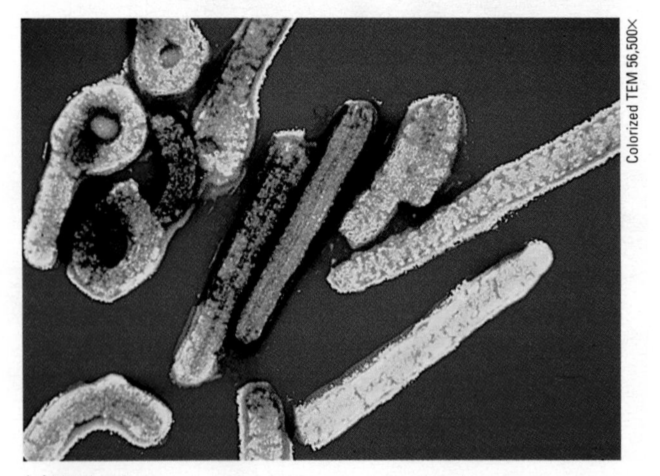

(a) Ebola virus

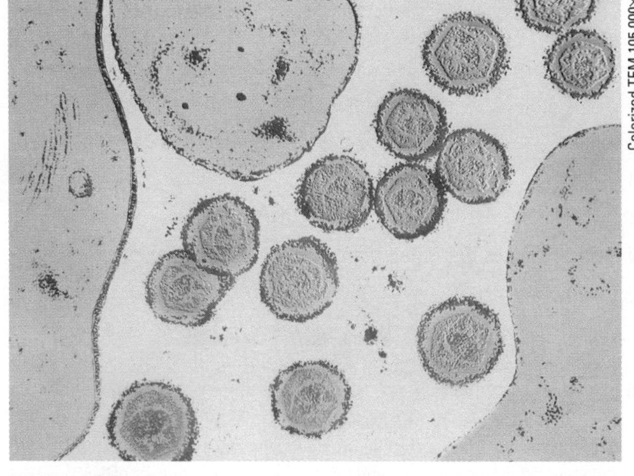

(b) West Nile virus

Figure 10.31 Emerging viruses. Emerging viruses are viruses that have recently appeared or recently come to the attention of medical scientists.

Figure 10.32 SARS Virus. Young girls in Hong Kong attempt to protect themselves from exposure to SARS virus (inset).

viruses to evolve into new strains (genetic varieties) that can cause disease in individuals who have developed resistance to the ancestral virus. Flu epidemics, for instance, are caused by new influenza virus strains that are different enough from earlier strains that people have little immunity to them.

New viral diseases often arise from the spread of existing viruses from one host species to another. Scientists estimate that about three-quarters of new human diseases have originated in other animals. For example, hantavirus is common in rodents, especially deer mice. The population of deer mice in the southwestern United States exploded in 1993 after unusually wet weather increased the rodents' food supply. Many people who inhaled dust containing traces of urine and feces from infected mice became infected with hantavirus, and dozens died. The source of the SARS-causing virus is still undetermined, although candidates include exotic animals found in livestock markets in China. And early 2004 brought reports of the first cases of people in Southeast Asia infected with a flu virus previously seen only in birds (the "avian flu"). As of early 2006, over 140 people in six Asian countries had become infected, most as a result of direct contact with sick poultry **(Figure 10.33)**. If this virus evolves so that it can spread easily from person to person, the potential for a major human outbreak is significant. Indeed, evidence is strong that the flu pandemic of 1918–1919, which killed about 40 million people, originated in birds.

The spread of a viral disease from a small, isolated population can lead to widespread epidemics. For instance, AIDS went unnamed and virtually unnoticed for decades before it began to spread around the world. In this case, technological and social factors, including affordable international travel, blood transfusions, sexual promiscuity, and the abuse of intravenous drugs, allowed a previously rare human disease to become a global scourge. It is likely that when we do find the means to control HIV and other deadly viruses, genetic research—in particular, molecular biology—will be responsible for the discovery. ■

Figure 10.33 Ducks in Vietnam being checked for infection by the Avian flu virus (inset).

Chapter Review

SUMMARY OF KEY CONCEPTS

For study help and activities, go to campbellbiology.com or the student CD-ROM.

Biology and Society: Sabotaging HIV

- Molecular biology, the study of heredity at the molecular level, can provide insight into many areas of biology, including the action of anti-HIV drugs.

DNA: Structure and Replication

Activity *The Hershey-Chase Experiment*

- **DNA and RNA Structure**

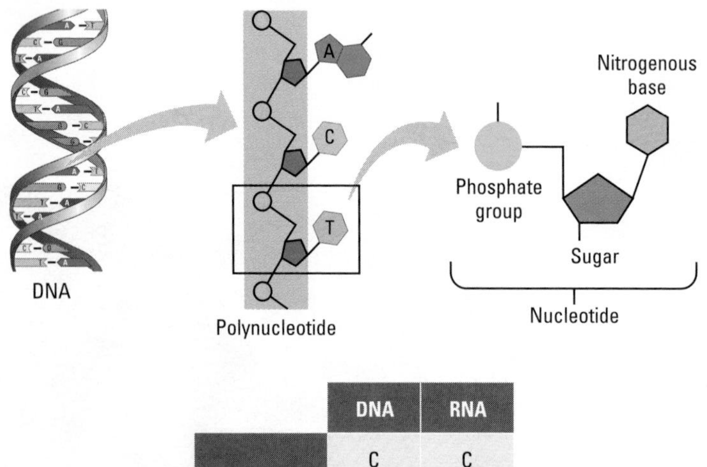

	DNA	RNA
Nitrogenous base	C G A T	C G A U
Sugar	Deoxy-ribose	Ribose
Number of strands	2	1

Activity *DNA and RNA Structure*

- **Watson and Crick's Discovery of the Double Helix** Watson and Crick worked out the three-dimensional structure of DNA: two polynucleotide strands wrapped around each other in a double helix. Hydrogen bonds between bases hold the strands together. Each base pairs with a complementary partner: A with T, and G with C.

Activity *DNA Double Helix*

- **DNA Replication**

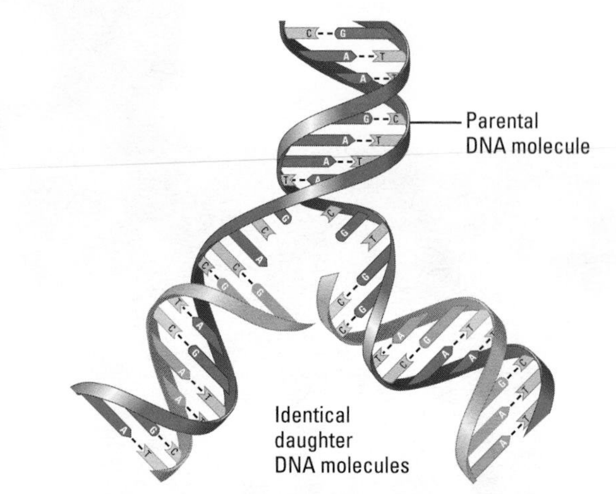

Case Study in the Process of Science *What Is the Correct Model for DNA Replication?*

Activity *DNA Replication*

The Flow of Genetic Information from DNA to RNA to Protein

eTutor *Protein Synthesis*

MP3 Tutor *DNA to RNA to Protein*

- **How an Organism's Genotype Produces Its Phenotype** The information constituting an organism's genotype is carried in the sequence of its DNA bases. Studies of inherited metabolic defects first suggested that phenotype is expressed through proteins. A particular gene—a linear sequence of many nucleotides—specifies a polypeptide. The DNA of the gene is transcribed into RNA, which is translated into the polypeptide.

Activity *Overview of Protein Synthesis*

Case Study in the Process of Science *How Is a Metabolic Pathway Analyzed?*

- **From Nucleotides to Amino Acids: An Overview** The DNA of a gene is transcribed into RNA using the usual base-pairing rules, except that an A in DNA pairs with U in RNA. In the translation of a genetic message, each triplet of nucleotide bases in the RNA, called a codon, specifies one amino acid in the polypeptide.

- **The Genetic Code** In addition to codons that specify amino acids, the genetic code has one codon that is a start signal and three that are stop signals for translation. The genetic code is redundant: There is more than one codon for most amino acids.

- **Transcription: From DNA to RNA** In transcription, RNA polymerase binds to the promoter of a gene, opens the DNA double helix there, and catalyzes the synthesis of an RNA molecule using one DNA strand as a template. As the single-stranded RNA transcript peels away from the gene, the DNA strands rejoin.

- **The Processing of Eukaryotic RNA** The RNA transcribed from a eukaryotic gene is processed before leaving the nucleus to serve as messenger RNA (mRNA). Introns are spliced out, and a cap and tail are added.

Activity *Transcription and RNA Processing*

- **Translation: The Players**

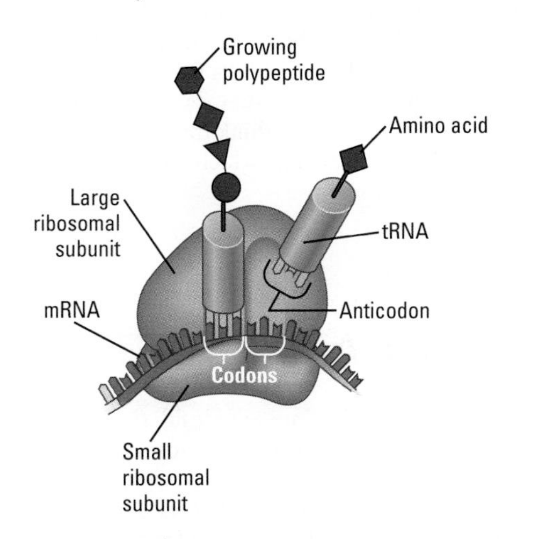

- **Translation: The Process** In initiation, a ribosome assembles with the mRNA and the initiator tRNA bearing the first amino acid. One by one, the codons of the mRNA are recognized by tRNAs bearing succeeding amino acids. The ribosome bonds the amino acids together. With each addition, the mRNA translocates by one codon through the ribosome. When a stop codon is reached, the completed polypeptide is released.

Activity *Translation*

- **Review: DNA → RNA → Protein** Figure 10.20 summarizes transcription, RNA processing, and translation. The sequence of codons in DNA, via the sequence of codons in mRNA, spells out the primary structure of a polypeptide.

- **Mutations** Mutations are changes in the DNA base sequence, caused by errors in DNA replication or by mutagens. Substituting, inserting, or deleting nucleotides in a gene has varying effects on the polypeptide and organism.

Case Study in the Process of Science *How Do You Diagnose a Genetic Disorder?*

Viruses: Genes in Packages

Activity *Simplified Reproductive Cycle of a Virus*

- Viruses can be regarded as genes packaged in protein.

- **Bacteriophages** When phage DNA enters a lytic cycle inside a bacterium, it is replicated, transcribed, and translated. The new viral DNA and protein molecules then assemble into new phages, which burst from the cell. In the lysogenic cycle, phage DNA inserts into the cell's chromosome and is passed on to generations of daughter cells. Much later, it may initiate phage production.

Activity *Phage Lytic Cycle*

Activity *Phage Lysogenic and Lytic Cycles*

- **Plant Viruses** Viruses that infect plants can be a serious agricultural problem. Most have RNA genomes. They enter plants via breaks in the plant's outer layers.

- **Animal Viruses** Many animal viruses, such as flu viruses, have RNA genomes; others, such as hepatitis viruses, have DNA. Some animal viruses "steal " a bit of cell membrane as a protective envelope. Some, such as the herpesviruses, can remain latent inside cells for long periods.

- **HIV, the AIDS Virus** HIV is a retrovirus. Inside a cell it uses its RNA as a template for making DNA, which is then inserted into a chromosome.

Activity *HIV Reproductive Cycle*

Case Study in the Process of Science *What Causes Infections in AIDS Patients?*

Case Study in the Process of Science *Why Do AIDS Rates Differ across the United States?*

Evolution Connection: Emerging Viruses

- Emerging viruses, ones that have appeared recently, result from mutations, contact between species, and spreading from previously isolated populations.

SELF-QUIZ

1. A molecule of DNA contains two polymer strands called __________, made by bonding many monomers called __________ together. Each monomer contains three parts: __________, __________, and __________.

2. Which of the following correctly ranks nucleic acid structures in order of size, from largest to smallest?
   a. gene, chromosome, nucleotide, codon
   b. chromosome, gene, codon, nucleotide
   c. nucleotide, chromosome, gene, codon
   d. chromosome, nucleotide, gene, codon

3. A scientist inserts a radioactively labeled DNA molecule into a bacterium. The bacterium replicates this DNA molecule and distributes one daughter molecule (double helix) to each of two daughter cells. How much radioactivity will the DNA in each of the two daughter cells contain? Why?

4. The nucleotide sequence of a DNA codon is GTA. In an mRNA molecule transcribed from this DNA, the codon has the sequence __________. In the process of protein synthesis, a tRNA pairs with the mRNA codon. The nucleotide sequence of the tRNA anticodon is __________. The amino acid attached to the tRNA is __________. (See Figure 10.11.)

5. Describe the process by which the information in a gene is transcribed and translated into a protein. Correctly use these terms in your description: tRNA, amino acid, start codon, transcription, mRNA, gene, codon, RNA polymerase, ribosome, translation, anticodon, peptide bond, stop codon.

6. Match the following molecules with the cellular process or processes in which they are primarily involved.

a. ribosomes	1. DNA replication
b. tRNA	2. transcription
c. DNA polymerases	3. translation
d. RNA polymerase	
e. mRNA	

7. A geneticist found that a particular mutation had no effect on the polypeptide encoded by the gene. This mutation probably involved
   a. deletion of one nucleotide.
   b. alteration of the start codon.
   c. insertion of one nucleotide.
   d. substitution of one nucleotide.

8. Scientists have discovered how to put together a bacteriophage with the protein coat of phage A and the DNA of phage B. If this composite phage were allowed to infect a bacterium, the phages produced in the cell would have
   a. the protein of A and the DNA of B.
   b. the protein of B and the DNA of A.
   c. the protein and DNA of A.
   d. the protein and DNA of B.

9. HIV requires an enzyme called __________ to convert its RNA genome to a DNA version.

10. Why is reverse transcriptase a particularly good target for anti-AIDS drugs? (*Hint:* Would you expect such a drug to harm the human host?)

Answers to the Self-Quiz questions can be found in Appendix D.

Go to the website or CD-ROM for more Self-Quiz questions.

THE PROCESS OF SCIENCE

11. A cell containing a single chromosome is placed in a medium containing radioactive phosphate, making any new DNA strands formed by DNA replication radioactive. The cell replicates its DNA and divides. Then the daughter cells (still in the radioactive medium) replicate their DNA and divide, resulting in a total of four cells. Sketch the DNA molecules in all four cells, showing a normal (nonradioactive) DNA strand as a solid line and a radioactive DNA strand as a dashed line.

12. In a classic 1952 experiment, biologists Alfred Hershey and Martha Chase labeled two batches of bacteriophage, one with radioactive sulfur (which only tags protein) and the other with radioactive phosphorus (which only tags DNA). In separate test tubes, they allowed each batch of phage to bind to nonradioactive bacteria and inject its DNA. After a few minutes, they separated the bacterial cells from the viral parts that remained outside the bacterial cells and measured the radioactivity of both portions. What results do you think they obtained? How would these results help them to determine which viral component—DNA or protein—was the infectious portion?

BIOLOGY AND SOCIETY

13. Scientists at the National Institutes of Health (NIH) have worked out thousands of sequences of genes and the proteins they encode, and similar analysis is being carried out at universities and private companies. Knowledge of the nucleotide sequences of genes might be used to treat genetic defects or produce lifesaving medicines. The NIH and some U.S. biotechnology companies have applied for patents on their discoveries. In Britain, the courts have ruled that a naturally occurring gene cannot be patented. Do you think individuals and companies should be able to patent genes and gene products? Before answering, consider the following: What are the purposes of a patent? How might the discoverer of a gene benefit from a patent? How might the public benefit? What negative effects might result from patenting genes?

14. AZT and other anti-HIV drugs have the potential to prevent the transmission of HIV to the children of infected mothers. But anti-AIDS drugs are too expensive for most people in developing nations. Some researchers have proposed a series of trials in which HIV-infected pregnant women are given very small doses of anti-HIV medications. Such research has the potential to determine the smallest effective dose of medication required to prevent transmission, and this could save many lives by reducing the cost of drug treatment. But such a trial would also result in many HIV-infected babies being born—a situation that could potentially have been prevented. Do you think such trials should be permitted? What are the potential benefits? The potential costs? How do you measure the relative value of each?

11 How Genes Are Controlled

Some researchers focus their **cloning efforts** on the most endangered species.

Shortly after you eat ice cream, bacteria living in your large intestine **turn on certain genes.**

Cancer-causing genes were first discovered in a chicken virus.

Lung cancer causes **more deaths** than any other kind of cancer.

BIOLOGY AND SOCIETY

Cloning at the Edge of Extinction

The animals in **Figure 11.1** all share two unusual features. First, they are all members of endangered species. Second, they are all **clones,** organisms created by asexual reproduction and thus genetically identical to a single parent.

First accomplished in the 1950s with frogs, animal cloning became much more commonplace after 1997. That year, Scottish researchers announced the first successful cloning of a mammal—the celebrated Dolly the sheep, who was cloned using an udder cell from an adult ewe. In the years since Dolly's landmark birth, researchers have cloned a variety of other mammals, including horses, cats, cows, pigs, monkeys, and, most recently, dogs.

Some researchers are now concentrating their efforts on cloning members of endangered species. Among the rare animals that have been cloned are a gaur (an Asian ox) and a wild mouflon (a small European sheep). One remarkable case was the 2003 cloning of a banteng, a Javanese bovine whose numbers have dwindled to just a few in the wild. Using frozen cells from a zoo-raised banteng that had died 23 years prior, scientists transplanted nuclei from the frozen cells into nucleus-free eggs from ordinary cows. The resulting embryos were implanted into surrogate cows, leading to the birth of a healthy baby banteng. This success shows that it is possible to produce a baby even when a female of the donor species is unavailable. Scientists may someday be able to use similar cross-species methods to clone an animal from a recently extinct species.

The use of cloning to repopulate endangered species holds tremendous promise. However, cloning may also create new problems. Conservationists object that cloning may detract from efforts to preserve natural habitats. They correctly point out that cloning does not increase genetic diversity and is therefore not as beneficial to endangered species as natural reproduction. Cloned animals are also less healthy than those arising from a fertilized egg. In 2003, Dolly was euthanized after suffering complications from a lung disease normally seen only in much older sheep. Other cloned animals have exhibited defects as well, such as susceptibility to obesity, pneumonia, liver failure, and premature death.

The ability to clone an animal using a transplanted nucleus demonstrates an important point: The nucleus of an adult body cell contains a complete genome capable of directing the production of an entire organism. The development of a multicellular organism, with many different kinds of cells, depends on **gene regulation,** the turning on and off of genes. Different genes are active in different kinds of cells. The health problems seen in cloned animals underscore the fact that proper gene regulation is important to the well-being of all organisms.

How genes are controlled and the role of gene regulation in the lives of organisms are the subjects of this chapter. As you will see, these subjects touch on some of the most interesting topics in all of biology. ■

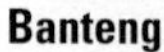

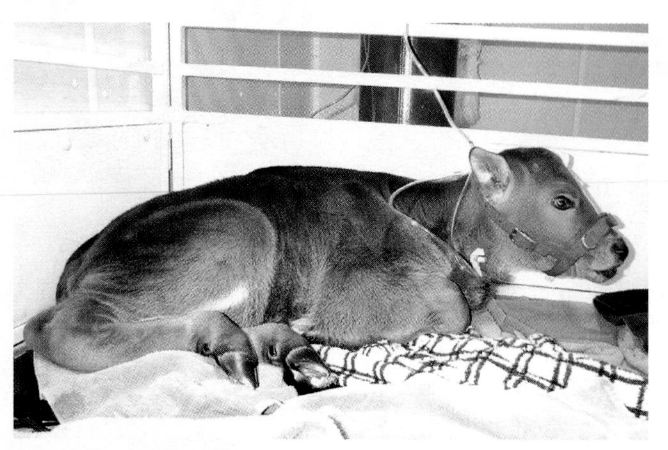

Gaur

Mouflon calf with mother

Banteng

Figure 11.1 Clones of endangered species.

How and Why Genes Are Regulated

MP3 Tutor Control of Gene Expression

The micrographs in **Figure 11.2** show four of the many different types of human cells. Each type of cell has a unique structure appropriate for carrying out its function. For example, a sperm cell has a powerful flagellum that can propel it through a woman's reproductive tract to meet an egg, and a nerve cell has long extensions for conducting nerve signals between widely separated parts of the body. Together, the various types of human cells enable a person to function as a whole.

What makes each of these types of cells different? Do they have different genes? Remember that every somatic cell in your body (that is, every cell except your gametes) was produced by repeated rounds of mitosis that started with a zygote. Since mitosis duplicates the genome exactly (or very nearly so), each of the many different kinds of cells in your body has the same DNA as the zygote.

If they all contain identical genetic instructions, then how do cells become different from one another? The only way that cells with the same genetic information can develop into cells with different structures and functions is if gene activity is regulated. In some way, control mechanisms must determine that certain genes will be turned *on* while others remain turned *off* in a particular cell. In other words, individual cells must undergo **cellular differentiation**—that is, they must become specialized in structure and function. It is the regulation of genes that leads to this specialization.

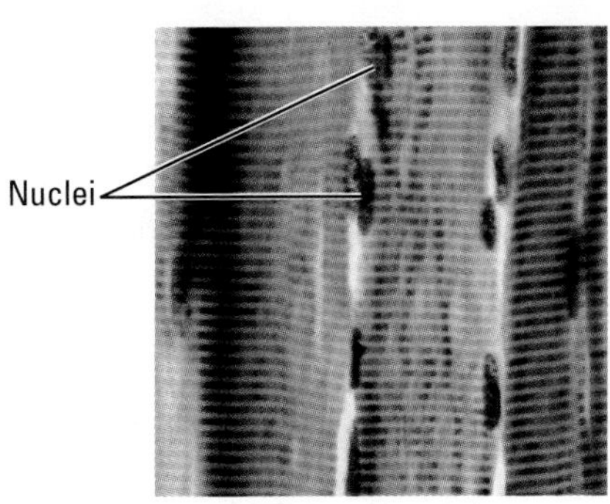

(a) Three muscle cells (partial). This micrograph shows short segments of three muscle cells, which are long fibers with multiple nuclei. The horizontal stripes result from the arrangement of the proteins that enable the cells to contract.

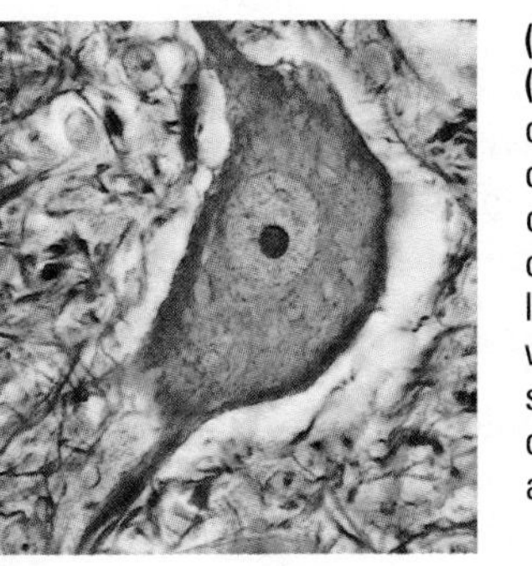

(b) Nerve cell (partial). The large orange shape is part of a nerve cell. You can see only a little of this cell's three long extensions, which carry nerve signals from one part of the body to another.

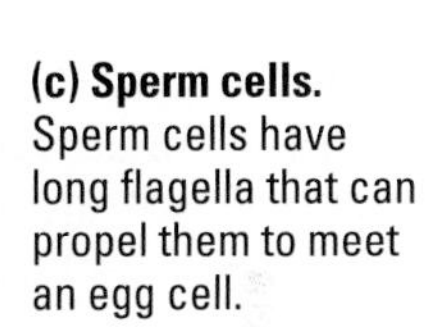

(c) Sperm cells. Sperm cells have long flagella that can propel them to meet an egg cell.

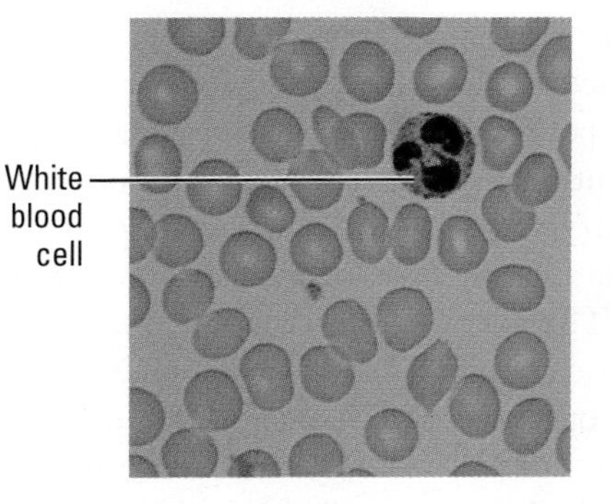

(d) Blood cells. A single white blood cell is surrounded by red blood cells. The small size and disklike shape of red blood cells help them efficiently transport oxygen molecules through the narrowest of blood vessels.

Figure 11.2 Four types of human cells (each shown at 750× magnification).

Patterns of Gene Expression in Differentiated Cells

What does it mean to say that genes are active or inactive, turned on or off? As we discussed in Chapter 10, genes determine the nucleotide sequences of specific mRNA molecules, and mRNA in turn determines the sequences of amino acids in protein molecules (DNA→RNA→protein). A gene that is turned on is being transcribed into mRNA, and that message is being translated into specific protein molecules. The overall process by which genetic information flows from genes to proteins—that is, from genotype to phenotype—is called **gene expression.**

Because all the differentiated cells in an individual organism contain the same genes and all the genes have the potential of being expressed, the great differences among cells in an organism must result from the selective expression of genes—that is, from the pattern of genes turned on in a given cell at a given time. Such regulation of gene expression plays a central role in the development of a unicellular zygote into a multicellular organism. During embryonic development, groups of cells follow diverging developmental pathways, and each group develops into a particular kind of tissue. In the mature organism, each cell type—nerve or pancreas, for instance—has a different pattern of turned-on genes.

Consider the patterns of gene expression for four genes in the three different specialized cells of an adult human shown in **Figure 11.3**. Note that the genes for the enzymes of the metabolic pathway of glycolysis are "on" in all the cells. However, the genes for specialized proteins, such as insulin and hemoglobin, are expressed only by particular kinds of cells.

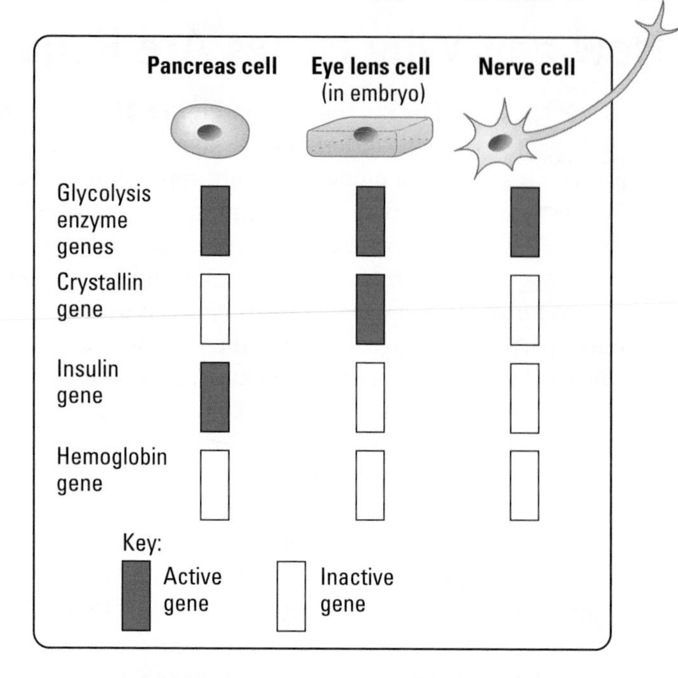

Figure 11.3 Patterns of gene expression. Shown here are the activity states of four genes in three types of human cells. Different types of cells express different combinations of genes. The genes for the enzymes of glycolysis are "housekeeping" genes; they are active in all metabolizing cells. The specialized proteins whose genes are represented here are the transparent protein crystallin, which forms the lens of the eye; insulin, a hormone made in the pancreas; and the oxygen transport protein hemoglobin. Notice that the hemoglobin genes are not active in any of the cell types shown here. They are turned on only in cells that are developing into red blood cells. Insulin genes are activated only in the cells of the pancreas that produce that hormone. Nerve cells express genes for other specialized proteins not indicated here.

Gene Regulation in Bacteria

Selective gene expression accounts for how different cell types arise. But exactly how is gene expression regulated in a cell? To examine how genes can be regulated, let's consider the relatively simple case of bacteria.

Bacteria are single-celled creatures that do not exhibit cellular differentiation in the eukaryotic sense. But in the course of their lives, they regulate their genes in response to environmental changes. Imagine an *Escherichia coli* bacterium living in your intestine. It will be bathed in varying nutrients, depending on what you eat. If you drink a milk shake, for example, there will be a sudden rush of lactose. In response, *E. coli* will express three genes for enzymes that enable it to absorb and digest this sugar. Once the lactose is gone, *E. coli* does not waste its energy continuing to produce these enzymes. Thus, the bacterium is able to adjust its gene expression to changes in the environment.

How does the presence or absence of lactose influence the activity of the genes responsible for lactose metabolism? The key is the way the three genes are organized: They are adjacent in the DNA and regulated (turned on and off) as a single unit. This regulation is achieved through control sequences, stretches of DNA that help turn all three genes on and off, coordinating their expression. Such a cluster of genes with related functions, along with the control sequences, is called an **operon.** The operon considered here, the *lac* (short for lactose) operon, was first described in the 1960s by French biologists François Jacob and Jacques Monod. The *lac* operon illustrates principles of gene regulation that apply to a wide variety of prokaryotic genes.

How are DNA control sequences able to cause genes to turn on or off? One control sequence, called a **promoter,** is the site where the transcription enzyme, RNA polymerase, attaches and initiates transcription—in our example, transcription of the three lactose enzyme genes. Between the promoter and the enzyme genes, a DNA segment called an **operator** acts as a switch that is turned on or off, depending on whether a specific protein is bound there. The operator and protein together determine whether RNA polymerase can attach to the promoter and start transcribing the genes. In the *lac* operon, when the operator switch is turned on, all the enzymes needed to metabolize lactose are made at once.

Figure 11.4a shows the *lac* operon in "off" mode, its status when there is no lactose around. Transcription is turned off when a protein called a **repressor** ( ) ❶ binds to the operator and ❷ physically blocks the attachment of RNA polymerase ( ) to the promoter.

Figure 11.4b shows the operon in "on" mode, when lactose is present. Lactose ( ) interferes with the attachment of the *lac* repressor to the operator by ❶ binding to the repressor and ❷ changing its shape. In its new shape, the repressor is inactive; it cannot bind to the operator, and the operator switch remains on. ❸ RNA polymerase can now bind to the promoter and from there ❹ transcribe the genes of the operon into mRNA. ❺ Translation produces all three enzymes needed for lactose utilization.

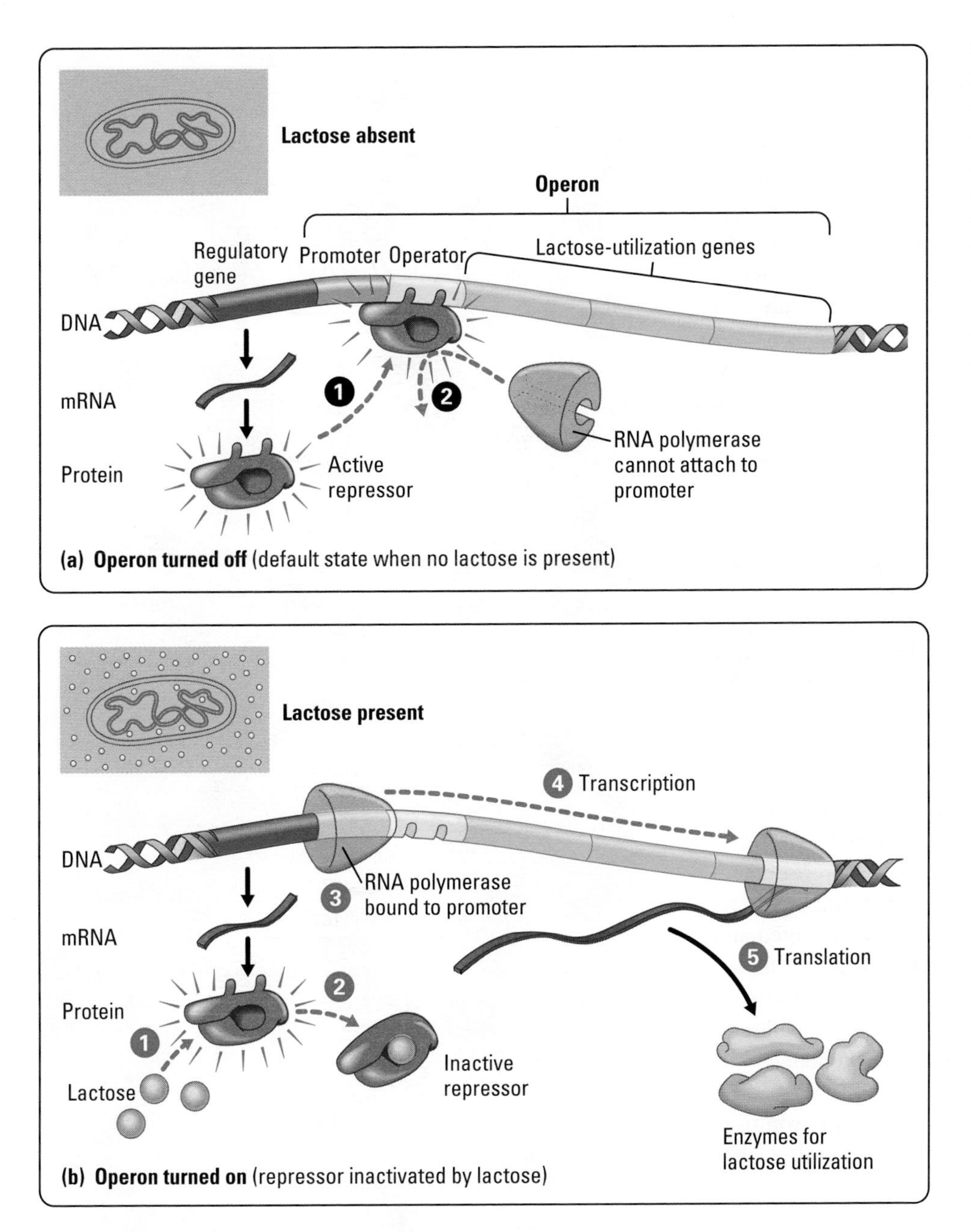

Figure 11.4 The *lac* operon of *E. coli*. In this bacterial chromosome segment, the three genes that code for the lactose-utilization enzymes are next to each other ( ). These genes and two control sequences, a promoter and an operator, make up the *lac* operon. A regulatory gene coding for a repressor ( ) is nearby. **(a)** In the absence of lactose, the repressor ( ) binds to the operator ( ), preventing transcription of the operon. **(b)** Lactose reverses the repression of the operon by inactivating the repressor. The result is the production of the three enzymes for lactose utilization.

Many operons have been identified in bacteria. Some are quite similar to the *lac* operon, while others have somewhat different mechanisms of control. For example, operons that control amino acid synthesis cause bacteria to stop making these molecules when they are already present in the environment, saving materials and energy for the cells. In these cases, the amino acid *activates* the repressor. Armed with a variety of operons, *E. coli* and other prokaryotes can thrive in frequently changing environments.

Gene Regulation in Eukaryotic Cells

While operons are a common method of gene regulation in bacteria, they generally do not exist in eukaryotes. Eukaryotic cells have more sophisticated mechanisms than bacteria for regulating the expression of their genes. This is not surprising because, as a single cell, a prokaryote does not require the elaborate regulation of gene expression that leads to cell specialization in multicellular eukaryotic organisms.

The pathway from gene to functioning protein in eukaryotic cells is a long one, providing a number of points where the process can be regulated—turned on or off, speeded up or slowed down. Picture the series of pipes that carry water from your local water supply to a faucet in your home. At various points, valves control the flow of water. We use this analogy in **Figure 11.5** to illustrate the flow of genetic information from a eukaryotic chromosome—a reservoir of genetic information—to an active protein that has been made in the cell's cytoplasm. The multiple mechanisms that control gene expression are analogous to the control valves in your water pipes. In the figure, a control knob indicates a gene expression "valve." All these knobs represent possible control points, although only one or a few control points are likely to be important for a typical protein.

Using a reduced version of Figure 11.5 as a guide to the flow of genetic information through the cell, we will explore several ways that eukaryotes can control gene expression, starting with what goes on in the cell nucleus.

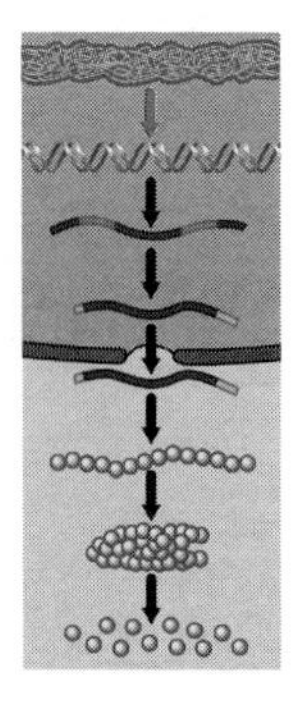

The Regulation of DNA Packing Recall from Chapter 8 that eukaryotic chromosomes may be in a more or less condensed state, with the DNA and accompanying proteins more or less tightly wrapped together. DNA packing tends to prevent gene expression, presumably by preventing RNA polymerase and other transcription proteins from contacting the DNA. Highly compacted chromatin, which is found in mitotic chromosomes and also in some regions of interphase chromosomes, is generally not expressed at all. For a gene to be transcribed, DNA packing must be loosened.

Cells may use DNA packing for the long-term inactivation of genes. One intriguing case is seen in female mammals, where one entire X chromosome in each somatic cell is highly compacted and almost entirely inactive. This **X chromosome inactivation** first takes place early in embryonic development, when one of the two X chromosomes in each cell is inactivated at random. The inactivation is subsequently inherited by a cell's descendants. Consequently, a female heterozygous for genes on the X chromosome has populations of cells that express different X-linked alleles. A striking effect of X chromosome inactivation is the tortoiseshell cat, which has orange and black patches of fur **(Figure 11.6)**.

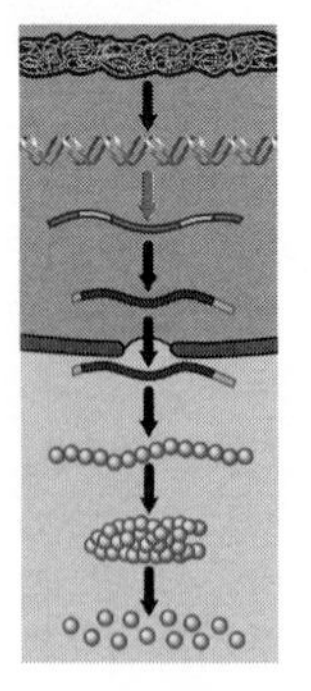

The Initiation of Transcription In both prokaryotes and eukaryotes, the most important stage for regulating gene expression is transcription. As with prokaryotes, this control is achieved in eukaryotic cells through regulatory proteins that bind to specific segments of DNA. There are, however, important differences in transcriptional regulation between prokaryotes and eukaryotes. For example, most eukaryotic genes have individual promoters and other control sequences—that is, eukaryotic genes are rarely organized into groups as operons.

As can be seen in **Figure 11.7**, transcriptional regulation in eukaryotes is complex, typically involving many proteins (collectively called **transcription factors**) acting in concert to bind to DNA sequences called **enhancers** and to the promoter. The DNA-protein assembly promotes the binding of RNA polymerase to the promoter. Genes coding for related enzymes, such as those in a metabolic pathway, may share a specific kind of enhancer (or collection of enhancers), allowing these genes to be activated at

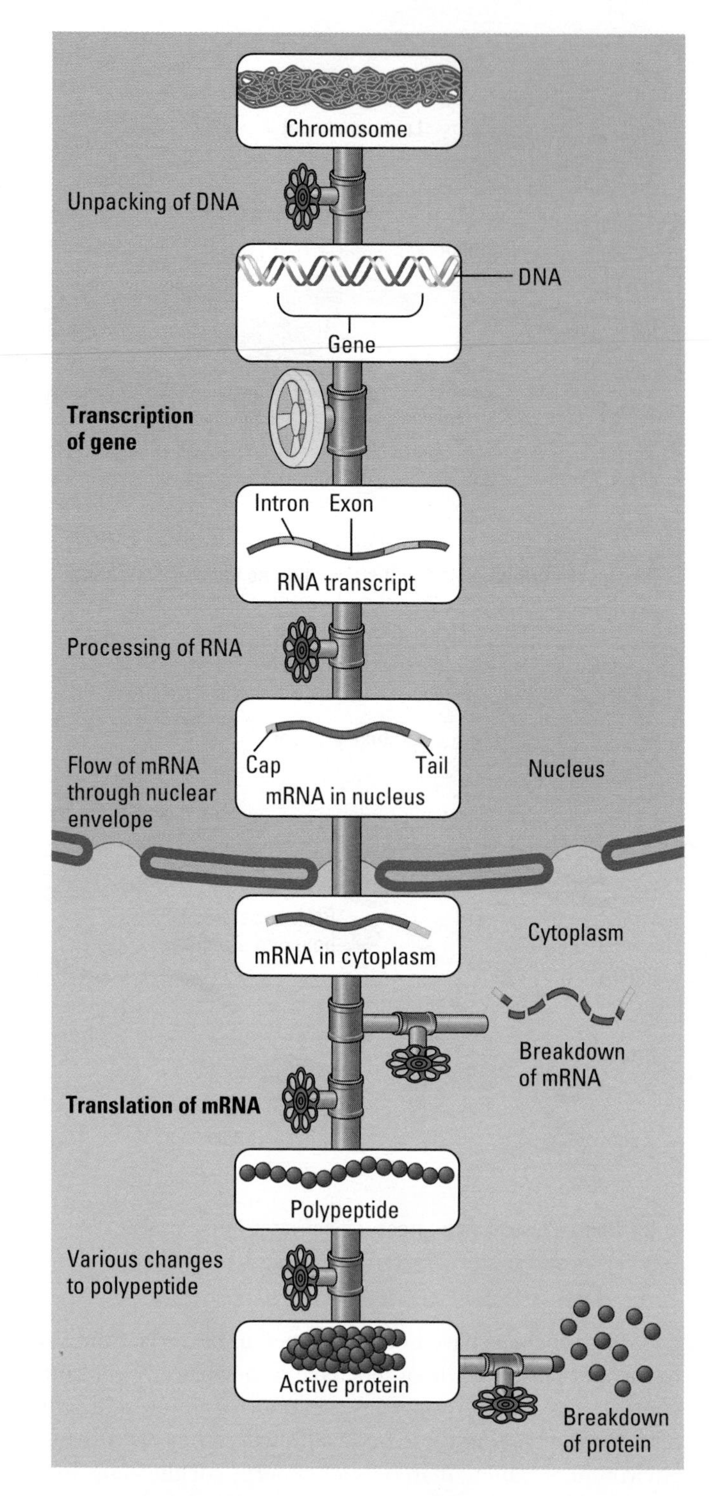

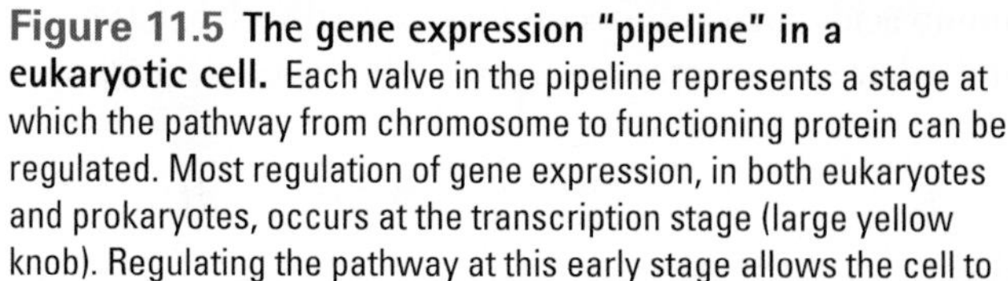
Figure 11.5 The gene expression "pipeline" in a eukaryotic cell. Each valve in the pipeline represents a stage at which the pathway from chromosome to functioning protein can be regulated. Most regulation of gene expression, in both eukaryotes and prokaryotes, occurs at the transcription stage (large yellow knob). Regulating the pathway at this early stage allows the cell to avoid wasting valuable resources.

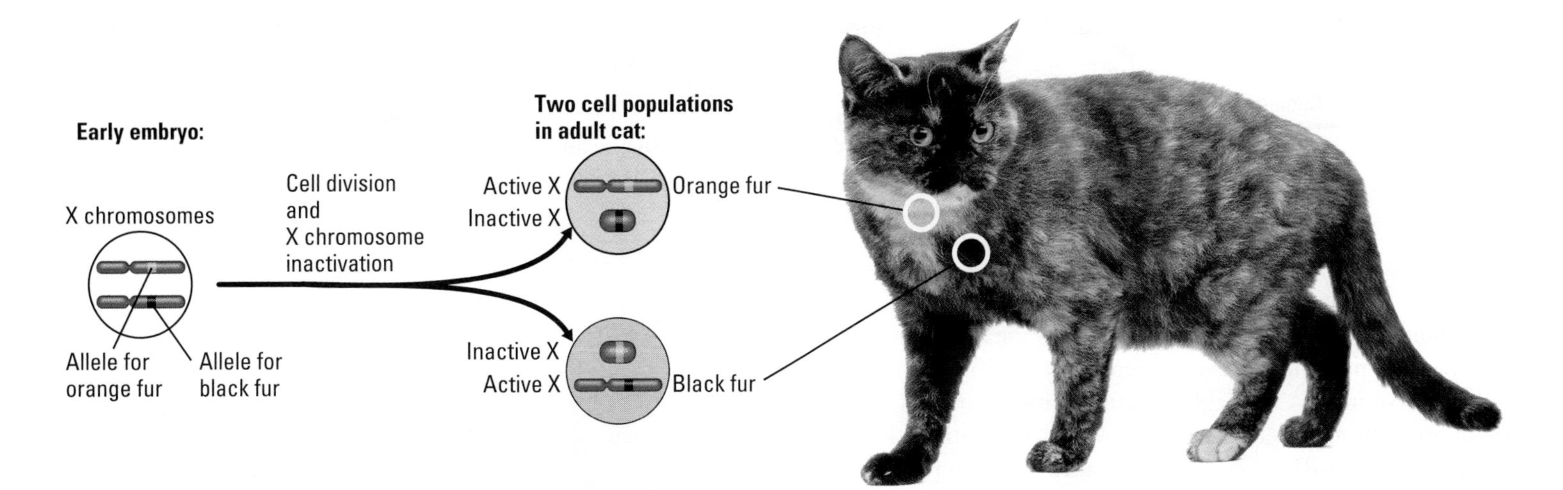

Figure 11.6 X chromosome inactivation: the tortoiseshell pattern on a cat. The tortoiseshell gene is on the X chromosome, and the tortoiseshell phenotype requires the presence of two different alleles, one for orange fur and one for nonorange (black) fur. Normally, only females can have both alleles, because only they have two X chromosomes. If a female is heterozygous for the tortoiseshell gene, she is tortoiseshell. Orange patches are formed by populations of cells in which the X chromosome with the orange allele is active; black patches have cells in which the X chromosome with the nonorange allele is active. (Calico cats also have white areas, which are determined by another gene.)

the same time. Not shown are repressor proteins, which may bind to DNA sequences called **silencers,** inhibiting the start of transcription.

In fact, repressors, which turn genes off (and play a prominent role in prokaryotic regulation), are less common in eukaryotes than **activators,** proteins that turn genes on by binding to DNA. (Activators act by making it easier for RNA polymerase to bind to the promoter.) The use of activators is efficient because a typical animal or plant cell needs to turn on (transcribe) only a small percentage of its genes, those required for the cell's specialized structure and function. The "default" state for most genes in multicellular eukaryotes seems to be "off," with the exception of "housekeeping" genes for routine activities such as glucose metabolism.

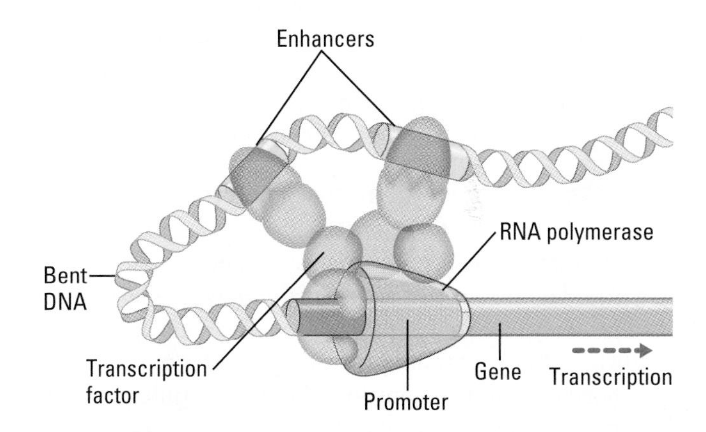

Figure 11.7 A model for the turning on of a eukaryotic gene. A large assembly of proteins and several control sequences in the DNA are involved in initiating the transcription of a eukaryotic gene.

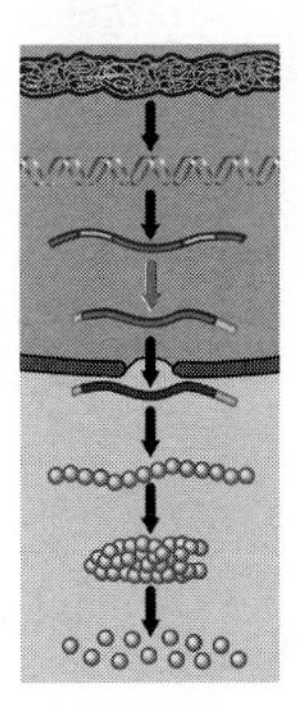

RNA Processing and Breakdown Within a eukaryotic cell, transcription is localized in the nucleus. There, RNA transcripts are modified before they move to the cytoplasm for translation by the ribosomes (see Figure 10.20). RNA processing includes the addition of a cap and a tail to the RNA, as well as the removal of any introns—noncoding DNA segments that interrupt the genetic message—and the splicing together of the remaining exons. Such processing creates additional opportunities for regulating gene expression.

In some cases, a cell can carry out exon splicing in more than one way, generating different mRNA molecules from the same starting RNA molecule. Notice in **Figure 11.8**, for example, that one mRNA molecule ends up with the green exon and the other with the brown exon. With this sort of **alternative RNA splicing,** an organism can get more than one type of polypeptide from a single gene. Results from the Human Genome Project suggest that alternative RNA splicing is very

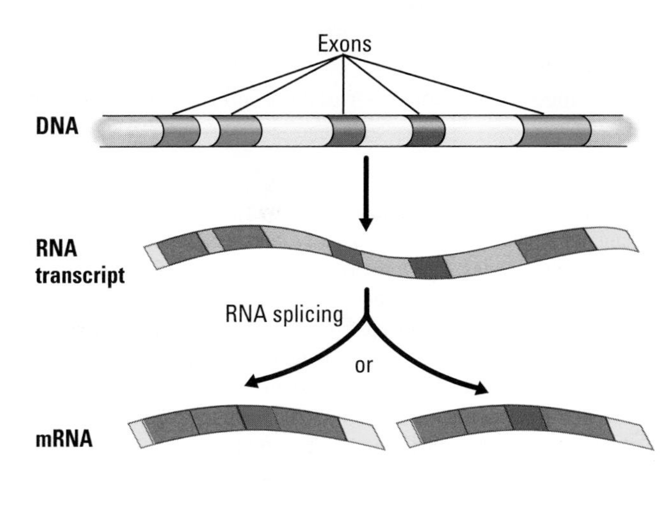

Figure 11.8 Producing two different mRNAs from the same gene. Two different cells can use a given DNA gene to synthesize somewhat different mRNAs and proteins. The cells might be cells of different tissues or the same type of cell at different times in the organism's life. Alternative RNA splicing is a means by which the cell can exert differential control over the expression of certain genes.

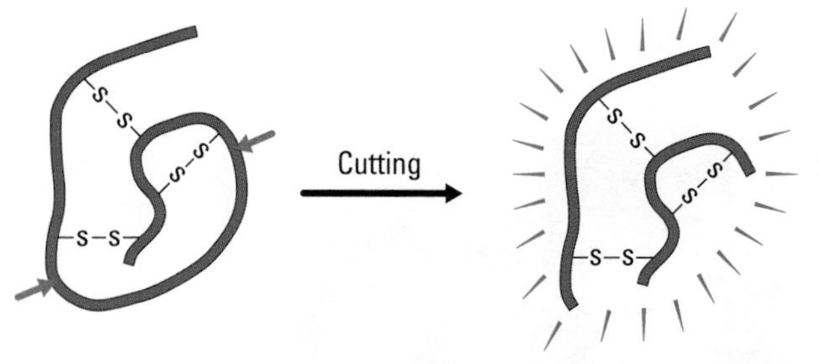

Figure 11.9 The formation of an active insulin molecule. One type of post-translational control is the cutting of a polypeptide to convert it to its active form. The hormone insulin, for example, is synthesized as one polypeptide. A special enzyme then removes an interior section. The remaining polypeptide chains are held together by covalent bonds between the sulfur atoms of sulfur-containing amino acids. Only in this form does the protein act as a hormone.

common in humans. In one well-documented case, the RNA transcript from a gene can be spliced to encode seven different versions of a cellular protein.

The lifetime of mRNA molecules is another important factor regulating protein production. Eukaryotic mRNAs can have lifetimes from hours to weeks. Long-lived mRNAs get translated into many more protein molecules than do short-lived ones. But eventually, all mRNAs are broken down and their parts recycled.

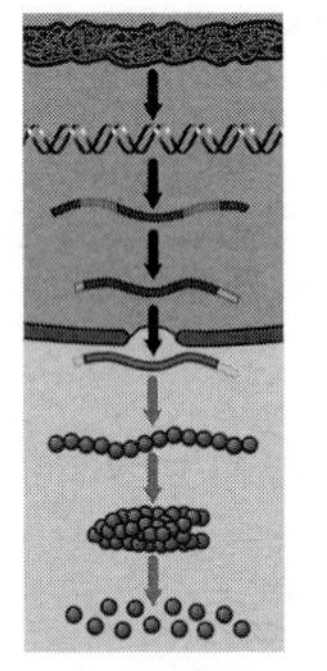

The Regulation of Translation After transcription is completed, the process of translation offers additional opportunities for regulation. Among the molecules involved in translation are a great many proteins that have regulatory functions. Red blood cells, for instance, have a protein that prevents the translation of hemoglobin mRNA unless the cell has a supply of heme, an iron-containing chemical group essential for hemoglobin function.

Protein Alterations and Breakdown The final opportunities for regulating gene expression occur after translation. Post-translational control mechanisms in eukaryotes often involve cutting polypeptides into smaller, active final products. The hormone insulin, for example, is synthesized in pancreatic cells as one long inactive polypeptide **(Figure 11.9)**. After translation, a large center portion is cut away, leaving two shorter chains that constitute the active insulin molecule.

Another control mechanism operating after translation is the selective breakdown of proteins. Some of the proteins that trigger metabolic changes in cells are broken down within a few minutes or hours. This regulation allows a cell to adjust the kinds and amounts of its proteins in response to changes in its environment.

Cell Signaling

So far, we have considered gene regulation only within a single cell. In a multicellular organism, the process of gene regulation can cross cell boundaries. A cell can produce and secrete chemicals, such as hormones, that induce a neighboring or distant cell to be regulated in a certain way.

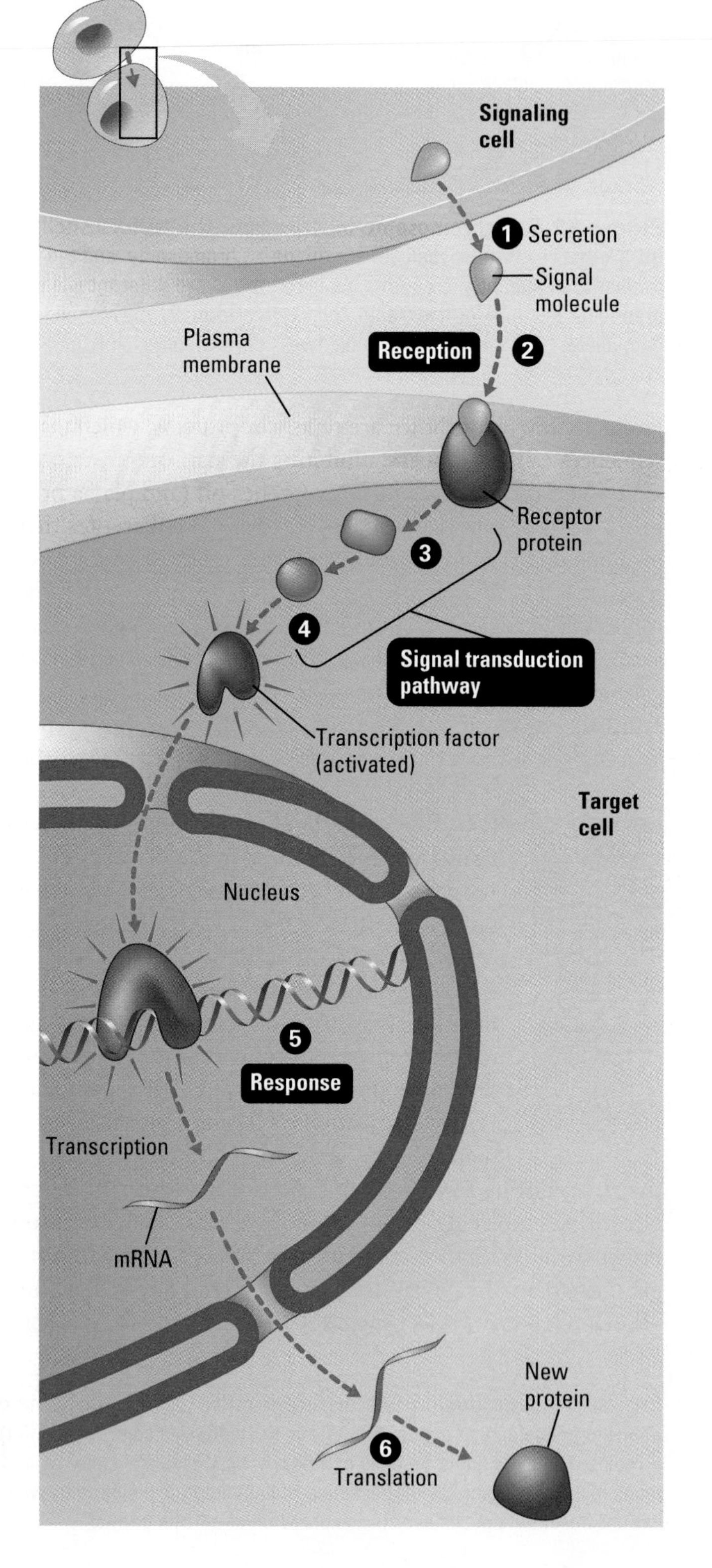

Figure 11.10 A cell-signaling pathway that turns on a gene. The coordination of cellular activities in a multicellular organism depends on cell-to-cell signaling that helps regulate genes. ❶ First, the signaling cell secretes the signal molecule. ❷ This molecule binds to a specific receptor protein embedded in the target cell's plasma membrane—the reception step. ❸ The binding activates the first in a series of relay proteins (●) within the target cell. Each relay molecule activates another. This is a signal transduction pathway. ❹ The last relay molecule in the series activates a transcription factor that ❺ triggers the transcription of a specific gene, the response to the signal. ❻ Translation of the mRNA produces a protein.

Cell-to-cell signaling, with proteins or other kinds of molecules carrying messages from signaling cells to receiving (target) cells, is a key mechanism in the development of a multicellular organism from a fertilized egg, as well as in the coordination of cellular activities in the mature organism. As you saw in Figure 5.20, a signal molecule usually acts by binding to a receptor protein in the plasma membrane of the target cell and initiating a signal transduction pathway, a series of molecular changes that converts a signal received on a target cell's surface to a specific response inside the cell. **Figure 11.10** shows the main elements of a signal transduction pathway in which the target cell's response is the transcription (turning on) of a gene.

DNA Microarrays: Visualizing Gene Expression

Scientists who study gene regulation can use a **DNA microarray** to visualize patterns of gene expression **(Figure 11.11)**. A DNA microarray is a glass slide carrying thousands of different kinds of single-stranded DNA fragments arranged in an array (grid). Each DNA fragment is prepared from a particular gene; a single microarray thus carries DNA from thousands of genes.

To use DNA a microarray, ❶ a researcher collects all of the mRNA transcribed from genes in a particular type of cell. This mRNA is mixed with reverse transcriptase, a viral enzyme that ❷ catalyzes the synthesis of DNA molecules that are complementary to each mRNA sequence. This **complementary DNA (cDNA)** is produced in the presence of nucleotides that have been modified to fluoresce (glow). ❸ A small amount of the fluorescently labeled cDNA mixture is added to each of the thousands of kinds of single-stranded DNA fragments in the microarray. If a molecule in the cDNA mixture is complementary to a DNA fragment at a particular location, the cDNA molecule binds to it, becoming fixed there. ❹ After nonbinding cDNA is rinsed away, the remaining cDNA produces a detectable glow in the microarray. The pattern of glowing spots enables the researcher to determine which genes are turned on or off in the starting cells.

Researchers can use microarrays to learn which genes are active in different tissues or in tissues from individuals in different states of health. This powerful new tool is giving researchers new insight into gene regulation.

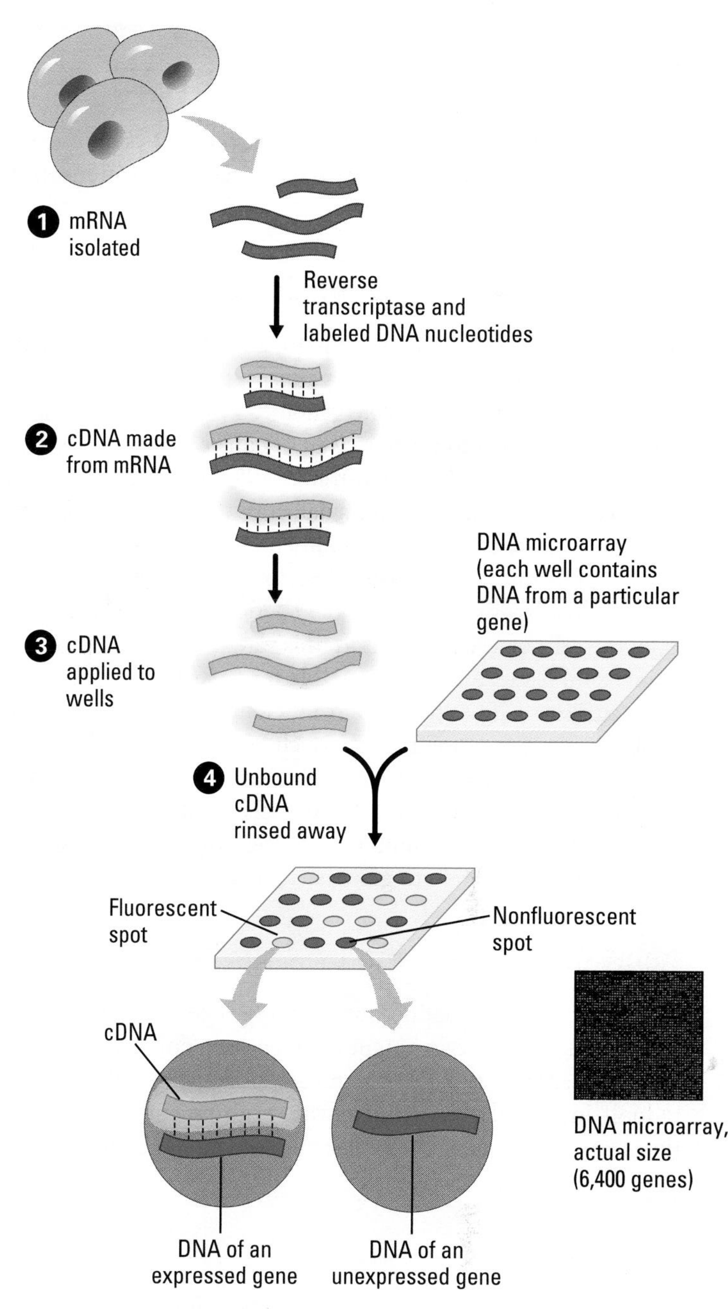

Figure 11.11 A DNA microarray. To visualize patterns of gene expression, researchers use a collection of mRNA from one type of cell to make fluorescently labeled complementary DNA (cDNA). The cDNA is applied to thousands of single-stranded DNA fragments from a large number of different genes fixed to the DNA microarray. After unbound cDNA is rinsed away, any remaining fluorescent spots represent active gene in the original cells.

CHECKPOINT

1. If your nerve cells and skin cells have exactly the same genes, how can they be so different?
2. A certain mutation in *E. coli* makes the *lac* operator unable to bind the active repressor. How would this affect the cell?
3. How does dense packing of DNA in chromosomes prevent gene expression?
4. Once mRNA encoding a particular protein reaches the cytoplasm, what are four mechanisms that can regulate the amount of the active protein in the cell?
5. How can a signal molecule from one cell alter gene expression in a target cell without entering the target cell?

Answers: 1. Each cell type must be expressing certain genes that are not expressed in the other cell type. **2.** The cell would wastefully produce the enzymes for lactose metabolism continuously, even in the absence of lactose. **3.** RNA polymerase and other proteins required for transcription do not have access to tightly packed DNA. **4.** Breakdown of the mRNA; regulation of translation; alteration of the protein (by polypeptide cutting, for example); and breakdown of the protein **5.** By binding to a receptor protein in the membrane of the target cell and triggering a signal transduction pathway that activates transcription factors

Cloning Plants and Animals

Now that we have examined how gene expression is regulated, we will devote the rest of this chapter to two important applications of this subject, organismal cloning and cancer.

The Genetic Potential of Cells

As already discussed, different cells are capable of performing different functions because they have different patterns of gene expression. And yet, each type of cell contains a complete genome. This suggests that a differentiated cell has the *potential* to express *all* an organism's genes. This sort of genetic potential can be readily demonstrated, at least for plants. For example, if you have ever grown a plant from a small cutting, you've seen evidence that differentiated plant cells have the ability to develop into a whole new organism. On a larger scale, the technique described in **Figure 11.12** can be used to produce hundreds or thousands of genetically identical organisms—clones—from the cells of a single plant. In this way, growers can propagate large numbers of plants that have desirable traits, such as high fruit yield or resistance to disease. The success of plant cloning shows that cell differentiation in plants does not cause irreversible changes in the DNA.

A similar, naturally occurring process in animals is **regeneration,** the regrowth of lost body parts. When a salamander loses a leg, for example, certain cells in the leg stump reverse their differentiated state, divide, and then differentiate again to give rise to a new leg.

Reproductive Cloning of Animals

Cells from animals that do not naturally regenerate major body parts also retain their full genetic potential. This was demonstrated in the 1950s when researchers replaced the nuclei of frog egg cells with nuclei from tadpole intestinal cells. Some of the resulting embryos developed into tadpoles and then frogs. The replacement of an egg's nucleus with one from

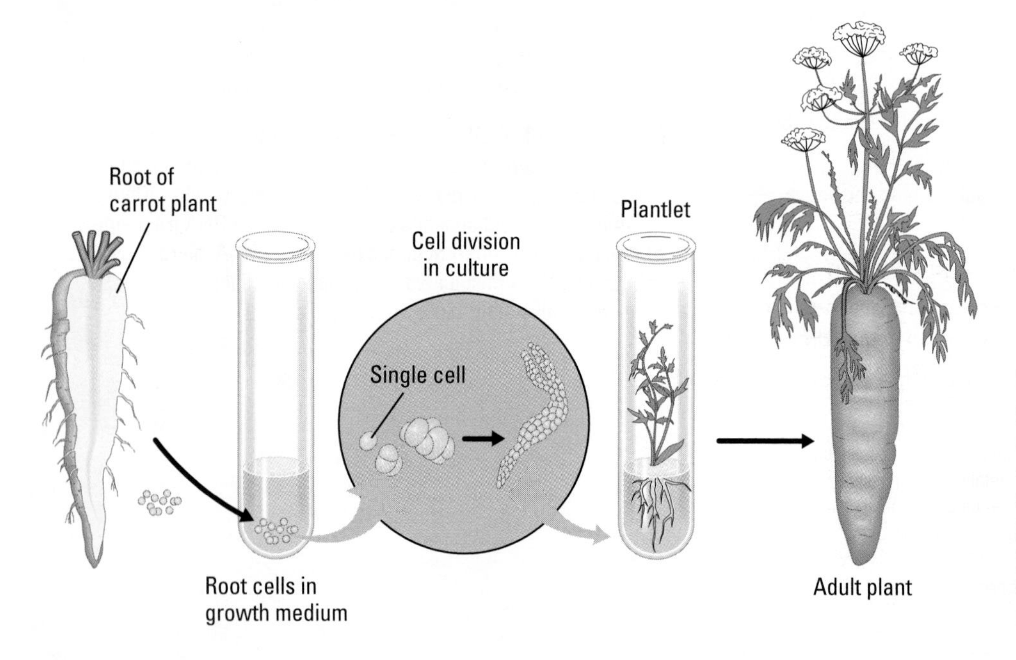

Figure 11.12 Test-tube cloning of a carrot plant. An entire carrot plant can be grown from a differentiated root cell. A single cell removed from the carrot's root and placed in growth medium may begin dividing and eventually grow into an adult plant. The new plant is a genetic duplicate of the parent plant. This process proves that mature plant cells can reverse their differentiation and develop into all the specialized cells of an adult plant.

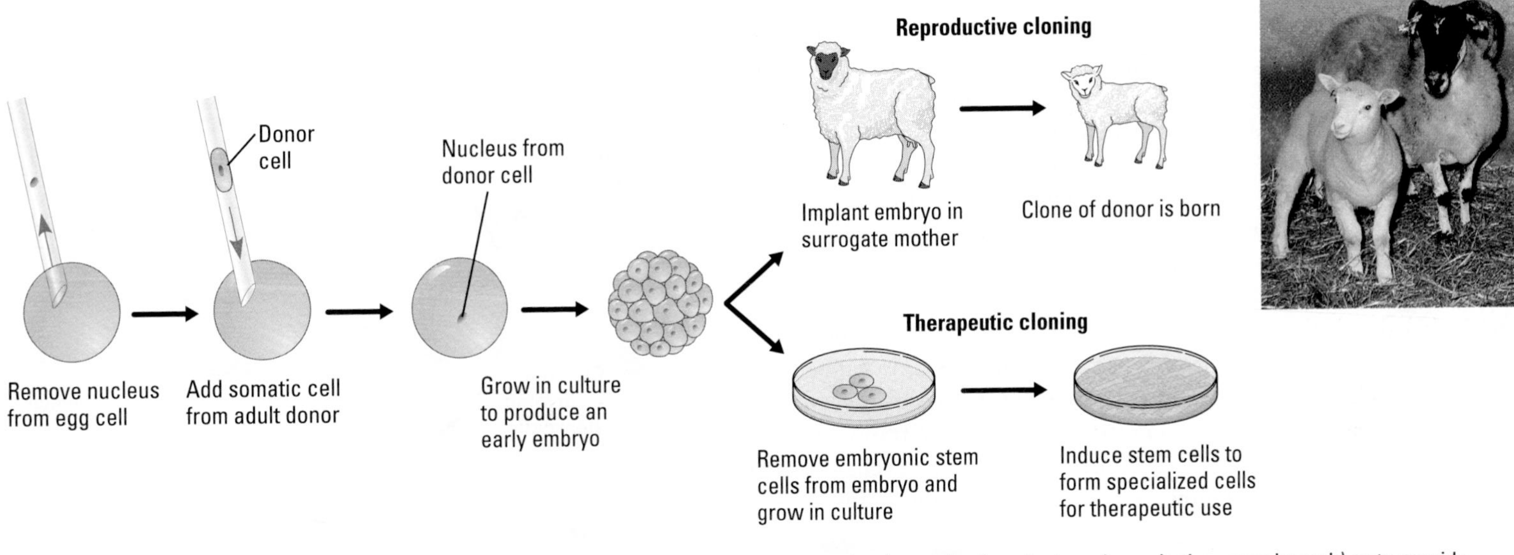
Reproductive cloning
Donor cell
Nucleus from donor cell
Implant embryo in surrogate mother
Clone of donor is born
Therapeutic cloning
Remove nucleus from egg cell
Add somatic cell from adult donor
Grow in culture to produce an early embryo
Remove embryonic stem cells from embryo and grow in culture
Induce stem cells to form specialized cells for therapeutic use

Figure 11.13 Cloning by nuclear transplantation. In this method of nuclear transplantation, an adult somatic nucleus is injected into a nucleus-free egg cell. The resulting embryo may then be used to produce a new organism (reproductive cloning, shown in the upper branch) or to provide stem cells (therapeutic cloning, lower branch). The lamb in the photograph is the famous Dolly, shown with her surrogate mother.

an adult cell, called **nuclear transplantation,** has been used to clone a variety of animals **(Figure 11.13)**.

As discussed in this chapter's opening essay, a breakthrough in animal cloning came in 1997 with the first cloning of a mammal using a nucleus from an adult cell (the upper branch of Figure 11.13). In this case, researchers in Scotland used an electric shock to fuse specially treated sheep udder cells with 277 eggs from which they had removed the nuclei. After several days of growth, 29 of the resulting embryos were implanted in the uteruses of unrelated ewes (surrogate mothers). One of the embryos developed into the world-famous Dolly. As expected, Dolly resembled her genetic parent, the nucleus donor, not the egg donor or the surrogate ewe.

The procedure depicted in the upper branch of Figure 11.13 is called **reproductive cloning** because it results in the creation of a new animal. Why would anyone want to do this? In agriculture, farm animals with specific sets of desirable traits might be cloned to produce herds of animals with these traits. In research, genetically identical animals may provide researchers with perfect "control animals" for many kinds of experiments. The pharmaceutical industry is experimenting with cloning animals for potential medical use. For example, the pigs in **Figure 11.14** are clones that lack a gene for a protein that can cause immune system rejection in humans. Organs from such pigs may one day be used in human patients who need transplants. And, as we saw in the chapter opener, reproductive cloning has the potential to restock populations of endangered animals.

Figure 11.14 Reproductive cloning of mammals. These piglets are clones of a pig that was genetically modified to lack a protein that causes transplant rejection in humans.

Therapeutic Cloning and Stem Cells

The lower branch of Figure 11.13 depicts an alternative cloning outcome called therapeutic cloning. The purpose of **therapeutic cloning** is not to produce a viable organism but to produce special cells called embryonic stem cells.

Embryonic Stem Cells Mammalian **embryonic stem cells (ES cells)** are derived from an early embryonic state called the blastocyst, a partially hollow mass of about 100 cells. A blastocyst is just a few days old and has not yet implanted into the uterus. During development, embryonic stem cells in the blastocyst differentiate to give rise to all the specialized cells in the body. When removed from the early embryo and grown in laboratory culture, ES cells can divide indefinitely. The right conditions—such as the presence of certain growth factors—can induce changes in gene expression that cause the cells to develop into a particular cell type **(Figure 11.15)**. If scientists can discover the right conditions, they may be able to grow cells for the repair of injured or diseased organs. In the future, embryos may be created using a cell nucleus from a patient so that ES cells can be harvested and induced to develop into replacement tissues or organs. Presumably, the patient's body would readily accept such a transplant because it would be a genetic match. The use of embryonic stem cells in therapeutic cloning is controversial, however, because the removal of ES cells destroys the embryo.

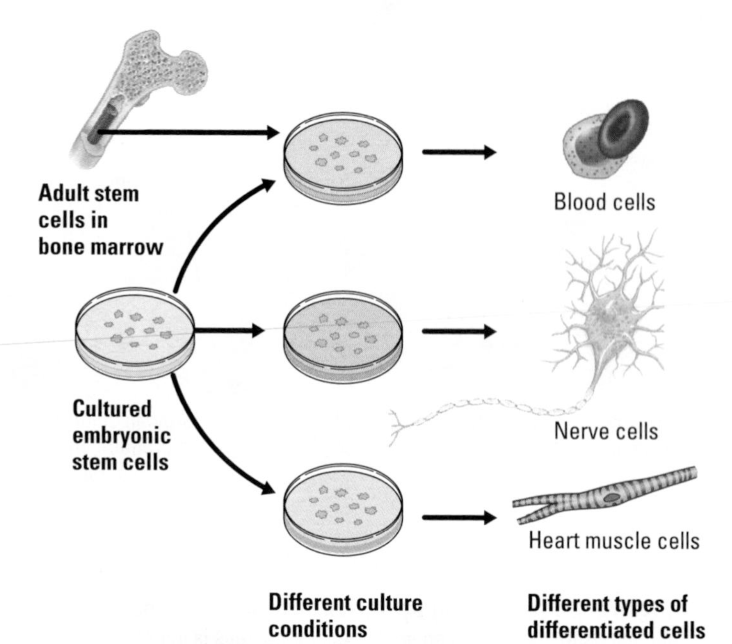

Figure 11.15 Differentiation of embryonic stem cells in culture. Scientists hope to discover growth conditions that will stimulate cultured embryonic and adult stem cells to differentiate into specialized cells of various types.

Adult Stem Cells Embryonic stem cells are not the only stem cells available to researchers. **Adult stem cells** are cells in adult tissues that generate replacements for nondividing differentiated cells. Unlike ES cells, adult stem cells are partway along the road to differentiation; in the body, they usually give rise to only a few related types of specialized cells. For example, stem cells in bone marrow generate the different kinds of blood cells. Adult stem cells are much more difficult than ES cells to grow in culture, but researchers have had some success. Because no embryonic tissue is involved in their harvest, adult stem cells may provide an ethically less problematic route for human tissue and organ replacement than ES cells. However, some researchers think that ES cells are the only types of cells likely to lead to groundbreaking advances in human health.

Umbilical Cord Blood Banking Another source of stem cells is blood collected from the umbilical cord and placenta at birth. Such stem cells appear to be partially differentiated, less so than adult stem cells, more so than embryonic stem cells. To obtain these cells, a physician inserts a needle into the umbilical cord and extracts $\frac{1}{4}$ to $\frac{1}{2}$ cup of blood **(Figure 11.16)**. The cells are then frozen until needed for later medical treatment. In 2005, doctors reported that an infusion of umbilical cord blood stem cells from a compatible (but unrelated) donor appeared to cure some babies of Krabbe's disease, a fatal inherited disorder of the nervous system. Other people have received cord blood as a treatment for leukemia. To date, however, most attempts at cord blood therapy have not been successful, and the American Academy of Pediatrics recommends cord blood banking only for babies born into families with a known genetic risk. So far, the promise of cord blood banking vastly exceeds the accomplishments.

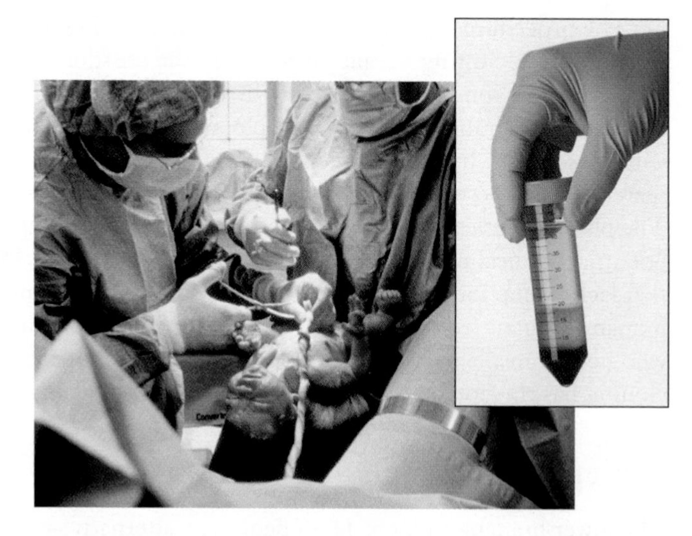

Figure 11.16 Umbilical cord blood banking. Just after birth, a doctor may remove blood from a newborn's umbilical cord. The umbilical cord blood (inset), rich in stem cells, is stored for possible future use.

Human Therapeutic Cloning In 2001, a biotechnology company in Massachusetts announced that it had created the first human embryo by cloning **(Figure 11.17)**. However, their most successful embryo stopped growing at about six cells. Although these embryos were not allowed to develop and the scientists' intentions were only to create ES cells, this achievement brought us one step closer to the possibility of human reproductive cloning.

Critics point out many obstacles—both practical and ethical—to human cloning. Practically, mammalian cloning is extremely difficult and inefficient. Only a small percentage of cloned embryos develop normally. Ethically, creating human embryos for such purposes raises many troubling questions. A consensus on the ethical use of human embryos is unlikely any time soon. Meanwhile, the research—and the debate—continues.

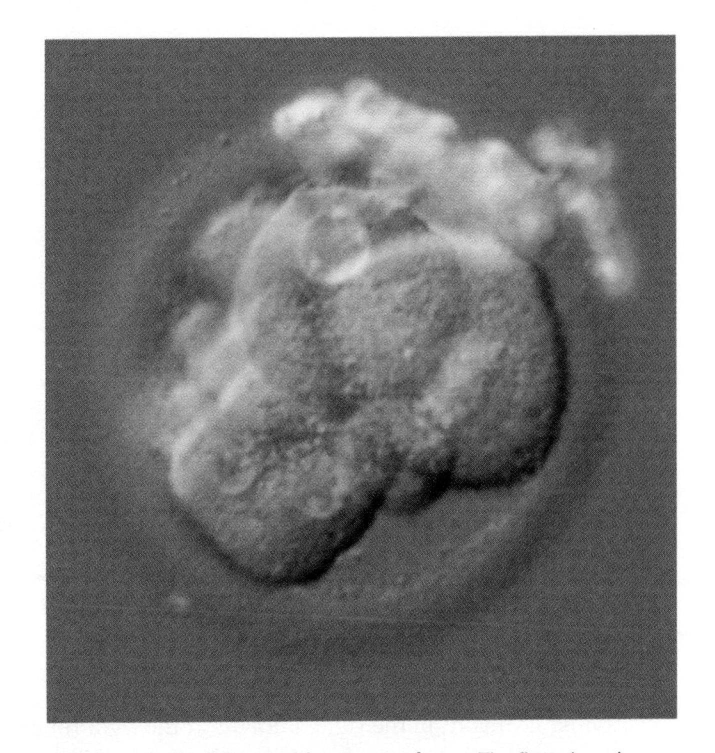

Figure 11.17 A cloned human embryo. The first cloned human embryo, created in 2001 by a biotechnology company, stopped growing at the six-cell stage. It was created for the purpose of therapeutic cloning, the production of embryonic stem cells for medical use.

CHECKPOINT

1. What is the difference between reproductive cloning and therapeutic cloning?
2. Name three potential sources of stem cells.

Answers: **1.** Reproductive cloning results in the production of a live individual, whereas therapeutic cloning produces stem cells. **2.** Embryonic tissue (ES cells), umbilical cord blood, and bone marrow (adult stem cells)

The Genetic Basis of Cancer

In Chapter 8, we introduced cancer as a variety of diseases in which cells escape from the control mechanisms that normally limit their growth and division. In recent years, scientists have learned that this escape from normal controls is due to changes in some of the cells' genes or to changes in the way certain genes are expressed.

Genes That Cause Cancer

The abnormal behavior of cancer cells was observed years before anything was known about the cell cycle, its control, or the role genes play in making cells cancerous. One of the earliest clues to the cancer puzzle was the discovery, in 1911, of a virus that causes cancer in chickens. Recall that viruses are simply "genes in a box," molecules of DNA or RNA surrounded by protein and in some cases a membranous envelope. Viruses that cause cancer can become permanent residents in host cells by inserting their nucleic acid into the DNA of host chromosomes. It is now known that a number of viruses that can produce cancer carry specific cancer-causing genes in their nucleic acid. When inserted into a host cell, these genes can make the host cell cancerous. A gene that causes cancer is called an **oncogene** ("tumor gene").

Oncogenes and Tumor-Suppressor Genes In 1976, American molecular biologists J. Michael Bishop, Harold Varmus, and their colleagues made a startling discovery. They found that the virus that causes cancer in chickens contains an oncogene that is an altered version of a normal chicken gene. Subsequent research has shown that the chromosomes of many animals, including humans, contain genes that can be converted to oncogenes. A normal gene with the potential to become an oncogene is called a **proto-oncogene.** (These terms can be confusing, so they bear repeating: a *proto-oncogene* is a normal gene that, if changed, can become a cancer-causing *oncogene*.) A cell can acquire an oncogene from a virus or from the conversion of one of its own proto-oncogenes.

How can a change in a gene cause cancer? Searching for the normal roles of proto-oncogenes in the cell, researchers found that many of these genes code for **growth factors**—proteins that stimulate cell division—or

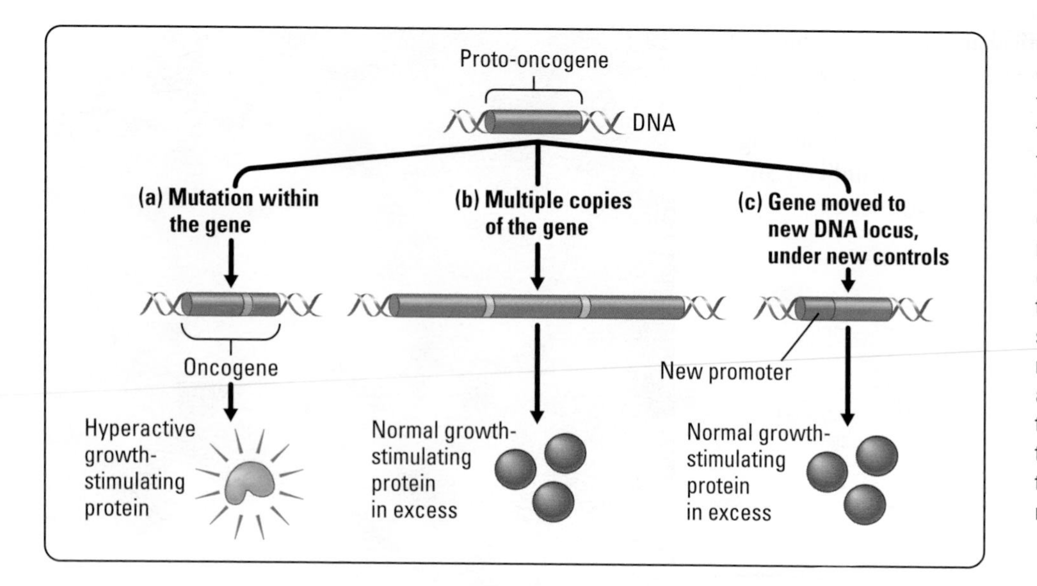

Figure 11.18 Alternative ways to make an oncogene from a proto-oncogene. Let's assume that the starting proto-oncogene codes for a protein that stimulates cell division. **(a)** A mutation (green) in the proto-oncogene itself may create an oncogene that codes for a hyperactive protein, one whose stimulating effect is stronger than normal. **(b)** An error in DNA replication or recombination may generate multiple copies of the gene, which are all transcribed and translated. The result would be an excess of the normal stimulatory protein. **(c)** The proto-oncogene may be moved from its normal location in the cell's DNA to another location. At its new site, the gene may be under the control of a different promoter and other transcriptional control sequences that cause it to be transcribed more often than normal. In that case, the normal protein would be made in excess.

for other proteins that affect the cell cycle. When all these proteins are functioning normally, in the right amounts at the right times, they help keep the rate of cell division at an appropriate level. When they malfunction, cancer—uncontrolled cell growth—may result.

For a proto-oncogene to become an oncogene, a mutation must occur in the cell's DNA. **Figure 11.18** illustrates three kinds of changes in DNA that can produce active oncogenes. In all three cases, normal gene expression is changed, and the cell is stimulated to divide excessively.

Changes in genes whose products inhibit cell division are also involved in cancer. These genes are called **tumor-suppressor genes** because the proteins they encode normally help prevent uncontrolled cell growth **(Figure 11.19)**. Any mutation that keeps a normal tumor-suppressor protein from being made or from functioning may contribute to the development of cancer.

The Effects of Cancer Genes on Cell-Signaling Pathways

Researchers are now working out the details of how oncogenes and defective tumor-suppressor genes contribute to uncontrolled cell growth. Normal proto-oncogenes and tumor-suppressor genes often code for proteins involved in signal transduction pathways leading to gene expression, pathways similar to the one shown in Figure 11.10. Let's consider the case of a proto-oncogene named *ras*, which in mutant form is an oncogene involved in many kinds of cancer. The protein encoded by *ras* is a component of signal transduction pathways leading to growth-stimulating responses. Therefore, if an oncogene-creating mutation causes the *ras* protein to be hyperactive, for example, the pathway may be overstimulated, and excessive cell division may result. To produce cancer, however, further abnormalities are usually needed, as we discuss next.

The Progression of a Cancer

Nearly 150,000 Americans were stricken by cancer of the colon (large intestine) or rectum in 2005. One of the best-understood types of human cancer, colon cancer illustrates an important principle about

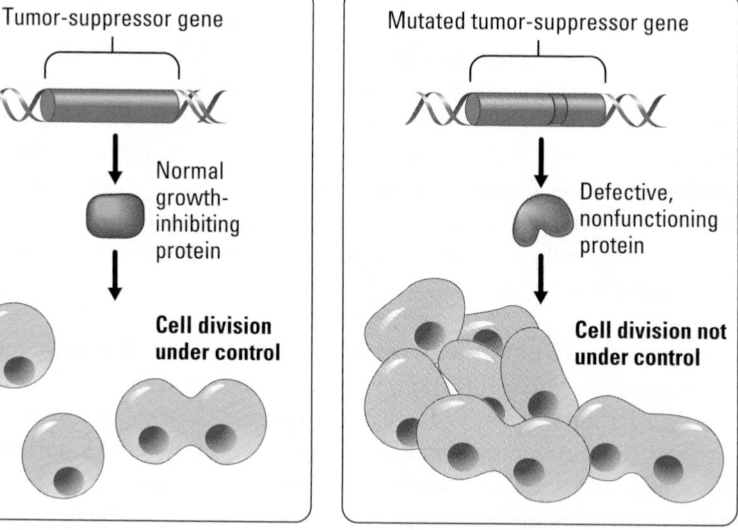

(a) Normal cell growth. A tumor-suppressor gene normally codes for a protein that inhibits cell growth and division. In this way, the gene helps prevent cancerous tumors from arising.

(b) Uncontrolled cell growth (cancer). When a mutation in a tumor-suppressor gene makes its protein defective, cells that are usually under the control of the normal protein may divide excessively, eventually forming a tumor.

Figure 11.19 Tumor-suppressor genes.

Figure 11.20 **Colon cancer develops after a series of mutations.**

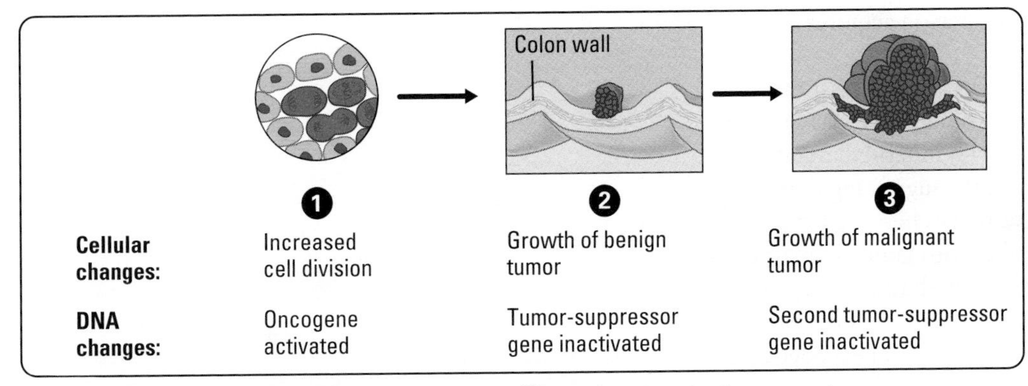

(a) Stepwise development of a typical colon cancer. ❶ The first sign of colon cancer is the unusually frequent division of apparently normal cells in the colon lining. ❷ Later, a benign tumor appears in the colon wall. ❸ Eventually, a malignant tumor develops.

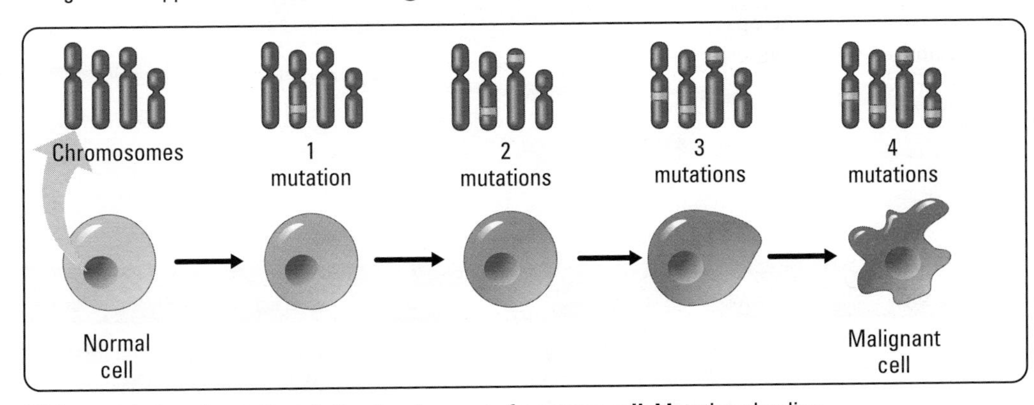

(b) Accumulation of mutations in the development of a cancer cell. Mutations leading to cancer accumulate in a lineage of cells. Colors distinguish the normal cell from cells with one or more mutations, leading to increased cell division and cancer. Once a cancer-promoting mutation occurs (orange band on chromosome), it is passed to all the descendants of the cell carrying it.

how cancer develops: More than one mutation is needed to produce a full-fledged cancer cell. As in many cancers, the development of colon cancer that metastasizes (spreads) is gradual.

As shown in **Figure 11.20a**, colon cancer begins as an unusually frequent division of normal-looking cells in the colon lining. This and later cellular changes parallel changes at the DNA level, including the activation of a cellular oncogene and the inactivation of two tumor-suppressor genes. These genetic changes (mutations) result in altered signal transduction pathways. The requirement for several mutations—the total number is usually four or more—explains why cancers can take a long time to develop. **Figure 11.20b** shows how mutations leading to cancer may accumulate in a lineage of cells. By the time a fourth mutation has occurred, the structure of this cell and its descendants is grossly altered.

"Inherited" Cancer Cancer is always a genetic disease in the sense that it always results from changes in DNA. Most mutations that lead to cancer arise in the organ where the cancer starts—the colon, for example. Because these mutations do not affect the cells that give rise to eggs or sperm, they are not passed from parent to child. Sometimes, however, a cancer-causing mutation occurs in gametes and is passed on from generation to generation. Such cases result in inborn mutations that predispose the people who inherit them to developing cancer. We say that this cancer is familial or inherited, even though it doesn't appear unless the person acquires additional mutations in the susceptible tissue.

One example of inherited cancer genes is associated with breast cancer, a disease that strikes one out of every ten American women. The vast majority of breast cancer cases seem to have nothing to do with inherited mutations. But in some families, breast cancer appears frequently, suggesting that an inherited trait might be involved. In 1994, cancer researchers identified a gene called *BRCA1* that is mutated in many families with familial breast cancer. Some mutations in gene *BRCA1* put a woman at high risk for breast cancer (and ovarian cancer as well)—a more than 80% risk of developing cancer during her lifetime. Research suggests that the protein encoded by the normal version of *BRCA1* acts as a tumor suppressor.

In recent years, clinical tests have been developed for the presence of mutations affecting *BRCA1* and *BRCA2*, another breast cancer gene. These tests offer women, particularly those with a family history of breast cancer, the option of being tested. Unfortunately, this knowledge is of limited use, because surgical removal of the breasts and/or ovaries is the only preventive option currently available to women who carry the mutant genes.

Cancer Risk and Prevention

Cancer is now one of the leading causes of death in most developed countries, including the United States. Death rates due to certain forms of cancer (including stomach, cervical, and uterine cancers) have decreased, but the overall cancer death rate is still on the rise, currently increasing at about 1% per decade.

Cancer-causing agents are called **carcinogens.** Most mutagens (substances that promote mutations) are carcinogens capable of bringing about cancer-causing DNA changes like the ones in Figure 11.20. In many cases, these changes result from decades of exposure to the mutagenic effects of carcinogens. One of the most potent carcinogens is ultraviolet (UV) radiation. Excessive exposure to UV radiation from the sun can cause skin cancer, including a deadly type called melanoma.

The one substance known to cause more cases and types of cancer than any other is tobacco. In 1900, lung cancer was a rare disease. Since then, largely because of an increase in cigarette smoking, rates of lung cancer have steadily increased, and today more people die of lung cancer (over 170,000 Americans in 2005) than of any other form of cancer. Most tobacco-related cancers come from smoking, but the passive inhalation of secondhand smoke also poses a risk. As **Table 11.1** indicates, tobacco use, sometimes in combination with alcohol consumption, causes a number of types of cancer. In nearly all cases, cigarettes are the

Table 11.1 Cancer in the United State (Ranked by Number of Cases)

Cancer	Known or Likely Carcinogen or Factor	Estimated Cases (2006)	Estimated Deaths (2006)
Prostate	Testosterone; possibly dietary fat	234,500	27,400
Breast	Estrogen; possibly dietary fat	214,600	41,400
Lung	Cigarette smoke	174,500	162,500
Colon and rectum	High dietary fat; low dietary fiber	148,600	55,200
Skin	Ultraviolet light	68,800	10,700
Lymphomas	Viruses (for some types)	66,700	20,300
Bladder	Cigarette smoke	61,400	13,100
Uterus	Estrogen	41,200	7,400
Kidney	Cigarette smoke	38,900	12,800
Leukemias	X-rays; benzene; viruses (for some types)	35,100	22,300
Pancreas	Cigarette smoke	33,700	32,300
Mouth and throat	Tobacco in various forms; alcohol	31,000	7,400
Stomach	Table salt; cigarette smoke	22,300	11,400
Ovary	Large number of ovulation cycles	20,200	15,300
Cervix	Viruses; cigarette smoke	19,700	3,700
Brain and nerve	Trauma; X-rays	18,800	12,800
Liver	Alcohol; hepatitis viruses	18,500	16,200
All others		151,300	92,600
Total		1,399,800	564,800

Source: *Cancer Facts and Figures 2006* (American Cancer Society Inc.)

main culprit, but smokeless tobacco products (snuff and chewing tobacco) are linked to cancer of the mouth and throat. Exposure to some of the most lethal carcinogens is often a matter of individual choice: Tobacco use, the consumption of alcohol, and excessive time spent in the sun are all avoidable behaviors that affect cancer risk.

Avoiding carcinogens is not the whole story. There is increasing evidence that some food choices significantly reduce a person's cancer risk. For instance, eating 20–30 g of plant fiber daily (about twice the amount the average American consumes) while eating less animal fat may help prevent colon cancer. Evidence also indicates that other substances in fruits and vegetables, including vitamins C and E and certain compounds related to vitamin A, may offer protection against a variety of cancers. Cabbage and its relatives, such as broccoli and cauliflower, are thought to be especially rich in substances that help prevent cancer, although the identities of all these substances are not yet established. Determining how diet influences cancer has become an important focus of research.

The battle against cancer is being waged on many fronts, and there is reason for optimism in the progress being made. It is especially encouraging that we can help reduce our risk of acquiring some of the most common forms of cancer by the choices we make in daily life.

CHECKPOINT

1. How can a mutation in a tumor-suppressor gene contribute to the development of cancer?
2. Why are most cases of breast cancer considered nonhereditary?
3. Of all known behavioral factors, which one causes the most cancer cases and deaths?

Answers: 1. A mutated tumor-suppressor gene may produce a defective protein unable to function in a pathway that normally inhibits cell division (that is, normally suppresses tumors). **2.** Most breast cancers are associated with mutations in body cells, not inherited mutations that are passed from parents to offspring via gametes. **3.** Tobacco use

Homeotic Genes

Homeotic genes are master control genes that regulate batteries of other genes involved in embryonic development. For example, one set of homeotic genes in fruit flies instructs cells in the embryonic head and thorax (midbody) to form antennae and legs, respectively. Elsewhere, these homeotic genes remain turned off, while others are turned on. Mutations in homeotic genes can produce bizarre effects. For example, homeotic fruit fly mutants may have extra sets of wings or legs growing from their head **(Figure 11.21)**.

Among the most exciting biological discoveries in recent years is that homeotic genes help direct embryonic development in a wide variety of organisms. Researchers studying homeotic genes in fruit flies found a common

Normal fruit fly

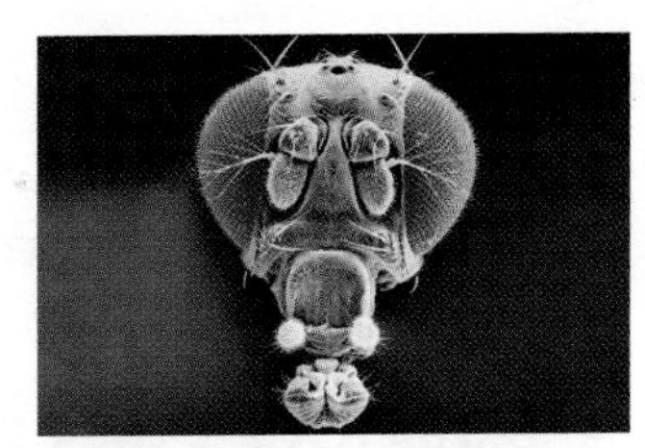

Normal head

Mutant fly with extra wings

Mutant fly with extra legs growing from head

Figure 11.21 The effect of homeotic genes. These strange mutant fruit flies result from mutations in homeotic (master control) genes.

structural feature: Every homeotic gene they looked at contained a common sequence of 180 nucleotides. Very similar sequences have since been found in virtually every eukaryotic organism examined so far, including yeasts, plants, earthworms, frogs, chickens, mice, and humans. These nucleotide sequences are called **homeoboxes,** and each is translated into a segment (60 amino acids long) of the protein product of the homeotic gene. The polypeptide segment encoded by the homeobox binds to specific sequences in DNA, enabling homeotic proteins that contain it to turn groups of genes on or off during development.

Figure 11.22 highlights some striking similarities in the chromosomal locations and developmental roles of homeobox-containing homeotic genes in two quite different animals. Notice that the order of genes on the fly chromosome is the same as on the four mouse chromosomes and that the gene order on the chromosomes corresponds to analogous body regions in both animals. These similarities suggest that the original version of these homeotic genes arose very early in the history of life and that the genes have remained remarkably unchanged over eons of animal evolution.

By their presence in such diverse creatures, homeotic genes illustrate one of the central themes of biology: unity in diversity. The fact that these key genes are control genes underscores the importance of regulation in the lives of organisms. ■

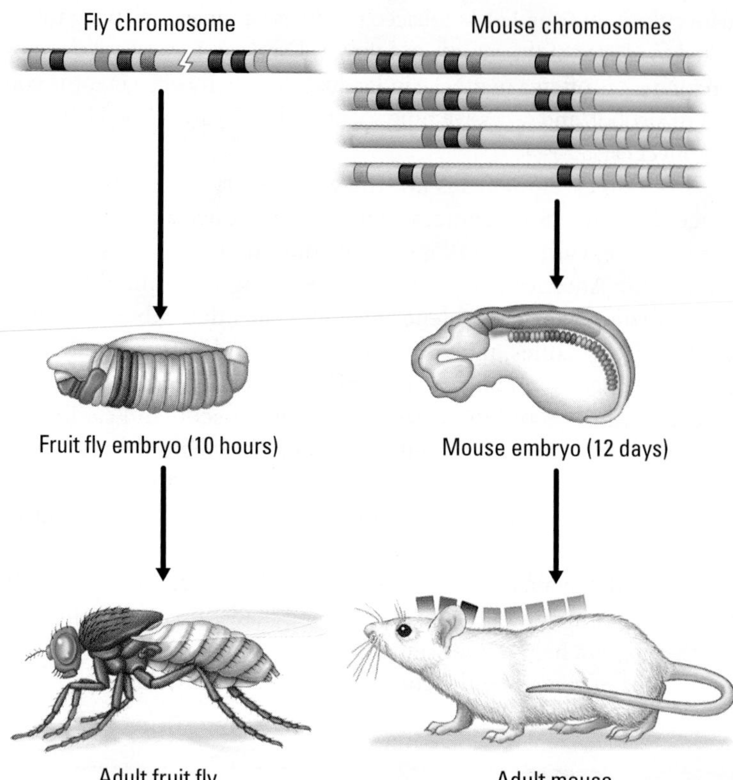

Figure 11.22 Homeotic genes in two different animals. At the top of the figure are portions of chromosomes that carry homeotic genes in the fruit fly and the mouse. The colored boxes represent homeotic genes that are very similar (have especially close DNA sequences) in flies and mice. The same color coding identifies the parts of the animals that are affected by these genes.

Chapter Review

SUMMARY OF KEY CONCEPTS

For study help and activities, go to campbellbiology.com or the student CD-ROM.

Biology and Society: Cloning at the Edge of Extinction

• Some scientists focus their efforts on making clones (genetically identical organisms created by asexual reproduction) of endangered species.

How and Why Genes Are Regulated

MP3 Tutor *Control of Gene Expression*

• **Patterns of Gene Expression in Differentiated Cells** The various cell types of a multicellular organism are different because different combinations of genes are turned on and off via gene regulation.

• **Gene Regulation in Bacteria** An operon is a cluster of genes with related functions together with their promoter and other sequences for controlling their transcription. The *lac* operon produces enzymes that break down lactose only when it is present.

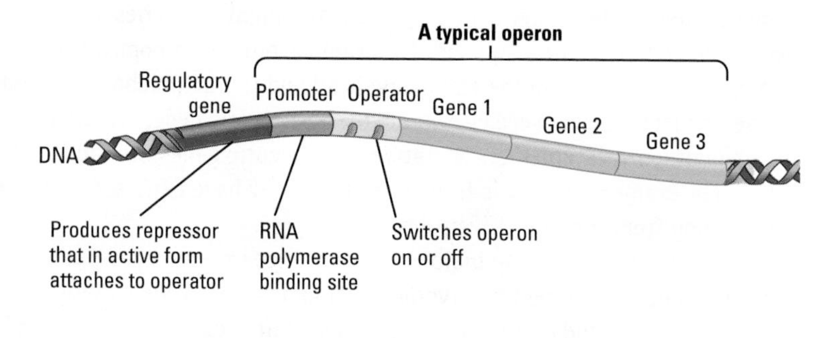

Activity *The* lac *Operon in* E. coli

• **Gene Regulation in Eukaryotic Cells** In the nucleus of eukaryotic cells, there are multiple possible control points in the pathway of gene expression. DNA packing tends to block gene expression, presumably by preventing access of transcription proteins to the DNA. An extreme example is X chromosome inactivation in the cells of female mammals. The most important control point in both eukaryotes and prokaryotes is at gene transcription. A variety of regulatory proteins interact with DNA and with each other to turn the transcription of eukaryotic genes on or off. There are also opportunities for the control of eukaryotic gene expression after transcription, when introns are cut out of the RNA and a cap and tail are added.

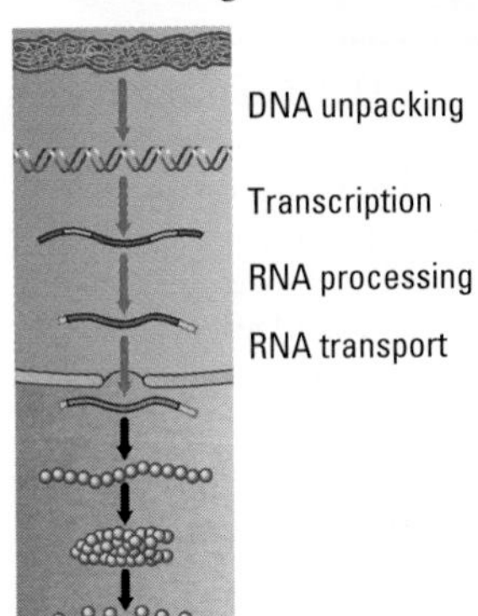

• The lifetime of an mRNA molecule helps determine how much protein is made, as do factors involved in translation. Finally, the cell may activate the finished protein in various ways (for instance, by cutting out portions) and later break it down.

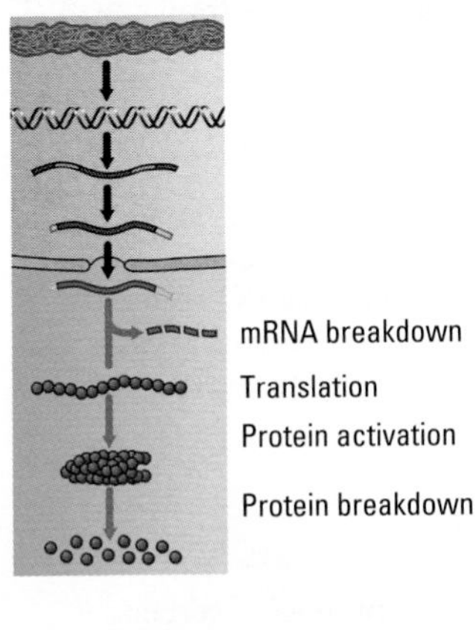

Case Study in the Process of Science *How Do You Design a Gene Expression System?*

Activity *Gene Regulation in Eukaryotes*

Activity *Review: Gene Regulation in Eukaryotes*

• **Cell Signaling** Cell-to-cell signaling is key to the development and functioning of multicellular organisms. Signal transduction pathways convert molecular messages to cell responses, often the transcription of particular genes.

Activity *Signal Transduction Pathway*

• **DNA Microarrays: Visualizing Gene Expression** DNA microarrays can be used to determine which of many genes are turned on in a particular cell type.

Cloning Plants and Animals

• **The Genetic Potential of Cells** Most differentiated cells retain a complete set of genes, so a carrot plant, for example, can be made to grow from a single carrot cell. Under special conditions, animals can also be cloned.

• **Reproductive Cloning of Animals** Nuclear transplantation is a procedure whereby a donor cell nucleus is inserted into a nucleus-free egg. First demonstrated in frogs in the 1950s, reproductive cloning was used in 1997 to clone a sheep from an adult mammary cell and has since been used to create many other cloned animals.

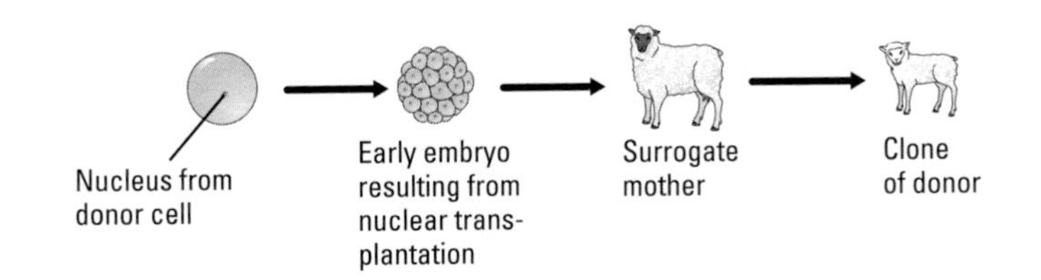

• **Therapeutic Cloning and Stem Cells** The purpose of therapeutic cloning is to produce embryonic stem cells for medical uses. Both embryonic and adult stem cells show promise for future therapeutic uses.

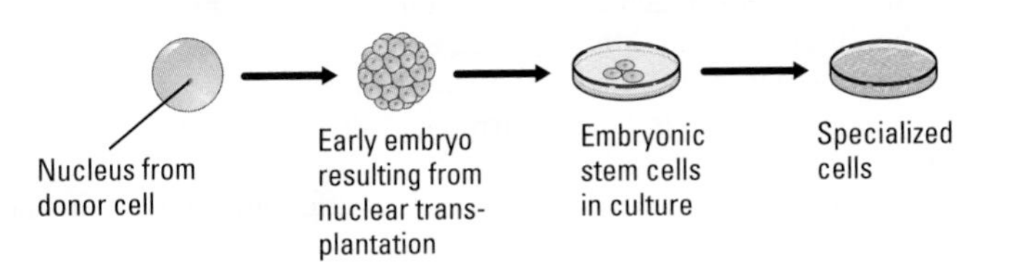

The Genetic Basis of Cancer

• **Genes That Cause Cancer** Cancer cells, which divide uncontrollably, can result from mutations in genes whose protein products regulate the cell cycle.

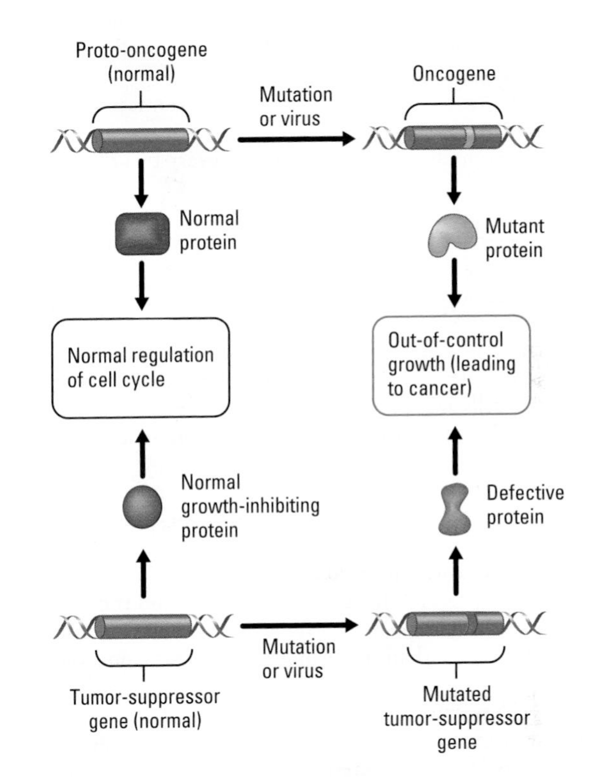

Many proto-oncogenes and tumor-suppressor genes code for proteins active in signal transduction pathways regulating cell division. Mutations of these genes cause malfunction of the pathways. Cancers result from a series of genetic changes in a cell lineage. Researchers have gained insight into the genetic basis of breast cancer by studying families in which a disease-predisposing mutation is inherited.

• **Cancer Risk and Prevention** Reducing exposure to carcinogens (which induce cancer-causing mutations) and other lifestyle choices can help reduce cancer risk.

Activity *Causes of Cancer*

Evolution Connection: Homeotic Genes

• Evidence for the evolutionary importance of gene regulation is apparent in homeotic genes, master genes that regulate groups of other genes that in turn control embryonic development.

SELF-QUIZ

1. Your bone cells, muscle cells, and skin cells look different because
   a. different kinds of genes are present in each kind of cell.
   b. they are present in different organs.
   c. different genes are active in each kind of cell.
   d. different mutations have occurred in each kind of cell.

2. What kinds of evidence demonstrate that differentiated cells in a plant or animal retain their full genetic potential?

3. The most common procedure for cloning an animal is called __________.

4. What is learned from a DNA microarray?

5. Which of the following is a valid difference between embryonic stem cells and the stem cells found in adult tissues?
   a. In laboratory culture, only adult stem cells are immortal.
   b. In nature, only embryonic stem cells give rise to all the different types of cells in the organism.
   c. Only adult stem cells can be made to differentiate in the laboratory.
   d. Only embryonic stem cells are found in every tissue of the adult body.

6. A group of prokaryotic genes with related functions that are regulated as a single unit, along with the control sequences that perform this regulation, is called a(n) __________.

7. The regulation of gene expression must be more complex in multicellular eukaryotes than in prokaryotes because
   a. eukaryotic cells are much smaller.
   b. in a multicellular eukaryote, different cells are specialized for different functions.
   c. prokaryotes are restricted to stable environments.
   d. eukaryotes have fewer genes, so each gene must do several jobs.

8. A eukaryotic gene was inserted into the DNA of a bacterium. The bacterium then transcribed this gene into mRNA and translated the mRNA into protein. The protein produced was useless; it contained many more amino acids than the protein made by the eukaryotic cell. Why?
   a. The mRNA was not spliced as it is in eukaryotes.
   b. Eukaryotes and prokaryotes use different genetic codes.
   c. Repressor proteins interfered with transcription and translation.
   d. Ribosomes were not able to bind to tRNA.

9. What is the difference between oncogenes and proto-oncogenes? How can one turn into the other? What purpose do proto-oncogenes serve?

10. A mutation in a single gene may cause a major change in the body of a fruit fly, such as an extra pair of legs or wings. Yet it takes the combined action of many genes to produce a wing or leg. How can a change in just one gene cause such a big change in the body? What are such genes called?

Answers to the Self-Quiz questions can be found in Appendix D.

Go to the website or CD-ROM for more Self-Quiz questions.

THE PROCESS OF SCIENCE

11. Study the depiction of the *lac* operon in Figure 11.4. Normally, the genes are turned off when lactose is not present. Lactose activates the genes, which code for enzymes that enable the cell to use lactose. Mutations can alter the function of this operon; in fact, it was the effects of various mutations that enabled Jacob and Monod to figure out how the operon works. Predict how the following mutations would affect the function of the operon in the presence and absence of lactose:
   a. Mutation of regulatory gene; repressor will not bind to lactose.
   b. Mutation of operator; repressor will not bind to operator.
   c. Mutation of regulatory gene; repressor will not bind to operator.
   d. Mutation of promoter; RNA polymerase will not attach to promoter.

12. One of the surprising results from the Human Genome Project is that there are many fewer genes than there are proteins in the human body. This result seems to highlight the importance of alternative RNA splicing, which allows several different mRNAs to be made from a single gene. Suppose you have samples of two types of adult cells from one person. These cell types are different in structure and function. Design an experiment using microarrays to determine whether or not the different gene expression in the two cell types is due to alternative RNA splicing.

13. Because a cat must have both orange and nonorange alleles to be tortoiseshell (see Figure 11.6), we would expect only female cats, which have two X chromosomes, to be tortoiseshell. Normal male cats (XY) can carry only one of the two alleles. Male tortoiseshell cats are occasionally seen, although they are usually sterile. What might you guess their genotype to be?

BIOLOGY AND SOCIETY

14. A chemical called dioxin is produced as a by-product of certain chemical manufacturing processes. Trace amounts of this substance were present in Agent Orange, a defoliant sprayed on vegetation during the Vietnam War. There has been a continuing controversy over its effects on soldiers exposed to Agent Orange during the war. Animal tests have suggested that dioxin can cause cancer, liver and thymus damage, immune system suppression, and birth defects; at high dosage it can be lethal. But such animal tests are inconclusive; a hamster is not affected by a dose that can kill a much larger guinea pig, for example. Researchers have discovered that dioxin enters a cell and binds to a protein that in turn attaches to the cell's DNA. How might this mechanism help explain the variety of dioxin's effects on different body systems and in different animals? How might you determine whether a particular individual became ill as a result of exposure to dioxin? Do you think this information is relevant in the lawsuits of soldiers suing over exposure to Agent Orange? Why or why not?

15. There are genetic tests available for several types of "inherited cancer." The results from these tests cannot usually predict that someone will get cancer within a particular amount of time. Rather, they indicate only that a person has an increased risk of developing cancer. For many of the cancers involved, there are no lifestyle changes that can decrease a person's risk of actually getting the disease. Some people feel that this makes the tests useless. If your close family had a history of cancer and a test were available, would you want to get screened? Why or why not? What would you do with this information? If a sibling decided to get screened, would you want to know the results?

12 DNA Technology

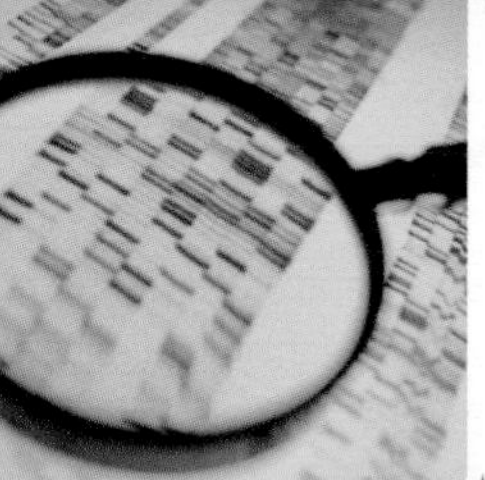

The first use of **DNA fingerprinting** proved one man innocent and another guilty of murder.

Chances are you ate a **genetically modified food** today.

The DNA of two people of the same sex is **99.9% identical.**

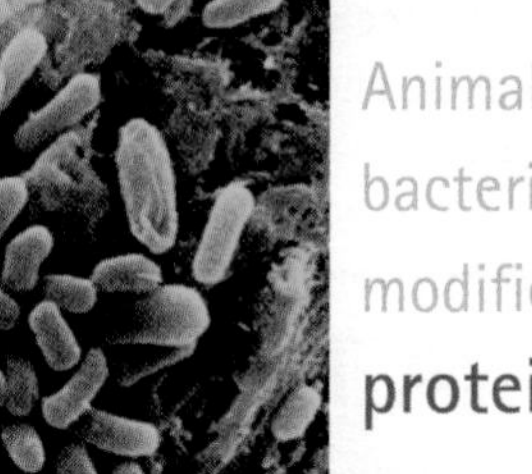

Animals, plants, and even bacteria can be genetically modified to produce **human proteins.**

BIOLOGY AND SOCIETY

Crime Scene Investigations: Murders in a Small Town

On November 22, 1983, the sleepy English village of Narborough awoke to news of a horrific crime: A 15-year-old girl had been raped and murdered on a quiet country lane. The killer left behind few clues, except for semen on the victim's body and clothes. Despite extensive investigation, the crime went unsolved. Three years later, another 15-year-old girl was raped and murdered, and police believed that a double murderer was at large **(Figure 12.1)**. Within days, a local man was arrested and charged with both crimes. Under considerable pressure from police, the suspect confessed to the second murder, but denied committing the first.

In an attempt to pin both murders on the suspect, investigators turned to Alec Jeffreys, a biology professor who had recently developed the first system for matching a DNA sample to a specific person. Because the DNA sequence of every person is unique (except for identical twins), DNA fingerprinting can be used to determine with near certainty whether two samples of genetic material come from the same individual. Jeffreys compared DNA from the 1983 and 1986 semen samples. As the police suspected, the DNA analysis indicated that the same person had committed both crimes. However, when Jeffreys analyzed the suspect's DNA, the results were shocking: DNA from the suspect did not match either crime scene sample. The suspect, for reasons unknown, had falsely confessed. Proved innocent, the man was released and entered legal history as the first person to be exonerated by DNA evidence.

The police then collected DNA samples from nearly 5,000 other local men, but none matched the evidence from the crime scenes. The case finally broke when a pub-goer described how a local named Colin Pitchfork had bullied him into submitting blood on Pitchfork's behalf. After the arrest of Pitchfork (Figure 12.1, inset), tests confirmed that his DNA matched the samples from the two murders. Pitchfork pled guilty to both crimes; the case again entered legal history, this time as the first ever to be solved by DNA evidence.

Since its introduction in 1986, DNA fingerprinting has become a standard law enforcement tool and has provided crucial evidence in many famous cases. **DNA technology**—methods for studying and manipulating genetic material—has thus quickly revolutionized the field of **forensics,** the scientific analysis of evidence for legal investigations.

Beyond the courtroom, DNA technology is responsible for some of the most remarkable scientific advances in recent years: Corn has been genetically modified to produce its own insecticide; the study of whole genomes has begun to reveal what it means to be human; and significant advances have been made toward curing fatal genetic diseases. In this chapter, you'll learn about these and other uses of DNA technology. Along the way, we'll consider the specific techniques involved, how they are applied, and some of the social, legal, and ethical issues these new technologies raise. ■

Figure 12.1 The Narborough murders. In the Narborough murder case, the first ever to be solved by DNA evidence, one man was proved innocent and another guilty.

Recombinant DNA Technology

Over 50 years ago, American geneticists Joshua Lederberg and Edward Tatum performed a series of experiments with *E. coli* that demonstrated that two individual bacteria can combine genes—a phenomenon that was previously thought to be limited to sexually reproducing eukaryotic organisms. With this work, they pioneered bacterial genetics, a field that within 20 years made *E. coli* the most thoroughly studied and understood organism at the molecular level. In the 1970s, research on *E. coli* led to the development of **recombinant DNA technology,** a set of laboratory techniques for combining genes from different sources—even different species—into a single DNA molecule.

Today, recombinant DNA technology is widely used to alter the genes of many types of cells for practical purposes. Scientists have genetically engineered bacteria to mass-produce a variety of useful chemicals, from cancer drugs to pesticides. Genes have also been transferred from bacteria into plants and from humans into farm animals. These applications are the latest developments in **biotechnology,** the use of organisms to perform practical tasks. Biotechnology is not new; in fact, it actually dates back thousands of years to the first uses of yeast to make bread and wine.

Figure 12.2 presents an overview of how recombinant DNA technology can be used to produce useful products. DNA carrying a gene of interest is taken from a cell of one organism and inserted into the DNA of a host cell. The gene of interest can be a human gene encoding a protein of medical value or perhaps a plant gene conferring resistance to pest insects. The host cell now contains **recombinant DNA,** a molecule carrying DNA from more than one source. A host that carries recombinant DNA (such as the corn plants shown in Figure 12.2) is called a **transgenic organism** or a **genetically modified organism (GMO).** As a genetically modified host multiplies its cells, it copies the gene of interest along with the rest of its DNA. Furthermore, the host cell may express the foreign gene and produce the desired protein.

DNA technology has revolutionized biotechnology. As shown at the bottom of Figure 12.2, the methods of DNA technology can be used to create many useful products. In the examples on the left, copies of the gene itself are the immediate product. In the examples on the right, the protein product of the gene is harvested.

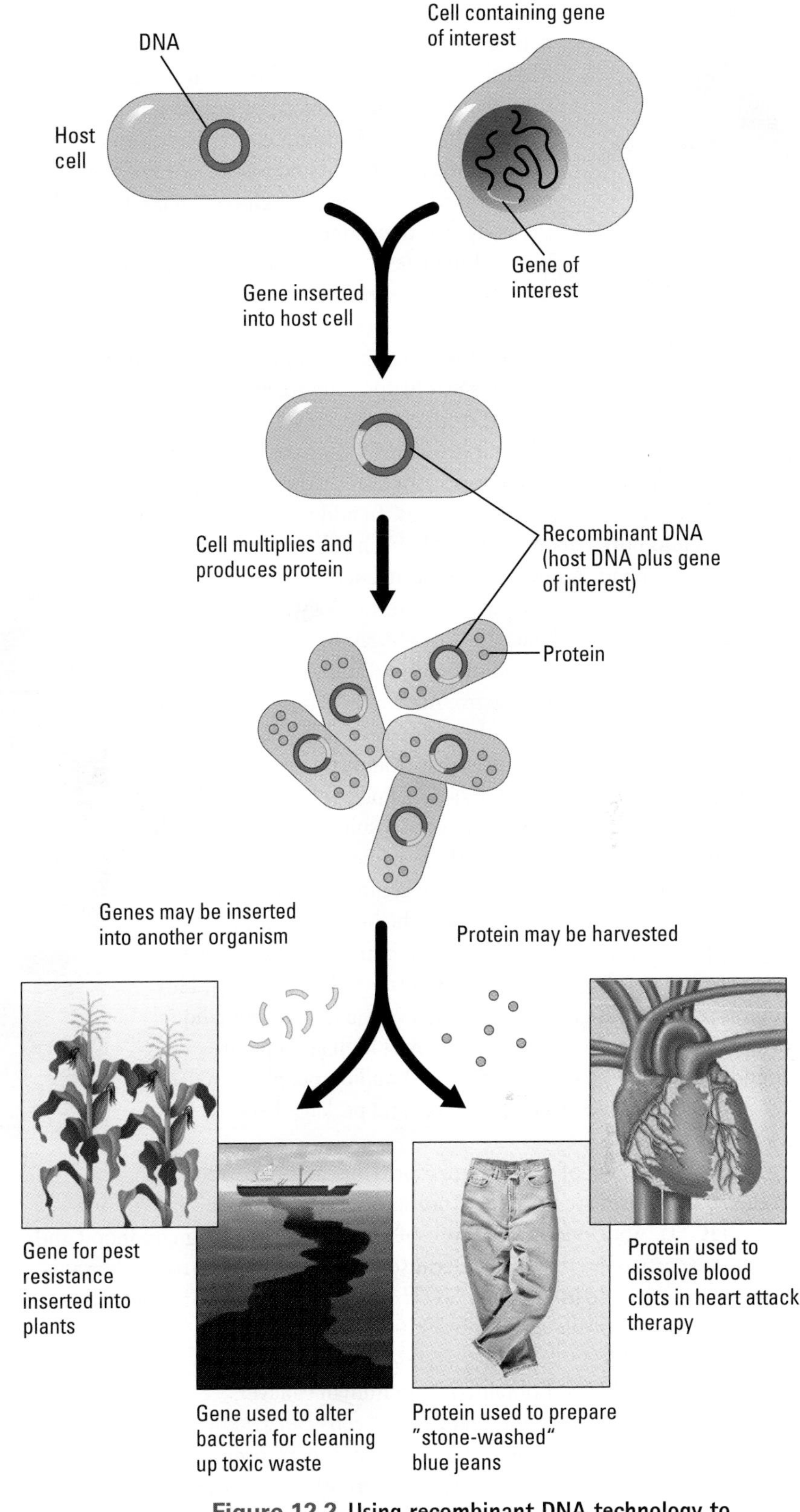

Figure 12.2 Using recombinant DNA technology to produce useful products.

From Humulin to Genetically Modified Foods

By transferring the gene for a desired protein product into a bacterium, yeast, or other kind of cell that is easy to grow in culture, proteins that are present naturally in only small amounts can be produced in large quantities. In this section, you'll learn about a few specific examples of this application of recombinant DNA technology.

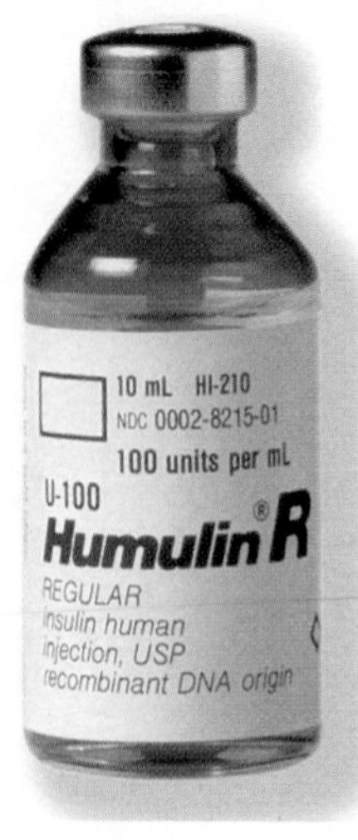

Making Humulin Humulin, the world's first genetically engineered pharmaceutical product, hit the market nearly 25 years ago. Humulin is human insulin that has been produced by genetically modified bacteria. In humans, insulin is a protein normally made by the pancreas. Functioning as a hormone, insulin regulates the level of glucose in the blood. If the body fails to produce enough insulin, the result is type 1 diabetes. There is no cure, so people with this disease must inject themselves with daily doses of insulin for the rest of their lives.

Because human insulin is not readily available, diabetes was historically treated using insulin extracted from cows and pigs. This treatment was not without problems, however. Because the chemical structure of pig and cow insulins are not exactly the same as that of human insulin, they can cause allergic reactions. In addition, by the 1970s, the supply of beef and pork pancreas available for insulin extraction could not keep up with the demand. A new source was needed.

In 1978, scientists working at the biotechnology company Genentech took on the challenge. They began by chemically synthesizing genes for human insulin. Since the amino acid sequences of the two polypeptides of the active human insulin protein were already known, it was easy to use the genetic code (see Figure 10.11) to determine nucleotide sequences that would code for them. Researchers arduously synthesized individual pieces of DNA and linked them together to form the desired insulin genes. In 1979, the researchers succeeded in inserting these artificial human genes into *E. coli* host cells. Under proper growing conditions, the transgenic bacteria cranked out large quantities of the human protein.

In 1982, Humulin became the first recombinant DNA drug approved by the Food and Drug Administration (FDA). Today, Humulin is produced in gigantic fermentation vats filled with a liquid culture of bacteria; the vats are four stories high and operate around the clock. Each day, more than 4 million people with diabetes use the insulin collected, purified, and packaged at such facilities **(Figure 12.3)**.

Figure 12.3 The world's largest plant for producing insulin, located in Denmark.

Insulin is just one of many human proteins that have been produced by transgenic bacteria. Another example is human growth hormone (HGH). Abnormally low levels of this hormone during childhood and adolescence can cause dwarfism. Because growth hormones from other animals are not effective in humans, HGH was an early target of genetic engineers. Before genetically engineered HGH became available in 1985, children with an HGH deficiency could only be treated with scarce and expensive supplies of HGH obtained from human cadavers.

Besides bacteria, yeast and mammalian cells can also be used to produce medically valuable human proteins. For example, transgenic mammalian cells growing in laboratory cultures are currently used to produce erythropoietin (EPO), a hormone that stimulates the production of red blood cells. EPO is used as a treatment for anemia; unfortunately, the drug is also abused by some athletes who seek the advantage of artificially high levels of oxygen-carrying red blood cells (called "blood doping").

DNA technology is also helping medical researchers develop vaccines. A **vaccine** is a harmless variant or derivative of a pathogen (a disease-causing microbe, such as a bacterium or virus) that is used to prevent an infectious disease. When a person is inoculated, the vaccine stimulates the immune system to develop lasting defenses against the pathogen. For the many viral diseases for which there is no effective drug treatment, prevention by vaccination is virtually the only medical way to fight the disease. One approach to vaccine production is to use genetically engineered cells to make large amounts of a protein molecule that is found on the pathogen's outside surface. Such a method, using transgenic yeast, is used to produce a vaccine against hepatitis B, a disabling and sometimes fatal liver disease.

Figure 12.4 Genetically modified corn. The corn plants in this field carry a bacterial gene that helps prevent infestation by the European corn borer (inset).

Genetically Modified (GM) Foods Since ancient times, humans have selectively bred agricultural crops to make them more useful (see the discussion of artificial selection in Chapter 1). Today, DNA technology is quickly replacing traditional breeding programs as scientists work to improve the productivity of plants and animals important to agriculture. In the United States today, roughly half the corn crop and over three-quarters of the soybean and cotton crops are genetically modified. **Figure 12.4** shows a field of corn that has been genetically engineered to resist attack by an insect called the European corn borer. Growing insect-resistant plants reduces the need for chemical insecticides. In another example, modified strawberries produce bacterial proteins that act as a natural antifreeze, providing protection from cold weather, which can harm this delicate crop. In an experimental setting, genetically engineered potatoes produce harmless proteins derived from the cholera bacterium; researchers hope that these modified potatoes will one day serve as an edible vaccine against cholera, a disease that kills thousands of children in developing nations every year. Scientists are also using genetic engineering to improve the nutritional value of crop plants. One example is transgenic "golden rice" that produces yellow rice grains containing beta-carotene, which our body uses to make vitamin A **(Figure 12.5)**. This rice could help prevent vitamin A deficiency among people who depend on rice as their staple food—half the world's population. However, controversy surrounds the use of GM foods, as we'll discuss at the end of the chapter.

Figure 12.5 Genetically modified rice. "Golden rice," shown here alongside ordinary rice, has been genetically modified to produce high levels of beta-carotene, a precursor to vitamin A.

Farm Animals and "Pharm" Animals **Figure 12.6** shows a herd of transgenic sheep that carry a gene for a human blood protein. The human protein can be harvested from the sheep's milk and is being tested as a treatment for cystic fibrosis. Because transgenic animals are difficult to produce, researchers may create a single transgenic animal and then clone it. The resulting herd of genetically identical transgenic animals, all carrying a recombinant human gene, could then serve as a grazing pharmaceutical factory—"pharm" animals.

Unlike transgenic plants, transgenic animals are currently used only to produce potentially useful proteins (not food). It is possible that DNA technology will eventually replace traditional animal breeding. Scientists might, for example, identify a gene that causes the development of larger muscles (which make up most of the meat we eat) in one variety of cattle and transfer it to other cattle or even to chickens. In 2006, University of Pittsburgh researchers genetically engineered pigs to carry a roundworm gene whose protein produces omega-3 fatty acids. Meat from the modified pigs contains 4–5 times as much healthy omega-3 fat as regular pork.

Recombinant DNA technology serves many roles today and will certainly play an even larger part in our future. In the next section, you'll learn more about the methods that scientists use to create and manipulate recombinant DNA.

Figure 12.6 Genetically modified sheep. A human blood protein, a potential treatment for cystic fibrosis and other chronic respiratory diseases, can be collected and purified from the milk of these transgenic sheep.

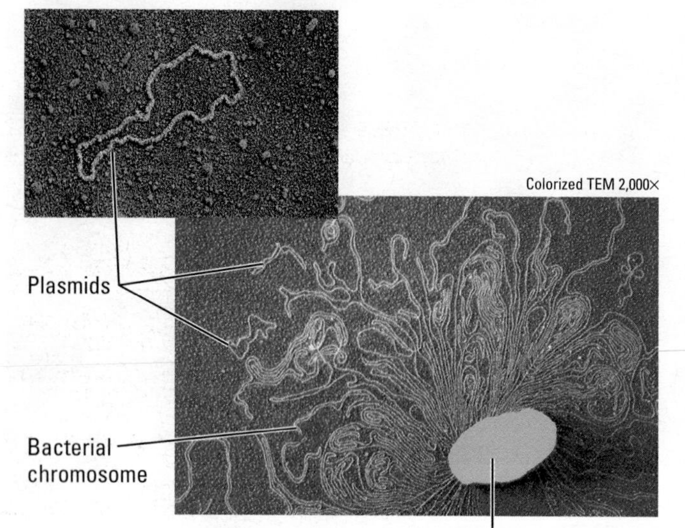

Figure 12.7 Bacterial plasmids. The large, light blue oval shape is the remnant of a bacterium that has ruptured and released all its DNA. Most of the DNA is the bacterial chromosome, which extends in loops from the cell. Two plasmids are also present. The inset shows an enlarged view of a single plasmid.

Recombinant DNA Techniques

Although recombinant DNA techniques can use a variety of cell types, bacteria are the workhorses of modern biotechnology. To manipulate genes in the laboratory, biologists often use bacterial **plasmids,** which are small, circular DNA molecules separate from the much larger bacterial chromosome. In **Figure 12.7**, you can see a bacterial cell that has been ruptured, revealing one long chromosome and several smaller plasmids. Like the main bacterial chromosome, plasmids carry genes. Their small size and the ease with which they are taken up by bacterial cells make plasmids useful to biologists. When they are taken up, plasmids act as **vectors,** DNA carriers that move genes from one cell to another. When the bacterial cell divides, the cell's replication machinery copies the plasmid, just as it would the main bacterial chromosome. Any foreign DNA that has been inserted will be replicated along with the rest of the plasmid.

Making recombinant DNA in large enough quantities to be useful requires several steps. Consider a typical genetic engineering challenge: A molecular biologist at a pharmaceutical company has identified a human gene *V* that codes for a valuable product—a hypothetical substance called protein V that kills certain human viruses. The biologist wants to set up a system for manufacturing the protein on a large scale. **Figure 12.8** illustrates a way to accomplish this using recombinant DNA techniques.

❶ First, the biologist isolates two kinds of DNA: many copies of a bacterial plasmid (to serve as a vector) and human DNA containing many genes, including gene *V*, the gene of interest. ❷ The researcher cuts both the plasmids and the human DNA. Each plasmid is cut in only one place; the

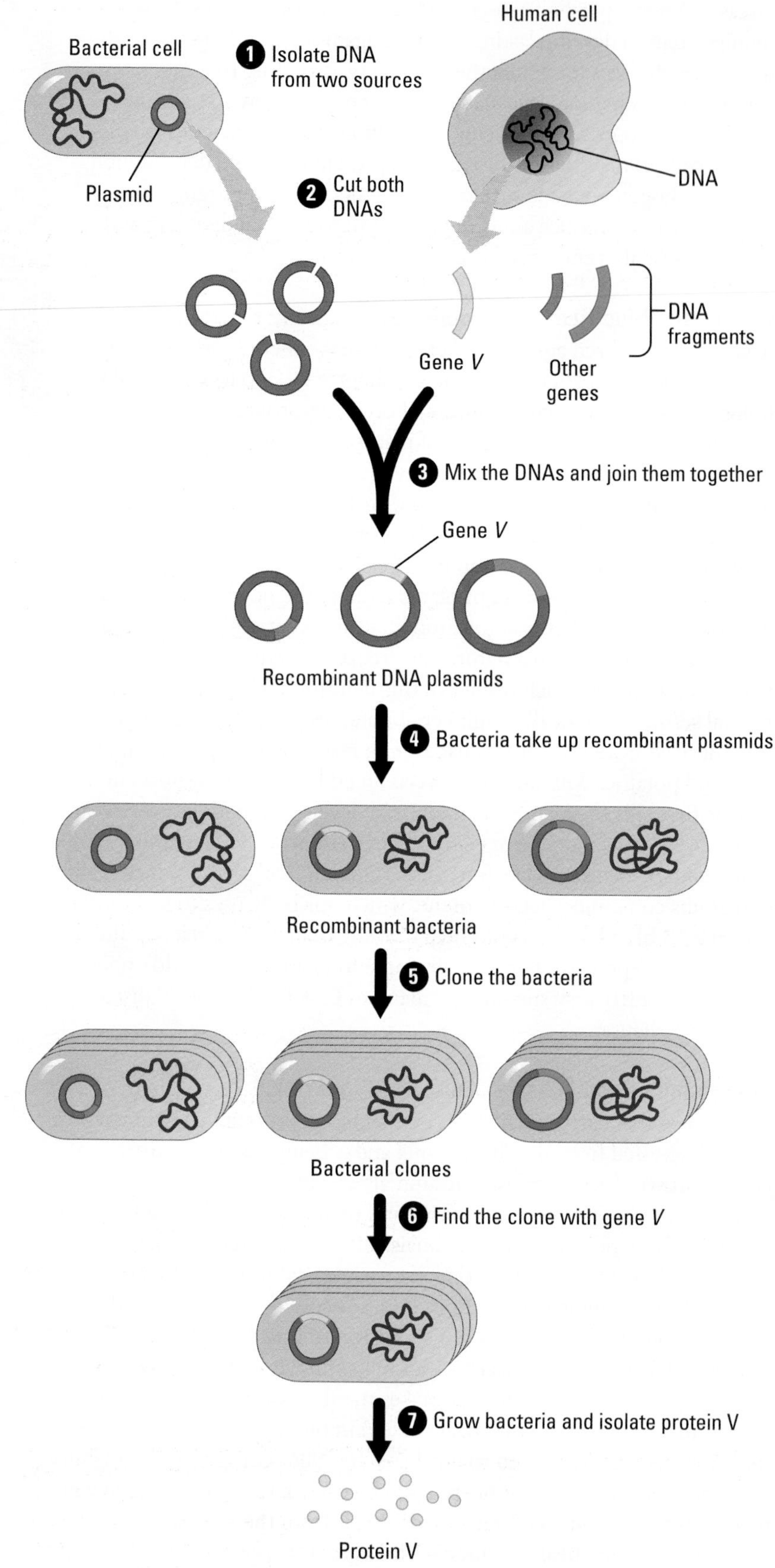

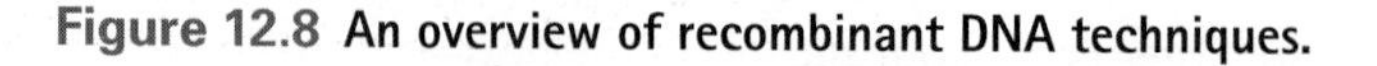
Figure 12.8 **An overview of recombinant DNA techniques.**

human DNA is cut into many fragments, one of which carries gene *V*. The figure shows the processing of just three human DNA fragments and three plasmids, but actually millions of plasmids and human DNA fragments (most of which do not contain gene *V*) are treated simultaneously. ❸ Next, the human DNA fragments are mixed with the cut plasmids. The plasmid and human DNA join together, resulting in recombinant DNA plasmids, some of which contain gene *V*. ❹ The recombinant plasmids are then mixed with bacteria, and, under the right conditions, the bacteria take up the recombinant plasmids. ❺ Each bacterium, with its recombinant plasmid, is allowed to reproduce. This step is the actual **gene cloning,** the production of multiple copies of the gene. As the bacterium forms a clone (a group of identical cells descended asexually from a single ancestral cell), any genes carried by the recombinant plasmid are also cloned (copied). ❻ The molecular biologist finds those few bacterial clones that contain gene *V*. ❼ The transgenic bacteria with gene *V* can then be grown in large tanks, producing protein V in marketable quantities.

A Closer Look: Cutting and Pasting DNA with Restriction Enzymes

As you saw in Figure 12.8, recombinant DNA results from the combination of two ingredients: a bacterial plasmid and the gene of interest (see steps 2 and 3). To understand how these DNA molecules are spliced together, you need to learn about the enzymes used to cut and paste DNA.

The cutting tools used for making recombinant DNA are bacterial enzymes called **restriction enzymes.** Most restriction enzymes recognize short nucleotide sequences (usually four to eight nucleotides long) in DNA molecules and cut at specific points within these recognition sequences. Hundreds of different restriction enzymes have been isolated, each of which recognizes and cuts a different DNA sequence. The top of **Figure 12.9** shows a piece of DNA that contains a recognition sequence for a particular restriction enzyme. ❶ The restriction enzyme cuts the DNA strands between the bases A and G within the recognition sequence, producing pieces of DNA called **restriction fragments.** The staggered cuts yield two double-stranded DNA fragments with single-stranded ends, called "sticky ends." Sticky ends are the key to joining DNA restriction fragments originating from different sources. ❷ Next, a piece of DNA from another source (orange) is added. Notice that the orange DNA has single-stranded ends identical in base sequence to the sticky ends on the blue DNA because the same restriction enzyme was used to cut both types of DNA. ❸ The complementary ends on the blue and orange fragments bond together by base pairing. ❹ This union is then made permanent by the "pasting" enzyme **DNA ligase.** This enzyme, which is one of the proteins the cell normally uses in DNA replication, connects the DNA pieces into continuous strands by forming covalent bonds between adjacent nucleotides. The final outcome is recombinant DNA, a molecule that contains DNA from two different sources.

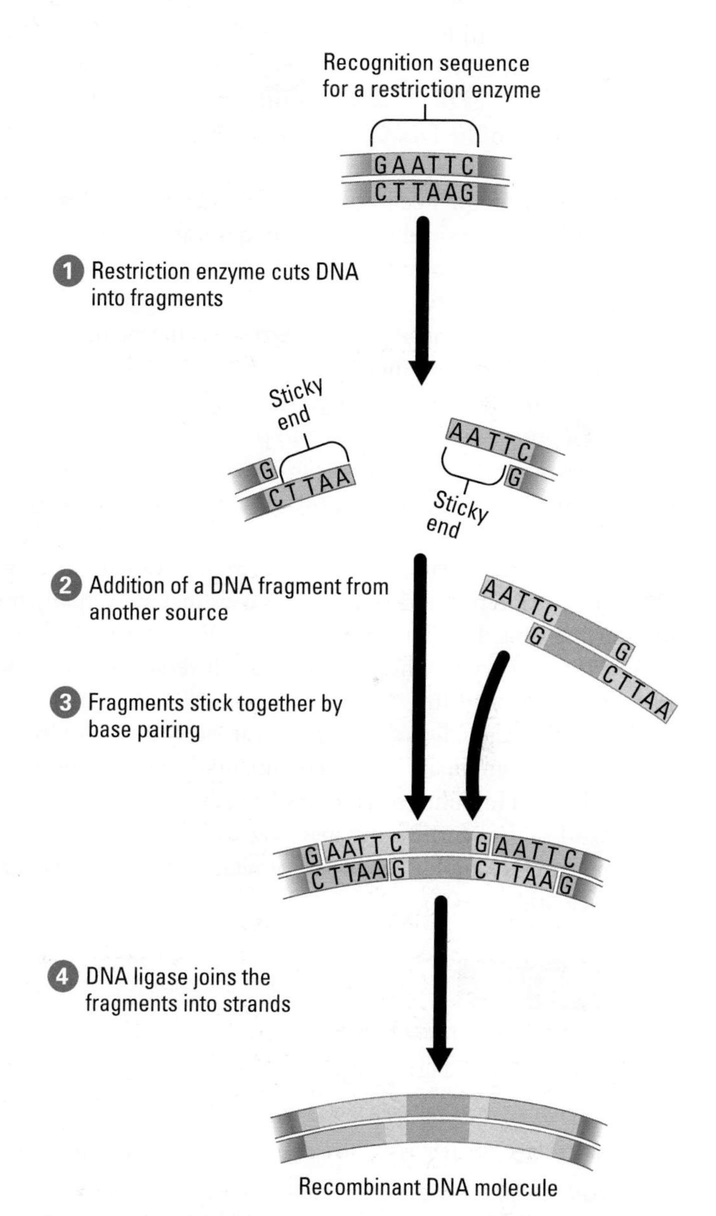

Figure 12.9 Cutting and pasting DNA. The production of recombinant DNA requires two enzymes: a restriction enzyme, which cuts the original DNA molecules into pieces, and DNA ligase, which pastes them together.

A Closer Look: Obtaining the Gene of Interest

The procedure shown in Figure 12.8 can yield millions of recombinant plasmids carrying many different segments of foreign DNA. Such a procedure is called a "shotgun" approach to gene cloning because it "hits" an enormous number of different pieces of DNA. The entire collection of cloned DNA fragments from a shotgun experiment, in which the starting material is bulk DNA from whole cells, is called a **genomic library.** A typical cloned DNA fragment is big enough to carry one or a few genes, and together the collection of fragments includes the entire genome of the organism from which the DNA was derived.

Once you've created a genomic library, you have to find the right "book"—that is, you must identify the bacterial clone containing a desired gene (step 6 in Figure 12.8). Methods for detecting a gene depend on base pairing between the gene and a complementary sequence on another nucleic acid molecule, either DNA or RNA. When at least part of the nucleotide sequence of a gene is already known, this information can be used to advantage. For example, if we know that a gene contains the sequence TAGGCT, a genetic engineer can use nucleotides labeled with a radioactive isotope to synthesize a short single strand of DNA with a complementary sequence (ATCCGA). This sort of labeled nucleic acid molecule is called a **nucleic acid probe** because it is used to find a specific gene or other nucleotide sequence within a mass of DNA. (In actual practice, a probe molecule would be considerably longer than six nucleotides.) When a radioactive DNA probe is added to the DNA of various clones, it tags the correct molecule—finds the right book in the library—by base-pairing to the complementary sequence in the gene of interest **(Figure 12.10)**. Once a probe detects the desired clone within a library, the cells can be grown further and the gene of interest produced in large amounts.

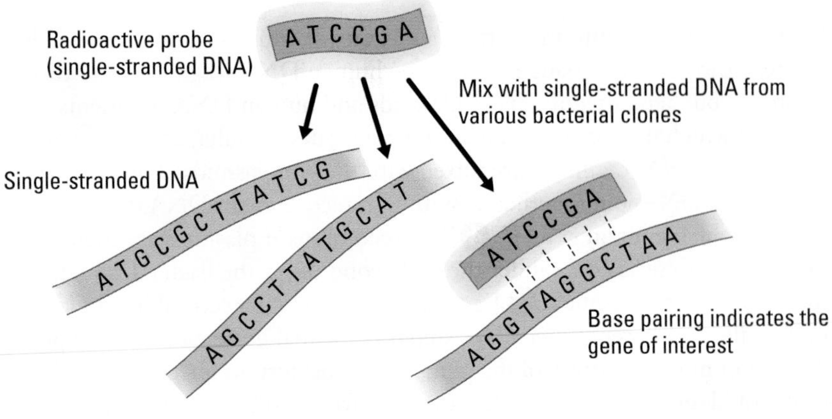

Figure 12.10 How a DNA probe tags a gene. The probe is a short, radioactive, single-stranded molecule of DNA or RNA. When it is mixed with single-stranded DNA from a gene with a complementary sequence, it attaches by hydrogen bonds, labeling the gene.

Another approach to obtain a gene of interest is to synthesize it. One method uses reverse transcriptase, a retroviral enzyme that can produce a molecule of DNA from a molecule of mRNA (see Chapter 10). **Figure 12.11** shows the steps involved. A eukaryotic cell ❶ transcribes the gene of interest and ❷ processes the transcript to produce mRNA. A researcher then ❸ isolates the mRNA and ❹ makes single-stranded DNA from it using reverse transcriptase. ❺ DNA polymerase is then used to synthesize a second DNA strand.

The DNA that results from this procedure, called complementary DNA (cDNA), represents only those genes that are actually transcribed in the starting cells. Furthermore, because cDNA strands lack introns (see Figure 10.14), they are shorter than the full version of the genes and therefore easier to work with.

When the desired gene is small, it can be synthesized from scratch. In the earliest recombinant DNA experiments, including those leading to the production of Humulin, researchers laboriously synthesized genes. Today, automated DNA-synthesizing machines can accurately and rapidly produce customized DNA molecules of any sequence up to lengths of a few hundred nucleotides.

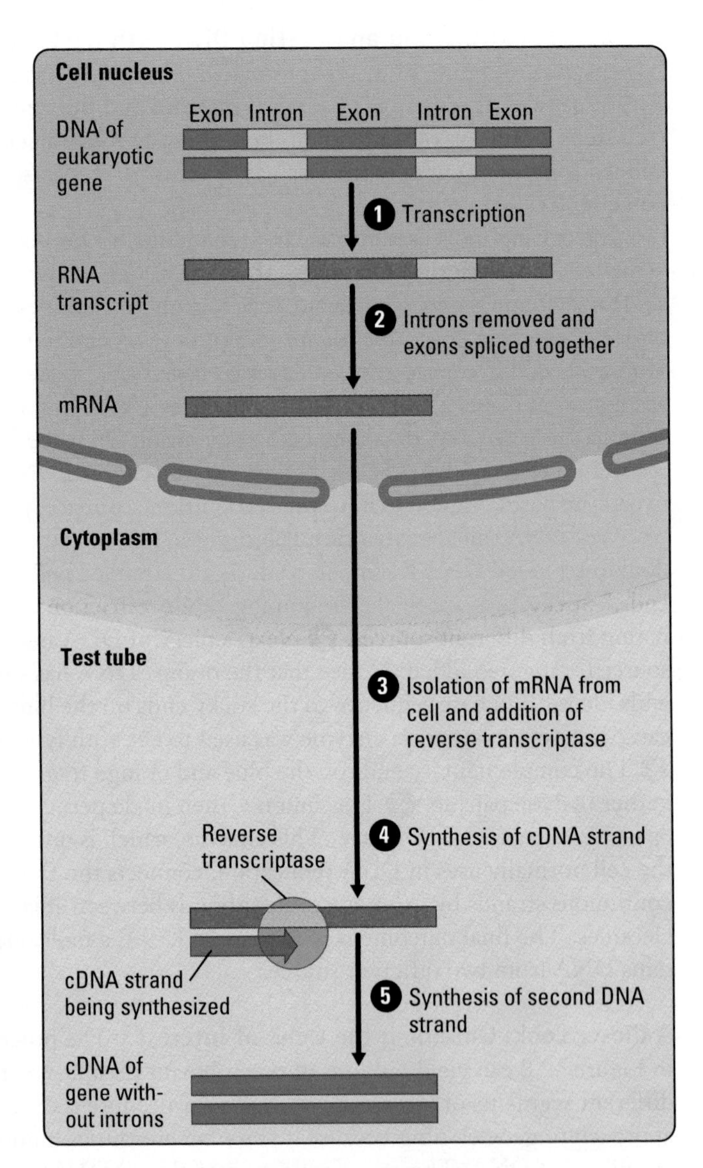

Figure 12.11 Making a gene from eukaryotic mRNA. Using the enzyme reverse transcriptase, a researcher can produce an artificial DNA gene (cDNA) from a molecule of messenger RNA (mRNA).

CHECKPOINT

1. What is recombinant DNA technology?
2. Corn that carries a bacterial gene is an example of a __________ organism.
3. Why are plasmids valuable tools for producing recombinant DNA?
4. Name three different ways that a gene of interest can be obtained.

Answers: 1. A set of methods for creating a DNA molecule that carries DNA originating from different sources **2.** transgenic (or genetically modified) **3.** Plasmids can carry virtually any foreign gene, are small enough to be taken up easily by bacteria, and are replicated by their bacterial host cells. **4.** The gene can be isolated from a genomic library created by a shotgun approach, produced from mRNA using reverse transcriptase, or synthesized from scratch.

DNA Fingerprinting and Forensic Science

DNA technology has rapidly transformed the field of forensics, the scientific analysis of evidence for crime scene investigations and other legal proceedings. The most important application of biology to the science of forensics is **DNA fingerprinting,** the analysis of DNA fragments to determine whether they come from a particular individual.

Figure 12.12 presents an overview of a typical investigation using DNA fingerprinting. After a crime has occurred, ❶ DNA samples are collected from different sources; they may be from the present crime scene, from an old crime scene, from a suspect, or from a victim. ❷ The DNA samples are then amplified (copied many times) to produce a large sample of DNA fragments. ❸ The fragments from the different sources can then be compared with each other. The data from DNA fingerprinting show which samples are from the same individual and which samples are unique.

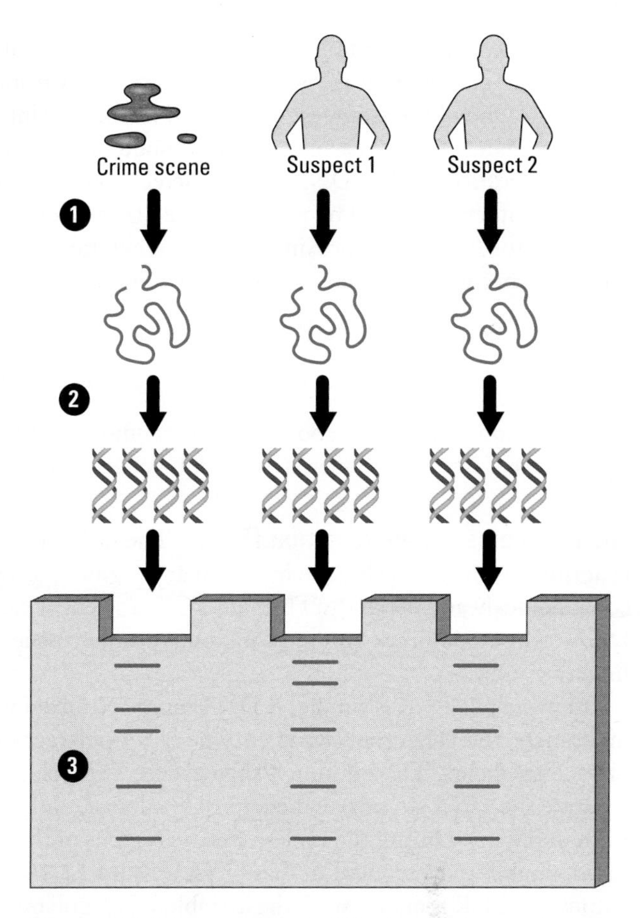

Figure 12.12 Overview of DNA fingerprinting. In this example, DNA from suspect 2 matches DNA found at the crime scene, but DNA from suspect 1 does not match.

Murder, Paternity, and Ancient DNA

Since its introduction in 1986, DNA fingerprinting has become a standard criminology tool and has provided crucial evidence in many famous cases. In the O. J. Simpson murder trial, DNA analysis proved that blood in Simpson's car belonged to the victims and that blood at the crime scene belonged to Simpson. (The jury in this case did not find the DNA evidence alone to be sufficient to convict the suspect, and Simpson was found not guilty.) During the investigation that led up to his impeachment, President Bill Clinton repeatedly denied that he had sexual relations with Monica Lewinsky—until DNA fingerprinting proved that his semen was on her dress. Of course, DNA evidence can prove innocence as well as guilt. The Innocence Project, headquartered in New York City, has used DNA technology and legal work to exonerate over 180 convicted criminals, including several on death row. DNA fingerprinting can also be used to identify crime victims. The largest such effort in history took place after the World Trade Center attack on September 11, 2001. Forensic scientists worked for years to match over 10,000 samples of victims' remains to DNA from items provided by families, such as toothbrushes and cigarette butts. When no sample of a victim's DNA was available, blood samples from close relatives were used to find near matches that confirmed identity.

The use of DNA fingerprinting extends beyond crimes. For instance, comparing the DNA of a mother, her child, and the purported father can conclusively settle a question of paternity. Sometimes paternity is of historical interest: DNA fingerprinting proved that Thomas Jefferson or a close male relative fathered a child with his slave Sally Hemings. DNA analysis can also be used to probe the origin of nonhuman materials. In 1998, the U.S. Fish and Wildlife Service began testing the DNA in caviar to determine if the fish eggs originated from the species claimed on the label. DNA fingerprinting can also help protect endangered species by conclusively proving the origin of contraband animal products.

Modern methods of DNA fingerprinting are so specific and powerful that the starting DNA material can be in a partially degraded state. This allows DNA analysis to be applied in a great number of ways. In evolution research, the technique has been used to study DNA pieces recovered from an ancient mummified human and from a 30-million-year-old plant fossil. A 2005 study determined that DNA extracted from a 27,000-year-old Siberian mammoth was 98.6% identical to DNA from modern African elephants. One of the strangest cases of DNA fingerprinting is that of Cheddar Man, a 9,000-year-old skeleton found in a cave near Cheddar, England **(Figure 12.13)**. DNA was extracted from his tooth and analyzed by DNA fingerprinting. The results showed that Cheddar Man was a direct ancestor (through approximately 300 generations) of a present-day schoolteacher who lived only a half mile from the cave!

Figure 12.13 Cheddar Man. Analysis of DNA extracted from "Cheddar Man"—a 9,000-year-old skeleton found in an English cave—showed that it was a direct ancestor of a local schoolteacher (right).

DNA Fingerprinting Techniques

In this section, you'll learn about current techniques for making a DNA fingerprint. These techniques are used for steps 2 and 3 in Figure 12.12.

The Polymerase Chain Reaction (PCR) The **polymerase chain reaction (PCR)** is a technique by which any segment of DNA can be copied quickly and precisely. Through PCR, a scientist can obtain enough DNA from even minute amounts of blood or other tissue to allow DNA fingerprinting.

In principle, PCR is simple. A DNA sample is mixed with nucleotide monomers, the DNA replication enzyme DNA polymerase, and a few other ingredients. The solution is then exposed to cycles of heating (to separate the DNA strands) and cooling (to allow the DNA strands to reform duplexes). During the cycles, specific regions of the DNA are replicated, doubling the amount of that DNA **(Figure 12.14)**. The key to automated PCR is an unusually heat-stable DNA polymerase, first isolated from prokaryotes living in hot springs. Unlike most proteins, this enzyme can withstand the heat at the start of each cycle. Beginning with a single DNA molecule, automated PCR can generate hundreds of billions of copies in a few hours.

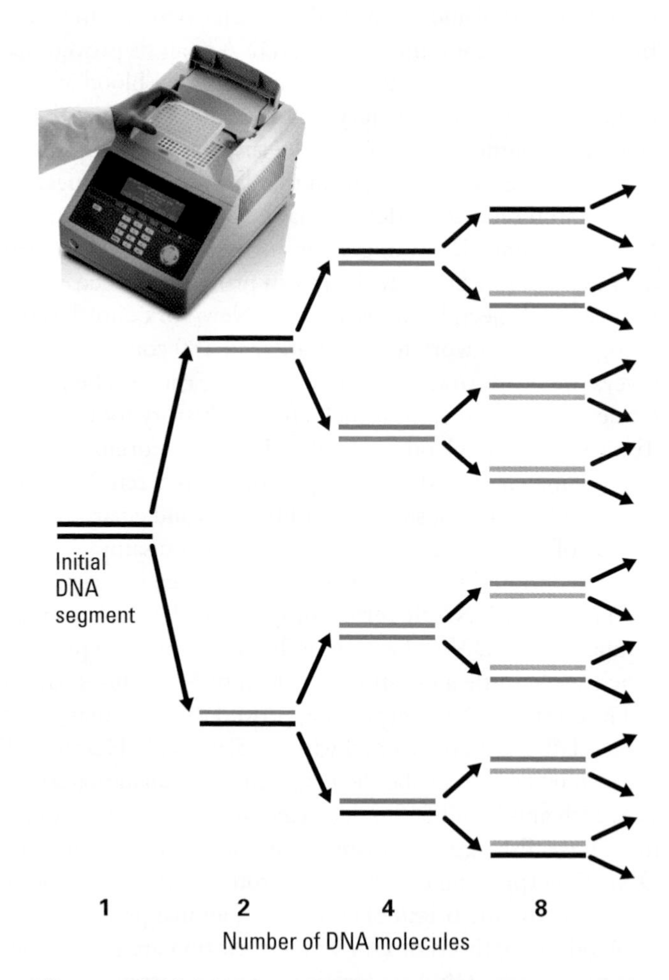

Figure 12.14 DNA amplification by PCR. The polymerase chain reaction (PCR) is a method for making many copies of a specific segment of DNA. Each round of PCR, performed on tabletop thermal cyclers (shown at top), doubles the total quantity of DNA.

Short Tandem Repeat (STR) Analysis How do you prove that two samples of DNA come from the same person? You could compare the entire genomes found in the two samples. But such an approach would be extremely impractical, requiring a lot of time and money. Instead, scientists compare a series of short noncoding DNA segments that vary from person to person. Such a variable region in the genome is a **genetic marker,** a chromosomal landmark whose inheritance can be studied. And just like a gene (which is also a type of genetic marker), a noncoding genetic marker is more likely to be an exact match between relatives than between unrelated individuals.

The genetic markers most often used in DNA fingerprinting are inherited variations in the lengths of repetitive DNA segments. **Repetitive DNA,** which makes up much of the DNA that lies between genes in humans, consists of nucleotide sequences that are present in multiple copies in the genome. Some of this DNA consists of short sequences repeated many times in a row (tandemly); such sequences are called **short tandem repeats (STRs).** For example, one person might have the sequence AGAT repeated 12 times at one place in the genome, the sequence GATA repeated 15 times at a second place, and so on; another person is likely to have the same sequences at the same places but with a different number of repeats.

STR analysis is a method for producing and comparing DNA fingerprints that reflect the lengths of STR sequences at specific sites in the genome. Most commonly, STR analysis compares the number of repeats of 13 specific four-nucleotide DNA sequences scattered throughout the genome. Each repeat site, which typically contains from 3 to 50 four-nucleotide repeats in a row, varies widely from person to person. In fact, some of the short tandem repeats used in the standard procedure can be found in up to 80 different forms in the human population.

Consider the two samples of DNA shown in **Figure 12.15**. Imagine that the top DNA segment was obtained at a crime scene and the bottom from a suspect's blood. The two segments have the same number of repeats at the first site: 7 repeats of the four-nucleotide DNA sequence AGAT (shown in orange). Notice, however, that they differ in the number of repeats at the second site: 8 repeats of GATA (shown in purple) in the crime scene DNA, compared to 13 repeats in the suspect's DNA. To create a DNA fingerprint, a scientist uses PCR to specifically amplify the regions of DNA that include these STR sites. The resulting fragments are then compared. The details of how this comparison is made is our next topic.

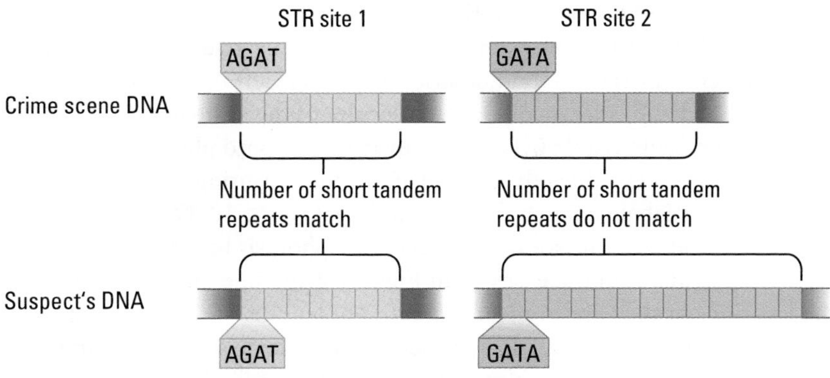

Figure 12.15 Short tandem repeat (STR) sites. Scattered throughout the genome, STR sites contain tandem repeats of four-nucleotide sequences. The number of repetitions at each site can vary from individual to individual. In this figure, both DNA samples have the same number of repeats (7) at the first STR site, but different numbers (8 versus 13) at the second.

Gel Electrophoresis Once a set of DNA fragments is prepared, the next step in STR analysis is to determine the lengths of these fragments. An essential tool of DNA technology, **gel electrophoresis** is a method for sorting macromolecules—usually proteins or nucleic acids—primarily on the basis of their electrical charge and length. **Figure 12.16** shows how gel electrophoresis can be used to separate the DNA fragments obtained from three different sources. A sample of each mixture is placed in a well (hole) at one end of a flat, rectangular gel, a thin slab of jellylike material that acts as a molecular sieve. A negatively charged electrode is then attached to the DNA-containing end of the gel and a positive electrode to the other end. Because phosphate (PO_4^-) groups give DNA fragments a negative charge, the fragments move through the gel toward the positive pole. However, the longer DNA fragments are held back by a

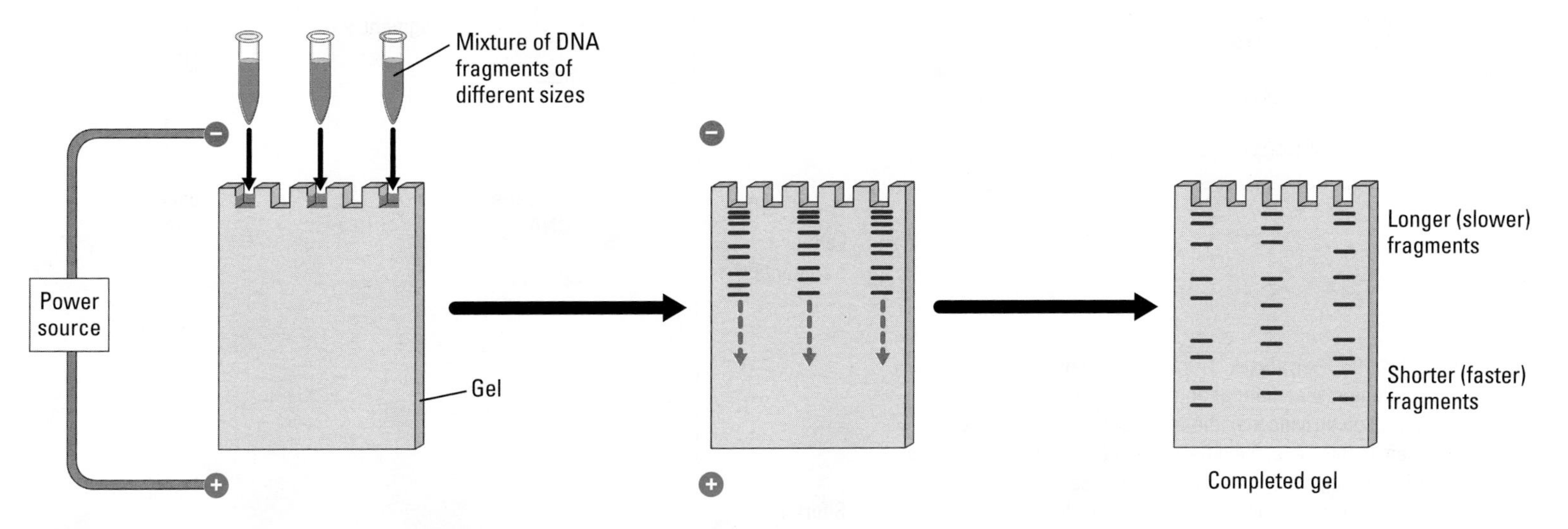

Figure 12.16 Gel electrophoresis of DNA molecules.

thicket of polymer fibers within the gel, so they move more slowly than the shorter DNA fragments—and thus not as far in a given time period. When the current is turned off, a series of bands is left in each "lane" of the gel. Each band consists of DNA molecules of the same length. The bands can be made visible by staining, by exposure onto photographic film (if the DNA is radioactively labeled), or by measuring fluorescence (if the DNA is labeled with a fluorescent dye). **Figure 12.17** shows the bands that would result from using gel electrophoresis to separate the DNA fragments from the example in Figure 12.15. The differences in the locations of the bands reflect the different lengths of the DNA fragments. This gel would provide evidence that the crime scene DNA did not come from the suspect. Notice that electrophoresis allows us to see similarities as well as differences between mixtures of restriction fragments, reflecting similarities as well as differences between the nucleotide sequences from two DNA samples.

Gel electrophoresis has many uses besides STR analysis. One common application is RFLP analysis. RFLP (pronounced "rif-lip") stands for restriction fragment length polymorphism. In this method, the molecules of DNA to be compared are exposed to a series of restriction enzymes **(Figure 12.18)**. The resulting restriction fragments are separated and made visible on a gel. In this case, both the number and location of bands indicate whether the original samples of DNA had identical DNA sequences at the restriction cut sites.

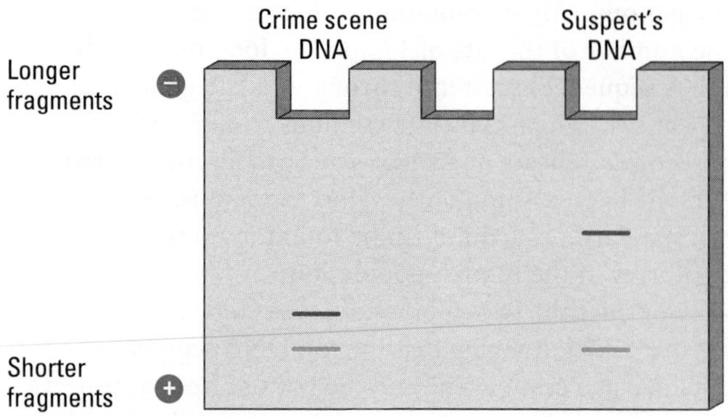

Crime scene DNA
Suspect's DNA
Longer fragments
Shorter fragments

Figure 12.17 Visualizing STR fragment patterns. This figure shows the bands that would result from gel electrophoresis of the STR sites created in Figure 12.15. Notice that the crime scene DNA has fragments of length 28 and 32, while the suspect DNA has fragments of length 28 and 52.

CHECKPOINT

1. Why is only the slightest trace of DNA at a crime scene often sufficient for forensic analysis?
2. What are STRs, and why are they useful for DNA fingerprinting?
3. You use a restriction enzyme to cut a DNA molecule. The base sequence of this DNA is known, and the molecule has a total of three restriction-enzyme cutting sites clustered close together near one end. When you separate the restriction fragments by electrophoresis, how do you expect the bands to be distributed in the electrophoresis lane?

Answers: 1. Because PCR can be used to produce enough molecules for analysis **2.** STRs (short tandem repeats) are nucleotide sequences repeated many times in a row within the human genome. STRs are valuable for DNA fingerprint identification because different people have different numbers of repeats at the various STR sites. **3.** Three bands near the positive pole at the bottom of the gel (small fragments) and one band near the negative pole at the top of the gel (large fragment)

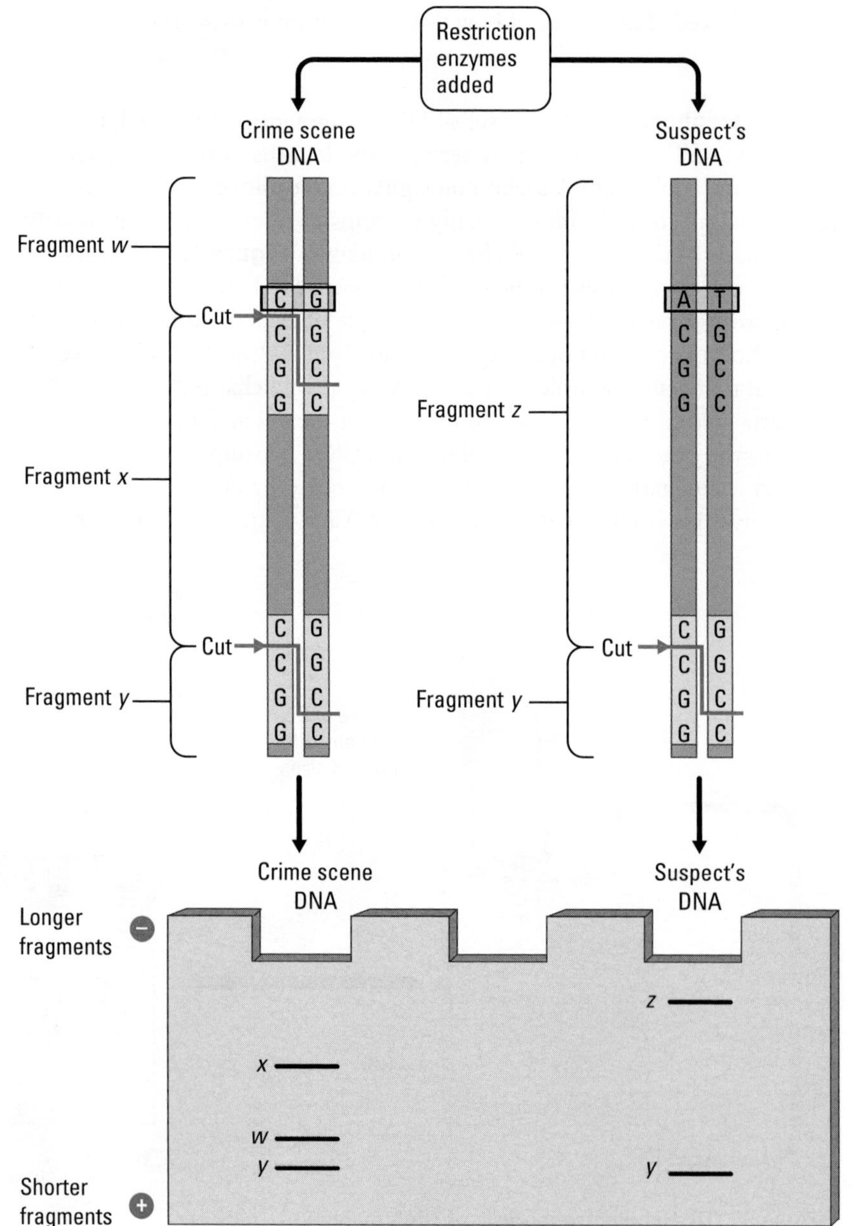

Figure 12.18 RFLP analysis. The DNA segments shown have nucleotide sequences that differ at one base pair (yellow boxes). A particular restriction enzyme may therefore cut the segments at different places. In this case, the difference in DNA sequence results in three restriction fragments from the first DNA sample and two from the second. This difference, revealed by gel electrophoresis, indicates that the two DNA samples come from different individuals.

Genomics and Proteomics

By the 1980s, the nucleotide sequences of many important genes from humans and other organisms had been determined. It didn't take long for biologists to think on a larger scale. In 1995, a team of scientists announced that they had determined the nucleotide sequence of the entire genome of *Haemophilus influenzae*, a bacterium that can cause several human diseases, including pneumonia and meningitis. **Genomics,** the science of studying whole genomes, was born.

The first targets of genomics research were bacteria, which have relatively little DNA. The genome of *H. influenzae*, for example, contains 1.8 million nucleotides and 1,709 genes. But soon the attention of genomics researchers turned toward more complex organisms with much larger genomes **(Table 12.1)**. The majority of organisms sequenced to date are prokaryotes, including *E. coli* and a few hundred other bacteria (some of medical importance) and a few dozen archaea. Several dozen eukaryotic genomes have been completed. Baker's yeast (*Saccharomyces cerevisiae*) was the first eukaryote to have its full sequence determined, and the roundworm *Caenorhabditis elegans* was the first multicellular organism. Other sequenced animals include the fruit fly (*Drosophila melanogaster*) and the lab mouse, both model organisms for genetics. Plants, such as *Arabidopsis thaliana* (another important research organism) and rice (one of the world's most economically important crops), have also been completed. Recently, the genomes of the mosquito *Anopheles gambiae* and the parasite *Plasmodium falciparum* were sequenced. These organisms are important to study because together they transmit malaria, a disease that kills over 1 million children worldwide each year. Another recently sequenced pathogen is *Trypanosoma brucei*, a protozoan parasite that causes African sleeping sickness. But of all the genomes sequenced by researchers so far, none is larger or more complex than the human genome.

Table 12.1 Some Important Sequenced Genomes

Organism	Date Completed	Size of Genome (in base pairs)	Approximate Number of Genes
Haemophilus influenzae (bacterium)	1995	1.8 million	1,700
Saccharomyces cerevisiae (yeast)	1996	12 million	5,800
Escherichia coli (bacterium)	1997	4.6 million	4,400
Caenorhabditis elegans (roundworm)	1998	97 million	19,100
Drosophila melanogaster (fruit fly)	2000	180 million	13,700
Arabidopsis thaliana (mustard plant)	2000	120 million	25,500
Homo sapiens (human)	2001	2.9 billion	25,000
Oryza sativa (rice)	2002	430 million	40,000
Rattus norvegicus (lab rat)	2004	2.8 billion	25,000
Trypanosoma brucei (pathogen)	2005	26 million	9,068

The Human Genome Project

The **Human Genome Project (HGP)** was a massive scientific endeavor with the goals of determining the nucleotide sequence of all DNA in the human genome and identifying the location and sequence of every gene. The HGP began in 1990 as an effort by an international consortium of government-funded researchers. Several years into the project, private companies, chiefly Celera Genomics in the United States, joined the effort. The original sequencing deadline was 2005, but fierce competition between the public consortium and Celera spurred both groups on. On April 14, 2003, both groups jointly announced the successful completion of the first draft of the sequence—more than two years ahead of schedule (**Figure 12.19**). At the completion of the project in 2004, it was announced that over 99% of the genome had been determined to 99.999% accuracy. (As of 2006, there remain a few hundred gaps of unknown sequence within the human genome that will require special methods to figure out.) This ambitious project has provided a wealth of data that may illuminate the genetic basis of what it means to be human.

The 24 chromosomes in the human genome (22 autosomes plus the X and Y sex chromosomes) contain approximately 2.9 billion nucleotide pairs of DNA and about 25,000 genes. To try to get a sense of this quantity of DNA, imagine that its nucleotide sequence is printed in letters (A, T, C, and G) like the letters in this book. At this size, the sequence would fill a stack of books 18 stories high!

Our genome presents a major challenge not only because of its size, but also because, like that of most complex eukaryotes, only a small amount of our total DNA is contained in genes that code for proteins, tRNAs, or rRNAs. Most complex eukaryotes have a huge amount of noncoding DNA—about 97% of human DNA is of this type. Some noncoding DNA is made up of gene control sequences such as promoters and enhancers (see Chapter 11). The remaining noncoding DNA has been dubbed "junk DNA," a tongue-in-cheek way of saying that scientists don't understand its functions. Junk DNA includes introns (whose total length in a gene may be ten times greater than the total length of the exons), repetitive DNA, and other noncoding DNA located between genes and at the ends of chromosomes. (Some of this noncoding DNA is used to create DNA fingerprints, as discussed in the section on STR analysis.)

The potential benefits of having a complete map of the human genome are enormous. For instance, hundreds of disease-associated genes have already been identified. One example is the gene that is mutated in an inherited type of Parkinson's disease, a debilitating brain disorder that causes tremors of increasing severity. Half a million Americans suffer from this disorder (**Figure 12.20**). Until recently, Parkinson's disease was thought to have only an environmental basis; there was no evidence of a hereditary component. But data from the Human Genome Project mapped a very small number of cases of Parkinson's disease to a specific gene. Interestingly, an altered version of the protein encoded by this gene has also been tied to Alzheimer's disease, suggesting a link between these two brain disorders. Moreover, the same gene is also found in rats, where it plays a role in the sense of smell, and in zebra finches, where it is thought to be involved in song learning. Cross-species comparisons such as these may uncover clues about the role played by the normal version of the protein in the brain. And such knowledge could eventually lead to treatment for the half a million Americans with Parkinson's disease.

Figure 12.19 Completion of the first draft of the human genome. Yoshiyuki Sakaki (left), director of the Japanese branch of the Human Genome Project, presents Japan's Prime Minister Junichiro Koizumi with a set of CDs that contain data from the completed first draft.

Figure 12.20 The fight against Parkinson's disease. Actor Michael J. Fox and boxer Muhammad Ali—both of whom have Parkinson's disease—testify before the Senate on the status of federal funding for Parkinson's research.

Tracking the Anthrax Killer

In October 2001, a 63-year-old Florida man died from inhalation anthrax, a disease caused by breathing spores of the bacterium *Bacillus anthracis*. As the first victim of this disease in the United States since 1976, his death was immediately suspicious. By the end of the year, four more people had died from anthrax. Law enforcement officials realized that someone was sending anthrax spores through the mail. The United States found itself in the grip of an unprecedented bioterrorist attack.

In the investigation that followed, one of the most helpful clues turned out to be the anthrax spores themselves. Investigators compared the genomes of the mailed anthrax spores—which were 3 million nucleotides long—with several laboratory strains. They quickly established that all of the mailed spores were genetically identical, suggesting that a single perpetrator was behind all the attacks. Furthermore, they were able to match the deadly spores with a laboratory subtype called the Ames strain. The Ames strain was first isolated from a dead Texas cow in 1981. From there, the strain was sent to the U.S. Army Medical Research Institute in Fort Detrick, Maryland. And from there, it was sent to at least 14 other labs for use in experiments. Unfortunately, the data from the comparisons were not detailed enough to tie the mailed anthrax spores to any particular laboratory.

The anthrax investigation is a prominent example of the new field of comparative genomics, the comparison of whole genomes. Other investigations have also taken advantage of this approach. In 1991, sequence data provided strong evidence that a Florida dentist transmitted HIV to several patients. In 1993, after the Aum Shinrikyo cult released anthrax spores in downtown Tokyo, genomic analysis showed why their attack didn't kill anyone: They had used a harmless veterinary vaccine strain. And investigation of the West Nile virus outbreak in 1999 proved that a single natural strain of virus was infecting both birds and humans.

In addition to aiding in the study of disease-causing microorganisms, comparative genomics can reveal similarities and differences among organisms more closely related to humans. In 2005, researchers completed a rough draft of the chimpanzee genome. Comparisons with human DNA revealed that we share 96% of our genome with our closest animal relatives. Genomic scientists are currently finding and studying the important differences, shedding scientific light on the age-old question of what makes us human.

Genome-Mapping Techniques

The early work on the Human Genome Project included obtaining preliminary maps with "road markers" that helped researchers find their way through a huge amount of data. These early stages relied on genetic markers that had been previously identified through pedigree analysis (see Chapter 9). The preliminary maps helped the researchers determine the locations of nucleotide sequences.

Today, genomic researchers largely rely on the whole-genome shotgun method, a technique pioneered by the researchers at Celera. In this method, an entire genome is chopped by restriction enzymes into fragments that are cloned and sequenced in just one stage. High-performance computers running specialized mapping software can reassemble the millions of partial sequences into an entire genome **(Figure 12.21)**.

The DNA sequences determined by the Human Genome Project have been deposited in a database that is available via the Internet. (You can

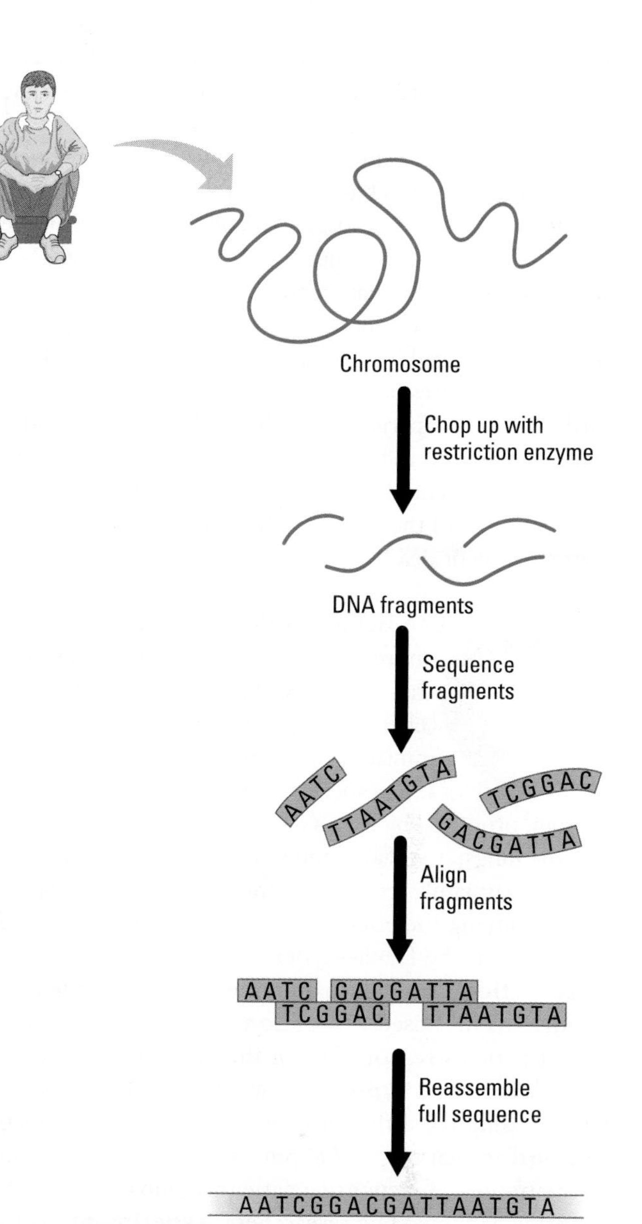

(a) The whole-genome shotgun method

(b) A genome-sequencing center

Figure 12.21 Genome mapping.

browse it yourself at the website for the National Center for Biotechnology Information.) Scientists use software to scan and analyze the sequences for genes, control elements, and other features. Now comes the most exciting challenge: figuring out the functions of the genes and other sequences and how they work together to direct the structure and function of a living organism. This challenge and the applications of the new knowledge should keep scientists busy well into the twenty-first century.

One interesting question to ask about the Human Genome Project is whose genome was sequenced. The answer is no one's—or at least not any one person's. The human genome sequenced by the public consortium was actually a reference genome compiled from a group of individuals. The genome sequenced by Celera consisted primarily of DNA sampled from the company's president. These representative sequences will serve as standards so that comparisons of individual differences and similarities can be made. Eventually, as the amount of sequence data multiplies, the small differences that account for individual variation within our species will come to light. Some of these seemingly minuscule differences can actually be a matter of life or death, as we'll see next.

Can Genomics Cure Cancer? Lung cancer, which kills more Americans every year than any other type of cancer, has long been the target of searches for effective chemotherapy drugs. In 2003, the Food and Drug Administration approved the drug gefitinib for the treatment of lung cancer. Gefitinib targets a protein called EGFR. This protein, produced by the *EGFR* gene, is found on the outside of cells that line the lungs; it is also found in lung cancer tumors.

Unfortunately, gefitinib is ineffective for many patients. However, while studying the effectiveness of gefitinib, Matthew Meyerson and his colleagues at the Dana-Farber Cancer Institute in Boston made the **observation** that a few patients actually responded quite dramatically to the drug. This posed a **question**: Are genetic differences among lung cancer patients responsible for the differences in gefitinib's effectiveness? Meyerson's **hypothesis** was that mutations in the *EGFR* gene were causing the different responses to gefitinib. His team made the **prediction** that the *EGFR* gene would have a different DNA sequence in the tumors of responsive patients compared to the tumors of unresponsive patients. The researchers' **experiment** involved sequencing the *EGFR* gene in cells extracted from the tumors of five patients who responded to the drug and four who did not.

The **results**, published in 2004, were quite striking: All five tumors from gefitinib-responsive patients harbored mutations in *EGFR*, while none of the other four tumors did (**Figure 12.22**). These results suggest that doctors can screen lung cancer patients for those who are most likely to benefit from treatment with this drug. In broader terms, this work suggests that genomics may bring about a revolution in the treatment of disease by allowing therapies to be custom-tailored to the genetic makeup of each patient. ■

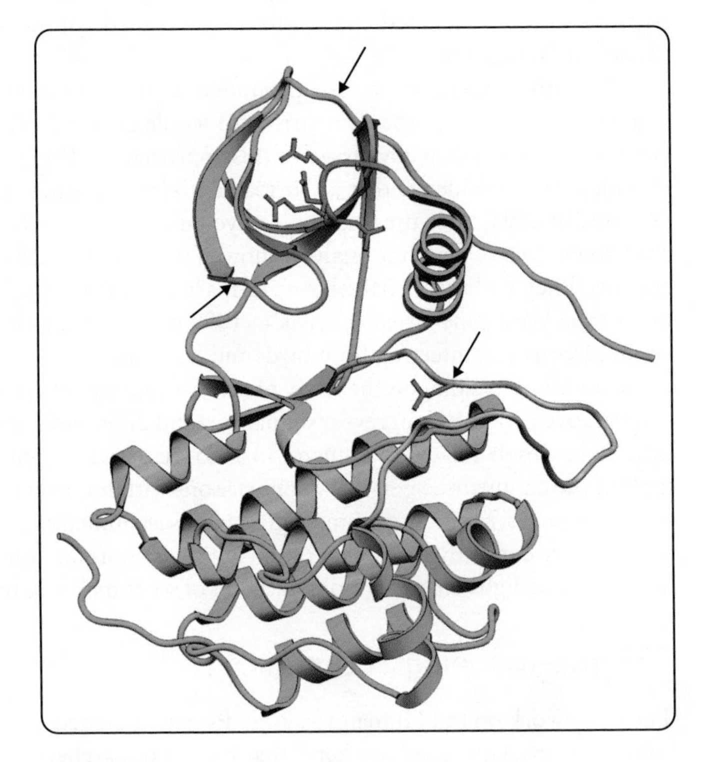

Figure 12.22 The *EGFR* protein: Fighting cancer with genomics. A 2004 study by Matthew Meyerson and colleagues showed that mutations (located at sites indicated by arrows) in a protein encoded by the *EGFR* gene can affect the ability of a cancer-fighting drug to destroy lung tumors.

Proteomics

The success in studying whole genomes is encouraging scientists to attempt similar systematic studies of the full protein sets (proteomes) that genomes encode, an approach called **proteomics.** The number of proteins in humans far exceeds the number of genes. And since proteins, not genes, actually carry out the activities of the cell, scientists must study when and where proteins are produced and how they interact in order to understand the functioning of cells and organisms.

Genomics and proteomics are enabling biologists to approach the study of life from an increasingly holistic perspective. Biologists are now in a position to compile catalogs of genes and proteins—that is, a listing of all the "parts" that contribute to the operation of cells, tissues, and organisms. With such catalogs in hand, researchers are shifting their attention from the individual parts to how these parts function as a whole in biological systems.

CHECKPOINT

1. Approximately how many nucleotides and genes are contained in the human genome?
2. Name three types of DNA that do not code for another molecule.
3. What is the difference between genomics and proteomics?

Answers: 1. About 3 billion nucleotides and 25,000 genes **2.** Introns, STR and other repeated DNA, and control elements (such as promoters and enhancers) **3.** Genomics studies the complete set of an organism's genes, while proteomics studies the complete set of proteins.

Human Gene Therapy

Human gene therapy is a recombinant DNA procedure intended to treat disease by altering an afflicted person's genes. In some cases, a mutant version of a gene may be replaced or supplemented with the normal allele. This could potentially correct a genetic disorder, perhaps permanently. In other cases, genes are inserted and expressed only long enough to treat a medical problem.

Figure 12.23 summarizes one approach to human gene therapy. ❶ A gene from a normal individual is isolated and cloned by recombinant DNA techniques. ❷ The gene is inserted into a vector, such as a nonharmful virus. ❸ The virus is then injected into the patient. The virus inserts a copy of its genome, including the human gene, into the DNA of the patient's cells. The normal gene is then transcribed and translated within the patient's body, producing the desired protein. Ideally, the nonmutant version of the gene would be inserted into cells that multiply throughout a person's life. Bone marrow cells, which include the stem cells that give rise to all the types of blood cells, are prime candidates. If the procedure succeeds, the cells will multiply throughout the patient's life and express the normal gene. The engineered cells will supply the missing protein, and the patient will be cured.

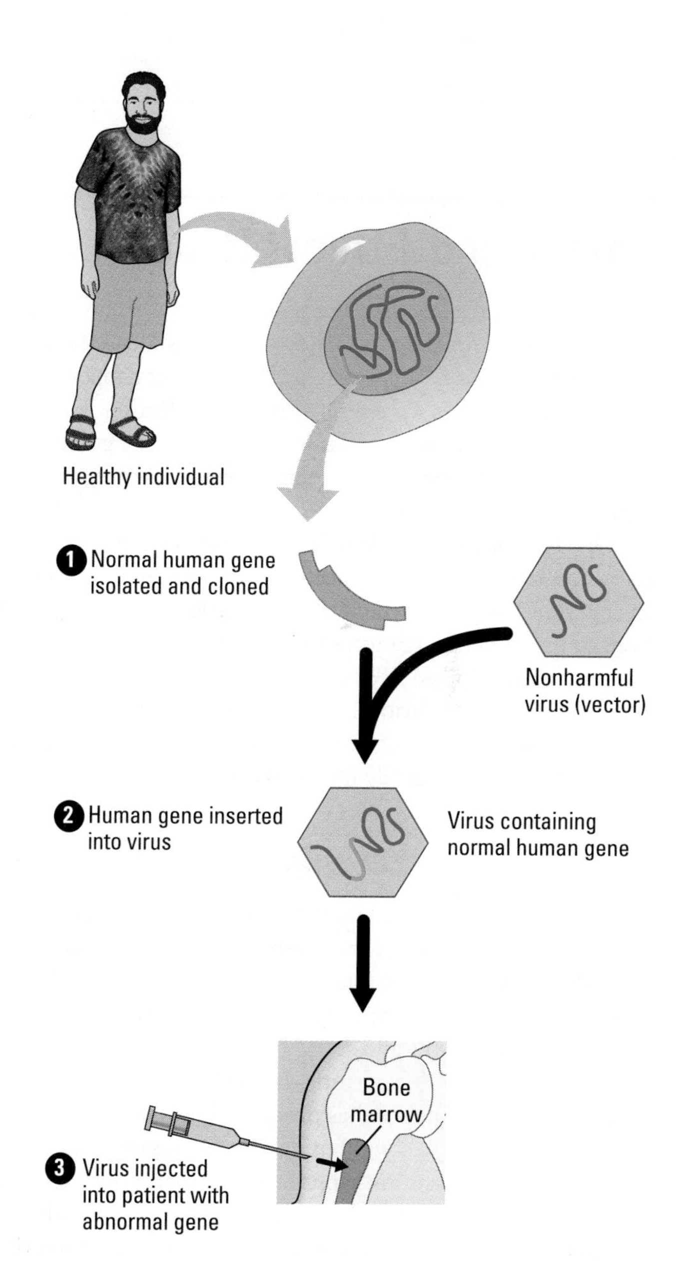

Healthy individual
❶ Normal human gene isolated and cloned
Nonharmful virus (vector)
❷ Human gene inserted into virus
Virus containing normal human gene
Bone marrow
❸ Virus injected into patient with abnormal gene

Figure 12.23 One approach to human gene therapy.

Treating Severe Combined Immunodeficiency

Severe combined immunodeficiency (SCID) is a fatal inherited disease caused by a single defective gene. Absence of the enzyme encoded by this gene prevents the development of the immune system, requiring patients to remain isolated within protective "bubbles." Unless treated with a bone marrow transplant (effective just 60% of the time), SCID patients quickly die from infections by ever-present microbes that most of us easily fend off.

Since the year 2000, gene therapy has successfully cured 22 children with inborn SCID, providing the first scientifically strong evidence of the effectiveness of gene therapy. As part of this treatment, researchers periodically removed immune system cells from the patients' blood, infected them with a virus engineered to carry the normal allele of the defective gene, then reinjected the blood back into the patient. The celebrations of this medical breakthrough were short-lived, however; three of the children who received gene therapy developed leukemia, and one of them died. Apparently, the retrovirus used as a vector activated an oncogene (see Chapter 8), creating cancerous blood cells. Thus, although the concept of gene therapy remains promising, very little evidence of safe and effective gene therapy has yet appeared. Active research into human gene therapy, with new, tougher safety guidelines, continues.

CHECKPOINT

1. Why are bone marrow stem cells ideally suited as targets for gene therapy?
2. How can viruses be used in gene therapy to treat human disorders?

Answers: 1. Because bone marrow stem cells multiply throughout a person's life **2.** Viruses can be used as vectors to introduce normal genes into human cells.

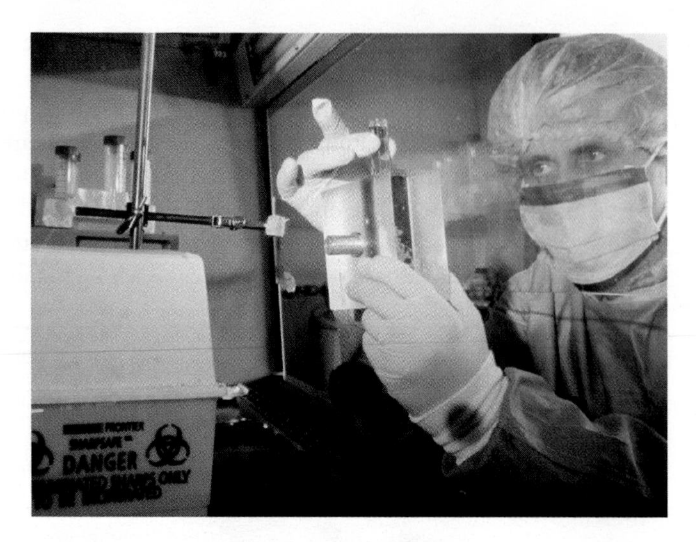

Figure 12.24 Maximum-security laboratory. A scientist in a high-containment laboratory uses a "hood" for working with dangerous microorganisms. Air flow through the hood prevents the microbes from escaping into the environment.

Safety and Ethical Issues

As soon as scientists realized the power of DNA technology, they began to worry about potential dangers. Early concerns focused on the possibility of creating hazardous new pathogens. What might happen, for instance, if cancer-causing genes were transferred into infectious bacteria or viruses? To address such concerns, scientists developed a set of guidelines that in the United States and some other countries have become formal government regulations.

One type of safety measure is a set of strict laboratory procedures designed to protect researchers from infection by engineered microbes and to prevent the microbes from accidentally leaving the laboratory **(Figure 12.24)**. In addition, strains of microorganisms to be used in recombinant DNA experiments are genetically crippled to ensure that they cannot survive outside the laboratory. Finally, certain obviously dangerous experiments have been banned. Today, most public concern about possible hazards centers not on recombinant microbes but on genetically modified (GM) foods.

The Controversy over Genetically Modified Foods

GM strains account for a significant percentage of several agricultural crops. Controversy about the safety of these foods is an important political issue throughout Europe and other parts of the world **(Figure 12.25)**. In response to these concerns, the European Union suspended the introduction of new GM crops and started considering the possibility of banning the import of all GM foodstuffs. In the United States and other countries where the GM revolution has proceeded more quietly, the labeling of GM foods is now being debated but has not yet become law.

Advocates of a cautious approach fear that crops carrying genes from other species might harm the environment or be hazardous to human health (by, for example, transferring allergens, molecules that can cause allergic reactions). A major concern is that transgenic plants might pass their

Figure 12.25 Opposition to genetically modified organisms (GMOs). In Europe, there is strong opposition to GM foods. This photo shows tractors destroying a field of genetically modified rape plants in Belgium. The rape plant is a member of the mustard family and is grown as a forage crop for sheep and hogs. The seeds are used to make cooking oil and as bird feed. The destruction was ordered by the Belgian Minister for Consumer Protection. The sign reads "GMO, no thanks! Yes to biodiversity."

new genes to close relatives in nearby wild areas. We know that lawn and crop grasses, for example, commonly exchange genes with wild relatives via pollen transfer. If domestic plants carrying genes for resistance to herbicides, diseases, or insect pests pollinated wild plants, the offspring might become "superweeds" that would be very difficult to control. However, researchers may be able to prevent the escape of such plant genes by engineering plants so that they cannot hybridize. Concern has also been raised that the widespread use of GM seeds may reduce natural genetic diversity, leaving crops susceptible to catastrophic die-offs in the event of a sudden change to the environment or introduction of a new pest.

The U.S. National Academy of Sciences released a study finding no scientific evidence that transgenic crops pose any special health or environmental risks. But the authors of the study also recommended more stringent long-term monitoring to watch for unanticipated environmental impacts.

Negotiators from 130 countries (including the United States) have agreed on a Biosafety Protocol that requires exporters to identify GM organisms present in bulk food shipments and allows importing countries to decide whether the shipments pose environmental or health risks. This agreement has been hailed as a breakthrough by environmentalists.

Today, governments and regulatory agencies throughout the world are grappling with how to facilitate the use of biotechnology in agriculture, industry, and medicine while ensuring that new products and procedures are safe. In the United States, all projects are evaluated for potential risks by a number of regulatory agencies, including the Food and Drug Administration, the Environmental Protection Agency, the National Institutes of Health, and the Department of Agriculture.

Ethical Questions Raised by DNA Technology

DNA technology raises moral, legal, and ethical questions—few of which have clear answers. Consider, for example, the child in **Figure 12.26**. She is growing at a normal rate, thanks to regular injections of human growth hormone (HGH) produced by genetically engineered cells. Should this type of therapy be reserved for treating only serious conditions? Should parents of short but hormonally normal children be able to seek HGH treatment to make their kids taller? If not, who decides which children are "tall enough" to be excluded from treatment?

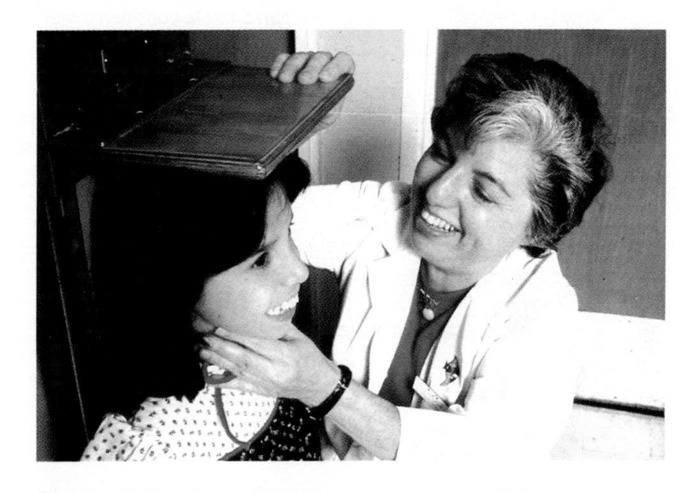

Figure 12.26 Treatment with human growth hormone. This child has been treated with human growth hormone made by bacteria.

Genetic engineering of gametes (sperm or ova) and zygotes has been accomplished in lab animals. It has not been attempted in humans because such a procedure would raise very difficult ethical questions. Should we try to eliminate genetic defects in our children and their descendants? Should we interfere with evolution in this way? From a long-term perspective, the elimination of unwanted versions of genes from the gene pool could backfire. Genetic variety is a necessary ingredient for the adaptation of a species as environmental conditions change with time. Genes that are damaging under some conditions may be advantageous under others (one example is the sickle-cell allele, discussed in Chapter 9). Are we willing to risk making genetic changes that could be detrimental to our species in the future? We may have to face such questions soon.

Advances in genetic fingerprinting raise privacy issues. If we were to create a DNA fingerprint of every person at birth, then we could theoretically match every rape to a perpetrator, since it is nearly impossible for someone to commit such a violent crime without leaving behind DNA evidence. But are we, as a society, prepared to sacrifice our genetic privacy, even for such a worthwhile goal?

And what of the information from the Human Genome Project? There is a danger that information about disease-associated genes could be abused. One issue is the possibility of discrimination and stigmatization. Would you, as an employer, want to know if a potential employee had an increased risk of schizophrenia? Should you be able to find out? Should insurance companies have the right to screen applicants for disease genes? People might be coerced into taking a DNA test in order to be considered for a job or an insurance policy. How do we prevent genetic information from being used in a discriminatory manner?

A much broader ethical question is how do we really feel about wielding one of nature's singular powers—the ability to make new microorganisms, plants, and animals? Some might ask if we have any right to alter an organism's genes—or to add our new creations to an already beleaguered environment.

Such questions must be weighed against the apparent benefits to humans and the environment that can be brought about by DNA technology. For example, bacteria are being engineered to clean up mining wastes and other pollutants that threaten our soil, water, and air. These organisms may be the only feasible solutions to some of our most pressing environmental problems.

DNA technologies raise many complex issues that have no easy answers. It is up to you, as a "citizen scientist," to educate yourself about these issues so that you can make informed choices.

CHECKPOINT

1. What is the main concern about adding genes for herbicide resistance to crop plants?
2. Why is genetically modifying a human gamete considered morally different from genetically modifying a human somatic (body) cell?

Answers: 1. The possibility that the genes could escape, via cross-pollination, to weeds that are closely related to the crop species **2.** A genetically modified somatic cell will affect only the patient. Modifying a gamete will affect an unborn individual as well as all of his or her descendants.

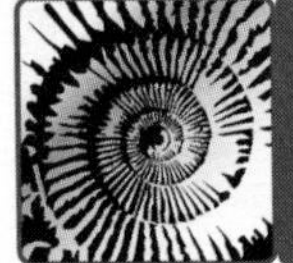

EVOLUTION CONNECTION

Genomes Hold Clues to Evolution

Genome data obtained to date confirm the evolutionary connections between even distantly related organisms and the relevance of research on simpler organisms to understanding human biology. Yeast, for example, have a number of genes close enough to the human versions that they can substitute for them in a human cell. In fact, researchers can sometimes work out what a human disease gene does by studying its counterpart in yeast. Many genes of disparate organisms are turning out to be astonishingly similar, to the point that one researcher has joked that he now views fruit flies as "little people with wings." On a grander scale, comparisons of the completed genome sequences of bacteria, archaea, and eukaryotes strongly support the theory that these are the three fundamental domains of life—a topic we discuss further in the next unit, "Evolution and Diversity." ■

Chapter Review

SUMMARY OF KEY CONCEPTS

For study help and activities, go to campbellbiology.com or the student CD-ROM.

Biology and Society: Crime Scene Investigations: Murders in a Small Town

- Forensics, the scientific analysis of legal evidence, has been revolutionized by DNA technology.

Recombinant DNA Technology

- Recombinant DNA technology is a set of laboratory procedures for combining DNA from different sources—even different species—into a single DNA molecule.

Activity *Applications of DNA Technology*

- **From Humulin to Genetically Modified Foods** Recombinant DNA techniques have been used to create nonhuman cells that produce human proteins, genetically modified food crops, and transgenic farm animals.

Activity *DNA Technology and Golden Rice*

- **Recombinant DNA Techniques**

Case Study in the Process of Science *How Are Plasmids Introduced into Bacterial Cells?*

Activity *Cloning a Gene in Bacteria*

Activity *Restriction Enzymes*

DNA Fingerprinting and Forensic Science

- DNA fingerprinting is used to determine whether two DNA samples come from the same individual.

Activity *DNA Fingerprinting*

- **Murder, Paternity, and Ancient DNA** DNA fingerprinting can be used to establish innocence or guilt of a criminal suspect, identify victims, determine paternity, and contribute to basic research.

- **DNA Fingerprinting Techniques** Short tandem repeat (STR) analysis compares DNA fragments using PCR and gel electrophoresis.

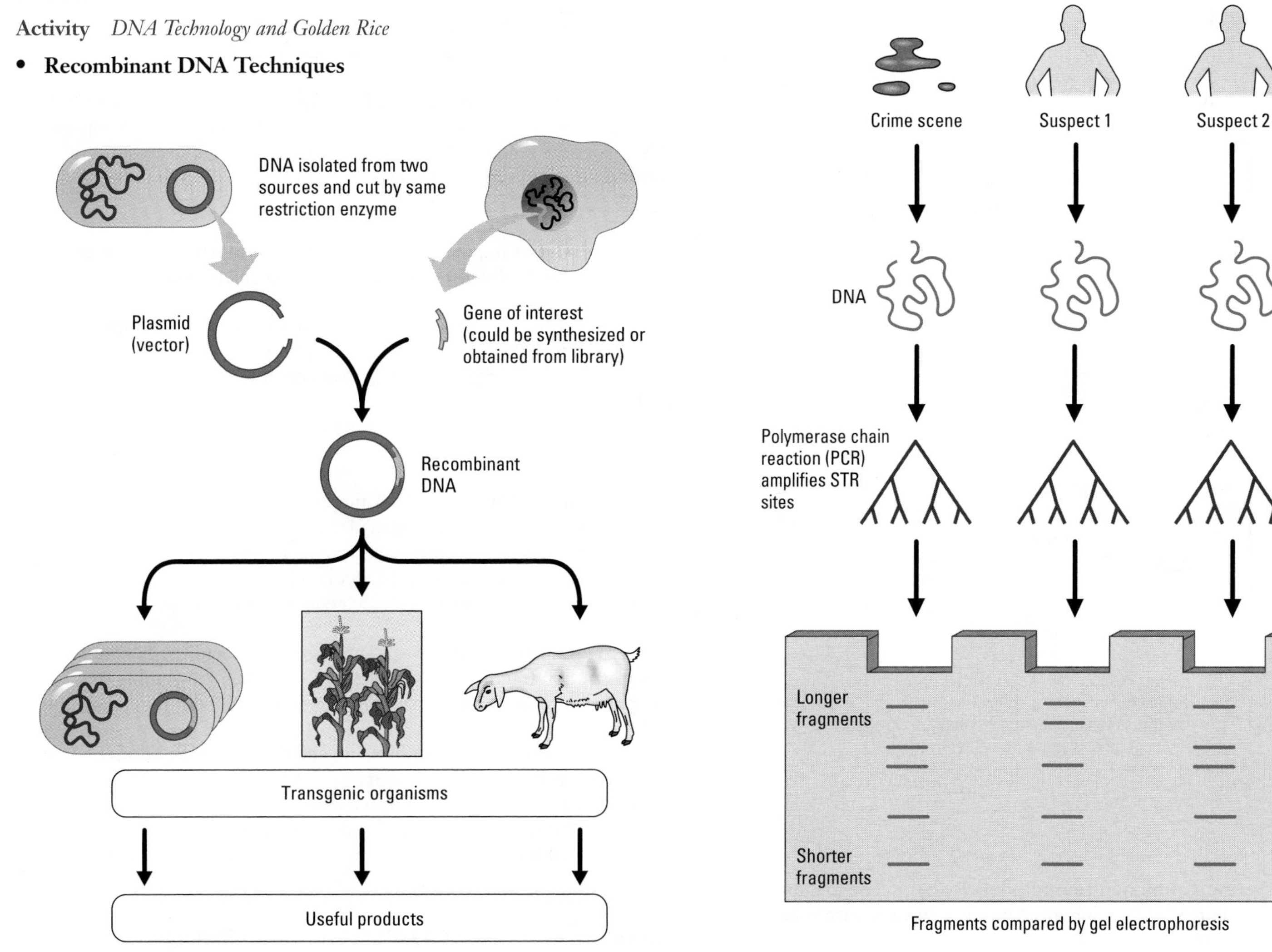

Fragments compared by gel electrophoresis

Activity *Gel Electrophoresis of DNA*

Activity *Analyzing DNA Fragments Using Gel Electrophoresis*

Case Study in the Process of Science *How Can Gel Electrophoresis Be Used to Analyze DNA?*

Genomics and Proteomics

- **The Human Genome Project** Started in 1990 and largely completed in 2003, the nucleotide sequence of the human genome is providing a wealth of useful data. The 24 different chromosomes of the human genome contain about 2.9 billion nucleotide pairs and 25,000 genes. The majority of the genome consists of noncoding DNA.

Activity *The Human Genome Project: Human Chromosome 17*

- **Tracking the Anthrax Killer** Comparing genomes can aid criminal investigations and basic research.

- **Genome-Mapping Techniques** The Human Genome Project proceeded through several stages during which preliminary maps were created and refined. The whole-genome shotgun method involves sequencing DNA fragments from an entire genome and reassembling them in a single stage.

- **Proteomics** Success in genomics has given rise to proteomics, the systematic study of the full set of proteins found in organisms.

Human Gene Therapy

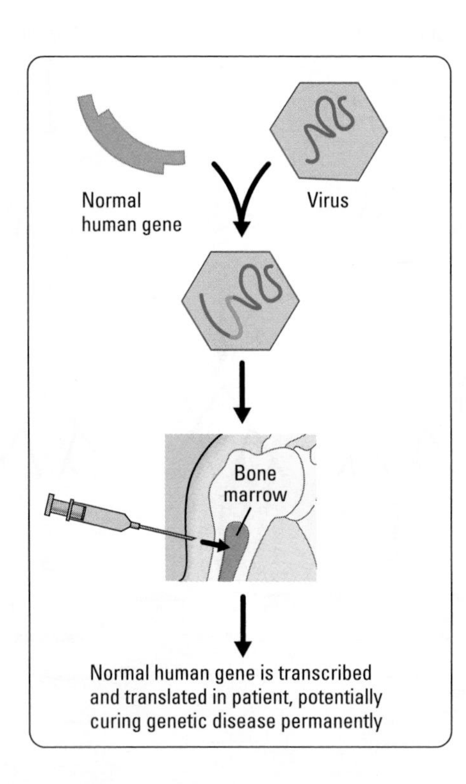

Normal human gene
Virus
Bone marrow
Normal human gene is transcribed and translated in patient, potentially curing genetic disease permanently

- **Treating Severe Combined Immunodeficiency** Gene therapy trials have focused on SCID, an inherited immune disease, with some success and some setbacks.

Safety and Ethical Issues

- **The Controversy over Genetically Modified Foods** The debate about genetically modified crops centers on whether they might harm humans or damage the environment by transferring genes through cross-pollination with other species.

- **Ethical Questions Raised by DNA Technology** We as a society and as individuals must become educated about DNA technologies to address the ethical questions raised by their use.

SELF-QUIZ

1. Suppose you wish to create a large batch of the protein lactase using recombinant DNA. Place the following steps in the order you would have to perform them.
   a. Find the clone with the gene for lactase.
   b. Insert the plasmids into bacteria and grow the bacteria into clones.
   c. Isolate the gene for lactase.
   d. Create recombinant plasmids, including one that carries the gene for lactase.

2. Why is an artificial gene that is made using reverse transcriptase often shorter than the natural form of the gene?

3. A carrier that moves DNA from one cell to another, such as a plasmid, is called a __________.

4. In making recombinant DNA, what is the benefit of using a restriction enzyme that cuts DNA in a staggered fashion?

5. A paleontologist has recovered a bit of organic material from the 400-year-old preserved skin of an extinct dodo. She would like to compare DNA from the sample with DNA from living birds. The most useful method for increasing the amount of dodo DNA available for testing is __________.

6. Why do DNA fragments containing STR sites from different people tend to migrate to different locations during gel electrophoresis?

7. What feature of a DNA fragment causes it to move through a gel during electrophoresis?
   a. the electrical charges of its phosphate groups
   b. its nucleotide sequence
   c. the hydrogen bonds between its base pairs
   d. its double helix shape

8. The pattern of bars in a DNA fingerprint shows
   a. the order of bases in a particular gene.
   b. the presence of various-sized fragments of DNA.
   c. the order of genes along particular chromosomes.
   d. the exact location of a specific gene in a genomic library.

9. If you wanted to sequence an entire genome rapidly, you could chop it into restriction fragments, sequence each one, then reassemble the sequences in their proper order. This method is called __________.

10. Put the following steps of human gene therapy in the correct order.
    a. Virus is injected into patient.
    b. Human gene is inserted into a virus.
    c. Normal human gene is isolated and cloned.
    d. Normal human gene is transcribed and translated in the patient.

Answers to the Self-Quiz questions can be found in Appendix D.

Go to the website or CD-ROM for more Self-Quiz questions.

THE PROCESS OF SCIENCE

11. A biochemist hopes to find a gene in human liver cells that codes for an important blood-clotting protein. She knows that the nucleotide sequence of a small part of the gene is CTGGACTGACA. Briefly explain how to obtain the desired gene.

12. Some scientists once joked that when the DNA sequence of the human genome was complete, "we can all go home" because there would be nothing left for genetic researchers to discover. Why haven't they all "gone home"?

BIOLOGY AND SOCIETY

13. In the not-too-distant future, gene therapy may be an option for the treatment and cure of many inherited disorders. What do you think are the most serious ethical issues that must be dealt with before human gene therapy is used on a large scale? Why do you think these issues are important?

14. Today, it is fairly easy to make transgenic plants and animals. What are some important safety and ethical issues raised by this use of recombinant DNA technology? What are some of the possible dangers of introducing genetically engineered organisms into the environment? What are some reasons for and against leaving decisions in these areas to scientists? Who do you think should make these decisions?

15. In October 2002, the government of the African nation of Zambia announced that it was refusing to distribute 15,000 tons of corn donated by the United States, enough corn to feed 2.5 million Zambians for three weeks. The government rejected the corn because it was likely to contain genetically modified kernels. The government made the decision after its scientific advisers concluded that the studies of the health risks posed by GM crops "are inconclusive." Do you agree with this assessment? Do you think that it is a good justification for refusing the donated corn? At the time of the government's decision, Zambia was facing food shortages, and 35,000 Zambians were expected to die from starvation over the next six months. In light of this, do you think it was morally acceptable for the government to refuse the food? How do the relative risks posed by GM crops compare with the relative risks posed by starvation?

16. In 1983, a 10-year-old girl was kidnapped from her home, raped, and murdered. A jury convicted a local teenager of the crimes and sentenced him to death for the brutal killing. In 1995, DNA analysis proved that semen found near the scene could not have come from the man accused. After 12 years on death row, he was exonerated and released from prison. His case, which took place in Illinois, was far from unique. From 1977 to 2000, 12 convicts were executed and 13 exonerated. In 2000, the governor of Illinois declared a moratorium on all executions in his state because the death penalty system was "fraught with errors." To date, no other states have followed Illinois in declaring a ban on the death penalty, even though the case described here is just one of many similar exonerations. Do you support the Illinois governor's decision? Why do you think no other states have done the same? What rights should death penalty inmates have with regard to DNA testing of old evidence? Who should pay for this additional testing?

UNIT 3

Evolution and Diversity

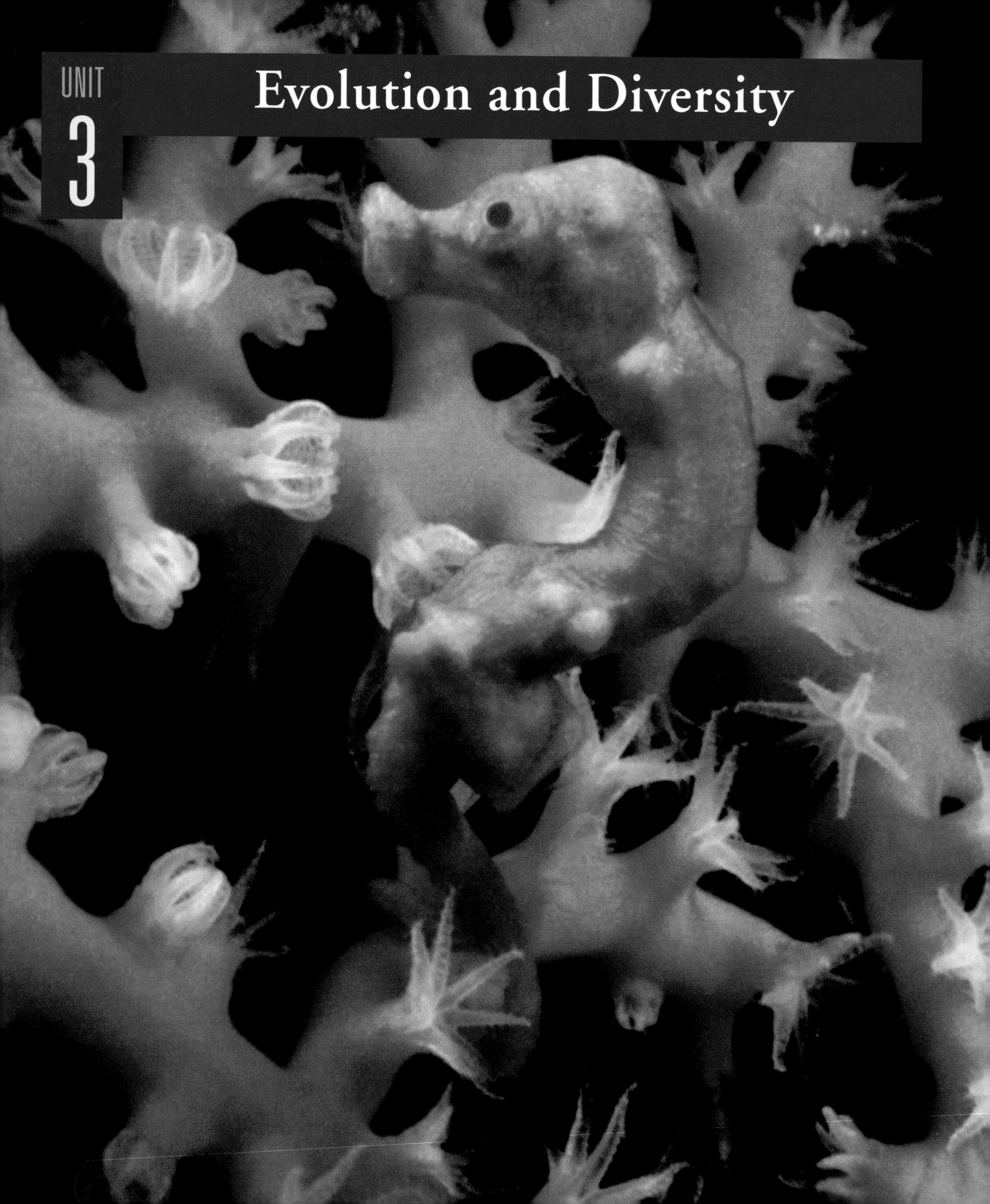

13 How Populations Evolve

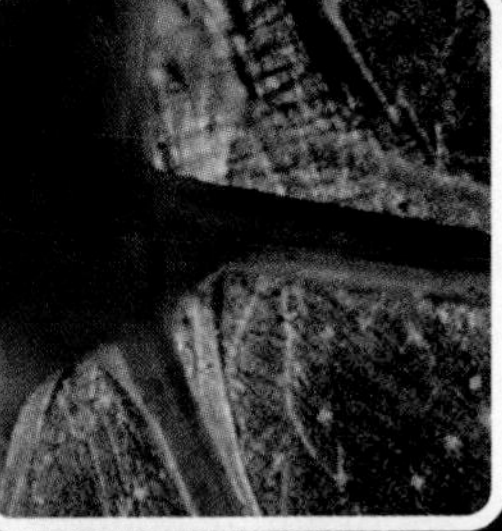

The **same types of bones** make up the forelimbs of humans, cats, whales, and bats.

All humans are connected by descent from African ancestors.

Use of pesticides has hastened the evolution of pesticide-resistant insects.

The A&E channel's *Biography* ranked **Charles Darwin** as the fourth most influential person of the past 1,000 years.

BIOLOGY AND SOCIETY

Persistent Pests

In the 1960s, the World Health Organization (WHO) began a campaign to eradicate the mosquitoes that transmit the disease malaria. It was a noble goal, since malaria kills an estimated 3 million people each year in the world's tropical regions. WHO led an effort to spray the mosquitoes' habitat with a chemical pesticide—an insect-killing poison—called DDT. Early results were promising, and the mosquito was eliminated from the edge of its native range. Progress soon stalled, however, and the eradication plan was dropped. How could a tiny mosquito thwart the best efforts of a large group of well-funded scientists?

Scenarios like this one have occurred dozens of times in the last several decades. When a new type of pesticide is introduced to control agricultural pests, it often produces encouraging early results. A relatively small amount of the poison dusted onto a crop may kill 99% of the insects. The few survivors of the first pesticide wave are insects with genes that somehow enable them to resist the chemical attack, just as the bacteria we described in Chapter 1 developed resistance to antibiotics. For example, the lucky few may carry genes coding for enzymes that destroy the pesticide. After the poison kills most members of the insect population, the survivors are free to reproduce. And when they do, the offspring of resistant insects may inherit the genes for pesticide resistance **(Figure 13.1)**. In each generation, the proportion of pesticide-resistant individuals in the insect population increases, making subsequent sprayings less and less effective.

Since the widespread use of chemical pesticides began in the 1940s, scientists have documented pesticide resistance in more than 500 species of insects. The problems such insects pose through their impact on agriculture and human health are just some of the many ways that evolution has a direct connection to our daily lives; we'll discuss other examples throughout this chapter. Everywhere, all the time, populations of organisms are fine-tuning adaptations to local environments through the evolutionary process of natural selection. Even the kinds of organisms on the planet—the world's catalog of species—have changed over time.

But while change characterizes life, so does continuity. All members of your family are connected by shared ancestry. In fact, all humans are connected by descent from our African ancestors. And all of life, with its dazzling diversity of millions of species, is united by descent from the first microbes that populated the primordial planet. It is this duality of life's unity and diversity that defines modern biology.

An understanding of evolution informs every field of biology, from exploring life's molecules to analyzing ecosystems. And applications of evolutionary biology are transforming medicine, agriculture, biotechnology, and conservation biology. Because evolution integrates all of biology, it is the thematic thread woven throughout this book. This unit of chapters features mechanisms of evolution and traces the history of life on Earth. This chapter starts with the story of how Charles Darwin came to formulate his ideas. Next, some of the lines of evidence in support of evolution are presented, followed by a closer look at Darwin's theory of natural selection. The chapter then focuses on the genetic basis of evolution and the ways it proceeds. ■

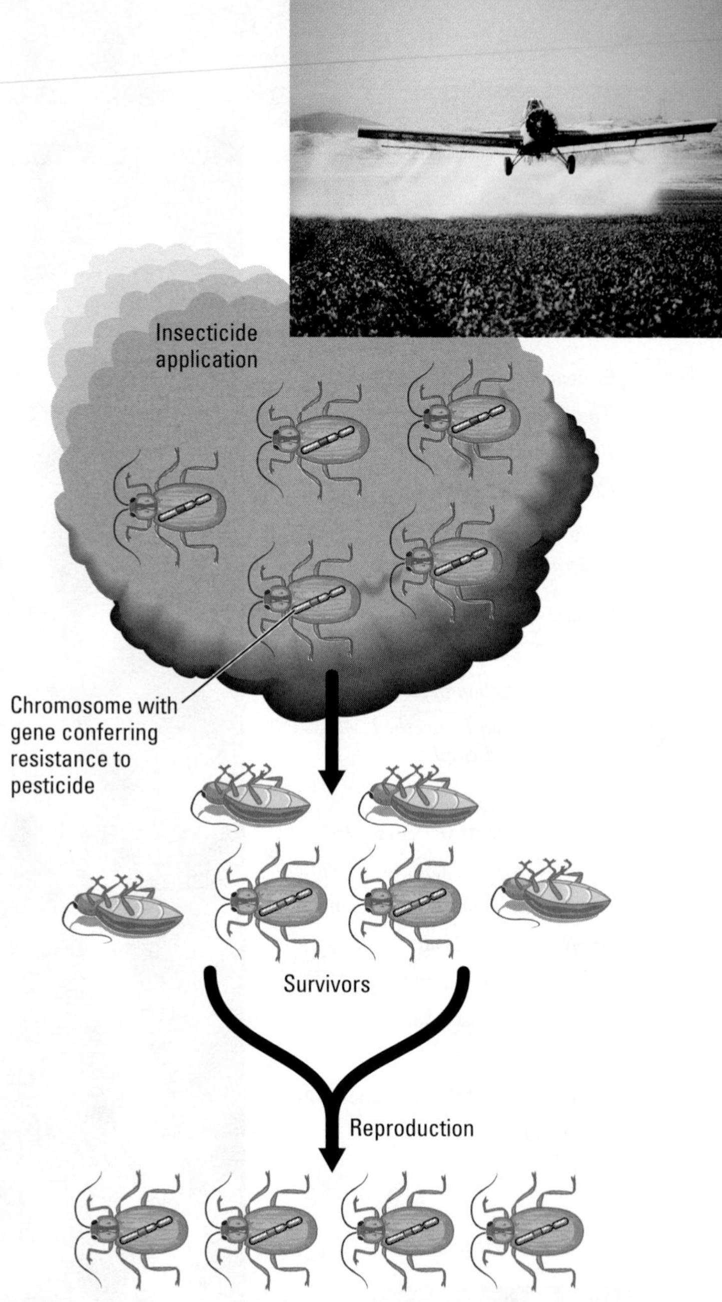

Insecticide application
Chromosome with gene conferring resistance to pesticide
Survivors
Reproduction
Additional applications of the same pesticide will be less effective, and the frequency of resistant insects in the population will grow.

Figure 13.1 Evolution of pesticide resistance in insect populations. By spraying crops with poisons to kill insect pests, humans have unwittingly favored the reproductive success of insects with inherent resistance to the poisons.

Charles Darwin and *The Origin of Species*

Biology came of age November 24, 1859, the day Charles Darwin published *On the Origin of Species by Means of Natural Selection*. Darwin's book presented two main concepts. First, Darwin argued convincingly from several lines of evidence that contemporary species arose from a succession of ancestors through a process of "descent with modification," his phrase for evolution. Darwin's second concept in *The Origin of Species* was his mechanism for how life evolves: natural selection.

The basic idea of **natural selection,** which we introduced in Chapter 1, is that a population of organisms can change over the generations if individuals having certain heritable traits leave more offspring than other individuals. The result of natural selection is **evolutionary adaptation,** a population's increase in the frequency of traits that are suited to the environment **(Figure 13.2)**. In modern terms, we would say that the genetic composition of the population has changed over time, and that is one way of defining **evolution.** But we can also use the term *evolution* on a much grander scale to mean all of biological history, from the earliest microbes to the enormous diversity of modern organisms.

Darwin's book drew a cohesive picture of life by connecting the dots between a bewildering array of seemingly unrelated facts. *The Origin of Species* focused biologists' attention on the great diversity of organisms—their origins and relationships, their similarities and differences, their geographic distribution, and their adaptations to surrounding environments. Before we

(a) A Trinidad tree mantid that mimics dead leaves

(b) A leaf mantid in Costa Rica

(c) A flower mantid in Malaysia

Figure 13.2 Camouflage as an example of evolutionary adaptation. Related species of insects called mantids have diverse shapes and colors that evolved in different environments.

examine how natural selection works and how Darwin derived the idea, let's place the Darwinian revolution in its historical context **(Figure 13.3)**.

Darwin's Cultural and Scientific Context

The view of life developed in *The Origin of Species* contrasts sharply with the view that prevailed during Darwin's lifetime. Many scientists of his day thought Earth was relatively young and populated by a huge number of unrelated species. *The Origin of Species* challenged that widely held notion. It was truly radical for its time. Not only did it challenge prevailing scientific views; it also shook the deepest roots of Western culture.

The Idea of Fixed Species Like many concepts in science, the basic idea of biological evolution can be traced back to the ancient Greeks. About 2,500 years ago, the Greek philosopher Anaximander promoted the idea that life arose in water and that simpler forms of life preceded more complex ones. However, the Greek philosopher Aristotle, whose views had an enormous impact on Western culture, generally held that species are fixed, or permanent, and do not evolve. Judeo-Christian culture fortified this idea with a literal interpretation of the biblical book of Genesis, which tells the story of each form of life being individually created in its present-day form. The idea that all living species are static in form and inhabit an Earth that is only about 6,000 years old dominated the intellectual climate of the Western world for centuries.

Lamarck and Evolutionary Adaptations In the mid-1700s, the study of **fossils**—imprints or remains of organisms that lived in the past—led French naturalist Georges Buffon to suggest that Earth might be much older than 6,000 years. He also observed some telling similarities between particular fossils and living animals. In 1766, Buffon proposed that certain fossil forms might be ancient versions of similar living species. Then, in the early 1800s, French naturalist Jean-Baptiste de Lamarck suggested that the best explanation for this relationship of fossils to current organisms is that life evolves. Lamarck explained evolution as a process of adaptation, the refinement of characteristics that equip organisms to perform successfully in their environments. An example of evolutionary adaptation is the powerful beak that enables some birds to crack tough seeds.

Today, we remember Lamarck mainly for his erroneous view of *how* species evolve. He proposed that by using or not using its body parts, an individual may develop certain traits that it passes on to its offspring. In other words, Lamarck proposed that acquired characteristics are inherited. He suggested, for example, that the strong beaks of seed-cracking birds are the cumulative result of ancestors exercising their beaks during feeding and passing that acquired beak power on to offspring. However, simple observations provide evidence against the inheritance of acquired characteristics: A carpenter who builds up strength and stamina through a lifetime of pounding nails with a heavy hammer will not pass enhanced biceps on to children. Lamarck's mistaken idea obscures the important fact that he helped set the stage for Darwin by proposing that species evolve as a result of interactions between organisms and their environments.

The Voyage of the *Beagle* Charles Darwin was born in 1809, on the same day as Abraham Lincoln. Even as a boy, Darwin's consuming interest in nature was evident. When he was not reading nature books, he was in the fields and forests fishing, hunting, and collecting insects. His father, an eminent physician, could see no future for a naturalist and sent Charles to the University of

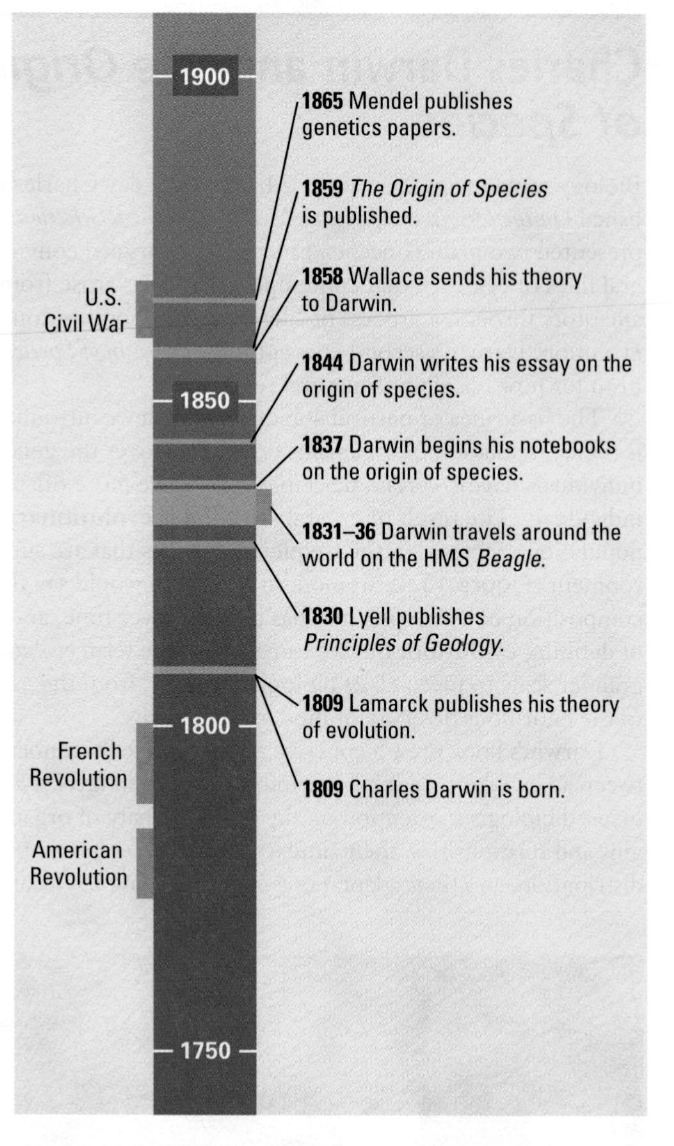
1900
1865 Mendel publishes genetics papers.
1859 *The Origin of Species* is published.
1858 Wallace sends his theory to Darwin.
U.S. Civil War
1844 Darwin writes his essay on the origin of species.
1850
1837 Darwin begins his notebooks on the origin of species.
1831–36 Darwin travels around the world on the HMS *Beagle.*
1830 Lyell publishes *Principles of Geology.*
1809 Lamarck publishes his theory of evolution.
1800
French Revolution
1809 Charles Darwin is born.
American Revolution
1750

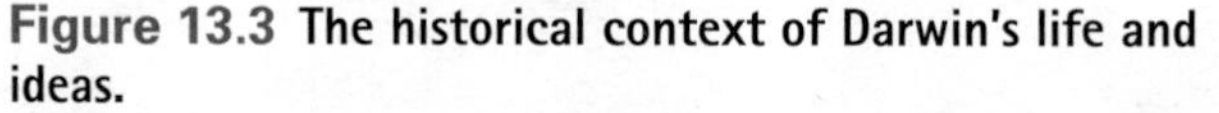
Figure 13.3 The historical context of Darwin's life and ideas.

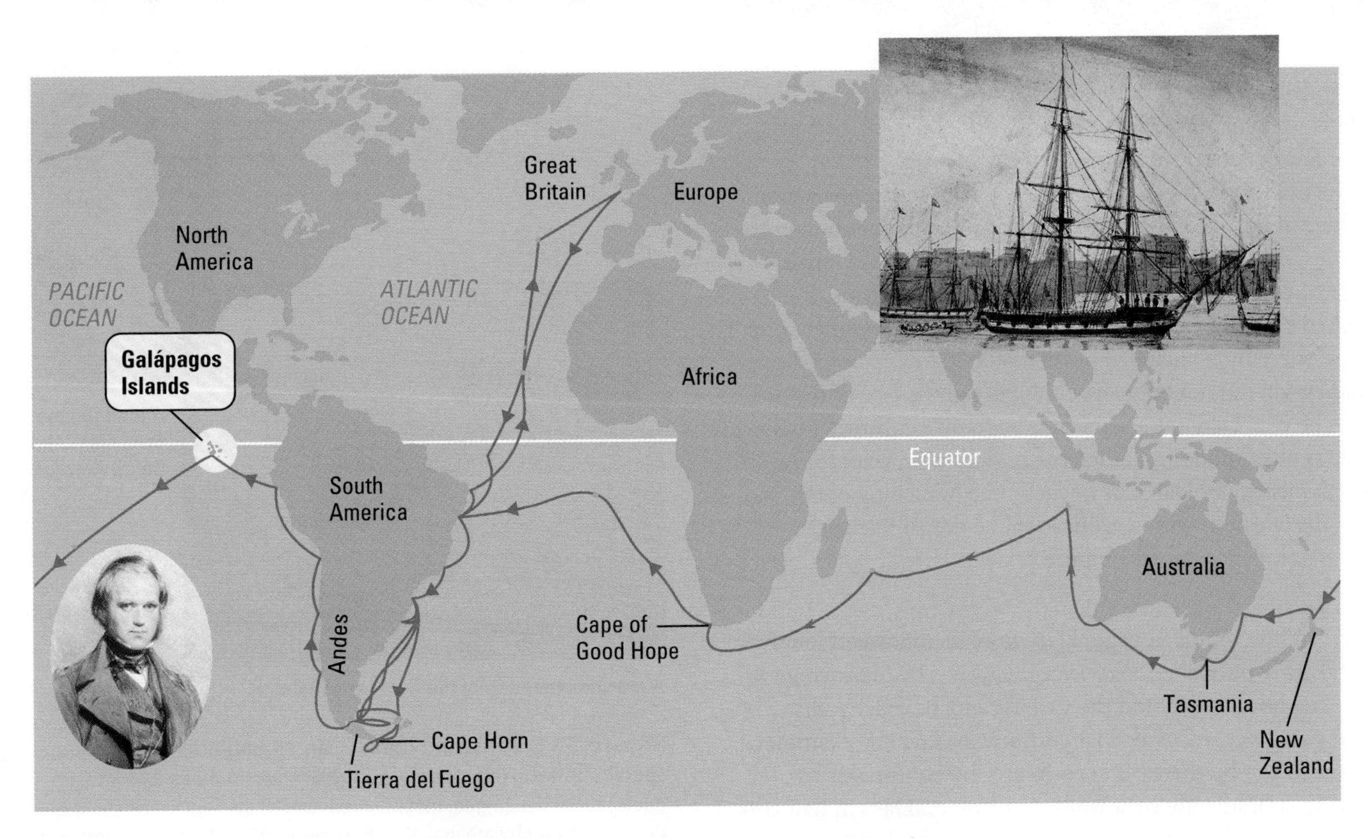

Figure 13.4 The voyage of the *Beagle*. The two insets show the ship and a young Darwin.

Edinburgh to study medicine. But Charles, only 16 years old at the time, found medical school boring and distasteful. He left Edinburgh without a degree and then enrolled at Christ College at Cambridge University, intending to become a minister. At Cambridge, Darwin became the protégé of the Reverend John Henslow, a professor of botany. Soon after Darwin received his B.A. degree in 1831, Professor Henslow recommended the young graduate to Captain Robert FitzRoy, who was preparing the survey ship HMS *Beagle* for a voyage around the world. It was a tour that would have a profound effect on Darwin's thinking and eventually on the thinking of the entire world.

Darwin was 22 years old when he sailed from Great Britain on the *Beagle* in December 1831. The main mission of the voyage was to chart poorly known stretches of the South American coastline **(Figure 13.4)**. Darwin spent most of his time on shore collecting thousands of specimens of fossils and living plants and animals. He noted the various adaptations of organisms that inhabited such diverse environments as the Brazilian jungles, the grasslands of the Argentine pampas, and the desolate and frigid lands at the southern tip of South America.

In spite of their unique adaptations, the plants and animals throughout the continent all had a definite South American stamp, very distinct from the life-forms of Europe. That in itself may not have surprised Darwin. But the plants and animals living in temperate regions of South America seemed more closely related to species living in tropical regions of that continent than to species living in temperate regions of Europe. And the South American fossils Darwin found, though clearly different species from modern ones, were distinctly South American in their resemblance to the living plants and animals of that continent. This led Darwin to wonder if contemporary South American species owed their features to descent from ancestral species on that continent.

Darwin was particularly intrigued by the geographic distribution of organisms on the Galápagos Islands. These are relatively young volcanic islands about 900 km (540 miles) off the Pacific coast of South America. Most of the animals of the Galápagos live nowhere else in the world, but they resemble species living on the South American mainland (**Figure 13.5**).

It is as though the islands had been colonized by plants and animals that strayed from elsewhere and then diversified as they adapted to environments on the different islands. Among the birds Darwin collected on the Galápagos were several types of finches. Some were unique to individual islands, while others were distributed on two or more islands that were close together. The unique adaptations of these birds included beaks modified for feeding on certain kinds of foods. Darwin did not appreciate the full significance of the diversity of the finches he collected until years after he returned to Britain. Since then, biologists have applied modern methods of comparing species to reconstruct the evolutionary history of Darwin's finches. The branching of this evolutionary tree traces the "descent with modification" of the 14 finch species from a common immigrant ancestor (see Figure 1.13).

A land iguana from mainland South America

A marine iguana from the Galápagos Islands

Figure 13.5 A marine iguana, an example of the unique species inhabiting the Galápagos. These reptiles dive into the ocean to feed on algae, making them one of the few lizard species that spend much time in water. Partially webbed feet and a flattened tail are two of the adaptations that make marine iguanas such good swimmers. Darwin noticed that the Galápagos marine iguanas are similar to, but distinct from, land-dwelling iguanas on the islands and on the South American mainland (inset).

The New Geology While on his voyage, Darwin read and was strongly influenced by the recently published *Principles of Geology*, by Scottish geologist Charles Lyell. The book presented the case for an ancient Earth sculpted by gradual geologic processes that continue today. For example, a mighty mountain range can be thrust up centimeter by centimeter by earthquakes occurring sporadically over millions of years. Darwin had collected fossils of sea snails in the Andes Mountains. Perhaps, he reasoned, earthquakes gradually lifted the rock bearing those marine fossils from the seafloor. Darwin would eventually apply this principle of *gradualism* to the evolution of Earth's life.

Descent with Modification

By the early 1840s, Darwin had composed a long essay describing the major features of his theory of evolution. He realized that his ideas would cause a social furor, however, and he delayed publishing his essay. Then, in the mid-1850s, Alfred Wallace, a British naturalist doing fieldwork in Indonesia, developed a theory almost identical to Darwin's. When Wallace sent Darwin a manuscript describing his own ideas on natural selection, Darwin wrote, "All my originality . . . will be smashed." However, in 1858, two of Darwin's colleagues presented Wallace's paper and excerpts from Darwin's earlier essay together to the scientific community. With the publication in 1859 of *The Origin of Species*, Darwin presented the world with an avalanche of evidence and a strong, logical argument for evolution. He also explained natural selection as the mechanism by which evolution occurs.

Darwin made two main points in *The Origin of Species*. First, he argued from evidence that the species of organisms inhabiting Earth today descended from ancestral species. In the first edition of his book, Darwin did not actually use the word "evolution." He referred instead to "descent with modification." Darwin postulated that as the descendants of the earliest organisms spread into various habitats over millions of years, they accumulated different modifications, or adaptations, to diverse ways of life. In Darwin's view, the history of life is analogous to a tree. Patterns of descent are like the branching and rebranching from a common trunk, the first organism, to the tips of millions of twigs representing the species living today. At each fork of the evolutionary tree is an ancestor common to all evolutionary branches extending from that fork. Closely related species, such as Asian

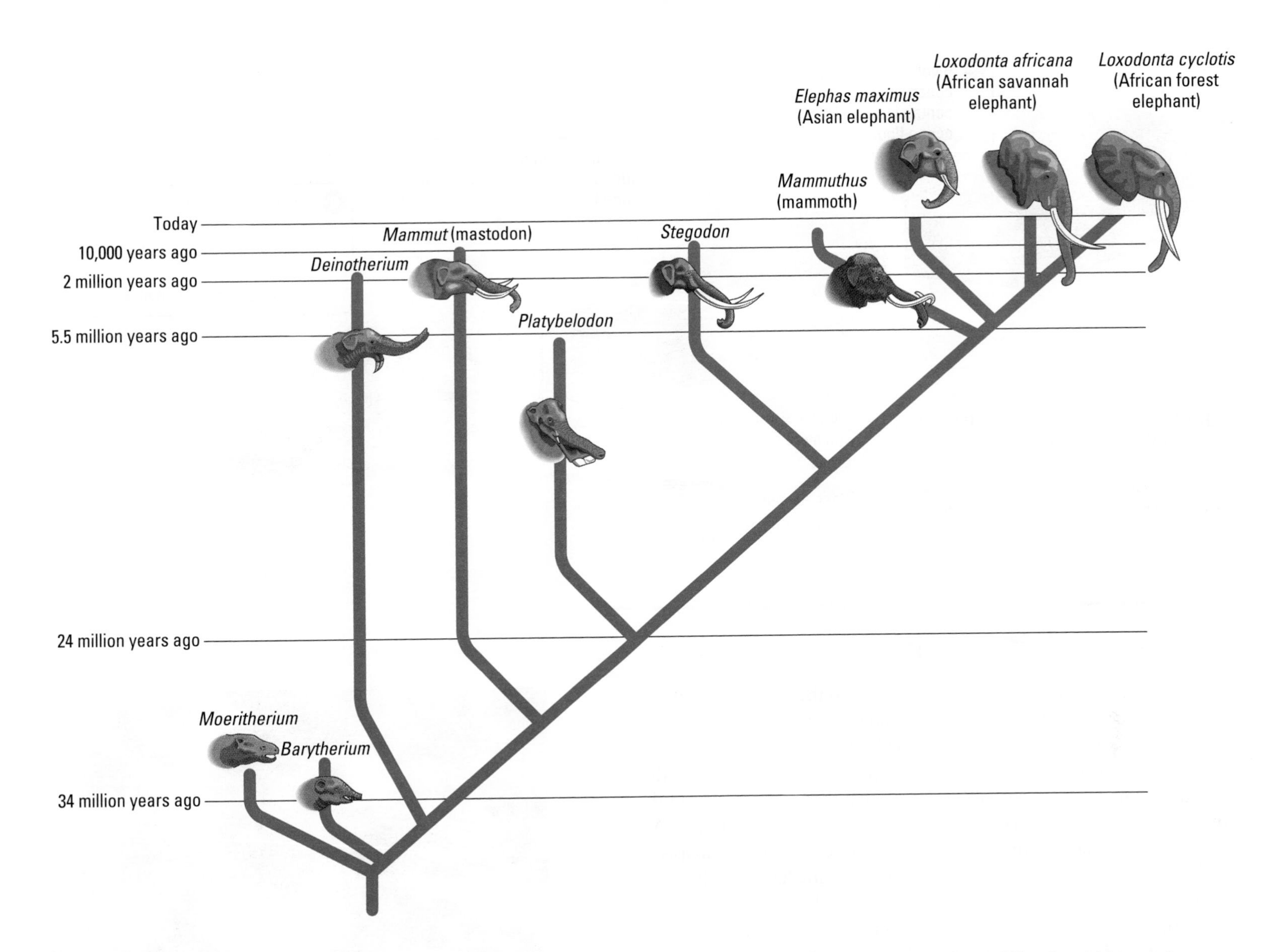

and African elephants, share many characteristics because their lineage of common descent traces to a recent fork of the tree of life **(Figure 13.6)**.

Darwin's second main point in *The Origin of Species* was his argument for natural selection as the mechanism for descent with modification. When biologists speak of Darwin's theory of evolution, they mean natural selection as a cause of evolution, not the phenomenon of evolution itself. In the next sections of this chapter, we will examine the evidence for evolution and then look more closely at Darwin's theory of natural selection.

Figure 13.6 Descent with modification. This evolutionary tree of the elephant family is based mainly on evidence from fossils—their anatomy, their order of appearance in geologic time, and their geographic distribution. Genetic analyses suggest that African elephants—long thought of as a single species—are actually two separate species.

CHECKPOINT

1. What is gradualism? How did Darwin apply that idea to the evolution of life?
2. What were the two main points in Darwin's *The Origin of Species*?
3. Darwin's phrase for evolution, __________ with__________, captured the idea that an ancestral species could diversify into many descendant species by the accumulation of different __________ to various environments.

Answers: 1. Gradualism is the idea that large changes on Earth can result from the accumulation of small changes over a very long time. Darwin applied this idea to suggest that species evolve through the slow accumulation of small changes over time. **2.** Descent of diverse species from common ancestors and natural selection as the mechanism of evolution **3.** descent; modification; adaptations

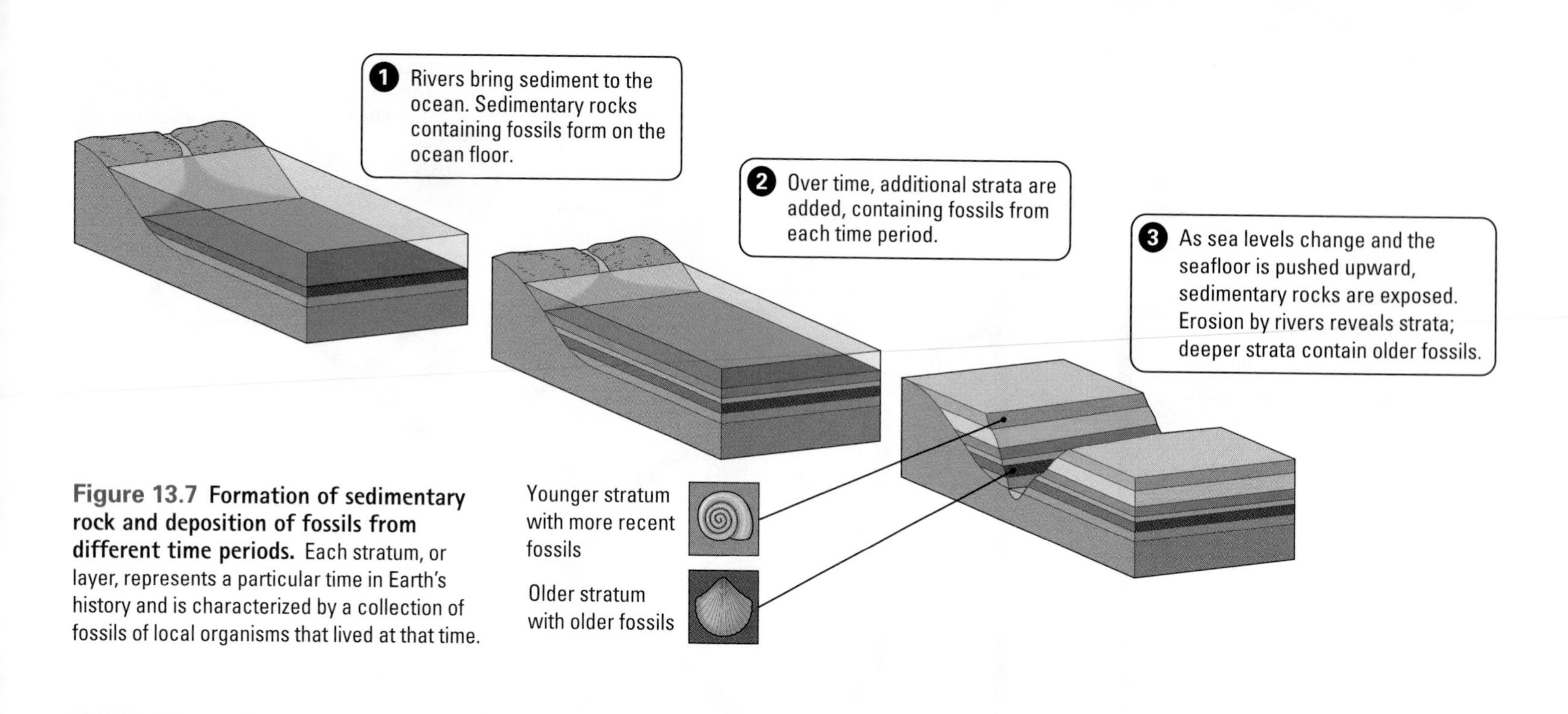

1 Rivers bring sediment to the ocean. Sedimentary rocks containing fossils form on the ocean floor.
2 Over time, additional strata are added, containing fossils from each time period.
3 As sea levels change and the seafloor is pushed upward, sedimentary rocks are exposed. Erosion by rivers reveals strata; deeper strata contain older fossils.
Younger stratum with more recent fossils
Older stratum with older fossils

Figure 13.7 Formation of sedimentary rock and deposition of fossils from different time periods. Each stratum, or layer, represents a particular time in Earth's history and is characterized by a collection of fossils of local organisms that lived at that time.

Evidence of Evolution

Evolution leaves observable signs. Such clues to the past are essential to any historical science. Historians of human civilization can study written records from earlier times. But they can also piece together the evolution of societies by recognizing vestiges of the past in modern cultures. Even if we did not know from written documents that Spaniards colonized the Americas, we would deduce this from the Hispanic stamp on Latin American culture. Similarly, biological evolution has left observable marks.

In this section, we will examine five of the many lines of evidence in support of evolution. One of them—fossils—is a historical record. The other four—biogeography, comparative anatomy, comparative embryology, and molecular biology—encompass historical vestiges of evolution evident in modern life.

The Fossil Record

Most fossils are found in sedimentary rocks. Sand and silt eroded from the land are carried by rivers to seas and swamps, where the particles settle to the bottom. Over millions of years, deposits pile up and compress the older sediments below into rock. Varying rates of sedimentation and types of particles lead to identifiable layers, or strata, of rock **(Figure 13.7)**. When aquatic organisms or land organisms that have been swept into the water die, they settle along with the sediments and may leave imprints in the rocks. Organisms that remain on land when they die may first be covered by windblown silt and then buried in waterborne sediments when sea levels rise over them. Thus, each rock layer bears a unique set of fossils representing a local sampling of the organisms that lived when that sediment was deposited. Younger strata are on top of older ones so the positions of fossils in the strata reveal their relative age. (The ages of fossils can be confirmed using radiometric dating—see Figure 14.17.) The **fossil record** is this ordered sequence of fossils as they appear in the rock layers, marking the passing of geologic time **(Figure 13.8)**.

Figure 13.8 Strata of sedimentary rock at the Grand Canyon. The Colorado River has cut through over 2,000 m of rock, exposing sedimentary layers that are like huge pages from the book of life. Scan the canyon wall from rim to floor, and you look back through hundreds of millions of years. Each layer entombs fossils that represent some of the organisms from that period of Earth's history.

The fossil record reveals the appearance of organisms in a historical sequence. The oldest known fossils, dating from about 3.5 billion years ago, are prokaryotes. This fits with the molecular and cellular evidence that prokaryotes are the ancestors of all life. Fossils in younger layers of rock reveal the evolution of various groups of eukaryotic organisms. One example is the successive appearance of the different classes of vertebrates (animals with backbones). Fishlike fossils are the oldest vertebrates in the fossil record. Amphibians are next, followed by reptiles, then mammals.

Paleontologists (scientists who study fossils) have discovered many transitional forms that link past and present. For example, a series of fossils provides evidence that birds descended from one branch of dinosaurs. Another example is a series of fossilized whales that connect these aquatic mammals to their land-dwelling ancestors **(Figure 13.9)**.

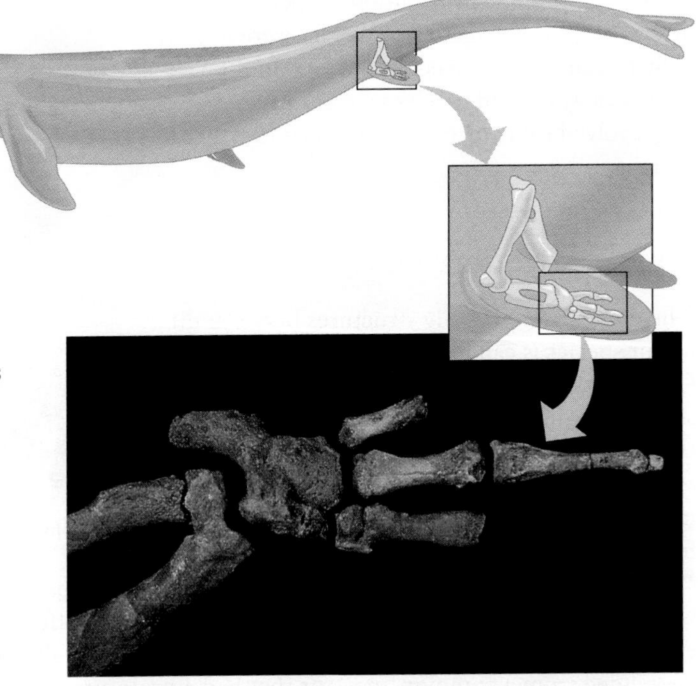

Figure 13.9 A transitional fossil linking past and present. The hypothesis that whales evolved from terrestrial (land-dwelling) ancestors predicts a four-limbed beginning for whales. Paleontologists digging in Egypt and Pakistan have identified extinct whales that had hind limbs. Shown here are fossilized leg bones of *Basilosaurus*, one of those ancient whales. These whales were already aquatic animals that no longer used their legs to support their weight.

Biogeography

It was the geographic distribution of species, called **biogeography,** that first suggested to Darwin that today's organisms evolved from ancestral forms. Consider, for example, Darwin's visit to the Galápagos Islands (see Figure 13.4). Darwin noted that the Galápagos animals resembled species of the South American mainland more than they resembled animals on similar but distant islands. This is what we should expect if the Galápagos species evolved from South American immigrants.

There are many other examples from biogeography that seem baffling without an evolutionary perspective. Why, for example, is Australia home to so many kinds of pouched mammals—marsupials **(Figure 13.10)**—but relatively few placental mammals (those in which embryonic development is completed in the uterus)? It is not because Australia is inhospitable to placental mammals. Humans have introduced rabbits, foxes, and many other placental mammals to Australia, where these introduced species have thrived to the point of becoming ecological and economic nuisances. The prevailing hypothesis is that the unique Australian wildlife evolved on that island continent in isolation from regions where early placental mammals diversified.

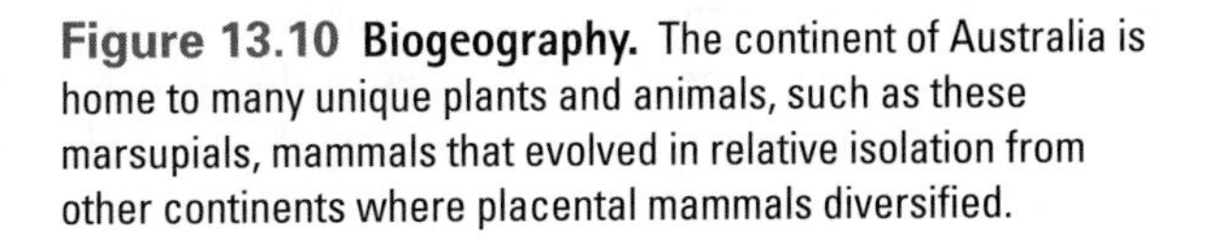

Figure 13.10 Biogeography. The continent of Australia is home to many unique plants and animals, such as these marsupials, mammals that evolved in relative isolation from other continents where placental mammals diversified.

The geographic distribution of species makes little sense if we imagine that species were individually placed in suitable environments. In the Darwinian view, we find species where they are because they evolved from ancestors that inhabited those regions.

Comparative Anatomy

The comparison of body structures between different species is called **comparative anatomy.** Certain anatomical similarities among species show signs of evolutionary history. For example, the same skeletal elements make up the forelimbs of humans, cats, whales, and bats, all of which are mammals (**Figure 13.11**). The functions of these forelimbs differ. A whale's flipper does not do the same job as a bat's wing. If these limbs had completely separate origins, we would expect that their basic designs would be very different. However, their structural similarity would not be surprising if all mammals descended from a common ancestor that had a prototype forelimb. Arms, forelegs, flippers, and wings of different mammals are variations on a common anatomical theme that has become adapted to different functions. Such similarity due to common ancestry is called **homology.** The forelimbs of diverse mammals are homologous structures.

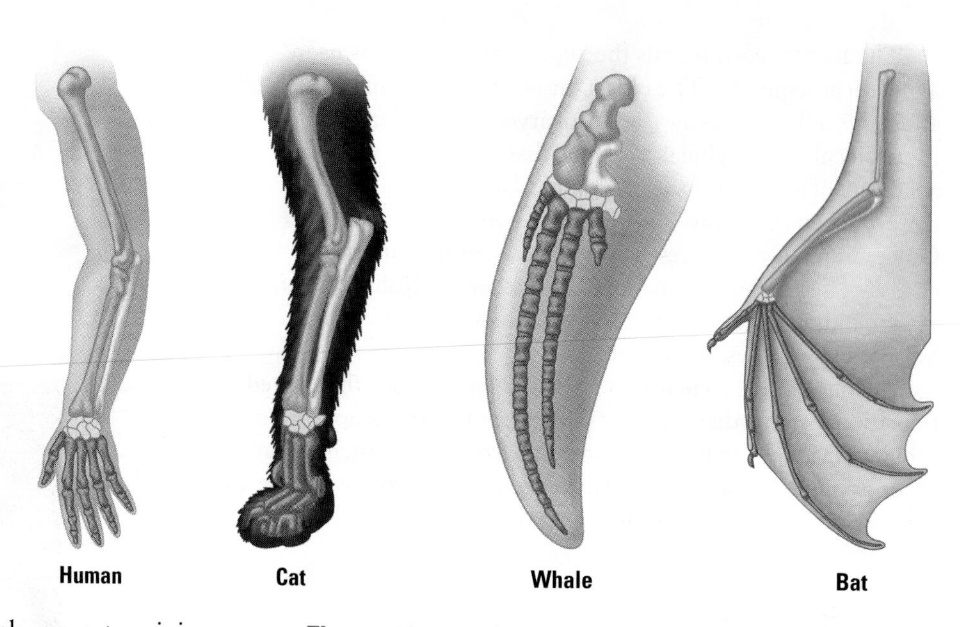

Figure 13.11 Homologous structures: anatomical signs of descent with modification. The forelimbs of all mammals are constructed from the same skeletal elements. (Homologous bones in each of these four mammals are colored the same.) The hypothesis that all mammals descended from a common ancestor predicts that their forelimbs, though diversely adapted, would be variations on a common anatomical theme.

Comparative anatomy attests that evolution is a remodeling process in which ancestral structures that originally functioned in one capacity become modified as they take on new functions—descent with modification. The historical constraints of this retrofitting are evident in anatomical imperfections. For example, the human spine and knee joint were derived from ancestral structures that supported four-legged mammals. Almost none of us will reach old age without experiencing knee or back problems. If these structures had first taken form specifically to support our bipedal posture, we would expect them to be less subject to sprains, spasms, and other common injuries.

Some of the most interesting homologous structures are vestigial organs, structures of marginal, if any, importance to the organism. **Vestigial organs** are remnants of structures that served important functions in the organism's ancestors, such as the rear leg bones evident in ancient whale fossils (see Figure 13.9). In another example, the skeletons of some snakes retain vestiges of the pelvis and leg bones of walking ancestors. If limbs were a hindrance to ancient snakes' way of life, natural selection would favor snake descendants with successively smaller limbs.

Comparative Embryology

Comparative embryology is the comparison of anatomical structures that appear during the early stages of development of different organisms. Closely related organisms have many similar stages in their embryonic development. One sign that vertebrates evolved from a common ancestor is that all of them have an embryonic stage in which structures called gill pouches appear on the sides of the throat. At this stage, the embryos of fishes, frogs, snakes, birds, apes—indeed, all vertebrates—

look more alike than different **(Figure 13.12)**. The different classes of vertebrates take on more and more distinctive features as development progresses. For example, gill pouches develop into gills in fishes but into parts of the ear and throat in humans.

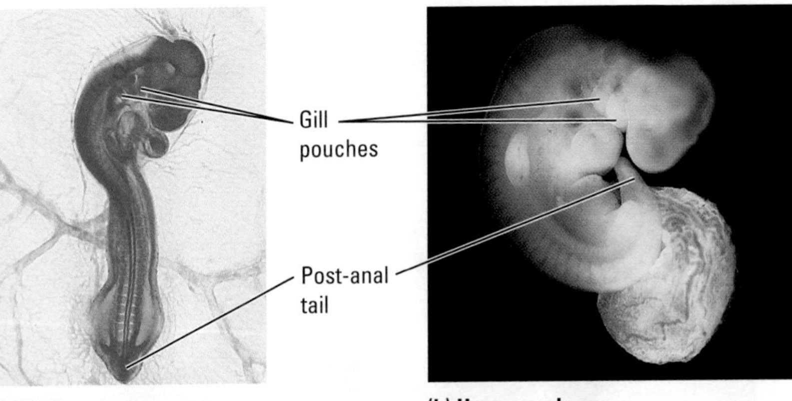

Figure 13.12 Evolutionary signs from comparative embryology. At this early stage of development, the kinship of vertebrates is unmistakable. Notice, for example, the gill pouches and tails in both **(a)** the bird (chick) embryo and **(b)** the human embryo. Comparative embryology helps biologists identify anatomical homology that is less apparent in adults.

Molecular Biology

As we saw in Chapter 10, the hereditary background of an organism is documented in its DNA and in the proteins encoded by the DNA. If two species have genes with nucleotide sequences that match closely (and therefore proteins with amino acid sequences that match closely), biologists conclude that these sequences must have been inherited from a relatively recent common ancestor **(Figure 13.13)**. In contrast, the greater the number of sequence differences between species, the less likely they share a close common ancestor. Molecular comparisons between diverse organisms have allowed biologists to develop hypotheses about the evolutionary divergence of branches on the tree of life. For example, genetic analyses suggests that what appeared to be a single species of African elephant is actually two separate species (see Figure 13.6).

Darwin's boldest hypothesis was that *all* forms of life are related to some extent through branching evolution from the earliest organisms. About 100 years after Darwin made this claim, molecular biology has provided strong evidence for it: All forms of life use DNA and RNA, and the genetic code (how RNA triplets are translated into amino acids) is nearly universal. This genetic language has been passed along through all the branches of evolution since its beginnings in an early form of life.

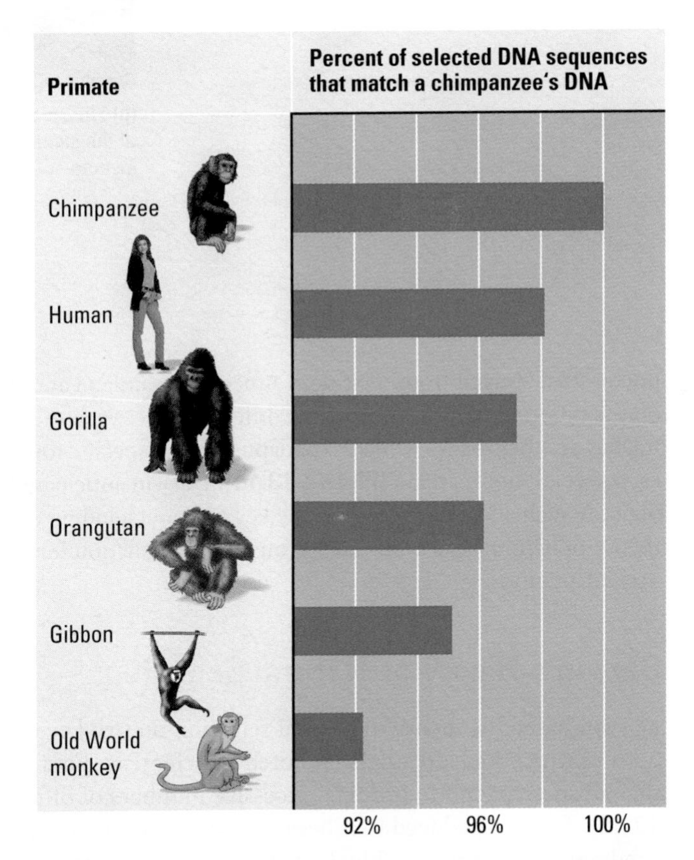

Figure 13.13 Genetic relationships among some primates. The bars in this diagram show the percent of selected DNA sequences that match between a chimpanzee and other primates, the animal group that includes monkeys, apes, and humans. For example, note that the DNA sequences of chimps and humans are better than a 98% match within the chosen regions. In contrast, the DNA sequences of chimps and Old World monkeys (macaques, madrills, baboons, and rhesus monkeys) have less than a 93% match.

CHECKPOINT

1. Why are older fossils generally in deeper rock layers than younger fossils?
2. What is homology?

Answers: 1. Sedimentation adds younger rock layers on top of older ones. **2.** Similarity between species that is due to shared ancestry

Natural Selection

MP3 Tutor
Natural Selection

Darwin perceived adaptation to the environment and the origin of new species as closely related processes. Imagine, for example, that an animal species from a mainland colonizes a chain of distant, relatively isolated islands. In the Darwinian view, populations on the different islands may diverge more and more in appearance as each population adapts to its local environment. Over many generations, the populations on different islands could become dissimilar enough to be designated as separate species. Evolution of the finches on the Galápagos Islands is an example (see Figure 1.13). It is a reasonable hypothesis that the islands were colonized by

(a) The large ground finch. This species of Galápagos finch has a large beak specialized for cracking seeds that fall from plants to the ground.

(b) The small tree finch. The smaller beak of the small tree finch is used to grasp insects.

(c) The woodpecker finch. The long, narrow beak of the woodpecker finch allows it to hold tools such as cactus spines to probe for termites and other wood-boring insects.

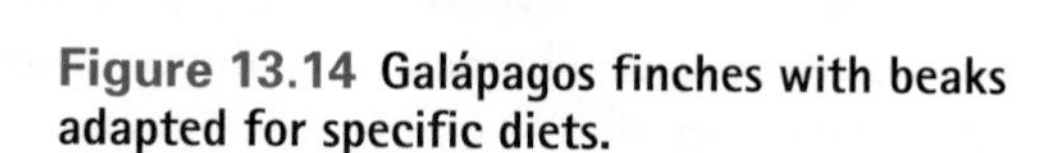

Figure 13.14 Galápagos finches with beaks adapted for specific diets.

finches that strayed from the South American mainland and then diversified on the different islands. Among the differences between the Galápagos finches are their beaks, which are adapted to the specific foods available on each species' home island **(Figure 13.14)**. Darwin anticipated that explaining how such adaptations arise is the key to understanding evolution. And his theory of natural selection remains our best explanation for the formation of new adaptations.

Darwin's Theory of Natural Selection

Darwin based his theory of natural selection on two key observations, both of which had already been noted by scientists. First, he recognized that all species tend to produce excessive numbers of offspring **(Figure 13.15)**. Darwin deduced that because natural resources are limited, the production of more individuals than the environment can support leads to a struggle for existence among the individuals of a population. In most cases, only a small percentage of offspring will survive in each generation. Many eggs are laid, young born, and seeds spread, but only a tiny fraction complete their development and leave offspring of their own. The rest are starved, eaten, frozen, diseased, unmated, or unable to reproduce for other reasons.

The second key observation that led Darwin to natural selection was his awareness of variation among individuals of a population. Just as no two people in a human population are alike, individual variation abounds in all

Figure 13.15 Overproduction of offspring. A cloud of millions of spores is exploding from these puffballs, a type of fungus. (Each puffball in this photo is about 2 cm across.) The wind will disperse the spores far and wide. Each spore, if it lands in a suitable environment, has the potential to grow and develop into a new fungus. Only a tiny fraction of the spores will actually give rise to offspring that survive and reproduce.

species **(Figure 13.16)**. Much of this variation is heritable. Siblings share more traits with each other and with their parents than they do with less closely related members of the population.

From these two observations, Darwin arrived at the conclusion that defines natural selection: Individuals whose inherited traits are best suited to the local environment are more likely than less fit individuals to survive and reproduce. In other words, the individuals that function best should leave the most surviving offspring. Darwin's genius was in connecting two observations that anyone could make and drawing an inference that could explain how adaptations evolve.

- **Observation 1:** *Overproduction.* Populations of all species have the potential to produce many more offspring than the environment can possibly support with food, space, and other resources. This overproduction makes a struggle for existence among individuals inevitable.
- **Observation 2:** *Individual variation.* Individuals in a population vary in many heritable traits.

→ **Inference:** *Differential reproductive success (natural selection).* Those individuals with traits best suited to the local environment generally leave a larger share of surviving, fertile offspring.

Darwin's insight was both simple and profound. The environment screens a population's inherent variability. Differential success in reproduction (natural selection) leads to an accumulation of the favored traits in the population over generations (evolution). In other words, natural selection promotes evolutionary adaptations.

Figure 13.16 Color variations in a population of Asian lady beetles.

Natural Selection in Action

Natural selection and evolution are observable phenomena. You learned about one classic and unsettling example at the start of the chapter: the evolution of pesticide resistance in hundreds of insect species. We have used pesticides to control insects that eat our crops, transmit diseases such as malaria, or just annoy us around the house or campground. But widespread use of these poisons has led to the unintended development of pesticide-resistant insect populations. Another example of natural selection in action was described in Chapter 1: the development of antibiotic-resistant bacteria. More recently, doctors have documented an increase in drug-resistant strains of HIV, the virus that causes AIDS.

These examples of evolution highlight two key points about natural selection. First, natural selection is more of an editing process than a creative mechanism. A pesticide does not create resistant individuals, but selects for resistant insects that were already present in the population (see Figure 13.1). Second, natural selection depends on time and place: It favors those characteristics in a varying population that fit the current, local environment. Environmental factors vary from place to place and from time to time. An adaptation in one situation may be useless or even detrimental in different circumstances. For example, some genetic mutations that happen to endow houseflies with resistance to the pesticide DDT also reduce a fly's growth rate. Before DDT was introduced, the gene for resistance was a handicap. But the appearance of DDT changed the rules in the environmental arena and favored pesticide-resistant individuals in the reproduction sweepstakes. Such are the dynamics of evolution by natural selection.

Does Predation Drive the Evolution of Lizard Horn Length? One recent and particularly elegant demonstration of natural selection in action involved the flat-tailed horned lizard *Phrynosoma mcalli*, a desert inhabitant of the American Southwest **(Figure 13.17a)**. The lizard's main predator, birds called shrikes, attack by biting a lizard's neck just behind the skull, severing the spine. The shrike then carries the dead prey to a convenient place, such as a fence or branch, impales it, and eats it **(Figure 13.17b)**.

Kevin Young and colleagues at Utah State University and Indiana University made the **observation** that flat-tailed horned lizards defend against attack by thrusting their heads backward, stabbing the shrike with the spiked horns that protrude from the rear of the skull. This led the researchers to **question** whether longer horn length represented a survival advantage. Young's **hypothesis** was that it did, and he formed the **prediction** that live horned lizards would have longer horn lengths than dead ones.

Young's **experiment** was simple: He measured the length of rear horns and the tip-to-tip spread distance of side horns from the skulls of 29 dead lizards (found where they had been impaled) and 155 live lizards. His **results**, shown in **Figure 13.17c**, indicate that the average horn length of live lizards is about 10% longer than that of dead lizards. The researchers concluded that defensive behavior against predators is one factor driving natural selection of horn length among this lizard species. ■

(a)

(b)

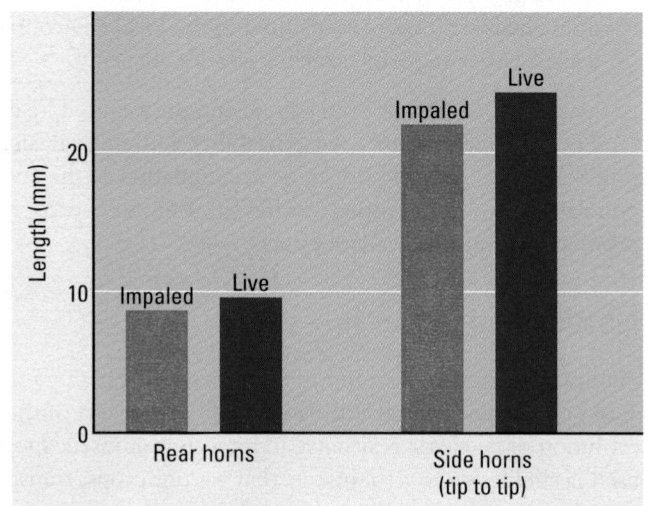

(c)

Figure 13.17 Does Predation Drive the Evolution of Lizard Horn Length? (a) A flat-tailed horned lizard. The lizards use the spiky horns that protrude from the back and side of the skull to ward off attacks. **(b)** Shrikes prey on the lizards. After killing a lizard, the bird often impales the lizard on a fence or branch. **(c)** Researchers measured the horns of impaled lizards and compared them with the horns of live lizards. The fact that both rear and side horns are significantly longer on live lizards than on dead lizards suggests that horn length is an adaptation that is evolving in response to predation by birds.

CHECKPOINT

1. Define natural selection.
2. Explain why the following statement is incorrect: "Pesticides cause pesticide resistance in insects."

Answers: 1. Natural selection is the differential reproductive success among a population's varying individuals. **2.** An environmental factor does not create new traits such as pesticide resistance, but rather selects among the traits that are already represented in the population.

The Modern Synthesis: Darwinism Meets Genetics

Natural selection requires hereditary processes that Darwin could not explain. How do the variations that are the raw material for natural selection arise in a population? And how are these variations passed along from parents to offspring? Darwin and Gregor Mendel lived and worked at the same time. In fact, by breeding peas in his abbey garden, Mendel illuminated the very hereditary processes required for natural selection to work. However, Mendel's discoveries went unappreciated by the scientific community during his lifetime (see Chapter 9). Mendelism and Darwinism finally came together in the mid-1900s, decades after both scientists had died. This fusion of genetics with evolutionary biology came to be known as the **modern synthesis** (here, "synthesis" means "combination"). One of its key elements is an emphasis on the biology of populations.

Populations as the Units of Evolution

We have already used the term *population* several times in this chapter. Now it's time for a biological definition: A **population** is a group of individuals of the same species living in the same place at the same time. One population may be isolated from other populations of the same species, with little interbreeding and thus little exchange of genes between them. Such isolation is common for populations confined to widely separated islands, unconnected lakes, or mountain ranges separated by lowlands **(Figure 13.18a)**. However, populations are not usually so isolated, and they rarely have sharp boundaries. One population center may blur into another in a region of overlap, where members of both populations are present but less numerous. Nevertheless, individuals are more concentrated in the population centers and are more likely to breed with other locals **(Figure 13.18b)**. Therefore, organisms of a population are generally more closely related to one another than to members of other populations.

A population is the smallest biological unit that can evolve. A common misconception is that individual organisms evolve during their lifetimes. It is true that natural selection acts on individuals; inherited characteristics affect their survival and reproductive success. But, the evolutionary impact of this natural selection is only apparent when we track how a population changes over time. In our example of pesticide resistance in insects, we measured evolution by the change in the relative numbers of resistant individuals over a span of generations, not by the survivability of any individual insect.

A focus on populations as the evolutionary units led to a new field in science called population genetics. **Population genetics** tracks the genetic makeup of populations over time.

(a) Trees

(b) Humans

Figure 13.18 Populations. **(a)** Two dense populations of trees are separated by a lake. The two populations are not totally isolated. Interbreeding occurs when wind blows pollen between the populations. Nevertheless, trees are more likely to breed with members of the same population than with trees on the other side of the lake. **(b)** This nighttime satellite view of North America shows the lights of human population centers, or cities. People move around the country, of course, and there are suburban and rural communities between cities, but people are more likely to choose mates locally.

Genetic Variation in Populations

You have no trouble recognizing your friends in a crowd. Each person has a unique genome, reflected in individual variations in appearance and temperament. Individual variation abounds in populations of all species that reproduce sexually. In addition to differences we can see, most populations have a great deal of variation that can be detected only by biochemical means. For example, you cannot tell a person's ABO blood group (A, B, AB, or O) just by looking at her or him.

Not all variation in a population is heritable. Phenotype results from a combination of the genotype, which is inherited, and many environmental influences. For instance, a strength-training program can build up your muscle mass beyond what would naturally occur from your genetic makeup. However, you would not pass this environmentally induced physique on to your offspring. Only the genetic component of variation is relevant to natural selection.

Many of the variable traits in a population result from the combined effect of several genes (see Chapter 9). This polygenic inheritance produces traits that vary more or less continuously—in human height, for instance, from very short individuals to very tall ones. By contrast, other features, such as human ABO blood group, are determined by a single gene locus, with different alleles producing one of only a few distinct phenotypes; there are no in-between types. In such cases, when a population includes two or more forms of a phenotypic characteristic, the contrasting forms are called morphs. A population is said to be **polymorphic** for a characteristic if two or more morphs are present in noticeable numbers—that is, if neither morph is extremely rare **(Figure 13.19)**.

Figure 13.19 Polymorphism in a garter snake population. These four garter snakes, which belong to the same species, were all captured in one Oregon field. The behavior of each morph (form) is correlated with its coloration. When approached, spotted snakes, which blend in with their background, generally freeze. In contrast, snakes with stripes, which make it difficult to judge the speed of motion, usually flee rapidly when approached.

Sources of Genetic Variation Mutations and sexual recombination, which are both random processes, produce genetic variation. Let's examine each of these processes in detail.

An organism is a refined product of thousands of generations of past selection. And yet mutations, those random changes in genetic material, can actually create new alleles (alternative forms of genes). For example, a mutation in a gene may substitute one nucleotide for another. This type of change will be harmless if it does not affect the function of the protein the DNA encodes. If it does affect the protein's function, the mutation will probably be harmful. A random mutation is like a shot in the dark; it is not likely to improve a genome any more than shooting a bullet through the hood of a car is likely to improve engine performance.

On rare occasions, however, a mutant allele may actually enhance reproductive success. This kind of effect is more likely when the environment is changing in such a way that alleles that were once disadvantageous are favorable under the new conditions. We already considered one example, the changing fortunes of the DDT resistance allele in insect populations.

Organisms with very short generation spans, such as bacteria, can evolve rapidly with mutations as the only source of genetic variation. Bacteria multiply so quickly that natural selection can increase a population's frequency of a beneficial mutation in just hours or days. For most animals and plants, however, their long generation times prevent new mutations from significantly affecting overall genetic variation in the short term. Consequently, animals and plants depend mainly on sexual recombination for the genetic variation that makes adaptation possible. The two sexual processes of meiosis and random fertilization shuffle alleles and deal them out to offspring in fresh combinations (see Chapter 8).

While the processes that generate genetic variation—mutation and sexual recombination—are random, natural selection (and hence evolution) is not. The environment selectively promotes the propagation of those genetic combinations that enhance survival and reproductive success.

Analyzing Gene Pools

A key concept of population genetics is the gene pool. The **gene pool** consists of all alleles in a population at any one time. The gene pool is the reservoir from which the next generation draws its genes.

Imagine a wildflower population with two varieties (morphs) contrasting in flower color. An allele for red flowers, which we will symbolize by *R*, is dominant to an allele for white flowers, symbolized by *r*. These are the only two alleles for flower color in the gene pool of this plant population. Now, let's say that 80%, or 0.8, of all flower-color loci in the gene pool have the *R* allele. We'll use the letter p to represent the relative frequency of the dominant allele in the population. Thus, $p = 0.8$. Because there are only two alleles in this example, the *r* allele must be present at the other 20% (0.2) of the gene pool's flower-color loci. Let's use the letter q for the frequency of the recessive allele in the population. For the wildflower population, $q = 0.2$. And since there are only two alleles for flower color, we know that

$$p + q = 1$$

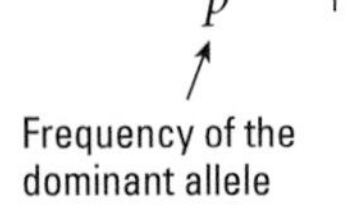

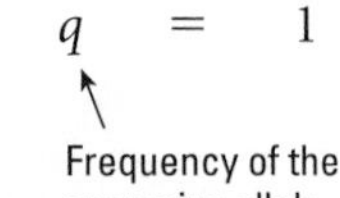

Notice that if we know the frequency of either allele in the gene pool, we can subtract it from 1 to calculate the frequency of the other allele.

From the frequencies of alleles, we can also calculate the frequencies of different genotypes in the population if the gene pool is completely stable (not evolving). In the wildflower population, what is the probability of producing an *RR* individual by "drawing" two *R* alleles from the pool of gametes? Here we apply the rule of multiplication that you learned in Chapter 9. The probability of drawing an *R* sperm multiplied by the probability of drawing an *R* egg is $p \times p = p^2$, or $0.8 \times 0.8 = 0.64$. In other words, 64% of the plants in the population will have the *RR* genotype. Applying the same math, we also know the frequency of *rr* individuals in the population: $q^2 = 0.2 \times 0.2 = 0.04$. Thus, 4% of the plants are *rr*, giving them white flowers. Calculating the frequency of heterozygous individuals, *Rr*, is trickier. That's because the heterozygous genotype can form in two ways, depending on whether the sperm or egg supplies the dominant allele. So the frequency of the *Rr* genotype is $2pq$, which is $2 \times 0.8 \times 0.2 = 0.32$. In our imaginary wildflower population, 32% of the plants are *Rr*, with red flowers. **Figure 13.20** reviews these calculations graphically.

Now we can write a general formula for calculating the frequencies of genotypes in a gene pool from the frequencies of alleles, and vice versa:

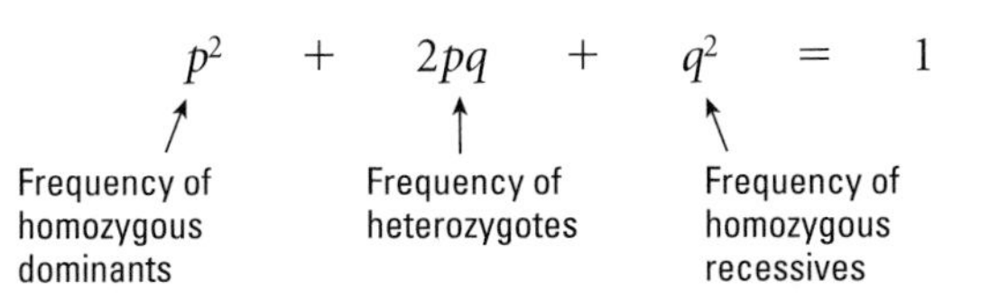

Notice that the frequencies of all genotypes in the gene pool must add up to 1. This formula is called the **Hardy-Weinberg formula,** named for the two scientists who derived it in 1908.

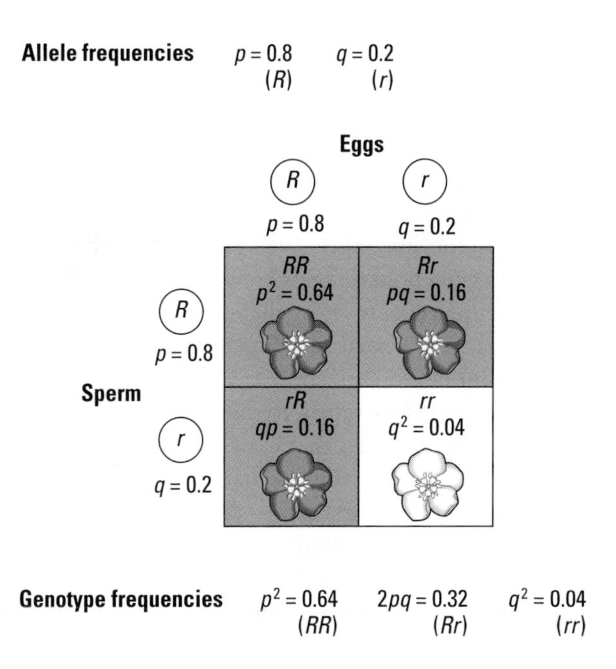

Figure 13.20 A mathematical swim in the gene pool. Each of the four boxes in the grid corresponds to an equally probable "draw" of alleles from the gene pool.

Population Genetics and Health Science

We can use the Hardy-Weinberg formula to calculate the percentage of a human population that carries the allele for a particular inherited disease. Consider phenylketonuria (PKU), which is an inherited inability to break down the amino acid phenylalanine. If untreated, the disorder causes severe mental retardation. PKU occurs in about one out of 10,000 babies born in the United States. Newborn babies are now routinely tested for PKU, and symptoms can be prevented if individuals living with the disease follow a strict diet **(Figure 13.21)**.

PKU is due to a recessive allele. Thus, the frequency of individuals in the U.S. population born with PKU corresponds to the double recessive q^2 term in the Hardy-Weinberg formula. For one PKU occurrence per 10,000 births, $q^2 = 0.0001$. Therefore, q, the frequency of the recessive allele in the population, equals the square root of 0.0001, or 0.01. And p, the frequency of the dominant allele, equals $1 - q$, or 0.99.

Now let's calculate the frequency of heterozygous individuals, those who carry the PKU allele in a single dosage. These carriers are free of the disorder but may pass the PKU allele on to offspring. Carriers are represented in the Hardy-Weinberg formula by $2pq$. And that's $2 \times 0.99 \times 0.01$, or 0.0198. Thus, the formula tells us that about 2% (actually, 1.98%) of the U.S. population carry the PKU allele. Estimating the frequency of a harmful allele is essential for any public health program dealing with genetic diseases.

Diet
INGREDIENTS: SORBITOL, MAGNESIUM STEARATE, ARTIFICIAL FLAVOR,
ASPARTAME† (SWEETENER),
ARTIFICIAL COLOR (YELLOW 5 LAKE, BLUE 1 LAKE), ZINC GLUCONATE.
†PHENYLKETONURICS: CONTAINS PHENYLALANINE

Figure 13.21 A warning to individuals with PKU. People with PKU (phenylketonurics) must strictly regulate their dietary intake of the amino acid phenylalanine. In addition to natural sources, phenylalanine is found in aspartame, a common artificial sweetener. The frequency of the PKU allele is high enough to warrant a public health program that includes warnings on foods that contain phenylalanine.

Microevolution as Change in a Gene Pool

How can we tell if a population is evolving? As stated earlier, evolution can be regarded as changes in the genetic composition of a population over time. How can we measure that? It helps, as a basis of comparison, to know what to expect if a population is *not* evolving. A nonevolving population is in genetic equilibrium, also called **Hardy-Weinberg equilibrium.** The population's gene pool remains constant over time. From generation to generation, the frequencies of alleles (p and q) and genotypes (p^2, $2pq$, and q^2) are unchanged. Sexual shuffling of genes cannot by itself change a large gene pool. But natural selection can.

As an example, let's return to our wildflower population. Imagine the arrival of an insect species that is a vigorous pollinator of plants. If this insect is attracted to white flowers, its presence could enhance the reproductive success of white-flowered plants. Over the generations, this selection factor would increase the frequency of the *r* allele at the expense of the *R* allele. In contrast to the Hardy-Weinberg equilibrium of a nonevolving population, we would now have the changing gene pool of an evolving population.

One of the products of the modern synthesis was a definition of evolution that is based on population genetics: Evolution is a generation-to-generation change in a population's frequencies of alleles. Because this describes evolution on the smallest scale, it is sometimes referred to more specifically as **microevolution.**

CHECKPOINT

1. What is the smallest biological unit that can evolve?
2. Define microevolution.
3. Which term in the Hardy-Weinberg formula ($p^2 + 2pq + q^2 = 1$) corresponds to the frequency of individuals who have *no* alleles for the recessive disease PKU?
4. Which process, mutation or sexual recombination, results in most of the generation-to-generation variability in human populations?

Answers: 1. A population **2.** Microevolution is a change in a population's frequencies of alleles. **3.** p^2 **4.** Sexual recombination

Mechanisms of Microevolution

Now that we understand microevolution to be the changes in a population's genetic makeup from generation to generation, we come to an obvious question: Aside from natural selection, what other mechanisms can change a gene pool? The other main causes of microevolution are genetic drift, gene flow, and mutation.

Genetic Drift

Flip a coin a thousand times, and a result of 700 heads and 300 tails would make you very suspicious about that coin. But flip a coin ten times, and an outcome of seven heads and three tails would seem within reason. The smaller the sample, the greater the chance of deviation from an idealized result—in this case, an equal number of heads and tails.

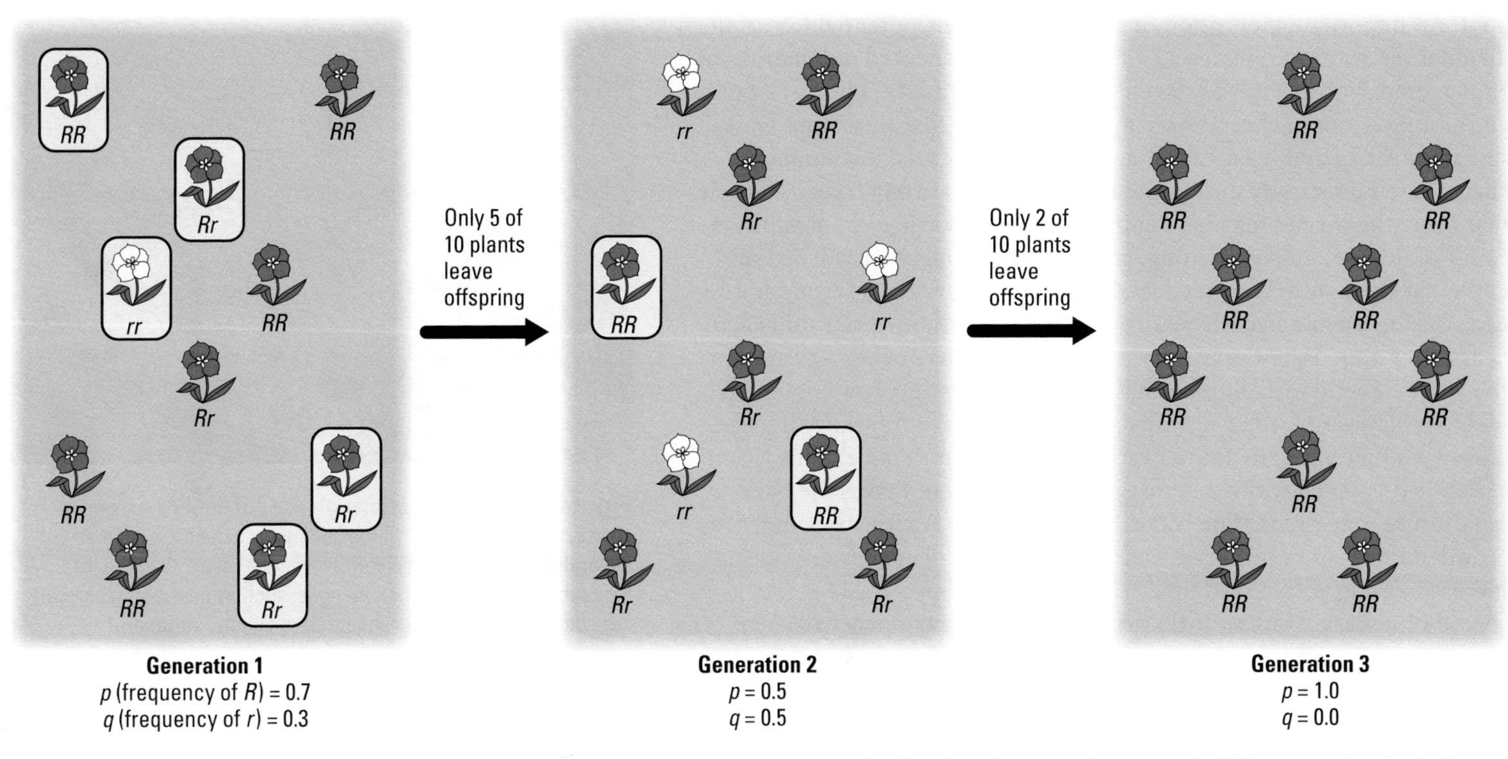

Figure 13.22 Genetic drift. This small wildflower population has a stable size of only about ten plants. For generation 1, only the five boxed plants produce fertile offspring. Only two plants of generation 2 manage to leave fertile offspring. Over the generations, genetic drift can completely eliminate some alleles, as is the case for the *r* allele in generation 3 of this imaginary population.

Let's apply coin-toss logic to a population's gene pool. If a new generation draws its alleles at random from the previous generation, then the larger the population (the sample size), the better the new generation will represent the gene pool of the previous generation. Thus, one requirement for a gene pool to maintain the status quo is a large population size. The gene pool of a small population may not be accurately represented in the next generation because of sampling error. The changed gene pool is analogous to the erratic outcome from a small sample of coin tosses.

Figure 13.22 applies this concept of sampling error to a small population of wildflowers. Chance causes the frequencies of the alleles for red (*R*) and white (*r*) flowers to change over the generations. And that fits our definition of microevolution. This evolutionary mechanism, a change in the gene pool of a small population due to chance, is called **genetic drift.** But what would cause a population to shrink down to a size where genetic drift occurs? Two ways this can occur are the bottleneck effect and the founder effect.

The Bottleneck Effect Disasters such as earthquakes, floods, and fires may kill large numbers of individuals, producing a small surviving population that is unlikely to have the same genetic makeup as the original population. By chance, certain alleles may be overrepresented among the survivors. Other alleles may be underrepresented. And some alleles may be eliminated altogether. Chance may continue to change the gene pool for many generations until the population is again large enough for sampling errors to be insignificant. The analogy in **Figure 13.23** illustrates why genetic drift due to a drastic reduction in population size is called the **bottleneck effect.**

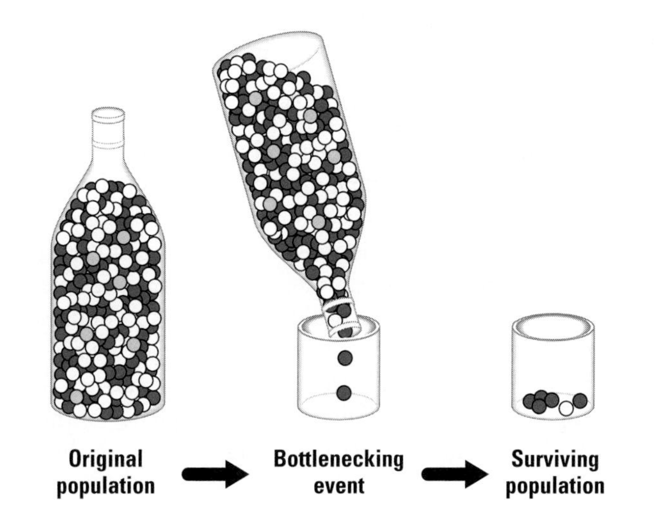

Figure 13.23 The bottleneck effect. The colored marbles in this analogy represent three morphs in an imaginary population. Shaking just a few of the marbles through the bottleneck is like drastically reducing the size of a population struck by some environmental disaster. By chance, blue marbles are overrepresented in the new population, white marbles are underrepresented, and gold marbles are absent. Similarly, bottlenecking a population of organisms tends to reduce variability.

Bottlenecking usually reduces the overall genetic variability in a population because at least some alleles are likely to be lost from the gene pool. An important application of this concept is the potential loss of individual variation, and hence adaptability, in bottlenecked populations of endangered species, such as the cheetah **(Figure 13.24)**. The fastest of all running animals, cheetahs are magnificent cats that were once widespread in Africa and Asia. Like many African mammals, the number of cheetahs fell drastically during the last ice age some 10,000 years ago. At that time, the species may have suffered a severe bottleneck, possibly as a result of disease, human hunting, and periodic droughts. Evidence suggests that the South African cheetah population suffered a second bottleneck during the nineteenth century when farmers hunted the animals to near extinction. Today, only a few small populations of cheetahs exist in the wild. Genetic variability in these populations is very low compared with populations of other mammals. In fact, genetic uniformity in cheetahs rivals that of highly inbred varieties of laboratory mice! This lack of variability, coupled with an increasing loss of habitat, makes the cheetah's future precarious. The cheetahs remaining in Africa are being crowded into nature preserves and parks as human demands on the land increase. Along with crowding comes an increased potential for the spread of disease. With so little variability, the cheetah may have a reduced capacity to adapt to such environmental challenges. Captive breeding programs are already under way and may be required for the cheetah's long-term survival.

Figure 13.24 Implications of the bottleneck effect in conservation biology. Some endangered species, such as this cheetah, have low genetic variability. As a result, they are less adaptable to environmental changes, such as new diseases, than are species with a greater resource of genetic variation.

The Founder Effect Genetic drift is also likely when a few individuals colonize an isolated island, lake, or other new habitat. The smaller the colony, the less its genetic makeup will match the gene pool of the larger population from which the colonists emigrated. If the colony succeeds, random drift will continue to affect the frequency of alleles until the population is large enough for genetic drift to be minimal.

The establishment of a new population whose gene pool differs from the parent population and the subsequent genetic drift in the new colony is known as the **founder effect.** The effect undoubtedly contributed to the evolutionary divergence of the finches and other organisms that arrived as strays on the remote Galápagos Islands that Darwin visited.

Genetic Drift and Hereditary Disorders in Human Populations The founder effect explains the relatively high frequency of certain inherited disorders in some human populations established by small numbers of colonists. In 1814, 15 people founded a British colony on Tristan da Cunha, a group of small islands in the middle of the Atlantic Ocean **(Figure 13.25)**. Apparently, one of the colonists carried a recessive allele for retinitis pigmentosa, a progressive form of blindness. Of the 240 descendants who still lived on the islands in the 1960s, 4 had retinitis pigmentosa. At least 9 others were known to be carriers (heterozygous, with one copy of the recessive allele). That frequency of the retinitis pigmentosa allele was much higher than in Great Britain, the source of the colonists.

Figure 13.25 Residents of Tristan da Cunha in the early 1900s. The island of Tristan da Cunha, located in the middle of the Atlantic Ocean, is listed in the *Guinness Book of Worlds Records* as the world's most remote inhabited island. Such genetic isolation resulted in a disproportionately high rate of hereditary blindness.

Gene Flow

In addition to genetic drift, another source of evolutionary change is **gene flow.** A population may gain or lose alleles when fertile individuals move into or out of a population or when gametes (such as plant pollen) are transferred between populations. Gene flow tends to reduce differences between populations. For example, because humans today move more freely about the world than in the past, gene flow has become an important agent of evolutionary change in previously isolated human populations **(Figure 13.26)**.

Mutation

As discussed in Unit 2, **mutations** are changes in an organism's DNA. A new mutation that is transmitted in gametes can immediately change the gene pool of a population by substituting one allele for another. For example, a mutation that causes a white-flowered plant (*rr*) in our hypothetical wildflower population to produce gametes bearing the dominant allele for red flowers (*R*) would decrease the frequency of the *r* allele in the population and increase the frequency of the *R* allele.

For any one gene locus, however, mutation alone does not have much quantitative effect on a large population in a single generation. This is because a mutation at any given gene locus is a very rare event. If some new allele increases its frequency by a significant amount in a population, it is not because mutation is generating the allele in abundance, but because individuals carrying the mutant allele are producing a disproportionate number of offspring as a result of natural selection or genetic drift.

Although mutations at a particular gene locus are rare, the cumulative impact of mutations across the entire genome can be significant. This is because each individual has thousands of genes, and many populations have thousands or millions of individuals. Certainly over the long term, mutation is, in itself, essential to evolution because it is the original source of the genetic variation that serves as raw material for natural selection.

Figure 13.26 Gene flow and human evolution. The migration of people throughout the world is transferring alleles between populations that were once isolated. This magazine cover celebrates our changing gene pools and culture with a computer-generated image blending facial features from several races.

Natural Selection: A Closer Look

Genetic drift, gene flow, and mutation can cause microevolution, but they do not necessarily lead to adaptation. In fact, only blind luck could result in random drift, migrant alleles, or shot-in-the-dark mutations improving a population's fit to its environment. Of all causes of microevolution, only natural selection promotes adaptation. And such evolutionary adaptation, remember, is a blend of chance and sorting—chance in the random generation of genetic variability, and sorting in the differential reproductive success among the varying individuals. Darwin explained the basics of natural selection. But it took the modern synthesis to fill in the details.

The Hardy-Weinberg equilibrium, which defines a nonevolving population, demands that all individuals in a population be equal in their ability to survive and reproduce. This condition is probably never completely met. On average, those individuals that function best in the environment leave the most offspring and therefore have a disproportionate impact on the gene pool. When farmers began spraying their fields with pesticides, resistant pests started outreproducing other members of the insect populations. This increased the frequency of alleles for pesticide resistance in gene pools. Microevolution by natural selection occurred.

Darwinian Fitness The phrases "struggle for existence" and "survival of the fittest" are misleading if we take them to mean direct competitive contests between individuals. There *are* animal species in which individuals lock horns or otherwise do combat to determine mating privilege. But reproductive success is generally more subtle and passive. Plants in a wildflower population, for example, may differ in reproductive success because some attract more pollinators—perhaps the result of slight differences in flower color, shape, or fragrance **(Figure 13.27)**. In a varying population

Figure 13.27 Darwinian fitness of some flowering plants depends in part on competition in attracting pollinators.

of moths, certain individuals may survive and produce more offspring than others because their wing colors better hide them from predators. A frog may produce more eggs than her neighbors because she is better at catching insects for food. These examples point to a biological definition of **fitness,** the contribution an individual makes to the gene pool of the next generation relative to the contributions of other individuals.

Survival to sexual maturity, of course, is prerequisite to reproductive success. But the biggest, fastest, toughest frog in the pond has a fitness of zero if it is sterile. Production of fertile offspring is the only score that counts in natural selection.

Three General Outcomes of Natural Selection Imagine a population of mice. Individuals range in fur color from very light to very dark brown. If we graph the number of mice in each color category, we get a bell-shaped curve, like the grading curve some of your instructors draw after an exam. If natural selection favors certain fur-color phenotypes over others, the population of mice will change over the generations. Three general outcomes are possible, depending on which phenotypes are favored. These three modes of natural selection are called directional selection, disruptive selection, and stabilizing selection.

Directional selection shifts the phenotypic curve of a population by selecting in favor of some extreme phenotype—the darkest mice, for example (**Figure 13.28a**). Directional selection is most common when the local environment changes or when organisms migrate to a new environment. An actual example is the shift of insect populations toward a greater frequency of pesticide-resistant individuals.

Disruptive selection can lead to a balance between two or more contrasting

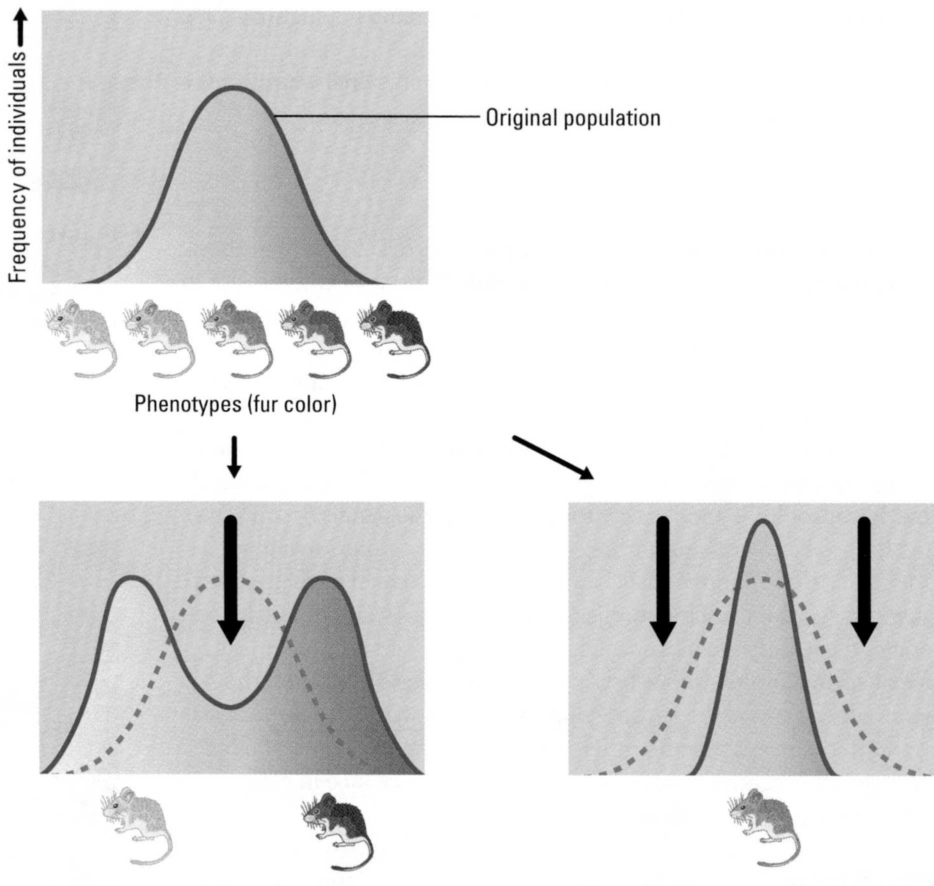

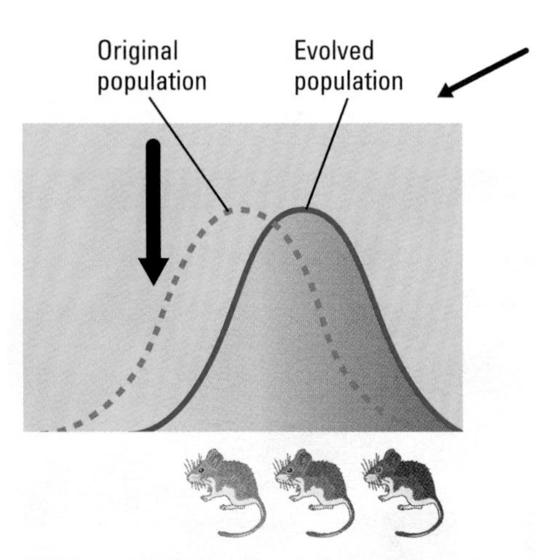

(a) Directional selection shifts the overall makeup of the population by favoring variants of one extreme. In this case, the trend is toward darker color, perhaps because the landscape has been shaded by the growth of trees.

(b) Disruptive selection favors variants of opposite extremes over intermediate individuals. Here, the relative frequencies of very light and very dark mice have increased. Perhaps the mice have colonized a patchy habitat where a background of light soil is studded with dark rocks.

(c) Stabilizing selection culls extreme variants from the population, in this case eliminating individuals that are unusually light or dark. The trend is toward reduced phenotypic variation and maintenance of the status quo.

Figure 13.28 Three general effects of natural selection on a phenotypic character. Here are three possible outcomes for selection working on fur color in imaginary populations of mice. The large downward arrows symbolize the pressure of natural selection working against certain phenotypes.

morphs (phenotypic forms) in a population **(Figure 13.28b)**. A patchy environment, which favors different phenotypes in different patches, is one situation associated with disruptive selection. The polymorphic snake population in Figure 13.19 is an example of disruptive selection.

Stabilizing selection maintains variation for a particular trait within a narrow range **(Figure 13.28c)**. It typically occurs in a relatively stable environment to which populations are already well adapted. This evolutionary conservatism works by selecting against the more extreme phenotypes. For example, stabilizing selection keeps the majority of human birth weights between 3 and 4 kg (approximately 6.5–9 pounds). For babies much lighter or heavier than this, infant mortality is greater.

Of the three selection modes, stabilizing selection probably prevails most of the time, resisting change in well-adapted populations. Evolutionary spurts occur when a population is stressed by a change in the environment or by migration to a new place. When challenged with a new set of environmental problems, a population either adapts through natural selection or dies off in that locale. The fossil record tells us that extinction is the most common result. Those populations that do survive crises often change enough to be designated new species, as we will see in Chapter 14.

CHECKPOINT

1. Compare and contrast the bottleneck effect and the founder effect as causes of genetic drift.
2. Why might new diseases pose a greater threat to cheetah populations than to mammalian populations having more genetic variation?
3. Which mechanism of microevolution has been most affected by the ease of human travel resulting from new modes of transportation?
4. What is the best measure of Darwinian fitness?
5. The thickness of fur in a bear population increases over several generations as the climate in the region becomes colder. This is an example of which type of selection: directional, disruptive, or stabilizing?

Answers: 1. Both processes result in populations being small enough for significant sampling error in the gene pool. A bottleneck reduces the size of an existing population. The founder effect is a new, small population consisting of individuals from a larger population. **2.** Because cheetah populations have so little variation, there is the potential for some new disease against which no individuals are resistant. **3.** Gene flow **4.** The number of fertile offspring an individual leaves **5.** Directional

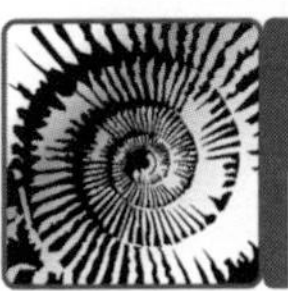

EVOLUTION CONNECTION

Population Genetics of the Sickle-Cell Allele

As the capstone of every chapter in this textbook, the Evolution Connection section reinforces biology's unifying theme. But this unit of chapters is all about evolution itself, so here we'll use this section to relate evolutionary biology to society. And for this chapter, our Connection is the study of two diseases in an evolutionary context.

About one out of every 400 African-Americans have sickle-cell disease, a genetic disorder in which oxygen delivery by the blood is impaired owing to abnormally shaped red blood cells (**Figure 13.29** inset). Red blood cell sickling causes periodic painful episodes as well as some potentially life-threatening complications, such as stroke. Sickle-cell disease is caused by a recessive allele. Only homozygous individuals, who inherit the recessive alleles from both parents, have the disorder. About one in ten African-Americans (and far fewer members of other ethnic groups) have a single copy of the sickle-cell allele. These heterozygous individuals do not have sickle-cell disease, but can pass the allele for the disorder on to their children. (You can review the biology of sickle-cell disease in Chapters 3 and 9.)

Why is the sickle-cell allele so much more common in African-Americans than in the general U.S. population? And how can we explain such a high frequency among African-Americans for an allele with the potential to shorten life (and hence reproductive success)? Evolutionary biology holds the answers.

In the African tropics, the sickle-cell allele is both boon and bane. It is true that when inherited in double dosage, the allele causes a life-threatening disease. But heterozygous individuals, who have just one copy of the sickle-cell allele, are relatively resistant to malaria. This is an important advantage in tropical regions where malaria, caused by a parasitic microorganism, is a major cause of death. The frequency of the sickle-cell allele in Africa is generally highest in areas where the malaria parasite is most common (see Figure 13.29).

How do we weigh the health impact of an allele that is both beneficial (in heterozygotes) and harmful (in homozygotes)? The Hardy-Weinberg formula is the tool we need. In some African populations, the sickle-cell allele has a frequency of 0.2, or 20%. This is q, the frequency of the recessive allele in the gene pool. And q^2, the frequency of homozygous recessives, is 0.2×0.2, or 0.04. So about 4% of individuals in these populations have sickle-cell disease. Now let's calculate the frequency of heterozygous individuals, who have enhanced resistance to malaria. That would be $2pq$ in the Hardy-Weinberg formula: $2 \times 0.8 \times 0.2$, or 0.32. So in this case, the sickle-cell allele benefits about 32% of the population, while it causes sickle-cell disease in only about 4% of the population. That explains the high frequency of the sickle-cell allele compared with other alleles that cause debilitating diseases. The representation of the allele among black Americans is a vestige of African roots.

This example of the intersection of evolutionary biology and health science is a reminder that biology is the foundation of all medicine. And evolution is the foundation of all biology. ■

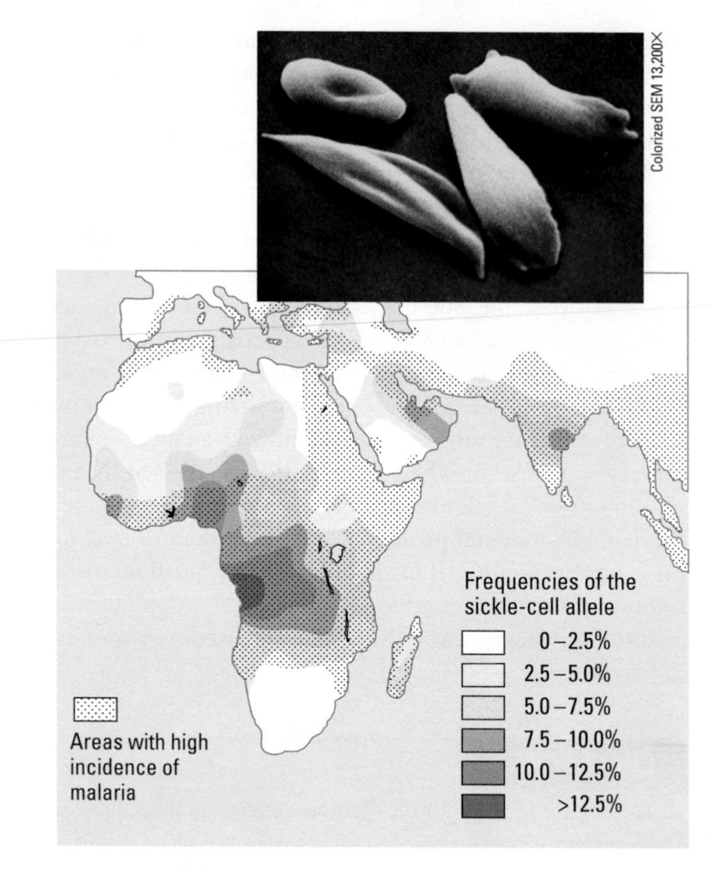

Figure 13.29 Mapping malaria and the sickle-cell allele. The inset shows sickled red blood cells.

Chapter Review

SUMMARY OF KEY CONCEPTS

For study help and activities, go to campbellbiology.com or the student CD-ROM.

Charles Darwin and *The Origin of Species*

• Charles Darwin established the ideas of evolution and natural selection in his 1859 publication *On the Origin of Species by Means of Natural Selection*.

• **Darwin's Cultural and Scientific Context** During his around-the-world voyage on the *Beagle*, Darwin observed adaptations of organisms that inhabited diverse environments. In particular, Darwin was struck by the geographic distribution of organisms on the Galápagos Islands, off the South American coast. When Darwin considered his observations in light of new evidence for a very old Earth that changed slowly, he arrived at ideas that were at odds with the long-held notion of a young Earth populated by unrelated and unchanging species.

Activity *The Voyage of the* Beagle: *Darwin's Trip Around the World*

Activity *Darwin and the Galápagos Islands*

• **Descent with Modification** Darwin made two proposals in *The Origin of Species:* (1) Modern species descended from ancestral species, and (2) natural selection is the mechanism of evolution.

Evidence of Evolution

• **The Fossil Record** The fossil record shows that organisms have appeared in a historical sequence, and many fossils link ancestral species with those living today.

• **Biogeography** Biogeography, the study of the geographic distribution of species, suggests that species evolved from ancestors that inhabited the same region.

• **Comparative Anatomy** Homologous structures among species and vestigial organs provide evidence of evolutionary history.

Activity *Reconstructing Forelimbs*

• **Comparative Embryology** Closely related species often have similar stages in their embryonic development.

• **Molecular Biology** All species share a common genetic code, suggesting that all forms of life are related through branching evolution from the earliest organisms. Comparisons of DNA and proteins provide evidence of evolutionary relationships.

Natural Selection

MP3 Tutor *Natural Selection*

• **Darwin's Theory of Natural Selection** Individuals best suited for a particular environment are more likely to survive and reproduce than less fit individuals.

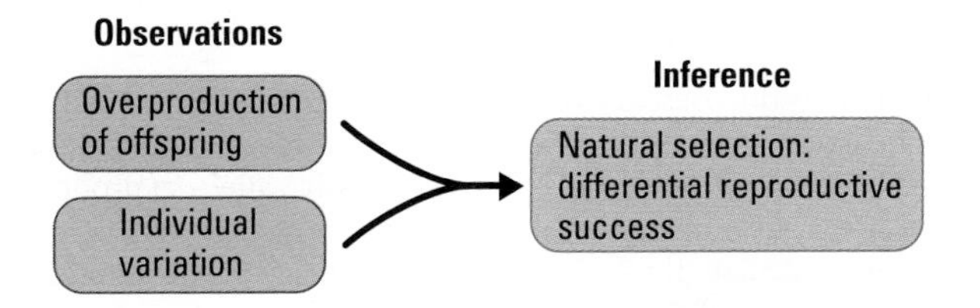

• **Natural Selection in Action** Natural selection can be observed in the evolution of pesticide-resistant insects, drug-resistant microbes, and horned lizards, among many other organisms.

Case Study in the Process of Science *What Are the Patterns of Antibiotic Resistance?*

Case Study in the Process of Science *How Do Environmental Changes Affect a Population of Leafhoppers?*

The Modern Synthesis: Darwinism Meets Genetics

• The modern synthesis fused genetics (Mendelism) and evolutionary biology (Darwinism) in the mid-1900s.

• **Populations as the Units of Evolution** A population, members of the same species living in the same time and place, is the smallest biological unit that can evolve. Population genetics emphasizes the extensive genetic variation within populations and tracks the genetic makeup of populations over time.

• **Genetic Variation in Populations** Polygenic ("many gene") inheritance produces traits that vary continuously, whereas traits that are determined by one genetic locus may be polymorphic, with two or more distinct forms. Mutation and sexual recombination produce genetic variation.

Activity *Genetic Variation from Sexual Recombination*

• **Analyzing Gene Pools** The gene pool consists of all alleles in all the individuals making up a population. The Hardy-Weinberg formula can be used to calculate the frequencies of genotypes in a gene pool from the frequencies of alleles, and vice versa:

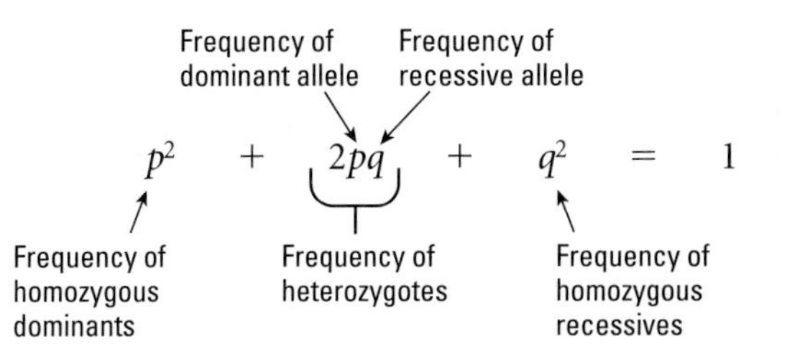

Case Study in the Process of Science *How Can Frequency of Alleles Be Calculated?*

• **Population Genetics and Health Science** The Hardy-Weinberg formula can be used to estimate the frequency of a harmful allele, which is useful information for public health programs dealing with genetic diseases.

• **Microevolution as Change in a Gene Pool** Microevolution is a generation-to-generation change in a population's frequencies of alleles.

Mechanisms of Microevolution

• **Genetic Drift** Genetic drift is a change in the gene pool of a small population due to chance. Bottlenecking (a drastic reduction in population size) and the founder effect (a new population started by a few individuals) are two situations leading to genetic drift.

• **Gene Flow** A population may gain or lose alleles by gene flow, which is genetic exchange with another population.

• **Mutation** Individual mutations have relatively little short-term effect on a large gene pool. In the long term, mutation is the source of genetic variation.

• **Natural Selection: A Closer Look** Of all causes of microevolution, only natural selection promotes evolutionary adaptations. Darwinian fitness is the contribution an individual makes to the gene pool of the next generation relative to the contributions of other individuals. The outcome of natural selection may be directional, disruptive, or stabilizing.

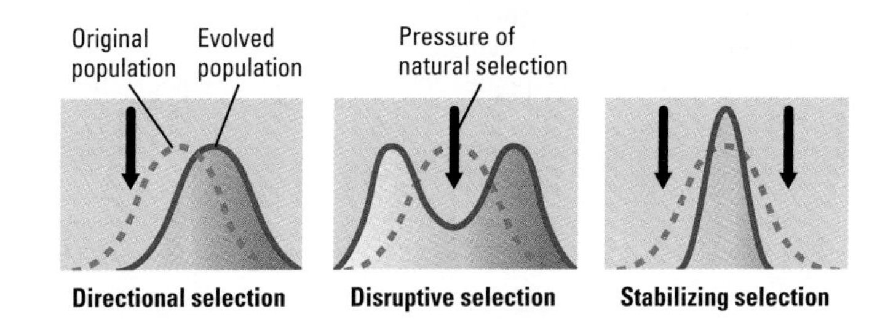

Activity *Causes of Microevolution*

SELF-QUIZ

1. Which of the following is *not* an observation or inference on which Darwin's theory of natural selection is based?
   a. There is heritable variation among individuals.
   b. Poorly adapted individuals never produce offspring.
   c. Because excessive numbers of offspring are produced, there is a struggle for limited resources.
   d. Individuals whose inherited characteristics best fit them to the environment will generally produce more offspring.

2. Which of the following is a true statement about Charles Darwin?
   a. He was the first to discover that living things can change, or evolve.
   b. He based his theory on the inheritance of acquired characteristics.
   c. He proposed natural selection as the mechanism of evolution.
   d. He was the first to realize that Earth is more than 6,000 years old.

3. In a population with two alleles for a particular genetic locus, B and b, the allele frequency of B is 0.7. If this population is in Hardy-Weinberg equilibrium, the frequency of heterozygotes is __________, the frequency of homozygous dominants is __________, and the frequency of homozygous recessives is __________.

4. Define fitness from an evolutionary perspective.

5. The processes of __________ and __________ generate variation, and __________ produces adaptation to the environment.
   a. sexual recombination . . . natural selection . . . mutation
   b. mutation . . . sexual recombination . . . genetic drift
   c. genetic drift . . . mutation . . . sexual recombination
   d. mutation . . . natural selection . . . sexual recombination
   e. mutation . . . sexual recombination . . . natural selection

6. As a mechanism of microevolution, natural selection can be most closely equated with
   a. random mating.
   b. genetic drift.
   c. unequal reproductive success.
   d. gene flow.

7. Why does a founder event favor microevolution in the founding population?

8. In a particular bird species, individuals with average-sized wings survive severe storms more successfully than other birds in the same population with longer or shorter wings. Of the three general outcomes of natural selection (directional, disruptive, or stabilizing), this example illustrates __________.

9. Which of the following statements is (are) true about a population in Hardy-Weinberg equilibrium? (More than one may be true.)
   a. The population is quite small.
   b. The population is not evolving.
   c. Gene flow between the population and surrounding populations does not occur.
   d. Natural selection is not occurring.

10. What environmental factor accounts for the relatively high frequency of the sickle-cell allele in tropical Africa?

Answers to the Self-Quiz questions can be found in Appendix D.

Go to the website or CD-ROM for more Self-Quiz questions.

THE PROCESS OF SCIENCE

11. A population of snails has recently become established in a new region. The snails are preyed on by birds that break the snails open on rocks, eat the soft bodies, and leave the shells. The snails occur in both striped and unstriped forms. In one area, researchers counted both live snails and broken shells. Their data are summarized here:

	Striped	Unstriped
Living	264	296
Broken	486	377
Total	750	673

Based on these data, which snail form is more subject to predation by birds? Predict how the frequencies of striped and unstriped individuals might change over the generations.

12. Imagine that the presence or absence of stripes on the snails from the previous question is determined by a single gene locus, with the dominant allele (S) producing striped snails and the recessive allele (s) producing unstriped snails. Combining the data from both the living snails and broken shells, calculate the following: the frequency of the dominant allele, the frequency of the recessive allele, and the number of heterozygotes in the observed groups.

BIOLOGY AND SOCIETY

13. To what extent are humans in a technological society exempt from natural selection? Explain your answer.

14 How Biological Diversity Evolves

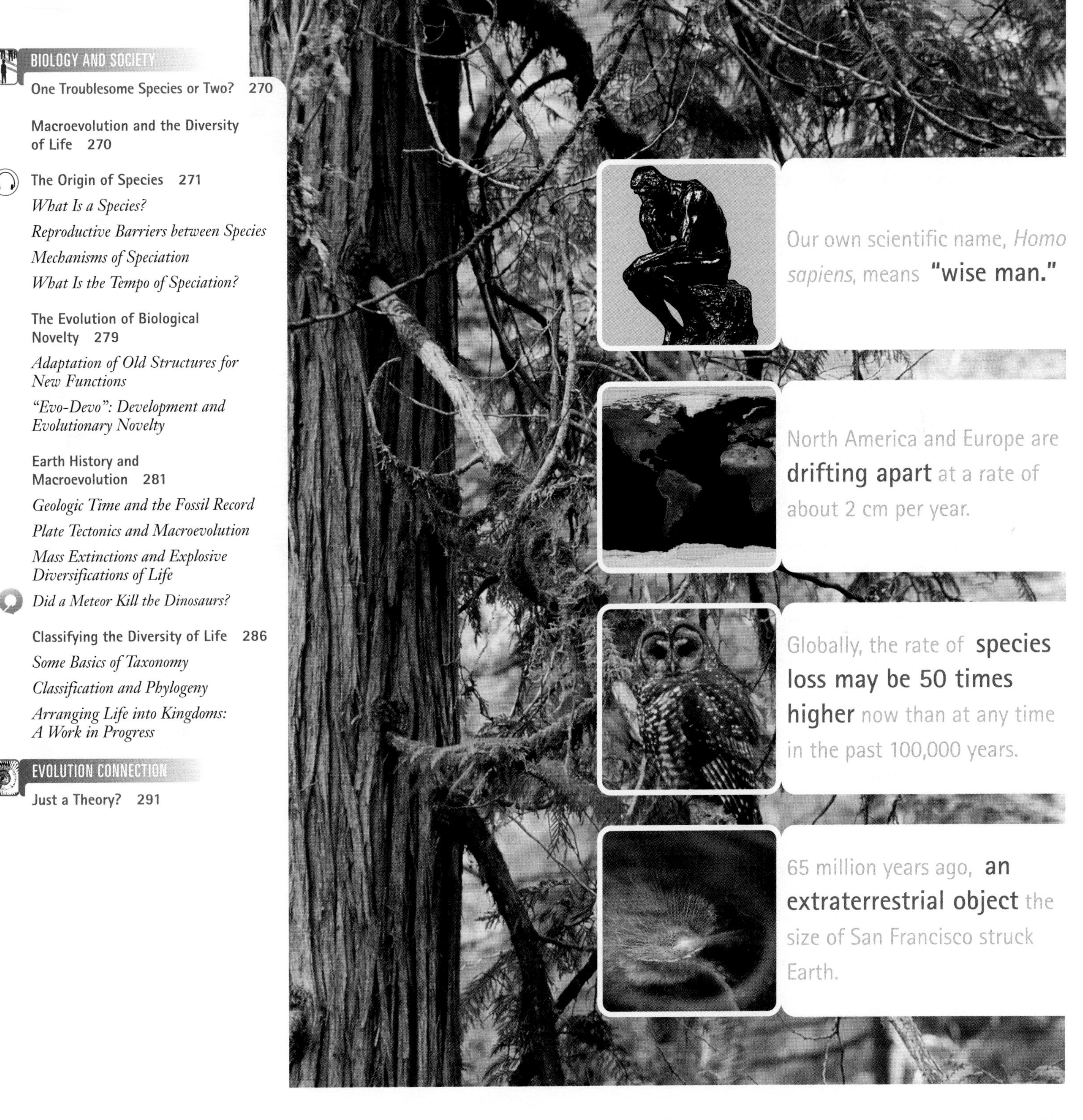

Our own scientific name, *Homo sapiens*, means **"wise man."**

North America and Europe are **drifting apart** at a rate of about 2 cm per year.

Globally, the rate of **species loss may be 50 times higher** now than at any time in the past 100,000 years.

65 million years ago, **an extraterrestrial object** the size of San Francisco struck Earth.

BIOLOGY AND SOCIETY

One Troublesome Species or Two?

Although you might think of them as just a pesky summertime nuisance, mosquitoes have assumed a more serious role in public health since the outbreak of West Nile virus **(Figure 14.1)**. This virus lives primarily in birds. If a mosquito bites an infected bird, the virus may travel from the bird's bloodstream to the mosquito's salivary glands. From there, the virus may be transmitted to a human. Most people infected with West Nile virus (WNV) show only mild symptoms or none at all. But a few (less than 1%) develop serious symptoms, including coma, paralysis, and permanent brain damage.

All mosquitoes may look roughly alike, but close examination reveals many distinct species. An important question facing public health officials is, Which mosquitoes transmit the virus? One tantalizing clue is that very few people are stricken with WNV in northern Europe. Perhaps the responsible mosquito species is one found in the United States but not there.

To find out, a group of researchers studied genetic variation of mosquito populations from around the world by compiling a catalog of mosquito DNA fingerprints. Their data showed that northern Europe is occupied by two distinct species of mosquito: one that lives above-ground and primarily bites birds, and a second that lives below-ground (in subway tunnels, for example) and primarily bites mammals, including humans. The second species was first discovered in the London Underground. Despite having nearby habitats and identical appearance, the two types of European mosquitoes do not interbreed, which fits one of the definitions of a species. The lack of interbreeding between the European above- and below-ground mosquitoes is therefore consistent with the DNA data, suggesting that the mosquitoes do represent two distinct species.

The DNA fingerprints of North American mosquitoes revealed a different story: Many of the insects have DNA in common with both of the European species. Apparently, on the North American continent, the two types of mosquitoes can interbreed. Because these hybrids bite both birds and people, they are capable of spreading WNV between the two. The researchers speculated that this is why WNV has spread so quickly in the United States but not in Europe: The mosquitoes that spread WNV in North America are a blend of two species that remain distinct in Europe and thus do not bite both birds and people there.

As the spread of West Nile virus shows, geography, behavior, and genetics can all keep species separate and influence the emergence of new ones. This chapter begins with the question of what a species is and then discusses the birth of new species. After that, we'll examine the methods biologists use to trace the evolution of biological diversity. ■

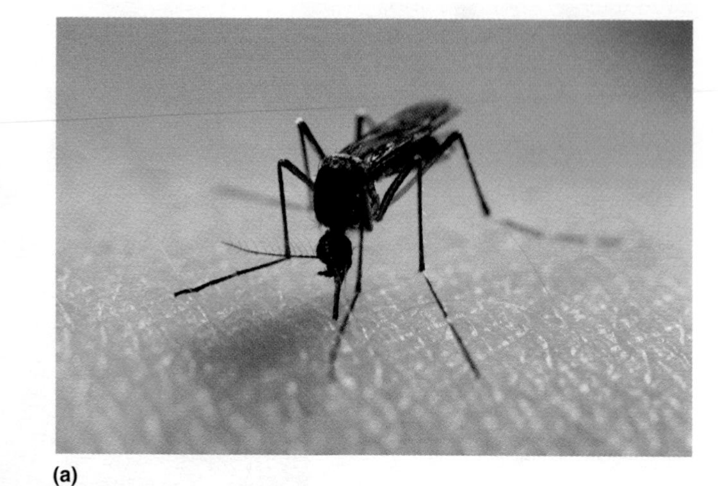

(a)

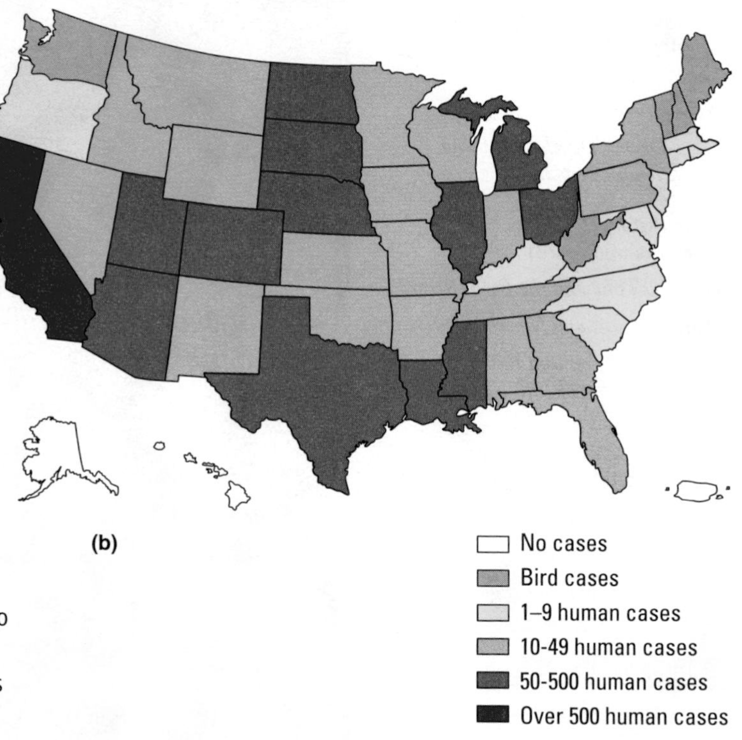

(b)

Figure 14.1 The threat of West Nile virus. (a) West Nile virus—named for the region of Africa where it was originally discovered—is spread when a mosquito feeds on a bird and then a human. **(b)** This map shows the distribution of cases of West Nile virus in the United States during the year 2005.

Macroevolution and the Diversity of Life

When Darwin traveled to the Galápagos Islands, he realized that he was visiting a place of origins. Though the volcanic islands were geologically young, they were already home to many plants and animals known nowhere else in the world. Among the islands' unique inhabitants were its giant tortoises, for which the Galápagos are named (the word *galapago* is Spanish for "tortoise"). After visiting the Galápagos, Darwin wrote in his diary: "Both in space and time, we seem to be brought somewhat near to that great fact—that mystery of mysteries—the first appearance of new beings on this Earth."

To understand the "appearance of new beings," as Darwin put it, it is not enough to explain microevolution and the adaptation of populations, which you learned about in Chapter 13. If that were all that ever happened, then Earth would be populated only by a highly adapted version of the first form of life. Evolutionary theory must also explain **macroevolution,** the major changes in the history of life, which are usually evident in the fossil record. Macroevolution includes the origin of new species, which generates biological diversity; the origin of evolutionary novelty, such as the wings and feathers of birds and the big brains of humans; the explosive diversification that follows some evolutionary breakthrough, such as the origin of thousands of plant species after the flower evolved; and mass extinctions, which clear the way for new adaptive explosions, such as the diversification of mammals that followed the disappearance of dinosaurs.

The beginning of new forms of life—the origin of species—is at the focal point of our study of macroevolution. Sometimes a population may change significantly through adaptation to a changing environment. Such linear evolution does not create a new species **(Figure 14.2a)**. The formation of new species, called **speciation,** occurs when one or more new species branch from a parent species, which may continue to exist **(Figure 14.2b)**. Speciation generates biological diversity by increasing the number of species.

How did evolution produce over 250,000 species of flowering plants? Or 35,000 species of fishes? Or over a million insect species? To explain such diversification, we must understand *how* new species emerge. This is our next topic.

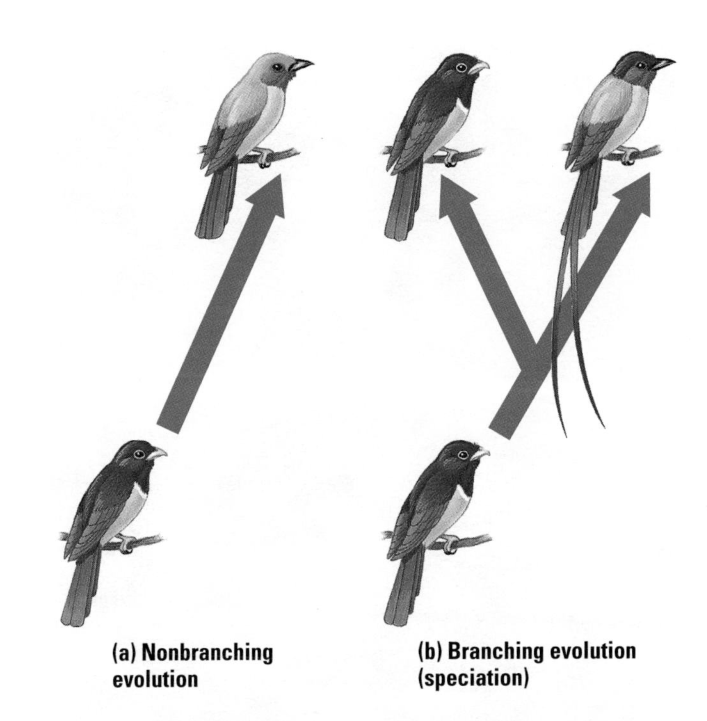

Figure 14.2 Two patterns of evolution. (a) Nonbranching evolution can transform a population significantly but does not create a new species. **(b)** Branching evolution splits a lineage into two or more species, thereby increasing the total number of species.

CHECKPOINT

1. Contrast microevolution with macroevolution.
2. What is speciation?

Answers: 1. Microevolution is a change in the gene pool of a population, often associated with adaptation; macroevolution includes major changes in the history of life, such as the origin of a new species, which is often noticeable enough to be evident in the fossil record. **2.** The formation of new species through branching evolution from previously existing species

The Origin of Species

MP3 Tutor **Speciation**

Species is a Latin word meaning "kind" or "appearance." Indeed, we learn to distinguish between the kinds of plants and animals—between dogs and cats, for example—from differences in their appearance. Although the basic idea of species as distinct life-forms seems intuitive, devising a more formal definition is not so easy.

What Is a Species?

One way of defining a species is called the **biological species concept.** It defines a **species** as a population or group of populations whose members have the potential to interbreed with one another in nature to produce

(a) Similarity between different species **(b) Diversity within one species**

Figure 14.3 The biological species concept is based on reproductive compatibility rather than physical similarity. **(a)** The eastern spotted skunk (top) and the western spotted skunk (bottom) are very similar in appearance, but they are separate species and cannot interbreed. **(b)** In contrast, humans, as diverse in appearance as we are, belong to a single species (*Homo sapiens*) and can interbreed.

fertile offspring (offspring who themselves can reproduce) **(Figure 14.3)**. Geography and culture may conspire to keep a Manhattan businesswoman and a Mongolian dairyman apart. But if the two did get together, they could have viable babies that develop into fertile adults because all humans belong to the same species. In contrast, humans and chimpanzees, as similar and closely related in their evolution as they are, remain distinct species because they do not interbreed. Such reproductive isolation blocks the exchange of genes between species and keeps their gene pools separate.

We cannot apply the biological species concept to all situations. For example, basing the definition of species on reproductive compatibility excludes organisms that only reproduce asexually (producing offspring from a single parent rather than from a mating pair). Fossils aren't doing much sexual reproduction either, so they cannot be evaluated by the biological species concept. The puzzling case of the mosquito species we described in the chapter-opening essay—where noninterbreeding European species produce hybrids in America—further illustrates the limitations of the biological species concept.

In response to such challenges, biologists have developed several other ways to define species. In practice, classification of most organisms is based mainly on observable and measurable physical traits. For example, biologists distinguish fossil species mainly by differences in their appearance. Another approach identifies species in terms of their ecological niches, focusing on unique adaptations to particular roles in a biological community. Yet another classification scheme defines a species as a set of organisms with a unique genetic history—that is, as one tip on the branching tree of life.

Each species concept is useful, depending on the situation and the questions being asked. The biological species concept, however, is particularly useful when focusing on how discrete groups of organisms may arise and be maintained by reproductive isolation. Because reproductive isolation is an essential factor in the evolution of many species, we look at it more closely next.

Reproductive Barriers between Species

Clearly, a fly will not mate with a frog or a fern. But what prevents biological species that are closely related from interbreeding? What, for example, maintains the species boundary between the western spotted skunk and the eastern spotted skunk (shown in Figure 14.3)? Their geographic ranges overlap in the Great Plains region, and they are so similar that only expert zoologists can tell them apart. And yet, these two skunk species do not interbreed.

Let's examine the reproductive barriers that isolate the gene pools of species **(Figure 14.4)**. We can classify reproductive barriers as either prezygotic or postzygotic, depending on whether they block interbreeding before or after the formation of zygotes (fertilized eggs).

Prezygotic barriers prevent mating or fertilization between species. The barrier may be time based (temporal isolation). For example, western spotted skunks breed in the fall, but the eastern species breeds in late winter. Temporal isolation keeps the species from mating even where they coexist on the Great Plains. In other cases, species live in the same region but not in the same habitats (habitat isolation). For example, one species of North American garter snake lives mainly in water, while a closely related species lives on land. Traits that enable individuals to recognize potential mates, such as odor, coloration, or courtship ritual, can also function as reproductive barriers (behavioral isolation). In many bird species, for example, courtship behavior is so elaborate that individuals are unlikely to mistake a bird of a different species as one of their kind **(Figure 14.5)**. In still other cases, the male and female sex organs of different species are anatomically incompatible (mechanical isolation). For example, even if insects of closely related species attempt to mate, the male and female sex organs may not fit together correctly, and no

Figure 14.4 Reproductive barriers between closely related species.

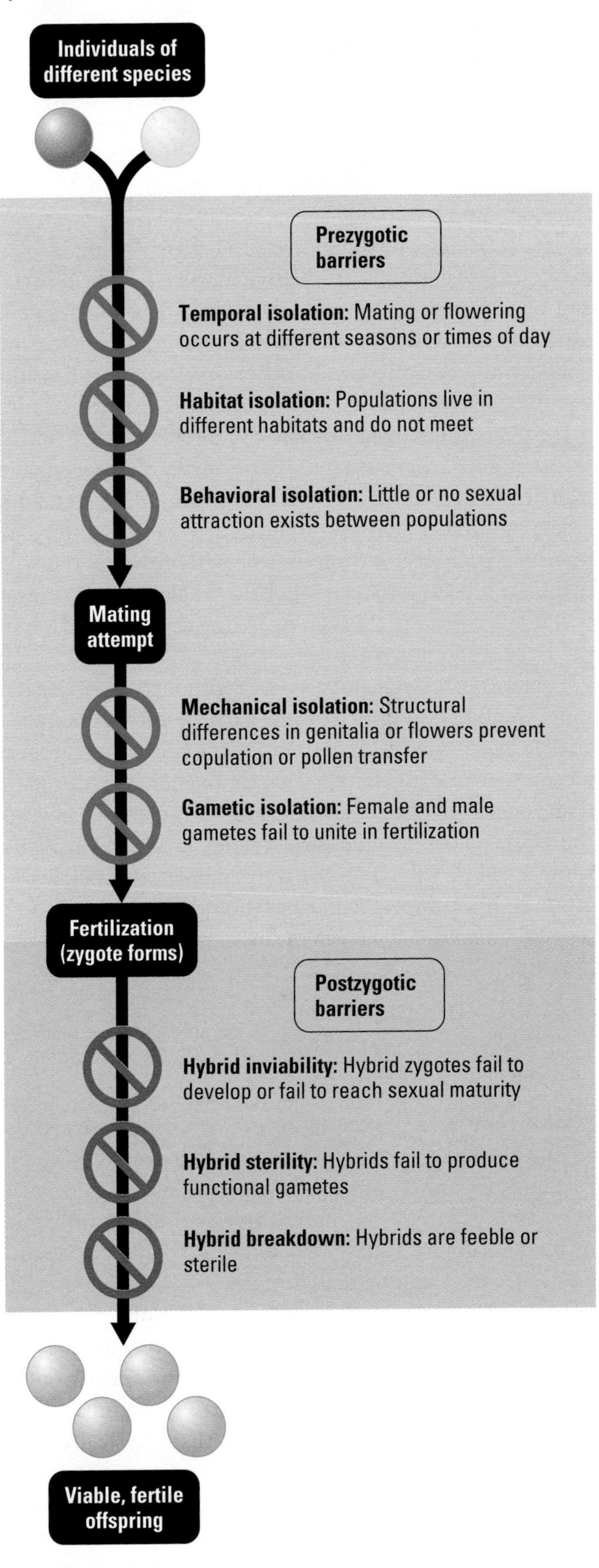

Individuals of different species
Prezygotic barriers
Temporal isolation: Mating or flowering occurs at different seasons or times of day
Habitat isolation: Populations live in different habitats and do not meet
Behavioral isolation: Little or no sexual attraction exists between populations
Mating attempt
Mechanical isolation: Structural differences in genitalia or flowers prevent copulation or pollen transfer
Gametic isolation: Female and male gametes fail to unite in fertilization
Fertilization (zygote forms)
Postzygotic barriers
Hybrid inviability: Hybrid zygotes fail to develop or fail to reach sexual maturity
Hybrid sterility: Hybrids fail to produce functional gametes
Hybrid breakdown: Hybrids are feeble or sterile
Viable, fertile offspring

Figure 14.5 Courtship ritual as a behavioral barrier between species. These blue-footed boobies, inhabitants of the Galápagos Islands, will mate only after a specific ritual of courtship displays. Part of the "script" calls for the male to high-step, a dance that advertises the bright blue feet characteristic of the species.

sperm is transferred. In still other cases, individuals of different species may actually copulate, but their gametes are incompatible or do not survive to meet, and fertilization does not occur (gametic isolation). In many mammals, for example, sperm does not survive within the reproductive tract of females of a different species.

Postzygotic barriers are mechanisms that operate should interspecies mating actually occur and form hybrid zygotes ("hybrid" here meaning that the egg comes from one species and the sperm from another species). In some cases, hybrid offspring die before reaching reproductive maturity (hybrid inviability). For example, although certain closely related frog species will hybridize, the offspring fail to develop normally because of genetic incompatibilities between the two species. In other cases of hybridization, offspring may become vigorous adults, but are infertile (hybrid sterility). A mule, for example, is the hybrid offspring of a female horse and a male donkey **(Figure 14.6)**. Because mules are sterile, there is no avenue for gene transfer between the two parental species, horse and donkey. In other cases, the first-generation hybrids are viable and fertile, but when these hybrids mate with one another or with either parent species, the offspring are feeble or sterile (hybrid breakdown). For example, different species of cotton plants can produce fertile hybrids, but the offspring of the hybrids do not survive.

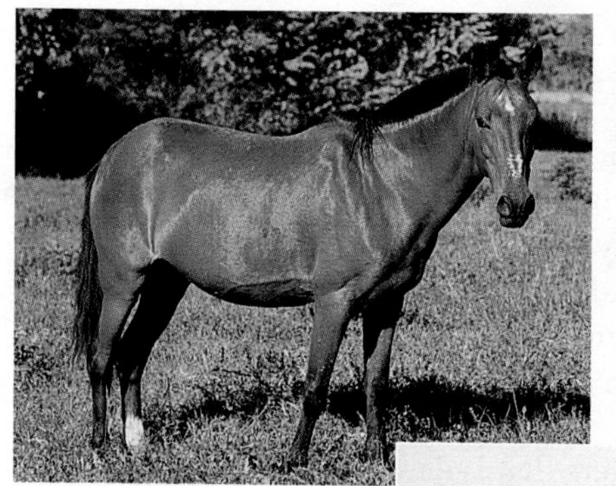

Horse

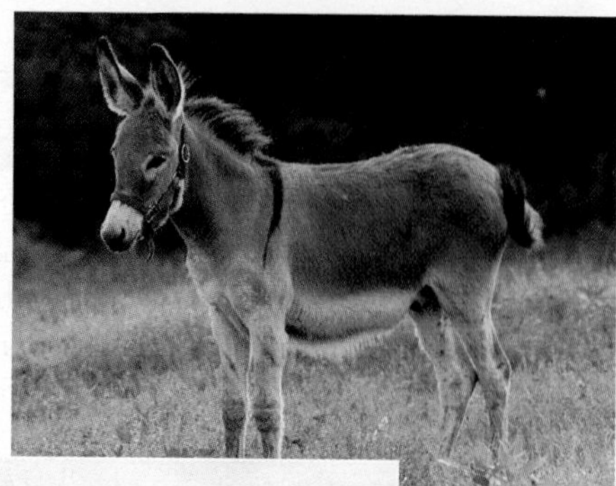

Donkey

Mule (hybrid)

Figure 14.6 Hybrid sterility, a postzygotic barrier. Horses and donkeys remain separate species because their hybrid offspring, mules, are sterile.

In most cases, it is not a single reproductive barrier but some combination of two or more that reinforces boundaries between species. For example, the northern European mosquitoes described in the chapter opener are isolated both temporally (the below-ground variety breeds all year round; the above-ground variety does not breed during the winter) and by habitat (above the ground versus below the ground). If it is reproductive isolation that keeps species separate, then the evolution of these barriers is the key to the origin of new species.

Mechanisms of Speciation

A key event in the potential origin of a species occurs when the gene pool of a population is somehow severed from other populations of the parent species. With its gene pool isolated, the splinter population can follow its own evolutionary course. Changes in its allele frequencies caused by genetic drift and natural selection are unaffected by gene flow from other populations. Such reproductive isolation can result from two general scenarios: allopatric speciation and sympatric speciation **(Figure 14.7)**. In **allopatric speciation,** the initial block to gene flow is a geographic barrier that physically isolates the splinter population. In contrast, **sympatric speciation** is the origin of a new species without geographic isolation. The splinter population becomes reproductively isolated right in the midst of the parent population.

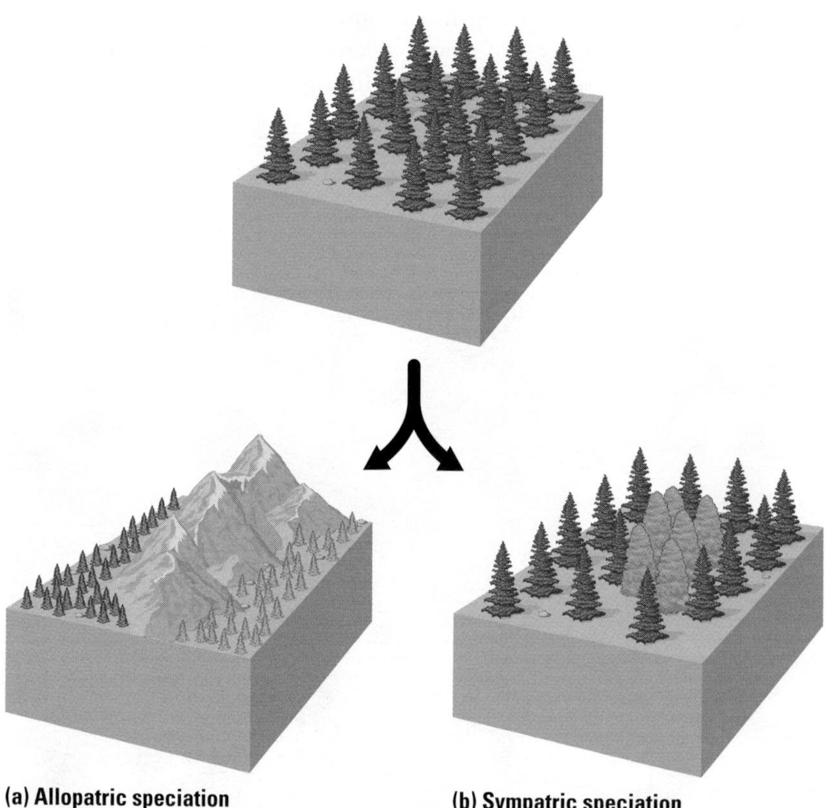

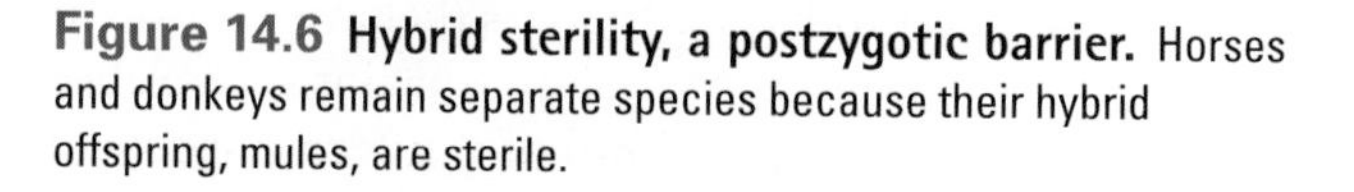

(a) Allopatric speciation

(b) Sympatric speciation

Figure 14.7 Two modes of speciation. **(a)** In allopatric speciation, a population forms a new species after being geographically isolated from its parent population. **(b)** In sympatric speciation, a part of a population becomes a new species while in the midst of its parent population.

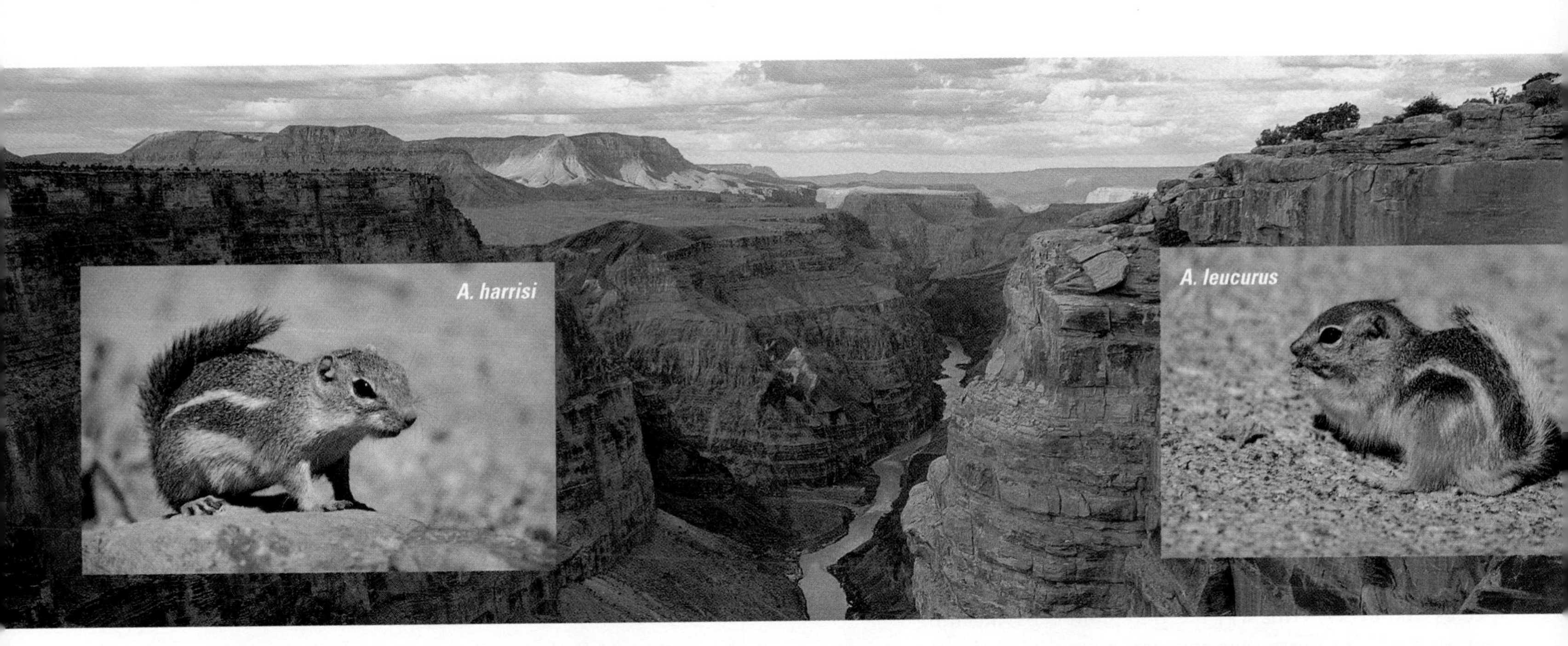

Figure 14.8 Allopatric speciation of antelope squirrels on opposite rims of the Grand Canyon. Harris's antelope squirrel (*Ammospermophilus harrisi*) is found on the south rim of the Grand Canyon. Just a few miles away on the north rim is the closely related white-tailed antelope squirrel (*Ammospermophilus leucurus*). Birds and other organisms that can disperse easily across the canyon have not diverged into different species on opposite rims.

Allopatric Speciation Several kinds of geologic processes can fragment a population into two or more isolated populations. A mountain may emerge and gradually split a population of organisms that can inhabit only lowlands. A land bridge, such as the Isthmus of Panama, may form and separate the marine life on either side. A large lake may subside until there are several smaller lakes, with their populations now isolated. Even without such geologic remodeling, geographic isolation and allopatric speciation can occur if individuals colonize a new, geographically remote area and become isolated from the parent population. An example is the speciation that occurred on the Galápagos Islands following colonization by immigrant organisms.

How formidable must a geographic barrier be to keep allopatric populations apart? The answer depends partly on the ability of the organisms to move about. Birds, mountain lions, and coyotes can cross mountain ranges, rivers, and canyons. Nor do such barriers hinder the windblown pollen of pine trees, and the seeds of many plants may be carried back and forth on animals. In contrast, small rodents may find a deep canyon or a wide river a formidable barrier **(Figure 14.8)**.

The likelihood of allopatric speciation increases when a population is both small and isolated. A small, isolated population is more likely than a large population to have its gene pool changed substantially by both genetic drift and natural selection. For example, in less than 2 million years, the few animals and plants that successfully colonized the Galápagos Islands gave rise to all the species now found there. But for each small, isolated population that becomes a new species, many more simply perish in their new environment. Life on the frontier is harsh, and most pioneer populations probably become extinct.

Even if a small, isolated population survives, it does not necessarily evolve into a new species. The population may adapt to its local environment and begin to look very different from the ancestral population, but that doesn't make it a new species. Speciation occurs with the evolution of reproductive barriers between the isolated population and its parent population. In other words, if speciation occurs during geographic separation, the new species will not breed with its ancestral population, even if the two populations should come back into contact **(Figure 14.9)**.

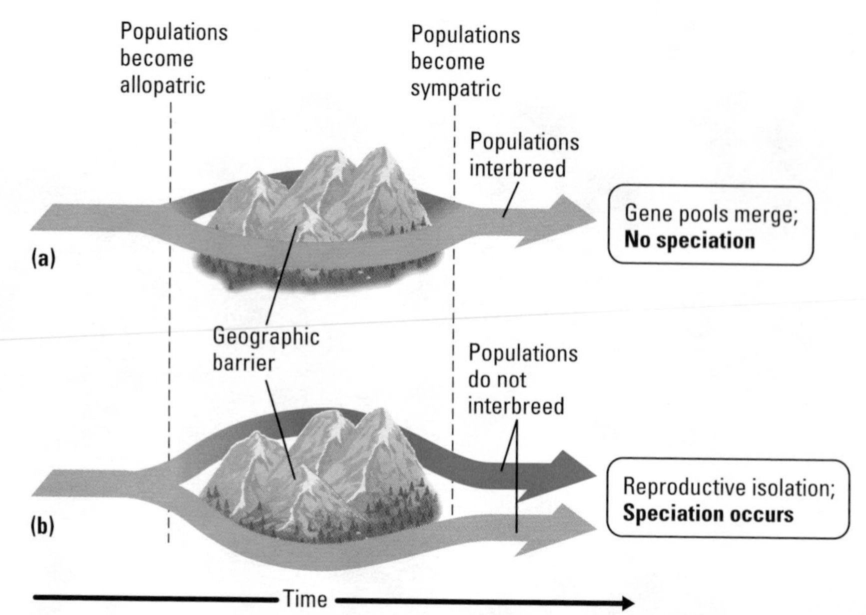

Figure 14.9 Has speciation occurred during geographic isolation? In this analogy, the arrows track populations over time. The mountain symbolizes a period of geographic isolation. The two parts of the figure show the two possibilities when populations come back together after a long period of separation.

Sympatric Speciation How can a subpopulation become reproductively isolated while in the midst of its parent population? This can occur in a single generation if a genetic change produces a reproductive barrier between mutants and the parent population. Sympatric speciation does not seem to be widespread among animals but has been important in plant evolution.

Many plant species have originated from accidents during cell division that resulted in extra sets of chromosomes (see Evolution Connection in Chapter 8). A new species that evolves this way has polyploid cells, meaning that each cell has more than two complete sets of chromosomes. The new species cannot produce fertile hybrids with its parent species, so reproductive isolation and speciation have occurred in a single generation without geographic isolation. This mechanism of sympatric speciation was first discovered in the early 1900s by Dutch botanist Hugo de Vries. During his experiments, he identified a new species of primrose that arose in this way **(Figure 14.10)**.

Polyploids do not always come from a single parent species. In fact, most polyploid species arise from the hybridization of two parent species. This mechanism of sympatric speciation accounts for many of the plant species we grow for food, including oats, potatoes, bananas, peanuts,

Figure 14.10 Botanist Hugo de Vries and his new primrose species. Working in the early 1900s, de Vries studied variation in evening primroses. During his breeding experiments, de Vries observed a new primrose variety that arose through sympatric speciation in just a single generation. The new species, named *Oenothera gigas* for its large size (lower right), could not interbreed with its parent species (upper left).

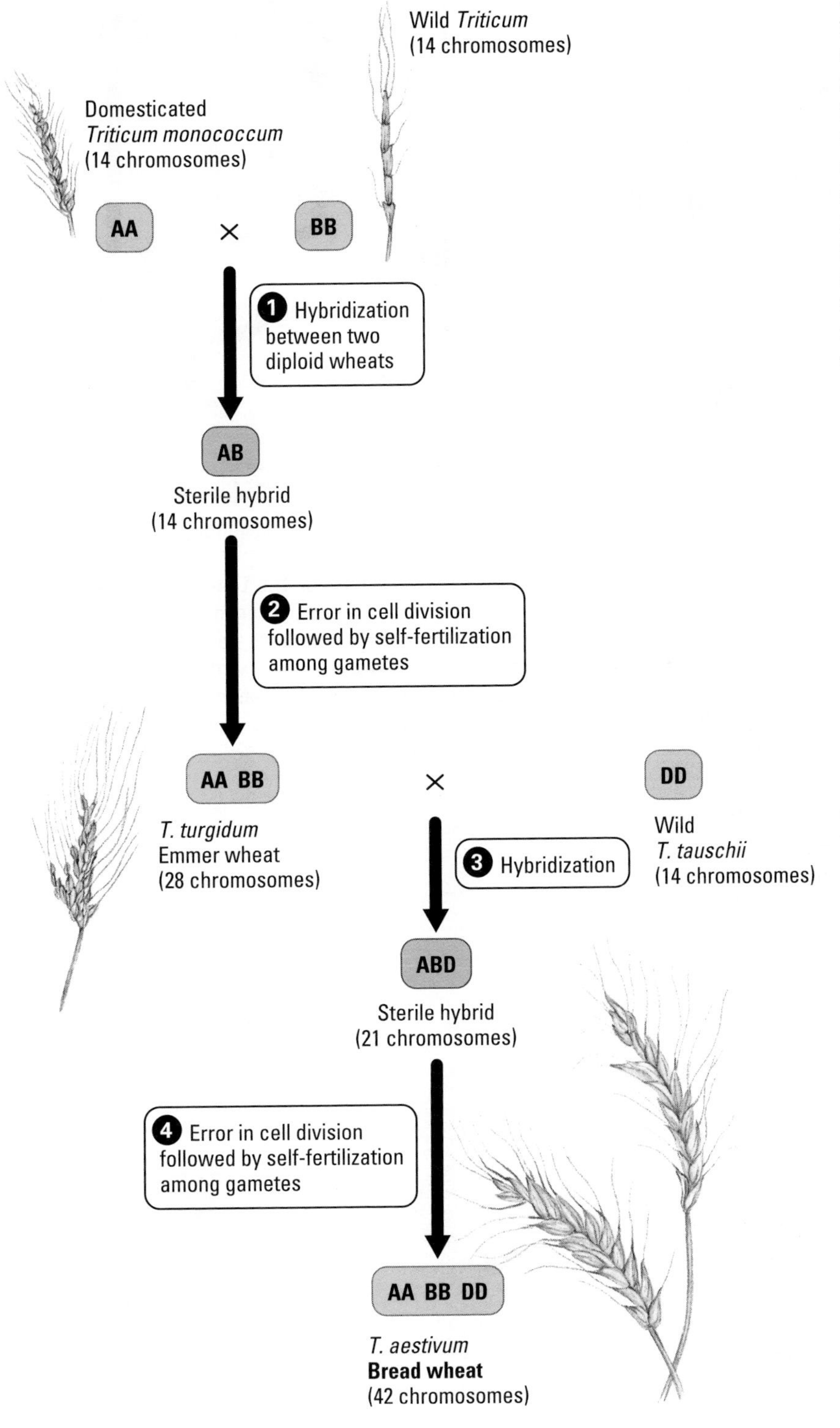

Figure 14.11 The evolution of wheat. The uppercase letters in this diagram represent not alleles but sets of chromosomes that we are tracing in the evolution of wheat, or *Triticum*. 1 The evolutionary descent of wheat began when a domesticated species hybridized with a wild relative that probably grew as a weed at the edges of cultivated fields. 2 Chromosome sets A and B of the two species would not have been able to pair at meiosis, making the AB hybrid sterile. However, an error in cell division, followed by self-fertilization, produced a new species, AABB. This hybrid, emmer wheat (*T. turgidum*), is still grown widely in Eurasia and western North America and is used mainly for making pasta products. 3 Over 8,000 years ago, cultivated emmer wheat hybridized with another closely related wild species. 4 This sterile hybrid gave rise to bread wheat, with two each of the three ancestral sets of chromosomes (AABBDD).

barley, plums, apples, sugarcane, coffee, and wheat. Wheat makes a good case study. The most widely cultivated plant in the world, what we call wheat is actually represented by 20 different species. Humans began domesticating wheat from wild grasses at least 11,000 years ago in the Middle East. **Figure 14.11** traces the evolutionary path from that first cultivated wheat species to bread wheat, the most important wheat species today.

What Is the Tempo of Speciation?

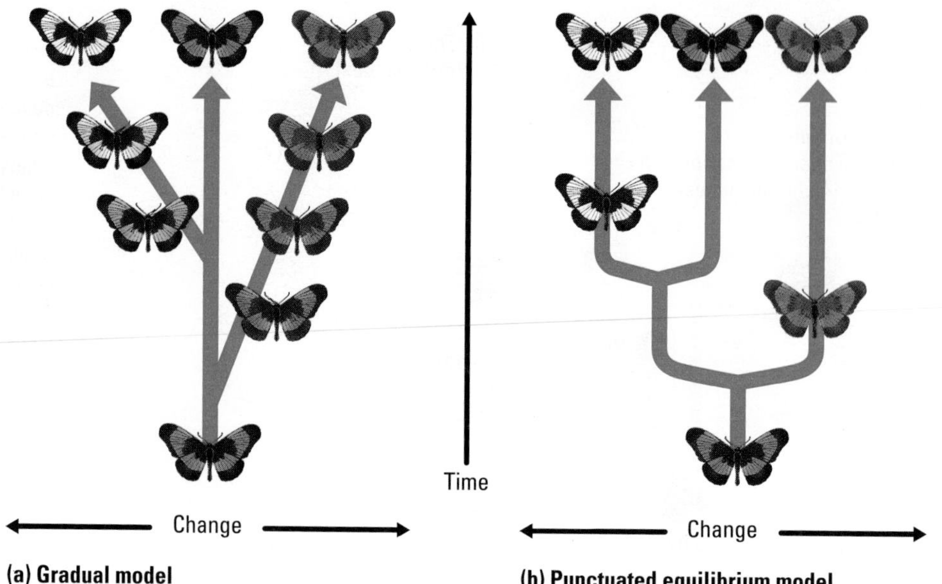

Figure 14.12 Two models for the tempo of evolution. **(a)** In the traditional model, species descended from a common ancestor diverge gradually in form as they acquire unique adaptations. **(b)** According to the punctuated equilibrium model, a new species changes most as it first branches from a parent species. After this speciation episode, there is little change for the rest of the species' existence.

Traditional evolutionary trees that diagram the descent of species sprout branches that diverge gradually **(Figure 14.12a)**. Such trees are based on the idea that the changes significant enough to distinguish a new species are an accumulation of many smaller changes occurring over vast spans of time. This concept applies the processes of microevolution you learned about in Chapter 13 to the multiplication of species. To Darwin, the origin of species was an extension of adaptation by natural selection, with isolated populations from common ancestral stock evolving differences gradually as they adapted to their local environments.

On the time scale of the fossil record, new species often appear more abruptly than the traditional model would predict. Paleontologists do find slow and steady transitions of fossil forms in some cases. However, the more common observation is that species appear as new forms rather suddenly (in geologic terms), persist essentially unchanged for their tenure on Earth, and then disappear from the fossil record as suddenly as they appeared. Darwin himself was bewildered by the dearth of slow and steady change in fossil lineages, as when he wrote: "Although each species must have passed through numerous transitional stages, it is probable that the periods during which each underwent modification, though many and long as measured by years, have been short in comparison with the periods during which each remained in an unchanged condition."

A model known as **punctuated equilibrium** addresses the nongradual appearance of species. According to this model, species most often diverge in spurts of relatively sudden change, instead of diverging slowly and gradually **(Figure 14.12b)**. In other words, species undergo most of their modification in appearance as they first branch from parent species; after that, little change occurs. The term punctuated equilibrium is derived from the idea of long periods of stasis (equilibrium) punctuated by episodes of speciation (relatively rapid change).

One cause of sudden speciation is the polyploidy you learned about as a mechanism of sympatric speciation in plants. Proponents of punctuated equilibrium point out that allopatric speciation can also be quite rapid. In just a few hundred to a few thousand generations, genetic drift and natural selection can cause significant change in the gene pool of a small population cloistered in a challenging new environment.

How can speciation in a few thousand generations, which may require several thousand years, be called an abrupt episode? The fossil record indicates that successful species last for a few million years, on average. Suppose that a particular species survives for 5 million years, but most of its evolutionary changes in anatomy occur during the first 50,000 years of its existence. In this case, the evolution of the species-defining characteristics is compressed into just 1% of the lifetime of the species. In the fossil record, the species will appear suddenly in rocks of a certain age and then linger with little or no change before becoming extinct. During its formative millennia, the species may accumulate its modifications gradually, but relative to the overall history of the species, its inception is abrupt. This scenario of an evolutionary spurt preceding a much longer period of morphological stasis would help explain why paleontologists find relatively few smooth transitions in the fossil record of species. Also, since the best candidates for speciation are small populations, we are less likely to have fossils of transitional stages than if they occurred in a large population.

Once it is acknowledged that "sudden" may be many thousands of years on the vast scale of geologic time, the debate over the tempo of speciation is muted somewhat. The degree to which a species changes after its origin is another issue. If the species is adapted to an environment that stays the same, then natural selection would not favor changes in the gene pool. In this view, stabilizing selection tends to hold a population in a long period of stasis. But some evolutionary biologists argue that stasis is an illusion. They propose that species may continue to change after they come into existence, but in nonanatomical ways that cannot be detected from fossils.

CHECKPOINT

1. By defining a species by its reproductive ___________ from other populations, the biological species concept can only be applied to organisms that reproduce ___________.
2. Why would allopatric speciation be less common on an island close to a mainland than on a more isolated island of the same size?
3. Each speciation episode in the evolution of bread wheat is an example of___________ speciation, which is the origin of a new species without geographic isolation from the parent species.
4. How does the punctuated equilibrium model account for the relative rarity of transitional fossils linking newer species to older ones?

Answers: 1. isolation; sexually **2.** Continued gene flow between mainland populations and those on nearby islands reduces the chance of enough genetic divergence for speciation. **3.** sympatric **4.** According to this model, the time required for speciation in most cases is relatively short compared with the overall duration of the species' existence. Thus, on the vast geologic time scale of the fossil record, the transition of one species to another seems abrupt.

The Evolution of Biological Novelty

The two squirrels in Figure 14.8 are different species, but they are, after all, very similar animals that live very much the same way. When most people think of evolution, they envision much more dramatic transformation. How can we account for such evolutionary products as flight in birds and braininess in humans?

Adaptation of Old Structures for New Functions

Birds are derived from a lineage of earthbound reptiles **(Figure 14.13)**. How could flying vertebrates evolve from flightless ancestors? More generally, how do major novelties of biological structure and function evolve? One mechanism is the gradual refinement of existing structures that take on new functions.

Most biological structures have an evolutionary plasticity that makes alternative functions possible. Biologists use the term **exaptation** for a structure that evolves in one context and later becomes adapted for other functions. This concept does not imply that a structure somehow evolves in anticipation of future use. Natural selection cannot predict the future and can only refine a structure in the context of its current utility. Birds have lightweight skeletons with honeycombed bones; such

Figure 14.13 An artist's reconstruction of an extinct bird. Called *Archaeopteryx* ("ancient wing"), this animal lived near tropical lagoons in central Europe about 150 million years ago. Like modern birds, it had flight feathers, but otherwise it was more like some small bipedal dinosaurs of its era. *Archaeopteryx* probably relied mainly on gliding from trees. Despite its feathers, *Archaeopteryx* is not considered an ancestor of modern birds. Instead, it probably represents an extinct side branch of the bird lineage.

a feature is also found in the dinosaur ancestors of birds. Since light bones predated flight, as is clearly indicated by the fossil record, then they must have had some function on the ground. The ancestors of birds were small, agile, bipedal dinosaurs, and they, too, would have benefited from a light frame. Moreover, a winglike form with feathers would have increased the surface area of the forelimbs. These enlarged forelimbs were adapted for flight after functioning in some other capacity, such as thermal regulation, courtship displays, or camouflage. The first flights may have been only glides or extended hops in an effort to pursue prey or escape from a predator. Once flight itself became an advantage, feathers and wings remodeled by natural selection to better fit their new function would have had a selective advantage.

Exaptation is the process by which novel features arise gradually through a series of intermediate stages, each of which has some function in the organism's current context. Harvard zoologist Karel Liem puts it this way: "Evolution is like modifying a machine while it's running." The concept that biological novelties can evolve by the remodeling of old structures for new functions is in the Darwinian tradition of large changes being an accumulation of many small changes crafted by natural selection.

Figure 14.14 Paedomorphosis. Some species retain as adults features that were solely juvenile in ancestors. The axolotl, a salamander, becomes an adult and reproduces while retaining certain tadpole characteristics, including gills. It is an example of how changes in the genes controlling development can produce a very different organism.

"Evo-Devo": Development and Evolutionary Novelty

Gradual evolutionary remodeling, such as the accumulation of flight adaptations in birds, probably involves a large number of genetic changes in populations. In other cases of macroevolution, relatively few genetic changes can cause major structural modifications. How can slight genetic divergence become magnified into major differences between organisms? Scientists working in the field of "evo-devo"—the intersection of evolutionary biology and the study of embryonic development—are finding some answers to this question.

Genes that program development control the rate, timing, and spatial pattern of changes in an organism's form as it is transfigured from a zygote into an adult. (Chapter 11 provides a closer look at how genes control development.) A subtle change in a species' developmental program can have profound effects. For example, the animal in **Figure 14.14**, a salamander called an axolotl, illustrates a phenomenon called **paedomorphosis,** which is the retention into adulthood of features that were solely juvenile in its ancestors. The axolotl grows to full size and reproduces without losing its external gills, a juvenile feature in most species of salamanders.

Paedomorphosis has also been important in human evolution. Humans and chimpanzees are even more alike in body form as fetuses than they are as adults. In the fetuses of both species, the skulls are rounded and the jaws are small, making the face rather flat **(Figure 14.15)**. As development proceeds, uneven bone growth makes the chimpanzee skull sharply angular, with heavy browridges and massive jaws. The adult chimpanzee has much greater jaw strength than we have, and its teeth are proportionately larger. In contrast, the adult human has a skull with decidedly rounded, more fetus-like contours. Put another way, our skull is paedomorphic; it retains fetal features even after we are mature. Our large skull is one of our most distinctive features. Our large, complex brain, which fills that bulbous skull, is another. The human brain is proportionately larger than the chimpanzee

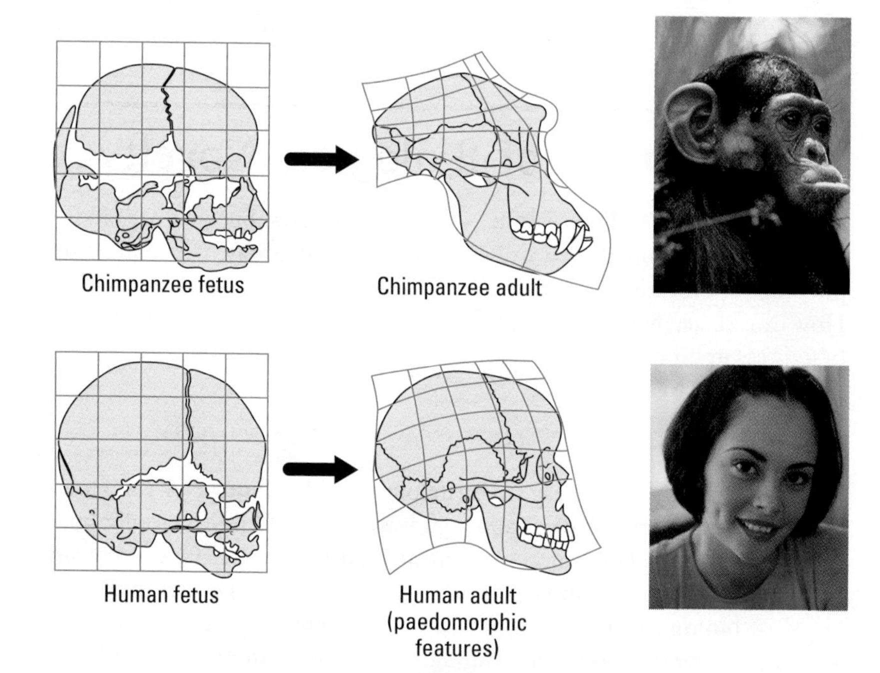

Figure 14.15 Comparison of human and chimpanzee skull development. Starting with fetal skulls that are very similar (left), the differential growth rates of different bones making up the skulls produce adult heads with very different proportions. The grid lines will help you relate the fetal skulls to the adult.

brain because growth of the organ is switched off much later in human development. Compared with the brain of a chimpanzee, our brain continues to grow for several more years, which can be interpreted as the prolonging of a juvenile process. We owe our culture to this evolutionary novelty, coupled with an extended childhood during which parents and teachers can influence what the young, growing brain stores.

CHECKPOINT

1. Explain why the concept of exaptation does not imply that a structure evolves in anticipation of some future environmental change.
2. How is the development of our large, roundish skulls similar to the development of the gills of axolotls?

Answers: 1. Although an exaptation is adapted for new or additional functions in a new environment, it existed earlier because it worked as an adaptation to the old environment. **2.** Both structures are paedomorphic (juvenile anatomies retained into adulthood).

Earth History and Macroevolution

Having examined how new species and evolutionary adaptations arise, we are ready to turn our attention to the history of biological diversity. Macroevolution is closely tied to the history of Earth.

Geologic Time and the Fossil Record

The fossil record is an archive of macroevolution. **Figure 14.16** surveys some of the diverse ways that organisms can fossilize. Sedimentary rocks are the richest sources of fossils and provide a record of life on Earth in their layers, or strata (see Chapter 13). The fossils in each stratum of sedimentary rock are a local sample of the organisms that existed at the time the sediment was deposited.

(c) Some fossils form when minerals seep into and replace organic matter. These petrified (stone) trees in the Petrified Forest National Park in Arizona are about 190 million years old.

(d) This 30-million-year-old insect is embedded in amber (hardened resin from a tree).

(a) Sedimentary rocks are the richest hunting grounds for paleontologists, scientists who study the fossil record. This researcher is excavating a fossilized dinosaur skeleton from sandstone in Dinosaur National Monument, located in Utah and Colorado.

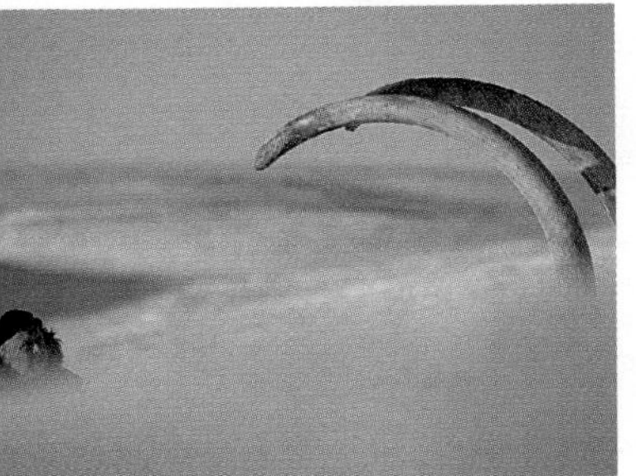

(b) These tusks belong to a whole 23,000-year-old mammoth, which scientists discovered in Siberian ice in 1999.

(e) Trace fossils are footprints, burrows, and other remnants of an ancient organism's behavior. A dinosaur left these footprints in a creek bed in what is now Oklahoma.

Figure 14.16 A gallery of fossils.

Table 14.1 The Geologic Time Scale

Relative Time Span	Geologic Time	Period	Epoch	Age (millions of years ago)	Some Important Events in the History of Life
Cenozoic	Cenozoic era	Quaternary	Recent		Historical time
				0.01	
			Pleistocene		Ice ages; humans appear
				1.8	
		Tertiary	Pliocene		Apelike ancestors of humans appear
				5	
			Miocene		Continued radiation of mammals and angiosperms
				23	
			Oligocene		Origins of many primate groups, including apes
				34	
			Eocene		Angiosperm dominance increases; origins of most modern mammalian orders
				56	
			Paleocene		Major radiation of mammals, birds, and pollinating insects
				65	
Mesozoic	Mesozoic era	Cretaceous			Flowering plants (angiosperms) appear; many groups of organisms, including most dinosaur lineages, become extinct at end of period (Cretaceous extinctions)
				145	
		Jurassic			Gymnosperms continue as dominant plants; dinosaurs dominant
				200	
		Triassic			Cone-bearing plants (gymnosperms) dominate landscape; radiation of dinosaurs, early mammals, and birds
				251	
Paleozoic	Paleozoic era	Permian			Extinction of many marine and terrestrial organisms (Permian extinctions); radiation of reptiles; origins of mammal-like reptiles and most modern orders of insects
				299	
		Carboniferous			Extensive forests of vascular plants; first seed plants; origin of reptiles; amphibians dominant
				359	
		Devonian			Diversification of bony fishes; first amphibians and insects
				416	
		Silurian			Diversity of jawless fishes; first jawed fishes; colonization of land by vascular plants and arthropods
				444	
		Ordovician			Origin of plants; marine algae abundant
				488	
		Cambrian			Origin of most modern animal phyla (Cambrian explosion)
				542	
Pre-cambrian	Precambrian			600	Diverse soft-bodied invertebrate animals; diverse algae
				700	Oldest animal fossils
				2,200	Oldest eukaryotic fossils
				2,700	Oxygen begins accumulating in atmosphere
				3,500	Oldest fossils known (prokaryotes)
				4,600	Approximate time of origin of Earth

By studying many different sites over the past two centuries, geologists have established a **geologic time scale,** reflecting a consistent sequence of geologic periods **(Table 14.1)**. Notice that the time line presented in Table 14.1 is separated into four comprehensive divisions: the Precambrian (a general term for the time before about 540 million years ago), followed by the Paleozoic, Mesozoic, and Cenozoic eras. Each of these divisions represents a distinct age in the history of Earth and its life. The boundaries between eras are marked by mass extinctions, when many forms of life disappeared from the fossil record and were replaced by species that diversified from the survivors. For example, the beginning of the Cambrian period is delineated by a great diversity of fossilized animals that are absent in rocks of the late Precambrian. And most of the animals that lived during the late Precambrian became extinct at the end of that era.

Fossils are reliable chronological documents only if we can determine their ages. The record of the rocks directly shows only the *relative* ages of fossils. It tells us the order in which groups of species evolved. However, the series of sedimentary layers alone does not tell the *absolute* ages of the embedded fossils. The layers of rock are analogous to the layers of wallpaper you might peel from the walls of a very old house. You could determine the sequence in which the wallpapers had been applied, but not the year that each layer was added. Geologists use a variety of methods to determine the ages of rocks and the fossils they contain. The most common method is **radiometric dating (Figure 14.17)**, which is based on the decay of radioactive isotopes (recall that isotopes are alternative forms of elements; see Table 2.1). The dates you see on the geologic time scale in Table 14.1 were established by radiometric dating.

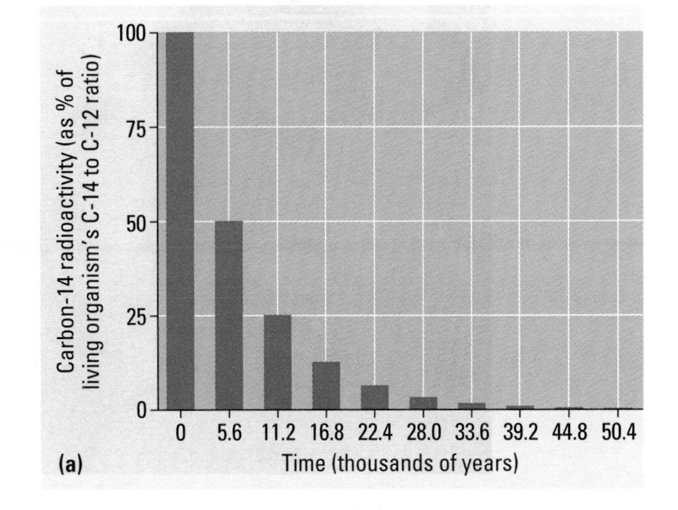

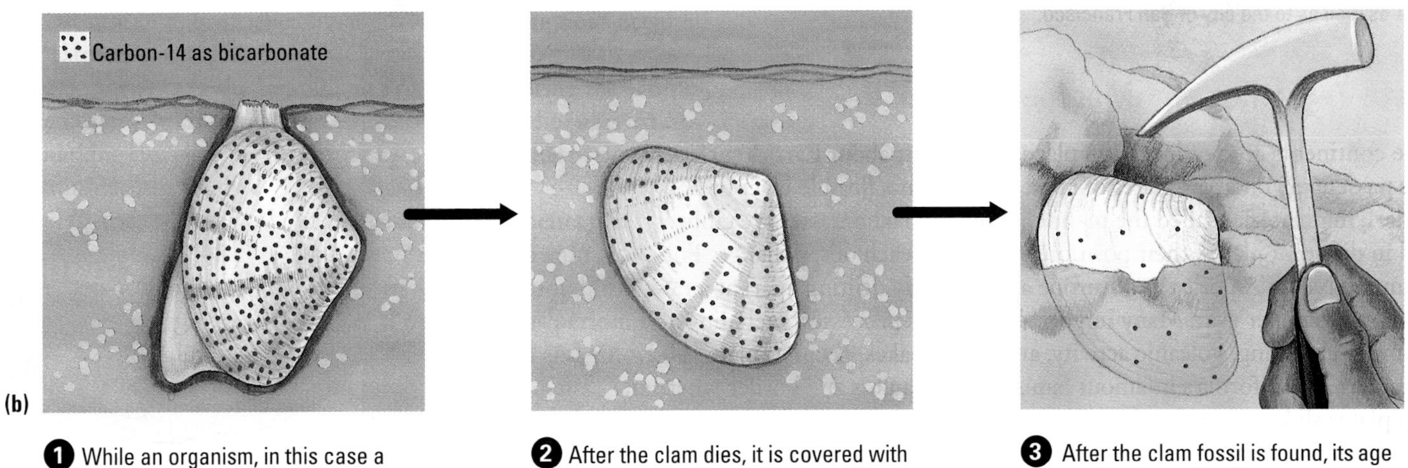

Carbon-14 as bicarbonate
(b)
1 While an organism, in this case a clam, is alive, it assimilates the different isotopes of each element in proportions determined by their relative abundances in the environment. Carbon-14 is taken up in trace quantities, along with much larger quantities of the more common carbon-12.
2 After the clam dies, it is covered with sediment, and its shell eventually becomes consolidated into a layer of rock as the sediment is compressed. From the time the clam dies and ceases to assimilate carbon, the amount of carbon-14 relative to carbon-12 in the fossil declines due to radioactive decay.
3 After the clam fossil is found, its age can be determined by measuring the ratio of the two isotopes to learn how many half-life reductions have occurred since it died. For example, if the ratio of carbon-14 to carbon-12 in this fossil clam was found to be 25% that of a living organism, this fossil would be about 11,200 years old (see graph in part a).

Figure 14.17 Radiometric dating. Amounts of radioactive isotopes can be measured by the radiation they emit as they decompose to more stable atoms (see Chapter 2). Paleontologists use this clocklike decay to date fossils. **(a)** Carbon-14 is a radioactive isotope with a half-life of 5,600 years. From the time an organism dies, it takes 5,600 years for half of the radioactive carbon-14 to decay; half of the remainder is present after another 5,600 years; and so on. Because the half-life of carbon-14 is relatively short, this isotope is only useful for dating fossils less than about 50,000 years old. To date older fossils, paleontologists use radioactive isotopes with longer half-lives, or they use other methods. Other methods for dating fossils generally confirm the ages determined by radiometric dating. **(b)** How carbon-14 dating is used to determine the vintage of a fossilized clam shell.

Did a Meteor Kill the Dinosaurs? For decades, scientists have been debating the cause of the rapid dinosaur die-off that occurred 65 million years ago. Many **observations** provide clues. The fossil record shows that the climate had cooled and that shallow seas were receding from continental lowlands. It also shows that many plant species died out. Perhaps the most telling evidence was discovered by physicist Luis Alvarez and his geologist son Walter Alvarez, both of the University of California. In 1980, they found that rock deposited around 65 million years ago contains a thin layer of clay rich in iridium, an element very rare on Earth but common in meteors and other extraterrestrial material that occasionally falls to Earth. This led the Alvarez team to ask the **question**, Is the iridium layer the result of fallout from a huge cloud of dust that billowed into the atmosphere when a large meteor or asteroid hit Earth?

The father and son formed the **hypothesis** that the mass extinction 65 million years ago was caused by the impact of an extraterrestrial object. This hypothesis makes a clear **prediction**: A huge impact crater of the right age should be found somewhere on Earth's surface. (This is a good example of discovery science, which relies on verifiable observations rather than a direct **experiment**; see Chapter 1.) In 1981, two petroleum geologists found the **results** predicted by Alvarez's hypothesis: the Chicxulub crater, located near Mexico's Yucatán Peninsula in the Caribbean Sea **(Figure 14.20)**. This impact site, 150–300 km wide and dating from the predicted time, was created when a meteor or asteroid the size of San Francisco slammed into Earth, releasing thousands of times more energy than is stored in the world's combined stockpile of nuclear weapons. Such a cloud could have blocked sunlight and disturbed climate severely for months, perhaps killing off many plant species and, later, the animals that depended on those plants for food.

Debate continues about whether this impact alone caused the dinosaurs to die out or whether other factors—such as continental movements or volcanic activity—also contributed. Most scientists agree, however, that the collision that created the Chicxulub crater could indeed have been a major factor in global climatic changes and mass extinctions. ■

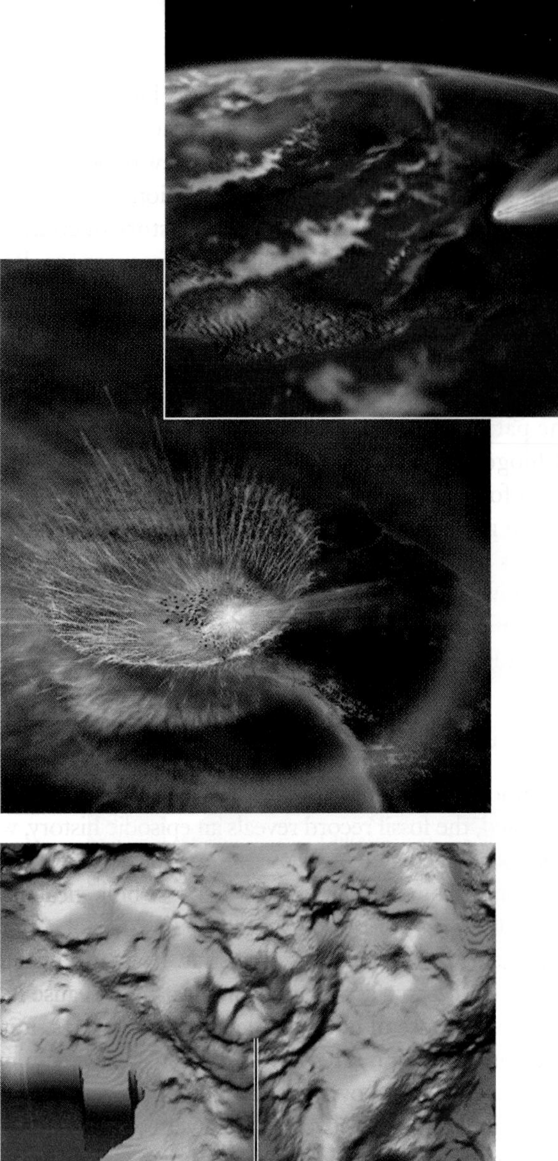

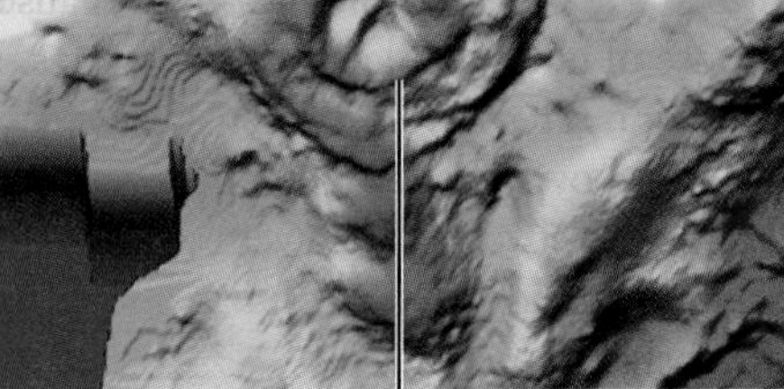

Figure 14.20 Trauma for planet Earth and its Cretaceous life. The 65-million-year-old Chicxulub impact crater (shown in blue in the sonar image at the bottom) is located in the Caribbean Sea near the Yucatán Peninsula of Mexico. The horseshoe shape of the crater and the pattern of debris in sedimentary rocks indicate that an asteroid or comet struck at a low angle from the southeast. The artist's interpretation in the top two images represents the impact and its immediate effect—a cloud of hot vapor and debris that could have killed most of the plants and animals in North America within minutes. That would explain the higher extinction rates of land animals and plants in North America than elsewhere around the globe.

CHECKPOINT

1. Use Table 14.1 to estimate how long prokaryotes inhabited Earth before eukaryotes evolved.
2. Imagine you unearth a skull with a carbon-14 to carbon-12 ratio about one-sixteenth that of a living organism. What is the approximate age of the skull? (Refer to Figure 14.17a.)
3. How many continents did Earth have at the time of Pangaea?

Answers: 1. About 1,800 million years, or 1.8 billion years **2.** 22,400 years (four half-life reductions) **3.** One

Classifying the Diversity of Life

Reconstructing evolutionary history is part of the science of **systematics,** the study of the diversity and relationships of organisms, both past and present. Systematics includes **taxonomy,** which is the identification, naming, and classification of species.

Some Basics of Taxonomy

Assigning scientific names to species is an essential part of systematics. Common names, such as monkey, fly, and pea, may work well in everyday communication, but they can be ambiguous because there are many species of each of these organisms. And some common names are misleading. Consider these three "fishes": jellyfish (a cnidarian), crayfish (a crustacean), and silverfish (an insect).

Using an agreed-upon formal naming system eases communication among scientists, allows researchers to unambiguously identify an organism, and makes it easier to recognize when a new species is discovered. The formal taxonomic system used by biologists today dates back to Carolus Linnaeus (1707–1778), a Swedish physician and botanist (plant specialist). Linnaeus's system has two main characteristics: a two-part name for each species and a hierarchical classification of species into broader and broader groups of organisms.

Naming Species Linnaeus's system assigns to each species a two-part latinized name, or **binomial.** The first part of a binomial is the **genus** (plural, *genera*) to which the species belongs. The second part of a binomial refers to one particular species within the genus. For example, the scientific name for the domestic cat is *Felis catus*. Notice that the first letter of the genus is capitalized and that the whole binomial is italicized and latinized. (You can name a bug you discover after a friend, but you must add the appropriate Latin ending.) The scientific name of our own species, *Homo sapiens*, which Linnaeus assigned in a show of optimism, means "wise man."

Hierarchical Classification In addition to defining and naming species, a major objective of systematics is to group species into an ordered list of categories. The first step of such a hierarchical classification is built into the binomial. We group species that are closely related into the same genus. For example, the domestic cat, *Felis catus*, belongs to a genus of relatively small cats that also includes the cougar (*Felis concolor*) and the ocelot (*Felis pardalis*). Grouping species is natural for us, at least in concept. We lump together several trees we know as oaks and distinguish them from several other species of trees we call maples. Indeed, oaks and maples belong to separate genera. Biology's taxonomic scheme formalizes our tendency to group related objects as a way of structuring our view of the world.

Beyond the grouping of species within genera, taxonomy extends to progressively broader categories of classification. It places similar genera in the same **family,** puts families into **orders,** orders into **classes,** classes into **phyla** (singular, *phylum*), phyla into **kingdoms,** and kingdoms into **domains.** (See Chapter 1 to review the three domains of life.) **Figure 14.21** places the domestic cat in this taxonomic scheme of groups within groups. Classifying a species by kingdom, phylum, and so on, is analogous to sorting mail, first by zip code and then by street, house number, and specific member of the household.

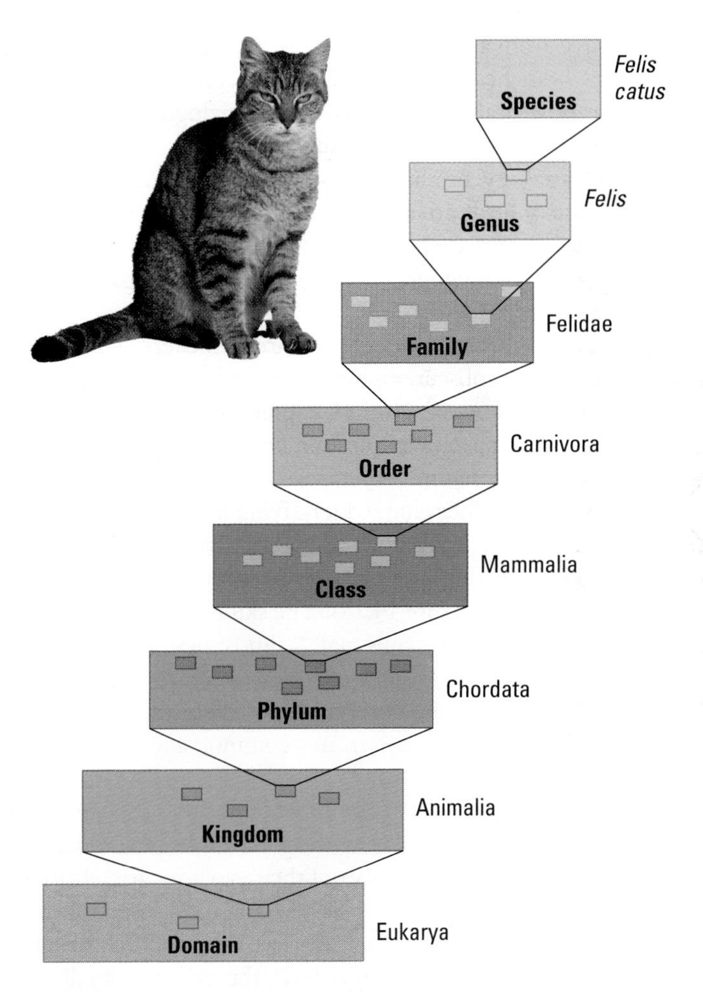

Figure 14.21 Hierarchical classification. Taxonomy classifies species into increasingly comprehensive groups.

Classification and Phylogeny

Ever since Darwin, systematics has had a goal beyond simple organization: to have classification reflect **phylogeny,** which is the evolutionary history of a species. In other words, how an organism is named and classified should reflect its place within the evolutionary tree of life. As a systematist classifies species into groups subordinate to other groups in the taxonomic hierarchy, the final product takes on the branching pattern of a **phylogenetic tree.** The principle of common descent is reflected in the branches of the tree **(Figure 14.22)**.

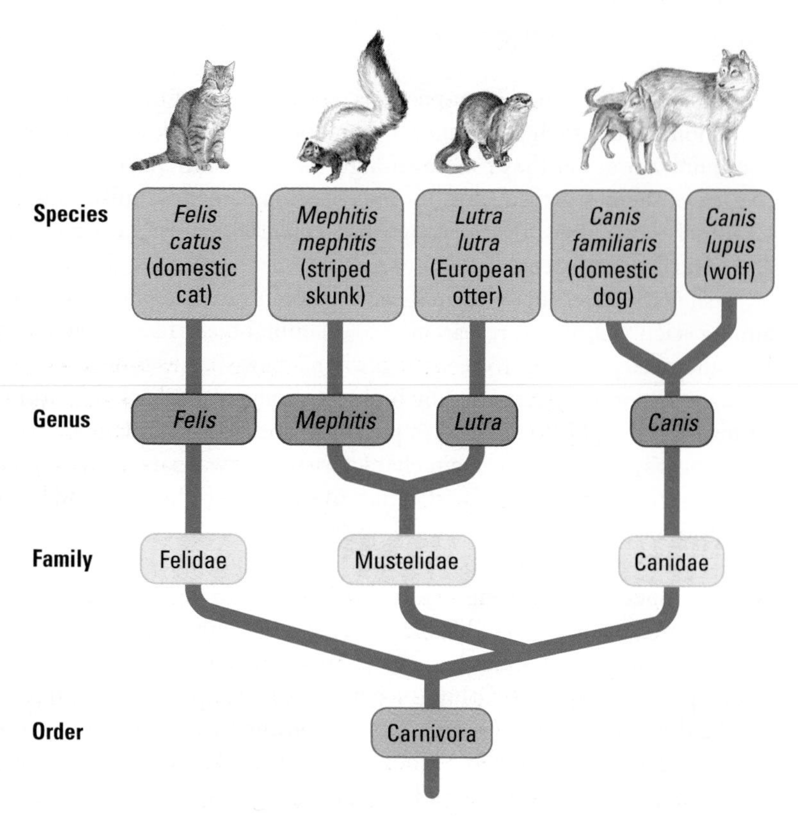

Figure 14.22 The relationship of classification and phylogeny for some members of the order Carnivora. The hierarchical classification is reflected in the finer and finer branching of the phylogenetic tree. Each branch point in the tree represents an ancestor common to species above that branch point.

Sorting Homology from Analogy Homologous structures are one of the best sources of information about phylogenetic relationships. Recall from Chapter 13 that homologous structures may look different and function very differently in different species, but they exhibit fundamental similarities because they evolved from the same structure that existed in a common ancestor. Among the vertebrates, for instance, the whale limb is adapted for steering in the water; the bat wing is adapted for flight. Nonetheless, there are many basic similarities in the bones supporting these two structures (see Figure 13.11). The greater the number of homologous structures between two species, the more closely the species are related.

There are pitfalls in the search for homology: Not all likeness is inherited from a common ancestor. Species from different evolutionary branches may have certain structures that are superficially similar if natural selection has shaped analogous adaptations. This is called **convergent evolution.** Similarity due to convergence is termed **analogy,** not homology. For example, the wings of insects and those of birds are analogous flight equipment: They evolved independently and are built from entirely different structures.

To develop phylogenetic trees and classify organisms according to evolutionary history, we must use only homologous similarities. This guideline is generally straightforward, but there can be complications. Adaptation can obscure homologies, and convergence can create misleading similarities. As we saw in Chapter 13, comparing the embryonic development of the species in question can often expose homology that is not apparent in the mature structures.

There is another clue to identifying homology and sorting it from analogy: The more complex two similar structures are, the less likely it is they evolved independently. Compare the skulls of a human and a chimpanzee, for example (see Figure 14.15). The skulls are not single bones, but a fusion of many, and the two skulls match almost perfectly, bone for bone. It is highly improbable that such complex structures matching in so many details could have separate origins. Most likely, the genes required to build these skulls were inherited from a common ancestor.

Molecular Biology as a Tool in Systematics If homology is about common ancestry, then comparing the genes and gene products (proteins) of organisms gets right to the heart of their evolutionary relationships. Sequences of nucleotides in DNA are inherited, and they program corresponding sequences of amino acids in proteins (see the Evolution Connection at the end of Chapter 3). At the molecular level, the evolutionary divergence of species parallels the accumulation of differences in their genomes. The more recently two species have branched from a common ancestor, the more similar their DNA and amino acid sequences should be.

Today, both the amino acid sequences for many proteins and the nucleotide sequences for a rapidly increasing number of genomes are in databases that are available via the Internet (see Chapter 12). This has catalyzed a boom in systematics as researchers use the databases to compare the hereditary information of different species in search of homology at its most basic level.

Molecular systematics provides a new way to test hypotheses about the phylogeny of species. The strongest support for any such hypothesis is agreement between molecular data and other means of tracing phylogeny, such as evaluating anatomical homology and analyzing the fossil record. And speaking of fossils, some are preserved in such a way that it is possible to extract DNA fragments for comparison with modern organisms **(Figure 14.23)**.

Figure 14.23 Studying ancient DNA. Some sedimentary fossils, such as this 40-million-year-old leaf, retain organic material, including DNA, which scientists can analyze.

The Cladistic Revolution Systematics entered a vigorous new era in the 1960s. Just as molecular methods became readily available for comparing species, computer technology that could crunch a new wealth of data helped usher in a new approach called cladistics.

Cladistics is the scientific search for clades (from the Greek word for "branch"). A **clade** consists of an ancestral species and all its descendants—a distinctive branch in the tree of life. Identifying clades involves identifying homologies unique to each group, where a group may be a species or a higher taxonomic cluster, such as a class or phylum **(Figure 14.24)**. In other words, cladistics focuses on the evolutionary innovations that define the branch points in evolution. Identifying clades makes it possible to construct classification schemes that reflect phylogeny.

Cladistics has become the most widely used method in systematics. This approach clarifies the degree of evolutionary relationship between groups that was not necessarily apparent in other taxonomic classifications. For instance, biologists traditionally placed birds and reptiles in separate classes of vertebrate animals (class Aves and class Reptilia, respectively). This classification, however, is inconsistent with cladistics. An inventory of

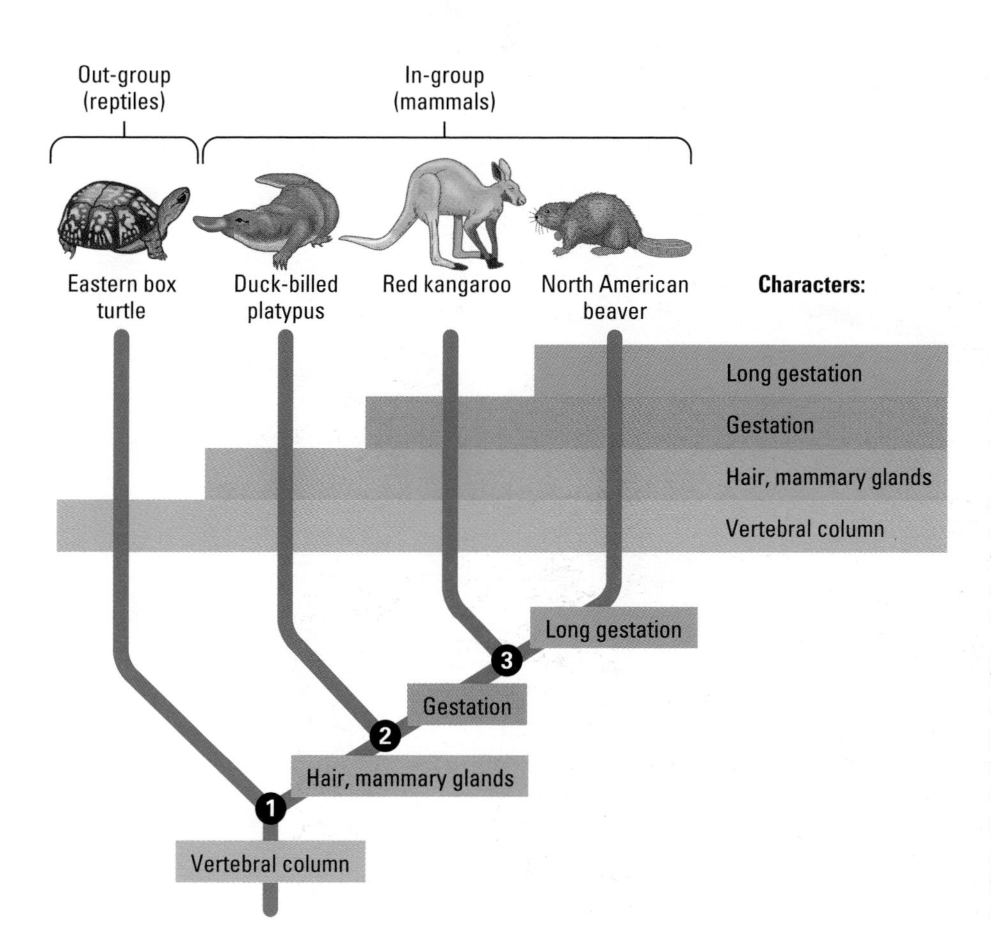

Figure 14.24 A simplified example of cladistics. Cladistics requires a comparison between a so-called in-group and an out-group. In our example, the three mammals make up the in-group, while the turtle, a reptile, is the out-group. This provides a reference point for distinguishing primitive characters from derived characters. A primitive character is a homology present in all the organisms being compared and so must also have been present in the common ancestor. All four animals in our example have vertebral columns (backbones), which is a primitive character. The derived characters, such as hair and gestation, are the evolutionary innovations that define the sequence of branch points (numbered) in the phylogeny of the in-group. ❶ Hair and mammary glands are among the derived characters that distinguish mammals from reptiles. ❷ Among the derived characters defining the next branch point is gestation, the carrying of offspring in the womb of the female parent. The duck-billed platypus, though a mammal, lays eggs, as do turtles and other reptiles. We can infer that gestation evolved in an ancestor common to the kangaroo and beaver that is more recent than the ancestor shared by all mammals, including the platypus. ❸ The last branch point in this cladistic analysis is defined by the much longer gestation of the beaver compared with that of the kangaroo (in which the embryo emerges from the womb very early and completes its development within the mother's pouch).

homologies indicates that crocodiles are more closely related to birds than they are to lizards and snakes **(Figure 14.25)**.

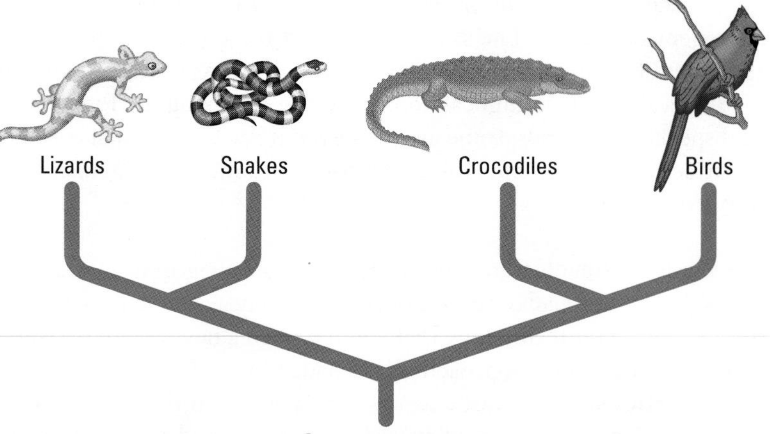

Figure 14.25 How cladistics is shaking phylogenetic trees. Strict application of cladistics sometimes produces phylogenetic trees that conflict with classical taxonomy. Traditionally, most biologists placed reptiles and birds in separate vertebrate classes. But based on derived characters, birds and crocodiles make up one clade, and lizards and snakes make up another. Or if we go back as far as the ancestor that crocodiles share with lizards and snakes to make up a clade, then the class Reptilia must also include birds. The tree you see here is thus more consistent with cladistics than with traditional classifications.

Arranging Life into Kingdoms: A Work in Progress

Phylogenetic trees are hypotheses about evolutionary history. Like all hypotheses, they are revised (or in some cases completely rejected) in accordance with new evidence. Molecular systematics and cladistics are combining to remodel phylogenetic trees and challenge conventional classifications.

Linnaeus divided all known forms of life between the plant and animal kingdoms, and the two-kingdom system prevailed in biology for over 200 years. In the mid-20th century, the two-kingdom system was replaced by a five-kingdom system that placed all prokaryotes in one kingdom and divided the eukaryotes among four other kingdoms.

In the late 20th century, molecular studies and cladistics led to the development of a **three-domain system (Figure 14.26)**. This current

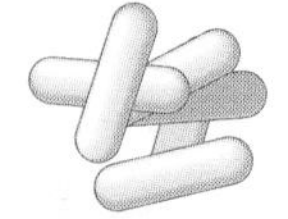

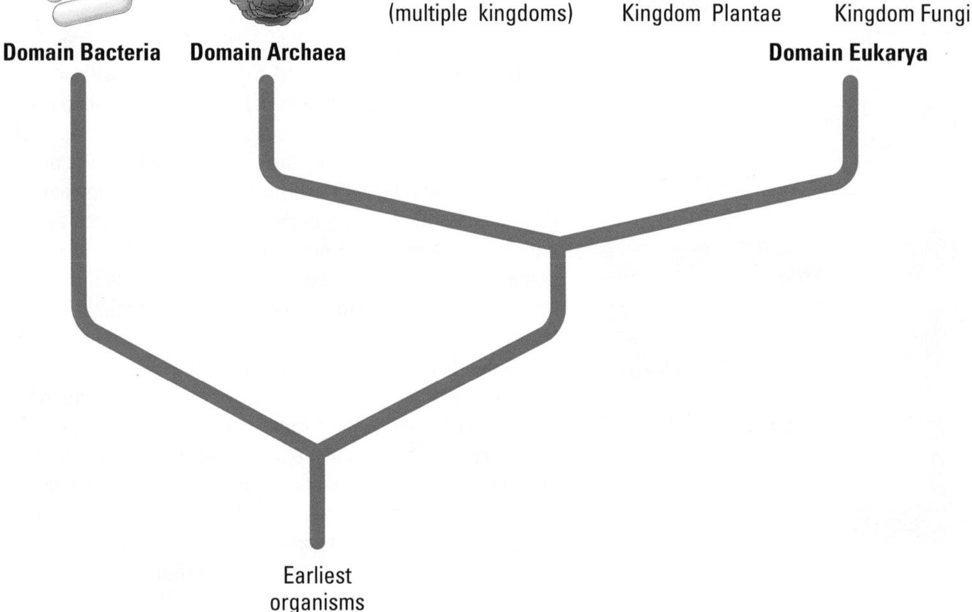

Figure 14.26 The three-domain classification system. Molecular and cellular evidence supports the phylogenetic hypothesis that two lineages of prokaryotes, domains Bacteria and Archaea, diverged very early in the history of life. Molecular evidence also suggests that domain Archaea is more closely related to domain Eukarya than to domain Bacteria.

scheme recognizes three basic groups: two domains of prokaryotes—Bacteria and Archaea—and one domain of eukaryotes, called Eukarya. The domains Bacteria and Archaea differ in a number of important structural, biochemical, and functional features, which we will discuss in Chapter 15.

The domain Eukarya is currently divided into kingdoms, but the exact number of kingdoms is still under debate. Biologists generally agree on the three kingdoms Plantae, Fungi, and Animalia. These consist of multicellular eukaryotes that differ in structure, development, and modes of nutrition. Plants make their own food by photosynthesis. Fungi live by decomposing the remains of other organisms and absorbing small organic molecules. Most animals live by ingesting food and digesting it within their bodies.

The remaining eukaryotes, the protists, include all those that do not fit the definition of plant, fungus, or animal—effectively, a taxonomic grab bag. Most protists are unicellular (amoebas, for example). But the protists also include certain large, multicellular organisms that are believed to be direct descendants of unicellular protists. For example, many biologists now classify the seaweeds as protists because they are more closely related to certain single-celled algae than they are to true plants.

It is important to understand that classifying Earth's diverse species of life is a work in progress as we learn more about organisms and their evolution. Charles Darwin envisioned the goals of modern systematics when he wrote in *The Origin of Species*, "Our classifications will come to be, as far as they can be so made, genealogies."

CHECKPOINT

1. How much of the classification in Figure 14.21 do humans share with the cat?
2. Our forearms and the wings of a bat are derived from the same ancestral prototype; thus, they are ___________. In contrast, the wings of a bat and the wings of a bee are derived from totally unrelated structures; thus, they are ___________.
3. The study of the relationships between diverse organisms is called ___________. Such studies help identify ___________, each of which contains an ancestral species and all its descendants.
4. The current classification scheme favored by most biologists places life into ___________ domains. The domain ___________ includes humans in the kingdom called ___________.

Answers: 1. We are classified the same down to the class level; both cat and human are mammals. We do not belong to the same order. **2.** homologous; analogous **3.** systematics; clades **4.** three; Eukarya; Animalia

EVOLUTION CONNECTION

Just a Theory?

So far in our study of evolution, we have come across several important debates. Classification schemes (the number of kingdoms in Eukarya) and the tempo of evolution (gradualism versus punctuated equilibrium) are two examples. These and other controversies stimulate research and lead to refinements in evolutionary theory. But such debates among evolutionary biologists also seem to fuel the "just a theory" argument that certain groups use in an attempt to discredit evolutionary theory and eliminate the study of evolution from the science curricula of our public schools.

The "just a theory" tactic for nullifying biology's unifying theme has two flaws. First, it fails to separate Darwin's two main points: the evidence that modern species *evolved* from ancestral forms and the theory that natural selection is the main *mechanism* for this evolution. To biologists, Darwin's "theory of evolution" is the idea that changes in species over time are due to natural selection—the mechanism Darwin proposed to explain the historical record of evolution documented by fossils, biogeography, and other types of evidence.

This brings us to the second flaw in the "just a theory" argument. The term *theory* has a very different meaning in science than in general use. The everyday use of "theory" comes close to what scientists mean by "hypothesis"—a tentative explanation. In science, however, a theory is more comprehensive than a hypothesis (see Chapter 1). A theory, such as Newton's theory of gravity, Einstein's theory of relativity, or Darwin's theory of natural selection, accounts for many facts and attempts to explain a great variety of phenomena. Such a unifying theory does not become widely accepted in science unless its predictions stand up to thorough and continual testing by experiments and observations. In other words, a comprehensive theory, to be accepted, is held to a higher standard of evidence than a hypothesis. Even then, science does not allow theories to become dogma. For example, most evolutionary biologists now doubt that natural selection alone accounts for the evolutionary history observed in the fossil record. Unpredictable events, such as asteroids crashing into Earth, are also thought to be important in shaping the patterns of biological diversity we observe.

The study of evolution is more robust than ever as a branch of science, and questions about how life evolves in no way imply that most biologists consider evolution itself to be "just a theory." Debates about evolutionary theory are like arguments over competing theories about gravity; we know that objects keep right on falling while physicists debate the details of possible mechanisms.

By attributing the diversity of life to natural causes rather than to supernatural forces, Darwin gave biology a sound, scientific basis. Nevertheless, the diverse products of evolution are elegant and inspiring. As Darwin said in the closing paragraph of *The Origin of Species*, "There is grandeur in this view of life." ■

Chapter Review

SUMMARY OF KEY CONCEPTS

For study help and activities, go to campbellbiology.com or the student CD-ROM.

Macroevolution and the Diversity of Life

- Whereas microevolution is a change in the gene pool of a population, macroevolution comprises changes at the species level, often evident in the fossil record. Macroevolution includes the appearance of new species, the origins of evolutionary novelties, and the explosive diversification that follows some evolutionary breakthroughs and mass extinctions.

Activity *Mechanisms of Macroevolution*

The Origin of Species

MP3 Tutor *Speciation*

- **What Is a Species?** According to one definition of a species, the biological species concept, a species is a population or group of populations whose members have the potential to interbreed with one another in nature to produce fertile offspring.

• **Reproductive Barriers between Species**

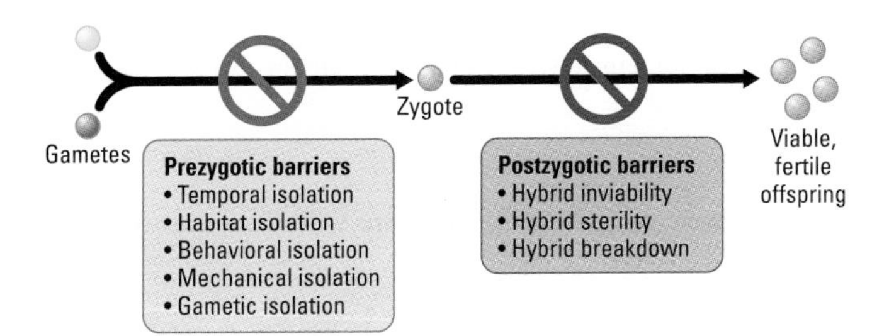

• **Mechanisms of Speciation** When the gene pool of a population is severed from other populations of the parent species, the splinter population can follow its own evolutionary course.

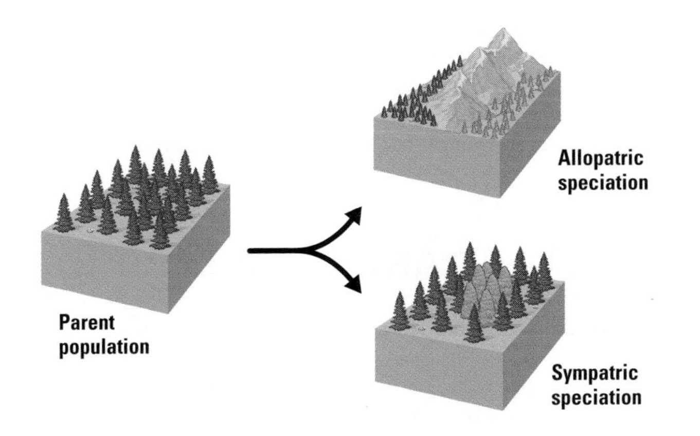

Hybridization leading to polyploids is a common mechanism of sympatric speciation in plants.

Activity *Exploring Speciation on Islands*

Activity *Polyploid Plants*

Case Study in the Process of Science *How Do New Species Arise by Genetic Isolation?*

• **What Is the Tempo of Speciation?** According to the punctuated equilibrium model, the time required for speciation in most cases is relatively short compared with the overall duration of the species' existence. This accounts for the relative rarity of transitional fossils linking newer species to older ones.

The Evolution of Biological Novelty

• **Adaptation of Old Structures for New Functions** An exaptation is a structure that evolves in one context and gradually becomes adapted for other functions.

• **"Evo-Devo": Development and Evolutionary Novelty** A subtle change in the genes that control a species' development can have profound effects. In paedomorphosis, for example, the adult retains juvenile body features.

Activity *Paedomorphosis: Morphing Chimps and Humans*

Earth History and Macroevolution

• **Geologic Time and the Fossil Record** Geologists have established a geologic time scale with four broad divisions: Precambrian, Paleozoic, Mesozoic, and Cenozoic. The most common method for determining the ages of fossils is radiometric dating.

• **Plate Tectonics and Macroevolution** About 250 million years ago, plate movements brought all the landmasses together into the supercontinent Pangaea, causing extinctions and providing new opportunities for the survivors to diversify. About 180 million years ago, Pangaea began to break up, causing geographic isolation.

• **Mass Extinctions and Explosive Diversifications of Life** The fossil record reveals long, relatively stable periods punctuated by mass extinctions followed by explosive diversification of certain survivors. For example, during the Cretaceous extinctions, about 65 million years ago, the world lost an enormous number of species, including dinosaurs. Mammals greatly increased in diversity after the Cretaceous.

Activity *The Geologic Time Scale*

Classifying the Diversity of Life

• Systematics, the study of biological diversity, includes taxonomy, which is the identification, naming, and classification of species.

• **Some Basics of Taxonomy** Each species is assigned a two-part name consisting of the genus and the species. In the taxonomic hierarchy, domain > kingdom > phylum > class > order > family > genus > species.

• **Classification and Phylogeny** The goal of classification is to reflect phylogeny, which is the evolutionary history of a species. Classification is based on the fossil record, homologous structures, and comparisons of DNA and amino acid sequences. Cladistics uses shared characteristics to group related organisms into clades.

Case Study in the Process of Science *How Is Phylogeny Determined Using Protein Comparisons?*

• **Arranging Life into Kingdoms: A Work in Progress** Biologists currently classify life into a three-domain system:

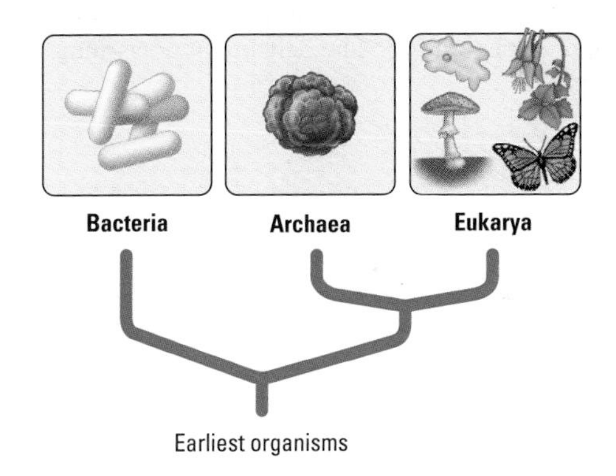

Activity *Classification Schemes*

SELF-QUIZ

1. Bird guides once listed the myrtle warbler and Audubon's warbler as distinct species that lived side by side in parts of their ranges. However, recent books show them as eastern and western forms of a single species, the yellow-rumped warbler. Apparently, it has been found that the two kinds of warblers
   a. live in the same areas.
   b. successfully interbreed.
   c. are almost identical in appearance.
   d. are merging to form a single species.

2. Label each of the following reproductive barriers as prezygotic or postzygotic:
 a. One lilac species lives on acid soil, another on basic soil.
 b. Mallard and pintail ducks mate at different times of year.
 c. Two species of leopard frogs have different mating calls.
 d. Hybrid offspring of two species of jimsonweed always die before reproducing.
 e. Pollen of one kind of tobacco cannot fertilize another kind.

3. Why is a small, isolated population more likely to undergo speciation than a large one?

4. Many species of plants and animals adapted to desert conditions probably did not arise there. Their success in living in deserts could be due to __________, structures that evolved in one context but became adapted for different functions.

5. Mass extinctions that occurred in the past
 a. cut the number of species to the few survivors left today.
 b. resulted mainly from the separation of the continents.
 c. occurred regularly, about every million years.
 d. were followed by diversification of the survivors.

6. The animals and plants of India are almost completely different from the species in nearby Southeast Asia. Why might this be true?
 a. They have become separated by convergent evolution.
 b. The climates of the two regions are completely different.
 c. India is in the process of separating from the rest of Asia.
 d. India was a separate continent until relatively recently.

7. Place these levels of classification in order from least inclusive to most inclusive: class, domain, family, genus, kingdom, order, phylum, species.

8. A paleontologist estimates that when a particular rock formed, it contained 12 mg of the radioactive isotope potassium-40. The rock now contains 3 mg of potassium-40. The half-life of potassium-40 is 1.3 billion years. From this information, you can conclude that the rock is approximately __________ billion years old.

9. In the three-domain system, which two domains contain prokaryotic organisms?

Answers to the Self-Quiz questions can be found in Appendix D.

Go to the website or CD-ROM for more Self-Quiz questions.

THE PROCESS OF SCIENCE

10. Distinguish "hypothesis" from "theory" in the vocabulary of science.

11. Imagine you are conducting fieldwork and discover two groups of mice living on opposite sides of a river. Assuming that you will not disturb the mice, design a study to determine whether or not these two groups belong to the same species. If you could capture some of the mice and bring them to the lab, would that affect your experimental design?

BIOLOGY AND SOCIETY

12. Experts estimate that human activities cause the extinction of hundreds or thousands of species every year. The natural rate of extinction is thought to be a few species per year. As we continue to alter the global environment, especially by cutting down tropical rain forests, the resulting extinction will probably rival that at the end of the Cretaceous period. Most biologists are alarmed at this prospect. What are some reasons for their concern? Considering that life has endured numerous mass extinctions and has always bounced back, how is the present mass extinction different? What might be some of the consequences for the surviving species?

15 The Evolution of Microbial Life

More than half of our antibiotics come from one genus of **soil bacterium.**

You may be eating a **protist with your sushi.**

Bacterial fermentation is used to produce cheese, yogurt, buttermilk, and sausage.

The number of **bacteria in one human's mouth** is greater than the total number of people who ever lived.

Bioterrorism

During the fall of 2001, five Americans died from the disease anthrax in a presumed terrorist attack **(Figure 15.1)**. Unfortunately, this act of bioterrorism was hardly unprecedented; there is a long and ugly history of humans using organisms as weapons. Animals, plants, fungi, and viruses have all served this purpose, but the most frequently employed biowarfare agents have been bacteria.

During the Middle Ages, the bacterium *Yersinia pestis* (the cause of bubonic plague) played a role in battle when armies hurled the bodies of plague victims into enemy ranks. Early conquerors, settlers, and warring armies in South and North America gave native peoples items purposely contaminated with infectious bacteria. In the 1930s, the Japanese government instituted a biowarfare program that killed tens of thousands of Chinese soldiers and civilians using bacteria that cause plague, anthrax, and cholera. In 1984, members of a cult in Oregon contaminated restaurant salad bars with *Salmonella* bacteria (which can cause food poisoning); over 700 people became sick, and 45 were hospitalized. During the 1990s, another cult tried to start an anthrax epidemic in Tokyo, and the Iraqi army loaded missiles with harmful bacteria. Luckily, neither of these latter two attempts resulted in casualties.

Figure 15.1 Cleaning up after a bioterrorist attack. This 2001 photo shows a hazardous-material worker spraying his colleagues after they had searched the Senate Office Building in Washington, DC, for anthrax spores.

The United States opened its first biological weapons research facility in 1943 at Fort Detrick, Maryland. There, the military studied and bred new strains of bacteria that cause such illnesses as anthrax, botulism, and tularemia. To "weaponize" naturally occurring pathogens, researchers selected highly virulent strains, made them resistant to antibiotic medications, and developed formulations for effective dispersal. But the practical difficulties of controlling such weapons—and a measure of moral repugnance—led President Richard Nixon to end the U.S. bioweapons program in 1969 and to order its products destroyed. In 1975, the Untied States signed the Biological Weapons Convention, pledging never to develop or store biological weapons. Eventually, 103 nations joined the ban, although not every signatory has honored it.

Of course, not all bacteria are harmful to humans. In fact, nearly all life on Earth depends on bacteria and other microbial life in one way or another. In this chapter, we'll examine the origins, structures, and diversity of microbes, starting with the prokaryotes and then moving on to the protists. We'll begin by examining some of the key events in the history of life on Earth. ■

Major Episodes in the History of Life

Life began when Earth was young. The planet formed about 4.5 billion years ago, and its crust began to solidify about 4 billion years ago (see Evolution Connection at the end of Chapter 2). A few hundred million years later, by 3.5 billion years ago, Earth was already inhabited by a diversity of

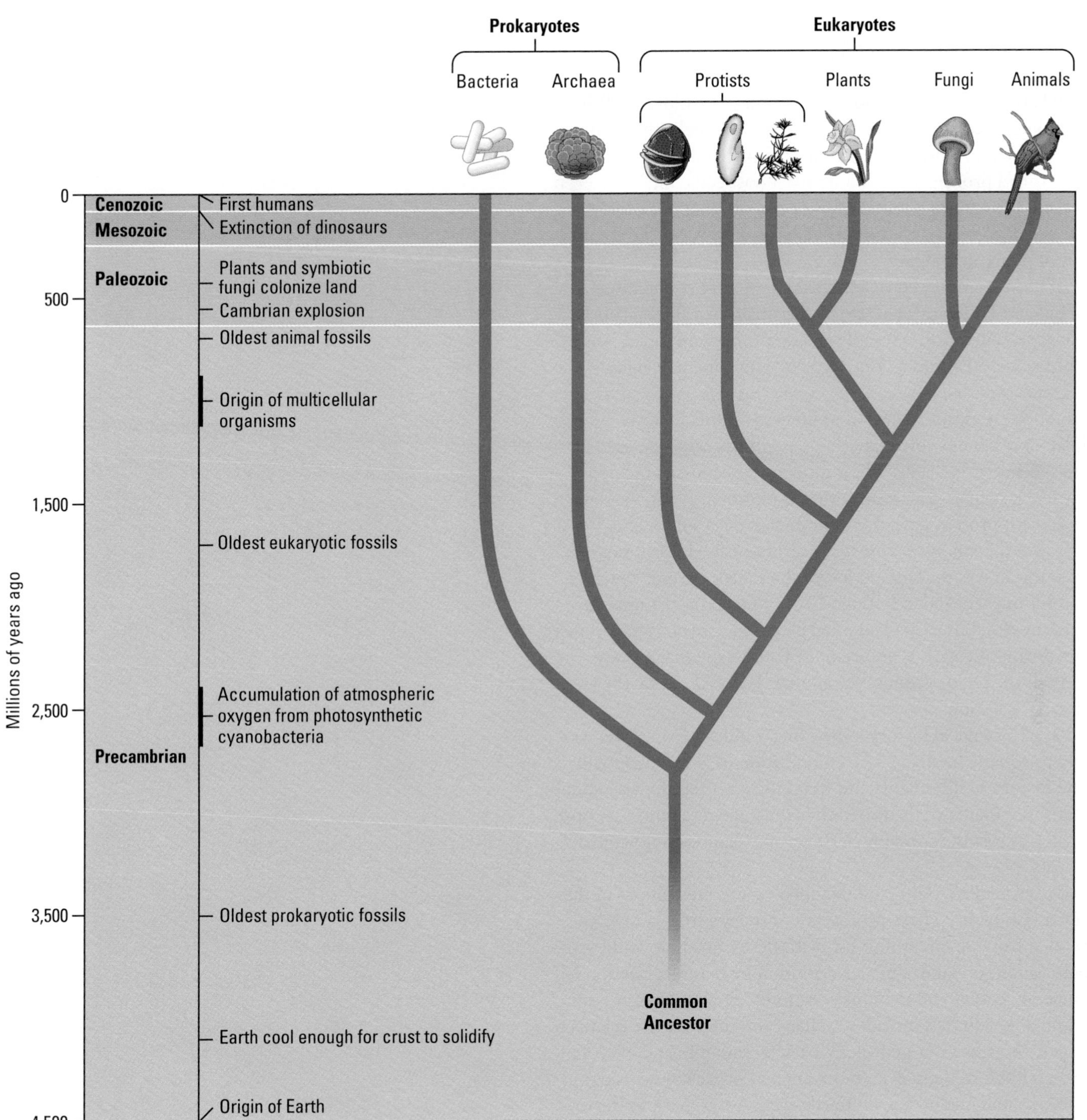

Figure 15.2 Some major episodes in the history of life. The timing of events is based on fossil evidence and molecular analysis.

organisms **(Figure 15.2)**. Those earliest organisms were all **prokaryotes,** their cells lacking true nuclei (see Chapter 4). Within the next billion years, two distinct groups of prokaryotes—bacteria and archaea—diverged.

An oxygen revolution began about 2.5 billion years ago. Photosynthetic prokaryotes that split water molecules released oxygen gas, changing Earth's atmosphere profoundly (see Evolution Connection at the end of Chapter 7). The corrosive O_2 doomed many prokaryotic groups. Among the survivors, a diversity of metabolic modes evolved, including cellular respiration, which uses O_2 to extract energy from food (see Chapter 6). All of this metabolic evolution occurred during the almost 2 billion years that prokaryotes had Earth to themselves.

The oldest eukaryotic fossils are about 2.2 billion years old. Recall from Chapter 4 that **eukaryotes** are composed of one or more cells that contain nuclei and many other organelles absent in prokaryotic cells. The eukaryotic cell evolved from a prokaryotic community, a host cell containing even smaller prokaryotes. The mitochondria of our cells and those of every other eukaryote are descendants of those smaller prokaryotes. And so are the chloroplasts of plants and algae.

The origin of more complex cells launched a period of tremendous diversification of eukaryotic forms. These new organisms were the protists. Represented today by a great diversity of species, protists are mostly microscopic and unicellular. Examples you may recognize include algae, amoebas, and *Paramecium*.

The next great event in the evolution of life was multicellularity. The first multicellular eukaryotes evolved, perhaps a billion years ago, as colonies of single-celled organisms. Their modern descendants include multicellular protists, such as seaweeds. Other evolutionary branches stemming from the ancient protists gave rise to animals, fungi, and plants.

The greatest diversification of animals was the so-called Cambrian explosion. The Cambrian was the first period of the Paleozoic era, which began about 540 million years ago (see Table 14.1). The earliest animals lived in late-Precambrian seas, but they diversified extensively over a span of just 10 million years during the early Cambrian. In fact, all the major body plans—as well as all the major groups—of animals had evolved by the end of that evolutionary eruption.

For over 85% of biological history—life's first 3 billion years—life was confined mostly to aquatic habitats. The colonization of land was a major milestone in the history of life. Plants and fungi together led the way about 475 million years ago. Plants transformed the landscape, creating new opportunities for all life-forms, especially herbivorous (plant-eating) animals and their predators.

The evolutionary venture onto land included vertebrate animals in the form of the first amphibians. These prototypes of today's frogs and salamanders descended from air-breathing fish with fleshy fins that could support the animal's weight on land. Further evolution by natural selection led to the appearance of reptiles and mammals. Among the mammals are the primates, the animal group that includes humans and their closest relatives, apes and monkeys. But trace our genealogy back far enough, and we count ancient protists, and before them even more ancient prokaryotes, as our distant ancestors. To understand life on Earth, we must go back to the origin and diversification of microbes.

CHECKPOINT

Put the following events in order, from the earliest to the most recent (it may help to refer to Table 14.1): diversification of animals (Cambrian explosion), evolution of eukaryotic cells, first humans, colonization of land by plants and fungi, origin of prokaryotes, evolution of land animals, evolution of multicellular organisms.

Answer: Origin of prokaryotes, evolution of eukaryotic cells, evolution of multicellular organisms, diversification of animals (Cambrian explosion), colonization of land by plants and fungi, evolution of land animals, first humans

The Origin of Life

We will never know for sure, of course, how life on Earth began. But as with any scientific investigation, we start with the assumption that science seeks natural causes for natural phenomena.

Resolving the Biogenesis Paradox

From the time of the ancient Greeks until less than 200 years ago, it was commonly believed that life could arise from nonliving matter. This idea of life emerging from inanimate material, called **spontaneous generation,** persisted well into the 19th century as an explanation for the rapid appearance of organisms around spoiled foods. Then in 1862, a series of experiments by Louis Pasteur confirmed what many others had suspected: All life today, including microbes, arises only by the reproduction of preexisting life. This "life-from-life" principle is called **biogenesis.**

But wait! If life always arises from previous life, then how could the first organisms arise? Although there is no evidence that spontaneous generation occurs today, it could have early in Earth's history, when conditions were very different. For instance, there was little corrosive atmospheric oxygen to tear apart complex molecules. And such energy sources as lightning, volcanic activity, and ultraviolet sunlight were all more intense than what we experience today (see Evolution Connection at the end of Chapter 2). The resolution to the biogenesis paradox is that life did not begin on a planet anything like the modern Earth, but on a young Earth that was a very different world **(Figure 15.3)**. Most biologists now think that it is at least a reasonable hypothesis that chemical and physical processes in Earth's primordial environment could have eventually produced very simple cells through a sequence of stages. Debate abounds about the nature of those stages.

Figure 15.3 An artist's rendition of Earth about 3 billion years ago. The pad-like objects in the scene represent colonies of prokaryotes known from the fossil record.

A Four-Stage Hypothesis for the Origin of Life

According to one hypothesis, the first organisms were products of chemical evolution in four stages: (1) the abiotic (nonliving) synthesis of small organic molecules, such as amino acid and nucleotide monomers; (2) the joining of these small molecules into polymers, including proteins and nucleic acids; (3) the origin of self-replicating molecules that eventually made inheritance possible; and (4) the packaging of all these molecules into pre-cells, droplets with membranes that maintained an internal chemistry different from the surroundings. This is all speculative, of course, but what makes it a valid scientific hypothesis is that it leads to predictions that can be tested in the laboratory.

Stage 1: Abiotic Synthesis of Organic Monomers Of the four stages proposed here for the origin of life, the first stage has been the most extensively studied by scientists. We'll begin our investigation of chemical evolution with a breakthrough experiment that first brought this idea to the forefront of scientific thinking.

Can Biological Monomers Form Spontaneously? In 1953, University of Chicago scientist Harold Urey and his 23-year-old graduate student Stanley Miller conducted a classic experiment. They began with the **observation** that modern biological macromolecules (DNA, protein, carbohydrates, and so on) are all composed of elements (primarily oxygen, hydrogen, carbon, and nitrogen) that were present in abundance on the early Earth. This led to the **question**, Could biological molecules arise spontaneously under conditions like those on the early Earth? Miller and Urey began with the **hypothesis** that a closed system designed in the laboratory to simulate such conditions could produce biologically important organic molecules from inorganic ingredients.

Figure 15.4 shows the apparatus they created to test their hypothesis. A warmed flask of water simulated the primordial sea. An "atmosphere"—in the form of gases added to a reaction chamber—contained hydrogen gas (H_2), methane (CH_4), ammonia (NH_3), and water vapor (H_2O). To mimic the prevalent lightning of the early Earth, sparks were discharged into the chamber. A condenser cooled the atmosphere, causing water and any dissolved compounds to "rain" into the miniature sea. Miller and Urey's **prediction** was that organic molecules would form and accumulate during this **experiment**.

Miller and Urey's **results** made front-page news. After the apparatus had run for a week, an abundance of organic molecules essential for life, including amino acids, the monomers of proteins, had collected in the "sea."

Since Miller and Urey's seminal experiment, other scientists have repeated and extended the research, varying such conditions as the composition of the ancient "atmosphere" and "sea." These laboratory analogues of the primeval Earth have produced all 20 amino acids, several sugars, lipids, nucleotides, and even ATP, the molecule that powers most biological work. The abiotic synthesis of organic molecules on the early Earth is certainly a plausible scenario. ■

(a) Stanley Miller with his experimental apparatus.

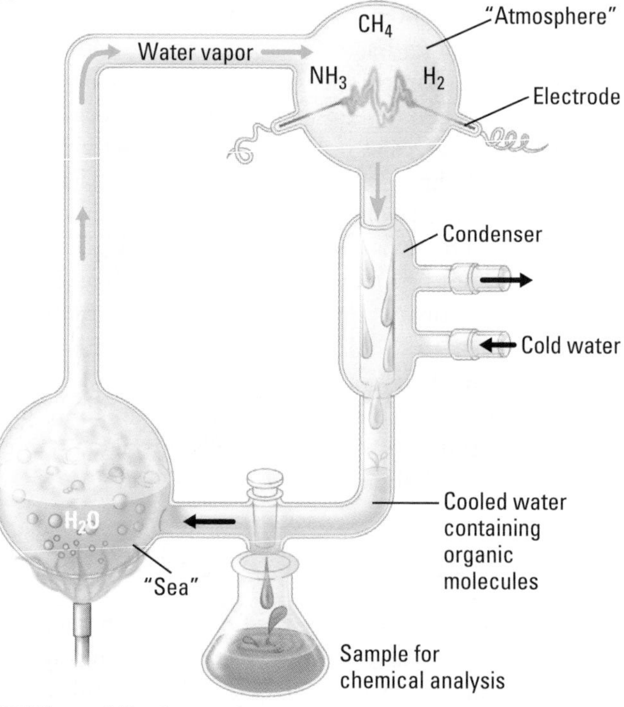

(b) Miller and Urey's experiment.

Figure 15.4 Making organic molecules in a laboratory simulation of early-Earth chemistry.

Stage 2: Abiotic Synthesis of Polymers If the hypothesis of an abiotic origin of life is correct, then it should be possible to link organic monomers to form polymers such as proteins and nucleic acids without the help of enzymes and other cellular equipment. Researchers have brought about such polymerization by dripping solutions of organic monomers onto hot sand, clay, or rock. The heat vaporizes the water in the solutions and concentrates the monomers on the underlying material. Some of the monomers then spontaneously bond together to form polymers. On the early Earth, raindrops or waves may have splashed dilute solutions of organic monomers onto fresh lava or other hot rocks and then washed polypeptides and other polymers back into the sea. Alternatively, submerged volcanoes and deep-sea vents, where gases and superheated water with dissolved minerals escape from Earth's interior, may have been locales for the abiotic synthesis of organic monomers and polymers. (You'll learn about these unique ecosystems in Chapter 19.)

Stage 3: Origin of Self-Replicating Molecules Life is defined partly by the process of inheritance, which is based on self-replicating molecules. Today's cells store their genetic information as DNA. They transcribe the information into RNA and then translate RNA messages into specific enzymes and other proteins (see Chapter 10). This mechanism of information flow

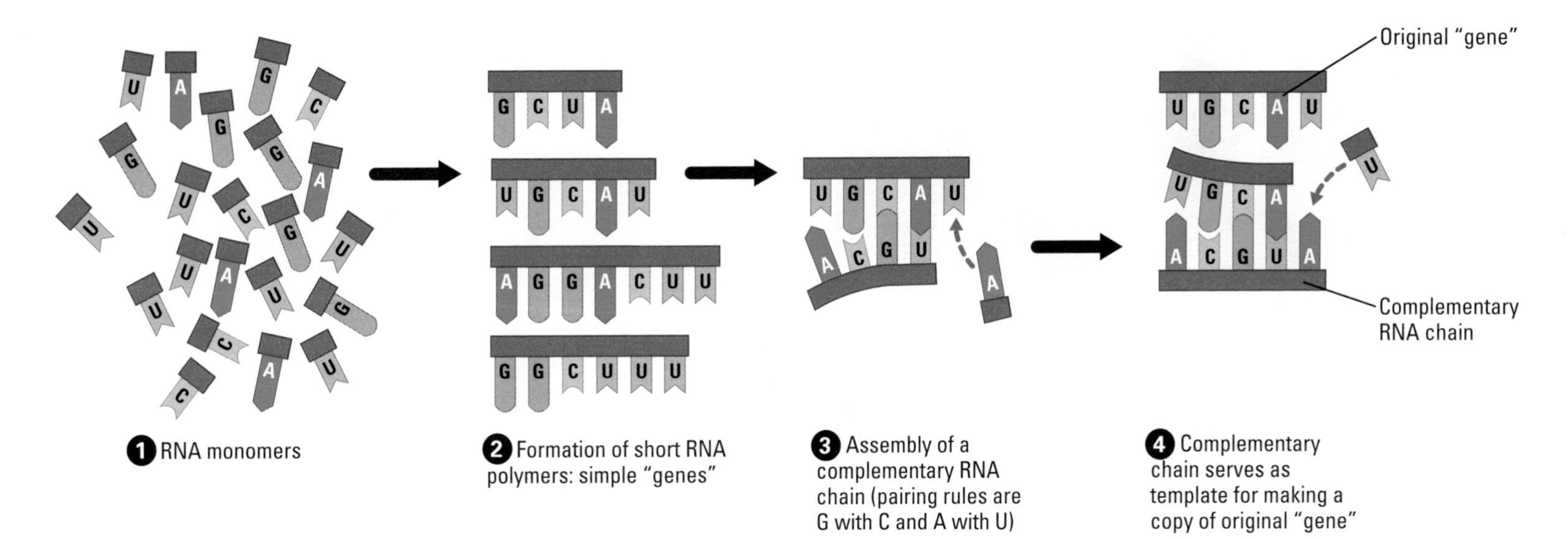

Figure 15.5 Abiotic replication of RNA "genes."

probably emerged gradually through a series of refinements to much simpler processes. What were the first genes like?

One hypothesis is that the first genes were short strands of RNA that replicated themselves without the assistance of proteins. In laboratory experiments, short RNA molecules can assemble spontaneously from nucleotide monomers in the absence of cells or enzymes **(Figure 15.5)**. The result is a population of RNA molecules, each with a random sequence of monomers. Some of the molecules self-replicate, but their success at this reproduction varies. What happens can be described as molecular evolution: The RNA varieties that replicate fastest increase their frequency in the population.

In addition to the experimental evidence, there is another reason the idea of RNA genes in the primordial world is plausible. Cells actually have RNAs that can act like enzymes; they are called **ribozymes.** Perhaps early ribozymes catalyzed their own replication. That would help with the "chicken and egg" paradox of which came first, enzymes or genes. Maybe the "chicken and egg" came together in the same RNA molecules. The molecular biology of today may have been preceded by an ancient "RNA world."

Stage 4: Formation of Pre-Cells The properties of life emerge from an interaction of molecules organized into higher levels of order. For example, **Figure 15.6a** diagrams one way that RNA and polypeptides could have cooperated in the prebiotic world. But such molecular teams would be much more efficient if they were packaged within membranes that kept the molecules close together. We'll call these molecular aggregates pre-cells—not really cells, but molecular packages with some of the properties of life **(Figure 15.6b)**.

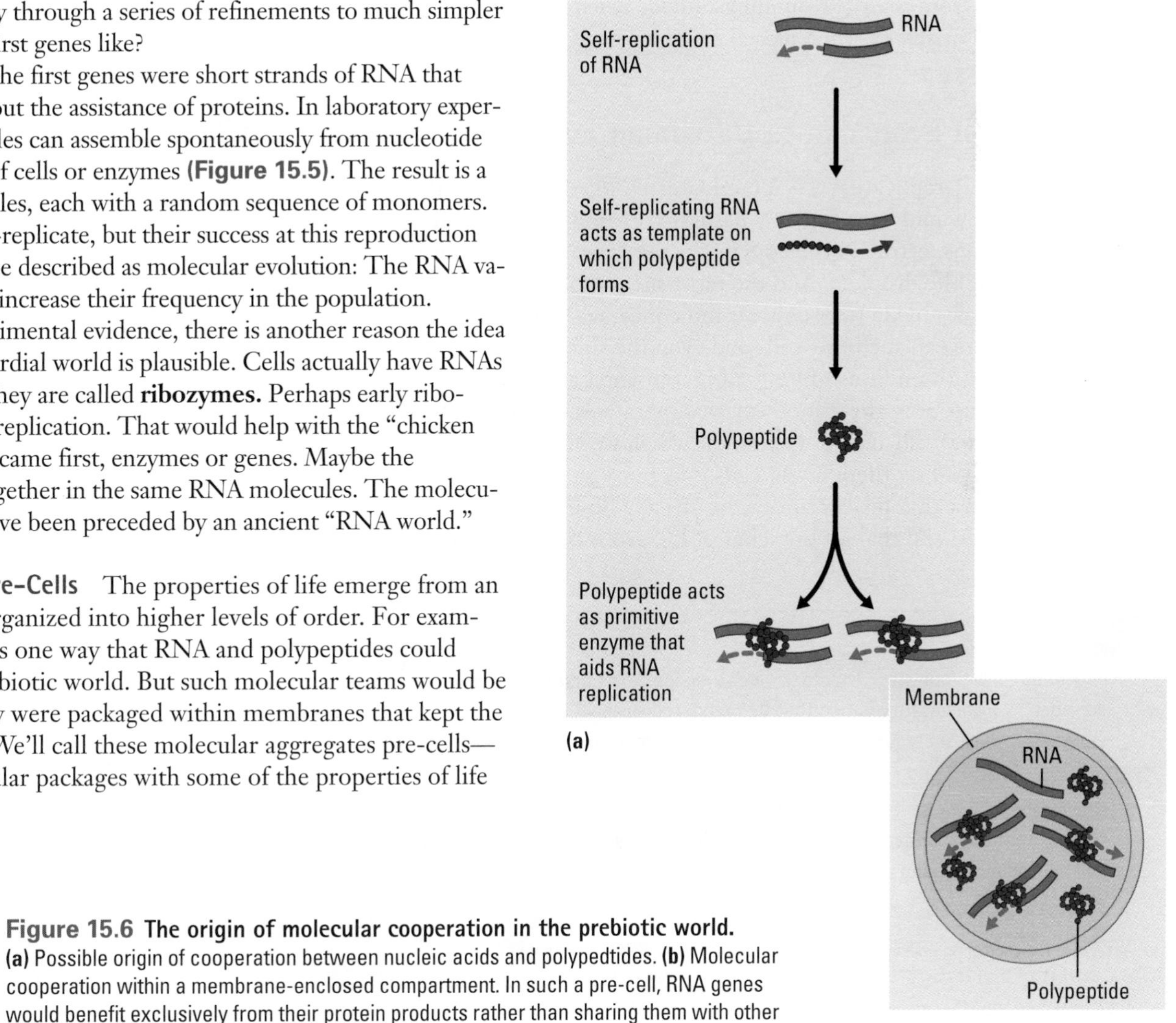

Figure 15.6 The origin of molecular cooperation in the prebiotic world. **(a)** Possible origin of cooperation between nucleic acids and polypedtides. **(b)** Molecular cooperation within a membrane-enclosed compartment. In such a pre-cell, RNA genes would benefit exclusively from their protein products rather than sharing them with other molecular complexes.

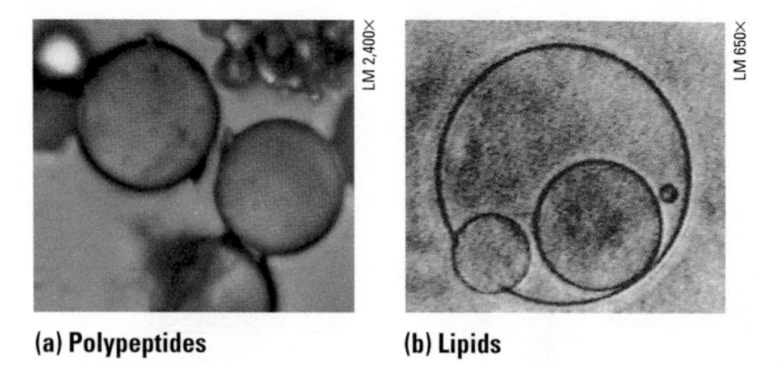

Figure 15.7 Laboratory versions of pre-cells. **(a)** These tiny spheres are made by cooling solutions of polypeptides that are produced abiotically. The spheres grow by absorbing more polypeptides until they reach an unstable size, at which time they split to form "daughter" spheres. **(b)** Lipids mixed with water self-assemble into these membrane-enclosed droplets (see Evolution Connection in Chapter 4). In some cases, large droplets bud to give rise to smaller "offspring."

Laboratory experiments demonstrate that pre-cells could have formed spontaneously from abiotically produced organic compounds **(Figure 15.7)**. Although pre-cells are not alive, they display some of the properties of living cells. They have a selectively permeable surface, can grow by absorbing molecules from their surroundings, divide when they reach a certain size, and swell or shrink when placed in solutions of different salt concentrations.

From Chemical Evolution to Darwinian Evolution

If pre-cells with self-replicating RNA (and later DNA) did form on the young Earth, they would be refined by natural selection—Darwinian evolution. Mutations, errors in the copying of the "genes," would result in variation among the droplets. And the most successful of these droplets would grow, divide (reproduce), and continue to evolve. Of course, the gap between such pre-cells and even the simplest of true cells is enormous. But with millions of years of incremental refinement through natural selection, these molecular cooperatives could have become more and more cell-like. The point at which we stop calling them pre-cells and start calling them living cells is as fuzzy as our definition of life. But we do know that prokaryotes were already flourishing at least 3.5 billion years ago and that all branches of life arose from those ancient prokaryotes.

CHECKPOINT

1. According to the four-stage hypothesis we've explored, the first cells were preceded by a chemical evolution that first produced small ___________ molecules, which subsequently joined to form larger molecules called ___________.
2. One reason why the spontaneous generation of life on Earth could not occur today is that the abundance of ___________ in our modern atmosphere would destroy complex organic molecules that are not inside organisms.
3. What is a ribozyme?

Answers: 1. monomer (or organic); polymers **2.** oxygen (O_2) **3.** An RNA molecule that functions as a catalyst

Prokaryotes

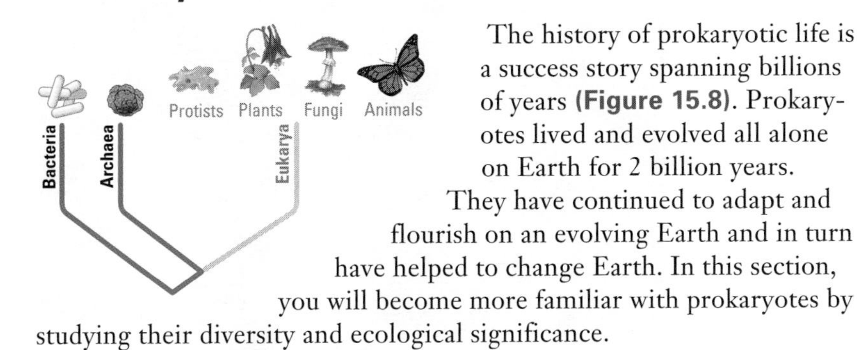

The history of prokaryotic life is a success story spanning billions of years **(Figure 15.8)**. Prokaryotes lived and evolved all alone on Earth for 2 billion years. They have continued to adapt and flourish on an evolving Earth and in turn have helped to change Earth. In this section, you will become more familiar with prokaryotes by studying their diversity and ecological significance.

They're Everywhere!

Today, prokaryotes are found wherever there is life, and they far outnumber the eukaryotes. In fact, more prokaryotes inhabit your mouth than the total number of people who have ever lived! Prokaryotes also thrive in habitats too cold, too hot, too salty, too acidic, or too alkaline for any eukaryote. In 1999, biologists even discovered prokaryotes living on the walls of a gold mine 2 miles below Earth's surface.

Though individual prokaryotes are small organisms, they are giants in their collective impact on Earth and its life. We hear most about a few species that cause serious illness. During the 14th century, Black Death—bubonic plague, a bacterial disease—spread across Europe, killing an estimated 25% of the human population. Tuberculosis, cholera, many sexually transmissible diseases, and certain types of food poisoning are also caused by bacteria.

However, prokaryotic life is much more than just a rogues' gallery. Benign or beneficial prokaryotes are far more common than harmful ones. Bacteria in our intestines provide us with important vitamins, and others living in our mouth prevent harmful fungi from growing there. Prokaryotes also recycle carbon and other vital chemical elements back and forth between organic matter and the soil and atmosphere. For example, some prokaryotes decompose dead organisms. Found in soil and at the bottom of lakes, rivers, and oceans, these decomposers return chemical elements to the environment in the form of inorganic compounds that can be used by plants, which in turn feed animals. If prokaryotic decomposers were to disappear, the chemical cycles that sustain life would come to a halt. All forms of eukaryotic life would also be doomed. In contrast, prokaryotic life would undoubtedly persist in the absence of eukaryotes, as it once did for 2 billion years.

The Two Main Branches of Prokaryotic Evolution: Bacteria and Archaea

Prokaryotes have a cellular organization fundamentally different from that of eukaryotes. Whereas eukaryotic cells have a membrane-enclosed nucleus and numerous other membrane-enclosed organelles, prokaryotic cells lack these structural features (see Chapter 4). By comparing diverse prokaryotes at the molecular level, biologists have identified two major branches of prokaryotic evolution: **bacteria** and **archaea.** Though they

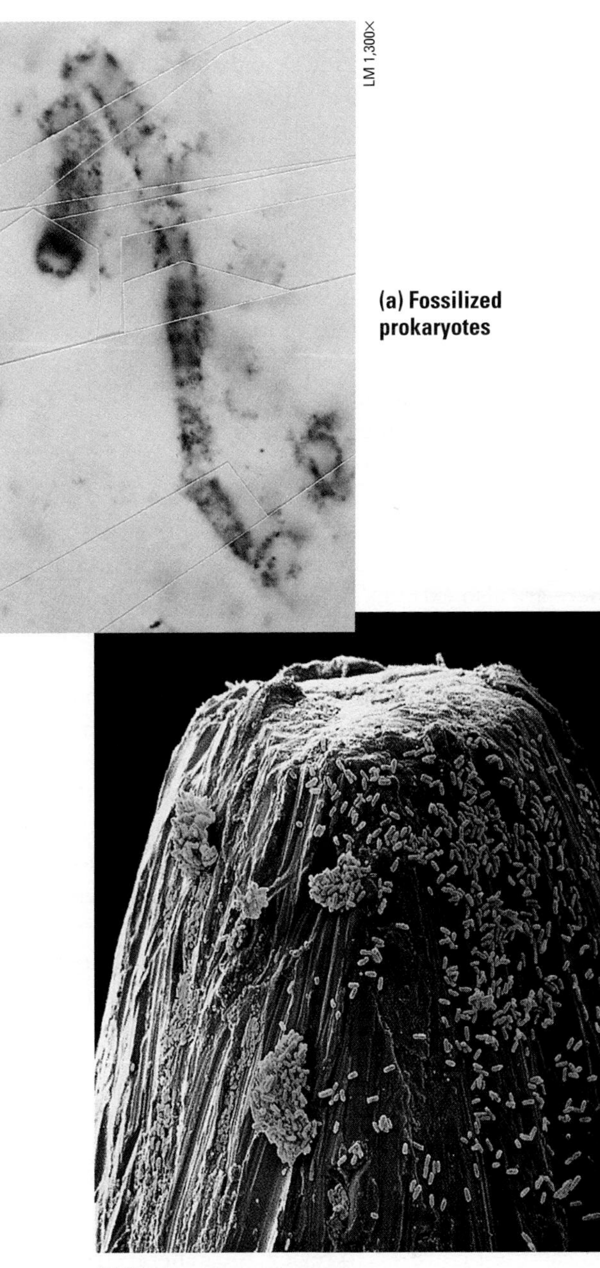

(a) Fossilized prokaryotes

(b) Modern-day prokaryotes

Figure 15.8 Over 3 billion years of prokaryotes. **(a)** This microscopic, 3.5-billion-year-old Australian fossil is the remains of a chain of prokaryotic cells. **(b)** The orange rods are individual modern bacteria, each about 5 μm long, on the head of a pin. This micrograph will help you understand why a pin prick can cause infection.

have prokaryotic cell organization in common, bacteria and archaea differ in many structural, biochemical, and physiological characteristics. Some of these suggest that archaea are more closely related to eukaryotes than they are to bacteria. It was these discoveries that prompted the three-domain classification—domains Bacteria, Archaea, and Eukarya—which you can review in Figure 14.26. The majority of known prokaryotes are bacteria, but the archaea are worth studying for their evolutionary and ecological significance.

The term archaea ("ancient") refers to the antiquity of the group's origin from the earliest cells. Even today, many species of archaea inhabit extreme environments, such as hot springs and salt ponds. Few other modern organisms (if any) can survive in some of these environments, which may resemble habitats on the early Earth.

Biologists refer to some archaea as "extremophiles," meaning "lovers of the extreme." There are extreme halophiles ("salt lovers"), archaea that thrive in such environments as Utah's Great Salt Lake, the Dead Sea, and seawater-evaporating ponds used to produce salt **(Figure 15.9)**. There are also extreme thermophiles ("heat lovers") that live in very hot water; some even populate the deep-ocean vents that gush water hotter than 100°C. Also among the archaea are the methanogens, which live in anaerobic (oxygen-free) environments and give off methane as a waste product. They are abundant in the mud at the bottom of lakes and swamps. You may have seen methane, also called marsh gas, bubbling up from a swamp. Great numbers of methanogens also inhabit the digestive tracts of animals. In humans, intestinal gas is largely the result of their metabolism. More importantly, methanogens aid digestion in cattle, deer, and other animals that depend heavily on cellulose for their nutrition. Normally, bloating does not occur in these animals because they regularly expel large volumes of gas produced by the methanogens and other microorganisms that enable them to utilize cellulose. (And that may be more than you wanted to know about these gas-producing microbes!)

Figure 15.9 Extreme halophiles ("salt-loving" archaea). These are seawater-evaporating ponds at the edge of San Francisco Bay. The colors of the ponds result from dense growth of the prokaryotes that thrive when the salinity of the water reaches 15–20% (before evaporation, seawater has a salt concentration of about 3%). The ponds are used for commercial salt production; the halophilic archaea are harmless.

The Structure, Function, and Reproduction of Prokaryotes

You can use Figure 4.5 to review the general structure of prokaryotic cells. Note the absence of a true nucleus and other membrane-enclosed organelles characteristic of the much more complex eukaryotic cells. And nearly all species of prokaryotes have cell walls exterior to their plasma membranes. These walls are chemically different from the cellulose walls of plant cells. Some antibiotics, including the penicillins, kill certain bacteria by incapacitating an enzyme the microbes use to make their walls.

Determining cell shape by microscopic examination is an important step in identifying bacteria **(Figure 15.10)**. Spherical species are called **cocci** (singular, *coccus*). Cocci that occur in clusters are called staphylococci. Other cocci occur in chains; they are called streptococci. The bacterium that causes strep throat in humans is a streptococcus. Rod-shaped prokaryotes are called **bacilli** (singular, *bacillus*). A third group of bacteria are curved or spiral-shaped. These include **spirochetes,** different species of which cause syphilis and Lyme disease.

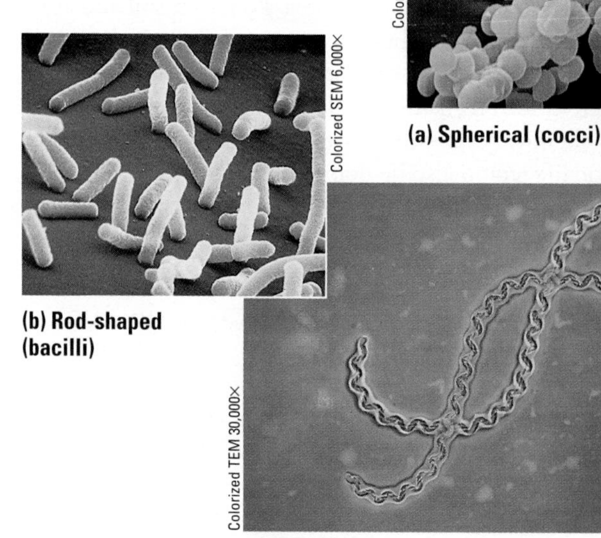

Figure 15.10 Three common shapes of bacteria.

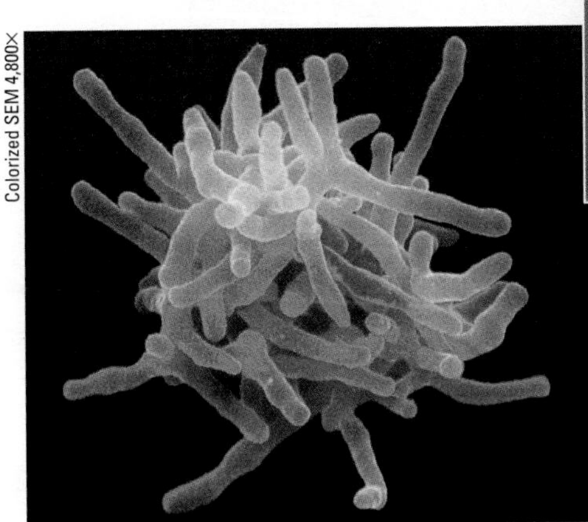

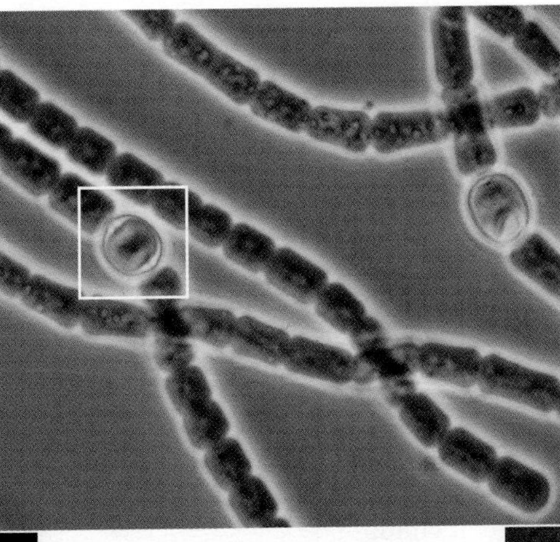

(a) Actinomycete. An actinomycete is a mass of branching chains of rod-shaped cells. These bacteria are very common in soil, where they break down organic substances. Most species secrete antibiotics, which inhibit the growth of competing bacteria. Pharmaceutical companies use various species of actinomycetes to produce antibiotic drugs, such as streptomycin.

(b) Cyanobacteria. These photosynthetic cyanobacteria are truly multicellular in having a division of labor among their cells. The box highlights a cell that converts atmospheric nitrogen to ammonia, which can then be incorporated into amino acids and other organic compounds.

(c) Giant bacterium. The bright ball in this photo is the marine bacterium *Thiomargarita namibiensis,* discovered in 1997 (the two smaller spheres above it are dead cells). This prokaryotic cell is over 0.5 mm in diameter, about the size of a fruit fly's head.

Figure 15.11 Some examples of bacterial diversity.

Most prokaryotes are unicellular and very small, but there are exceptions to both of these generalizations. Some species tend to aggregate transiently into groups of two or more cells, such as the streptococci already mentioned. Others form true colonies, which are permanent aggregates of identical cells **(Figure 15.11a)**. And some species even exhibit a simple multicellular organization, with a division of labor between specialized types of cells **(Figure 15.11b)**. Among unicellular species, there are some giants that actually dwarf most eukaryotic cells **(Figure 15.11c)**.

About half of all prokaryotic species are mobile. Many of those travelers have one or more flagella that propel the cells away from unfavorable places or toward more favorable places, such as nutrient-rich locales **(Figure 15.12)**.

Although few bacteria can thrive in the extreme environments favored by many archaea, some bacteria can survive extended periods of very harsh

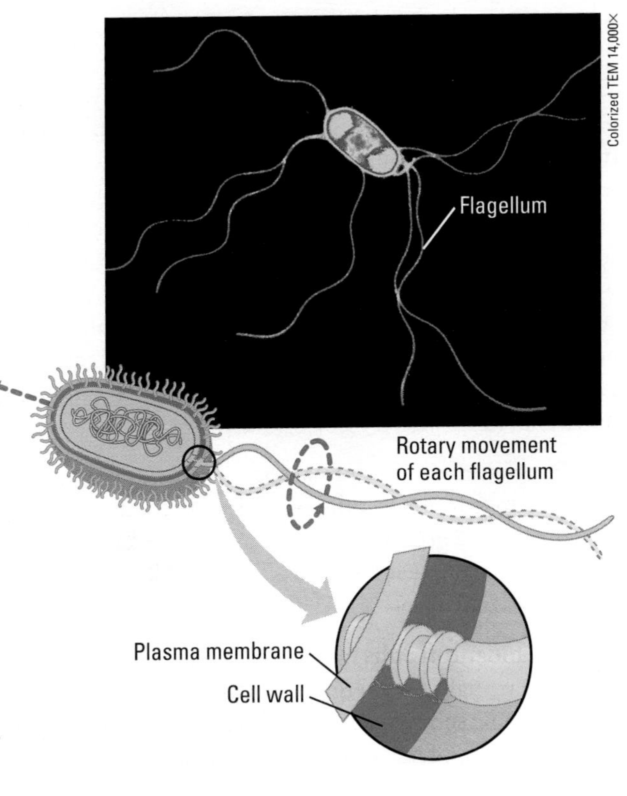

Figure 15.12 Prokaryotic flagella. These locomotor appendages are entirely different in structure and mechanics from the eukaryotic flagella discussed in Chapter 4. At the base of the prokaryotic version is a motor and set of rings embedded in the plasma membrane and cell wall. This machinery actually spins like a wheel, rotating the filament of the flagellum.

conditions by forming specialized "resting" cells, or **endospores (Figure 15.13)**. Some endospores can remain dormant for centuries. Not even boiling water kills most of these resistant cells. When anthrax is used as a biological weapon, it is the spores, not the bacteria themselves, that are deployed. To ensure that all cells including endospores are killed when laboratory equipment is sterilized, microbiologists use an appliance called an autoclave, a pressure cooker that heats to a temperature of 121°C (250°F) with high-pressure steam. The food-canning industry uses similar methods to kill endospores of dangerous soil bacteria such as *Clostridium botulinum*, which produces a toxin that causes the potentially fatal disease botulism.

Many prokaryotes can reproduce at a phenomenal rate if conditions are favorable. The cells copy their DNA almost continuously and divide again and again by the process called **binary fission.** To understand how this makes explosive population growth possible, flash back to that childhood numbers game: "Would you rather have a million dollars or start out with just a penny and have it doubled every day for a month?" If you opt for the penny, you feel like a loser at mid-month, when you have only a few hundred dollars. But by the end of the month, you've bagged about 10 million bucks. This is the exponential growth that repeated doublings make possible. Now apply that concept to prokaryotic reproduction, except double the number every 20 minutes. That's the rate at which some bacteria can divide if there is plenty of food and space. In just 24 hours, a single tiny cell could give rise to a bacterial colony equivalent in mass to about 15,000 humans!

Fortunately, few prokaryotic populations can sustain exponential growth for long. Environments are usually limiting in resources such as food and space. Prokaryotes also produce metabolic waste products that may eventually pollute the colony's environment. Still, you can understand why certain bacteria can make you sick so soon after infection or why food can spoil so rapidly. Refrigeration retards food spoilage not because the cold kills the bacteria on food but because most microorganisms reproduce very slowly at such low temperatures.

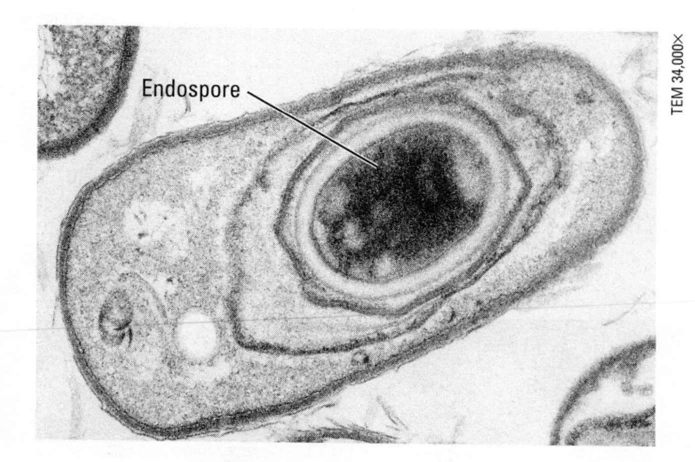

Figure 15.13 An endospore in an anthrax bacterium. This prokaryote is *Bacillus anthracis,* the notorious bacterium that produces the deadly disease called anthrax in cattle, sheep, and humans. There are actually two cells here, one inside the other. The outer cell produced the specialized dormant inner cell, called an endospore. With its thick, protective coat, the endospore can survive all sorts of trauma, including lack of water and nutrients, extreme heat or cold, and most poisons. When the environment becomes more hospitable, the endospore can absorb water and resume growth.

The Nutritional Diversity of Prokaryotes

When classifying diverse organisms, biologists often use the phrase *mode of nutrition* to describe how an organism obtains two main resources: carbon (for synthesizing organic compounds) and energy. Species that obtain energy from light are termed phototrophs, while species that obtain energy from environmental chemicals are called chemotrophs. Species that obtain carbon from the inorganic compound carbon dioxide (CO_2) are called autotrophs, while species that obtain carbon from at least one organic nutrient—the sugar glucose, for instance—are called heterotrophs. We can combine energy source (phototroph versus chemotroph) and carbon source (autotroph versus heterotroph) to group all organisms according to four major modes of nutrition **(Table 15.1)**:

Table 15.1 Nutritional Classification of Organisms

Nutritional Type	Energy Source	Carbon Source
Photoautotroph (photosynthesizer)	Sunlight	CO_2
Chemoautotroph	Inorganic chemicals	CO_2
Photoheterotroph	Sunlignt	Organic compound
Chemoheterotroph	Organic compounds	Organic compounds

1. **Photoautotrophs** are photosynthetic organisms that harness light energy to drive the synthesis of organic compounds from CO_2. Among the diverse groups of photosynthetic prokaryotes are the cyanobacteria, such as the species in Figure 15.11b. All photosynthetic eukaryotes—plants and algae—also fit into this nutritional category.

2. **Chemoautotrophs** need only CO_2 as a carbon source. However, instead of using light for energy, these prokaryotes extract energy from certain inorganic substances, such as hydrogen sulfide (H_2S) or ammonia (NH_3). This mode of nutrition is unique to certain prokaryotes. For example,

prokaryotic species living around the hot-water vents deep in the seas are the main food producers in those bizarre ecosystems (see Chapter 19).

3. **Photoheterotrophs** can use light to generate ATP but must obtain their carbon in organic form. This mode of nutrition is restricted to certain prokaryotes.
4. **Chemoheterotrophs** must consume organic molecules for both energy and carbon. This nutritional mode is found widely among prokaryotes, certain protists, and even some plants. All fungi and animals are chemoheterotrophs.

The Ecological Impact of Prokaryotes

Organisms as pervasive, abundant, and diverse as prokaryotes are guaranteed to have tremendous impact on Earth and all its inhabitants. Here we survey just a few examples of prokaryotic clout.

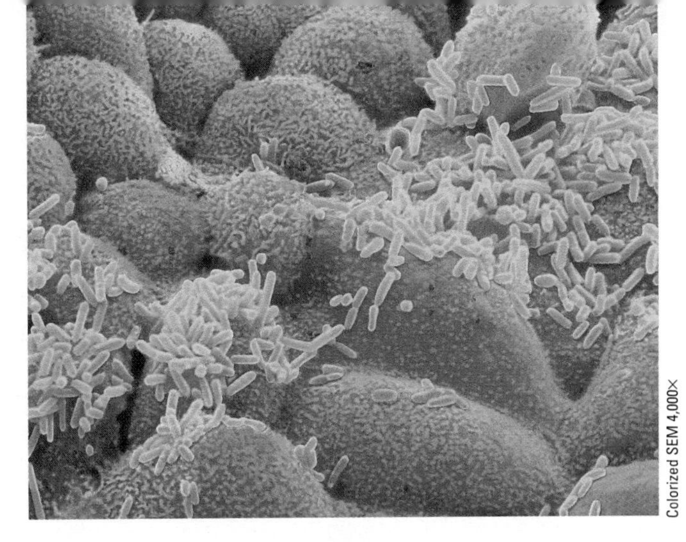

Figure 15.14 Really bad bacteria. The yellow rods are *Haemophilus influenzae* bacteria on cells lining the interior of a human nose. These pathogens are transmitted through the air. Not to be confused with influenza (flu) viruses, *H. influenzae* causes pneumonia and other lung infections that kill about 4 million people worldwide per year. Most victims are children in developing countries, where malnutrition lowers resistance to all pathogens.

Bacteria That Cause Disease We are constantly exposed to bacteria, some of which are potentially harmful **(Figure 15.14)**. Bacteria and other organisms that cause disease are called **pathogens.** We're healthy most of the time because our body's defenses check the growth of pathogens. Occasionally, the balance shifts in favor of a pathogen, and we become ill. Even some of the bacteria that are normal residents of the human body can make us sick when our defenses have been weakened by poor nutrition, medical treatment (especially cancer therapy), or a viral infection.

Most pathogenic bacteria cause disease by producing poisons. There are two classes of these poisons: exotoxins and endotoxins. **Exotoxins** are poisonous proteins secreted by bacterial cells. A single gram of the exotoxin that causes botulism could kill a million people. Another exotoxin producer is *Staphylococcus aureus* (abbreviated *S. aureus*). It is a common, usually harmless resident of our skin surface. However, if *S. aureus* enters the body through a cut or other wound or is swallowed in contaminated food, it can cause serious diseases collectively called staph infections. One type of *S. aureus* produces exotoxins that cause layers of skin to slough off; another can cause vomiting and severe diarrhea; yet another can produce a potentially deadly disease called toxic shock syndrome.

In contrast to exotoxins, **endotoxins** are not cell secretions but are chemical components of the cell walls of certain bacteria. All endotoxins induce the same general symptoms: fever, aches, and sometimes a dangerous drop in blood pressure (shock). The severity of symptoms varies with the host's condition and with the bacterium. Different species of *Salmonella*, for example, produce endotoxins that cause food poisoning and typhoid fever.

During the last 100 years, following the 19th-century discovery that "germs" cause disease, the incidence of bacterial infections has declined, particularly in developed nations. Sanitation is generally the most effective way to prevent bacterial disease. The installation of water treatment and sewage systems continues to be a public health priority throughout the world. Antibiotics have been discovered that can cure most bacterial diseases. However, resistance to widely used antibiotics has evolved in many of these pathogens (see Chapter 1).

In addition to sanitation and antibiotics, a third defense against bacterial disease is education. A case in point is Lyme disease, the most widespread pest-carried disease in the United States. The disease is caused by a spirochete bacterium carried by ticks that live on deer and field mice **(Figure 15.15)**.

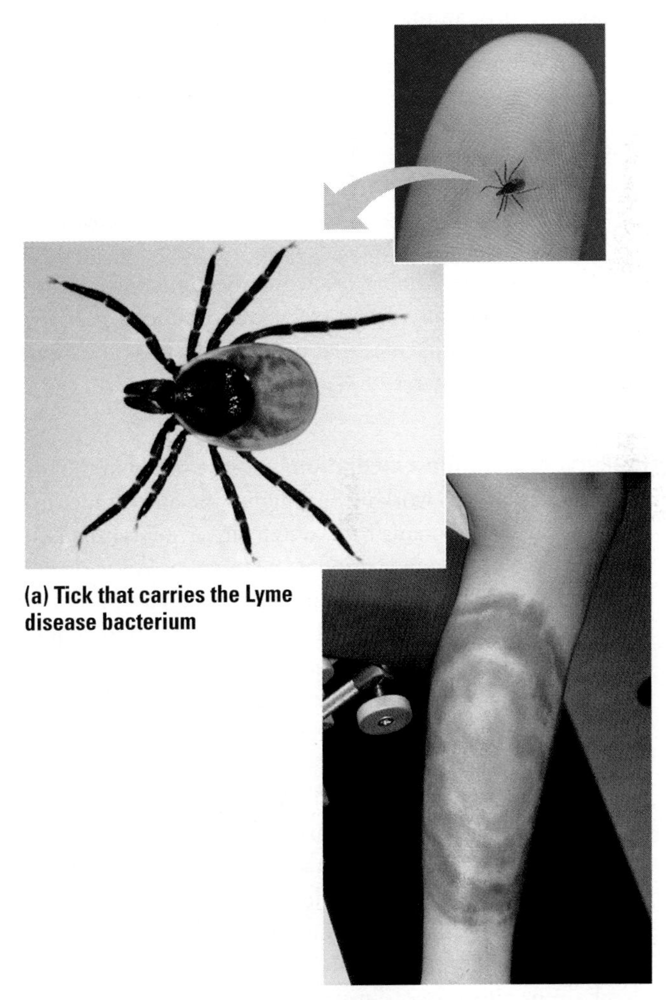

(a) Tick that carries the Lyme disease bacterium

(b) "Bull's eye" rash

Figure 15.15 Lyme disease, a bacterial disease transmitted by ticks.

Lyme disease usually starts as a red rash shaped like a bull's-eye around a tick bite. Antibiotics can cure the disease if administered within about a month of exposure. If untreated, Lyme disease can cause debilitating arthritis, heart disease, and nervous system disorders. So far, the best defense is public education about avoiding tick bites and the importance of seeking treatment if a rash develops. The Centers for Disease Control and Prevention recommends that you avoid vegetation in tick-infested areas, wear light-colored clothing when walking through brush so that you can easily see any ticks, and use an appropriate insect repellent.

As we've said, pathogenic bacteria are in the minority among prokaryotes. Far more common are species that are essential to our well-being, either directly or indirectly. Let's turn our attention now to the vital role that prokaryotes play in sustaining the biosphere.

Prokaryotes and Chemical Recycling Not too long ago, the atoms of the organic molecules in your body were parts of the inorganic compounds of soil, air, and water, as they will be again. Life depends on the recycling of chemical elements between the biological and physical components of ecosystems. Prokaryotes play essential roles in these chemical cycles. For example, all the nitrogen that plants use to make proteins and nucleic acids comes from prokaryotic metabolism in the soil. In turn, animals get their nitrogen compounds from plants.

Another vital function of prokaryotes, mentioned earlier in the chapter, is the breakdown of organic wastes and dead organisms. Prokaryotes decompose organic matter and, in the process, return elements to the environment in inorganic forms that can be used by other organisms. If it were not for such decomposers, carbon, nitrogen, and other elements essential to life would become locked in the organic molecules of corpses and waste products. You'll learn more about the roles that prokaryotes play in chemical cycling in Chapter 19.

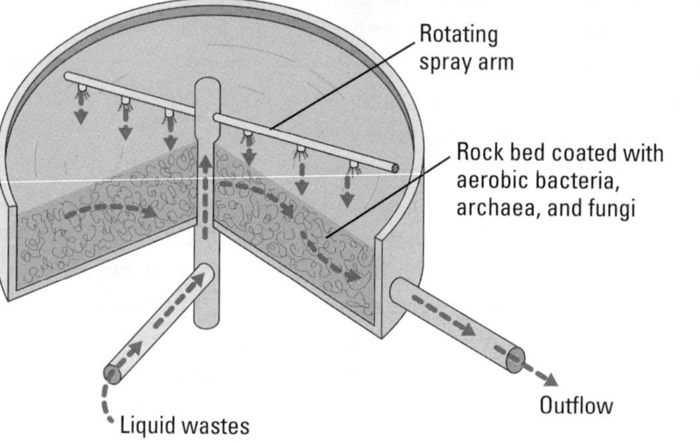

Figure 15.16 Putting prokaryotes to work in sewage treatment facilities. This is a trickling filter system, one type of mechanism for treating liquid wastes after sludge is removed. The long horizontal pipes rotate slowly, spraying liquid wastes onto a thick bed of rocks. Bacteria, archaea, and fungi growing on the rocks remove much of the organic material dissolved in the waste. Outflow from the rock bed is sterilized and then released, usually into a river or ocean.

Prokaryotes and Bioremediation People have put the metabolically diverse prokaryotes to work in cleaning up the environment. The use of organisms to remove pollutants from water, air, or soil is called **bioremediation.** One example of bioremediation is the use of prokaryotic decomposers to treat our sewage. Raw sewage is first passed through a series of screens and shredders, and solid matter is allowed to settle out from the liquid waste. This solid matter, called sludge, is then gradually added to a culture of anaerobic prokaryotes, including both bacteria and archaea. The microbes decompose the organic matter in the sludge, converting it to material that can be used as landfill or fertilizer after chemical sterilization. Liquid wastes are treated separately from the sludge **(Figure 15.16)**.

We are just beginning to explore the great potential that prokaryotes offer for bioremediation. Certain bacteria that occur naturally on ocean beaches can decompose petroleum and are useful in cleaning up oil spills **(Figure 15.17)**. Genetically engineered bacteria may be able to degrade oil more rapidly than the naturally occurring oil-eaters. Bacteria may also help us clean up old mining sites. The water that drains from mines is highly acidic and is also laced with poisons—often compounds of arsenic, copper, zinc, and the heavy metals lead, mercury, and cadmium. Contamination of our soils and groundwater by these toxic substances poses a widespread threat, and cleaning up the mess is extremely expensive. Although there are no simple solutions to the problem, prokaryotes may be able to help. Bacteria called *Thiobacillus* thrive in the acidic waters that drain from mines. Some mining companies use these microbes to extract copper and other valuable metals from low-grade ores. While obtaining energy by oxidizing sulfur or sulfur-containing compounds, the bacteria also accumulate metals from the mine waters. Unfortunately, their use in cleaning up mine wastes is limited because their metabolism also adds sulfuric acid to the water. If this problem is solved, perhaps through genetic engineering, *Thiobacillus* and other prokaryotes may help us overcome some environmental dilemmas that seem intractable today. One current research focus is a bacterium that tolerates radiation doses thousands of times stronger than those that would kill a person. This species may help clean up toxic dump sites that include radioactive waste.

Figure 15.17 Treatment of an oil spill in Alaska. These workers are spraying fertilizers onto an oil-soaked beach. The fertilizers stimulate growth of naturally occurring bacteria that initiate the breakdown of the oil. This technique is the fastest and least expensive way yet devised to clean up spills on beaches. Of course, it would be much better to keep oil off the beaches in the first place!

It is the nutritional diversity of prokaryotes that makes such benefits as chemical recycling and bioremediation possible. The various modes of nutrition and metabolic pathways we find in organisms living today are all variations on themes that prokaryotes "invented" during their long reign as Earth's exclusive inhabitants. Prokaryotes are at the foundation of life in both the ecological sense and the evolutionary sense. The subsequent breakthroughs in evolution were mostly structural, including the origin of the eukaryotic cell and the diversification of the protists.

CHECKPOINT

1. As different as archaea and bacteria are, both groups are characterized by __________ cells, which lack nuclei and other membrane-bounded organelles.
2. Upon microscopic examination, how do you think you could distinguish the cocci that cause staph infections from those that cause strep throat?
3. A species of bacterium requires only the amino acid methionine as an organic nutrient and lives in very deep caves where no light penetrates. Based on its mode of nutrition, this species would be classified as a __________.
4. Why are some archaea referred to as "extremophiles"?
5. Why do microbiologists autoclave their laboratory instruments and glassware rather than just washing them in very hot water?
6. How do bacteria help restore the atmospheric CO_2 required by plants for photosynthesis?

Answers: 1. prokaryotic **2.** By the arrangement of the cell aggregates: grapelike clusters for staphylococcus and chains of cells for streptococcus **3.** chemoheterotroph **4.** Because they thrive in extreme environments too hot, too salty, or too acidic for other organisms **5.** To kill bacterial endospores, which can survive boiling water **6.** By decomposing the organic molecules of dead organisms and organic refuse such as leaf litter, bacteria release carbon from the organic matter in the form of CO_2.

Protists

Anton van Leeuwenhoek, an early Dutch microscopist, was the first person to describe the microbial world: "No more pleasant sight has met my eye than this of so many thousands of living creatures in one small drop of water," he wrote more than three centuries ago. It is a world every biology student should have the opportunity to rediscover by peering through a microscope into a droplet of pond water filled with diverse creatures called protists.

Protists are eukaryotic, and thus even the simplest are much more complex than the prokaryotes. The first eukaryotes to evolve from prokaryotic ancestors were protists. The very word implies great antiquity (from the Greek *protos*, first). The primal eukaryotes were not only the predecessors of the great variety of modern protists; they were also ancestral to all other eukaryotes—plants, fungi, and animals. Two of the most significant chapters in the history of life—the origin of the eukaryotic cell and the subsequent emergence of multicellular eukaryotes—both occurred during the evolution of protists.

The Origin of Eukaryotic Cells

The differences between prokaryotic and eukaryotic cells far outnumber the differences between plant and animal cells. The fossil record indicates that eukaryotes evolved from prokaryotes more than 2 billion years ago. One of biology's most engaging questions is how this happened—in particular, how the membrane-enclosed organelles of eukaryotic cells arose. A widely accepted theory is that eukaryotic cells evolved through a combination of two processes. In one process, the eukaryotic cell's endomembrane system—all the membrane-enclosed organelles except mitochondria and chloroplasts (see Chapter 4)—evolved from inward folds of the plasma membrane of a prokaryotic cell **(Figure 15.18a)**.

A second, very different process, called endosymbiosis, generated mitochondria and chloroplasts. **Symbiosis** ("living together") is a close association between organisms of two or more species, and **endosymbiosis** refers to one species living inside another host species. Chloroplasts and mitochondria seem to have evolved from small symbiotic prokaryotes that established residence within other, larger host prokaryotes **(Figure 15.18b)**. The ancestors of mitochondria may have been aerobic bacteria that were able to use oxygen to release large amounts of energy from organic molecules by cellular respiration. At some point, such a prokaryote might have been an internal parasite of a larger heterotroph, or an ancestral host cell may have ingested some of these aerobic

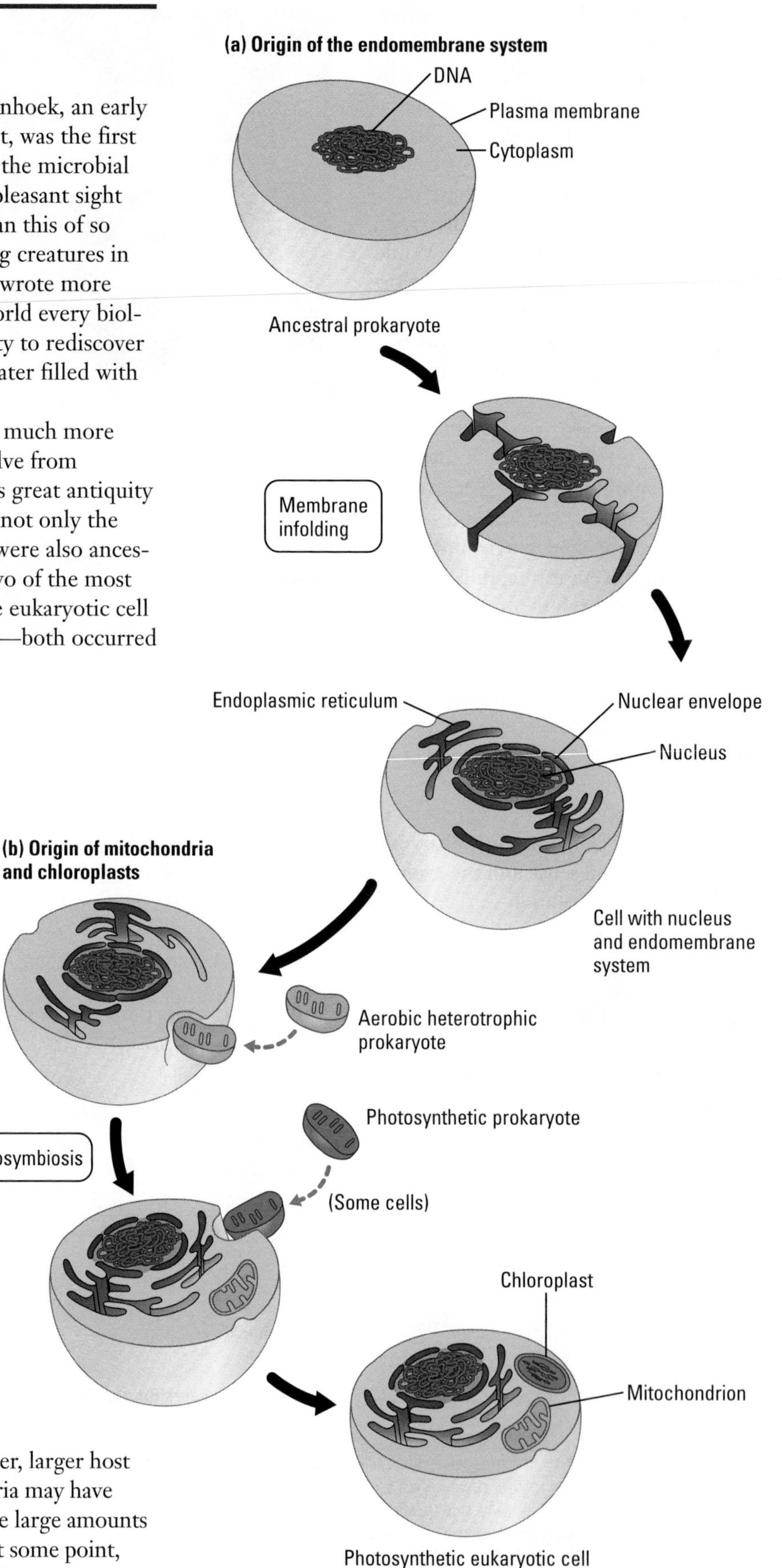

Figure 15.18 How did eukaryotic cells evolve?

cells for food. If some of the smaller cells were indigestible, they might have remained alive and continued to perform respiration in the host cell. In a similar way, photosynthetic bacteria ancestral to chloroplasts may have come to live inside a larger host cell. Because almost all eukaryotes have mitochondria but only some have chloroplasts, it is logical to suppose that mitochondria evolved first.

By whatever means the relationships began, it is not hard to imagine how a symbiosis between engulfed aerobic or photosynthetic cells and a larger host cell might have become mutually beneficial. In a world that was becoming increasingly aerobic, a cell that was itself an anaerobe would have benefited from aerobic endosymbionts that turned the oxygen to advantage. And a heterotrophic host could derive nourishment from photosynthetic endosymbionts. In the process of becoming more interdependent, the host and endosymbionts would have become a single organism, its parts inseparable.

Developed primarily by Lynn Margulis of the University of Massachusetts, the endosymbiosis model is supported by extensive evidence. Present-day mitochondria and chloroplasts are similar to prokaryotic cells in a number of ways. For example, both types of organelles contain small amounts of DNA, RNA, and ribosomes that resemble prokaryotic versions more than eukaryotic ones. These components enable chloroplasts and mitochondria to exhibit some autonomy in their activities. The organelles transcribe and translate their DNA into polypeptides, contributing to some of their own enzymes. They also replicate their own DNA and reproduce within the cell by a process resembling the binary fission of prokaryotes.

The origin of the eukaryotic cell made more complex organisms possible, and a vast variety of protists evolved.

The Diversity of Protists

All protists are eukaryotes, but they are so diverse that few other general characteristics can be cited. In fact, biologists use the term *protist* to refer to those eukaryotes that are not plants, animals, or fungi. At present, there is not yet consensus among biologists about how to divide the protists into kingdoms. This catch-all category will almost certainly be reorganized in the coming years.

Given their diversity, it is not surprising that protists vary in structure and function more than any other group of organisms. Most protists are unicellular, but there are some colonial and multicellular species. Because most protists are unicellular, they are justifiably considered the simplest eukaryotic organisms. But at the cellular level, many protists are exceedingly complex—the most elaborate of all cells. We should expect this of organisms that must carry out, within the boundaries of a single cell, all the basic functions performed by the collective of specialized cells that make up the bodies of plants and animals. Each unicellular protist is not analogous to a single cell from a human, but is itself an organism as complete as any whole animal or plant.

For our survey of these diverse organisms, we'll look at four major categories of protists, grouped more by lifestyle than by their evolutionary relationships: protozoans, slime molds, unicellular algae, and seaweeds.

Protozoans Protists that live primarily by ingesting food, a mode of nutrition that is heterotrophic (see Table 15.1) and animal-like, are called **protozoans** ("first animal"). Protozoans thrive in all types of aquatic environments, including wet soil and the watery environment inside animals. Most species eat bacteria or other protozoans, but some can absorb nutrients dissolved in the water. Protozoans that live as parasites in animals, though in the minority,

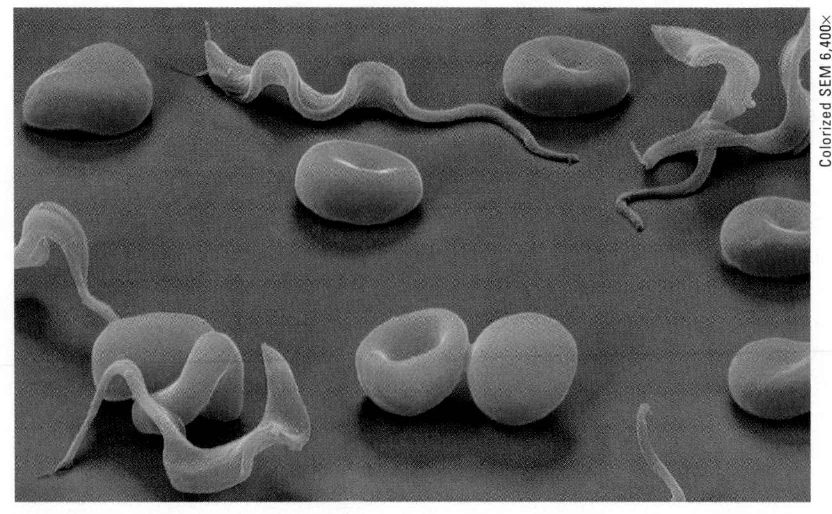

(a) A flagellate. The squiggly forms among these human red blood cells are trypanosomes, parasitic flagellates that cause sleeping sickness, a debilitating disease common in parts of Africa. (The flagellum borders the undulating membrane.) Trypanosomes alter the molecular structure of their coats frequently and thus prevent immunity from developing in the host.

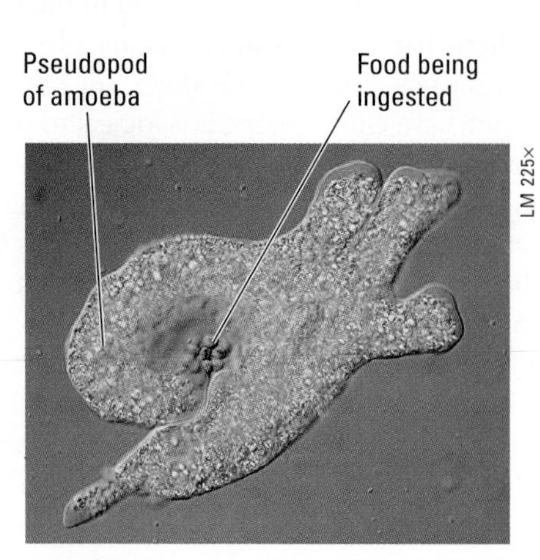

(b) An amoeba. This amoeba is ingesting a smaller protozoan as food. The amoeba's pseudopodium arches around the prey and engulfs it into a food vacuole (also see Chapter 5).

(c) A foram. Forams are almost all marine. The foram cell secretes a porous, multi-chambered shell made of organic material hardened with calcium carbonate. Thin strands of cytoplasm (pseudopodia) extend through the pores, functioning in swimming, shell formation, and feeding. The shells of fossilized forms are major components of the limestone rocks that are now land formations.

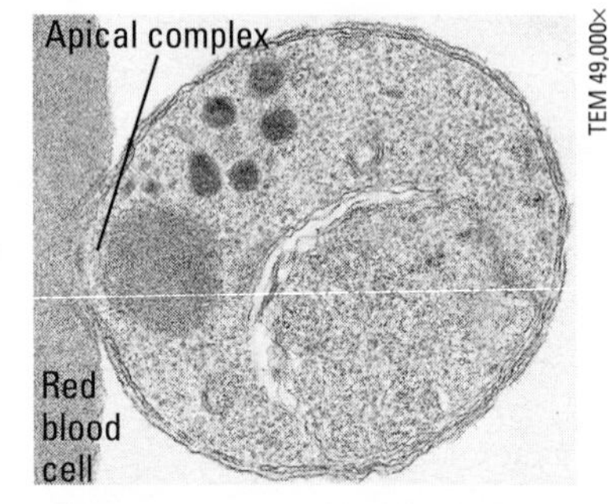

(d) An apicomplexan. *Plasmodium,* the apicomplexan that causes malaria, uses its apical complex to enter red blood cells of its human host. The parasite feeds on the host cell from within, eventually destroying it.

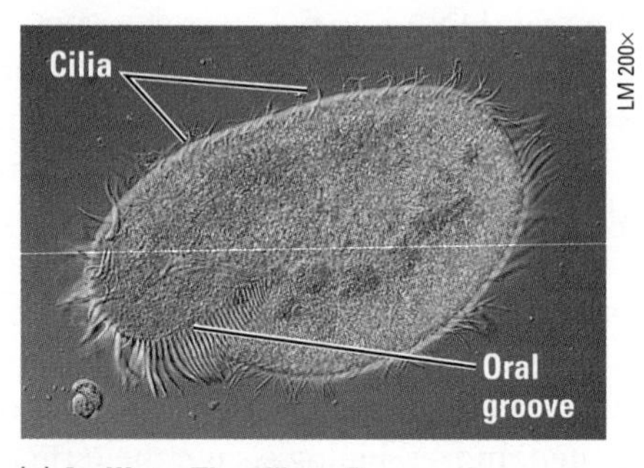

(e) A ciliate. The ciliate *Paramecium* uses its cilia to move through pond water. The beating of cilia lining the oral groove keeps a current of water containing bacteria and small protists moving toward the cell "mouth" at the base of the groove.

Figure 15.19 A gallery of protozoans.

cause some of the world's most harmful human diseases. We'll examine five groups of protozoans: flagellates, amoebas, forams, apicomplexans, and ciliates.

Flagellates are protozoans that move by means of one or more flagella. Most species are free-living (nonparasitic). However, there are also some nasty parasites that make humans sick. An example is *Giardia,* a flagellate that infects the human intestine and can cause abdominal cramps and severe diarrhea. People become infected mainly by drinking water contaminated with feces from infected animals. Another group of dangerous flagellates are the trypanosomes, including a species that causes sleeping sickness, a serious illness prevalent in tropical Africa and transmitted by the tsetse fly **(Figure 15.19a)**.

Amoebas are characterized by great flexibility and the absence of permanent locomotor organelles. Most species move and feed by means of **pseudopodia** (singular, *pseudopodium*), temporary extensions of the cell **(Figure 15.19b)**. Amoebas can assume virtually any shape as they creep over rocks, sticks, or mud at the bottom of a pond or ocean. One species of parasitic amoeba causes amebic dysentery, responsible for up to 100,000 deaths worldwide every year. Other protozoans with pseudopodia include the **forams** **(Figure 15.19c)**, which also have shells.

Apicomplexans are all parasitic, and some cause serious human diseases. They are named for an apparatus at their apex (tip) that is specialized for penetrating host cells and tissues. This group of protozoans includes *Plasmodium*, the parasite that causes malaria (**Figure 15.19d;** also see Biology and Society in Chapter 13). As part of the effort to combat malaria, scientists determined the complete sequence of the *Plasmodium* genome in 2002.

Ciliates are protozoans that use locomotor structures called cilia to move and feed. Nearly all ciliates are free-living (nonparasitic). The best-known example is the freshwater ciliate *Paramecium* **(Figure 15.19e)**.

Figure 15.20 A plasmodial slime mold. The weblike form of the slime mold's feeding stage is an adaptation that enlarges the organism's surface area, increasing its contact with food, water, and oxygen. Within the fine channels of the plasmodium, cytoplasm streams first one way and then the other in pulses that are beautiful to watch with a microscope. The cytoplasmic streaming helps to distribute nutrients and oxygen within the giant cell.

Slime Molds

These protists are more attractive than their name. Slime molds resemble fungi in appearance and lifestyle, but the similarities are due to convergent evolution; slime molds and fungi are not at all closely related. The filamentous body of a slime mold, like that of a fungus, is an adaptation that increases exposure to the environment. This suits the role of these organisms as decomposers. The two main groups of these protists are plasmodial slime molds and cellular slime molds.

Plasmodial slime molds are named for the feeding stage in their life cycle, an amoeboid mass called a plasmodium (not to be confused with *Plasmodium*, the apicomplexan parasite that causes malaria). You can find plasmodial slime molds among the leaf litter and other decaying material on a forest floor, and you won't need a microscope to see them. A plasmodium can measure several centimeters across, with its network of fine filaments taking in bacteria and bits of dead organic matter amoeboid style. Large as it is, the plasmodium is actually a single cell with many nuclei **(Figure 15.20)**.

Cellular slime molds raise a question about what it means to be an individual organism **(Figure 15.21)**. The feeding stage in the life cycle of a cellular slime mold consists of ❶ solitary amoeboid cells. They function individually, using their pseudopodia to creep through organic matter and engulf bacteria. But when food is in short supply, the amoeboid cells swarm together to form ❷ a sluglike colony that moves and functions as a single unit. After a brief period of mobility, the colony ❸ extends a stalk and develops into a multicellular reproductive structure.

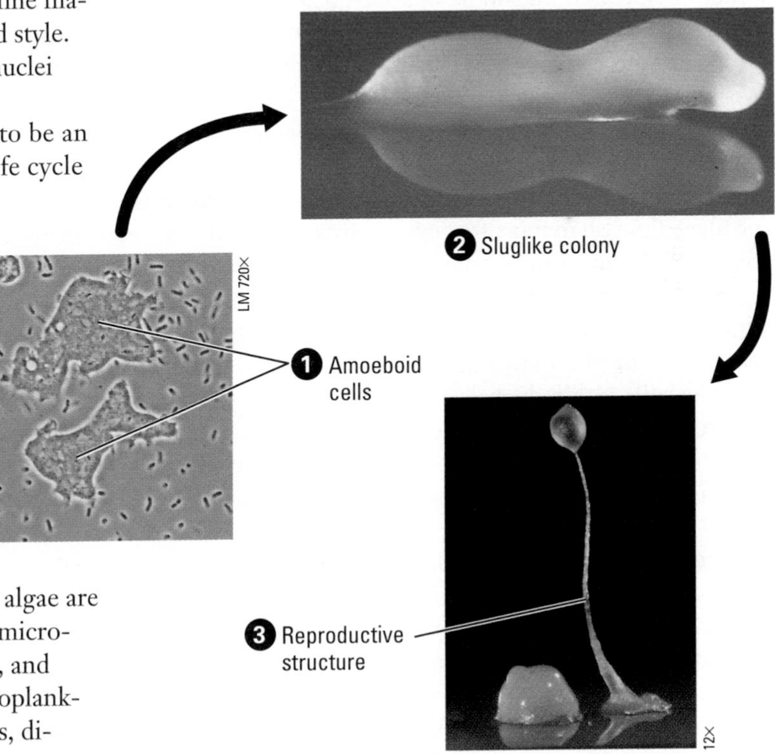

Figure 15.21 Stages in the life cycle of a cellular slime mold.

Unicellular Algae

Photosynthetic protists are called **algae** (singular, *alga*). Their chloroplasts support food chains in freshwater and marine ecosystems. Algae come in unicellular, colonial, and multicellular forms. Many unicellular algae are components of **plankton,** the communities of organisms, mostly microscopic, that drift or swim weakly near the surfaces of ponds, lakes, and oceans. More specifically, planktonic algae are referred to as phytoplankton. We'll look at three groups of unicellular algae: dinoflagellates, diatoms, and green algae.

Dinoflagellates are abundant in the vast aquatic pastures of phytoplankton. Each dinoflagellate species has a characteristic shape reinforced

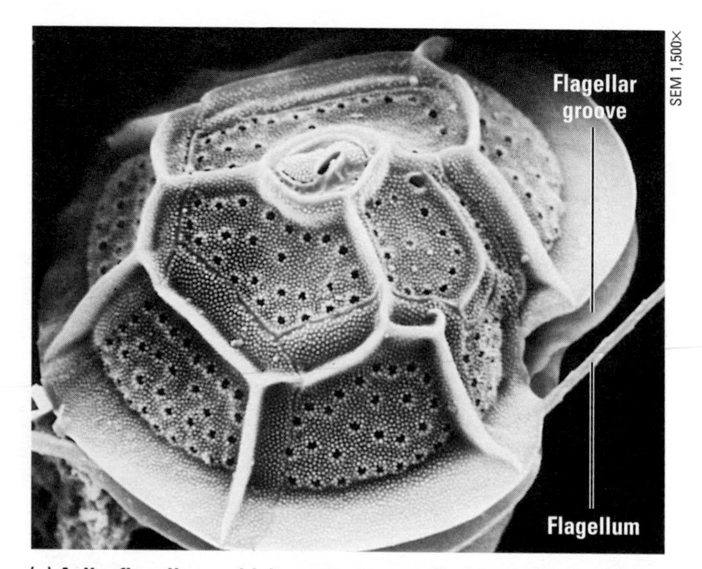

(a) A dinoflagellate, with its wall of protective plates

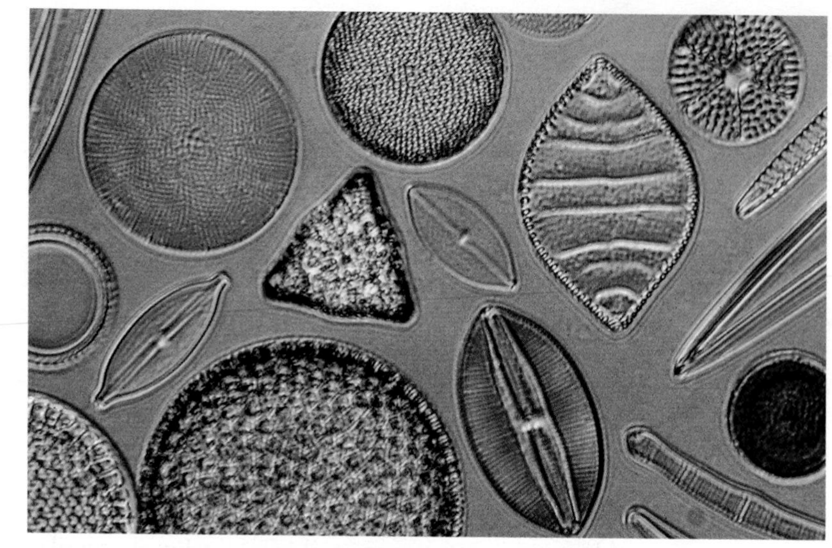

(b) A sample of diverse diatoms, which have glassy walls

by external plates made of cellulose **(Figure 15.22a)**. The beating of two flagella in perpendicular grooves produces a spinning movement. Dinoflagellate blooms—population explosions—sometimes cause warm coastal waters to turn pinkish orange, a phenomenon known as a red tide. Toxins produced by some red-tide dinoflagellates have caused massive fish kills, especially in the tropics, and are poisonous to humans as well.

Diatoms have glassy cell walls containing silica, the mineral used to make glass **(Figure 15.22b)**. The cell wall consists of two halves that fit together like the bottom and lid of a shoe box. Diatoms store their food reserves in the form of an oil that provides buoyancy, keeping diatoms floating as phytoplankton near the sunlit surface. Massive accumulations of fossilized diatoms make up thick sediments known as diatomaceous earth, which is mined for its use as a filtering material, an abrasive, and a natural insecticide.

Green algae are named for their grass-green chloroplasts. Unicellular green algae flourish in most freshwater lakes and ponds, as well as many home pools and aquariums. Some species are flagellated **(Figure 15.22c)**. The green algal group also includes colonial forms, such as *Volvox*, shown in **Figure 15.22d**. Each *Volvox* colony is a ball of flagellated cells (the small green dots in the photo) that are very similar to certain unicellular green algae. The balls within the balls in Figure 15.22d are daughter colonies that will be released when the parent colonies rupture. Of all photosynthetic protists, green algae are the most closely related to true plants. (We'll examine the evidence of this evolutionary relationship in the next chapter.)

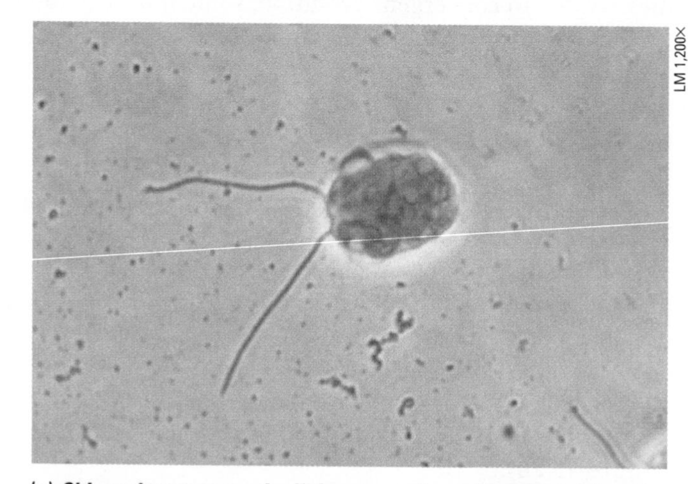

(c) *Chlamydomonas*, a unicellular green alga with a pair of flagella

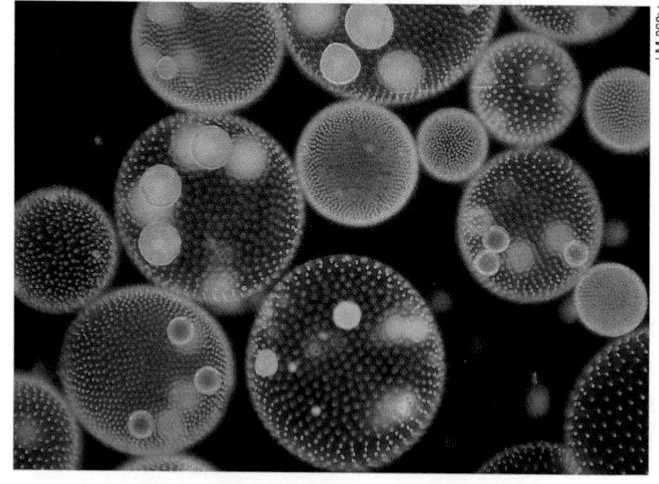

(d) *Volvox*, a colonial green alga

Figure 15.22 **Unicellular and colonial algae.**

Seaweeds

Defined as large, multicellular marine algae, **seaweeds** grow on rocky shores and just offshore beyond the zone of the pounding surf. Their cell walls have slimy and rubbery substances that cushion their bodies against the agitation of the waves. Some seaweeds are as large and complex as many plants. And though the word *seaweed* implies a plantlike appearance, the similarities between these algae and true plants are a consequence of convergent evolution. In fact, the closest

(a) Green algae. This sea lettuce is an edible species that inhabits the intertidal zone where the land meets the ocean. In addition to seaweeds, the green algal group includes unicellular and colonial species, such as those in Figures 15.22c and d.

(b) Red algae. These seaweeds are most abundant in the warm coastal waters of the tropics. Because their chloroplasts have special pigments that absorb the blue and green light that penetrates best through water, red algae can generally live in the deepest water. The species in this photo is an example of corraline algae, whose mineral-hardened cell walls contribute to the architecture of some coral reefs.

(c) Brown algae. This group includes the largest seaweeds, known as kelp, which grow as marine "forests" in relatively deep water beyond the intertidal zone. Some species grow to a height of over 60 m in a single season. Kelp is a renewable resource, and, more importantly, kelp forests provide habitat for many animals, including a great diversity of fishes. If you have walked on a beach covered with kelp washed ashore during a storm, you may have noticed the bubble-like organs called floats; these keep the photosynthetic blades of the kelp in the light near the water's surface.

Figure 15.23 The three major groups of seaweeds.

relatives of seaweeds are certain unicellular algae, which is why many biologists include seaweeds with the protists. Seaweeds are classified into three different groups, based partly on the types of pigments present in their chloroplasts: green algae, red algae, and brown algae (some of which are known as kelp) **(Figure 15.23)**.

Many coastal people, particularly in Asia, harvest seaweeds for food. For example, in Japan and Korea, some seaweed species (such as brown algae called kombu) are ingredients in soups. Other seaweeds (such as red algae called nori) are used to wrap sushi. Marine algae are rich in iodine and other essential minerals. However, much of their organic material consists of unusual polysaccharides that humans cannot digest, which prevents seaweeds from becoming staple foods. They are ingested mostly for their rich tastes and unusual textures. The gel-forming substances in the cell walls of seaweeds are widely used as thickeners for such processed foods as puddings, ice cream, and salad dressing. The seaweed extract called agar provides the gel-forming base for the media microbiologists use to culture bacteria in petri dishes and for a type of gel commonly used to perform gel electrophoresis.

CHECKPOINT

1. Which organelles of eukaryotic cells probably descended from endosymbiotic bacteria?
2. Why are protists especially important to biologists investigating the evolution of eukaryotic life?
3. What three modes of locomotion occur among protozoans?
4. Which protozoans are most similar in their movement to the unicellular stage of slime molds?
5. What metabolic process mainly distinguishes algae from protozoans?
6. Are seaweeds plants?

Answers: 1. Mitochondria and chloroplasts **2.** Because the first eukaryotes were protists, and these ancient protists were ancestral to all other eukaryotes, including plants, fungi, animals, and modern protists **3.** Movement using flagella, cilia, and pseudopodia **4.** Amoebas **5.** Photosynthesis **6.** No, they are large, multicellular marine algae. Most biologists do not consider them plants.

EVOLUTION CONNECTION

The Origin of Multicellular Life

An orchestra can play a greater variety of musical compositions than a violin soloist can. Put simply, increased complexity makes more variations possible. Thus, the origin of the eukaryotic cell led to an evolutionary radiation of new forms of life. Unicellular protists, which are organized on the complex eukaryotic plan, are much more diverse in form than the simpler prokaryotes. The evolution of multicellular bodies crossed another threshold in structural organization.

Multicellular organisms are fundamentally different from unicellular ones. In a unicellular organism, all of life's activities occur within a single cell. In contrast, a multicellular organism has various specialized cells that perform different functions and are dependent on each other. For example, some cells procure food, while others transport materials or provide movement.

The evolutionary links between unicellular and multicellular life were probably colonial forms, in which unicellular protists stuck together as loose federations of independent cells **(Figure 15.24)**. The gradual transition from colonies to truly multicellular organisms involved the cells becoming increasingly interdependent as a division of labor evolved. We can see one level of specialization and cooperation in the colonial green alga *Volvox* (see Figure 15.22d). *Volvox* produces gametes (sperm and ova), which depend on nonreproductive (somatic) cells while developing. Cells in truly multicellular organisms are specialized for many more nonreproductive functions, including feeding, waste disposal, gas exchange, and protection, to name a few.

Multicellularity evolved many times among the ancestral stock of protists, leading to new waves of biological diversification. The diverse seaweeds are examples of the descendants, and so are plants, fungi, and animals. In the next chapter, we'll trace the long evolutionary movement of plants and fungi onto land. Then we'll follow the threads of animal evolution in Chapter 17. ■

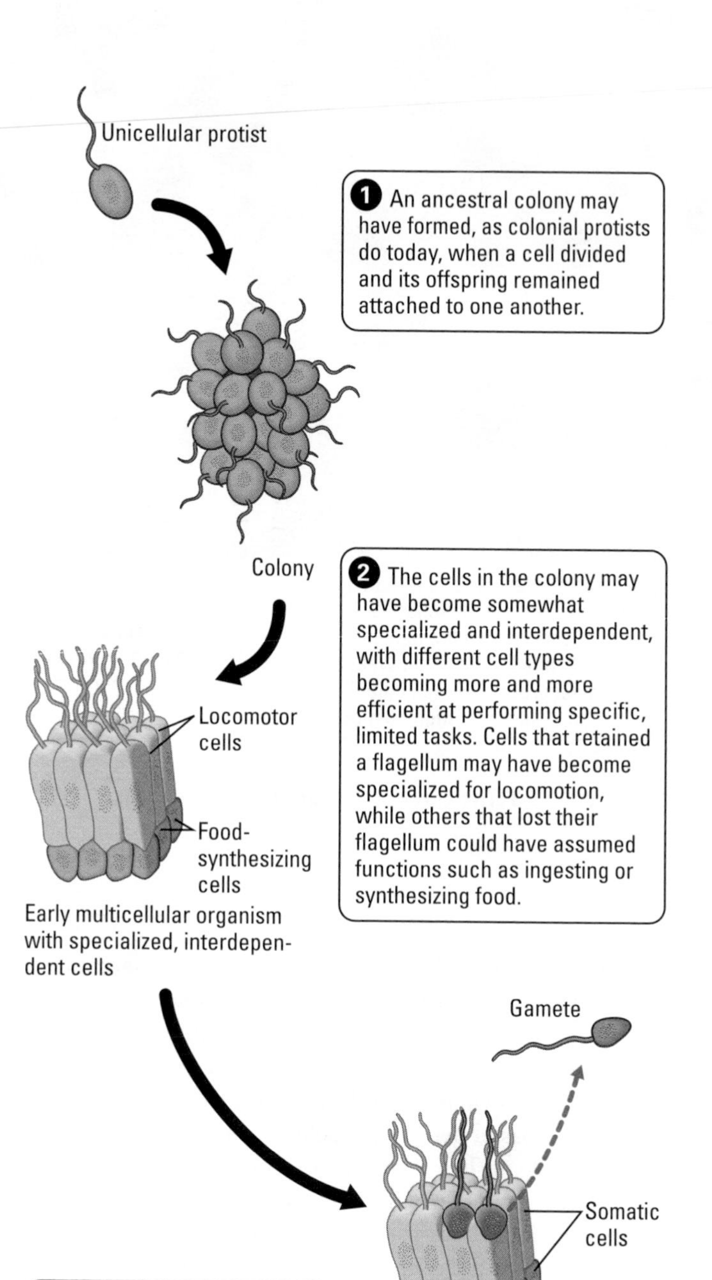

Unicellular protist
1 An ancestral colony may have formed, as colonial protists do today, when a cell divided and its offspring remained attached to one another.
Colony
2 The cells in the colony may have become somewhat specialized and interdependent, with different cell types becoming more and more efficient at performing specific, limited tasks. Cells that retained a flagellum may have become specialized for locomotion, while others that lost their flagellum could have assumed functions such as ingesting or synthesizing food.
Locomotor cells
Food-synthesizing cells
Early multicellular organism with specialized, interdependent cells
Gamete
Somatic cells
3 Additional specialization among the cells in the colony may have led to distinctions between sex cells (gametes) and nonreproductive cells (somatic cells).
Later organism with gametes and somatic cells

Figure 15.24 A model for the evolution of multicellular organisms from unicellular protists.

Chapter Review

SUMMARY OF KEY CONCEPTS

For study help and activities, go to campbellbiology.com or the student CD-ROM.

Major Episodes in the History of Life

Millions of years ago	Major episode
475	Plants and fungi colonize land
530	All major animal phyla established
1,000	First multicellular organisms
2,200	Oldest eukaryotic fossils
2,700	Accumulation of atmospheric O_2
3,500	Oldest prokaryotic fossils
4,600	Origin of Earth

Activity *The History of Life*

The Origin of Life

- **Resolving the Biogenesis Paradox** All life today arises only by the reproduction of preexisting life. However, most biologists think it possible that chemical and physical processes in Earth's primordial environment produced the first cells through a series of stages.

- **A Four-Stage Hypothesis for the Origin of Life** One scenario suggests that the first organisms were products of chemical evolution in four stages:

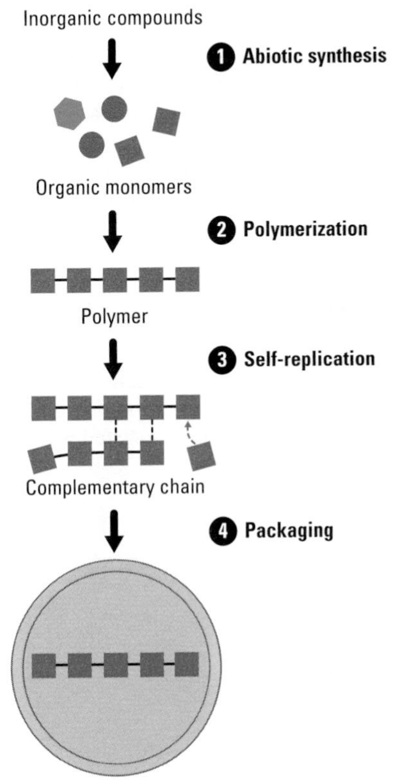

Case Study in the Process of Science *How Did Life Begin on Early Earth?*

- **From Chemical Evolution to Darwinian Evolution** Over millions of years, natural selection favored the most efficient pre-cells, which evolved into the first prokaryotic cells.

Prokaryotes

- **They're Everywhere!** Prokaryotes are found wherever there is life and greatly outnumber eukaryotes. Prokaryotes thrive in habitats where eukaryotes cannot live. A few prokaryotic species cause serious diseases, but most are either benign or beneficial to other forms of life.

- **The Two Main Branches of Prokaryotic Evolution:** Bacteria and Archaea

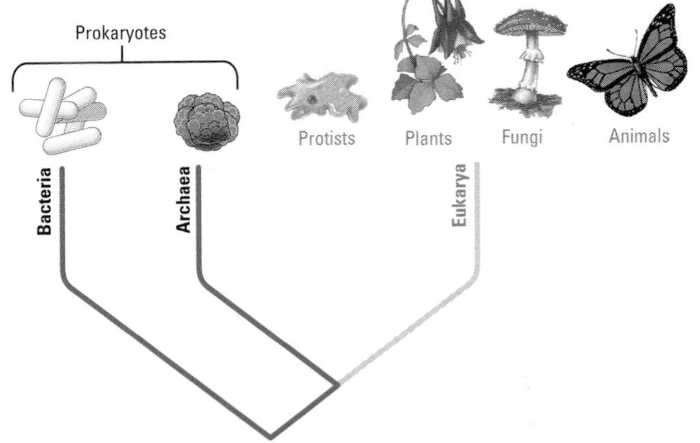

- **The Structure, Function, and Reproduction of Prokaryotes** Prokaryotic cells lack nuclei and other membrane-enclosed organelles. Most have cell walls. Some of the most common shapes of prokaryotes are:

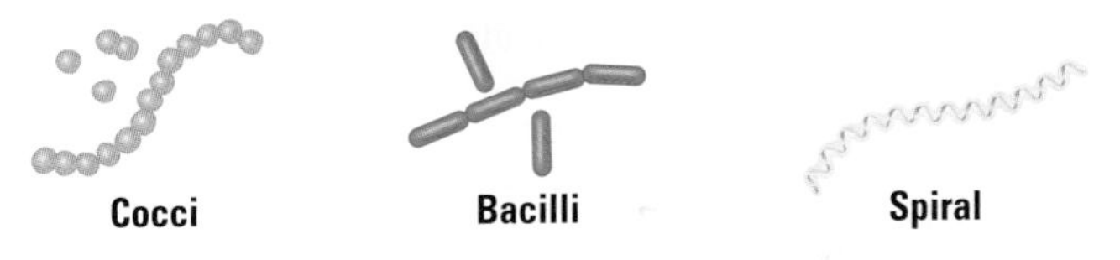

About half of all prokaryotic species are mobile, most of these using flagella to move. Some prokaryotes can survive extended periods of harsh conditions by forming endospores. Many prokaryotes can reproduce by binary fission at high rates if conditions are favorable, but growth is usually restricted by limited resources.

Activity *Prokaryotic Cell Structure and Function*

- **The Nutritional Diversity of Prokaryotes** Prokaryotes exhibit four major modes of nutrition:

Nutritional Mode	Energy Source	Carbon Source
Photoautotroph	Sunlight	CO_2
Chemoautotroph	Inorganic chemicals	
Photoheterotroph	Sunlight	Organic compounds
Chemoheterotroph	Organic compounds	

Case Study in the Process of Science *What Are the Modes of Nutrition in Prokaryotes?*

• **The Ecological Impact of Prokaryotes** Most pathogenic bacteria cause disease by producing exotoxins or endotoxins. Sanitation, antibiotics, and education are the best defenses against bacterial disease. Prokaryotes help recycle chemical elements between the biological and physical components of ecosystems. Humans can use prokaryotes to remove pollutants from water, air, and soil in the process called bioremediation.

Activity *Diversity of Prokaryotes*

Protists

• **The Origin of Eukaryotic Cells** The nucleus and endomembrane system of eukaryotes probably evolved from infoldings of the plasma membrane of ancestral prokaryotes. Mitochondria and chloroplasts probably evolved from symbiotic prokaryotes that took up residence inside larger cells.

• **The Diversity of Protists** Protists are unicellular eukaryotes and their closest multicellular relatives.

- Protozoans (including flagellates, amoebas, apicomplexans, and ciliates) primarily live in aquatic environments and ingest their food.
- Slime molds (including plasmodial slime molds and cellular slime molds) resemble fungi in appearance and lifestyle as decomposers, but are not at all closely related.
- Unicellular algae (including dinoflagellates, diatoms, and unicellular green algae) are photosynthetic protists that support food chains in freshwater and marine ecosystems.
- Seaweeds—which include green, red, and brown algae—are large, multicellular marine algae that grow on and near rocky shores.

Case Study in the Process of Science *What Kinds of Protists Are Found in Various Habitats?*

SELF-QUIZ

1. Place these events in the history of life on Earth in the order that they occurred.
   a. origin of multicellular organisms
   b. colonization of land by plants and fungi
   c. origin of eukaryotes
   d. origin of prokaryotes
   e. colonization of land by animals

2. Place the following steps in the origin of life in the order that they are hypothesized to have occurred.
   a. integration of self-replicating molecules into membrane-enclosed pre-cells
   b. origin of the first molecules capable of self-replication
   c. abiotic joining of organic monomers into polymers
   d. abiotic synthesis of organic monomers
   e. natural selection among pre-cells

3. DNA replication relies on the enzyme DNA polymerase. Why does this suggest that the earliest genes were made from RNA?

4. The two main evolutionary branches of prokaryotic life are __________ and __________. Which is more likely to be found on your table top?

5. Why do penicillins kill certain bacteria but not the human host?

6. Contrast exotoxins with endotoxins.

7. What is the difference between autotrophs and heterotrophs in terms of the source of their organic compounds?

8. The bacteria that cause tetanus can be killed only by prolonged heating at temperatures considerably above boiling. What does this suggest about tetanus bacteria?

9. To what nutritional classification do you belong? (*Hint:* Review Table 15.1.)

10. Of the following, which describes protists most inclusively?
    a. multicellular eukaryotes
    b. protozoans
    c. eukaryotes that are not plants, fungi, or animals
    d. single-celled organisms closely related to bacteria

11. Which algal group is most closely related to plants?
    a. diatoms
    b. green algae
    c. dinoflagellates
    d. seaweeds

Answers to the Self-Quiz questions can be found in Appendix D.

Go to the website or CD-ROM for more Self-Quiz questions.

THE PROCESS OF SCIENCE

12. Imagine you are on a team designing a moon base that will be self-contained and self-sustaining. Once supplied with building materials, equipment, and organisms from Earth, the base will be expected to function indefinitely. One of the members of your team has suggested that everything sent to the base be chemically treated or irradiated so that no bacteria of any kind are present. Do you think this is a good idea? Predict some of the consequences of eliminating all bacteria from an environment.

13. Your classmate says that organisms that require oxygen existed before photosynthetic organisms. Do you support this idea? Explain why or why not.

BIOLOGY AND SOCIETY

14. Many local newspapers publish a weekly list of restaurants that have been cited by inspectors for poor sanitation. Locate such a report and highlight the cases that are probably associated with food contamination by pathogenic prokaryotes.

15. What do you think should be done to prevent bioterrorism?

16 Plants, Fungi, and the Move onto Land

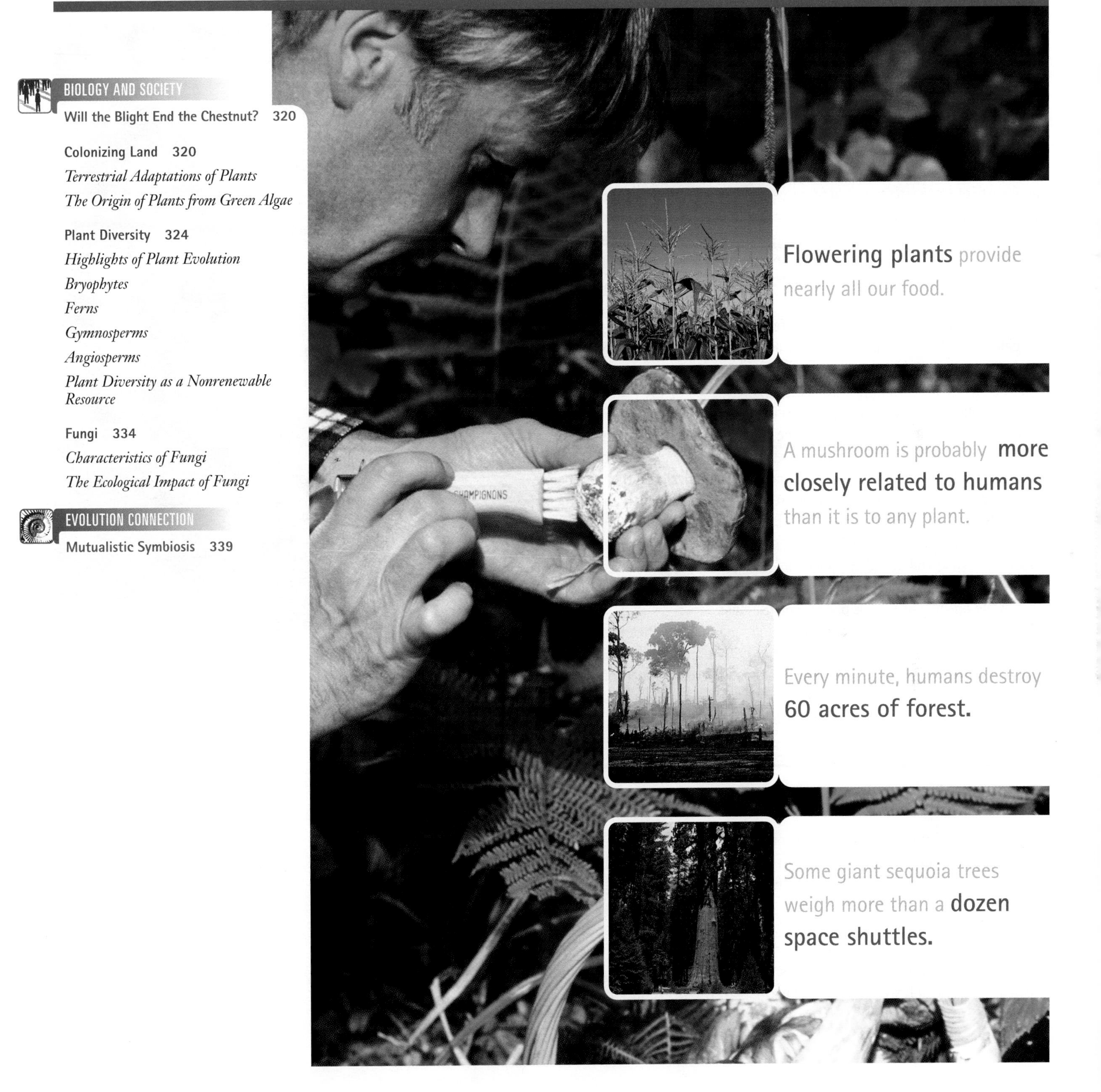

Flowering plants provide nearly all our food.
A mushroom is probably **more closely related to humans** than it is to any plant.
Every minute, humans destroy **60 acres of forest.**
Some giant sequoia trees weigh more than a **dozen space shuttles.**

BIOLOGY AND SOCIETY

Will the Blight End the Chestnut?

Will the blight end the chestnut?
The farmers rather guess not.
It keeps smoldering at the roots
And sending up new shoots
Till another parasite
Shall come to end the blight.

—Robert Frost, "Evil Tendencies" (1930)

The forests of the eastern United States, from Maine to Georgia, were once dominated by the American chestnut tree (*Castanea dentate*) **(Figure 16.1)**. Prized for their rapid growth, huge size, rot-resistant wood (which made them ideal for log cabin foundations), and bountiful harvest of edible nuts (traditionally roasted and eaten during the Thanksgiving-to-Christmas holiday season), American chestnuts were a mainstay of rural life.

Tragically, all this changed in just a few decades. Around 1900, an Asian fungus called *Cryphonectria parasitica* was accidentally introduced from China into North America. While many Asian trees had evolved defenses against the fungus, American trees had not. In just 25 years, blight caused by the fungus killed virtually every one of the estimated 3.5 billion adult American chestnut trees.

Despite the decimation of trees by the blight, the American chestnut is not extinct. You can still find small chestnut trees sprouting from old roots or stumps in many forests. Unfortunately, the blight fungus kills practically all the trees before they reach sexual maturity, so the trees are not propagating and restoring the population. However, forest researchers are working with the wealth of genetic material contained in the young sprouts in an attempt to develop a blight-resistant strain through breeding and genetic engineering. Perhaps this magnificent tree will once again be a presence within American forests. But the question posed by Robert Frost in the opening line of his poem remains unanswered.

The harmful interaction between plant and fungus exemplified by the chestnut blight is an unusual case. As we explore the diversity of plants and fungi in this chapter, you will see that it is much more common for the members of these two kingdoms to benefit from each other's presence. Indeed, as we'll discuss first, aquatic plants probably never could have adapted to land without the aid of fungi. ■

Figure 16.1 An American chestnut tree, circa 1920.

Colonizing Land

Plants are terrestrial (land-dwelling) organisms. True, some, such as water lilies, have returned to the water, but they evolved from terrestrial ancestors (just as several species of aquatic mammals, such as whales, evolved from terrestrial mammals).

What exactly is a plant? A **plant** is a multicellular eukaryote that makes organic molecules by photosynthesis. Photosynthesis distinguishes plants from the animal and fungal kingdoms. But what about large algae, including seaweeds, which we classified as protists in the preceding chapter? They, too, are multicellular, eukaryotic, and photosynthetic. What distinguishes plants from algae is a set of terrestrial adaptations.

Terrestrial Adaptations of Plants

Structural Adaptations Living on land poses very different problems from living in water **(Figure 16.2)**. In terrestrial habitats, the resources that a photosynthetic organism needs are found in two very different places. Light and carbon dioxide are mainly available above-ground, while water and mineral nutrients are found mainly in the soil. Thus, the complex bodies of plants show varying degrees of structural specialization, exhibiting subterranean organs, the **roots,** and aerial organs, the leaf-bearing **shoots.**

Most plants have symbiotic fungi associated with their roots. These root-fungus combinations are called **mycorrhizae** ("fungus root"). For their part, the fungi absorb water and essential minerals from the soil and provide these materials to the plant. The sugars produced by the plant nourish the fungi. Mycorrhizae are evident on some of the oldest plant fossils. They are key adaptations that made it possible to live on land **(Figure 16.3)**.

Leaves are the main photosynthetic organs of most plants. Exchange of carbon dioxide and oxygen between the atmosphere and the photosynthetic interior of a leaf occurs via **stomata,** the microscopic pores through the leaf's surface (see Figure 7.3). A waxy layer called the **cuticle** coats the leaves and other aerial parts of most plants, helping the plant body retain water. (Think of the waxy surface of some houseplant leaves or a cucumber.)

Differentiation of the plant body into root and shoot systems solved one problem but created new ones. For the shoot system to stand up straight in the air, it must have support. This is not a problem in the water: Huge seaweeds do not need skeletons because the surrounding water buoys them. An important terrestrial adaptation of plants is **lignin,** a chemical that hardens cell walls. Imagine what would happen to you if your skeleton were to suddenly turn mushy. A tree would also collapse if it were not for its "skeleton," its framework of lignin-rich cell walls.

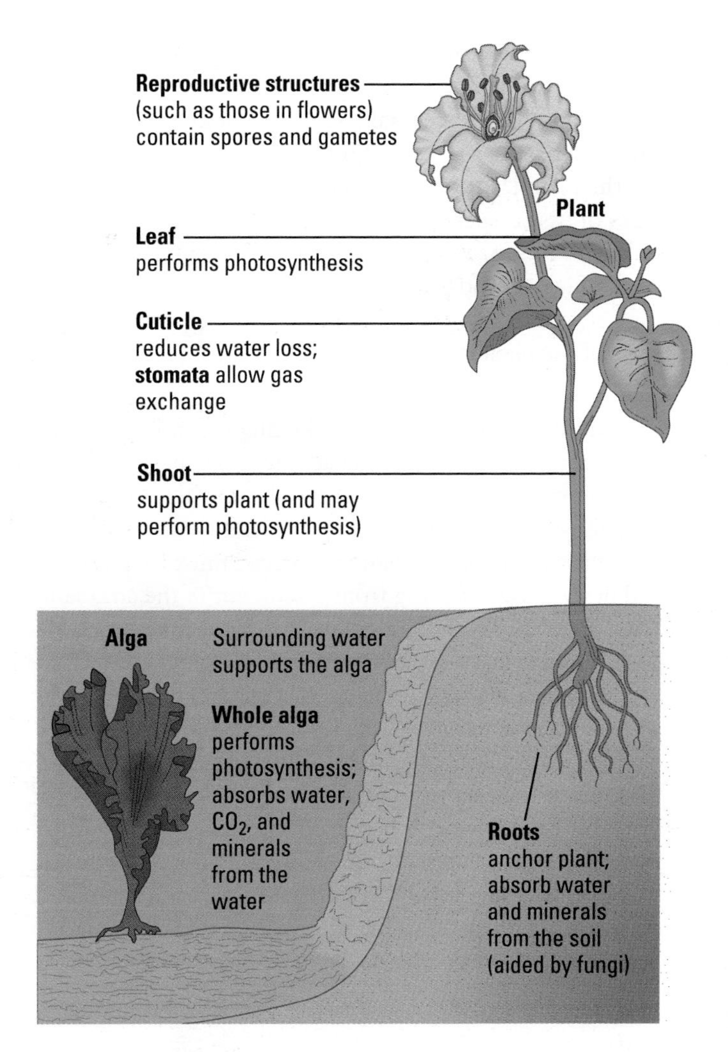

Figure 16.2 Contrasting environments for algae and plants.

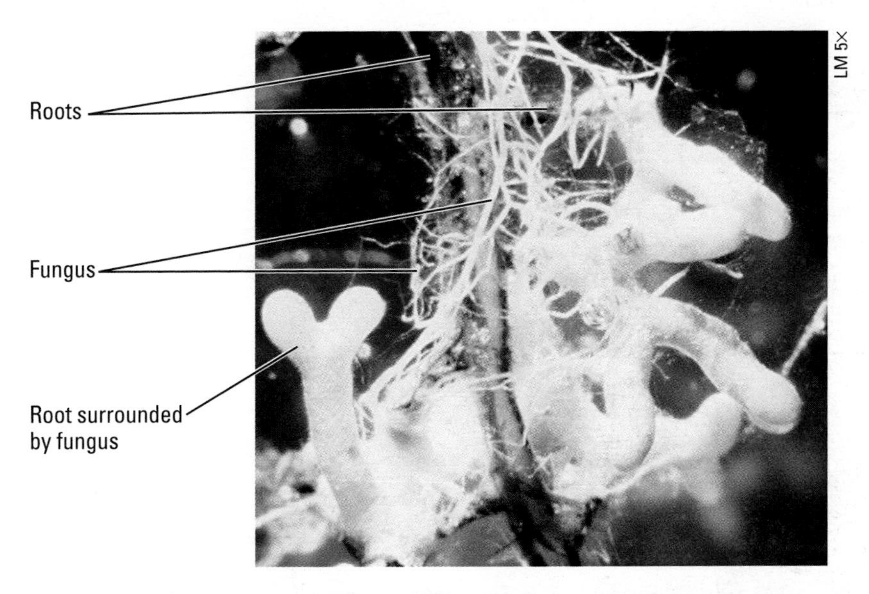

Figure 16.3 Mycorrhizae: symbiotic associations of fungi and roots. The finely branched filaments of the fungus (white in the photo) provide an extensive surface area for absorption of water and minerals from the soil. The fungus provides some of those materials to the plant and benefits in turn by receiving sugars and other organic products of the plant's photosynthesis.

Specialization of the plant body into roots and shoots also introduced the problem of transporting vital materials between the distant organs. The terrestrial equipment of most plants includes **vascular tissue,** a system of tube-shaped cells that branch throughout the plant **(Figure 16.4)**. The vascular tissue actually has two types of tissues specialized for transport: **xylem,** consisting of dead cells with tubular cavities for transporting water and minerals from roots to leaves; and **phloem,** consisting of living cells that distribute sugars from the leaves to the roots and other nonphotosynthetic parts of the plant.

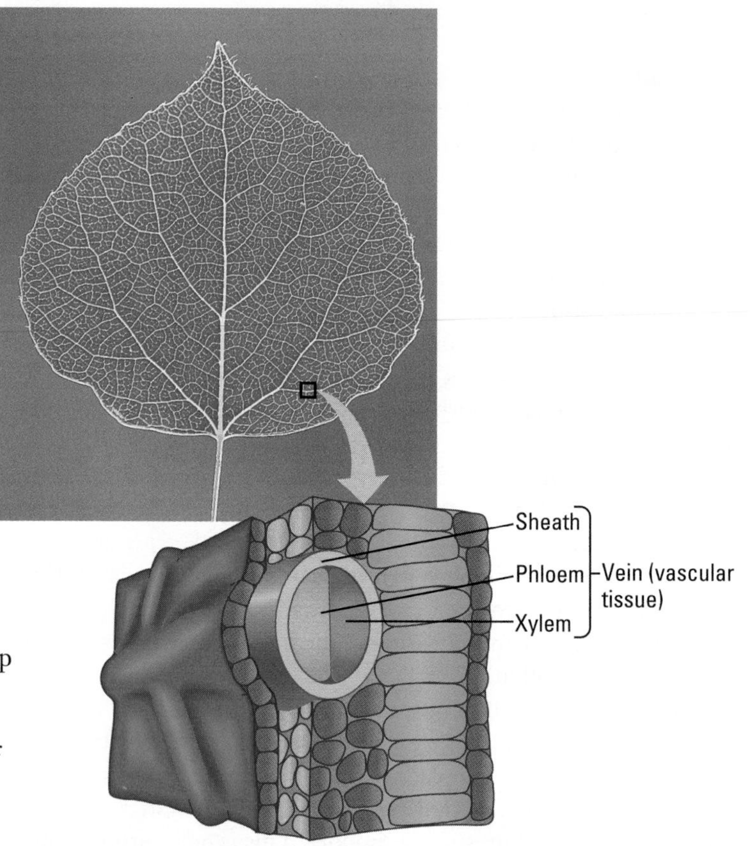

Figure 16.4 Network of veins in a leaf. The vascular tissue of a plant, visible in the photograph of the leaf as yellow veins, delivers water and minerals absorbed by the roots and carries away the sugars produced in the leaves.

Reproductive Adaptations Adapting to land also required a new mode of reproduction. For algae, the surrounding water ensures that gametes (sperm and eggs) and developing offspring stay moist. The aquatic environment also provides a means of dispersing the gametes and offspring. Plants, however, must keep their gametes and developing offspring from drying out in the air. Plants (and some algae) produce their gametes in protective structures called **gametangia** (singular, *gametangium*). A gametangium has a jacket of protective cells surrounding a moist chamber where gametes can develop without dehydrating.

For most plants, sperm reach the eggs by traveling within pollen, which is carried by wind or animals. The egg remains within tissues of the mother plant and is fertilized there. In plants, but not algae, the zygote (fertilized egg) develops into an embryo while still contained within the female parent, which protects the embryo and keeps it from dehydrating **(Figure 16.5)**. Most plants rely on wind or animals, such as fruit-eating birds or mammals, to disperse their offspring, which are in the form of embryos contained in seeds.

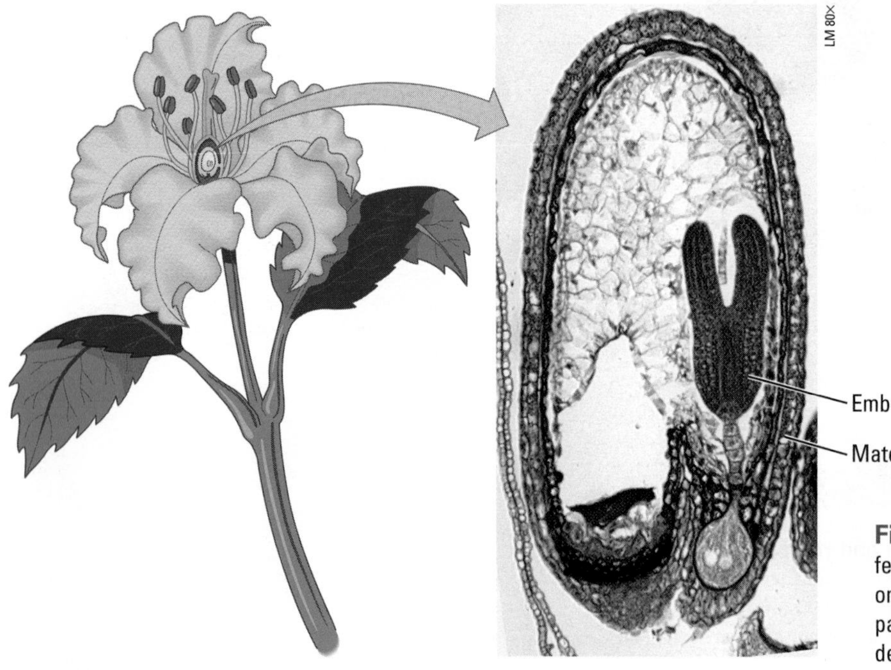

Figure 16.5 The protected embryo of a plant. Internal fertilization, with sperm and egg combining within a moist chamber on the mother plant, is an adaptation for living on land. The female parent continues to nurture and protect the plant embryo, which develops from the zygote.

The Origin of Plants from Green Algae

The move onto land and the spread of plants to diverse terrestrial environments was incremental. It paralleled the gradual accumulation of terrestrial adaptations, beginning with populations that descended from algae.

Green algae are the protists most closely related to plants. More specifically, molecular comparisons and other evidence place a group of multicellular green algae called **charophyceans** closest to plants **(Figure 16.6)**. Many species of modern charophyceans are found in shallow water around the edges of ponds and lakes. Some of the ancient charophyceans that lived about the time that land was first colonized may have inhabited shallow-water habitats subject to occasional drying. Natural selection would have favored individual algae that could survive through periods when they were not submerged. The protection of developing gametes and embryos within jacketed organs (gametangia) on the parent is one adaptation to living in shallow water that would also prove essential on land. We know that by about 475 million years ago, the vintage of the oldest plant fossils, an accumulation of adaptations allowed permanent residence above water. The plants that color our world today diversified from those early descendants of green algae.

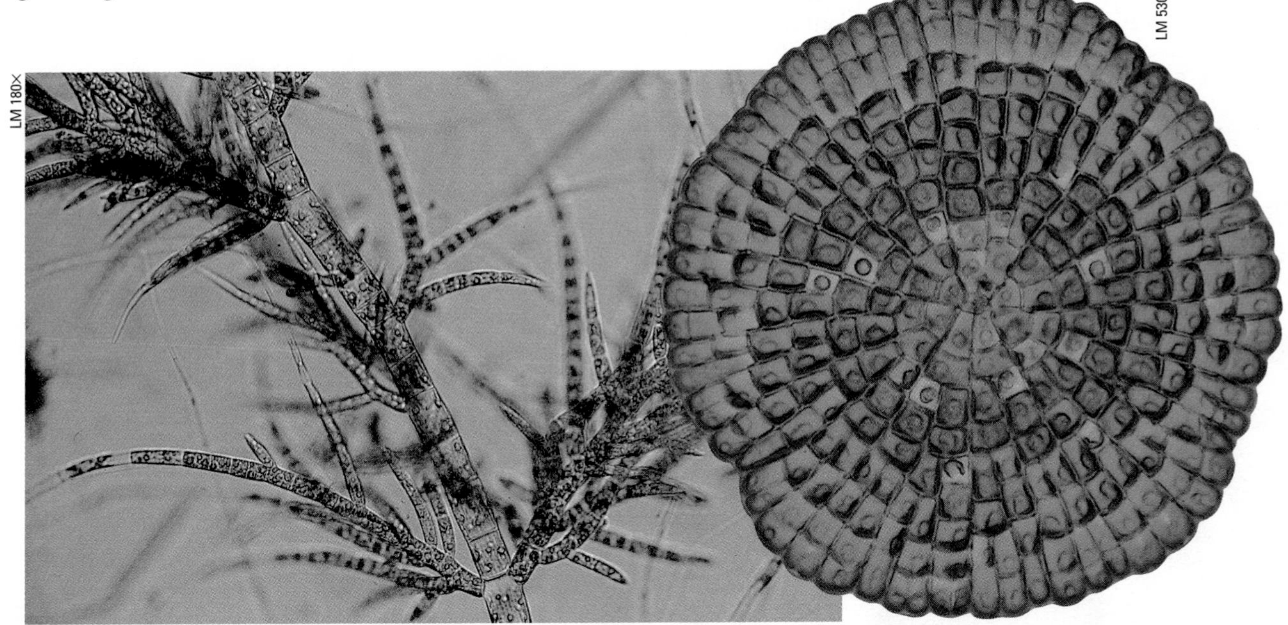

(a) ***Chara***

(b) ***Coleochaete***

Figure 16.6 Two species of charophyceans, the closest algal relatives of plants. (a) *Chara* is a particularly elaborate green alga. **(b)** *Coleochaete,* though less plantlike than *Chara* in appearance, is actually more closely related to plants.

CHECKPOINT

1. Name some adaptations of plants for living on land.
2. What mode of nutrition is used by both plants and algae?

Answers: 1. Any of the following: cuticle; stomata; vascular tissue; lignin-hardened cell walls; gametangia, which protect gametes; protected embryos; and differentiation of the body into a subterranean root system and above-ground stems and leaves. **2.** Both plants and algae obtain energy via photosynthesis.

Plant Diversity

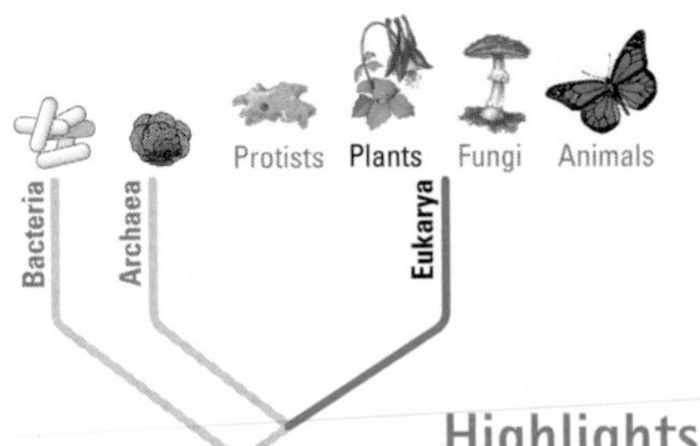

As we survey the diversity of modern plants, remember that the past is the key to the present. The history of the plant kingdom is a story of adaptation to diverse terrestrial habitats.

Highlights of Plant Evolution

The fossil record chronicles four major periods of plant evolution, which are also evident in the diversity of modern plants **(Figure 16.7)**. Each stage is marked by the evolution of structures that opened new opportunities on land.

The first period of evolution was the origin of plants from their aquatic ancestors, the green algae called charophyceans. The first terrestrial adaptations included gametangia, which protected the gametes and embryos. Gametangia made it possible for the plants known as **bryophytes,** which include mosses, to diversify from the earliest plants. Vascular tissue that conducted water and nutrients also evolved relatively early in plant history.

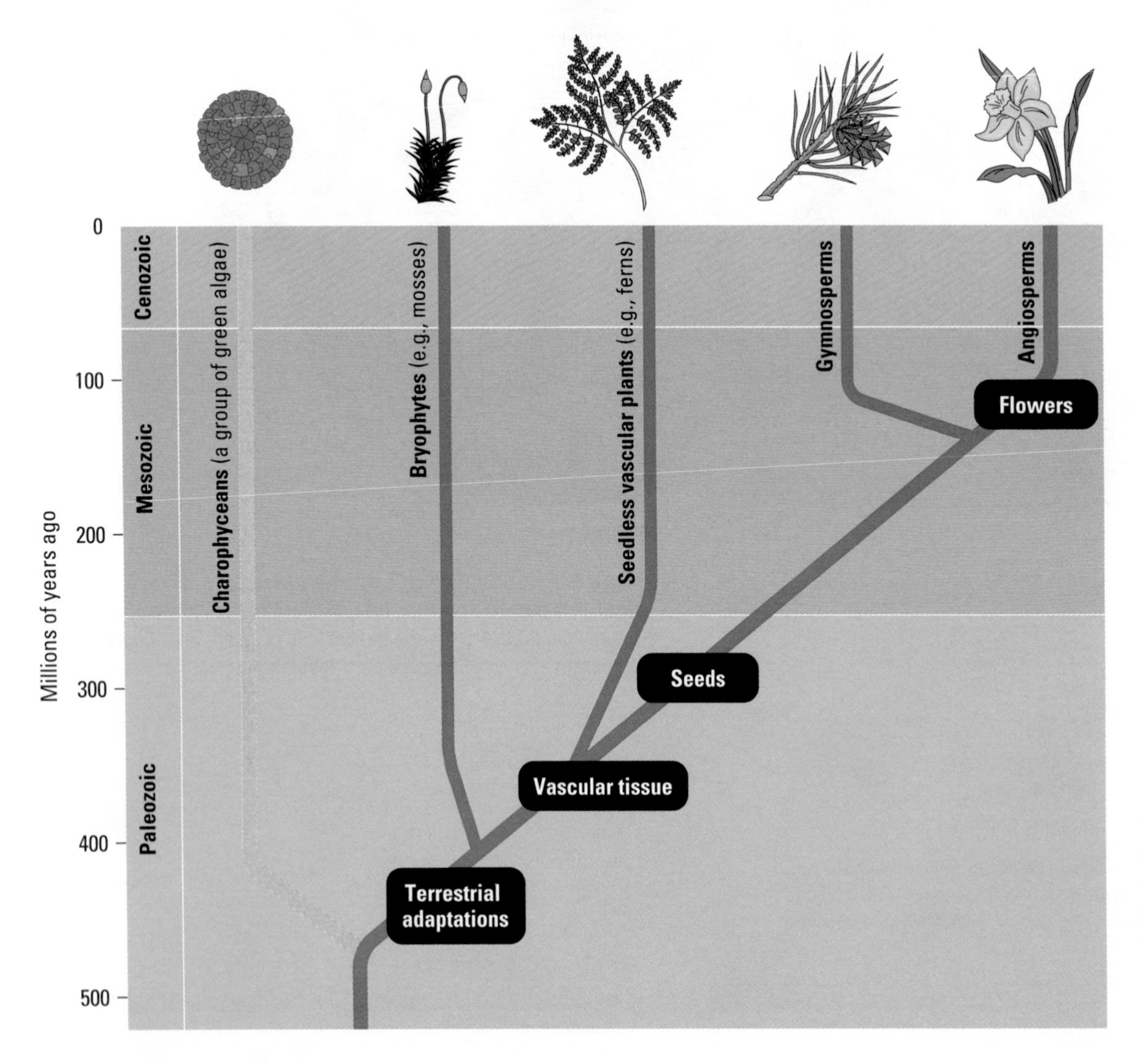

Figure 16.7 Highlights of plant evolution. Modern representatives of the major evolutionary branches are illustrated at the top of this phylogenetic tree. As we survey the diversity of plants, miniature versions of this tree will help you place each plant group in its evolutionary context.

However, most bryophytes lack vascular tissue, and they are categorized as nonvascular plants.

The second period of plant evolution was the diversification of plants with vascular tissue. The presence of conducting tissues allowed vascular plants to grow much larger, rising above the ground to achieve significant height (whereas bryophytes tend to grow in low mats). The earliest vascular plants lacked seeds. Today, this seedless condition is retained by **ferns** and a few other groups of vascular plants.

The third major period of plant evolution began with the origin of the seed. Seeds advanced the colonization of land by further protecting plant embryos from drying and other hazards. A seed consists of an embryo packaged along with a store of food within a protective covering. The seeds of early seed plants were not enclosed in any specialized chambers. These plants gave rise to many types of **gymnosperms** ("naked seed"). Today, the most widespread and diverse gymnosperms are the conifers, which are the pines and other plants with cones.

The fourth major episode in the evolutionary history of plants was the emergence of flowering plants, or **angiosperms** ("contained seed"). The flower is a complex reproductive structure that bears seeds within protective chambers called ovaries. This contrasts with the bearing of naked seeds by gymnosperms. The great majority of modern-day plants are angiosperms.

With these highlights as our framework, we are now ready to survey the four major groups of modern plants: bryophytes, ferns, gymnosperms, and angiosperms.

Figure 16.8 A peat moss bog in Norway. Peat mosses, or *Sphagnum,* carpet at least 3% of Earth's terrestrial surface, with greatest density in high northern latitudes. The accumulation of peat, the thick mat of living and dead plants in wetlands, ties up an enormous amount of organic carbon because peat has an abundance of chemical materials that are not easily degraded by microbes. The carbon storage by peat bogs plays an important role in stabilizing Earth's atmospheric carbon dioxide concentrations, and hence climate, through the CO_2-related greenhouse effect (see Chapter 7). The ability of peat moss to absorb and retain water makes it an excellent addition to garden soil.

Bryophytes

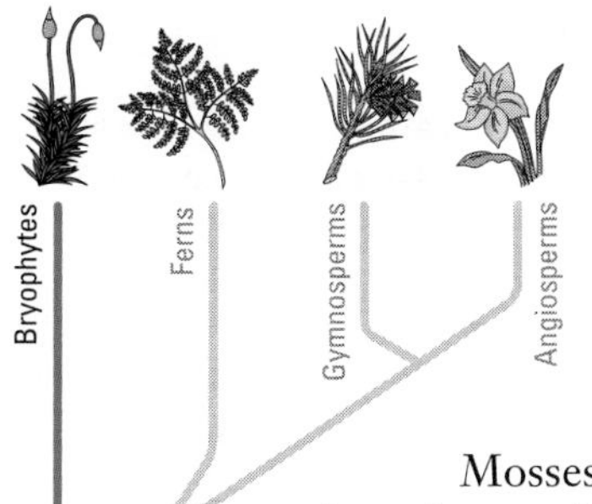

The most familiar bryophytes are **mosses.** (Other bryophytes include liverworts and hornworts.) A mat of moss actually consists of many plants growing in a tight pack, helping to hold one another up **(Figure 16.8)**. The mat has a spongy quality that enables it to absorb and retain water.

Mosses display two of the key terrestrial adaptations that made the move onto land possible: (1) a waxy cuticle that helps prevent dehydration and (2) the retention of developing embryos within the mother plant's gametangium. However, they are not totally liberated from their ancestral aquatic habitat. Mosses need water to reproduce. Their sperm have flagella, like those of most green algae. These sperm must swim through water to reach eggs. (A film of rainwater or dew is enough moisture for the sperm to travel.) In addition, most mosses have no vascular tissue to carry water from soil to aerial parts of the plant. This explains why damp, shady places are the most common habitats of mosses. These plants also lack lignin, the wall-hardening material that enables other plants to stand tall. Mosses may sprawl as mats over acres, but they always have a low profile.

If you look closely at some moss growing on your campus, you may actually see two distinct forms of the plant. The greener, sponge-like plant that is the more obvious is called the **gametophyte.** Careful examination will reveal the other form of the moss, called a **sporophyte,** growing out of a gametophyte as a stalk with a capsule at its tip **(Figure 16.9)**. The cells of the gametophyte are haploid (they have one set of chromosomes;

Figure 16.9 The two forms of a moss. The feathery plant we generally know as a moss is the gametophyte. The stalk with the capsule at its tip is the sporophyte. This photo shows the capsule releasing its tiny spores.

see Chapter 8). In contrast, the sporophyte is made up of diploid cells (with two chromosome sets). These two different stages of the plant life cycle are named for the types of reproductive cells they produce. Gametophytes produce gametes (sperm and eggs), while sporophytes produce spores. As reproductive cells, **spores** differ from gametes in two ways: A spore can develop into a new organism without fusing with another cell (two gametes must fuse to form a zygote); and spores usually have tough coats that enable them to resist harsh environments (whereas gametes must stay moist).

The gametophyte and sporophyte are alternating generations that take turns producing each other. Gametophytes produce gametes that unite to form zygotes, which develop into new sporophytes. And sporophytes produce spores that give rise to new gametophytes. This type of life cycle, called **alternation of generations,** occurs only in plants and certain algae **(Figure 16.10)**. Among plants, mosses and other bryophytes are unique in having the gametophyte as the dominant generation—the larger, more obvious plant. As we continue our survey of plants, we'll see an increasing dominance of the sporophyte as the more highly developed generation.

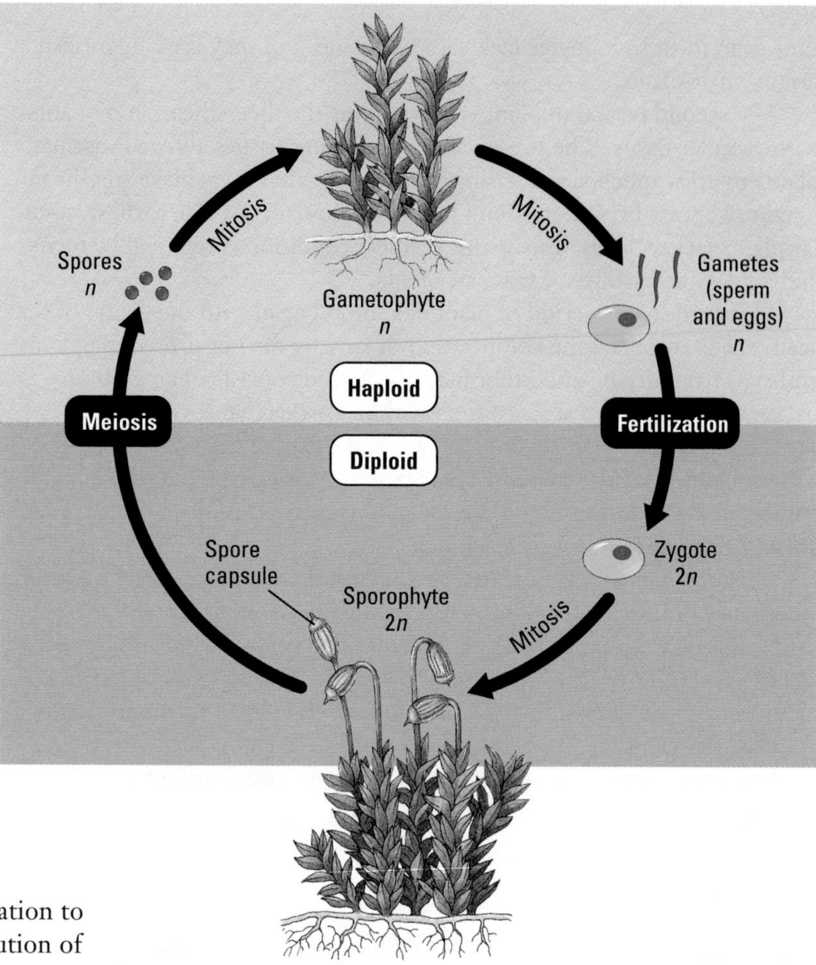

Figure 16.10 Alternation of generations. Plants have life cycles very different from ours. Each of us is a diploid individual; the only haploid stages in the human life cycle, as for nearly all animals, are sperm and eggs. By contrast, plants have alternating generations: Diploid ($2n$) individuals (sporophytes) and haploid (n) individuals (gametophytes) generate each other in the life cycle. In the case of mosses, the sporophyte remains attached to the gametophyte, depending on its parent for water and nutrients.

Ferns

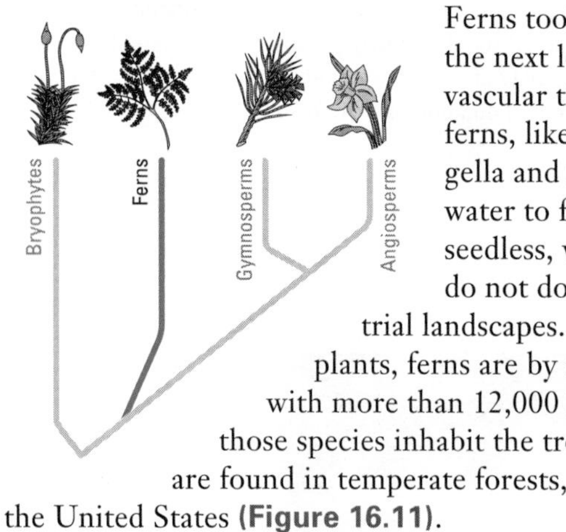

Ferns took terrestrial adaptation to the next level with the evolution of vascular tissue. However, the sperm of ferns, like those of mosses, have flagella and must swim through a film of water to fertilize eggs. Ferns are also seedless, which helps explain why they do not dominate most modern terrestrial landscapes. Still, of all seedless vascular plants, ferns are by far the most diverse today, with more than 12,000 known species. Most of those species inhabit the tropics, although many species are found in temperate forests, such as many woodlands in the United States **(Figure 16.11)**.

Figure 16.11 Ferns (seedless vascular plants). This fern species grows on the forest floor in the eastern United States. The "fiddleheads" in the inset on the right are young fronds (leaves) ready to unfurl. The fern generation familiar to us is the sporophyte generation. The inset on the left is the underside of a sporophyte leaf specialized for reproduction. The yellow dots are spore capsules that can release numerous tiny spores. The spores develop into gametophytes. However, you would have to crawl on the forest floor and explore with careful hands and sharp eyes to find fern gametophytes, tiny plants growing on or just below the soil surface.

Figure 16.12 A "coal forest" of the Carboniferous period. This painting, based on fossil evidence, reconstructs one of the great seedless forests. Most of the large trees with straight trunks are seedless plants called lycophytes. On the left, the tree with numerous feathery branches is another type of seedless plant called a horsetail. The plants near the base of the trees are ferns. Note the giant bird-sized dragonfly, which must have made quite a buzz.

During the Carboniferous period, from about 360 to 300 million years ago, ancient ferns were part of a much greater diversity of seedless plants that formed vast, swampy forests over much of what is now Eurasia and North America **(Figure 16.12)**. At that time, these continents were close to the equator and had tropical swamp forests that generated great quantities of organic matter. As the plants died, they fell into stagnant wetlands and did not decay completely. Their remains formed thick deposits of organic debris, or peat. Later, seawater flooded the swamps, marine sediments covered the peat, and pressure and heat gradually converted the peat to coal. Coal is black sedimentary rock made up of fossilized plant material. It formed during several geologic periods, but the most extensive coal beds are derived from Carboniferous deposits. Like coal, oil and natural gas also formed from the remains of extinct organisms; thus, all three are known as **fossil fuels.** We burn fossil fuels to generate much of our electricity. As we deplete our oil and gas reserves, the use of coal is likely to increase.

Gymnosperms

Bryophytes Ferns Gymnosperms Angiosperms

"Coal forests" dominated the North American and Eurasian landscapes until near the end of the Carboniferous period. At that time, global climate turned drier and colder, and the vast swamps began to disappear. This climatic change provided an opportunity for seed plants, which can complete their life cycles on dry land and withstand long, harsh winters. Of the earliest seed plants, the most successful were the gymnosperms, and several kinds grew along with the seedless plants in the Carboniferous swamps. Their descendants include the **conifers,** or cone-bearing plants.

Conifers Perhaps you have had the fun of hiking or skiing through a forest of conifers, the most common gymnosperms. Pines, firs, spruces, junipers, cedars, and redwoods are all conifers. A broad band of coniferous forests covers much of northern Eurasia and North America and extends southward in mountainous regions **(Figure 16.13)**. Today, about 190 million acres of coniferous forests in the United States, mostly in the western states and Alaska, are designated national forests.

Conifers are among the tallest, largest, and oldest organisms on Earth. Redwoods, native to the northern California coast, grow to heights of more than 110 m (360 feet), only certain eucalyptus trees in Australia are taller. The most massive organisms alive are the giant sequoias, relatives of redwoods that grow in the Sierra Nevada mountains of California. One, known as the General Sherman tree, is about 84 m (275 feet) high and weighs more than the combined weight of a dozen space shuttles. Bristlecone pines, another species of California conifer, are among the oldest organisms alive. One bristlecone, named Methuselah, is more than 4,600 years old; it was a young tree when humans invented writing.

Figure 16.13 A coniferous forest in Jasper National Park, in the Canadian Rockies. Coniferous forests are widespread in northern Eurasia and North America; conifers also grow in the Southern Hemisphere, though they are less numerous there.

Nearly all conifers are evergreens, meaning they retain leaves throughout the year. Even during winter, a limited amount of photosynthesis occurs on sunny days. And when spring comes, conifers already have fully developed leaves that can take advantage of the sunnier days. The needle-shaped leaves of pines and firs are also adapted to survive dry seasons. A thick cuticle covers the leaf, and the stomata are located in pits, further reducing water loss.

Coniferous forests are highly productive; you probably use products harvested from them every day. For example, conifers provide much of our lumber for building and wood pulp for paper production. What we call wood is actually an accumulation of vascular tissue with lignin, which gives the tree structural support.

Terrestrial Adaptations of Seed Plants Compared to ferns, conifers and most other gymnosperms have three additional adaptations that make survival in diverse terrestrial habitats possible: (1) further reduction of the gametophyte, (2) the evolution of pollen, and (3) the advent of the seed.

The first adaptation is an even greater development of the diploid sporophyte compared to the haploid gametophyte generation **(Figure 16.14)**. A

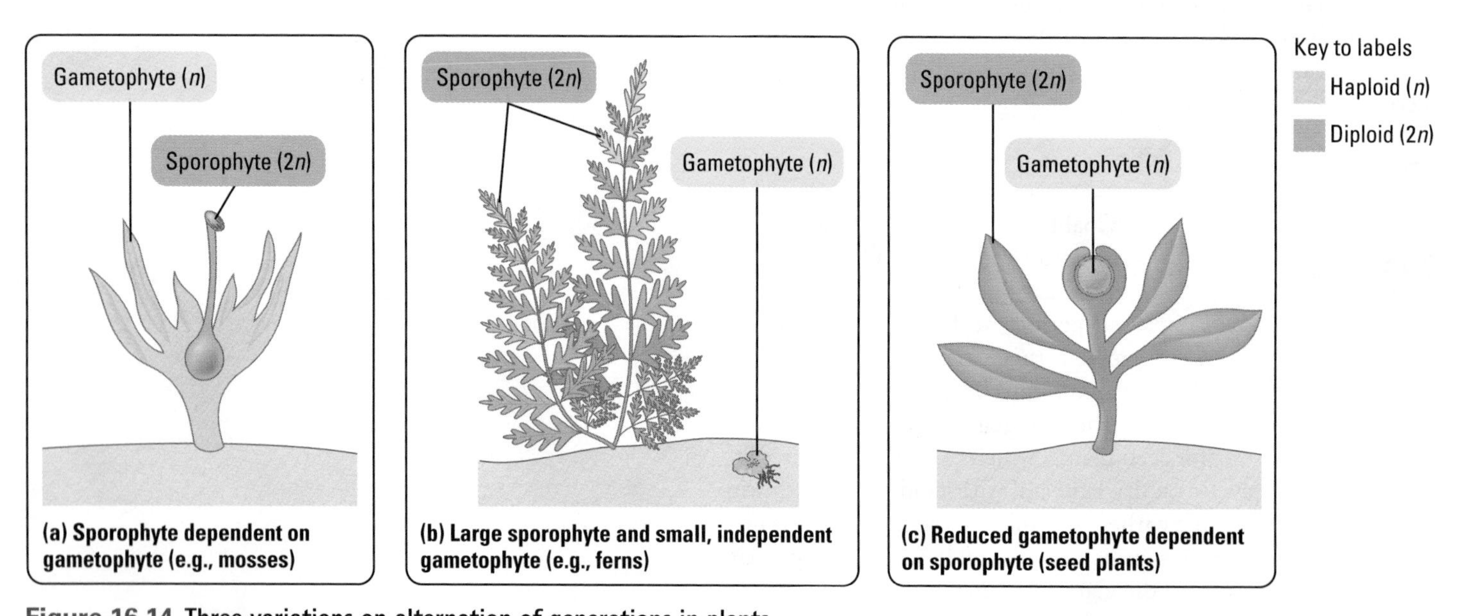

Figure 16.14 Three variations on alternation of generations in plants.

pine tree or other conifer is actually a sporophyte with tiny gametophytes living in its cones **(Figure 16.15)**. In contrast to bryophytes and ferns, gymnosperm gametophytes are totally dependent on and protected by the tissues of the parent sporophyte. Some plant biologists speculate that the shift toward diploidy in land plants was related to the harmful impact of the sun's ionizing radiation, which causes mutations. This damaging radiation is more intense on land than in aquatic habitats, where organisms are somewhat protected by the light-filtering properties of water. Of the two generations of land plants, the diploid form (sporophyte) may cope better with mutagenic radiation. A diploid organism homozygous for a particular essential allele has a "spare tire" in the sense that one copy of the allele may be sufficient for survival if the other is damaged.

A second adaptation of seed plants to dry land was the evolution of **pollen.** A pollen grain is actually the much-reduced male gametophyte. It houses cells that will develop into sperm. In the case of conifers, wind carries the pollen from male to female cones, where eggs develop within female gametophytes (see Figure 16.15). This mechanism for sperm transfer contrasts with the swimming sperm of mosses and ferns. In seed plants, this use of tough, airborne pollen that carries sperm to egg is a terrestrial adaptation that led to even greater success and diversity of plants on land.

The third important terrestrial adaptation of seed plants is the seed itself. A **seed** consists of a plant embryo packaged along with a food supply within a protective coat. Seeds develop from structures called **ovules** **(Figure 16.16)**. In conifers, the ovules are located on the scales of female cones. Conifers and other gymnosperms, lacking ovaries, bear their seeds "naked" on the cone scales (though the seeds do have protective coats).

Scale (contains
female gametophyte)
Female
cones
Male cones
(produce
pollen, male
gametophyte)

Figure 16.15 A pine tree, a conifer. The tree bears two types of cones. The hard, woody ones we usually notice are female cones (upper left inset). Each scale of the female cone is actually a modified leaf bearing a pair of structures called ovules. Smaller, softer mature male cones release clouds of millions of pollen grains, some of which land on female cones on trees of the same species. The female cones generally develop on the higher branches, where they are unlikely to be dusted with pollen from the same tree. Sperm released by pollen fertilizes eggs in the ovules of the female cones. The ovules eventually develop into seeds.

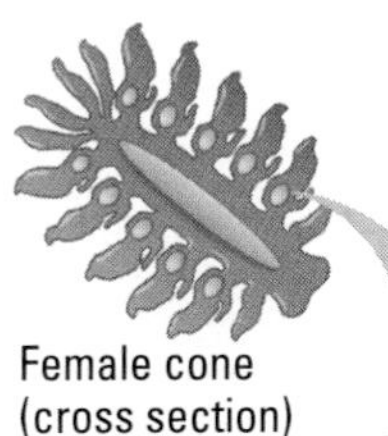
Female cone
(cross section)

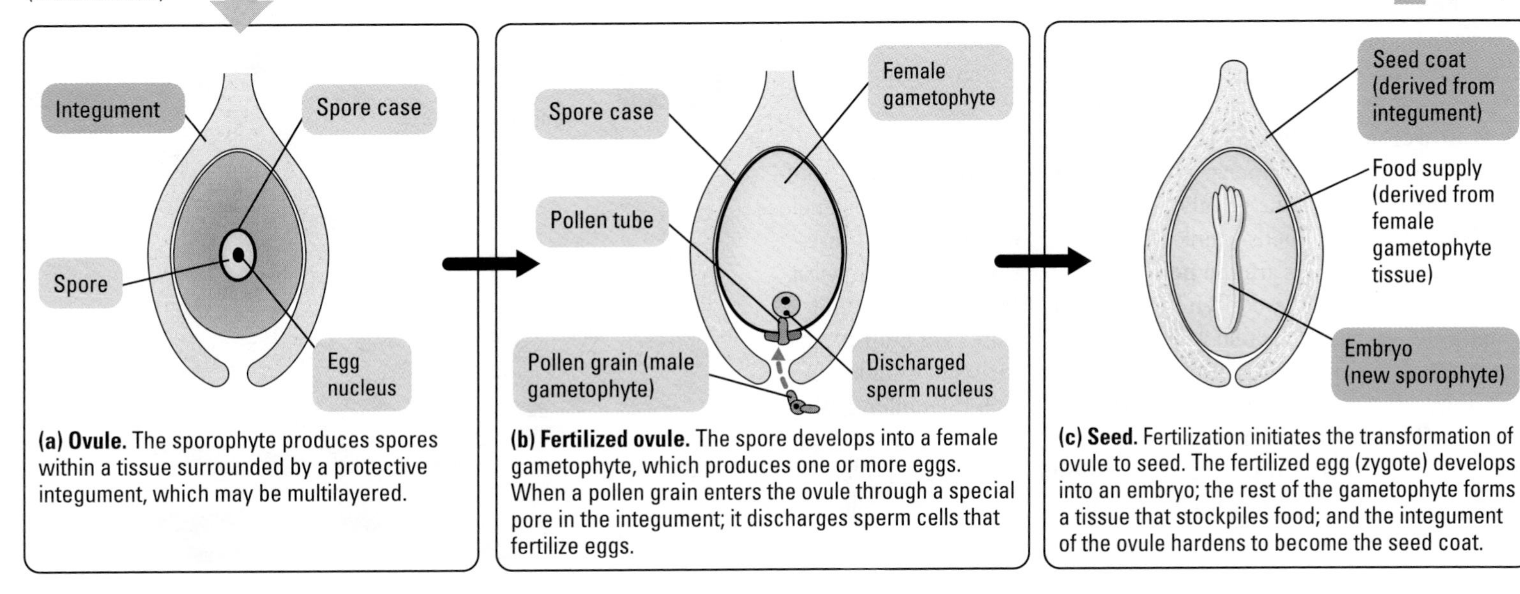
Key to labels
Haploid (n)
Diploid (2n)
Integument
Spore case
Spore
Egg
nucleus
(a) Ovule. The sporophyte produces spores within a tissue surrounded by a protective integument, which may be multilayered.
Spore case
Female
gametophyte
Pollen tube
Pollen grain (male
gametophyte)
Discharged
sperm nucleus
(b) Fertilized ovule. The spore develops into a female gametophyte, which produces one or more eggs. When a pollen grain enters the ovule through a special pore in the integument; it discharges sperm cells that fertilize eggs.
Seed coat
(derived from
integument)
Food supply
(derived from
female
gametophyte
tissue)
Embryo
(new sporophyte)
(c) Seed. Fertilization initiates the transformation of ovule to seed. The fertilized egg (zygote) develops into an embryo; the rest of the gametophyte forms a tissue that stockpiles food; and the integument of the ovule hardens to become the seed coat.

Figure 16.16 From ovule to seed.

Once released from the parent plant, the seed can remain dormant for days, months, or even years. Under favorable conditions, the seed can then **germinate:** Its embryo emerges through the seed coat as a seedling. Some seeds drop close to their parents, while others are carried far by the wind or animals.

Angiosperms

The photograph of the coniferous forest in Figure 16.13 could give us a somewhat distorted view of today's plant life. Conifers do cover much land in the northern parts of the globe, but it is the angiosperms, or flowering plants, that dominate most other regions. There are about 250,000 angiosperm species versus about 700 species of conifers and other gymnosperms. Whereas gymnosperms supply most of our lumber and paper, angiosperms supply nearly all our food and much of our fiber for textiles. Over 90% of the plant kingdom are angiosperms, including cereal grains such as wheat and corn, citrus and other fruit trees, garden vegetables, and cotton. Fine hardwoods from flowering plants such as oak, cherry, and walnut trees supplement the lumber we get from conifers.

Several unique adaptations account for the success of angiosperms. For example, refinements in vascular tissue make water transport even more efficient in angiosperms than in gymnosperms. Of all terrestrial adaptations, however, it is the flower that accounts for the unparalleled success of the angiosperms.

Flowers, Fruits, and the Angiosperm Life Cycle No organisms make a showier display of their sex lives than angiosperms. From roses to dandelions, flowers display a plant's sex organs. For many angiosperms, this showiness helps to attract insects and other animals that transfer pollen from the sperm-bearing organs of one flower to the egg-bearing organs of another. This dependance on animals for pollen transfer targets the pollen to other plants of the same species, rather than relying on uncertain winds to blow the pollen around.

A **flower** is actually a short stem with four whorls of modified leaves: sepals, petals, stamens, and carpels **(Figure 16.17)**. At the bottom of the flower are the **sepals,** which are usually green. They enclose the flower before it opens (think of the green "wrapping" on a rosebud). Above the sepals are the **petals,** which are usually the showiest part of the flower and are often important in attracting insects and other pollinators. The actual reproductive structures are multiple stamens and one or more carpels. Each **stamen** consists of a stalk—the **filament**—bearing a sac called an **anther,** in which the pollen grains develop. The carpel consists of a stalk—the **style**—with an ovary at the base and a sticky tip known as the **stigma,** which traps pollen. The **ovary** is a protective chamber containing one or more ovules, in which the eggs develop.

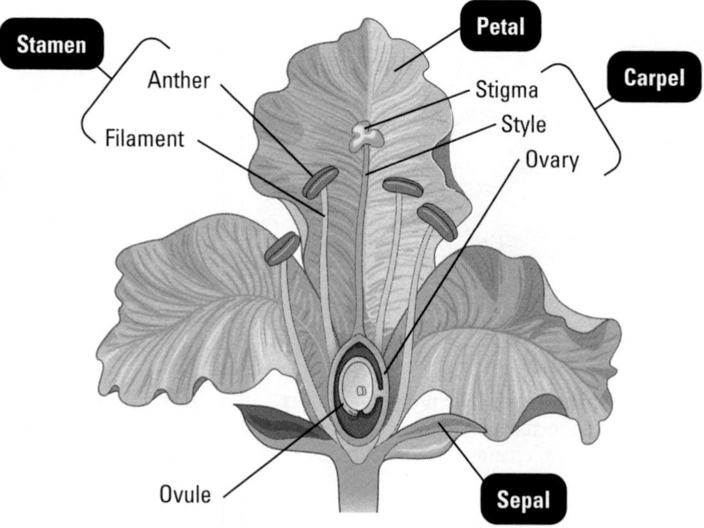

Figure 16.17 Structure of a flower.

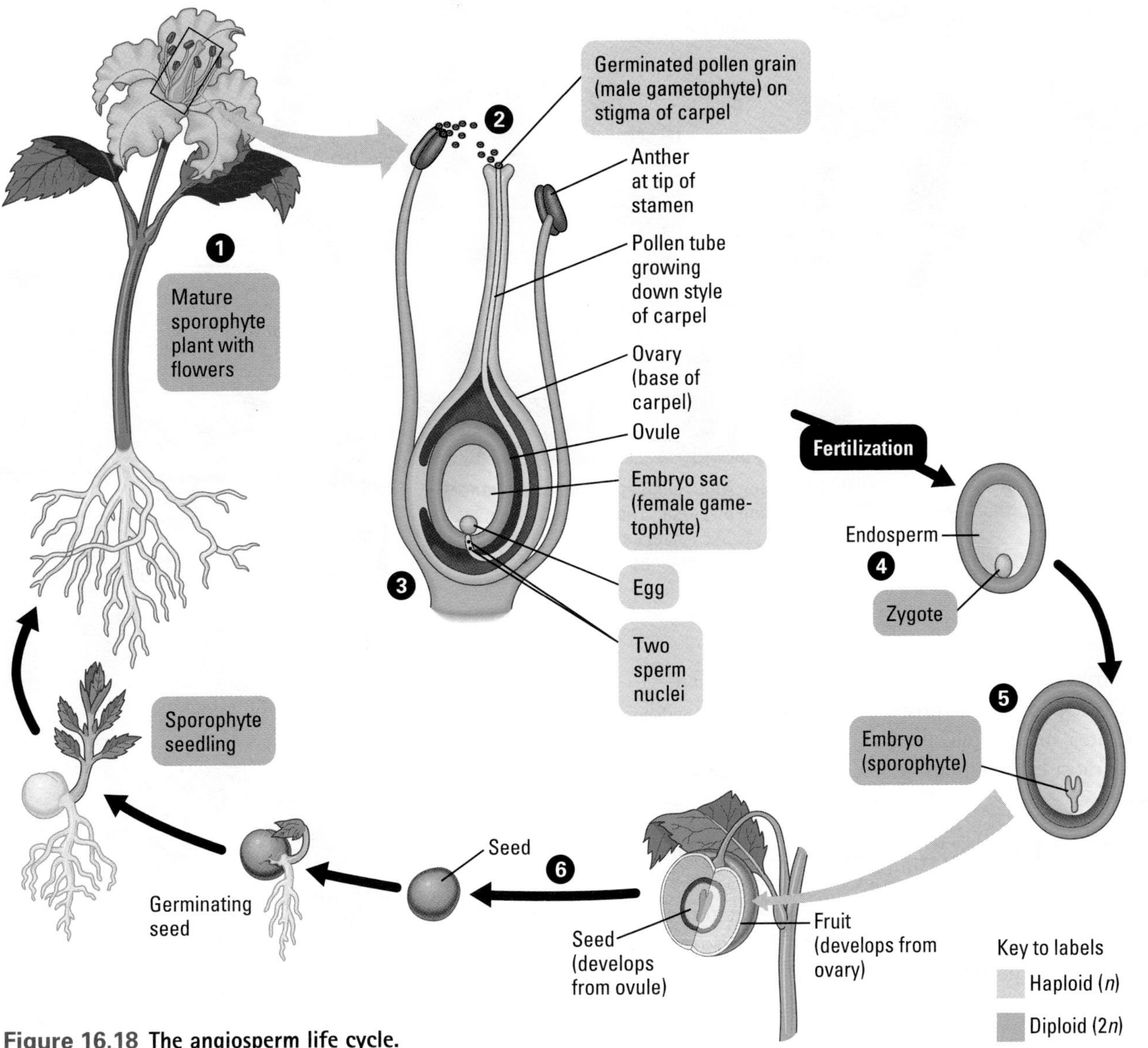

Figure 16.18 The angiosperm life cycle.

Figure 16.18 highlights key stages in the angiosperm life cycle. ❶ The flower we see is part of the sporophyte plant. As in gymnosperms, the pollen grain is the male gametophyte of angiosperms. The female gametophyte—the embryo sac—is located within an ovule, which in turn resides within a chamber of the ovary. ❷ A pollen grain that lands on the sticky stigma of a carpel extends a tube down to the ovule and ❸ deposits two sperm nuclei within the embryo sac. This **double fertilization** is an angiosperm characteristic. One sperm cell fertilizes an egg in the embryo sac. ❹ This produces a zygote, which ❺ develops into an embryo. The second sperm cell fertilizes another female gametophyte cell, which then develops into a nutrient-storing tissue called **endosperm,** which nourishes the embryo. Double fertilization thus synchronizes the development of the embryo and food reserves within an ovule. ❻ The whole ovule develops into a seed. The seed's enclosure within an ovary is what distinguishes angiosperms from the naked-seed condition of gymnosperms.

A **fruit** is the ripened ovary of a flower. As seeds are developing from ovules, the ovary wall thickens, forming the fruit that encloses the seeds. A

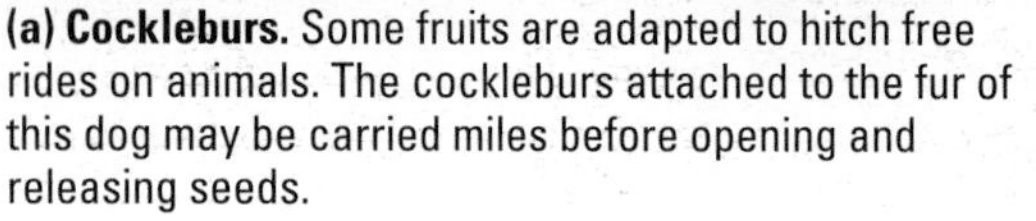

(a) Cockleburs. Some fruits are adapted to hitch free rides on animals. The cockleburs attached to the fur of this dog may be carried miles before opening and releasing seeds.

(b) Dandelion fruit. Some angiosperms depend on wind for seed dispersal. The dandelion fruit, for example, acts like a kite, carrying a tiny seed far away from its parent plant.

(c) Berries. Many angiosperms produce fleshy, edible fruits that are attractive to animals as food. When a mouse eats a berry, it digests the fleshy part of the fruit, but most of the tough seeds pass unharmed through the mouse's digestive tract. The mouse later deposits the seeds, along with a fertilizer supply, some distance from where it ate the fruit.

Figure 16.19 Fruits and seed dispersal. Many types of garden produce—tomatoes, squash, strawberries, and oranges, to name just a few—are the edible fruits from plants we have domesticated. Not all fruits, however, look like something you'd eat for lunch.

pea pod is an example of a fruit, with seeds (mature ovules, the peas) encased in the ripened ovary (the pod). Fruits protect and help disperse seeds. As **Figure 16.19** demonstrates, many angiosperms depend on animals to disperse seeds. Conversely, most land animals, including humans, rely on angiosperms as a food source.

Angiosperms and Agriculture Flowering plants provide nearly all our food. All of our fruit and vegetable crops are angiosperms. Corn, rice, wheat, and the other grains are grass fruits. Grains are also the main food source for domesticated animals, such as cows and chickens. We also grow angiosperms for fiber, medications, perfumes, and decoration.

Early humans probably collected wild seeds and fruits. Agriculture gradually developed as people began sowing seeds and cultivating plants to have a more dependable food source. And as they domesticated certain plants, people began to intervene in plant evolution by selective breeding designed to improve the quantity and quality of the foods. Agriculture is a unique kind of evolutionary relationship between plants and animals.

Plant Diversity as a Nonrenewable Resource

The exploding human population, with its demand for space and natural resources, is extinguishing plant species at an unprecedented rate. The problem is especially critical in the tropics, where more than half the human population lives and population growth is fastest. Tropical rain forests are being destroyed at a frightening pace. The most common cause of this destruction is large-scale slash-and-burn clearing of forest for agricultural use. Fifty million acres, an area about the size of the state of Washington, are cleared each year, a rate that could completely eliminate Earth' s tropical forests within 25 years. As the forest disappears, so do thousands of plant species. Insects and other rain forest animals that depend on these plants are also vanishing. In all, researchers estimate that the destruction of habitat in the rain forest and other ecosystems is

Table 16.1	A Sampling of Medicines Derived from Plants	
Compound	**Source**	**Example of Use**
Atropine	Belladonna plant	Pupil dilator in eye exams
Digitalin	Foxglove	Heart medication
Menthol	Eucalyptus tree	Ingredient in cough medicines
Morphine	Opium poppy	Pain reliever
Quinine	Quinine tree	Malaria preventive
Taxol	Pacific yew	Ovarian cancer drug
Tubocurarine	Curare tree	Muscle relaxant during surgery
Vinblastine	Periwinkle	Leukemia drug

Source: Adapted from Randy Moore et al., *Botany*, 2nd ed. Dubuque, IA: Brown, 1998. Table 2.2, p. 37.

claiming hundreds of species each year. The toll is greatest in the tropics because that is where most species live; but environmental assault is a generically human tendency. Europeans eliminated most of their forests centuries ago, and habitat destruction is now endangering many species in North America. Extinction is irrevocable; plant diversity is a nonrenewable resource.

Many people have ethical concerns about contributing to the extinction of living forms. But there are also practical reasons to be concerned about the loss of plant diversity. As already mentioned, we depend on plants for thousands of products, including food, building materials, and medicines **(Table 16.1)**. More than 120 prescription drugs are extracted from plants. However, researchers have investigated fewer than 5,000 of the 300,000 known plant species as potential sources of medicine. Pharmaceutical companies were led to most of these species by local peoples who use the plants in preparing their traditional medicines.

The tropical rain forest may be a medicine chest of healing plants that could become extinct before we even know they exist. This is only one reason to value what is left of plant diversity and to search for ways to slow the loss. The solutions we propose must be economically realistic. If the goal is only profit for the short term, then we will continue to slash and burn until the forests are gone. If, however, we begin to see rain forests and other ecosystems as living treasures that can regenerate only slowly, we may learn to harvest their products at sustainable rates.

We have seen in our survey of plants, especially the angiosperms, how entangled the botanical world is with other terrestrial life. We switch our attention now to that other group of organisms that moved onto land with plants, the kingdom Fungi.

CHECKPOINT

1. Which of the following structures is common to all four major plant groups: vascular tissue, flowers, seeds, cuticle, pollen?
2. Gametophyte is to __________ as __________ is to diploid.
3. How does the evergreen nature of pines and other conifers adapt the plants for living where the growing season is very short?
4. Contrast the mode of sperm delivery in ferns with sperm delivery in conifers.
5. (a) In what way are forests renewable resources? (b) Under what conditions does reality belie the "renewable" designation?
6. __________ are to conifers as flowers are to __________.
7. What are the four main types of organs of a flower?
8. What is a fruit?

Answers: 1. Cuticle **2.** haploid; sporophyte **3.** Because the plants do not lose their leaves during autumn and winter, the leaves are already fully developed for photosynthesis when the short growing season begins in spring. **4.** The flagellated sperm of ferns must swim through water to reach eggs. In contrast, the airborne pollen of conifers delivers sperm without need of an aqueous route; the pollen releases sperm near the egg after pollination occurs. **5.** (a) Forests are renewable in the sense that new trees can grow where old growth has been removed by logging. (b) The current situation is that wood is harvested at a much faster rate than new growth can replace it, and thus "renewable" is an unrealistic designation for forest resources. **6.** Cones; angiosperms **7.** Sepals, petals, stamens, and carpels **8.** A ripened ovary of a flower that protects and aids in the dispersal of seeds contained in the fruit

Fungi

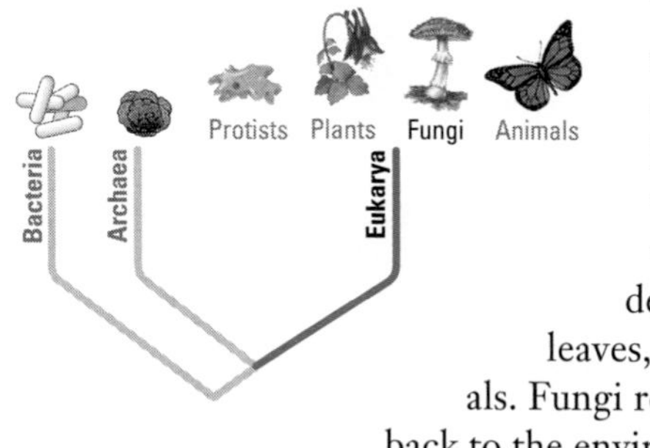

The word *fungus* often evokes some unpleasant images. Fungi rot timbers, spoil food, and afflict humans with athlete's foot and worse. However, ecosystems would collapse without fungi to decompose dead organisms, fallen leaves, feces, and other organic materials. Fungi recycle vital chemical elements back to the environment in forms other organisms can assimilate. And you have already learned that nearly all plants have mycorrhizae, fungus-root associations that help plants absorb minerals and water from the soil. In addition to these ecological roles, fungi have been used by humans in various ways for centuries. We eat some fungi (mushrooms and truffles, for instance), culture fungi to produce antibiotics and other drugs, add them to dough to make bread rise, culture them in milk to produce a variety of cheeses, and use them to ferment beer and wine.

Fungi are eukaryotes, and most are multicellular. Molecular studies indicate that fungi and animals probably arose from a common ancestor. In

(a) Fly agaric mushrooms. These are the reproductive structures of a fungus that absorbs nutrients as it decomposes compost on a forest floor.

(b) A "fairy ring." Some mushroom-producing fungi poke up "fairy rings," which can appear on a lawn overnight. A ring develops at the edge of the main body of the fungus, which consists of an underground mass of tiny filaments (hyphae) within the ring. As the underground fungal mass grows outward from its center, the diameter of the fairy rings produced at its expanding perimeter increases annually.

(c) *Pilobolus*. The bulbs at the tips of the stalks are sacs of reproductive spores. *Pilobolus* stalks bend toward light and then shoot their spore sacs like cannonballs. Grazing animals eat the spore sacs and scatter the spores in feces, where the spores grow into new fungi.

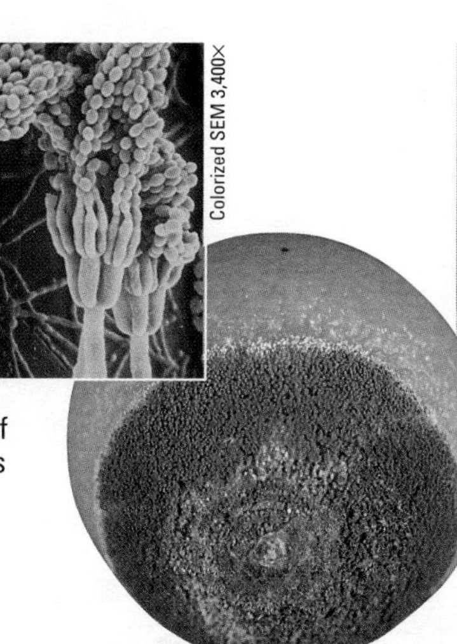

(d) Mold. Molds grow rapidly on their food sources, which are often *our* food sources as well. The mold on this orange reproduces asexually by producing chains of microscopic spores (inset) that are dispersed via air currents.

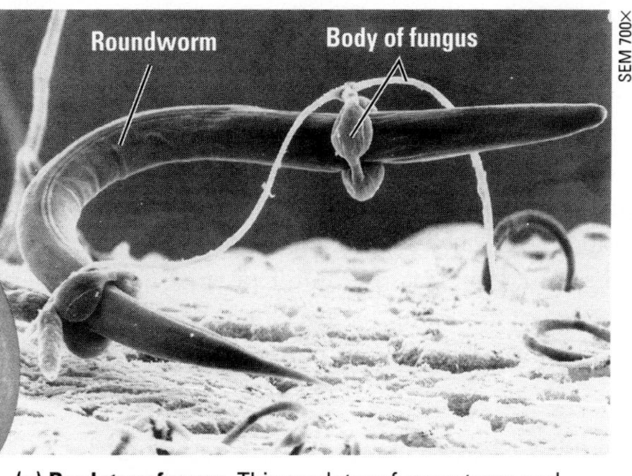

(e) Predatory fungus. This predatory fungus traps and feeds on tiny roundworms in the soil. The fungus is equipped with hoops that can constrict around a worm in less than a second.

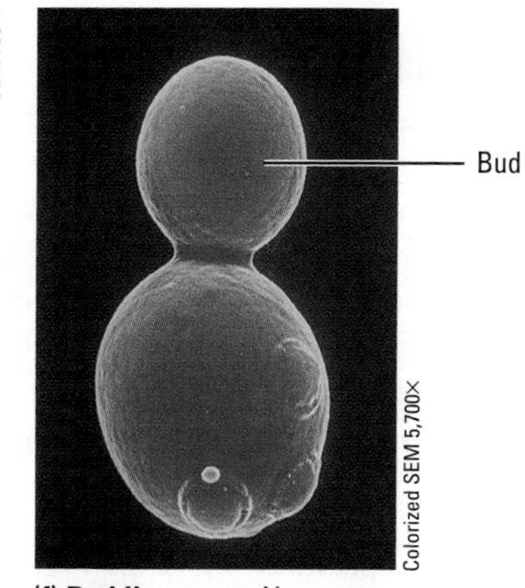

(f) Budding yeast. Yeasts are unicellular fungi. This yeast cell is reproducing asexually by a process called budding.

Figure 16.20 A gallery of diverse fungi.

other words, a mushroom is probably more closely related to you than it is to any plant! However, fungi are actually a form of life so distinctive that they are accorded their own kingdom, the kingdom Fungi **(Figure 16.20)**.

Characteristics of Fungi

In this section, we'll examine the structure and function of fungi, beginning with an overview of how fungi obtain nutrients.

Fungal Nutrition Fungi are heterotrophs that acquire their nutrients by **absorption.** In this mode of nutrition, small organic molecules are absorbed from the surrounding medium. A fungus digests food outside its body by secreting powerful digestive enzymes into the food. The enzymes decompose complex molecules to the simpler compounds that the fungus can absorb. For example, fungi that are decomposers absorb nutrients from such nonliving organic material as fallen logs, animal corpses, or the wastes of live organisms. Parasitic fungi absorb nutrients from the cells or body fluids of living hosts. Some of these fungi, such as certain species infecting the lungs of humans, cause disease. In other cases, such as mycorrhizae, the relationships between fungi and their hosts are mutually beneficial.

Fungal Structure Fungi are structurally adapted for their absorptive nutrition. The bodies of most fungi are constructed of structures called **hyphae** (singular, *hypha*). Fungal hyphae are minute threads of cytoplasm surrounded by a plasma membrane and cell wall. The cell walls of fungi differ from the cellulose walls of plants. Fungal cell walls are usually built mainly of chitin, a strong but flexible polysaccharide that is also found in the external skeletons of insects. Most fungi have multicellular hyphae, which consist of chains of cells separated by cross-walls with pores large enough to allow ribosomes, mitochondria, and even nuclei to flow from cell to cell.

Fungal hyphae associate into an interwoven mat called a **mycelium** (plural, *mycelia*), which is the feeding network of the fungus (**Figure 16.21**). Fungal mycelia can be huge, but they usually escape our notice because they are often subterranean. In 2000, scientists discovered the mycelium of one humongous fungus in Oregon that is 5.5 kilometers (km)—that's 3.4 miles!—in diameter and spreads through 2,200 acres of forest. This fungus is at least 2,400 years old and hundreds of tons in weight, qualifying it as one of Earth's oldest and largest organisms.

A mycelium maximizes contact with its food source by mingling with the organic matter it is decomposing and absorbing. Ten cubic centimeters (cm^3) of rich organic soil may contain as much as a kilometer of hyphae. And a fungal mycelium grows rapidly, adding as much as a kilometer of hyphae each day as it branches within its food. Fungi are nonmotile organisms; they cannot run, swim, or fly in search of food. But the mycelium makes up for the lack of mobility by swiftly extending the tips of its hyphae into new territory.

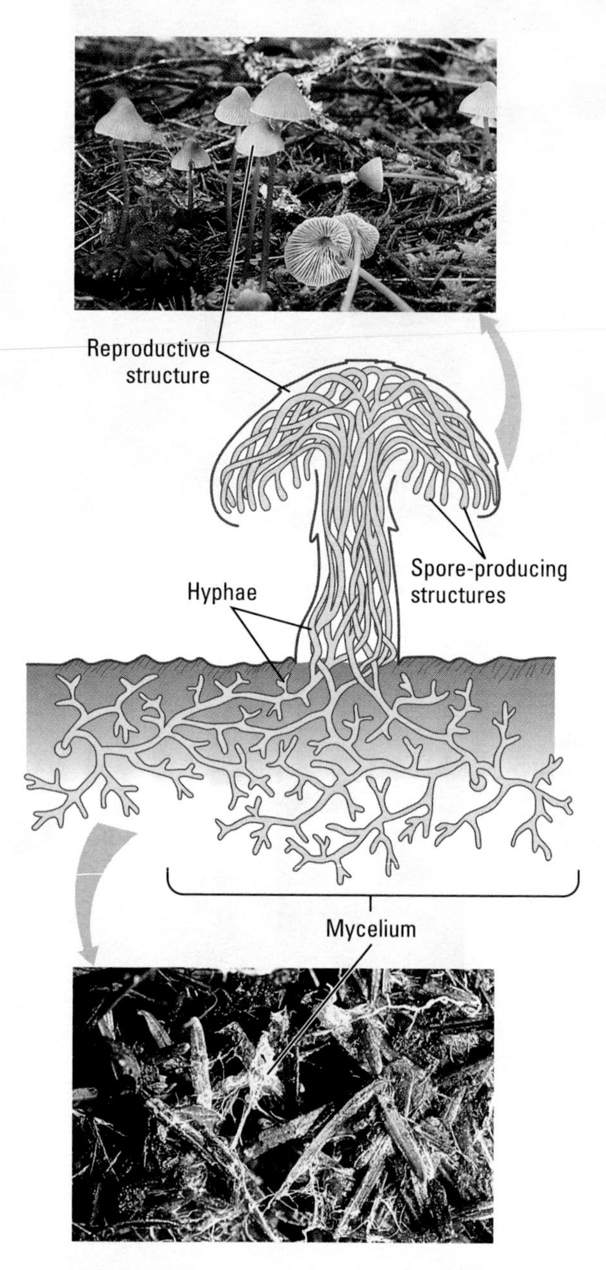

Figure 16.21 The fungal mycelium. The mushroom we see is like the tip of an iceberg. It is a reproductive structure consisting of tightly packed hyphae that extend upward from a much more massive mycelium of hyphae growing underground. The photos show mushrooms and the mycelium of cottony threads that decompose organic litter.

Fungal Reproduction Fungi reproduce by releasing spores that are produced either sexually or asexually. The output of spores is mind-boggling. For example, puffballs, which are the reproductive structures of certain fungi, can puff out clouds containing trillions of spores (see Figure 13.15). Carried by wind or water, spores germinate to produce mycelia if they land in a moist place where there is food. Spores thus function in dispersal and account for the wide geographic distribution of many species of fungi. The airborne spores of fungi have been found more than 160 km (100 miles) above Earth. Closer to home, try leaving a slice of bread out for a week or two and you will observe the furry mycelia that grow from the invisible spores raining down from the surrounding air.

The Ecological Impact of Fungi

Fungi have been major players in terrestrial communities ever since plants and fungi together moved onto land. Let's examine a few examples of how fungi continue to have an enormous ecological impact.

Fungi as Decomposers Fungi and bacteria are the principal decomposers that keep ecosystems stocked with the inorganic nutrients essential for plant growth. Without decomposers, carbon, nitrogen, and other elements would accumulate in organic matter. Plants and the animals they feed would starve because elements taken from the soil would not be returned (see Chapter 19).

Fungi are well adapted as decomposers of organic refuse. Their invasive hyphae enter the tissues and cells of dead organisms and digest polymers, including the cellulose of plant cell walls. A succession of fungi, in concert with bacteria and, in some environments, invertebrate animals, are responsible for the complete breakdown of organic litter. The air is so loaded with fungal spores that as soon as a leaf falls or an insect dies, it is covered with spores, and it is soon infiltrated by fungal hyphae.

We may applaud fungi that decompose forest litter or dung, but it is a different story when molds attack our fruit or our shower curtains. Between 10% and 50% of the world's fruit harvest is lost each year to fungal attack. And a wood-digesting fungus does not distinguish between a fallen oak limb and the oak planks of a boat. During the Revolutionary War, the British lost more ships to fungal rot than to enemy attack. What's more, soldiers stationed in the tropics during World War II watched as their tents, clothing, boots, and binoculars were destroyed by molds. Some fungi can even decompose certain plastics.

Parasitic Fungi Parasitism is a relationship in which two species live in contact and one organism benefits while the other is harmed. Of the 100,000 known species of fungi, about 30% make their living as parasites, mostly on or in plants. In some cases, fungi that infect plants have literally changed landscapes. In the chapter-opening essay, you learned how a fungal blight devastated the American chestnut. A related fungus killed most American elm trees **(Figure 16.22a)**. Fungi are also serious agricultural pests. Some species infect grain crops and cause tremendous economic losses each year **(Figure 16.22b)**.

(a) American elm trees killed by Dutch elm disease fungus

(b) Ergots (dark structures) on rye seed heads

Figure 16.22 Parasitic fungi that cause plant disease. **(a)** The parasitic fungus that causes Dutch elm disease evolved with European species of elm trees, and it is relatively harmless to them. But it is deadly to American elms. The fungus was accidentally introduced into the United States on logs sent from Europe to pay World War I debts. Insects called bark beetles carried the fungus from tree to tree. Since then, the disease has destroyed elm trees all across North America. **(b)** The seeds of some kinds of grain, including rye, wheat, and oats, are sometimes infected with fungal growths called ergots. Consumption of flour made from ergot-infested grain can cause ergotism, a disease with extremely severe symptoms that may be fatal. Both medicines and toxins have been isolated from ergots. The latter include lysergic acid, the raw material from which the hallucinogenic drug LSD is made.

Figure 16.23 Feeding on fungi. (a) Truffles (the fungal kind, not the chocolates) are the reproductive structures of fungi that grow with tree roots as mycorrhizae. Truffles release strong odors that attract certain mammals and insects, which excavate the fungi and disperse their spores. Truffle hunters use pigs or dogs to locate their prizes, which may command prices of hundreds of dollars per pound. Gourmets describe the complex flavors of truffles as nutty, musky, cheesy, or some combination of those tastes. **(b)** The turquoise streaks in blue cheese are the mycelia of a specific fungus.

Animals are much less susceptible to parasitic fungi than are plants. Only about 50 species of fungi are known to be parasitic in humans and other animals. However, their effects are significant enough to make us take them seriously. Among the diseases that fungi cause in humans are yeast infections of the lungs, some of which can be fatal, and vaginal yeast infections. Other fungal parasites produce a skin disease called ringworm, so named because it appears as circular red areas on the skin. Ringworm fungi can infect virtually any skin surface. Most commonly, they attack the feet and cause intense itching and sometimes blisters. This condition, known as athlete's foot, is highly contagious but can be treated with various fungicidal medications.

Commercial Uses of Fungi It would not be fair to fungi to end our discussion with an account of diseases. In addition to their positive global impact as decomposers, fungi also have a number of practical uses for humans.

Most of us have eaten mushrooms, although we may not have realized that we were ingesting the reproductive extensions of subterranean fungi. Mushrooms are often cultivated commercially in artificial caves in which cow manure is piled (be sure to wash your store-bought mushrooms thoroughly). Edible mushrooms also grow wild in fields, forests, and backyards, but so do poisonous ones. There are no simple rules to help the novice distinguish edible from deadly mushrooms. Only experts should dare to collect wild fungi for eating.

Mushrooms are not the only fungi we eat. The fungi called truffles are highly prized by gourmets **(Figure 16.23a)**. And the distinctive flavors of certain kinds of cheeses come from the fungi used to ripen them **(Figure 16.23b)**. Particularly important in food production are unicellular fungi, the yeasts. As discussed in Chapter 6, yeasts are used in baking, brewing, and winemaking.

Fungi are medically valuable as well. Some fungi produce antibiotics that are used to treat bacterial diseases. In fact, the first antibiotic discovered was penicillin, made by the common mold *Penicillium* **(Figure 16.24)**.

As sources of antibiotics and food, as decomposers, and as partners with plants in mycorrhizae, fungi play vital roles in life on Earth.

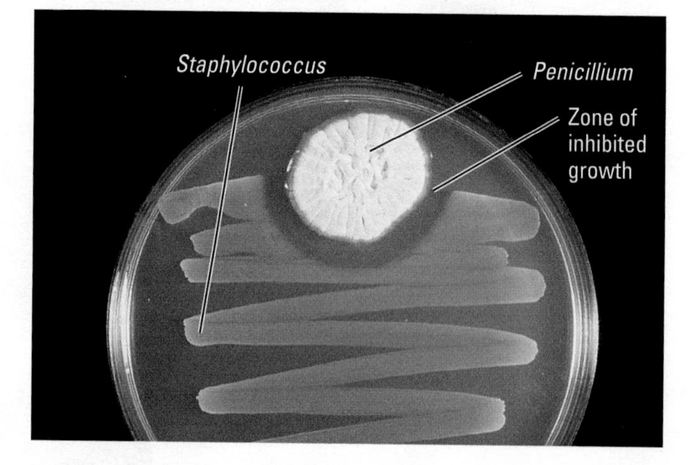

Figure 16.24 Fungal production of an antibiotic. Penicillin is made by the common mold *Penicillium*. In this petri dish, the clear area between the mold and the growing *Staphylococcus* bacteria is where the antibiotic produced by the *Penicillium* inhibits the growth of the bacteria.

CHECKPOINT

1. What are mycorrhizae?
2. Contrast the heterotrophic nutrition of a fungus with your own heterotrophic nutrition.
3. Describe how the structure of a fungal mycelium reflects its function.
4. What is athlete's foot?
5. What do you think is the natural function of the antibiotics that fungi produce in their native environments?

Answers: 1. Root-fungus symbiotic associations that enhance the uptake of water and minerals by the plant and provide organic nutrients to the fungus **2.** A fungus digests its food externally by secreting digestive juices into the food and then absorbing the small nutrients that result from digestion. In contrast, humans and most other animals ingest relatively large pieces of food and digest the food within their bodies. **3.** The extensive network of hyphae puts a large surface area in touch with the food source. **4.** Infection of the foot's skin with ringworm fungus **5.** The antibiotics block the growth of microorganisms, especially bacteria, that compete with the fungi for nutrients and other resources.

EVOLUTION CONNECTION

Mutualistic Symbiosis

Evolution is not just about the origin and adaptation of individual species. Relationships between species are also an evolutionary product. **Symbiosis** is the term used to describe ecological relationships between organisms of different species that are in direct physical contact. As mentioned earlier, **parasitism** is a symbiotic relationship in which one species, the parasite, benefits while harming its host in the process. Our focus here, however, is on **mutualism,** symbiosis that benefits both species.

We have seen many examples of mutualism over the past two chapters. Eukaryotic cells evolved from mutualistic symbiosis among prokaryotes. And today, bacteria living in the roots of certain plants provide nitrogen compounds to their host and receive food in exchange. We have our own mutually symbiotic bacteria that help keep our skin healthy and produce certain vitamins in our intestines. Particularly relevant to this chapter is the symbiotic association of fungi and plant roots—mycorrhizae—that made life's move onto land possible.

Lichens, symbiotic associations of fungi and algae, are striking examples of how two species can become so merged that the cooperative is essentially a new life-form. At a distance, it is easy to mistake lichens for mosses or other simple plants growing on rocks, rotting logs, trees, roofs, or gravestones **(Figure 16.25)**. In fact, lichens are not mosses or any other kind of plant, nor are they even individual organisms. A lichen is a symbiotic association of millions of tiny algae embraced by a mesh of fungal hyphae. The photosynthetic algae feed the fungi. The fungal mycelium, in turn, provides a suitable habitat for the algae, helping to absorb and retain water and minerals. The mutualistic merger of partners is so complete that lichens are actually named as species, as though they are individual organisms. Mutualisms such as lichens and mycorrhizae showcase the web of life that has evolved on Earth. ■

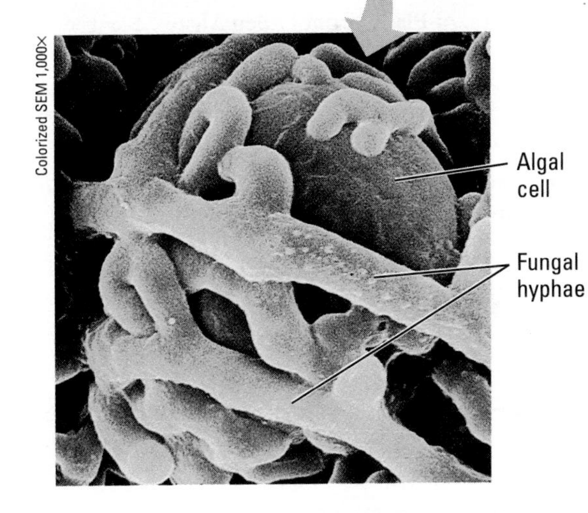

Figure 16.25 Lichens: symbiotic associations of fungi and algae. Lichens generally grow very slowly, sometimes less than a millimeter per year. Some lichens are thousands of years old, rivaling the oldest plants as Earth's elders. The close relationship between the fungal and algal partners is evident in the microscopic blowup of a lichen.

Chapter Review

SUMMARY OF KEY CONCEPTS

For study help and activities, go to campbellbiology.com or the student CD-ROM.

Colonizing Land

• **Terrestrial Adaptations of Plants** Plants are multicellular photosynthetic eukaryotes with adaptations for living on land.

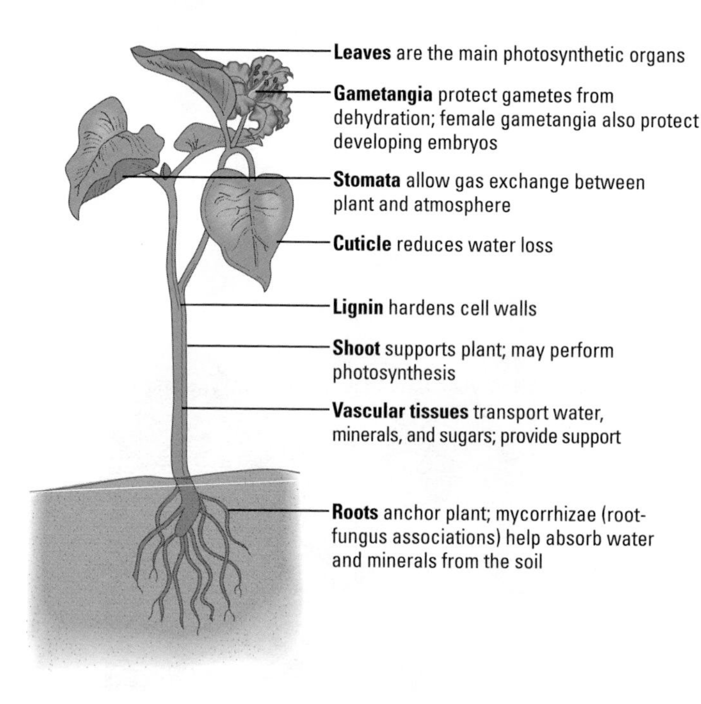

Leaves are the main photosynthetic organs
Gametangia protect gametes from dehydration; female gametangia also protect developing embryos
Stomata allow gas exchange between plant and atmosphere
Cuticle reduces water loss
Lignin hardens cell walls
Shoot supports plant; may perform photosynthesis
Vascular tissues transport water, minerals, and sugars; provide support
Roots anchor plant; mycorrhizae (root-fungus associations) help absorb water and minerals from the soil

Activity *Terrestrial Adaptations of Plants*

• **The Origin of Plants from Green Algae** Plants evolved from a group of multicellular green algae called charophyceans.

Plant Diversity

• **Highlights of Plant Evolution** Four major periods of plant evolution are marked by terrestrial adaptations.

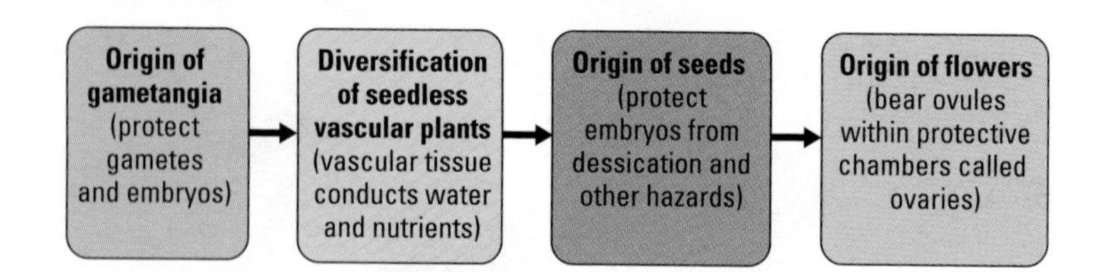

Origin of gametangia (protect gametes and embryos)
Diversification of seedless vascular plants (vascular tissue conducts water and nutrients)
Origin of seeds (protect embryos from dessication and other hazards)
Origin of flowers (bear ovules within protective chambers called ovaries)

Activity *Highlights of Plant Evolution*

• **Bryophytes** The most familiar bryophytes are mosses. Mosses display two key terrestrial adaptations: a waxy cuticle that prevents dehydration and the retention of developing embryos within the mother plant's gametangia. Mosses are most common in moist environments because their sperm must swim to the eggs and because they lack lignin in their cell walls and thus cannot stand tall. Bryophytes are unique among plants in having the gametophyte as the dominant generation in the life cycle.

Activity *Moss Life Cycle*

• **Ferns** Ferns are seedless plants that have vascular tissues but still use flagellated sperm to fertilize eggs. During the Carboniferous period, giant ferns were among the plants that decayed to thick deposits of organic matter, which were gradually converted to coal.

Activity *Fern Life Cycle*

Case Study in the Process of Science *What Are the Different Stages of a Fern Life Cycle?*

• **Gymnosperms** A drier and colder global climate near the end of the Carboniferous favored the evolution of the first seed plants. The most successful were the gymnosperms, represented by conifers. Needle-shaped leaves with thick cuticles and sunken stomata are adaptations to dry conditions. Conifers and most other gymnosperms have three additional terrestrial adaptations: (1) further reduction of the gametophyte generation and greater development of the diploid sporophyte, (2) the evolution of pollen, which doesn't require water for transport, and (3) the advent of the seed, which consists of a plant embryo packaged along with a food supply within a protective coat.

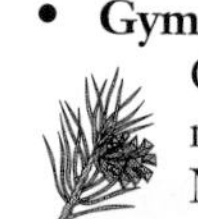

Activity *Pine Life Cycle*

• **Angiosperms** Angiosperms supply nearly all our food and much of our fiber for textiles. The evolution of the flower and more efficient water transport help account for the success of the angiosperms. The dominant stage is a sporophyte with gametophytes in its flowers. The female gametophyte is located within an ovule, which in turn resides within a chamber of the ovary. Fertilization of an egg in the female gametophyte produces a zygote, which develops into an embryo. The whole ovule develops into a seed. The seed's enclosure within an ovary is what distinguishes angiosperms from the naked-seed condition of gymnosperms. A fruit is the ripened ovary of a flower. Fruits protect and help disperse seeds. Angiosperms are a major food source for animals, while animals aid plants in pollination and seed dispersal. Agriculture constitutes a unique kind of evolutionary relationship among plants, humans, and other animals.

Case Study in the Process of Science *How Are Trees Identified by Their Leaves?*

Activity *Angiosperm Life Cycle*

• **Plant Diversity as a Nonrenewable Resource** The exploding human population, with its demand for space and natural resources, is causing the extinction of plant species at an unprecedented rate.

Activity *Madagascar and the Biodiversity Crisis*

Fungi

• **Characteristics of Fungi** Fungi are unicellular or multicellular eukaryotes that are more closely related to animals than to plants. Fungi are heterotrophs that digest their food externally and absorb the nutrients. A fungus usually consists of a mass of threadlike hyphae, forming a mycelium. The cell walls of fungi are mainly composed of chitin. Although most fungi are nonmotile, the mycelium can grow very quickly, extending the tips of its hyphae into new territory. Fungi reproduce and disperse by releasing spores that are produced either sexually or asexually.

Case Study in the Process of Science *How Does the Fungus* Pilobolus *Succeed as a Decomposer?*

Activity *Fungal Reproduction and Nutrition*

- **The Ecological Impact of Fungi** Fungi and bacteria are the principal decomposers of ecosystems. Many molds destroy fruit, wood, and human-made materials. About 50 species of fungi are known to be parasitic in humans and other animals. Fungi are also commercially important as food and in baking, beer and wine production, and the manufacture of antibiotics.

Evolution Connection: Mutualistic Symbiosis

- Lichens, in which algae are surrounded by fungal hyphae, are an example of mutualism, a mutually beneficial symbiosis.

SELF-QUIZ

1. Angiosperms are different from all other plants because only angiosperms have __________.

2. Ovule is to seed as ovary is to __________.

3. Under a microscope, a piece of a mushroom would look most like
   a. jelly.
   b. a tangle of string.
   c. grains of sand.
   d. a sponge.

4. During the Carboniferous period, the dominant plants, which later formed the great coal beds, were mainly
   a. mosses and other bryophytes.
   b. ferns and other seedless vascular plants.
   c. charophyceans and other green algae.
   d. conifers and other gymnosperms.

5. You discover a new species of plant. Under the microscope, you find that it produces flagellated sperm. A genetic analysis shows that its dominant generation has diploid cells. What kind of plant do you have?

6. Which of the following terms includes all others in the list: angiosperm, fern, vascular plant, gymnosperm, seed plant.

7. Plant diversity is greatest in
   a. tropical forests.
   b. the temperate forests of Europe.
   c. deserts.
   d. the oceans.

8. Name five products you've used today that come from angiosperms.

9. Lichens are symbionts of photosynthetic __________ with __________.

10. Fungi acquire nutrients by __________.

Answers to the Self-Quiz questions can be found in Appendix D.

Go to the website or CD-ROM for more Self-Quiz questions.

THE PROCESS OF SCIENCE

11. In April 1986, an accident at a nuclear power plant in Chernobyl, Ukraine, scattered radioactive fallout for hundreds of miles. In assessing the biological effects of the radiation, researchers found mosses to be especially valuable as organisms for monitoring the damage. Radiation damages organisms by causing mutations. Explain why it is faster to observe the genetic effects of radiation on mosses than on other types of plants. Imagine that you are conducting tests shortly after a nuclear accident. Using potted moss plants as your experimental organisms, design an experiment to test the hypothesis that the frequency of mutations decreases with the organism's distance from the source of radiation.

12. You discover what you think may be one extremely large underground fungal mycelium living beneath your campus. How could you prove that it is, in fact, one individual organism spread across a very large area, as opposed to a group of separate organisms?

BIOLOGY AND SOCIETY

13. Why are tropical rain forests being destroyed at such an alarming rate? What kinds of social, technological, and economic factors are responsible? Most forests in developed Northern Hemisphere countries have already been cut. Do the developed nations have a right to pressure the developing nations in the Southern Hemisphere to slow or stop the destruction of their forests? Defend your answer. What kinds of benefits, incentives, or programs might slow the assault on the rain forests?

14. Imagine you were charged with the task of managing a coniferous forest. How would you balance the need for productive use of the forest (to provide lumber, for example) with preservation of its diversity? What activities would you allow or prohibit in the forest (for example, snowmobiling, logging, hiking, mushroom harvesting, mining, camping, grazing, making campfires)? How would you defend your choices to the public?

17 The Evolution of Animals

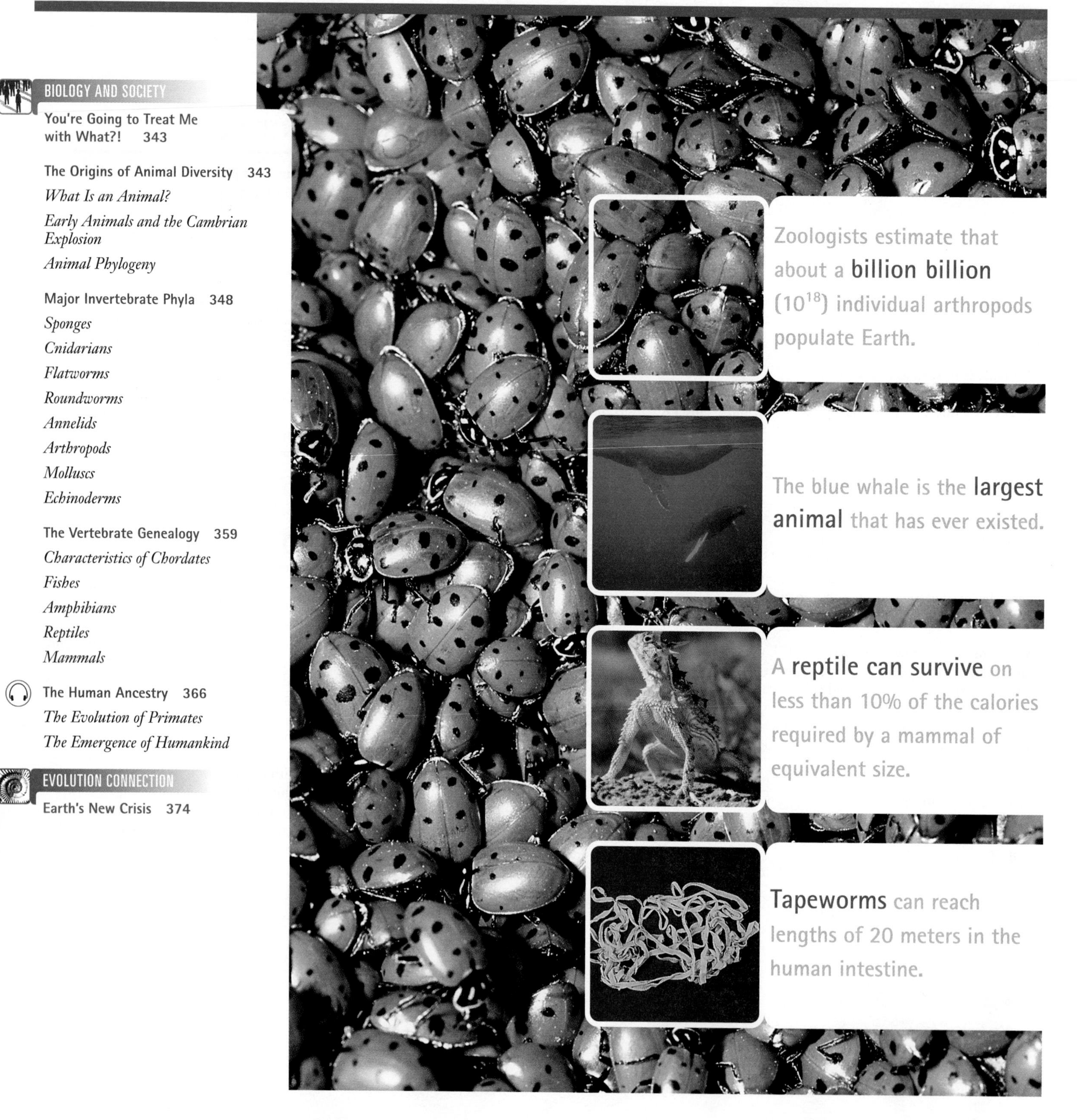

Zoologists estimate that about a **billion billion** (10^{18}) individual arthropods populate Earth.
The blue whale is the **largest animal** that has ever existed.
A **reptile can survive** on less than 10% of the calories required by a mammal of equivalent size.
Tapeworms can reach lengths of 20 meters in the human intestine.

BIOLOGY AND SOCIETY

You're Going to Treat Me with What?!

In January 2004, the U.S. Food and Drug Administration (FDA) did something it had never done before: It approved the use of a live animal as a medical device. And not just any animal; the FDA gave permission for doctors to use maggots—specifically, juvenile blowflies (*Phaenicia sericata*)—to cleanse infected wounds **(Figure 17.1a)**. Called maggot debridement therapy, this practice is becoming increasingly common for the treatment of bedsores, burns, surgical wounds, and traumatic injuries. It is safe and effective because the enzymes secreted by the immature flies break down only dead or dying tissue, leaving healthy skin and bone unharmed.

Later in 2004, the FDA approved a second animal for medical treatment: leeches—specifically, a species of European freshwater leech called *Hirudo medicinalis* **(Figure 17.1b)**. The FDA approved leeches for the treatment of circulatory complications. Most commonly, leeches are applied after reconstructive microsurgery in which limbs or digits are reattached. Because arteries (which transport blood into a reattached area) are easier to reconnect than veins (which transport blood out), blood can pool in the reattached area and stagnate, starving the healing tissue of oxygen. Medicinal leeches have razor-like jaws with hundreds of tiny teeth that cut through the skin. They secrete saliva containing a strong anesthetic and an anticoagulant into the wound. The anesthetic makes the bite virtually painless, and the anticoagulant prevents clotting as the leech drains excess blood from the wound.

As of today, maggots and leeches are the only animals approved for use in medical treatment in the United States, but many animal species fill our world. In fact, of the 1.7 million species of organisms known to science, over two-thirds are animals. This amazing diversity arose through hundreds of millions of years of evolution as natural selection shaped animal adaptations to Earth's many environments. In this chapter, we'll look at 9 of the roughly 35 phyla (major groups) in the kingdom Animalia. These major phyla contain the greatest number of species and are the most abundant and widespread. Along the way, we'll give special attention to the major milestones in animal evolution. ■

(a) Blowfly maggots. A nurse holds a dish of young blowflies (*Phaenicia sericata*) that can be applied to infected wounds.

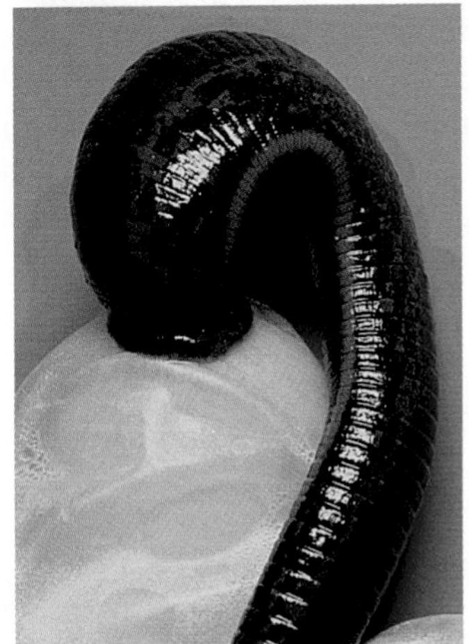

(b) A medicinal leech. A nurse applied this leech (*Hirudo medicinalis*) to a patient to drain blood from a hematoma (abnormal accumulation of blood around an internal injury).

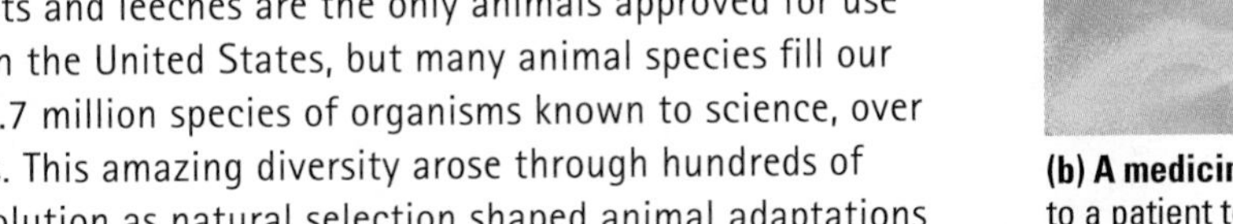

Figure 17.1 Animals as medical devices.

The Origins of Animal Diversity

Animal life began in Precambrian seas with the evolution of multicellular creatures that ate other organisms. We are among their descendants.

What Is an Animal?

Animals are eukaryotic, multicellular, heterotrophic organisms that obtain nutrients by ingestion. That's a mouthful. And speaking of mouthfuls, **ingestion** means eating food. This mode of nutrition contrasts animals with fungi, which obtain nutrients by absorption after digesting the food outside the body (see Chapter 16). Animals digest their food within their bodies after ingesting other organisms, dead or alive, whole or by the piece **(Figure 17.2)**.

A few key features of life history also distinguish animals. Most animals reproduce sexually. Once an egg is fertilized, the zygote develops into an early embryonic stage called a **blastula,** which is usually a hollow ball of cells **(Figure 17.3)**. The next embryonic stage in most animals is a gastrula, which has layers of cells that will eventually form the adult body parts. The gastrula also has a primitive gut, which will develop into the animal's digestive compartment. Continued development, growth, and maturation transform some animals directly from the embryo into an adult. However, the life histories of many animals include a sexually immature form called a **larva.** A larva is anatomically distinct from the adult form, usually eats different foods, and may even have a different habitat. Think how different a frog is from its larval form, a tadpole. A change in body form, called **metamorphosis,** eventually remodels the larva into the adult form.

Most animals have muscle cells, as well as nerve cells that control the muscles. The evolution of this equipment for coordinated movement enabled some animals to search for or chase their food. The most complex animals, of course, can use their muscular and nervous systems for many functions other than eating. Some species even use massive networks of nerve cells called brains to think.

Figure 17.2 Nutrition by ingestion, the animal way of life. Most animals ingest relatively large pieces of food, though rarely as large as the prey in this case. In this amazing scene, a rock python is beginning to ingest a gazelle. The snake will spend two weeks or more digesting its meal.

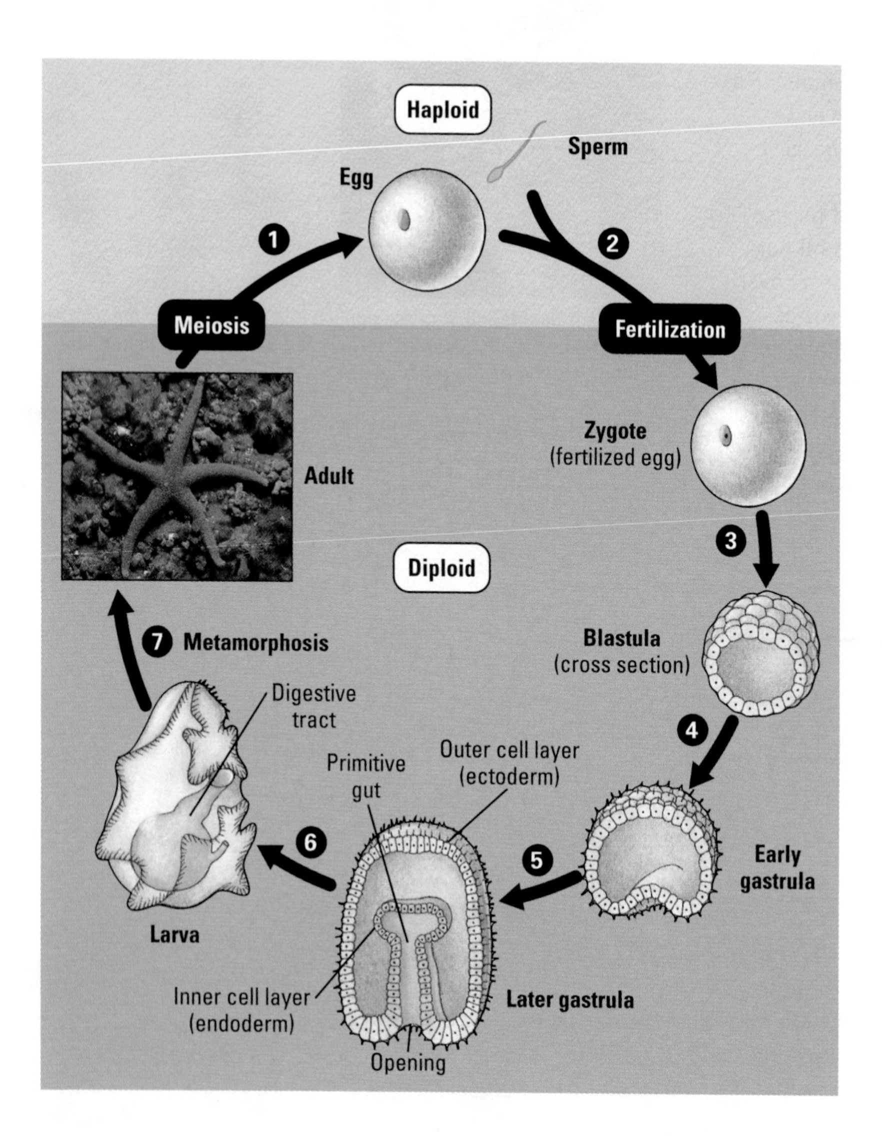

Figure 17.3 Life cycle of a sea star as an example of animal development. ❶ Male and female adult animals produce haploid gametes (eggs and sperm) by meiosis. ❷ An egg and a sperm fuse to produce a diploid zygote. ❸ Early mitotic divisions lead to an embryonic stage called a blastula, common to all animals. Typically, the blastula consists of a ball of cells surrounding a hollow cavity. ❹ Later, in the sea star and many other animals, one side of the blastula cups inward, forming an embryonic stage called a gastrula. ❺ The gastrula develops into a saclike embryo with a two-layered wall and an opening at one end. Eventually, the outer layer (ectoderm) develops into the animal's epidermis (skin) and nervous system. The inner layer (endoderm) forms the digestive tract. Still later in development, in most animals, a third layer (mesoderm) forms between the other two and develops into most of the other internal organs (not shown in the figure). ❻ Following the gastrula, many animals continue to develop and then mature directly into adults. But others, including the sea star, develop into one or more larval stages first. ❼ The larva undergoes a major change of body form, called metamorphosis, in becoming an adult—a mature animal capable of reproducing sexually.

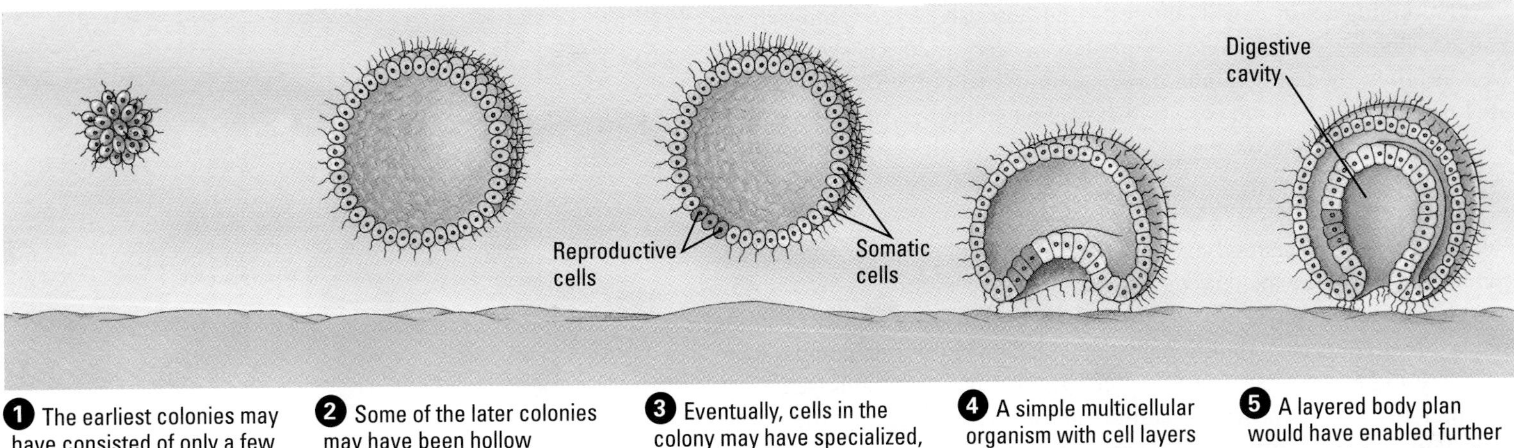

1 The earliest colonies may have consisted of only a few cells, all of which were flagellated and basically identical.

2 Some of the later colonies may have been hollow spheres—floating aggregates of heterotrophic cells—that ingested organic nutrients from the water.

3 Eventually, cells in the colony may have specialized, with some cells adapted for reproduction and others for somatic (nonreproductive) functions, such as locomotion and feeding.

4 A simple multicellular organism with cell layers may have evolved from a hollow colony, with cells on one side of the colony cupping inward, the way they do in the gastrula of an animal embryo (see Figure 17.3).

5 A layered body plan would have enabled further division of labor among the cells. The outer flagellated cells would have provided locomotion and some protection, while the inner cells could have specialized in reproduction or feeding.

Figure 17.4 One hypothesis for a sequence of stages in the origin of animals from a colonial protist. With its specialized cells and a simple digestive compartment, the proto-animal shown at the end of the process could have fed on organic matter on the seafloor.

Early Animals and the Cambrian Explosion

Animals probably evolved from a colonial, flagellated protist that lived in Precambrian seas **(Figure 17.4)**. By the late Precambrian, about 600–700 million years ago, a diversity of animals had already evolved. Then came the Cambrian explosion. At the beginning of the Cambrian period, 542 million years ago, animals underwent a relatively rapid diversification. In fact, during a span of only about 15 million years, all the major animal body plans we see today evolved. It is an evolutionary episode so boldly marked in the fossil record that geologists use the dawn of the Cambrian period as the beginning of the Paleozoic era (see Table 14.1). Many of the Cambrian animals seem bizarre compared with the versions we see today, but most zoologists now agree that Cambrian fossils can be classified as ancient representatives of modern animal phyla **(Figure 17.5)**.

What ignited the Cambrian explosion? Hypotheses abound. Most researchers now believe that the Cambrian explosion simply extended animal diversification that was already well under way during the late Precambrian. But what caused the radiation of animal forms to accelerate so dramatically during the early Cambrian? One hypothesis emphasizes increasingly complex predator-prey relationships that led to diverse adaptations for feeding, motility, and protection. This would help explain why most Cambrian animals had shells or hard outer skeletons, in contrast to Precambrian animals, which were mostly soft-bodied. Another hypothesis focuses on the evolution of genes that control the development of animal form, such as the placement of body parts in embryos. At least some of these genes are common to diverse animal phyla. However, variation in how, when, and where these genes are expressed in an embryo can produce some of the major differences in body form that distinguish the phyla. Perhaps such changes in gene expression were partly responsible for the relatively rapid diversification of animals during the early Cambrian.

Figure 17.5 A Cambrian seascape. This drawing is based on fossils (such as the *Olenoides Serratus*, inset) collected at a site called the Burgess Shale in British Columbia, Canada.

Continuing research will help test hypotheses about the Cambrian explosion. But as the explosion becomes less mysterious, it will seem no less wonderful. In the last half billion years, animal evolution has mainly generated new variations of old themes that originated in the Cambrian seas.

Animal Phylogeny

Because animals diversified so rapidly on the scale of geologic time, it is difficult, using only the fossil record, to sort out the evolutionary relationships among the various phyla. To reconstruct the evolutionary history of animal phyla, researchers have traditionally depended on clues from comparative anatomy and embryology (see Chapter 13). Molecular methods (see Chapter 14) are now providing additional tools for testing hypotheses about animal phylogeny. **Figure 17.6** represents a set of hypotheses about the evolutionary relationships among nine major animal phyla based on body

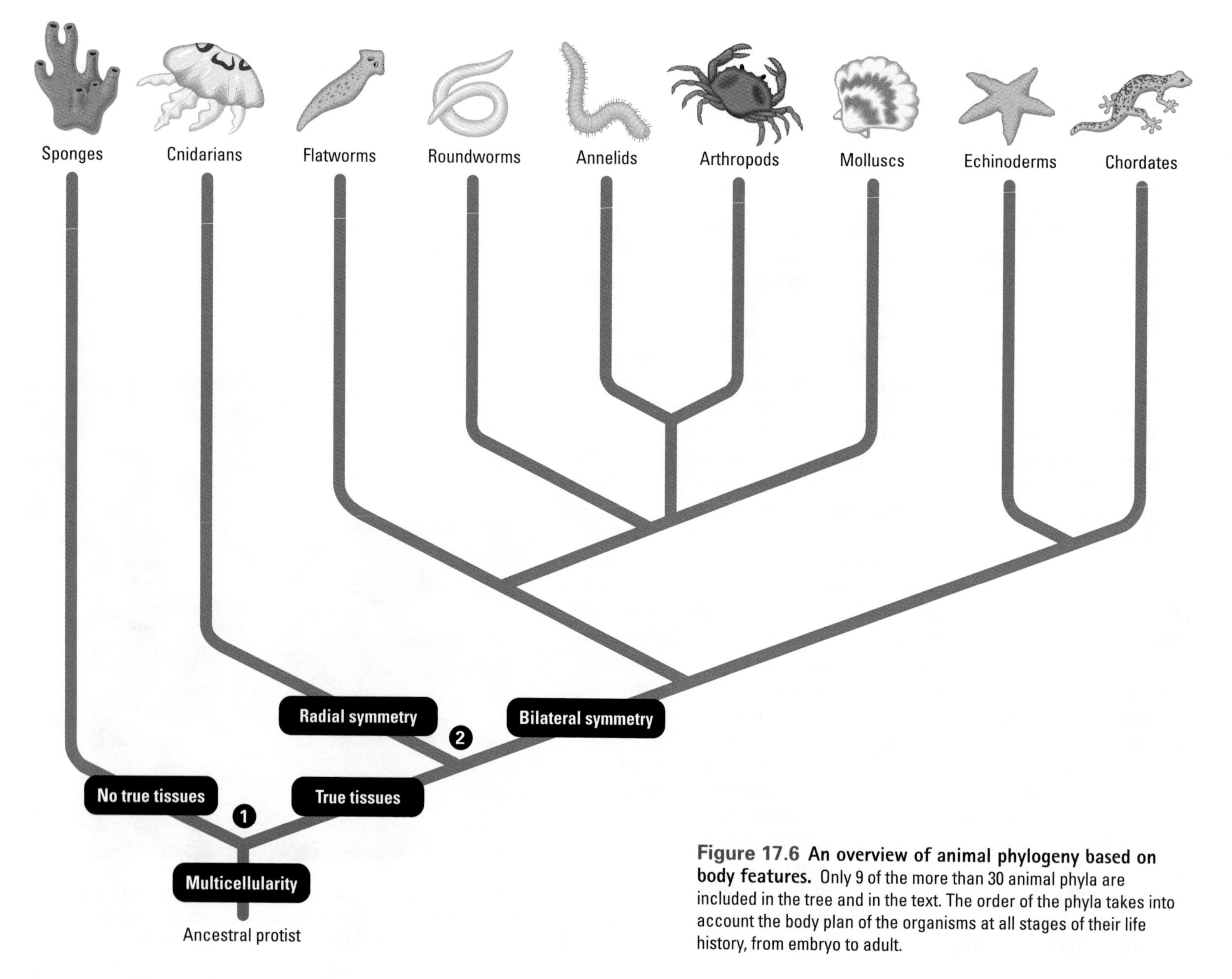

Figure 17.6 An overview of animal phylogeny based on body features. Only 9 of the more than 30 animal phyla are included in the tree and in the text. The order of the phyla takes into account the body plan of the organisms at all stages of their life history, from embryo to adult.

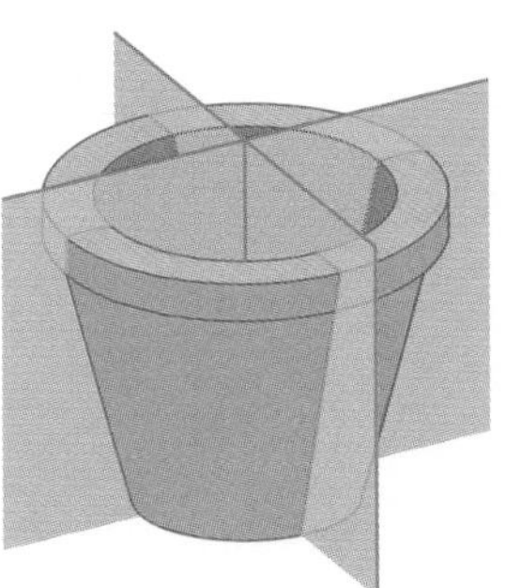

(a) Radial symmetry

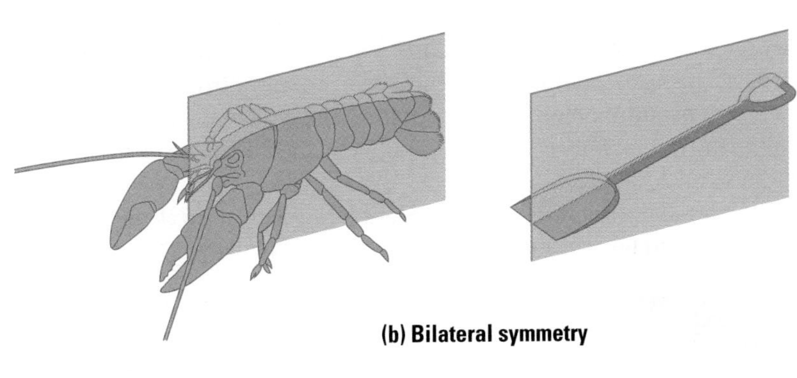

(b) Bilateral symmetry

Figure 17.7 Body symmetry. **(a)** The parts of a radial animal, such as this sea anemone, radiate from the center. Any imaginary slice through the central axis would divide the animal into mirror images. **(b)** A bilateral animal, such as this lobster, has a left and right side. Only one imaginary cut would divide the animal into mirror-image halves.

plan and embryonic development. The circled numbers on the tree highlight two important evolutionary branch points, and these numbers are keyed to the following discussion.

❶ The first branch point distinguishes sponges from all other animals based on structural complexity. Sponges, though multicellular, lack the true tissues, such as nervous tissue, that characterize more complex animals.

❷ The second major evolutionary split is based on body symmetry: radial versus bilateral. To understand this difference, imagine a pail and shovel. The pail has **radial symmetry,** identical all around a central axis. The shovel has **bilateral symmetry,** which means there's only one way to split it into two equal halves—right down the midline. **Figure 17.7** contrasts radial and bilateral symmetry.

The symmetry of an animal generally fits its lifestyle. Many radial animals are sessile forms (attached to a larger object) or drifting or weakly swimming organisms (plankton, for example). Their symmetry equips them to meet the environment equally well from all sides. Most animals that move actively from place to place are bilateral. A bilateral animal has a definite "head end" that encounters food, danger, and other stimuli first when the animal is traveling. In most bilateral animals, a nerve center in the form of a brain is at the head end, near a concentration of sense organs such as eyes. Thus, bilateral symmetry is an adaptation that aids movement, such as crawling, burrowing, or swimming.

The evolution of body cavities, which occurred multiple times in various lineages, also helped lead to more complex animals. A **body cavity** is a fluid-filled space separating the digestive tract from the outer body wall. A body cavity has many functions. Its fluid cushions internal organs, helping to prevent injury. The cavity also enables the internal organs to grow and move independently of the outer body wall. If it were not for your body cavity, exercise would be very hard on your internal organs. And every beat of your heart or ripple of your intestine would deform your body surface. It would be a scary sight. In soft-bodied animals such as earthworms, the noncompressible fluid of the body cavity is under pressure and functions as a hydrostatic skeleton against which muscles can work. In fact, body cavities may have first evolved as adaptations for burrowing. In Figure 17.6, only sponges, cnidarians, and flatworms lack a body cavity.

Among animals with a body cavity, there are differences in how the cavity develops **(Figure 17.8)**. In all cases, the cavity is at least partly lined by

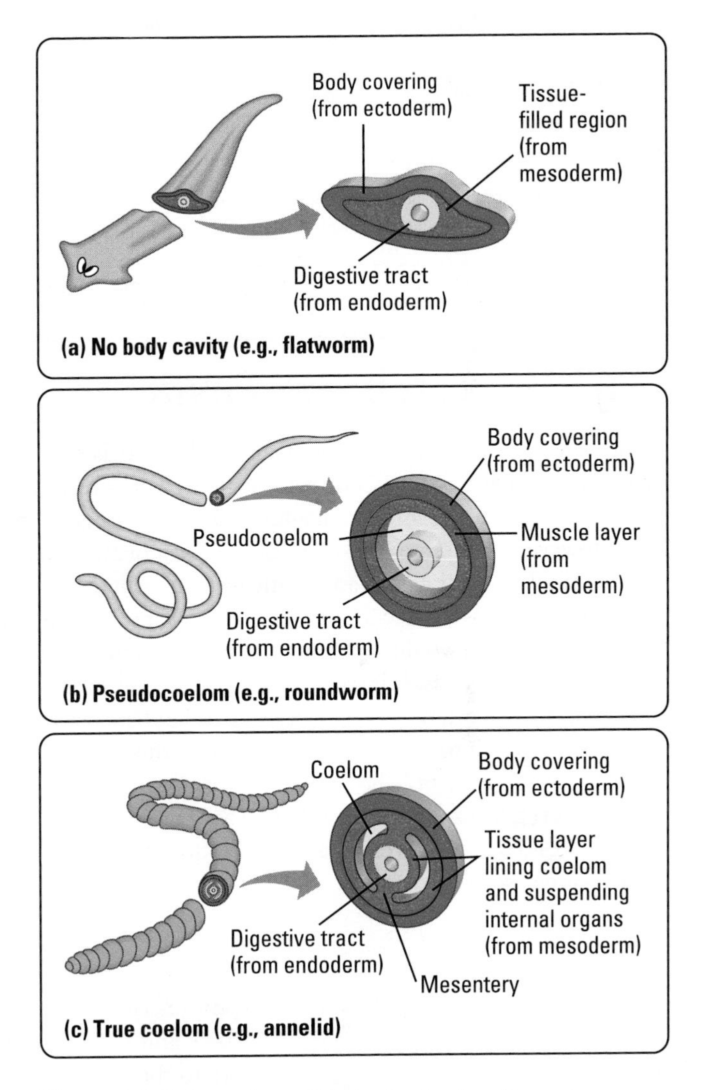

Figure 17.8 Body plans of bilateral animals. The various organ systems of these animals develop from the three tissue layers that form in the embryo. **(a)** Flatworms are examples of animals that lack a body cavity. **(b)** Roundworms have a pseudocoelom, a body cavity only partially lined by mesoderm, the middle tissue layer. **(c)** Earthworms and other annelids are examples of animals with a true coelom. A coelom is a fluid-filled body cavity, completely lined by mesoderm, within which organs are suspended.

a middle layer of tissue, called mesoderm, which develops between the inner (endoderm) and outer (ectoderm) layers of the gastrula embryo. If the body cavity is not completely lined by tissue derived from mesoderm, it is termed a **pseudocoelom.** A true **coelom,** the type of body cavity humans and many other animals have, is completely lined by tissue derived from mesoderm.

With the overview of animal evolution in Figure 17.6 as our guide, we're ready to take a closer look at some animal phyla.

CHECKPOINT

1. In terms of key body features, chordates are most like which other animal phylum?
2. What mode of nutrition distinguishes animals from fungi, both of which are heterotrophs?
3. Why is animal evolution during the early Cambrian referred to as an "explosion"?
4. A round pizza displays ___________ symmetry, while a fork displays ___________ symmetry.
5. The fully lined cavity between your outer body wall and your digestive tract is an example of a true ___________. Roundworms have a ___________, a body cavity not completely lined by tissue.

Answers: 1. Echinoderms **2.** Ingestion **3.** Because a great diversity of animals evolved in a relatively short time span **4.** radial; bilateral **5.** coelom; pseudocoelem

Major Invertebrate Phyla

Living as we do on land, our sense of animal diversity is biased in favor of vertebrates, which are animals with a backbone. Vertebrates are well represented on land in the form of such animals as amphibians, reptiles, and mammals. However, vertebrates make up less than 5% of all animal species. If we were to sample the animals in an aquatic habitat, such as a pond, tide pool, or coral reef, or if we were to consider the millions of insects that share our terrestrial world, we would find ourselves in the realm of **invertebrates,** animals without backbones. It is traditional to divide the animal kingdom into vertebrates and invertebrates, but this makes about as much zoological sense as sorting animals into flatworms and nonflatworms. We give special attention to the vertebrates only because we humans are among the backboned ones. However, by exploring the other 95% of the animal kingdom—the invertebrates—we'll discover an astonishing diversity of beautiful creatures that too often escape our notice.

Sponges

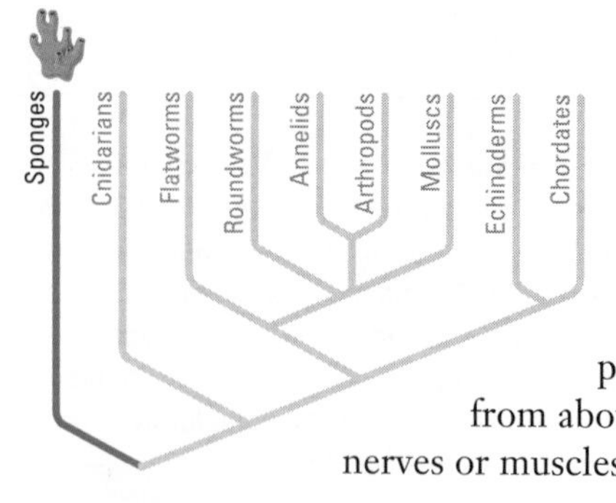

Sponges (phylum Porifera) are sessile animals that appear so sedate to the human eye that the ancient Greeks believed them to be plants **(Figure 17.9)** The simplest of all animals, sponges probably evolved very early from colonial protists. Sponges range in height from about 1 cm to 2 m. Sponges have no nerves or muscles, but the individual cells can sense

Figure 17.9 A sponge, a member of the phylum Porifera.

and react to changes in the environment. The cell layers of sponges are loose federations of cells, not really tissues, because the cells are relatively unspecialized. Of the 9,000 or so species of sponges, only about 100 live in fresh water; the rest are marine.

The body of a sponge resembles a sac perforated with holes. Water is drawn through the pores into a central cavity, then flows out of the sponge through a larger opening **(Figure 17.10)**. Most sponges feed by collecting bacteria from the water that streams through their porous bodies. Flagellated cells called choanocytes trap bacteria in mucus and then engulf the food by phagocytosis (see Chapter 4). Cells called amoebocytes pick up food from the choanocytes, digest it, and carry the nutrients to other cells. Amoebocytes are the "do-all" cells of sponges. Moving about by means of pseudopodia, they digest and distribute food, transport oxygen, and dispose of wastes. Amoebocytes also manufacture the fibers that make up a sponge's skeleton. In some sponges, these fibers are sharp and spur-like. Other sponges have softer, more flexible skeletons; these pliant, honeycombed skeletons are the "natural sponges" we may use in the bath and to wash our cars.

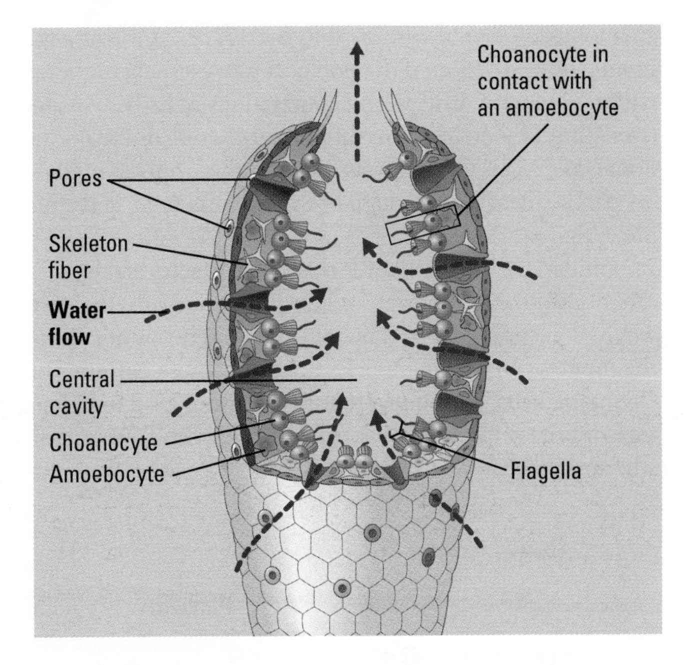

Figure 17.10 Anatomy of a sponge. Feeding cells called choanocytes have flagella that sweep water through the sponge's body. Choanocytes trap bacteria and other food particles, and amoebocytes distribute the food to other cells. To obtain enough food to grow by 100 g (about 3 ounces), a sponge must filter 1,000 kg (about 275 gallons) of seawater.

Cnidarians

Cnidarians (phylum Cnidaria) are characterized by the presence of body tissues (as are all the remaining animals we will discuss) as well as by radial symmetry and tentacles with stinging cells. Jellies, sea anemones, hydras, and coral animals are all cnidarians. Most of the 10,000 cnidarian species are marine.

The basic body plan of a cnidarian is a sac with a central digestive compartment, the **gastrovascular cavity.** A single opening to this cavity functions as both mouth and anus. This basic body plan has two variations: the sessile **polyp** and the floating **medusa** (plural, *medusae*) **(Figure 17.11)**. Polyps adhere to larger objects and extend their tentacles, waiting for prey. Examples of the polyp form are hydras, sea

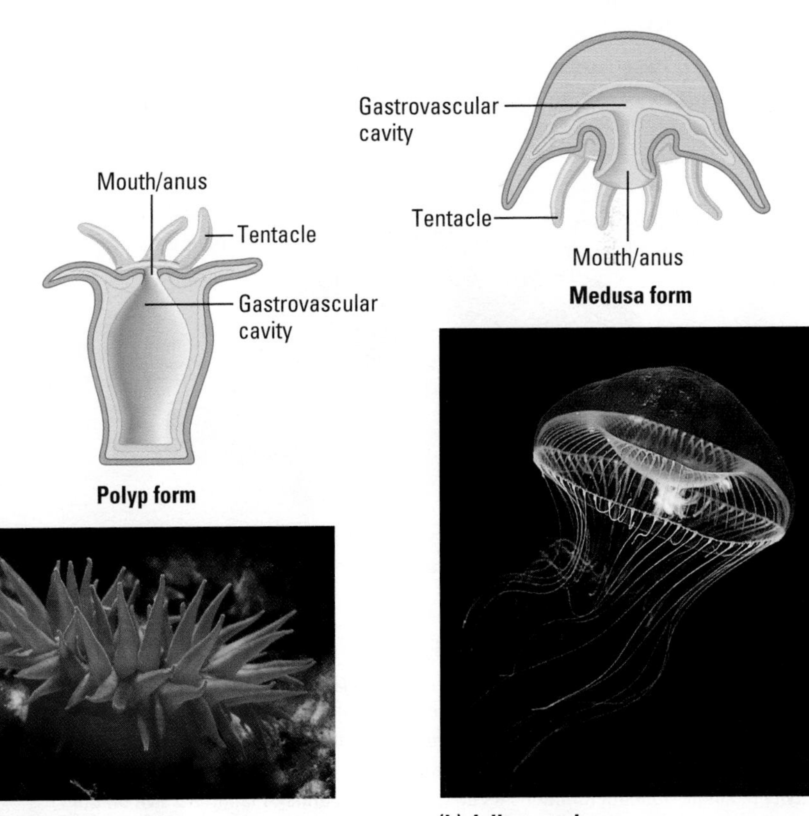

Figure 17.11 Polyp and medusa forms of cnidarians. Note that cnidarians have two tissue layers, distinguished in the diagrams by blue and yellow. The gastrovascular cavity has only one opening, which functions as both mouth and anus.

anemones, and coral animals **(Figure 17.12)**. A medusa is a flattened, mouth-down version of the polyp. It moves freely in the water by a combination of passive drifting and contractions of its bell-shaped body. Jellies (sometimes called jellyfish, though these animals are not fish) are medusae. Some cnidarians exist only as polyps, others only as medusae, and still others pass sequentially through both a medusa stage and a polyp stage in their life cycle.

Cnidarians are carnivores that use tentacles arranged in a ring around the mouth to capture prey and push the food into the gastrovascular cavity, where digestion begins. The undigested remains are eliminated through the mouth/anus. The tentacles are armed with batteries of cnidocytes ("stinging cells"), unique structures that function in defense and in the capture of prey **(Figure 17.13)**. The phylum Cnidaria is named for these stinging cells.

Figure 17.12 Coral animals. Each polyp in this colony is about 3 mm in diameter. Coral animals secrete hard external skeletons of calcium carbonate (limestone). Each polyp builds on the skeletal remains of earlier generations to construct the "rocks" we call coral. Though individual coral animals are small, their collective construction accounts for such biological wonders as Australia's Great Barrier Reef. Tropical coral reefs are home to an enormous variety of invertebrates and fishes.

Flatworms

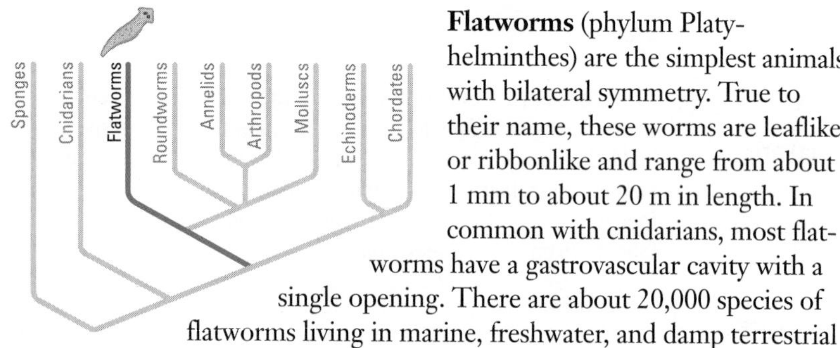

Flatworms (phylum Platyhelminthes) are the simplest animals with bilateral symmetry. True to their name, these worms are leaflike or ribbonlike and range from about 1 mm to about 20 m in length. In common with cnidarians, most flatworms have a gastrovascular cavity with a single opening. There are about 20,000 species of flatworms living in marine, freshwater, and damp terrestrial habitats. Planarians are examples of free-living (nonparasitic) flatworms **(Figure 17.14)**. The phylum also includes many parasitic species, such as flukes and tapeworms.

Parasitic flatworms called blood flukes are a major health problem in the tropics. These worms have suckers that attach to the inside of the blood vessels near the human host's intestines. Infection by flatworms

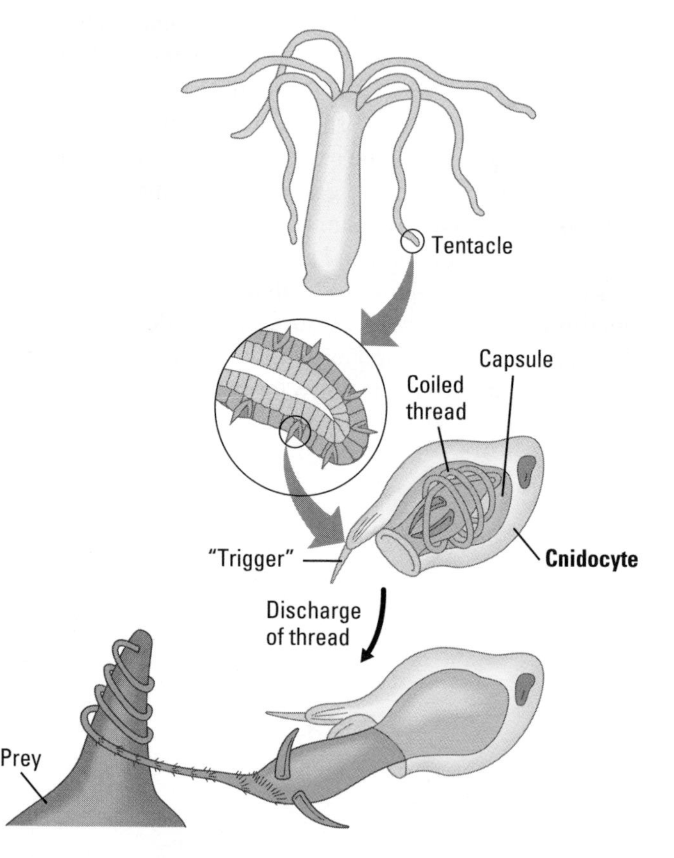

Figure 17.13 Cnidocyte action. Each cnidocyte contains a fine thread coiled within a capsule. When a trigger is stimulated by touch, the thread shoots out. Some cnidocyte threads entangle prey, while others puncture the prey and inject a poison.

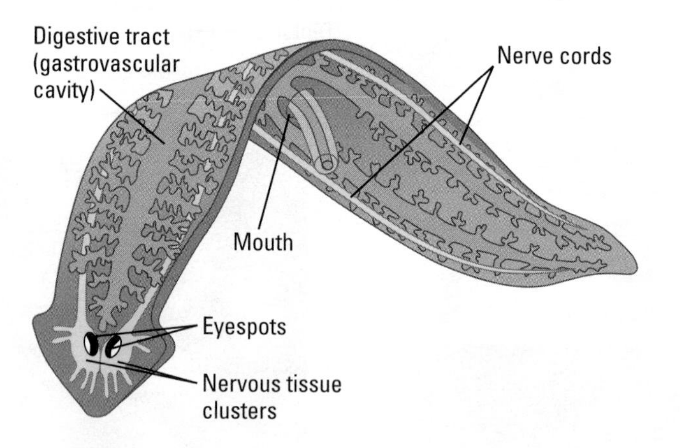

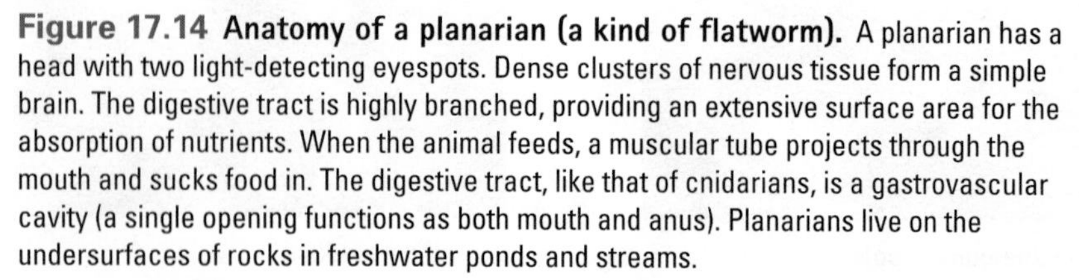

Figure 17.14 Anatomy of a planarian (a kind of flatworm). A planarian has a head with two light-detecting eyespots. Dense clusters of nervous tissue form a simple brain. The digestive tract is highly branched, providing an extensive surface area for the absorption of nutrients. When the animal feeds, a muscular tube projects through the mouth and sucks food in. The digestive tract, like that of cnidarians, is a gastrovascular cavity (a single opening functions as both mouth and anus). Planarians live on the undersurfaces of rocks in freshwater ponds and streams.

causes a long-lasting disease with such symptoms as severe abdominal pain, anemia, and dysentery. About 250 million people in 70 countries suffer from blood fluke disease.

Tapeworms parasitize many vertebrates, including humans. Most tapeworms have a very long, ribbonlike body with repeated parts **(Figure 17.15)**. They also differ from other flatworms in not having any digestive tract at all. Living in partially digested food in the intestines of their hosts, tapeworms simply absorb nutrients across their body surface. Tapeworms have a complex life cycle, usually involving more than one host. Humans can become infected with tapeworms by eating rare beef containing the worm's larvae. The larvae are microscopic, but the adults can reach lengths of 20 m in the human intestine. Such large tapeworms can cause intestinal blockage and rob enough nutrients from the human host to cause nutritional deficiencies. Fortunately, an orally administered drug can kill the adult worms.

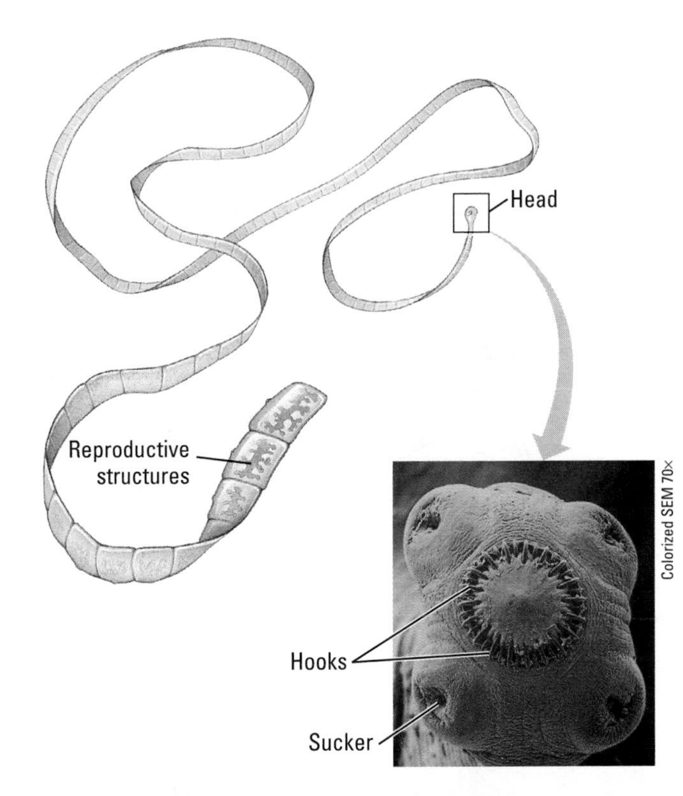

Figure 17.15 Anatomy of a tapeworm. The head of a tapeworm is armed with suckers and hooks that lock the worm to the intestinal lining of the host. Behind the head is a long ribbon of units that are little more than sacs of sex organs. At the back of the worm, mature units containing thousands of eggs break off and leave the host's body with the feces.

Roundworms

Roundworms (also called **nematodes,** members of the phylum Nematoda) get their common name from their cylindrical body, which is usually tapered at both ends **(Figure 17.16a)**. Roundworms are among the most diverse (in species number) and widespread of all animals. About 90,000 species of roundworms are known, and perhaps ten times that number actually exist. Roundworms range in length from about a millimeter to a meter. They are found in most aquatic habitats, in wet soil, and as parasites in the body fluids and tissues of plants and animals. Free-living roundworms in the soil are important decomposers. Other species are major agricultural pests that attack the roots of plants. At least 50 parasitic roundworm species infect humans, including pinworms, hookworms, and the parasite that causes trichinosis **(Figure 17.16b)**.

Roundworms exhibit two innovations not found in flatworms. First, roundworms (and all the remaining animals we will discuss) have a **complete digestive tract,** which is a digestive tube with two openings, a mouth and an anus. This anatomy contrasts with the gastrovascular cavity of cnidarians and flatworms, which uses a single opening as both mouth and anus. A complete digestive tract can process food and absorb nutrients as a meal moves in one direction from one specialized digestive organ to the next. In humans, for example, the mouth, stomach, and intestines act as digestive organs. A second evolutionary innovation we see for the first time in roundworms is a body cavity, which in this case is a pseudocoelom (that is, a body cavity not completely lined by mesoderm-derived tissue; see Figure 17.8b).

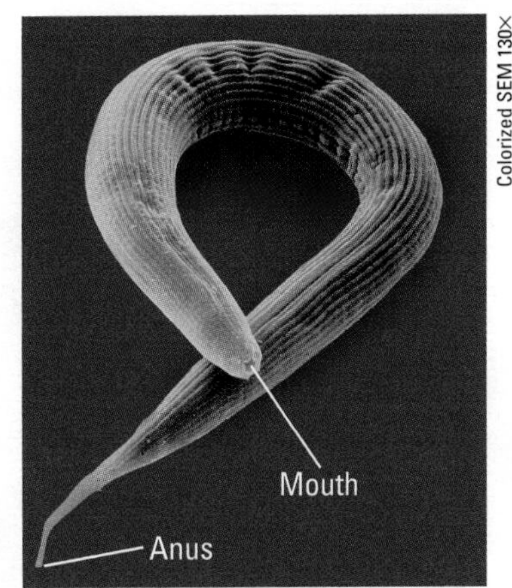

(a) A free-living roundworm. This species has the classic roundworm shape: cylindrical with tapered ends. This worm looks like it's wearing a corduroy coat, but the ridges actually indicate muscles that run the length of the body.

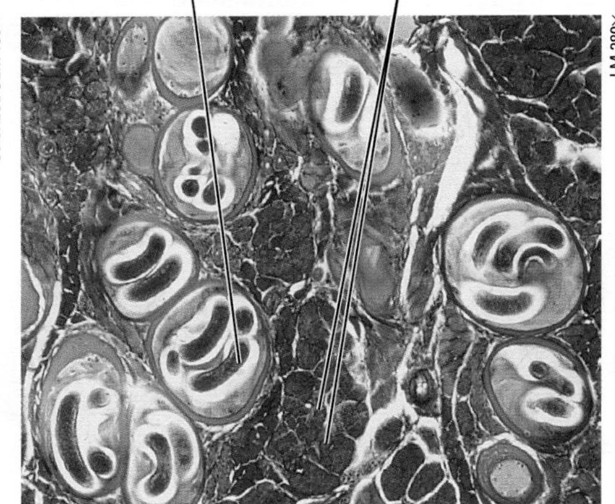

(b) Parasitic roundworms. The disease called trichinosis is caused by the *Trichinella* roundworms shown here in human muscle tissue. Humans may acquire the parasite by eating infected, undercooked pork. The worms burrow into the human intestine and eventually travel to other parts of the body, invading muscles and other organs. Symptoms of trichinosis include fatigue, diarrhea, and fever. In severe cases, trichinosis can be fatal.

Figure 17.16 Roundworms.

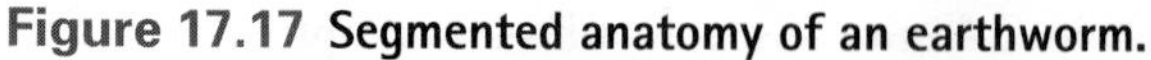

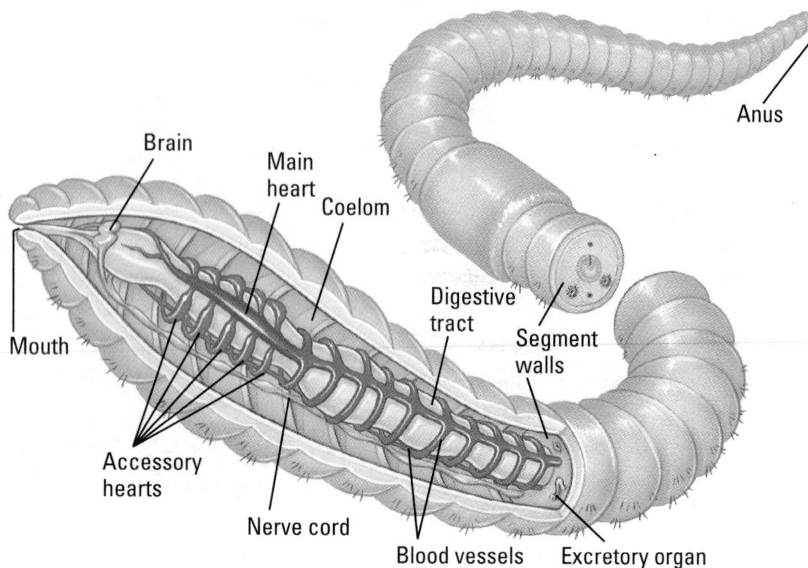

Figure 17.17 Segmented anatomy of an earthworm. Annelids are segmented both externally and internally. Many of the internal structures are repeated, segment by segment. The coelom (body cavity) is partitioned by walls (only two segment walls are fully shown here). The nervous system (yellow) includes a nerve cord with a cluster of nerve cells in each segment. Excretory organs (green), which dispose of fluid wastes, are also repeated in each segment. The digestive tract, however, is not segmented; it passes through the segment walls from the mouth to the anus. Segmental blood vessels connect continuous vessels that run along the top (dorsal location) and bottom (ventral location) of the worm. The segmental vessels include five pairs of accessory hearts. The main heart is simply an enlarged region of the dorsal blood vessel near the head end of the worm.

Annelids

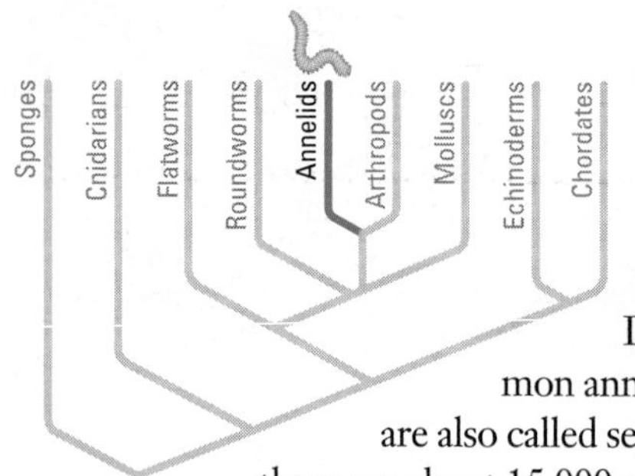

Annelids (phylum Annelida) are worms with body **segmentation,** which is the subdivision of the body along its length into a series of repeated segments. In annelids, the segments look like a set of fused rings. Look closely at an earthworm, a common annelid, and you'll see why these creatures are also called segmented worms **(Figure 17.17)**. In all, there are about 15,000 annelid species, ranging in length from less than 1 mm to the 3-m giant Australian earthworm **(Figure 17.18a)**. Annelids live in the sea, most freshwater habitats, and damp soil. There are three main groups of annelids: earthworms, polychaetes **(Figure 17.18b)**, and leeches (see Figure 17.1b).

(a) A giant Australian earthworm. These worms are bigger than most snakes. Perhaps you've slipped on slimy worms, but imagine actually *tripping* over one!

(b) Polychaetes. These worms have segmental appendages that function in movement and as gills. On the left is a sandworm. The beautiful polychaete on the right is an example of a fan worm, which lives in a tube it constructs by mixing mucus with bits of sand and broken shells. Fan worms use their feathery headdresses as gills and to extract food particles from the seawater. This species is called a Christmas tree worm.

Figure 17.18 Annelids.

Earthworms eat their way through the soil, extracting nutrients as the soil passes through the digestive tract. Undigested material, mixed with mucus secreted into the digestive tract, is eliminated as castings through the anus. Farmers and gardeners value earthworms because the animals till the earth, and the castings improve the texture of the soil. Charles Darwin estimated that each acre of British farmland contained about 50,000 earthworms that produced 18 tons of castings per year.

In contrast to earthworms, most polychaetes are marine, mainly crawling on or burrowing in the seafloor. Segmental appendages with hard bristles help the worm wriggle about in search of small invertebrates to eat. The appendages also increase the animal's surface area for taking up oxygen and disposing of metabolic wastes, including carbon dioxide.

The third group of annelids, leeches, are notorious for the bloodsucking habits of some species (as described in the chapter-opening essay). However, most species are free-living carnivores that eat small invertebrates such as snails and insects. The majority of leeches live in fresh water, but a few terrestrial species inhabit moist vegetation in the tropics.

Arthropods

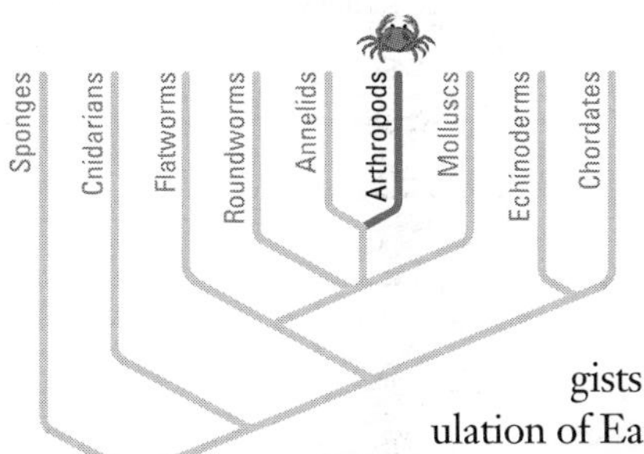

Arthropods (phylum Arthropoda) are named for their jointed appendages. Crustaceans (such as crabs and lobsters), arachnids (such as spiders and scorpions), and insects (such as grasshoppers and moths) are all examples of arthropods. Zoologists estimate that the total arthropod population of Earth numbers about a billion billion (10^{18}) individuals (that's about 150 million arthropods for each person!). Researchers have identified over a million arthropod species, mostly insects. In fact, two out of every three species of life that have been scientifically described are arthropods. And arthropods are represented in nearly all habitats of the biosphere. On the criteria of species diversity, distribution, and sheer numbers, arthropods must be regarded as the most successful animal phylum.

General Characteristics of Arthropods Arthropods are segmented animals. In contrast to the repeating similar segments of annelids, however, arthropod segments and their appendages have become specialized for a great variety of functions. This evolutionary flexibility contributed to the great diversification of arthropods. Specialization of segments (or of fused groups of segments) provides for an efficient division of labor among body regions. For example, the appendages of different segments are variously adapted for walking, feeding, sensory reception, swimming, and defense **(Figure 17.19)**.

The body of an arthropod is completely covered by an **exoskeleton** (external skeleton). This coat is constructed from layers of protein and a polysaccharide called chitin. The exoskeleton can be a thick, hard armor over some parts of the body (such as the head), yet paper-thin and flexible in other locations (such as the joints). The exoskeleton protects the animal and provides points of

Figure 17.19 Anatomy of a lobster, an arthropod. The whole body, including the appendages, is covered by an exoskeleton. The two distinct regions of the body are the cephalothorax (consisting of the head and thorax) and the abdomen. The head bears a pair of eyes, each situated on a movable stalk. The body is segmented, but this characteristic is obvious only in the abdomen. The animal has a tool kit of specialized appendages, including pincers, walking legs, swimming appendages, and two pairs of sensory antennae. Even the multiple mouthparts are modified legs, which is why they work from side to side rather than up and down (as our jaws do).

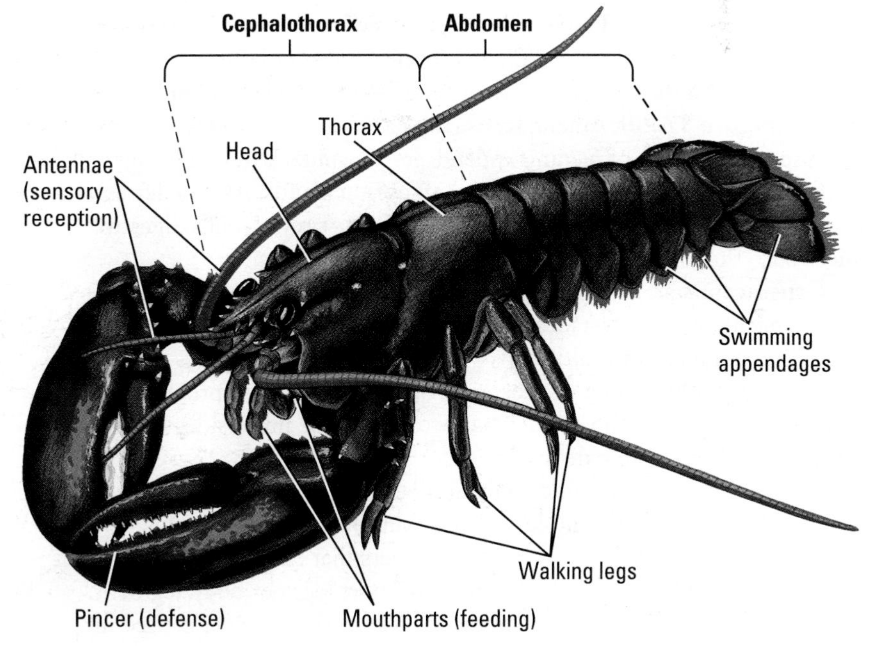

(a) A scorpion (about 8 cm long). Scorpions are nocturnal hunters. Their ancestors were among the first terrestrial carnivores, preying on arthropods that fed on the early land plants. Scorpions have a pair of appendages modified as large pincers that function in defense and food capture. The tip of the tail bears a poisonous stinger. Scorpions eat mainly insects and spiders. They will sting people only when prodded or stepped on.

(b) A black widow spider (about 3 cm long). Spiders are usually most active during the daytime, hunting insects or trapping them in webs. Spiders spin their webs of liquid silk, which solidifies as it comes out of specialized glands. Each spider engineers a style of web that is characteristic of its species. Spiders also use silk as droplines for rapid escape; as a covering for their eggs; and even as "gift wrapping" for food that certain male spiders offer to potential mates.

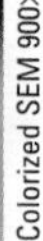

Colorized SEM 900×

(c) A microscopic dust mite. This magnified house dust mite is a ubiquitous scavenger in our homes. Each square inch of carpet and every one of those dust balls under a bed is like a city of thousands of dust mites. Unlike some mites that carry pathogenic bacteria, dust mites are harmless except to people who are allergic to the mites' feces.

Figure 17.20 **Arachnids.**

attachment for the muscles that move the appendages. There are, of course, advantages to wearing hard parts on the outside. Our own skeleton is interior to most of our soft tissues, an arrangement that doesn't provide much protection from injury. But our skeleton does offer the advantage of being able to grow along with the rest of our body. In contrast, a growing arthropod must occasionally shed its old exoskeleton and secrete a larger one. This process, called **molting,** leaves the animal temporarily vulnerable to predators and other dangers.

Arthropod Diversity The four main groups of arthropods are the arachnids, the crustaceans, the millipedes and centipedes, and the insects.

Most **arachnids** live on land. Scorpions, spiders, ticks, and mites are examples **(Figure 17.20)**. Arachnids usually have four pairs of walking legs and a specialized pair of feeding appendages. In spiders, these feeding appendages are fang-like and equipped with poison glands. As a spider uses these appendages to immobilize and dismantle its prey, it spills digestive juices onto the torn tissues and sucks up its liquid meal.

Crustaceans are nearly all aquatic. Crabs, lobsters, crayfish, shrimps, and barnacles are all crustaceans **(Figure 17.21)**. They all exhibit the arthropod hallmark of multiple pairs of specialized appendages (see Figure 17.19). One group of crustaceans, the isopods, is represented on land by pill bugs, which you may have found on the undersides of moist leaves and other organic debris.

Millipedes and **centipedes** have similar segments over most of the body and superficially resemble annelids, but their jointed legs give them away as arthropods. Millipedes are landlubbers that eat decaying plant matter **(Figure 17.22)**. They have two pairs of short legs per body segment. Centipedes are terrestrial carnivores, with a pair of poison claws used in

(a) A grass shrimp

(b) Barnacles

Figure 17.21 Crustaceans. **(a)** Grass shrimp, also known as popcorn shrimp, are translucent crustaceans that live in shallow marine waters, where they scavenge on algae. **(b)** Easily confused with bivalve molluscs, barnacles are actually sessile crustaceans with exoskeletons hardened into shells by calcium carbonate (lime). The jointed appendages projecting from the shell capture small plankton.

defense and to paralyze prey, such as cockroaches and flies. Each of their body segments bears a single pair of long legs.

In species diversity, **insects** outnumber all other forms of life combined. They live in almost every terrestrial habitat and in fresh water, and flying insects fill the air. Insects are rare, though not absent, in the seas, where crustaceans are the dominant arthropods. **Entomology** is the branch of biology that specializes in the study of insects.

The oldest insect fossils date back to about 400 million years ago, during the Paleozoic era (see Table 14.1). Later, the evolution of flight sparked an explosion in insect variety **(Figure 17.23)**.

Figure 17.22 A millipede.

(a) Black-banded mantis

(b) Bee hawkmoth

(c) Longhorn beetle

Figure 17.23 A small sample of insect diversity.

Like the grasshopper in **Figure 17.24**, most insects have a three-part body: head, thorax, and abdomen. The head usually bears a pair of sensory antennae and a pair of eyes. Several pairs of mouthparts are adapted for particular kinds of eating—for example, for biting and chewing plant material in grasshoppers; for lapping up fluids in houseflies; and for piercing skin and sucking blood in mosquitoes. Most adult insects have three pairs of legs and one or two pairs of wings, all borne on the thorax.

Flight is obviously one key to the great success of insects. An animal that can fly can escape many predators, find food and mates, and disperse to new habitats much faster than an animal that must crawl about on the ground. Because their wings are extensions of the exoskeleton and not true appendages, insects can fly without sacrificing any walking legs. By contrast, the flying vertebrates—birds and bats—have one of their two pairs of walking legs modified for wings, which explains why these vertebrates are generally not very swift on the ground.

Many insects undergo metamorphosis in their development. In the case of grasshoppers and some other insect groups, the young resemble adults but are smaller and have different body proportions. The animal goes through a series of molts, each time looking more like an adult, until it reaches full size. In other cases, insects have distinctive larval stages specialized for eating and growing that are known by such names as maggots (fly larvae) or caterpillars (larvae of moths and butterflies). The larval stage looks entirely different from the adult stage, which is specialized for dispersal and reproduction. Metamorphosis from the larva to the adult occurs during a pupal stage **(Figure 17.25)**.

Animals so numerous, diverse, and widespread as insects are bound to affect the lives of all other terrestrial organisms, including humans. On one hand, we depend on bees, flies, and many other insects to pollinate our crops and orchards. On the other hand, insects are carriers of the microorganisms that cause many diseases, including malaria and West Nile virus. Insects also compete with humans for food. In parts of Africa, for instance, insects claim about 75% of the crops. Trying to minimize their losses, farmers in the United States spend billions of dollars each year on pesticides, spraying crops with massive doses of some of the deadliest poisons ever invented. Try as they may, not even humans have significantly challenged the preeminence of insects and their arthropod kin. As Cornell University's Thomas Eisner puts it: "Bugs are not going to inherit the Earth. They own it now. So we might as well make peace with the landlord."

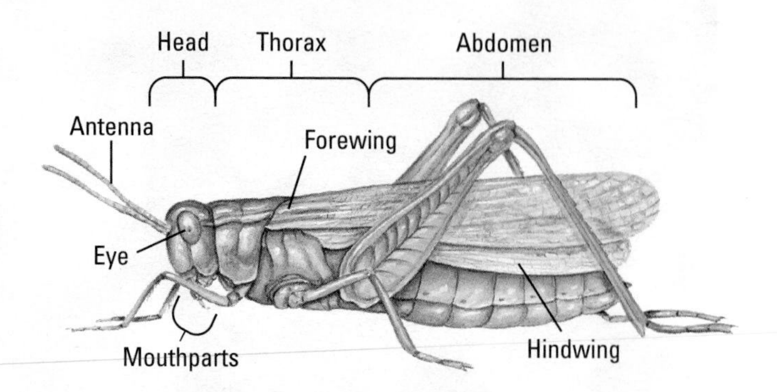

Figure 17.24 Anatomy of a grasshopper.

(a) Larva (caterpillar)

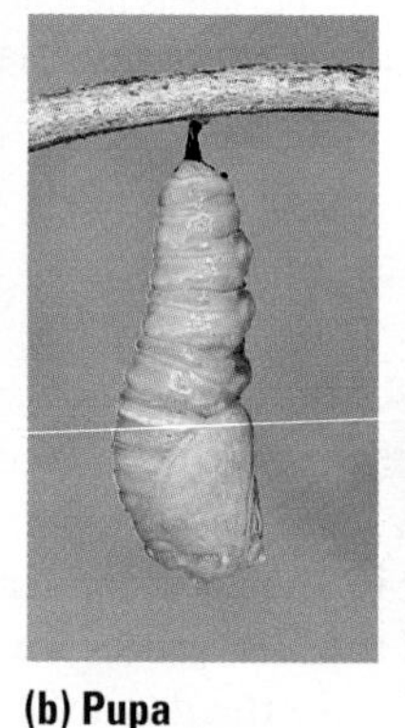
(b) Pupa

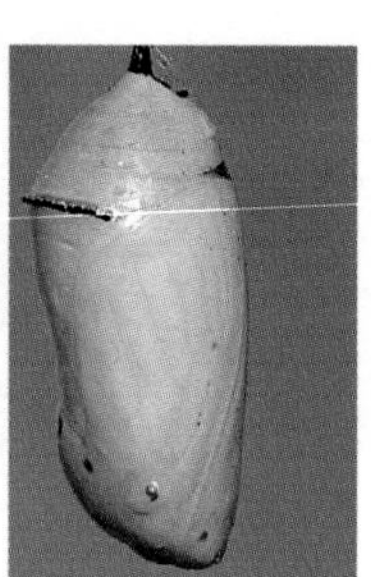
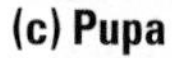
(c) Pupa

(d) Emerging adult

(e) Adult

Figure 17.25 Metamorphosis of a monarch butterfly. **(a)** The larva (caterpillar) spends its time eating and growing, molting as it grows. **(b)** After several molts, the larva encases itself in a cocoon and becomes a pupa. **(c)** Within the pupa, the larval organs break down and adult organs develop from cells that were dormant in the larva. **(d)** Finally, the adult emerges from the cocoon. **(e)** The butterfly flies off and reproduces, nourished mainly from the calories it stored when it was a caterpillar.

Molluscs

Snails and slugs, oysters and clams, and octopuses and squids are all **molluscs** (phylum Mollusca). Molluscs are soft-bodied animals, but most are protected by a hard shell. Slugs, squids, and octopuses have reduced shells, most of which are internal, or they have lost their shells completely during their evolution. Many molluscs feed by using a straplike rasping organ called a **radula** to scrape up food. Garden snails use their radulas like tiny saws to cut pieces out of leaves. Most of the 150,000 known species of molluscs are marine, though some inhabit fresh water, and there are land-dwelling molluscs in the form of snails and slugs.

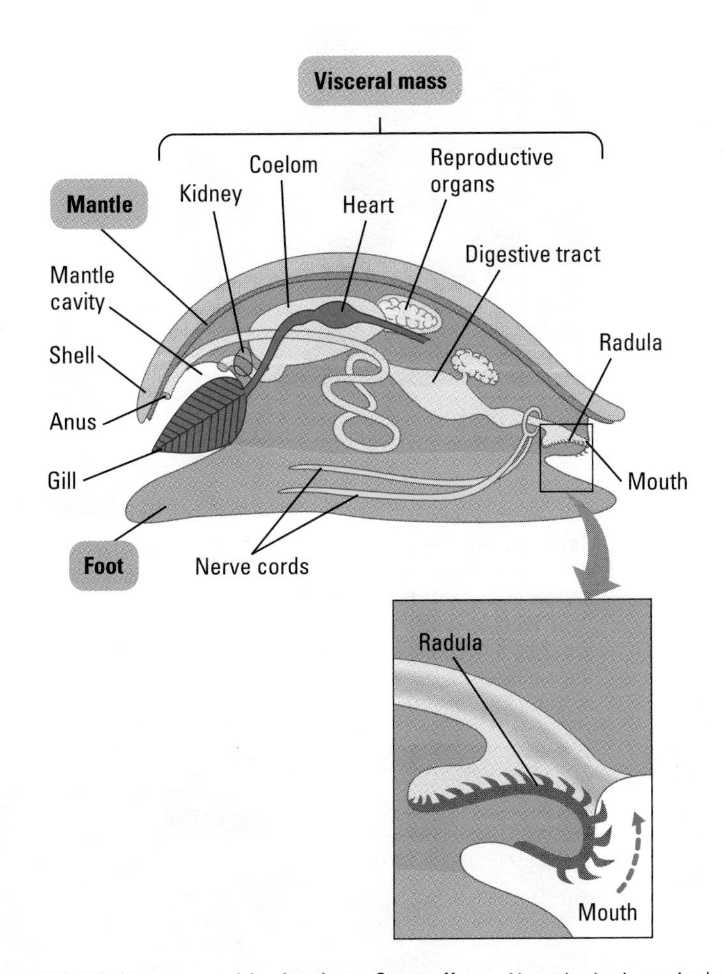

Figure 17.26 The general body plan of a mollusc. Note the body cavity (a true coelom, though a small one) and the complete digestive tract, with both mouth and anus.

Despite their apparent differences, all molluscs have a similar body plan **(Figure 17.26)**. The body has three main parts: a muscular foot, usually used for movement; a visceral mass containing most of the internal organs; and a fold of tissue called the mantle. The **mantle** drapes over the visceral mass and secretes the shell (if one is present).

The three major groups of molluscs are gastropods (including snails and slugs), bivalves (including clams and oysters), and cephalopods (including squids and octopuses). Most **gastropods** are protected by a single spiraled shell into which the animal can retreat when threatened **(Figure 17.27a)**. Many gastropods have a distinct head with eyes at the tips of tentacles (think of a garden snail). **Bivalves,** including numerous species of clams, oysters, mussels, and scallops, have shells divided into two halves hinged together **(Figure 17.27b)**. Most bivalves are sedentary, living in sand or mud in marine and freshwater environments. They use their muscular foot for digging and anchoring. **Cephalopods** generally differ from gastropods and sedentary bivalves in being built for speed and agility. A few cephalopods have large, heavy shells, but in most the shell is small and internal (as in squids) or missing altogether (as in octopuses). Cephalopods have large brains and sophisticated sense organs, which contribute to the success of these animals as mobile predators. Cephalopods use beak-like jaws and a radula to crush or rip prey apart. The cephalopod mouth is at the base of the foot, which is drawn out into several long tentacles for catching and holding prey **(Figure 17.27c)**.

(a) Gastropod shells. Shell collectors are delighted by the variety of gastropods.

(b) A bivalve. This scallop has many eyes (small round structures) peering out between the two halves of the hinged shell.

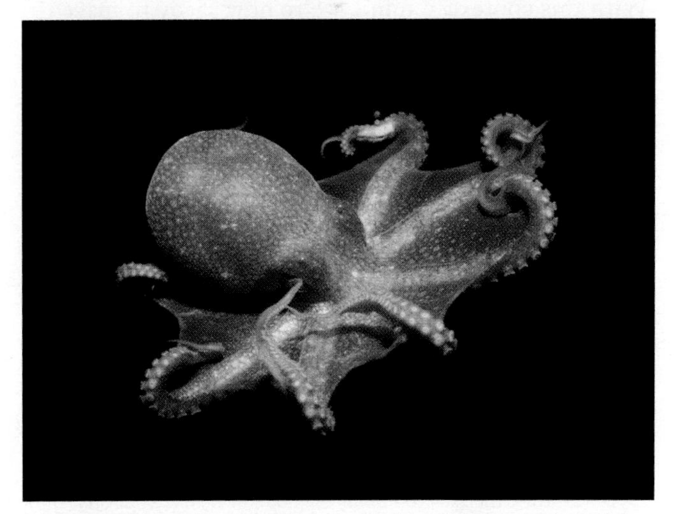

(c) A cephalopod. An octopus is a cephalopod without a shell. This octopus lives on the seafloor, where it scurries about in search of crabs and other food. Its brain is larger and more complex, proportionate to body size, than that of any other invertebrate.

Figure 17.27 Molluscs.

Echinoderms

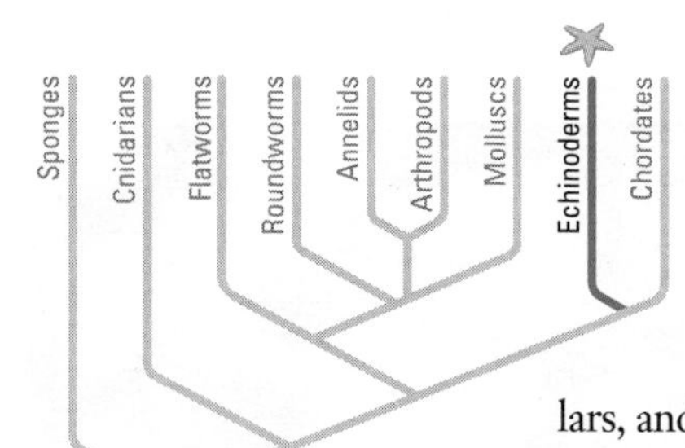

The **echinoderms** (phylum Echinodermata) are named for their spiny surfaces (*echin* is Greek for "spiny"). Sea urchins, the porcupines of the invertebrates, certainly live up to the phylum name. Among the other echinoderms are sea stars, sand dollars, and sea cucumbers **(Figure 17.28)**.

Echinoderms are all marine. Most are sessile or slow moving. Echinoderms lack body segments, and most have radial symmetry as adults. Both the external and the internal parts of a sea star, for instance, radiate from the center like spokes of a wheel. In contrast to the adult, the larval stage of echinoderms is bilaterally symmetrical. This supports other evidence that echinoderms are not closely related to other radial animals, such as cnidarians, that never show bilateral symmetry. Most echinoderms have an **endoskeleton** (interior skeleton) constructed from hard plates just beneath the skin. Bumps and spines of this endoskeleton account for the animal's rough or prickly surface. Unique to echinoderms is the **water vascular system,** a network of water-filled canals that circulate water throughout the echinoderm's body, facilitating gas exchange (the entry of O_2 and the removal of CO_2) and waste disposal. The water vascular system also branches into extensions called tube feet. A sea star or sea urchin pulls itself slowly over the seafloor using its suction-cup-like tube feet. Sea stars also use their tube feet to grip prey during feeding (see Figure 17.28a).

Looking at sea stars and other adult echinoderms, you may think they have little in common with humans and other vertebrates. But if you return to Figure 17.6, you'll see that echinoderms share an evolutionary branch with chordates, the phylum that includes vertebrates. Analysis of embryonic development reveals this relationship. The mechanism of coelom formation and many other details of embryology differentiate the echinoderms and chordates from the evolutionary branch that includes roundworms, molluscs, annelids, and arthropods. With this context in mind, we're now ready to make the transition from invertebrates to vertebrates.

CHECKPOINT

1. How does the digestive tract of a jelly differ from that of a roundworm?
2. In what fundamental way does the structure of a sponge differ from that of all other animals?
3. A sea anemone is a member of the phylum __________ while a blood fluke is a member of the phylum __________.
4. Which major arthropod group is mainly aquatic?
5. Classify the following molluscs: a garden snail is an example of a __________; a clam is an example of a __________; and a squid is an example of a __________.
6. Contrast the skeleton of an echinoderm with that of an arthropod.

Answers: 1. The digestive tract of a jelly is a sac with one opening; the roundworm tract is a tube with a separate mouth and anus. **2.** A sponge has no true tissues. **3.** Cnidaria; Platyhelminthes **4.** Crustaceans. **5.** gastropod; bivalve; cephalopod **6.** An echinoderm has an endoskeleton; an arthropod has an exoskeleton.

(a) A sea star. The mouth of a sea star, not visible here, is located in the center of the undersurface. When a sea star encounters an oyster or clam, its favorite foods, it grips the mollusc's shell with its tube feet (see inset) and positions its mouth next to the narrow opening between the two halves of the prey's shell. The sea star then pushes its stomach out through its mouth and through the crack in the mollusc's shell.

(b) A sea urchin. In contrast to sea stars, sea urchins are spherical and have no arms. If you look closely, you can see the long tube feet projecting among the spines. Unlike sea stars, which are mostly carnivorous, sea urchins mainly graze on seaweed and other algae.

(c) A sea cucumber. On casual inspection, sea cucumbers do not look much like other echinoderms. Sea cucumbers lack spines, and the hard endoskeleton is much reduced. However, closer inspection reveals many echinoderm traits, including five rows of tube feet.

Figure 17.28 Echinoderms.

The Vertebrate Genealogy

Most of us are curious about our genealogy. On the personal level, we wonder about our family ancestry. As biology students, we are interested in tracing human ancestry within the broader scope of the entire animal kingdom. In this quest, we ask three questions: What were our ancestors like? How are we related to other animals? and What are our closest relatives?

In this section, we trace the evolution of the vertebrates, the group that includes humans and their closest relatives. Mammals, reptiles, amphibians, and fishes are all classified as vertebrates. Among the unique vertebrate features are the cranium (skull) and the backbone, a series of bones called vertebrae, for which the group is named **(Figure 17.29)**. Our first step in tracing the vertebrate genealogy is to determine where vertebrates fit in the animal kingdom.

Figure 17.29 Backbone, extra long. Vertebrates are named for their backbone, which consists of a series of bones called vertebrae. The vertebrate hallmark is quite apparent in this snake skeleton. You can also see the skull, the bony case protecting the brain. The backbone and skull are parts of an endoskeleton, a skeleton inside the animal rather than covering it.

Characteristics of Chordates

The last phylum in our survey of the animal kingdom is Chordata. **Chordates** all share four key features that appear in the embryo and sometimes in the adult. These four chordate hallmarks are (1) a **dorsal, hollow nerve cord;** (2) a **notochord,** which is a flexible, longitudinal rod located between the digestive tract and the nerve cord; (3) **pharyngeal slits,** which are gill structures in the pharynx, the region of the digestive tube just behind the mouth; and (4) a **post-anal tail,** which is a tail to the rear of the anus **(Figure 17.30)**. Though these chordate characteristics are often difficult to recognize in the adult animal, they are always present in chordate embryos. For example, the notochord, for which our phylum is named, persists in adult humans only in the form of the cartilage disks that function as cushions between the vertebrae. (Back injuries described as "ruptured disks" or "slipped disks" refer to these structures.)

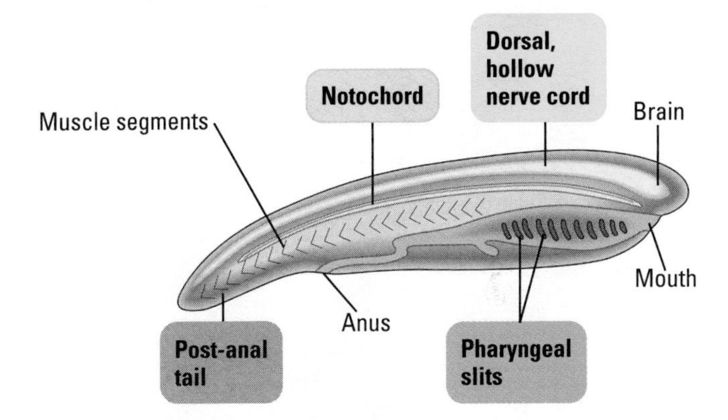

Figure 17.30 Chordate characteristics.

Body segmentation is another chordate characteristic, though not a unique one. The chordate version of segmentation probably evolved independently of the segmentation we observe in annelids and arthropods. Chordate segmentation is apparent in the backbone of vertebrates (see Figure 17.29) and in the segmental muscles of all chordates (see the chevron-shaped—>>>>—muscles in the lancelet of Figure 17.31a). Segmental musculature is not so obvious in adult humans unless one is motivated enough to sculpt those "washboard abs."

The phylum Chordata is divided into three subphyla. The first two subphyla contain **lancelets** and **tunicates,** invertebrate chordates **(Figure 17.31)**. The third subphylum, **vertebrates,** retains the basic chordate characteristics but

(a) A lancelet. This marine invertebrate owes its name to its bladelike shape. Only a few centimeters long, lancelets wiggle backward into the sand, leaving their head exposed, and filter tiny food particles from the seawater.

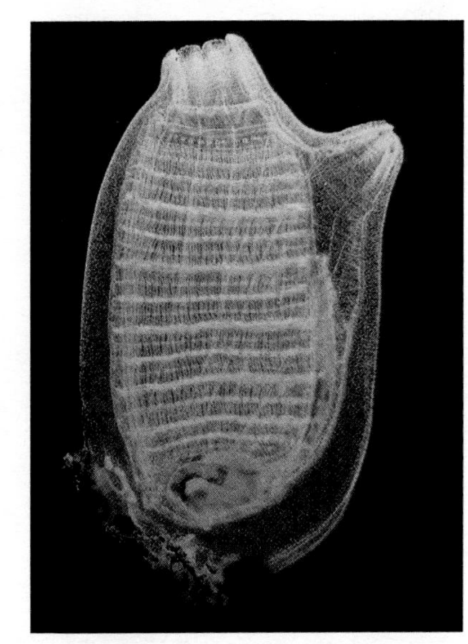

(b) A tunicate. This 3-cm-high adult tunicate, or sea squirt, is a sessile filter feeder that bears little resemblance to other chordates. However, a tunicate goes through a larval stage that is unmistakably chordate.

Figure 17.31 Invertebrate chordates.

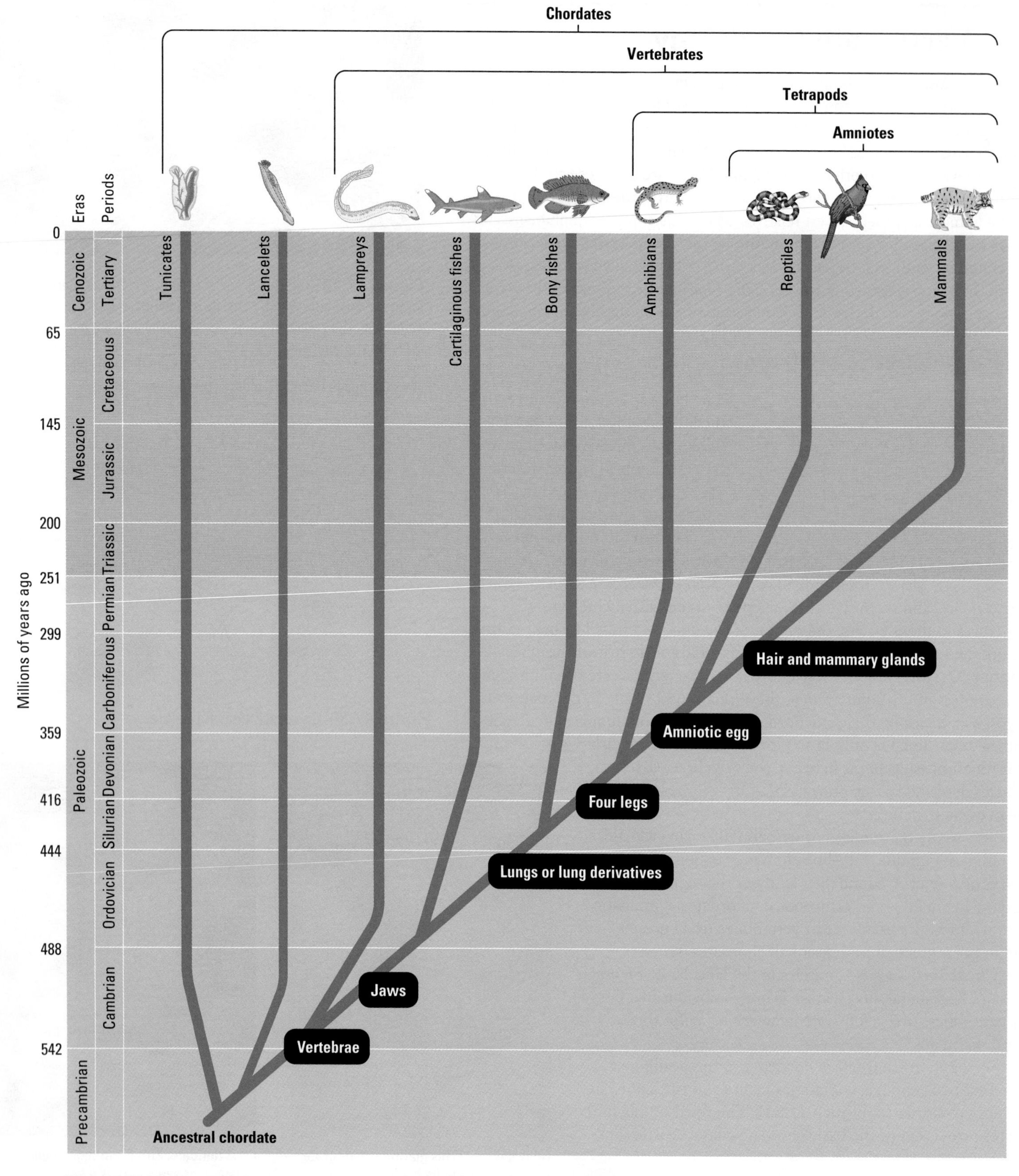

Figure 17.32 The vertebrate genealogy. "Tetrapods" refers to the four legs of terrestrial vertebrates. "Amniotes" refers to the evolution of the amniotic egg, a shelled egg that made it possible for vertebrates to reproduce on land (a chicken egg is one example).

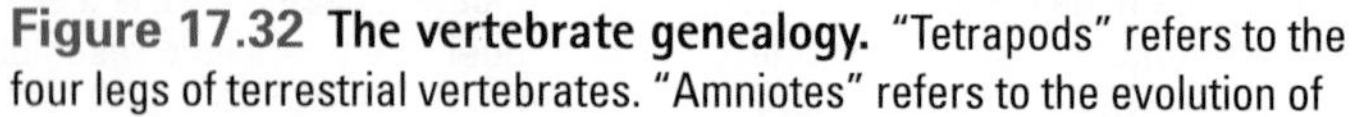

has additional features that are unique, including, of course, the backbone (see Figure 17.29). **Figure 17.32** is an overview of chordate and vertebrate evolution that will provide a context for our survey of the vertebrates.

Fishes

The first vertebrates were aquatic and probably evolved during the early Cambrian period about 542 million years ago. These early vertebrates lacked jaws. The jawless vertebrates are represented today by the fishes called lampreys **(Figure 17.33a)**. Some lampreys are parasites that use their jawless mouths as suckers to attach to the sides of large fishes and draw blood. In contrast, most vertebrates—including the other fishes, amphibians, reptiles, and mammals—have jaws, which are hinged skeletons that work the mouth. We know from the fossil record that the first jawed vertebrates were fishes that evolved by about 444 million years ago. In addition to jaws, these fishes had two pairs of fins, which made them maneuverable swimmers. Some of those fishes were more than 10 m long. Some early fishes were active predators that could chase prey and bite off chunks of flesh. Even today, most fishes are carnivores. The two major groups of living fishes are the **chondrichthyans** (cartilaginous fishes—the sharks and rays) and the **osteichthyans** (bony fishes—including such familiar groups as tuna, trout, and goldfish).

Cartilaginous fishes have a flexible skeleton made of cartilage. Most sharks are adept predators—fast swimmers with streamlined bodies, acute senses, and powerful jaws **(Figure 17.33b)**. A shark does not have keen eyesight, but its sense of smell is very sharp. In addition, special electrosensors on the head can detect minute electrical fields produced by muscle contractions in nearby animals. Sharks also have a **lateral line system,** a row of sensory organs running along each side of the body. Sensitive to changes in water pressure, the lateral line system enables a shark to detect minor vibrations caused by animals swimming in its neighborhood. There are fewer than 1,000 living species of cartilaginous fishes, nearly all of them marine.

Bony fishes (Figure 17.33c) have skeletons that are reinforced by calcium. They also have a lateral line system, a keen sense of smell, and excellent eyesight. On each side of the head, a protective flap called the **operculum** (plural, *opercula*) covers a chamber housing the gills. Movement of the operculum allows the fish to breathe without swimming. By contrast, sharks lack opercula and must swim to pass water over their gills—hence the idea that a shark must keep moving to stay alive. Also unlike sharks, bony fishes have a specialized organ that helps keep them buoyant—the **swim bladder,** a gas-filled sac. Thus, many bony fishes can conserve energy by remaining almost motionless, in contrast to sharks, which sink to the bottom if they stop swimming. Some bony fishes have a connection between the swim bladder and the digestive tract that enables

(a) A sea lamprey (the inset shows its rasping mouth)

(b) A cartilaginous fish (a sand bar shark)

(c) A bony fish (a yellow perch)

Figure 17.33 Three kinds of fishes.

them to gulp air and extract oxygen from it when the dissolved oxygen level in the water gets too low.

Bony fishes are the largest group of vertebrates, with about 30,000 species. They are common in the seas and in freshwater habitats. Most bony fishes, including trout, bass, perch, and tuna, are **ray-finned fishes.** Their fins are supported by thin, flexible skeletal rays (see Figure 17.33c). A second evolutionary branch of bony fishes includes the **lobe-finned fishes.** The lobe-fins are named for their muscular fins supported by stout bones. One lineage of lobe-finned fish, **lungfishes,** lives today in the Southern Hemisphere. They inhabit stagnant ponds and swamps, surfacing to gulp air into their lungs. A second lineage of lobe-finned fish is represented today by the coelacanth, a deep-sea dweller that was believed to have been extinct for millions of years before being "rediscovered" in the 20th century. Coelacanths may use their fins to waddle along the seafloor. The third lineage of lobe-finned fish migrated out of fresh water and adapted to life on land, playing a key role in the evolution of amphibians, the first terrestrial vertebrates.

Amphibians

In Greek, the word *amphibios* means "living a double life." Most **amphibians** exhibit a mixture of aquatic and terrestrial adaptations. Most species are tied to water because their eggs, lacking shells, dry out quickly in the air. The frog in **Figure 17.34** spends much of its time on land, but it lays its eggs in water. An egg develops into a larva called a tadpole, a legless, aquatic algae-eater with gills, a lateral line system resembling that of fishes, and a long finned tail. In changing into a frog, the tadpole undergoes a radical metamorphosis. When a young frog crawls onto shore and begins life as a terrestrial insect-eater, it has four legs, air-breathing lungs instead of gills, a pair of external eardrums, and no lateral line system. Because of metamorphosis, many amphibians truly live a double life. But even as adults, amphibians are most abundant in damp habitats, such as swamps and rain forests. This is partly because amphibians depend on their moist skin to supplement lung function in exchanging gases with the environment. Thus, even those frogs that are adapted to relatively dry habitats spend much of their time in humid burrows or under piles of moist leaves. The amphibians of today, including frogs and salamanders, account for only about 8% of all living vertebrates, or about 4,000 species.

(a) Tadpole

(b) Tadpole undergoing metamorphosis

(c) Adult frog

Figure 17.34 The "dual life" of an amphibian. This group of vertebrates is named for the fact that most of its members have adaptations for living both on land and in the water.

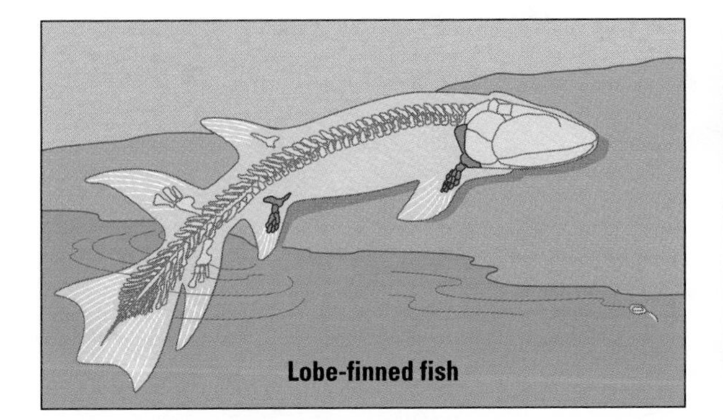

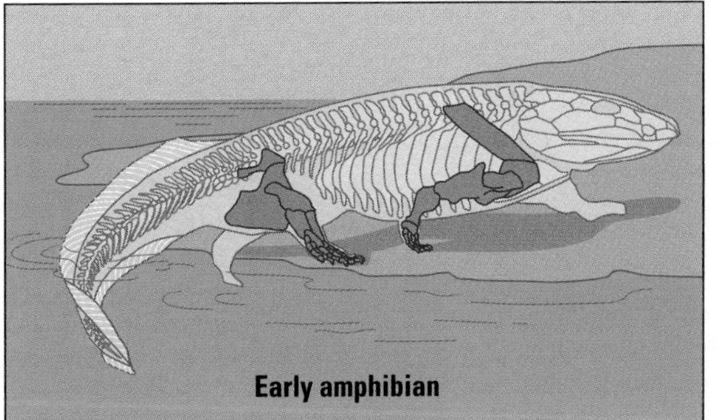

Figure 17.35 The origin of tetrapods. Fossils of some lobe-finned fishes have skeletal supports extending into their fins. Fossils of early amphibians have limb skeletons that probably functioned in helping early amphibians move on land.

Amphibians were the first vertebrates to colonize land. They descended from fishes that had lungs and fins with muscles and skeletal supports strong enough to enable some movement, however clumsy, on land **(Figure 17.35)**. The fossil record chronicles the evolution of four-limbed amphibians from fishlike ancestors. Terrestrial vertebrates—amphibians, reptiles, and mammals—are collectively called **tetrapods,** which means "four feet." Had our amphibian ancestors had three pairs of legs on their undersides instead of just two, we might be hexapods. This image seems silly, but serves to reinforce the point that evolution, as descent with modification, is constrained by history.

Reptiles

Tunicates
Lancelets
Lampreys
Cartilaginous fishes
Bony fishes
Amphibians
Reptiles
Mammals

Reptiles (including birds) and mammals are **amniotes.** The evolution of amniotes from an amphibian ancestor included many additional adaptations for living on land. One such adaptation, which gives the group its name, is the **amniotic egg,** a fluid-filled egg enclosed in a shell inside of which the embryo develops **(Figure 17.36)**. The amniotic egg functions as a "self-contained pond" that enables amniotes to complete their life cycle on land.

The **reptiles** include snakes, lizards, turtles, crocodiles, alligators, and birds, along with a number of extinct groups, including most of the dinosaurs. In addition to an amniotic egg, adaptations such as waterproof skin allowed reptiles to break their ancestral ties to aquatic habitats. Scales, which contain a protein called keratin (also found in human hair), help prevent dehydration in dry air. Reptiles cannot breathe through their dry skin and so obtain most of their oxygen with their lungs.

Figure 17.36 Terrestrial equipment of reptiles. This bull snake displays two reptilian adaptations to living on land: a waterproof skin with keratinized scales; and amniotic eggs, with shells that protect a watery, nutritious internal environment where the embryo can develop on land. Snakes evolved from lizards that adapted to a burrowing lifestyle.

Nonbird reptiles are sometimes referred to as "cold-blooded" animals because they do not use their metabolism extensively to control body temperature. Reptiles do regulate body temperature, but largely through behavioral adaptations. For example, many lizards regulate their internal temperature by basking in the sun when the air is cool and seeking shade when the air is too warm. Because they absorb external heat rather than generating much of their own, lizards are said to be **ectotherms,** a term more accurate than cold-blooded. By heating directly with solar energy rather than through the metabolic breakdown of food, a lizard can survive on less than 10% of the calories required by a mammal of equivalent size.

As successful as reptiles are today, they were far more widespread, numerous, and diverse during the Mesozoic era, which is sometimes called the "age of reptiles." Reptiles diversified extensively during that era, producing a dynasty that lasted until about 65 million years ago. Dinosaurs, the most diverse reptile group, included the largest animals ever to inhabit land. Some were gentle giants that lumbered about while browsing vegetation. Others were voracious carnivores that chased their larger prey on two legs (**Figure 17.37**).

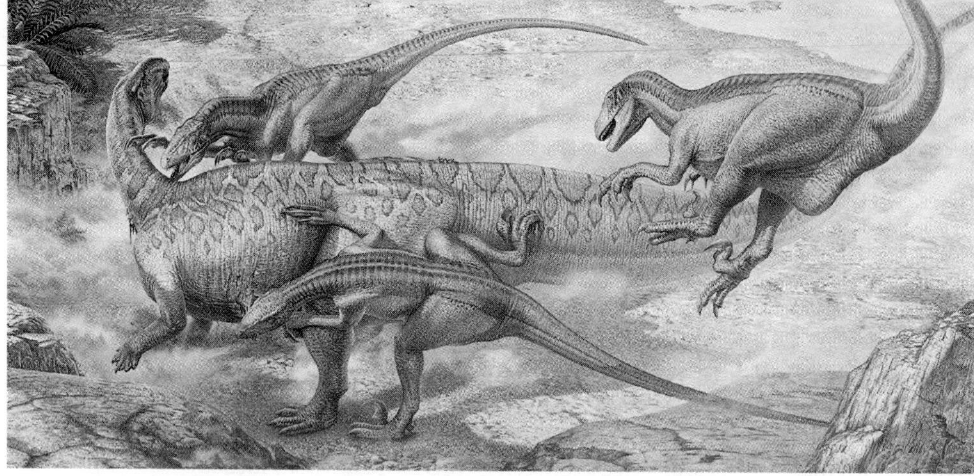

Figure 17.37 A Mesozoic feeding frenzy. Hunting in packs, *Deinonychus* (meaning "terrible claw") probably used its sickle-shaped claws to slash at larger prey.

The age of reptiles began to wane about 70 million years ago. During the Cretaceous, the last period of the Mesozoic era, global climate became cooler and more variable (see Chapter 14). This was a period of mass extinctions that claimed all the dinosaurs by about 65 million years ago, except for one lineage. That lone surviving lineage is represented today by the reptilian group we know as birds.

Birds **Birds** evolved from a lineage of small, two-legged dinosaurs during the great reptilian radiation of the Mesozoic era. But modern birds look quite different from modern reptiles because of their feathers and other distinctive flight equipment. Almost all of the 8,600 living bird species are airborne. The few flightless species, including the ostrich and the penguin, evolved from flying ancestors. Appreciating the avian world is all about understanding flight.

Almost every element of bird anatomy is modified in some way that enhances flight. The bones have a honeycombed structure that makes them strong but light (the wings of airplanes have the same basic construction). For example, a huge seagoing species called the frigate bird has a wingspan of more than 2 m, but its whole skeleton weighs only about 113 g (4 ounces). Another adaptation that reduces the weight of birds is the absence of some internal organs found in other vertebrates. Female birds, for instance, have only one ovary instead of a pair. Also, modern birds are toothless, an adaptation that trims the weight of the head (no uncontrolled nosedives). Birds do not chew food in the mouth but grind it in the gizzard, a chamber of the digestive tract near the stomach.

Flying requires a great expenditure of energy and an active metabolism. Unlike other reptiles, birds are **endotherms,** meaning they use their own metabolic heat to maintain a warm, constant body temperature.

A bird's most obvious flight equipment is its wings. Bird wings are airfoils that illustrate the same principles of aerodynamics as the wings of an airplane (**Figure 17.38**). A bird's flight motors are its powerful breast muscles, which are anchored to a keel-like breastbone. It is mainly these flight

Figure 17.38 The aerodynamics of a bald eagle in flight. Bird wings are airfoils, which have shapes that create lift by altering air currents. Air passing over a wing must travel farther in the same amount of time than air passing under the wing. This expands the air above the wing relative to the air below the wing. And this makes the air pressure pushing upward against the lower wing surface greater than the pressure of the expanded air pushing downward on the wing. The wings of birds and airplanes owe their "lift" to this pressure differential.

muscles that we call "white meat" on a turkey or chicken. Some birds, such as eagles and hawks, have wings adapted for soaring on air currents and flap their wings only occasionally. Other birds, including hummingbirds, excel at maneuvering but must flap continuously to stay aloft. In either case, it is the shape and arrangement of the feathers that form the wing into an airfoil. Feathers are made of keratin, the same protein that forms the scales of reptiles. Feathers may have functioned first as insulation, helping birds retain body heat, only later being co-opted as flight gear.

Mammals

There are two major lineages of amniotes: one that led to the reptiles and one that produced the mammals. The first true **mammals** arose about 200 million years ago and were probably small, nocturnal insect-eaters. Mammals became much more diverse after the downfall of the dinosaurs. Most mammals are terrestrial. However, there are nearly 1,000 species of winged mammals, the bats. And about 80 species of dolphins, porpoises, and whales are totally aquatic. The blue whale, an endangered mammal that grows to lengths of nearly 30 m, is the largest animal that has ever existed. Two features—hair and mammary glands that produce milk that nourishes the young—are mammalian hallmarks. The main function of hair is to insulate the body and help maintain a warm, constant internal temperature; mammals, like birds, are endotherms.

There are three major groups of mammals: the monotremes, the marsupials, and the eutherians. The duck-billed platypus and the echidna are the only existing species of **monotremes,** egg-laying mammals **(Figure 17.39a)**. The platypus lives along rivers in eastern Australia and on the nearby island of Tasmania. The female usually lays two eggs and incubates them in a leaf nest. After hatching, the young nurse by licking up milk secreted onto the mother's fur.

Most mammals are born rather than hatched. During gestation in marsupials and eutherians, the embryos are nurtured inside the mother by an organ called the **placenta.** Consisting of both embryonic and maternal tissues, the placenta joins the embryo to the mother within the uterus. The embryo is nurtured by maternal blood that flows close to the embryonic blood system in the placenta.

Marsupials, the so-called pouched mammals, include kangaroos, koalas, and opossums. These mammals have a brief gestation and give birth to tiny embryonic offspring that complete development while attached to the mother's nipples. The nursing young are usually housed in an external pouch on the mother's abdomen **(Figure 17.39b)**. Nearly all marsupials live in Australia, New Zealand, and North and South America. Australia has been a marsupial sanctuary for much of the past 60 million years. Australian marsupials have diversified extensively, filling terrestrial habitats that on other continents are occupied by eutherian mammals.

Eutherians are also called **placental mammals** because their placentas provide a more intimate and longer-lasting association between the mother and her developing young than do marsupial placentas **(Figure 17.39c)**.

(a) A monotreme. Monotremes, such as this echidna, are the only mammals that lay eggs (inset).

(b) A marsupial. The young of marsupials are born very early in their development. This brushtail opossum will finish its growth while nursing from a nipple in its mother's pouch.

(c) Eutherians. In eutherians (placental mammals), young develop within the uterus of the mother. There they are nurtured by the flow of blood through the dense network of vessels in the placenta. The reddish portion of the afterbirth clinging to the newborn zebra in this photograph is the placenta.

Figure 17.39 The three major groups of mammals.

Eutherians make up almost 95% of the 4,500 species of living mammals. Dogs, cats, cows, rodents, rabbits, bats, and whales are all examples of eutherian mammals. One of the eutherian groups is the primates, which includes monkeys, apes, and humans.

CHECKPOINT

1. What four features do we share with invertebrate chordates such as lancelets?
2. What feature of your skeleton is an example of segmentation?
3. A shark has a __________ skeleton, while a tuna has a __________ skeleton.
4. The oldest group of tetrapods is the __________.
5. What is an amniotic egg?
6. Birds differ from other reptiles in their main source of body heat, with birds being __________ and reptiles being __________.
7. What are two main hallmarks of mammals?

Answers: 1. (1) Dorsal, hollow nerve cord; (2) notochord; (3) gill structures at some time during development; (4) post-anal tail at some time during development **2.** The vertebrae of your backbone **3.** cartilaginous; bony **4.** amphibians **5.** A shelled egg with the embryo contained in a fluid-filled sac **6.** endotherms; ectotherms **7.** Hair and mammary glands

The Human Ancestry

MP3 Tutor
Human Evolution

We have now traced animal genealogy to the **primates,** the mammalian group that includes *Homo sapiens* and its closest kin. To understand what that means, we must trace our ancestry back to the trees, where some of our most treasured traits originated.

The Evolution of Primates

Primate evolution provides a context for understanding human origins. The fossil record supports the hypothesis that primates evolved from insect-eating mammals during the late Cretaceous period, about 65 million years ago. Those early primates were small, arboreal (tree-dwelling) mammals. Thus, primates were first distinguished by characteristics that were shaped, through natural selection, by the demands of living in the trees. For example, primates have limber shoulder joints, which make it possible to brachiate (use arms to swing from branch to branch). The dexterous hands of primates can hang on to branches and manipulate food. Nails have replaced claws in many primate species, and the fingers are very sensitive. The eyes of primates are close together on the front of the face. The overlapping fields of vision of the two eyes enhance depth perception, an obvious advantage when brachiating. Excellent eye-hand coordination is also important for arboreal maneuvering **(Figure 17.40)**. Parental care is essential for young animals in the trees. Mammals devote more energy to caring for their young than most other vertebrates,

Figure 17.40 The arboreal athleticism of primates. This orangutan displays primate adaptations for living in the trees: limber shoulder joints, manual dexterity, and stereo vision due to eyes on the front of the face.

and primates are among the most attentive parents of all mammals. Most primates have single births and nurture their offspring for a long time. Though humans do not live in trees, we retain in modified form many of the traits that originated there.

Taxonomists divide the primates into three main groups **(Figure 17.41)**. The first includes lorises, pottos, and lemurs. These primates live in

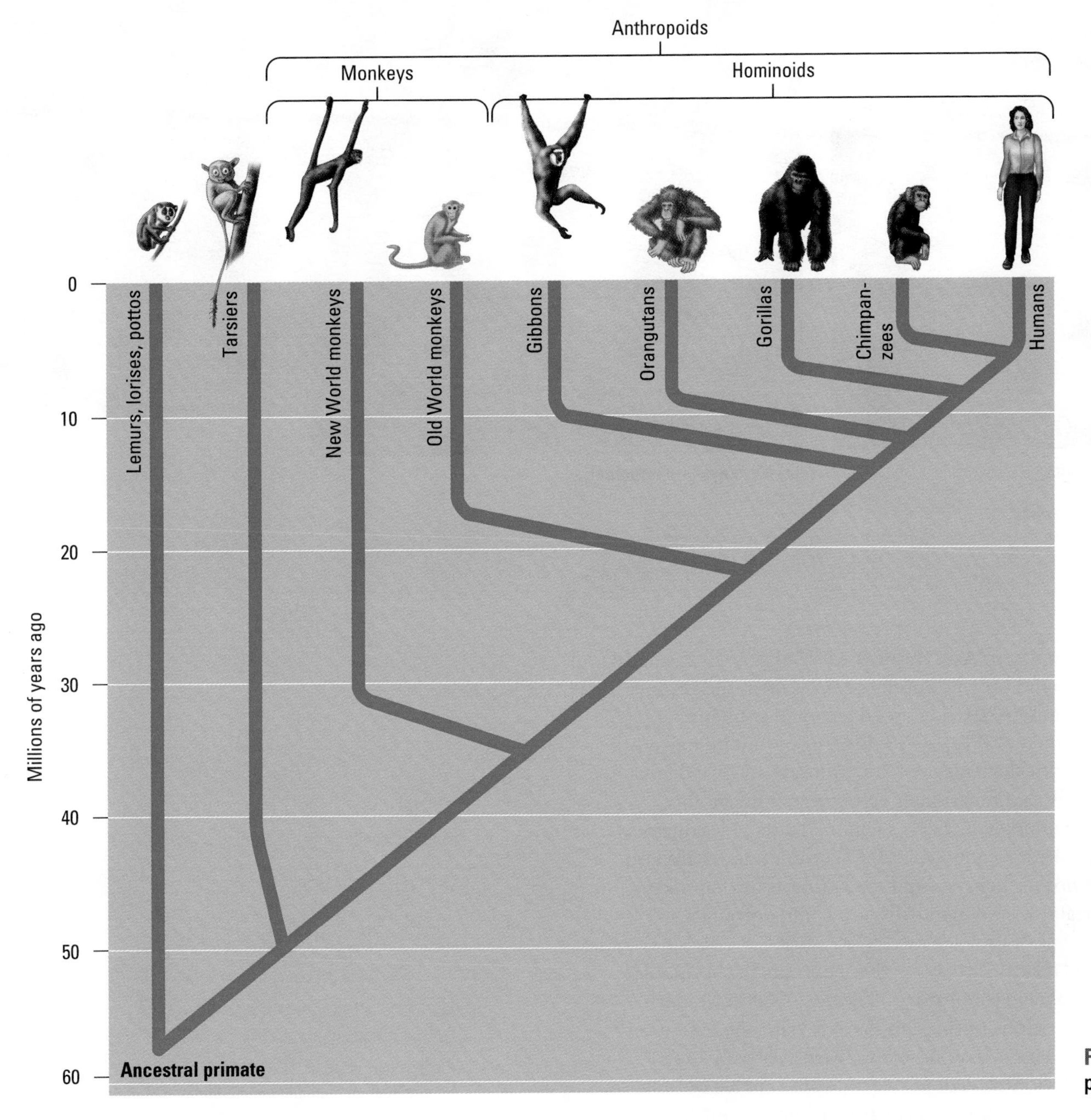

Figure 17.41 Primate phylogeny.

(a) A slender loris

(b) A tarsier

A wooly spider monkey
with young (New
World monkeys)

A pig-tail macaque
with young
(Old World monkeys)

(c) Monkeys (anthropoids)

Figure 17.42 Primate diversity.

Madagascar, Africa, and southern Asia **(Figure 17.42a)**. Tarsiers, small nocturnal tree-dwellers found only in Southeast Asia, form the second group of primates **(Figure 17.42b)**. The third group of primates, **anthropoids,** includes monkeys **(Figure 17.42c)**, apes, and humans. All monkeys in the New World (the Americas) are arboreal and are distinguished by prehensile tails that function as an extra appendage for brachiating. (If you see a monkey in a zoo swinging by its tail, you know it's from the New World.) Although some Old World monkeys are also arboreal, their tails are not prehensile. And many Old World monkeys, including baboons, macaques, and mandrills, are mainly ground-dwellers.

Our closest anthropoid relatives are within the ape group, the **hominoids:** gibbons, orangutans, gorillas, and chimpanzees **(Figure 17.42d)**. Modern apes live only in tropical regions of the Old World. With the exception of some gibbons, apes are larger than monkeys, with relatively long arms, short legs, and no tail. Although all the apes are capable of brachiation, only gibbons and orangutans are primarily arboreal. Gorillas and chimpanzees are highly social. Apes have larger brains proportionate to body size than monkeys, and their behavior is more adaptable. And, of course, the hominoids include humans **(Figure 17.42e)**.

(d) Apes (hominoids)

(e) A human with young (hominoids)

Figure 17.42 Primate diversity, continued.

The Emergence of Humankind

Humanity is one very young twig on the vertebrate branch, just one of many branches on the tree of life. In the continuum of life spanning 3.5 billion years, humans and chimpanzees have shared a common ancestry for all but the last 5–7 million years (see Figure 17.41). Put another way, if we compressed the history of life to a year, humans and chimpanzees would diverge from a common ancestor at about 6 A.M on December 31st. The fossil record and molecular systematics concur in that vintage for the human lineage. **Paleoanthropology,** the study of human evolution, focuses on this very thin slice of biological history.

Some Common Misconceptions

Certain misconceptions about human evolution that were generated during the early part of the 20th century still persist today in the minds of many, long after these myths have been debunked by the fossil evidence.

Let's first dispose of the myth that our ancestors were chimpanzees or any other modern ape. Chimpanzees and humans represent two divergent branches of the anthropoid tree that evolved from a common, less specialized ancestor. Chimps are not our parent species, but more like our phylogenetic siblings or cousins.

Another misconception envisions human evolution as a ladder with a series of steps leading directly from an ancestral anthropoid to *Homo sapiens*.

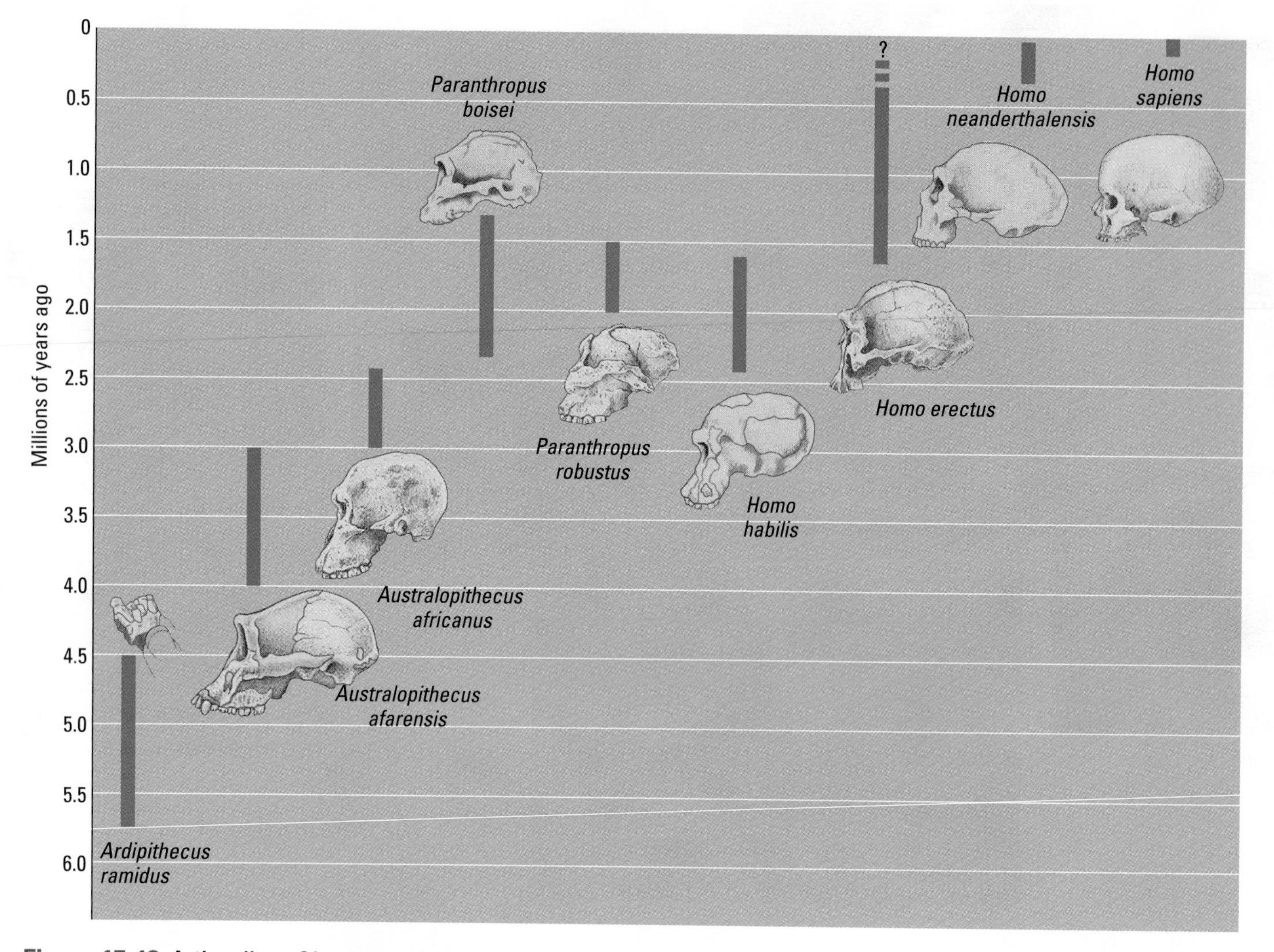

Figure 17.43 A time line of human evolution. Notice that there have been times when two or more hominid species coexisted. The skulls are all drawn to the same scale to enable you to compare the sizes of craniums and hence brains.

This is often illustrated as a parade of fossil **hominids** (members of the human family) becoming progressively more modern as they march across the page. If human evolution is a parade, then it is a disorderly one, with many splinter groups having traveled down dead ends. At times in hominid history, several different human species coexisted **(Figure 17.43)**. Human phylogeny is more like a multibranched bush than a ladder, with our species being the tip of the only twig that still lives.

One more myth we must bury is the notion that various human characteristics, such as upright posture and an enlarged brain, evolved in unison. A popular image is of early humans as half-stooped, half-witted cave-dwellers. In fact, we know from the fossil record that different human features evolved at different rates, with erect posture, or bipedalism, leading the way. Our pedigree includes ancestors who walked upright but had ape-sized brains.

After dismissing some of the folklore on human evolution, however, we must admit that many questions about our ancestry have not yet been resolved.

***Australopithecus* and the Antiquity of Bipedalism** Before there was *Homo*, several hominid species of the genus *Australopithecus* walked the

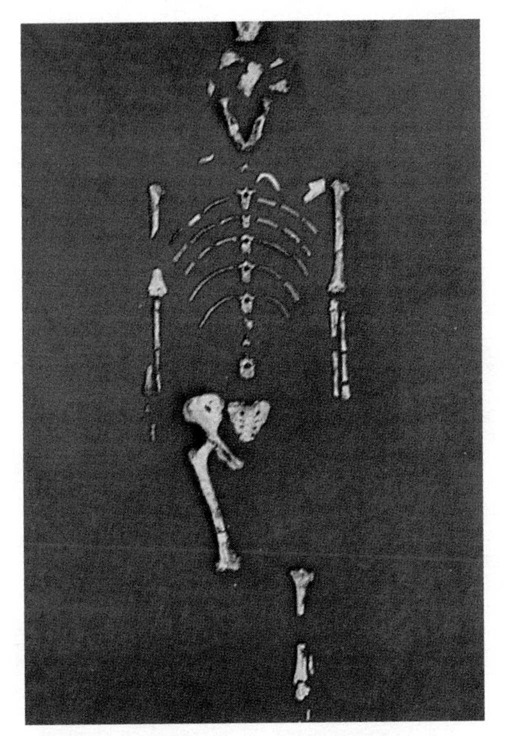

(a) *Australopithecus afarensis* skeleton. Lucy, a 3.2-million-year-old skeleton, represents the hominid species *Australopithecus afarensis*. Fragments of the pelvis and skull indicate that *A. afarensis* walked on two feet.

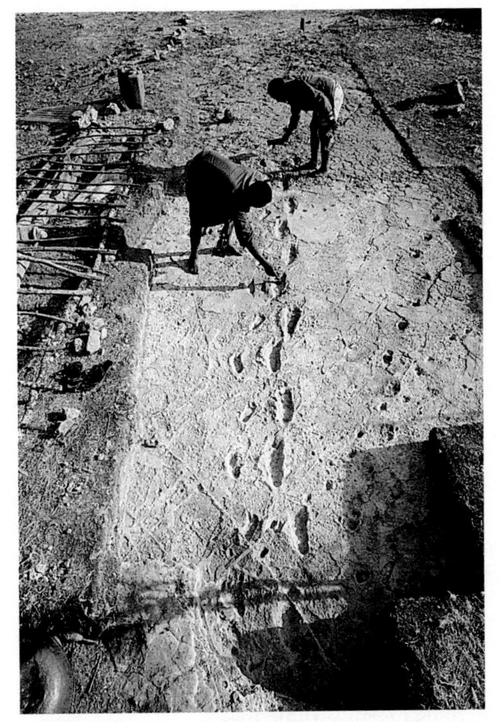

(b) Ancient footprints. Some 3.7 million years ago, several bipedal (upright-walking) hominids left footprints in damp volcanic ash in what is now Tanzania in East Africa. The fossilized prints were discovered by British anthropologist Mary Leakey in 1978. The footprints are part of the strong evidence that bipedalism is a very old human trait.

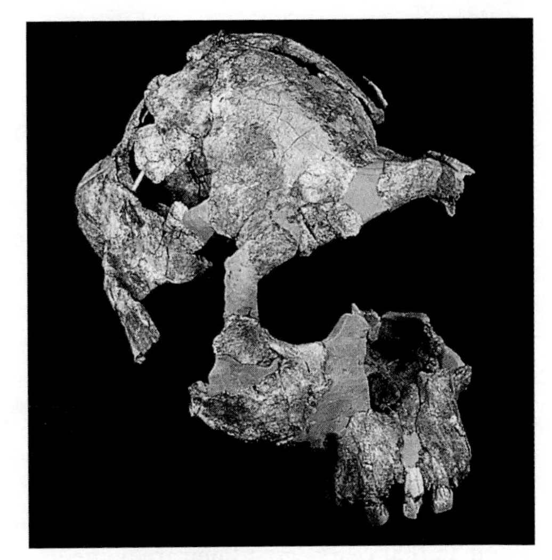

(c) Early hominid skull. This 3.9 million year old *A. afarensis* skull shows evidence of being attached to a vertical backbone. Thus, the upright posture of humans is at least that old.

Figure 17.44 The antiquity of upright posture.

African savanna (grasslands with clumps of trees). Paleoanthropologists have focused much of their attention on *A. afarensis*, an early species of *Australopithecus*. Fossil evidence now pushes bipedalism in *Australopithecus* back to at least 4 million years ago **(Figure 17.44)**.

One of the most complete fossil skeletons of *A. afarensis* dates to about 3.2 million years ago in East Africa. Nicknamed Lucy by her discoverers, the individual was a female, only about 3 feet tall, with a head about the size of a softball. Lucy and her kind lived in savanna areas and may have subsisted on nuts and seeds, bird eggs, and whatever animals they could catch or scavenge from kills made by more efficient predators such as large cats and dogs.

Homo Habilis and the Evolution of Inventive Minds Enlargement of the human brain is first evident in fossils from East Africa dating to about 2.4 million years ago. Thus, the fundamental human trait of an enlarged brain evolved a few million years after the other major human trait—bipedalism. As evolutionary biologist Stephen Jay Gould put it, "Mankind stood up first and got smart later."

Anthropologists have found skulls with brain capacities intermediate in size between those of the latest *Australopithecus* species and those of *Homo sapiens*. Simple handmade stone tools are sometimes found with the larger-brained fossils, which have been dubbed *Homo habilis* ("handy man"). After walking upright for about 2 million years, humans were finally beginning to use their manual dexterity and big brains to invent tools that enhanced their hunting, gathering, and scavenging on the African savanna.

***Homo Erectus* and the Global Dispersal of Humanity** The first species to extend humanity's range from its birthplace in Africa to other continents was *Homo erectus*, perhaps a descendant of *H. habilis*. Skeletons of *H. erectus* dating to 1.8 million years ago represent the oldest known fossils of hominids outside Africa. *Homo erectus* migrated to populate many regions of Asia and Europe, eventually moving as far as Indonesia.

Homo erectus was taller than *H. habilis* and had a larger brain capacity. During the 1.5 million years the species existed, the *H. erectus* brain increased to as large as 1,200 cubic centimeters (cm^3), a brain capacity that overlaps the normal range for modern humans. Intelligence enabled humans to continue succeeding in Africa and also to survive in the colder climates of the north. *Homo erectus* resided in huts or caves, built fires, made clothes from animal skins, and designed stone tools. In anatomical and physiological adaptations, *H. erectus* was poorly equipped for life outside the tropics, but made up for the deficiencies with cleverness and social cooperation.

Some African, Asian, European, and Australasian (from Indonesia, New Guinea, and Australia) populations of *H. erectus* gave rise to regionally diverse descendants that had even larger brains. Among these descendants of *H. erectus* were *Homo neanderthalensis* (the Neanderthals), who lived in Europe, the Middle East, and parts of Asia from about 200,000 years ago to about 35,000 years ago. (They are named Neanderthals because their fossils were first found in the Neander Valley of Germany.) Compared with us, Neanderthals had slightly heavier browridges and less pronounced chins, but their brains, on average, were slightly larger than ours. Neanderthals were skilled toolmakers, and they participated in burials and other rituals that required abstract thought.

The Origin and Dispersal of *Homo Sapiens* The oldest known fossils of our own species, *H. sapiens*, were discovered in Ethiopia and date from 160,000 to 195,000 years ago. These early humans lacked the heavy browridges of *H. erectus* and *H. neanderthalensis* and were more slender, suggesting that they belong to a distinct lineage.

The Ethiopian fossils support molecular evidence about the origin of humans. DNA studies indicate that Europeans and Asians share a relatively recent common ancestor and that many African lineages represent earlier branches on the human tree. These findings strongly suggest that all living humans have ancestors that originated as *Homo sapiens* in Africa.

Some of the evidence for this conclusion comes from analyses of mitochondrial DNA, which is maternally inherited, and Y chromosomes, which are transmitted from fathers to sons. These studies suggest that all living humans inherited their mitochondrial DNA from a common ancestral woman who lived 150,000 to 200,000 years ago. Mutations on the Y chromosome can serve as markers for tracing the ancestry and relationships among males alive today. By comparing the Y chromosomes of males from various geographic regions, researchers were able to infer divergence from a common African male ancestor.

Fossil evidence suggests that our species emerged from Africa in one or more waves, spreading first into Asia and then to Europe and Australia. The oldest fossils of *H. sapiens* outside Africa date back about 50,000 years. The date of the arrival of humans in the New World is uncertain, although the oldest generally accepted evidence puts it at 15,000 years ago.

Cultural Evolution An erect stance was the most radical anatomical change in our evolution; it required major remodeling of the foot, pelvis, and vertebral column. Enlargement of the brain was a secondary alteration made possible by prolonging the growth period of the skull and its contents (see Figure 14.15). The primate brain continues to grow after birth, and the period of growth is longer for a human than for any other primate. The extended period of human development also lengthens the time parents care for their offspring, which contributes to the child's ability to benefit from the experiences of earlier generations. This is the basis of **culture**—the social transmission of accumulated knowledge, customs, beliefs, and art over generations. The major means of this transmission is language, written and spoken.

Cultural evolution is continuous, but there have been three major stages. The first stage began with nomads who hunted and gathered food on the African grasslands 2 million years ago. They made tools, organized communal activities, and divided labor. Beautiful ancient art, such as the 30,000-year-old cave paintings shown in **Figure 17.45**, is just one example of our cultural roots in early societies. The second main stage of cultural evolution came with the development of agriculture in Africa, Eurasia, and the Americas about 10,000 to 15,000 years ago. Along with agriculture came permanent settlements and the first cities. The third major stage in our cultural evolution was the Industrial Revolution, which began in the 18th century. Since then, the development of new technology has escalated exponentially; a single generation spanned the flight of the Wright brothers and Neil Armstrong's walk on the moon. It took less than a decade for the Internet to transform commerce, communication, and education. Through all this cultural evolution, from simple hunter-gatherers to high-tech societies, we have not changed biologically in any significant way. We are probably no more intelligent than our cave-dwelling ancestors. The same toolmaker who chipped away at stones now designs microchips and software. The know-how to build skyscrapers, computers, and spaceships is stored not in our genes but in the cumulative product of hundreds of generations of human experience, passed along by parents, teachers, books, and electronic media.

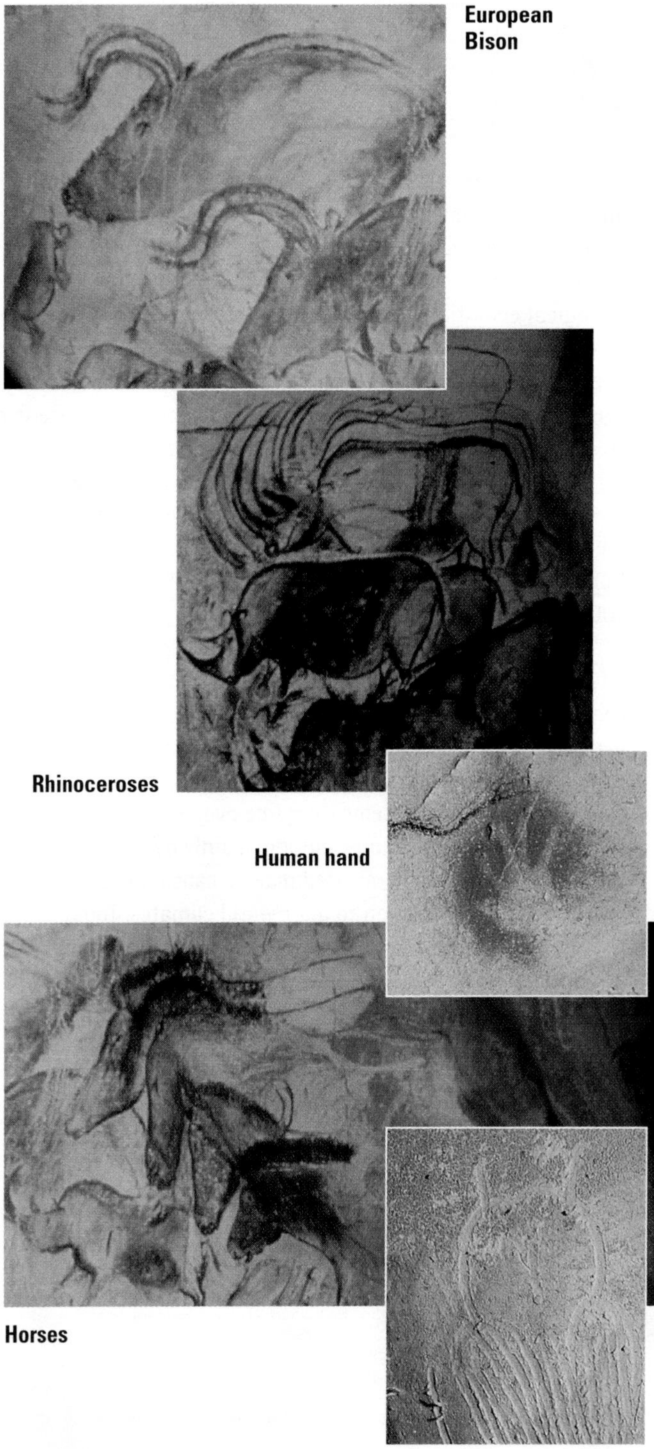

European Bison

Rhinoceroses

Human hand

Horses

Figure 17.45 Art history goes way back—and so does our fascination with and dependence on animal diversity. Early artists created these remarkable images beginning about 30,000 years ago. Three cave explorers found this prehistoric art gallery on Christmas eve, 1994, when they ventured into a cavern near Vallon-Pont d'Arc, in southern France.

CHECKPOINT

1. To which group of mammals do we belong? To which subgroup do we belong?
2. Based on the fossil evidence in Figure 17.43, how many hominid species existed 1.7 million years ago?
3. Humans first evolved on which continent?
4. What role does a long period of parental care play in culture?
5. Why is *Homo habilis* known as "handy man"?
6. Is it possible that a *Homo sapiens* individual ever met a Neanderthal?

Answers: 1. Primates; anthropoids (or hominoids) **2.** Four **3.** Africa **4.** It extends the opportunity for parents to transmit the lessons of the past to offspring. **5.** Because it was the earliest hominid that definitely created and used stone tools **6.** Yes. Neanderthals lived in Europe and Asia until about 35,000 years ago, and *H. sapiens* first migrated out of Africa about 50,000 years ago.

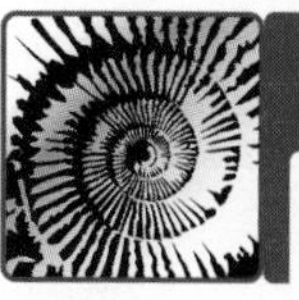

EVOLUTION CONNECTION

Earth's New Crisis

Evolution of the human brain may have been a more subtle process than acquiring an upright stance, but the global consequences of cerebral expansion have been enormous. Cultural evolution made *Homo sapiens* a new force in the history of life—a species that could defy its physical limitations and shortcut biological evolution. We do not have to wait to adapt to an environment through natural selection; we simply change the environment to meet our needs. We are the most numerous and widespread of all large animals, and everywhere we go, we bring environmental change. There is nothing new about environmental change. The history of life is the story of biological evolution on a changing planet. But it is unlikely that change has ever been as rapid as in the age of humans. Cultural evolution vastly outpaces biological evolution. We are changing the world faster than many species can adapt; the rate of extinctions in the 20th century was 50 times greater than the average for the past 100,000 years.

This rapid rate of extinction is mainly a result of habitat destruction, which is a function of human cultural changes and overpopulation **(Figure 17.46)**. Feeding, clothing, and housing 6 billion people impose an enormous strain on Earth's capacity to sustain life. If all these people suddenly assumed the high standard of living enjoyed by many people in developed nations, it is very likely that Earth's support systems would be overwhelmed. Already, for example, current rates of fossil fuel consumption, mainly by developed nations, are so great that the carbon dioxide released may be causing the temperature of the atmosphere to increase enough to alter world climates. Today, it is not just individual species that are endangered, but entire ecosystems, the global atmosphere, and the oceans. Tropical rain forests, which play a vital role in moderating global weather, are being cut down at a startling rate. Scientists have hardly begun to study these ecosystems, and many species in them may become extinct before they are even discovered.

Of the many crises in the history of life, the impact of one species, *Homo sapiens*, is the latest and potentially the most devastating. We will examine the interactions of humans—as well as other species—with the environment in our next unit on ecology. ■

Figure 17.46 A clear-cut forest, an example of habitat destruction caused by *Homo sapiens*.

Chapter Review

SUMMARY OF KEY CONCEPTS

For study help and activities, go to campbellbiology.com or the student CD-ROM.

The Origins of Animal Diversity

• **What Is an Animal?** Animals are eukaryotic, multicellular, heterotrophic organisms that obtain nutrients by ingestion. Most animals reproduce sexually and develop from a zygote to a blastula, gastrula, and perhaps a larval stage before becoming an adult.

• **Early Animals and the Cambrian Explosion** Animals probably evolved from a colonial, flagellated protist more than 700 million years ago. At the beginning of the Cambrian period, 542 million years ago, animal diversity increased rapidly.

• **Animal Phylogeny** Major branches of animal evolution are defined by two key evolutionary differences: the presence or absence of tissues and radial versus bilateral body symmetry. A body cavity at least partly lined by mesoderm evolved in a number of later branches.

Activity *Overview of Animal Phylogeny*

Major Invertebrate Phyla

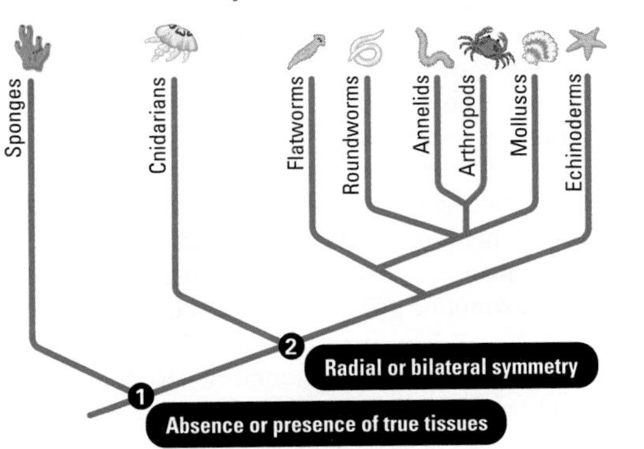

- **Sponges** Sponges (phylum Porifera) are sessile animals with porous bodies but no true tissues. They filter-feed by drawing water through pores in the sides of the body.
- **Cnidarians** Cnidarians (phylum Cnidaria) have radial symmetry, a gastrovascular cavity, and tentacles with cnidocytes. The body is either a sessile polyp or a floating medusa.
- **Flatworms** Flatworms (phylum Platyhelminthes) are the simplest bilateral animals. They may be free-living or parasitic.
- **Roundworms** Roundworms, also called nematodes (phylum Nematoda), are unsegmented and cylindrical with tapered ends. They have a complete digestive tract and a pseudocoelom. They may be free-living or parasitic.
- **Annelids** Annelids (phylum Annelida) are segmented worms. They may be free-living or parasitic on other animals.
- **Arthropods** Arthropods (phylum Arthropoda) are segmented animals with an exoskeleton and specialized, jointed appendages. Arthropods consist of four main groups: arachnids, crustaceans, millipedes and centipedes, and insects. In species diversity, insects outnumber all other forms of life combined.

Case Study in the Process of Science *How Are Insect Species Identified?*

- **Molluscs** Molluscs (phylum Mollusca) are soft-bodied animals often protected by a hard shell. The body has three main parts: a muscular foot, a visceral mass, and a fold of tissue called the mantle.
- **Echinoderms** Echinoderms (phylum Echinodermata) are sessile or slow-moving marine animals that lack body segments and possess a unique water vascular system. Bilaterally symmetrical larvae usually change to radially symmetrical adults.

Activity *Characteristics of Invertebrates*

The Vertebrate Genealogy

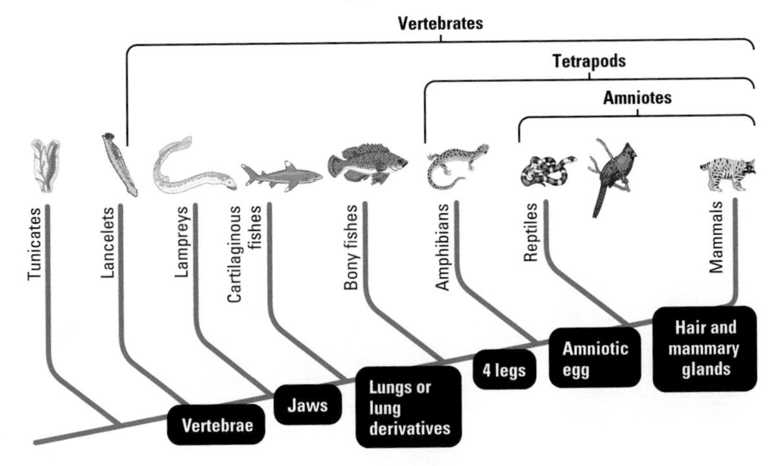

- **Characteristics of Chordates** Chordates (phylum Chordata) are defined by a dorsal, hollow nerve cord; a flexible notochord; pharyngeal slits; and a post-anal tail. Tunicates and lancelets are invertebrate chordates. All other chordates are vertebrates, possessing a cranium and backbone.
- **Fishes** Lampreys are jawless vertebrates. Cartilaginous fishes (chondrichthyans), such as sharks, are mostly predators with powerful jaws and a flexible skeleton made of cartilage. Bony fishes (osteichthyans) have a stiff skeleton reinforced by calcium. Bony fishes are further classified into ray-finned fishes and lobe-finned fishes (including lungfishes).
- **Amphibians** Amphibians are tetrapod vertebrates that usually deposit their eggs (lacking shells) in water. Aquatic larvae typically undergo a radical metamorphosis into the adult stage. Their moist skin requires that amphibians spend much of their adult life in moist environments.
- **Reptiles** Reptiles are amniotes, vertebrates that develop in a fluid-filled egg enclosed by a shell. Reptiles include terrestrial ectotherms with lungs and waterproof skin covered by scales. Scales and amniotic eggs enhanced reproduction on land. Birds are endothermic reptiles with wings, feathers, and other adaptations for flight.

Case Study in the Process of Science *How Does Bone Structure Shed Light on the Origin of Birds?*

- **Mammals** Mammals are endothermic vertebrates with hair and mammary glands. There are three major groups: Monotremes lay eggs; marsupials use a placenta but give birth to tiny offspring that usually complete development while attached to nipples inside the mother's pouch; and eutherians, or placental mammals, use their placenta in a longer-lasting association between the mother and her developing young.

Activity *Characteristics of Chordates*

The Human Ancestry

MP3 Tutor *Human Evolution*

- **The Evolution of Primates** The first primates were small, arboreal mammals that evolved from insect-eating mammals about 65 million years ago. Anthropoids consist of New World monkeys (with prehensile tails), Old World monkeys (without prehensile tails), apes, and humans. Modern apes are confined to tropical regions of the Old World.

Activity *Primate Diversity*

- **The Emergence of Humankind** Chimpanzees and humans evolved from a common ancestor about 5–7 million years ago. Upright posture evolved in several hominid species of the genus *Australopithecus* at least 4 million years ago. Enlargement of the human brain in *Homo habilis* came later, about 2.4 million years ago. *Homo erectus* was the first species to extend humanity's range from its birthplace in Africa to other continents. *Homo erectus* gave rise to regionally diverse descendants, such as the Neanderthals (*H. neanderthalensis*). Current data indicate a relatively recent dispersal of modern Africans that gave rise to today's human diversity. Human cultural evolution began in Africa with wandering hunter-gatherers, progressed to the development of agriculture and the Industrial Revolution, and continues today with accelerating technological change.

Activity *Human Evolution*

SELF-QUIZ

1. Bilateral symmetry in the animal kingdom is best correlated with
   a. an ability to sense equally in all directions.
   b. the presence of a skeleton.
   c. motility and active predation and escape.
   d. development of a true coelom.

2. Members of a small group of worms classified in the phylum Gnathostomulida (not discussed in this chapter) live between sand grains on ocean beaches. These animals are unsegmented and lack a body cavity, but some have a complete digestive tract. These characteristics suggest that gnathostomulid worms might branch from the main trunk of the animal phylogenetic tree between which two phyla?

3. Which of the following categories includes all others in the list: arthropod, arachnid, insect, butterfly, crustacean, millipede?

4. Reptiles are much more extensively adapted to life on land than amphibians in that reptiles
   a. have a complete digestive tract.
   b. lay eggs that are enclosed in shells.
   c. are endothermic.
   d. go through a larval stage.

5. What is the name of the phylum to which humans belong? For what anatomical structure is the phylum named? Where in your body is a derivative of this anatomical structure found?

6. Fossils suggest that the first major trait distinguishing human primates from other primates was __________.

7. Which of the following types of animals is not included in the human ancestry? (*Hint*: See Figure 17.32.)
   a. a bird
   b. a bony fish
   c. an amphibian
   d. a primate

8. Put the following list of species in order, from the oldest to the most recent: *Homo erectus*, *Australopithecus* species, *Homo habilis*, *Homo sapiens*.

9. Match each of the following animals to its phylum:

a. Human	1. Porifera
b. Leech	2. Arthropoda
c. Sponge	3. Cnidaria
d. Lobster	4. Chordata
e. Sea anemone	5. Annelida

Answers to the Self-Quiz questions can be found in Appendix D.

Go to the website or CD-ROM for more Self-Quiz questions.

THE PROCESS OF SCIENCE

10. Imagine that you are a marine biologist. As part of your exploration, you dredge up an unknown animal from the seafloor. Describe some of the characteristics you should look at to determine the phylum to which the creature should be assigned.

11. Many people describe themselves as vegetarians. Strictly speaking, vegetarians eat only plant products. Most vegetarians are not, in fact, that strict. Interview acquaintances who describe themselves as vegetarians and determine which taxonomic groups they avoid eating (see Figures 17.6 and 17.32). Try to generalize about their diet. For example, do they avoid eating vertebrates but eat some invertebrates? Do they avoid only birds and mammals? Do they eat dairy products or eggs?

12. Some researchers think that drying and cooling of the climate caused expansion of the African savanna and that this environment favored upright walking in early humans. Why might bipedalism be advantageous in the savanna? How might an erect posture relate to the evolution of a larger brain?

BIOLOGY AND SOCIETY

13. Coral reefs harbor a greater diversity of animals than any other environment in the sea. Australia's Great Barrier Reef has been protected as a marine reserve and is a mecca for scientists and nature enthusiasts. Elsewhere, such as in Indonesia and the Philippines, coral reefs are in danger. Many reefs have been depleted of fish, and runoff from the shore has covered coral with sediment. Nearly all the changes in the reefs can be traced back to human activities. What kinds of activities do you think might be contributing to the decline of the reefs? What are some reasons to be concerned about this decline? Do you think the situation is likely to improve or worsen in the future? Why? What might the local people do to halt the decline? Should the developed countries help? Why or why not?

14. The human body has not changed much in the last 100,000 years, but human culture has changed a great deal. As a result of our culture, we change the environment at a rate far greater than many species, including our own, can evolve. What evidence of rapid environmental change do you see regularly? What aspects of human culture are responsible for these changes? Do you see any evidence of a decrease in the rate of human-caused environmental changes?

UNIT 4

Ecology

18 The Ecology of Organisms and Populations

Every four years, the **world's population increases** by the population equivalent of the United States.

Some ecological communities depend on **periodic fires.**

After several days at high altitude, your body will begin to produce **more red blood cells.**

The biosphere is currently undergoing **a mass extinction.**

BIOLOGY AND SOCIETY

The Human Population Explosion

Earth's human population currently stands at over 6.5 billion—and keeps right on growing. The population has more than doubled from 3 billion in just 40 years. At the present growth rate of 74 million additional people per year, it takes about four years to add the population equivalent of the United States. We are by far the most abundant large animals, and given our technological prowess, we have a disproportionately high impact on the environment **(Figure 18.1)**.

Every day, we hear about local and global problems that threaten our well-being or provoke disputes between individuals or nations: global warming, toxic waste, conflicts over oil, the declining health of the oceans, civil strife aggravated by depressed economics. Contributing to all these apparently unrelated problems is a common factor: the continued growth of the human population in the face of limited resources. The human population explosion is now Earth's most significant biological phenomenon. Our species requires vast amounts of materials and space, including places to live, land to grow our food, and places to dump our waste. Incessantly expanding our presence on Earth, we have devastated the environment for many other species and now threaten to make it unfit for ourselves.

Every 20 minutes, our population increases by nearly 3,000 individuals. In the same 20 minutes, some researchers estimate, one species of plant or animal becomes extinct. Most ecologists believe that the two trends—exploding human population and the mass extinction that the biosphere is now experiencing—are related, at least over the long term. The most important factor is our encroachment on and destruction of habitat as we spread out and "tame" more land to satisfy our growing demand for space, food, shelter, fuel, water, and other resources.

To understand the problem of human population growth, we must understand the principles of ecology that apply to all species. This chapter therefore begins with an overview of ecology as a scientific discipline. Next, we'll consider how organisms adapt to Earth's varied ecological arenas. We will then focus on the study of populations of individual species, eventually returning to our discussion of the human population. ■

Figure 18.1 The impact of the ever-growing human population on the environment.

An Overview of Ecology

The scientific study of the interactions between organisms and their environments is called **ecology.** All organisms, including humans, interact continuously with their environments. We breathe, exchanging gases with the atmosphere. We eat other organisms, taking in secondhand energy that entered plants as sunlight. We add urea and other wastes to water and soil. We absorb and radiate heat. We are inextricably connected to the world outside our own bodies, as are all of Earth's creatures.

This outside world, the environment, can be divided into two major components. The **abiotic component** consists of nonliving chemical and physical factors, such as temperature, light, water, minerals, and air. The

biotic component includes the living factors—all the other organisms that are part of an individual's environment. Other organisms may compete with an individual for food and other resources, prey upon it, or change its physical and chemical environment.

In this section, we'll take a closer look at three key words in our definition of ecology: the *scientific* study of the *interactions* between organisms and their *environments*. This straightforward definition masks an enormously complex and exciting area of biology that is also of crucial practical importance. The science of ecology provides a basic understanding of how natural processes and organisms interact, giving us the tools we need to manage the planet's limited resources over the long term.

Ecology as Scientific Study

Humans have always had an interest in other organisms and their environments. As hunters and gatherers, prehistoric people had to learn where game and edible plants could be found in greatest abundance. Naturalists, from Aristotle to Darwin and beyond, made the process of observing and describing organisms in their natural habitats an end in itself rather than simply a means of survival. Extraordinary insight can still be gained from this descriptive approach of watching nature and recording its structure and processes. (As baseball-player-cum-philosopher Yogi Berra once put it, "You can observe a lot just by watching.") Thus, natural history as a "discovery science" (see Chapter 1) remains fundamental to ecology **(Figure 18.2)**. But in the past few decades, ecology has become increasingly experimental. In spite of the difficulty of conducting experiments that often involve large amounts of time and space, many ecologists are testing hypotheses in the laboratory and in the field. Ecologists also complement descriptive and experimental studies with computer simulations of large-scale experiments that might be impossible to conduct in the field. Whatever the approach, ecology employs that most basic of scientific processes, the posing of hypotheses and the use of observations and experiments to test those hypotheses. One such experiment is described near the end of this chapter.

(a) Establishing a canopy research station. This hot-air dirigible is placing a giant rubber raft on the treetops of a tropical rain forest in French Guiana, in northeastern South America.

(b) Studying the canopy. Living and working on this field station 30 m above the forest floor, an international research team studies the canopy environment. French biologist Pierre Grard and two other scientists are cataloging plants and the insects that feed on those plants. Among Earth's diverse environments, the canopy (upper tier) of tropical rain forests is especially rich in its diversity of insects, birds, and other animals.

Figure 18.2 Discovery science in a rain forest canopy.

A Hierarchy of Interactions

When we study the interactions between organisms and their environments, it is convenient to divide ecology into four increasingly comprehensive levels: organismal ecology, population ecology, community ecology, and ecosystem ecology.

Organismal ecology is concerned with the evolutionary adaptations that enable individual organisms to meet the challenges posed by their abiotic environments **(Figure 18.3a)**. The distribution of organisms is limited by the abiotic conditions they can tolerate. Earthworms, for instance, are restricted to moist environments because their skin does not prevent dehydration, as is apparent the day after a rain, when many desiccated earthworms can be found on pavement that was soaked by rain not long before.

The next level of organization in ecology is the **population,** a group of individuals of the same species living in a particular geographic area. **Population ecology** concentrates mainly on factors that affect population density and growth **(Figure 18.3b)**. The latter part of this chapter focuses on population ecology.

A **community** consists of all the organisms that inhabit a particular area; it is an assemblage of populations of different species. Questions in

community ecology focus on how interactions between species, such as predation, competition, and symbiosis, affect community structure and organization **(Figure 18.3c)**.

An **ecosystem** includes all the abiotic factors in addition to the community of species in a certain area. For example, a forest ecosystem includes not only the organisms, such as diverse plants and animals, but also the soil, water sources, sunlight, and other abiotic factors of the environment. In **ecosystem ecology,** questions concern energy flow and the cycling of chemicals among the various biotic and abiotic factors **(Figure 18.3d)**. (Chapter 19 focuses on community and ecosystem ecology.)

The **biosphere** is the global ecosystem—the sum of all the planet's ecosystems, or all of life and where it lives. The most complex level in ecology, the biosphere includes the atmosphere to an altitude of several kilometers, the land down to water-bearing rocks about 1,500 m deep, lakes and streams, caves, and the oceans to a depth of several kilometers. Isolated in space, the biosphere is self-contained, or closed, except that its photosynthetic producers derive energy from sunlight and it loses heat to space.

Ecology and Environmentalism

The science of ecology should be distinguished from the informal use of the word *ecology* to refer to environmental concerns. At the same time, we need to understand the complicated and delicate relationships between organisms and their environments to address environmental problems.

Our current awareness of the biosphere's limits stems mainly from the 1960s, a time of growing disillusionment with environmental practices of the past. In the 1950s, technology seemed poised to free humankind from several age-old bonds. New chemical fertilizers and pesticides, for example, showed great promise for increasing agricultural productivity and eliminating insect-borne diseases. Fertilizers were applied extensively, and pests were attacked with massive aerial spraying of pesticides, including a poison called DDT. The immediate results were astonishing: Increases in farm productivity enabled developed nations such as the United States to grow surplus food and sell it overseas, and the worldwide incidence of malaria and several other insect-borne diseases was markedly reduced. The pesticide DDT was hailed as a miracle weapon to wield anywhere insects caused problems.

Our enthusiasm for chemical fertilizers and pesticides began to wane as some of the side effects of DDT and other widely used poisons began appearing in the late 1950s. One of the first to perceive the

(a) Organismal ecology: How do diving whales stay under water for so long?

(b) Population ecology: What factors limit the number of striped mice that can inhabit a particular area?

(c) Community ecology: What factors influence the diversity of tree species that make up a particular forest?

(d) Ecosystem ecology: What processes recycle vital chemical elements such as nitrogen within a savanna ecosystem?

Figure 18.3 Examples of questions at different levels of ecology.

global dangers of pesticide abuse was Rachel Carson (**Figure 18.4**). We can trace our current environmental awareness to Carson's 1962 book *Silent Spring*. Her warnings were underscored when scientists began reporting that DDT was threatening the survival of predatory birds and was showing up in human milk. Another serious problem was the genetic resistance to pesticides that evolved in an increasing number of pest populations (see Chapter 13). By the early 1970s, disillusionment with the overuse of chemicals and a realization that our finite biosphere could not tolerate unlimited exploitation had developed into widespread concern about the environment. The environmental movement that Carson helped catalyze continues, with the students of today sustaining a tradition of activism on behalf of the biosphere's future health (**Figure 18.5**).

Today, it's clear that no part of the biosphere is untouched by the abusive impact of human populations and their technology. Depletion of natural resources, localized famine aggravated by land misuse and expanding population, the growing list of species extinguished or endangered by loss of habitat, the poisoning of soil and streams with toxic wastes, and global warming caused by deforestation and combustion of fossil fuels—these are just a few of the problems that we have created and now must solve. Analyzing environmental issues and planning for better practices start with a basic understanding of ecology, which should be part of every student's education.

Figure 18.4 Rachel Carson. Although her book *Silent Spring* focused on the biosphere's hangover from the pesticide DDT, Carson's message was much broader: "The 'control of nature' is a phrase conceived in arrogance, born of the Neanderthal age of biology and philosophy, when it was supposed that nature exists for the convenience of man." *Silent Spring* was seminal to the modern environmental movement.

CHECKPOINT

1. A (an) __________ consists of a biological __________, or all the biotic factors in the area, along with the nonliving, or __________, factors.
2. Why is it more accurate to define the biosphere as the global ecosystem rather than the global community?
3. Rachel Carson's *Silent Spring* focused on the destructive consequences of toxic pollutants, especially __________.

Answers: 1. ecosystem; community; abiotic **2.** Because the biosphere includes both abiotic and biotic factors **3.** the pesticide DDT

The Evolutionary Adaptations of Organisms

The fields of ecology and evolutionary biology are tightly linked. Charles Darwin was an ecologist (although he predated the word *ecology*). It was the geographic distribution of organisms and their exquisite adaptations to specific environments that provided Darwin with evidence for evolution. Evolutionary adaptation via natural selection results from the interaction of organisms with their environments, which brings us back to our definition of ecology. Thus, events that occur in the short term—or what is sometimes called ecological time—translate into effects over the longer scale of evolutionary time. For instance, hawks feeding on field mice have an impact on the gene pool of the prey population by curtailing the reproductive success of certain individuals. One long-term effect of such a predator-prey interaction may be the prevalence in the mouse population of fur coloration that camouflages the animals.

In our brief survey of organismal ecology, we'll focus on three types of adaptations—physiological, anatomical, and behavioral—that enable plants

Figure 18.5 Environmental activism. Students in Vancouver, British Columbia, continue the tradition of environmental activism by advocating for preservation of an old-growth forest.

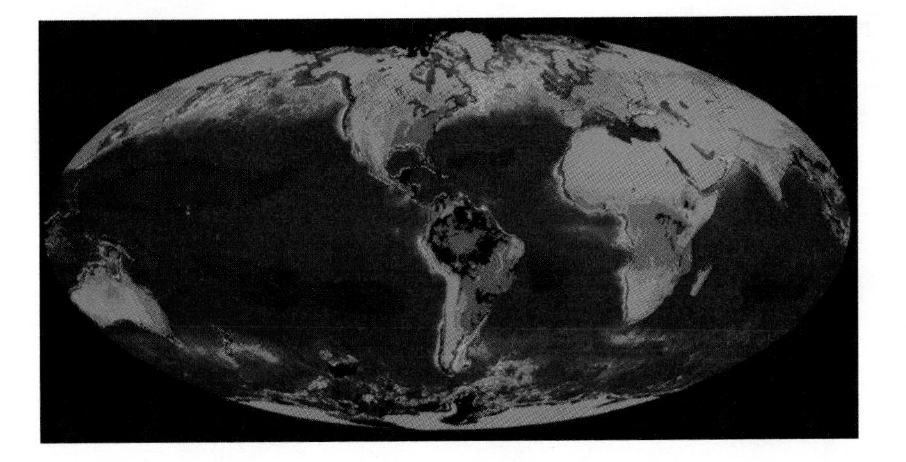

Figure 18.6 Regional distribution of life in the biosphere. In this image of Earth, colors are keyed to the relative abundance of chlorophyll, which correlates with the regional densities of life. Green areas on land are dense forests, including the tropical forests of South America and Africa. Orange areas on land are relatively barren regions, such as the Sahara of Africa. The patchy productivity of the oceans is also evident, with green regions having a greater abundance of phytoplankton (algae) than darker regions. The patchiness of the biosphere is mainly due to regional variations in abiotic factors such as temperature and the availability of water and mineral nutrients.

and animals to adjust to changes in their environments. Note that these changes occur during the lifetime of an individual, so they do not qualify as evolution, which is change in a population over time (see Chapter 13). But an individual organism's abilities to adjust to environmental change during ecological time are themselves adaptations refined by natural selection. We begin by looking at some of the most important abiotic factors and how they might affect the organisms that interact with them.

Abiotic Factors of the Biosphere

The biosphere is patchy. We can see this environmental patchwork on several levels. On a global scale, ecologists have long recognized striking regional patterns in the distribution of terrestrial and aquatic life **(Figure 18.6)**. These patterns mainly reflect regional differences in climate and other abiotic factors. **Figure 18.7** shows patchiness on a local scale; we can see a mixture of forest, small lakes, a meandering river, and open meadows. If we moved even closer, into any one of these different environments, we would find patchiness on yet a smaller scale. For example, we would find that each lake has several different **habitats**—specific environments in which organisms live. And each habitat has a characteristic community of organisms. As with the global patchiness of the biosphere, smaller-scale environmental variation is based mainly on differences in abiotic factors.

Figure 18.7 Patchiness of the environment in an Alaskan wilderness.

Sunlight Solar energy powers nearly all ecosystems. In aquatic environments, the availability of sunlight has a significant effect on the growth and distribution of algae. Because water itself and the microorganisms in it absorb light and keep it from penetrating very far, most photosynthesis occurs near the surface of the water. In terrestrial environments, light is often not the most important factor limiting plant growth. In many forests, however, shading by trees creates intense competition for light at ground level.

Water Aquatic organisms have a seemingly unlimited supply of water, but they face problems of water balance if their own solute concentration does not match that of their surroundings (see Figures 5.14 and 5.15). For a terrestrial organism, the main water problem is the threat of drying out. Many land species have watertight coverings that reduce water loss. For example, a waxy coating on the leaves and other aerial parts of most plants helps prevent dehydration. And humans and other mammals have a layer

Figure 18.8 Home of hot prokaryotes. "Heat-loving" archaea thrive in this pool at Yellowstone National Park, where temperatures may exceed 80°C (176°F).

of dead outer skin containing a waterproofing protein. Moreover, the ability of our kidneys to excrete very concentrated urine is an evolutionary adaptation that enables us to rid our body of urea, a waste product, with minimal water loss.

Temperature Environmental temperature is an important abiotic factor because of its effect on metabolism. For example, few organisms can maintain a sufficiently active metabolism at ambient temperatures near or below 0°C (32°F). At the other extreme, temperatures above 50°C (122°F) destroy the enzymes of most organisms. Still, extraordinary adaptations enable some species to live outside this temperature range. For example, some North American frogs and turtles can freeze during the winter months and still survive, and archaea (one domain of prokaryotes) living in hot springs have enzymes that function optimally at extremely high temperatures **(Figure 18.8)**. Mammals and birds can remain considerably warmer than their surroundings and can be active in a fairly wide range of temperatures, but even these animals cannot tolerate a very hot or very cold environment and function best at moderate temperatures.

Wind Wind affects organisms in many different ways. Some organisms—for example, bacteria, protists, and many insects that live on nutrient-poor snow-covered mountain peaks—depend on nutrients blown to them by wind. Many plants depend on wind to disperse their pollen and seeds. Local wind damage often creates openings in forests, contributing to patchiness in ecosystems. Wind also increases an organism's rate of water loss by evaporation. The consequent increase in evaporative cooling can be advantageous on a hot summer day but can cause dangerous wind chill in the winter. Wind can also affect the pattern of a plant's growth **(Figure 18.9)**.

Figure 18.9 Wind as an abiotic factor that shapes trees. The prevailing winds gradually caused the "flagging" of these fir trees on a ridge on Oregon's Mount Hood. The mechanical disturbance of the wind inhibits limb growth on the windward side of the trees, while limbs on the leeward side grow normally. This growth response is an evolutionary adaptation that reduces the number of limbs that are broken during strong winds.

Rocks and Soil The physical structure and chemical composition of rocks and soil limit the distribution of plants and of the animals that feed on the vegetation. Soil variation contributes to the patchiness we see in terrestrial landscapes such as that in Figure 18.7. In streams and rivers, the composition of the underlying rock can affect water chemistry, which in turn influences the resident plants and animals. In marine environments, the structure of the seafloor determines the types of organisms that can attach to or burrow in those habitats.

Periodic Disturbances Catastrophic disturbances, such as fires, hurricanes, tornadoes, and volcanic eruptions, can devastate biological communities.

After the disturbance, the area is recolonized by organisms or repopulated by survivors, but the structure of the community undergoes a succession of changes during the rebound **(Figure 18.10)**. Some disturbances, such as volcanic eruptions, are so infrequent and irregular over space and time that organisms have not acquired evolutionary adaptations to them. Fire, on the other hand, although unpredictable over the short term, is a recurrent phenomenon in some communities, and many plants have adapted to this periodic disturbance. For example, manzanita, an evergreen shrub that grows in western North America, actually requires the heating and chemical effects of fire to stimulate the germination of its seeds. In fact, several communities—such as the chaparral in which manzanita grows—require periodic fires to maintain them.

Figure 18.10 Recovery after a forest fire. Just a few months after a forest fire swept through Yellowstone National Park (left), wildflowers and other small plants were already colonizing the area (right). The increased sunlight and soil nutrients released from the trees that burned were among the abiotic factors contributing to the regreening of the scorched land.

Physiological Responses

You may have seen a cat's fur fluff up on a cold day, a response that helps insulate the animal's body. The mechanism for this response is the contraction of tiny muscles attached to the hairs. (Our own muscles do this, too, but we just get "goose bumps" instead of a furry insulation.) The blood vessels in the cat's skin also constrict, which slows the loss of body heat. (This works for humans, too.) These are examples of physiological responses to environmental change. In these mechanisms of temperature regulation (thermoregulation), the response occurs in just seconds.

Physiological response that is longer term, though still reversible, is called **acclimation.** For example, suppose you moved from Boston, which is essentially at sea level, to the mile-high city of Denver. One physiological response to the lower oxygen supply in your new environment would be a gradual increase in the number of your red blood cells, which transport O_2 from your lungs to other parts of your body. Acclimation can take days or weeks. This is why high-altitude climbers, such as those attempting to scale Mount Everest, need extended stays at a high-elevation base camp before proceeding to the summit.

The ability to acclimate is generally related to the range of environmental conditions the species naturally experiences. Species that live in very warm climates, for example, usually do not acclimate to extreme cold. Among vertebrates, birds and mammals can generally tolerate the greatest temperature extremes because, as endotherms, they use their metabolism to regulate internal temperature (see Chapter 17). In contrast, ectothermic reptiles can only tolerate more limited climates **(Figure 18.11)**.

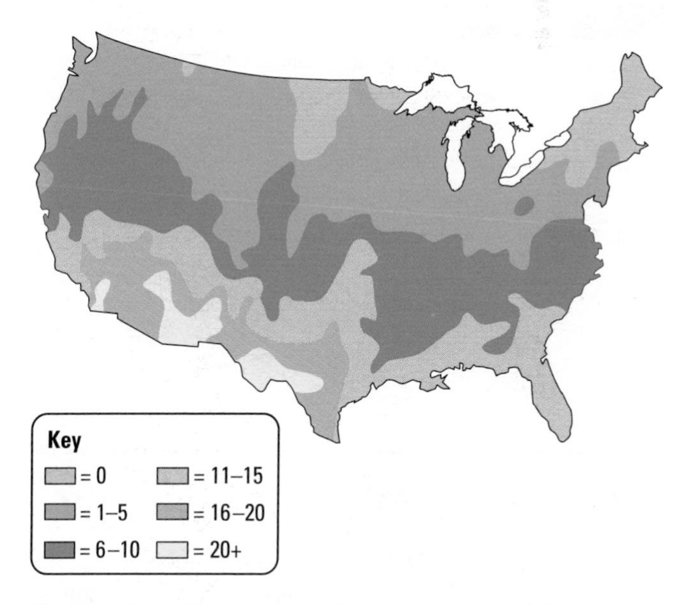

Figure 18.11 The number of lizard species in different regions of the contiguous United States. Notice that there are fewer and fewer lizard species in more northern regions. This reflects lizards' ectothermic physiology, which depends on environmental heat for keeping the body warm enough for the animal to be active.

Anatomical Responses

Many organisms respond to environmental challenge with some type of change in body shape or anatomy (structure). When the change is reversible, the response is an example of acclimation. Many mammals and birds, for example, grow a heavier coat of fur or feathers in winter; sometimes coat color changes seasonally as well, camouflaging the animal against winter snow and summer vegetation.

Other anatomical changes are irreversible over the lifetime of an individual. Environmental variation can affect growth and development so much that there may be remarkable differences in body shape within a population. You can see an example in Figure 18.9, which shows the "flagging" that wind causes in certain trees. In general, plants are more anatomically plastic than animals. Rooted and unable to move to a better location, plants rely entirely on their anatomical and physiological responses to survive environmental fluctuations.

Behavioral Responses

In contrast to plants, most animals can respond to an unfavorable change in the environment by moving to a new location. Such movement may be fairly localized. For example, many desert ectotherms, including reptiles, maintain a reasonably constant body temperature by shuttling between sun and shade. Some animals, however, are capable of migrating great distances in response to such environmental cues as the changing seasons. Many migratory birds overwinter in Central and South America, returning to northern latitudes to breed during summer. And humans, with their large brains and available technology, have an especially rich range of behavioral responses available to them **(Figure 18.12)**.

Figure 18.12 Behavioral responses have expanded the geographic range of humans. Dressing for the weather is a thermoregulatory behavior unique to humans.

CHECKPOINT

1. __________ energy is such an important abiotic factor because __________ provides most of the organic fuel and building material for the organisms of most ecosystems.
2. Why does the set of evolutionary adaptations characterizing a species tend to limit the geographic distribution of that species?
3. What is acclimation?

Answers: 1. Solar (or light); photosynthesis **2.** To the extent that the species is adapted to particular environmental conditions, it is not so well equipped to survive and reproduce where there are different conditions to which the species is not adapted. **3.** A gradual, reversible change in anatomy or physiology in response to an environmental change

What Is Population Ecology?

Now that you have a basic understanding of the factors that make up the environment and how organisms respond to them, we will turn our attention to another of the four hierarchical levels of ecology: population ecology.

No population can grow indefinitely. Species sometimes exhibit population explosions, but their populations inevitably crash. In contrast to these radical booms and busts, many populations are relatively stable over time, with only minor increases or decreases in population size. Population ecology focuses on the factors that influence a population's size (number of individuals), growth rate (rate of change in population size), density (number of individuals per unit area or volume), and structure (such as relative numbers of individuals of different ages).

Biologists define a population as a group of individuals of the same species living in a given area at a given time (see Chapter 13). A population's geographic boundaries may be natural, as with certain species of trout in an isolated lake. But ecologists often define a population's boundaries in more arbitrary ways that fit their research questions. For example, an ecologist studying the contribution of asexual reproduction to the population growth of sea anemones might define a population as all the anemones of one species in a particular tide pool. Another researcher studying the effects of hunting on deer might define a population as all the deer within a particular state. Yet another researcher, attempting to determine which segment of the human population will be most affected by the AIDS epidemic, might study the HIV infection rate of the human population in one nation or throughout the world. Whatever the scale of the population we're studying, there are some common principles of population structure and growth that will guide our analysis.

Figure 18.13 An indirect census of a prairie dog population. We could estimate the number of prairie dogs in this colony in South Dakota by counting the number of mounds constructed by the rodents. The estimate is rough because the animals are social, and the number of individuals that cohabit the system of tunnels under each mound varies.

Population Density

Population density is the number of individuals of a species per unit area or volume—the number of oak trees per square kilometer (km^2) in a forest, for example, or the number of earthworms per cubic meter (m^3) in the forest's soil.

How do we measure population density? In rare cases, it is possible to actually count all individuals within the boundaries of the population. For example, we could count the total number of oak trees (say, 200) in a forest covering 50 km^2. The population density would be the total number of trees divided by the area, or 4 trees per square kilometer ($4/km^2$).

In most cases, it is impractical or impossible to count all individuals in a population. Instead, ecologists use a variety of sampling techniques to estimate population density. For example, they might estimate the density of alligators in the Florida Everglades based on a count of individuals in a few sample plots of 1 km^2 each. Generally speaking, the larger the number and size of sample plots, the more accurate the estimates. In some cases, population densities are estimated not by counts of organisms but by indirect indicators, such as number of bird nests or rodent burrows **(Figure 18.13)**.

Another sampling technique commonly used to estimate wildlife populations is the **mark-recapture method (Figure 18.14)**. The researcher places traps within the boundaries of the population under study. The researcher then marks the captured animals with tags, collars, bands, or spots of dye. The marked animals are released. After a few days or weeks—enough time for the marked individuals to mix randomly with unmarked members of the population—traps are set again. This second capture will yield both marked and unmarked individuals. From these data, the total number of individuals in the population can be estimated. The mark-recapture method assumes that each marked individual has the same probability of being trapped as each unmarked individual. This is not always a safe assumption, because an animal that has been trapped once may be wary of traps in the future.

Figure 18.14 A mark-recapture estimate of population size. This biologist is using the mark-recapture method to estimate the number of individuals in a population of birds called sanderlings. Let's say that he has captured 50 sanderlings in a harmless trap called a mist net. He marks the birds with leg bands (inset) and then releases them. A second capture two weeks later yields a total of 100 sanderlings, of which 10 are marked birds that have been recaptured. We can estimate that 10% of the total sanderling population is marked. Since the biologist marked 50 birds, our estimate for the entire population is about 500 birds.

Patterns of Dispersion

The **dispersion pattern** of a population is the way individuals are spaced within the population's geographic range. A clumped pattern, in which individuals are aggregated in patches, is the most common in nature. Clumping often results from an unequal distribution of resources in the environment. For instance, cottonwood trees are usually clumped along a streamside in patches of moist and sandy soil. Clumping of animals is often associated with uneven food distribution, mating or other social behavior, or predator avoidance. For instance, mosquitoes often swarm in great numbers, which increases their chances for mating. Schooling fishes are another example of clumped dispersion **(Figure 18.15a)**.

A uniform pattern of dispersion often results from interactions among the individuals of a population. For instance, creosote bushes in the desert tend to be uniformly spaced because their roots compete for water and dissolved nutrients. Animals often exhibit uniform dispersion as a result of social interactions. Examples are birds nesting in large numbers on small islands **(Figure 18.15b)**.

In a third type of dispersion, random dispersion, individuals in a population are spaced in a patternless, unpredictable way. This only occurs in the absence of strong attractions or repulsions among individuals in a population. Clams living in a coastal mudflat, for instance, might be randomly dispersed at times of the year when they are not breeding and when resources are plentiful and do not affect their distribution. Forest trees are also randomly distributed in some cases **(Figure 18.15c)**. However, environmental conditions and social interactions make random dispersion rare.

Some populations exhibit both clumped and uniform dispersion patterns, but on different scales. For instance, if you studied dispersion patterns of the human population in the northeastern United States **(Figure 18.16)**, you would find most of the population clumped in metropolitan areas, such as New York City and Boston. Within each clump, however, individuals or family groups might be more or less uniformly dispersed—for example, in housing subdivisions of fixed lot size.

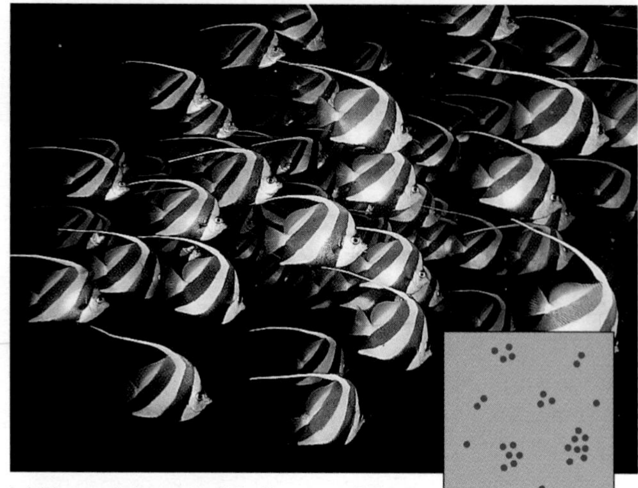

(a) Clumped. Butterfly fish, like many fishes, often clump into schools. Schooling may enhance the hydrodynamic efficiency of swimming, reduce predation ("safety in numbers"), and increase feeding efficiency.

(b) Uniform. Birds nesting on small islands, such as these king penguins on South Georgia Island in the South Atlantic, often exhibit uniform spacing. Territorial behavior helps stabilize the distance between individuals.

(c) Random. Trees of one species are often randomly distributed among trees of other species in tropical rain forests.

Figure 18.15 Patterns of dispersion within a population's geographic range.

Population Growth Models

To appreciate the explosive potential for population increase, consider a single bacterium that can reproduce by fission every 20 minutes under ideal laboratory conditions. After 20 minutes, there would be two bacteria, four after 40 minutes, and so on. If this continued for only a day and a half—a mere 36 hours—there would be bacteria enough to form a layer a foot deep over the entire Earth! At the other extreme, elephants may produce only six young in a 100-year life span. Still, it would take only 750 years for a single mating pair of elephants to give rise to a population of 19 million. However, indefinite increase does not occur, either in the laboratory or in nature. A small population in a favorable environment may increase rapidly for a while, but eventually the numbers must, as a result of limited resources and other factors, stop growing.

Ecologists use several measures to calculate changes in population size over time. The growth of a population is equal to the number of births minus the number of deaths. The **growth rate** is the change in population size per time interval. In the chapter-opening essay, for example, we mention that the growth rate for the world's human population is about 74 million additional people per year. Now let's take a look at two models that will help us understand the raw potential and limitations for the growth rates of populations.

The Exponential Growth Model: The Ideal of an Unlimited Environment

The rate of expansion of a population under ideal, unregulated conditions is described by the **exponential growth model,** in which the whole population

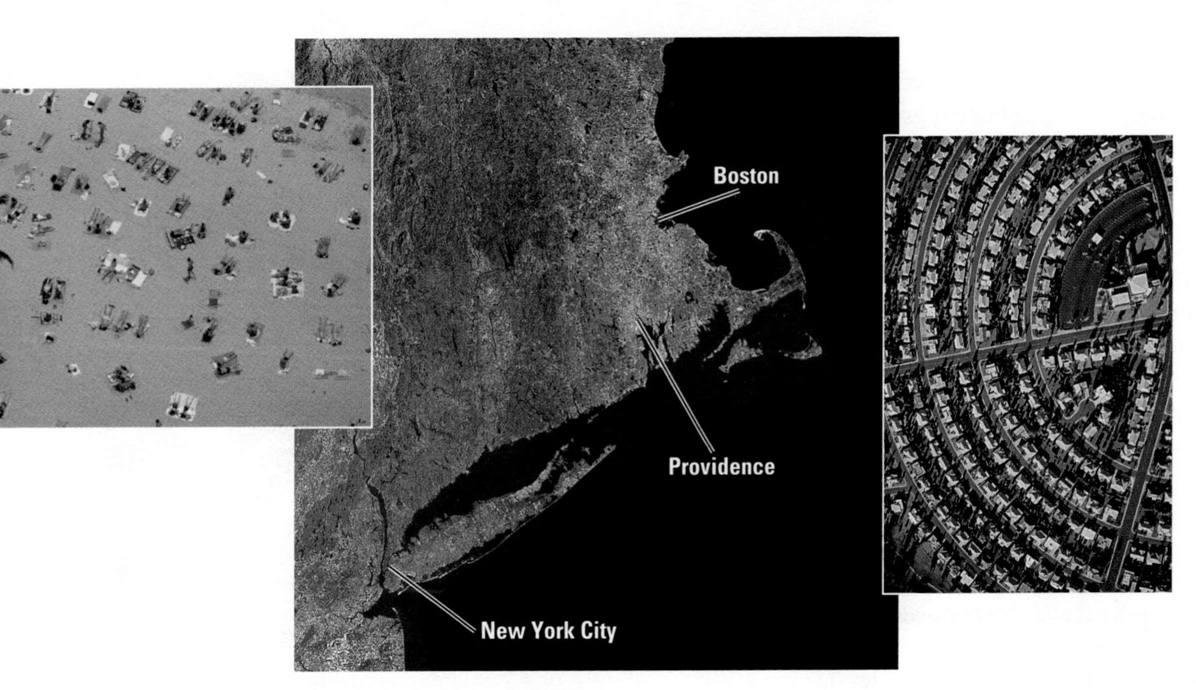

Figure 18.16 Different scales, different spacing patterns. On the large-scale view of the northeastern United States (center), the human population is mostly clumped into cities and towns (purple areas). But on a more local scale, humans tend to be territorial enough to distribute themselves uniformly.

multiplies by a constant factor during constant time intervals (the generation time). For example, the constant factor for the bacterial population represented in **Figure 18.17** is 2, because each parent cell splits to produce two daughter cells. The generation time for these bacteria is 20 minutes. The progression for bacterial growth—2, 4, 8, 16, and so on—is the number 2 raised to a successively higher power (exponent) each generation (that is, 2^1, 2^2, 2^3, 2^4, and so on).

Here's a key feature of the exponential growth model: The rate at which a population grows depends on the number of individuals already in the population. It's like compound interest on a long-term savings account at a fixed annual yield. At 7% per annum, you make only $70 the first year on a $1,000 deposit. But leave it in the bank for 50 years, and you'll be earning about $2,200 per year in interest on an account that has grown to over $30,000. Similarly, in our model for exponential population growth, the bigger the population size, the faster the population increases. As time goes by, the population gets bigger faster and faster.

The exponential growth model gives an idealized picture of the unregulated growth of a population. For bacteria, unregulated growth means there is no restriction on the ability of the cells to live, grow, and reproduce. Given a few days of unregulated growth, bacteria would smother every other living thing. Obviously, long periods of exponential increases are not common in the real world, or life could not continue on Earth. Where we do observe exponential growth in nature, it is generally a short-lived consequence of organisms being introduced to a new or underexploited environment.

The Logistic Growth Model: The Reality of a Limited Environment

In nature, a population may grow exponentially for a while, but eventually, one or more environmental factors will limit its growth. Population size then stops increasing or may even crash. Environmental factors that restrict population growth are called **population-limiting factors.** The

Time	Number of Cells	
0 minutes	1	$= 2^0$
20	2	$= 2^1$
40	4	$= 2^2$
60	8	$= 2^3$
80	16	$= 2^4$
100	32	$= 2^5$
120 (= 2 hours)	64	$= 2^6$
3 hours	512	$= 2^9$
4 hours	4096	$= 2^{12}$
8 hours	16,777,216	$= 2^{24}$
12 hours	68,719,476,736	$= 2^{36}$

(a) Exponential growth. Note that after 120 minutes the time is expressed in hours. The exponents reflect the number of 20-minute generations.

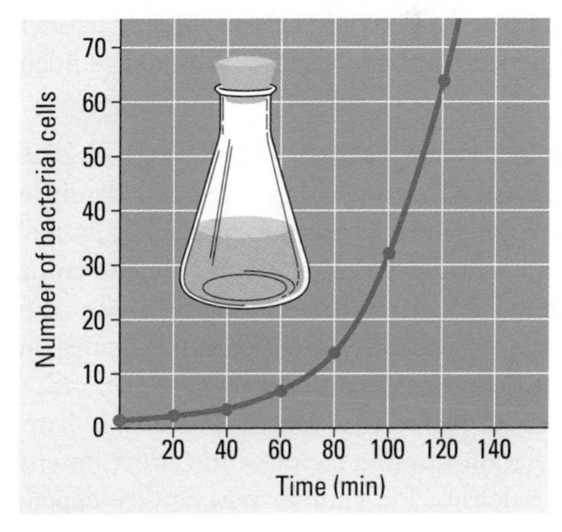

(b) Plotting the data.

Figure 18.17 Exponential growth of a bacterial colony.

graph for the seal population in **Figure 18.18** resembles what is called the **logistic growth model,** a description of idealized population growth that is slowed by limiting factors. **Figure 18.19** contrasts the logistic growth model with the exponential growth model.

Carrying capacity is the number of individuals in a population that the environment can just maintain ("carry") with no net increase or decrease. For the fur seal population in Figure 18.18, for instance, the carrying capacity is about 10,000 mated males. This value varies, depending on species and habitat. Carrying capacity might be considerably less than 10,000 for a fur seal population on a smaller island with fewer breeding sites, for example. At carrying capacity, the population is as big as it can theoretically get in its environment, and the population growth rate is zero.

The logistic model predicts that a population's growth rate will be low when the population size is either small or large, and highest when the population is at an intermediate level relative to the carrying capacity. At a low population level, resources are abundant, and the population is able to grow nearly exponentially. However, at this point the increase is small because the population is small. In contrast, at a high population level, population-limiting factors strongly oppose the population's potential to increase. In nature, there might be less food available per individual or fewer breeding sites, nest sites, or shelters. The limiting factors make the birth rate decrease, the death rate increase, or both. Eventually, the population stabilizes at the carrying capacity, when the birth rate equals the death rate.

Both the logistic growth model and the exponential growth model are theoretical ideals. No natural population fits either one perfectly. However, these models are useful starting points for studying population growth. Ecologists use them to predict how populations will grow in certain environments and as a basis for constructing more complex models. The models have stimulated research leading to an improved understanding of populations in nature.

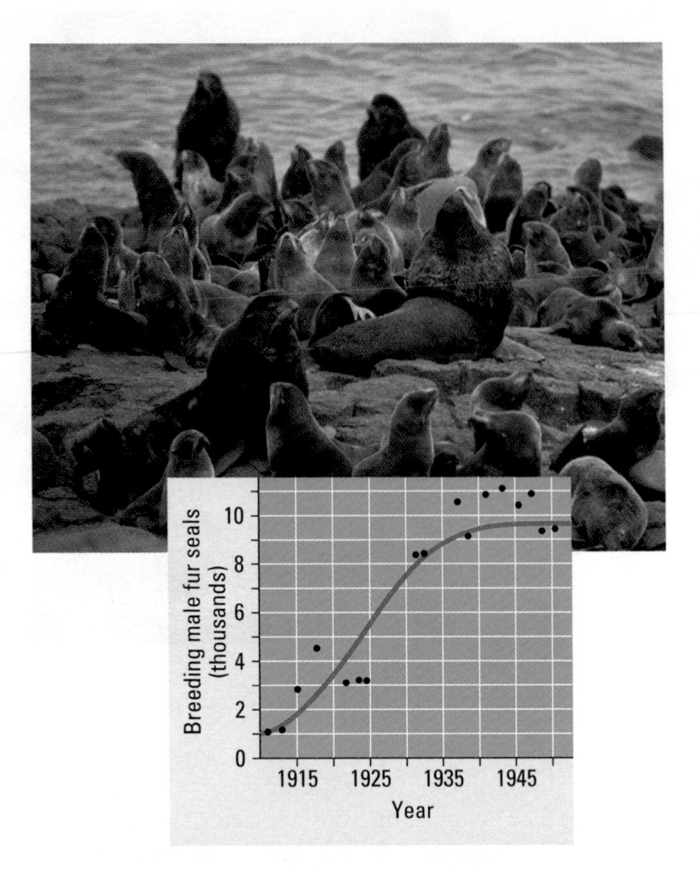

Figure 18.18 Effect of population-limiting factors on the growth of an animal population: the logistic growth model. The graph profiles the growth of a population of fur seals on Saint Paul Island, off the coast of Alaska. (For simplicity, only the mated bulls were counted. Each has a harem of females, as shown in the photograph.) Before 1915, the seal population on the island remained low because of uncontrolled hunting, although it changed from year to year. After hunting was controlled, the population increased rapidly until about 1935. At this point, a number of population-limiting factors, including some hunting and the amount of space suitable for breeding, restricted growth, and the population leveled off and began fluctuating around a size of about 10,000 bull seals.

Regulation of Population Growth

Let's take a closer look at the population-limiting factors that contribute to carrying capacity. Ecologists classify these factors into two categories: density-dependent factors and density-independent factors.

Density-Dependent Factors The major biological assumption of the logistic model is that increasing population density reduces the resources available for individual organisms, ultimately limiting population growth. The logistic model is a description of **intraspecific competition**—competition between individuals of the same species for the same limited resources. As population size increases, competition becomes more intense, and the growth rate declines in proportion to the intensity of competition. Thus, population growth rate is density dependent. A **density-dependent factor** is a population-limiting factor whose effects intensify as the population increases in density. Put another way, density-dependent factors affect a greater percentage of individuals in a population as the number of individuals increases. Limited food supply and the buildup of poisonous wastes are examples of density-dependent factors. Such factors depress a population's

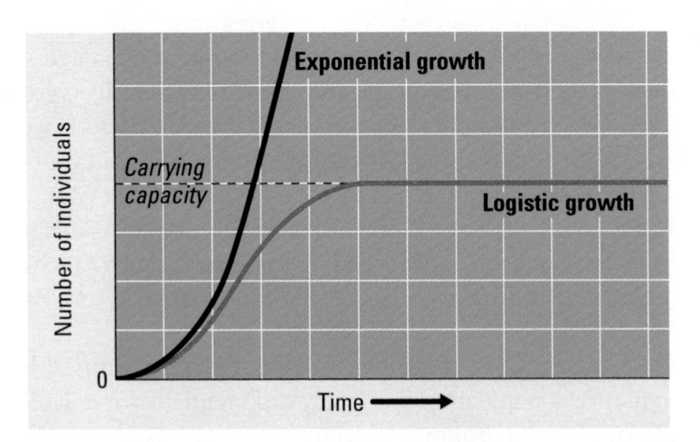

Figure 18.19 Logistic growth and exponential growth contrasted.

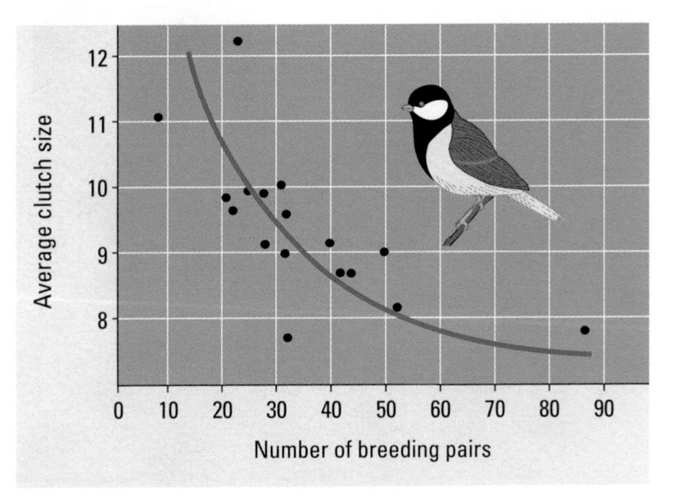

(a) Decreasing birth rate with increasing density

(b) Increasing death rate with increasing density

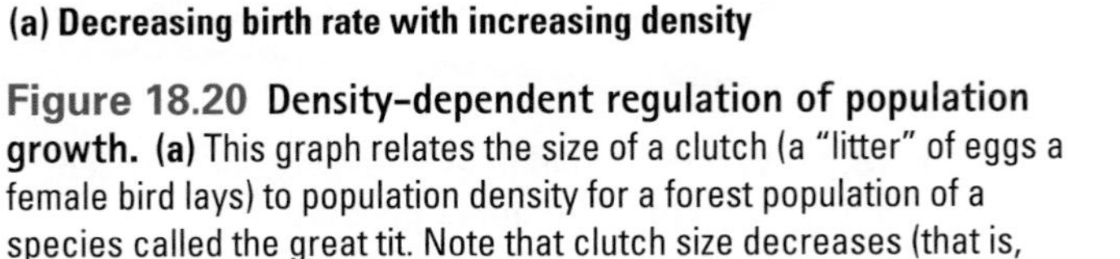
Figure 18.20 Density-dependent regulation of population growth. **(a)** This graph relates the size of a clutch (a "litter" of eggs a female bird lays) to population density for a forest population of a species called the great tit. Note that clutch size decreases (that is, birth rate decreases) as population density increases. **(b)** In this laboratory culture of flour beetles, the percentage of individuals that survive from the egg stage to reproductive maturity decreases (that is, death rate increases) as population density increases.

growth rate by increasing the death rate, decreasing the birth rate, or both **(Figure 18.20)**.

We often see density-dependent regulation in laboratory populations. For example, when a pair of fruit flies are placed in a jar with a constant amount of food added each day, population growth fits the logistic model. After a rapid increase, population growth levels off as the flies become so numerous that they outstrip their limited food supply. Each individual in a large population has a smaller share of the limited food than it would in a small population. Also, the more flies, the more concentrated the poisonous wastes become in the jar.

Laboratory populations are one thing; natural populations are another. We do not often see clear-cut cases of density-dependent factors regulating populations in nature. To test whether such factors are operating, it is necessary to change the density of individuals in the population while keeping other factors constant. This is sometimes done in the management of game populations. For instance, state agencies may allow hunters to reduce populations of white-tailed deer to levels that keep the animals from permanently damaging the plants they use for food. White-tailed deer are browsers, preferring the highly nutritious parts of woody shrubs—young stems, leaves, and buds. When deer populations are kept low and high-quality food is therefore abundant, a high percentage of females become pregnant and bear offspring; in fact, many of them produce twins **(Figure 18.21)**. On the other hand, when populations are high and food quality is poor, many females fail to reproduce at all. These observations support the hypothesis that food supply is a density-dependent factor regulating white-tailed deer populations.

Figure 18.21 Increased birth rates in times of plenty. In deer populations, the birth of twin fawns is much more common when population densities are low.

Density-Independent Factors A population-limiting factor whose intensity is unrelated to population density is called a **density-independent factor.** Examples are such abiotic factors as unfavorable changes in the weather. A freeze in the fall, for example, may kill a certain percentage of insects in a population. The date and severity of the first freeze obviously are not affected by the density of the insect population. Density-independent factors such as a killing frost affect the same percentage of individuals regardless of population size. (In larger populations, of course, greater numbers will die.)

In many natural populations, density-independent factors limit population size well before resources or other density-dependent factors become important. In such cases, the population may decline suddenly. If we look at the growth curve of such a population, we see something like exponential growth followed by a rapid decline rather than a leveling off. Ecologists have observed such history, for example, in certain populations of aphids, insects that feed on the sap of plants (**Figure 18.22**). Aphids and many other insects often show virtually exponential growth in the spring and then rapid die-offs when it becomes hot and dry in the summer. A few individuals may remain, and these may allow population growth to resume again if favorable conditions return. Some insect populations—many mosquitoes and grasshoppers, for instance—will die off entirely, leaving only eggs, which will initiate population growth the following year. In addition to seasonal changes in the weather, abrupt environmental trauma, such as fire, floods, storms, and habitat disruption by human activity, can affect populations in a density-independent manner.

Over the long term, most populations are probably regulated by a mixture of density-independent and density-dependent factors. Many populations remain fairly stable in size and are presumably close to a carrying capacity that is determined by density-dependent factors. In addition, however, many show short-term fluctuations due to density-independent factors. In some cases, the distinction between density-dependent and density-independent factors is not clear. In the case of white-tailed deer, for example, many individuals may starve to death in very cold, snowy areas. The severity of this effect is related to the harshness of the winter; cold temperatures increase energy requirements (and therefore the need for food), while deeper snow makes it harder to find food. But the severity of the effect is also density dependent, because the larger the population, the less food available per individual.

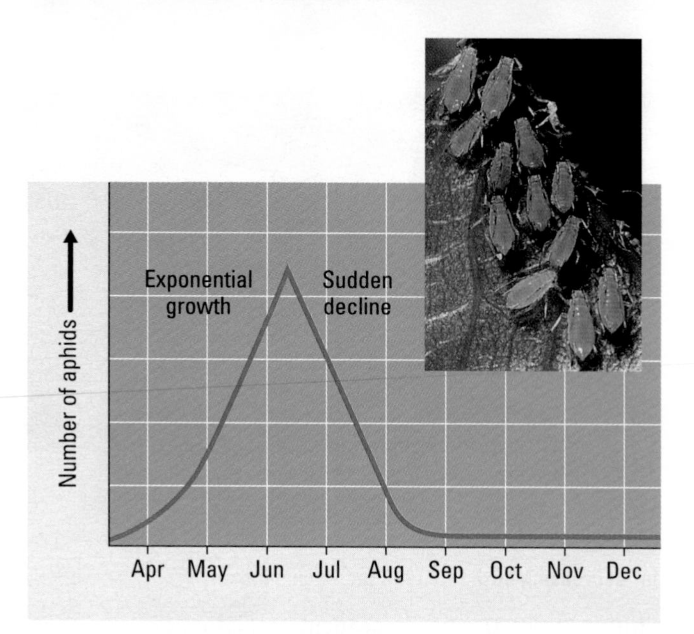

Figure 18.22 Weather change as a density-independent factor limiting growth of an aphid population.

Population Cycles Some populations of insects, birds, and mammals have regular boom-and-bust cycles. Perhaps the most striking is that of the periodic cicadas, grasshopper-like insects that complete their life cycle every 13 or 17 years, emerging from the ground at phenomenal densities (as high as 600 individuals per square meter). This long life cycle may be an adaptation that reduces predation; few predators can wait 13 or 17 years for their prey to appear.

As a case study of population cycles, let's examine the boom-and-bust cycles of the snowshoe hare and one of its predators, the lynx. Both animals inhabit the northern coniferous forests of North America (**Figure 18.23**). About every ten years, both hare and lynx populations have a rapid increase (a "boom") followed by a sharp decline (a "bust"). What causes these boom-and-bust cycles? The ups and downs in the two populations seem to almost match each other on the graph. Does this mean that changes in one directly affect the other? In other words, does predation by the lynx make the hare population fluctuate, and do the ups and downs of the hare population cause the changes in the lynx population? For the lynx and many other predators that depend heavily on a single species of prey, the availability of prey can influence population changes. Thus, the ten-year cycle in the lynx population probably results at least in part from the ten-year cycle in the hare

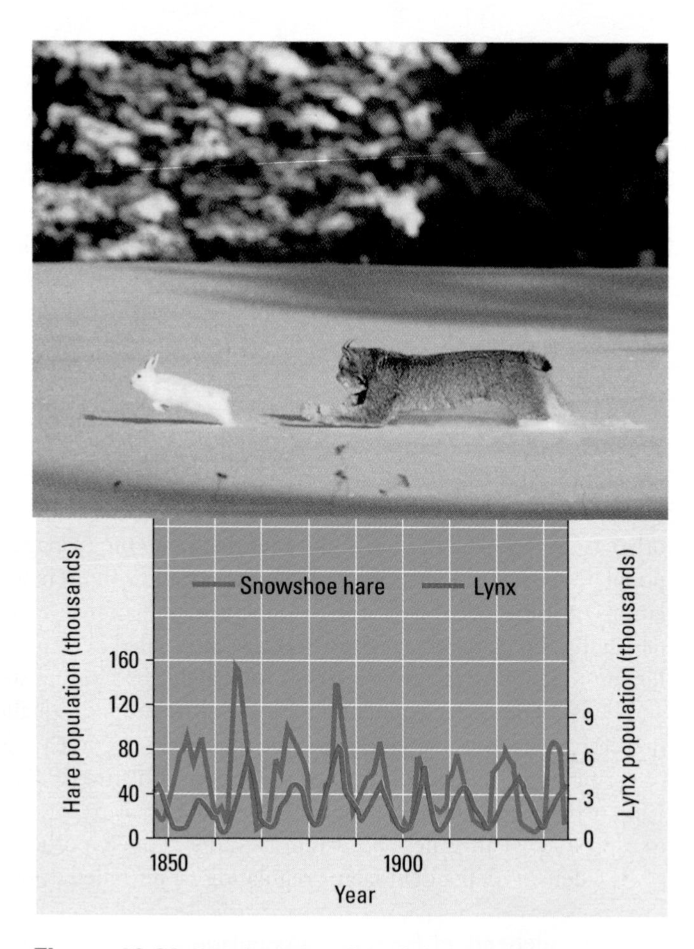

Figure 18.23 Population cycles of the snowshoe hare and the lynx. Research has suggested that the parallel nature of the hare and lynx cycles is the result of several factors, including the availability of plants and the predator-prey relationship between the lynx and the hare.

population. But we cannot conclude from the graph that lynx predation alone causes the cyclic changes in the hare population. In fact, researchers have discovered that hare populations will cycle about every ten years, whether or not lynx are present. Another hypothesis is that fluctuations in hare populations are tied to fluctuations in the populations of plants that are food for the hares. There is evidence that when certain plants are damaged by herbivores, the nutrient content of the plants decreases. Experimental studies performed in the field support the hypothesis that the ten-year cycle of snowshoe hares results from the combined effects of predation and fluctuations in the hare's food sources.

Populations of many rodents also exhibit boom-and-bust changes, often cycling every three to five years. For such short-term cycles, some researchers postulate that either predation or food supply alone may be the underlying cause. Another hypothesis, based on laboratory studies of mice and other small rodents, is that stress from crowding may alter hormonal balance and reduce fertility. The causes of cycles probably vary among species and maybe even among populations of the same species.

Human Population Growth

Now that we have examined some general concepts of population dynamics, let's return to the specific case of the human population.

The History of Global Population Growth The human population has been growing almost exponentially for centuries. In fact, if we compare the history of human population growth in **Figure 18.24** with the exponential growth model in Figure 18.17b, it almost looks as if we've been multiplying like bacteria. Of course, our generation span is about 20 years instead of the mere 20 minutes for some bacteria, so our population explosion has been stretched out in time. Still, when we compare the two graphs, it seems as though we've been proliferating into the space and resources of the biosphere as though it were an enormous petri dish. And like bacterial growth in a real petri dish, human exponential growth cannot continue forever.

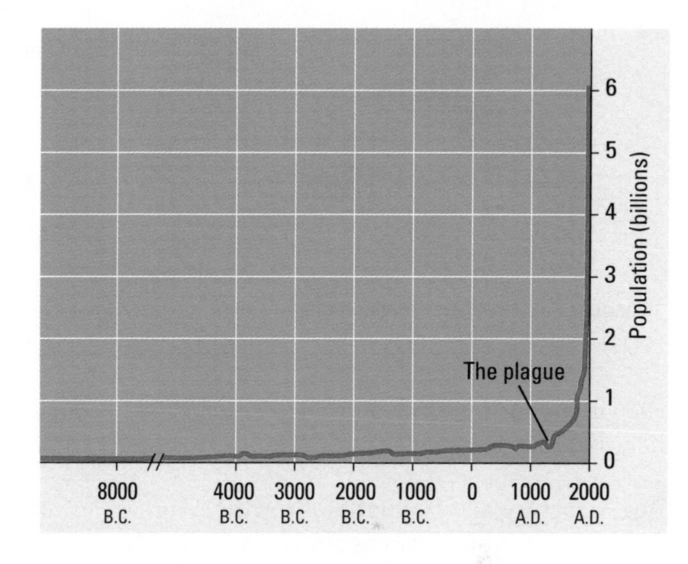

Figure 18.24 The history of human population growth.

You can see in Figure 18.24 that the human population increased relatively slowly until about the mid-1600s, when approximately 500 million people inhabited Earth. The population doubled to 1 billion within the next two centuries, doubled again to 2 billion between 1850 and 1930, and doubled still again by 1975 to more than 4 billion. Projections of future trends in population growth vary, but most predict that the total human population will reach 7–8 billion by 2025.

Human population growth is based on the same two general parameters that affect other animal and plant populations: birth rates and death rates. Birth rates increased and death rates decreased when agricultural societies replaced a lifestyle of hunting and gathering about 10,000 years ago. Since the Industrial Revolution (which began in the mid-1700s), virtually exponential growth has resulted mainly from a drop in death rates, especially infant mortality, even in the least developed countries. Improved nutrition, better medical care, and sanitation have all contributed to an increased percentage of newborns that survive long enough to leave offspring of their own. A decrease in mortality coupled with birth rates that are still relatively high in most developing countries results in an actual increase in population growth rates. In Sri Lanka, for example, the birth rate has been decreasing over the past 60 years, but it has never declined

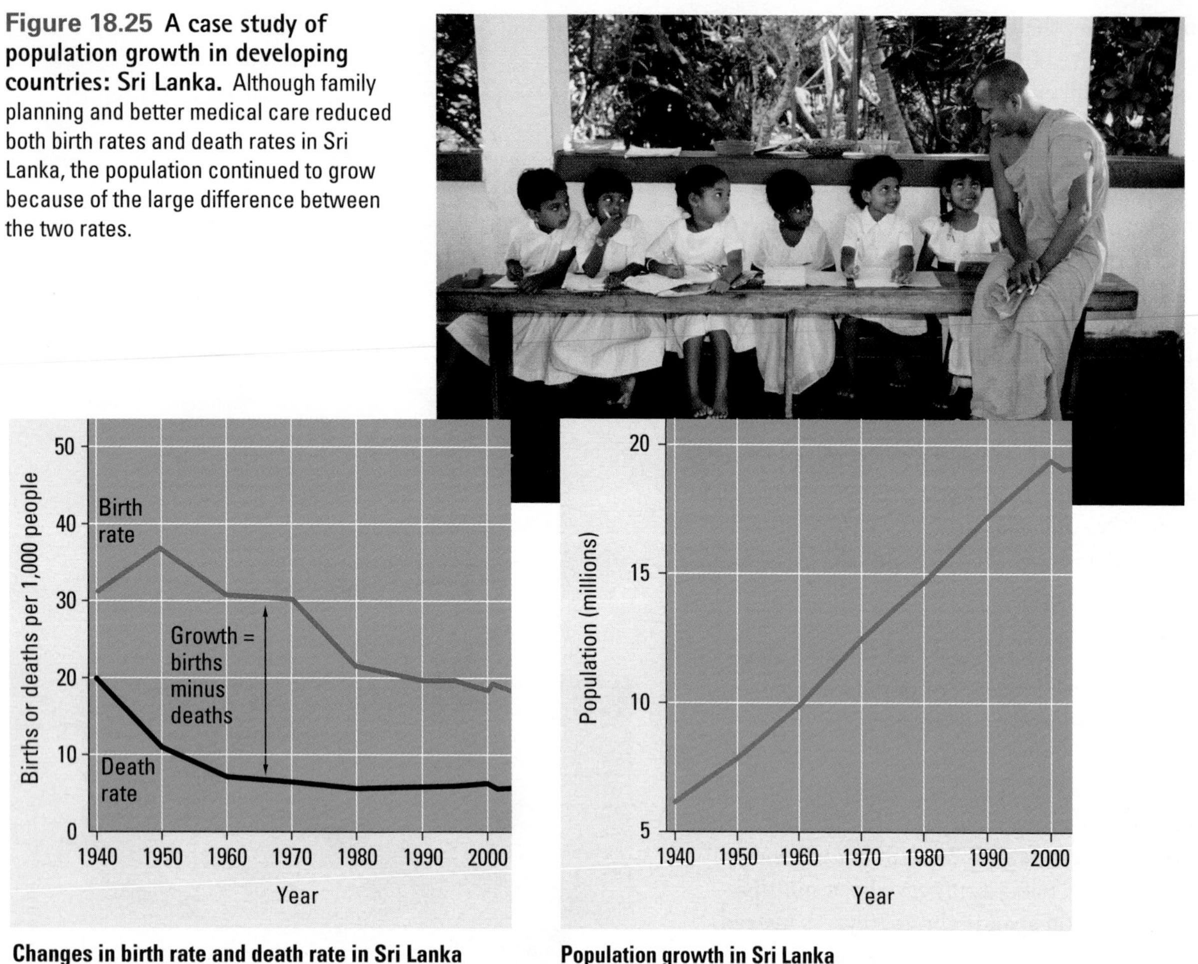

Figure 18.25 A case study of population growth in developing countries: Sri Lanka. Although family planning and better medical care reduced both birth rates and death rates in Sri Lanka, the population continued to grow because of the large difference between the two rates.

Changes in birth rate and death rate in Sri Lanka

Population growth in Sri Lanka

enough to offset the drop in the death rate. Thus, the Sri Lankan population continues to grow **(Figure 18.25)**.

Age Structure and Population Growth Worldwide, human population growth is a mosaic of various rates of growth in different countries. Some developed countries, such as Sweden, have stable populations because birth rates and death rates balance. In sharp contrast to Sweden, most developing nations have burgeoning populations in which birth rates greatly exceed death rates. Partly as a result of such unchecked growth, many people in such countries face serious housing, water, and food shortages, as well as severe pollution problems.

A population characteristic called age structure can help us predict the future growth of populations in different countries. The **age structure** of a population is the proportion of individuals in different age-groups. In Italy, for instance, individuals younger than reproductive age are relatively underrepresented in the population **(Figure 18.26)**. This contributes to that country's slightly negative population growth. In contrast, Afghanistan has an age structure that is bottom-heavy, skewed toward young individuals who will grow up and sustain the explosive growth (currently 2.4% per year) with their own reproduction. The United States is growing slowly, at a rate of about 0.6% per year.

Notice in Figure 18.26 that the age structure for the United States has a bulge that corresponds to the "baby boom" that lasted for about two decades after the end of World War II. Even though couples born

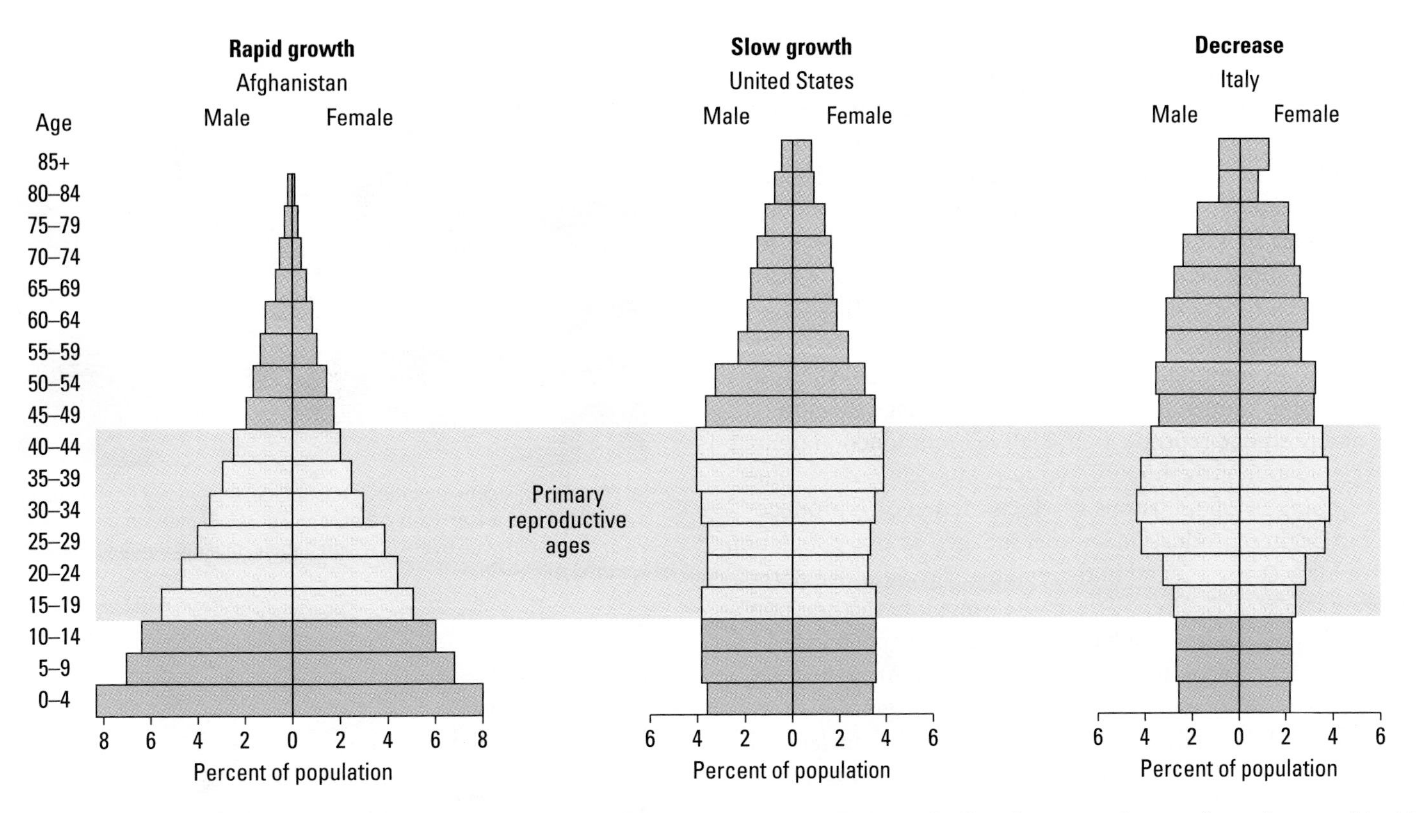

Figure 18.26 Age structures of three nations. The proportion of individuals in different age groups has a significant impact on the potential for future population growth. Afghanistan, like many developing countries, has a disproportionately large number of young people, and their future reproduction will sustain a steep population increase for many years. In contrast, Italy's population is distributed more evenly over all age classes, with a high proportion of individuals past their prime reproductive years. The United States has a fairly even age distribution up to age 55 except for the bulge corresponding to the post–World War II "baby boom." (The data shown are from 2004.)

during those years had an average of fewer than two children the population continued to increase during the past few decades because there were so many "boomers" of reproductive age. Because the boomers are now in their 50s and early 60s, they are having less impact on population growth. The U.S. population is still growing, however, because the combination of birth rate and immigration exceeds the death rate. Immigration (legal and illegal combined) now contributes about 40% of the current growth of the U.S. population. And while the average woman in the United States tends to have only two children, the total number of women having children remains high enough for the birth rate to exceed the death rate. Population researchers predict that the U.S. population will continue to grow well into the 21st century, perhaps increasing from about 300 million today to about 390 million by the year 2050.

The Sociology, Economics, and Politics of Population Growth Age-structure diagrams not only reveal a population's growth trends; they also relate to social conditions. Based on the diagrams in Figure 18.26, we can predict, for instance, that employment for an increasing number of working-age people will continue to be a significant problem for Afghanistan in the foreseeable future. For Italy and the United States, a decreasing proportion of working-age people—mostly those of college age today—will be supporting an increasing proportion of retired people. Programs such as the U.S. Social Security system and Medicare, which are crucial to many older citizens, will become severely strained as the proportion of senior citizens swells.

Predictions of future trends in global human population growth vary widely. The gloomiest models predict a continuing high rate of growth through the 21st century, with a doubling to about 12 billion as early as 2050. Perhaps more realistic is a computer model based on the almost-global trend toward smaller families. This model predicts that by about 2080, the human population will peak at about 10.6 billion and then begin a slight decline to about 10.4 billion by the end of the 21st century. Either way, there will be a lot more people consuming resources and dumping pollutants, bruising a biosphere that already ails **(Figure 18.27)**.

A unique feature of human population growth is our ability to control it with contraception. In addition, social change and the rising educational and career aspirations of women in many cultures encourage them to delay marriage and postpone reproduction. Delayed reproduction dramatically decreases population growth rates. You can get a sense of this phenomenon by imagining two populations in which women each produce three children but begin reproduction at different ages. In one population, females first give birth at age 15, and in the other, at age 30. If we start with a group of newborn girls, after 30 years the women in the first population will already begin to have grandchildren, whereas women in the second population will be giving birth to their first child. After 60 years, women in the first population will have a large number of great-great-grandchildren (who will themselves begin to reproduce 15 years later), but women in the second population will just begin to see their grandchildren being born.

There is a great deal of disagreement among leaders in many countries about how much support the government should provide for family planning. The issue is certainly socially charged, but there are also political and economic threads to the debate. For example, the United States and many other developed countries, as well as some developing ones, currently face labor shortages, a problem cited by some opponents as stemming from policies that encourage family planning. But the debate is also related in part to the difficulty of answering the question, How many people are too many? The problem of defining carrying capacity for humans is confounded by the observation that carrying capacity has changed with human cultural evolution. The advent of agricultural and industrial technology has significantly increased carrying capacity at least twice during human history, and opponents of population control are counting on new, as yet unidentified, technological breakthroughs that will allow our population to grow and plateau at some higher level.

Technology has undoubtedly increased Earth's carrying capacity for humans, but, as we have seen, no population can continue to grow indefinitely. Ideally, human populations will reach carrying capacity smoothly and then level off. This will occur when birth rates and death rates are equal, and a decrease in birth rate is more desirable than an increase in death rate. If, however, the population fluctuates about carrying capacity, we can expect periods of increase followed by mass death, as has occurred during plagues, localized famines, and international military conflicts. In any case, the human population must eventually stop growing. Unlike other organisms, we can decide whether zero population growth will be attained through social changes involving individual choice or government intervention or through increased mortality due to resource limitation and environmental degradation. For better or worse, we have the unique responsibility to decide the fate of our species and that of the rest of the biosphere.

(a) Manila. This scene captures the stifling, unsanitary conditions in a slum that has arisen within a garbage dump. Manila, in the Philippines, has well over 10 million people, making it one of the world's largest cities.

(b) Los Angeles. On a per capita basis, people in developed countries have much more impact on the environment than people in developing countries. Americans, such as these commuters on a Los Angeles freeway, consume ten times as much energy per person (mostly as fossil fuels) as people in developing countries.

Figure 18.27 **Two very crowded places.**

CHECKPOINT

1. What is the relationship between a population and a species?
2. What is the approximate current size of Earth's human population?
3. An aquarium population of guppies has reached a stable population size. We decide to add twice as much guppy food per day to the aquarium, but this turns out to have no effect on population size. What is the most likely explanation for this observation?
4. Of the following factors with the potential to limit the growth of a human population, which one is most density independent? (a) lowering of fertility due to hormonal changes in very crowded conditions; (b) a famine; (c) mass drowning caused by hurricane floods; (d) epidemic of a highly contagious disease; (e) freezing deaths due to a shortage of housing.
5. What new research question arises from evidence that the population cycle of snowshoe hares is caused at least partly by a cycle in the availability or nutritional value of the plants eaten by the hares?
6. If a population's changing growth rate approximates the logistic model, what causes the population's size to level off?
7. (a) How does the age structure of the U.S. population explain the current surplus in the Social Security fund? (b) If the system is not changed, why will the surplus give way to a deficit sometime in the next few decades?
8. From the graph in Figure 18.25, what was the approximate percentage increase in the Sri Lankan population over the 20-year span from 1960 to 1980?

Answers: 1. A population is a localized subset of a species. **2.** 6.5 billion **3.** The population was already at carrying capacity before we increased food supply, and the key limiting factor was something other than food availability. **4.** c **5.** What causes the cyclic changes in the plants? **6.** The population size reaches the environment's carrying capacity. **7.** (a) The largest population segment, the baby boomers, are currently in the work force in their peak earning years, paying into the system. (b) The boomers will retire over the next few decades and begin drawing from the system at a time when there will be fewer employees paying into Social Security. **8.** 50%, an increase from about 10 million to 15 million people.

Life Histories and Their Evolution

Earlier in this chapter, we looked at how certain physiological, anatomical, and behavioral responses to environmental variation can increase an organism's chances of survival. However, natural selection does not act only on traits that increase survival; organisms that survive but do not reproduce are not at all "fit" in the Darwinian sense (see Chapter 13). Clearly, an organism can pass along its genes only if it survives long enough to reproduce. But how long is long enough? In many cases, there are trade-offs between survival traits and traits that enhance reproductive output—traits such as frequency of reproduction, investment in parental care, and the number of offspring per reproductive episode (usually called seed crop for seed plants and litter size or clutch size for animals).

The traits that affect an organism's schedule of reproduction and death make up part of its **life history**, the series of events from birth through reproduction and death. Of course, a particular life history, like most characteristics of an organism, is the result of natural selection operating on a species over evolutionary time. In this section, we'll see how life history traits affect population growth.

Table 18.1	Life Table for the U.S. Population in 2003			
Age Interval	**Number Living at Start of Age Interval (*N*)**	**Number Dying During Interval (*D*)**	**Mortality (Death Rate) During Interval (*D/N*)**	**Chance of Surviving Interval (1 − *D/N*)**
0–10	100,000	884	0.009	0.991
10–20	99,116	423	0.004	0.996
20–30	98,693	941	0.010	0.990
30–40	97,752	1,308	0.013	0.987
40–50	96,444	2,859	0.030	0.970
50–60	93,585	5,825	0.062	0.938
60–70	87,760	12,225	0.139	0.861
70–80	75,535	22,794	0.302	0.698
80–90	52,741	31,401	0.595	0.405
90+	21,340	21,340	1.000	0.000

Life Tables and Survivorship Curves

When the life insurance industry was established about a century ago, insurance companies set their rates based on early scientific studies of human populations. Needing to determine how long, on average, an individual of a given age could be expected to live, the insurance companies began using what are called life tables.

A **life table** tracks survivorship and mortality (death) in a population. For example, **Table 18.1** arranges the survivorship/mortality data for a sample of 100,000 U.S. citizens over a ten-year period, ending in 2003. Using this table, an insurance agent could predict that a 21-year-old has about a 0.99 (99%) chance of surviving to age 30. Borrowing the basic idea from the insurance industry, population ecologists construct life tables for plants and nonhuman animals.

A graphic way of representing some of the data in a life table is to draw a **survivorship curve,** a plot of the number of people still alive at each age **(Figure 18.28)**. We can classify survivorship curves for diverse organisms into three general types. A Type I curve is relatively flat at the start, reflecting low death rates during early and middle life, and dropping steeply as death rates increase among older age-groups. Humans and many other large mammals that produce relatively few offspring but provide them with good care often exhibit this kind of curve. In contrast, a Type III curve indicates high death rates for the very young and then a period when death rates are much lower for those few individuals who survive to a certain age. Species with this type of survivorship curve usually produce very large numbers of offspring but provide little or no care for them. An oyster, for instance, may release millions of eggs, but most offspring die as larvae from predation or other causes. A Type II curve is intermediate, with mortality more constant over the life span. This type of survivorship has been observed in several invertebrates, including hydras, and in certain rodents, such as the gray squirrel.

A population's pattern of mortality is certainly a key feature of life history. But we have defined life history as the series of events from birth

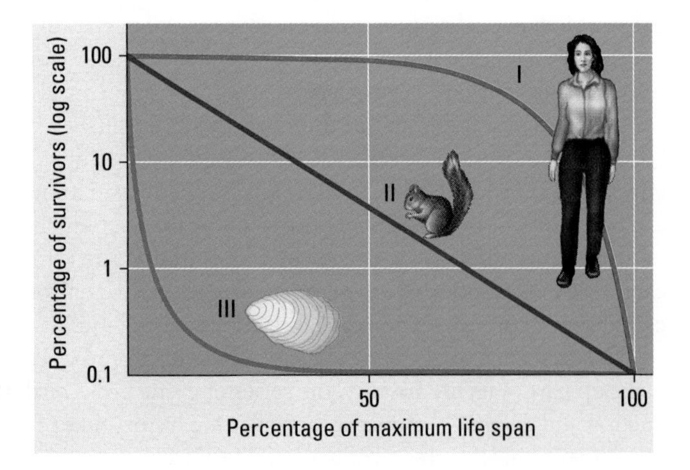

Figure 18.28 **Three idealized types of survivorship curves.**

through reproduction and death. Let's take a closer look now at how natural selection affects the reproductive strategies that evolve in populations.

Life History Traits as Evolutionary Adaptations

Some key life history traits affecting population growth are the age at which reproduction first occurs, the number of offspring, the amount of parental care committed to offspring, and the overall energy cost of reproduction. For a given population in a particular environment, natural selection will favor the combination of life history traits that maximizes an individual's output of viable, fertile offspring. In other words, life history traits, like anatomical features, are shaped by evolutionary adaptation.

Figure 18.29 illustrates one type of life history, called "big-bang" reproduction, also called an **opportunistic life history.** The agave, or century plant, grows in arid climates with sparse and unpredictable rainfall. It may grow for decades (a century in some cases) without flowering or reproducing. Then one rainy spring, it grows a floral stalk that may be as tall as a telephone pole, produces many seeds, then withers and dies, its food reserves and water spent in the formation of its massive bloom. By growing and storing nutrients until an unusually wet year and then putting all its resources into one grand burst of seed production, the plant maximizes its reproductive success. Populations that exhibit a big-bang reproductive strategy typically live in an unpredictable environment that is controlled by density-independent factors, such as the weather.

Figure 18.29 The "big-bang" reproduction of century plants. The century plant's life history is an example of what ecologists call the "big-bang" strategy of reproduction. It's an evolutionary adaptation to the organism's environmental situation.

Other populations with similar life histories include dandelions and many other annual weeds that grow quickly in open, disturbed areas, producing a large number of seeds in a brief time when the weather is favorable. Although most of the seeds will not produce mature plants, their large number and ability to disperse to new habitats ensure that at least some will grow and eventually produce seeds themselves. Many insects, including locusts, also exhibit an opportunistic life history, maximizing reproductive output whenever environmental opportunity knocks. For such species, natural selection has reinforced quantity of reproduction more than individual survivorship. In general, populations with a big-bang life history exhibit a Type III survivorship curve (see Figure 18.28).

In contrast, some populations, typically larger-bodied and longer-lived species, exhibit an **equilibrial life history,** which generally results in a Type I survivorship curve. Individuals usually mature later and produce few offspring but care for their young. The population size may be quite stable, held near the carrying capacity by density-dependent factors (stabilization around carrying capacity accounts for the term equilibrial). Natural selection has resulted in the production of better-endowed offspring that can become established in the well-adapted population into which they are born. The life histories of many large terrestrial vertebrates fit this model. A female polar bear, for instance, has only one or two offspring every three years, but the cubs remain in her protective custody for over two years. In the plant kingdom, the coconut palm tree also fits the equilibrial model. Compared with most other trees, it produces relatively few, very large seeds, which provide ample nutrients (including the coconut "milk") for the embryo; this is a plant's version of parental care.

Table 18.2 Some Life History Characteristics of Opportunistic and Equilibrial Populations

Characteristic	Opportunistic Populations (e.g., many wildflowers)	Equilibrial Populations (e.g., many large mammals)
Climate	Relatively unpredictable	Relatively predictable
Maturation time	Short	Long
Life span	Short	Long
Death rate	Often high	Usually low
Number of offspring produced per reproductive episode	Many	Few
Number of reproductions per lifetime	Usually one	Often several
Timing of first reproduction	Early in life	Later in life
Size of offspring or eggs	Small	Large
Parental care	None	Often extensive

Source: Adapted from E. R. Pianka, *Evolutionary Ecology,* 6th ed. (San Francisco, CA: Benjamin/Cummings, 2000), p. 186.

Table 18.2 contrasts some life history traits between opportunistic and equilibrial strategies. Most populations probably fall between the opportunistic and equilibrial extremes. Thus, these two life history strategies are only hypothetical models, starting points for studying the complex interplay of the forces of natural selection on reproductive characteristics.

Does the Environment Shape Guppy Life Histories? As mentioned at the start of the chapter, ecologists conduct hypothesis-driven scientific investigations to probe many aspects of their discipline. For many years, David Reznick of the University of California, Riverside, and John Endler of the University of California, Santa Barbara, have been investigating the life histories of guppy populations in Trinidad, a Caribbean island. Guppies are small freshwater fish you probably recognize as popular aquarium pets. In the Aripo River system of Trinidad, guppies live as isolated populations in small pools. In some cases, two populations inhabiting the same stream live less than 100 m apart, but they are separated by a waterfall that impedes the migration of guppies between the two ponds.

Early in their research, Reznick and Endler made the **observation** that certain life history traits among guppy populations correlated with the main type of predator in a stream pool. Some guppy populations live in pools where the predator is the killifish, which eats mainly small, immature guppies. Other guppy populations live where larger fish, called pike-cichlids, eat mostly large, mature guppies. Where preyed on by pike-cichlids, guppies tend to be smaller, mature earlier, and produce more offspring each time they give birth than those in areas without pike-cichlids.

Reznick and Endler asked the **question**: Do the differences between the populations result from natural selection? A reasonable **hypothesis** is that the selective predation of larger versus smaller guppies results in the observed life history adaptations. They made the **prediction** that artificially manipulating the predator-prey relationship could result in a measurable evolutionary change in the population.

Reznick and Endler tested their hypothesis with field **experiments (Figure 18.30)** in which they transplanted guppies from locations with pike-cichlids (predators of mature guppies) to guppy-free sites inhabited by killifish (predators of juvenile guppies). Over 11 years, they measured the age, size at maturity, and other life history traits of the populations. Their **results** are summarized in **Figure 18.31**. Over 30 to 60 guppy generations, the average weight at maturity for guppies in the transplanted (experimental) populations increased by about 14% compared with control populations. Other life history traits also changed in the direction predicted by the hypothesis of evolutionary adaptation due to selective predation. Reznick and Endler had in fact tested an evolutionary hypothesis and documented evolution in a natural setting over a relatively short time. ■

Figure 18.30 David Reznick conducting field experiments on guppy evolution in Trinidad.

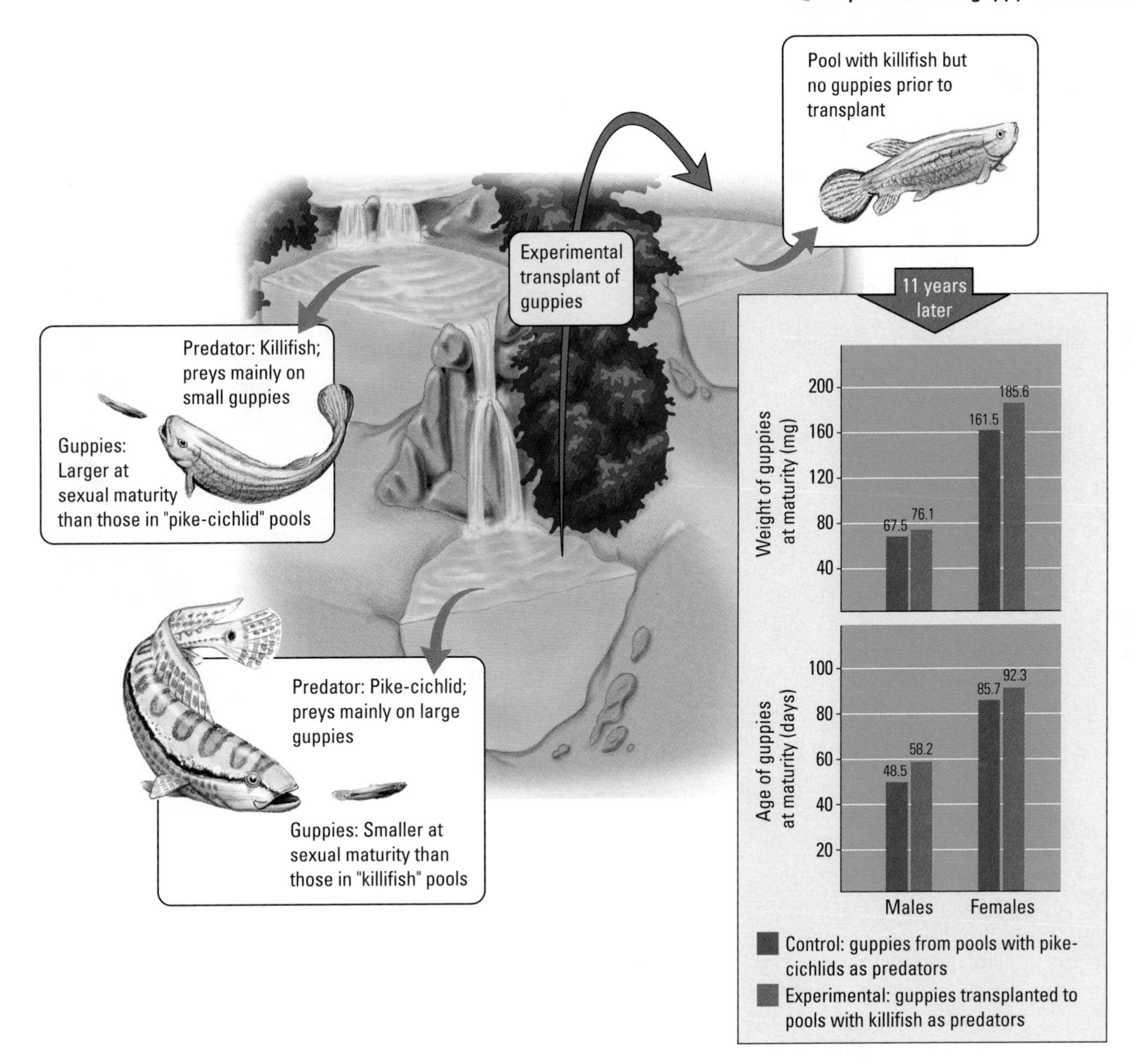

Figure 18.31 Testing a Darwinian hypothesis. Researchers tested the hypothesis that selective predation accounts for the life history differences between guppy populations. As shown in the graphs, the researchers observed that the average weight and age at sexual maturity of the transplanted experimental populations increased significantly as compared to the control populations.

CHECKPOINT

1. How do the terms *opportunistic* and *equilibrial* contrast the key characteristics of those life history strategies?
2. What is the key feature of the mortality column in the life table for a population exhibiting a Type II survivorship curve?

Answers: **1.** An opportunistic life history is characterized by an ability to produce a large number of offspring very rapidly when the environment affords a temporary opportunity for a burst of reproduction. An equilibrial life history is characterized by a population size that fluctuates only slightly from carrying capacity. **2.** The mortality is about the same for every age interval.

EVOLUTION CONNECTION

Natural Selection on the American Plains

Natural selection adapts species to the mix of abiotic and biotic factors they encounter. As a case study, consider the pronghorn antelope, pictured in **Figure 18.32**. The pronghorn first evolved on the open plains and shrub deserts of North America over 1 million years ago, and it is found nowhere else in the world. Let's see how some of the unique adaptations of the pronghorn fit the environmental conditions in which it evolved.

First, what about the major abiotic factors? The pronghorn's habitat is arid, windswept, and subject to extreme temperature fluctuations. With a thick coat that traps air and provides insulation, the pronghorn is superbly adapted to these conditions. If you drive through Wyoming or Colorado in the winter, you may see herds of these animals foraging when temperatures are well below 0°C. Water is rarely a problem for a pronghorn because it obtains a great deal of moisture from the vegetation it eats.

Figure 18.32 **Pronghorn antelopes *(Antilocapra Americana).***

What about the pronghorn's adaptations to the biotic components of its habitat? The pronghorn's main foods are small broadleaf plants, grasses, and woody shrubs, and its teeth are adapted for biting and chewing these plants. Also, like a cow, it has a stomach containing cellulose-digesting bacteria. As the pronghorn eats plants, the bacteria digest cellulose, and the animal obtains most of its nutrients from the bacteria. The pronghorn's main predators are wolves, coyotes, and cougars. As the pronghorn evolved, adaptations—primarily great speed and endurance—enabled it to escape its predators. Capable of sprinting about 95 km/hr (60 mph) on flat ground, it is one of the fastest mammals. An adult pronghorn can also keep up a pace of about 65 km/hr for at least 30 minutes—a definite benefit when being chased by long-distance runners such as wolves. Other adaptations that help the pronghorn foil predators include its tan and white coat, which often camouflages the animal on the open plains, and its keen eyes, which can detect movement at great distances.

The pronghorn is highly successful in its habitat. In a different environment, the pronghorn's adaptations for escaping predators might not be as effective. This suggests that an organism can usually tolerate environmental fluctuations only within the set of conditions to which it is adapted. Outside that set, the organism may not survive long enough to reproduce. Thus, in adapting populations to local environmental conditions, natural selection may limit the distribution of organisms. We'll keep these principles in mind as we begin, in the next chapter, to examine higher levels of ecological organization: communities and whole ecosystems. ■

Chapter Review

SUMMARY OF KEY CONCEPTS

For study help and activities, go to campbellbiology.com or the student CD-ROM.

An Overview of Ecology

• Ecology is the scientific study of interactions between organisms and their environments. The environment includes abiotic (nonliving) and biotic (living) components.

• **Ecology as Scientific Study** Ecologists use observation, experiments, and computer models to test hypothetical explanations of these interactions.

• **A Hierarchy of Interactions** Ecologists study interactions at four increasingly complex levels.

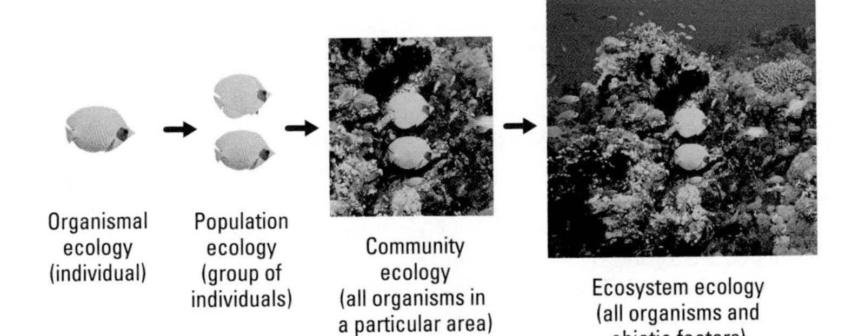

• **Ecology and Environmentalism** Human activities have had an impact on all parts of the biosphere. Ecology provides the basis for understanding and addressing these environmental problems.

Activity *DDT and the Environment*

The Evolutionary Adaptations of Organisms

• **Abiotic Factors of the Biosphere** The biosphere is an environmental patchwork in which abiotic factors affect the distribution and abundance of organisms. These include the availability of sunlight and water, temperature, wind, rock and soil characteristics, and catastrophic disturbances such as fires, hurricanes, tornadoes, and volcanic eruptions.

Case Study in the Process of Science *Do Pillbugs Prefer Wet or Dry Environments?*

Case Study in the Process of Science *How Do Abiotic Factors Affect the Distribution of Organisms?*

• **Physiological Responses** Most organisms adjust their physiological conditions in response to changes in the environment. Acclimation is a longer-term response that can take days or weeks. The ability to acclimate is generally related to the range of environmental conditions that a species naturally experiences.

• **Anatomical Responses** Many organisms respond to environmental change with reversible or irreversible changes in anatomy. Unable to move to a better location, plants are generally more anatomically plastic than animals.

• **Behavioral Responses** Able to travel about, animals frequently adjust to poor environmental conditions by moving to a new location. These may be small adjustments such as shuttling between sun and shade or seasonal migrations to new regions.

Activity *Evolutionary Adaptations*

What Is Population Ecology?

• Population ecology focuses on the factors that influence a population's size, growth rate, density, and structure. A population consists of members of a species living in the same place at the same time.

• **Population Density** Population density, the number of individuals of a species per unit area or volume, can be estimated by a variety of sampling techniques. These include counting the number of individuals in sample plots and the mark-recapture method.

Activity *Techniques for Estimating Population Density and Size*

• **Patterns of Dispersion** Dispersion patterns of a population are determined by various environmental or social factors.

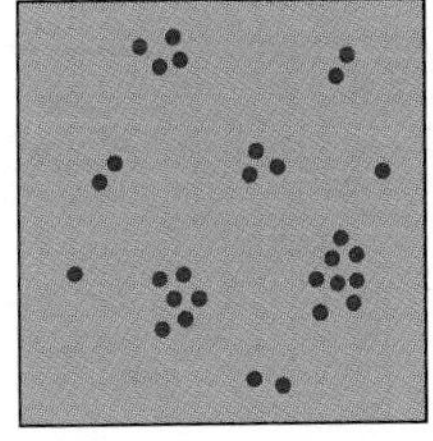

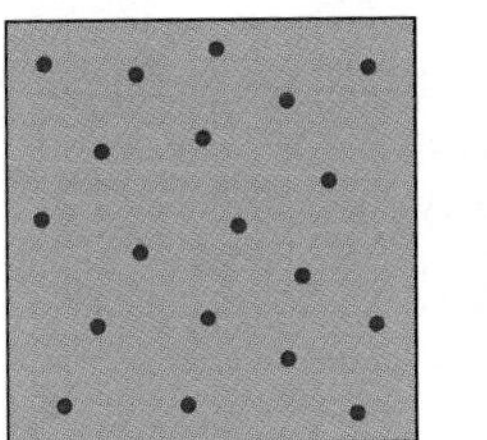

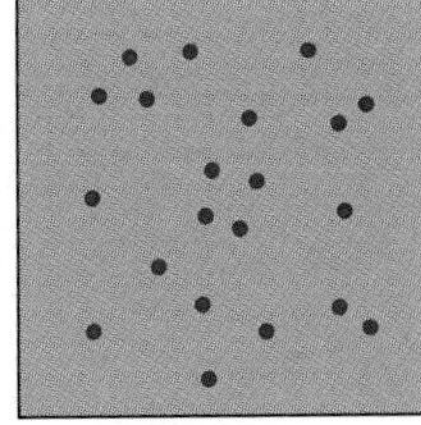

• **Population Growth Models**

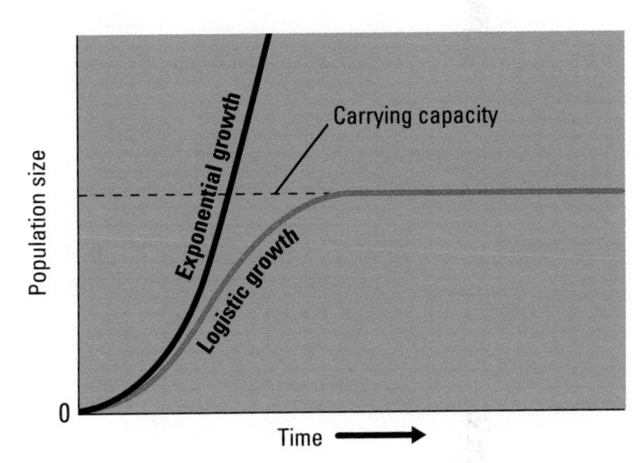

The exponential model of population growth describes an idealized population in an unlimited environment. This model predicts that the larger a population becomes, the faster it grows. Exponential growth in nature is generally a short-lived consequence of organisms being introduced to a new or underexploited environment. The logistic model of population growth describes an idealized population that is slowed by limiting factors. This model predicts that a population's growth rate will be small when the population size is either small or large, and highest when the population is at an intermediate level relative to the carrying capacity.

• **Regulation of Population Growth** Over the long term, most population growth is limited by a mixture of density-independent and density-dependent factors. Density-dependent factors intensify as a population increases in density, increasing the death rate, decreasing the birth rate, or both. Density-independent factors affect the same percentage of individuals regardless of population size. Some populations have regular boom-and-bust cycles.

• **Human Population Growth** The human population has been growing almost exponentially for centuries. Human population growth is based on the same two general parameters that affect other animal and plant populations:

birth rates and death rates. Birth rates increased and death rates decreased when agricultural societies replaced a lifestyle of hunting and gathering. The age structure of the population is a major factor in the different growth rates of different countries. Population researchers predict that the U.S. population will continue to grow well into the 21st century. Age-structure diagrams also predict social predicaments. Predictions of future trends in global human population growth vary widely. The human species is unique in having the ability to consciously control its own population growth, the fate of our species, and the fate of the rest of the biosphere.

Activity *Human Population Growth*

Activity *Analyzing Age-Structure Diagrams*

Life Histories and Their Evolution

- **Life Tables and Survivorship Curves** A population's pattern of mortality is a key feature of life history. A life table tracks survivorship and mortality in a population. Survivorship curves can be classified into three general types, depending on the rate of mortality over the entire life span.

Activity *Investigating Survivorship Curves*

- **Life History Traits as Evolutionary Adaptations** Life history traits are shaped by evolutionary adaptation; they may vary within a species and may change as the environmental context changes. Most populations probably fall between the extreme opportunistic strategies of many insects and equilibrial strategies of many larger-bodied species.

SELF-QUIZ

1. Place these levels of ecological study in order, from the least to the most comprehensive: community ecology, ecosystem ecology, organismal ecology, population ecology.

2. Name several abiotic factors that might affect the community of organisms living inside a home fish tank.

3. The formation of goose bumps on your skin in cold weather is an example of a (an) __________ response, while seasonal migration is an example of a (an) __________ response.

4. What two values would you need to know to figure out the human population density of your community?

5. Pine trees in a forest tend to shade and kill pine seedlings that sprout nearby. What pattern of growth will this produce?

6. A uniform dispersion pattern for a population may indicate that
   a. the population is spreading out and increasing its range.
   b. resources are heterogeneously distributed.
   c. individuals of the population are competing for some resource, such as water and minerals for plants or nesting sites for animals.
   d. there is an absence of strong attractions or repulsions among individuals.

7. With regard to its percent increase, a population that is growing logistically grows fastest when its density is __________ compared to the carrying capacity.
   a. low b. intermediate c. high

8. Which of the following shows the effects of a density-dependent limiting factor?
   a. A forest fire kills all the pine trees in a patch of forest.
   b. Early rainfall triggers the explosion of a locust population.
   c. Drought decimates a wheat crop.
   d. Rabbits multiply, and their food supply begins to dwindle.

9. Skyrocketing growth of the human population since the beginning of the Industrial Revolution appears to be mainly a result of
   a. migration to thinly settled regions of the globe.
   b. better nutrition boosting the birth rate.
   c. a drop in the death rate due to better nutrition and health care.
   d. the concentration of humans in cities.

10. If members of a species produce a large number of offspring but provide minimal parent care, then a Type __________ survivorship curve is expected. In contrast, if members of a species produce few offspring and provide them with long-standing care, then a Type __________ survivorship curve is expected. (*Hint:* Refer to Figure 18.28.)

Answers to the Self-Quiz questions can be found in Appendix D.

Go to the website or CD-ROM for more Self-Quiz questions.

THE PROCESS OF SCIENCE

11. Design a laboratory procedure to measure the effect of water temperature on the population growth of a certain phytoplankton species from a pond.

12. We estimate the size of a population of small mice in a particular field by the mark-recapture method (see Figure 18.14). Our estimate is 350. Later, we learn from experiments on the behavior of these mice that they can locate a baited trap faster if they have already been rewarded with food by visiting that trap once before. Does this mean that our original estimate of 350 individuals was too low or too high? Explain your answer in terms of the principles that underlie the mark-recapture method.

BIOLOGY AND SOCIETY

13. During the summer of 1988, lightning ignited huge forest fires that burned a large portion of Yellowstone National Park (see Figure 18.10). The National Park Service has a natural-burn policy: Fires that start naturally are allowed to burn unless they endanger human settlements. The fires were allowed to spread and burn while firefighters primarily protected people. The public accused the park service of letting a national treasure go up in flames. Park service scientists stuck with the natural-burn policy. Do you think this was the best decision? Support your position.

14. In the spring of 2000, national park officials authorized a controlled burn of a forested area near Los Alamos, New Mexico, with the intention of clearing away brush and dead wood to reduce the severity of future fires. Unfortunately, a weather warning was ignored and the fire escaped control, claiming over 200 homes and thousands of acres of forest. Assuming that the basic concept of controlled burning is scientifically sound, how would you justify future burns to a public that remembers the Los Alamos fiasco?

15. Many people regard the rapid population growth of developing countries as our most serious environmental problem. Others think that the population growth in developed countries, though smaller, is actually a greater threat to the environment. What kinds of problems result from population growth in developing countries and in industrialized countries? Which do you think is the greater threat, and why?

19 Communities and Ecosystems

The oceans **cover about 75%** of Earth's surface.

Only a **tiny fraction** of the sunlight that shines on Earth is converted to chemical energy.

Sunken ships, army tanks, and subway cars are being used as **artificial reefs.**

Some plants **attract parasitic wasps** that lay their eggs in caterpillars that eat the plants.

BIOLOGY AND SOCIETY

Reefs, Coral and Artificial

Found in warm tropical waters, coral reefs are distinctive and complex ecosystems. Coral reefs are dominated by the structure of the coral itself, formed mainly by cnidarians that secrete hard external skeletons made of calcium carbonate. These skeletons vary in shape, forming an underlying surface on which other corals, sponges, and algae grow. The coral animals themselves (see Figure 17.12) feed on microscopic organisms and particles of organic debris. They also obtain organic molecules from the photosynthesis of symbiotic algae that live in their tissues. Reefs are important because they support a diversity of invertebrates and fishes, which also makes them a popular destination for people who enjoy diving and fishing **(Figure 19.1)**.

Some coral reefs cover enormous expanses of shallow ocean, but these delicate ecosystems fail to thrive in polluted waters. While no part of the biosphere is safe from human intrusion, it is particularly difficult for coral reef ecosystems to recover because corals grow very slowly. Corals are also subject to damage from both native and introduced predators, such as the crown-of-thorns sea star, which has destroyed coral reefs in parts of the western Pacific Ocean.

But there is a flip side to human intrusion: the possibility of creating new, artificial reefs. Artificial reefs are formed by sinking large, stable objects to the ocean bottom. Some reefs, for example, have been constructed from steel and concrete, while others consist of sunken ships, army tanks, and even subway cars. Though artificial reefs contain no coral, they can serve as a base for new reef ecosystems. Within just a few years, artificial reefs teem with marine organisms that live in or on nearly every square inch. One large artificial reef program is run by the state of South Carolina. Since 1973, the state has established 44 offshore reefs. Two of them serve as research stations, while the rest are open to the public for fishing and diving, generating revenue for the state while also providing new marine habitats.

In this chapter, we'll examine the diverse interactions among organisms—in reefs and all other parts of the biosphere—and how those relationships determine the species composition and other features of communities. On a larger scale, we'll explore the dynamics of ecosystems, such as forests, ponds, and oceanic zones. Ultimately, we'll widen our scope to view the biosphere as the global ecosystem, the sum of all ecosystems. Throughout the chapter, biology's core theme of evolution will be apparent in the adaptations that fit organisms to their ecological roles in communities. ■

Figure 19.1 A coral reef community. This underwater scene shows the stunning diversity of life in one coral reef community near the Fiji Islands.

Key Properties of Communities

On your next walk through a field or woodland, or even across campus or through your own backyard, try to observe some of the interactions among the species present. You may see birds using trees as nesting sites, bees pollinating flowers, caterpillars feeding on leaves, spiders trapping insects in their webs, cats stalking rodents, or ferns growing in the shade provided by trees. These examples are but a sample of the many interactions that occur in any ecological theater. In addition to the physical and chemical factors (abiotic factors) we discussed in Chapter 18, an organism's environment includes other individuals in its population and populations of other species living in the same area. Such an assemblage of species living close enough together for potential interaction is called a **community.** In **Figure 19.2**, the lion, the zebra, the hyena, the vultures, and the grasses and other plants are all members of a community in Kenya. In this section, we'll examine four key properties of a community: its diversity, its prevalent form of vegetation, its stability, and its trophic structure.

Figure 19.2 Diverse species interacting in a Kenyan savanna community.

Diversity

The diversity of a community—the variety of different kinds of organisms that make up the community—has two components. The first component is **species richness,** or the total number of different species in the community. The other component is the relative abundance of the different species. For example, imagine two forest communities, each with 100 individuals distributed among four different tree species (A, B, C, and D) as follows:

Community 1: 25A, 25B, 25C, 25D

Community 2: 80A, 10B, 5C, 5D

The species richness is the same for both communities because they both contain four species, but the relative abundance is different **(Figure 19.3)**. You would easily notice the four different types of trees in community 1, but without looking carefully, you might see only the abundant species A in the second forest. Most observers would intuitively describe community 1 as the more diverse of the two communities. Indeed, as used by ecologists, the term **species diversity** considers both diversity factors: species richness and relative abundance. As we'll discuss in Chapter 20, species diversity is a key factor in maintaining the health of all ecosystems and, indeed, the health of the whole biosphere.

Figure 19.3 Which forest is more diverse? Because both forests include the same four tree species, these two communities are equal in their species richness of trees. But if we factor in the relative abundance of species, then community 1 is the more diverse because no one species predominates. Ecologists take both factors into account when evaluating species diversity.

Prevalent Form of Vegetation

The second property of a community, its prevalent form of vegetation, applies mainly to terrestrial communities. For example, deciduous trees are the prevalent components in a temperate deciduous forest. When we look more closely at such a community, we see not only which plants are dominant, but also how the plants are arranged, or "structured." For instance, a deciduous forest has a pronounced vertical structure: The treetops form a top layer, or canopy, under which there is a subcanopy of lower branches; and below the subcanopy, small shrubs and herbs carpet the forest floor. The types and structural features of plants largely determine the kinds of animals that live in a community.

Stability

The third property of a community, **stability,** refers to the community's ability to resist change and return to its original species composition after being disturbed. Stability depends on both the type of community and the nature of disturbances. For example, a forest dominated by cedar and hemlock trees is a highly stable community in that it may last for thousands of years with little change in species composition. Large cedars and hemlocks even withstand most lightning-caused fires, which kill small trees and shrubs. However, when a fire does kill the dominant trees, a cedar/hemlock forest might seem less stable than, say, a grassland, because it will take much longer for the forest to return to its original species composition.

Trophic Structure

The fourth property of a community is its **trophic structure,** the feeding relationships among the various species making up the community. A community's trophic structure determines the passage of energy and nutrients from plants and other photosynthetic organisms to herbivores and then to carnivores. A trophic relationship is certainly implied in the scene in **Figure 19.4**, and we'll see many more examples as we take a closer look at community interactions.

Figure 19.4 Trophic structure: feeding relationships. After spending four to six years feeding in the open ocean, this salmon is attempting to return to the stream of its birth. Along the way, however, it may become a meal for a grizzly bear.

CHECKPOINT

1. How could a community appear to have relatively little diversity even though it is rich in species?
2. A community's feeding relationships of producers and consumers is referred to as the community's __________ structure.

Answers: 1. If one or a few of the diverse species accounted for almost all the organisms in the community, with the other species being rare **2.** trophic

Interspecific Interactions in Communities

With the four main properties of a community in mind, we turn next to the various kinds of interactions *between* species—what ecologists call **interspecific interactions.** The three main types of interspecific interactions we'll explore are competition, predation, and symbiosis. In each case, we'll see how these community relationships function as environmental factors in the evolutionary adaptation of organisms through natural selection.

Competition Between Species

When populations of two or more species in a community rely on similar limiting resources, they may be subject to **interspecific competition.** You already learned in Chapter 18 about intraspecific, or *within*-species, competition: As a population's density increases and nears carrying capacity, every individual has access to a smaller share of some limiting resource, such as food. As a result, mortality rates increase, birth rates decrease, and population growth is curtailed. In interspecific competition, the population growth of a species may be limited by the density of competing species as well as by the density of its own population. For example, if several bird species in a forest feed on a limited population of insects, the density of each species may have a negative impact on population growth in the other species. Similarly, species may compete for nesting sites, shelters, or any other resource that is in short supply.

What Happens When Species Compete? In 1934, Russian ecologist G. F. Gause studied the effects of interspecific competition in laboratory experiments with two closely related species of protists, *Paramecium aurelia* and *Paramecium caudatum*. Gause cultured the protists under stable conditions with a constant amount of food added every

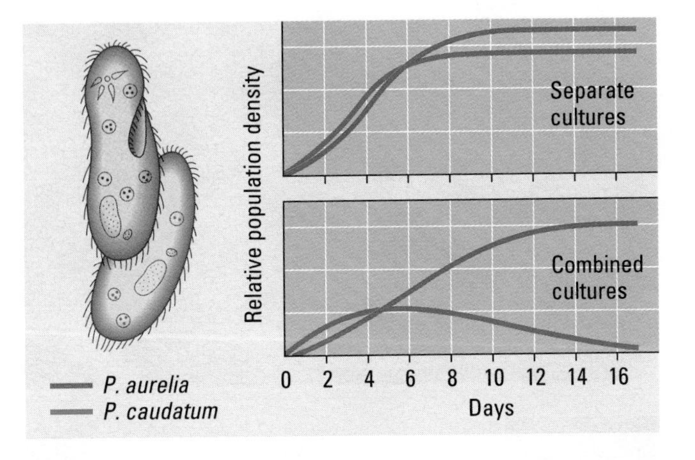

Figure 19.5 Competitive exclusion in laboratory populations of *Paramecium*. When cultured separately (upper graph) with constant amounts of food added daily, populations of each of the two *Paramecium* species grow to carrying capacity. But when the two species are cultured together (lower graph), *P. aurelia* has a competitive edge in obtaining food, an advantage that drives *P. caudatum* to extinction in the microcosm of the culture jar.

day. When he grew the two species in separate cultures, he made the **observation** that each population grew rapidly and then leveled off at what was apparently the carrying capacity of the culture. This caused Guase to **question** what would happen when the two species were grown in one culture. Gause's **hypothesis** was that the two species had such similar roles in the environment that they would compete for the same limiting resources and thus could not coexist in the same place. This led him to the **prediction** that one species would outcompete the other for the available resources.

In his **experiment**, Gause cultured *P. caudatum* and *P. aurelia* together while adding a constant supply of food daily. His **results** are shown in **Figure 19.5**. *P. caudatum* soon became extinct. Apparently, *P. aurelia* had a competitive edge in obtaining food and drove the other species to extinction in the culture by using the resources more efficiently and thus reproducing more rapidly. Even a slight reproductive advantage will eventually lead to local elimination of the inferior competitor.

Ecologists called Gause's concept the **competitive exclusion principle.** As another example of this principle, consider the classic field experiments with two species of barnacles that attach to intertidal rocks on the North Atlantic coast, shown in **Figure 19.6.** ■

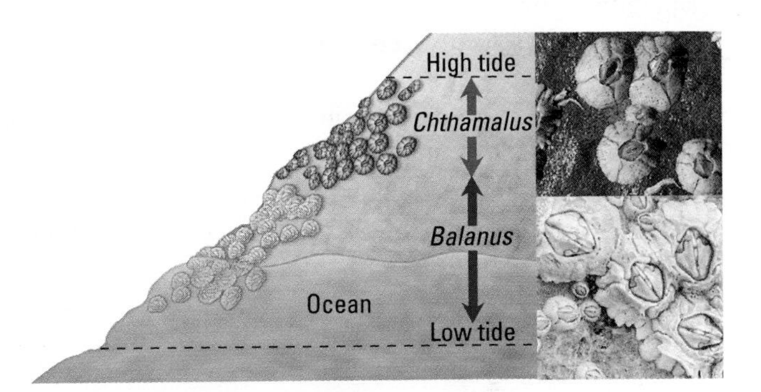

Figure 19.6 Testing the hypothesis of competitive exclusion in the field. These two types of barnacles, *Balanus* and *Chthamalus*, grow on rocks that are exposed during low tide. *Chthamalus* occupies the upper parts of the rocks, which are out of the water longer during low tides. *Balanus* fails to survive as high on the rocks as *Chthamalus*, apparently because *Balanus* dries out in the air. When ecologists experimented by removing *Balanus* from the lower rocks, *Chthamalus* spread lower, colonizing the unoccupied rocks. However, when both species colonize the same rock, *Balanus* eventually displaces *Chthamalus* on the lower part of the rock. The researchers concluded that the upper limit of *Balanus*'s distribution is set mainly by the availability of water, whereas the lower limit of *Chthamalus*'s distribution is set by competition.

The Ecological Niche The sum total of a species' use of the biotic and abiotic resources in its environment is called the species' **niche.** An organism's niche is its ecological role—how it "fits into" an ecosystem. The niche of a population of tropical tree lizards, for example, consists of, among many other components, the temperature range it tolerates, the size of trees on which it perches, the time of day in which it is active, and the size and type of insects it eats.

We can now restate the competitive exclusion principle to say that two species cannot coexist in a community if their niches are identical. However, ecologically similar species can coexist in a community if there are one or more significant differences in their niches.

Resource Partitioning There are two possible outcomes of competition between species having identical niches: Either the less competitive species will be driven to local extinction, or one of the species may evolve enough to use a different set of resources. This differentiation of niches that enables similar species to coexist in a community is called **resource partitioning.**

Figure 19.7 illustrates an example. We can think of resource partitioning within a community as "the ghost of competition past"—circumstantial evidence of earlier interspecific competition resolved by the evolution of niche differences.

Predation

In everyday usage, the term *community* is benign, maybe even implying the warmness of cooperation, what we think of as "community spirit." In contrast, an ecological community exhibits the Darwinian realities of competition and predation, where organisms eat other organisms. In the interspecific interaction called **predation,** the consumer is the **predator** and the food species is the **prey.** Most ecologists include herbivory, the eating of plants by animals, as a form of predation, even in cases such as grazing, where the animal does not kill the whole plant.

It shouldn't surprise you that predation is a potent factor in evolutionary adaptation. Eating and avoiding being eaten are prerequisite to reproductive success. Natural selection refines the adaptations of both predators and prey.

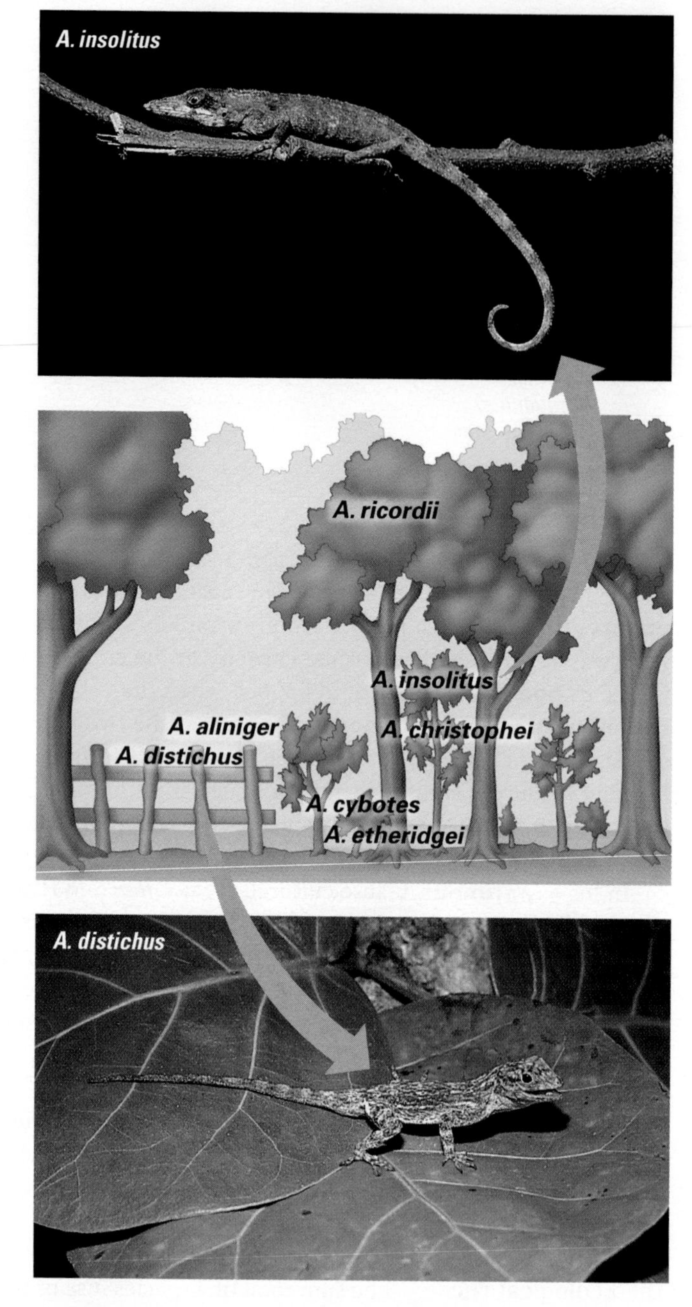

Figure 19.7 Resource partitioning in a group of lizards. Seven species of *Anolis* lizards live in close proximity in La Palma, in the Dominican Republic. The lizards all feed on insects and other small arthropods. However, competition for food is minimized because each lizard species perches in a certain microhabitat. *Anolis distichus* (bottom), for example, perches on fence posts and other sunny surfaces, such as leaves, whereas *A. insolitus* (top) usually perches on shady branches.

Predator Adaptations Many important feeding adaptations of predators are both obvious and familiar. Most predators have acute senses that enable them to locate and identify potential prey. In addition, many predators have adaptations such as claws, teeth, fangs, stingers, or poison that help catch and subdue the organisms on which they feed. Rattlesnakes and other pit vipers, for example, locate their prey with special heat-sensing organs located between each eye and nostril, and they kill small birds and mammals by injecting them with toxins through their fangs. Similarly, many herbivorous insects locate appropriate food plants through chemical sensors on their feet, and their mouthparts are adapted for shredding tough vegetation. Predators that pursue their prey are generally fast and agile, whereas those that lie in ambush are often camouflaged in their environments. In our own case, perhaps it is valid to view our capacity for learning and teaching to be adaptations contributing to our success as predators; the invention and refinement of agriculture over the centuries has certainly increased Earth's carrying capacity for humans (see Chapter 18).

Plant Defenses Against Herbivores Plants cannot run away from herbivores. Chemical toxins, often in combination with various kinds of spines or thorns, are plants' main defenses against being eaten to extinction. Among such chemical weapons are the poison strychnine, produced by a tropical vine called *Strychnos toxifera*; morphine, from the opium poppy; nicotine, produced by the tobacco plant; mescaline, from peyote cactus; and tannins (such as those found in tea and wine), from a variety of plant species. Other defensive compounds that are not toxic to humans but may be distasteful to herbivores are responsible for the familiar flavors of cinnamon, cloves, and peppermint. Some plants even produce chemicals that imitate insect hormones and cause abnormal development in some insects that eat them.

Animal Defenses Against Predators Animals can avoid being eaten by using passive defenses, such as hiding, or active defenses, such as escaping or defending themselves against predators. Fleeing is a common antipredator response, though it can be very costly in terms of energy. Many animals flee into a shelter and avoid being caught without expending the energy required for a prolonged flight. Active self-defense is less common, though some large grazing mammals will vigorously defend their young from predators such as

Figure 19.8 Mobbing, a behavioral defense against predators. Many prey species turn the tables and attack their predators. Here, two crows mob a barn owl, a predator that sometimes kills and eats crow eggs and nestlings.

Figure 19.9 Protecting offspring by faking an injury. This killdeer uses deception to defend her nest against predators and human disturbance. When danger threatens, she leaves the nest and fakes a broken wing. This behavior distracts and draws a potential predator away from the nest, and then the mother just flies away. The trickery often saves the lives of a killdeer's offspring.

lions. Other behavioral defenses include alarm calls, which can result in many individuals of the prey species mobbing the predator **(Figure 19.8)**. Distraction displays direct the attention of the predator away from a vulnerable prey, such as a bird chick, to another potential prey that is more likely to escape, such as the chick's parent **(Figure 19.9)**.

Many other defenses rely on adaptive coloration, which has evolved repeatedly among animals. Camouflage, called **cryptic coloration,** is a passive defense that makes potential prey difficult to spot against its background **(Figure 19.10)**.

Some animals have mechanical or chemical defenses against would-be predators. Most predators are strongly discouraged by the familiar defenses of skunks and porcupines. Some animals, such as poisonous toads and frogs, can synthesize toxins. Others acquire chemical defense passively by accumulating toxins from the plants they eat. For example, monarch butterflies store poisons from the milkweed plants they eat as larvae, making the butterflies distasteful to some potential predators. Animals with chemical defenses are often brightly colored, a caution to predators known as **warning coloration (Figure 19.11)**.

Figure 19.10 Camouflage. A dark brown Asian leaf frog blends in among the dead leaves of a forest floor.

Figure 19.11 Warning coloration of the poison arrow frog. The vivid markings of this pair of tree frogs, inhabitants of Costa Rican rain forests, warn of noxious chemicals in the frog's skin; predators learn about this as soon as they touch the frog. In some parts of South America, human hunters in the rain forest tip their arrows with poisons from similar frogs to bring down large mammals.

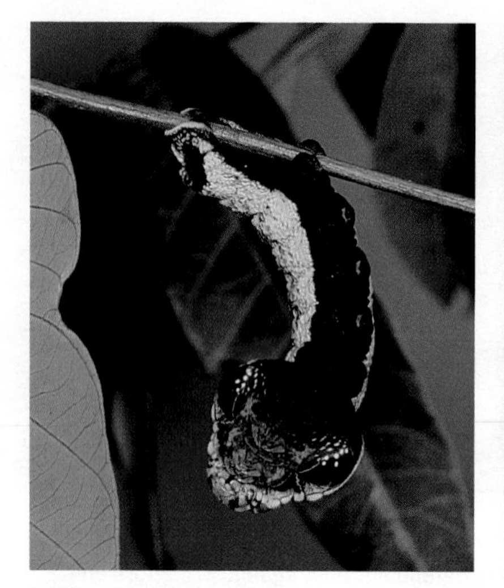

(a) Hawk moth larva

(b) Snake

Figure 19.12 Batesian mimicry. When disturbed, **(a)** the hawk moth larva resembles **(b)** a snake.

(a) Cuckoo bee

(b) Yellow jacket

Figure 19.13 Müllerian mimicry. Both **(a)** the cuckoo bee and **(b)** the yellow jacket wasp have stingers that release toxins. The cross-mimicry in appearance presumably benefits both species because predators learn more quickly to avoid any prey with these distinctive markings.

A species of prey may gain significant protection through mimicry, a "copycat" adaptation in which one species mimics another. In **Batesian mimicry,** a palatable or harmless species mimics an unpalatable or harmful model. In one intriguing example, the larva of the hawk moth puffs up its head and thorax when disturbed, looking like the head of a small poisonous snake, complete with eyes **(Figure 19.12)**. The mimicry even involves behavior; the larva weaves its head back and forth and hisses like a snake. In **Müllerian mimicry,** two or more unpalatable species resemble each other. Presumably, each species gains an additional advantage, because the pooling of numbers causes predators to learn more quickly to avoid any prey with a particular appearance **(Figure 19.13)**.

Predators also use mimicry in a variety of ways. For example, some snapping turtles have tongues that resemble a wriggling worm, thus luring small fish; any fish that tries to eat the "bait" is itself quickly consumed as the turtle's strong jaws snap closed.

Predation and Species Diversity in Communities

You might think that organisms eating other organisms would always reduce species diversity, but field experiments reveal that predator-prey relationships can actually preserve diversity. Experiments by American ecologist Robert Paine in the 1960s were among the first to provide such evidence. Paine removed the dominant predator, a sea star of the genus *Pisaster*, from experimental areas within the intertidal zone of the Washington coast **(Figure 19.14)**. The result was that *Pisaster*'s main prey, a mussel of the genus *Mytilus*, outcompeted many of the other shoreline

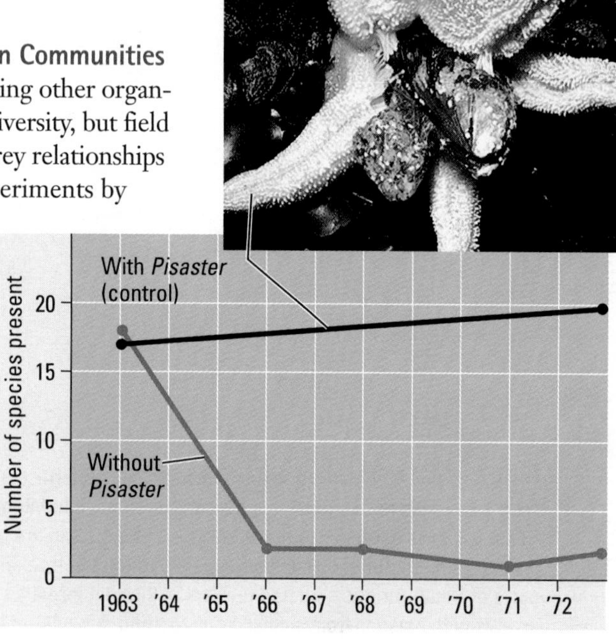

Figure 19.14 Effect of a keystone predator on species diversity. Ecologist Robert Paine removed the sea star *Pisaster* (inset, eating its favorite food, a mussel) from a tide pool in Washington State. Mussels then gradually took over the rock space and eliminated other invertebrates and algae.

organisms (barnacles and snails, for instance) for the important resource of space on the rocks. The number of species dropped.

The experiments of Paine and others led to the concept of keystone predators. A **keystone predator** is a species, such as *Pisaster*, that reduces the density of the strongest competitors in a community. In so doing, the predator helps maintain species diversity by preventing competitive exclusion of weaker competitors. By checking the population growth of mussels, the most successful (and most abundant) species, the sea star *Pisaster* helped maintain populations of several less competitive species in Paine's experimental tide pools. Predation has its constructive side.

Symbiotic Relationships

A **symbiotic relationship** is an interaction between two or more species that live together in direct contact. The two types of symbiotic relationships most important in structuring ecological communities are parasitism and mutualism.

Parasitism A symbiotic relationship is called **parasitism** if one organism benefits while the other is harmed. The **parasite,** usually the smaller of the two organisms, obtains its nutrients by living on or in its **host** organism. We can actually think of parasitism as a specialized form of predation. Tapeworms and the protozoa that cause malaria are examples of internal parasites (see Figures 17.15 and 15.19d). External parasites include mosquitoes, which suck blood from animals, and aphids, which tap into the sap of plants.

Natural selection has refined the relationships between parasites and their hosts. Many parasites, particularly microorganisms, have adapted to specific hosts, often a single species. In any parasite population, reproductive success is greatest for individuals that are best at locating and feeding on their hosts. For example, some aquatic leeches first locate a host by detecting movement in the water and then confirm its identity based on temperature and chemical cues on the host's skin. Natural selection has also favored the evolution of host defenses. In humans and other vertebrates, an elaborate immune system helps defend the body against specific internal parasites. With natural selection working on both host and parasite, the eventual outcome is usually a relatively stable relationship that does not kill the host quickly, an excess that would eliminate the parasite as well. An example of how rapidly natural selection can temper a host-parasite relationship is the evolution of resistance to a viral parasite in Australian rabbit populations **(Figure 19.15)**.

Figure 19.15 A case study in the evolution of host-parasite relations. In the 1940s, Australia was overrun by hundreds of millions of rabbits, all descended from just 12 pairs imported a century earlier for sport. The rabbits destroyed vegetation in huge expanses of Australia and threatened the sheep and cattle industries. In 1950, in an effort to control the exploding rabbit population, biologists deliberately introduced a myxoma virus that specifically parasitizes rabbits. The virus spread rapidly and killed 99.8% of the infected rabbits. However, a second exposure to the virus killed only 90% of the rabbit population derived from survivors of the first application. And a third exposure eliminated only about 50% of the surviving rabbit population. Apparently, viral infection selected for host genotypes that were better able to resist the parasite. Over the generations, the myxoma virus had less and less effect, and the rabbit population rebounded. The Australian government is now having more success with a different virus, which was introduced into the rabbit population in 1995.

Mutualism In contrast to the one-sidedness of parasitism, **mutualism** is a symbiosis that benefits both partners. Just two of the examples we encountered in earlier chapters are the root-fungus associations called mycorrhizae and the symbiotic association of fungi and algae in lichens (see Figures 16.3 and 16.25). **Figure 19.16** illustrates another case of mutual symbiosis, the relationship between acacia trees and the ants that protect these trees from herbivorous insects and competing plants.

Figure 19.16 Mutualism between acacia trees and ants. Certain species of Central and South American acacia trees have hollow thorns that house stinging ants. The ants feed on sugar and proteins from specialized glands on the trees (the orange swellings at the tips of the leaflets). The acacia also benefits from housing and feeding this population of ants, for the ants will attack anything that touches the tree. The pugnacious ants sting other insects, remove fungal spores, and clip surrounding vegetation that happens to grow close to the foliage of the acacia.

Many mutualistic relationships may have evolved from predator-prey or host-parasite interactions. Certain flowering plants, for example, have adaptations that attract animals that function in pollination or seed dispersal; these adaptations may represent an evolutionary response to a history of the herbivores' feeding on pollen and seeds. In many cases, pollen is spared when the pollinator is able to consume nectar instead. Indigestible seeds within fruits are dispersed by animals, which are nourished by the fruits. Any plants that could derive some benefit by sacrificing organic materials other than pollen or seeds would increase their reproductive success, and the adaptations for mutualistic interactions would spread through the plant population over the generations.

The Complexity of Community Networks

So far, we have reduced the networks known as biological communities to interactions between species: competition, predator-prey interactions, and symbiosis. It is the branching of these interactions that makes communities so complex. For example, some plants attract parasitic wasps that lay their eggs in caterpillars that eat the plants, thereby protecting the plant from extensive caterpillar damage (**Figure 19.17**). Both the wasps and caterpillars are also eaten by predators such as spiders and birds, and all of these organisms are hosts to specific parasites. Ecologists are only beginning to sort out the complex networks of just a few biological communities. Contributing to the challenges of community ecology is the fact that the structure of a community may change, sometimes over relatively short periods of time, owing to a variety of disturbances. The next section examines some of these changes.

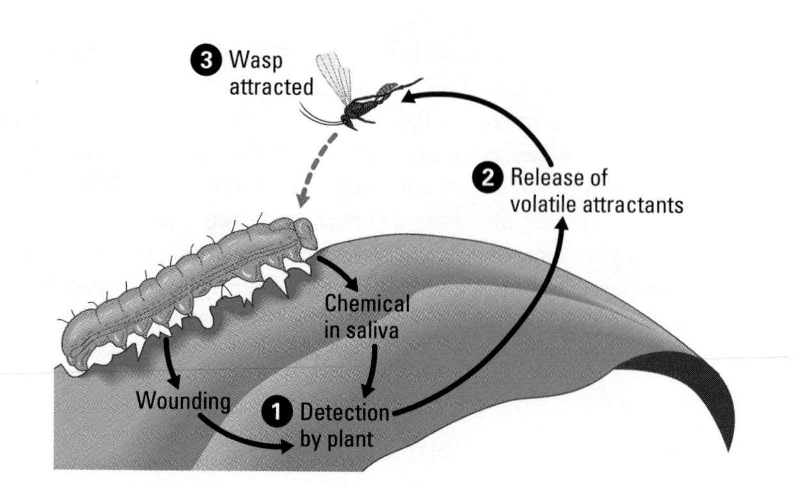

Figure 19.17 A three-way community interaction. Some plants attract wasps that lay their eggs in caterpillars feeding on the plants. After a caterpillar injures the plant's cells, ❶ the saliva from the caterpillar stimulates the leaf to ❷ release a wasp-attracting vapor. ❸ The wasp stings the leaf-munching caterpillar and then lays its eggs within the caterpillar. The wasp larvae that hatch from these eggs will eat their way out of the caterpillar. (Adapted with permission from Edward Farmer, "Plant Biology: New Fatty Acid-Based Signals: A Lesson from the Plant World," *Science* vol. 276, page 912 (May 9, 1997). Copyright © 1997 American Association for the Advancement of Science.)

CHECKPOINT

1. Reexamine the distribution of the two barnacle species in Figure 19.6. What experiment could you perform to test the hypothesis that it is mainly susceptibility to drying, not competitive exclusion, that keeps *Balanus* from populating the upper parts of the rock?
2. A ___________ is a species that helps maintain diversity in a community by preying on the strongest competitor.
3. Which type of symbiotic relationship is represented by bacteria in the human colon (large intestine) that produce vitamin K, a nutrient that can be utilized by the human host?

Answers: 1. You could remove *Chthamalus* from the upper parts of the rock to see whether *Balanus* still failed to spread upward, even in the absence of a competitor. **2.** keystone predator **3.** Mutualism

Disturbance of Communities

Disturbances are episodes that damage biological communities, at least temporarily, by destroying organisms and altering the availability of resources such as mineral nutrients and water. Examples of disturbances are storms, fires, floods and the severe erosion they can cause, droughts, and human activities such as deforestation and the introduction of domesticated grazing animals.

Disturbances affect all communities, which rarely have time to settle into a constant state before another disturbance occurs. Put another way, most communities spend much of their time in various stages of recovery from various disturbances. Change seems to characterize most communities more than constancy and balance.

Ecological Succession

Communities may change drastically after a flood, fire, glacial advance and retreat, or volcanic eruption strips away their vegetation. A variety of species may colonize the disturbed area. Later these may be replaced as yet other species colonize the area. This process of community change is called **ecological succession.**

When a community arises in a virtually lifeless area with no soil, the change is called **primary succession. Figure 19.18** shows different stages of primary successions, starting with ❶ the rubble left by a retreating glacier. ❷ Often the only life-forms initially present are autotrophic microorganisms. ❸ Lichens and mosses, which grow from windblown spores, are commonly the first large photosynthetic organisms to colonize the barren ground. Soil develops gradually as organic matter accumulates from the decomposed remains of the early colonizers. ❹ Once soil is present, the lichens and mosses are overgrown by grasses, shrubs, and trees that sprout from seeds blown in from nearby areas or carried in by animals. ❺ Eventually, the area may be colonized by plants that will become the community's prevalent form of vegetation. ❻ Primary succession from barren soil to a community such as a deciduous forest can take hundreds or thousands of years.

❶ Retreating glacier

❷ Barren landscape after glacial retreat

❸ Moss and lichen stage

❹ Moss, grass, shrubs, and trees (alders and cottonwoods) covering newly formed soil

❺ Spruce coming into the alder and cottonwood forest

❻ Spruce and hemlock forest

Figure 19.18 Succession after retreat of glaciers. Ecologists sometimes deduce the process of succession by studying several locations that are at different successional stages. For example, these photographs, taken at different sites where glaciers are retreating (melting), represent about 200 years of succession.

Secondary succession occurs where a disturbance has destroyed an existing community but left the soil intact. For instance, forested areas that have been cleared for farming will, if abandoned, undergo secondary succession and may eventually return to forest. The earliest plants to recolonize an area are often species with opportunistic life histories (see Chapter 18). Many are herbaceous (nonwoody) species that grow from windblown or animal-borne seeds. These plants thrive where there is little competition from other plants. Woody shrubs may eventually replace most of the herbaceous species. Later yet, trees may replace most of the shrubs.

A Dynamic View of Community Structure

Disturbances keep communities in a continual flux, making them mosaics of patches at various successional stages and preventing them from ever reaching a state of equilibrium or complete balance. We tend to think of a disturbance as having a negative impact, but this is only part of the story, at least in the case of natural disturbances. Small-scale disturbances often have positive effects, such as creating new opportunities for species. For example, when a tree falls, it disturbs the immediate surroundings. However, the fallen tree fosters new habitats, and the depression left by its roots may fill with water and be used as egg-laying sites by frogs, salamanders, and numerous insects **(Figure 19.19)**.

Figure 19.19 A small-scale disturbance. When this tree fell during a windstorm in a forest in Michigan, its root system and the surrounding soil uplifted, resulting in a depression that filled with water. The dead tree, the root mound, and the water-filled depression created new habitats.

If most communities never really stabilize, what is the effect of this constant change on species diversity? When disturbance is severe and relatively frequent, the community may include only those species capable of colonizing at early stages of succession. At the other end of the disturbance scale, when disruptions are mild and rare in a particular location, then the late-successional species that are most competitive will make up the community to the exclusion of other species. Between these two extremes, in an area where disturbance is moderate in both severity and frequency, species diversity may be greatest. Studies of species diversity in tropical rain forests support this intermediate-disturbance hypothesis. Scattered throughout these forests are gaps where trees have fallen. In these disturbed areas, species of various successional stages coexist within a relatively small space.

CHECKPOINT

1. What is the main abiotic factor that distinguishes primary from secondary succession?
2. Why are trees unlikely to be among the first colonizers of a landscape cleared by the advance and retreat of a glacier?

Answers: 1. Absence of soil (primary succession) versus presence of soil (secondary succession) at the onset of succession **2.** Trees can't grow until earlier colonizers form soil.

An Overview of Ecosystem Dynamics

To this point in the chapter, we have focused on the study of communities. We will now broaden our scope to ecological systems, or ecosystems, the highest level of ecological complexity. An **ecosystem** consists of all the organisms in a given area plus the physical environment, including soil, water, and air. In other words, an ecosystem is a biological community plus the abiotic factors with which the community interacts.

Perhaps you have seen or even made a terrarium such as the one in **Figure 19.20**. Such a microcosm qualifies as an ecosystem because it exhibits the two major processes that sustain all ecosystems: energy flow and chemical cycling. **Energy flow** is the passage of energy through the components of the ecosystem. **Chemical cycling** is the use and reuse of chemical elements such as carbon and nitrogen within the ecosystem.

Energy reaches most ecosystems in the form of sunlight. Plants and other photosynthetic producers convert the light energy to the chemical energy of food. Energy continues its flow through an ecosystem when animals and other consumers acquire organic matter by feeding on plants and other producers. Especially important consumers are the soil bacteria and fungi that obtain their energy by decomposing the dead remains of plants, animals, and other organisms. In using their chemical energy for work, all organisms dissipate heat energy to their surroundings. Thus, energy cannot be recycled within an ecosystem, but must flow continuously through the ecosystem, entering as light and exiting as heat (see Figure 19.20).

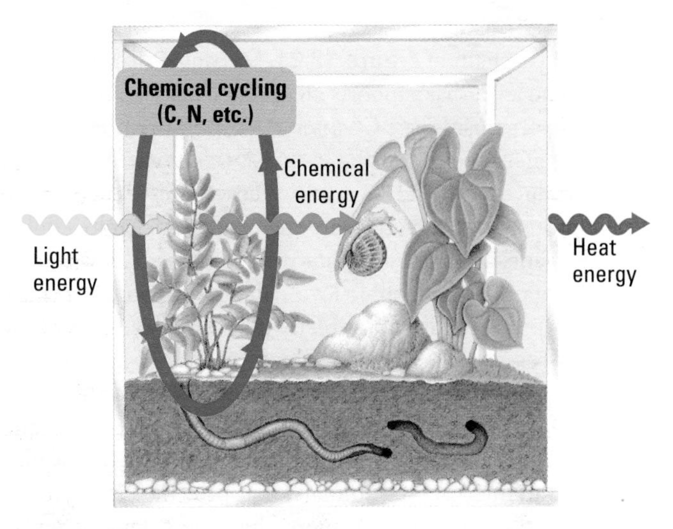

Figure 19.20 A terrarium ecosystem. Small and artificial as it is, this terrarium illustrates the two major ecosystem processes: energy flow and chemical cycling.

In contrast to the flow of energy through an ecosystem, chemical elements can be recycled between an ecosystem's living community and the abiotic environment. The plants and other producers acquire their carbon, nitrogen, and other chemical elements in inorganic form from the air and soil. Photosynthesis then enables plants to incorporate these elements into organic compounds such as carbohydrates and proteins. Animals acquire these elements in organic form by eating. The metabolism of all organisms returns some of the chemical elements to the abiotic environment in inorganic form. Cellular respiration, for example, breaks organic molecules down to carbon dioxide and water. This job of recycling is finished by microorganisms that decompose dead organisms and their organic wastes, such as feces and leaf litter. These decomposers restock the soil, water, and air with chemical elements in the inorganic form that plants and other producers can again build into organic matter. And the chemical cycles continue.

Note again the key distinction between energy *flow* and chemical *cycling*. Because energy, unlike matter, cannot be recycled, an ecosystem must be powered by a continuous influx of energy from an external source, usually the sun. Notice also that an ecosystem's energy flow and chemical cycling are closely related. Both depend on transfer of substances in the feeding relationships, or trophic structure, of the ecosystem.

Trophic relationships determine an ecosystem's routes of energy flow and chemical cycling. In analyzing these feeding relationships, ecologists divide the species of an ecosystem into different **trophic levels** based on their main sources of nutrition. Let's see how these trophic levels connect as the living components of an ecosystem's energy flow and chemical cycling.

Trophic Levels and Food Chains

The sequence of food transfer from trophic level to trophic level is called a **food chain.** You can see a comparison of a terrestrial food chain and a marine food chain in **Figure 19.21**. Starting at the bottom of such diagrams, the trophic level that supports all others consists of the **producers,** the autotrophic organisms (see Chapter 6). As photosynthetic organisms, producers use light energy to power the synthesis of organic compounds. Plants are the main producers on land. In aquatic ecosystems, **phytoplankton**—photosynthetic protists and bacteria—are the producers. Multicellular algae and aquatic plants are also important producers in shallow waters.

All organisms in trophic levels above the producers are consumers, the heterotrophs directly or indirectly dependent on the output of producers. **Herbivores,** which eat plants, algae, or autotrophic bacteria, are the **primary consumers** of an ecosystem. Primary consumers on land include insects, snails, and certain vertebrates, such as grazing mammals. In aquatic ecosystems, primary consumers include a variety of **zooplankton** (mainly protists and microscopic animals) that prey on the phytoplankton.

Above the primary consumers, the trophic levels are made up of **carnivores,** which eat the consumers from the levels below. On land, **secondary consumers** include many small mammals, such as rodents that eat herbivorous insects, and a great variety of small birds, frogs, and spiders, as well as lions and other large carnivores that eat grazers. (We humans qualify as secondary consumers, too, because we eat grazers, such as cows, and their products, such as milk.) In aquatic ecosystems, secondary consumers are mainly small fishes that eat zooplankton and bottom-dwelling invertebrates. Higher trophic levels include **tertiary consumers,** such as snakes that eat mice and other secondary consumers. Some ecosystems even have **quaternary consumers,** such as hawks in terrestrial ecosystems and killer whales in marine environments (see Figure 19.21).

Most diagrams of food chains are incomplete because they do not show a critical trophic level of consumers called **detritivores,** another name for **decomposers.** Detritivores derive their energy from **detritus,** the dead material left by all trophic levels. Detritus includes animal wastes, plant litter, and dead organisms. Most organic matter eventually becomes detritus and is consumed by detritivores. A great variety of animals, often called scavengers, eat detritus. For instance, earthworms, many rodents, and insects eat fallen leaves and other detritus. Other scavengers include crayfish, catfish, and vultures. But an ecosystem's main detritivores are the prokaryotes and fungi **(Figure 19.22)**. Enormous numbers of microorganisms in the soil and in mud at the bottoms of lakes and oceans recycle most of the ecosystem's organic materials to inorganic compounds that plants and other producers can reuse.

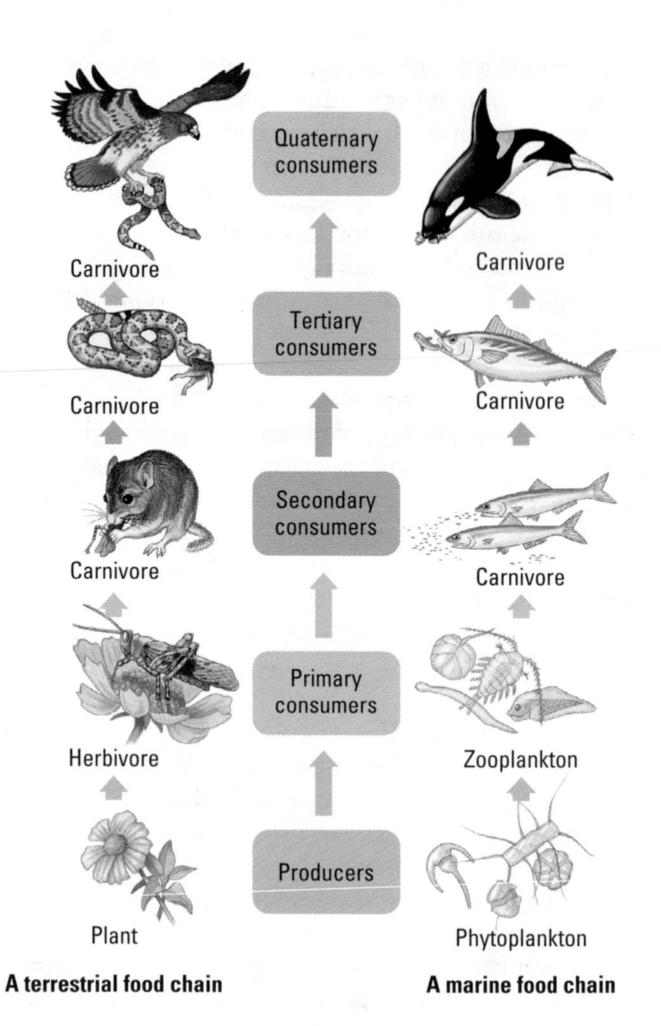

Figure 19.21 Examples of food chains. The arrows trace the transfer of food from producers through various levels of consumers. Detritivores (decomposers), important consumers in all ecosystems, are not included in these simplified diagrams of terrestrial and marine ecosystems.

Figure 19.22 Fungi decomposing a dead log.

Food Webs

Actually, few ecosystems are so simple that they are characterized by a single unbranched food chain. Several types of primary consumers usually feed on the same plant species, and one species of primary consumer may eat several different plants. Such branching of food chains occurs at the other trophic levels as well. For example, adult frogs, which are secondary consumers, eat several insect species that may also be eaten by various birds. In addition, some consumers feed at several different trophic levels. An owl, for instance, may eat mice, which are mainly primary consumers that may also eat certain invertebrates. But an owl may also feed on snakes, which are strictly carnivorous. **Omnivores,** including humans, eat producers as well as consumers of different levels. Thus, the feeding relationships in an ecosystem are usually woven into elaborate **food webs.**

Though more realistic than a food chain, a food web diagram, such as the one in **Figure 19.23**, is still a highly simplified model of the feeding

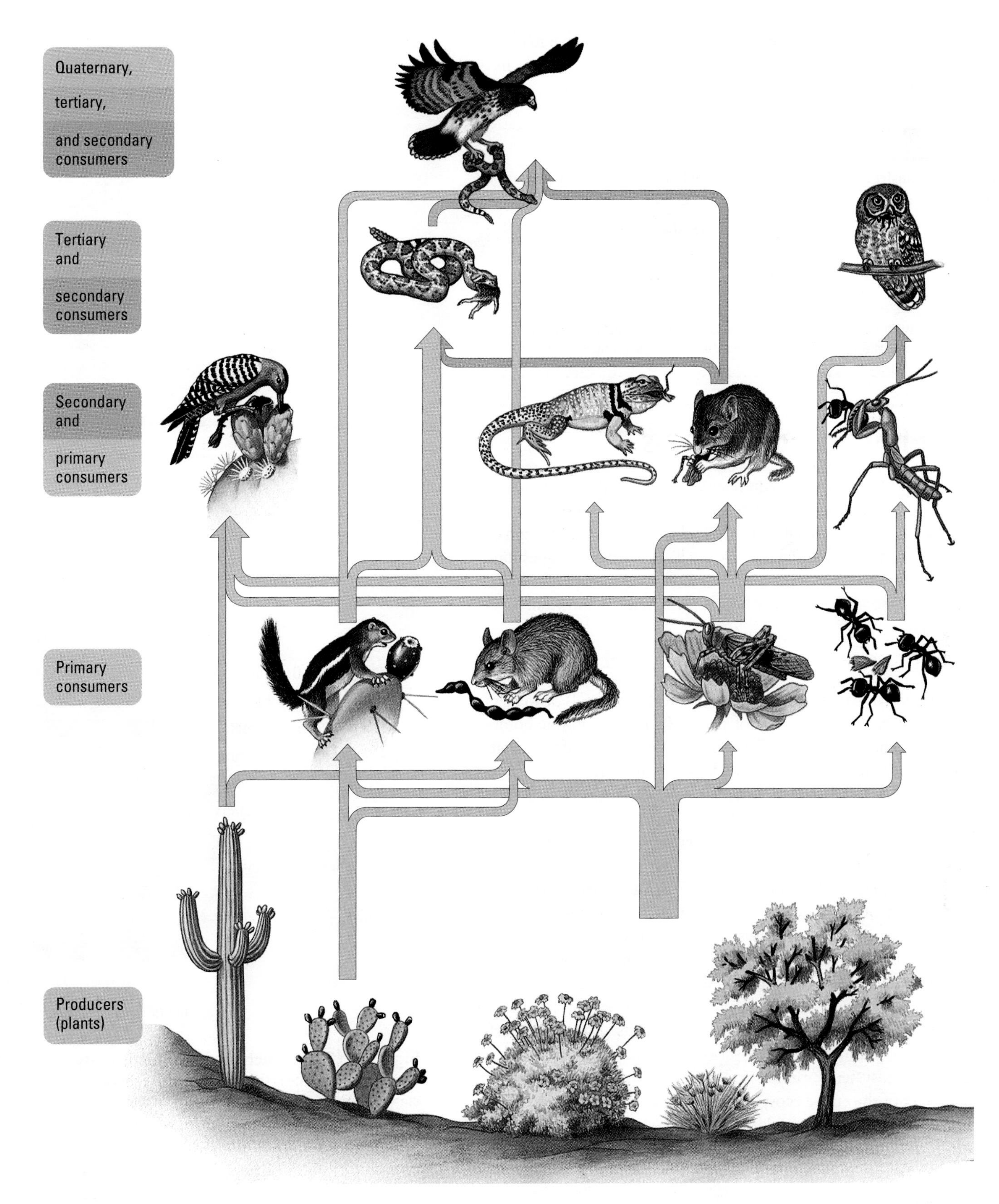

Figure 19.23 A simplified food web. As in the food chains of Figure 19.21, the arrows in this web indicate "who eats whom," the direction of nutrient transfers. We also continue the color coding introduced in Figure 19.21 for the trophic levels and food transfers. Note that in a food web, such as this one for a Sonoran Desert ecosystem, some species feed at more than one trophic level. The other branching factor that turns food chains into webs is the consumption of the same species by more than one consumer.

relationships in an ecosystem. An actual food web would involve many more organisms at each trophic level, and most of the animals would have a more diverse diet than shown in the figure.

CHECKPOINT

1. Why is the transfer of energy in an ecosystem referred to as energy flow, not energy cycling?
2. I'm eating a cheese pizza. At which trophic level(s) am I feeding?
3. In the "who eats whom" dynamics of a food web, even consumers of the highest level in the ecosystem eventually become food for ________.

Answers: 1. Because energy passes through an ecosystem, entering as sunlight and leaving as heat. It is not recycled within the ecosystem. **2.** I'm a primary consumer when I eat flour and tomato sauce, and I'm a secondary consumer when I eat cheese (or meat) on the pizza. **3.** detritivores, or decomposers

Energy Flow in Ecosystems

MP3 Tutor
Energy Flow in Ecosystems

All organisms require energy for growth, maintenance, reproduction, and, in some species, locomotion. In this section, we take a closer look at energy flow through ecosystems. Along the way, we'll answer two key questions: What limits the length of food chains? and How do lessons about energy flow apply to human nutrition?

Productivity and the Energy Budgets of Ecosystems

Each day, Earth is bombarded by about 10^{19} kilocalories (kcal) of solar radiation. This is the energy equivalent of about 100 million atomic bombs the size of the one that devastated Hiroshima in 1945. Most of this solar energy is absorbed, scattered, or reflected by the atmosphere or by Earth's surface. Of the visible light that reaches plants and other producers, only about 1% is converted to chemical energy by photosynthesis.

Ecologists call the amount, or mass, of organic material in an ecosystem the **biomass.** And the rate at which plants and other producers build biomass, or organic matter, is called the ecosystem's **primary productivity.** The primary productivity of the entire biosphere is about 170 billion tons of organic material per year. Different ecosystems vary considerably in their primary productivity **(Figure 19.24)** as well as in their contribution to the total productivity of the biosphere. Open ocean, for example, has a low productivity per unit area. But its huge volume results in a large total contribution to the planet's productivity. In contrast, tropical rain forests and tropical reefs have a high primary productivity but cover relatively little area.

Whatever the ecosystem, primary productivity sets the spending limit for the energy budget of the entire ecosystem because consumers must acquire their organic fuels from producers. Now let's see how this energy budget is divided among the different trophic levels in an ecosystem's food web.

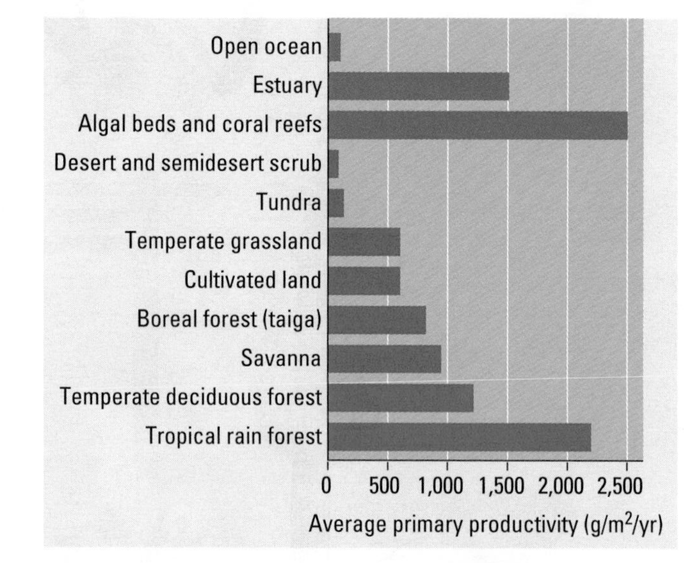

Figure 19.24 Productivity of different ecosystems. Primary productivity is the rate at which plants and other producers store chemical energy in biomass over a unit of time, in this case a year. Aquatic ecosystems are color-coded blue in these histograms; terrestrial ecosystems are green.

Energy Pyramids

When energy flows as organic matter through the trophic levels of an ecosystem, much of it is lost at each link in the food chain. Consider the transfer of organic matter from plants (producers) to herbivores (primary consumers). In most ecosystems, herbivores manage to eat only a fraction of the plant material produced, and they cannot digest all the organic compounds they do ingest. Of the organic compounds an herbivore can use, much of it is consumed as fuel for cellular respiration. Only the remainder is available as raw material for growth of the herbivore. For example, in the case of a caterpillar consuming leaves, only about 15% of the biomass of the food is transformed into caterpillar biomass **(Figure 19.25)**. On average, only about 10% of energy in the form of organic matter at each trophic level is stored as biomass in the next level of the food chain. The rest of the energy winds up in feces or is released as heat by the working organisms in the ecosystem.

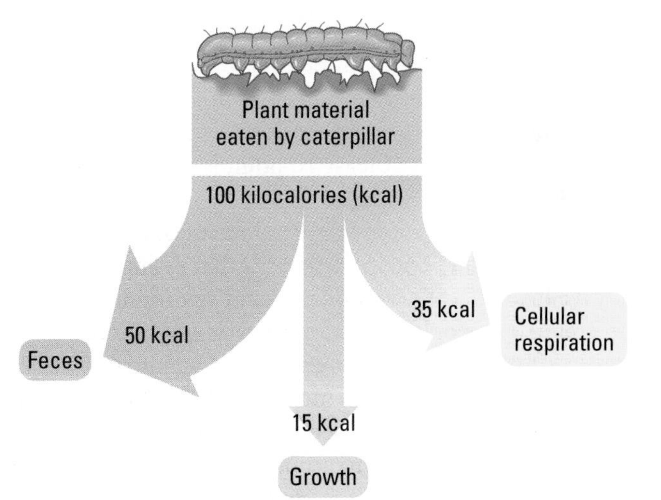

Figure 19.25 What becomes of a caterpillar's food? Only about 15% of the calories of plant material this herbivore consumes becomes stored as biomass available to the next link in the food chain.

The cumulative loss of energy from a food chain can be represented in a diagram called an **energy pyramid.** The trophic levels are stacked in blocks, with the block representing producers forming the foundation of the pyramid **(Figure 19.26)**. The size of each block is proportional to the biomass in each trophic level. The pyramid owes its steep shape to the loss of 90% or so of the energy with each food transfer in the chain.

An important implication of this stepwise decline of energy along a food chain is that the amount of energy available to top-level consumers is small compared with that available to lower-level consumers. Only a tiny fraction of the energy stored by photosynthesis flows through a food chain to a tertiary consumer, such as a snake feeding on a mouse. This explains why top-level consumers such as killer whales and hawks require so much geographic territory; it takes a lot of vegetation to support trophic levels so many steps removed from photosynthetic production. You can also understand now why most food chains are limited to three to five levels; there is simply not enough energy at the very top of an energy pyramid to support another trophic level. There are, for example, no nonhuman predators of lions, eagles, and killer whales; the biomass in populations of these top-level consumers is insufficient to support yet another trophic level with a reliable source of nutrition.

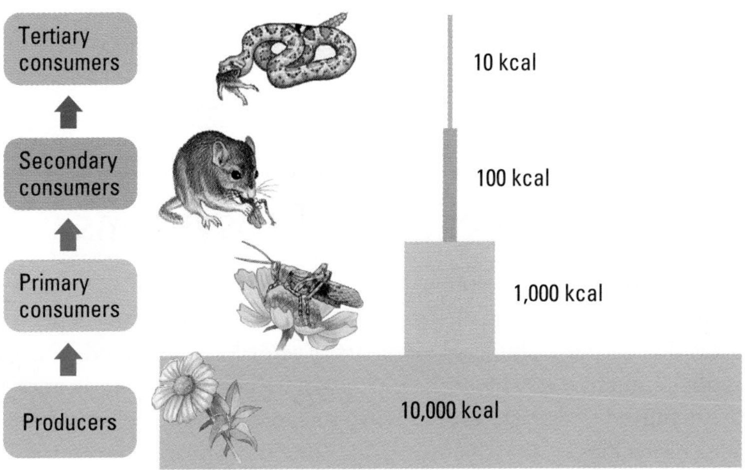

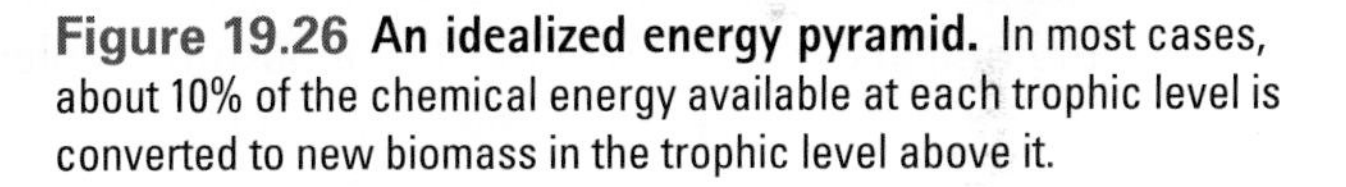

Figure 19.26 An idealized energy pyramid. In most cases, about 10% of the chemical energy available at each trophic level is converted to new biomass in the trophic level above it.

Ecosystem Energetics and Human Nutrition

The dynamics of energy flow apply to the human population as much as to other organisms. Like other consumers, we depend entirely on productivity by plants for our food. As omnivores, most of us eat both plant material and meat. When we eat grain or fruit, we are primary consumers; when we eat beef or other meat cut from herbivores, we are secondary consumers. When we eat fish like salmon (which eat insects and other small animals) and tuna (which eat smaller fish), we are tertiary or quaternary consumers.

Figure 19.27 applies the concept of energy pyramids to show why consuming a high proportion of our calories in the form of meat is a

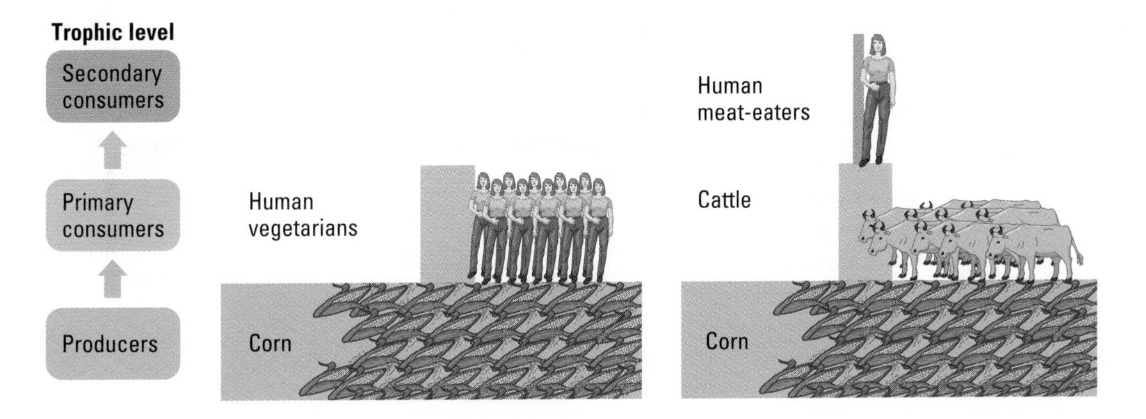

Figure 19.27 Food energy available to the human population at different trophic levels. Most humans have a diet between these two extremes. The point here is that a greater proportion of photosynthetic energy reaches us when we feed as primary consumers directly on plants than when we are nourished by photosynthesis indirectly by feeding as secondary consumers on animals.

relatively inefficient way of tapping photosynthetic productivity. About 90% of the food energy in a crop such as corn is lost to us if that corn is first "processed" by primary consumers such as hogs or cattle. Put still another way, it takes at least 10 pounds of feed corn to produce 1 pound of steak.

This analysis of energy flow to human consumers is not meant as a plea for vegetarianism. The main point is that eating meat of any kind is an expensive luxury, both economically and environmentally. In many developing countries, people are mainly vegetarian by necessity because they cannot afford to buy much meat or their country cannot afford to produce it. Wherever meat is on the menu, producing it requires that more land be cultivated, more water be used for irrigation, and more chemical fertilizers and pesticides be applied to croplands used for growing grain. It is likely that as the human population continues to expand, meat consumption will become even more of a luxury than it is today.

CHECKPOINT

1. In a shrubland ecosystem, which is likely to have the greatest total biomass: the sum of all insects or the sum of all birds that feed on the insects?
2. Why is a pound of bacon so much more expensive than a pound of corn?

Answers: 1. The insects **2.** Because it took about 10 pounds of feed corn to produce that pound of bacon

Chemical Cycling in Ecosystems

The sun supplies ecosystems with energy, but there are no extraterrestrial sources of the chemical elements life requires. Ecosystems depend on a recycling of these chemical elements. Even while an individual organism is alive, much of its chemical stock is rotated continuously as nutrients are acquired and waste products are released. Atoms present in the complex molecules of an organism at the time of its death are returned in simpler compounds to the atmosphere, water, or soil by the action of bacterial and fungal detritivores. This decomposition replenishes the pools of inorganic nutrients that plants and other producers use to build new organic matter. In a sense, we only borrow an ecosystem's chemical elements, returning what is left in our bodies after we die. "Ashes to ashes, dust to dust" is one metaphor for this fact of life. Let's take a closer look at how chemicals cycle between organisms and the abiotic (nonliving) components of ecosystems.

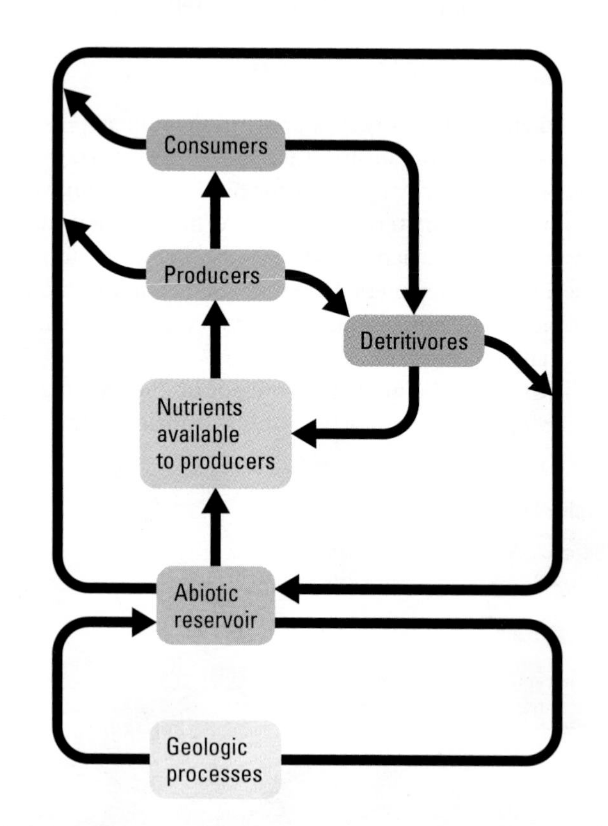

Figure 19.28 Generalized scheme for biogeochemical cycles.

The General Scheme of Chemical Cycling

Because chemical cycles in an ecosystem involve both biotic and abiotic components, they are called **biogeochemical cycles. Figure 19.28** is a general scheme for these cycles. Here are three key points to note:

- Each circuit, whether for carbon, nitrogen, or some other chemical material required for life, has an **abiotic reservoir** through which the chemical cycles. Carbon's main abiotic reservoir, for example, is the atmosphere, where carbon atoms are stocked mainly in the form of CO_2 gas (carbon dioxide). Carbon makes the abiotic → biotic transition when it is incorporated from CO_2 into food by plants and other producers. And the carbon is returned to the abiotic reservoir by the cellular respiration of the ecosystem's organisms, including detritivores.

- A portion of chemical cycling can bypass the biotic components and rely completely on geologic processes. Water, for example, can exit lakes and oceans by evaporation and recycle to those abiotic reservoirs by precipitation, such as rain.
- Some chemical elements require "processing" by certain microorganisms before they are available to plants as inorganic nutrients. The element nitrogen is an example. Its main abiotic reservoir is the atmosphere, which is almost 80% nitrogen in the form of N_2 gas. However, plants can utilize nitrogen only in the forms of ammonium (NH_4^+) and nitrate (NO_3^-), which roots absorb from the soil. Certain soil bacteria convert atmospheric N_2 to ammonium and nitrate in the soil, making nitrogen available to plants (see Chapter 15). In fact, microorganisms, mainly prokaryotes, play a critical role in all biogeochemical cycles.

Examples of Biogeochemical Cycles

Like most generalizations, the basic scheme of Figure 19.28 can go only so far in helping us understand biogeochemical cycles. A chemical's specific route through an ecosystem varies with the particular element and the trophic structure of the ecosystem. **Figure 19.29**, on the next two pages, will help you apply the basic principles of chemical cycling to four key materials: carbon, nitrogen, phosphorus, and water (the only one that's a compound, not an element).

The carbon, nitrogen, phosphorus, and water cycles (and others) can be classified into two main categories based on how mobile the chemicals are within environments. Some chemicals, including phosphorus, are not very mobile and thus cycle almost entirely locally, at least over the short term. Soil is the main reservoir for these nutrients, which are absorbed by plant roots and eventually returned to the soil by detritivores living in the same vicinity. In contrast, for those materials that spend part of their time in gaseous form—carbon, nitrogen, and water are examples—the cycling is essentially global. For instance, some of the carbon atoms a plant acquires from the air as CO_2 may have been released to the atmosphere by the respiration of some animal living far away.

CHECKPOINT

1. What is the main abiotic reservoir for carbon?
2. What would happen to the carbon cycle if all the detritivores suddenly "went on strike" and stopped working?
3. Over the short term, why does phosphorus cycling tend to be more localized than carbon, nitrogen, or water cycling?

Answers: 1. The atmospheric stock of CO_2 **2.** Carbon would accumulate in organic mass, the atmospheric reservoir of carbon would decline, plants would eventually be starved for CO_2, and photosynthesis would cease. **3.** Because phosphorus is cycled almost entirely within the soil rather than being transferred over long distances via the atmosphere

Figure 19.29 Examples of biogeochemical cycles.

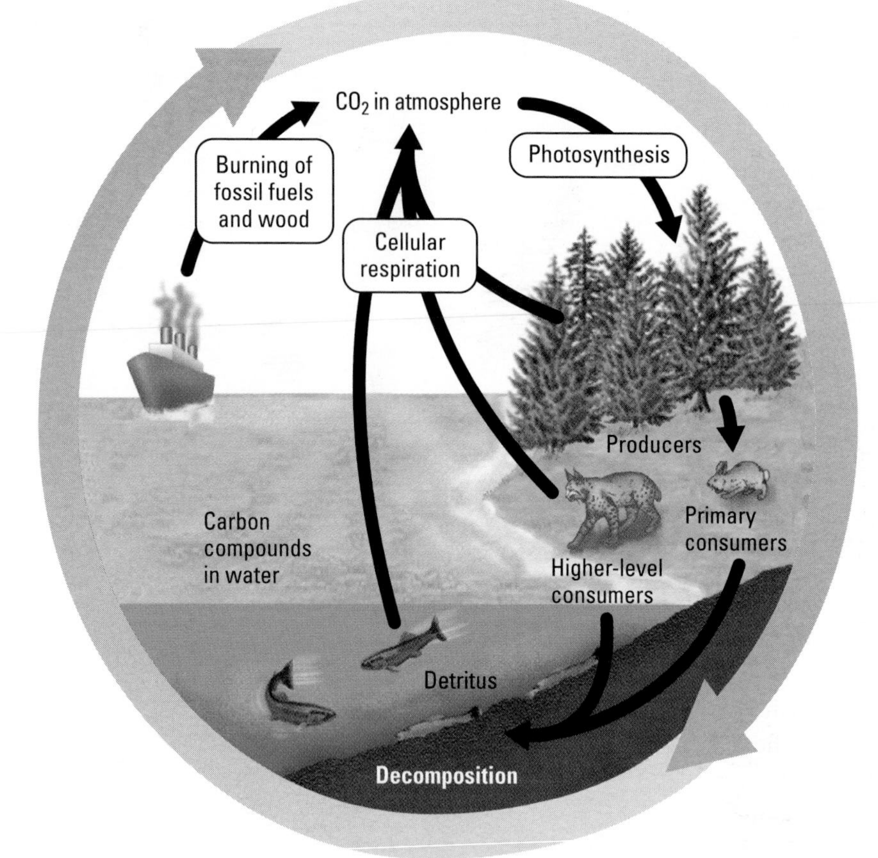

(a) The carbon cycle. Carbon is a major ingredient of all organic molecules. With the atmosphere as its main abiotic reservoir, carbon cycles globally. The reciprocal metabolic processes of photosynthesis and cellular respiration are mainly responsible for the cycling of carbon between the biotic and abiotic world. On a global scale, the return of CO_2 to the atmosphere by respiration closely balances its removal by photosynthesis. However, the increased burning of wood (during slash-and-burn deforestation) and fossil fuels (coal and petroleum) is steadily raising the level of CO_2 in the atmosphere, causing significant environmental problems, including global warming.

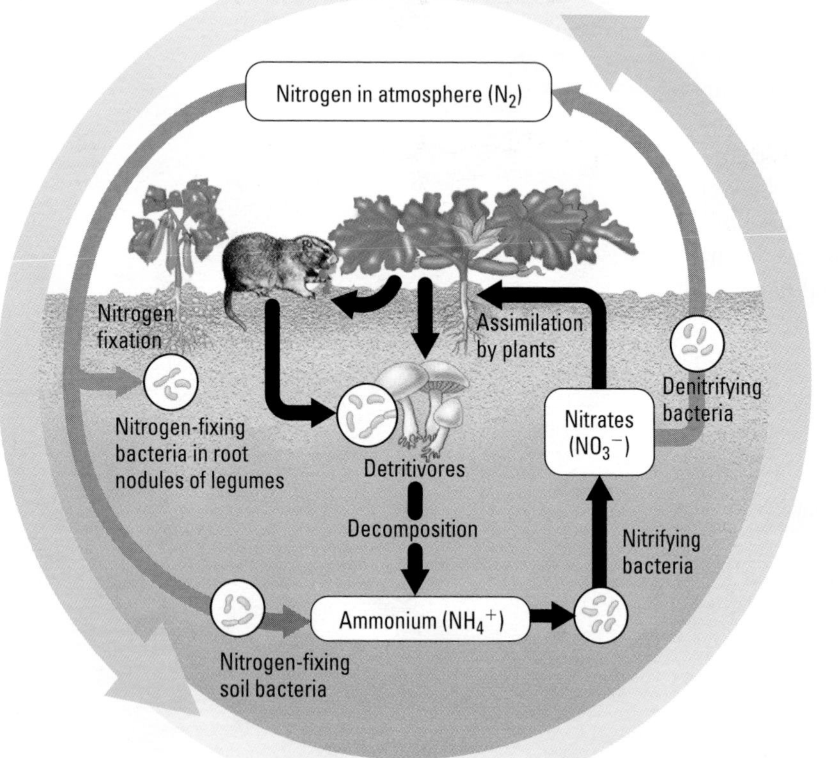

(b) The nitrogen cycle. As an ingredient of amino acids and other organic compounds essential to life, nitrogen is a key chemical element in all ecosystems. Earth's atmosphere is almost 80% N_2 gas, but that form of nitrogen cannot be used by plants. Soil bacteria called nitrogen-fixing bacteria convert the N_2 to ammonium, and nitrifying bacteria convert the ammonium to nitrates. Ammonium and nitrates are the soil minerals that the roots of plants absorb as nitrogen sources for synthesis of amino acids and other organic molecules. Notice in the diagram that there are two groups of nitrogen-fixing bacteria: free-living bacteria and those that live as symbiotic organisms in root nodules of beans and other legumes (see Chapter 15). Other types of bacteria also play key roles at various junctures in the nitrogen cycle. For example, denitrifying bacteria complete the nitrogen cycle by converting soil nitrate to atmospheric N_2. Much of the nitrogen cycling by bacteria in many ecosystems involves the pathway shown by black arrows in the diagram. The pathway shown by gray arrows often moves only a tiny fraction of nitrogen into and out of natural ecosystems.

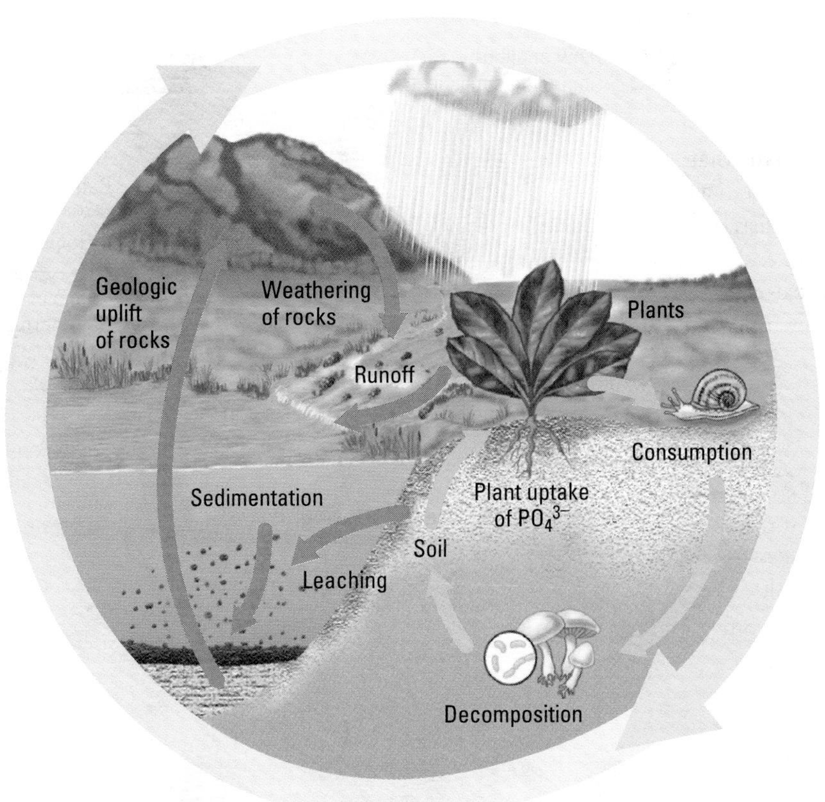

(c) The phosphorus cycle. Organisms require phosphorus as an ingredient of ATP and nucleic acids (such as DNA) and, in vertebrates, as a major component of bones and teeth. In contrast to carbon and nitrogen, phosphorus does not have an atmospheric presence and tends to recycle only locally. The main abiotic reservoir is rock, which, upon weathering, releases phosphorus mainly in the form of the mineral phosphate (PO_4^{3-}). Plants absorb the dissolved phosphate ions in the soil and build them into organic compounds. Consumers obtain phosphorus in organic form from plants. Detritivores return phosphates to the soil. Some phosphates also precipitate out of solution at the bottom of deep lakes and oceans. The phosphates in this form may eventually become part of new rocks and will not cycle back into living organisms until geologic processes uplift the rocks and expose them to weathering. Thus, phosphorus actually cycles on two time scales: through a local ecosystem on the scale of ecological time (yellow arrows in diagram) and through strictly geologic processes on the much longer scale of geologic time (gray arrows).

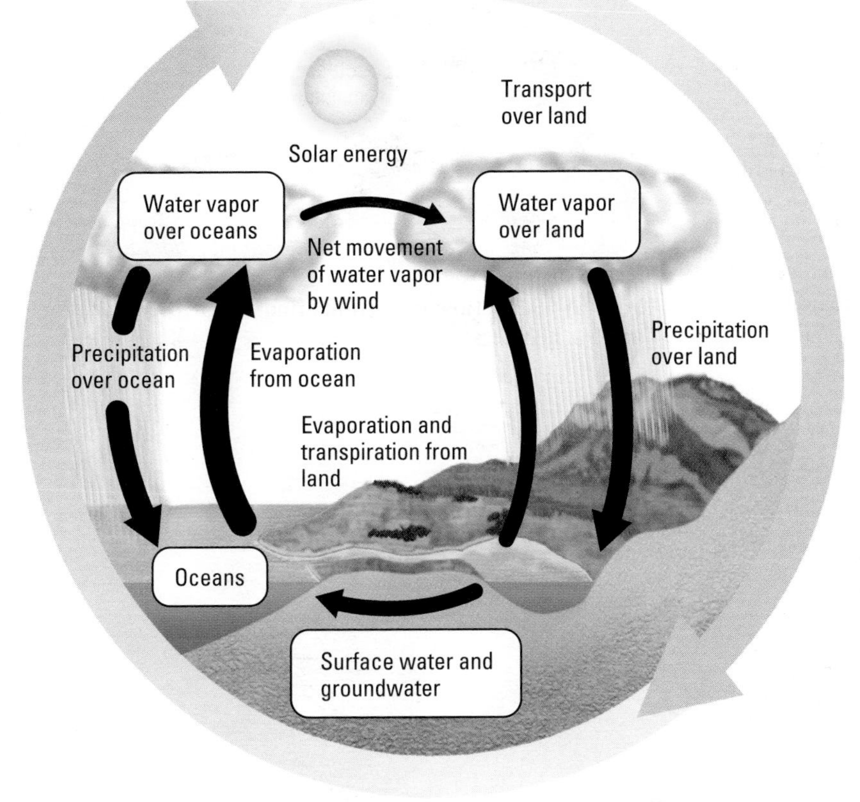

(d) The water cycle. Water is essential to life, both because organisms are made mostly of water and because water's unique properties make environments on Earth habitable (see Chapter 2). Three major processes driven by solar energy—precipitation, evaporation, and transpiration from plants (evaporation from leaves)—continuously move water between the land, the oceans, and the atmosphere. In this diagram, the widths of the arrows indicate the relative quantities of water moving through various parts of the global water cycle. Note that over the oceans, evaporation exceeds precipitation. The result is a net movement of this excess amount of water vapor in clouds that are carried by winds from the oceans across the land. On land, precipitation exceeds evaporation and transpiration. The excess precipitation forms systems of surface water (such as lakes and rivers) and groundwater, all of which flow back to the sea, completing the water cycle. The water cycle has a global character because there is a large reservoir of water in the atmosphere. Thus, water molecules that have evaporated from the Pacific Ocean, for instance, may appear in a lake or in an animal's body far inland in North America.

Biomes

Now that we've examined how ecosystems are organized, we can turn our attention to the ecosystems found on Earth. Major types of ecosystems that cover large geographic regions are called **biomes.** You can think of them as types of landscapes. Tropical forests and deserts are examples. The distribution of the biomes largely depends on climate, with temperature and rainfall often the key factors determining the kind of biome that exists in a particular region. Let's look at how climate is determined and how it helps explain the locations of various types of biomes.

How Climate Affects Biome Distribution

Because of its curvature, Earth receives an uneven distribution of solar energy **(Figure 19.30)**. The equator receives the greatest intensity of solar radiation. Heated by the direct rays of the sun, air at the equator rises. It then cools, forms clouds, and drops rain **(Figure 19.31)**. This largely explains why rain forests are concentrated in the **tropics**—the region from the Tropic of Cancer to the Tropic of Capricorn.

After losing moisture over equatorial zones, dry high-altitude air masses spread away from the equator until they cool and descend again at latitudes of about 30° north and south. Many of the world's great deserts—the Sahara in North Africa and the Arabian on the Arabian Peninsula, for example—are centered at these latitudes because of the dry air they receive.

Latitudes between the tropics and the Arctic Circle in the north and the Antarctic Circle in the south are called **temperate zones.** Generally, these regions have milder climates than the tropics or the polar regions. Notice in Figure 19.31 that some of the descending dry air heads into the latitudes above 30°. At first these air masses pick up moisture, but they tend to drop it as they cool at higher latitudes. This is why the north and south temperate zones tend to be relatively wet. Coniferous forests dominate the landscape at the wet but cool latitudes around 60° north.

Proximity to large bodies of water and the presence of landforms such as mountain ranges also affect climate. Oceans and large lakes moderate climate by absorbing heat when the air is warm and releasing heat to cold air. Mountains affect climate in two major ways. First, air temperature declines as elevation increases. Second, mountains can block the flow of cool, moist air from a coast, causing radically different climates on opposite sides of a mountain range **(Figure 19.32)**. Now let's survey Earth's major ecosystems, noting the influence of climate.

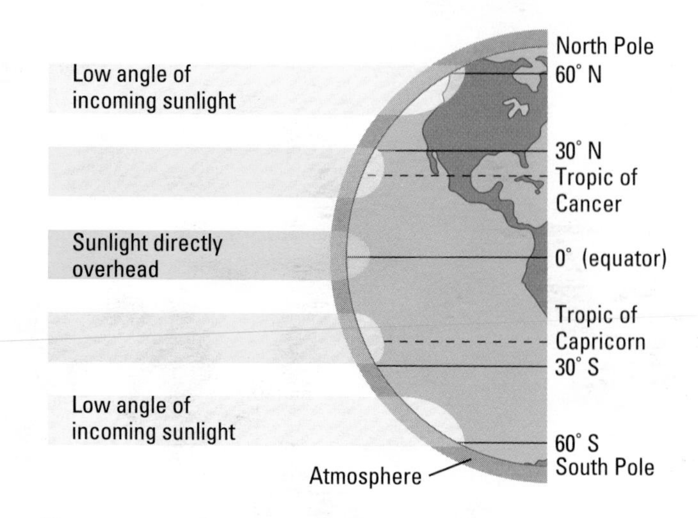

Figure 19.30 Uneven heating of Earth.

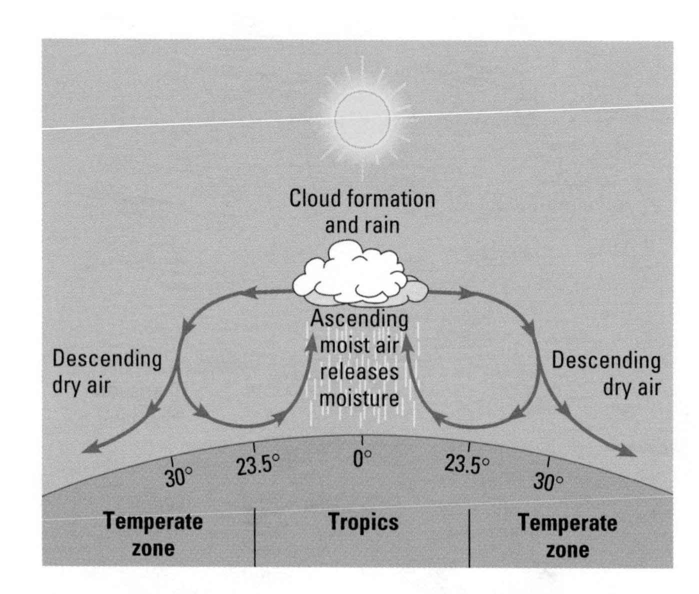

Figure 19.31 How uneven heating of Earth produces various climates.

Terrestrial Biomes

Figure 19.33 shows a map of Earth's major terrestrial biomes. Because there are latitudinal patterns of climate over Earth's surface, there are also latitudinal patterns of biome distribution. If the climate in two geographically separate areas is similar, the same type of biome may occur in both places. Coniferous forests, for instance, extend in a broad band across North America, Europe, and Asia. Note, however, that a biome is characterized by a type of biological community, not a collection of certain species of organisms. For example, the groups of species living in the Sahara Desert of Africa and the Gobi Desert of eastern Asia are different, but the species in both regions are adapted to desert conditions. Organisms in widely separated biomes may look alike because of convergence, the evolution of similar adaptations in independently evolved species living in similar environments (see Chapter 14).

Most biomes are named for major physical or climatic features and for their predominant vegetation. For example, temperate grasslands are dominated by various grass species and are generally found in the temperate zones. Each biome is also characterized by microorganisms, fungi, and animals adapted to that particular environment. Temperate grasslands, for example, are more likely than forests to be populated by large grazing mammals.

In most biomes today, extensive human activities have radically altered the natural patterns. Most of the eastern United States, for example, is classified as temperate deciduous forest, but human activity has eliminated all but a tiny percentage of the original forest. In fact, humans have altered much of Earth's surface, replacing original biomes with urban and agricultural ones.

Figure 19.34, spread over the next four pages, surveys the major terrestrial biomes, beginning near the equator and generally approaching the poles.

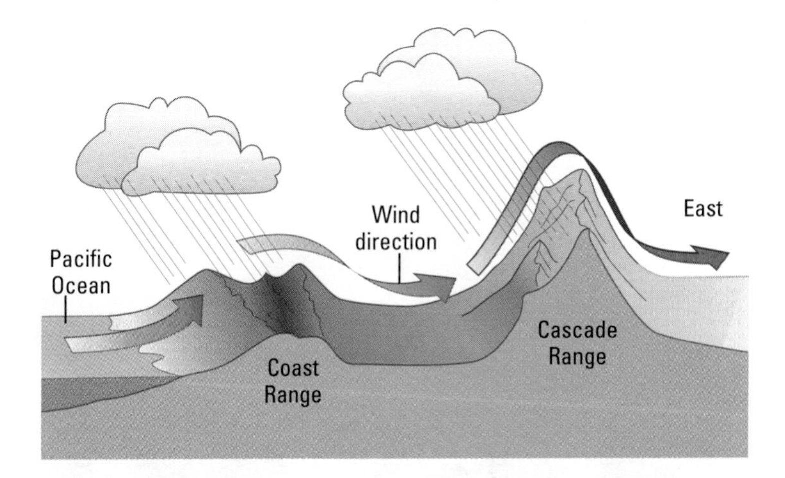

Figure 19.32 How mountains affect rainfall. In this example, moist air moves in off the Pacific Ocean and encounters the Coast Range in the state of Washington. Air flows upward, cools at higher altitudes, and drops a large amount of water. The biological community in this wet region is a temperate rain forest. Farther inland, precipitation increases again as the air moves up and over the Cascade Range. On the eastern side of the Cascades, there is little precipitation. As a result of this so-called rain shadow, much of central Washington is very arid, almost qualifying as a desert.

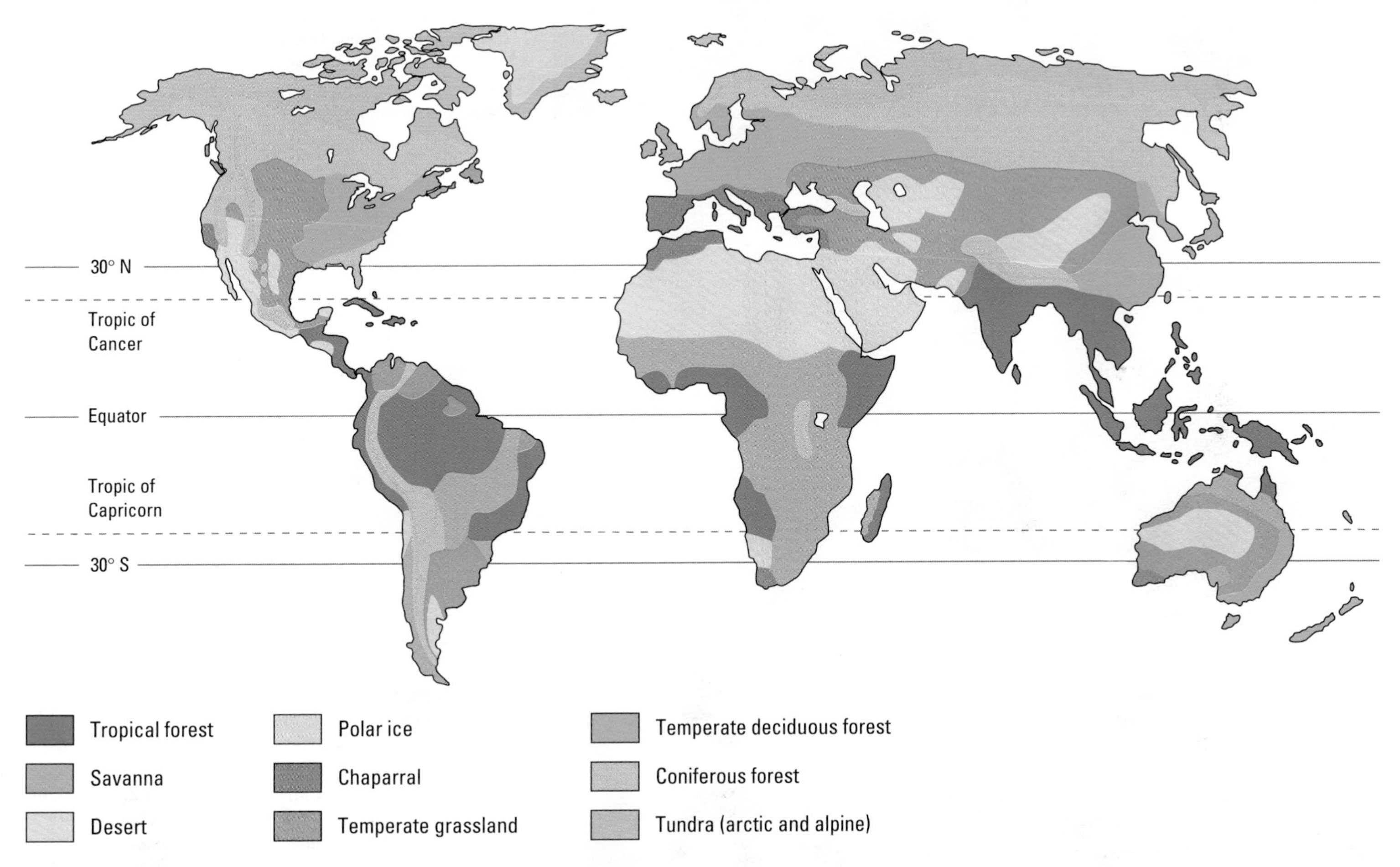

Figure 19.33 Map of the major terrestrial biomes. Although this map has sharp boundaries, biomes actually grade into one another. We'll use smaller versions of this map, highlighted by color coding, during our closer look at the terrestrial biomes in Figure 19.34.

Figure 19.34 Major terrestrial biomes.

(a) Tropical forest. Rainfall, which is quite variable in the tropics, is the prime determinant of the vegetation growing in a tropical forest. This photograph shows a tropical rain forest in Costa Rica. Tropical rain forests have pronounced vertical stratification, or layering. Trees in the canopy make up the topmost stratum. The canopy is often so dense that little light reaches the ground below. When an opening does occur, perhaps because of a fallen tree, other trees and large woody vines grow rapidly, competing for light and space as they fill the gap. Many of the trees are covered with epiphytes (plants that grow on other plants rather than in soil), such as orchids and bromeliads. In lowland areas that have a prolonged dry season or scarce rainfall at any time, tropical dry forests predominate. The plants found there are a mixture of thorny shrubs and trees and succulents. In regions with distinct wet and dry seasons, tropical deciduous trees are common.

(b) Savanna. Savannas are grasslands with scattered trees. Savannas such as this one in Kenya are home to many large herbivores and their predators. Actually, the dominant herbivores here and in other savannas are insects, especially ants and termites. Fire is an important abiotic component, and the dominant plant species are fire adapted. The luxuriant growth of grasses and forbs (small broadleaf plants) during the rainy season provides a rich food source for animals. However, large grazing mammals must migrate to greener pastures and scattered watering holes during regular periods of seasonal drought.

(c) Desert. Sparse rainfall (less than 30 cm per year) largely determines that an area will be a desert. Some deserts have soil surface temperatures above 60°C (140°F) during the day. Other deserts, such as those west of the Rocky Mountains and in central Asia, are relatively cold. The Sonoran Desert of southern Arizona, shown here, is characterized by giant saguaro cacti and deeply rooted shrubs. Evolutionary adaptations of desert plants and animals include a remarkable array of mechanisms that store water. The "pleated" structure of saguaro cacti enables the plants to expand when they absorb water during wet periods. Some desert mice never drink, deriving all their water from the metabolic breakdown of the seeds they eat. Protective adaptations that deter feeding by mammals and insects, such as spines on cacti and poisons in the leaves of shrubs, are common in desert plants.

(d) Chaparral. This biome occurs in midlatitude coastal areas with mild, rainy winters and long, hot, dry summers. Dense, spiny, evergreen shrubs dominate chaparral. Plants of the chaparral, such as those in this California scrubland, are adapted to and dependent on periodic fires. The dry, woody shrubs are frequently ignited by lightning and by careless human activities, creating summer and autumn brushfires in the densely populated canyons of Southern California and elsewhere. Some of the shrubs produce seeds that will germinate only after a hot fire. Food reserves stored in their fire-resistant roots enable them to resprout quickly and use mineral nutrients released by fires.

Figure 19.34 Major terrestrial biomes, continued.

(e) Temperate grassland. The puszta of Hungary, the pampas of Argentina and Uruguay, the steppes of Russia, and the plains and prairies of central North America are all temperate grasslands. The key to the persistence of grasslands is seasonal drought, occasional fires, and grazing by large mammals, all of which prevent establishment of woody shrubs and trees. Temperate grasslands, such as the tallgrass prairie in South Dakota shown here, once covered much of central North America. Because grassland soil is both deep and rich in nutrients, these habitats provide fertile land for agriculture. Most grassland in the United States has been converted to farmland, and very little natural prairie exists today.

(f) Temperate deciduous forest. Dense stands of deciduous trees are trademarks of temperate deciduous forests, such as this one in Great Smoky Mountains National Park in North Carolina. Temperate deciduous forests occur throughout midlatitudes where there is sufficient moisture to support the growth of large trees. Deciduous forest trees drop their leaves before winter, when temperatures are too low for effective photosynthesis and water lost by evaporation is not easily replaced from frozen soil. Many temperate deciduous forest mammals also enter a dormant winter state called hibernation, and some bird species migrate to warmer climates. Virtually all the original deciduous forests in North America were destroyed by logging and land clearing for agriculture and urban development. In contrast to drier biomes, these forests tend to recover after disturbance, and today we see deciduous trees dominating undeveloped areas over much of their former range.

(g) Coniferous forest. Cone-bearing evergreen trees such as pine, spruce, fir, and hemlock dominate coniferous forests, such as this one in Olympic National Park in western Washington. Warm, moist air from the Pacific Ocean supports these unique communities, which, like most coniferous forests, are dominated by one or a few tree species. Extending in a broad band across northern North America and Eurasia to the southern border of the arctic tundra, the northern coniferous forest, or taiga, is the largest terrestrial biome on Earth (see Figure 19.33). Taiga receives heavy snowfall during winter. The conical shape of many conifers prevents too much snow from accumulating on and breaking their branches. Coniferous forests are being logged at an alarming rate, and the old-growth stands of these trees may soon disappear.

(h) Tundra. Permafrost (permanently frozen subsoil), bitterly cold temperatures, and high winds are responsible for the absence of trees and other tall plants in this arctic tundra in central Alaska (photographed in autumn). The arctic tundra receives very little annual precipitation, and what water does fall cannot penetrate the underlying permafrost and accumulates in pools on the shallow topsoil during the short summer. Tundra covers expansive areas of the Arctic, amounting to 20% of Earth's land surface. High winds and cold temperatures create similar plant communities, called alpine tundra, on very high mountaintops at all latitudes, including the tropics.

Freshwater Biomes

Aquatic biomes, consisting of freshwater and marine ecosystems, occupy the largest part of the biosphere.

Lakes and Ponds Standing bodies of water range from small ponds only a few square meters in area to large lakes, such as North America's Great Lakes, that are thousands of square kilometers **(Figure 19.35a)**.

In lakes and large ponds, the communities of plants, algae, and animals are distributed according to the depth of the water and its distance from shore. Shallow water near shore and the upper stratum of water away from shore make up the **photic zone,** so named because light is available for photosynthesis. Phytoplankton (microscopic algae and cyanobacteria) grow in the photic zone, joined by rooted plants and floating plants such as water lilies in photic regions near shore. If a lake or pond is deep enough or murky enough, it has an **aphotic zone,** where light levels are too low to support photosynthesis.

At the bottom of all aquatic biomes is the **benthic zone.** Made up of sand and organic and inorganic sediments, the benthic zone is occupied by communities of organisms, including a diversity of bacteria, collectively called benthos. (If you have ever waded barefoot into a pond or lake, you have felt the benthos squish between your toes.) A major source of food for the benthos is detritus that "rains" down from the productive surface waters of the photic zone.

Nitrogen and phosphorus are the mineral nutrients that usually limit the amount of phytoplankton growth in a lake or pond. Many lakes and ponds are affected by large inputs of nitrogen and phosphorus from sewage and runoff from fertilized lawns and agricultural fields. These nutrients often produce blooms, or population explosions, of algae. Heavy algal growth reduces light penetration into the water, and when the algae die and decompose, a pond or lake can suffer serious oxygen depletion. Fish and other aerobic organisms may then die.

Rivers and Streams Bodies of water flowing in one direction, rivers and streams generally support quite different communities of organisms than lakes and ponds **(Figure 19.35b)**. A river or stream changes greatly between its source (perhaps a spring or snowmelt) and the point at which it empties into a lake or the ocean. Near a source, the water is usually cold, low in nutrients, and clear. The channel is often narrow, with a swift current that does not allow much silt to accumulate on the bottom. The current also inhibits the growth of phytoplankton; most of the organisms found here are supported by the photosynthesis of algae attached to rocks or organic material (such as leaves) carried into the stream from the surrounding land. The most abundant benthic animals are usually insects that eat algae, leaves, or one another. Trout are often the predominant fishes, locating their food, including insects, mainly by sight in the clear water.

Downstream, a river or stream generally widens and slows. Marshes, ponds, and other wetlands are common in downstream areas **(Figure 19.35c)**. There the water is usually warmer and may be murkier because of sediments and phytoplankton suspended in it. Worms and insects that burrow into mud are often abundant, as are waterfowl, frogs, and catfish and other fishes that find food more by scent and taste than by sight.

Many streams and rivers have been affected by pollution from human activities, by the construction of stream channels to speed water flow, and by dams that hold water. For centuries, humans used streams and rivers as

(a) Satellite view of the Great lakes

(b) A stream in the Great Smoky Mountains, Tennessee

(c) A freshwater wetland in Georgia

Figure 19.35 Freshwater biomes.

depositories of waste, thinking that these materials would be diluted and carried downstream. While some pollutants are carried far from their source, many settle to the bottom, where they can be taken up by aquatic organisms. Even the pollutants that are carried away contribute to ocean and lake pollution. In many cases, dams have completely changed the downstream ecosystems, altering the intensity and volume of water flow and affecting fish and invertebrate populations **(Figure 19.36)**.

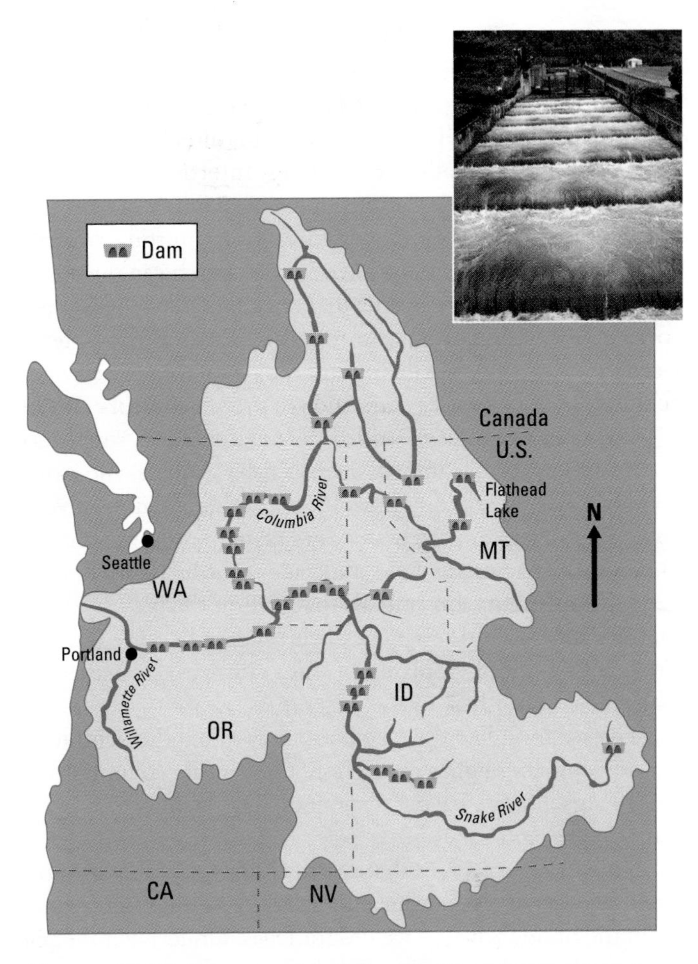

Figure 19.36 Damming the Columbia River Basin. If Lewis and Clark lived today, they would have a hard time navigating the Columbia. This map shows only the largest of the 250 dams that have altered freshwater ecosystems throughout the Pacific Northwest. The great concrete obstacles make it difficult for salmon to swim upriver to their breeding streams, though many dams now have "fish ladders" that provide detours (inset).

Marine Biomes

Life originated in the sea and evolved there for almost 3 billion years before plants and animals began colonizing land. Covering about 75% of the planet's surface, oceans have always had an enormous impact on the biosphere. Their evaporation provides most of Earth's rainfall, and photosynthesis by marine algae supplies a substantial portion of the biosphere's oxygen.

Estuaries The area where a freshwater stream or river merges with the ocean is called an **estuary.** Most estuaries are bordered by extensive coastal wetlands called mudflats and saltmarshes **(Figure 19.37)**. Salinity (salt concentration) varies spatially within estuaries, from nearly that of freshwater to that of the ocean; it also varies over the course of a day with the rise and fall of the tides. Nutrients from the river enrich estuarine waters, making estuaries one of the most biologically productive environments on Earth.

Saltmarsh grasses and algae are the major producers in estuaries. This environment also supports a variety of worms, oysters, crabs, and many of the fish species that humans consume. A diversity of marine invertebrates and fishes use estuaries as a breeding ground or migrate through them to freshwater habitats upstream. Estuaries are also crucial feeding areas for a great diversity of birds.

Although estuaries support a wide variety of commercially valuable species, areas around estuaries are also prime locations for commercial and residential developments. In addition, estuaries are unfortunately at the receiving end for pollutants dumped upstream. Very little undisturbed estuarine habitat remains, and a large percentage has been totally eliminated by landfill and development. Many states have now—rather belatedly—taken steps to preserve their remaining estuaries.

Figure 19.37 An estuary on the edge of Chesapeake Bay, Maryland.

Major Oceanic Zones Marine life is distributed according to depth of the water, degree of light penetration, distance from shore, and open water versus bottom **(Figure 19.38)**. The area where land meets sea is called the **intertidal zone.** This environment is alternately submerged and exposed by the twice-daily cycle of tides. Intertidal organisms are therefore subject to huge daily variations in the availability of seawater (and the nutrients it carries) and in temperature. Also, intertidal organisms are subject to the mechanical forces of wave action, which can dislodge them from their habitats. The organisms that are adapted to this agitation attach to rocks or vegetation or burrow into mud or sand. Examples are certain seaweeds (large algae), sessile (nonmotile) crustaceans such as barnacles, and echinoderms such as sea stars and sea urchins **(Figure 19.39)**.

The open ocean itself, called the **pelagic zone,** supports communities dominated by motile animals such as fishes, squids, and marine mammals, including whales and dolphins. Phytoplankton drift passively in the pelagic zone. Phytoplankton are the ocean's main photosynthetic producers, making most of the organic food molecules on which other ocean-dwellers depend. Zooplankton are animals that drift in the pelagic zone either because they are too small to resist ocean currents or because they don't swim. Zooplankton eat phytoplankton and in turn are consumed by other animals, including fishes.

The seafloor, like the lake bottom, is called the benthic zone. Depending on depth and light penetration, the benthic community consists of attached algae, fungi, bacteria, sponges, burrowing worms, sea anemones, clams, crabs, and fishes.

Note in Figure 19.38 that marine biologists often group the illuminated regions of the benthic and pelagic communities together, calling them the photic zone, as we did for lakes. Underlying the photic zone is the vast, dark aphotic zone. This is the most extensive part of the biosphere. Without light, there are no photosynthetic organisms, but life is still diverse in the aphotic zone. Many kinds of invertebrates, such as sea urchins and polychaete worms, and some fishes scavenge organic matter that sinks from the lighted waters above. The seafloor also includes **hydrothermal vent communities,** which are powered by chemical energy from Earth's interior rather than by sunlight **(Figure 19.40)**.

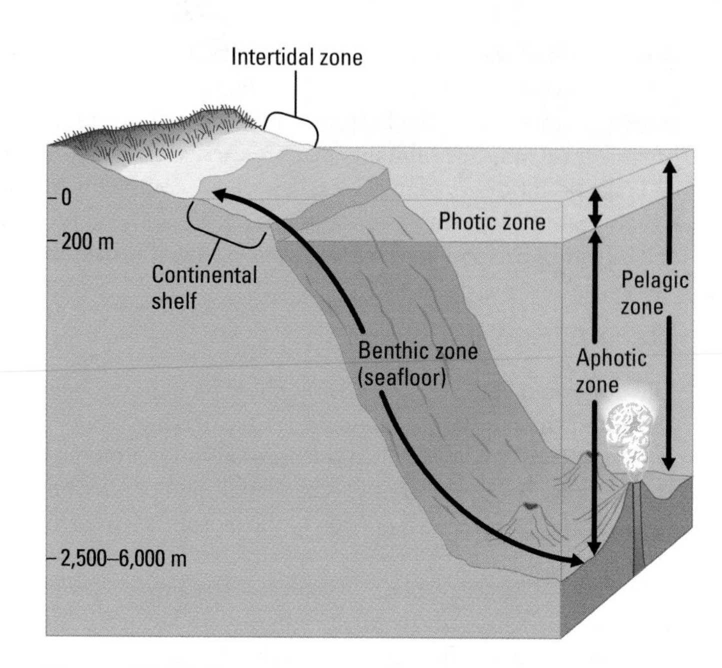

Figure 19.38 Oceanic zones.

Figure 19.39 A sampling of organisms in a tide pool.

CHECKPOINT

1. Why is homeowner's insurance relatively expensive for people who live in the chaparral?
2. How do humans now use most of the North American land that was once temperate grasslands?
3. How does the loss of leaves function as an adaptation of deciduous trees to cold winters?
4. What two abiotic factors account for the rarity of trees in arctic tundra?
5. The __________, small photosynthetic organisms inhabiting the __________ zone of the pelagic zone, provide most of the food for oceanic life.
6. Why does sewage cause algal blooms in lakes?
7. What is the energy source for hydrothermal vent communities?

Answers: 1. High fire risk **2.** For farming **3.** By reducing loss of water from the trees when that water cannot be replaced because of frozen soil **4.** Long, very cold winters (short growing season) and permafrost **5.** phytoplankton; photic **6.** The sewage adds minerals that stimulate growth of the algae. **7.** Chemicals that spew up from Earth's interior

Figure 19.40 Exploring hydrothermal vent communities. Research vessels such as *Alvin* (shown here) have enabled scientists to explore the seafloor to depths of 2,500 m, more than a mile deeper than sunlight penetrates. In the late 1970s, researchers first encountered a diversity of organisms living around hydrothermal vents, where molten rock and hot gases escape from Earth's interior. Nearby lives a community of sea anemones, giant clams, shrimps, worms, and a few fish species. The food chain is supported by sulfur bacteria, which use the oxidation of hydrogen sulfide (H_2S) in the vent's effluent to power the synthesis of organic molecules from CO_2. Some of the animals eat the bacteria. Others, such as the meter-tall tube worms in the photo at the right, harbor the food-producing bacteria as symbionts. Unlike all the other ecosystems we know, life at the hydrothermal vents on the seafloor is driven by energy from Earth itself rather than from the sun.

EVOLUTION CONNECTION

Coevolution in Biological Communities

We have seen many examples in this chapter of adaptations that evolved in populations as a result of interactions with other species in a community. Defenses of plants against herbivores, animal defenses against predators, and parasite-host relationships are all examples. Ecologists use the term **coevolution** when the adaptations of two species are closely connected—that is, when an adaptation in one species leads to a counteradaptation in a second species. Let's examine one such case of coevolution, the reciprocal adaptations between certain butterfly species and passionflowers, plant species on which the caterpillars of the butterflies feed.

Passionflower vines of the genus *Passiflora* are protected against most herbivorous insects by their production of toxic compounds in young leaves and shoots. However, the larvae (caterpillars) of butterflies of the genus *Heliconius* can tolerate these defensive chemicals. This counteradaptation has enabled *Heliconius* larvae to become specialized feeders on plants that few other insects can eat **(Figure 19.41)**. Survival of the larvae is further enhanced by a behavioral adaptation of the butterflies. The eggs that female *Heliconius* butterflies lay on the leaves of passionflower vines are bright yellow, and other *Heliconius* females generally avoid laying eggs on leaves marked by these yellow dots. This behavior presumably reduces competition among the larvae for food.

An infestation of *Heliconius* larvae can devastate a passionflower vine, and these poison-resistant insects are likely to be a strong selection force favoring the evolution of more defenses in the plants. In some species of *Passiflora*, the leaves have conspicuous yellow spots that mimic *Heliconius* eggs, an adaptation that may divert the butterflies to other plants in their search for egg-laying sites.

On closer inspection, such seemingly clear-cut examples of coevolution between two species usually turn out to be more complicated. The yellow "spots" on the passionflower vine are actually nectar-secreting glands, which attract ants and wasps that prey on *Heliconius* eggs or larvae. And there is evidence that the mere presence of ants on a leaf will discourage a *Heliconius* butterfly from laying eggs there. (We saw a similar example in the mutualism of ants and acacia trees; see Figure 19.16.) The relationship of *Passiflora* and *Heliconius* is another example of the complexity—shaped by evolution—that characterizes Earth's communities and ecosystems. ■

(a) *Heliconius* caterpillar

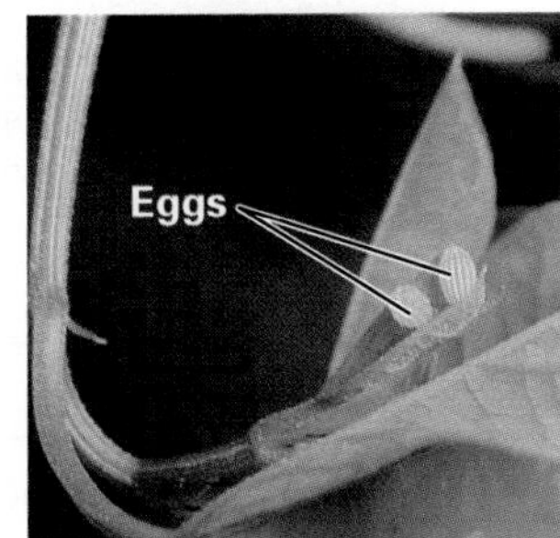

(b) *Heliconius* eggs

(c) Passionflower leaves with sugar-secreting nectar glands

Figure 19.41 Coevolution of a plant and insect. Passionflower vines produce toxic chemicals that help protect leaves from most herbivorous insects. **(a)** A counteradaptation has evolved in *Heliconius* butterflies; their larva can feed on passionflower leaves because they have digestive enzymes that break down the plant's toxins. **(b)** The female butterflies avoid laying eggs on passionflower leaves with egg clusters deposited by other *Heliconius* females. This behavior ensures an adequate food supply when eggs hatch and the larvae begin feeding on the leaves. **(c)** Some passionflower species have yellow nectaries (sugar-secreting glands) that mimic *Heliconius* eggs. The butterflies avoid laying eggs on leaves with these yellow spots. The nectaries also attract ants that prey on the butterfly eggs.

Chapter Review

SUMMARY OF KEY CONCEPTS

For study help and activities, go to campbellbiology.com or the student CD-ROM.

Key Properties of Communities

- **Diversity** Community diversity includes the species richness and relative abundance of different species.

Case Study in the Process of Science *How Are Impacts on Community Diversity Measured?*

- **Prevalent Form of Vegetation** The types and structural features of plants largely determine the kinds of animals that live in a community.
- **Stability** A community's stability is its ability to resist change and return to its original species composition after being disturbed. Stability depends on the type of community and the nature of the disturbances.
- **Trophic Structure** Trophic structure consists of the feeding relationships among the various species making up a community.

Interspecific Interactions in Communities

- **Competition Between Species** When populations of two or more species in a community rely on similar limiting resources, they may be subject to interspecific competition. Two species cannot coexist in a community if their niches are identical. Resource partitioning is the differentiation of niches that enables similar species to coexist in a local community.
- **Predation** Natural selection refines the adaptations of both predators and prey. Most predators have acute senses that enable them to locate and identify potential prey. Plants mainly use chemical toxins, spines, and thorns to defend against predators. Animals may defend themselves by using passive defenses, such as cryptic coloration (camouflage), or active defenses, such as escaping, alarm calls, mobbing, or distraction displays. Animals with chemical defenses are often brightly colored, a warning to potential predators. A species of prey may also gain protection through mimicry. In Batesian mimicry, a palatable species mimics an unpalatable model. In Müllerian mimicry, two or more unpalatable species resemble each other. Predator-prey relationships can preserve diversity. A keystone predator is a species that reduces the density of the strongest competitors in a community. This predator helps maintain species diversity by preventing competitive exclusion of weaker competitors.
- **Symbiotic Relationships** A symbiotic relationship is an interspecific interaction in which two species live in direct contact. Parasitism is a one-sided relationship in which the parasite benefits at the expense of the host. Natural selection has refined the relationships between parasites and their hosts. Mutualism is a symbiosis that benefits both partners. Many mutualistic relationships may have evolved from predator-prey or host-parasite interactions.
- **The Complexity of Community Networks** The branching of interactions between species makes communities complex.

Activity *Interspecific Interactions*

Disturbance of Communities

- Disturbances are episodes that damage biological communities, at least temporarily, by destroying organisms and altering the availability of resources such as mineral nutrients and water. Most communities spend much of their time in various stages of recovery from disturbances.
- **Ecological Succession** The sequence of changes in a community after a disturbance is called ecological succession. Primary succession occurs where a community arises in a virtually lifeless area with no soil. Secondary succession occurs where a disturbance has destroyed an existing community but left the soil intact.

Activity *Primary Succession*

- **A Dynamic View of Community Structure** In general, disturbances keep communities in a continual flux, making them mosaics of patches at various successional stages and preventing them from reaching a completely stable state. Small-scale disturbances often have positive effects, such as creating new opportunities for species. Species diversity may be greatest in places where disturbances are moderate in severity and frequency.

An Overview of Ecosystem Dynamics

- An ecosystem is a biological community and the abiotic factors with which the community interacts. Energy must flow continuously through an ecosystem, from producers to consumers and decomposers. Chemical elements can be recycled between an ecosystem's living community and the abiotic environment.

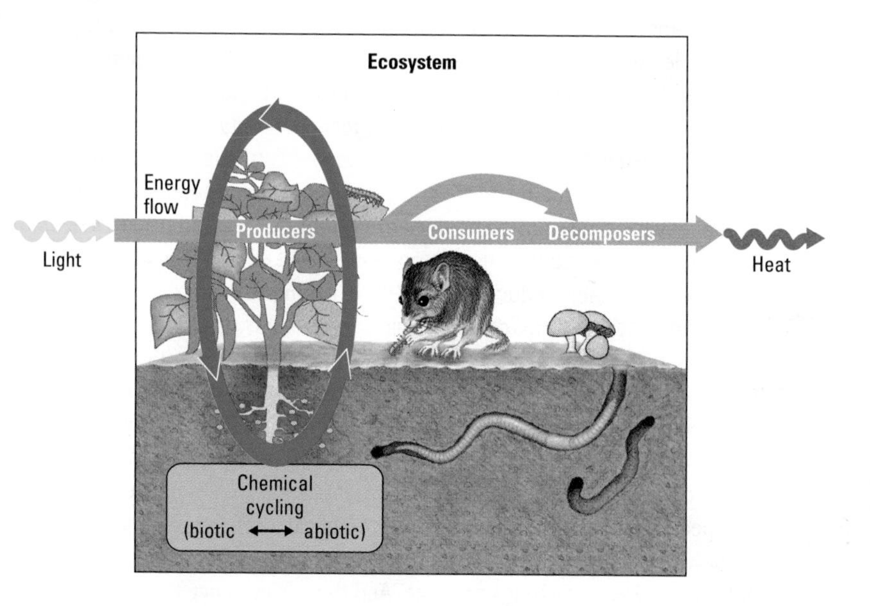

Activity *Energy Flow and Chemical Cycling*

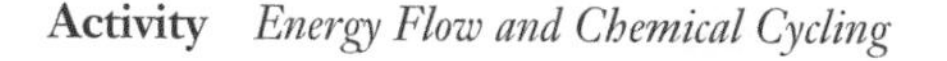

• **Trophic Levels and Food Chains** Trophic relationships determine an ecosystem's routes of energy flow and chemical cycling.

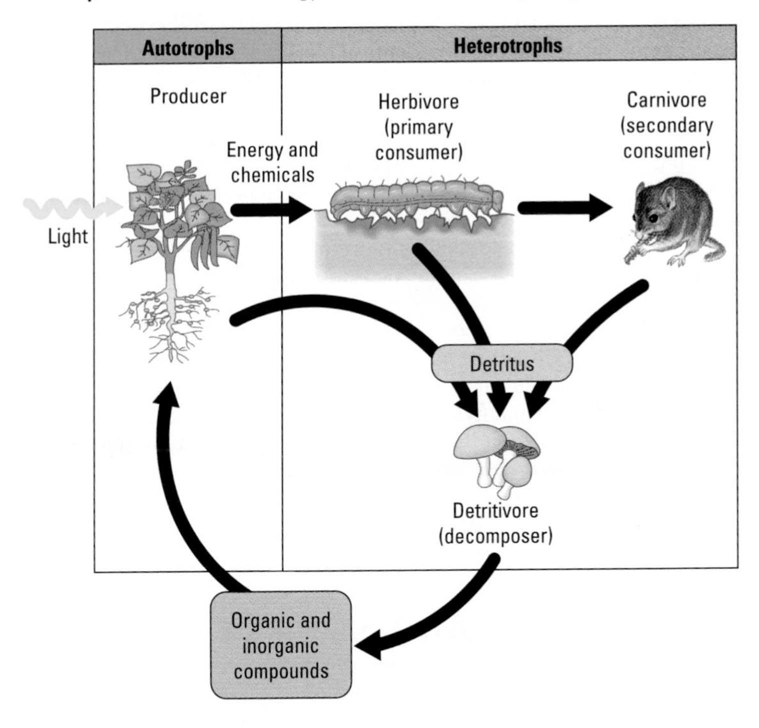

• **Food Webs** The feeding relationships in an ecosystem are usually woven into elaborate food webs.

Activity *Food Webs*

Energy Flow in Ecosystems

MP3 Tutor *Energy Flow in Ecosystems*

• **Productivity and the Energy Budgets of Ecosystems** The rate at which plants and other producers build biomass is the ecosystem's primary productivity. Ecosystems vary considerably in their productivity. Primary productivity sets the spending limit for the energy budget of the entire ecosystem because consumers must acquire their organic fuels from producers.

Case Study in the Process of Science *How Does Light Affect Primary Productivity?*

• **Energy Pyramids** In a food chain, only about 10% of the biomass at one trophic level is available to the next:

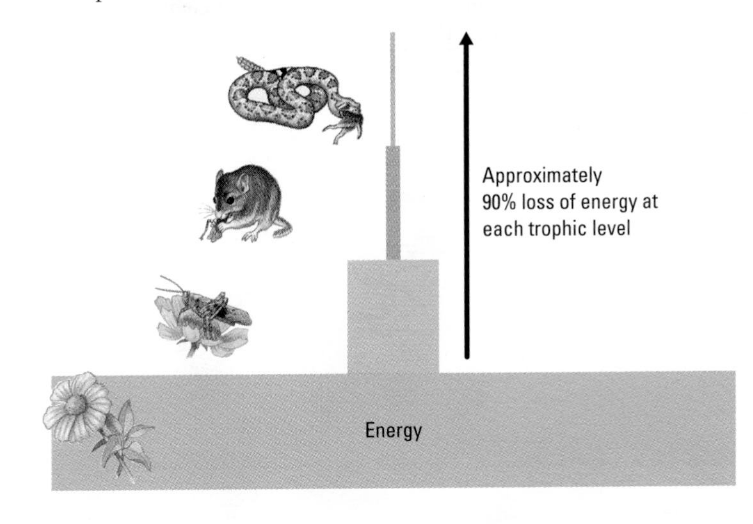

Activity *Energy Pyramids*

• **Ecosystem Energetics and Human Nutrition** Eating producers instead of consumers requires less photosynthetic productivity and reduces the impact to the environment.

Chemical Cycling in Ecosystems

• **The General Scheme of Chemical Cycling** Biogeochemical cycles involve biotic and abiotic components. Each circuit has an abiotic reservoir through which the chemical cycles. Some chemical elements require "processing" by certain microorganisms before they are available to plants as inorganic nutrients.

• **Examples of Biogeochemical Cycles** A chemical's specific route through an ecosystem varies with the element and the trophic structure of the ecosystem. Phosphorus is not very mobile and is cycled locally. Carbon, nitrogen, and water spend part of their time in gaseous form and are cycled globally.

Activity *The Carbon Cycle*

Activity *The Nitrogen Cycle*

Biomes

• **How Climate Affects Biome Distribution** The geographic distribution of terrestrial biomes is based mainly on regional variations in climate. Climate is largely determined by the uneven distribution of solar energy on Earth. Proximity to large bodies of water and the presence of landforms such as mountains also affect climate.

• **Terrestrial Biomes** If the climate in two geographically separate areas is similar, the same type of biome may occur in both. Most biomes are named for major physical or climatic features and for their predominant vegetation. The major terrestrial biomes include tropical forest, savanna, desert, chaparral, temperate grassland, temperate deciduous forest, coniferous forest, and tundra.

Activity *Terrestrial Biomes*

• **Freshwater Biomes** Freshwater biomes include lakes, ponds, rivers, streams, and wetlands. Lakes vary, depending on depth, with regard to light penetration, temperature, nutrients, oxygen levels, and community structure. Rivers change greatly from their source to the point at which they empty into a lake or an ocean.

• **Marine Biomes** Estuaries, located where a freshwater river or stream merges with the ocean, are one of the most biologically productive environments on Earth. As in freshwater environments, marine life is distributed into distinct zones (intertidal, pelagic, and benthic) according to the depth of the water, degree of light penetration, distance from shore, and open water versus bottom. Hydrothermal vent communities are a marine deepwater biome powered by chemical energy from Earth's interior instead of by sunlight.

Activity *Aquatic Biomes*

Evolution Connection: Coevolution in Biological Communities

• The interaction of *Passiflora* passionflowers and *Heliconius* butterflies is an example of coevolution, when an adaptation in one species leads to a counteradaptation in a second species.

SELF-QUIZ

1. The concept of trophic structure of a community emphasizes the
   a. prevalent form of vegetation.
   b. keystone predator.
   c. feeding relationships within a community.
   d. species richness of the community.

2. Match each organism with its trophic level (you may choose a level more than once).

a. alga	1. detritivore
b. grasshopper	2. producer
c. zooplankton	3. tertiary consumer
d. eagle	4. secondary consumer
e. fungi	5. primary consumer

3. According to the concept of competitive exclusion,
   a. two species cannot coexist in the same habitat.
   b. extinction or emigration is the only possible result of competitive interactions.
   c. intraspecific competition results in the success of the best-adapted individuals.
   d. two species cannot share the same niche in a community.

4. How can a keystone predator help maintain species diversity within a community?

5. Match the defense mechanism with the term that describes it.

a. a harmless beetle that resembles a scorpion	1. camouflage coloration
b. the bright markings of a poisonous tropical frog	2. warning coloration
c. the mottled coloring of moths that rest on lichens	3. Batesian mimicry
d. two poisonous frogs that resemble each other in coloration	4. Müllerian mimicry

6. Over a period of many years, grass grows on a sand dune, then shrubs grow, and then eventually trees grow. This is an example of ecological __________.

7. According to the energy pyramid model, why is eating grain-fed beef a relatively inefficient means of obtaining the energy trapped by photosynthesis?

8. Local conditions, such as heavy rainfall or the removal of plants, may limit the amount of nitrogen, phosphorus, or calcium available to a particular ecosystem, but the amount of carbon available to the system is seldom a problem. Why?

9. We are on a coastal hillside on a hot, dry summer day among evergreen shrubs that are adapted to fire. We are most likely standing in a __________ biome.

10. In volume, which biome is largest?
   a. deserts
   b. the intertidal zone
   c. the photic zone of the oceans
   d. the pelagic zone of the oceans

Answers to the Self-Quiz questions can be found in Appendix D.

Go to the website or CD-ROM for more Self-Quiz questions.

THE PROCESS OF SCIENCE

11. An ecologist studying desert plants performed the following experiment. She staked out two identical plots that included a few sagebrush plants and numerous small annual wildflowers. She found the same five wildflower species in similar numbers in both plots. Then she enclosed one of the plots with a fence to keep out kangaroo rats, the most common herbivores in the area. After two years, four species of wildflowers were no longer present in the fenced plot, but one wildflower species had increased dramatically. The unfenced control plot had not changed significantly in species composition. Using the concepts discussed in the chapter, what do you think happened?

12. Imagine that you have been chosen as the biologist for the design team implementing a self-contained space station to be assembled in orbit. It will be stocked with organisms you choose, creating an ecosystem that will support you and five other people for two years. Describe the main functions you expect the organisms to perform. List the types of organisms you would select, and explain why you chose them.

BIOLOGY AND SOCIETY

13. Sometime in 1986, near Detroit, a freighter pumped out water ballast containing larvae of European zebra mussels. The molluscs have multiplied wildly, spreading through Lake Erie and entering Lake Ontario. In some places in the Great Lakes region, they have become so numerous that they have blocked the intake pipes of power plants and water treatment plants, fouled boat hulls, and sunk buoys. What makes this type of population explosion occur? What might happen to native organisms that suddenly must share the Great Lakes ecosystem with zebra mussels? Why is this a problem? How would you suggest trying to solve the mussel population problem?

14. Near Lawrence, Kansas, there was a rare patch of the original North American temperate grassland that had never been plowed. It was home to numerous native grasses, annual plants, and grassland animals. Among the species present were two endangered plants. Environmental activists thought the area should be set aside as a nature preserve, and they started to raise money to save the patch of land. In 1990, the owner of the land plowed it, stating that there are no federal laws protecting endangered plants on private grasslands and that he did not want to be told what he could do with his property. What issues and values are in conflict in this situation? How could this story have had a more satisfactory ending for all concerned? What would you have done if you were an environmental activist? If you were the farmer?

20 Human Impact on the Environment

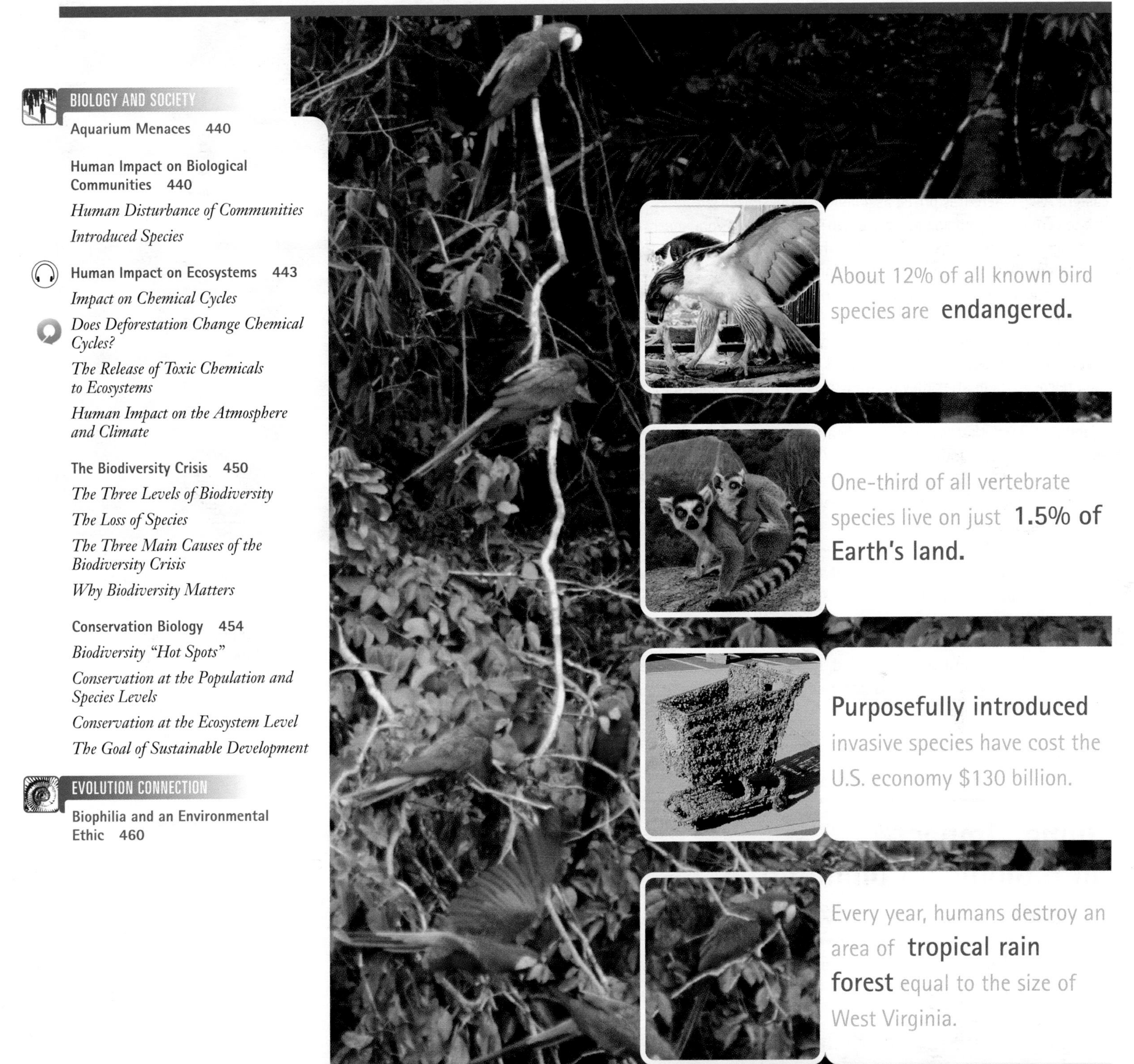

About 12% of all known bird species are **endangered.**
One-third of all vertebrate species live on just **1.5% of Earth's land.**
Purposefully introduced invasive species have cost the U.S. economy $130 billion.
Every year, humans destroy an area of **tropical rain forest** equal to the size of West Virginia.

BIOLOGY AND SOCIETY

Aquarium Menaces

In May 2002, in a pond near Crofton, Maryland, a fisherman caught, photographed, and released an 18-inch exotic-looking fish. A month later, another fisherman caught and froze a 26-inch specimen, which was identified as *Channa argus*, the northern snakehead. State officials found hundreds of juvenile snakeheads living in the pond.

The fish soon caught the attention of the press, which dubbed it "Frankenfish" because of its odd anatomy and physiology. Up to 3 feet long, it has a snakelike head with large sharp teeth **(Figure 20.1a)**. In addition, it is able to gulp air and wiggle across land for several days (if it can stay moist). Native to eastern Asia, where it is prized as a delicacy, the snakehead poses a significant risk to ecosystems outside its native habitat because it voraciously eats smaller fishes, crustaceans, frogs, and insects, while having no predators to keep its numbers in check. By the end of the summer, the state successfully eradicated the snakehead—but only by killing all the fish in the pond (and in two neighboring ponds) with huge doses of poison.

How did an Asian fish get into a Maryland pond? Investigators discovered that a local man bought two snakeheads in New York, raised them as pets, and then released them after they outgrew his home aquarium. This situation is hardly unique. In 2000, an alga called *Caulerpa*—a Caribbean species bred as an aquarium "plant" that is particularly hearty and resistant to disease, predators, and herbicides—was found growing in a California lagoon, probably after someone dumped a home saltwater aquarium **(Figure 20.1b)**. An earlier invasion of the Mediterranean Sea by this "super seaweed" is displacing many of the native algae there, and the same thing could happen now all along the Pacific coast of North America.

These stories of human-introduced species disrupting native ecosystems make two important points: First, humans have a disproportionately high impact on the environment. This is why we begin the chapter with an exploration of how human activities are putting communities, ecosystems, and biological diversity at risk. And second, humans continue to be fascinated with the natural world—which is, after all, why we keep aquariums. It's also why we end the chapter (and the unit) with a discussion of conservation biology and the challenge facing us as individuals and as a society to preserve biodiversity. ■

(a) The northern snakehead. Non-native fish like this one have the potential to disrupt local ecosystems.

(b) Aquarium-bred algae. In this underwater photo, you can see the fast-growing algae *Caulerpa* (foreground) crowding out the native eelgrass.

Figure 20.1 Non-native species released from your aquarium can cause significant damage to local ecosystems.

Human Impact on Biological Communities

In Chapter 19, you learned that many natural disturbances are actually constructive in enhancing species diversity in a community. Unfortunately, human disturbance of biological communities is almost always destructive, reducing species diversity.

Human Disturbance of Communities

Of all animals, humans have the greatest impact on communities worldwide **(Figure 20.2)**. Logging and clearing for urban development, mining, and farming have reduced large tracts of forests to small patches of disconnected woodlots in many parts of the United States and throughout Europe. Similarly, agricultural development has disrupted what were once the vast grasslands of the North American prairie.

Much of the United States is now a hodgepodge of early successional growth where more mature communities once prevailed. After forests are clear-cut and abandoned, weedy and shrubby vegetation often colonizes the area and dominates it for many years. This type of vegetation is also found extensively in agricultural fields that are no longer under cultivation and in vacant lots and construction sites that are periodically cleared.

Human disturbance of communities is by no means limited to the United States; nor is it a recent problem. Tropical rain forests are quickly disappearing as a result of clear-cutting for lumber and pastureland. Centuries of overgrazing and agricultural disturbance have contributed to the recurrent famines in parts of Africa by turning seasonal grasslands into great barren areas.

Human disturbance usually reduces species diversity in communities. We currently use about 60% of Earth's land in one way or another, mostly as cropland and rangeland. Most crops are grown in **monocultures,** intensive cultivations of a single plant variety over large areas. For example, forests that are cut to produce pulpwood and lumber are often replanted in single-species stands. And intensive grazing on rangelands often results in the removal of several native plant species that may be replaced by only a few invasive introduced species.

Figure 20.2 A large-scale human disturbance. This open-pit mine in Butte, Montana, known as the Berkeley Pit, is nearly 2,000 feet deep and covers a land area of 1.5 by 1 miles. To get a sense of its scale, notice the trees growing on several plateaus. The mine has not operated since 1982, when the last of 1 billion tons of high-quality copper ore was removed. Water in the pit, which is highly acidic and loaded with heavy-metal ions, threatens groundwater, soils, adjacent rivers, and communities for hundreds of miles downstream. The Berkeley Pit is one of the largest and most troublesome cleanup sites in the United States, and the federal government has spent millions to initiate waste removal at this site.

Introduced Species

Sometimes called exotic species, **introduced species** are those that humans move from the species' native locations to new geographic regions (as with the fish and alga discussed in the chapter-opening essay). In some cases, the introductions are intentional; Australia's rabbit population explosion is an example (see Figure 19.15). In other cases, humans transplant species accidentally. For instance, the fungus that causes chestnut blight was accidentally introduced into the United States from China (see the opening essay in Chapter 16). Whether purposeful or unintentional, introduced species that gain a foothold usually disrupt their adopted community, often by preying on native organisms or outcompeting native species that use some of the same resources.

Throughout human history, moving plants and animals from place to place was considered "improving on nature." This was especially true of domesticated species. As Europeans began colonizing the Americas in the 15th and 16th centuries, plants and animals that provide food were transported back and forth across the Atlantic. Corn, potatoes, tomatoes, beans, and certain kinds of squash are just a few examples of New World species that became European staples.

Humans have also introduced many nonfood species with the best of intentions. For example, the U.S. Department of Agriculture encouraged the import of a Japanese plant called kudzu to the American South in the 1930s to help control erosion, especially along irrigation canals. The government even paid farmers to plant kudzu vines. The enthusiasm for the new vines led to kudzu festivals in southern towns, complete with the crowning of kudzu queens. But kudzu celebrations ended decades ago as the invasive plant took over vast expanses of the southern

(a) Kudzu

(b) European starlings

landscape **(Figure 20.3a)**. Another introduced plant called purple loosestrife is claiming over 200,000 acres of wetlands per year, crowding out native plants and the animals that feed on the native flora. The story is similar for the introduction into the United States of a bird called the European starling. A citizens group intent on introducing all plants and animals mentioned in Shakespeare's plays imported 120 starlings to New York's Central Park in 1890 (the starling is mentioned in one line of Shakespeare's Henry IV). From that foothold, starlings spread rapidly across North America. In less than a century, the population increased to about 100 million, displacing many of the native songbird species in the United States and Canada **(Figure 20.3b)**.

The ease of travel by ships and airplanes has accelerated the transplant of species, especially unintentional introductions. For example, fire ants, which can inflict very painful beelike stings, reached the southeastern United States in the early 1900s from South America, probably in the hold of a produce ship. Fire ants have been extending their range northward and westward ever since. In Texas, fire ants have eliminated about two-thirds of the native ant species. Another accidentally introduced ant species, the Argentine ant, is decimating populations of native ants in California **(Figure 20.3c)**. Still another recent troublesome invader in the United States is the zebra mussel, a mollusc that entered the St. Lawrence Seaway in the mid-1980s in ballast water released by a cargo ship that had traveled from the mussels' native Caspian Sea **(Figure 20.3d)**. By the summer of 1993, zebra mussels were found throughout the Great Lakes and in the Mississippi River as far south as New Orleans. Of the many problems caused by the mussels, those that have provoked the most uproar have resulted from their competition with humans—by clogging reservoir intake pipes, for example. However, zebra mussels also compete with native shellfish for space and with fish for the plankton used for nourishment. The full extent to which this new competitor is altering community structure has yet to be determined.

All told, the United States has at least 50,000 introduced species, with a cost to the economy of over $130 billion in damage and control efforts. And that does not include the priceless loss of native species. Fortunately, most introduced species fail to thrive in their new homes. But for those exotic species that succeed, elimination of native species is a common consequence. In fact, introduced species rank second only to habitat destruction as a cause of extinctions and loss of Earth's biodiversity.

Why should some introduced species be able to outcompete native organisms? In many cases, the answer is that there are relatively few

(c) Argentine ants ganging up on a native red ant

(d) Zebra mussels The shopping cart, covered with the mussels (inset), was retrieved from a lake.

Figure 20.3 A few of America's 50,000 introduced species.

pathogens, parasites, and predators to hold population explosions of the introduced species in check once the globetrotters arrive. In contrast, the native species of a biological community fit into a network of interactions shaped by the evolutionary adaptations of multiple species.

CHECKPOINT

1. What is an "exotic" or "introduced" species?
2. If an introduced species gains a foothold in its new home, why does its success often contribute to loss of native species?

Answers: 1. A species that humans have transferred—intentionally or not—from the species' native location to another location **2.** The introduced species may prey on some native species and outcompete others owing to the absence of the introduced species' natural pathogens, predators, and parasites.

Human Impact on Ecosystems

MP3 Tutor
Global Warming

Human population growth plus technology add up to a badly bruised biosphere. We have managed to intrude in one way or another into the dynamics of most ecosystems. Even where we have not completely destroyed a natural system, our actions have disrupted the trophic structure, energy flow, and chemical cycling of ecosystems in most areas of the world. The effects are sometimes local or regional, but the ecological impact of humans can be far-reaching or even global. For example, acid precipitation may be carried by prevailing winds and fall as rain hundreds or thousands of miles from the smokestacks emitting the chemicals that produce it (see Chapter 2). And the carbon dioxide (CO_2) exhaust of our machinery is probably causing a global warming that will affect all life on Earth, including humans **(Figure 20.4)**. In this section, we'll examine a few of the ways that humans affect local ecosystems and the biosphere, the global ecosystem.

Figure 20.4 Carbon dioxide producers. Burning fossil fuels, such as gasoline, natural gas, and coal, produces CO_2. The rising level of carbon dioxide in the atmosphere is probably contributing to an increase in the average global temperature.

Impact on Chemical Cycles

Human activities often intrude in biogeochemical cycles by removing nutrients from one location and adding them to another. This may result in the depletion of key nutrients in one area, excesses in another place, and the disruption of the natural chemical cycling in both locations. For example, nutrients in the soil of croplands or rangelands make their way into the waste products of humans and livestock and then appear in streams and lakes through discharge as sewage and runoff from stockyards. Someone eating a salad in Washington, DC, is consuming nutrients that only days before might have been in the soil in California. And a short time later, some of these nutrients will be in the Potomac River, having passed through an individual's digestive system and the local sewage facilities.

Humans have altered chemical cycles to such an extent that it is no longer possible to understand any cycle without taking the human impact into account. Let's consider just a few examples.

Impact on the Carbon Cycle The increased burning of fossil fuels (coal and petroleum) as well as wood from deforested areas is steadily raising the level of CO_2 in the atmosphere. This is leading to significant environmental problems, such as global warming (discussed later).

Figure 20.5 Eutrophication causing an algal bloom. To discover the effects of added nutrients on an aquatic ecosystem, researchers used a plastic curtain to partition a lake into an "experimental lake" (top) and a "control lake" (bottom). The team added certain mineral nutrients to the experimental lake, but not to the control, which served as a basis of comparison. Within two months after phosphorus was added, a bloom (population explosion) of algae gave the experimental lake the cloudy, whitish appearance you see in the photograph. In such overfertilized lakes, bacteria and other decomposers of the algal mass sometimes use up all the oxygen, leading to the death of fish and other aerobic organisms. (Reprinted with permission from D. W. Schindler, "Eutrophication and Recovery in Experimental Lakes: Implications for Lake Management," *Science* vol. 184, page 897, Figure 1.49 (1974). Copyright © 1974 American Association for the Advancement of Science.)

Impact on the Nitrogen Cycle Sewage treatment facilities typically empty large amounts of dissolved inorganic nitrogen compounds into rivers and streams. Additionally, farmers routinely apply large amounts of inorganic nitrogen fertilizers, mainly ammonium and nitrates, to croplands. Lawns and golf courses also receive sizable doses of fertilizer. Crop and lawn plants take up some of the nitrogen compounds, and soil bacteria convert some to atmospheric nitrogen (N_2) (see Figure 19.29b). However, chemical fertilizers usually exceed the soil's natural recycling capacity, and the excess nitrogen compounds often enter streams, lakes, and groundwater. While nitrogen is a beneficial nutrient to plants, too much of it can create an imbalance. In lakes and streams, nitrogen compounds from runoff continue to fertilize, causing heavy growth of algae. Groundwater pollution by nitrogen fertilizers is also a serious problem in many agricultural areas. In the human digestive tract, nitrates in drinking water are converted to nitrites, which can be toxic.

Impact on the Phosphorus Cycle Like nitrogen compounds, phosphates are a major component of sewage outflow. They are also used extensively in agricultural fertilizers and are a common ingredient in pesticides. Phosphate pollution of lakes and rivers, like nitrate pollution, stimulates a heavy algal growth. Such overfertilization, or **eutrophication,** leads to population explosions of algae and cyanobacteria that can eventually remove so much oxygen from lakes or ponds that aerobic life suffocates **(Figure 20.5)**.

Impact on the Water Cycle One of the main sources of atmospheric water is transpiration (evaporation) from the dense vegetation of tropical rain forests. The destruction of these forests, which is occurring rapidly today, will change the amount of water vapor in the air **(Figure 20.6)**. This, in turn, will most likely alter local, and perhaps global, weather patterns. Another change in the water cycle caused by humans results from pumping large amounts of groundwater to the surface to use for crop irrigation. This practice can increase the rate of evaporation from soil, and unless this loss is balanced by increased rainfall over land, groundwater supplies are depleted. Large areas in the midwestern United States, the southwestern American desert, parts of California, and areas bordering the Gulf of Mexico currently face this problem.

Figure 20.6 Deforestation and the water cycle. Destruction of tropical forests, such as this one in Belize, reduces return of water to the atmosphere via transpiration.

Does Deforestation Change Chemical Cycles? A case study of chemical cycling has been ongoing since 1955 at the Hubbard Brook Experimental Forest in the White Mountains of New Hampshire. The research area consists of a deciduous forest with several valleys, each drained by a small creek that is a tributary of Hubbard Brook. Bedrock impenetrable to water is close to the surface of the soil, so each valley constitutes a watershed that can drain only through its creek.

The research team first determined the mineral budget for each of six valleys by measuring the inflow and outflow of several key nutrients. They collected rainfall at several sites to measure the amount of water and dissolved minerals added to the ecosystem. To monitor the loss of water and minerals, the scientists constructed small concrete dams, each with a V-shaped spillway, across the creek at the bottom of each valley **(Figure 20.7a)**. About 60% of the water added to the ecosystem as rainfall and snow exits through the streams, and the remaining 40% is lost by transpiration from plants and evaporation from the soil.

The researchers **observed** that local cycling within each watershed conserved most of the mineral nutrients. That is, mineral inflow balanced outflow and both were relatively small compared with the quantity of minerals being recycled within the forest ecosystem.

The researchers posed the **question**: What would happen to the chemical cycles if a region was deforested? They made the **hypothesis** that a lack of plant life would have a drastic effect on the flow of chemicals through an experimental valley. Their **experiment** involved completely deforesting one valley and then comparing the inflow and outflow of water and minerals in the experimentally altered watershed with the inflow and outflow in a control watershed for three years **(Figure 20.7b)**.

The **results** of the experiment showed that water runoff from the altered watershed increased by 30–40%, apparently because there were no plants to absorb and transpire water from the soil. Net losses of minerals from the altered watershed were huge. Most remarkable was the loss of nitrate, which increased in concentration in the creek 60-fold **(Figure 20.7c)**. Not only was this vital mineral nutrient drained from the ecosystem, but nitrate in the creek reached a level considered unsafe for drinking water.

The Hubbard Brook researchers concluded that the amount of nutrients leaving an intact forest ecosystem is controlled mainly by plants. Nutrients were lost from the system when plants were not present to retain them. These effects were almost immediate, occurring within a few months of

(a) A dam at the Hubbard Brook study site

(b) Deforested watersheds in the Hubbard Brook Forest

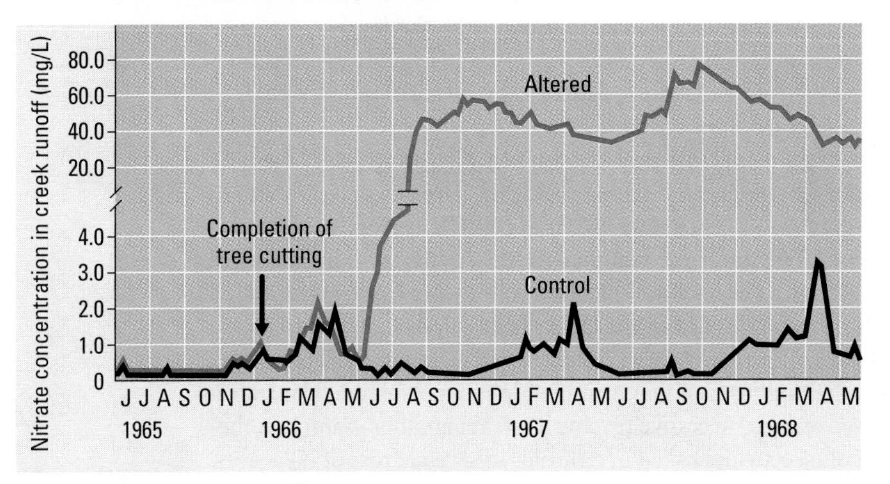

(c) The loss of nitrate from a deforested watershed

Figure 20.7 Chemical cycling in an experimental forest: the Hubbard Brook study. **(a)** A dam at a water-sampling station in the Hubbard Brook Forest. **(b)** "Experimental" (deforested, seen as snow-covered here) and "control" watersheds in the Hubbard Brook Experimental Forest. **(c)** Nitrate levels in the runoff from the deforested area began to rise markedly about five months after the plants were killed. Within eight months, the nitrate loss was about 60 times greater in the altered watershed than in the control area.

logging, and continuing as long as plants were absent. As new plants grew in the treatment area, transpiration increased and runoff decreased. Though scientists designed the Hubbard Brook experiments to assess natural ecosystem dynamics, the results also provided important insights into the mechanisms by which human activities such as deforestation affect these processes. ■

The Release of Toxic Chemicals to Ecosystems

In addition to transporting vital chemical elements from one location to another, humans have added entirely new materials, many of them toxic, to ecosystems. We produce an immense variety of these toxic chemicals, including thousands of synthetics previously unknown in nature. Many of these poisons cannot be degraded by microorganisms and consequently persist in the environment for years or even decades. In other cases, chemicals released into the environment may be relatively harmless but are converted to more toxic products by reaction with other substances or by the metabolism of microorganisms. For example, mercury, a by-product of plastic production, was once routinely expelled into rivers and the sea in an insoluble form. Bacteria in the bottom mud converted the waste to methyl mercury, an extremely toxic soluble compound that then accumulated in the tissues of organisms, including humans who consumed fish from the contaminated waters.

Organisms acquire toxic substances from the environment along with nutrients and water. Some of the poisons are metabolized or excreted, but others accumulate in specific tissues, especially fat. Examples of industrially synthesized compounds that act in this manner are the chlorinated hydrocarbons (which include many pesticides, such as DDT) as well as the industrial chemicals called PCBs (polychlorinated biphenols).

One of the reasons the toxins we add to ecosystems are such ecological disasters is that they become more concentrated in successive trophic levels of a food web, a process called **biological magnification.** Magnification occurs because the biomass at any given trophic level is produced from a much larger toxin-containing biomass ingested from the level below (see Figure 19.26 to review energy pyramids). Thus, top-level carnivores are usually the organisms most severely damaged by toxic compounds that have been released into the environment.

One classic example of biological magnification involves DDT, the poisonous pollutant that pioneering ecologist Rachel Carson warned about more than 40 years ago (see Chapter 18). In the 1960s, researchers began finding traces of DDT in marine mammals in the Arctic, far from any places DDT had been used. The chemical had been transported and concentrated as it passed through food webs. In the Great Lakes food chain shown in **Figure 20.8**, the concentration of PCBs measured in herring gull eggs was almost 5,000 times higher than that measured in phytoplankton. The concentration increased at each successive trophic level. Yet another example is the 1999 spraying of insecticides called pyrethroids in several areas of New York City. The spraying was a precaution against West Nile virus, which is carried by mosquitoes (see the opening essay in Chapter 14). Within months, there was a massive die-off of lobsters in Long Island Sound, likely caused by pyrethroid magnification within the local food chain. Scientists hypothesize that the poisons reached the ocean in rainstorm runoff from the city.

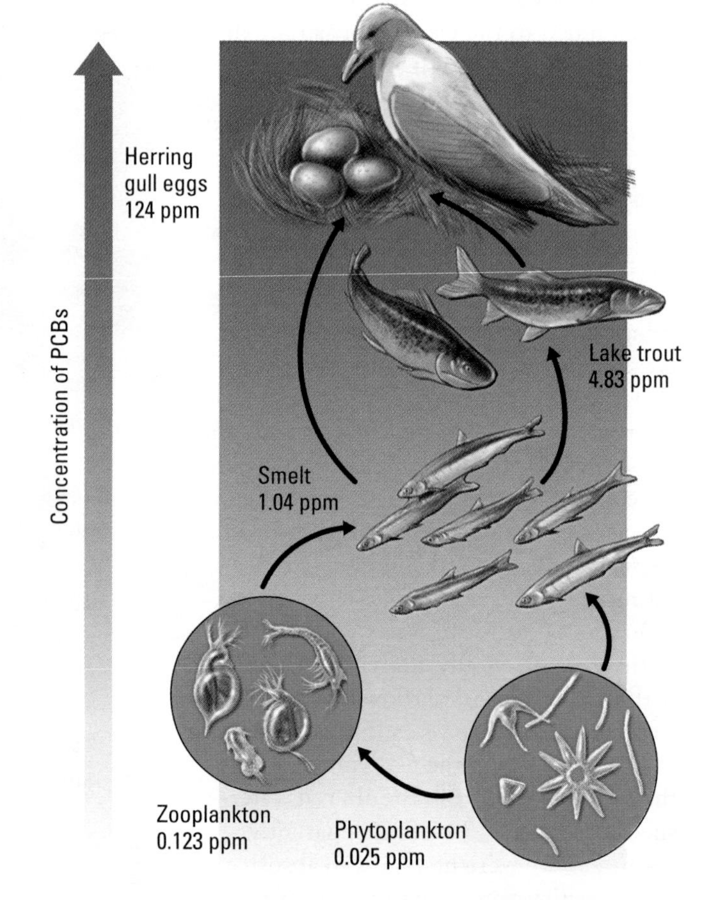

Figure 20.8 Biological magnification of PCBs in a food chain. The concentration of PCBs (polychlorinated biphenols) in a Great Lakes food chain was magnified by a factor of 5,000, from 0.025 ppm in phytoplankton to 124 ppm in the eggs of fish-eating birds. (The abbreviation ppm stands for "parts per million," a unit of measurement commonly used for toxins.)

Human Impact on the Atmosphere and Climate

We are causing radical changes in the composition of the atmosphere and, consequently, in the global climate. Our activities release a variety

Figure 20.9 An earthrise photographed from the moon. Apollo astronaut Rusty Schweickart, reflecting on such a view, once remarked: "On that small blue-and-white planet below is everything that means anything to you. National boundaries and human artifacts no longer seem real. Only the biosphere, whole and home of life."

of gaseous waste products. We once thought that the vastness of the atmosphere could absorb these materials without significant consequences, but an astronaut's view of our little planet squashes such naive notions **(Figure 20.9)**. One pressing problem that relates directly to one of the chemical cycles we examined is the rising level of carbon dioxide in the atmosphere.

Carbon Dioxide Emissions, the Greenhouse Effect, and Global Warming Since the Industrial Revolution, the concentration of CO_2 in the atmosphere has been increasing as a result of the combustion of fossil fuels and the burning of enormous quantities of wood removed by deforestation. Various methods have estimated that the average CO_2 concentration in the atmosphere before 1850 was about 274 parts per million (ppm). When a monitoring station on Hawaii's Mauna Loa peak began making very accurate measurements in 1958, the CO_2 concentration was 316 ppm **(Figure 20.10)**. Today, the concentration of CO_2 in the atmosphere exceeds 370 ppm, an increase of about 17% since the measurements began just over 45 years ago. If CO_2 emissions continue to increase at the present rate, by the year 2075 the atmospheric concentration of this gas will be double what it was at the start of the Industrial Revolution. It is difficult to predict the multiple ways this intrusion in the carbon cycle will affect the biosphere and its various ecosystems.

One factor that complicates predictions about the long-term effects of rising atmospheric CO_2 concentration is its possible influence on Earth's heat budget. Much of the solar radiation that strikes the planet is reflected back into space. Although CO_2 and water vapor in the atmosphere are transparent to visible light, they intercept and absorb much of the reflected heat radiation, bouncing it back toward Earth. This process, called the

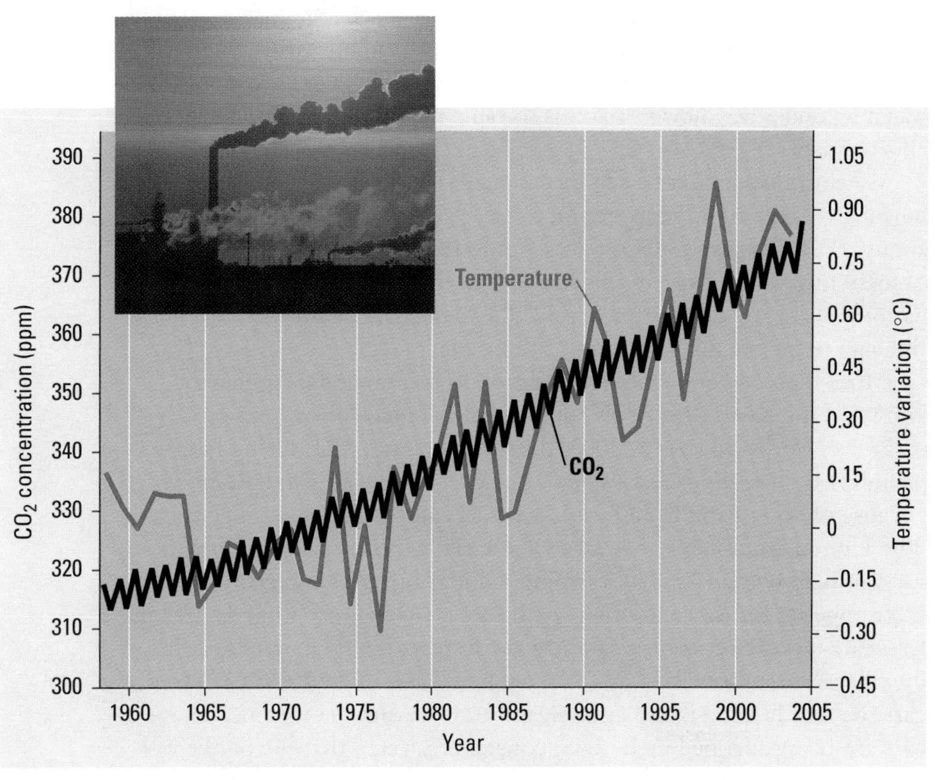

Figure 20.10 Increase in atmospheric CO_2 and temperature variation since 1958. The data were collected at Mauna Loa, Hawaii. The burning of fossil fuels (photo) was the main contributor to the increases observed.

greenhouse effect, retains some of the solar heat. **Figure 20.11** adds some details to the illustration of the greenhouse effect you saw in Figure 7.18.

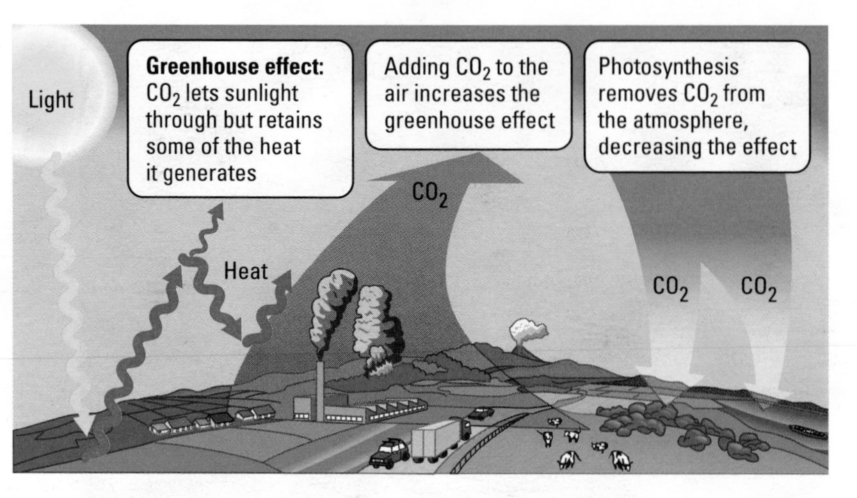

Figure 20.11 Factors influencing the greenhouse effect.

The marked increase in atmospheric CO_2 concentrations during the last 150 years concerns ecologists because of its potential effect on global temperature through the greenhouse effect. The black line in Figure 20.10 plots the warming trend during the period of increasing CO_2 concentration. A number of studies predict that a doubling of CO_2 concentration by the end of the 21st century will cause an average global temperature increase of about 2°C. Such an increase would make the world warmer than at any time in the past 100,000 years. A worst-case scenario suggests that the warming would be greatest near the poles. Melting of polar ice might raise sea level by an estimated 100 m, gradually flooding areas 150 km (or more) inland from the current coastline. New York, Miami, Los Angeles, and many other cities would then be underwater. A warming trend would also alter the geographic distribution of precipitation, making major agricultural areas of the central United States much drier. Most of Earth's natural ecosystems would also be affected, with boundaries between systems such as forests and grasslands shifting. However, the various mathematical models that have been used to study this question disagree about the details of how warming on a global level will change the climate in each region.

By studying how prehistoric periods of global warming and cooling affected plant communities, ecologists are using another strategy to help predict the consequences of future temperature changes. Records from fossilized pollen provide evidence that plant communities are altered dramatically by climate change. However, past climate changes occurred gradually, and plant and animal populations could spread into areas where conditions allowed them to survive. A major concern about the global warming under way now is that it is so rapid that many species may not be able to adapt.

What can be done to lessen the chances of greenhouse disaster? The burning of trees after deforestation to clear land in the tropics accounts for about 20% of the excess CO_2 released into the atmosphere. The burning of fossil fuels is the cause of the other 80%. International cooperation and national and individual action are needed to decrease fossil fuel consumption and to reduce the destruction of forests.

More than 189 countries have signed an international agreement known as the Kyoto Protocol that intends to reduce greenhouse gas emissions worldwide. As of 2006, the United States, the world's largest producer of greenhouse gases, had not accepted the agreement, mostly because of opposition from business leaders and global-warming skeptics. The United States and other large industrial nations are searching for ways to help fend off global warming. China, an economic giant with a huge appetite for cars and other fossil fuel–powered machines, is working to create new power sources and transit systems that generate less carbon dioxide. A few automakers are offering new types of "hybrid" gas/electric cars that use less fossil fuel and release fewer harmful emissions. These cars are being snapped up by U.S. consumers, a sign that the public welcomes a shift away from fossil fuels, but so far their share of the U.S. auto market is quite small. Another "green technology"—solar systems for heating water and generating electricity—is becoming an increasingly popular option for homes.

Depletion of Atmospheric Ozone Life on Earth is protected from the damaging effects of ultraviolet (UV) radiation (see Chapter 10) by a very thin protective layer of ozone molecules (O_3) located in the atmosphere between 17 and 25 km above Earth's surface. This **ozone layer** absorbs UV radiation, preventing much of it from reaching organisms in the biosphere. Measurements by atmospheric scientists document that the ozone layer has been gradually thinning since the middle of the 20th century **(Figure 20.12)**.

The destruction of atmospheric ozone probably results mainly from the accumulation of chlorofluorocarbons, chemicals used in refrigeration, as propellants in aerosol cans, and in certain manufacturing processes. When the breakdown products from these chemicals rise in the atmosphere, the chlorine they contain reacts with ozone, converting it to O_2. Subsequent chemical reactions liberate the chlorine, allowing it to react with other ozone molecules in a chain reaction. The effect is most apparent over Antarctica, where cold winter temperatures facilitate these atmospheric reactions. Scientists first described an "ozone hole" over Antarctica in 1985. Since then, the size of the ozone hole has increased (see Figure 20.12a), sometimes extending as far as the southernmost portions of Australia, New Zealand, and South America. And at the more heavily populated middle latitudes, ozone levels have decreased 2–10% during the past 20 years.

The consequences of ozone depletion may be quite severe for all life on Earth, including humans. Some scientists expect the growing intensity of UV radiation to increase the incidence of skin cancer and cataracts among humans. It is likely that there will also be damaging effects on crops and natural communities, especially the phytoplankton that are responsible for a large proportion of the biosphere's primary productivity. The danger posed by ozone depletion is so great that many nations agreed in 1987 to end the production of chlorofluorocarbons by the year 2010. (The United States and other industrialized nations have already substituted safer compounds for chlorofluorocarbons, but a grace period was allowed for developing countries.) As a result of such action, ozone depletion has slowed. Unfortunately, even if all chlorofluorocarbons were banned today, the chlorine molecules already in the atmosphere will continue to degrade atmospheric ozone for at least a century. It is just one more example of how far our technological tentacles reach in disrupting the dynamics of ecosystems and the entire biosphere.

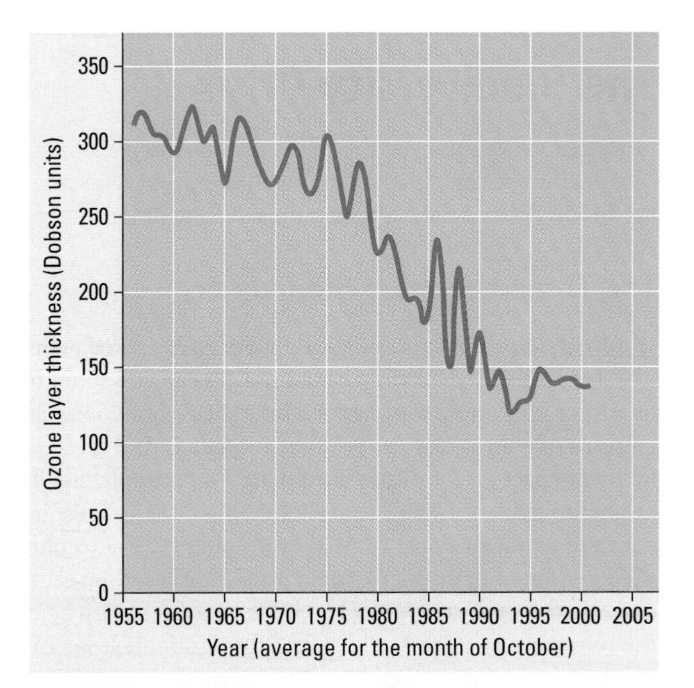

(a) Thickness of ozone layer

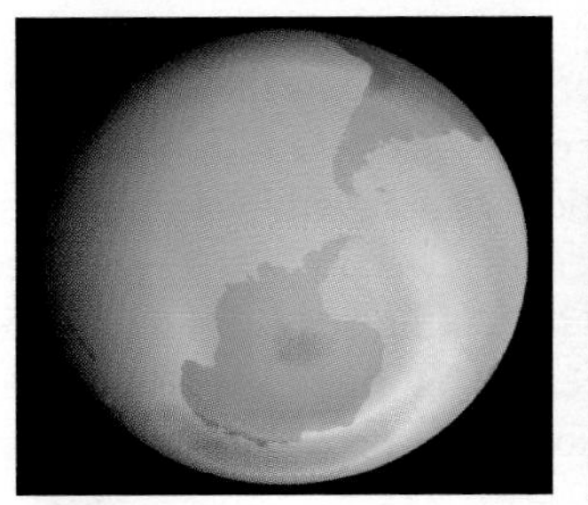

1979

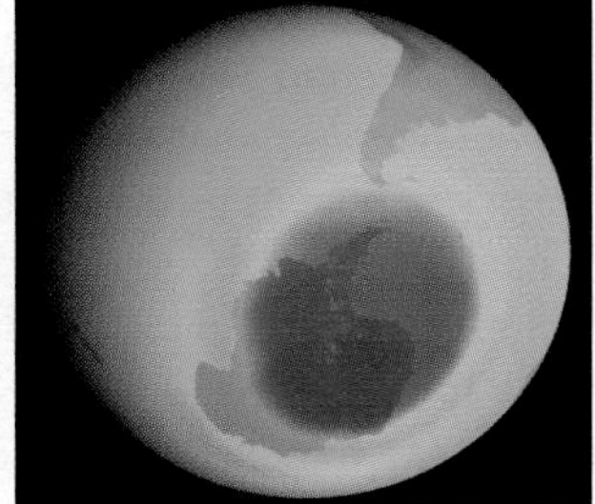

2000

(b) Ozone hole

Figure 20.12 Erosion of Earth's ozone shield. (a) This graph tracks the thickness of the ozone layer in units called Dobsons. **(b)** The ozone hole over Antarctica is visible as the blue patch in these images based on atmospheric data.

CHECKPOINT

1. How can clear-cutting a forest damage the water quality of nearby lakes?
2. How does the excessive addition of mineral nutrients to a lake eventually result in the loss of most fish in the lake?
3. How is biological magnification relevant to the health of most humans in developed countries?
4. From the data in Figure 20.10, what was the approximate percentage increase in atmospheric CO_2 between 1960 and 1990?

Answers: 1. Without the growing trees to assimilate minerals from the soil, more of the minerals run off and end up polluting water resources. **2.** The eutrophication (overfertilization) initially causes population explosions of algae and the organisms that feed on them. The respiration of so much life, including the detritivores decomposing all the organic refuse, consumes most of the lake's oxygen, which the fish require. **3.** People in developed countries generally eat more meat than do people in developing countries. As secondary or tertiary consumers in a food chain, meat-eaters acquire a greater dose of certain toxic chemicals than if they fed exclusively on plants as primary consumers. **4.** About 12%, from 315 ppm to about 352 ppm; $352 - 315 = 37$, and $(37/315) \times 100 = 11.7\%$

The Biodiversity Crisis

Now that we've surveyed the ways that humans can affect communities and ecosystems, we can understand one of the primary consequences: an alarming **biodiversity crisis,** a precipitous decline in Earth's great variety of life.

The Three Levels of Biodiversity

Biodiversity, short for biological diversity, has three main components. The first is the diversity of ecosystems. Each ecosystem, be it a rain forest, desert, or coral reef, has a unique biological community and characteristic patterns of energy flow and chemical cycling. And each ecosystem has a unique impact on the entire biosphere. For example, the floating "pastures" of phytoplankton in the oceans help moderate the greenhouse effect by consuming massive quantities of atmospheric CO_2 for photosynthesis and shell building (many microscopic protists in plankton secrete shells of bicarbonate, a derivative of CO_2). Some ecosystems are being erased from the biosphere at an astonishing rate. For example, over 25% of the world's coral reefs were destroyed by 1998 as a result of overfishing, coastal development, pollution, and global environmental change; that number is predicted to rise to 40% by 2010 and 60% by 2030.

The second component of biodiversity is the variety of species that make up the biological community of any ecosystem **(Figure 20.13)**. And the third component is the genetic variation within each species. You learned in Chapter 13 that the loss of genetic diversity—by a severe reduction in population size, for example—can hasten the demise of a dwindling species.

Though human impact reaches all three levels of biodiversity, most of the research focus so far has been on species extinction.

Figure 20.13 A tropical rain forest is a showcase of biodiversity.

The Loss of Species

The seventh mass extinction in the history of life is well under way. Previous episodes, including the Cretaceous extinctions that claimed the dinosaurs and many other groups, pale by comparison. The current mass extinction is both broader and faster, extinguishing species at a rate at least 50 times faster than just a few centuries ago. And unlike past poundings of biodiversity, which were triggered mainly by physical processes such as climate change caused by volcanism or asteroid crashes, this latest mass extinction is due to the evolution of a single species—a big-brained, manually dexterous, environment-manipulating toolmaker that has named itself *Homo sapien*—"wise man."

We do not know the full scale of the biodiversity crisis in terms of a species "body count," for we are undoubtedly losing species that we didn't even know existed; taxonomists estimate that the 1.8 million species that have been identified represent less than 10% of the true number of species. However, there are already enough signs to know that the biosphere is in deep trouble. For example:

- About 12% of the 9,946 known bird species in the world and 24% of the 4,763 known mammal species are threatened with extinction.
- Of the approximately 20,000 known plant species in the United States, 200 have become extinct since good records have been kept and 730 species are endangered or threatened.
- About 20% of the known freshwater fishes in the world either have become extinct during historical times or are seriously threatened. The toll on amphibians and reptiles has been almost as great.
- Harvard biologist Edward O. Wilson, a renowned scholar of biodiversity, has compiled what he grimly calls the Hundred Heartbeat Club. The species that belong are those animals that number fewer than 100 individuals on Earth and so are only that many heartbeats away from extinction **(Figure 20.14)**.
- Several researchers estimate that at the current rate of destruction, over half of all plant and animal species will be gone by the end of the 21st century.

The modern mass extinction is different from earlier biodiversity shakeouts in still another important way. The prehistoric crashes were all followed by rebounds in diversity as the survivors radiated and adapted to ecological niches left vacant by the extinctions. But as long as we humans continue to destroy habitats and degrade biodiversity at the ecosystem level, there can be no rebound in the evolutionary diversification of life. In fact, the trend is toward increased geographic range and prevalence of "disaster species," those life-forms such as house mice, kudzu and other weeds, cockroaches, and fire ants that seem to thrive in environments disrupted by human activities. Unless we can reverse the current trend of increasing biodiversity loss, we will leave our children and grandchildren a biosphere that is much less interesting and much more biologically impoverished.

(a) Philippine eagle

(b) Chinese river dolphin

(c) Javan rhinoceros

Figure 20.14 A hundred heartbeats from extinction. These are just three of the many members of what E. O. Wilson calls the Hundred Heartbeat Club, species with fewer than 100 individuals remaining on Earth.

The Three Main Causes of the Biodiversity Crisis

Before we can begin to address how to slow or reverse the biodiversity crisis, we must examine its root causes.

Habitat Destruction Human alteration of habitats poses the single greatest threat to biodiversity throughout the biosphere **(Figure 20.15a)**. Assaults on diversity at the ecosystem level result from the expansion of agriculture to feed the burgeoning human population, urban development, forestry, mining, and environmental pollution. The amount of human-altered land surface is approaching 50%, and we use over half of all accessible surface fresh water. Some of the most productive aquatic habitats in estuaries and intertidal wetlands are also prime locations for commercial and residential developments. The loss of marine habitats is also severe, especially in coastal areas and coral reefs.

Introduced Species Ranking second behind habitat loss as a cause of the biodiversity crisis is human introduction of exotic species that eliminate native species through predation or competition. For example, if your campus is in an urban setting, there is a good chance that the birds you see most often as you walk between classes are starlings, rock doves (often called "pigeons"), and house sparrows—all introduced species that have replaced native birds in many areas of North America. One of the largest rapid-extinction events yet recorded is the loss of freshwater fishes in Lake Victoria in East Africa. About 200 of the 300 species of native fishes, found nowhere else but in this lake, have become extinct since Europeans introduced a non-native predator, the Nile perch, in the 1960s **(Figure 20.15b)**.

Overexploitation As a third major threat to biodiversity, overexploitation of wildlife often compounds problems of shrinking habitat and introduced species. Animal species whose numbers have been drastically reduced by excessive commercial harvest or sport hunting include whales, the American bison, Galápagos tortoises, and numerous fishes. Many fish stocks in the ocean have been overfished to levels that cannot sustain further human exploitation **(Figure 20.15c)**. In addition to the commercially important species, members of many other species are often killed unintentionally by harvesting methods; for example, dolphins, marine turtles, and seabirds are often caught in fishing nets, and countless numbers of invertebrates are killed by marine trawls (big nets). An expanding, often illegal world trade in wildlife products, including rhinoceros horns, elephant tusks, and grizzly bear gallbladders, also threatens many species.

(a) Habitat destruction. This is an all-too-common scene: the clearing of a forest for lumber, agriculture, housing projects, or roads.

(b) Introduced species. One of the largest freshwater fishes (up to 2 m long and weighing up to 450 kg), the Nile perch was introduced to Lake Victoria in East Africa to provide high-protein food for the growing human population. Unfortunately, the perch's main effect has been to wipe out about 200 smaller native species.

(c) Overexploitation. Beginning in the 1980s, wholesalers began airfreighting fresh, iced bluefin tuna to Japan for sushi and sashimi. In auctions like the one shown here, the fish now brings up to $100 per pound! It took just ten years to reduce the North Atlantic bluefin population to less than 20% of its 1980 size. In spite of fishing quotas, the high price that bluefin tuna brings probably dooms the species to extinction.

Figure 20.15 The three main causes of the biodiversity crisis.

Why Biodiversity Matters

Why should we care about the loss of biodiversity? Perhaps the purest reason is our sense of connection to nature and other forms of life. We are naturally curious about our world and wish to preserve it so that we may learn from it. Additionally, many people share a moral belief that other species have an inherent right to life.

In addition to ethical and aesthetic reasons for preserving biodiversity, there are practical ones as well. First of all, we depend on many other species for food, clothing, shelter, oxygen, soil fertility—the list goes on and on. In the United States, 25% of all prescriptions dispensed from pharmacies contain substances derived from plants. For instance, two drugs effective against Hodgkin's disease and other cancers come from the rosy periwinkle, a flowering plant native to the island of Madagascar **(Figure 20.16)**. Madagascar alone harbors some 8,000 species of flowering plants, 80% of which are **endemic species,** meaning they occur only there. Among these unique plants are several species of wild coffee trees, some of which yield beans lacking caffeine ("naturally decaffeinated"). With an estimated 200,000 species of plants and animals, Madagascar is among the top five most biologically diverse countries in the world. Unfortunately, most of Madagascar's species are in serious trouble. People have lived on the island for only about 2,000 years, but in that time, Madagascar has lost 80% of its forests and about 50% of its native species. Madagascar's dilemma represents that of much of the developing world. The island is home to over 10 million people, most of whom are desperately poor and hardly in a position to be concerned with environmental conservation. Yet the people of Madagascar, as well as others around the globe, could derive vital benefits from the biodiversity that is being destroyed.

Who knows which of the species disappearing from the biosphere could provide new sources of food or medicine? And who understands the dynamics of ecosystems well enough to know which species are least essential to biological communities? The great American naturalist Aldo Leopold explained it this way: "If the biota, in the course of aeons, has built something we like but do not understand, then who but a fool would discard seemingly useless parts? To keep every cog and wheel is the first precaution of intelligent tinkering."

Another reason to be concerned about the changes that underlie the biodiversity crisis is that the human population itself is threatened by large-scale alterations in the biosphere. Like all other species, we evolved in Earth's ecosystems, and we are dependent on the living and nonliving components of these systems. By allowing the extinction of species and the degradation of habitats to continue, we are taking a risk with our own species' survival.

In an attempt to counter what they see as a tendency of policymakers and governments to undervalue life-sustaining features of the biosphere, a team of ecologists and economists recently estimated the cost of losing ecosystem "services." For example, they estimated part of the value of a wetland from the cost of flood damage that occurred because of the loss of the wetland's ability to hold floodwater. For the year 1997, the scientists estimated the average annual value of ecosystem dynamics in the biosphere at 33 trillion U.S. dollars. In contrast, the global gross national product for the same year was 18 trillion U.S. dollars. Although rough, these estimates help make the important point that we cannot afford to continue to take ecosystems for granted.

Figure 20.16 Madagascar's rosy periwinkle, a source of anticancer drugs.

CHECKPOINT

1. What are the three main levels of biodiversity?
2. What are the three main causes of the biodiversity crisis?

Answers: 1. Ecosystem diversity, species diversity, and the genetic diversity of populations
2. Habitat destruction, introduced species, and overexploitation

Conservation Biology

Conservation biology is a goal-oriented science that seeks to counter the loss of biodiversity. It began to take form in the late 1970s as an interdisciplinary collaboration of scientists working toward the long-term maintenance of functional ecosystems and a reduction in the rate of species extinction. Conservation biologists recognize that biodiversity can be sustained only if the evolutionary mechanisms that have given rise to species and communities of organisms continue to operate. Thus, the goal is not simply to preserve individual species but to sustain ecosystems, where natural selection can continue to function, and to maintain the genetic variability on which natural selection acts. The front lines for conservation biology are geographic areas that are especially rich in endangered biodiversity.

Biodiversity "Hot Spots"

A **biodiversity hot spot** is a relatively small area with an exceptional concentration of species **(Figure 20.17)**. Many of the organisms in biodiversity hot spots are endangered or threatened endemic species. Together, the biodiversity hotspots shown in Figure 20.17 total less than 1.5% of Earth's land but are home to one-third of all species of plants and vertebrates. Conservation biologists have also identified aquatic ecosystems, including certain river systems and coral reefs, that are biodiversity hot spots.

Because endemic species are limited to specific areas, they are highly sensitive to habitat degradation. Some biologists estimate that at the current rate of human development, loss of habitat will cause the extinction of about half of the species in terrestrial biodiversity hot spots in the next 10 to 15 years. Thus, biodiversity hot spots are also hot spots of extinction and rank high on the list of areas demanding strong global conservation efforts. In the United States, the greatest numbers of endangered species (terrestrial and freshwater) occur in Hawaii, southern California, the southern Appalachians, and the southeastern coastal states (especially Florida), areas with the highest numbers of endemic species.

With so much biodiversity concentrated in a relatively small portion of the biosphere, there is reason for optimism that conservation biologists can accomplish much by focusing on these hot spots. The goal is to keep as many of these hot spots as wild as possible. On the other hand, the biodiversity crisis is a global problem, and focusing on hot spots should not detract from efforts to conserve habitats and species diversity in other areas.

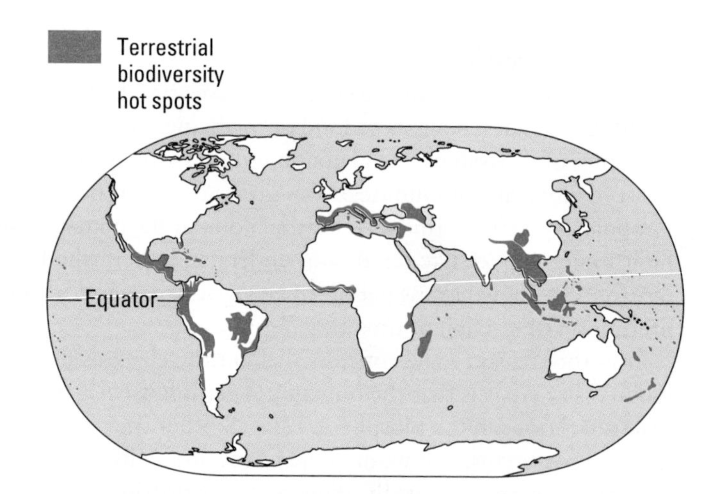

Figure 20.17 Earth's terrestrial biodiversity hot spots.

Conservation at the Population and Species Levels

Much of the popular and political discussion of the biodiversity crisis centers on species. The U.S. Endangered Species Act (ESA) defines an **endangered species** as one that is "in danger of extinction throughout all or a significant portion of its range." Also defined for protection by the ESA, **threatened species** are those that are likely to become endangered in the foreseeable future throughout all or a significant portion of their geographic range.

Habitat Fragmentation and Subdivided Populations What can be done to protect endangered or threatened species? Much current research focuses on the dynamics of populations that have been reduced in numbers and fragmented by human activities. Because habitats are naturally patchy, the populations of many species were subdivided into groups with varying degrees of isolation from one another before humans began altering habitats significantly. Gene flow among population subgroups varied according to the degree of isolation (see Chapter 13). Today, severe **population fragmentation,** the splitting and consequent isolation of portions of populations by habitat degradation, is one of the most harmful effects of habitat loss due to human activities **(Figure 20.18)**. We'll refer to these fragments of a population as subpopulations. A decrease in the overall size of populations and a reduction in gene flow among subpopulations usually accompany fragmentation.

In most cases, subpopulations are separated into habitat patches that vary in quality. For instance, in a fragmented forest, the presence of some large, dead trees can make the difference between a high-quality patch and a low-quality one for squirrels, owls, and many birds that require rot cavities for their nests. Patches with abundant high-quality resources tend to have stable, persistent subpopulations. Reproductive individuals in a high-quality patch tend to produce more offspring than the patch can sustain. Such a habitat where a subpopulation's reproductive success exceeds its death rate is called a **source habitat.** Source habitats produce enough new individuals that some disperse to other areas, often in search of food or a place to reproduce. In contrast, a habitat where a subpopulation's death rate exceeds its reproductive success is called a **sink habitat.** The persistence of many subpopulations in sink habitats depends on individuals entering it from source habitats.

Today, because of habitat loss, the number of sink habitats is increasing. For example, the remaining source habitats of the northern spotted owl are relatively small fragments of old-growth rain forests. The owls tend to disperse from such fragments of source habitat into larger, neighboring areas where trees are regrowing after logging. Unfortunately, such areas are sink habitats for the owls. Sustaining the owl populations depends in part on keeping enough reproductive individuals as robust breeding stock in the source habitats. Some researchers have suggested surrounding old-growth fragments with distinct boundaries—such as clear-cut areas—that the owls will not enter. The spotted owl situation illustrates the importance of identifying source and sink habitats and of protecting source habitats.

Northern spotted owl

Figure 20.18 Fragmentation of a forest ecosystem. This aerial photograph illustrates fragmentation of a coniferous forest in the Mount Hood National Forest in northwestern Oregon. The forest was originally contiguous. The open snowy areas in the photo were logged, creating forest fragments, some of which are islands within clear-cut areas. A common result of human activities, this kind of habitat alteration has reduced and fragmented the populations of many species. An example is the northern spotted owl (inset), which inhabits coniferous forests of the U.S. Pacific Northwest. Owl populations declined markedly and were fragmented after these forests were logged.

What Makes a Good Habitat? What factors make some habitats sources and others sinks? Identifying the specific combination of habitat factors that is critical for a species is fundamental in conservation biology.

As a case study in identifying critical habitat factors, we'll consider the red-cockaded woodpecker (*Picoides borealis*), an endangered, endemic species originally found throughout the southeastern United States. This species requires mature pine forests, preferably ones dominated by the longleaf pine. Most woodpeckers nest in dead trees, but the red-cockaded woodpecker drills its nest holes in mature, living pine trees

(Figure 20.19a). It also drills small holes around the entrance to their nest cavity, which causes resin from the tree to ooze down the trunk. The resin seems to repel certain predators, such as corn snakes, that eat bird eggs and nestlings. Another critical habitat factor for this woodpecker is a low growth of plants among the mature pine trees **(Figure 20.19b)**. Biologists have found that a habitat becomes a sink, with breeding birds tending to abandon nests, when vegetation among the pines is thick and higher than about 15 feet **(Figure 20.19c)**. Apparently, the birds require a clear flight path between their home trees and the neighboring feeding grounds. Historically, periodic fires swept through longleaf pine forests, keeping the undergrowth low.

(a) A red-cockaded woodpecker perches at the entrance to its nest site in a longleaf pine tree

(b) Forest habitat with low undergrowth sustains red-cockaded woodpeckers

(c) High, dense undergrowth impedes the woodpeckers' access to feeding grounds

Figure 20.19 Habitat requirements of the red-cockaded woodpecker.

The recent recovery of the red-cockaded woodpecker from near extinction to sustainable populations is largely due to recognition of the bird's key habitat factors and protection of some longleaf pine forests that support viable numbers of the birds. The use of controlled fires to reduce forest undergrowth helps maintain mature pine trees and thus the woodpeckers as well.

Conserving Species amid Conflicting Demands Determining habitat requirements is only one aspect of the effort to save species. It is usually necessary to weigh a species' biological and ecological needs against the conflicting demands of our complex culture. Thus, conservation biology often highlights the relationships between biology and society. For example, an ongoing, often bitter debate in the U.S. Pacific Northwest pits saving habitats for populations of the northern spotted owl, timber wolf, grizzly bear, and bull trout against demands for jobs in the timber, mining, and other resource extraction industries. Programs to restock wolves and to bolster the populations of grizzly bears and other large carnivores are opposed by some recreationists concerned about safety and by many ranchers concerned with the potential loss of livestock.

Large, high-profile vertebrates are not always the focal point in conflicts involving conservation biology. It is habitat use that is almost always at issue. Should work proceed on a new highway bridge if it destroys the only remaining habitat of a species of freshwater mussel? If you were the owner of a coffee plantation growing varieties that thrive in bright sunlight, do you think you would be willing to change to shade-tolerant coffee varieties that are less productive and less profitable but support large numbers of songbirds?

Conservation at the Ecosystem Level

Most conservation efforts in the past have focused on saving individual species, and this work continues. More and more, however, conservation biology aims at sustaining the biodiversity of entire communities and ecosystems. On an even broader scale, conservation biology considers the biodiversity of whole landscapes. Ecologically, a **landscape** is a regional assemblage of interacting ecosystems, such as an area with forest, adjacent fields, wetlands, streams, and streamside habitats. **Landscape ecology** is the application of ecological principles to the study of land-use patterns. Its goal is to make ecosystem conservation a functional part of the planning for land use.

Edges and Corridors Edges between ecosystems are prominent features of landscapes, whether natural or altered by humans **(Figure 20.20)**. Such edges have their own sets of physical conditions, such as soil type and surface features, that differ from the ecosystems on either side of them. Edges also may have their own type and amount of disturbance. For instance, the edge of a forest often has more blown-down trees than a forest interior because the edge is less protected from strong winds. Because of their specific physical features, edges also have their own communities of organisms. Some organisms thrive in edges because they require resources found only there. For instance, whitetail deer browse on woody shrubs found in edge areas between woods and fields, and their populations often expand when forests are logged or interrupted by development.

Edges can have both positive and negative effects on biodiversity. A recent study in a tropical rain forest in western Africa indicated that natural edge communities are important sites of speciation (the origin of new species; see Chapter 13). On the other hand, landscapes where human activities have produced edges often have fewer species and are dominated by species that are adapted to edges. An example is the brown-headed cowbird, an edge-adapted species that is currently expanding its populations in many areas of North America. Cowbirds forage in open fields on insects disturbed by or attracted to cattle and other large herbivores. But the cowbirds also need forests, where they lay their eggs in the nests of other bird species. The "host" bird feeds the cowbird young as though they were the host's own offspring. (This is a form of parasitism.) Cowbird numbers are burgeoning where forests are being heavily cut and fragmented, creating more forest-edge habitats and open land for cattle, horses, and sheep. Increasing cowbird populations and loss of habitats are correlated with declining populations of several songbirds, such as warblers, that the cowbirds parasitize.

Another important landscape feature, especially where habitats have been severely fragmented, is the **movement corridor,** a narrow strip or series of small clumps of quality habitat connecting otherwise isolated patches. Streamside habitats often serve as natural corridors, and government policy in some nations prohibits destruction of these areas. In places where there is extremely heavy human impact, government agencies sometimes construct artificial corridors **(Figure 20.21)**. Corridors can promote dispersal and help sustain populations, and they are especially important to species that migrate between different habitats seasonally. On the other hand, a corridor can be harmful—as, for example, in the spread of diseases, especially among small subpopulations in closely situated habitat patches. A movement corridor that connects a source habitat to a sink habitat may also reduce population numbers in the source. The effects of movement corridors have not yet been thoroughly studied, and researchers tend to evaluate the potential effects of corridors on a case-by-case basis.

(a) Natural edges between ecosystems

(b) Edges created by human activities

Figure 20.20 Landscape edges between ecosystems. **(a)** This landscape in Australia includes several edges that lie between a dry forest, a rocky area with grassy islands, and a flat, grass-covered lakeshore. **(b)** Human activities, such as logging and road building, often create edges that are more abrupt than those delineating natural landscapes. Sharp edges surround clear-cuts in this photograph of a heavily logged rain forest in Malaysia.

Zoned Reserves In an attempt to slow the disruption of ecosystems, a number of countries are setting up what they call zoned reserves. A **zoned reserve** is an extensive region of land that includes one or more areas undisturbed by humans. The undisturbed areas are surrounded by lands that have been changed by human activity and are used for economic gain. The key factor of the zoned reserve concept is the development of a social and economic climate in the surrounding lands that is compatible with ecosystem conservation. These surrounding areas continue to be used to support the human population, but they are protected from extensive alteration. As a result, they serve as buffer zones against further intrusion into the undisturbed areas.

The small Central American nation of Costa Rica has become a world leader in establishing zoned reserves. In exchange for reducing its international debt, the Costa Rican government established eight zoned reserves, called

Figure 20.21 An artificial corridor. This highway underpass allows movement between protected areas for the few remaining Florida panthers. High fences along the highway reduce road kills of panthers and other species.

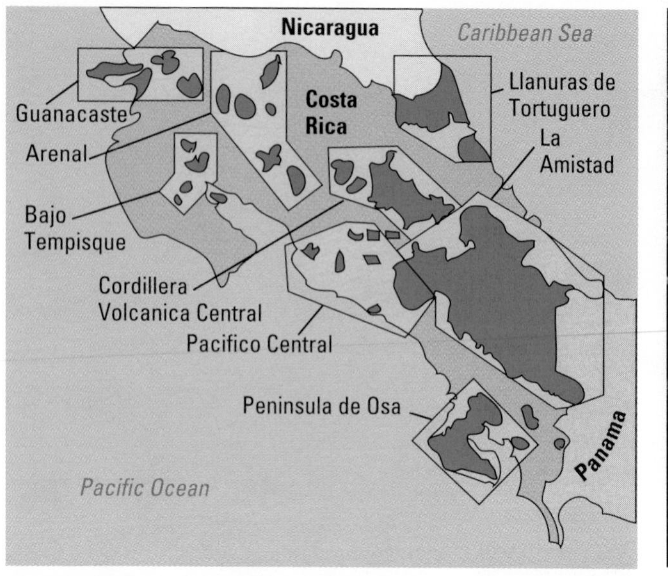

(a)

(b)

Figure 20.22 Zoned reserves in Costa Rica. (a) The green areas on the map are national park lands, core areas where human disruption is minimized. Surrounding these conservation cores, gold areas indicate buffer zones. These are transition areas, mainly privately owned, where most of the human population live and work. Ideally, the most destructive practices—industries such as mining, large-scale monoculture (growth of a single type of crop over a large area), and urban development—are confined to the outermost fringes of the buffer zones. Within the buffer zones, the trend is toward sustainable agriculture and forestry, activities that can provide comfortable economic support for local residents without drastically altering habitats. **(b)** Local students marvel at the diversity of life in one of Costa Rica's reserves.

"conservation areas" **(Figure 20.22)**. Costa Rica is making progress toward managing its zoned reserves so that the buffer zones provide a steady, lasting supply of forest products, water, and hydroelectric power and also support sustainable agriculture and tourism. An important goal is providing a stable economic base for the people living there. Destructive practices that are not compatible with long-term ecosystem conservation (such as massive logging, large-scale single-crop agriculture, and extensive mining), and from which there is often little local profit, are gradually being discouraged. Costa Rica looks to its zoned reserve system to maintain at least 80% of its native species.

The Goal of Sustainable Development

With conservation progress in countries such as Costa Rica as a model, many nations, scientific organizations, and private foundations are embracing the concept of sustainable development. Balancing human needs with the health of the biosphere, **sustainable development** has the goal of long-term prosperity of human societies and the ecosystems that support them. The significance of that responsibility was nicely phrased by former Norwegian Prime Minister G. H. Brundtland: "We must consider our planet to be on loan from our children, rather than being a gift from our ancestors."

Sustainable development will depend on the continued research and applications of basic ecology and conservation biology. It will also require a cultural commitment to conserve ecosystem processes and biodiversity—a priority that relatively few nations have placed at the top of their political, economic, and social agendas. Those of us living in affluent developed nations are responsible for the greatest amount of environmental degradation. Reality demands that we rearrange some of our priorities, learn to revere the natural processes that sustain us, and temper our orientation toward

Table 20.1	Some Important Ways You Can Promote Sustainability

Reduce consumption
- Buy less.
- Avoid excess packaging.
- Avoid products made from nonrenewable resources.
- Fix things rather than discarding them.

Be more energy efficient
- Bike, walk, or take public transportation instead of driving.
- Share rides.
- Purchase efficient appliances and vehicles.

Promote recycling
- Recycle at home and at work.
- Purchase products made from recycled materials.

Take political action
- Vote for pro-environment policymakers.
- Approve ecologically sound ballot measures.
- Join an environmental advocacy group.
- Write letters in support of environmental causes.
- Run for political office yourself.

Promote research and education
- Talk about environmental issues with friends and family to raise their awareness.
- Sponsor environmental initiatives on your campus or in your workplace.
- Serve as a role model through your actions.

Think long term
- Realize that addressing environmental problems sometimes has short-term costs but long-term benefits.
- Support policymakers and businesses that promote long-term environmental thinking.

short-term personal gain. What can you do to help? **Table 20.1** lists some simple changes we can all make to improve the health of the biosphere.

The current state of the biosphere demonstrates that we are treading precariously on uncharted ecological ground, entering a time of unprecedented environmental change. But despite the uncertainties, now is not the time for gloom and doom, but a time to meet the challenges to pursue more knowledge about life and to work as individuals and a society toward long-term sustainability. Along the way, we will reap the bonus of appreciating our connections to the biosphere and its diversity of life.

CHECKPOINT

1. What is a biodiversity hot spot?
2. Why is the fragmentation of populations of endangered species increasing?
3. Are critical habitat factors generally more favorable in a source habitat or a sink habitat? Why?
4. How is a landscape different from an ecosystem?
5. How can "living on the edge" be a good thing for some species, such as white-tailed deer and cowbirds?

Answers: 1. A relatively small area with a disproportionately large number of species, including endangered species **2.** By destroying habitat, human activities are fragmenting the populations of many species into subpopulations that live in the separate patches of remaining habitat. **3.** Source habitat, because the population is increasing in a source habitat (but not in a sink habitat) **4.** A landscape is more inclusive in that it consists of several interacting ecosystems in the same region. **5.** They use a combination of resources from the two ecosystems on either side of the edge.

EVOLUTION CONNECTION

Biophilia and an Environmental Ethic

Not many people today live in truly wild environments or even visit such places often. Our modern lives are very different from those of early humans, who hunted and gathered and painted wildlife murals on cave walls. But our behavior reflects remnants of our ancestral attachment to nature and the diversity of life. People keep pets, nurture houseplants, invite avian visitors with backyard birdhouses, and visit zoos, gardens, and nature parks. These pleasures are examples of what biologist Edward O. Wilson calls **biophilia,** a human desire to affiliate with other life in its many forms. (You met Wilson earlier in the context of the Hundred Heartbeat Club.) Wilson extends biophilia to include our attraction to pristine landscapes with clean water and lush vegetation **(Figure 20.23)**. We evolved in natural environments rich in biodiversity, and we still have an affinity for such settings. Wilson makes the case that our biophilia is innate, an evolutionary product of natural selection acting on a brainy species whose survival depended on a close connection to the environment and a practical appreciation of plants and animals.

It will come as no surprise that most biologists have embraced the concept of biophilia **(Figure 20.24)**. After all, these are people who have turned their passion for nature into careers. But biophilia strikes a harmonic chord with biologists for another reason. If biophilia is evolutionarily embedded in our genomes, then there is hope that we can become better custodians of the biosphere. If we all pay more attention to our biophilia, a new environmental ethic could catch on among individuals and societies. And that ethic is a resolve never to knowingly allow a single species to become extinct or any ecosystem to be destroyed as long as there are reasonable ways to prevent such ecological violence. It is an environmental ethic that balances out another human trait—our tendency to subdue Earth. Yes, we should be motivated to preserve biodiversity because we depend on it for food, medicine, building materials, fertile soil, flood control, habitable climate, drinkable water, and breathable air. But maybe we can also work harder to prevent the extinction of other forms of life just because it is the ethical thing for us to do as the most thoughtful species in the biosphere. Again, Wilson sounds the call: "Right now, we're pushing the species of the world through a bottleneck. We've got to make it a major moral principle to get as many of them through this as possible. It's the challenge now and for the next century. And there's one good thing about our species: We like a challenge!"

Biophilia is a fitting capstone for this unit. Modern biology is the scientific extension of our human tendency to feel connected to and curious about all forms of life. We hope that our discussion of ecology has deepened your biophilia and broadened your education. ■

Figure 20.23 Doing what comes naturally. One of modern biology's greatest naturalists, Edward O. Wilson has helped teach scientists and the general public a greater respect for Earth's biodiversity (two of his books have won Pulitzer Prizes). He coined the term *biophilia* to denote humans' inherent passion for nature. This photograph finds biophiliac Wilson in the woods near Massachusetts's Walden Pond, a landscape immortalized by another great naturalist and writer, Henry David Thoreau.

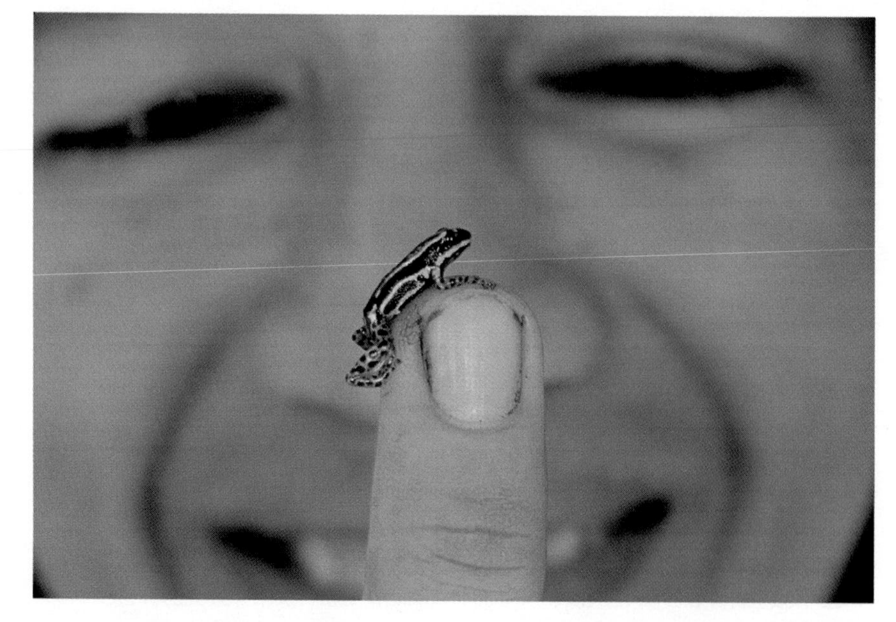

Figure 20.24 The face of biophilia. Biologist Carlos Rivera Gonzales, who is participating in a biodiversity survey in a remote region of Peru, could not resist a closer look at a tiny tree frog. You can also see biophilia in the faces of the children in Figure 20.22b.

Chapter Review

SUMMARY OF KEY CONCEPTS

For study help and activities, go to campbellbiology.com or the student CD-ROM.

Human Impact on Biological Communities

- **Human Disturbance of Communities** Human disturbance usually reduces species diversity in communities. Much of the United States is now a hodgepodge of early successional growth where more mature communities once prevailed.

- **Introduced Species** Introduced species are those that humans intentionally or accidentally move from the species' native locations to new geographic regions. Introduced species rank second only to habitat destruction as a cause of extinctions and loss of Earth's biodiversity.

Activity *Fire Ants: An Introduced Species*

Human Impact on Ecosystems

MP3 Tutor *Global Warming*

- **Impact on Chemical Cycles** Human activities often intrude in biogeochemical cycles by removing nutrients from one location and adding them to another. For example, the increased burning of fossil fuels is steadily raising CO_2 in the atmosphere, and sewage treatment facilities and fertilizers add large amounts of nitrogen and phosphorus to aquatic systems. Deforestation and extensive removal of groundwater change the water cycle.

Activity *Water Pollution from Nitrates*

- **The Release of Toxic Chemicals to Ecosystems** Humans have added entirely new materials, many of them toxic, to ecosystems. These toxins often become more concentrated through biological magnification.

Increasing PCB concentration

Activity *DDT and the Environment*

- **Human Impact on the Atmosphere and Climate** Deforestation and the burning of fossil fuels have increased concentrations of CO_2 in the atmosphere, contributing to the greenhouse effect and global warming, with potentially disastrous consequences. Developed countries, which have the greatest energy consumption, have the greatest responsibility to reduce their use of fossil fuels by conserving energy and developing alternative energy sources. In addition, the protective ozone layer has been gradually thinning since 1975 because of the accumulation of chlorofluorocarbons. Scientists expect that a thinning ozone layer will result in an increase in skin cancer and cataracts among humans.

Activity *The Greenhouse Effect*

The Biodiversity Crisis

- **The Three Levels of Biodiversity**

- **The Loss of Species** The current mass extinction, caused by human activities, is broader and faster than the Cretaceous extinction that claimed dinosaurs and many other groups. Species now go extinct at a rate at least 50 times faster than just a few centuries ago.

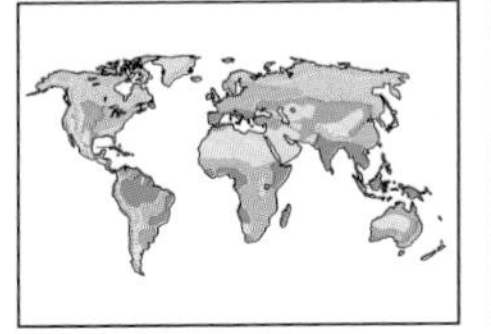

Diversity of ecosystems

Diversity of species within communities

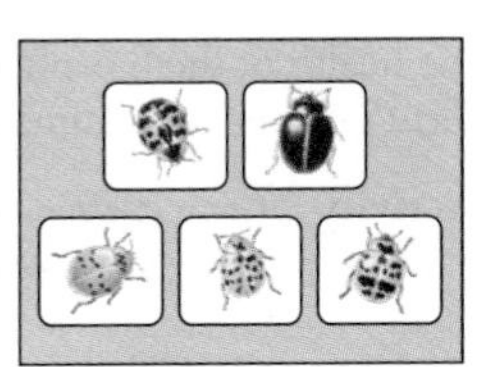

Diversity within species

- **The Three Main Causes of the Biodiversity Crisis**

- **Why Biodiversity Matters** Humans have relied on biodiversity for food, clothing, shelter, oxygen, soil fertility, and medicinal substances. The loss of diversity limits the potential for new discoveries and reflects large-scale changes in the biosphere that could have catastrophic consequences.

Habitat destruction

Introduction of non-native species

Overexploitation of wildlife

Activity *Madagascar and the Biodiversity Crisis*

Conservation Biology

- Conservation biology is a goal-oriented science that seeks to counter the loss of biodiversity.

- **Biodiversity "Hot Spots"** The front lines for conservation biology are relatively small geographic areas that are especially rich in endangered species.

- **Conservation at the Population and Species Levels** Severe population fragmentation is one of the most harmful effects of habitat loss due to human activities. Fragmentation usually results in a decrease in the overall size of populations and a reduction in gene flow among subpopulations. In addition, because of habitat loss, the survival of subpopulations in source habitats is threatened by the dispersal of individuals to an increasing number of sink habitats. Identifying the specific combination of habitat factors that is critical for a species is fundamental in conservation biology. Conservation biology often highlights the relationships between biology and society. Competing demands for habitat use is almost always at issue.

• **Conservation at the Ecosystem Level** Increasingly, conservation biology aims at sustaining the biodiversity of entire communities, ecosystems, and landscapes. Edges between ecosystems are prominent features of landscapes, with positive and negative effects on biodiversity. Natural edge communities are important sites of speciation. But human-produced edges often have fewer species and are dominated by species adapted to edges, such as cowbirds. Corridors can promote dispersal and help sustain populations. But corridors can also promote the spread of diseases and connect source and sink habitats. Zoned reserves are now used to slow the disruptions of ecosystems.

Case Study in the Process of Science *How Are Potential Prairie Restoration Sites Analyzed?*

• **The Goal of Sustainable Development** Balancing human needs with the health of the biosphere, sustainable development has the goal of long-term prosperity of human societies and the ecosystems that support them.

Activity *Conservation Biology Review*

SELF-QUIZ

1. What is an introduced species? Why are introduced species often able to outcompete native organisms?

2. The recent increase in atmospheric CO_2 concentration is mainly a result of an increase in
   a. primary productivity.
   b. the absorption of infrared radiation escaping from Earth.
   c. the burning of fossil fuels and wood.
   d. cellular respiration by the exploding human population.

3. The Hubbard Brook Experimental Forest study demonstrated all of the following except:
   a. Most minerals were recycled within the unaltered forest ecosystem.
   b. Mineral inflow and outflow within a natural watershed were nearly balanced.
   c. Deforestation resulted in an increase in water runoff.
   d. The nitrate concentration in waters draining from the deforested area decreased.

4. Why are falcons and other top predators in food chains most severely affected by pesticides such as DDT?

5. What is the greenhouse effect? How could the greenhouse effect be related to global warming?

6. The people in which region consume the most energy each day?
   a. the United States
   b. South America
   c. China
   d. India
   e. Africa

7. The ozone layer
   a. affects the nitrogen cycle.
   b. contributes to global warming.
   c. absorbs UV radiation, possibly protecting organisms from UV damage.
   d. causes cancer.

8. Which of the following statements most comprehensively addresses what conservation biologists mean by the "biodiversity crisis"?
   a. Worldwide extinction rates are currently 50 times greater than at any time during the past 100,000 years.
   b. Introduced species, such as house sparrows and starlings, have rapidly expanded their ranges.
   c. Harvests of marine fish, such as cod and bluefin tuna, are declining.
   d. Many pest species have developed resistance and are no longer effectively controlled by insecticide applications.

9. Currently, the number one cause of biodiversity loss is __________.

10. A __________ is a local grouping of interacting ecosystems with several adjacent habitats.

Answers to the Self-Quiz questions can be found in Appendix D.

Go to the website or CD-ROM for more Self-Quiz questions.

THE PROCESS OF SCIENCE

11. Biologists in the United States are concerned that populations of many migratory songbirds, such as warblers, are declining. Evidence suggests that some of these birds might be victims of pesticides. Most of the pesticides implicated in songbird mortality have not been used in the United States since the 1970s. Suggest a hypothesis to explain the current decline in songbird numbers. Design an experiment that could test your hypothesis.

BIOLOGY AND SOCIETY

12. By 1935, hunting and trapping had eliminated wolves from the continental United States. Since their protection as an endangered species, wolves have moved south from Canada and have become reestablished in the Rocky Mountains and northern Great Lakes region. Conservationists who would like to speed up this process have reintroduced wolves into Yellowstone National Park. Local ranchers are opposed to bringing back the wolves because they fear predation on their cattle and sheep. What are some reasons for reestablishing wolves in Yellowstone Park? What effects might the reintroduction of wolves have on the ecological communities in the region? What might be done to mitigate the conflicts between ranchers and wolves?

13. Some organizations are starting to envision a sustainable society—one in which each generation inherits sufficient natural and economic resources and a relatively stable environment. The Worldwatch Institute, an environmental policy organization, estimates that we must reach sustainability by the year 2030 to avoid economic and environmental collapse. To get there, we must begin shaping a sustainable society during the next decade or so. In what ways is our current system not sustainable? What might we do to work toward sustainability, and what are the major roadblocks to achieving it? How would your life be different in a sustainable society?

UNIT 5 Animal Structure and Function

21 Unifying Concepts of Animal Structure and Function

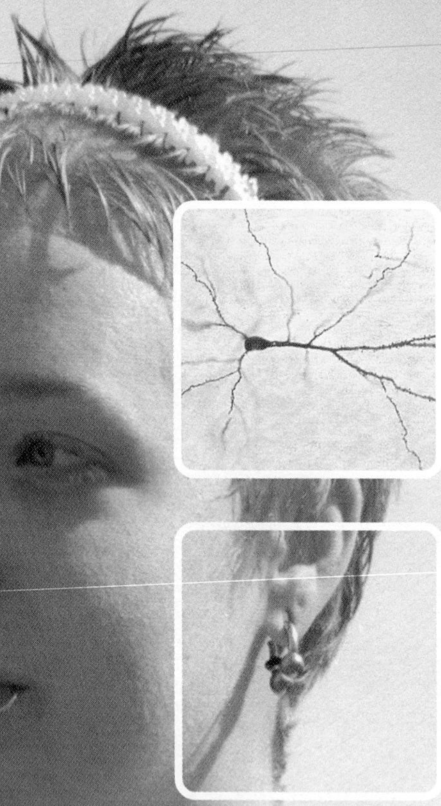

Some of your **nerve cells** are a meter long.

Your nose and earlobes are **good for piercing** because the cartilage in them contains no blood vessels.

During **hibernation,** the body temperature of a ground squirrel may drop 30°C.

Each day, **your kidneys** process enough blood to fill a hot tub.

BIOLOGY AND SOCIETY

Keeping Cool

The photo in **Figure 21.1** shows soccer star David Beckham just after a rigorous game on a warm, sunny day. During the game, the metabolic activity of his muscles produced a lot of heat, and the sun's rays warmed him. Yet his body temperature did not rise significantly. Instead, a temperature control center within his brain initiated several changes in his body that helped keep his internal temperature within its normal range. Sweating cooled his skin (by evaporation), and blood vessels near the surface of his body expanded, allowing more of his body heat to be released.

However, what if the game had gone into overtime? It's possible that an extended period of vigorous play could hinder his body's ability to regulate temperature, allowing his internal temperature to rise dangerously high. In response to such an increase in temperature, surface blood vessels widen to their maximum, and excessive sweating eventually causes dehydration. Both of these responses could lower blood pressure enough to cause him to faint. Losing consciousness would lower heat production by forcing him to stop moving. This condition, called **heat exhaustion,** is the body's final attempt to cool itself.

Under extremely severe conditions, his body could lose the ability to regulate its temperature completely. A highly elevated body temperature could depress his brain's control center, disrupting its functions. This would result in a sharply rising body temperature that further impairs the brain, creating a vicious cycle. His skin would become hot and dry (since sweating would be suspended), his organs would begin to fail, and brain damage and death would quickly follow unless his body were rapidly cooled and he were provided fluids. This condition, called **heat stroke,** represents a total breakdown of the body's internal temperature control system. Heat stroke is a life-threatening emergency and demonstrates how vital the ability to control the internal environment is to a properly functioning organism.

In this unit of chapters, you will learn about how your body works. The emphasis will be on humans, but comparisons with other animals will illustrate some of the ways in which different organisms "solve" a common set of problems. How, for example, do creatures as diverse as hydras, halibut, and humans obtain oxygen? How does your manner of extracting nutrients from food compare with an earthworm's? Is frog reproduction much different from lizard reproduction? By learning about different evolutionary adaptations that solve the general challenges of life, you will gain a richer context for understanding your own body. This chapter sets the stage by introducing some unifying concepts of structure and function that apply across the entire animal kingdom. ■

Figure 21.1 Regulating body heat. British soccer star David Beckham demonstrates some of the ways the human body can regulate its internal temperature, including sweating and expanding surface blood vessels (causing his face to redden).

The Structural Organization of Animals

Life is characterized by a hierarchy of organization. In animals, individual cells are grouped into tissues, which are in turn grouped into organs, which participate in organ systems, which together make up the entire organism **(Figure 21.2)**. Parts of the body at each level of the hierarchy act together to perform the functions of life, such as regulating an animal's internal environment. For example, the regulation of body temperature within David Beckham's body in the chapter-opening essay requires cooperation among several levels of body organization. The brain, an organ, sends signals via the nervous system that trigger specific changes in tissues and individual cells. The result is a change within the organism as a whole.

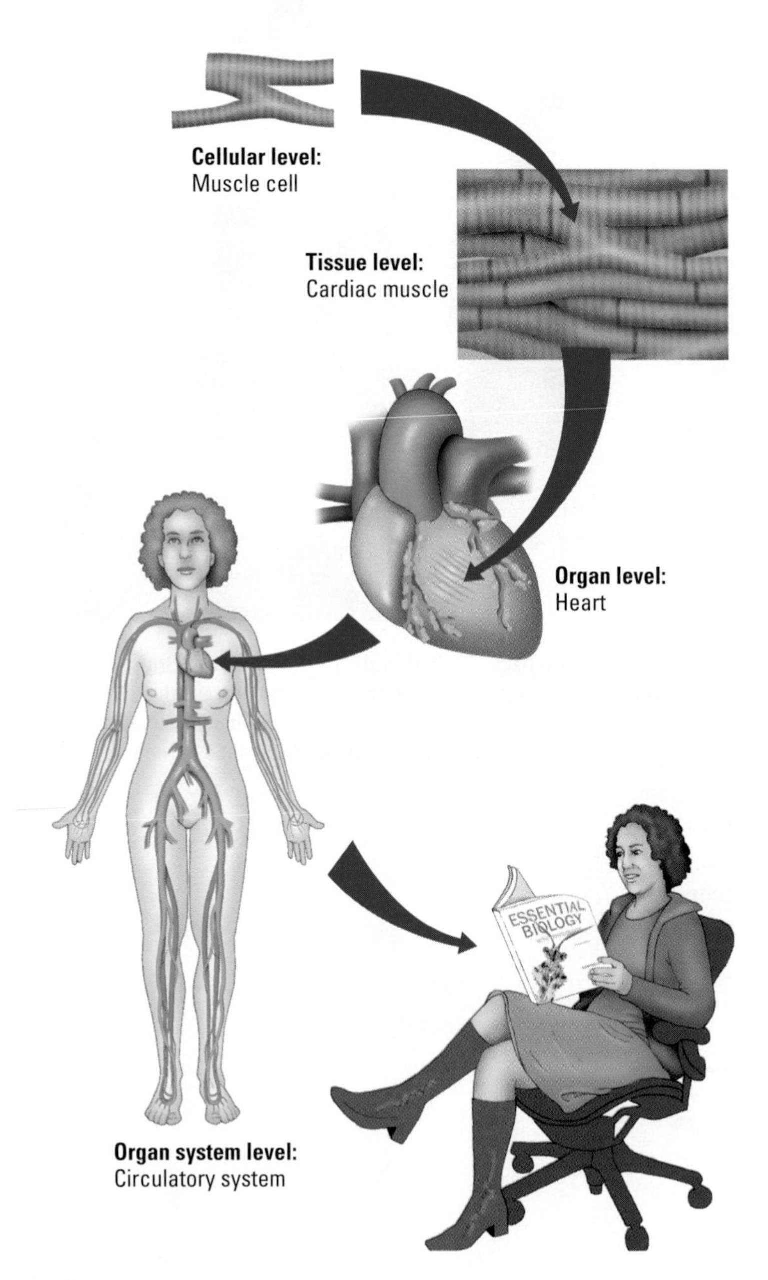

Figure 21.2 Structural hierarchy in a human.

Form Fits Function

Which is the better tool: a hammer or a screwdriver? The answer, of course, depends on what you're trying to do. Given a choice of tools, you would not use a hammer to loosen a screw or a screwdriver to pound a nail. How a device works is correlated with its structure: Form fits function. A chair, for example, must have a surface to sit on and a stable base to support that surface. These elements of design are dictated by the task that the chair must perform. The same principle applies to life at its many structural levels, from cells to organisms. Analyzing a biological structure gives us clues about what it does and how it works. Conversely, knowing the function of a structure provides insight about its construction. In exploring life at its many levels, we will discover functional elegance at every turn **(Figure 21.3)**.

When discussing form and function, biologists distinguish anatomy from physiology. **Anatomy** is the study of the structure of an organism and its parts. **Physiology** is the study of the function of an organism's structural equipment. For example, an anatomist might study the path of blood from the intestine to and through the liver, while a physiologist might study how the liver aids in digestion. Despite their different approaches, the two disciplines actually serve the same purpose: to better understand the connections between structure and function. For this reason, anatomy and physiology are often studied together.

The axiom "form fits function" will guide us throughout our study of animals. But this rule of "design" does not mean that such biological tools as a bird's wings, a fish's gills, or a mammal's teeth are products of purposeful invention. Here the analogy to household tools such as hammers and screwdrivers fails, because those tools were designed with a specific goal. It is natural selection that refines biological equipment. It does so by screening for the most effective variations among individuals of a population—those variations that are most advantageous in the local environment. Generations of selecting for what works best in a particular environment will fit form to function without any sort of goal-oriented plan. Thus, the form-function principle is just another face of biology's overarching theme: evolution.

(b) The structure-function principle also applies to organs within an animal. For example, the honeycombed construction of a bird's bones provides a lightweight skeleton that is very strong.

(a) A bird's form makes flight possible. The structure-function connection can apply to the shape of the whole organism, as you can see from this white tern in flight.

(c) At the cellular level, form again fits function. Nerve cells that control the muscles of a bird or other animal have long extensions that transmit signals.

Figure 21.3 Form fits function.

Tissues

As we learned in Chapter 4, the cell is the basic unit of all living organisms. In most multicellular animals, including humans, cells usually do not act alone, but instead are grouped into tissues. A **tissue** is an integrated group of similar cells (from dozens to billions) that perform a specific function. The cells composing a tissue are specialized; they have an overall structure that enables them to perform a specific task.

The term *tissue* is from a Latin word meaning "weave," and some tissues resemble woven cloth with their meshwork of nonliving fibers surrounding living cells. Other tissues are held together by a sticky substance that coats the cells or by special junctions that rivet the cells together. An animal has four main categories of tissue: epithelial tissue, connective tissue, muscle tissue, and nervous tissue.

Epithelial Tissue

Epithelial tissue, also called **epithelium,** covers the surface of the body and lines organs and cavities within the body **(Figure 21.4)**. The outer layer of skin and the linings of the heart, blood vessels, digestive tract, respiratory tract, and genitourinary tract are examples of epithelial tissue. Some epithelial tissue forms glands that secrete chemical substances, such as sweat glands in the skin and mammary glands in the breast. Epithelial tissue is made of sheets of tightly packed cells that are riveted together, forming a continuous layer. The architecture of an epithelium is an example of how form fits function at the tissue level. For example, the epidermis (the outer layer of skin) contains dense layers of epithelial cells bound tightly together, forming a protective barrier that surrounds the body and helps maintain its integrity.

The body continuously renews surface-lining epithelium, shedding old cells and growing new ones. Think of the turnover of cells necessary inside the mouth, where hot liquids and sharp-edged snacks frequently damage the epithelial layer. As a result of continuous shedding, the epidermis is entirely renewed about every two weeks. Such turnover requires rapid cell division, which increases the risk of cancer developing. Consequently, about 80% of all cancers arise in epithelial tissue. Cancers of the epithelium are called carcinomas and include skin, lung, and breast cancer.

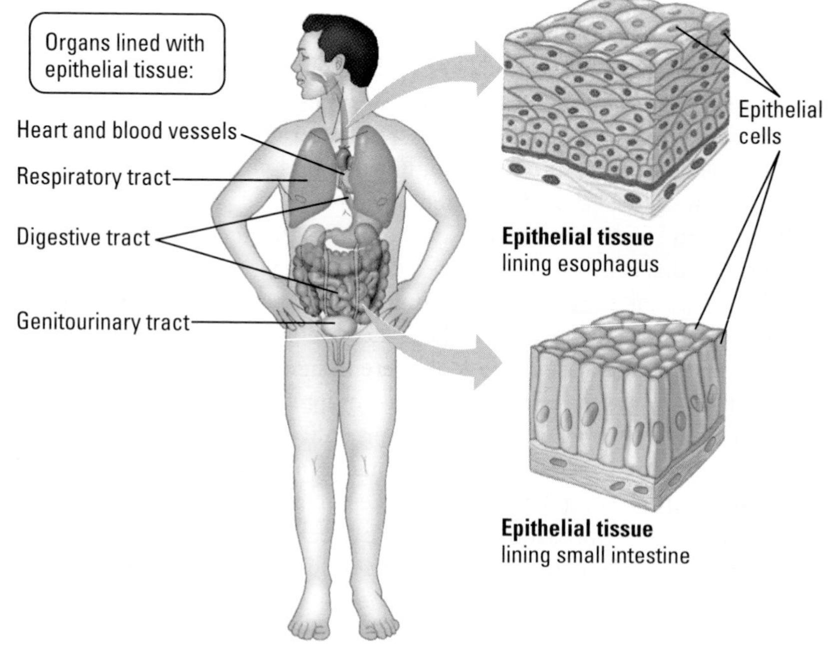

Figure 21.4 Epithelial tissue. Epithelial tissue (epithelium) is found in a variety of body structures, including the skin and the linings of most organs. The lining of the esophagus is made up of multiple layers of tightly packed epithelial cells. The lining of the intestine contains a single layer of epithelial cells.

Connective Tissue

In contrast to epithelium, with its tightly packed cells, **connective tissue** has a sparse population of cells scattered through an extracellular matrix (see Figure 4.7b). The matrix consists of a web of protein fibers embedded in a uniform foundation that may be liquid (as in blood), jellylike (as in adipose tissue), or solid (as in bone).

The structure of connective tissue is correlated with its function: to bind and support other tissues. In fact, connective tissue is so interwoven with your organs that if all your other tissues were to suddenly disappear, you would still be recognizable to your friends (though they might run away after they got a good look!). For example, the epithelial tissue of your skin (epidermis) lies on top of a layer of connective tissue called the dermis, which contains the blood vessels and extracellular fluids that nourish the epidermis.

Figure 21.5 illustrates the six major types of connective tissue. The most widespread connective tissue in the body of vertebrates (animals with backbones, including humans) is **loose connective tissue** (**Figure 21.5a**). It binds epithelia to underlying tissues and holds organs in place. The matrix of this connective tissue is a loose weave of protein fibers, including collagen fibers, which give connective tissue great strength. For example, if your aunt pinches your cheek, it is collagen fibers that keep your flesh from tearing away from the bone. The connective tissue also has elastic fibers, which enable your pinched skin to return quickly to its original shape.

Adipose tissue stores fat in closely packed cells in a sparse matrix of fibers (**Figure 21.5b**). The fat stockpiles energy and pads and insulates the body. Each adipose cell contains a large globule of fat (mostly triglycerides); the cell swells when fat is stored and shrinks when fat is used for energy.

Blood is a connective tissue with a matrix that is liquid (**Figure 21.5c**). Red and white blood cells are suspended in this liquid matrix, which is called plasma. Blood functions mainly in transporting substances from one part of the body to another and also plays a major role in immunity.

Figure 21.5 Types of connective tissue.

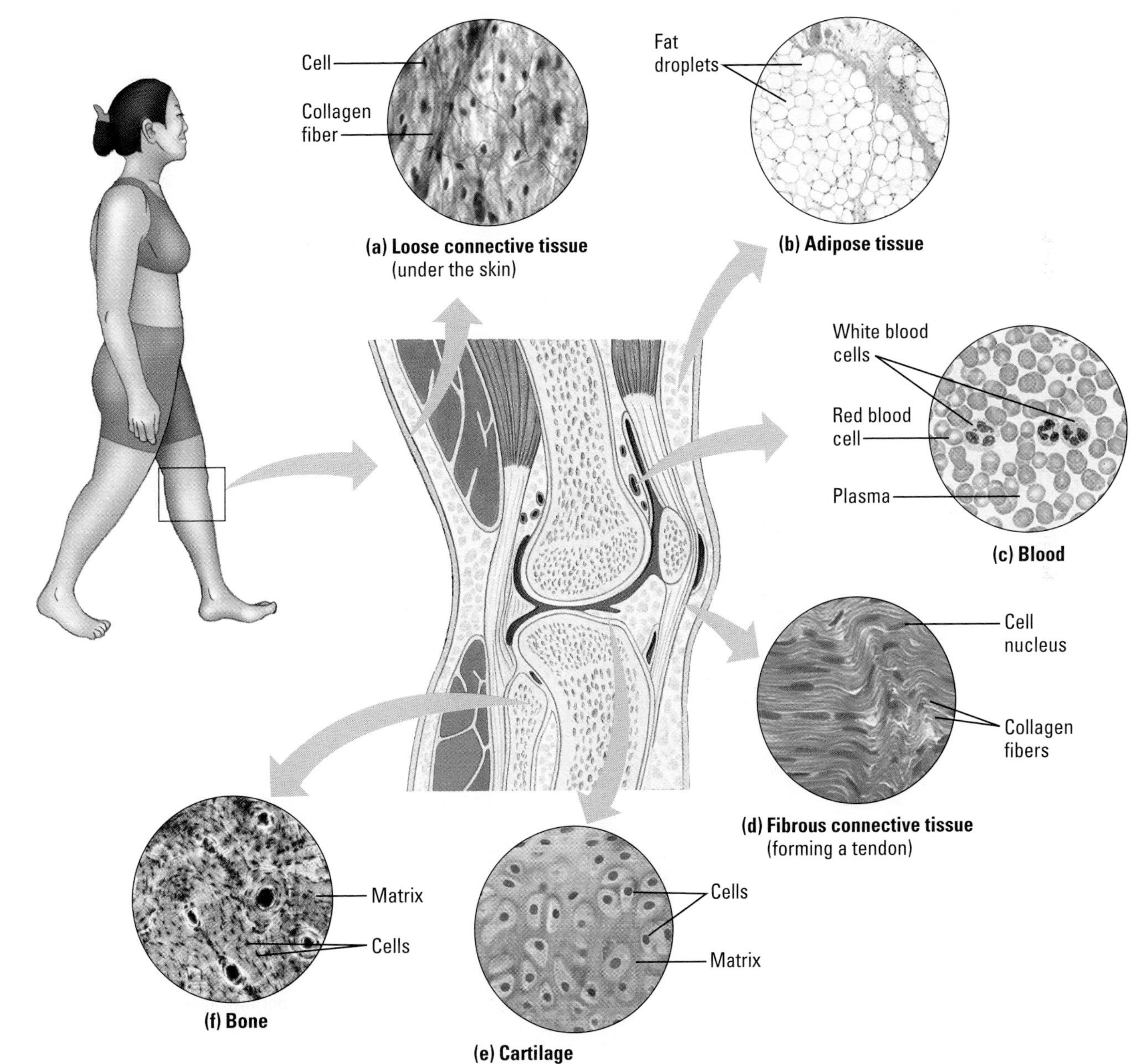

The other three types of connective tissue have dense matrices. **Fibrous connective tissue** has a dense matrix of collagen. It forms tendons, which attach muscles to bones, and ligaments, which join bones together at joints **(Figure 21.5d)**. The matrix of **cartilage** is strong but flexible **(Figure 21.5e)**. Flip your outer ear, and you'll get the idea of how cartilage functions as a flexible, boneless skeleton. Cartilage is often found at the end of bones and forms the shock-absorbing pads that cushion the bony vertebrae of the spinal column. Cartilage has no blood vessels, so injuries to it heal very slowly. **Bone** is a rigid connective tissue with a matrix of collagen fibers hardened with deposits of calcium salts. This combination makes bone hard without being brittle **(Figure 21.5f)**.

Muscle Tissue Muscle is the most abundant tissue in most animals. In fact, what we call "meat" in a grocery store is mostly the muscles of cattle, chicken, fish, and other animals. **Muscle tissue** consists of bundles of long, thin, cylindrical cells called muscle fibers. Each cell has specialized proteins arranged into a structure that contracts when the cell is stimulated by a signal from a nerve. Humans and other vertebrates have three types of this contractile tissue, each with a slightly different structure: skeletal muscle, cardiac (heart) muscle, and smooth muscle.

Skeletal muscle is attached to bones by tendons and is responsible for our voluntary movements, such as walking and talking **(Figure 21.6a)**. This type of muscle is said to be striated (striped) because the contractile apparatus forms a banded pattern in each muscle fiber, or cell. Adults have a fixed number of these cells. Weight lifting and other bodybuilding methods do not increase the number of muscle cells but rather enlarge those already present.

Cardiac muscle, also striated, is found only in heart tissue **(Figure 21.6b)**. Its contraction produces the heartbeat. Cardiac muscle cells are branched

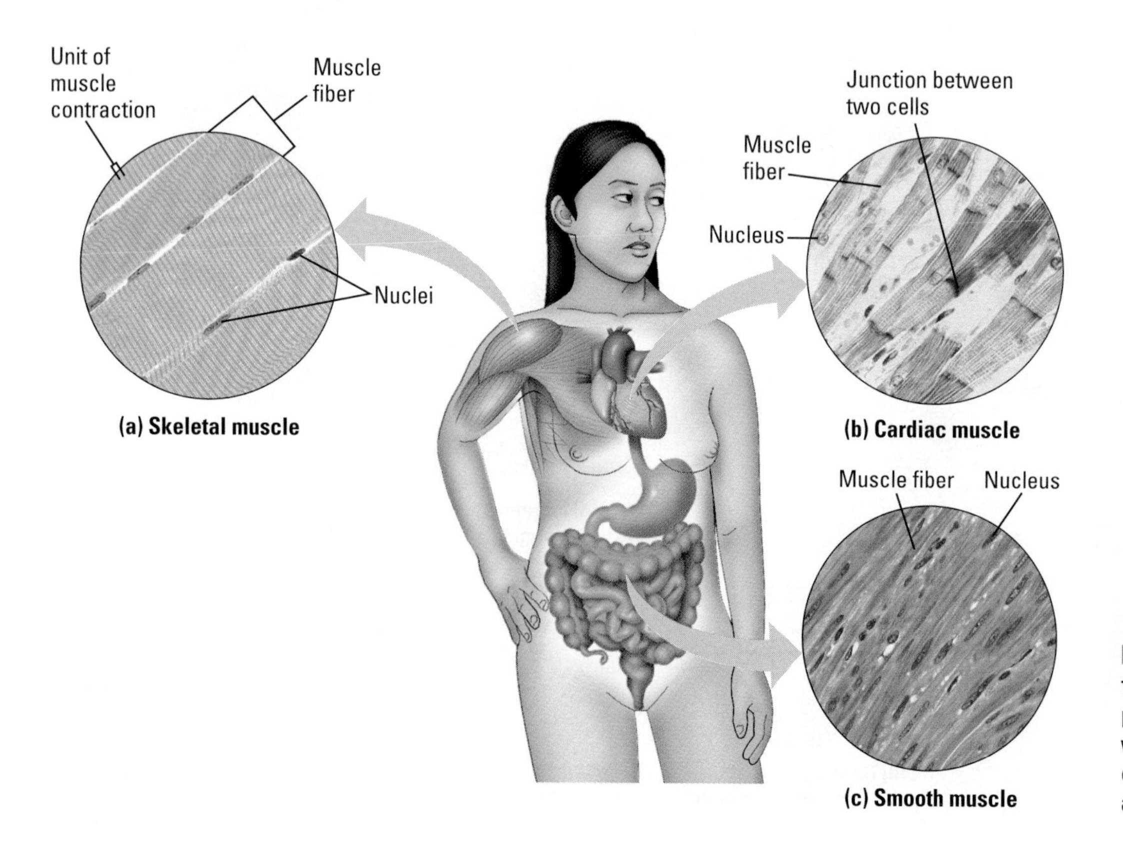

Figure 21.6 Three types of muscle tissue. Skeletal muscle, which helps us perform all intentional movements, such as walking and talking, is the only type that we can consciously control; the other two types act involuntarily.

and joined to one another, appearing like one large interconnected mass of muscle. This allows the contraction signal to be quickly propagated to all muscle cells at once, producing a coordinated beat. Cardiac muscle is an involuntary muscle, meaning that your heart contracts without any conscious control on your part.

In contrast to skeletal and cardiac muscle, **smooth muscle** is named for its lack of obvious striations **(Figure 21.6c)**. Found in the walls of such organs as the intestines and blood vessels, smooth muscle is involuntary. For example, we have no control over the contractions of our intestinal muscles that move food along the digestive tract. Smooth muscle contracts more slowly than skeletal muscle, but it can remain contracted for a long time.

Nervous Tissue Most animals are active creatures that respond rapidly to stimuli from their environment. For example, if you touch a very hot object, your arm quickly jerks your hand away. This requires sensory input, processing, and motor output—information that needs to be relayed from one part of the body to another. It is **nervous tissue** that makes such communication possible. Nervous tissue is found in the brain and spinal cord, as well as in the nerves that connect them to all other parts of the body.

The basic unit of nervous tissue is the **neuron,** or nerve cell (see Figure 21.3c). With their long extensions, neurons can transmit electrical signals called nerve impulses very rapidly over long distances. For example, neurons in the sciatic nerve of your leg may extend to 1 m (3.3 feet) long, running all the way from the base of your spinal cord to the tips of your toes. You are wired for action—and for thinking, since it is the network of neurons in your brain that functions as your mind.

Organs and Organ Systems

The next level in the structural hierarchy is the **organ.** An organ consists of two or more tissues packaged into one working unit that performs a specific function. At each level of the hierarchy, functions emerge from the collective interaction of the structures at lower levels. An organ, for example, performs functions that none of its component tissues can carry out alone.

Your heart, brain, liver, stomach, and lungs are all examples of organs. The heart is mostly muscle, but it also has epithelial, connective, and nervous tissues. Epithelial tissue lining the heart chambers prevents leakage and provides a smooth surface over which blood can flow. Connective tissue makes the heart elastic and strengthens its walls and valves. Neurons regulate the rhythmic contractions of cardiac muscles. To see another example, examine the layered arrangement of tissues in the wall of the small intestine in **Figure 21.7**.

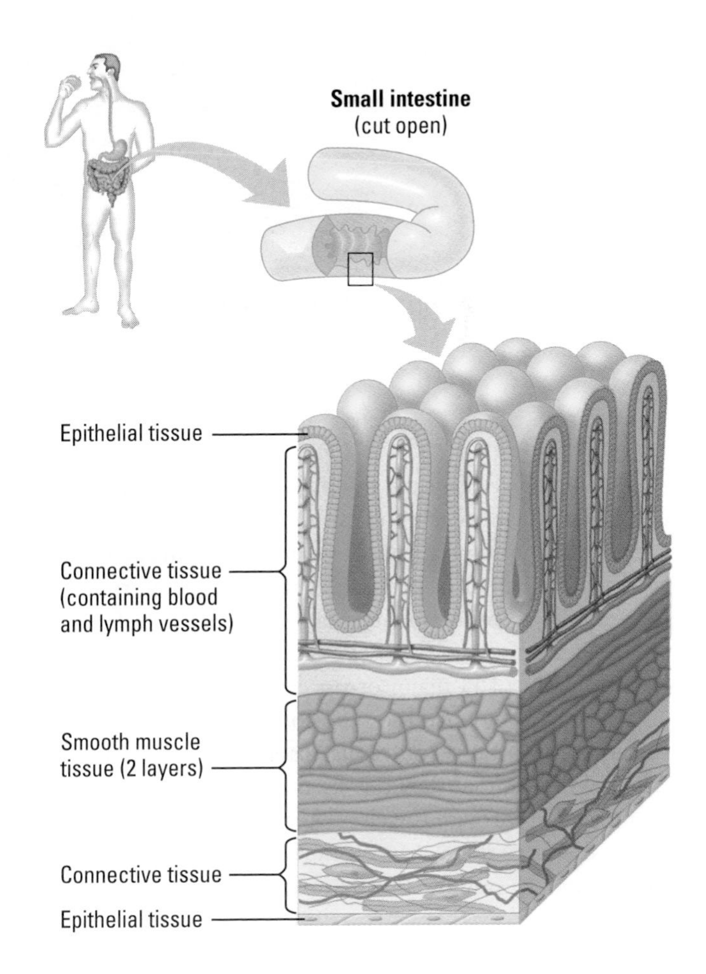

Figure 21.7 Tissue layers of the small intestine, an organ. The small intestine, an organ of the digestive system, consists mainly of three types of tissue arranged in layers. The inside wall of the small intestine is lined with epithelial tissue that secretes mucus and digestive enzymes. (Notice that the epithelium bends to form fingerlike projections, increasing its surface area.) Underneath this layer are two layers of connective tissue—one of which contains blood capillaries and lymph vessels—separated by two layers of smooth muscle. Beneath that, a final layer of epithelial tissue forms the outside surface.

Figure 21.8 Organ systems of a vertebrate.

The organs of humans and most other animals are organized into **organ systems,** teams of organs that work together to perform vital body functions. The components of an organ system can be either physically connected together or dispersed within the body. Your circulatory system, for example, transports materials throughout your body. Its main organs are the heart and blood vessels (arteries, veins, and capillaries). **Figure 21.8** is an overview of 11 major organ systems in vertebrates.

An organism depends on the coordination of all its organ systems for survival. For instance, nutrients absorbed from the digestive tract are distributed throughout the body by the circulatory system. But the heart that pumps blood through the circulatory system requires nutrients absorbed by the digestive tract and oxygen (O_2) obtained from the air or water by the respiratory system. The failure of any organ system jeopardizes the

Figure 21.8 Organ systems of a vertebrate, *continued*

entire animal because the systems are so interwoven and dependent on each other. Your body is a whole, living unit greater than the sum of its parts. In the next section, we'll look at how the body's organ systems interact with the external environment.

CHECKPOINT

1. What is the difference between anatomy and physiology?
2. Explain how a tennis racket illustrates the "form fits function" principle.
3. Name the four types of tissue found in most organs.

Answers: 1. Anatomy is the study of structures, while physiology is the study of functions. **2.** The function of a tennis racket is to hit a ball. The elements of its form that help with this include a handle for gripping, a large head to strike the ball, and strings that stretch and recoil to provide bounce. **3.** Epithelial, connective, muscle, and nervous

Exchanges with the External Environment

Most animals are covered with a protective layer that separates the external and internal chemical environments. This does not mean, however, that an animal barricades itself against the harsh world outside. On the contrary, every organism is an **open system.** This means that organisms continuously exchange chemicals and energy with their surroundings. In fact, they must do so to survive. You eat, breathe, defecate, urinate, sweat, and radiate heat—all examples of how you operate as an open system. And the exchange of materials extends down to the cellular level. Nutrients and oxygen must enter every living cell, and carbon dioxide and other wastes must exit.

An animal's size and shape affect how it exchanges energy and materials with its surrounding environment. In an animal body, every living cell must be bathed in a watery solution, partly because substances must be dissolved in water to cross cell membranes. An amoeba (a single-celled protozoan) lives in water and consists of only one cell **(Figure 21.9a)**. Thus, no part of its body is very far away from the outside plasma membrane, where exchange with its watery environment can occur.

Exchange with the environment is more complicated for multicellular animals. Each microscopic cell in a multicellular organism has a plasma membrane where exchange of materials can occur. But this exchange only works if all the cells of the animal have access to a suitable watery environment. A hydra (a simple pond-dwelling relative of jellies) has a body wall only two cell layers thick **(Figure 21.9b)**. Both outer and inner layers of cells are bathed in pond water, which enters the digestive sac through the mouth. Every cell of the hydra is thus able to exchange materials through direct contact with the watery environment. A flat body shape is another way of maximizing exposure to the environment. For instance, a tapeworm may be several meters long, but because it is very thin, most of its cells are bathed in the body fluid of the worm's host, from which it obtains nutrients.

As discussed in Chapter 17, simple body forms, such as two-layered sacs and flat shapes, do not allow for much complexity in internal organization. But animals with more complex body forms face the same basic problem: Every living cell in their body must be bathed in fluid and have access to oxygen, nutrients, and other resources from the outside environment. Complex animals have extensively folded or branched internal surfaces that maximize surface area for exchange with the immediate environment. For example, your lungs, which exchange oxygen and carbon dioxide with the air you breathe, are not shaped like big balloons, but like millions of tiny balloons at the tips of finely branched air tubes **(Figure 21.10)**. The epithelium of the lungs has a very large total surface area—about the size of a tennis court.

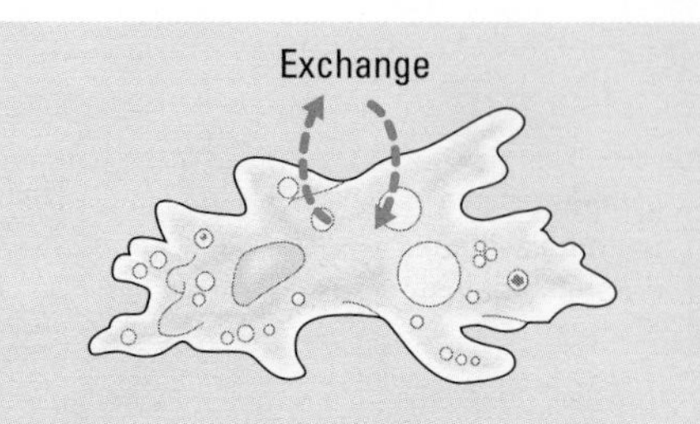

(a) Single cell. The entire surface area of a single-celled organism, such as this amoeba, contacts the environment. Because of its small size, the organism has a large surface area (relative to its volume) through which it exchanges materials with the external world. (The exchange is indicated by the red dashed arrows here.)

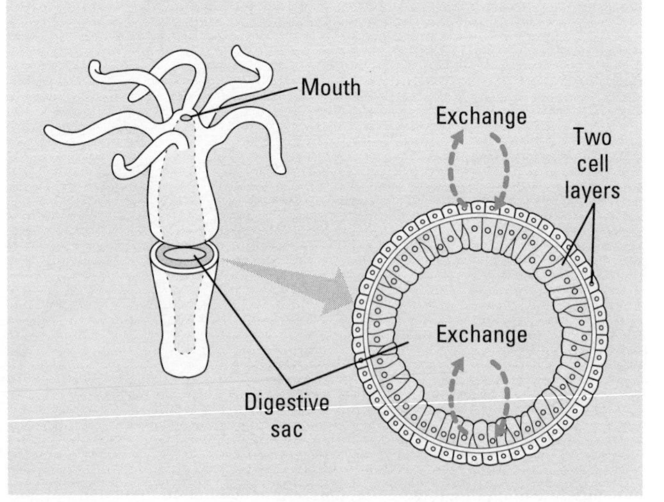

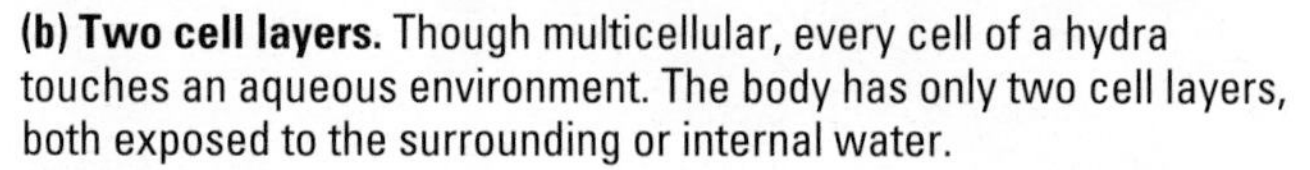

(b) Two cell layers. Though multicellular, every cell of a hydra touches an aqueous environment. The body has only two cell layers, both exposed to the surrounding or internal water.

Figure 21.9 Contact of simple organisms with the environment.

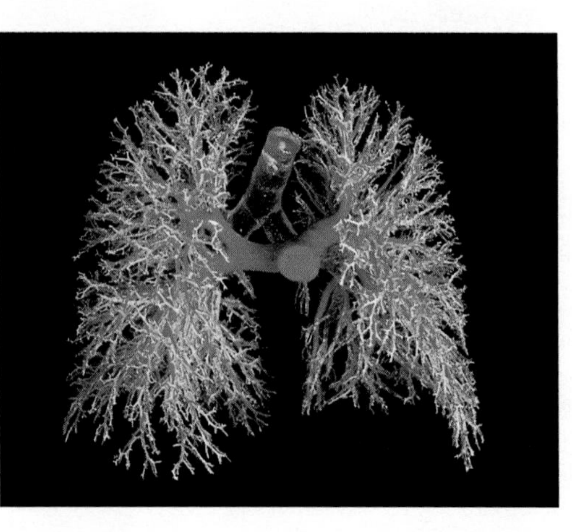

Figure 21.10 The branched surface area of the human lung. Animals with complex body shapes need internal structures that provide a lot of surface area for exchange with the external environment. The human lungs, for example, contain extensively branched structures to take in oxygen and release carbon dioxide. This model shows the tiny air tubes of the lungs (white) and the fine blood vessels (red) that transport gases between the heart and lungs.

Figure 21.11 shows three organ systems—digestive, respiratory, and urinary—that exchange materials with the external environment. Inside the body, materials are transported by the circulatory system from exchange surfaces to nearly every cell in the body, and wastes are transported away from cells to be disposed of at exchange surfaces.

CHECKPOINT

Why doesn't a tapeworm require a circulatory system?

Answer: Because most cells of a tapeworm are in direct contact with the environment

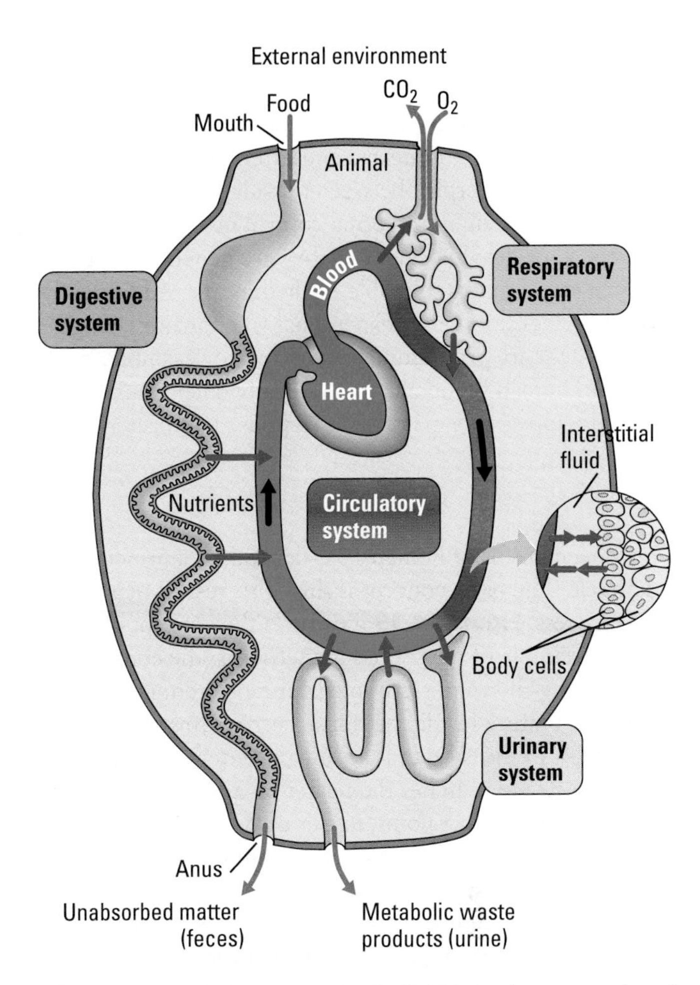

Figure 21.11 Indirect exchange between the external and internal environments in complex animals. This simplified model of an animal body shows three organ systems responsible for exchange with the external environment, connected by the circulatory system. The blue arrows indicate the exchange of materials between the circulatory system and the other three systems and between the circulatory system and body cells.

Regulating the Internal Environment

Every living organism has the ability to respond to its environment. Among the responses are ones that prevent changes in the outside world from adversely affecting the internal makeup of an organism. In this section, you'll learn how animals adjust to changing environmental conditions.

Homeostasis

One of the body's most important functions is to maintain its integrity even when the world around it changes. The internal environment of vertebrates is the **interstitial fluid** that fills the spaces between cells and exchanges nutrients and wastes with microscopic blood vessels. It is important that the composition of the interstitial fluid remain relatively constant no matter what occurs in the outside world.

Homeostasis, which literally means "steady state," is the body's tendency to maintain relatively constant conditions in the internal environment even when the external environment changes. Large changes in the environment outside the body normally cause only small changes inside the body **(Figure 21.12)**. For example, your body regulates the amount of sugar in your blood so that it does not fluctuate for long from a concentration of 0.1%, regardless of what you ate for lunch. And the water content

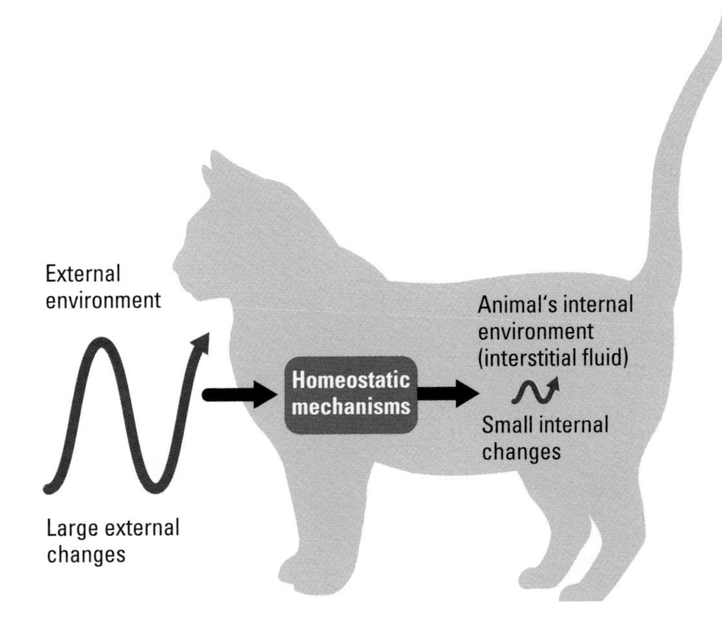

Figure 21.12 Homeostasis. An animal has mechanisms of homeostasis that regulate the fluid environment bathing its cells. In the face of large changes in the external environment, the homeostatic systems maintain internal conditions within a range that the animal's metabolism can tolerate.

of your cells stays about the same no matter what you drink or how dry the climate you live in.

Actually, the internal environment of an animal always fluctuates slightly in response to internal and external changes. Homeostasis is a dynamic state, an interplay between outside forces that tend to change the internal environment and internal control mechanisms that oppose such changes. Changes do occur, but they are generally moderated to stay within a range that is tolerable for living cells. Additionally, there are times during the development of an animal when major changes in the internal environment are programmed to occur. For example, the balance of hormones in human blood changes radically during puberty or pregnancy, and internal temperature may rise in response to a bacterial infection.

Negative and Positive Feedback

Most mechanisms of homeostasis depend on a principle called **negative feedback.** The basic concept is this: The results of a process inhibit that same process. **Figure 21.13** illustrates the concept of negative feedback with an example of household homeostasis, the control of room temperature. During the winter, a home heating system maintains a relatively constant temperature inside the house despite drastic changes outside. A key component of this system is a control center—a thermostat—that monitors temperature and switches the heater on and off. The thermostat needle oscillates around the set point, in this case the setting of the thermostat at 20°C (about 68°F). Whenever room temperature drops below the set point, the thermostat switches the heater on (top pathway in Figure 21.13). When the temperature rises above the set point, the thermostat switches the heater off (bottom pathway). This illustrates the basic principle of negative feedback: The result (increased room temperature, for example) of a process (heating of the air) inhibits that very process (by switching the heater off). Negative feedback is the most common mechanism of homeostatic control in animals.

Less commonly, organisms use **positive feedback,** where the results of a process intensify that same process. For example, during the labor that leads up to childbirth, hormones cause the muscles of the uterus to contract. This contraction stimulates the release of more hormone molecules, which cause more contractions, and so on, in a positive-feedback loop of increasing intensity. The result is climactic muscle contractions that push the baby from the womb.

In the rest of this section, we'll examine several specific examples of homeostasis. We'll first discuss the control of temperature and the control of water gain and loss in a variety of animals. We'll then focus on the urinary system—an organ system that performs several homeostatic roles.

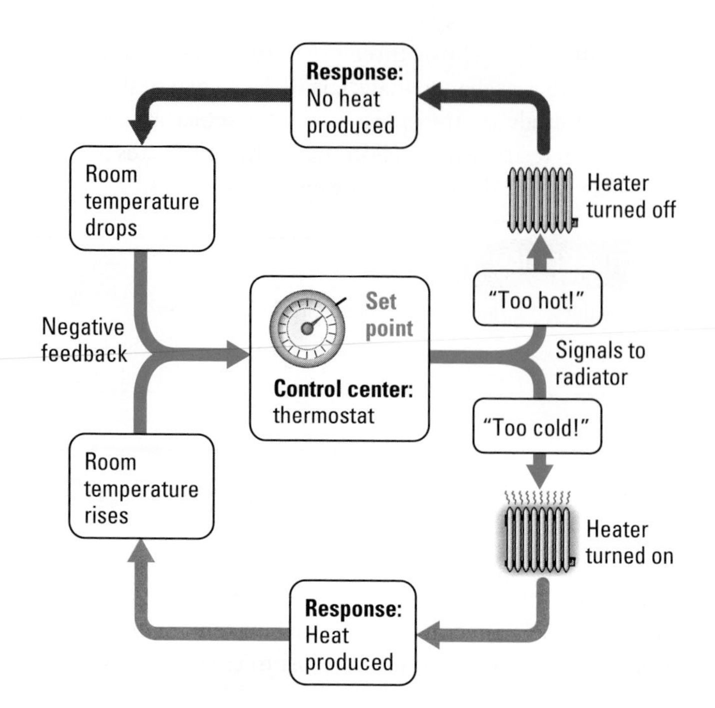

Figure 21.13 An example of negative feedback: control of temperature. Regulating the temperature in a room depends on a thermostat that acts as a control center. The thermostat detects temperature changes and activates a mechanism that reverses that change. This is an example of negative feedback because the result of the process (heat) shuts that process down.

Thermoregulation

As you saw in the example of David Beckham at the start of the chapter, **thermoregulation,** the maintenance of internal body temperature, is an important homeostatic mechanism in animals. As you walk from one warm building to the next on a cold winter day, your body's temperature will barely fluctuate, despite the drastic change in temperature in the environment around you. The ability to maintain a body temperature substantially warmer than the surrounding environment is characteristic of **endotherms,** animals that derive the majority of their body heat from

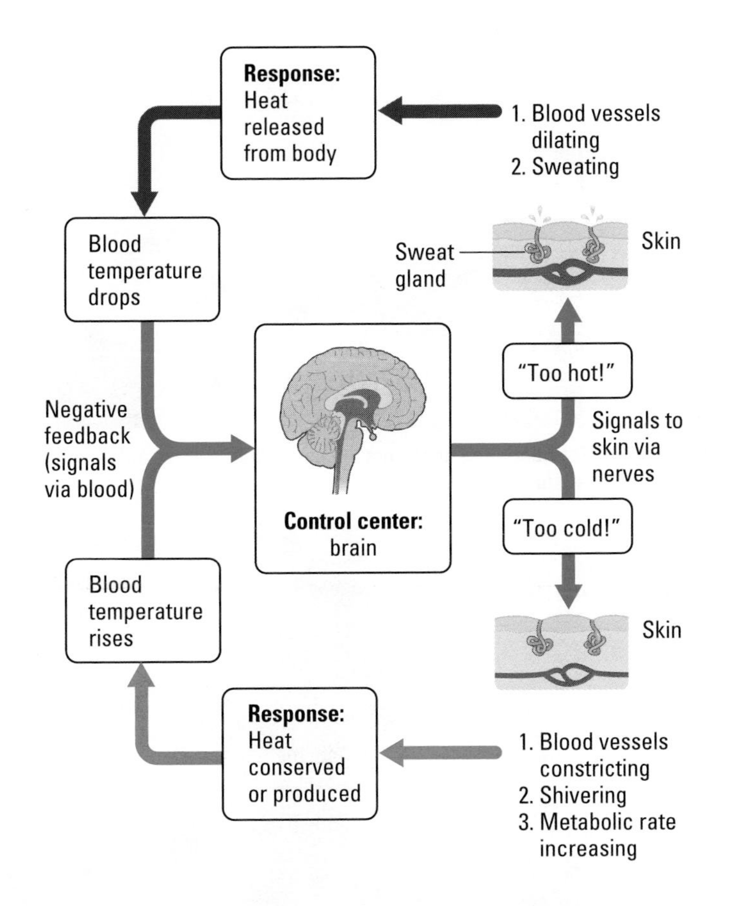

Figure 21.14 Thermoregulation in the human body. By comparing this figure with Figure 21.13, you can see how similar negative-feedback controls moderate the temperature of a room and your body. The control center for body temperature—our thermostatic control center—is located in the brain.

their metabolism. Mammals and birds are important groups of endotherms. In contrast, **ectotherms,** which include most invertebrates, fishes, amphibians, and nonbird reptiles, obtain body heat primarily by absorbing it from their surroundings.

Humans and other animals have a number of structures and mechanisms that aid in thermoregulation. As shown in **Figure 21.14**, our brain has a control center that maintains our body temperature near 37°C (98.6°F). When body temperature falls below normal (in frigid weather, for example), the brain's control center sends signals that trigger changes that will bring it back to normal: Blood vessels near the body's surface constrict and muscles contract, causing us to shiver. When body temperature gets too high, the control center sends signals to dilate the blood vessels near the skin and activate sweat glands, allowing excess heat to escape. Like a home controlled by a thermostat, the temperature of the body does not stay completely constant, but fluctuates up and down within an acceptable range.

Fever, an abnormally high internal temperature, is a body-wide response that usually indicates an ongoing fight against infection. When immune system cells encounter invading microorganisms, they release chemicals called pyrogens ("fire makers"). Pyrogens travel through the bloodstream to the brain's thermostatic control center. There, the pyrogens stimulate the control center to raise the body's internal temperature, producing a fever. Many people mistakenly believe that the invading microbes themselves cause a fever. In fact, the actual cause is usually the body's fight *against* the microbes. A fever over 40°C (104°F) may be life-threatening because a temperature that high can damage important body proteins. A moderate fever of 38–39°C (100–102°F), however, discourages bacterial growth and speeds the body's internal defenses. A moderate fever thus helps protect the body's internal environment against potentially harmful invaders from the external environment.

In addition to the brain's temperature control center, a number of other mechanisms help animals cope with heat gained from or lost to the environment (**Figure 21.15**). Heat loss in mammals is often partly regulated by hair: A thin coat of fur in the summer becomes a thick, insulating coat during the winter. Insulating fat helps protect aquatic mammals, such as seals, and aquatic birds, such as penguins. A wide variety of behavioral adaptations assist thermoregulation, such as migrating to suitable climates, huddling in the cold, and even dressing for the weather in the case of humans. When a lizard basks on a sunlit rock, for example, it is absorbing heat that warms its body. Other behaviors help some animals deal with seasonal changes in the weather. Ground squirrels and chipmunks, for example, hibernate in burrows during the winter, drastically reducing their body temperature and metabolic rate and thus conserving energy. When an animal's body gets too warm, heat can be lost by evaporation. Humans sweat, while other animals (such as lizards and dogs) lose heat as moisture evaporates from their nostrils and mouth.

(a) Absorbing heat from the sun. When a lizard basks in the sun, it absorbs heat.

(b) Releasing heat through evaporation. Humans can release heat by the evaporation of water—sweat, for example.

Figure 21.15 Methods of thermoregulation in animals.

Osmoregulation

Living cells depend on a precise balance of water and solutes. Whether an animal inhabits land, fresh water, or salt water, its cells cannot survive a *net* gain or loss of water. That is, a cell may exchange water with the environment as long as the total amount leaving and entering is the same. If too much water enters, animal cells will burst; if too much leaves, they will shrivel and die.

Osmoregulation is the control of the gain or loss of water and dissolved solutes. Osmoregulation is based largely on regulating solutes because water follows the movement of solutes by osmosis. Osmosis occurs whenever two solutions separated by a membrane differ in total solute concentration (see Chapter 5). There is always a net movement of water from the solution with lower solute concentration to the one with higher solute concentration.

A variety of mechanisms for maintaining water balance in aquatic animals have evolved. Most marine invertebrates—including lobsters, scallops, and jellies (sometimes called jellyfish)—are **osmoconformers,** meaning that their internal and external environments have similar water concentrations. Because their internal water concentration matches the external concentration, osmoconformers do not undergo a net gain or loss of water. In contrast, all freshwater animals and most marine vertebrates are **osmoregulators**: They must actively regulate water loss and gain.

As an example, let's compare osmoregulation in freshwater and saltwater fish. In fresh water, the external solute concentration is low. Thus, water enters the body by osmosis. To compensate, freshwater fish take up solutes (such as ions) via the digestive system and gills and expel excess water by producing large amounts of dilute urine. In osmoregulators that live in salt water, osmosis draws water out of the body. In response, saltwater fish take in water by drinking, pump out ions via the gills, and produce only small amounts of concentrated urine.

All land animals are also osmoregulators. Most land animals gain water from eating and drinking and lose it through urinating, defecating, breathing, and perspiring. The challenge facing land-dwelling osmoregulators is to avoid becoming dehydrated. In insects, this need is addressed with waxy exoskeletons, while terrestrial vertebrates (including humans) are covered with water-resistant layers of dead skin cells. As we'll see in the next section, our kidneys also play a major role in regulating water balance.

Homeostasis in Action: The Urinary System

Survival in any environment requires a precise balance between waste disposal and the animal's needs for water and solutes. The urinary system plays a central role in several kinds of homeostasis, forming and excreting waste-carrying urine while regulating the amount of water and solutes in body fluids.

In the human urinary system, the main processing centers are the two kidneys. Each is a compact organ, a bit smaller than a fist, located on either side of the abdomen. The kidneys contain many fine tubes called **tubules** and an intricate network of capillaries (tiny blood vessels). Every day, the total volume of blood in the body passes through the kidneys hundreds of times. As the blood circulates, a fraction of the plasma (the liquid portion of the blood) enters the kidney tubules. Once there, it is called **filtrate.** The filtrate contains valuable substances that need to be reclaimed (such as water, glucose, salts, and amino acids) and other substances that need to be disposed of. Chief among the waste products to be excreted is urea, a nitrogen-containing compound produced from the breakdown of proteins and nucleic acids. Humans must dispose of urea, but we cannot simply excrete all of the filtrate as urine; if we did, we would lose vital nutrients and dehydrate rapidly. Instead, our kidneys refine the filtrate, concentrating the wastes and returning most of the water and useful solutes to the blood.

The anatomy of the human urinary system is shown in **Figure 21.16**. Starting with the whole system in part (a), blood to be filtered enters each kidney via a blood vessel called the renal artery, shown in red; blood that

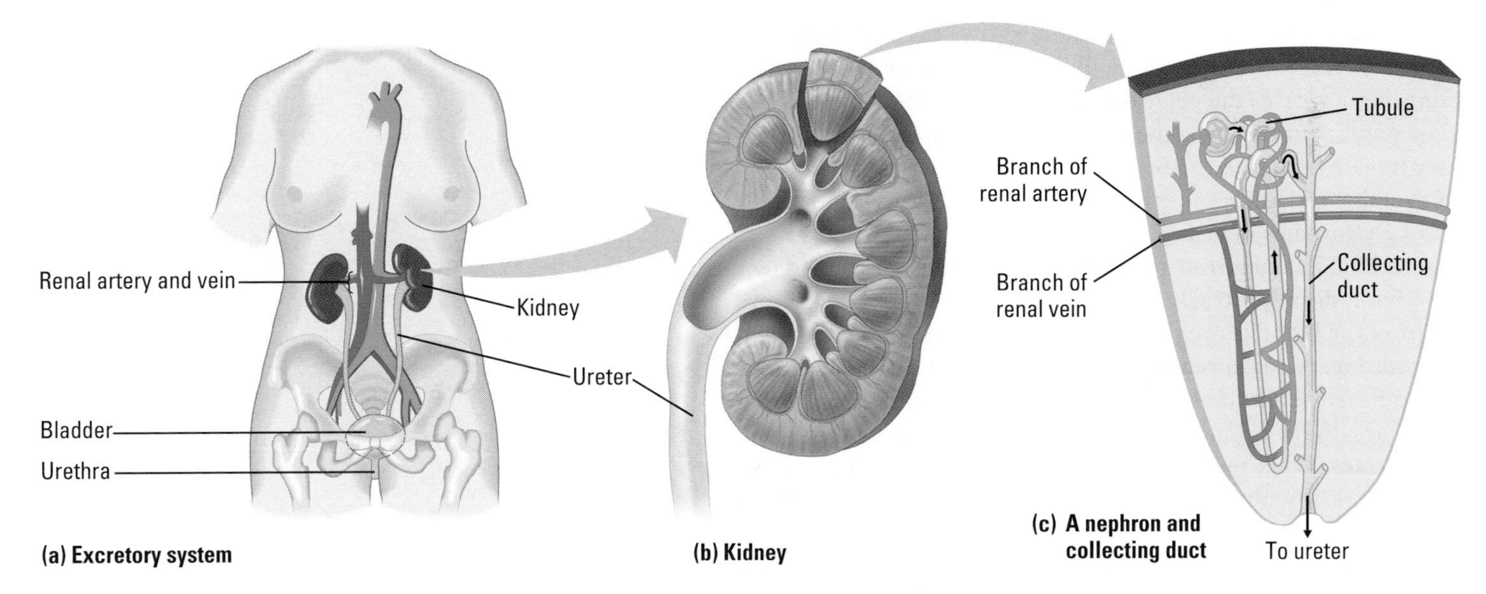

Figure 21.16 Anatomy of the human urinary system.
(a) The human urinary system filters 1,000–2,000 liters (L) of blood each day. **(b)** The main processing centers of the urinary system are the two kidneys, located on each side of the body. Renal arteries carry blood into the kidneys, and renal veins take blood away. Urine is collected in the ureter for transport to the urinary bladder. **(c)** Each kidney contains about a million nephrons. It is here that the functions of the urinary system are carried out.

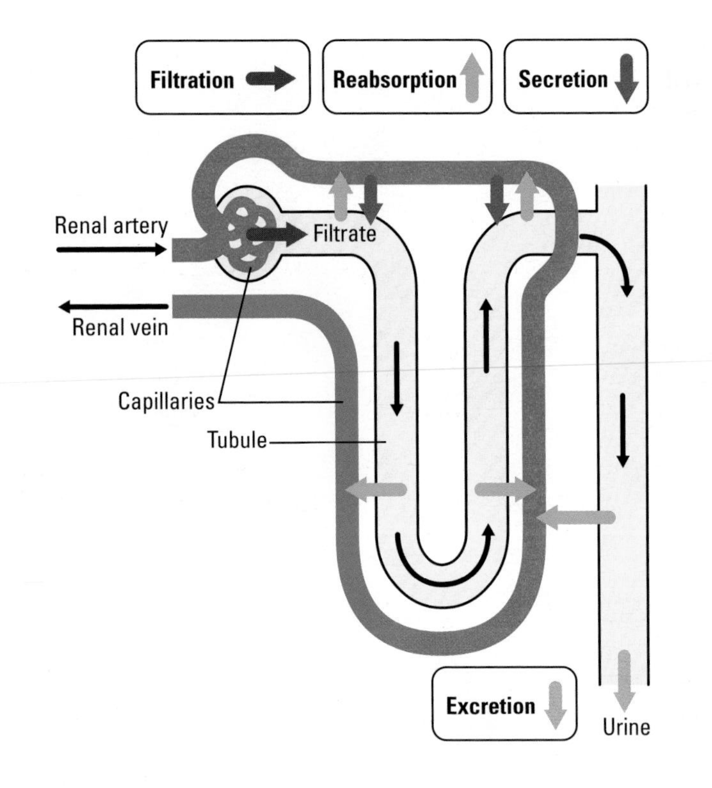

Figure 21.17 Major functions of the urinary system. The urinary system rids the body of wastes (such as urea) while keeping water and needed molecules. This is achieved through four different mechanisms: filtration, reabsorption, secretion, and excretion.

has been filtered leaves the kidney in the renal vein, shown in blue. The renal artery branches into millions of fine blood vessels in the kidney, as shown in a cutaway view in part (b). Part (c) shows one branch of the renal artery supplying blood via a network of capillaries to a **nephron,** the functional unit of the kidneys. A nephron consists of a tubule and its associated blood vessels. Blood pressure forces water and solutes from the blood into the nephron tubule, creating filtrate. As the filtrate passes through the tubule, water and needed nutrients are reabsorbed into the bloodstream and wastes are secreted into the filtrate. The filtrate becomes more and more concentrated, resulting in a small quantity of **urine.** At the end of the tubule, urine leaves the nephron via a collecting duct. Urine collects in the kidney and then leaves via the **ureter.** Urine is stored in the **urinary bladder** (visible in Figure 21.16a). Periodically, urine is expelled from the bladder via the **urethra,** a tube that empties near the vagina in females and through the penis in males.

Figure 21.17 summarizes the excretory functions performed by the kidneys. First, during **filtration,** water and other small molecules are forced from the blood through a capillary wall into a kidney tubule, forming filtrate. Next, in **reabsorption,** water and valuable solutes are reclaimed from the filtrate and returned to the blood. In **secretion,** certain specific substances, such as ions and some drugs, are transported into the filtrate. What remains after filtration, reabsorption, and secretion is the urine. Finally, in **excretion,** urine passes from the kidneys to the outside.

The body is able to control its internal concentration of water and dissolved molecules through hormonal control of the nephrons in the kidney. When the solute concentration of body fluids rises too high (indicating that not enough water is present), the brain increases levels of a hormone called ADH (antidiuretic hormone) in the blood. This hormone signals the nephrons to reabsorb more water from the filtrate, effectively increasing the body's water content. Conversely, when body fluids become too dilute, blood levels of ADH drop, water reabsorption in the kidneys decreases, and the excreted urine becomes much more dilute. This is why your urine is very clear when you have been drinking a lot of water. Diuretics, such as alcohol, are substances that inhibit the release of ADH and therefore cause excessive urinary water loss. Drinking alcohol will make you urinate more frequently, and the resulting dehydration contributes to the symptoms of a hangover.

Kidney failure, the inability of the kidneys to filter blood, can be caused by injury, illness (such as high blood pressure, diabetes, or an infection), or prolonged use of pain relievers, alcohol, or other drugs. Kidney failure can leave the body unable to rid itself of accumulated wastes. Patients with one functioning kidney can lead a normal life, but dual kidney failure is lethal if untreated. One option is to place the person on **dialysis,** filtration of the blood by a machine. A dialysis machine mimics the action of a nephron **(Figure 21.18)**. Blood from an artery enters a series of selectively permeable tubes. Urea and excess fluids pass from the blood into a dialyzing solution, while needed substances transfer the opposite way. The machine discards the used dialyzing solution as wastes accumulate. Dialysis treatment is life sustaining for people with kidney failure, but it is costly, time-consuming (about 4–6 hours three times a week), and must be continued for life. Alternatively, a kidney from a living compatible donor (usually a relative) or a deceased organ donor can be transplanted into a person with kidney failure. Unfortunately, the number of people who need kidneys is much greater than the number of kidneys available.

Throughout our study of animal form and function in this unit, you will see many other examples of how mechanisms of homeostasis moderate changes in the internal environment despite larger fluctuations in the external environment. You will also see many other examples of what happens when homeostatic controls fail.

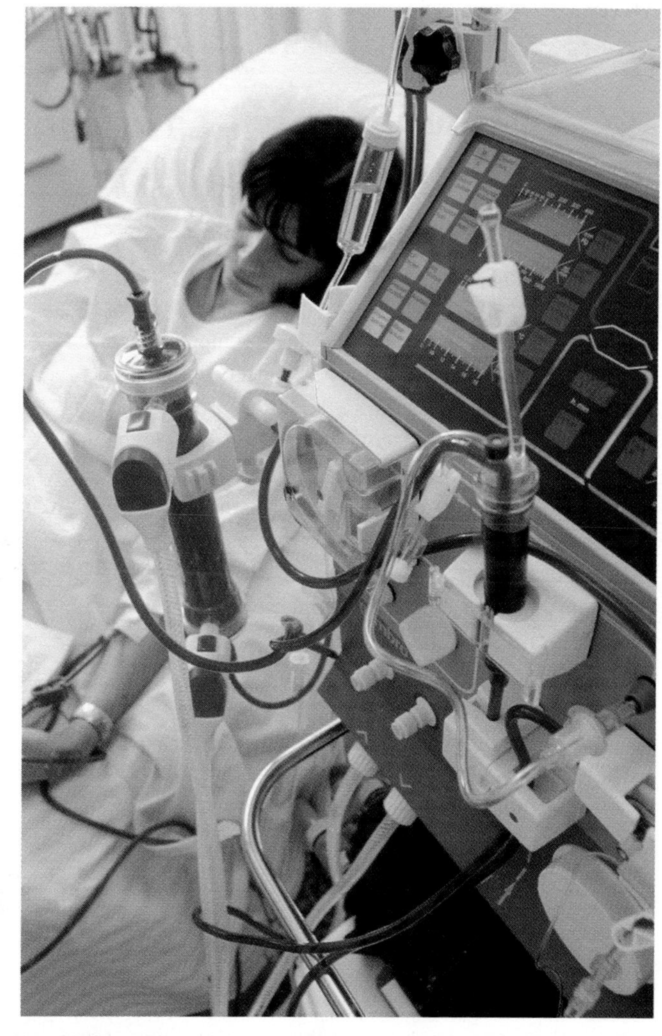

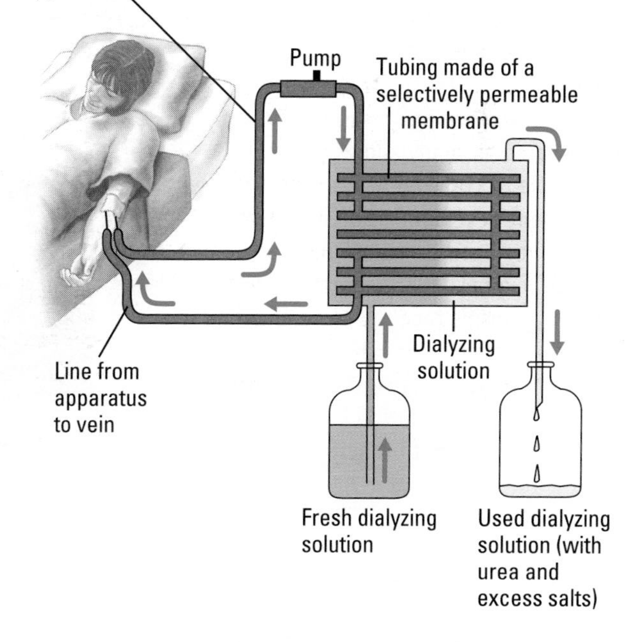

Figure 21.18 A dialysis machine.

CHECKPOINT

1. Through homeostasis, an animal's body tends to maintain a nearly constant ___________ despite changes in the ___________.
2. When you flush a toilet, water begins to refill the tank. The rising water level lifts a float. When the float reaches a certain height, it stops more water from entering the tank. This is an example of what kind of homeostatic control? Why?
3. Name two features of the human body that help release heat and three features that help retain heat.
4. Name two ways that you bring water into your body, and name four ways that water leaves your body.
5. Name the four processes that occur as kidney tubules process blood and create urine.

Answers: 1. internal environment; external environment **2.** This is an example of negative feedback because the result of a process (water filling the tank) inhibits that same process. **3.** Examples include sweating and dilation of blood vessels to release heat and shivering, body hair, and constriction of blood vessels to retain heat. **4.** Water enters via eating and drinking and leaves via breathing, sweating, urinating, and defecating. **5.** Filtration, reabsorption, secretion, and excretion

EVOLUTION CONNECTION

How Physical Laws Limit Animal Form

Imagine the horror of wading into a murky lake and feeling your legs engulfed by a squishy amoeba the size of a professional wrestler. Fortunately, you don't have to add this to your worry list. It will never happen. As we saw in our analysis of how body size affects an animal's ability to exchange materials with the environment, physical laws limit the size of single cells. An amoeba the size of a human could never move materials across its membrane fast enough to satisfy the metabolic needs of such a large blob of cytoplasm. This is just one example of how physical law affects the evolution of an organism's form. Physical requirements constrain what natural selection can "invent."

As another example, consider how the problem of moving quickly through water shapes a diversity of organisms. Tuna and other fast bony fishes can swim at speeds up to 80 km/hr (about 50 mph). Sharks, penguins, and aquatic mammals such as dolphins, seals, and whales are also fast swimmers. And they all have the same basic body shape **(Figure 21.19)**. This shape is called fusiform, which means tapered on both ends. The constraints of the physical world account for this similarity among speedy aquatic swimmers. Because water is about a thousand times denser than air, the slightest bump in the body contour causes immense drag, slowing a swimmer more than it would a runner. We would thus expect speedy fishes and marine mammals to have similar shapes because the laws of hydrodynamics are universal. The fusiform shape shared by many aquatic animals is an example of convergent evolution, the independent development of similar forms. Convergence occurs because natural selection favors similar adaptations when diverse organisms face the same environmental problem, such as the resistance of water to fast travel. We'll see many other examples of how evolution has resulted in adaptations of form to function in the chapters that follow. ■

(a) Dolphins

(b) Seal

(c) Shark

(d) Tuna

(e) Penguins

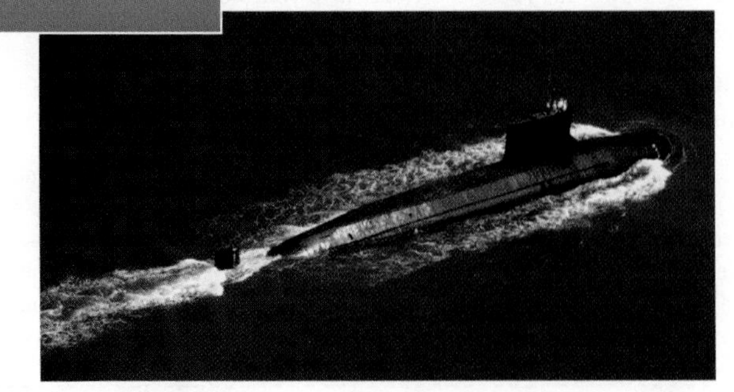

(f) Submarine

Figure 21.19 Convergent evolution of fast swimmers. The evolution of diverse animals and human-made vehicles has converged on a common shape because the laws of hydrodynamics are the same for fish, birds, mammals, and submarines.

Chapter Review

SUMMARY OF KEY CONCEPTS

For study help and activities, go to campbellbiology.com or the student CD-ROM.

Biology and Society: Keeping Cool

- A failure of the body's internal temperature control system can result in heat exhaustion or heat stroke.

The Structural Organization of Animals

Hierarchical Organization of Animals

Level	Description	Example
Cell	The basic unit of all living organisms	Muscle cell
Tissue	A collection of similar cells that perform a specific function	Cardiac muscle
Organ	Multiple tissues forming a structure that performs a specific function	Heart
Organ system	A team of organs that work together	Circulatory system
Organism	A living being, which depends on the coordination of all structural levels for homeostasis and survival	Person

- **Form Fits Function** At every level, the structure of a body part is correlated with the task it must perform. Anatomy is the study of the structure of organisms, while physiology is the study of the function of an organism's structures.

Activity *Correlating Structure and Function of Cells*

- **Tissues** Animals have four main kinds of tissue: epithelial (covers body surfaces), connective (supports other organs), muscle (contracts), and nervous (relays and integrates information).

Activity *The Levels of Life Card Game*

Activity *Overview of Animal Tissues*

Activity *Epithelial Tissue*

Activity *Connective Tissue*

Activity *Muscle Tissue*

Activity *Nervous Tissue*

- **Organs and Organ Systems** Figure 21.8 summarizes the 11 major human organ systems.

Exchanges with the External Environment

- All animals are open systems and must exchange chemicals and energy with the environment.
- Every cell of a simple organism can exchange materials through direct contact with the environment. Large and complex body shapes require indirect exchange between extensively branched internal structures and the environment, usually via a circulatory system.

Regulating the Internal Environment

- **Homeostasis** Homeostasis is the body's tendency to maintain relatively constant internal conditions despite large fluctuations in the external environment.

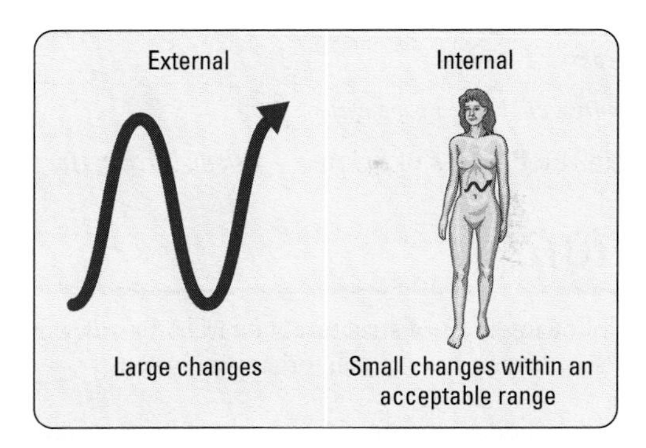

- **Negative and Positive Feedback** In negative feedback, the most common homeostatic mechanism, the results of a process inhibit that very process. Less common is positive feedback, in which the results of a process intensify that same process.

Activity *Regulation: Negative and Positive Feedback*

- **Thermoregulation** In thermoregulation, homeostatic mechanisms regulate internal body temperature. Body temperature may be decreased through dilation of surface blood vessels, sweating, and heat exhaustion (in severe cases). Body temperature may be increased through constriction of blood vessels, shivering, basking in the sun, growing thicker fur, and increasing metabolism.

Case Study in the Process of Science *How Does Temperature Affect Metabolic Rate in* Daphnia?

• **Osmoregulation** All organisms balance the gain or loss of water and dissolved solutes. Osmoconformers do not undergo a net gain or loss of water because their internal and external environments have similar solute concentrations. In contrast, osmoregulators must actively regulate their water balance—by drinking and urinating, for example.

• **Homeostasis in Action: The Urinary System** The urinary system expels wastes and regulates solute and water balance. The following diagram summarizes the structure and function of the nephron, the functional unit of the kidney.

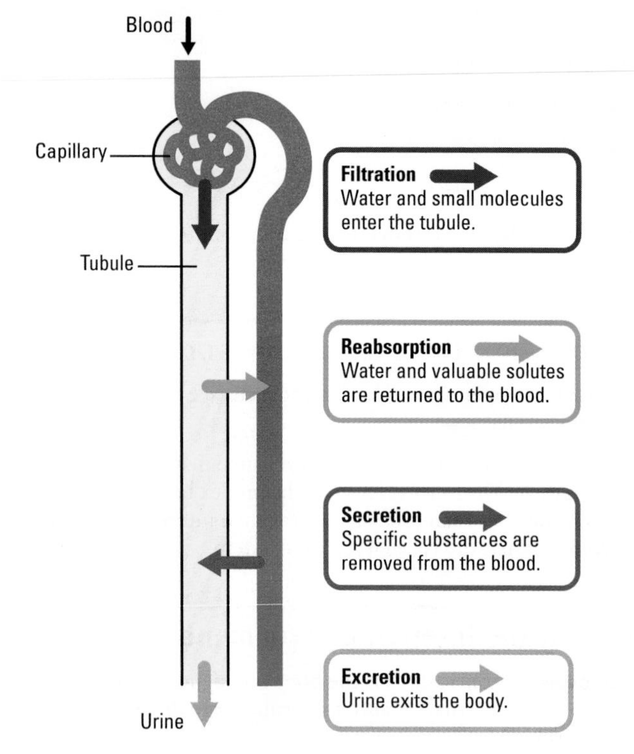

Activity *Human Urinary System*

Activity *Nephron Function*

Activity *Control of Water Reabsorption*

Case Study in the Process of Science *What Affects Urine Production?*

SELF-QUIZ

1. List the following units of structural hierarchy in order from largest to smallest: tissue, organ system, cell, organism, organ.

2. Is a roundworm or a flatworm more likely to have a circulatory system? Why?

3. Which of the following best illustrates homeostasis?
   a. Most adult human beings are between 5 and 6 feet tall.
   b. The lungs and intestines have large surface areas for exchange.
   c. When blood water concentration goes up, the kidneys expel more water.
   d. When oxygen in the blood decreases, you may feel light-headed.

4. Most human cells are surrounded by an aqueous solution called __________.

5. Stimulation of a nerve cell causes sodium ions to leak into the cell, and the sodium influx triggers the inward leaking of even more sodium. This is an example of __________ feedback. An increase in the concentration of glucose in the blood stimulates the pancreas to secrete insulin, a hormone that lowers blood glucose concentrations. This is an example of __________ feedback.

6. What is the main difference between endotherms and ectotherms?

7. __________ is a life-threatening condition caused by a breakdown of thermoregulation.

8. Which of the following is not one of the homeostatic functions of the kidney?
   a. reabsorption
   b. filtration
   c. excretion
   d. ingestion

9. Drinking alcohol makes you urinate more frequently because
   a. alcohol contains caffeine.
   b. alcohol inhibits the release of ADH, a hormone that increases water reabsorption in the kidneys.
   c. alcohol inhibits the release of ADH, a hormone that decreases water reabsorption in the kidneys.
   d. alcohol causes more water to filter from the blood into the kidneys.

10. What happens to most of the water that passes from the blood into the kidneys by filtration? What happens to the rest of it?

Answers to Self-Quiz questions can be found in Appendix D.

Go to the website or CD-ROM for more Self-Quiz questions.

THE PROCESS OF SCIENCE

11. Eastern tent caterpillars (*Malacosoma americanum*) live in sizable groups in silk nests, or tents, which they construct in cherry trees. They are among the first insects to become active in the spring, emerging very early in the season—a time when the caterpillars regularly contend with large daily temperature fluctuations, from freezing to hot conditions. Observing a colony over the course of a day, you notice striking changes in group behavior: Early in the morning, the black caterpillars rest in a tightly packed group on the east-facing surface of the tent. In midafternoon, the group is found on the tent undersurface, each caterpillar individually hanging from the tent by just a few of its legs. Propose a hypothesis to explain this behavior. How could you test your hypothesis?

BIOLOGY AND SOCIETY

12. The kidneys remove many drugs from the blood, and these substances show up in the urine. Some employers require a urine drug test at the time of hiring and/or at intervals during employment. Do you think an employee should have the right to refuse a urine test? Should every job be subject to mandatory drug testing or just certain jobs? Which jobs do *you* think should be subject to testing? Would you take a drug test to get or keep a job? Why or why not?

13. Kidneys were the first organs to be successfully transplanted. A donor can live a normal life with a single kidney, making it possible for individuals to donate a kidney to an ailing relative or even an unrelated individual. Should individuals be allowed to sell one of their kidneys to a wealthy person in need of a kidney? Why or why not? What are some of the ethical issues raised by organ commerce?

22 Nutrition and Digestion

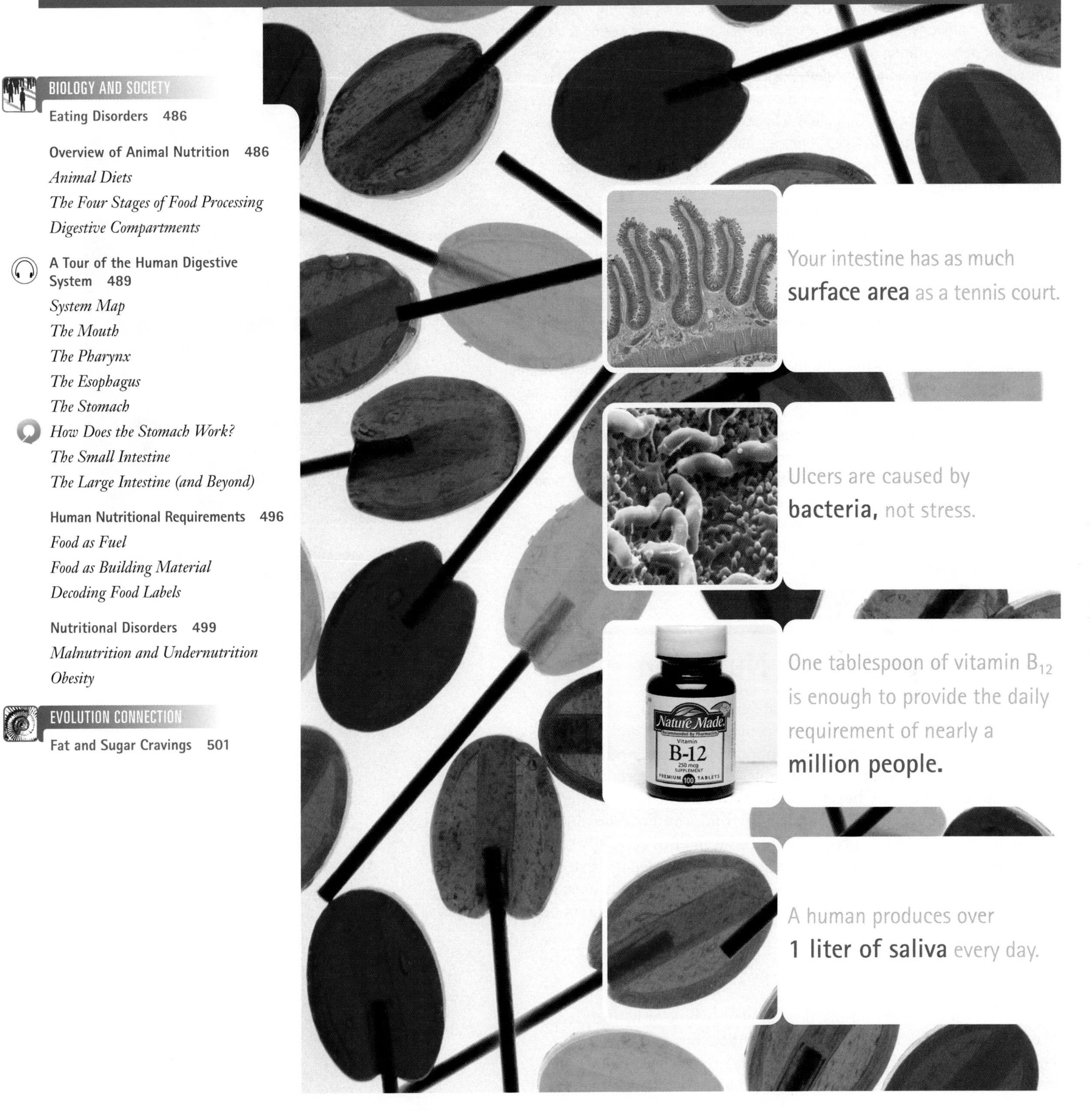

Your intestine has as much **surface area** as a tennis court.

Ulcers are caused by **bacteria,** not stress.

One tablespoon of vitamin B_{12} is enough to provide the daily requirement of nearly a **million people.**

A human produces over **1 liter of saliva** every day.

BIOLOGY AND SOCIETY

Eating Disorders

Imagine feeling ashamed and guilty every time you ate something, while at the same time feeling obsessively drawn to food. Imagine being severely underweight and malnourished, and yet, when you look in the mirror, seeing a grossly overweight person staring back at you. Such is the plight of millions of Americans who suffer from eating disorders **(Figure 22.1)**. **Anorexia** is characterized by self-starvation due to an intense fear of gaining weight, even when the person is actually underweight. **Bulimia** is a behavioral pattern of binge eating followed by purging through induced vomiting, abuse of laxatives, or excessive exercise. Both eating disorders are characterized by an obsession with body weight and shape and can result in serious health problems.

The causes of anorexia and bulimia are unknown, but there is evidence that certain individuals may be more prone to develop eating disorders than others because of differences in genes, psychology, and brain chemistry. Culture also seems to be a factor: Anorexia and bulimia occur almost exclusively in the most affluent industrialized countries, where food is plentiful but thinness is idealized. Some evidence suggests that popular culture and the media may promote unhealthy body images.

Treatment options for eating disorders include counseling and antidepressant medications. Some people with eating disorders eventually develop healthy eating habits without any treatment at all. For others, dysfunctional nutrition becomes a long-term problem that can impair health and even lead to death.

Eating disorders are a serious health threat for the simple reason that humans, like all animals, must eat to survive and thrive. Your body transforms the food you eat into building-block molecules and into the energy that powers your movements and thoughts. You are what you eat in the sense that your health and appearance depend on the quality of your diet. In this chapter, you'll learn essential concepts of nutrition and digestion, beginning with an overview of how animals eat. ■

Figure 22.1 Speaking out about eating disorders. The actress Portia de Rossi admitted in a 2006 interview that she has struggled with anorexia.

Overview of Animal Nutrition

Every mealtime is a reminder that we are animals, organisms that must feed on other organisms to survive. In contrast to plants and other autotrophs ("self-feeders"), animals cannot make sugars, proteins, and other organic matter from inorganic molecules such as carbon dioxide and water. All animals are **heterotrophs;** they must acquire nutrients in the form of organic material from their environment. Food provides the raw materials that animals need to build tissue and fuel cellular work. However, food primarily consists of large, complex molecules that are not in a form an animal's cells can use. Thus, the animal body must be able to break down the nutrients in food (that is, to digest them) in order for the nutrients to be useful.

Animal Diets

All animals eat other organisms, dead or alive, whole or by the piece. Beyond that generalization, however, animal diets vary extensively. **Herbivores,** such as cattle, gorillas, and sea urchins, feed mainly on plants or algae. **Carnivores,** such as lions, snakes, and frogs, mainly eat

Figure 22.2 Animal diets. Animals can be classified into three broad groups according to their diet. **(a)** This photo shows an herbivore (a caterpillar) eating its way through a leaf, leaving a trail of black feces in its wake. **(b)** In this scene from Kenya, a carnivore (cheetah) has begun to consume a Grant's gazelle. **(c)** Most humans are omnivores.

animals that eat plants (or animals that eat animals that eat plants). **Omnivores,** such as crows, cockroaches, and humans, eat both plants and animals **(Figure 22.2)**.

The Four Stages of Food Processing

Think of the fate of one of your meals—say, a slice of pizza—and you'll start to get a sense of the four stages of food processing: ingestion, digestion, absorption, and elimination. **Ingestion** is just another word for eating. You ingest pizza when you bite off a piece. **Digestion** is the breakdown of food to small nutrient molecules. The tomato sauce on a pizza, for example, is broken down to simple sugars and amino acids. **Absorption** is the uptake of the small nutrient molecules by cells lining the digestive tract. For example, amino acids made available by the breakdown of the cheese protein in our imaginary pizza are absorbed by cells lining the small intestine and transferred to the bloodstream, which distributes them throughout the body. **Elimination** is the disposal of undigested materials left over from the food we eat.

Digestion: A Closer Look

The process of digestion usually begins with **mechanical digestion,** physical processes such as chewing. Mechanical digestion breaks chunks of food into small pieces, exposing more of the food molecules to **chemical digestion,** the chemical breakdown of food by digestive enzymes. Most of the organic matter in the food we eat consists of polymers, large molecules made up of smaller building blocks called monomers (see Chapter 3). Chemical digestion breaks food polymers down into monomers. For instance, starch is digested to its component glucose monomers.

The digestive dismantling of large food molecules is necessary for two reasons. First, polymers are too large to cross the membranes of animal cells; they must be broken down into monomers, which are small enough for cells to absorb. Second, most of the polymers of food—the proteins in cheese, for example—are different from the polymers that make up an animal's body. Your body never uses the protein that you eat directly; it will dismantle it and then use the pieces (amino acids) to build its own new proteins **(Figure 22.3)**.

Chemical digestion proceeds via hydrolysis, chemical reactions that break down polymers into monomers using water molecules in the

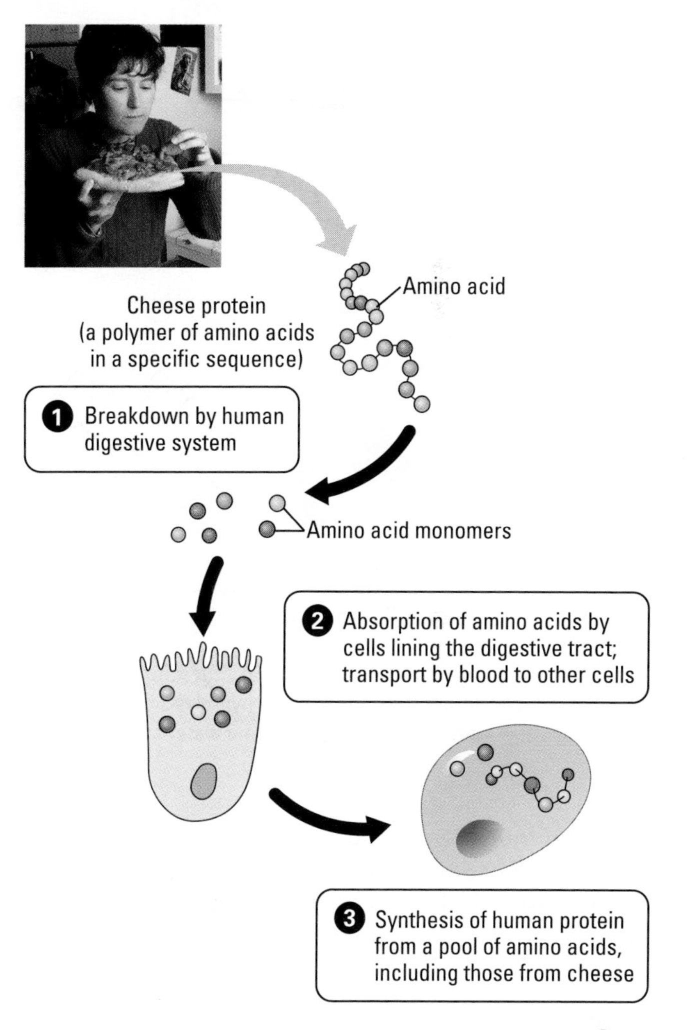

Figure 22.3 From cheese protein to human protein. ❶ A person ingests pizza and digests the protein from cheese. ❷ The amino acids from the cheese protein are absorbed by cells lining the digestive tract and travel through the bloodstream to other body cells. ❸ These cells use amino acids from the cheese and other food to produce a new protein. The human proteins have amino acid sequences different from those of the food proteins.

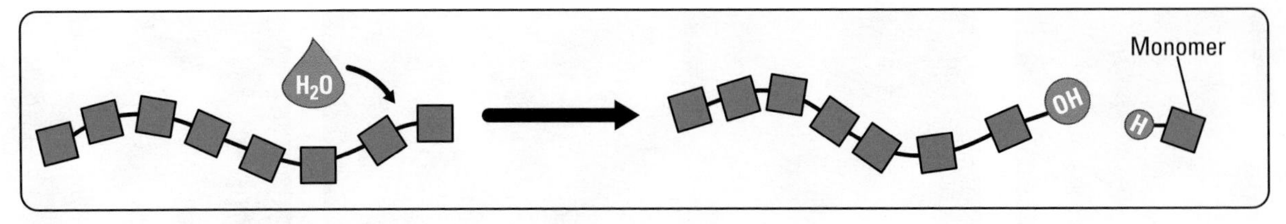

(a) Hydrolysis: general mechanism

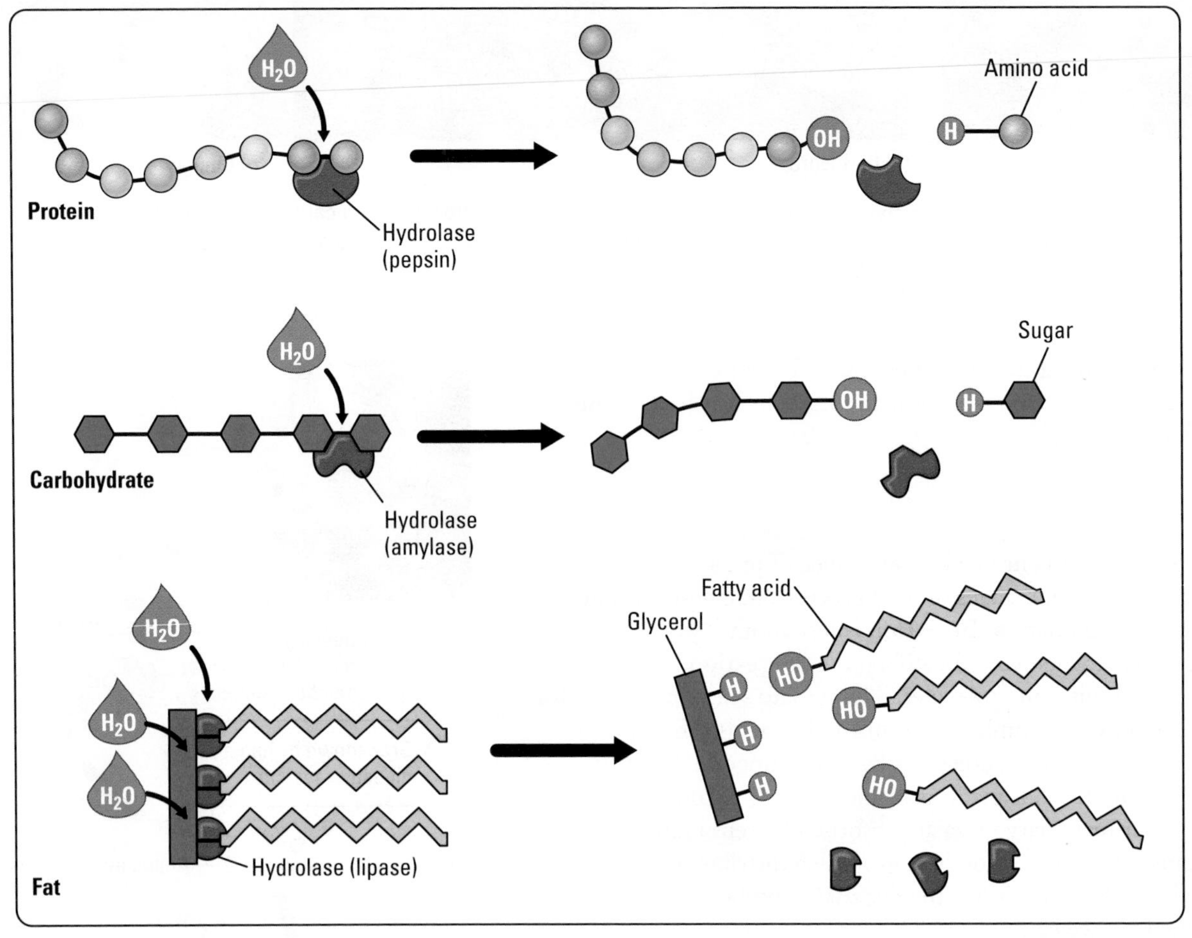

(b) Specific hydrolases

Figure 22.4 Chemical digestion: hydrolysis of food molecules. (a) Specific hydrolysis breaks a food molecule down by adding atoms from water molecules (H_2O) to the bonds between monomers. **(b)** Specific digestive enzymes called hydrolases catalyze hydrolysis.

process (**Figure 22.4a**, and see Figure 3.6b). Like most of life's chemical reactions, digestion requires enzymes. **Hydrolases** are enzymes that catalyze digestive hydrolysis reactions (**Figure 22.4b**). For example, lipases are hydrolases that digest fats.

Digestive Compartments

How do animals digest their food without digesting their own cells and tissues? After all, digestive enzymes hydrolyze the same kinds of biological molecules that make up the animal body, and it is obviously important to avoid digesting oneself! A common solution to this problem has evolved in animals: Chemical digestion proceeds safely within some kind of specialized compartment.

The simplest digestive compartments are **food vacuoles,** intracellular organelles filled with digestive enzymes. Digestion within food vacuoles begins after a cell engulfs food by phagocytosis (see Figure 5.18). As food is digested within the vacuole, the broken-down monomers pass through the membrane and nourish the cell. Sponges are the only animals that digest their food solely via food vacuoles.

All other animals have digestive compartments that are surrounded by, rather than within, cells. This allows for the digestion of food that is much larger than the size of a single cell. Simpler animals, including cnidarians (such as jellies and hydras) and flatworms (such as planarians and flukes), have **gastrovascular cavities.** These are compartments with a single opening that functions as both the entrance for food and the exit for undigested wastes **(Figure 22.5a)**. Digestion is initiated within the cavity, but it is completed within food vacuoles of the cells lining the cavity.

The vast majority of animals, including humans and other vertebrates, have **digestive tubes** with two separate openings, a mouth at one end and an anus at the other **(Figure 22.5b)**. Food enters the mouth and moves through specialized regions that digest and absorb nutrients in a stepwise fashion. (You know what an assembly line is; a digestive tube is a *dis*assembly line.) Undigested wastes are eliminated from the digestive tube as feces via the anus.

CHECKPOINT

1. Place the four stages of food processing in their proper order: absorption, digestion, elimination, ingestion.
2. Food polymers are broken down into the monomers that make them up through the action of a class of enzymes called __________ that catalyze __________ reactions.
3. What is the main difference between gastrovascular cavities and digestive tubes?

Answers: 1. Ingestion, digestion, absorption, elimination **2.** hydrolases; hydrolysis **3.** Gastrovascular cavities have just one opening; digestive tubes have two (mouth and anus).

A Tour of the Human Digestive System

MP3 Tutor
The Human Digestive System

We are now ready to follow our slice of pizza through the human digestive tube, from mouth to anus. It's important to have a good map so we don't get lost in there.

System Map

The human digestive system consists of a digestive tube, called the **alimentary canal** or gut, and several accessory organs (salivary glands, pancreas, liver, and gallbladder) that secrete digestive chemicals into the

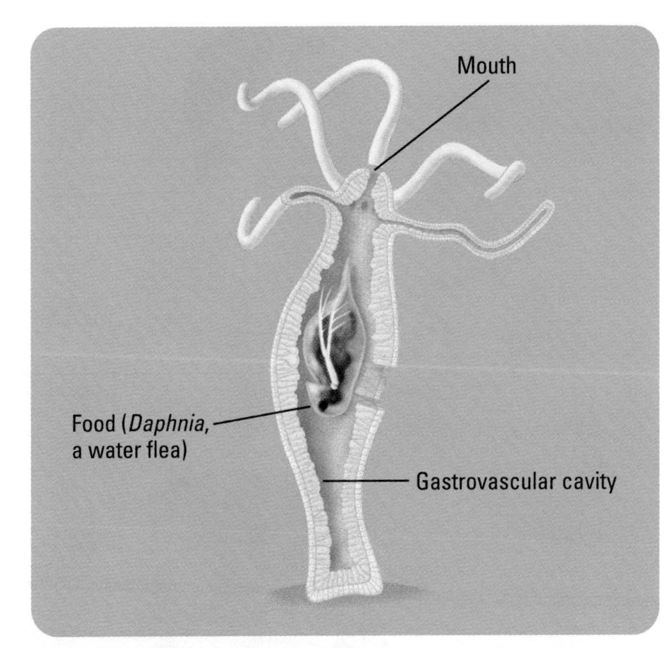

(a) Gastrovascular cavity in a hydra

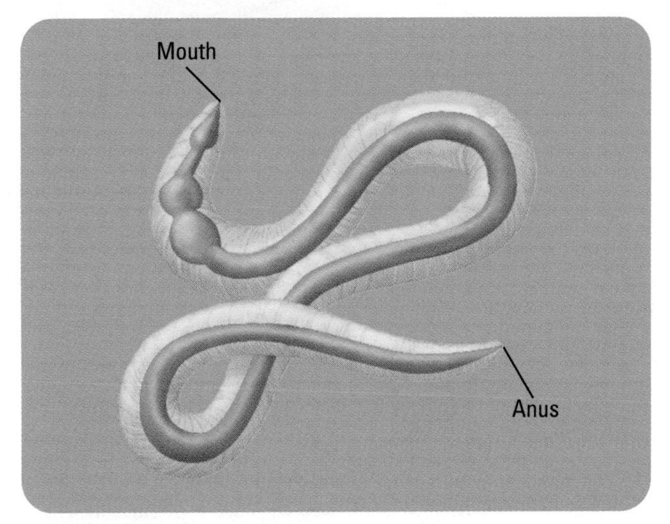

(b) Digestive tube in an earthworm

Figure 22.5 Digestive compartments. (a) A hydra's food is ingested into a gastrovascular cavity. Undigested wastes are eliminated through the same opening. **(b)** A digestive tube, such as this earthworm's, has both a mouth and an anus. Specialized organs along its length perform the four main functions of food processing: ingestion, digestion, absorption, and elimination.

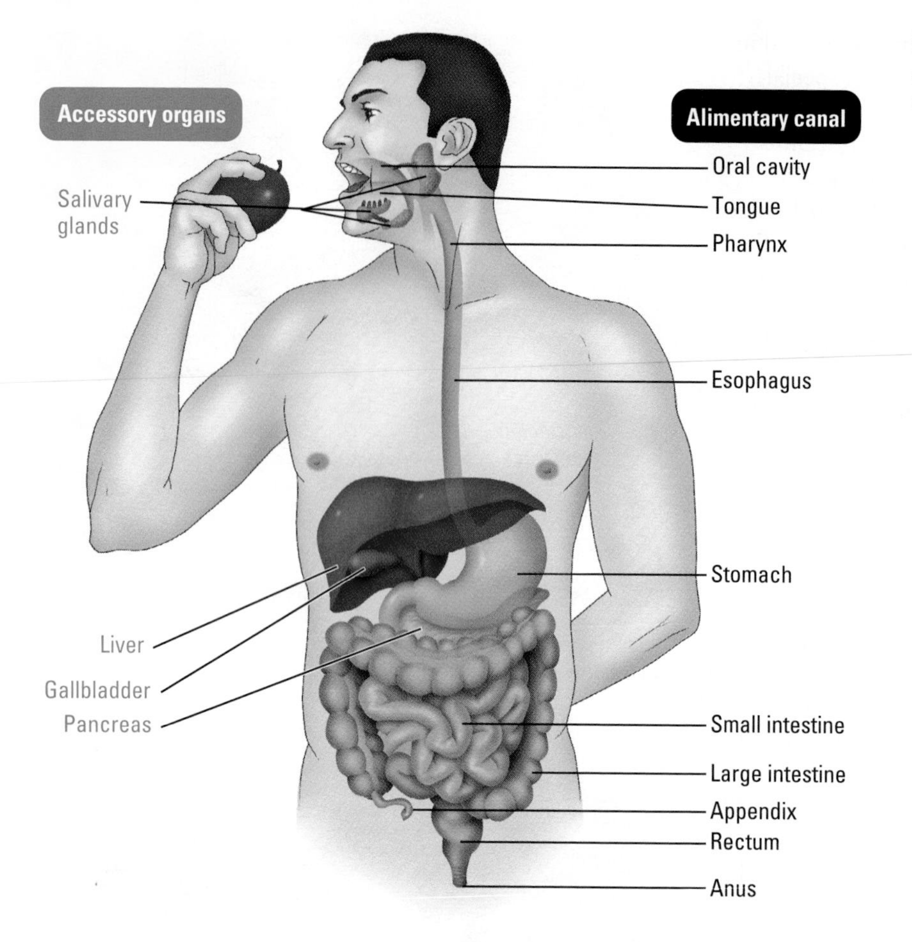

Figure 22.6 The human digestive system. The human digestive system consists of the alimentary canal (black labels) and accessory organs (blue labels).

canal via ducts **(Figure 22.6)**. The human alimentary canal is about 9 m (30 feet) in length. This long tube fits within a human body because it folds back and forth over itself with many switchbacks. The alimentary canal is divided into specialized digestive organs along its length: mouth (oral cavity) → pharynx → esophagus → stomach → small intestine → large intestine (colon and rectum) → anus. You'll get a closer look at the structure and function of the digestive organs on our journey through the alimentary canal. As you reach each organ, look back at Figure 22.6 to find exactly where you are. First stop: the mouth.

The Mouth

The **mouth** (also known as the **oral cavity**) functions in ingestion (food intake) and the preliminary steps of digestion **(Figure 22.7)**. Mechanical digestion begins before we even swallow as our teeth cut, smash, and grind the food. Chewing makes food easier to swallow and exposes more food surface to digestive juices. Chemical digestion also begins in the mouth with the secretion of saliva from three pairs of glands. Saliva contains the digestive enzyme **salivary amylase.** This enzyme hydrolyzes starch, a major ingredient of our pizza crust.

The muscular **tongue** is very busy during mealtime. Besides tasting the food, the tongue shapes it into a ball and pushes the food ball to the back of the mouth. Swallowing moves food into the pharynx.

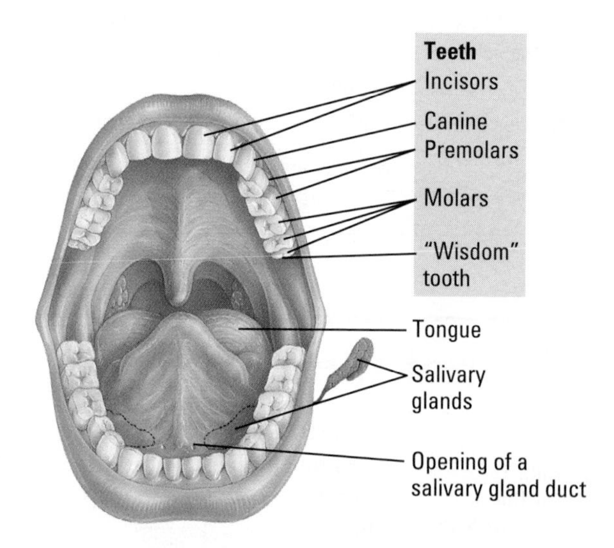

Figure 22.7 The human oral cavity. The process of mechanical digestion begins when large chunks of food are cut into smaller pieces by the teeth. Children acquire 20 immature (baby) teeth by about age 2, which are replaced by 32 adult teeth. The bladelike incisors are specialized for biting off pieces of food. The pointed canine teeth help with ingestion by ripping and tearing away pieces of food. The premolars and molars have broad surfaces that crush and grind food during chewing. The third set of molars is called the wisdom teeth. They do not appear in all people, and in some people they push against other teeth and must be removed.

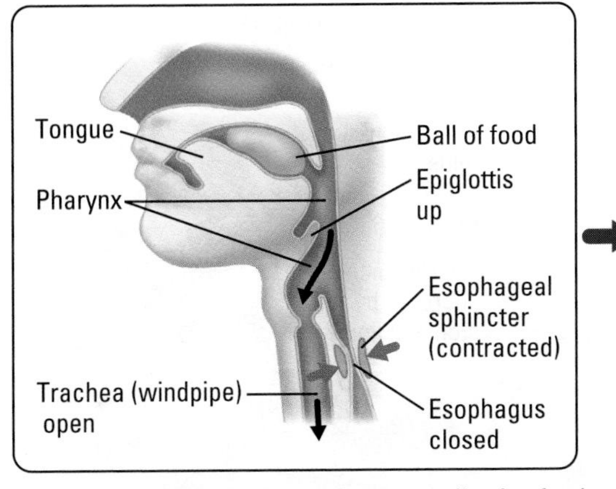

(a) Not swallowing. When you're not swallowing food or drink, air travels freely through the trachea (black arrows). The esophagus is closed because the esophageal sphincter, a ring of muscles, is contracted (blue arrows).

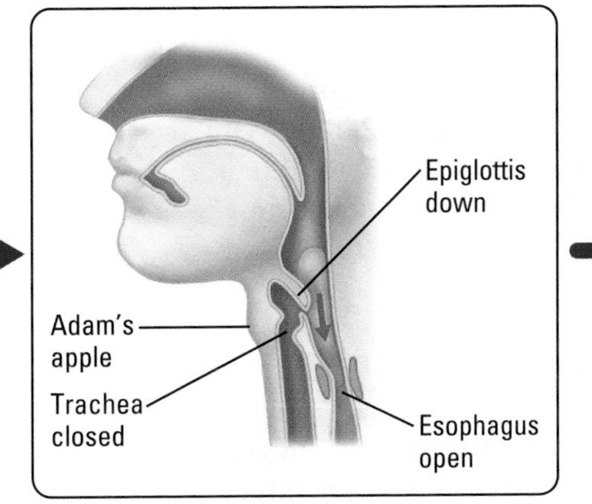

(b) Swallowing started. Once the food reaches the back of the mouth, a swallowing reflex is triggered. The top of the trachea rises against the epiglottis, closing the air passage, while the esophagus opens because the sphincter relaxes. Food travels down the esophagus (green arrow).

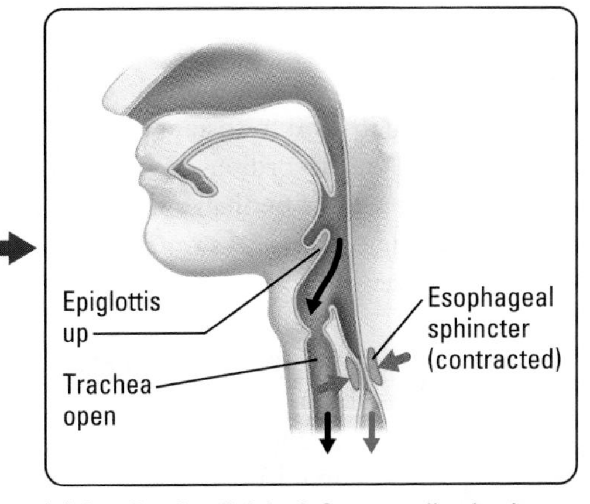

(c) Swallowing finished. Once swallowing is completed, the esophageal sphincter contracts, and air once again flows freely through the trachea.

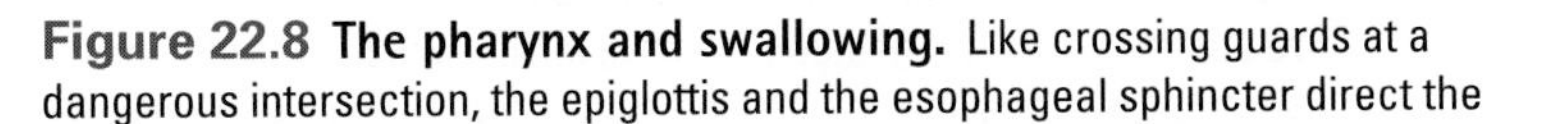

Figure 22.8 The pharynx and swallowing. Like crossing guards at a dangerous intersection, the epiglottis and the esophageal sphincter direct the flow of traffic through the pharynx. You can see this action in the bobbing of your Adam's apple every time you swallow.

The Pharynx

The chamber called the **pharynx,** located in your throat, is an intersection of the food and breathing pathways. It connects the mouth to the esophagus, but the pharynx also opens to the trachea, or windpipe, which leads to the lungs. When you're not swallowing, the trachea entrance is open, and you can breathe. When you swallow, a reflex moves the opening of the trachea upward and tips a door-like flap called the epiglottis to close the trachea entrance (**Figure 22.8**). The closing of the trachea ensures that the food will go down the esophagus. Occasionally, food begins to "go down the wrong pipe," triggering a strong coughing reflex that helps keep your airway clear of food.

The Esophagus

The **esophagus** is a muscular tube that connects the pharynx to the stomach. Imagine trying to move a tennis ball through a sock open at both ends. The ball will not just fall through, but you could move the ball along by pinching repeatedly just behind the ball until it pops out the other end. Your esophagus moves a ball of food by a similar mechanism. The action is called **peristalsis,** rhythmic waves of muscular contractions that squeeze the food ball along the esophagus (**Figure 22.9**). This muscular action ensures that you can swallow even when standing on your head. Peristalsis continues throughout the length of the alimentary canal, propelling food all the way along.

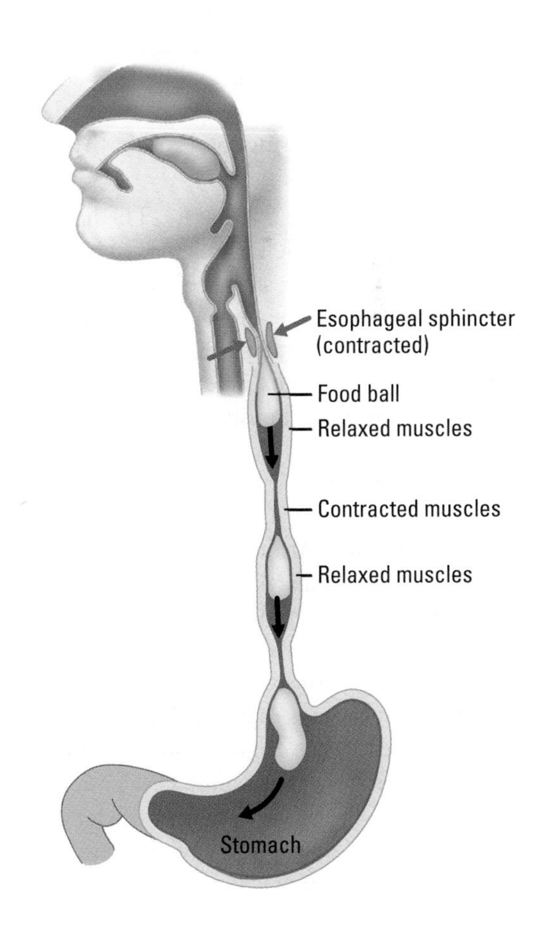

Figure 22.9 The esophagus and peristalsis. The wavelike muscular action called peristalsis squeezes balls of food through the esophagus to the stomach. Muscles of the esophageal wall contract just behind the food ball and relax just ahead of it.

The Stomach

We do not have to eat constantly because the **stomach** is a large organ that can store enough food to sustain us for several hours **(Figure 22.10)**. With its elastic wall and accordion-like folds, the stomach can stretch to accommodate about 2 L (more than a half gallon) of food and water. Once stored, how is food digested in the stomach? This question led to one of biology's most bizarre set of experiments.

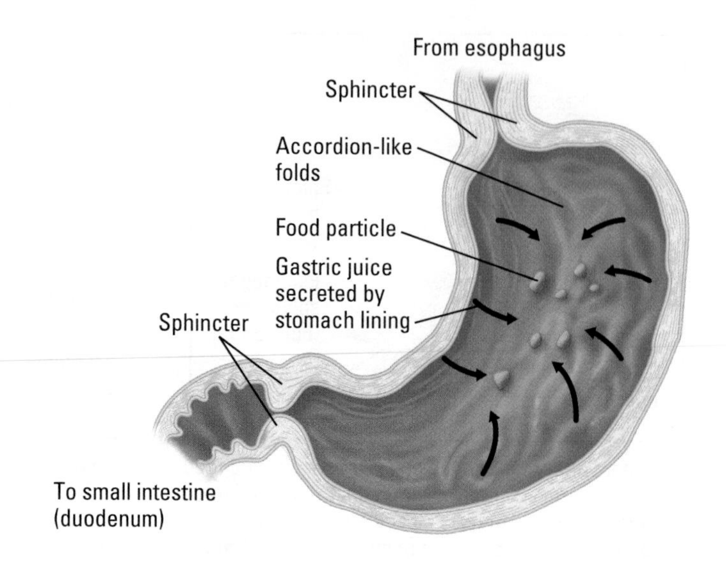

Figure 22.10 The stomach. The elastic stomach contains numerous accordion-like folds that allow it to expand during big meals. Sphincters control the flow of food into the stomach and the release of partially digested food from it. The wall of the stomach secretes gastric juices containing mucus, which helps to protect the stomach from self-digestion.

How Does the Stomach Work? In 1822, a 19-year-old trapper named Alexis St. Martin was accidentally shot in the upper abdomen at close range. A local army physician named William Beaumont performed emergency surgery **(Figure 22.11)**. The surgery saved St. Martin's life, but the wound healed improperly, leaving an opening to St. Martin's stomach.

St. Martin gradually regained his health and strength, but the hole in his stomach remained. At a time when no one had a clue how the stomach worked, Beaumont used this rare opportunity to perform experiments on St. Martin's digestive physiology. By pulling open a skin flap, Beaumont made many **observations** of the liquid within the stomach. This led Beaumont to **question** whether this fluid was responsible for the stomach's digestive action; Beaumont formed the **hypothesis** that digestion in the stomach was primarily a chemical process. He made the **prediction** that food wrapped in porous bags, tied to a string, and inserted through the hole into St. Martin's stomach would be degraded by chemicals in the stomach fluid.

In the ensuing decades, Beaumont performed a long series of such **experiments** on his unusual patient. After inserting a bag and waiting for a specified time interval, Beaumont would withdraw the string and record the condition of the food items. After 2 hours, for example, Beaumont noted that "cabbage, bread, pork, and boiled beef all cleanly digested." Over time, Beaumont's **results** helped confirm that chemical digestion plays a major role in processing food within the stomach. ■

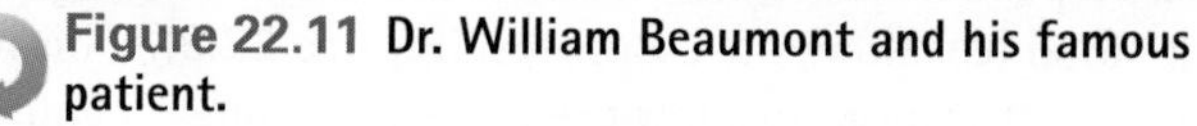
Figure 22.11 Dr. William Beaumont and his famous patient.

Chemical Digestion in the Stomach The liquid that Beaumont observed in St. Martin's stomach is called **gastric juice,** a digestive fluid secreted by the cells lining the stomach's interior. Gastric juice is made up of strong acid, digestive enzymes, and mucus. The acid in gastric juice is hydrochloric acid, and it is concentrated enough to dissolve iron nails. Gastric juice also contains **pepsin,** an enzyme that digests proteins, like those in the cheese on our pizza. Pepsin breaks the polypeptides of protein into smaller pieces.

When food passes from the esophagus into the stomach, the muscular stomach walls begin to churn, mixing the food and gastric juice into a thick soup called **acid chyme.** At the downstream end of the stomach, a sphincter (a ring of muscle) regulates transfer of acid chyme from the stomach to the small intestine. With the acid chyme leaving the stomach only a squirt at a time, it takes about 2–6 hours for the stomach to empty after a meal. (Continued contraction of stomach muscles after the stomach is empty causes the grumbling pangs that broadcast your hunger.)

With all that acid, what keeps the stomach from eating itself? Mucus coating the stomach lining helps protect it from gastric juices and from abrasive materials in the food. Timing is also a factor: Nerve signals and the chemical signals called hormones regulate secretion of gastric juice so that it is discharged only when food is in the stomach. Even with these safeguards, gastric juice still manages to erode the stomach lining, requiring the production of new cells by cell division. In fact, the stomach lining is completely replaced about once every three days.

Stomach Ailments Gastric juices, strong enough to digest a diverse array of foods, can be harmful. Occasional backflow of acid chyme into the esophagus causes heartburn (which, if we're going to be picky, should be called esophageal burn). Some people suffer this backflow frequently and severely enough to harm the lining of the esophagus, a condition called acid reflux or GERD (gastroesophageal reflux disease).

If the stomach lining is eroded by gastric juice faster than it can regenerate, painful open sores called gastric ulcers can form in the stomach wall. The cause of most ulcers is not stress, as was once thought, but infection of the stomach lining by a type of acid-tolerant bacterium named *Helicobacter pylori* **(Figure 22.12)**. The metabolism of these bacteria damages the mucous coat, making the lining more accessible to gastric juice. (The cause of ulcers was conclusively established in 1984 when biologist Barry Marshall experimented on himself by drinking beef soup laced with *H. pylori* bacteria; although Marshall eventually won a 2005 Nobel Prize for his work, we do not recommend this mode of experimentation!) Stomach ulcers are thus treated with antibiotics. Affected people can also get relief by taking medications that contain bismuth (such as Pepto Bismol), which helps reduce ulcer symptoms and may kill some bacteria.

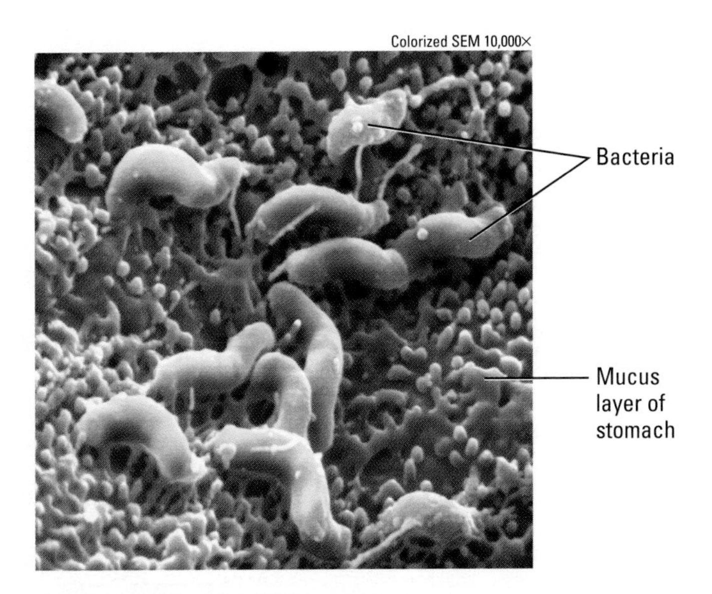

Figure 22.12 Ulcer-causing bacteria. The bacteria visible in this micrograph, *Helicobacter pylori*, initiate ulcers by destroying protective mucus and causing inflammation of the stomach lining. Then the serious damage starts, as the acidic gastric juice dissolves stomach tissue. In the most severe ulcers, the erosion can produce a hole in the stomach wall and cause life-threatening internal bleeding and infection.

The Small Intestine

The **small intestine,** at a length of about 6 m (about 20 feet), is the longest part of the alimentary canal. (The small intestine is so named because of its relatively small diameter: 2.5 cm across versus twice that for the large intestine.) It is also the major organ for chemical digestion and for absorption of nutrients into the bloodstream.

Chemical Digestion in the Small Intestine What kind of shape is our imaginary pizza in by the time it reaches the small intestine? So far, mechanical digestion has turned the meal into a thick, nutrient-rich soup. Chemical digestion by salivary amylase and gastric pepsin has initiated hydrolysis of the pizza's starch and proteins. Now the small intestine takes over with an arsenal of hydrolases that can break all the food macromolecules down to monomers. These enzymes are mixed with acid chyme in the first 25 cm or so (about a foot) of the small intestine, the region called the **duodenum.**

The duodenum receives digestive juices from the pancreas, liver, and gallbladder **(Figure 22.13)**. The **pancreas** is a large gland that secretes pancreatic juice into the duodenum via a duct. Pancreatic juice neutralizes the stomach acids that enter the duodenum and contains hydrolases that participate in the chemical digestion of carbohydrates, fats, proteins, and nucleic acids.

Bile is a juice produced by the **liver,** stored in the **gallbladder,** and secreted through a duct into the duodenum. Bile contains salts that bind to the droplets of fat that have been produced from larger globules by the contraction of muscles in the intestinal wall. The salts prevent the fat droplets from re-forming large globules. In droplet form, the fats are more accessible to lipase and therefore digested faster.

The intestinal lining itself aids in enzymatic digestion by producing a variety of hydrolases. The cumulative activities of all these hydrolytic enzymes break the different classes of food molecules completely down into monomers, which are now ready for absorption into the body.

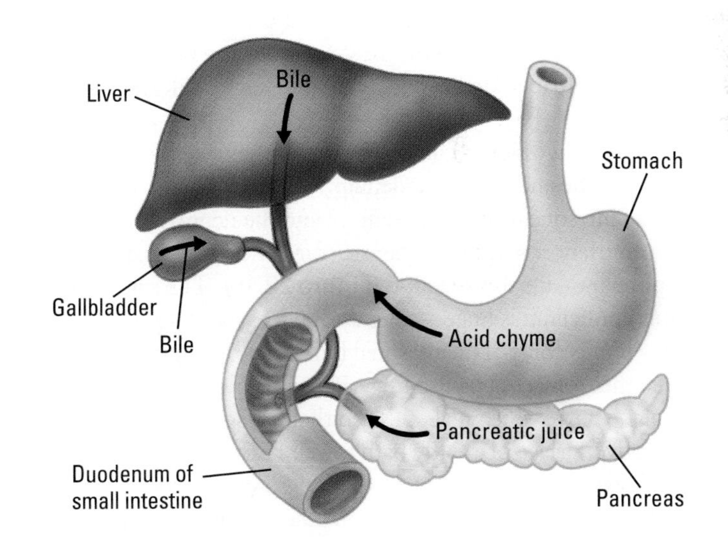

Figure 22.13 The duodenum. Acid chyme squirted from the stomach into the duodenum is mixed with pancreatic juice, bile from the liver and gallbladder, and intestinal juice produced in the duodenum itself. As peristalsis propels the mix along the small intestine, hydrolases break food macromolecules down to their monomers.

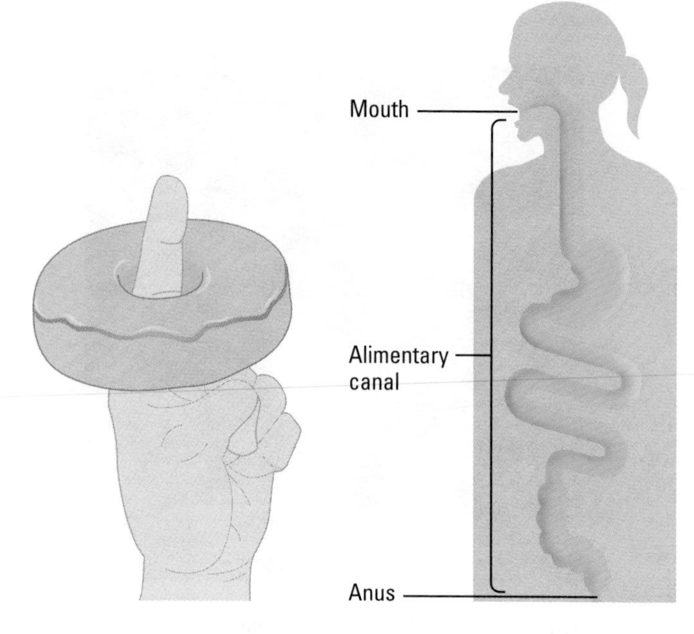

(a) A finger through a hole

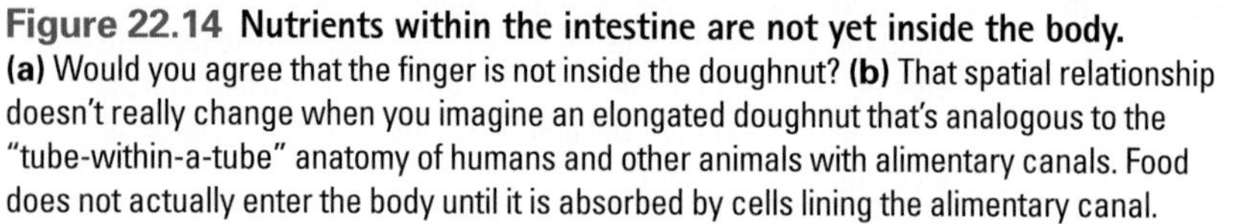

Figure 22.14 Nutrients within the intestine are not yet inside the body. **(a)** Would you agree that the finger is not inside the doughnut? **(b)** That spatial relationship doesn't really change when you imagine an elongated doughnut that's analogous to the "tube-within-a-tube" anatomy of humans and other animals with alimentary canals. Food does not actually enter the body until it is absorbed by cells lining the alimentary canal.

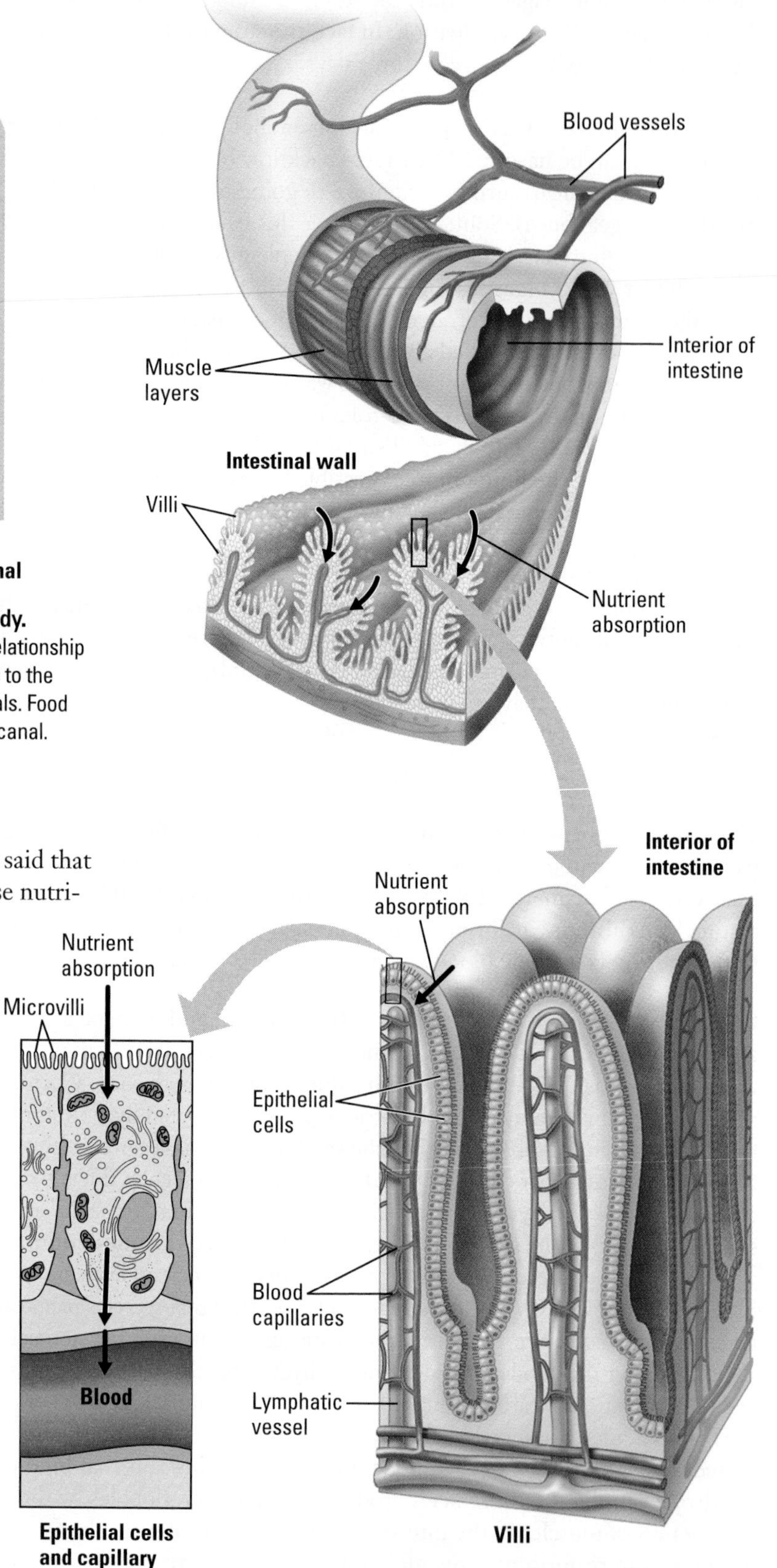

Absorption of Nutrients Wait a minute! The previous sentence said that nutrients "are now ready for absorption into the body." Aren't these nutrients already in the body? Not really. The alimentary canal is a tunnel running through the body, and its cavity is continuous with the great outdoors. The doughnut analogy shown in **Figure 22.14** should convince you that this is so. Until nutrients actually cross the tissue lining the alimentary canal to enter the bloodstream, they are still *outside* the body. If it were not for nutrient absorption, we could eat and digest huge meals but still starve.

Most digestion is complete by the time our pizza meal reaches the end of the duodenum. The next several meters of small intestine (called the jejunum and the ileum) are specialized for nutrient absorption. The structure of the intestinal lining, or epithelium, fits this function **(Figure 22.15)**. The surface area of this epithelium is huge—roughly 300 m^2, about equal to the size of a tennis court. The intestinal lining not only has large folds, like the stomach, but also fingerlike outgrowths called villi (singular, villus), which makes the epithelium something like the absorptive surface of a fluffy bath towel. Each cell of the epithelium adds even more surface by having microscopic projections called microvilli. Across this expansive surface of intestinal epithelium, nutrients are transported into the network of small blood vessels and lymphatic vessels in the core of each villus.

Once they have crossed the cell membranes of the microvilli, the nutrients from our pizza are finally *inside* the body, where the bloodstream carries them away to distant cells. But our tour of the tube is not yet over, for we still have to make it through the large intestine.

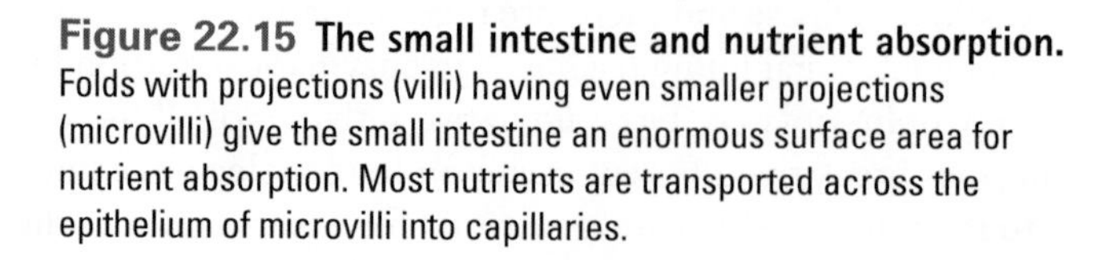

Figure 22.15 The small intestine and nutrient absorption. Folds with projections (villi) having even smaller projections (microvilli) give the small intestine an enormous surface area for nutrient absorption. Most nutrients are transported across the epithelium of microvilli into capillaries.

The Large Intestine (and Beyond)

At only 1.5 m in length, the **large intestine,** also called the **colon,** is shorter than the small intestine but almost twice as wide (about 5 cm). Where the two organs join, a sphincter controls the passage of what's left of a meal. Near this junction is a small fingerlike extension called the **appendix** (see Figures 22.6 and 24.7). If the junction between the appendix and the colon becomes blocked, appendicitis—a bacterial infection of the appendix—may result. Emergency surgery is usually required to remove the inflamed appendix and prevent the spread of infection.

The main function of the colon is to absorb water from the alimentary canal. About 7 L of water per day spill into your alimentary canal as the solvent of digestive juices. About 90% of this water is absorbed back into your blood and tissue fluids, with the small intestine reclaiming much of the water and the colon finishing the job. As water is absorbed, undigested materials from the meal become more solid as they are conveyed along the colon by peristalsis. The end product is **feces,** consisting mainly of indigestible plant fibers (cellulose from any vegetables on our pizza, for example). The feces also contain enormous numbers of intestinal prokaryotes, normal inhabitants of the colon. Some colon bacteria, such as *E. coli*, produce several B vitamins and vitamin K. These vitamins supplement your diet when they are absorbed into the bloodstream through the colon wall.

If the lining of the colon is irritated by a viral or bacterial infection (sometimes mistakenly called "stomach bugs"), diarrhea may result because the colon is not able to reabsorb water efficiently. Prolonged diarrhea can cause life-threatening dehydration, particularly among the very young and very old. The opposite problem, constipation, occurs when peristalsis moves feces along too slowly and the colon reabsorbs so much water that the feces become too compacted. Constipation can result from lack of exercise or from a diet that does not include enough plant fiber.

The **rectum,** the last 15 cm (6 inches) of the large intestine, stores feces until they can be eliminated. Contractions of the colon create the urge to defecate. Two rectal sphincters, one voluntary and one involuntary, regulate the opening of the **anus** (see Figure 22.6).

From entrance to exit, we have now followed a pizza slice all the way through the alimentary canal. **Figure 22.16** stretches the tube out to help you review food processing along its length.

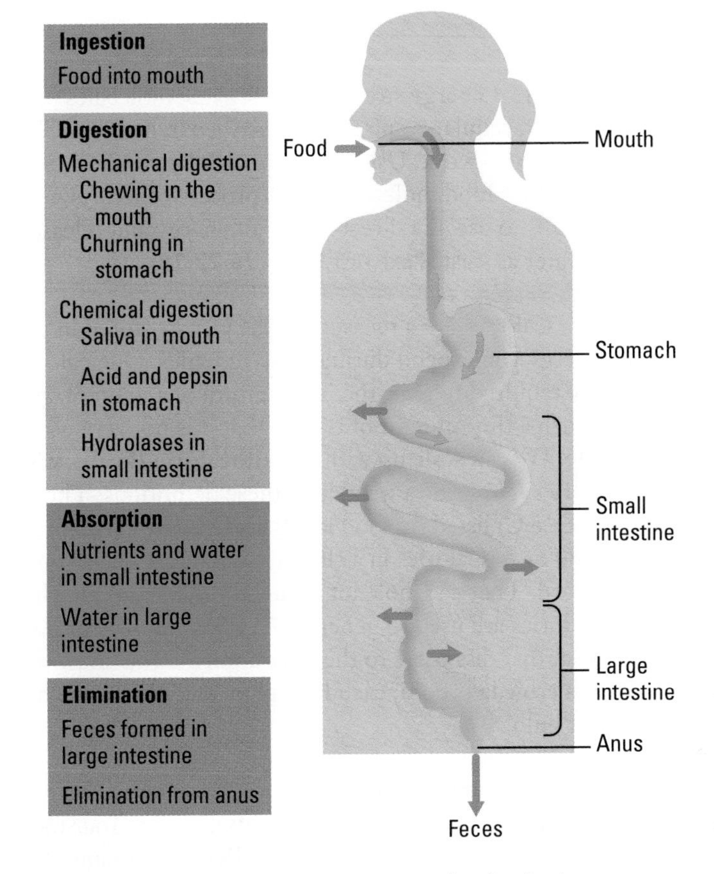

Figure 22.16 Review of food processing in the human alimentary canal.

CHECKPOINT

1. In the oral cavity, chewing functions in ___________ digestion and salivary amylase initiates the chemical digestion of ___________.
2. When we start coughing because food or drink "went down the wrong pipe," the material has entered the ___________ instead of the ___________.
3. Why don't astronauts in weightless environments have trouble swallowing?
4. How can antibiotics help treat gastric ulcers?
5. What is the major organ for nutrient absorption?
6. Amylase is to ___________ as ___________ is to fats.
7. What is the main function of the large intestine?
8. Explain how treatment of a chronic infection with antibiotics for an extended period of time can cause a vitamin K deficiency.

Answers: 1. mechanical; starch (or polysaccharides) **2.** trachea (windpipe); esophagus **3.** Because peristalsis pushes the food along **4.** By killing the bacteria that damage the stomach lining **5.** Small intestine **6.** Starch; lipase **7.** Water reabsorption **8.** By killing bacteria that synthesize vitamin K in the colon

Human Nutritional Requirements

For any animal, proper nutrition helps maintain homeostasis. A balanced diet provides the fuel for cellular work and the materials for building molecules. The food guide pyramid, most recently updated in 2005 by the U.S. Department of Agriculture, contains dietary guidelines for proper nutrition **(Figure 22.17)**.

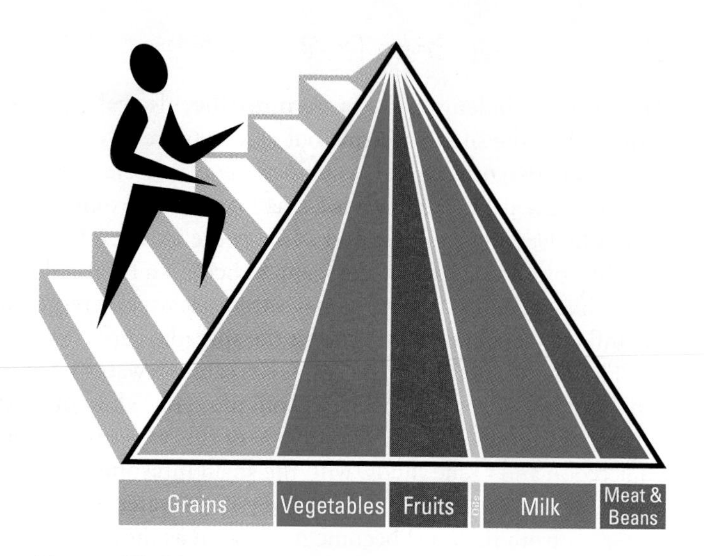

Figure 22.17 The official food guide pyramid. This food guide pyramid, released in 2005 by the U.S. Department of Agriculture, suggests a diet rich in whole grains, vegetables, fruits, and calcium-rich milk products, with less protein-rich meat and beans. Only moderate amounts of fats, sugars, and salt (not shown on the pyramid) are recommended. The width of each band in the pyramid reflects relative proportions of the different foods in a healthy diet. The left side of the pyramid reminds us that daily activity is important. The makeup of the pyramid undergoes periodic changes as we learn more about human nutrition.

Food as Fuel

Cells can extract energy stored in the organic molecules of food through the process of cellular respiration (described in detail in Chapter 6) and use that energy to do work. Using oxygen, cellular respiration breaks down sugar and other food molecules. This process generates many molecules of ATP for cells to use as a direct source of energy and releases carbon dioxide and water as waste "exhaust" **(Figure 22.18)**.

Calories Calories are a measure of the energy stored in your food as well as the energy you expend during daily activities. One **calorie** is the amount of energy required to raise the temperature of a gram of water by 1°C. That is such a tiny amount of energy that it is not very useful on the human scale. We can scale up with the **kilocalorie (kcal),** which is a thousand calories. Now, for a wrinkle in these definitions: The "Calories" (with an uppercase C) listed on food labels are actually kilocalories, not calories. So the 280 or so Calories in a slice of thick-crust pepperoni pizza are actually 280 kcal. That's a whole lot of fuel for making ATP. However, not all of the Calories are used to generate ATP. About 60% of our food energy is lost as heat that dissipates to the environment. You know this effect from being in a crowded room; each human produces as much heat as a 100-watt light bulb.

Metabolic Rate How fast do we "burn" our food? The rate of energy consumption by the body is called **metabolic rate.** Your metabolic rate is equal to your **basal metabolic rate (BMR),** the amount of energy it takes just to maintain your basic body functions, plus any additional energy consumption above that base rate. BMRs for humans average about 1,300–1,500 kcal per day for adult females and 1,600–1,800 kcal per day for adult males. The more active you are, the greater your actual metabolic rate and the greater the number of calories your body uses per day. Metabolic rate also depends on body size, age, stress level, and heredity. **Table 22.1** gives you an idea of the amount of activity it takes to use up the kilocalories in several common foods.

The maintenance of body weight is a simple exercise in caloric arithmetic. You add calories in the form of food and subtract them in the form of energy expended. If you take in more by eating than you burn by activity, you will gain weight. If you want to lose weight, you must decrease the number of calories coming in and/or increase the number of calories expended. Despite a multitude of fad diets, losing weight comes down to a simple formula: Eat less and exercise more! One pound of body fat is equivalent to about 3,500 kcal. So if you decrease your eating by about 300 kcal per day and work out to burn an additional 200 kcal each day, you should lose about 1 pound per week. Caloric balance, however, does not ensure good nutrition, and food is our source of substance as well as our source of energy.

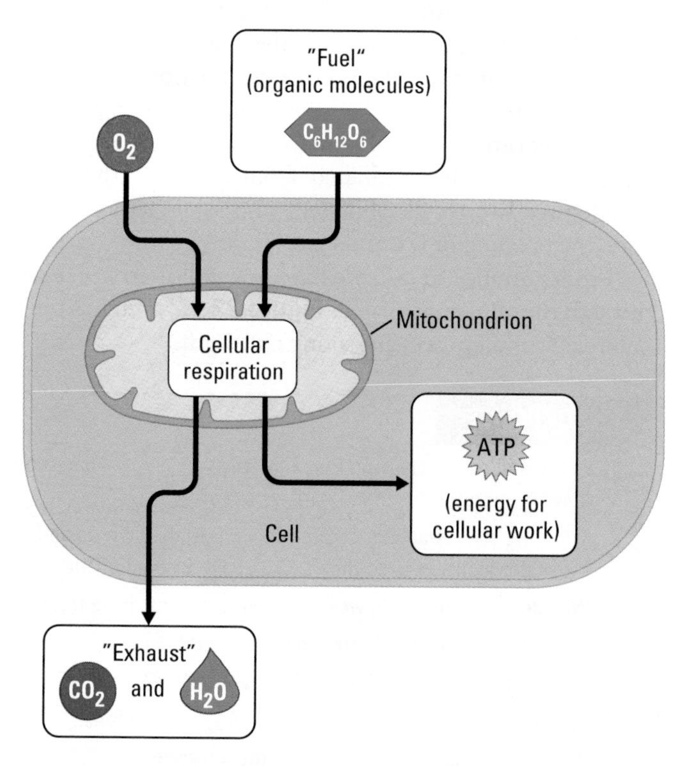

Figure 22.18 Review of cellular respiration. Within the mitochondria, a series of chemical reactions extracts energy from fuel molecules and generates ATP, which drives most cellular work.

Table 22.1	Exercise Required to Burn the Calories (kcal) in Common Foods		
	Jogging	**Swimming**	**Walking**
Speed	9 min/mi	30 min/mi	20 min/mi
kcal "burned" per hour	775	408	245
Cheeseburger (quarter-pound), 417 kcal	32 min	1 hr, 1 min	1 hr, 42 min
Pepperoni pizza (1 large slice), 280 kcal	22 min	42 min	1 hr, 8 min
Soft drink (12 oz), 152 kcal	12 min	22 min	37 min
Whole wheat bread (1 slice), 65 kcal	5 min	10 min	16 min

These data are for a person weighing 68 kg (150 lb).

Food as Building Material

Even if you have stopped growing, your health depends on continuous repair and maintenance of your tissues. The building materials required for such work are provided by the small organic monomers produced during the digestion of food polymers. Your cells can reassemble those universal monomers into various polymers, such as proteins and DNA of your own "brand."

Within limits, your metabolism can change organic material from one form to another and compensate for disparities in nutrient supply and demand. For instance, if a cell has a shortage of a particular amino acid, it may be able to make it from another amino acid that is present in excess. However, certain substances cannot be made from any other materials, so the body needs to receive them in preassembled form. These are called **essential nutrients,** and a healthful diet must include adequate amounts of all of them. The most important of these nutrients are essential amino acids, vitamins, minerals, and essential fatty acids.

Essential Amino Acids All proteins are built from 20 different kinds of amino acids (see Figure 3.20). Twelve of the 20 amino acids can be manufactured by the adult body from other compounds. The other eight are **essential amino acids:** They must be obtained from the diet because human cells cannot make them. (Infants also require a ninth, histidine.)

Different foods contain different proportions of amino acids. Animal products, such as meat, eggs, and milk, are said to be "complete" because they have all the essential amino acids in the proportions you need. In contrast, most plant proteins are incomplete, meaning they are deficient in one or more of the essential amino acids. If you are a vegetarian (by choice or, as for much of the world's population, by economic necessity), the key to good nutrition is to eat a variety of plants, each of which provides different essential amino acids. The combination of a grain and a legume (examples of legumes are peas, beans, and peanuts) often provides the right balance **(Figure 22.19)**. Most societies have a staple meal that includes such a combination.

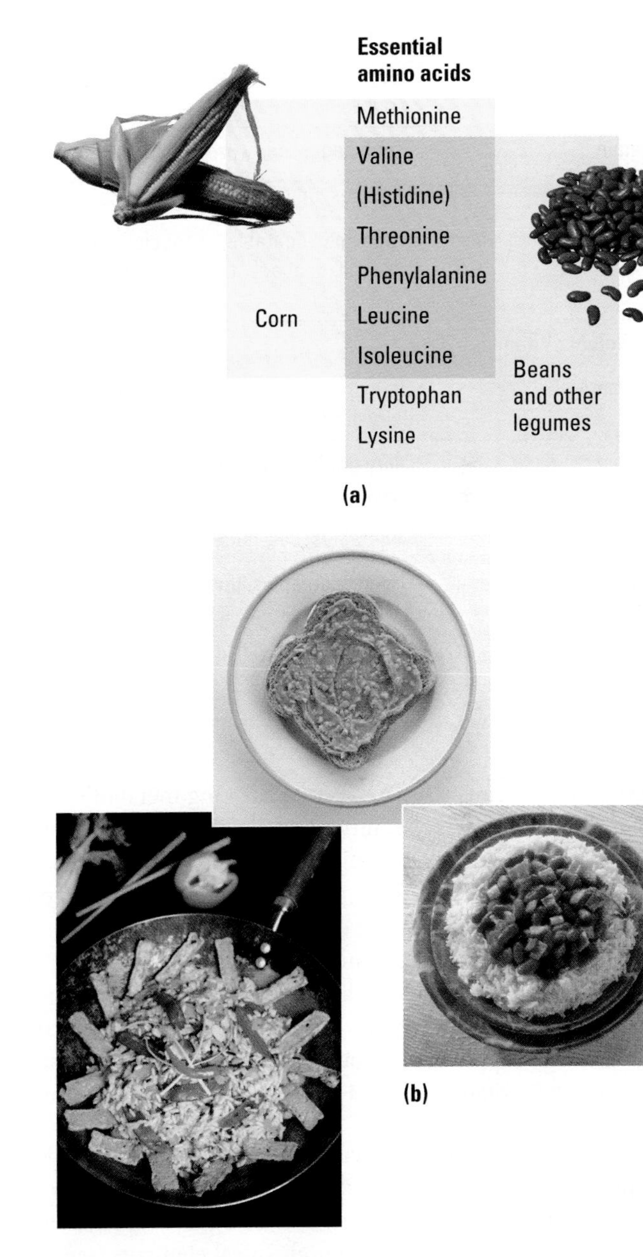

Figure 22.19 Essential amino acids from a vegetarian diet. (a) An adult human can obtain all eight essential amino acids by eating a meal of corn and beans. **(b)** Many societies have a staple meal that combines a grain with a legume, such as beans and rice, peanut butter and bread, or rice and tofu. Such a combination often provides a complete set of essential amino acids.

Vitamins Organic molecules required in the diet for good health are called **vitamins.** Vitamins are required in much smaller amounts than the essential amino acids. One tablespoon of vitamin B_{12}, for instance, can provide the daily requirement of nearly a million people. Most vitamins

Table 22.2	Vitamins	
Vitamin	**Major Dietary Sources**	**Symptoms of Deficiency or Extreme Excess***
Water-Soluble Vitamins		
Vitamin B_1 (thiamine)	Pork, legumes, peanuts, whole grains	Beriberi (nerve disorders, emaciation, anemia)
Vitamin B_2 (riboflavin)	Dairy products, meats, enriched grains, vegetables	Skin lesions such as cracks at corners of mouth
Niacin	Nuts, meats, grains	Skin and gastrointestinal lesions, nervous disorders Flushing of face and hands, liver damage
Vitamin B_6 (pyridoxine)	Meats, vegetables, whole grains	Irritability, convulsions, muscular twitching, anemia Unstable gait, numb feet, poor coordination
Pantothenic acid	Most foods: meats, dairy products, whole grains, etc.	Fatigue, numbness, tingling of hands and feet
Folic acid (folacin)	Green vegetables, oranges, nuts, legumes, whole grains	Anemia, gastrointestinal problems Masks deficiency of vitamin B_{12}
Vitamin B_{12}	Meats, eggs, dairy products	Anemia; nervous system disorders
Biotin	Legumes, other vegetables, meats	Scaly skin inflammation; neuromuscular disorders
Vitamin C (ascorbic acid)	Fruits and vegetables, especially citrus fruits, broccoli, cabbage, tomatoes, green peppers	Scurvy (degeneration of skin, teeth, blood vessels), weakness, delayed wound healing, impaired immunity Gastrointestinal upset
Fat-Soluble Vitamins		
Vitamin A	Dark green and orange vegetables and fruits, dairy products	Vision problems; dry, scaly skin Headache, irritability, vomiting, hair loss, blurred vision, liver and bone damage
Vitamin D	Dairy products, egg yolk; also made in human skin in presence of sunlight	Rickets (bone deformities) in children; bone softening in adults Brain, cardiovascular, and kidney damage
Vitamin E (tocopherol)	Vegetables, oils, nuts, seeds	None well documented
Vitamin K	Green vegetables, tea	Defective blood clotting Liver damage and anemia

**Symptoms of extreme excess are given in red type.*

function as assistants to enzymes in catalyzing metabolic reactions. Although the daily dietary requirements for the 13 vitamins essential to human health are quite small, deficiencies in any of them can cause serious health problems **(Table 22.2)**. A lack of vitamin C, for example, causes scurvy. (To avoid this problem during long sea voyages, British sailors ate citrus fruit, earning them the nickname "limeys.")

People who eat a balanced diet should be able to obtain enough of all needed nutrients in their food. For others, vitamin supplements can fill in the gaps, although supplements should not be used indiscriminately, since overdoses of certain vitamins (such as A, D, and K) can be harmful.

Minerals The organic molecules in our diet, such as carbohydrates, fats, and proteins, provide the four chemical elements most abundant in our body: carbon, oxygen, hydrogen, and nitrogen (see Chapter 2). We also require smaller amounts of 21 other chemical elements (see Figure 2.3) that are acquired mainly in the form of inorganic substances called **minerals.** Like vitamin deficiencies, mineral deficiencies can cause health problems. For example, the mineral calcium is needed as a building material for our bones and teeth and for the proper functioning of nerves and muscles. It can be obtained from dairy products, dark green vegetables such as spinach, and legumes. Too little calcium can result in the degenerative bone disease osteoporosis. Mineral excesses can also cause problems. For example, we require sodium for our nerves and muscles to function, but the average person in the United States, consumes about 20 times the required amount of sodium, mainly in the form of table salt (sodium chloride).

Essential Fatty Acids Our cells make fats and other lipids by combining fatty acids with other molecules, such as glycerol (see Chapter 3). We can make most of the required fatty acids, in turn, from simpler molecules; those we cannot make, called **essential fatty acids,** we must obtain in our diet. One essential fatty acid, linoleic acid (one of the omega-6 family of fatty acids), is especially important because it is needed to make some of the phospholipids of cell membranes. Most diets furnish ample amounts of essential fatty acids, and deficiencies are rare.

Decoding Food Labels

Have you ever read the label on a food product to pass the time while eating? Doing so can be quite enlightening. For example, you might notice that snack foods that proclaim "lite" on their packaging may actually have a significant fat content and contain more sugar than the ordinary version of the product.

To help you get past the hype and assess the nutritional value of a packaged food for yourself, the U.S. Food and Drug Administration (FDA) requires two blocks of information on labels **(Figure 22.20)**. One lists the ingredients, in order from greatest amount to least, by weight. The other lists key nutrition facts. The first item to pay attention to is the serving size; be sure to adjust the rest of the nutrition information to reflect your actual serving. Next, you'll see a wide variety of data, including Calories, fat, cholesterol, carbohydrates, fiber, protein, vitamins, and minerals. The FDA labeling regulations change from time to time; for example, in 2006, manufacturers were newly required to list trans fat levels (see Chapter 3). Reading food labels can't guarantee good nutrition, of course, but they do help us make more informed choices about what we put in our body.

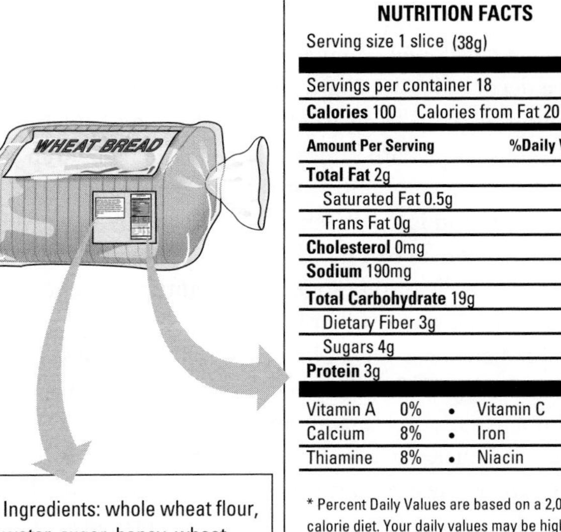

Figure 22.20 FDA-required food labels. The "Nutrition Facts" box defines a serving size and then indicates the number of total Calories per serving and the number of Calories in the form of fats. The box also lists selected nutrients as amounts per serving and as percentages of a daily value. Daily values are based on a 2,000-Calorie-per-day diet (actually 2,000 kilocalories, the "Calories" of food labels). For example, a slice of this bread has 1.5 g of fat, which is 2% of the daily fat allowance for a person needing 2,000 Calories per day. Fats, cholesterol, sugar, and sodium are included because there is evidence that excesses of these nutrients can contribute to certain health problems. The label also features food substances believed to enhance health, such as dietary fiber, protein, and certain vitamins and minerals. The lower part of the label prescribes some guidelines for healthful diets.

CHECKPOINT

1. Why is your actual metabolic rate much higher than your BMR?
2. In what sense is a stable body weight a matter of caloric accounting?
3. What is an "essential nutrient"?
4. How many slices of the bread in Figure 22.20 would you have to eat to obtain one day's recommended amount of fiber?

Answers: 1. Because BMR only accounts for an activity-free existence **2.** We maintain a steady weight when our caloric intake matches our caloric expenditure. **3.** A substance an organism requires but cannot make by its own metabolism **4.** Eight and a half slices

Nutritional Disorders

Considering the central role that nutrients play in a healthy body, it's not surprising that nutritional dysfunction can cause severe problems. In this section, we examine some common nutritional disorders.

Malnutrition and Undernutrition

Living in a developed country where food is plentiful and most people can afford a decent diet, it is hard to relate to starvation. But 800 million people around the world—nearly three times the population of the United States—must cope with hunger. About 11,000 children starve to death each day. The main type of nutritional deficiency is malnutrition.

Malnutrition is a deficiency of one or more of the essential nutrients. The deficiency may be caused by inadequate intake or medical problems (such as metabolic or digestive abnormalities). Due to lack of education about diet, even some affluent people eat badly enough to have vitamin or mineral deficiencies. On a global scale, however, it is protein deficiency—insufficient intake of one or more essential amino acids—that causes the most human suffering, mainly in nonindustrialized countries.

Protein deficiency is concentrated in geographic regions where there is a great gap between food supply and population size. The most reliable sources of essential amino acids are animal products, but these foods are expensive. People forced by economic necessity to get almost all their calories from a single plant staple, such as corn or potatoes, will suffer deficiencies of essential amino acids.

Most victims of protein deficiency are children, who, if they survive infancy, are likely to be retarded in mental and physical development. The resulting syndrome is called kwashiorkor from the Ghanaian word for "rejected one," a reference to the onset of the disease when a child is weaned from its mother's milk and placed on a starchy diet after a sibling is born **(Figure 22.21)**. The problem of protein deficiency in some nonindustrialized countries has been compounded by a trend away from breast-feeding altogether.

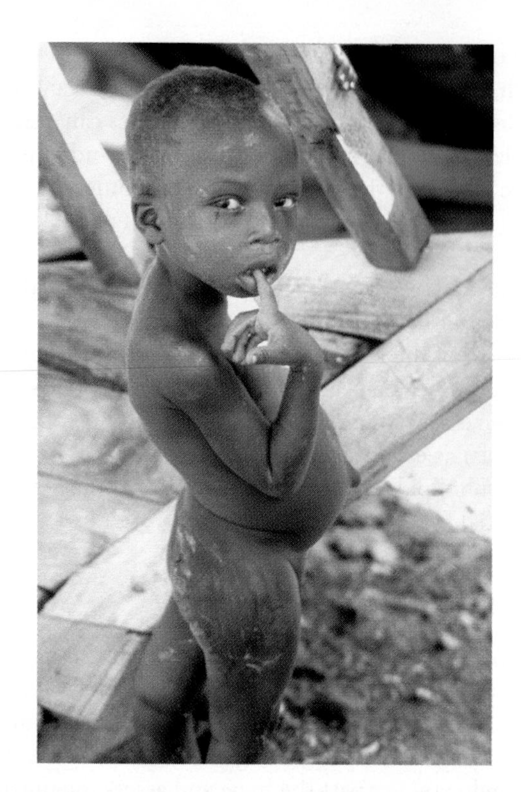

Figure 22.21 Kwashiorkor in a Haitian boy. A form of malnutrition caused by inadequate protein intake, kwashiorkor causes deficiency in blood proteins. This, in turn, causes swelling (edema) of the belly and limbs.

Undernutrition is caused by inadequate intake of calories. An undernourished person isn't getting enough food to supply basic body needs. To compensate, the body begins to break down its own molecules, beginning with stored carbohydrates and fats, then advancing to its own proteins as an energy source to stay alive. Muscles begin to atrophy, and even brain proteins are consumed for fuel. Since any food, even a diet of a single staple such as rice or corn, provides calories, undernourishment usually occurs when a crisis, such as drought or war, has cut off the food supply. As discussed in the chapter opening, eating disorders can also cause undernutrition.

Obesity

In the United States and many other industrialized countries, it is overnourishment that is the nutritional disorder of greatest concern. **Obesity** is defined as an inappropriately high **body mass index (BMI),** a ratio of weight to height. About one-third of all Americans are obese, and another one-third are overweight (a BMI that is between normal and obese). Overnourishment can be an advantage to some animals, such as hibernators. But in humans, obesity increases the risk of heart attack, diabetes, and several other diseases.

It is important to realize that not being as slim or as well toned as a magazine model does not mean that you are obese. Researchers continue to debate how heavy we can be before we are considered unhealthy. Reflecting this uncertainty, BMI charts show a range of acceptable values **(Figure 22.22)**.

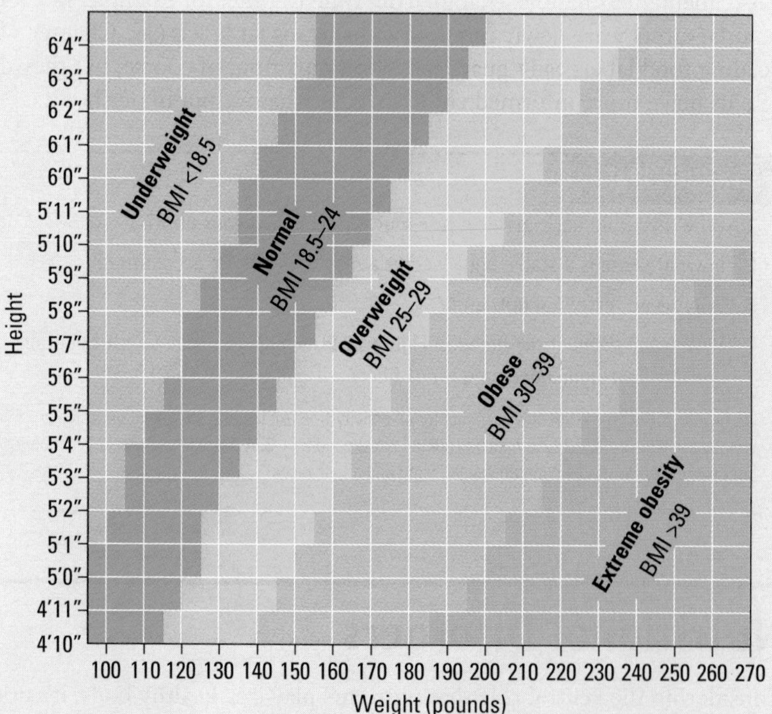

Figure 22.22 Body mass index (BMI): One measure of healthy weight. A range of healthy weights is defined by BMI, the ratio of body height to weight.

To at least some extent, a tendency toward obesity is inherited **(Figure 22.23)**. Researchers have identified over 100 genes that contribute to weight maintenance. Some of those genes affect metabolic rate, some affect the control mechanisms that tell our body when we've had enough to eat, while still others slow the use of stored fat as an energy source. Having a defect in one or more of these genes still can't produce obesity without environmental contribution in the form of lots of fattening foods. But it is partly genetics that explains why certain people have to fight so hard to control their weight, while others can eat all they want without gaining a pound. Remember, the best way to maintain a healthy weight is to eat a balanced diet and get plenty of exercise.

Figure 22.23 A ravenous rodent. The obese mouse on the left has a defect in a gene that normally produces an appetite-regulating protein. Several other genes function in weight management in mammals, including humans.

CHECKPOINT

Children with kwashiorkor may be very thin, but their bellies will be very swollen. Are these children undernourished or malnourished? Explain your answer.

Answer: These children are malnourished. Although they may be receiving enough calories, they are not receiving enough amino acids in their diet to make needed proteins.

EVOLUTION CONNECTION

Fat and Sugar Cravings

The majority of Americans consume too many high calorie foods, which contribute to obesity. Even though we all know that we need to limit the amount of fat in our diet, it sure is hard, isn't it? Most of us crave fatty foods: cheeseburgers, chips, fries, ice cream, candy. For many of us, such foods are "satisfying" in a way that no other foods are. We seem to be hardwired to crave foods that are bad for us.

The seemingly unhelpful trait of craving fat and sugar makes more sense from a broad evolutionary standpoint. It is only very recently, about the last 100 years, that large numbers of people have had access to a reliable supply of food. For most of human history, our ancestors were continually in danger of starvation. On the African savanna, where humans first evolved, people survived by gathering seeds and other edible plant products and hunting game or scavenging meat from animals killed by other predators. Foods that were fatty or sweet were probably hard to come by. In such a feast-or-famine existence, natural selection may have favored individuals who gorged themselves on rich, fatty foods on those rare occasions when such treats were available. Such individuals, with their ample reserves, were more likely than thinner friends to survive famines.

Perhaps our modern taste for fats and sugars reflects the selection advantage it conveyed in our evolutionary history. Of course, today most of us only hunt and gather in grocery stores, fast-food restaurants, and college cafeterias **(Figure 22.24)**. Although we know it is unhealthful, many of us find it difficult to overcome the ancient survival behavior of stockpiling for the next famine. ■

Figure 22.24 The modern hunter-gatherer.

Chapter Review

SUMMARY OF KEY CONCEPTS

For study help and activities, go to campbellbiology.com or the student CD-ROM.

Biology and Society: Eating Disorders

- Eating disorders include anorexia (self-starvation) and bulimia (cycles of bingeing and purging).

Overview of Animal Nutrition

- Animals must eat other organisms to obtain organic molecules for energy and building materials.
- **Animal Diets** Herbivores mainly eat plants and algae, carnivores mainly eat other animals, and omnivores eat both animals and plants.

Activity *How Animals Eat Food*

- **The Four Stages of Food Processing**

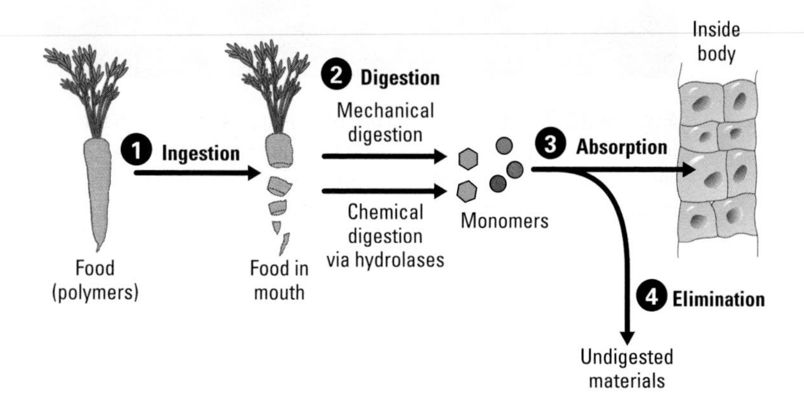

- **Digestive Compartments** In animals, hydrolysis of food occurs within food vacuoles (membrane-bounded compartments within cells), gastrovascular cavities (extracellular compartments with a single opening), or alimentary canals (internal compartments with two separate openings, a mouth and an anus).

A Tour of the Human Digestive System

MP3 Tutor *The Human Digestive System*

- **System Map**

Alimentary canal	Accessory organs	Digestion		Absorption
		Mechanical	Chemical	
Mouth (oral cavity)	Salivary glands	Chewing	Salivary amylase	
↓ Pharynx and esophagus				
↓ Stomach		Churning	Acid and pepsin (in gastric juice)	
↓ Small intestine	Liver, gallbladder, pancreas		Other hydrolases	Nutrients and water
↓ Large intestine				Water
↓ Anus				

- **The Mouth** Within the mouth (oral cavity), teeth function in mechanical digestion, breaking food into smaller pieces. A hydrolase in saliva, salivary amylase, initiates chemical digestion by starting the hydrolysis of starch.

Case Study in the Process of Science *What Role Does Salivary Amylase Play in Digestion?*

- **The Pharynx** This throat chamber leads to both the digestive and respiratory systems. A food-swallowing reflex closes the trachea and opens the esophagus.
- **The Esophagus** This muscular tube connects the pharynx to the stomach and moves food along via rhythmic muscular contractions called peristalsis.
- **The Stomach** This elastic organ stores food, functions in mechanical digestion by churning, and uses gastric juice for chemical digestion. Mucus secreted by the cells of the stomach lining helps protect it from self-digestion. Sphincters regulate the passage of food into and out of the stomach. In some people, a bacterium infects the stomach lining and causes ulcers.
- **The Small Intestine** Small in diameter but very long, the small intestine is the main organ of chemical digestion and nutrient absorption. Most chemical digestion occurs in the duodenum, the first part of the small intestine, where acid chyme from the stomach mixes with pancreatic juice, bile, and a digestive juice secreted by the intestinal lining. These digestive juices include all of the hydrolases necessary to complete the dismantling of food polymers to their monomers. The rest of the small intestine is specialized for absorption of these monomers. Across the large surface area of the intestinal lining, with its microvilli upon villi, intestinal cells transport nutrients into capillaries of the circulatory system.
- **The Large Intestine (and Beyond)** The colon, which is wider and shorter than the small intestine, makes up most of the large intestine. The large intestine completes the reclaiming of water that entered the alimentary canal as the solvent of digestive juices. By the time undigested wastes reach the rectum, most water has been reabsorbed into the blood, and the relatively solid feces can be stored in the rectum until eliminated via the anus.

Activity *Human Digestive System*

Human Nutritional Requirements

- **Food as Fuel** Cellular respiration extracts energy from food molecules and uses it to generate ATP. Our metabolic rate depends on our basal metabolic rate (energy expenditure at complete rest) plus additional energy expenditure due to our activity level.
- **Food as Building Material** In addition to providing organic compounds for building material, we require a set of specific nutrients in our diet. Essential amino acids, which we cannot make from other molecules, are required for protein production. Vitamins, organic molecules required in very small amounts, function mainly as enzyme helpers. Minerals are inorganic nutrients that provide essential chemical elements other than carbon, oxygen, hydrogen, and nitrogen.
- **Decoding Food Labels** The U.S. Food and Drug Administration requires labels on packaged foods that list ingredients in descending order of abundance and provide information about calories and specific nutrients.

Activity *Analyzing Food Labels*

Nutritional Disorders

- **Malnutrition** The most common dietary deficiency in nonindustrialized countries, malnutrition is a deficiency of one or more essential nutrients, most often essential amino acids (protein deficiency). Undernutrition is a caloric deficiency.

• **Obesity** Defined as an inappropriately high ratio of weight to height, obesity is the most common nutritional disorder in most industrialized countries. About one-third of all Americans are obese, putting them at higher risk for heart disease.

Activity *Case Studies of Nutritional Disorders*

SELF-QUIZ

1. A new species of large, multicellular animal has just been discovered. Most animals have a digestive tube, but at first, this animal appears to use food vacuoles. Why is it unlikely that this animal uses only food vacuoles for digestion?

2. A friend says, "It's not eating that causes you to gain weight, it's absorption." Is that statement true?

3. __________ is the chemical and mechanical breakdown of food into small molecules, while __________ is the uptake of these small molecules by the body's cells.

4. A patient enters the emergency room at a hospital with advanced stomach ulcers. The ulcers have damaged all tissue layers in her stomach, and she is bleeding into her abdominal cavity. Her condition must be immediately corrected by surgery. What type of organism is responsible for most stomach ulcers? If the ulcers had been detected earlier, how could they have been treated?

5. Most chemical digestion occurs in the __________ as a result of the activity of enzymes made by the __________.
   a. stomach; oral cavity
   b. pancreas; stomach
   c. large intestine; small intestine
   d. small intestine; pancreas

6. If you maintain a normal activity level and consume the same number of calories needed for your basal metabolic rate, you will
   a. gain weight.
   b. lose weight.
   c. stay the same weight.
   d. lose weight but then quickly regain it.

7. Why is the amount of oxygen you consume proportional to your metabolic rate?

8. Why are vitamins required in such small doses compared to essential amino acids?

9. Imagine that you and your roommate both eat according to FDA nutritional guidelines, but you are a vegetarian, while your roommate is not. Why do you have to be more concerned about the nutritional content of your food than your roommate does? What can you do to address this concern?

10. Your roommate informs you that she has no time to eat properly, so she is taking megadoses of vitamin supplements. You warn her that this might be dangerous. Which vitamin excess could pose the largest threat to her?
   a. A
   b. thiamin (B_1)
   c. C
   d. folic acid

Answers to the Self-Quiz questions can be found in Appendix D.

Go to the website or CD-ROM for more Self-Quiz questions.

THE PROCESS OF SCIENCE

11. Nutritional labels provide us with basic information concerning the food we eat, but often those labels don't tell us everything we need to know. For example, the FDA recommends that about 12% of our daily calories come from protein. How do we know if our diet is in accordance with those regulations? Labels tell us how many Calories are in the products, but often they do not directly tell us how many of those Calories come from protein. Proteins and carbohydrates have approximately 4 Calories per gram, while fats contain 9 Calories per gram. The nutritional label on a certain brand of cookies is shown below. Calculate how many Calories in this product are from fat, from carbohydrates, and from protein.

Total Fat 7g
Saturated Fat 5g
Cholesterol 0mg
Sodium 80mg
Total Carbohydrates 18g
Dietary Fiber 1g
Sugars 8g
Protein 1g

12. Sue has been obese for years. She has tried diet and exercise to lose weight, but nothing seems to work. If she doesn't lose weight, her health will suffer greatly. Sue has been reading about gastric bypass surgery as an option to lose large amounts of weight. In this major operation, the stomach is surgically divided into a top and bottom pouch. The top pouch is very tiny, so that only a small amount of food can enter. The duodenum of the small intestine is cut and attached to the small pouch of the stomach. Food from the small pouch automatically bypasses a portion of the small intestine so that less food can be absorbed. Using what you learned about the digestive process, discuss some of the problems Sue might have after the surgery concerning her nutrition and the functioning of her digestive system.

BIOLOGY AND SOCIETY

13. Obesity is at an all-time high in our country and increasing. Every day we are bombarded by advertisements promising to help us shed extra pounds the easy way with no exercise, no dieting, and no side effects. Some of the companies that market these products make false claims in their advertising. People spend billions of dollars on weight loss supplements each year in the hopes that the claims are true. Most people simply end up disappointed; but occasionally serious and even fatal consequences can occur. How could you test the claims of diet-marketing companies? Do you think that the makers of these products should be held accountable for any false claims? How would you propose that they be held accountable?

14. As described in question 12, some obese individuals elect to have gastric bypass surgery in an attempt to lose weight. Do you think that insurance companies should pay for such surgeries? Why or why not?

23 Circulation and Respiration

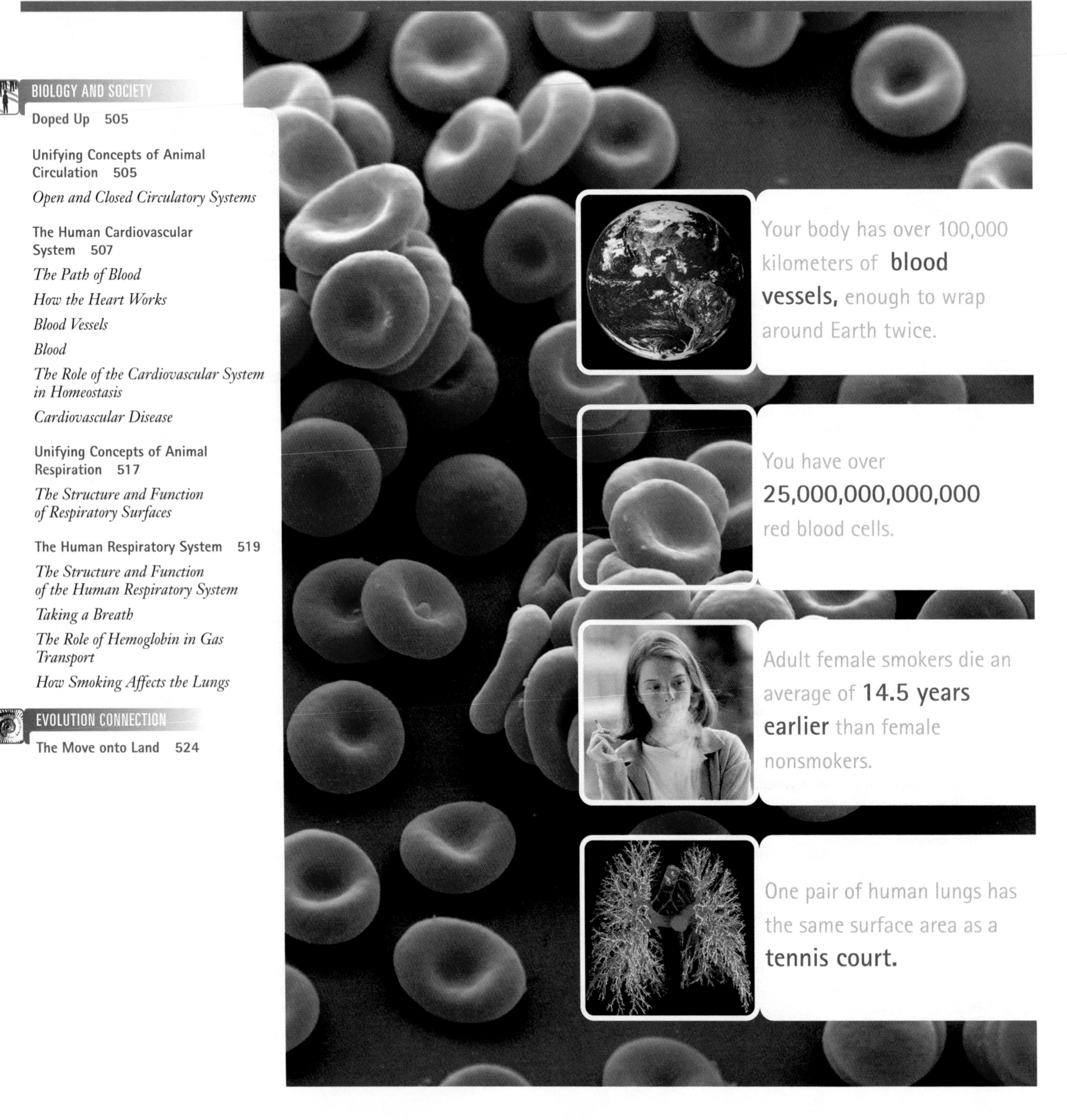

Your body has over 100,000 kilometers of **blood vessels,** enough to wrap around Earth twice.
You have over **25,000,000,000,000** red blood cells.
Adult female smokers die an average of **14.5 years earlier** than female nonsmokers.
One pair of human lungs has the same surface area as a **tennis court.**

BIOLOGY AND SOCIETY

Doped Up

The integrity of the Tour de France—the greatest event in the sport of cycling—began to unravel on July 8, 1998. Just three days before the start of the race, a masseur working for a French cycling team was caught by police with huge quantities of performance-enhancing drugs. By the end of the Tour, the scandal had rippled through the cycling community, resulting in the expulsion or withdrawal of 7 of the original 21 teams and nearly half the riders. The press dubbed that year's race "The Tour of Shame" **(Figure 23.1)**.

At the center of the controversy was the chemical erythropoietin (EPO). EPO is a hormone produced by the kidneys in response to low oxygen concentrations in body tissues. After traveling to the bone marrow, EPO boosts the production of oxygen-carrying red blood cells. Doctors prescribe synthetic EPO to relieve the symptoms of anemia—a condition caused by a shortage of red blood cells or their components—due to kidney disease, cancer, or AIDS. Because additional red blood cells boost the functions of the circulatory and respiratory systems, EPO increases athletic stamina, potentially offering a huge advantage in a race like the grueling 21-day, 2,000-mile Tour de France.

EPO abuse is difficult to detect because it is a hormone that is produced naturally by the body and because artificial EPO is rapidly cleared from the bloodstream. One way that athletic commissions test for cheaters is by measuring the percentage of red blood cells in the blood volume. A value over 50% is usually grounds for disqualification.

EPO is banned not just because it confers an unfair advantage, but also because it is dangerous. Artificially increasing the oxygen-carrying capacity of the blood can lead to medical problems such as abnormal clotting, heart disease, stroke, and even death. EPO has been blamed for the deaths of dozens of athletes, including eight professional European cyclists in 2003 and 2004 alone. (The disqualification of 2006 Tour de France winner Floyd Landis was the result of abnormal levels of testosterone, not EPO, in his blood.)

Proper functioning of the circulatory and respiratory systems is essential not only to endurance athletes but to all humans and most other animals. The two systems are such close partners that we will explore them in a single chapter. For each of the two systems, we'll start by considering general challenges that face all animals and survey the variety of ways these are addressed. We'll then take a closer look at each system in humans and end by examining what may happen when these systems fail. ■

Figure 23.1 1998's "Tour of Shame." These cyclists are protesting the blood doping scandals that marred the 1998 Tour de France.

Unifying Concepts of Animal Circulation

Every organism must exchange materials and energy with its environment. In simple animals, such as hydras and jellies (also known as jellyfishes), all the cells are in direct contact with the environment. Thus, every cell can easily exchange materials with its environment by **diffusion,** the movement of molecules along a concentration gradient, from an area of higher concentration to an area of lower concentration (see Figure 5.12). However, most animals are too large or too complex for exchange to take place by diffusion alone. In such animals, a

circulatory system facilitates the exchange of materials, providing a rapid, long-distance internal transport system that brings resources close enough to cells for diffusion to occur. Once there, the resources that cells need, such as nutrients and oxygen (O_2), can enter the cytoplasm through the plasma membrane. And metabolic wastes, such as carbon dioxide (CO_2), diffuse from the cells to the circulatory system for disposal.

Open and Closed Circulatory Systems

All but the simplest animals have a circulatory system with three main components: a central pump, a vascular system (a set of tubular blood vessels), and circulating fluid. Two main types of circulatory systems have evolved. Many invertebrates, including most molluscs and all arthropods, have what is called an **open circulatory system.** The system is termed "open" because the circulating fluid is pumped through open-ended vessels and flows out among the cells **(Figure 23.2a)**.

In a **closed circulatory system,** blood is confined to vessels and is distinct from the interstitial fluid, the fluid that fills the spaces around cells **(Figure 23.2b)**. One or more hearts pump blood into large vessels that branch into smaller ones coursing through the organs. Earthworms, octopuses, and vertebrates are examples of animals with a closed circulatory system.

The closed circulatory system in vertebrates, including humans, is called a **cardiovascular system,** which refers to both the heart (from the Greek *kardia*) and blood vessels (from the Latin *vas*). The fish in Figure 23.2b illustrates the main features of a cardiovascular system. The heart receives blood in a chamber called the **atrium** (plural, *atria*). A second chamber called the **ventricle** pumps blood away from the heart. The blood is confined within three main types of blood vessels: arteries, capillaries, and veins. **Arteries** carry blood away from the heart, branching into smaller **arterioles** as they approach the organs. Blood then flows from arterioles into capillary beds, networks of tiny vessels called **capillaries** that infiltrate nearly every organ and tissue in the body. The thin walls of capillaries allow exchange between the blood and interstitial fluid. In fish, blood is oxygenated in gill capillaries and then travels via arteries to other capillary beds, where oxygen enters tissue cells. Capillaries converge into **venules,** which in turn converge into larger **veins** that return blood back to the heart. Capillaries are the functional center of the circulatory system: This is where materials are transferred to and from surrounding tissues. Arteries and veins primarily transport blood back and forth between the heart and capillaries.

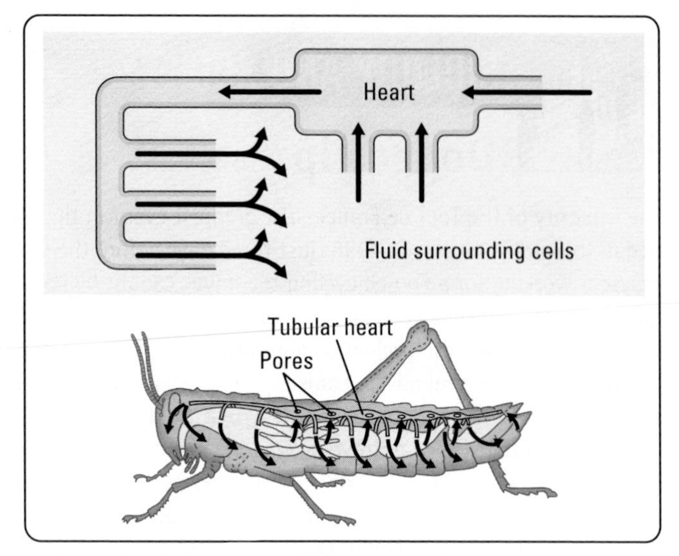

(a) Open circulatory system. In the open system of an insect, such as the grasshopper, the tubular heart pumps the circulatory fluid into the body tissues. Nutrients then diffuse from the fluid directly into the body cells. At the same time, contractions of body muscles move the fluid toward the tail. When the heart relaxes, the fluid returns to it through several pores.

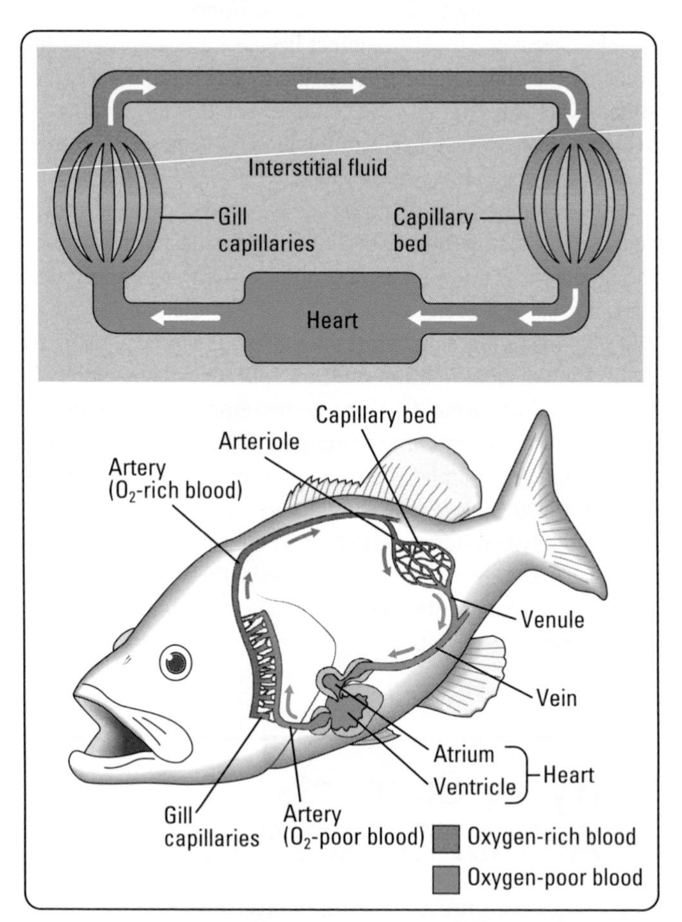

(b) Closed circulatory system. In the closed system of a fish, blood is always contained within vessels, separate from the interstitial fluid bathing the cells. The arrows indicate the direction of blood flow. Blood is pumped from the heart to the gills, to capillary beds in body tissues, and back to the heart.

Figure 23.2 Open and closed circulatory systems. In all figures in this chapter, the color red represents oxygen-rich blood, while blue represents oxygen-poor blood.

CHECKPOINT

1. Why can't the human body rely solely on diffusion to provide all needed chemicals to body cells?
2. Consider the different types of blood vessels in a cardiovascular system: capillaries, arteries, arterioles, venules, and veins. In what order does the blood travel through them as it moves from the heart to body cells and then back to the heart?

Answers: 1. Diffusion is only effective over short distances. Diffusion alone would take too long to convey materials throughout the large human body. **2.** Heart → arteries → arterioles → capillaries → venules → veins → heart

The Human Cardiovascular System

Recall that the circulatory system of most animals has three main components: a pump, a vascular system, and a circulating fluid. In the human cardiovascular system, the central pump is the heart, the vascular system is the blood vessels, and the circulating fluid is the blood.

The Path of Blood

In humans and other terrestrial vertebrates, the three components of the cardiovascular system are organized into a **double circulation system** with two distinct circuits of blood flow. As **Figure 23.3a** shows, the **pulmonary circuit** carries blood between the heart and the lungs. In the lungs, carbon dioxide diffuses one way (from the blood into the lungs) while oxygen diffuses the other way (from the lungs into the blood). The pulmonary circuit then returns the oxygen-rich blood to the heart. As shown in **Figure 23.3b**, the **systemic circuit** carries blood between the heart and the rest of the body. It supplies oxygen and nutrients to body tissues and organs and picks up carbon dioxide and other wastes from them. The oxygen-poor blood then returns to the heart via the systemic circuit.

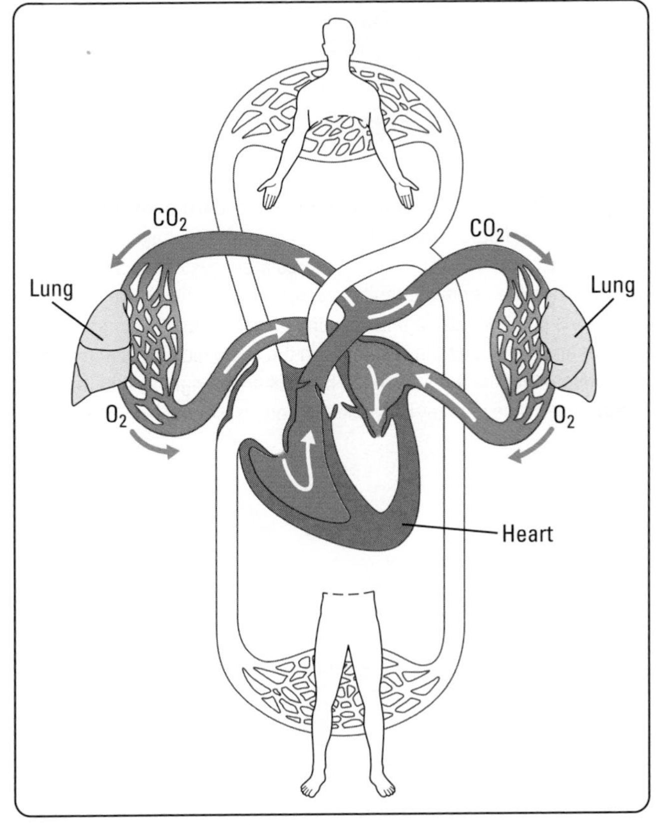

(a) Pulmonary circuit. In organisms with double circulation, the pulmonary circuit transports blood between the heart and lungs.

(b) Systemic circuit. The systemic circuit transports blood between the heart and body tissues.

Figure 23.3 Double circulation.

Figure 23.4 traces the path of blood in the human cardiovascular system as it makes one complete trip. It takes the blood in your body about 1 minute to travel through the circulatory system—less time than it will take you to read about it!

Let's start at ❶ the right atrium, where two large veins empty O_2-poor blood from the body into the heart. The blood is then pumped to ❷ the right ventricle, which pumps the O_2-poor blood to the lungs via ❸ two pulmonary arteries. ❹ As the blood flows through capillaries in the lungs, CO_2 diffuses out of the blood and O_2 diffuses into the blood. The newly O_2-rich blood then flows through ❺ the pulmonary veins to ❻ the left atrium of the heart, completing the pulmonary circuit. Next, the blood is pumped a short distance from the left atrium into ❼ the left ventricle. Blood leaves the left ventricle through ❽ the aorta, the largest blood vessel in the body, with a diameter about as big as a quarter. Branching from the aorta are several large arteries that lead to ❾ the head and arms, and the abdominal organs and legs. For simplicity, this diagram does not show the individual organs, but near each one, arteries lead to arterioles that in turn branch into capillaries that penetrate each organ. Diffusion to and from cells takes place across the thin walls of the capillaries. Downstream, the capillaries join into venules, which convey the blood back into veins. Oxygen-poor blood from the upper body and head is channeled into ❿ a large vein called the superior vena cava; ⓫ another large vein, the inferior vena cava, receives blood from the lower body. These two veins complete the circuit by returning blood to the heart.

In the section that follows, you'll learn about the structure and function of each component in the human cardiovascular system and how the components work together. You may find it useful to refer back to Figure 23.4 as you learn the details of each component.

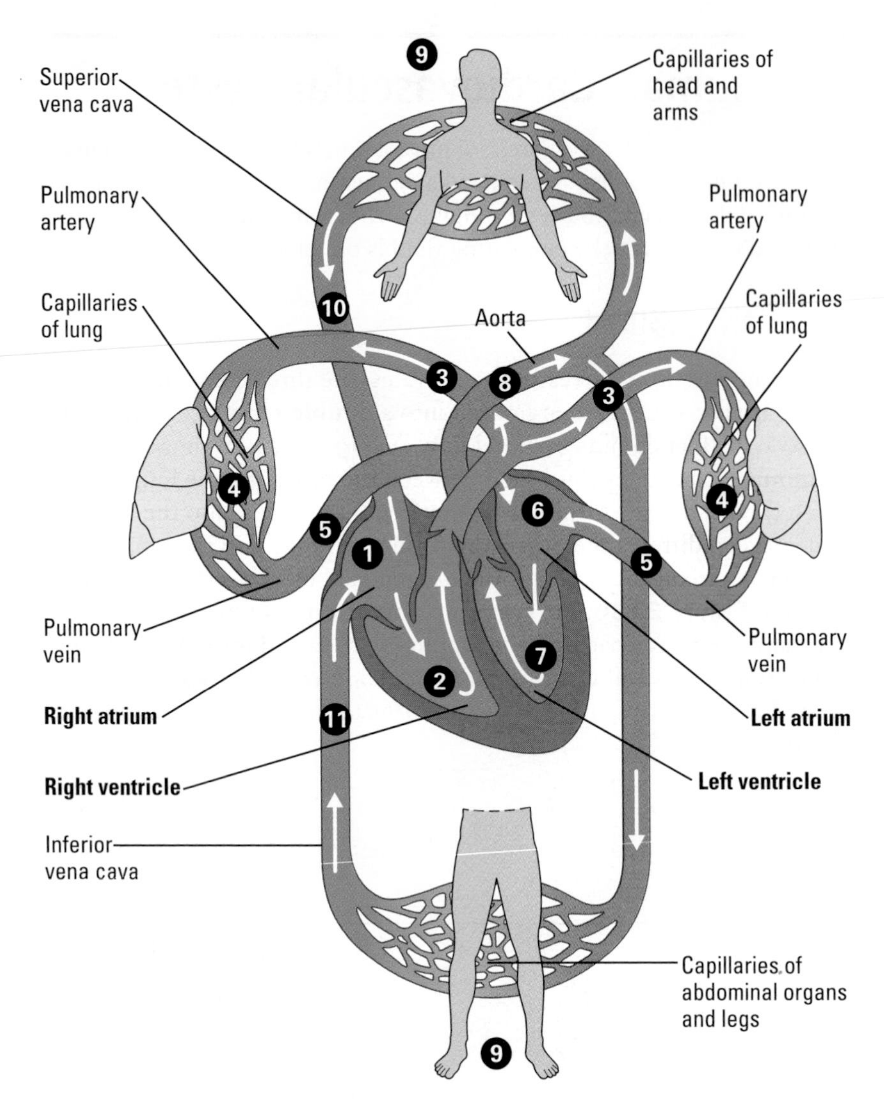

Figure 23.4 A trip through the human cardiovascular system. This diagram traces the path of blood as it makes one full trip around the body. If we consider the circulatory system as a whole, the pulmonary and systemic circuits operate simultaneously. The two ventricles pump almost in unison, sending some blood through the pulmonary circuit and the rest through the systemic circuit at the same time. (Note that the left side of the heart is on the right side of the figure—and vice versa—because the body is viewed from the front.)

How the Heart Works

The hub of the human cardiovascular system is the **heart,** a muscular organ about the size of a fist located under the breastbone. Remember that humans have a double circulation system, with the pulmonary circuit pumping blood between the heart and lungs, and the systemic circuit transporting blood to and from the rest of the body. This double circuit is supported by a four-chambered heart. **Figure 23.5** shows the flow of blood through these chambers.

The thin-walled atria collect blood returning to the heart and pump it into the ventricles. The thicker-walled ventricles, which pump blood out of the heart to the other body organs are much more powerful than the atria. Four valves in the heart prevent backflow and keep blood moving in the correct direction. Notice that the four-chambered structure of the heart prevents the oxygen-rich and oxygen-poor blood from mixing. The right atrium receives oxygen-poor blood from body tissues via two

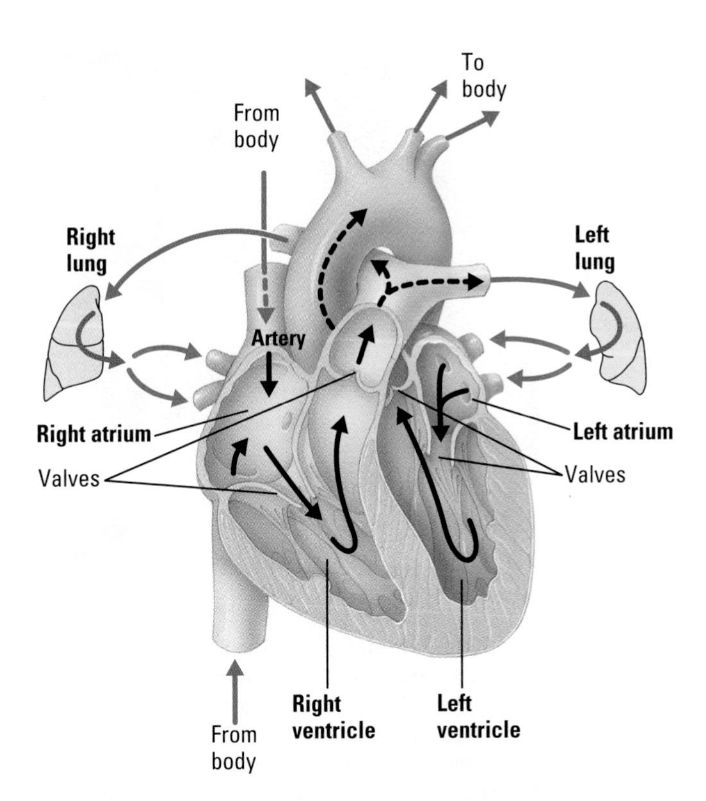

Figure 23.5 Path of blood flow through the human heart. The heart contains four chambers, with two atria located above two ventricles. Valves help maintain the direction of blood flow. Notice that oxygen-poor blood and oxygen-rich blood are kept separate.

large veins, and the right ventricle pumps it to the lungs. The left atrium receives oxygen-rich blood from the lungs, and the left ventricle pumps it out to body organs.

The Cardiac Cycle The heart relaxes and contracts rhythmically in what is called the **cardiac cycle (Figure 23.6)**. When the heart relaxes, the chambers fill with blood; when the heart contracts, it pumps blood. The relaxation phase of the heart cycle is known as **diastole;** the contraction phase is called **systole.** In a healthy adult person at rest, the number of beats per minute, or **heart rate,** ranges between 60 and 80.

During its cycle, the heart makes a distinctive "lub-dupp, lub-dupp" sound as the heart valves snap shut. You can hear this sound with a stethoscope or by pressing your ear against another person's chest. A trained ear can also detect the sound of a **heart murmur,** which may indicate a defect in one or more of the valves. A serious murmur sounds like a "hisssss" as a stream of blood squirts backward through a defective valve. Some people are born with murmurs, while others have valves damaged by infection (due to rheumatic fever, for instance). Severe murmurs indicate an excessive backwash of blood, resulting in inefficient blood flow that can lead to death. However, most cases of heart murmur are not serious, and those that are can be corrected by replacing the damaged valves with artificial ones or with valves taken from an organ donor (human or other animal, usually a pig).

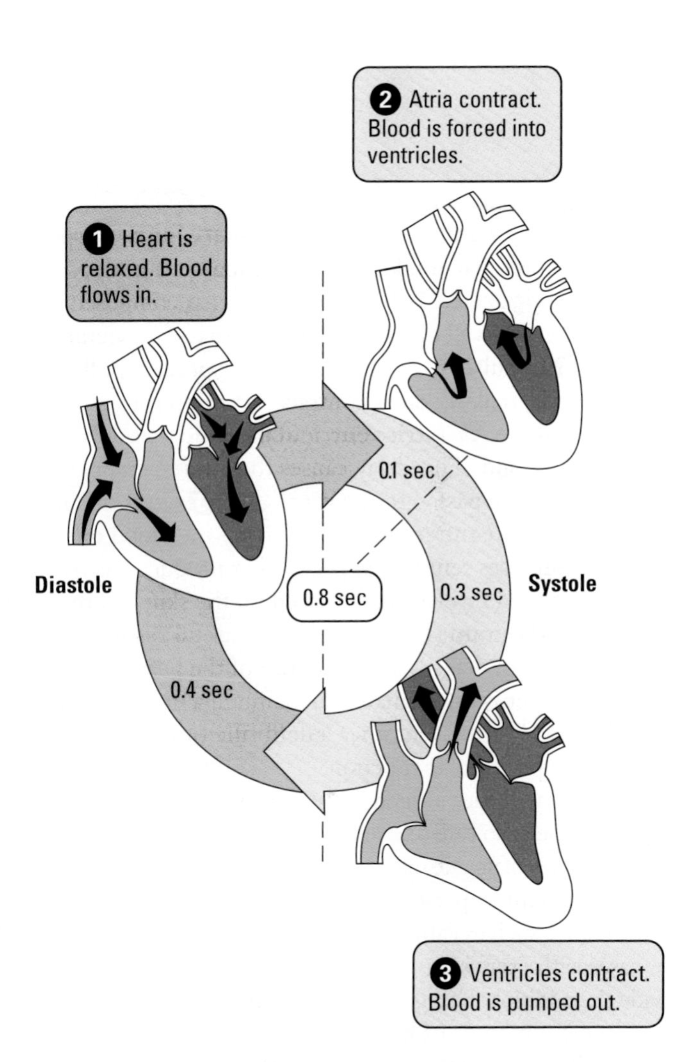

Figure 23.6 The cardiac cycle. The muscles of the heart relax (diastole) and contract (systole) in a rhythmic cycle. This diagram follows the heart through a cycle lasting 0.8 second. ❶ During diastole, which lasts about 0.4 second, blood returning to the heart via veins flows into all four chambers. ❷ During the first 0.1 second of systole, the atria contract, forcing all the blood into the ventricles. ❸ During the remaining 0.3 second of systole, the ventricles contract, pumping blood out of the heart and into the aorta and pulmonary arteries.

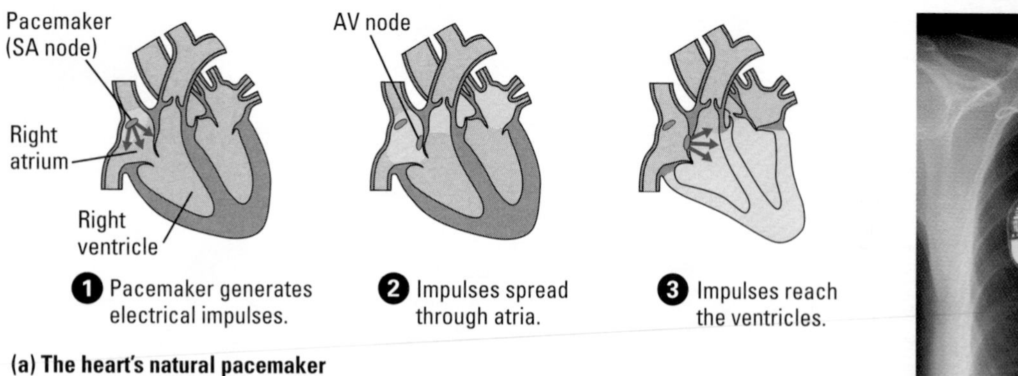
Pacemaker
(SA node)
AV node
Right
atrium
Right
ventricle
1 Pacemaker generates electrical impulses.
2 Impulses spread through atria.
3 Impulses reach the ventricles.

(a) The heart's natural pacemaker

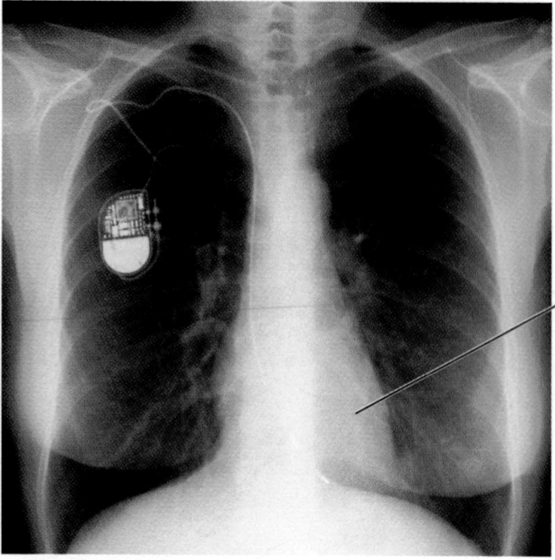
Heart

(b) Artificial pacemaker

Figure 23.7 Pacemakers. **(a)** The SA (sinoatrial) node is the heart's natural pacemaker. It is located within the muscle tissue in the wall of the right atrium. **(b)** An artificial pacemaker is a small electronic device that is surgically implanted into cardiac muscle or the chest cavity near the SA node. It helps to maintain the proper electrical rhythms of a defective heart that is unable to do so for itself.

The Pacemaker and the Control of Heart Rate A special region of the heart called the **pacemaker,** or **SA (sinoatrial) node,** sets the tempo of the heartbeat **(Figure 23.7a)**. The pacemaker is composed of specialized muscle tissue in the wall of the right atrium that ❶ generates electrical impulses. ❷ Impulses from the pacemaker spread rapidly through the walls of both atria, making them contract. The impulses then pass to a relay point called the **AV (atrioventricular) node** that delays the signals by about 0.1 second. This delay causes the atria to empty completely before the impulses are passed to the ventricles. ❸ Once the impulses reach the ventricles, they contract strongly, driving the blood out of the heart.

The impulses sent by the pacemaker produce electrical currents that can be detected by electrodes placed on the skin and recorded as an **electrocardiogram (ECG** or **EKG)**. Careful reading of an EKG can provide a wealth of data about the health of the heart. During a heart attack, the pacemaker is often unable to maintain a normal rhythm. Electrical shocks applied to the chest via a defibrillator may reset the pacemaker and restore proper cardiac function.

A variety of cues influence the signals sent by the pacemaker. For example, epinephrine (also called adrenaline), the "fight-or-flight" hormone released during times of stress, increases heart rate (see Chapter 25). Adrenaline is often injected into a person having a heart attack to help restart a stopped heart. Stimulants, such as caffeine, also make the heart beat faster. And heart rate increases with exercise, an adaptation that enables the circulatory system to provide the additional oxygen needed by muscles hard at work.

In certain kinds of heart disease, the heart's electrical control mechanism fails to maintain a normal rhythm. The remedy is an artificial pacemaker, a small electronic device surgically implanted near the SA node. This device emits electrical signals to maintain normal heart rhythms **(Figure 23.7b)**.

Blood Vessels

If we think of the heart as the body's "pump," then the system of arteries, veins, and capillaries connected to it can be thought of as the "plumbing." If you review Figure 23.4, you'll notice that arteries and veins are distinguished by the direction in which they carry blood: Arteries carry blood *away from* the heart, and veins carry blood *toward* the heart. Capillaries allow for exchange between the bloodstream and the tissue cells (via interstitial fluid).

All blood vessels are lined by a thin layer of tightly packed epithelial cells. Structural differences in the walls of the different kinds of blood vessels correlate with their different functions **(Figure 23.8)**. Capillaries have very thin walls—often just one cell thick—that allow the exchange of substances between the blood and the interstitial fluid that bathes tissue cells. The smallest capillaries are so small that blood cells must pass through them single file. The walls of arteries have two additional, thicker layers. An outer layer of elastic connective tissue allows the vessels to stretch and recoil. Between this layer and the epithelial cells is a middle layer of smooth muscle. The smooth muscle provides the strength and elasticity to accommodate the rapid flow of blood at high pressure produced by the beating heart. Veins convey blood back to the heart at low velocity and pressure after the blood has passed through capillary beds. Veins (but not arteries) also have one-way valves that prevent backflow, ensuring that blood always moves toward the heart.

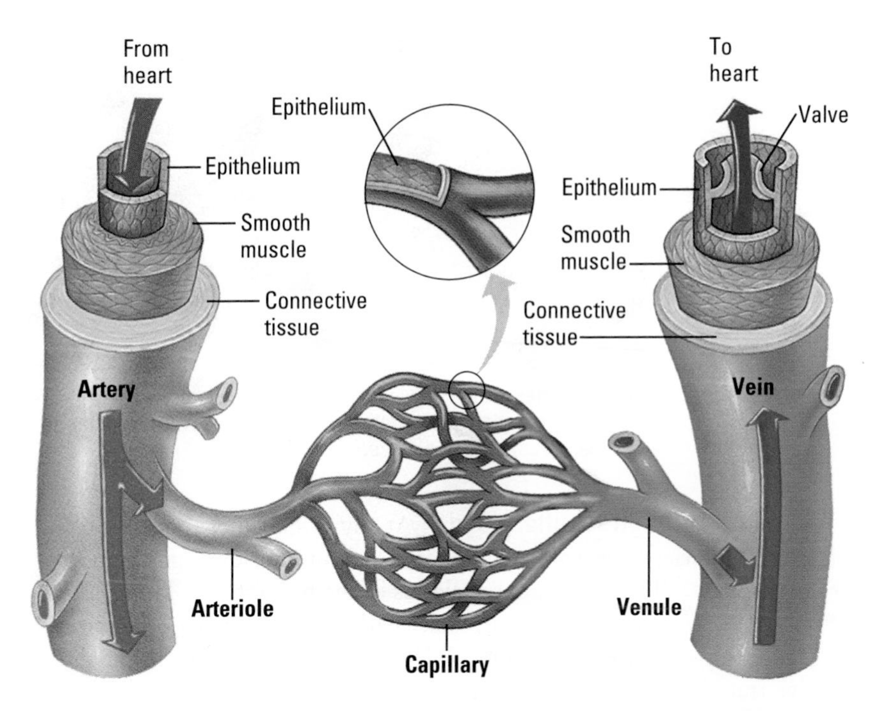

Figure 23.8 The structure of blood vessels. Arteries branch into smaller vessels called arterioles, which in turn branch into capillaries. Chemical exchange between blood and interstitial fluid occurs across the thin walls of the capillaries. The capillaries converge into venules, which deliver blood to veins. All of these vessels are lined by a thin, smooth epithelium. Arteries and veins also have two additional layers.

Blood Flow Through Arteries The force that blood exerts against the walls of your blood vessels is called **blood pressure.** Created by the beating of the heart, blood pressure is the main force driving the blood from the heart through the arteries and arterioles to the capillary beds. When the ventricles contract, blood is forced into the arteries faster than it can flow into the arterioles. This creates pressure that stretches the elastic walls of the arteries. You can feel this effect when you measure your heart rate by taking your pulse. A **pulse** is the rhythmic stretching of the arteries caused by the pressure of blood forced into the arteries during systole. The elastic walls of the arteries recoil during diastole, maintaining enough pressure on the remaining blood to sustain a constant flow into arterioles and capillaries. Thus, blood pressure is recorded as two numbers, such as 120/80 ("120 over 80"). The first number is blood pressure during systole (in millimeters of mercury, a standard pressure unit); the second number is the blood pressure that remains in the arteries during diastole.

Normal blood pressure falls within a range of values, but optimal blood pressure for adults is below 120 systolic and below 80 diastolic (as measured on the upper arm). Lower values are generally considered better, although very low blood pressure may lead to light-headedness and fainting. Blood pressure higher than the normal range may indicate a serious cardiovascular disorder.

High blood pressure, or **hypertension,** is persistent systolic blood pressure higher than 140 and/or diastolic blood pressure higher than 90. Hypertension affects approximately one-quarter of the adult population in the United States. It is sometimes called a "silent killer" because it often displays no outward symptoms for years while increasing the risk of heart attacks, heart disease, and strokes.

Lifestyle changes can help control hypertension: eating a heart-healthy diet, avoiding excess alcohol (more than two drinks per day), exercising regularly, and maintaining proper weight. If lifestyle changes don't work to decrease hypertension, there are several effective medications that can usually help.

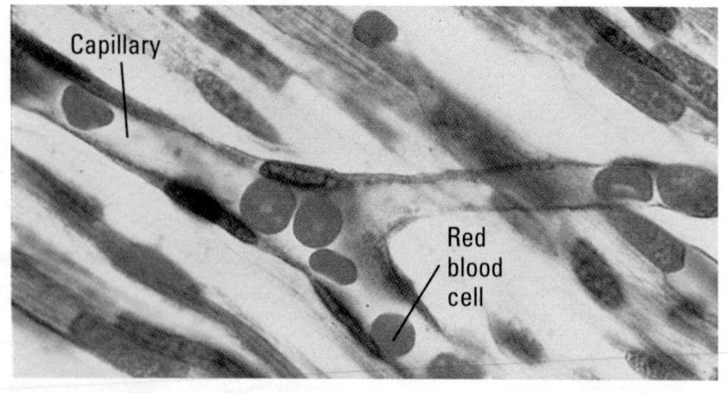

(a) Capillaries. Blood flowing through the circulatory system eventually reaches capillaries, the small vessels where exchange with cells actually takes place.

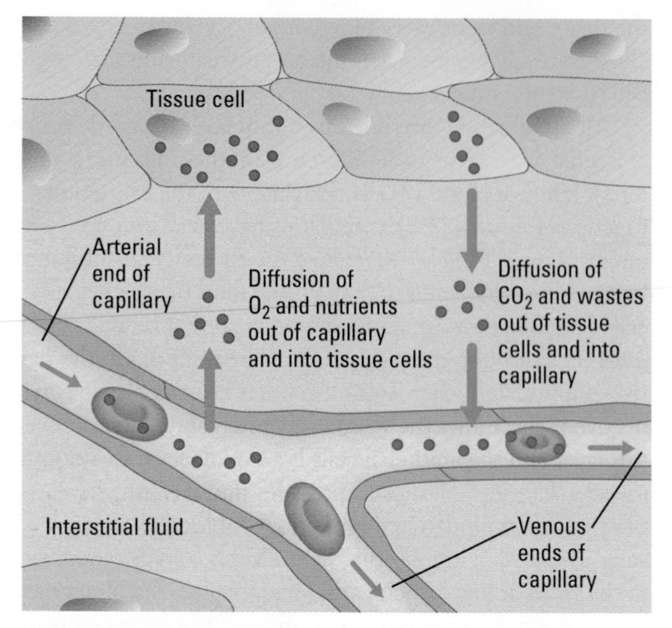

(b) Chemical exchange. Within the capillary beds, there is local exchange of molecules between the blood and interstitial fluid, which bathes the cells of tissues.

Figure 23.9 Chemical exchange between the blood and tissue cells.

Blood Flow Through Capillary Beds At any given time, about 5–10% of your capillaries have a steady flow of blood running through them **(Figure 23.9a)**. These include capillaries in the brain, heart, kidneys, and liver, which are usually filled to capacity with blood. In many other sites, the blood supply varies. After a meal, for instance, blood flow to the digestive tract increases. During strenuous exercise, blood is diverted from the digestive tract and supplied more generously to skeletal muscles and skin. This is one reason why heavy exercise immediately after eating may cause indigestion or muscle cramping (and why you shouldn't swim too soon after eating—just like mom always said!).

The walls of capillaries are thin and leaky. Consequently, as blood enters a capillary at the arterial end, blood pressure pushes fluid rich in O_2, nutrients, and other substances needed by the surrounding cells out of the capillary and into the interstitial fluid **(Figure 23.9b)**. These molecules then diffuse from the interstitial fluid into nearby tissue cells. Blood cells and other large components usually remain in the blood because they are too large to pass through the capillary walls. At the venous end of the capillary, CO_2 and other wastes diffuse from tissue cells, through the interstitial fluid, and into the capillary bloodstream. This local chemical exchange—between the blood and tissue cells within capillary beds—is the most important function of the circulatory system.

Blood Return Through Veins After chemicals are exchanged between the blood and body cells, blood returns to the heart via veins. From the capillaries, blood moves into small venules, then into larger veins, and finally to the inferior and superior venae cavae, the two large blood vessels that flow into the heart. By the time blood exits the capillaries and enters the veins, the pressure originating from the heart has dropped to near zero. The blood still moves through veins, even against the force of gravity, because veins are sandwiched between skeletal muscles **(Figure 23.10)**. As these muscles contract (when you walk, for example), they squeeze the blood along.

You may witness firsthand the importance of muscle contraction in conveying blood through your veins if you stand still for too long. After a while without contracting your muscles, you will start to become weak and dizzy and may even faint because gravity prevents blood from returning to your heart in sufficient amounts to supply your brain with oxygen.

Over time, leg veins may stretch and enlarge and the valves within them weaken. As a result, veins just under the skin can become visibly swollen, a condition called varicose veins. Besides being unsightly, varicose veins may cause serious problems, such as cramping, blood clots, and ulcerations (open breaks in the skin).

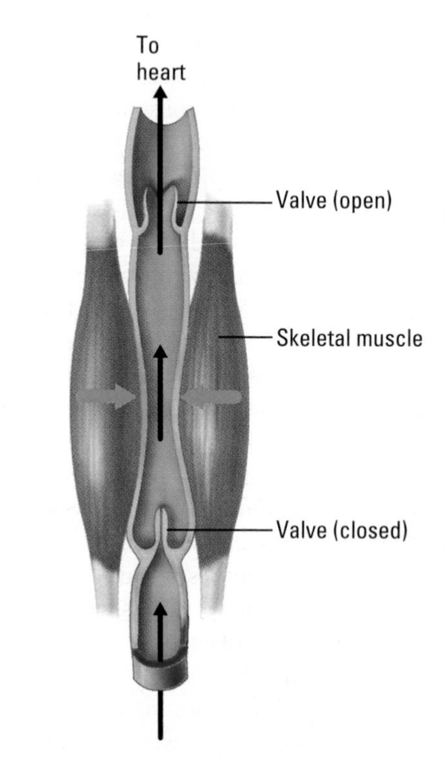

Figure 23.10 Blood flow in a vein. The contraction of muscles surrounding veins squeezes blood toward the heart. Flaps of tissue in the veins act as one-way valves, preventing backflow.

Blood

Now that we have examined the structures and functions of the heart and blood vessels, let's focus on the composition of blood itself **(Figure 23.11)**. The circulatory system of an adult human has about 5 L (11 pints) of blood. Just over half this volume consists of a yellowish liquid called **plasma**. Plasma is about 90% water. The other 10% of plasma is made up of dissolved salts, proteins, and various other molecules being transported by the blood, such as nutrients, wastes, and hormones. Suspended within the plasma are several types of cellular elements. Let's examine each of them in turn.

Red Blood Cells and Oxygen Transport **Red blood cells,** also called **erythrocytes**, are by far the most numerous type of blood cell. There are about 25 trillion of these tiny cells in the average person's bloodstream. Carbohydrate-containing molecules on the surface of red blood cells determine the blood type (such as A, B, AB, or O) of an individual (see Figure 9.18).

Each tiny red blood cell contains approximately 250 million molecules of **hemoglobin,** an iron-containing protein that transports oxygen. As red blood cells pass through the capillary beds of your lungs, oxygen diffuses into the red blood cells and binds to the hemoglobin. This process is reversed in the capillaries of the systemic circuit, where the hemoglobin unloads its cargo of oxygen to the cells. The structure of a red blood cell enhances its ability to carry and deliver oxygen. Human red blood cells are

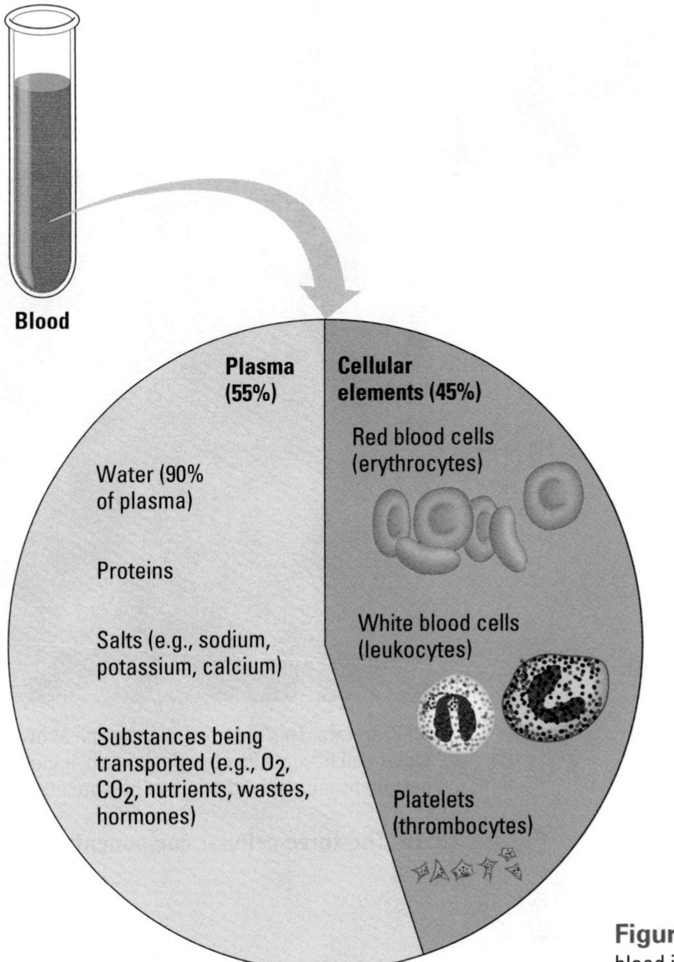

Figure 23.11 The composition of human blood. Just over half the volume of blood is liquid plasma. Blood cells and platelets account for the remaining blood volume.

shaped like disks with indentations in the middle, increasing the surface area available for gas exchange **(Figure 23.12a)**. In addition to their shape, the red blood cells of mammals have another feature that enhances their oxygen-carrying capacity: Red blood cells lack nuclei and other organelles, leaving more room to carry hemoglobin. Furthermore, red blood cells are very small relative to other cells. A population of smaller cells has a greater total surface area for gas exchange than the same volume of larger cells.

Adequate amounts of hemoglobin and red blood cells are essential to the normal functioning of the body. An abnormally low amount of hemoglobin or a low number of red blood cells can result in a condition called **anemia.** An anemic person feels constantly tired and run-down because the body cells do not get enough oxygen. Anemia can result from a variety of factors, including excessive blood loss, vitamin or mineral deficiencies, and certain cancers. A mutation in the gene for hemoglobin may result in sickle-cell disease, another cause of anemia (see Chapter 9).

White Blood Cells and Defense As a group, **white blood cells,** or **leukocytes**, fight infections and cancer **(Figure 23.12b)**. White blood cells are larger than red blood cells. And, unlike red blood cells, they lack hemoglobin but contain nuclei and a full complement of other organelles. There are about 700 times fewer white blood cells than red blood cells, although their numbers temporarily increase when the body is combating an infection. Some white blood cells actually spend most of their time outside the circulatory system, moving through interstitial fluid, where many of the battles against infection are waged. You'll learn more about white blood cells in Chapter 24.

Platelets and Blood Clotting We all get cuts and scrapes from time to time, yet we do not bleed to death because blood contains two components that help plug leaks by forming clots: platelets and fibrinogen. **Platelets,** or **thrombocytes,** are bits of cytoplasm pinched off from larger cells in the bone marrow. Clotting begins when the epithelium lining a blood vessel is damaged. Almost immediately, platelets adhere to the damaged tissue to form a sticky cluster that seals minor breaks. Platelets also release clotting factors, molecules that convert **fibrinogen,** a protein found in the plasma, into a threadlike protein called **fibrin.** Molecules of fibrin form a dense network to create a patch **(Figure 23.12c)**. On the skin, such clots are called scabs.

The clotting mechanism is so important that any defect in it can be life-threatening. In the inherited disease hemophilia (see Figure 9.32), excessive, sometimes fatal bleeding occurs from even minor cuts and bruises. Hemophilia is caused by a genetic mutation in one of several genes that produce clotting factors. The opposite problem—too much clotting—can also be fatal. A **thrombus** is a blood clot that forms in the absence of injury. An **embolus** is a thrombus that dislodges from its point of origin and travels in the blood. If an embolus lodges in an artery of the heart and

(a) Red blood cells. This electron micrograph shows the indented disk shape (called a biconcave disk) of human red blood cells.

Colorized SEM 2,600×

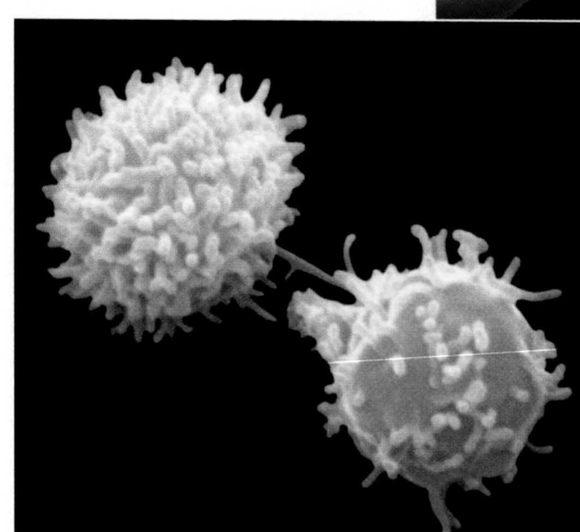

(b) White blood cells.

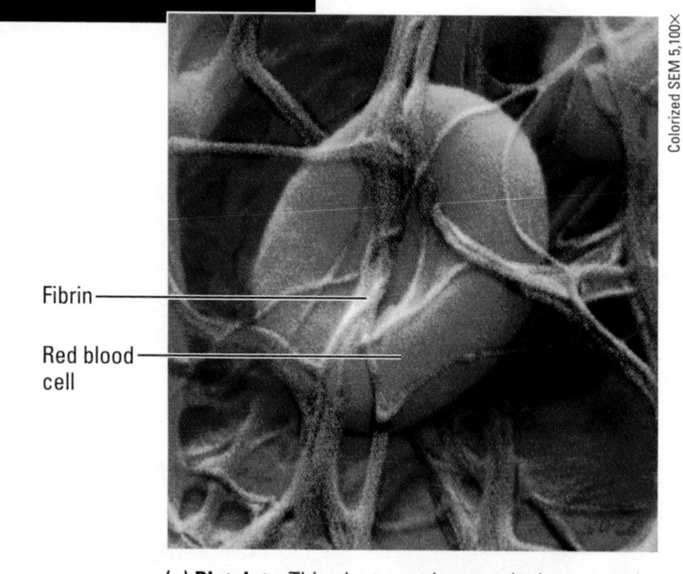

(c) Platelets. This electron micrograph shows a red blood cell trapped in a tangle of fibrin, a clotting protein produced by the action of platelets.

Figure 23.12 The three cellular components of blood.

is large enough to block it, a heart attack occurs. Similarly, an embolus in the head can cause a stroke, the death of brain tissue.

Stem Cells and the Treatment of Leukemia As we discussed in Chapter 11, stem cells have the potential to differentiate into other types of cells. The red marrow inside of bones such as the ribs, vertebrae, sternum, and pelvis contain stem cells that differentiate into red and white blood cells as well as into the cells that produce platelets. After forming in the early embryo, these stem cells continually reproduce themselves and create all the blood cells needed throughout life. Bone marrow stem cells show great promise for the treatment of disease, particularly leukemia.

Leukemia is cancer of the white blood cells (leukocytes). Because cancerous cells grow uncontrollably, a person with leukemia has an unusually high number of white blood cells, most of which are defective. The overabundance of these cells crowds out the bone marrow cells that produce red blood cells and platelets, causing severe anemia and impaired clotting.

Leukemia is usually fatal unless treated. Unfortunately, some cases do not respond to the standard cancer treatments of radiation and chemotherapy. An alternative treatment involves transplanting healthy bone marrow tissue into a patient whose own cancerous marrow has been purposely destroyed. Patients may be treated with their own bone marrow: Marrow from the patient is harvested, processed to remove as many of the cancerous cells as possible, and then reinjected. Alternatively, a donor, often a sibling, may provide marrow. Bone marrow stem cells are rare (only one in several thousand marrow cells) but very potent. Injection of as few as 30 of these cells can lead to complete repopulation of the patient's blood cells.

The Role of the Cardiovascular System in Homeostasis

The circulatory system performs several homeostatic functions that help maintain relatively constant internal conditions. First, by exchanging nutrients and wastes with the interstitial fluid, the circulatory system helps control the chemical balance of the fluid that surrounds cells. Second, the circulatory system helps control the composition of the blood by continuously moving it through organs, such as the lungs, liver, and kidneys, that regulate the blood's contents. The circulatory system is also involved in the body's temperature regulation, hormone distribution, and defense against foreign invaders—all topics that are discussed in other chapters.

Cardiovascular Disease

Cardiovascular disease, a set of diseases that affects the heart and blood vessels, accounts for 40% of all deaths in the United States, killing over 1 million people each year. The leading cause of death in the United States is heart attack. Like all of our cells, heart muscle cells require oxygen-rich blood to survive. When blood exits the heart via the aorta, several **coronary arteries** (shown in dark red in **Figure 23.13**) immediately branch off to supply the heart muscle. If one or more of these blood vessels become blocked, heart muscle cells quickly die from lack of O_2 (gray area in Figure 23.13). Such an event, and the subsequent failure of the heart to function properly, is called a **heart attack.** Approximately one-third of heart attack victims die almost immediately. For those who survive, the ability of the damaged heart to pump blood may be seriously impaired.

The suddenness of a heart attack belies the fact that the arteries of most victims became impaired gradually by a chronic cardiovascular disease called **atherosclerosis.** During the course of the disease, cholesterol

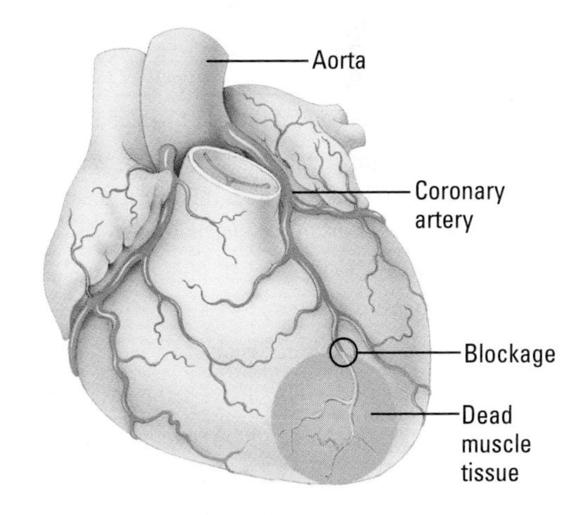

Figure 23.13 Blockage of a coronary artery, resulting in a heart attack. The coronary arteries supply oxygen to cardiac muscle cells. If one or more of these arteries becomes blocked, the heart muscle cells that it feeds will die from lack of oxygen. Such an event, called a heart attack, can lead to permanent damage of the heart muscle.

and other substances accumulate into buildups called plaque that form in the walls of arteries, narrowing the passages through which blood can flow **(Figure 23.14)**. Vessels narrowed by plaque increase blood pressure and can more easily form and trap clots. If a coronary artery becomes partially blocked, a person may feel occasional chest pain, a condition called angina.

How can you avoid becoming a heart disease victim? There are three everyday behaviors that have a significant impact on the risk of cardiovascular disease and heart attack. Smoking doubles the risk of heart attack and increases the severity if one does occur. Exercise (particularly "cardio" or "aerobic" workouts) can cut the risk of heart disease in half. Eating a heart-healthy diet, low in cholesterol, trans fat, and saturated fat (see Chapter 3) and high in fruits and vegetables, can reduce the risk of developing atherosclerosis.

If you already have cardiovascular disease, there are treatments available. Certain drugs can lower cholesterol. Angioplasty (inserting a tiny catheter with a balloon that is inflated to compress plaque and widen clogged coronary arteries) and stents (small wire mesh tubes that prop arteries open) can also help. Bypass surgery is a much more drastic remedy. In this procedure, blood vessels removed from a patient's legs are sewn onto the heart to shunt blood around clogged arteries. In an even more drastic measure, a defective heart can be replaced by a transplant or even with an artificial heart (although no one has survived long with the latter). Unfortunately, surgery of any kind only treats the disease symptoms, not the underlying causes, so problems will return if risk factors are not minimized.

On the positive side, the death rate from cardiovascular disease in the United States has been cut in half during the past 50 years. Health education, early diagnosis, and reduction of risk factors, particularly smoking, are mostly responsible.

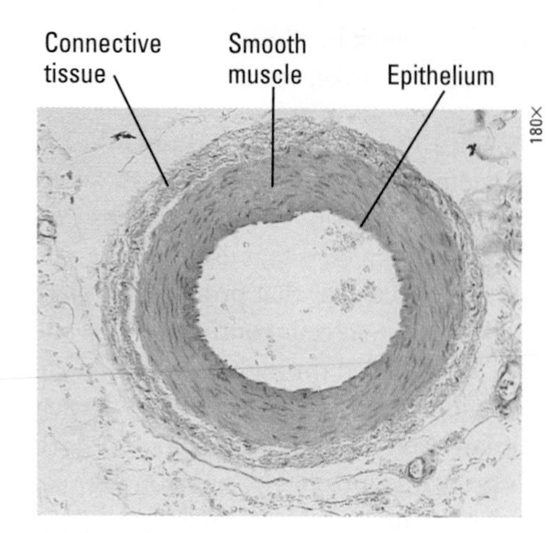

(a) Normal artery

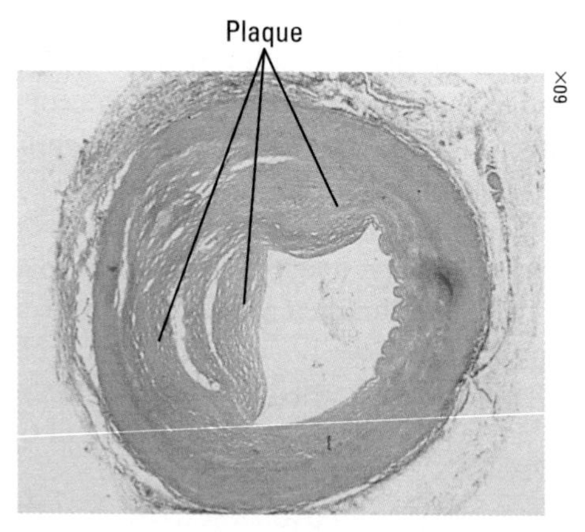

(b) Artery partially blocked by plaque

Figure 23.14 Atherosclerosis. These light micrographs contrast cross sections of **(a)** a normal artery with **(b)** an artery partially blocked by plaque, a condition called atherosclerosis.

CHECKPOINT

1. In the human cardiovascular system, what is the difference between the pulmonary circuit and the systemic circuit?
2. The heart chambers called the ___________ receive blood returning to the heart via blood vessels called ___________. The ___________ are the chambers where blood is pumped out of the heart to blood vessels called ___________.
3. Some babies are born with a small hole in the wall of muscle that separates the left and right ventricles. How does this affect the oxygen content of the blood pumped out of the heart in the systemic circuit?
4. List the three types of blood vessels in order of the pressure of blood within them, from highest pressure to lowest.
5. How does blood move through veins?
6. Why does eating saturated fat increase the risk of heart attack?
7. Name four things you can do to lower your risk of cardiovascular disease.

Answers: 1. The pulmonary circuit transports blood between the heart and lungs, while the systemic circuit transports blood between the heart and body tissues. **2.** atria; veins; ventricles; arteries **3.** Oxygen levels are reduced as oxygen-depleted blood mixes with oxygen-rich blood. **4.** Arteries, capillaries, veins **5.** Blood is squeezed along through veins by the contraction of surrounding muscles, and one-way valves prevent backflow. **6.** Saturated fat contributes to atherosclerosis, forming plaque that narrows arteries. **7.** Eat a heart-healthy diet, avoid excess alcohol, exercise regularly, and maintain a healthy weight.

Unifying Concepts of Animal Respiration

Do you know what your fate would be if you stopped breathing for more than a few minutes? Your body cells would begin to die from lack of oxygen. Without oxygen, the chemical reactions of cellular respiration—the series of metabolic steps that releases energy from food molecules (see Chapter 6)—cannot proceed. Recall that cellular respiration uses oxygen and glucose to produce water, carbon dioxide, and energy in the form of ATP. All working cells therefore require a steady supply of O_2 from the environment and must continuously dispose of CO_2:

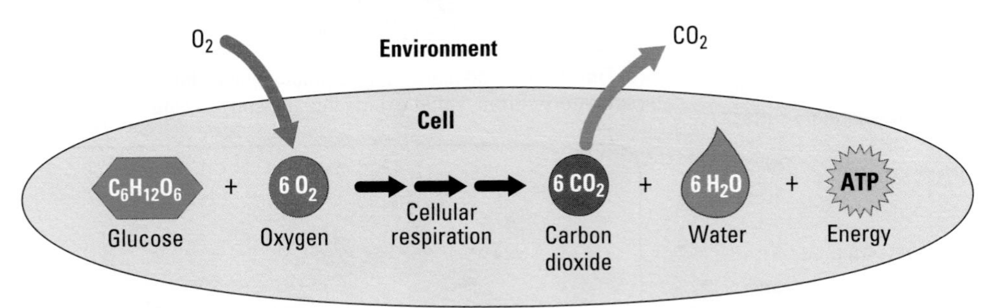

It is this exchange of O_2 and CO_2 between cells and the environment that makes it possible for animals to put to work the food molecules that the digestive system provides. The **respiratory system** consists of several organs that facilitate gas exchange. (Note that in the context of a whole organism, the word *respiratory* refers to the process of breathing, not to cellular respiration.) As you will see, gas exchange requires close cooperation between the circulatory and respiratory systems.

The Structure and Function of Respiratory Surfaces

Earth's main reservoir of oxygen is the atmosphere, which is about 21% O_2. Terrestrial animals access this oxygen by breathing air. Oceans, lakes, and other bodies of water also contain O_2 in the form of dissolved gas, but it makes up only 3–5% of that environment. Many aquatic animals obtain oxygen by passing water over their gills. The part of an animal where O_2 from the environment diffuses into living cells and CO_2 diffuses out to the surrounding environment is called the **respiratory surface.** The respiratory surface is usually covered with a single layer of living cells that is thin enough and moist enough to allow rapid diffusion between the body and the environment.

The respiratory surface of an animal must be large enough to take up O_2 for every cell in the body. A variety of solutions to the problem of providing a large enough respiratory surface have evolved, varying mainly with the size of the organism and whether it lives in water or on land. In the simplest cases, such as protozoans and other unicellular organisms, gas exchange occurs over the whole surface area of the organism. For some animals, such as sponges and flatworms, the plasma membrane of every cell in the body is close enough to the outside environment for gases to diffuse in and out. In many animals, however, the bulk of the body does not have direct access to the outer environment, and the respiratory surface is the interface between the environment and the blood, which transports gases to and from the rest of the body.

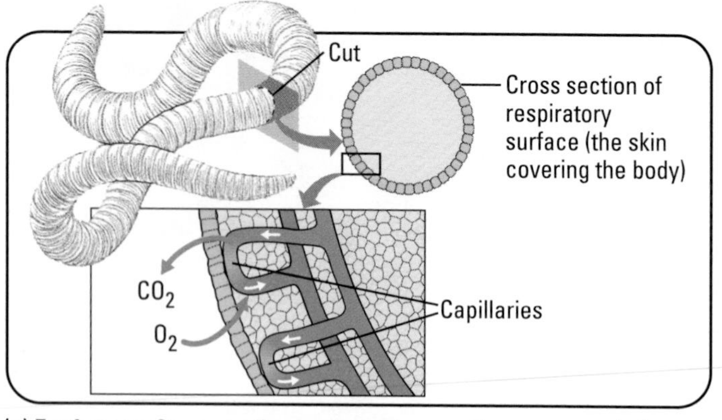

(a) Earthworm. Some small animals, such as earthworms, use their entire moist outer skin as a respiratory organ.

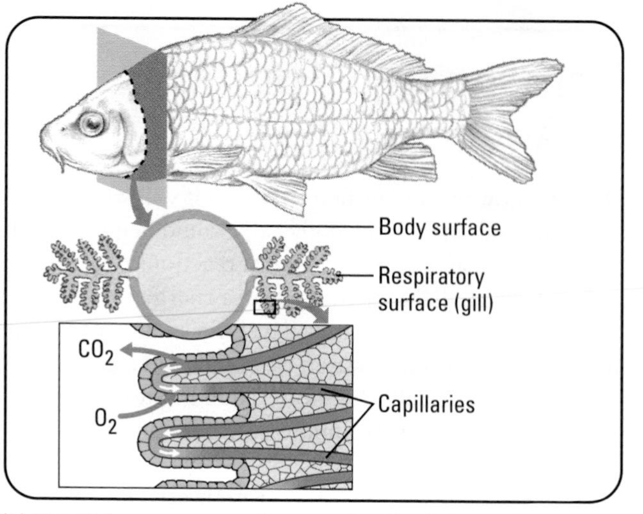

(b) Fish. Fishes and many other aquatic animals have gills, feathery respiratory surfaces that extend into the surrounding water.

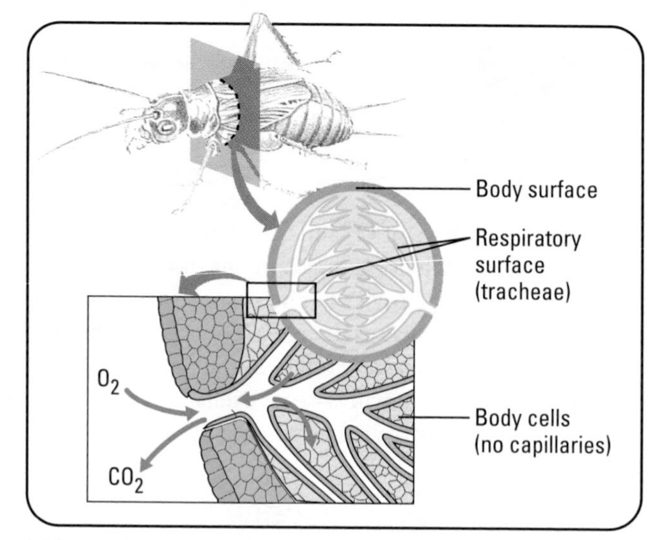

(c) Insect. Terrestrial insects have a pervasive system of internal tubes called tracheae that channel air directly to cells.

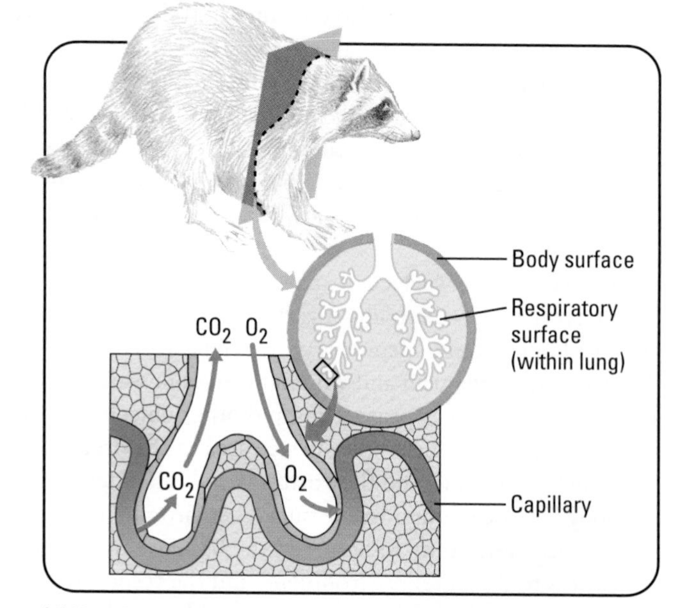

(d) Vertebrate. Most terrestrial vertebrates exchange gases across the thin lining of lungs, internal organs with a rich blood supply.

Figure 23.15 Four types of respiratory organs. Each of the four parts of this figure includes a cross section of the animal's body through the respiratory surface. The yellow areas represent respiratory surfaces; the green circles represent body surfaces with little or no role in respiration. The boxed enlargements show a portion of the respiratory surface in the process of diffusing O_2 and CO_2.

Some animals use their entire outer skin as a respiratory surface. An earthworm, for example, has moist skin, and gases diffuse across its general body surface. This results in an exchange of O_2 and CO_2 between the environment and a dense net of capillaries just below the earthworm's skin **(Figure 23.15a)**. For most animals, however, the outer body surface is either impermeable to gases or lacks sufficient area to exchange gases for the whole body. In such animals, specialized regions of the body have extensively folded or branched surfaces that provide a large surface area for gas exchange.

Within aquatic environments, the respiratory surfaces are found in extensions, or outfoldings, of the body surface called **gills.** Lobsters, which are invertebrates, and fishes, which are vertebrates, are examples of animals with gills. The feathery gills located on either side of a fish's head have a respiratory surface area much greater than the rest of the body surface. A fish obtains oxygen by continuously pumping water over its gills. As water passes over the respiratory surface of the gills, gases diffuse between the water and the blood. The blood then carries oxygen to the rest of the body **(Figure 23.15b)**. Gills—especially those of large, active fish—must be very efficient to obtain enough oxygen from water. Indeed, some aquatic animals can remove more than 80% of the oxygen from the water moving over their gills.

In most land-dwelling animals, the respiratory surfaces are folded into the body rather than extending from it. The infolded surfaces are open to the air only through narrow tubes. This allows the animals to breathe air while retaining the moisture needed to maintain the plasma membranes of respiratory surface cells. Insects breathe using **tracheae,** an extensive system of internal tubes that branch throughout the body **(Figure 23.15c)**. Tracheae begin near the body's surface and branch down to narrower tubes that extend to nearly every cell. There, gas exchange occurs via diffusion across the moist epithelium that lines the tips of the tubes. No circulatory system is required to transport oxygen in insects because virtually all body cells are within a very short distance of the respiratory surface.

Lungs are the most common respiratory surface among snails, some spiders, and terrestrial vertebrates such as amphibians, birds and other reptiles, and mammals **(Figure 23.15d)**. In contrast to the tracheae of insects, lungs are located in only one part of the body. The circulatory system transports oxygen from the respiratory surface to the rest of the body. We'll take a closer look at human lungs in the next section.

CHECKPOINT

1. Why do all human cells require a continuous supply of O_2 and disposal of CO_2?
2. What is the main difference between the gills of fishes and the lungs of humans in terms of where they are located in the body?
3. In what basic way is the process of gas exchange different in insects than in both fishes and humans?

Answers: 1. O_2 is required for cellular respiration, which produces energy and also produces CO_2 as a waste product. **2.** Gills extend outward from the body into the surrounding environment, while the lungs are internal sacs. **3.** The respiratory system of insects uses tracheae that bring O_2 directly to body cells; fishes and humans use a circulatory system that transports O_2 from the respiratory surface to body cells.

The Human Respiratory System

Figure 23.16 provides an overview of three phases of gas exchange in humans. ❶ The first step of gas exchange is breathing. When an animal with lungs breathes, a large, moist internal surface is exposed to air. Oxygen diffuses across the cells lining the lungs and into the surrounding blood vessels. Simultaneously, CO_2 passes out of the blood, into the lungs, and is exhaled. Most land animals require a lot of oxygen. The inner tubes of the lungs are extensively branched, providing a large respiratory surface. ❷ The second step in gas exchange is the transport of O_2 from the lungs to the rest of the body via the circulatory system. The blood also carries CO_2 from the tissues back to the lungs. ❸ The final step is for cells to take up O_2 from the blood and release CO_2 into the blood. The delivered O_2 is used by the body cells to obtain energy from food via the process of cellular respiration. This same process produces CO_2 as a waste product that cells pass to the blood. The circulatory system transports the CO_2 back to the lungs, where it is exhaled and released from the body.

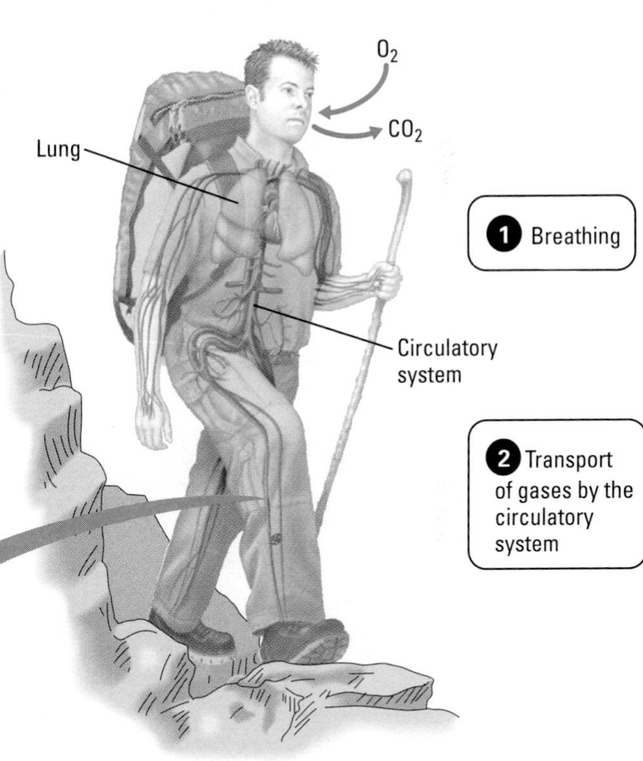

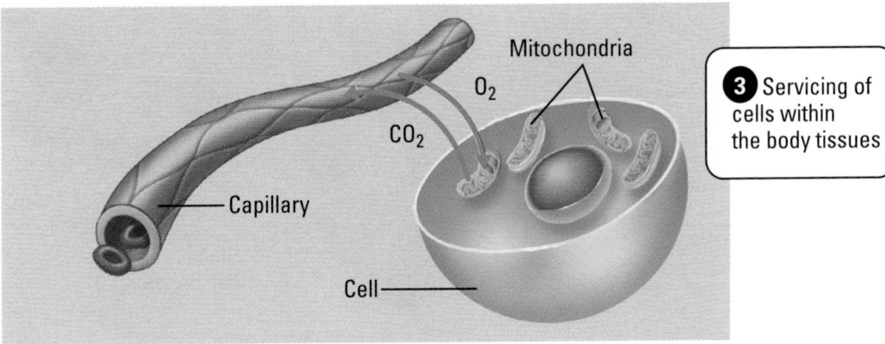

Figure 23.16 The three phases of gas exchange.

The Structure and Function of the Human Respiratory System

Figure 23.17 shows the human respiratory system (along with a few organs from other systems for orientation). Our lungs are located in the chest cavity, bordered along the bottom by the **diaphragm,** a sheet of muscle. Let's get an overview of the respiratory system by following the flow of air into the lungs.

Air enters the respiratory system through external openings: the nostrils and mouth. In the nasal cavity, the air is filtered by hairs, warmed, humidified, and sampled by smell receptors. The air passes to the **pharynx,** where the digestive and respiratory systems meet.

From the pharynx, air is inhaled into the **larynx** (voice box) and then into the **trachea** (windpipe). The trachea forks into two **bronchi** (singular, *bronchus*), one leading to each lung. Within the lungs, each bronchus branches repeatedly into finer and finer tubes called **bronchioles.** The system of branching tubes looks like an inverted tree, with the trachea as the trunk and the bronchioles as the smallest branches.

The bronchioles dead-end in grapelike clusters of air sacs called **alveoli** (singular, *alveolus*) **(Figure 23.18)**. Each of our lungs contains millions of these tiny sacs. The inner surface of each alveolus is lined with a layer of epithelial cells that forms the respiratory surface, where gases are actually exchanged. O_2 diffuses from the air into a web of blood capillaries that surrounds each alveolus, entering the bloodstream. CO_2 diffuses from the blood in the capillaries into the alveoli. It is then exhaled through the bronchioles to the bronchus, up through the trachea, and out of the body.

During exhalation, outgoing air moves over a pair of **vocal cords** within the larynx. Humans produce vocal sounds by flexing muscles in the

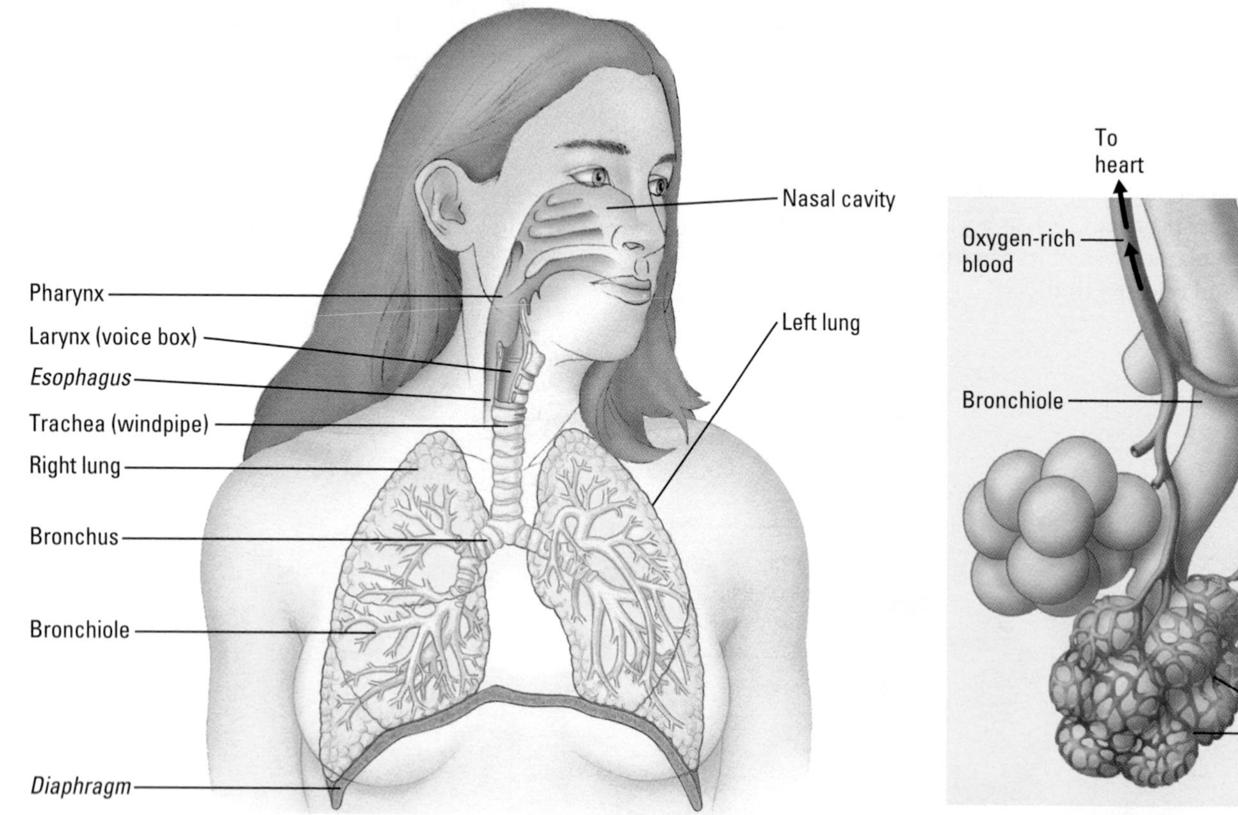

Figure 23.17 The human respiratory system.

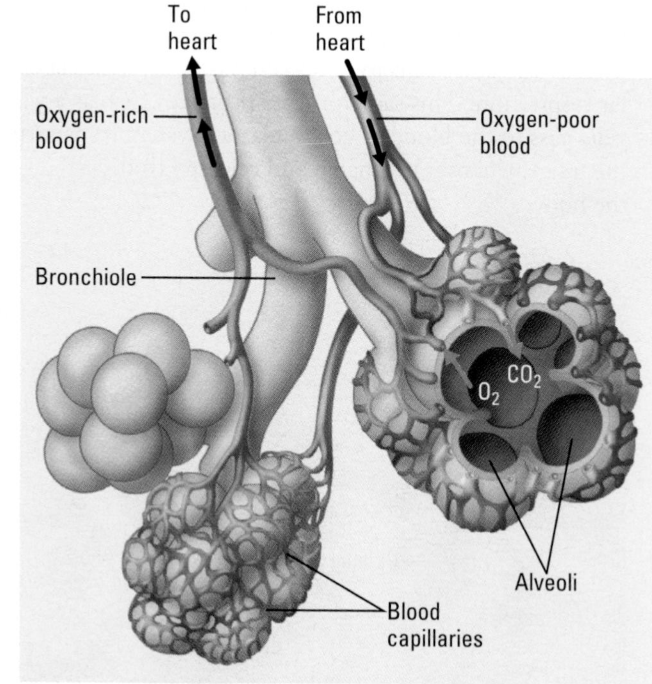

Figure 23.18 The structure of alveoli. Capillaries surround air sacs called alveoli. These are the sites of gas exchange in the lungs.

voice box as air rushes by, stretching the cords and making them vibrate. High-pitched sounds are produced by tensing the muscles and shortening the cords, while lower-pitched sounds are made by reducing the tension and lengthening the cords. During puberty, the voice box of males grows rapidly, resulting in a deeper voice. Pubescent boys often have trouble controlling their lengthening vocal cords, resulting in a "breaking" voice.

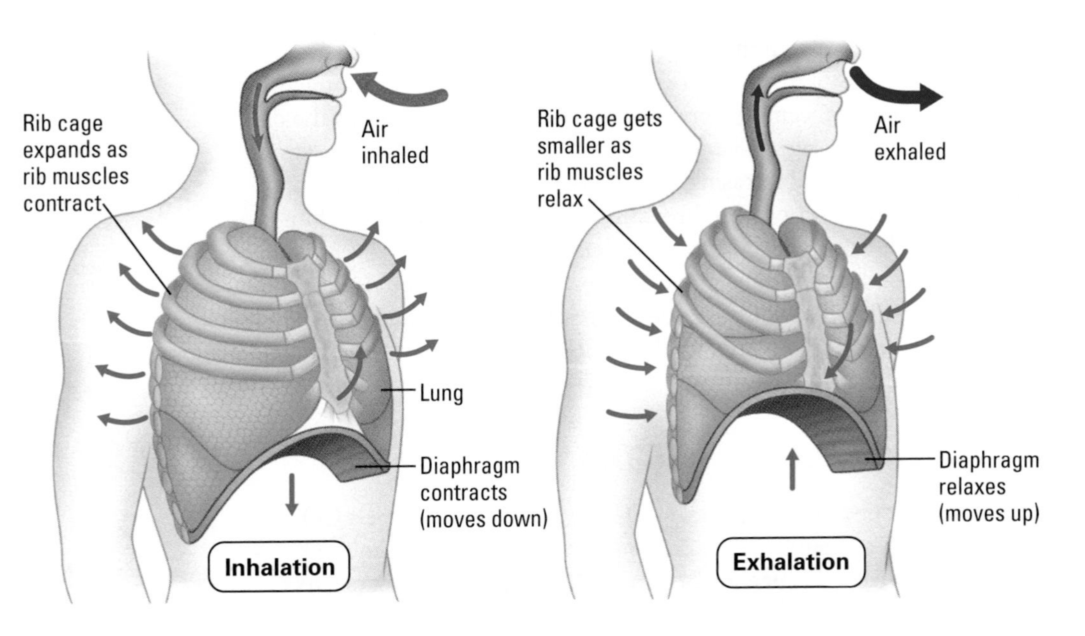

Rib cage expands as rib muscles contract
Air inhaled
Lung
Diaphragm contracts (moves down)
Inhalation
Rib cage gets smaller as rib muscles relax
Air exhaled
Diaphragm relaxes (moves up)
Exhalation

Figure 23.19 How a human breathes.

Taking a Breath

Breathing consists of the alternating processes of inhalation and exhalation. **Figure 23.19** shows the changes that occur during breathing. If you place your hands on your rib cage and inhale, you can feel your ribs move upward and spread out as muscles between them contract. Meanwhile, your diaphragm moves downward, expanding the chest cavity. All of this increases the volume of the lungs, dropping the air pressure in the lungs to below the air pressure of the atmosphere. The result is that air rushes in through the mouth and nostrils toward the region of lower pressure, filling the lungs. This is called **negative pressure breathing.** Although it seems that you actively suck in air when you inhale, air actually moves into your lungs passively after the air pressure in them drops.

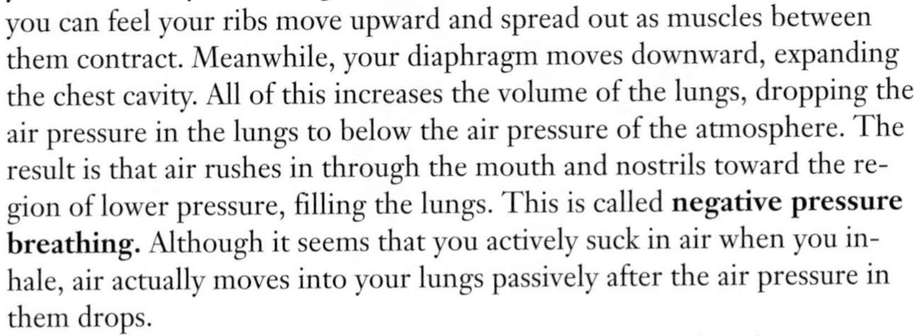

During exhalation, the rib and diaphragm muscles relax, decreasing the volume of the chest cavity. This increases the air pressure inside the lungs, forcing air to rush out of the respiratory system. Movement of the diaphragm is vital to normal breathing. That is why a punch to the diaphragm (the "breadbasket") can "knock the wind out of you." It temporarily shocks the diaphragm muscle, prevents movement of the chest cavity, and stops you from taking a normal breath. Other times, sudden involuntary contractions of the diaphragm force air through the voice box, causing hiccups.

You can consciously speed up or slow down your breathing. You can even hold your breath (although, despite the claims of angry children, you would pass out and return to normal breathing before you turned blue). Usually, however, you aren't aware of breathing; you certainly aren't aware of it when you're asleep. What, then, controls your breathing?

Most of the time, automatic control centers in the brain regulate breathing **(Figure 23.20)**. Nerves from the brain's breathing control centers signal the diaphragm and rib muscles to contract, making us inhale. Between inhalations, the muscles relax, and we exhale. While at rest, the control centers send out signals that maintain a respiratory rate of 10–14 inhalations per minute. And when we're not at rest? ❶ The breathing control centers increase or decrease breathing rate in response to levels of CO_2 (*not* O_2) in the blood. When you exercise, for example, cellular respiration kicks into high gear, producing more ATP for your muscles and raising the amount of CO_2 in the blood. ❷ When the brain senses the higher CO_2 level, ❸ the breathing control centers increase the breathing rate and depth. As a result, more CO_2 is eliminated in the exhaled air and more O_2 is provided to working muscle cells.

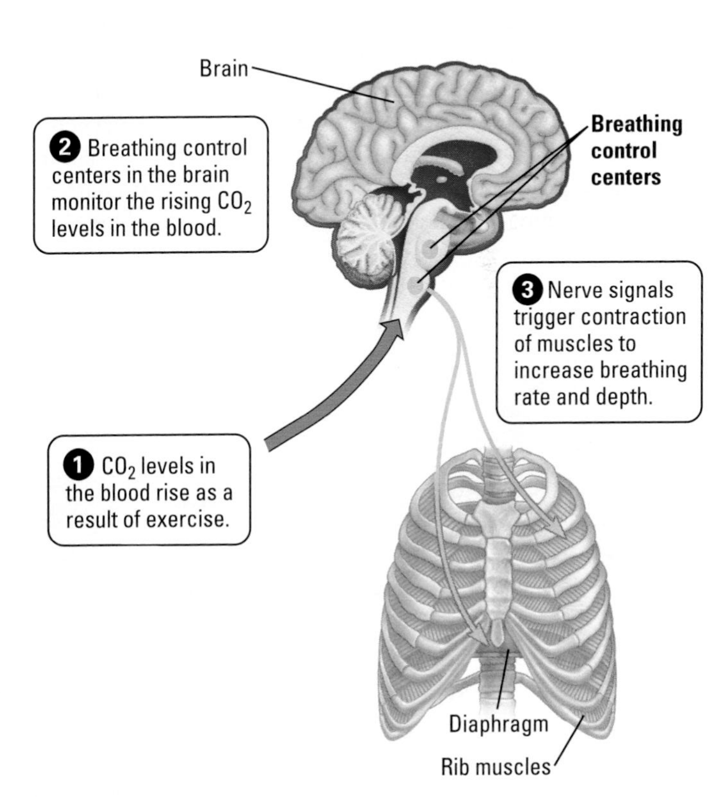

Brain
❷ Breathing control centers in the brain monitor the rising CO_2 levels in the blood.
Breathing control centers
❸ Nerve signals trigger contraction of muscles to increase breathing rate and depth.
❶ CO_2 levels in the blood rise as a result of exercise.
Diaphragm
Rib muscles

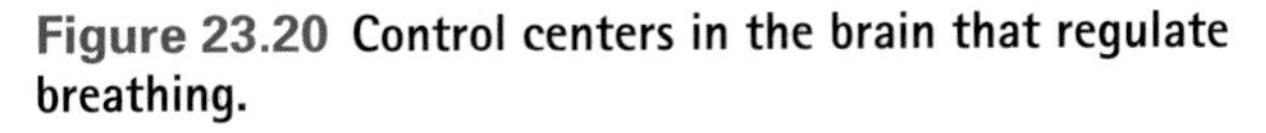

Figure 23.20 Control centers in the brain that regulate breathing.

You can make yourself dizzy by hyperventilating, taking excessively rapid, deep breaths. This rapid breathing purges the blood of so much CO_2 that the control centers temporarily cease sending signals to the diaphragm and rib muscles. Changes in blood chemistry cause arterioles in the brain to constrict, decreasing the brain's blood supply and causing dizziness. Automatic control of breathing stops until the CO_2 level increases enough to switch the breathing centers back on. If you begin to hyperventilate (during a panic attack, for example), you should breathe into a paper bag. This will cause you to breathe back air higher in CO_2 than normal air, increasing the level of CO_2 in the blood and helping to restore normal breathing.

The Role of Hemoglobin in Gas Transport

The human respiratory system takes O_2 into the body and expels CO_2, but it relies on the circulatory system to shuttle these gases between the lungs and the body's cells **(Figure 23.21)**. In the lungs, gas exchange occurs between blood in the capillaries surrounding the alveoli and the air within the alveoli. The gases exchange by diffusion. O_2 moves out of the air and into the blood because air is richer in O_2 than blood is. CO_2 also diffuses along a gradient, in this case from the blood to the air in the lungs.

There is a problem, however, with this simple scheme: Oxygen does not readily dissolve in blood, so oxygen will not tend to move from the air into the blood on its own. Solving this problem is the fact that most of the O_2 in blood is carried by hemoglobin molecules within red blood cells **(Figure 23.22)**. Each molecule of hemoglobin consists of four polypeptide chains. Attached to each polypeptide is a chemical group called a heme ( ) at the center of which is an atom of iron (shown in black). Each iron atom can hold one O_2 molecule, so one molecule of hemoglobin can carry a maximum of four molecules of O_2. Each red blood cell has about 250 million molecules of hemoglobin, so one tiny cell can carry 1 billion O_2 molecules. Hemoglobin picks up oxygen in the lungs, carries it to the body's cells, and releases it. When hemoglobin binds oxygen, it changes the color of blood to a bright cherry red. Oxygen-poor blood is a dark maroon that appears blue through the skin. This is why a person who has stopped breathing—such as a drowning victim—has blue skin.

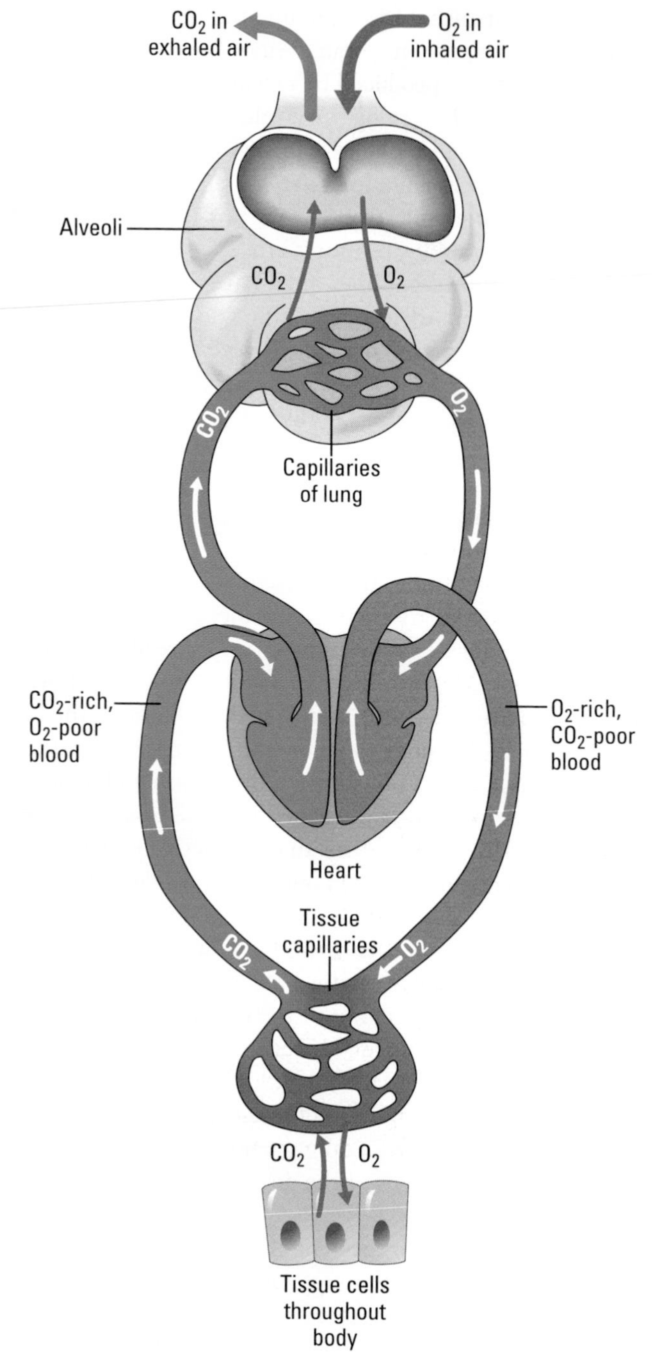

Figure 23.21 Gas transport and exchange in the body. The respiratory system exchanges gases through inhalation and exhalation, but it relies on the circulatory system to bring these gases to body cells. Within the lungs, O_2 moves from air into capillaries through the alveoli. At the same time, CO_2 exits. The O_2-rich blood is then sent to body tissue capillaries via the heart. In the body tissues, O_2 enters the tissue cells and CO_2 leaves the tissue cells and enters the blood. The O_2-poor blood is circulated back to the heart, then to the lungs.

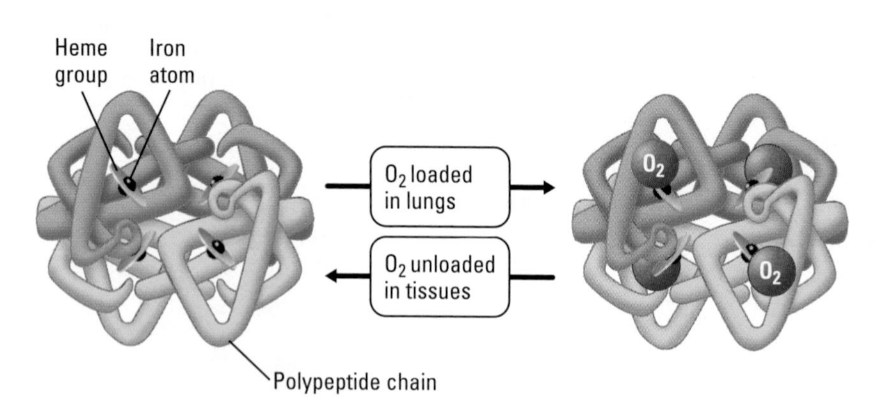

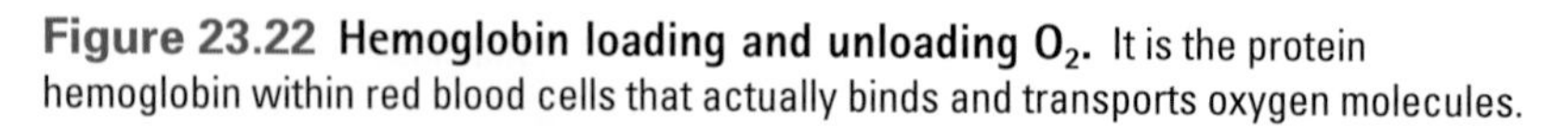

Figure 23.22 Hemoglobin loading and unloading O_2. It is the protein hemoglobin within red blood cells that actually binds and transports oxygen molecules.

Because iron (one atom per heme group) is so important in the structure of hemoglobin, a shortage of iron causes less hemoglobin to be produced by the body. In fact, iron deficiency is the most common cause of anemia. Women are more likely to develop iron deficiency than men because of blood lost during menstruation. Pregnant women generally benefit from iron supplements to support the developing fetus and placenta.

Besides binding O_2, hemoglobin can also bind carbon monoxide (CO), a colorless, odorless gas. In fact, CO binds to hemoglobin even more tightly than O_2 does. Breathing CO can therefore interfere with the delivery of O_2 to body cells. Because CO also interferes with cellular respiration (see Figure 6.12), it can cause rapid death. Sitting in an idling car in an enclosed space can be fatal because CO is emitted as an air pollutant in the exhaust from gasoline-powered engines. And despite its potentially deadly effects, millions of Americans willingly inhale CO in the form of cigarette smoke.

How Smoking Affects the Lungs

Every breath you take exposes your respiratory tissues to potentially damaging chemicals. Air pollutants such as sulfur dioxide, carbon monoxide, and ozone can all cause respiratory problems. One of the worst sources of airborne pollutants is tobacco smoke. The visible smoke from burning tobacco is mostly microscopic particles of carbon. Sticking to the carbon particles are over 4,000 different chemicals, many of which are known to be harmful and even potentially deadly.

The epithelial tissue lining our respiratory system is extremely delicate. Its main protection is the mucus covering the cells and the beating cilia that sweep dirt particles and microorganisms off their surfaces. Tobacco smoke irritates the cells that line the bronchi and trachea, inhibiting their ability to remove foreign substances from the airways. This interference with the normal cleansing mechanism of the respiratory system allows more toxin-laden smoke particles to reach and damage the lungs' delicate alveoli. Frequent coughing—common in heavy smokers—is the respiratory system's attempt to compensate and clean itself.

The health statistics associated with smoking are staggering. Every year, smoking kills about 440,000 Americans. One in two American smokers will die from their habit. On average, adults who smoke cigarettes die 13 to 14 years earlier than nonsmokers. Smokers account for 90% of all cases of lung cancer, one of the deadliest forms of cancer: Only 15% of people diagnosed with lung cancer are alive five years later, and lung cancer kills more Americans than any other form of cancer. **Figure 23.23a** shows a cutaway view of a pair of healthy human lungs (and heart). In contrast, **Figure 23.23b** shows the lungs of a smoker, turned black from the long-term buildup of smoke particles.

Besides lung cancer, smokers also have a markedly greater risk than nonsmokers of developing cancers of the mouth, throat, bladder, pancreas, and several other organs. Smoking can also cause emphysema, a disease that causes alveoli to disintegrate, reducing the lungs' ability to exchange gases and causing breathlessness and constant fatigue.

There is no lifestyle choice that can have a more positive impact on your long-term health than not smoking. After quitting, it takes about 15 years for a former smoker's risk to even out with that of a nonsmoker,

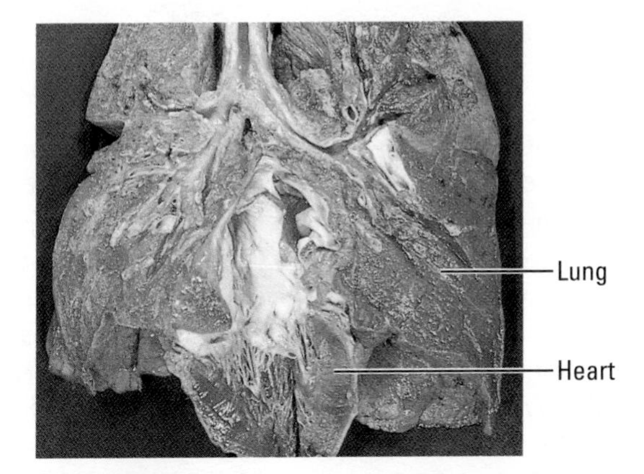

(a) Healthy lungs

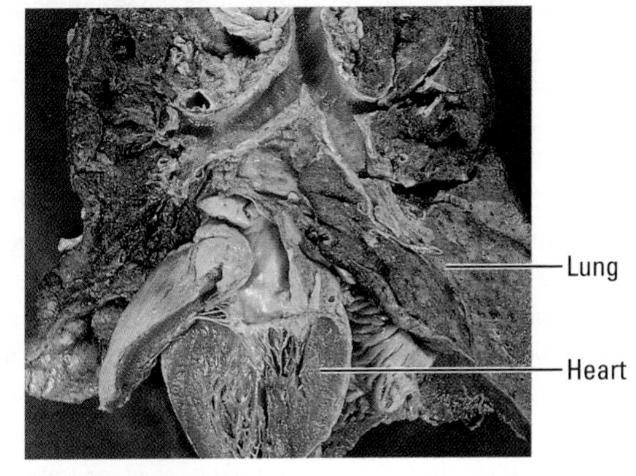

(b) Cancerous lungs

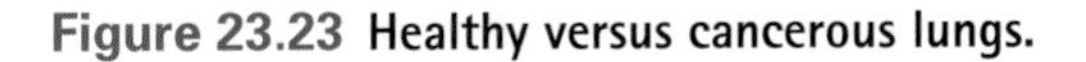

Figure 23.23 Healthy versus cancerous lungs.

so not starting is clearly the best option. Unfortunately, even nonsmokers may be affected by tobacco, as studies have found that secondhand cigarette smoke is a health hazard, particularly to newborns.

CHECKPOINT

1. Put the following parts of the human respiratory system in the order that an inhaled breath would encounter them: bronchi, larynx, nasal cavity, alveoli, pharynx, bronchioles, trachea.
2. Explain how hyperventilating disrupts the control of breathing.
3. Which type of cellular transport is responsible for moving respiratory gases between the blood and the body's cells?
4. Most oxygen in the blood is transported bound to the protein ___________ within ___________ cells.
5. What causes "smoker's cough"?

Answers: 1. Nasal cavity → pharynx → larynx → trachea → bronchi → bronchioles → alveoli **2.** Hyperventilating purges the blood of CO_2, causing the brain's breathing control centers to cease sending inhalation signals to the chest muscles. **3.** Diffusion **4.** hemoglobin; red blood **5.** Tobacco smoke damages the cells that line the airway and inhibits their ability to remove foreign substances. "Smoker's cough" is the body's way of compensating for this.

The Move onto Land

The colonization of land by vertebrates was one of the pivotal milestones in the history of life. The evolution of legs from fins may be the most obvious change in body plan, but the refinement of lung breathing was just as important. The gills that exchange gases in aquatic environments do not work as respiratory organs on land, where air is the oxygen source. But complex organs such as lungs do not evolve "on demand" as the environment changes. Natural selection generally refines existing equipment. Indeed, the fossil record indicates that lungs already functioned in one group of fishes before the first vertebrate moved onto land. In fact, this air-breathing ability has been retained in modern lungfishes **(Figure 23.24)**.

Figure 23.24 An African lungfish.

Equipped with both gills and lungs, lungfishes can obtain their oxygen from either water or air. These versatile creatures are found today on the southern continents of Africa, South America, and Australia. They generally inhabit stagnant, oxygen-poor ponds, rivers, and swamps, occasionally surfacing to gulp air into their lungs. When ponds and rivers shrink during dry seasons, some lungfishes can burrow into the mud and seal themselves in mucus-lined burrows. During this period of low activity similar to hibernation, the lungfish's metabolism slows drastically, and the animal obtains all the oxygen it needs by breathing.

Modern lungfishes are the descendants of a much greater diversity of lungfishes that lived in shallow water about 400 million years ago. Was one of those ancient animals the common ancestor of modern lungfishes and the terrestrial vertebrates—amphibians, reptiles, and mammals? If so, then lungfishes would be our closest fishy kin. In fact, DNA comparisons place lungfishes as the fishes most closely related to terrestrial vertebrates. This molecular evidence supports the hypothesis that our elaborate lungs evolved from much simpler versions that first functioned in shallow-water fishes—another example of Darwin's "descent with modification." ■

Chapter Review

SUMMARY OF KEY CONCEPTS

For study help and activities, go to campbellbiology.com or the student CD-ROM.

Unifying Concepts of Animal Circulation

- The circulatory system facilitates the exchange of materials and energy between the cells of an organism and its environment.
- **Open and Closed Circulatory Systems** In an open circulatory system, circulating fluid is pumped through open-ended vessels and circulates freely among cells. In a closed circulatory system, blood is confined within vessels.

The Human Cardiovascular System

- **The Path of Blood** Trace the path of blood in the following figure, noting where blood is oxygen-poor and where it is oxygen-rich.

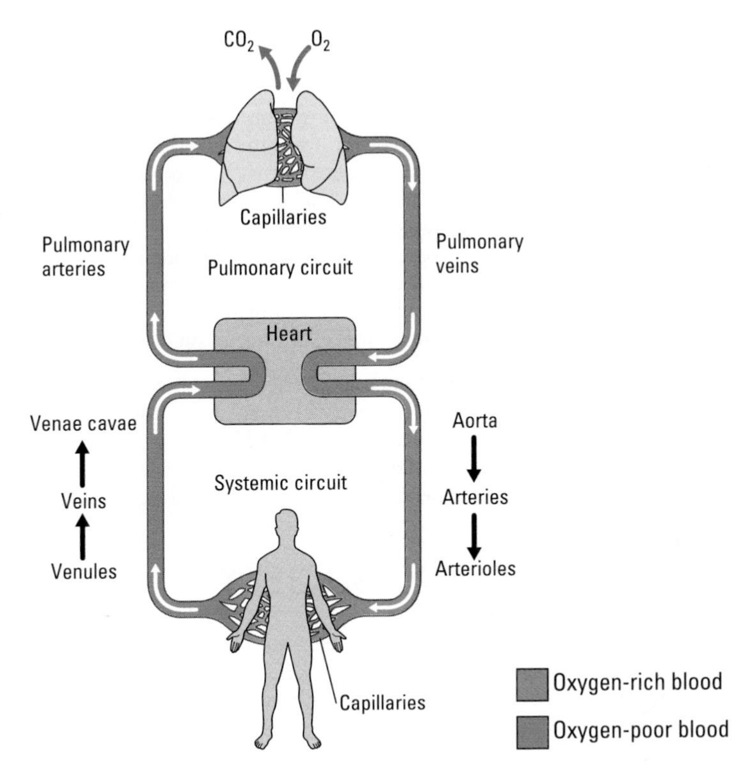

Activity *Path of Blood Flow*

Activity *Cardiovascular System Structure*

- **How the Heart Works** Review the structure of the heart by tracing the flow of blood in the following diagram.

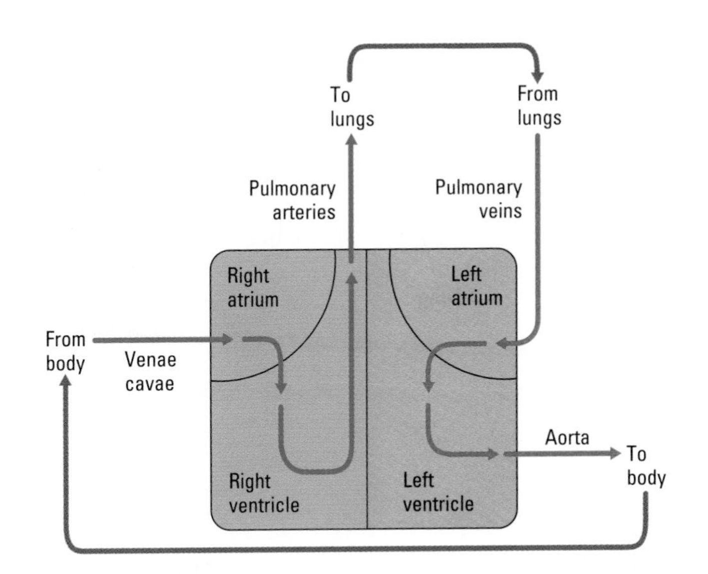

- The cardiac cycle is composed of two phases: systole (contraction) and diastole (relaxation). The pacemaker, which sets the tempo of the heartbeat, generates electrical impulses that stimulate the atria and ventricles to contract.
- **Blood Vessels** Muscular arteries carry blood away from the heart. Exchange between the blood and interstitial fluid occurs across the thin walls of capillaries. Valves in the veins and contractions of surrounding skeletal muscle keep blood moving back to the heart.

Activity *Cardiovascular System Function*

- **Blood** Blood consists of liquid plasma and cellular elements: red blood cells (erythrocytes), which transport oxygen; white blood cells (leukocytes), which aid in defense; and platelets (thrombocytes), which aid in clotting. New blood cells are continually formed from stem cells found in red bone marrow.
- **The Role of the Cardiovascular System in Homeostasis** The cardiovascular system is involved in fluid balance, temperature regulation, hormone distribution, and defense.
- **Cardiovascular Disease** Diseases of the heart and blood vessels—including heart attack and stroke—kill more Americans than any other type of disease. Atherosclerosis is the buildup of lipids and other substances in the inner walls of arteries.

Case Study in the Process of Science *How Is Cardiovascular Fitness Measured?*

Unifying Concepts of Animal Respiration

- The respiratory system facilitates gas exchange.
- **The Structure and Function of Respiratory Surfaces** A variety of respiratory surfaces have evolved in animals. The respiratory surface is the part of the body where gas exchange takes place: the entire body surface, gills, tracheae, or lungs.

The Human Respiratory System

- **The Structure and Function of the Human Respiratory System** When you take a breath, air moves sequentially from the nostrils and/or mouth to the pharynx, larynx, trachea, bronchi (which enter the lungs), bronchioles, and finally the alveoli (the actual respiratory surfaces).

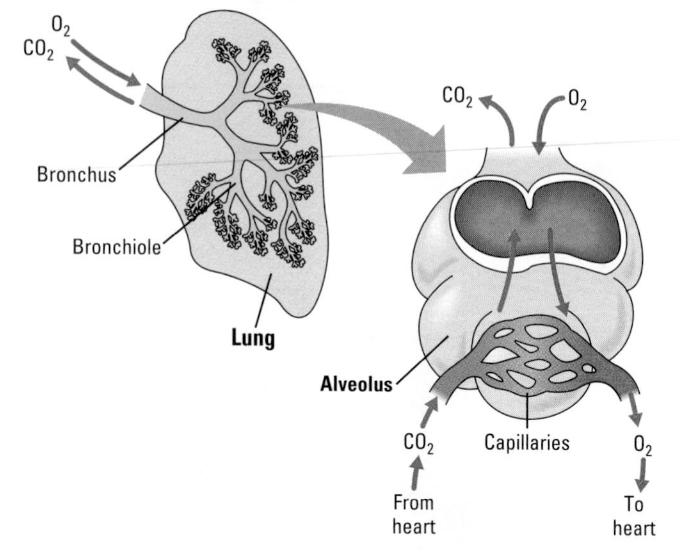

Activity *The Human Respiratory System*

- **Taking a Breath** During inhalation, the chest cavity expands and air pressure in the lungs decreases, causing air to rush into the lungs. During exhalation, air pressure in the lungs increases and air moves out of the lungs. Breathing rate is set by breathing control centers in the brain, which are influenced by CO_2 concentrations in the blood.

- **The Role of Hemoglobin in Gas Transport** After O_2 enters the lungs, it binds to hemoglobin in red blood cells and is transported to body tissue cells by the circulatory system.

Activity *Transport of Respiratory Gases*

- **How Smoking Affects the Lungs** Tobacco smoke irritates the respiratory surfaces of the lungs, impairing alveolar function and causing multiple health problems.

SELF-QUIZ

1. What is the difference between an open circulatory system and a closed circulatory system?

2. Why is the following statement false? "All arteries carry oxygen-rich blood, while all veins carry oxygen-poor blood."

3. Match each chamber of the heart with its function.

a. left atrium	1. receives blood from the body via the venae cavae
b. right atrium	2. sends blood to the lungs via the pulmonary arteries
c. left ventricle	3. receives blood from the lungs via the pulmonary veins
d. right ventricle	4. sends blood to the body via the aorta

4. People who do not get enough iron in their diet run the risk of developing ___________.

5. Match each element of the blood with its main function.

a. plasma	1. transport O_2
b. platelets	2. fight infections
c. red blood cells	3. carries dissolved elements
d. white blood cells	4. aid in clotting

6. Which of the following increases your risk of developing cardiovascular disease?
   a. taking cholesterol-lowering drugs
   b. exercise
   c. low-fat diet
   d. smoking

7. What is the primary difference between the tracheae of insects and the lungs of humans in terms of how they deliver oxygen to body cells?

8. The respiratory surface of vertebrate lungs consists of tiny sacs within the lungs called ___________.

9. During inhalation, when the diaphragm contracts and the ribs spread apart, air pressure inside the lungs
   a. increases.
   b. decreases.
   c. decreases, then increases.
   d. is unchanged.

10. Why is carbon monoxide deadly?
    a. It binds to hemoglobin in white blood cells in place of O_2.
    b. It binds to hemoglobin in red blood cells in place of O_2.
    c. It binds to hemoglobin in red blood cells in place of CO_2.
    d. It binds to hemoglobin in white blood cells in place of CO_2.

Answers to the Self-Quiz questions can be found in Appendix D.

Go to the website or CD-ROM for more Self-Quiz questions.

THE PROCESS OF SCIENCE

11. Birds and mammals have a four-chambered heart, with two ventricles and two atria, but other modern reptiles have a three-chambered heart, with just one ventricle. Paleontologists debate whether dinosaurs had a typical "reptile-like" heart or a "birdlike" heart. Long-necked sauropod dinosaurs could have had unusual circulatory demands because their head may have been raised far above their heart. The farther the head is above the heart, the greater the systolic pressure needs to be for blood to reach the brain. For example, the long-necked dinosaur *Brachiosaurus* may have carried its head as much as 6 m (20 feet) above its heart. It is estimated that such an anatomy demanded a systolic blood pressure of 500 mm of mercury for blood to reach the brain! Some paleontologists consider this to be evidence that dinosaurs must have had a four-chambered heart that supported a dual circulatory system similar to that of birds and mammals, rather than the three-chambered heart of nonbird reptiles. Can you explain why?

12. During a college swim meet, John told his friends that before a race, he likes to "charge up" on oxygen by hyperventilating so he can hold his breath longer. Although John is correct that this will allow him to hold his breath longer, it is not for the reason that he gave to his friends. How is John mistaken? Why is this use of hyperventilation actually very dangerous?

BIOLOGY AND SOCIETY

13. Recently, a 19-year-old woman received a bone marrow transplant from her 1-year-old sister. The woman was suffering from a deadly form of leukemia and was almost certain to die without a transplant. The woman's parents had decided to have another child in a final attempt to provide their daughter with a matching donor. Although the ethics of the parents' decision were criticized, doctors reported that this situation is not uncommon. In your opinion, is it acceptable to have a child to provide organ or tissue donation? Why or why not?

14. While administering first aid, emergency health-care workers are often exposed to a patient's body fluids. Considering that dangerous contagious diseases (such as hepatitis, tuberculosis, and AIDS) are not uncommon in society today, what do you think are the risks to a rescuer who is trying to save someone's life? What rights do you think a rescuer should have in such a situation? What about the rights of the patient? Should people be required to inform health-care workers that they have a contagious disease?

24 The Body's Defenses

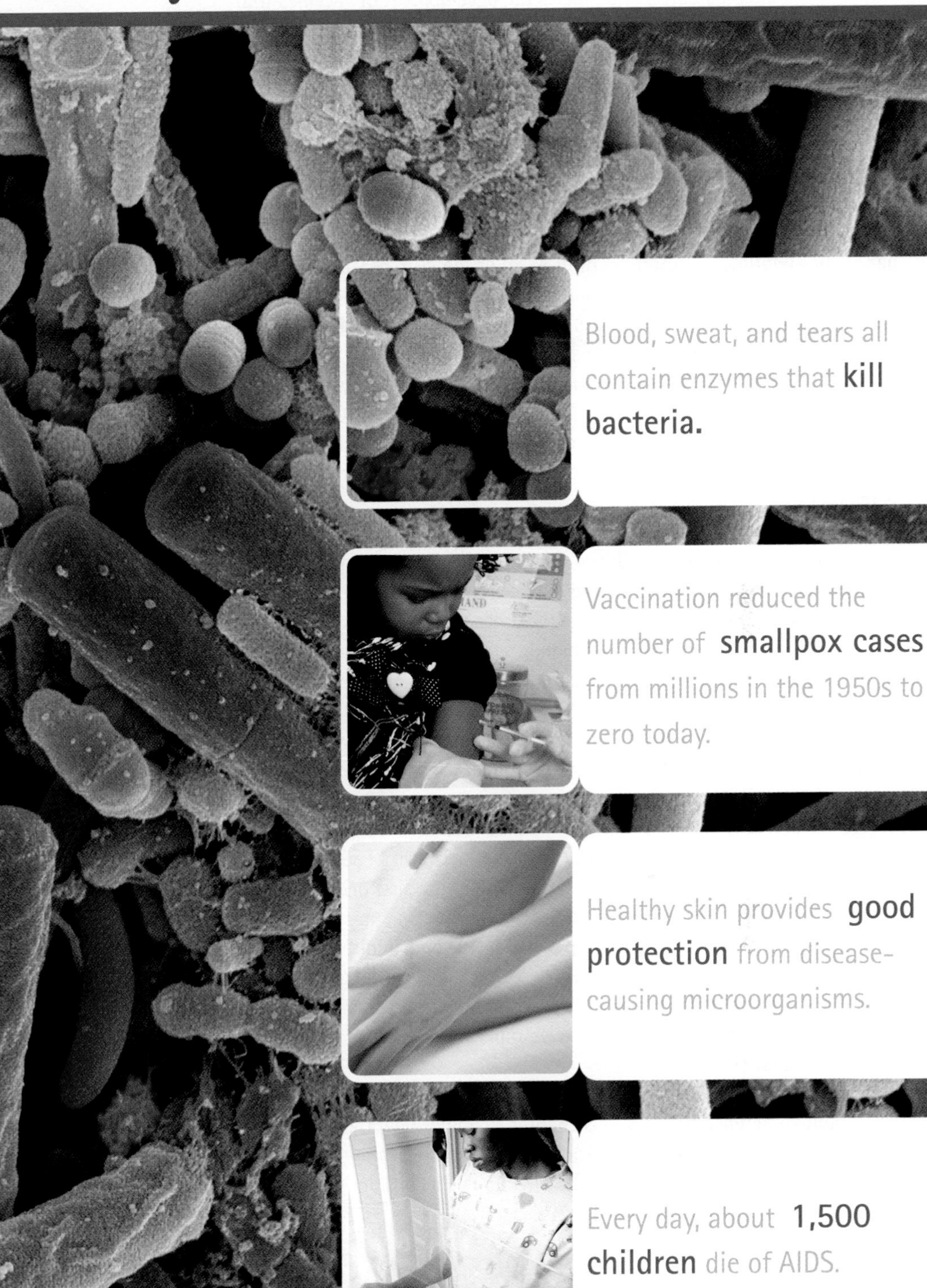

Blood, sweat, and tears all contain enzymes that **kill bacteria.**

Vaccination reduced the number of **smallpox cases** from millions in the 1950s to zero today.

Healthy skin provides **good protection** from disease-causing microorganisms.

Every day, about **1,500 children** die of AIDS.

BIOLOGY AND SOCIETY

The Next Pandemic?

One of medical science's greatest triumphs is the development of vaccines that offer protection from deadly viral diseases. A **vaccine** is a harmless derivative or variant of a disease-causing microbe. After injection, a vaccine can stimulate the immune system (an important part of our body's protective system) to mount a vigorous defense if the actual pathogen is ever encountered. In many industrialized nations, childhood vaccination programs have virtually eliminated previously feared viral diseases such as smallpox and polio.

So why can't we do the same for influenza (the "flu"), caused by the influenza virus **(Figure 24.1a)**? The flu virus is responsible for some of the deadliest **pandemics** (worldwide outbreaks of disease) of the 20th century. During an 18-month period from 1918 to 1919, "Spanish flu" killed at least 20 million people worldwide **(Figure 24.1b)**, about as many as have ever died of AIDS in the 25 years since it was first recognized. The "Asian flu" (1957–1958) and the "Hong Kong flu" (1968–1969) together killed 100,000 Americans. Even in an average nonpandemic year, influenza kills 20,000 people in the United States, mostly the elderly or people with chronic illnesses. The flu virus seems to make a mockery of our ability to develop vaccines against dangerous microbes.

The reason we can't be permanently vaccinated against the flu during childhood can be stated in one word: evolution. Our immune system learns to recognize and attack proteins found on the surface of the flu virus. But the RNA genome in the virus that encodes these proteins mutates rapidly. Because the genome has a high rate of mutation, new flu strains with new surface proteins are constantly emerging. Such changes mean that a vaccine against one flu strain will probably be ineffective against another. That is why a yearly vaccination against the flu is required; every year's batch of influenza vaccine is updated to include the latest strains.

Some medical researchers fear that a pandemic may start from a newly discovered strain of influenza virus called the "avian flu." Found primarily among poultry in Asia and Europe, human infections first occurred in southeast Asia and have since appeared farther west. So far, most human infections have occurred among people who have come in direct contact with infected birds. But health workers fear that the avian flu will mutate into a strain that can spread easily from person to person. If it does, and if a vaccine cannot be developed and deployed rapidly, the result may be the first pandemic of the 21st century.

The ongoing battle with the flu is one example of how much we depend on our body's built-in defense systems. This chapter examines how these defenses work to protect the body. We'll also look at how knowledge of the immune system has been applied to improve human health, and we'll consider what happens when the immune system malfunctions. ■

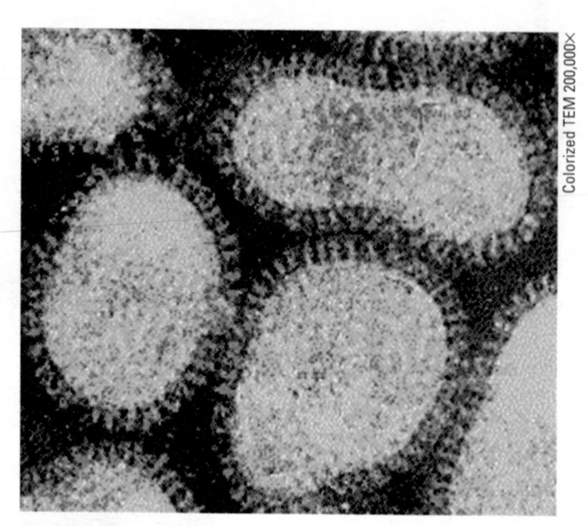

(a) The influenza virus.

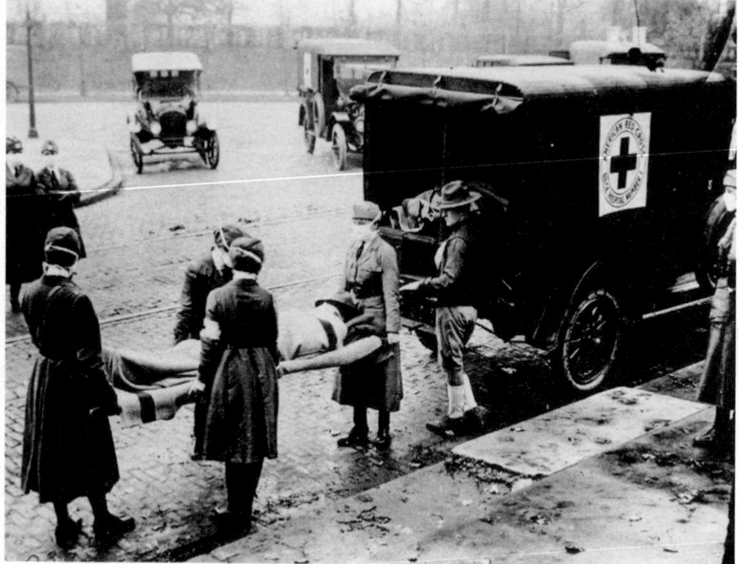

(b) The 1918–1919 influenza pandemic.

Figure 24.1 The world's deadliest virus.

Nonspecific Defenses

Nearly everything in the environment—water, food, and even the air—teems with microbes. The human body has three cooperative lines of

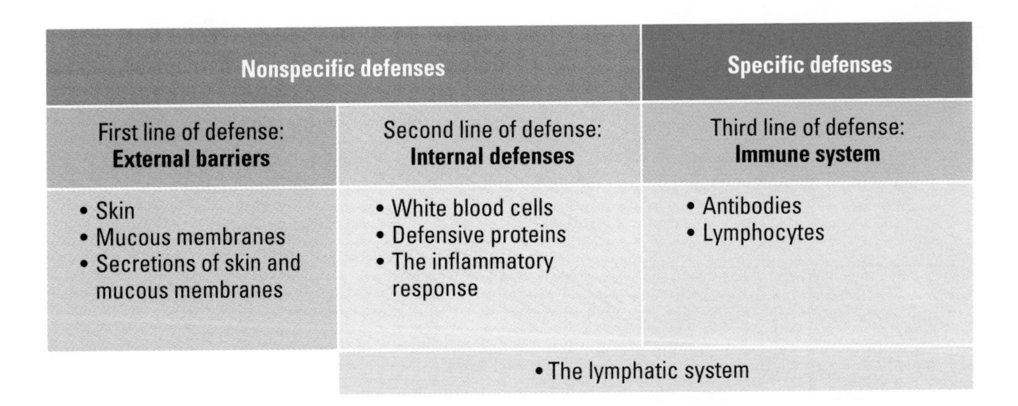

Nonspecific defenses		Specific defenses
First line of defense: **External barriers**	Second line of defense: **Internal defenses**	Third line of defense: **Immune system**
• Skin • Mucous membranes • Secretions of skin and mucous membranes	• White blood cells • Defensive proteins • The inflammatory response	• Antibodies • Lymphocytes
	• The lymphatic system	

Figure 24.2 Overview of the body's defenses. The nonspecific defenses do not distinguish one infectious microbe from another. The first line of defense prevents most infections from getting deep inside the body, while the second line deals with those that do. The specific defenses present a customized response to each particular kind of intruder. (Lymphocytes are a type of white blood cell.) The lymphatic system is involved in both nonspecific and specific defenses.

defense that protect us from the constant barrage of these and other potentially harmful substances **(Figure 24.2)**. The first line of defense is a set of physical barriers and chemical agents that prevent foreign invaders from getting deep inside the body. If that line fails, a variety of internal protections stand ready as a second line of defense. These first two lines of defense are nonspecific—that is, they do not distinguish one infectious microbe from another, but rather respond similarly to every infectious agent. Later in this chapter, we'll discuss the third line of defense: the immune system, which responds in a customized way to each particular microorganism.

External Barriers

For a foreign substance to get inside the body, it must first get by the body's external barriers. These barriers form an important protective front line because they actually *prevent* infection, as opposed to the body's other defenses, which come into play only *after* infection has already occurred.

The body has several physical barriers. Intact skin forms a tough outer layer that most bacteria and viruses cannot penetrate. Inside the nostrils, hairs filter many particles from the incoming air. Cells in the mucous membranes of the respiratory and digestive tracts secrete a sticky fluid called mucus. (Notice the different spellings: *mucus* is secreted by *mucous* membranes.) Mucus traps bacteria, dust, and other particles. Beating cilia extending from cells of the respiratory tract sweep mucus with the trapped particles outward until it is swallowed or expelled by sneezing, coughing, or blowing the nose **(Figure 24.3)**.

In addition to presenting physical barriers, the body also employs nonspecific chemical defenses against foreign microbes. Sweat, saliva, and tears contain antimicrobial chemicals such as lysozyme, an enzyme that disrupts the cell walls of some bacteria. Glands produce oils and acids that make the skin inhospitable to microorganisms. Concentrated stomach acid kills most of the bacteria swallowed with food, water, or saliva.

The flushing of some of the body's systems also protects the body from infection. Tears constantly wash the eyes with antibacterial proteins. The flow of urine moves foreign substances out of the urinary tract. Swallowing flushes bacteria into the highly acidic stomach, while vomiting rapidly expels harmful substances.

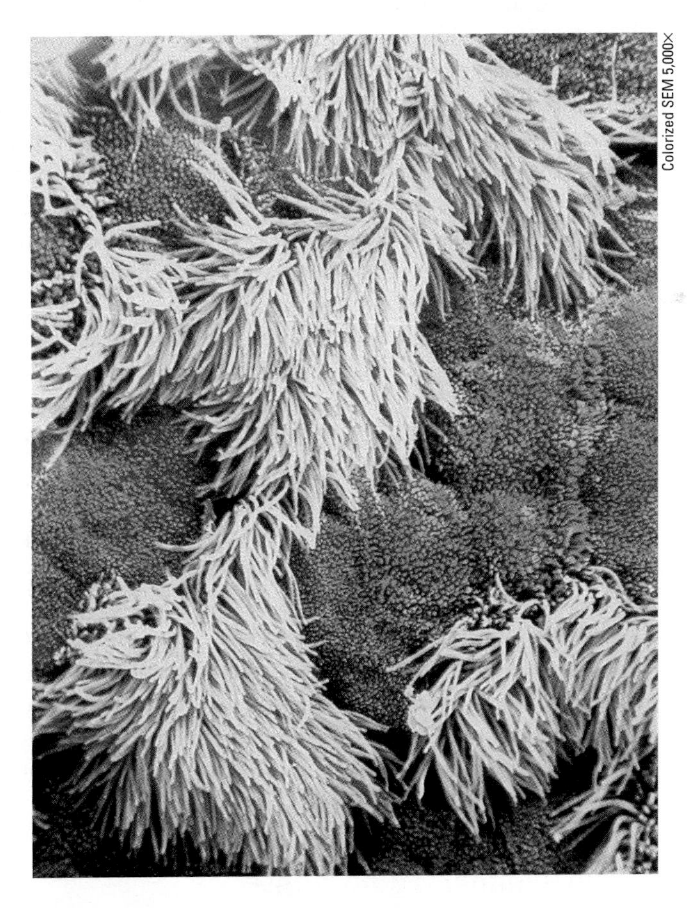

Figure 24.3 Cilia: a nonspecific external barrier. In this colorized electron micrograph of the lining of the trachea (windpipe), ciliated cells (yellow) are interspersed with mucus-producing cells (orange). The mucus helps trap foreign invaders, and the cilia help move them out of the respiratory tract.

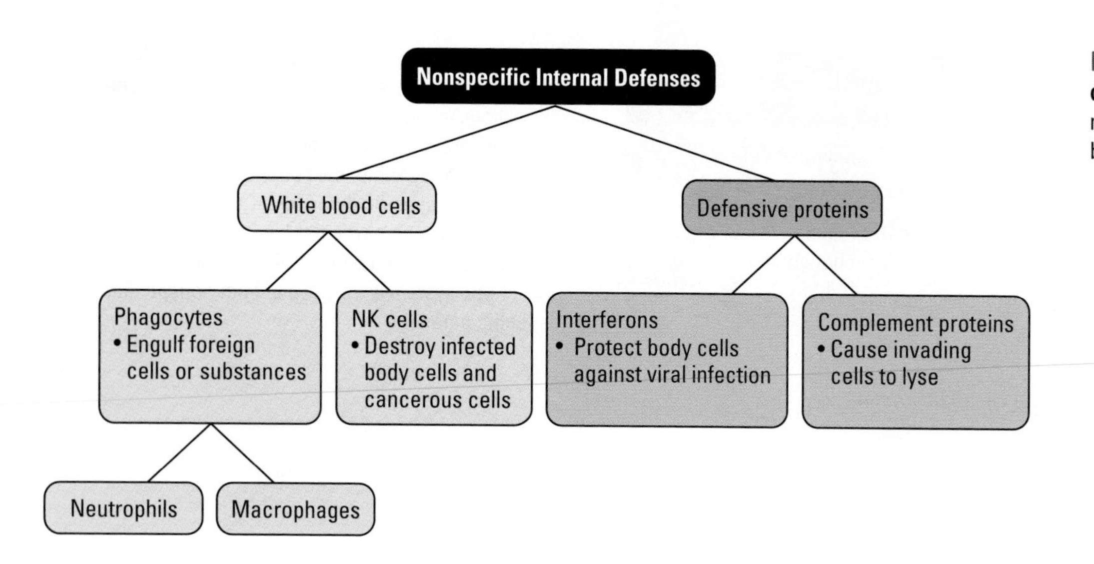

Figure 24.4 The body's second line of defense. When the external barriers fail, a set of nonspecific internal defenses stands ready as the body's next line of protection.

Internal Defenses

When an invader penetrates the body's external barriers, a set of nonspecific internal defenses acts as a second line of defense **(Figure 24.4)**. This line of defense depends mostly on white blood cells (leukocytes), which are produced in great numbers in the bone marrow. Several different types of white blood cells circulate through the bloodstream and interstitial fluid, destroying foreign cells and molecules, infected body cells, and debris from dead cells. **Neutrophils** and **macrophages** ("big eaters") are phagocytic white blood cells that engulf bacteria and viruses in infected tissues (see Figure 5.18). Other white blood cells, called **natural killer cells (NK cells),** attack virus-infected body cells and cancer cells. NK cells penetrate the plasma membrane of infected cells, causing them to burst.

Other nonspecific internal defenses include proteins that either attack microbes directly or impede their reproduction. **Interferons**, proteins produced by virus-infected body cells, help healthy cells resist viruses **(Figure 24.5)**. Infected cells ❶ release interferons that ❷ bind to plasma membrane receptors on nearby uninfected cells. This binding stimulates the healthy cells to ❸ produce proteins that inhibit viral reproduction. This is a good example of a nonspecific defense: The interferons produced by a virus-infected cell protect other cells from all kinds of viruses, not just the type that infected the first cell.

Defensive proteins of another sort, called **complement proteins,** circulate in the blood. Some of these proteins coat the surfaces of microbes, making them easier for macrophages to engulf. Other complement proteins cut lethal holes in microbial membranes, causing the cells to lyse. Also, complement proteins amplify another nonspecific defense, the inflammatory response, which we discuss next.

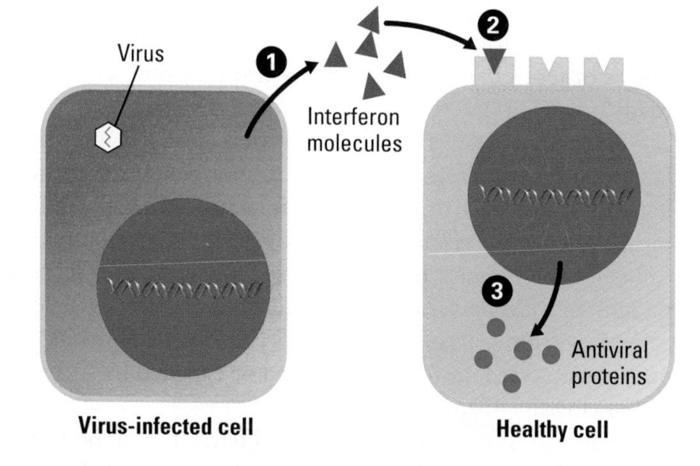

Figure 24.5 The action of interferon. When a virus infects a body cell, the cell produces and releases interferon molecules that bind to receptors on healthy cells, stimulating them to produce antiviral proteins that inhibit the growth of other viruses.

The Inflammatory Response

What happens if you get a bad cut on your finger? The next day, the finger is often red, swollen, painful, and warm to the touch. These are signs of the **inflammatory response,** a coordinated set of nonspecific defenses **(Figure 24.6)**. When tissue becomes damaged (from cuts, bug bites, or burns, for example), ❶ the injured cells and invading microorganisms

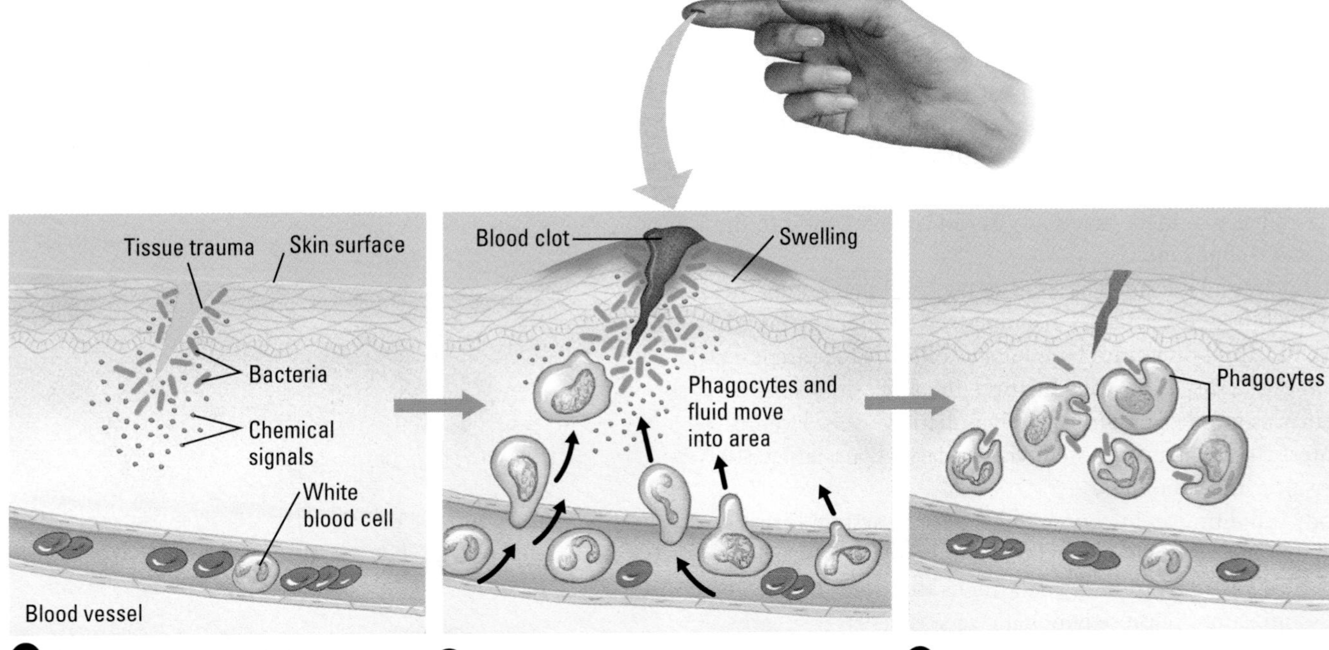

❶ Tissue injury; release of chemical signals such as histamine

❷ Dilation and increased leakiness of local blood vessels; migration of phagocytes to the area

❸ Phagocytes (macrophages and neutrophils) engulf bacteria and cell debris; tissue heals

Figure 24.6 The inflammatory response. Whenever tissue is damaged, the body responds with a coordinated set of nonspecific defenses called the inflammatory response.

release chemicals that trigger various nonspecific defenses. One such chemical signal, **histamine,** ❷ causes nearby blood vessels to dilate and leak fluid into the wounded tissue, causing it to swell. The swelling helps the tissue to heal by diluting toxins, bringing extra oxygen, and delivering platelets and clotting proteins that promote scabbing. The chemical signals also ❸ attract phagocytes and other white blood cells, which engulf bacteria and the remains of body cells killed by them or by the physical injury. The pus that often fills an infected injury consists of white blood cells, fluid that leaked from capillaries, and other tissue debris.

Damaged cells also release **prostaglandins,** chemical signals that increase blood flow to the damaged area, causing the wound to turn red and warm. (The word inflammation means "setting on fire.") Prostaglandins also stimulate nerves to send pain signals to the brain. While no one likes to feel pain, it has the benefit of changing our behavior to discourage additional damage. In response to severe tissue damage or infection, cells may also release **pyrogens** ("fire makers"), chemicals that can travel through the bloodstream to the hypothalamus—a control center in the brain—where they stimulate a fever, which may discourage bacterial growth.

Aspirin and ibuprofen are anti-inflammatory drugs; they dampen the body's normal inflammatory response and help reduce swelling and fever. Taking these drugs when you have an infection helps you feel better by blocking the production of the pain-causing prostaglandins. This is an example of a symptomatic cure, one that treats the outward symptoms of an illness (for example, pain and fever) without addressing the underlying cause (the infection).

All the defenses you've learned about so far respond the same way to any invader; that is, they are nonspecific. Soon you'll learn about the body's specific defenses—ones that are custom-tailored to each individual threat. But first let's examine a body system that contributes to both specific and nonspecific defenses.

The Lymphatic System

The **lymphatic system** consists of a branching network of vessels, numerous **lymph nodes** (saclike organs packed with macrophages and a type of white blood cells called lymphocytes), the bone marrow (where white blood cells develop), and several other organs **(Figure 24.7)**. The lymphatic vessels carry fluid called **lymph,** which is similar to interstitial fluid. The lymphatic system has two main functions: to return tissue fluid to the circulatory system and to fight infection.

As we noted in Chapter 23, a small amount of the fluid that enters the tissue spaces from the blood in a capillary bed does not reenter the blood capillaries. Instead, this fluid flows into lymphatic capillaries that are intermingled with blood capillaries. Now called lymph, this fluid drains from the lymphatic capillaries into larger and larger lymphatic vessels. Eventually, the fluid reenters the circulatory system via two large lymphatic vessels that fuse with veins near the shoulders.

When your body is fighting an infection, the lymphatic system is the main battleground. Lymph nodes fill with a huge number of lymphocytes, causing the "swollen glands" in your neck and armpits that your doctor feels for as a likely sign of infection. Because lymphatic vessels penetrate nearly every tissue, lymph can pick up microbes from infection sites just about anywhere in the body and deliver them to the lymphatic nodes and other organs. As the lymph passes through the lymphatic tissue, macrophages may engulf any invaders. Additionally, lymphocytes may be activated to mount a specific immune response, as we'll discuss in the next section.

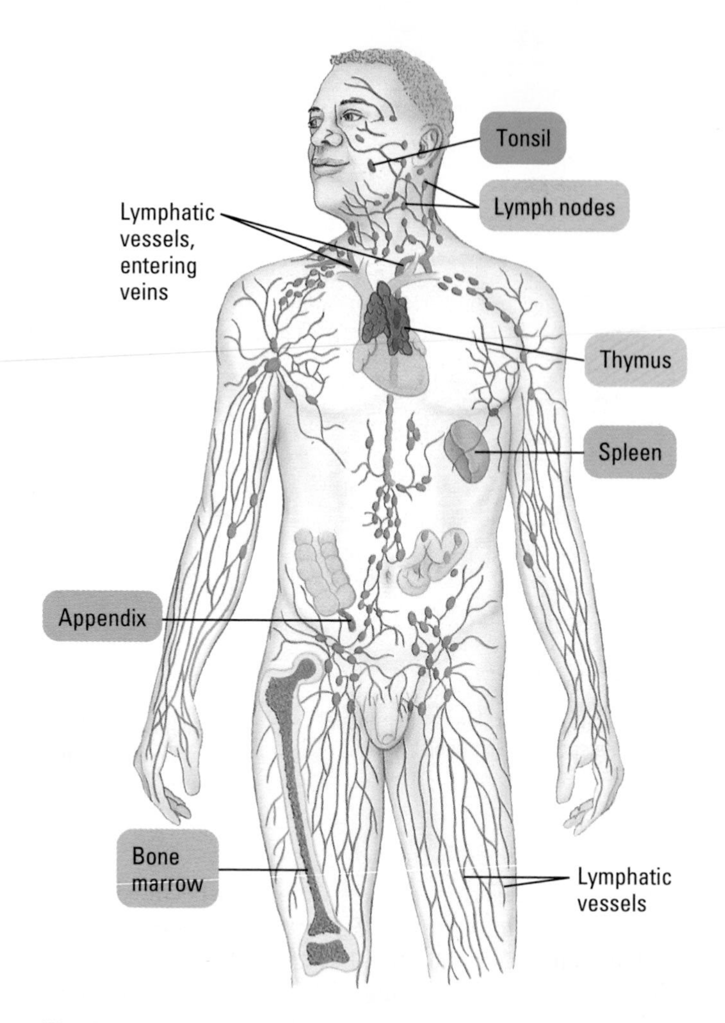

Figure 24.7 The lymphatic system. The lymphatic system transports fluid called lymph through various vessels and organs, including the spleen, thymus, tonsils, appendix, and many lymph nodes. White blood cells develop in the bone marrow. The lymphatic system plays a role in both the nonspecific and specific defenses.

CHECKPOINT

1. What makes the first and second lines of defense "nonspecific"?
2. What are phagocytes? Name two kinds of phagocytes.
3. Name the symptoms of the inflammatory response.
4. What are the two functions of the lymphatic system?

Answers: 1. They present the same response to every invader. **2.** White blood cells that engulf foreign cells and digest them; neutrophils and macrophages **3.** Swelling, redness, warmth, pain, and sometimes fever **4.** To collect fluid that leaks from blood capillaries within tissues and to help protect the body from foreign molecules and cells

Specific Defenses: The Immune System

MP3 Tutor
The Human Immune System

When the nonspecific defenses fail to ward off an infectious agent, the immune system provides a third line of defense. The **immune system** consists of a large collection of cells that work together to present a specific response to infection. The immune system recognizes and attacks specific kinds of invading microbes and cancer cells (which the body usually identifies as foreign). With its vast collection of over 2 trillion individual cells spread throughout the bloodstream and lymphatic system, the immune system presents an elaborately coordinated response to infection.

Whereas the nonspecific defenses are always ready to fight a variety of infections, the immune response must first be primed. An **antigen** is a foreign substance that elicits an immune response. Most antigens are molecules on the surfaces of viruses (such as the influenza surface proteins mentioned in the opening essay) or on the surfaces of other kinds of foreign cells, such as bacteria, protozoans, or parasitic worms. Antigens can also be molecules from mold spores, cancer cells, pollen, and house dust, as well as molecules on the cell surfaces of transplanted tissue.

When the immune system detects an antigen, it produces defensive proteins called antibodies. An **antibody** is a protein found in blood plasma that attaches to one particular kind of antigen and helps counter its effects. The antibodies produced against one antigen are usually ineffective against any other foreign substance—that is, they are specific.

In addition to being specific, the immune system has a remarkable "memory." It can "remember" antigens it has encountered before and react against them more promptly and vigorously on second and subsequent exposures. For example, if a person gets chicken pox, the immune system remembers certain molecules on the virus that causes this disease. Should the virus enter again, the immune system mounts a quick and decisive attack that usually destroys the virus before symptoms appear. Thus, the immune response, unlike nonspecific defenses, is adaptive; exposure to a particular foreign agent enhances future responses to that same agent.

The term **immunity** means resistance to specific invaders. Immunity is usually acquired by natural infection, but it can also be achieved by **vaccination** (also called immunization). In this procedure, the immune system is confronted with a vaccine composed of a harmless variant of a disease-causing microbe or one of its components. The vaccine stimulates the immune system to mount defenses against these antigens. These defenses will also be effective against the actual pathogen because it has similar antigens. Once a person has been successfully vaccinated, the immune system will respond quickly if it is exposed to the microbe.

In the United States, widespread vaccination of children has virtually eliminated viral diseases such as polio, mumps, and measles (**Figure 24.8**). Researchers are working hard to develop a vaccine against HIV, but so far success has been elusive. (See Evolution Connection at the end of this chapter for a major reason why.) As described in the opening essay, the influenza virus poses challenges to vaccine development as well. One of the major success stories of modern vaccination involves smallpox, a potentially fatal viral infection that affected over 50 million people a year worldwide in the 1950s. A massive vaccination effort has been so effective that there have been no cases of smallpox since 1977. Since 2001, however, the U.S. government has stockpiled hundreds of millions of doses of smallpox vaccine and has begun to vaccinate high-risk health-care workers in case the smallpox virus is used in a bioterrorist attack.

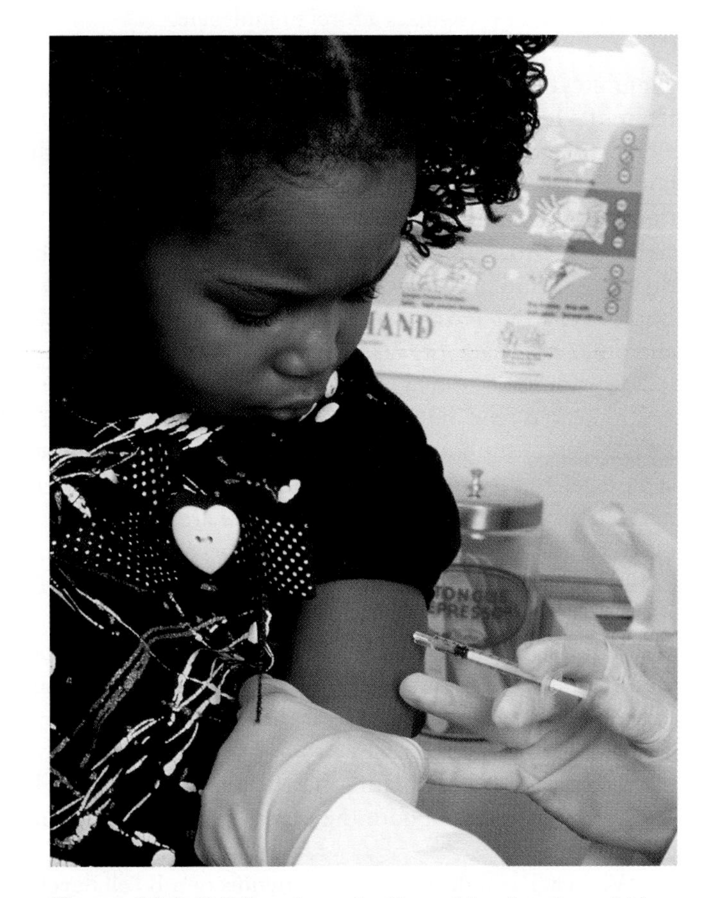

Figure 24.8 Childhood vaccinations. Most American children receive a series of vaccination shots starting soon after birth, including vaccinations against diphtheria/tetanus/pertussis ("diptet"), polio, hepatitis, chicken pox, and measles/mumps/rubella (MMR).

Whether antigens enter the body naturally (when you catch the flu, for example) or artificially (when you get a flu shot), the resulting immunity is called **active immunity** because the body is actively stimulated to produce antibodies in its own defense. It is also possible to acquire **passive immunity** by receiving premade antibodies. For example, a fetus obtains antibodies from its mother's bloodstream, and travelers sometimes get a shot containing antibodies to pathogens they are likely to encounter. In another example, the effects of a poisonous snakebite may be counteracted by injecting the victim with antivenin, which contains antibodies extracted from animals previously exposed to the venom. Passive immunity is temporary because the recipient's immune system is not stimulated by antigens; immunity lasts only as long as the antibodies do—usually a few weeks or months.

Recognizing the Invaders

To help you understand the process of immunity, we'll first look at the major players in the immune system and then see how the immune system recognizes a foreign molecule.

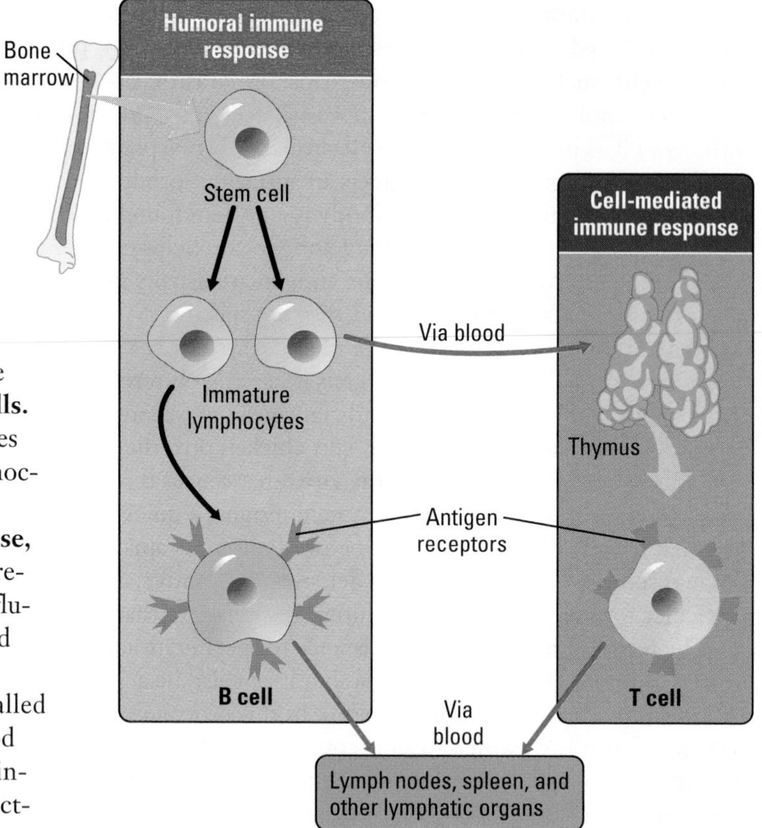

Figure 24.9 The development of lymphocytes. All lymphocytes are made in the bone marrow, but they mature in different locations: B cells in the bone marrow and T cells in the thymus. Once mature, lymphocytes are released into the blood and lymphatic system.

Lymphocytes **Lymphocytes,** white blood cells found most often in the lymphatic system, produce the immune response. Like all blood cells, lymphocytes originate from stem cells in the bone marrow **(Figure 24.9)**. Some immature lymphocytes continue developing in the bone marrow; these become specialized as **B cells.** Other immature lymphocytes migrate via the blood to the thymus, a gland in the chest, where they become specialized as **T cells.** Both B cells and T cells eventually make their way to the lymph nodes and other lymphatic organs. With these two different kinds of lymphoctyes, the immune system mounts a dual defense.

In the first type of immunity, called the **humoral immune response,** B cells secrete antibodies as a form of defense. The humoral immune response defends primarily against bacteria and viruses present in body fluids. The antibodies that the B cells secrete are carried in the lymph and blood to sites of infection wherever they occur in the body.

The second type of immune response, produced by T cells, is called the **cell-mediated immune response.** T cells circulate in the blood and lymph, attacking infected body cells. T cells also work against infections caused by fungi and protozoans and are important in protecting the body from its own cells if they become cancerous. In addition, T cells function indirectly by promoting phagocytosis by other white blood cells and by stimulating B cells to produce antibodies. Thus, T cells are involved in both the cell-mediated and humoral immune responses.

When a T cell develops in the thymus or a B cell develops in bone marrow, the cell synthesizes molecules of a specific protein and builds them into its plasma membrane. These protein molecules (purple in Figure 24.9) are antigen receptors that stick out from the cell's surface. All the antigen receptors on a lymphocyte are capable of binding to only one specific type of antigen. In the case of a B cell, the receptors are actually molecules of the particular antibody that the B cell will secrete. Once its surface proteins are in place, a B cell or T cell can recognize a specific antigen and mount an immune response against it. One cell may recognize an antigen on the mumps virus, for instance, while another detects a particular antigen on a tetanus-causing bacterium.

A great diversity of B cells and T cells develop in each individual—enough to recognize and bind to just about every possible antigen. A small population of each kind of lymphocyte lies in wait, ready to recognize and respond to a specific antigen. Only a tiny fraction of the immune system's lymphocytes will ever be used, but they are all available if needed. It is as if the immune system maintains a huge standing army of soldiers, each designed to recognize one particular kind of invader. The majority of soldiers never encounter their target and remain idle. But when an invader does appear, chances are good that some lymphocytes will be able to recognize and bind to it.

Figure 24.10 Structure of an antibody. Antibodies are immune system proteins that bind antigens. Every antibody is made from four polypeptides (two light chains and two heavy chains) that form a Y shape. The V (variable) regions at the ends of each arm of the Y form the antigen-binding site. The antigen-binding site comes in a tremendous variety of different shapes, each capable of binding to one kind of antigen in a highly specific lock-and-key manner. C stands for "constant" region.

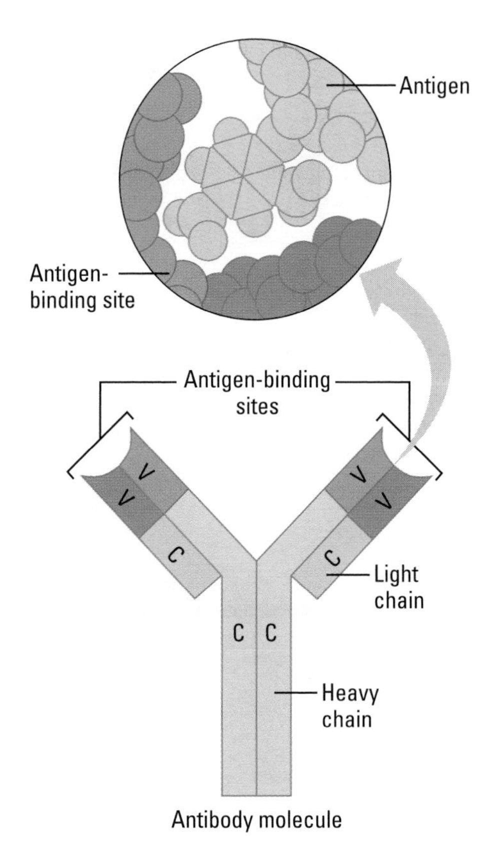

Antibodies The antibodies produced by B cells serve as molecular weapons of defense **(Figure 24.10)**. A typical antibody consists of four polypeptide chains, two longer ("heavy") chains (shown in purples) and two shorter ("light") chains (shown in blues). The four chains are joined together to form a Y shape.

In the humoral immune response, an antibody recognizes and binds to a certain antigen and in so doing helps counter its effect. The structure of an antibody allows it to perform these functions. Notice in Figure 24.10 that each of the four chains of the molecule has a C (constant) region and a V (variable) region. At the tip of each arm of the Y, a pair of V regions forms an antigen-binding site, a region of the molecule responsible for the antibody's recognition-and-binding function. An antigen-binding site and the antigen it binds have complementary shapes, like an enzyme and substrate or a lock and key. A huge variety in the shapes of the binding sites of different antibody molecules arises from a similarly large variety of amino acid sequences in the V regions. This structural variety gives the humoral immune system the ability to react to virtually any kind of antigen.

The main role of antibodies in eliminating invading microbes or molecules is to mark the invaders. An antibody marks an antigen by combining with it to form an antigen-antibody complex. Such binding triggers one of several mechanisms that help destroy the invader. Sometimes, the binding of antibody molecules physically blocks antigens, making them harmless. For example, if antibodies bind to virus surface proteins, that may be enough to inhibit the virus from entering body cells. Other times, antibody binding causes viruses, bacteria, or foreign eukaryotic cells to form large clumps that are easily captured by circulating phagocytes. Additionally, the antigen-antibody complex may activate complement proteins that destroy the marked antigen.

Responding to the Invaders

The immune system stops infections by destroying antigen-bearing invaders. When the immune system detects an antigen, it responds with an increase in the number of cells that either produce defensive proteins (B cells) or attack infected cells (T cells).

B Cells and Clonal Selection With so many different kinds of lymphocytes, how does the body marshall enough of the right one to fight a specific invading antigen? The key is a process called **clonal selection.** At first, an antigen activates only a tiny number of lymphocytes with receptors specific to that antigen. These "selected" cells then proliferate, forming a clone of cells (a population of genetically identical B cells or T cells) that are specific for the stimulating antigen.

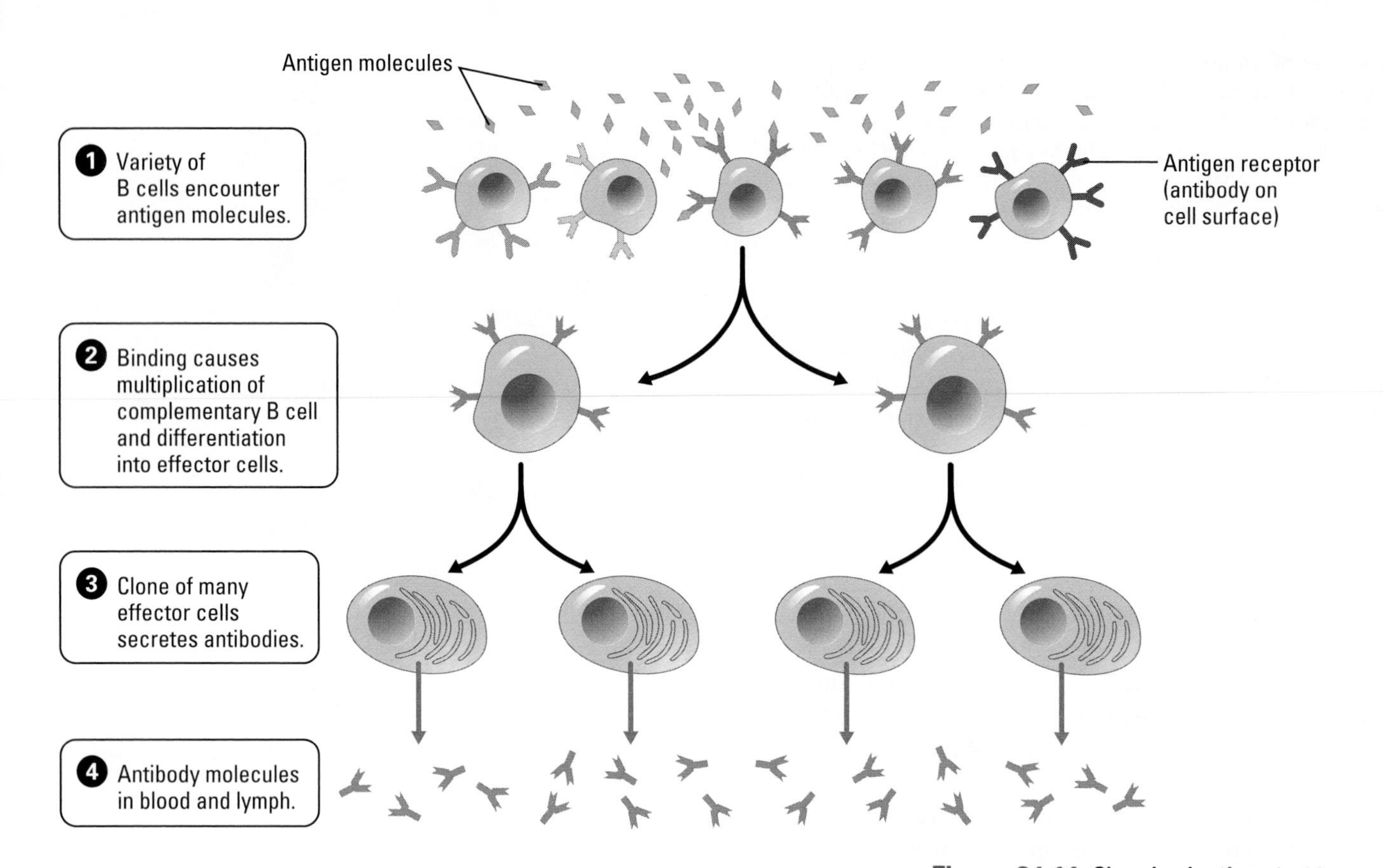
Antigen molecules
Antigen receptor (antibody on cell surface)
1 Variety of B cells encounter antigen molecules.
2 Binding causes multiplication of complementary B cell and differentiation into effector cells.
3 Clone of many effector cells secretes antibodies.
4 Antibody molecules in blood and lymph.

Figure 24.11 Clonal selection. In this example of clonal selection, an antigen "selects" those B cells that bind to it and causes their proliferation. A similar kind of clonal selection operates on T cells.

Figure 24.11 illustrates how clonal selection works for B cells. Each B cell has its own specific type of antigen receptor (antibody) embedded in its surface. ❶ Once an antigen enters the body, it binds with one of the few B cells that have complementary receptors. Lymphocytes without the appropriate binding sites are not affected. ❷ The selected cell is activated—it grows, divides, and differentiates further. ❸ The result is a clone of **effector cells** specialized for defending against the very antigen that triggered the response. Each effector cell makes as many as 2,000 copies of its antibody per second. ❹ These antibodies circulate in the blood and lymphatic fluid, binding to the invading antigen and contributing to its destruction or blocking its harmful effects. Clonal selection thus results in the production of many antibody molecules, all of which are specialized to recognize copies of the antigen that started the response.

The antibody molecules produced by a single clone of cells are called **monoclonal antibodies.** Using cell cultures, researchers can artificially produce monoclonal antibodies in large quantities. Such monoclonal antibodies have many practical uses, including home pregnancy tests. One type of home pregnancy test detects a hormone called human chorionic gonadotropin (HCG) that is present in the urine of pregnant women. When a testing strip is dipped into urine, monoclonal antibodies on it bind to any HCG that is present, causing a color change on the strip.

Immunological Memory The first exposure of lymphocytes to an antigen, called the **primary immune response,** takes several days to produce effector cells via clonal selection. The antibodies produced by these effector cells reach their peak levels about two weeks after first exposure and then start to decline. Each effector cell lives only 4 or 5 days, and the primary immune response subsides as the effector cells die out.

Clonal selection also produces **memory cells** (not shown in Figure 24.11). In contrast to short-lived effector cells, memory cells can last decades in the lymph nodes, ready to be activated by a second exposure to the antigen. When the antigen is encountered again, the memory cells that bind to it initiate a faster and stronger response called the **secondary immune response.** The activated memory cells multiply quickly, producing a large new clone of lymphocytes. This new clone in turn produces antibodies that are often more effective than those produced during the primary response. In some cases, memory cells seem to confer lifetime immunity, as in such childhood diseases as mumps and measles.

Although we have focused on B cells in this section, clonal selection, effector cells, and memory cells are features of the T cell system as well.

T Cells Whereas B cells respond to antigens circulating freely in body fluids (humoral immune response), T cells respond to pathogens that have already entered body cells (cell-mediated immune reponse). There are two main kinds of T cells: helper T cells and cytotoxic T cells.

Helper T cells bind to other white blood cells that have previously encountered an antigen. In the example in **Figure 24.12**, ❶ a macrophage ingests a microbe and breaks it into fragments—antigens that are foreign to the macrophage. Protein molecules belonging to the macrophage, which we will call self proteins (because they belong to the body itself), ❷ bind to the foreign antigens—nonself molecules—and ❸ display them on the cell's surface. ❹ Receptors embedded in the helper T cell's plasma membrane recognize and bind to the combination of self protein and nonself molecule displayed on the macrophage.

The binding of a T cell receptor to a self-nonself complex activates the helper T cell **(Figure 24.13)**. An activated helper T cell promotes the immune response in several ways. The helper T cell itself grows and divides, producing more activated helper T cells as well as memory cells. In addition, the helper T cell stimulates the activity of cytotoxic T cells and helps activate B cells, thus stimulating the humoral immune response.

Cytotoxic T cells are the only T cells that actually kill other body cells. Cytotoxic T cells identify infected body cells in the same way that helper

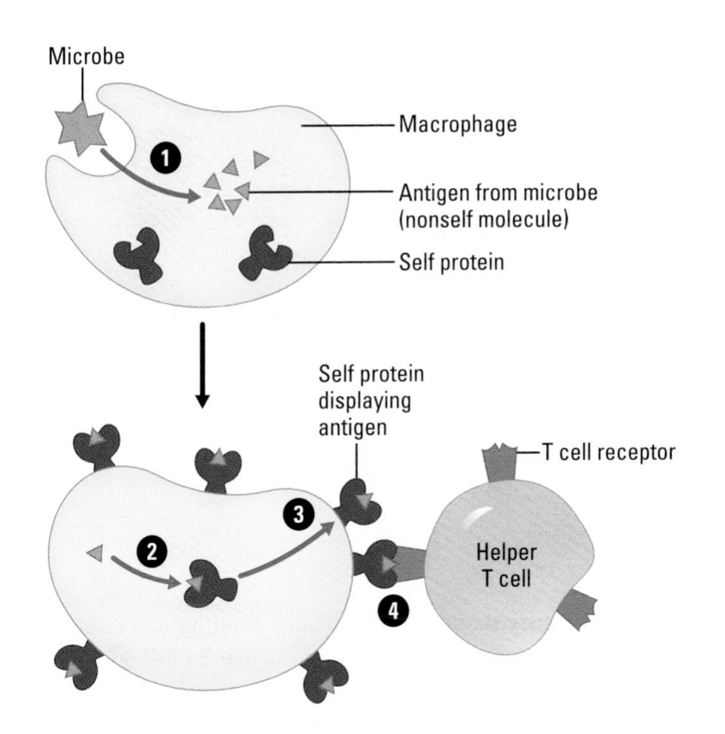

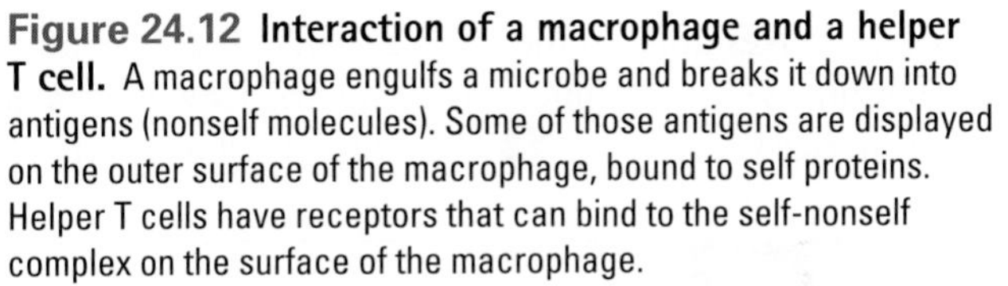
Figure 24.12 Interaction of a macrophage and a helper T cell. A macrophage engulfs a microbe and breaks it down into antigens (nonself molecules). Some of those antigens are displayed on the outer surface of the macrophage, bound to self proteins. Helper T cells have receptors that can bind to the self-nonself complex on the surface of the macrophage.

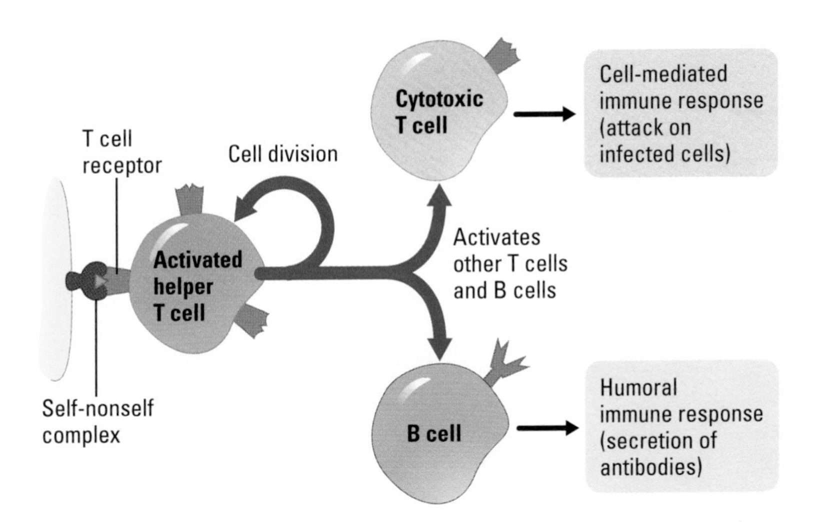

Figure 24.13 The roles of an activated helper T cell. Activated helper T cells promote both the cell-mediated and humoral immune responses by stimulating the activity of cytotoxic T cells and B cells.

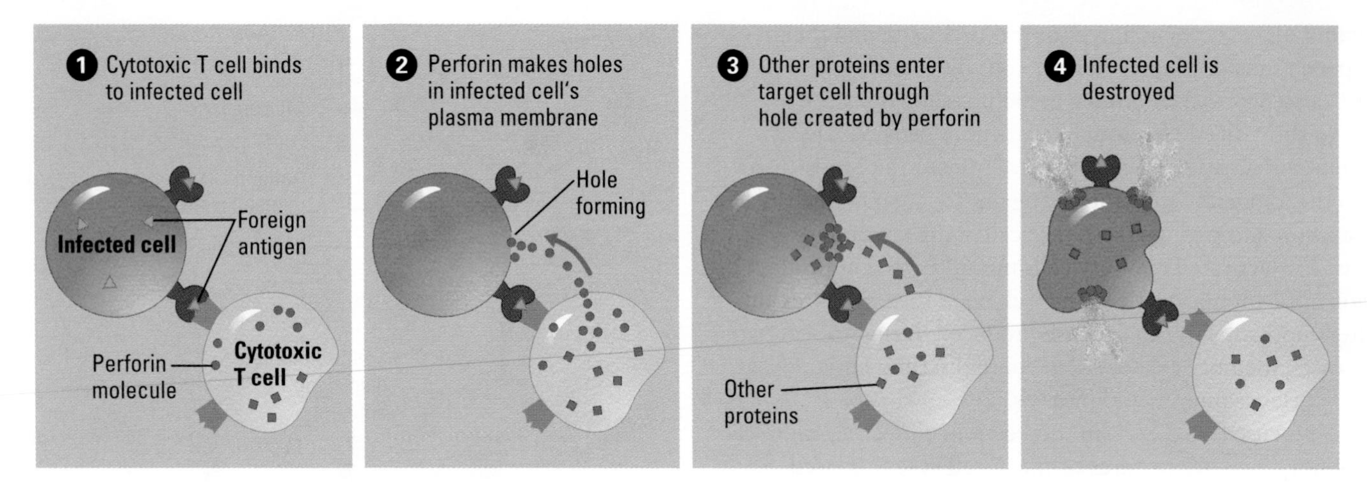

Figure 24.14 How a cytotoxic T cell kills an infected cell. When a cytotoxic T cell binds to an infected cell, the T cell produces a protein called perforin that forms holes in the infected cell. Other T cell proteins enter the cytoplasm of the target cell through these holes and trigger programmed cell death.

T cells identify their targets: through binding of a membrane receptor to a self-nonself complex. As shown in **Figure 24.14**, ❶ a cytotoxic T cell binds to an infected cell. The binding activates the T cell. ❷ The activated cell synthesizes and discharges several proteins, including one called perforin that makes holes in the infected cell's plasma membrane. ❸ Other T cell proteins enter the infected cell and trigger a process called programmed cell death (see Chapter 26). ❹ The infected cell dies and is destroyed.

CHECKPOINT

1. What is the immune system? What is its function?
2. What is an antigen? Name some examples.
3. What are the two main types of lymphocytes?
4. How does the body produce enough B cells to fight off a major infection?
5. What makes a secondary immune response faster than a primary immune response?
6. What are the two main types of T cells, and what does each do?

Answers: 1. The immune system is a large collection of cells that presents coordinated, specific responses to infection. **2.** An antigen is a molecule that elicits an immune response. Examples include proteins on the surfaces of microorganisms, pollen, and dust. **3.** B cells and T cells **4.** The process of clonal selection causes those few B cells that are complementary to the antigens of the infectious agent to selectively multiply. **5.** It takes a few weeks for a clone of effector cells to be formed during the primary response. In a secondary response, preexisting memory cells can respond more quickly. **6.** Helper T cells stimulate other body defenses. Cytotoxic T cells destroy infected cells.

Immune Disorders

If the intricate interplay of immune cells goes awry, problems can result that range from mild irritations to deadly diseases. In this section, we will examine some of the consequences of a malfunctioning immune system.

Allergies

Allergies are abnormal sensitivities to antigens in the environment. Antigens that cause allergies are called **allergens.** Common allergens include protein molecules on pollen grains, on the feces of tiny mites that live in

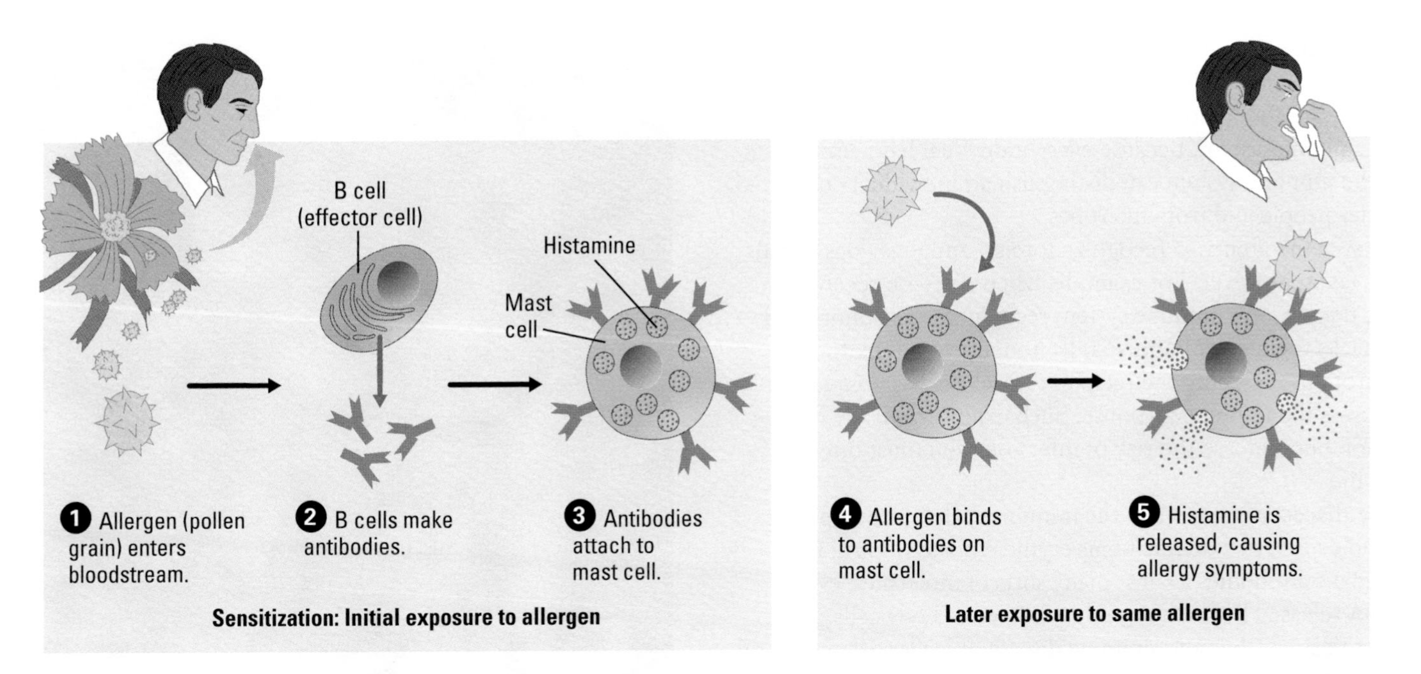
B cell
(effector cell)
Histamine
Mast
cell
1 Allergen (pollen grain) enters bloodstream.
2 B cells make antibodies.
3 Antibodies attach to mast cell.
Sensitization: Initial exposure to allergen
4 Allergen binds to antibodies on mast cell.
5 Histamine is released, causing allergy symptoms.
Later exposure to same allergen

Figure 24.15 How allergies develop. Allergies develop in two stages. The first exposure causes sensitization. Subsequent exposures to the same allergen produce allergy symptoms.

house dust, and in animal dander (shed skin cells). Allergic reactions typically occur very rapidly in response to tiny amounts—sometimes just a few molecules—of an allergen.

The symptoms of an allergy result from a two-stage reaction sequence outlined in **Figure 24.15**. The first stage, called sensitization, occurs when a person is first exposed to an allergen—pollen, for example. ❶ After an allergen enters the bloodstream, it binds to B cells with complementary receptors. ❷ The B cells then proliferate through clonal selection and secrete large amounts of antibodies to that allergen. ❸ Some of these antibodies attach to receptor proteins on the surface of mast cells, normal body cells that produce histamine and other chemicals that trigger the inflammatory response.

Allergy symptoms do not arise until the second stage, which begins when the same allergen reenters the body and ❹ binds to the antibodies on the mast cells. In response to this binding, ❺ the mast cells release histamine, which causes allergy symptoms such as sneezing, coughing, and itching. Because allergens usually enter the body through the nose and throat, symptoms are often most prominent there.

Allergic reactions range from seasonal nuisances to severe, life-threatening responses. **Anaphylactic shock** is an especially dangerous type of allergic reaction. Some people are extremely sensitive to certain allergens, such as the venom from a bee sting or allergens in peanuts or shellfish. Any contact with these allergens causes a sudden release of inflammatory chemicals. Blood vessels dilate abruptly, causing a precipitous and potentially fatal drop in blood pressure, a condition called shock. Fortunately, anaphylactic shock can be counteracted with injections of the hormone epinephrine **(Figure 24.16)**.

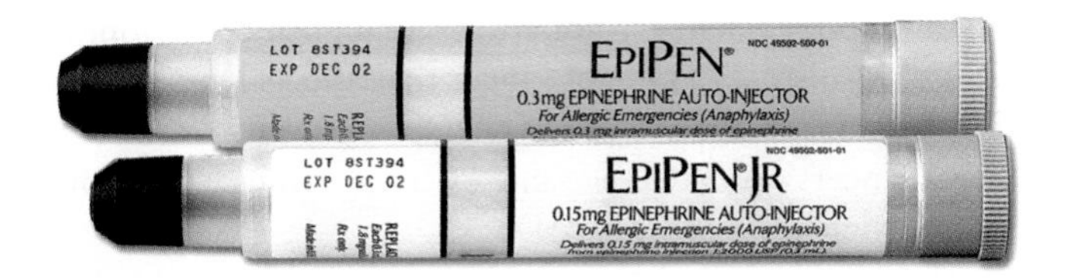
LOT 8ST394
EXP DEC 02
EPIPEN
0.3mg EPINEPHRINE AUTO-INJECTOR
For Allergic Emergencies (Anaphylaxis)
LOT 8ST394
EXP DEC 02
EPIPEN JR
0.15mg EPINEPHRINE AUTO-INJECTOR
For Allergic Emergencies (Anaphylaxis)

Figure 24.16 Single-use epinephrine syringes. People with known severe allergies can carry single-use syringes that contain the hormone epinephrine (also called adrenaline). An injection can quickly stop anaphylactic shock, a life-threatening allergic reaction.

Autoimmune Diseases

The immune system's ability to recognize the body's own molecules—that is, to distinguish self from nonself—enables it to battle foreign invaders without harming healthy body cells. Self proteins on cell surfaces are the key to this ability. Each person's cells have a particular collection of self proteins that provide the molecular "fingerprints" recognized by the immune system. The particular collections of proteins are specific to the individual in whom they are found, marking the body cells as "off-limits" to attacks from the immune system. Because every individual has a unique set of self proteins, the immune system can distinguish an individual's cells from those of other people and from microbes.

The immune system's ability to recognize foreign antigens does not always work in our favor, however. For example, when a person receives an organ transplant, the person's immune system recognizes the donor's cells as foreign and attacks them. To minimize rejection, doctors look for a donor with self proteins matching the recipient's as closely as possible and use drugs to suppress the immune response. Such drugs drastically reduce the risk of rejection but increase the risk of infections and must often be administered for life.

Autoimmune diseases result when the immune system turns against the body's own molecules. In systemic lupus erythematosus (lupus), for example, B cells make antibodies against many sorts of molecules, even histones and DNA released by the normal breakdown of body cells. Rheumatoid arthritis is another autoimmune disease; it leads to damage and painful inflammation of the cartilage and bone of joints (**Figure 24.17**). In insulin-dependent diabetes, the insulin-producing cells of the pancreas are the targets of autoimmune cell-mediated responses. In multiple sclerosis (MS), T cells wrongly attack proteins in neurons (see Figure 27.3); the resulting malfunctions can cause movement and vision problems.

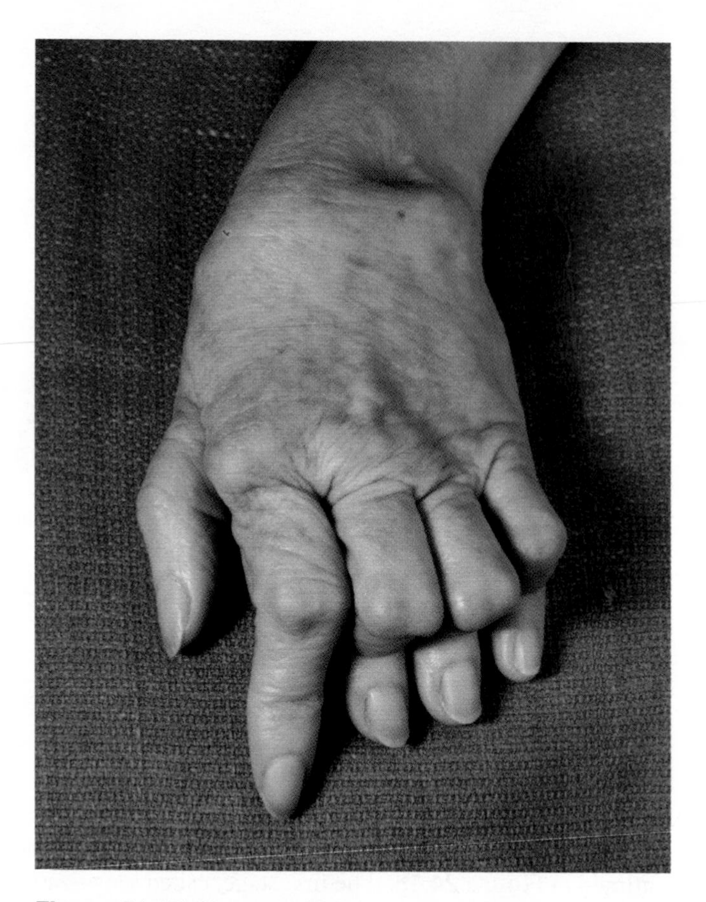

Figure 24.17 Rheumatoid arthritis. This photo shows the hand of a person with rheumatoid arthritis, an autoimmune disease in which the cartilage and bone of joints become damaged.

Immunodeficiency Diseases

In contrast to autoimmune diseases are a variety of defects called **immunodeficiency diseases.** Immunodeficient people lack one or more of the components of the immune system and as a result are susceptible to infections that would ordinarily not cause a problem. A rare inborn disease called severe combined immunodeficiency (SCID) causes a marked deficit in both T cells and B cells. People with SCID are extremely sensitive to even minor infections, forcing them to live behind protective barriers (providing inspiration for various "bubble boy" stories in the popular culture) or to receive bone marrow transplants. Recently, SCID has become the focus of several human gene therapy trials (see Chapter 12).

Immunodeficiency is not always an inborn condition. For instance, Hodgkin's disease, a type of cancer that affects lymphocytes, can depress the immune system. Radiation therapy and drug treatments used against many cancers can have the same effect. Another well-known acquired immunodeficiency disease is AIDS.

AIDS

AIDS (acquired immunodeficiency syndrome) has killed more than 25 million people worldwide since 1981, and more than 40 million people are currently living with the AIDS virus, HIV. In 2005, 5 million people were newly infected with HIV, and over 3 million died, including more than half a million children under 15. The vast majority of HIV infections and AIDS deaths occur in the nonindustrialized nations of Asia and sub-Saharan Africa.

HIV is deadly because it destroys the immune system, leaving the body defenseless against most invaders. HIV can infect a variety of cells, but it most often attacks helper T cells—the cells that activate other T cells and B cells. When HIV depletes the body of helper T cells, both the cell-mediated and humoral immune responses are severely impaired. Death usually results, not from HIV itself but from another infectious agent or from cancer. Hence the name of the disease: It is *acquired* through an infection that results in severe *immunodeficiency* due to a lack of helper T cells, and it presents as a *syndrome*, a combination of symptoms (see Figure 10.30 to review the course of cell infection by HIV).

Education is the best weapon against the spread of AIDS. Safe sex behaviors, such as reducing promiscuity and using condoms, can save many lives **(Figure 24.18)**. (See Chapter 26 for a more complete discussion.)

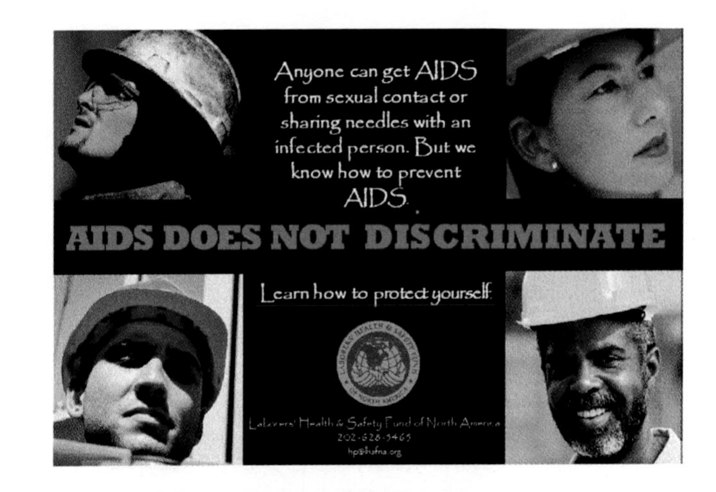

Figure 24.18 AIDS education.

CHECKPOINT

1. What is an autoimmune disease?
2. What is an immunodeficiency disease?

Answers: 1. A disease in which cells of the immune system attack certain of the body's own cells as foreign **2.** A disease in which one or more parts of the immune system are defective, resulting in an inability to fight infections

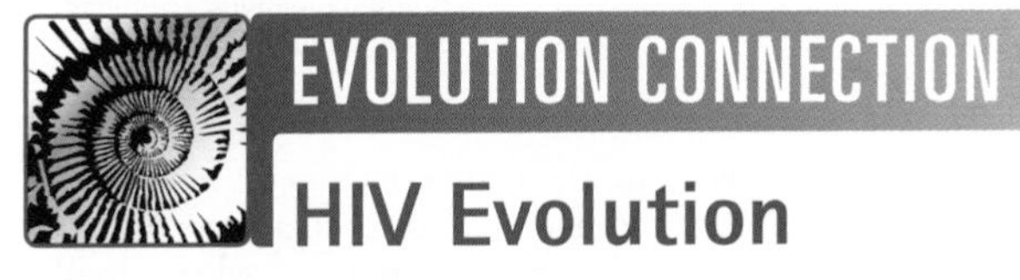

EVOLUTION CONNECTION

HIV Evolution

HIV has one of the fastest rates of mutation of any pathogen ever studied **(Figure 24.19)**. This startling fact has led researches to view the evolution of drug-resistant HIV strains as the number one obstacle to the eradication of AIDS. At one time, there was great hope that a "cocktail" of three anti-AIDS drugs, each of which attacks a different part of the HIV viral life cycle, could completely eliminate the virus in an infected person. Such hope greatly underestimated the ability of HIV to evolve. Although people with access to such drugs do survive much longer and have a greatly improved quality of life, the virus is usually not totally eliminated from a patient's immune system. Disturbingly, drug-resistant HIV strains are now being documented in newly infected patients. This demonstrates that HIV readily adapts through natural selection to a changing environment—one in which drug treatments are widely available. In other words, the presence of anti-AIDS drugs in the environment has created a selection pressure that favors the success and spread of drug-resistant strains. Thus, the battle continues, with medical science on one side and the constantly evolving HIV on the other. ■

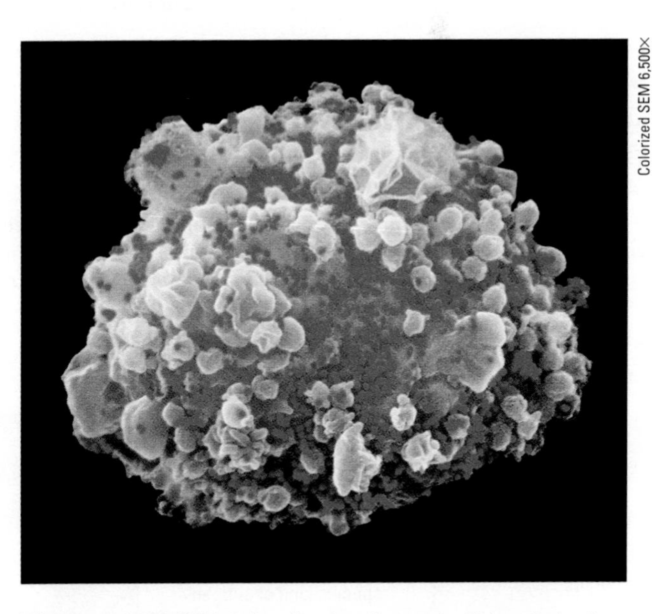

Figure 24.19 HIV doing its deadly work. This colorized electron micrograph shows a human immune cell under attack by HIV (small red spheres).

Chapter Review

SUMMARY OF KEY CONCEPTS

For study help and activities, go to campbellbiology.com or the student CD-ROM.

Biology and Society: The Next Pandemic?

- The influenza virus has caused several pandemics, but new vaccines must be developed every year because of the high mutation rate of the virus.

Nonspecific Defenses

- The human body contains two lines of defense (external barriers and internal nonspecific defenses) that respond identically to every type of invader.

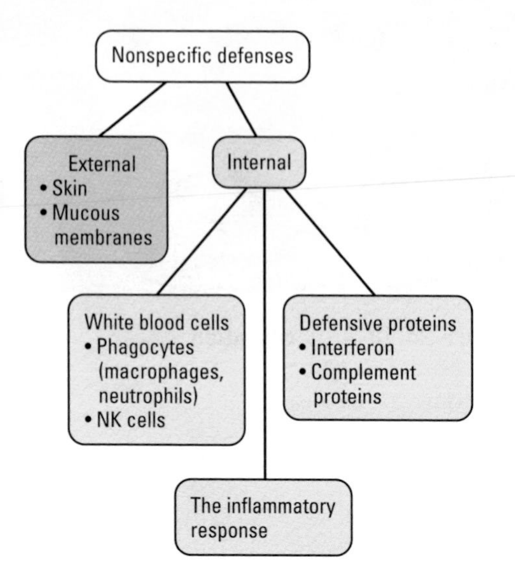

- **The Lymphatic System** The lymphatic system is a network of lymphatic vessels and organs. The vessels collect fluid from body tissues and return it as lymph to the blood. The organs, including the lymph nodes, are packed with white blood cells that fight infections.

Specific Defenses: The Immune System

MP3 Tutor *The Human Immune System*

- The immune system consists of a large collection of cells that work together to present a specific response to infection. Antigens are molecules that elicit immune reactions. The immune system reacts to antigens and "remembers" an invader. Infection or vaccination triggers active immunity, whereas passive immunity is obtained through transfer of antibodies.

- **Recognizing the Invaders** Two kinds of lymphocytes carry out the immune response.

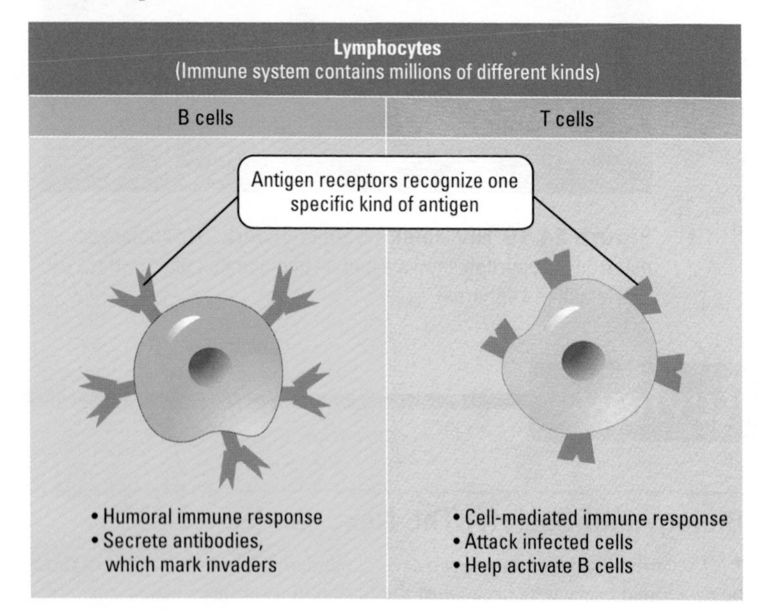

Antibodies recognize antigens through specific binding, and this binding can trigger destruction of the antigen-bearing molecule or cell through several different mechanisms. An antibody molecule contains four polypeptide chains arranged in a Y shape, with the antigen-binding sites at the tips of the arms of the Y.

- **Responding to the Invaders** When an antigen enters the body, it activates only lymphocytes with complementary receptors, a process called clonal selection.

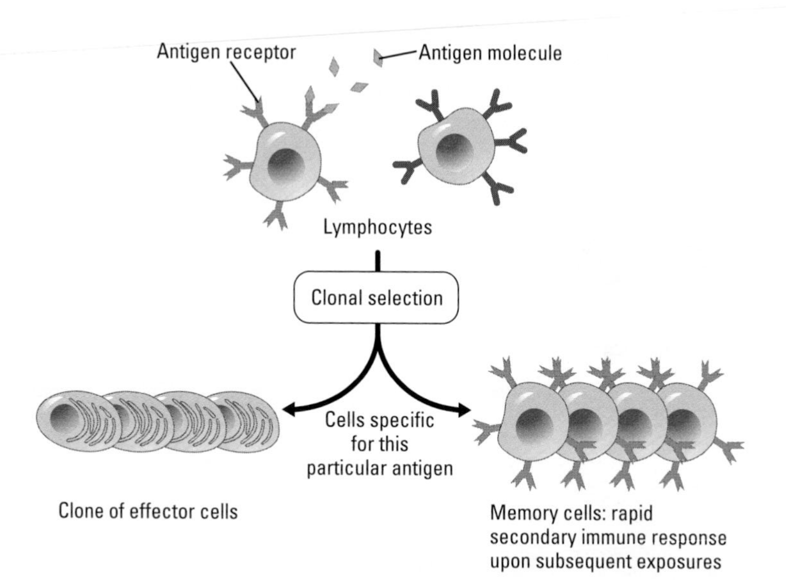

Helper T cells and cytotoxic T cells are the main effectors of the cell-mediated immune response, and helper T cells also stimulate the humoral immune response. In the cell-mediated immune response, a white blood cell, such as a macrophage, first displays a foreign antigen (a nonself molecule) and one of the body's own self proteins to a helper T cell. The helper T cell's receptors recognize the self-nonself complexes, and the interaction activates the helper T cell. In turn, the helper T cell can activate cytotoxic T cells with the same receptors (and can stimulate B cell activation as well). Cytotoxic T cells bind to infected body cells displaying both self proteins and foreign antigens and destroy them.

Activity *Immune System Responses*

Immune Disorders

- **Allergies** Allergies are abnormal sensitivities to antigens (allergens) in the surroundings. An allergic reaction produces inflammatory responses that result in uncomfortable symptoms.

- **Autoimmune Diseases** The immune system normally reacts only against nonself substances (foreign molecules and cells), not against self (the body's own macromolecules). It generally rejects transplanted organs, whose cells lack the recipient's unique "fingerprint" of self proteins. In autoimmune diseases, the system turns against some of the body's own molecules.

- **Immunodeficiency Diseases** In immunodeficiency diseases, immune components are lacking, and infections recur. Immunodeficiencies may arise through inborn genetic mutations or through disease.

- **AIDS** AIDS is a worldwide epidemic that kills millions of people each year. HIV, the AIDS virus, attacks helper T cells, crippling both the cell-mediated and humoral immune responses. Practicing safer sex could save many lives.

Case Study in the Process of Science *Why Do AIDS Rates Differ Across the U.S.?*

Activity *HIV Reproductive Cycle*

Case Study in the Process of Science *What Causes Infections in AIDS Patients?*

SELF-QUIZ

1. Foreign molecules that elicit an immune response are called __________.

2. Classify each of the following components of the immune system as belonging to either the specific or nonspecific defenses.
   a. Natural killer cells
   b. Complement proteins
   c. Antibodies
   d. Inflammation
   e. Interferons
   f. Cytotoxic T cells
   g. Helper T cells

3. Which of the following best describes the difference in the way B cells and T cells deal with invaders?
   a. B cells confer active immunity; T cells confer passive immunity.
   b. B cells send out antibodies that attack invaders; T cells themselves do the attacking.
   c. T cells handle the primary immune response; B cells handle the secondary response.
   d. B cells are responsible for cell-mediated immunity; T cells are responsible for humoral immunity.

4. The antigen-binding sites of an antibody molecule are formed from the molecule's variable (V) regions. Why are these regions called variable?
   a. They can change their shapes on command to fit different antigens.
   b. They change their shapes when they bind to an antigen.
   c. Their specific shapes are unimportant.
   d. They have different shapes on different antibody molecules.

5. Match each type of defensive cell with a function or description from the right column.

a. Lymphocyte	1. Attacks infected body cells
b. Cytotoxic T cell	2. Carries out humoral immunity
c. Helper T cell	3. Type of phagocytic white blood cell
d. Neutrophil	4. General name for B or T cell
e. B cell	5. Initiates the secondary immune response
f. Memory cell	6. Cell most commonly attacked by HIV

6. Explain how each of the following characteristics of the inflammatory response helps protect the body: swelling, pain, and fever.

7. Why is AIDS deadlier than most other viral diseases?

8. A baby has been born with an immunodeficiency disease. In trying to diagnose the problem, physicians discover that the child is not producing any antibodies. It is most likely that this child is missing the kind of immune cells called __________.

9. A person is diagnosed with the autoimmune disorder lupus. What has happened to this person's immune system?

10. Once vaccinated, you have had a primary exposure to specific antigens. If you ever encounter these antigens again, your body will mount a rapid immune response. The cells that account for this rapid secondary response are called __________. The process that produces these long-term cells is called __________.

Answers to the Self-Quiz questions can be found in Appendix D.

Go to the website or CD-ROM for more Self-Quiz questions.

THE PROCESS OF SCIENCE

11. Most biologists believe that the immune system's defense against infections largely rests on its ability to distinguish self molecules from nonself molecules. This concept seems central to our understanding of immune function. However, like all scientific ideas, it is not beyond question. Several immunologists have developed an alternative hypothesis: that the immune system's effectiveness rests mostly on its ability to recognize damage to body tissues caused by the invaders, not on the ability to recognize nonself. If you were going to test the "damage" hypothesis, what might you look for? Which type of cell would you expect to be directly affected by damaged tissues? Why? Some proponents argue that the "damage" hypothesis makes more sense from an evolutionary perspective, claiming that it is more advantageous for an organism's defense system to respond to tissue damage than to the mere presence of a foreign microbe. Do you agree? Why or why not?

12. Your roommate is rushed to the hospital after suffering a severe allergic reaction to a bee sting. After she is treated and released, she asks you (the local biology expert!) to explain what happened. She says, "I don't understand how this could have happened. I was stung by a bee before but didn't have a reaction." Suggest a hypothesis to explain what has happened to cause her severe allergic reaction and why she didn't have a reaction after her previous sting.

BIOLOGY AND SOCIETY

13. Organ donation saves many lives each year. Even though some transplanted organs are derived from living donors, the majority come from patients who die but still have healthy organs that can be of value to a transplant recipient whose cell surfaces have similar sets of proteins. Potential organ donors can fill out an organ donation card to specify their wishes. If the donor is in critical condition and dying, the donor's family is usually consulted to discuss the donation process. Generally, the next of kin must approve before donation can occur, regardless of whether the patient has completed a donor card. In some cases, the donor's wishes are overridden by a family member. Do you think that family members should be able to deny the stated intentions of the potential donor? Why or why not? Have you signed up to be an organ donor? Why or why not?

25 Hormones

Your body makes over **50 different kinds** of hormones.
All human embryos develop into females unless they produce **testosterone** early in development.
The hormone **epinephrine** is released during both stressful and pleasurable situations.
Abnormal levels of **human growth hormone** can cause either dwarfism or gigantism.

BIOLOGY AND SOCIETY

Of Hunger and Hormones

We all know it when we feel it: hunger, a craving for food. The urge can be powerful—nearly overwhelming—grabbing our attention and not letting go until satiated. But what causes hunger? Is it possible to turn it on and off? The answers to such questions could change the way we think about food and have a profound impact on human health. For example, appetite-suppressing drugs could be a boon in the treatment of obesity. At the other end of the spectrum, appetite-boosting treatments could help people who suffer from wasting syndromes associated with surgery, AIDS, and cancer.

A breakthrough in the quest to understand the nature of hunger came in 1999 when scientists discovered a protein fragment called ghrelin. Ghrelin is a **hormone,** a chemical that travels from its production site through body fluids (usually the blood) and affects cells at other sites in the body. In the case of ghrelin, production occurs mainly in the stomach. From there, ghrelin is released into the bloodstream. When it reaches the brain, it binds to receptors on the outside of certain brain cells.

Several lines of evidence suggest that ghrelin induces hunger and stimulates eating. Injecting rats with ghrelin prompts increased feeding and weight gain. In people, the amount of ghrelin in the blood rises as mealtime approaches but then quickly drops after a meal **(Figure 25.1)**. In one controlled study, people receiving a premeal dose of ghrelin consumed 30% more food than those who did not. In another study, 18 people with Prader-Willi syndrome, a genetic disorder that causes uncontrollable food cravings, were found to have unusually high levels of ghrelin.

But other evidence cautions against a simplistic view of the relationship between ghrelin and hunger. For example, genetically engineered mice that lack the ghrelin gene or the gene for its receptor in the brain do not eat less or lose weight. And notice in Figure 25.1 that there is a peak in ghrelin concentrations around midnight, but aside from the occasional late-night snack, most people do not feel hungry at that time. These facts coupled with other recent research suggest that ghrelin interacts with several other hormones to regulate the hunger sensation.

The overarching role of hormones like ghrelin is to coordinate cellular activities in different parts of the body, enabling the integrated functioning of organ systems. In addition to hunger, hormones regulate other basic body functions, such as energy use, metabolism, and growth. This chapter focuses on how hormones help maintain homeostasis within the human body. We begin with an overview of how hormones work and then turn to the major components of the human endocrine system. Along the way, we'll consider many examples of the effects of hormonal imbalance. ■

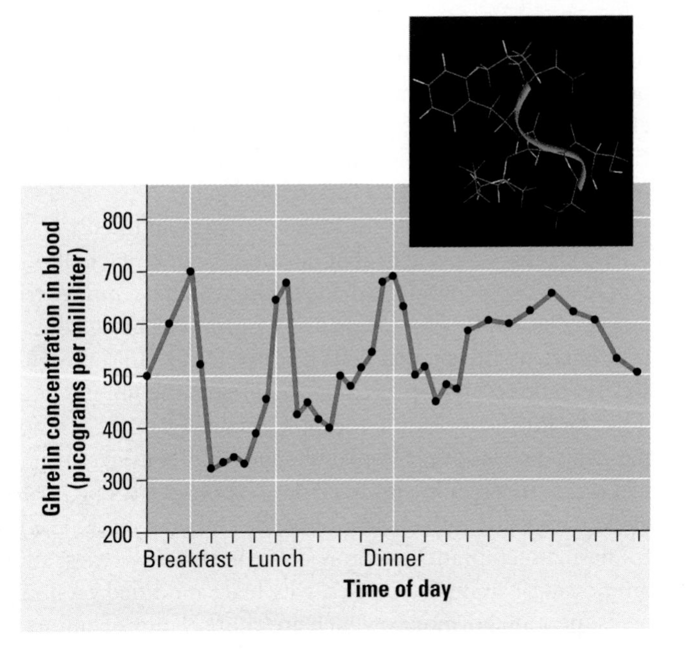

Figure 25.1 A hunger hormone? Researchers measured concentrations of the hormone ghrelin (shown at top) in the blood of ten test subjects over a 24-hour period. Ghrelin concentration rose before each meal and fell dramatically after each meal. (A picogram is a million-millionth of a gram.)

Hormones: An Overview

Vertebrates make over 50 different hormones. Hormones are made and secreted mainly by organs called **endocrine glands.** Collectively, all hormone-secreting cells constitute the **endocrine system,** the body's main system for internal chemical regulation.

Figure 25.2 shows the release of hormone molecules from secretory vesicles in an endocrine cell and their delivery, via the circulatory system, to a **target cell,** a cell with receptors for a specific hormone. Because

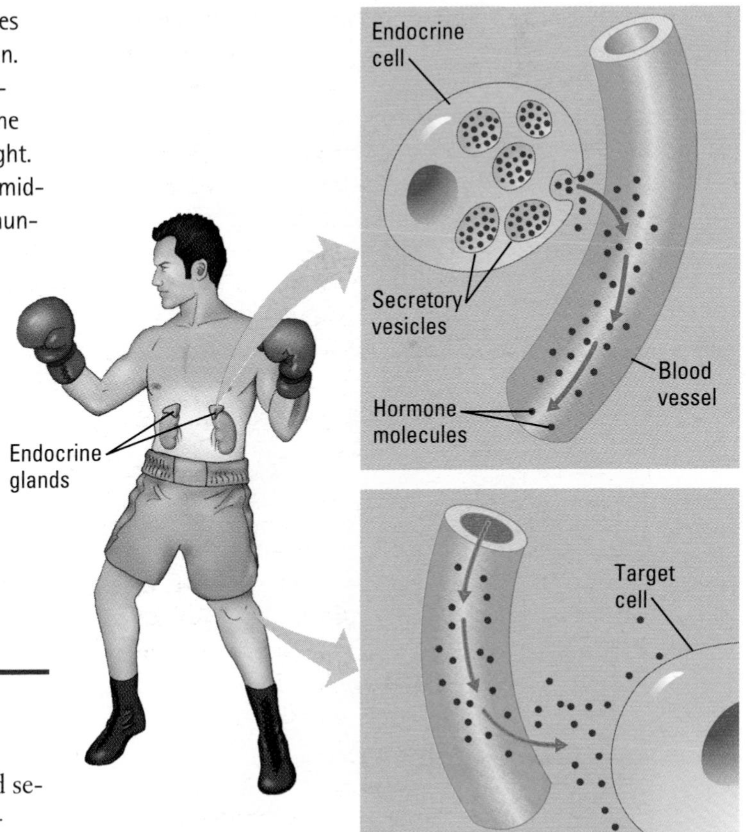

Figure 25.2 Hormone secretion from an endocrine cell. A cell within an endocrine gland secretes hormone molecules. The hormone is carried via the circulatory system to all cells of the body, but it only affects target cells.

hormones reach all parts of the body, the endocrine system is especially important in controlling whole-body activities. For example, hormones govern our metabolic rate, growth, maturation, and reproduction. In many cases, a tiny amount of a hormone influences the activities of an enormous number of target cells in a variety of organs.

A hormone may come into contact with all the tissues in the body, but only cells that respond to that hormone—its target cells—are affected by it. There are two general mechanisms by which hormones trigger changes in target cells. In the first mechanism, hormones bring about changes without ever entering their target cells **(Figure 25.3)**. To start, ❶ a hormone binds to a specific receptor protein in the plasma membrane of the target cell. The binding activates the receptor, which ❷ initiates a signal transduction pathway, a series of molecular changes that converts an extracellular chemical signal to a specific intracellular response. ❸ The final relay molecule activates a protein that carries out the cell's response. Hormones that bind to plasma membrane receptors come in three varieties, all derived from amino acids: amine hormones, which are modified versions of single amino acids; peptide hormones, which are short chains of amino acids; and protein hormones, made of polypeptides. Ghrelin, mentioned in the opening essay, is an example of a peptide hormone.

The second mechanism by which hormones trigger changes in target cells involves binding to receptors inside the cell. The **steroid hormones** work this way. Steroid hormones consist of small hydrophobic molecules derived from cholesterol. As shown in **Figure 25.4**, ① a steroid hormone enters a cell by diffusing through the plasma membrane. If the cell is a target cell, the hormone ② binds to a receptor protein in the cytoplasm or

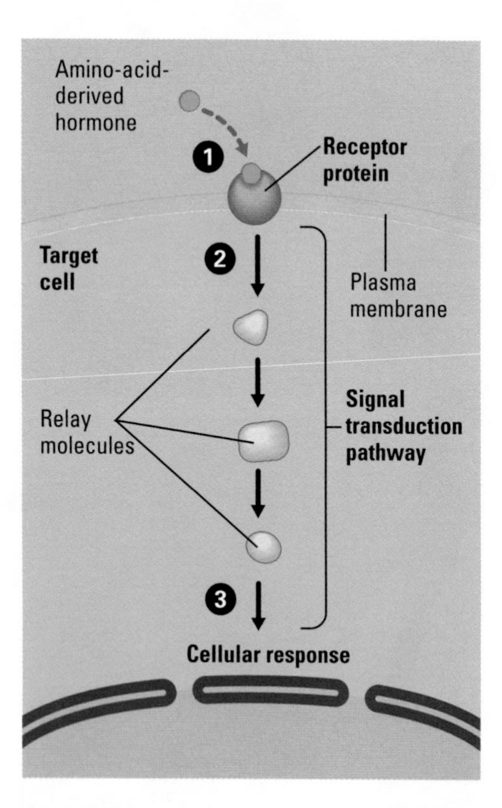

Figure 25.3 A hormone that binds to a plasma membrane receptor. Amino-acid-based hormones bind to receptors on the plasma membranes of target cells. Once bound, they affect the cell by activating a signal transduction pathway.

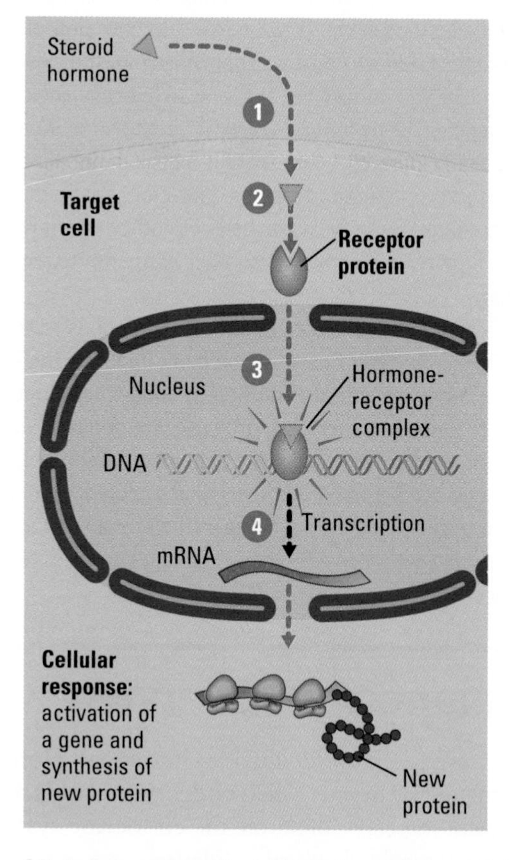

Figure 25.4 A hormone that binds to an intracellular receptor. Steroid hormones bind to receptors inside their target cells. Steroid hormones affect the cell by turning specific genes on or off.

nucleus. Rather than triggering a signal transduction pathway, the receptor itself carries the hormone's signal. ③ The hormone-receptor complex attaches to specific sites on the cell's DNA in the nucleus. ④ The binding to DNA may stimulate transcription of certain genes into mRNA molecules, which are translated into new proteins, or it may turn transcription off. The sex hormones estrogen and testosterone are examples of steroid hormones.

We've now completed an overview of how hormones work. In the next section, we'll take a closer look at the human endocrine system.

CHECKPOINT

1. How does a hormone get from the endocrine gland that secretes it to a distant target cell?
2. What are two major differences between the mechanisms of action of steroid and amino-acid-based hormones?

Answers: 1. Via body fluids, usually the bloodstream **2.** Steroid hormones bind to receptors inside the cell, while amino-acid-based hormones bind to plasma membrane receptors. Also, steroid hormones always affect gene expression directly; amino-acid-based hormones act by triggering a signal transduction pathway that leads to a cellular response.

The Human Endocrine System

The human endocrine system consists of about a dozen major glands. Some of these, such as the thyroid and the pituitary gland, are endocrine specialists; their primary function is to secrete hormones into the blood. Several other glands have both endocrine and nonendocrine functions. The pancreas, for example, secretes hormones that influence the level of glucose in the blood and also secretes digestive enzymes into the intestine. Still other organs, such as the stomach, are primarily nonendocrine but have some cells that secrete hormones (such as ghrelin).

Figure 25.5 shows the locations of the major human endocrine glands. **Table 25.1** (on page 548) summarizes the actions of the main hormones they produce. The table provides an overview of the human endocrine system, and you may wish to refer to it as we focus on the individual glands and their hormones.

Hormones have a wide range of targets. Some hormones, like the sex hormones, which promote male and female characteristics, affect most of the tissues of the body. Other hormones, such as glucagon from the pancreas, have only a few kinds of target cells (in this case, liver and fat cells). Some hormones have other endocrine glands as their targets. For example, the hypothalamus secretes many hormones that regulate other endocrine glands, especially the pituitary. In some cases, the same hormone may elicit different responses in different target cells, depending on the type of cell and its signal transduction pathway.

In the rest of this section, we'll look at the major organs of the human endocrine system. For each one, you'll learn about the hormones it secretes and how these chemicals help the body maintain homeostasis. Keep in mind that this chapter only covers the major endocrine glands and their hormones; there are other hormone-secreting structures—the heart and liver, for example—and dozens of other hormones that we will not discuss.

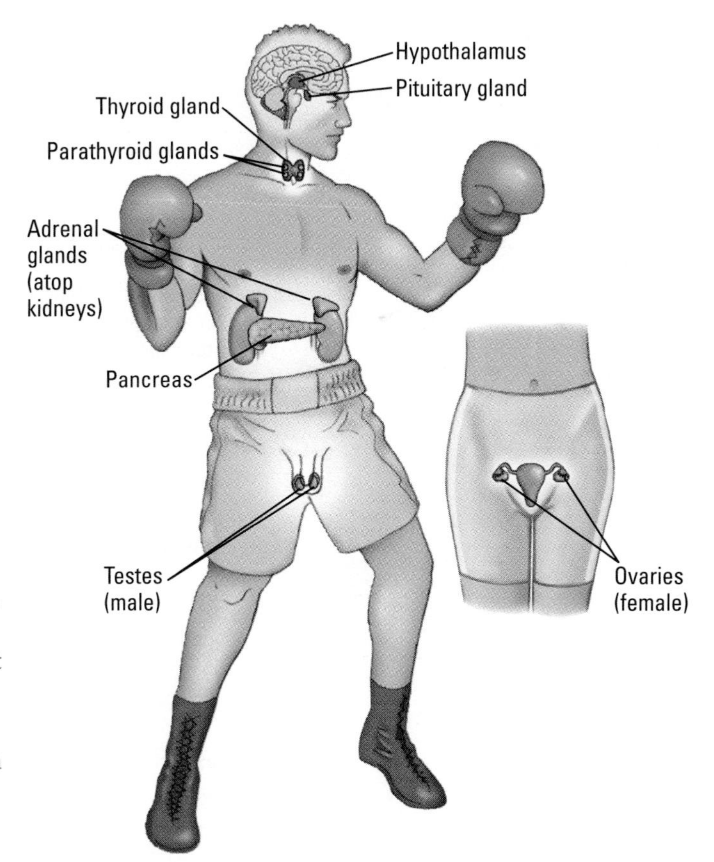

Figure 25.5 The major endocrine glands in humans. This figure shows only the major endocrine organs discussed in the text. Several other organs that have endocrine functions are not shown.

Table 25.1	Major Human Endocrine Glands and Some of Their Hormones			
Gland	**Hormone**	**Chemical Class**	**Representative Actions**	**Regulated By**
Hypothalamus	Hormones released by the posterior pituitary and hormones that regulate the anterior pituitary (see below)			
Pituitary gland				
Posterior pituitary (releases hormones made by hypothalamus)	Oxytocin	Peptide	Stimulate contraction of uterus and mammary gland cells	Nervous system
	Antidiuretic hormone (ADH)	Peptide	Promotes retention of water by kidneys	Water/salt balance
Anterior pituitary	Growth hormone (GH)	Protein	Stimulates growth (especially bones) and metabolic functions	Hypothalamic hormones
	Prolactin (PRL)	Protein	Stimulates milk production	Hypothalamic hormones
	Follicle-stimulating hormone (FSH)	Protein	Stimulates production of ova and sperm	Hypothalamic hormones
	Luteinizing hormone (LH)	Protein	Stimulates ovaries and testes	Hypothalamic hormones
	Thyroid-stimulating hormone (TSH)	Protein	Stimulates thyroid gland	Thyroxine in blood; hypothalamic hormones
	Adrenocorticotropic hormone (ACTH)	Peptide	Stimulates adrenal cortex to secrete glucocorticoids	Glucocorticoids; hypothalamic hormones
Thyroid gland	Triiodothyronine (T_3) and thyroxine (T_4)	Amine	Stimulate and maintain metabolic processes	TSH
	Calcitonin	Peptide	Lowers blood calcium level	Calcium in blood
Parathyroid glands	Parathyroid hormone (PTH)	Peptide	Raises blood calcium level	Calcium in blood
Pancreas	Insulin	Protein	Lowers blood glucose level	Glucose in blood
	Glucagon	Protein	Raises blood glucose level	Glucose in blood
Adrenal glands				
Adrenal medulla	Epinephrine (adrenaline) and norepinephrine (noradrenaline)	Amine	Raise blood glucose level; increases metabolic activities; constrict certain blood vessels	Nervous system
Adrenal cortex	Glucocorticoids	Steroid	Raise blood glucose level	ACTH
	Mineralocorticoids	Steroid	Promote reabsorption of Na^+ and excretion of K^+ in kidneys	K^+ (potassium) in blood
Gonads				
Testes	Androgens (testosterone)	Steroid	Support sperm formation; promote development and maintenance of male secondary sex characteristics	FSH and LH
Ovaries	Estrogens	Steroid	Stimulate uterine lining growth; promote development and maintenance of female secondary sex characteristics	FSH and LH
	Progesterone	Steroid	Promotes uterine lining growth	FSH and LH
Pineal gland (in brain)	Melatonin	Amine	Involved in biological rhythms	Light/dark cycles
Thymus	Thymosin	Peptide	Stimulates development of T cells (See chapter 24)	Not known

The Hypothalamus and Pituitary Gland

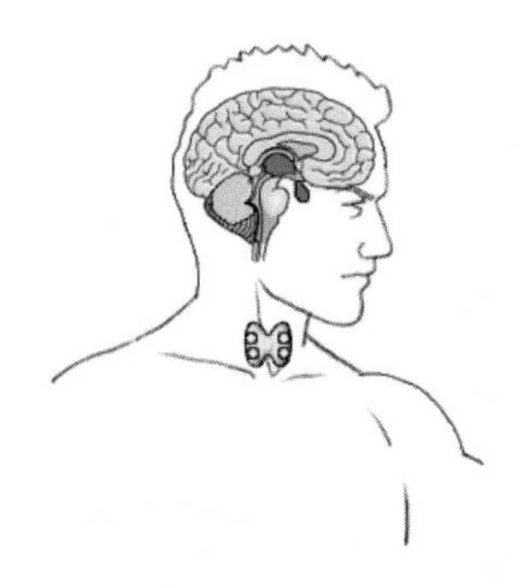

The **hypothalamus** is the main control center of the endocrine system **(Figure 25.6)**. As part of the brain, the hypothalamus receives information from nerves about the internal condition of the body and about the external environment. It then responds by sending out appropriate neural or endocrine signals. Its signals directly control the **pituitary gland**, a pea-sized structure that hangs down from the hypothalamus. In response to signals from the hypothalamus, the pituitary secretes hormones that influence numerous body functions. The hypothalamus thus exerts master control over the endocrine system by using the pituitary to relay directives to other glands.

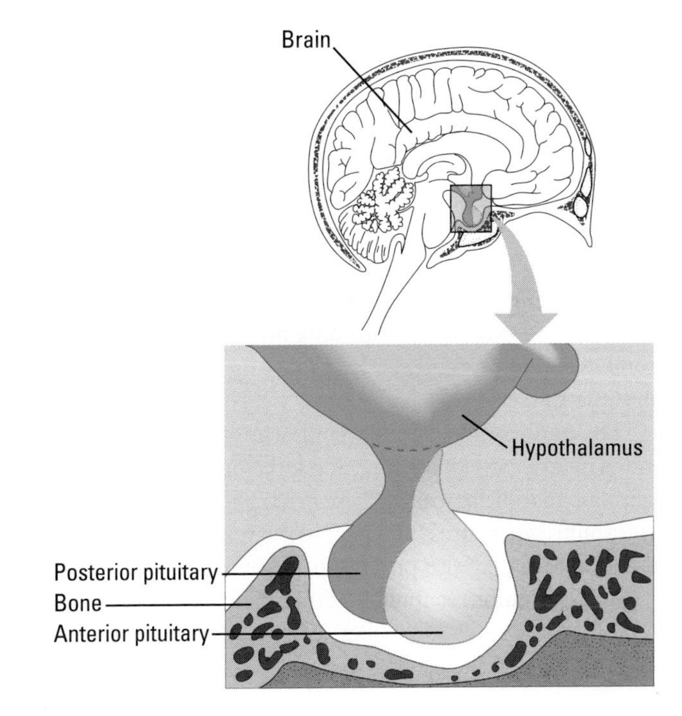

Figure 25.6 Location of the hypothalamus and pituitary.

As Figure 25.6 shows, the pituitary gland consists of two distinct parts: a posterior lobe and an anterior lobe, both situated in a pocket of skull bone just under the hypothalamus. The **posterior pituitary** is composed of nervous tissue and is actually an extension of the hypothalamus. It stores and secretes hormones made in the hypothalamus. In contrast, the **anterior pituitary** is composed of endocrine cells that synthesize and secrete numerous hormones directly into the blood. Several of these hormones control the activity of other endocrine glands. The hypothalamus exerts control over the anterior pituitary by secreting two kinds of hormones into short blood vessels that connect the two glands. Releasing hormones stimulate the anterior pituitary to secrete hormones, while inhibiting hormones induce the anterior pituitary to stop secreting hormones.

Figure 25.7 shows how the hypothalamus operates through the posterior pituitary to direct the activity of the kidneys. The hypothalamus makes antidiuretic hormone (ADH) (▲), which is stored and released by the posterior pituitary. As discussed in Chapter 21, ADH helps cells of the kidney reabsorb water, preventing dehydration by decreasing urine volume when the body needs to retain water. When the body has too much water, the hypothalamus responds by slowing the release of ADH from the pituitary. Some diuretics, such as alcohol, inhibit the release of ADH, which in turn increases the output of urine.

In response to releasing hormones secreted by the hypothalamus, the anterior pituitary synthesizes and releases many different hormones that influence a broad range of body activities. For example, the anterior pituitary releases **follicle-stimulating hormone (FSH)** and **luteinizing hormone (LH),** which, among other functions, help to regulate the female menstrual cycle (as we'll see in Chapter 26). Another anterior pituitary hormone, **prolactin (PRL),** stimulates mammary glands to produce milk.

Of all the anterior pituitary secretions, none has a broader effect than the protein called **growth hormone (GH).** During childhood and adolescence, high levels of GH promote the development and enlargement of all parts of the body. If too much GH is produced in a very young person,

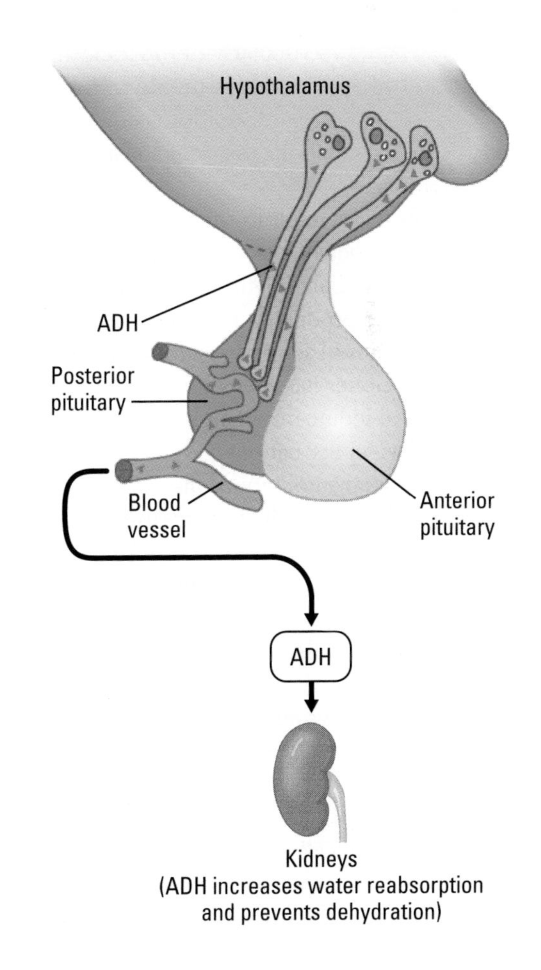

Figure 25.7 ADH and osmoregulation. This diagram shows how the hypothalamus, through signals sent via long cells (gray) to the posterior pituitary, regulates water homeostasis through the action of the hormone ADH (▲).

usually because of a pituitary tumor, gigantism can result **(Figure 25.8a)**. Overproduction of GH in adults results in a condition called acromegaly, characterized by enlarged bones in the hands, feet, and face **(Figure 25.8b)**. In contrast, too little GH during development can lead to dwarfism **(Figure 25.8c)**. Administering GH to children with GH deficiency can successfully prevent dwarfism. Once extracted in only minute quantities from the pituitary glands of cadavers, human GH is now produced in large quantities by bacteria modified to carry the human GH gene (see Chapter 12). Unfortunately, its increased availability has caused some athletes to abuse human GH in order to bulk up their muscles. Such abuse is extremely dangerous and can lead to disfigurement, heart failure, and multiple cancers.

The **endorphins,** another kind of anterior pituitary hormone, are the body's natural painkillers; they mask the perception of pain. Some researchers speculate that the so-called "runner's high" results partly from the release of endorphins when stress and pain in the body reach critical levels. The release of endorphins may also produce pleasant feelings during such diverse activities as deep meditation, acupuncture treatments, or even eating very spicy foods.

(a) Overproduction of GH during development. Prolonged overproduction of human growth hormone (GH) during development causes gigantism. André René Roussimoff (1946–1993), known as André the Giant, stood nearly 7 feet tall and weighed almost 500 pounds.

(b) Overproduction of GH during adulthood. Overproduction of GH later in life causes acromegaly. The most common symptom of this disease is enlargement of the hands and face.

(c) Underproduction of GH during development. The body of actor Verne Troyner, like those of other dwarfs, produced lower than average amounts of GH during childhood.

Figure 25.8 **Pituitary growth hormone disorders.**

The Thyroid and Parathyroid Glands

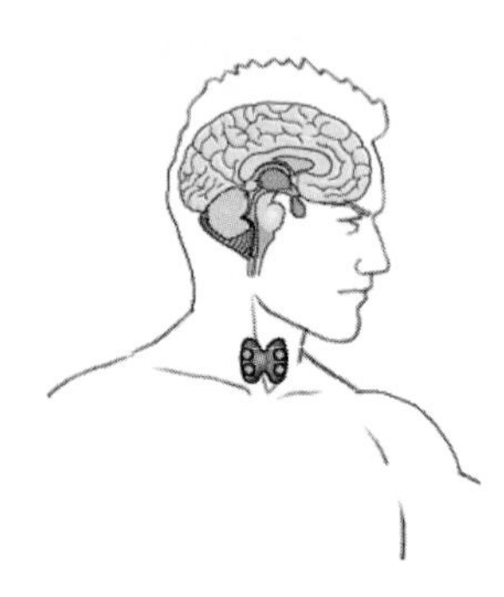

Your **thyroid gland** is located in your neck just under your larynx (voice box). The thyroid produces several hormones, some of which contain the element iodine. These thyroid hormones increase oxygen consumption and metabolic rate in all the cells of your body. They also play crucial roles in development and maturation, particularly of the bone and nerve cells. Insufficient levels of the thyroid hormones in the blood (hypothyroidism) or excess levels (hyperthyroidism) can result in serious metabolic disorders. Fortunately, thyroid disorders can be easily treated.

Hypothyroidism can result from dietary deficiencies or from a defective thyroid gland. One type of dietary hypothyroidism is goiter (see Figure 2.4). If too little iodine is available to the thyroid, then insufficient quantities of the thyroid hormones are produced. This underproduction interrupts feedback loops that control thyroid activity, resulting in overstimulation and swelling of the thyroid gland. Goiter has been reduced in many industrialized nations by the incorporation of iodine into table salt. Unfortunately, goiter still affects many people in developing nations.

Graves' disease is an autoimmune disease (seeChapter 24) that results in an overactive thyroid **(Figure 25.9)**. One treatment for Graves' disease takes advantage of the fact that the thyroid accumulates iodine: Patients drink a solution containing radioactive iodine, which travels to the thyroid and kills off some of the thyroid cells. By controlling the dose, the "radioactive cocktail" can kill off just enough cells to lower thyroid output and relieve symptoms.

Embedded within the thyroid are four disk-shaped **parathyroid glands** (see Figure 25.5). The thyroid and parathyroid glands function in calcium homeostasis, keeping the concentration of calcium ions (Ca^{2+}) within a narrow range. An appropriate level of calcium in the blood and interstitial fluid is essential for many body functions. Without calcium, nerve signals cannot be transmitted from cell to cell, muscles cannot function properly, blood cannot clot, and cells cannot transport molecules across their membranes. Two hormones, **calcitonin** from the thyroid gland and **parathyroid hormone (PTH)** from the parathyroids, regulate the blood calcium level. Calcitonin and PTH are said to be **antagonistic hormones** because they have opposite effects: Calcitonin lowers the calcium level in the blood, whereas PTH raises it.

As **Figure 25.10** indicates, these two antagonistic hormones operate by means of feedback systems that keep the calcium level near the homeostatic set point. ❶ A rise in the blood Ca^{2+} level above the homeostatic set point ❷ induces the thyroid gland to secrete calcitonin. Calcitonin, in turn, has two main effects: It causes more Ca^{2+} to be deposited in the bones, and it makes the kidneys reabsorb less Ca^{2+} as they form urine. ❸ The result is a lower Ca^{2+} level in the blood. ❹ When the blood Ca^{2+} level drops below the set point, ❺ the parathyroids release PTH into the blood. PTH stimulates the release of Ca^{2+} from bones and increases Ca^{2+} uptake by the kidneys and intestines, ❻ raising calcium levels.

In summary, a sensitive balancing system maintains calcium homeostasis. Failure of the system can have far-reaching effects in the body. For example, a shortage of PTH causes the blood calcium level to drop dramatically. Such a drop can lead to uncontrollable muscle contractions and potentially fatal convulsions known as tetany.

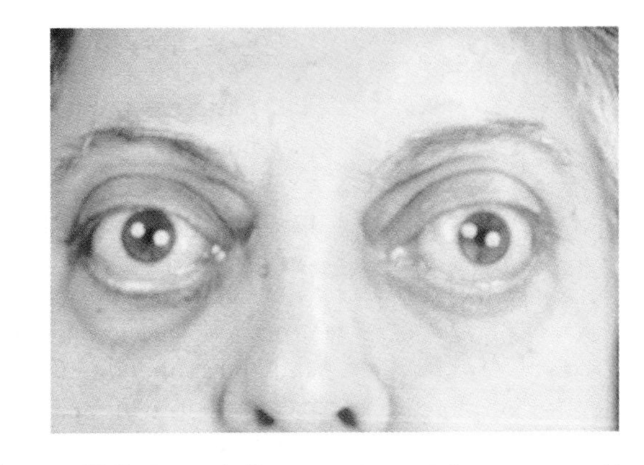

Figure 25.9 Graves' disease. Graves' disease, caused by hyperthyroidism, can make a person overheat, sweat profusely, develop high blood pressure, and produce protruding eyes.

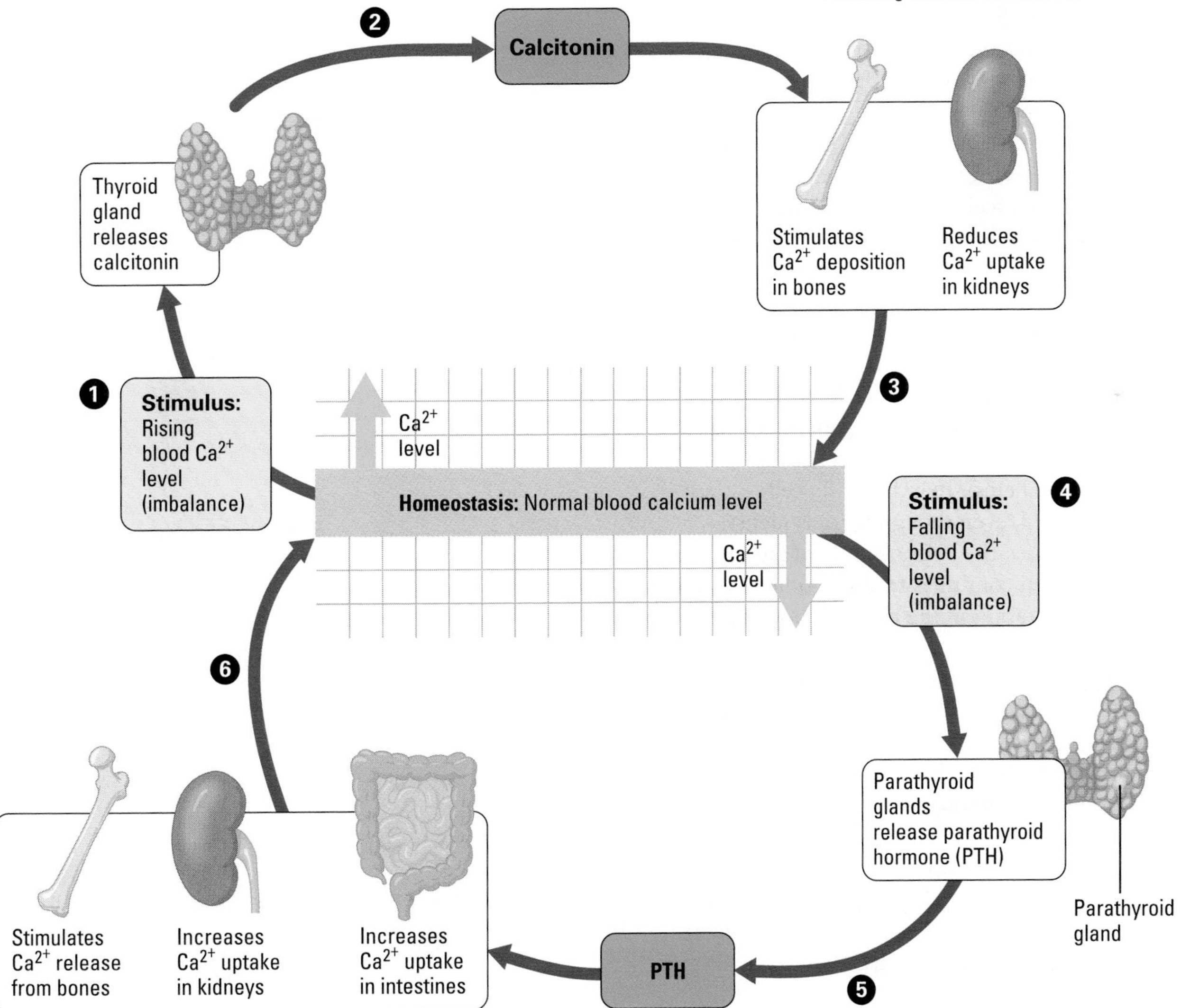

Figure 25.10 Calcium homeostasis. This diagram traces the regulation of blood calcium level by calcitonin and PTH, which act antagonistically. Note that in addition to their skeletal role in supporting the body, bones act as a calcium bank, receiving and releasing calcium as needed.

The Pancreas

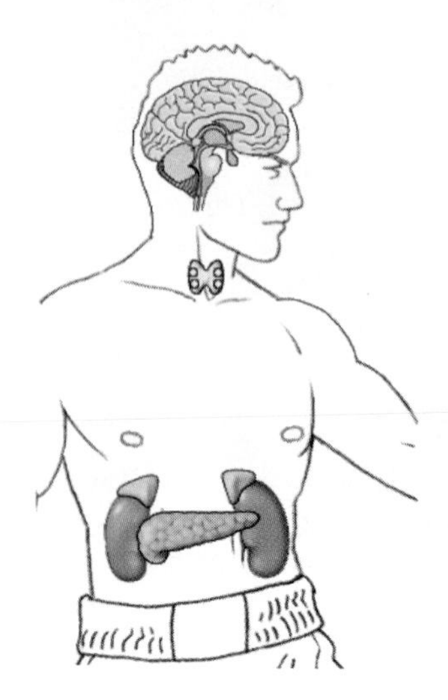

The **pancreas** produces two hormones that play important roles in managing the body's energy supplies. One of the hormones is **insulin,** a protein hormone produced by clusters of specialized pancreatic cells called **islet cells.** Other islet cells secrete another protein hormone called **glucagon.** Insulin and glucagon help maintain a homeostatic balance between the amount of glucose available in the blood and the amount of glucose stored as the polymer glycogen in body cells.

As shown in **Figure 25.11**, insulin and glucagon are antagonists, countering each other in a feedback circuit that precisely manages both glucose storage and glucose use by body cells. By negative feedback, the concentration of glucose in the blood determines the relative amounts of insulin and glucagon secreted by islet cells. ❶ Rising glucose concentration in the blood—shortly after you eat a carbohydrate-rich meal, for example—❷ stimulates the pancreas to secrete more insulin into the blood. Target cells for insulin take up more glucose from the blood, ❸ decreasing the blood glucose level. Liver and skeletal muscle cells take up much of the glucose and use it to form glycogen, which they store. When the blood glucose level falls to the set point, the cells of the pancreas lose their stimulus to secrete insulin.

❹ When the blood glucose level dips below the set point, as it may between meals or during strenuous exercise, ❺ pancreatic cells respond by secreting more glucagon. Glucagon is a fuel mobilizer, making liver cells break glycogen down into glucose and release it into the blood. ❻ Then, when the blood glucose level returns to the set point, the pancreas slows its secretion of glucagon.

Diabetes mellitus is a serious hormonal disease that affects up to 6% of Americans—about 18 million people. In diabetes, body cells are unable to absorb glucose from the blood, either because there isn't enough insulin in the blood (as in type 1, or insulin-dependent, diabetes) or because the target cells do not respond normally to the insulin in the blood (as in type 2, or non-insulin-dependent, diabetes). In either case, cells cannot obtain enough glucose from the blood, even though there is plenty. Starved for fuel, cells are forced to burn the body's supply of fats and proteins. Meanwhile, the digestive system continues to absorb glucose from the diet, causing the glucose concentration in the blood to become extremely high.

There are treatments for diabetes mellitus but no cure. Type 1 patients require regular injections of insulin, usually obtained from genetically modified organisms (see Chapter 12). Over 90% of American diabetics have type 2 diabetes. This form of the disease is almost always associated with being overweight and underactive, although whether obesity causes diabetes (and if so, how) remains unknown. This type of diabetes can often be managed by controlling sugar intake and by exercising and dieting to reduce weight. Although diabetes can be treated, there is no cure. Every year, some 350,000 Americans die from the disease or from its complications, which include severe dehydration, cardiovascular and kidney disease, and nerve damage.

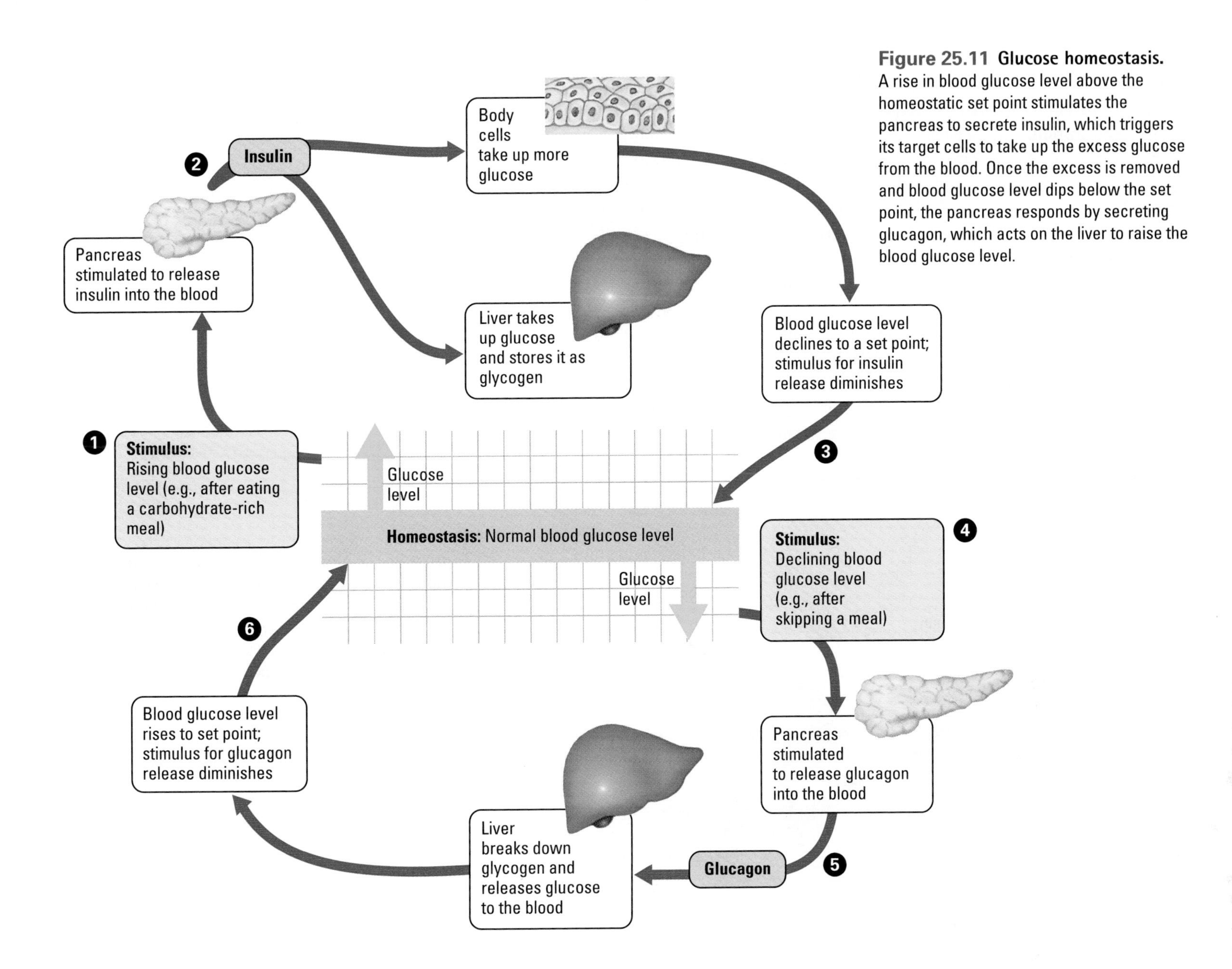
Body cells take up more glucose
Insulin
2
Pancreas stimulated to release insulin into the blood
Liver takes up glucose and stores it as glycogen
Blood glucose level declines to a set point; stimulus for insulin release diminishes
1
Stimulus:
Rising blood glucose level (e.g., after eating a carbohydrate-rich meal)
Glucose level
3
Homeostasis: Normal blood glucose level
Stimulus:
Declining blood glucose level (e.g., after skipping a meal)
4
Glucose level
6
Blood glucose level rises to set point; stimulus for glucagon release diminishes
Pancreas stimulated to release glucagon into the blood
Liver breaks down glycogen and releases glucose to the blood
Glucagon
5

Figure 25.11 Glucose homeostasis. A rise in blood glucose level above the homeostatic set point stimulates the pancreas to secrete insulin, which triggers its target cells to take up the excess glucose from the blood. Once the excess is removed and blood glucose level dips below the set point, the pancreas responds by secreting glucagon, which acts on the liver to raise the blood glucose level.

The Adrenal Glands

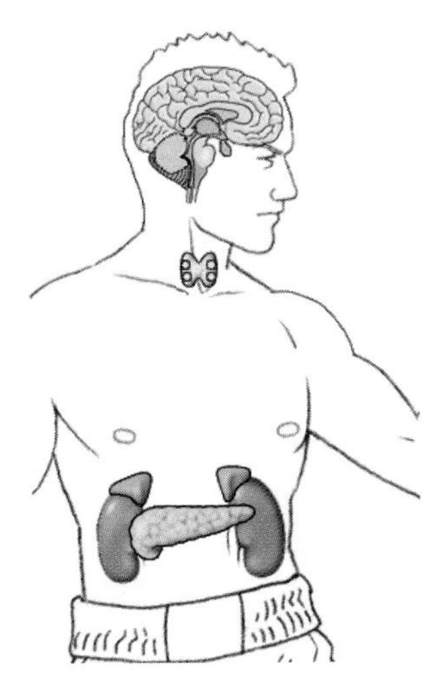

The human body has two **adrenal glands** sitting atop the kidneys. Each adrenal gland is actually two glands in one: a central portion called the **adrenal medulla** and an outer portion called the **adrenal cortex.** Though the cells they contain and the hormones they produce are different, both the medulla and the cortex secrete hormones that enable the body to respond to stress. You've probably felt your heart beat faster and your skin develop goose bumps when you've sensed danger or approached a stressful situation. Positive emotions—extreme pleasure, for instance—can produce the same effects. These reactions are triggered by two "fight-or-flight" hormones secreted by the adrenal medulla: **epinephrine** (also called adrenaline) and **norepinephrine**. These two hormones ensure a rapid, short-term response to stress.

Stressful stimuli, whether negative or positive, activate nerve cells in the hypothalamus. As indicated on the left side of **Figure 25.12**, these cells ❶ send signals that stimulate the adrenal medulla to ❷ secrete epinephrine and norepinephrine into the blood. Epinephrine and norepinephrine both contribute to the short-term stress response by stimulating liver cells to release glucose, making more fuel available for cellular work. They also prepare the body for action by raising blood pressure, breathing rate, heart rate, and metabolic rate. In addition, epinephrine and norepinephrine change blood-flow patterns to shuttle blood to where it is most needed: Blood vessels in the brain and skeletal muscles are dilated, increasing alertness and the muscles' ability to react to stress, while blood vessels elsewhere are constricted, reducing activities (such as digestion) that are not immediately involved in the stress response. The short-term stress response occurs and subsides rapidly.

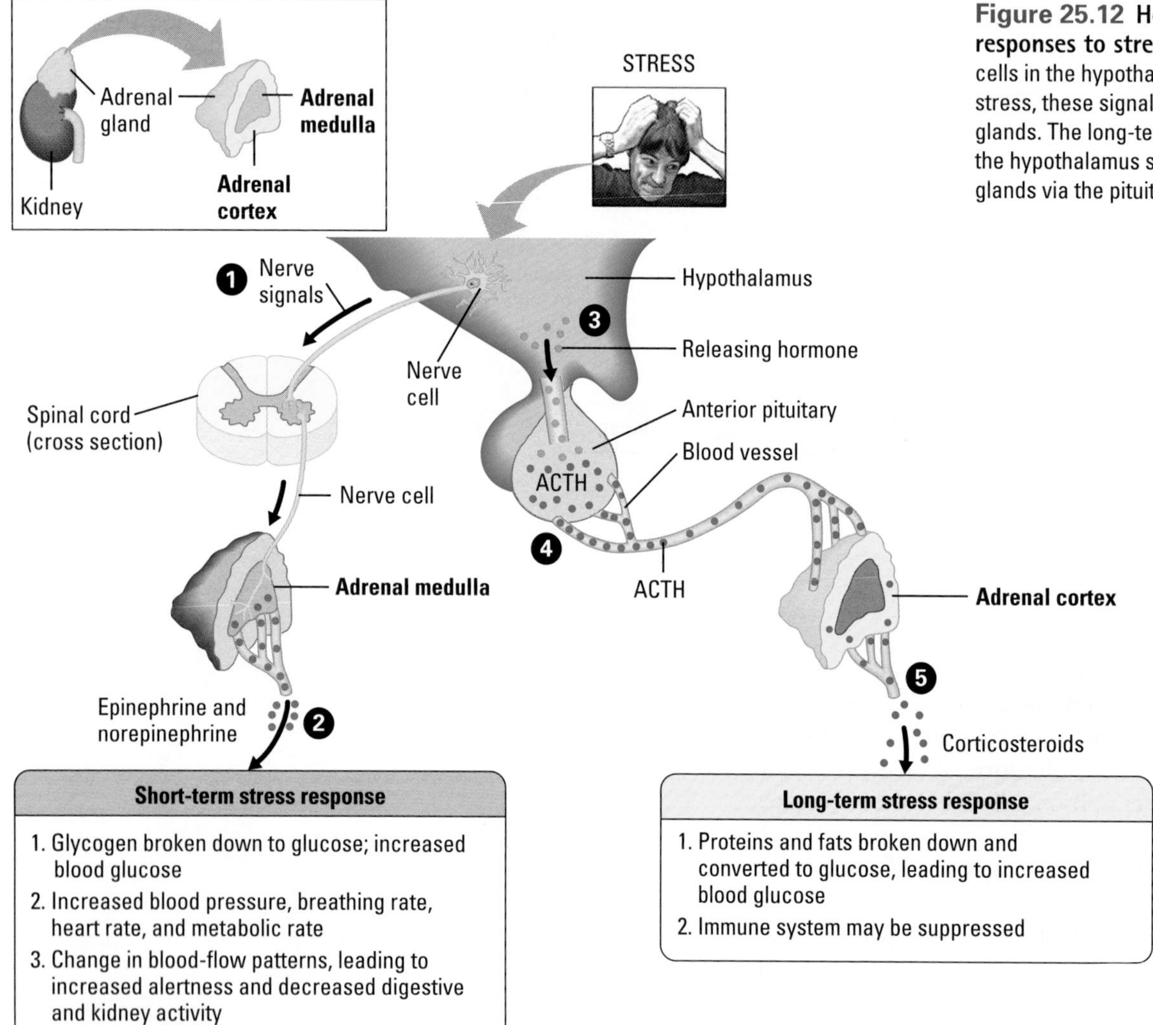

Adrenal gland
Adrenal medulla
Adrenal cortex
Kidney
STRESS
1 Nerve signals
Hypothalamus
3
Releasing hormone
Nerve cell
Spinal cord (cross section)
Anterior pituitary
Blood vessel
ACTH
Nerve cell
4
ACTH
Adrenal medulla
Adrenal cortex
5
Epinephrine and norepinephrine
2
Corticosteroids
Short-term stress response
1. Glycogen broken down to glucose; increased blood glucose
2. Increased blood pressure, breathing rate, heart rate, and metabolic rate
3. Change in blood-flow patterns, leading to increased alertness and decreased digestive and kidney activity
Long-term stress response
1. Proteins and fats broken down and converted to glucose, leading to increased blood glucose
2. Immune system may be suppressed

Figure 25.12 How the adrenal glands control our responses to stress. Stressful stimuli activate nerve cells in the hypothalamus. In the short-term response to stress, these signals are relayed onward to the adrenal glands. The long-term stress response is initiated when the hypothalamus sends hormonal signals to the adrenal glands via the pituitary.

In contrast to epinephrine and norepinephrine (secreted by the adrenal medulla), hormones secreted by the adrenal cortex can provide a slower, longer-lasting response to stress. As the right side of **Figure 25.12** indicates, the hypothalamus ❸ secretes a releasing hormone that stimulates the pituitary to ❹ secrete a hormone called ACTH. In turn, ACTH stimulates cells of the adrenal cortex to ❺ synthesize and secrete a family of steroid hormones called **corticosteroids**, which include the glucocorticoids. The **glucocorticoids** help promote the synthesis of glucose from noncarbohydrates, such as proteins and fats.

Very high levels of glucocorticoids in the blood can suppress the body's defense system, including the inflammatory response that occurs at infection sites (see Chapter 24). For this reason, physicians may use glucocorticoids to treat diseases in which excessive inflammation is a problem. Cortisone, for example, can be used to treat arthritis. But although cortisone and other glucocorticoids can relieve swelling and pain from inflammation, they also suppress immunity and so can make a person highly susceptible to infection.

Glucocorticoids can also be prescribed to relieve pain from athletic injuries. Professional baseball and football players often receive cortisone injections into injured knees or elbows. With this treatment, the pain usually subsides, but its underlying cause remains. Masking the pain covers up the pain's message—that tissue is damaged and may get worse if not allowed to heal. If an athlete exercises an injured site before the tissue has recovered, the added stress can cause more serious and longer lasting damage.

The Gonads

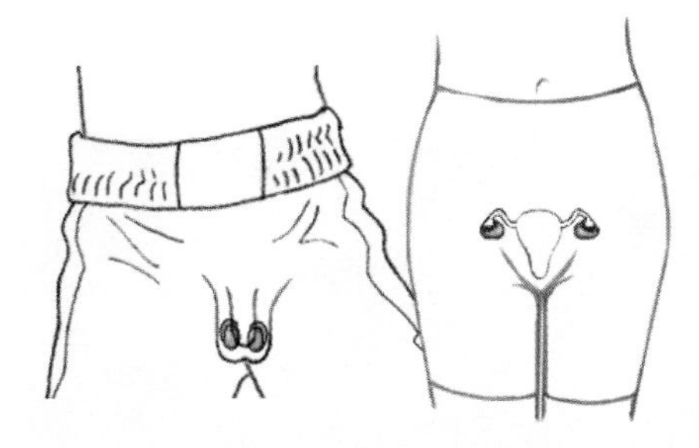

Sex hormones are steroid hormones that affect growth and development and regulate reproductive cycles and sexual behavior. Under the direction of the hypothalamus and pituitary gland, the **gonads,** or sex glands (ovaries in the female and testes in the male), secrete sex hormones in addition to producing gametes.

The gonads of humans produce three major categories of sex hormones: androgens, estrogens, and progestins. Both females and males have all three types, but in different proportions. Females have a high ratio of estrogens to androgens. **Estrogens** maintain the female reproductive system and promote the development of such female features as breasts and wider hips. **Progestins,** such as progesterone, are primarily involved in preparing the uterus to support a developing embryo.

Males have a high ratio of androgens to estrogens. **Androgens,** mainly testosterone, stimulate the development and maintenance of the male reproductive system. Androgens produced by male embryos during the seventh week of development stimulate the embryo to develop into a male rather than a female. During puberty, high concentrations of androgens trigger the development of male characteristics, such as a lower-pitched voice, facial hair, and large skeletal muscles. We'll discuss the male and female sex hormones further when we focus on human reproduction in Chapter 26.

CHECKPOINT

1. How does the relationship between the hypothalamus and the pituitary gland illustrate the close link between the nervous and endocrine systems?
2. Why does alcohol consumption lead to frequent urination?
3. Both calcitonin (from the thyroid) and parathyroid hormone (from the parathyroids) regulate blood calcium level. Why are these two hormones said to be antagonistic?
4. If someone with type 1 diabetes eats a big meal and does not take any medication, what will happen to that person's blood glucose level? Why?
5. The short-term stress response is regulated mainly by the hormones ___________ and ___________, while the long-term stress response is regulated mainly by a family of hormones called the ___________. All of these hormones are secreted by the ___________.
6. How could a hormonal imbalance result in a person who is genetically male but physically female?

Answers: 1. The hypothalamus, part of the brain and therefore the nervous system, controls the pituitary gland, part of the endocrine system. The pituitary releases various hormones into the bloodstream in response to signals from the hypothalamus. **2.** Alcohol slows the release of ADH from the pituitary. Decreased levels of ADH result in decreased levels of water reabsorption by the kidneys, increasing urine output. **3.** These two hormones have opposite effects: Calcitonin lowers blood calcium level, while PTH raises it. **4.** A person with type 1 diabetes has insufficient insulin in the bloodstream. Insulin normally stimulates cells to take up glucose, so an insufficient level of insulin results in a high level of glucose in the blood. **5.** epinephrine; norepinephrine; corticosteroids; adrenal glands **6.** If cells within a male embryo do not secrete testosterone at the proper time during development or the embryo's other cells lack testosterone receptors, the embryo will develop into a female despite being genetically male.

EVOLUTION CONNECTION

The Changing Roles of Hormones

Hormones play important roles in all vertebrates, and some of the same hormones can be found in vertebrates that are only distantly related. Interestingly, the same hormone can have different actions in different animals—a strong indication that hormonal regulation was an early evolutionary adaptation.

The hormone prolactin (PRL), produced by the anterior pituitary, is a good example. Prolactin produces a great diversity of effects in different vertebrate species. In mammals, for example, PRL stimulates mammary glands to grow and produce milk. Suckling by a newborn stimulates further release of PRL, which in turn increases the milk supply—an example of positive feedback (**Figure 25.13**).

In nonmammals, however, PRL performs a variety of other functions. In birds, PRL regulates fat metabolism and reproduction. In amphibians, it stimulates movement toward water and affects metamorphosis. In fish that migrate between salt and fresh water (salmon, for example), PRL helps regulate salt and water balance. These diverse effects suggest that prolactin is an ancient hormone whose functions diversified during vertebrate evolution. Over millions of years, the prolactin molecule stayed the same, but its role changed dramatically in response to varied environments. ■

Figure 25.13 Prolactin positive feedback in mammals. A suckling newborn stimulates the release of the hormone prolactin from the pituitary. This hormone stimulates the mammary glands to produce more milk.

Chapter Review

SUMMARY OF KEY CONCEPTS

For study help and activities, go to campbellbiology.com or the student CD-ROM.

Biology and Society: Of Hunger and Hormones

- Ghrelin is an example of a hormone, a regulatory chemical that travels via body fluids from its production site to target cells.

Hormones: An Overview

- The endocrine system consists of a collection of hormone-secreting cells and is the body's main system for internal chemical regulation, particularly of whole-body activities such as growth, reproduction, and control of metabolic rate. Endocrine glands are the primary sites of hormone production and secretion. Hormones trigger changes in target cells by two general mechanisms:

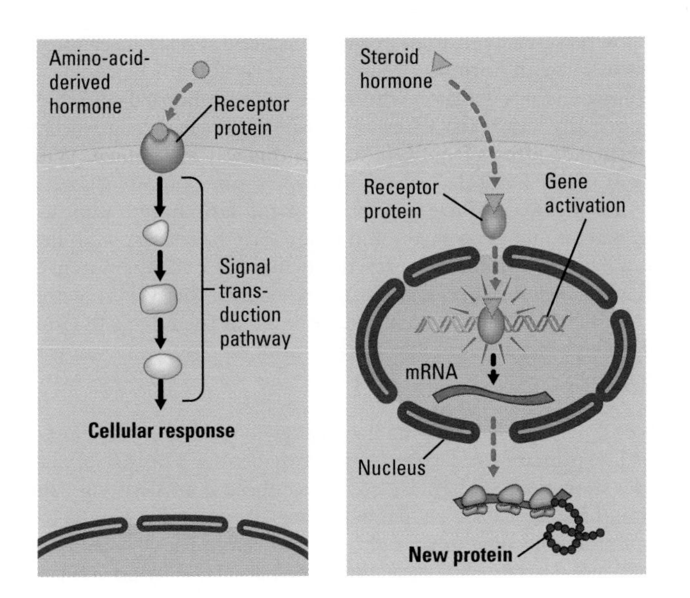

Activity *Overview of Cell Signaling*

Activity *Action of Amino-Acid-Based Hormones*

Activity *Action of Steroid Hormones*

The Human Endocrine System

- The human endocrine system consists of about a dozen glands that secrete several dozen hormones. These glands and hormones, summarized in Figure 25.5 and Table 25.1, vary widely in their functions, means of regulation, and targets.

- **The Hypothalamus and Pituitary Gland**

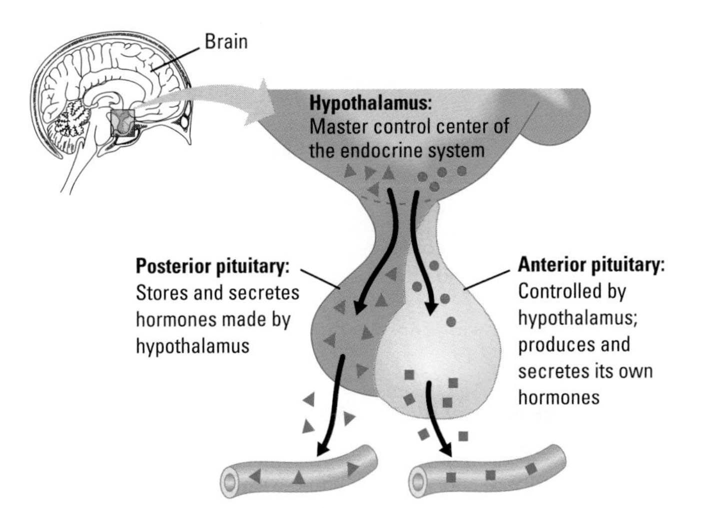

- **The Thyroid and Parathyroid Glands** Hormones from the thyroid gland regulate an animal's development and metabolism. Too little thyroid hormone in the blood (hypothyroidism) or too much (hyperthyroidism) can lead to metabolic disorders. Blood calcium level is regulated by the antagonistic hormones calcitonin from the thyroid (which lowers blood calcium level) and parathyroid hormone from the parathyroid glands (which raises blood calcium level).

Case Study in the Process of Science *How Does a Thyroid Hormone Affect Metabolism?*

- **The Pancreas** The pancreas secretes two hormones, insulin and glucagon, that control the blood glucose level. Insulin signals cells to take up glucose and signals the liver to store glucose. Glucagon causes the liver to release stored glucose into the blood. Diabetes mellitus results from a lack of insulin (type 1 diabetes) or a failure of cells to respond to it (type 2 diabetes).

- **The Adrenal Glands** Hormones from the adrenal glands help maintain homeostasis when the body is stressed. Nerve signals from the hypothalamus stimulate the adrenal medulla to secrete epinephrine and norepinephrine, which quickly trigger the fight-or-flight response. ACTH from the pituitary causes the adrenal cortex to secrete corticosteroids. Corticosteroids include glucocorticoids, which promote the synthesis of glucose. Glucocorticoids, such as cortisone, relieve inflammation and pain, but they can mask injury and suppress immunity.

- **The Gonads** Estrogens, progestins, and androgens are steroid sex hormones produced by the ovaries in females and the testes in males. Estrogens and progestins stimulate the development of female characteristics and maintain the female reproductive system. Androgens, such as testosterone, trigger the development of male characteristics. The secretion of sex hormones is controlled by the hypothalamus and pituitary.

Activity *Human Endocrine Glands and Hormones*

SELF-QUIZ

1. Which of the following is *not* true concerning homeostasis?
   a. It is the maintenance of constant internal conditions.
   b. It involves the regulation of body conditions such as calcium and glucose levels.
   c. It usually prevents major fluctuations in body conditions.
   d. Its maintenance is solely the responsibility of the endocrine system.

2. Unlike most amino-acid-based hormones, steroid hormones
   a. usually do not enter target cells.
   b. bind to receptors in the cytoplasm or nucleus of target cells.
   c. cause a cellular response.
   d. are made of amino acids.

3. Explain how the same hormone can have different effects on two different target cells and no effect on a third type of cell.

4. Which of the following controls the activity of all the others?
   a. thyroid gland
   b. hypothalamus
   c. pituitary gland
   d. adrenal cortex

5. A patient comes to a local health clinic with a large swelling in her neck that appears to be goiter. As the physician is taking the patient history, the patient mentions that she does not use iodized salt. The patient is overweight and has a slow metabolism. Does this patient have hyperthyroidism or hypothyroidism? Which endocrine gland is involved? How is this problem related to the lack of iodine in the diet?

6. The pancreas increases its output of insulin in response to
   a. an increase in body temperature.
   b. a decrease in blood glucose.
   c. a hormone secreted by the anterior pituitary.
   d. an increase in blood glucose.

7. Your roommate has just been diagnosed as hyperglycemic, which means he has an unusually high blood glucose level. Which of the following could be characteristic of this disorder?
   a. Too much insulin is produced after a high-sugar snack.
   b. The hormone glucagon is underactive.
   c. The receptors on his cells do not respond to insulin.
   d. any of the above

8. What sorts of problems might an athlete have if he or she chronically uses glucocorticoids to manage pain?

9. Testosterone belongs to a class of sex hormones called __________.

10. For each of the following situations, name the hormone that is most likely to account for the effect described. In some cases, there may be more than one answer.
    a. While running a marathon, athletes often report that they sense a lot of pain early on. The more they exert themselves, the less pain they report sensing. The hormones involved are __________.
    b. A 9-year-old child has been diagnosed with pituitary dwarfism. The hormone he is lacking is __________.
    c. A 30-year-old female has ovarian cancer and must have both ovaries removed. One class of hormones she will be lacking after her ovaries are removed is __________.
    d. After drinking several beers, your roommate wonders why he is urinating more frequently than normal. The technical reason is that alcohol inhibits release of a hormone that usually allows the kidneys to retain water for the body. The hormone being inhibited is __________.

Answers to the Self-Quiz questions can be found in Appendix D.

Go to the website or CD-ROM for more Self-Quiz questions.

THE PROCESS OF SCIENCE

11. Mice from a genetically modified strain remain healthy as long as you feed them regularly and do not let them exercise. After they eat, their blood glucose level rises slightly and then declines to a homeostatic level. However, if these mice fast or exercise at all, their blood glucose drops dangerously. Which hypothesis best explains their problem? Explain your choice.
    a. The mice have insulin-dependent diabetes.
    b. The mice lack insulin receptors on their cells.
    c. The mice lack glucagon receptors on their cells.
    d. The mice cannot synthesize glycogen from glucose.

BIOLOGY AND SOCIETY

12. A low rate of secretion of growth hormone (GH) causes pituitary dwarfism. Growth hormone made using recombinant DNA technology enables children who suffer from pituitary dwarfism to grow normally and reach a stature within the normal range. So far, no long-term side effects from GH used in this way are known. With GH readily available and relatively inexpensive, some parents who are afraid their normal children are not growing fast enough want to use GH to make them grow faster and taller. Are there reasons to hesitate treating a normal child with growth hormone, or are the potential benefits worth the risk? There is some evidence that GH injected into older adults may delay or even reverse some of the effects of aging. Should GH be freely available for any adult who wants to use it for that purpose?

13. Type 2 diabetes is becoming increasingly common in the United States. The primary risk factor for type 2 diabetes is a history of obesity. Statistics show that children are becoming obese at an alarmingly high rate, and this correlates with increased rates of type 2 diabetes. When young children become overweight, the problem is usually blamed on nutrition and exercise choices made by the parents. Why do you think today's parents have a harder time feeding their children nutritious meals and providing an exercise program than did parents of previous generations? What can be done to solve this problem? Recently, several families have filed legal action against a fast-food restaurant chain, claiming that their children are obese because they ate there frequently. These parents claim that the restaurant chain is liable because it made the parents think that the food was healthy and would not contribute to obesity. Do you believe that these parents have a case? Justify your response.

26 Reproduction and Development

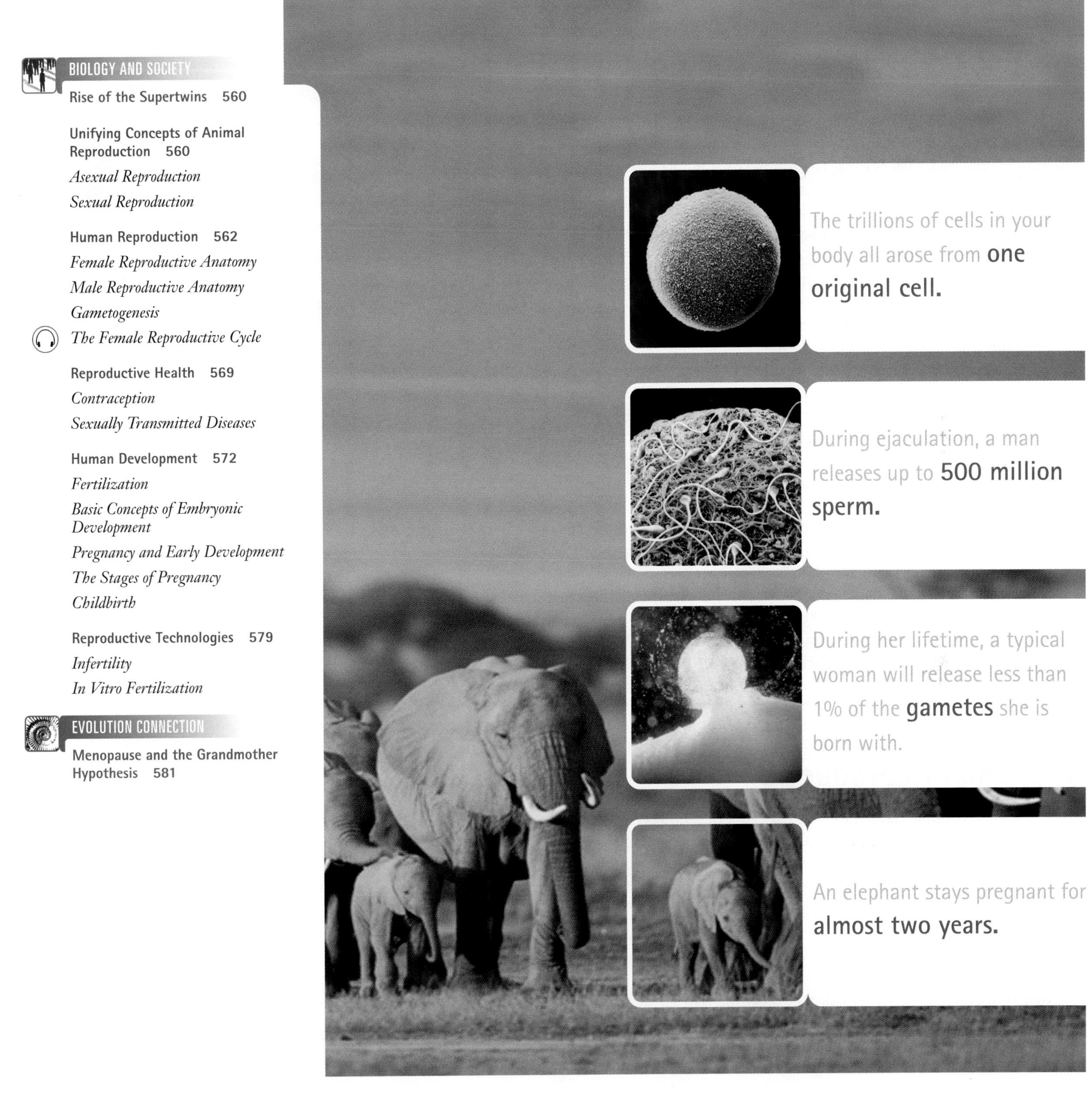

The trillions of cells in your body all arose from **one original cell.**
During ejaculation, a man releases up to **500 million sperm.**
During her lifetime, a typical woman will release less than 1% of the **gametes** she is born with.
An elephant stays pregnant for **almost two years.**

BIOLOGY AND SOCIETY

Rise of the Supertwins

On November 19, 1997, news reports heralded the arrival of "The Iowa Septuplets" **(Figure 26.1)**. Born to Kenny and Bobbi McCaughey, the four boys and three girls were heralded as the world's first set of surviving septuplets. Other incredible stories of multiple births soon followed. A Texas woman delivered octuplets in December 1998, and an Italian woman did the same in September 2000.

What accounted for the sudden glut of multiple births? All three of these women were taking fertility drugs because they couldn't become pregnant naturally. Fertility drugs stimulate the ovaries to release one or more eggs. If multiple eggs are released, each one may be fertilized (that is, fused with a sperm to form a zygote), resulting in multiple embryos. Over 10% of women taking fertility drugs become pregnant with more than one embryo—sometimes quite a few more. Increased use of fertility drugs has caused the multiple-birth rate in the United States to soar. In the last 25 years, the rate of twin births has risen more than 80%, and the rate of "supertwins"—triplets or higher—has risen more than 500%.

Fertility drugs have allowed thousands of infertile couples to have a baby. But there are also risks involved. Newborns from multiple births are more likely to be premature and more likely to have lower birth weights. They are also less likely to survive—the mortality rate is 5 times higher among twins and 12 times higher among supertwins—and less likely to be healthy if they do survive. The Iowa septuplets, for example, each weighed less than half the national average birth weight. Two of them developed cerebral palsy. One of the Texas octuplets died (weighing just 10 ounces), as did four of the Italian octuplets.

Couples turn to fertility drugs to overcome their natural reproductive limitations. That such treatments often have "unnatural" outcomes underscores the intricate nature of human reproduction. In this chapter, we explore the anatomy and physiology of animal reproduction, paying particular attention to the reproductive structures of humans and how human babies develop from a single cell. We also consider how modern health practices and technologies allow us to circumvent the natural process of reproduction. ■

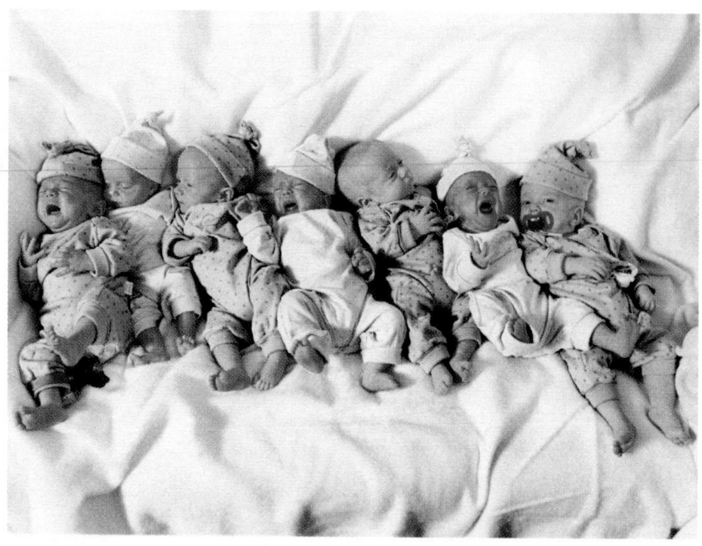

Figure 26.1 The Iowa septuplets. Born after their mother was treated with fertility drugs, these seven babies were the world's first surviving septuplets.

Unifying Concepts of Animal Reproduction

Every individual animal has a finite life span. Species, however, last much longer because of **reproduction,** the creation of new individuals from existing ones. All animals reproduce, but they do so in a great variety of ways. There are, however, two basic schemes of animal reproduction: asexual reproduction and sexual reproduction.

Asexual Reproduction

Asexual reproduction is the creation of offspring that are genetically identical to a lone parent. In the simplest type of asexual reproduction, bacteria and some single-celled eukaryotes such as amoebas reproduce by **binary fission** (splitting in two), in which a single cell divides into two genetically identical offspring cells (see Figure 8.2a). Some multicellular animals, including many invertebrates, such as the sea anemone shown in **Figure 26.2a**, reproduce by splitting into two or more individuals of roughly equal size. This process is called simply **fission** because more than two offspring may result. Another means of asexual reproduction is **fragmentation,** the breaking of a parent body into several pieces, followed by **regeneration,** the regrowth of two or more whole animals from the pieces. Some organisms have remarkable powers of fragmentation and regeneration. If the five arms of certain species of sea star are split off, for example, they can give rise to five offspring (see Figure 8.2b). Even more incredibly, if a sea sponge is pushed through a wire mesh, each of the resulting clumps of cells can regrow into a new sponge.

Another form of asexual reproduction is budding, the splitting off of new individuals from existing ones. Hydras, relatively simple multicellular animals related to jellies, reproduce this way. In **Figure 26.2b**, you can see a budding hydra that will soon detach from its parent.

Asexual reproduction has a number of advantages. Because it eliminates the need to find a mate, asexual reproduction allows a species to perpetuate itself even if its individual members are isolated from one another. Asexual reproduction also allows organisms to multiply quickly, without spending time or energy producing gametes (sperm and eggs). If an individual is very well suited to its environment, asexual reproduction allows it to reproduce rapidly and exploit available resources.

A potential disadvantage of asexual reproduction is that it produces genetically uniform populations. Genetically similar individuals may thrive in one particular environment; but if the environment changes and becomes less favorable to survival, all individuals may be affected equally, and the entire population may die out (see, for example, the story of the Irish potato famine that opens Chapter 28).

(a) Fission. This sea anemone is reproducing asexually through the process of fission.

(b) Budding. This hydra is in the process of reproducing by budding.

Figure 26.2 Different methods of asexual reproduction.

Sexual Reproduction

You know that you are a mix of traits from your mother and father. But you also know that you have a genome distinct from every other human (unless you happen to have an identical twin). You are the product of **sexual reproduction,** the creation of offspring by the fusion of two haploid sex cells called **gametes** to form a diploid **zygote.** The male gamete is the **sperm,** and the female gamete is the **ovum** (plural, *ova*). The zygote and the new individual it develops into contain a unique combination of genes carried from the parents via the egg (ovum) and sperm.

Unlike asexual reproduction, sexual reproduction increases genetic variability among offspring as a result of the huge variety of gametes produced by meiosis (see Chapter 8). In theory, when an environment changes suddenly or drastically, there is a better chance that some of the variant offspring will survive and reproduce than if all offspring are genetically very similar. Variation is the raw material of evolution by natural selection.

Some animals can reproduce both sexually and asexually. Rotifers, microscopic aquatic animals, reproduce asexually when food supplies are ample. When conditions become harsher, rotifers switch to sexual reproduction. Rotifers benefit from both modes of reproduction by increasing their numbers rapidly and efficiently when the living is easy and by creating a population of varying individuals when times get tough **(Figure 26.3a)**.

In some species, each individual is a **hermaphrodite,** meaning that it is equipped with both male and female reproductive systems. When earthworms mate, for example, each individual serves both as male and female, donating and receiving sperm **(Figure 26.3b)**.

The mechanics of fertilization play an important part in sexual reproduction. Many aquatic invertebrates and most fishes and amphibians use **external fertilization (Figure 26.3c)**. In this reproductive scheme, the parents discharge their gametes into the water, where fertilization occurs; the female and male don't necessarily have to touch to mate. In contrast, **internal fertilization** occurs when sperm are deposited in or near the female reproductive tract, the gametes fusing within the female's body. This requires **copulation,** or sexual intercourse, and complex reproductive systems. Internal fertilization is an evolutionary adaptation that allows terrestrial animals to reproduce in an environment where delicate gametes would otherwise dry out. Nearly all terrestrial animals reproduce this way. In the next section, we'll examine the reproductive anatomy that allows one particular terrestrial animal—namely, humans—to achieve internal fertilization.

CHECKPOINT

1. What is the most important difference in the genetic makeup of the offspring resulting from sexual and asexual reproduction?
2. Name three different types of asexual reproduction.
3. What is the primary advantage of sexual reproduction with respect to evolutionary adaptation?

Answers: 1. Asexual reproduction produces genetically identical offspring, while sexual reproduction produces genetically unique offspring. **2.** Methods of asexual reproduction include binary fission, fission, budding, and fragmentation. **3.** Sexual reproduction produces variation in a population, and variation is the raw material of evolution. Those individuals with traits best suited to the environment produce more offspring, gradually increasing a population's adaptation to the environment.

Human Reproduction

Although the differences between human male and female reproductive anatomies are obvious, there are also some similarities. Both sexes have a pair of **gonads,** the organs that produce gametes. And both sexes have ducts to store and deliver the gametes as well as structures to facilitate copulation. We'll now examine the anatomical features of human reproduction, beginning with female anatomy.

Female Reproductive Anatomy

Two views of the female reproductive system are shown in **Figure 26.4**. The outer features of the female reproductive anatomy are collectively called the **vulva.** The **vagina,** or birth canal, opens to the outside just behind the opening of the urethra, the tube through which urine is excreted. A pair of skin folds, the **labia minora,** borders the openings, and a pair of thick, fatty ridges, the **labia majora,** protects the entire genital region. Until sexual intercourse or other vigorous physical activity ruptures it, a thin

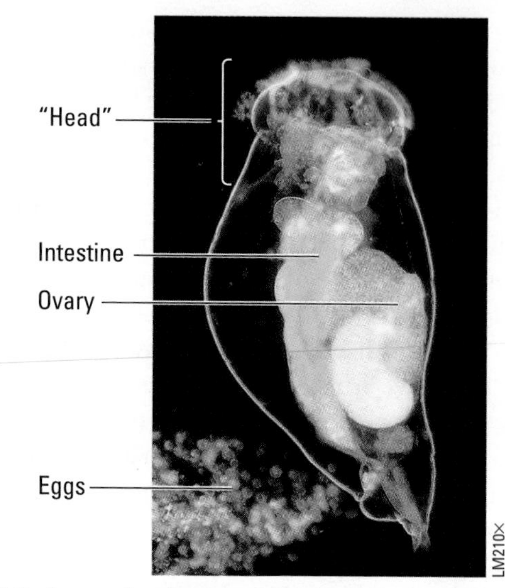

(a) Both sexual and asexual. This rotifer, a common aquatic invertebrate, can reproduce both sexually and asexually. In this photo, the rotifer is laying eggs.

(b) Hermaphroditism. When earthworms mate, each partner gives and receives sperm.

(c) External fertilization. Frogs release eggs and sperm (too small to be seen) into the water, where fertilization takes place. The embrace between the two frogs is a mating ritual that triggers the release of gametes.

Figure 26.3 Reproductive schemes.

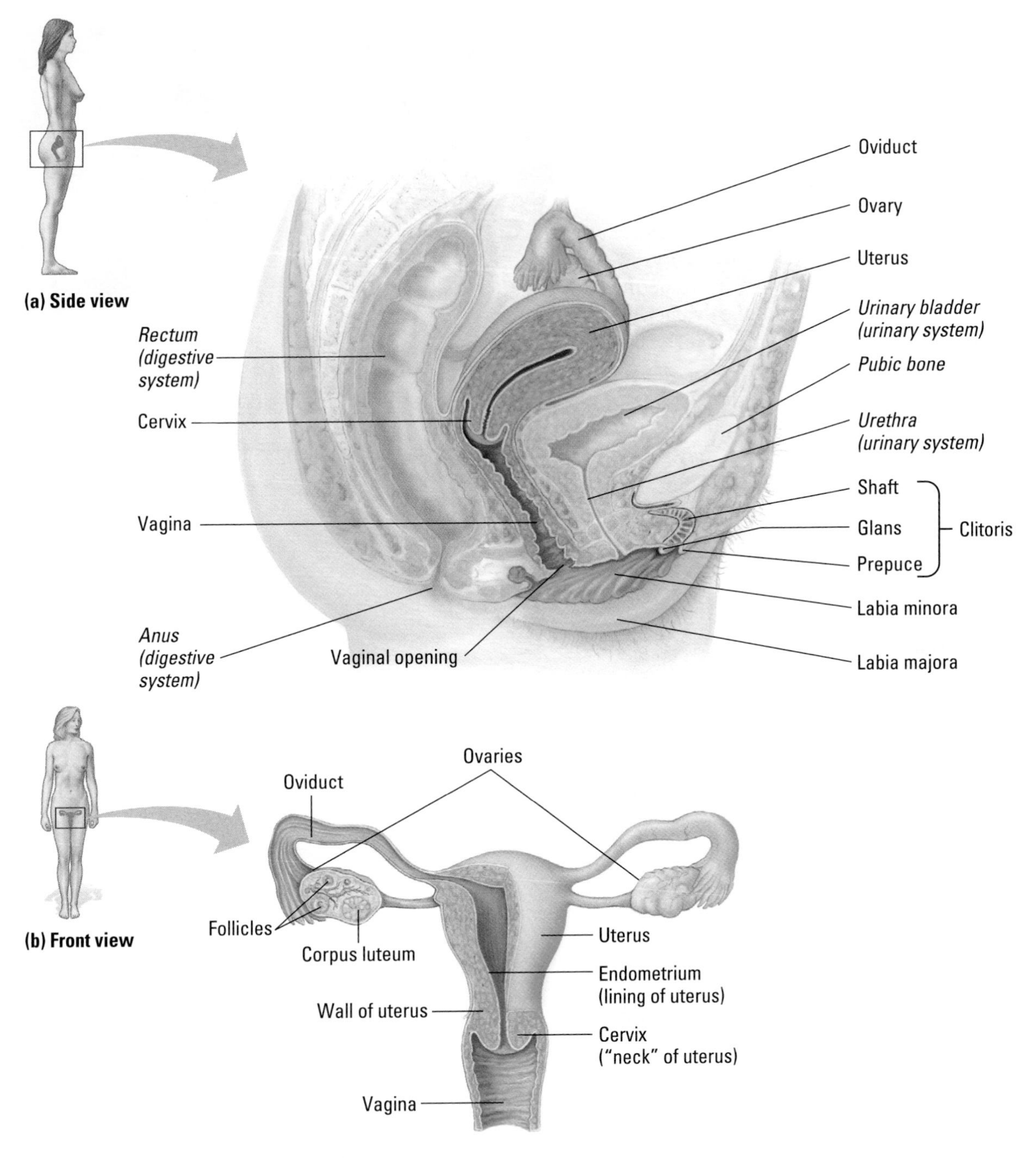

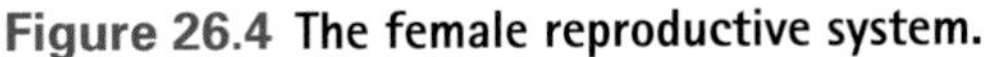

Figure 26.4 The female reproductive system.

membrane called the **hymen** partly covers the vaginal opening. During sexual arousal, the vagina, labia minora, and clitoris swell as they fill with blood. The **clitoris** consists of a short shaft supporting a rounded **glans,** or head, covered by a small hood of skin called the **prepuce.** In Figure 26.4, blue highlights the erectile tissue within the clitoris that fills with blood during arousal. The clitoris, especially the glans, has an enormous number of nerve endings and is very sensitive to touch.

The **ovaries** are the site of gamete production in human females. A woman's ovaries are each about an inch long and have a bumpy surface. The bumps are **follicles,** each consisting of a single developing egg cell surrounded by one or more layers of cells that nourish and protect it. The follicles also produce estrogen, the female sex hormone (see Chapter 25).

A female is born with over 1 million follicles, but only about a quarter million are viable at puberty, and only several hundred will release egg cells during her reproductive years. Starting at puberty and continuing until menopause, one follicle (or rarely two or more) matures and releases its egg cell about every 28 days. An egg cell is ejected from the follicle in the process of **ovulation** (**Figure 26.5**). After ovulation, what remains of the follicle grows to form a solid mass called the **corpus luteum,** which plays a crucial role in the release of hormones during the reproductive cycle (as you'll see later in the chapter). The released egg enters an **oviduct,** also known as a fallopian tube, where cilia sweep it toward the uterus. If sperm are present, fertilization may take place in the upper part of the oviduct. If the released egg is not fertilized, it is shed during menstruation, and a new follicle matures during the next cycle.

The **uterus,** also known as the womb, is the actual site of pregnancy. The uterus is about the size and shape of an upside-down pear, but can grow to several times that size to accommodate a growing fetus. The uterus has a thick muscular wall lined with a blood-rich layer of tissue called the **endometrium.** An embryo implants in the endometrium, and development is completed there. The term **embryo** is used for the stage in development from the first division of the zygote until body structures begin to appear, about the 9th week. From the 9th week until birth, a developing human is called a **fetus.**

The narrow neck at the bottom of the uterus is the **cervix.** It is recommended that a woman have a yearly Pap test in which cells are removed from around the cervix and examined under a microscope for signs of cervical cancer. Regular Pap smears greatly increase the chances of detecting cervical cancer early and therefore treating it successfully. The cervix opens into the vagina. During copulation, the vagina serves as a repository for sperm.

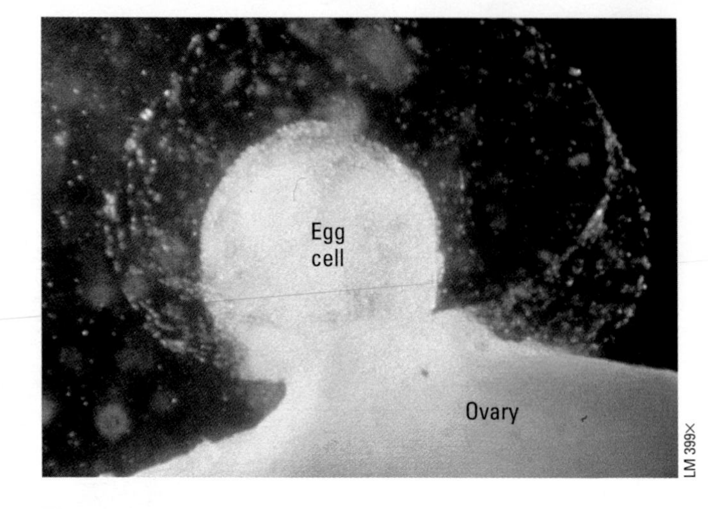

Figure 26.5 An egg cell being ejected from a follicle during ovulation.

Male Reproductive Anatomy

Figure 26.6 presents two views of the male reproductive system. The **penis** contains erectile tissue (shown in blue in the figures) that can fill with blood and cause an erection during sexual arousal. The penis consists of a shaft that supports the glans (head). The glans is richly supplied with nerve endings and is highly sensitive to stimulation. A prepuce, or foreskin, covers the glans. Circumcision, the surgical removal of the foreskin, is commonly performed for religious reasons or perceived health benefits. However, medical scientists have not reached consensus on whether circumcision has an overall positive or negative impact on a man's health.

The **testes** (singular, *testis*), the male gonads, are located outside the abdominal cavity in a sac called the **scrotum.** Sperm cannot develop at body temperature. By keeping the sperm-forming cells away from the body, the scrotum keeps them cool enough to function normally. In cold conditions, muscles around the scrotum contract, pulling the testes toward the body, maintaining the proper temperature.

From puberty well into old age, the testes normally produce hundreds of millions of sperm every day. From the testes, sperm pass into a coiled tube called the **epididymis,** which stores sperm while they complete development. During **ejaculation**—the expulsion of sperm-containing fluid from the penis—the sperm leave the epididymis and travel through a duct called the **vas deferens** (plural, *vasa deferentia*). The **seminal vesicles** and **prostate** gland add fluid that nourishes the sperm and provides protection

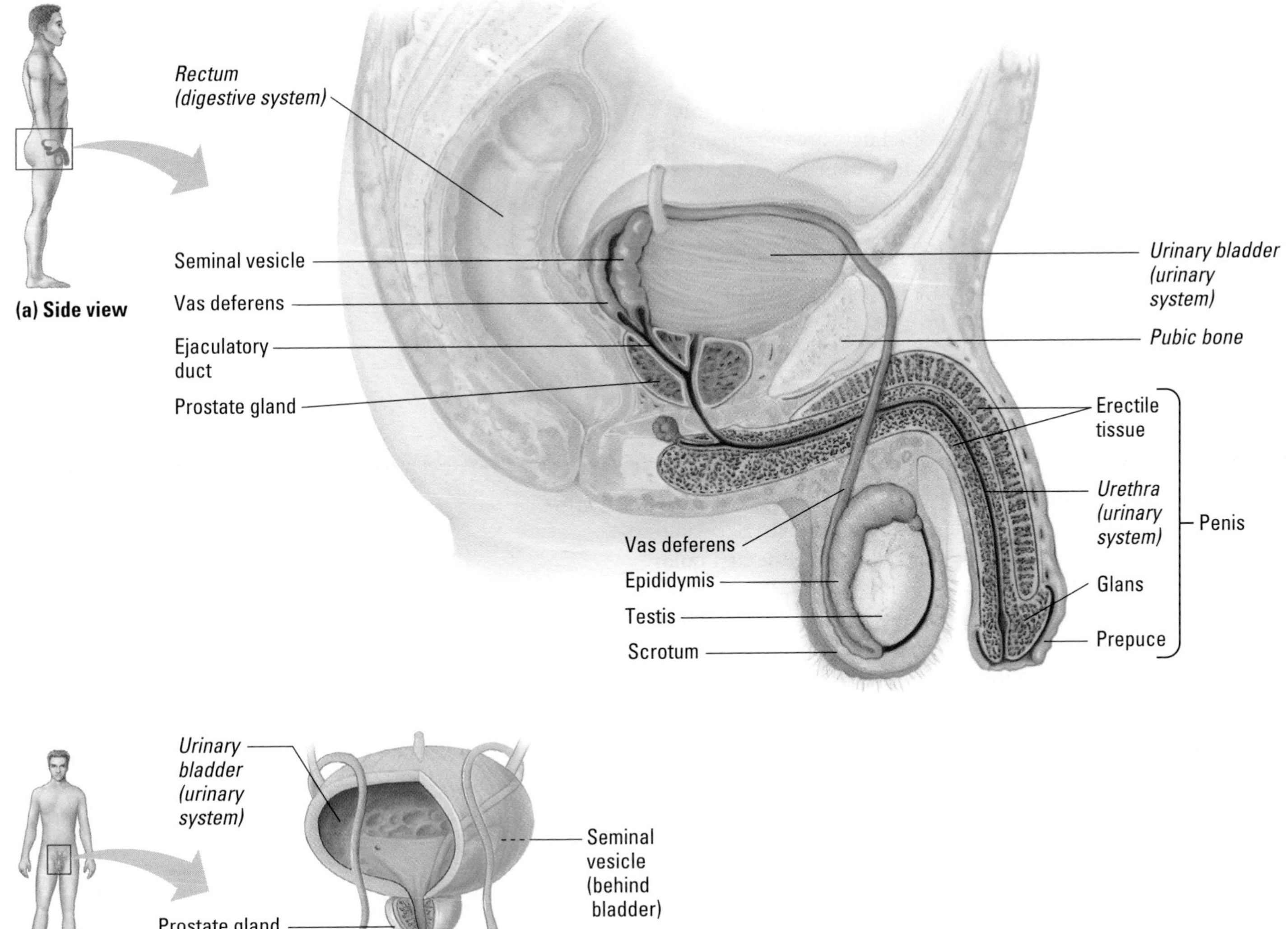

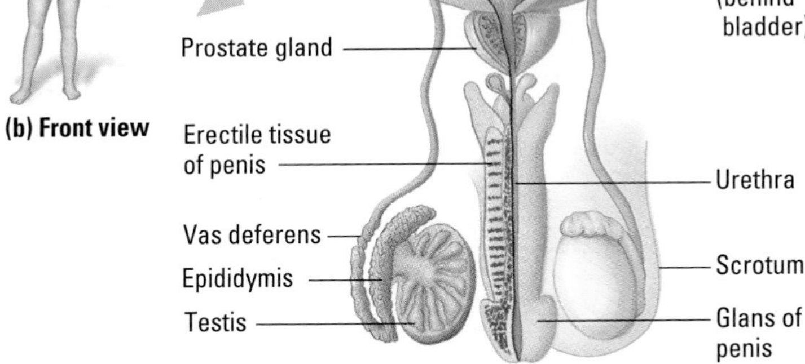

Figure 26.6 The male reproductive system.

from the natural acidity of the vagina. The prostate gland is the source of some of the most common medical problems in men over 40. Prostate cancer, for example, is the second most commonly diagnosed cancer in the United States (after skin cancer). The two vasa deferentia, one from each of the testes, empty into the **urethra.** The urethra conveys, at different times, both sperm and urine out through the penis.

Ejaculation, caused by the contraction of muscles along the sperm ducts, releases about 5 mL (1 teaspoonful) of **semen.** About 95% of semen is fluid secreted by the various glands, and about 5% is made up of 200–500 million sperm, only one of which may fertilize an egg.

Gametogenesis

The production of gametes is called **gametogenesis.** Human gametes—ova and sperm—are haploid cells with 23 chromosomes that develop by meiosis from diploid cells with 46 chromosomes. (You may find it helpful to review the discussion of meiosis in Chapter 8.) There are significant differences in gametogenesis between human females and males, so we'll examine them separately.

Oogenesis **Figure 26.7** summarizes **oogenesis,** the development of eggs within the ovaries. At birth, each ovary contains many thousands of follicles, and each follicle contains one dormant **primary oocyte,** a diploid cell that has paused its cell cycle in prophase of meiosis I.

A primary oocyte can be triggered to develop further by the hormone FSH (follicle-stimulating hormone). After puberty and until menopause, about every 28 days, FSH from the pituitary gland stimulates one of the dormant follicles to develop. The follicle enlarges, and the primary oocyte within it completes meiosis I and begins meiosis II. The division of the cytoplasm in meiosis I is unequal, with a single **secondary oocyte** receiving almost all of it. The smaller of the two daughter cells, called the first **polar body,** receives almost no cytoplasm.

About the time the secondary oocyte forms, the pituitary gland secretes LH (luteinizing hormone), which causes ovulation. The ripening follicle bursts, releasing its secondary oocyte from the ovary. The ruptured follicle then develops into a corpus luteum. The secondary oocyte enters the oviduct, and if a sperm cell fuses with it, the secondary oocyte completes

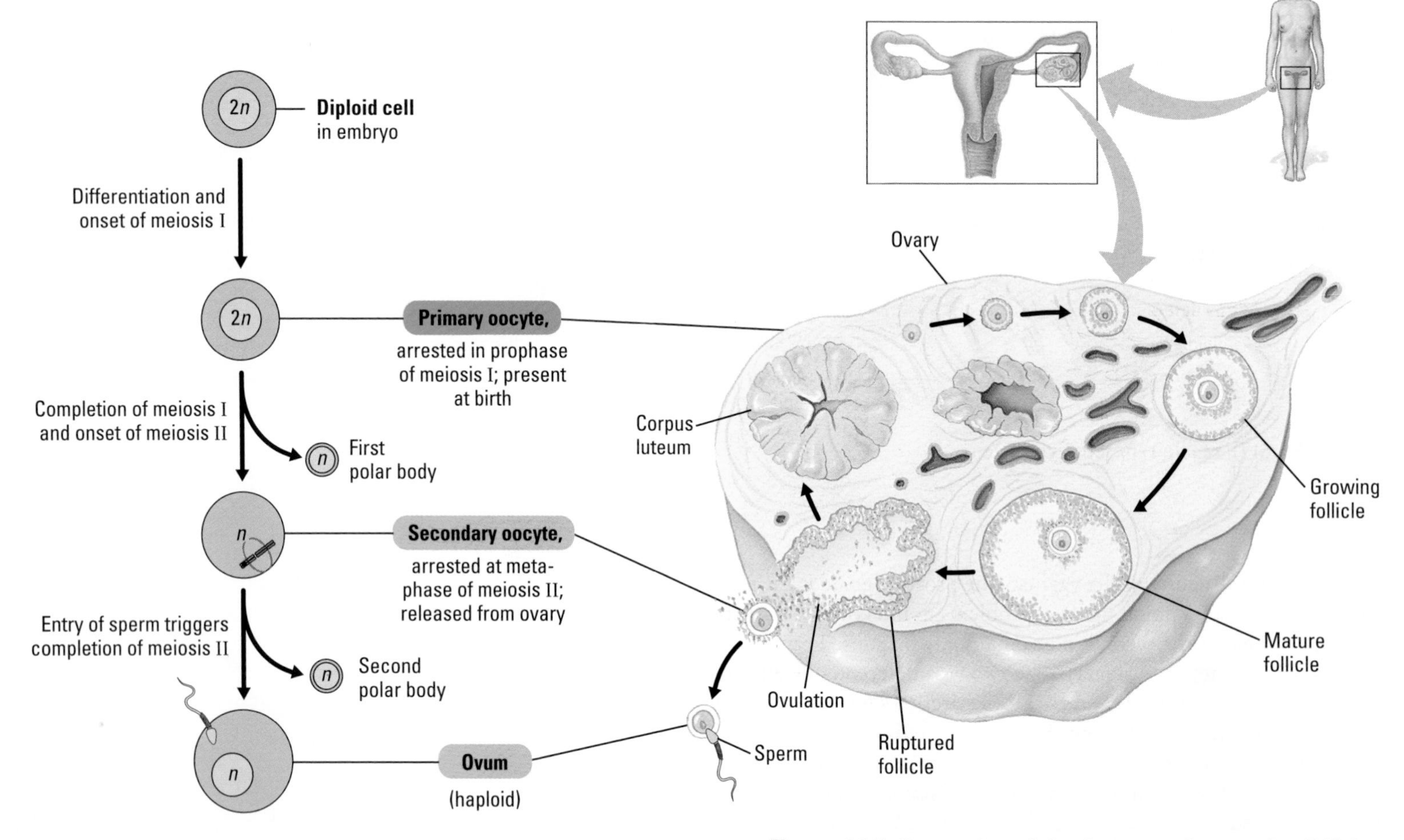

Figure 26.7 Oogenesis and development of an ovarian follicle.

meiosis II. Meiosis II is also unequal, yielding a small polar body and the actual ovum. The chromosomes of the ovum can then fuse with the chromosomes of the sperm cell, producing a diploid zygote. The polar bodies, which are quite small because they received almost no cytoplasm, degenerate. The ovum has acquired nearly all the cytoplasm and thus the bulk of the nutrients and organelles contained in the original cell.

Spermatogenesis The formation of sperm cells is called **spermatogenesis** (**Figure 26.8**). Sperm cells develop inside the testes in coiled tubes called the **seminiferous tubules.** Cells near the outer walls of the tubules multiply constantly by mitosis. Each day, about 3 million of them differentiate into **primary spermatocytes,** the cells that undergo meiosis. Meiosis I of a primary spermatocyte produces two **secondary spermatocytes.** Meiosis II then forms four cells, each with the haploid number of chromosomes. The sperm cells that develop from each of these haploid cells are gradually pushed toward the center of the seminiferous tubules. From there, the sperm cells pass into the epididymis, where they mature, become motile, and are stored until ejaculation.

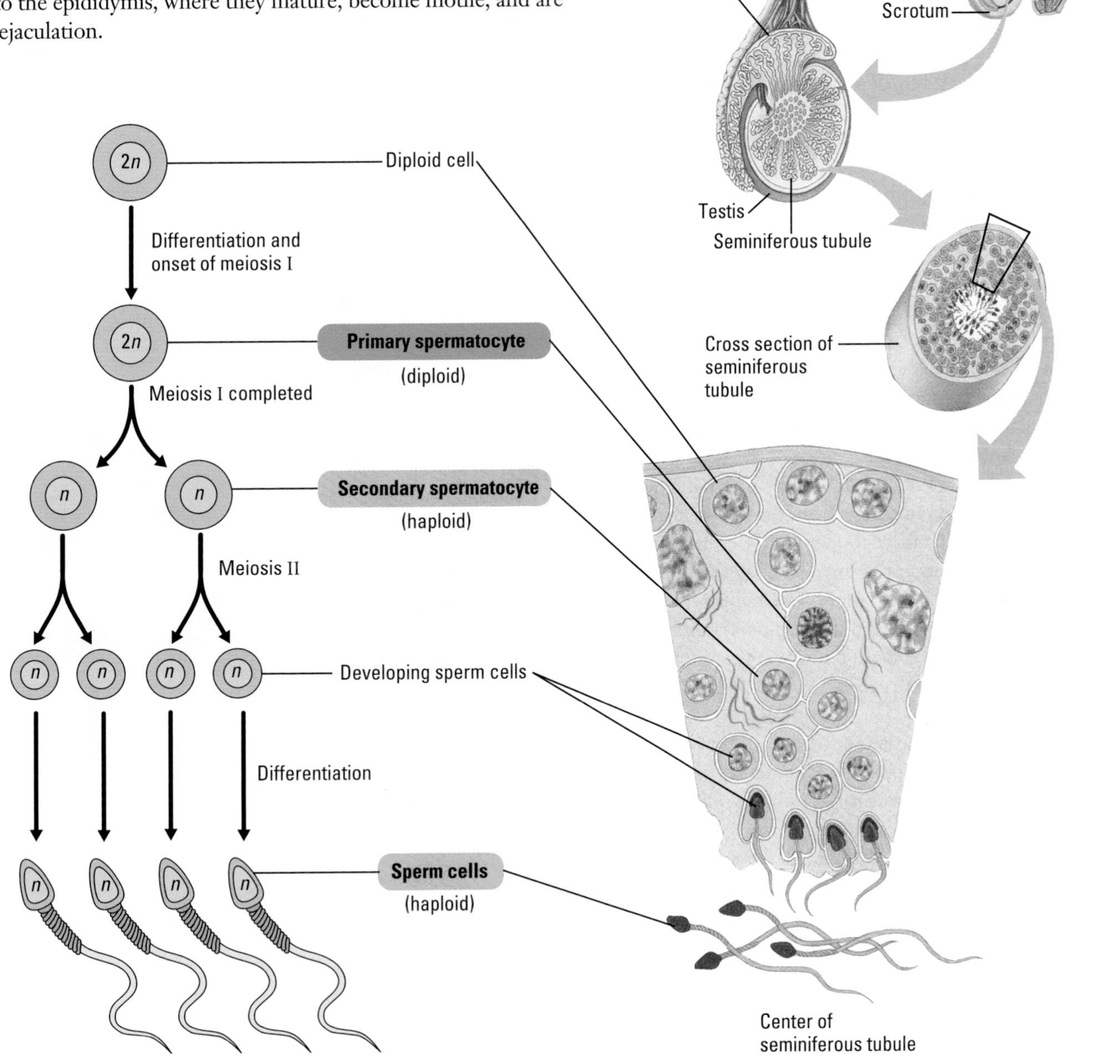

Figure 26.8 Spermatogenesis. This process takes about 65–75 days in the human male.

Both oogenesis and spermatogenesis produce haploid gametes, but there are several important differences between the two processes. One obvious difference is location: testes in the male and ovaries in the female. Furthermore, human males create new sperm every day from puberty through old age. Human females, on the other hand, create primary oocytes only during fetal development; a woman is thought to be born with all the primary oocytes she will ever have. Another difference is that four gametes result from each diploid parent cell during spermatogenesis, whereas oogenesis results in only one gamete from each parent cell. Moreover, there are significant differences in the cells produced by meiosis: Sperm are small and motile and contain relatively few nutrients; eggs are large, nonmotile, and well stocked with organelles. Finally, oogenesis cannot be completed without stimulation from a sperm cell, whereas spermatogenesis is completed before the sperm leave the testis.

The Female Reproductive Cycle

MP3 Tutor **Female Reproductive Cycle**

Human females have a **reproductive cycle,** a recurring series of events that produces gametes, makes them available for fertilization, and prepares the body for pregnancy **(Figure 26.9)**. The reproductive cycle repeats itself once every 28 days, on average, but cycles from 20 to 40 days are not uncommon. The reproductive cycle is actually two cycles in one. The **ovarian cycle** (Figure 26.9c) controls the growth and release of an egg. During the **menstrual cycle** (Figure 26.9e), the uterus is prepared for possible implantation of an embryo.

By convention, the first day of menstruation, a woman's "period," is designated as day 1 of the menstrual cycle. **Menstruation** is uterine bleeding caused by the breakdown of the endometrium, the blood-rich inner lining of the uterus. If an embryo implants in the uterine wall, it will obtain nutrients from the endometrium and the thickened lining will not be discharged. Menstruation is thus a sign that pregnancy has not occurred during the previous cycle. Menstruation usually lasts 3–5 days. The menstrual discharge, which leaves the body through the vagina, consists of blood, clusters of endometrial cells, and mucus. After menstruation, the endometrium regrows, reaching its maximum thickness in 20–25 days.

The hormones shown in parts (a), (b), and (d) of Figure 26.9 regulate the ovarian and menstrual cycles, synchronizing ovulation with preparation of the uterus for possible implantation of an embryo. At the start of the ovarian cycle, the hypothalamus secretes a releasing hormone that stimulates the anterior pituitary gland to ❶ increase its output of FSH and LH. True to its name, FSH ❷ stimulates the growth of an ovarian follicle. As

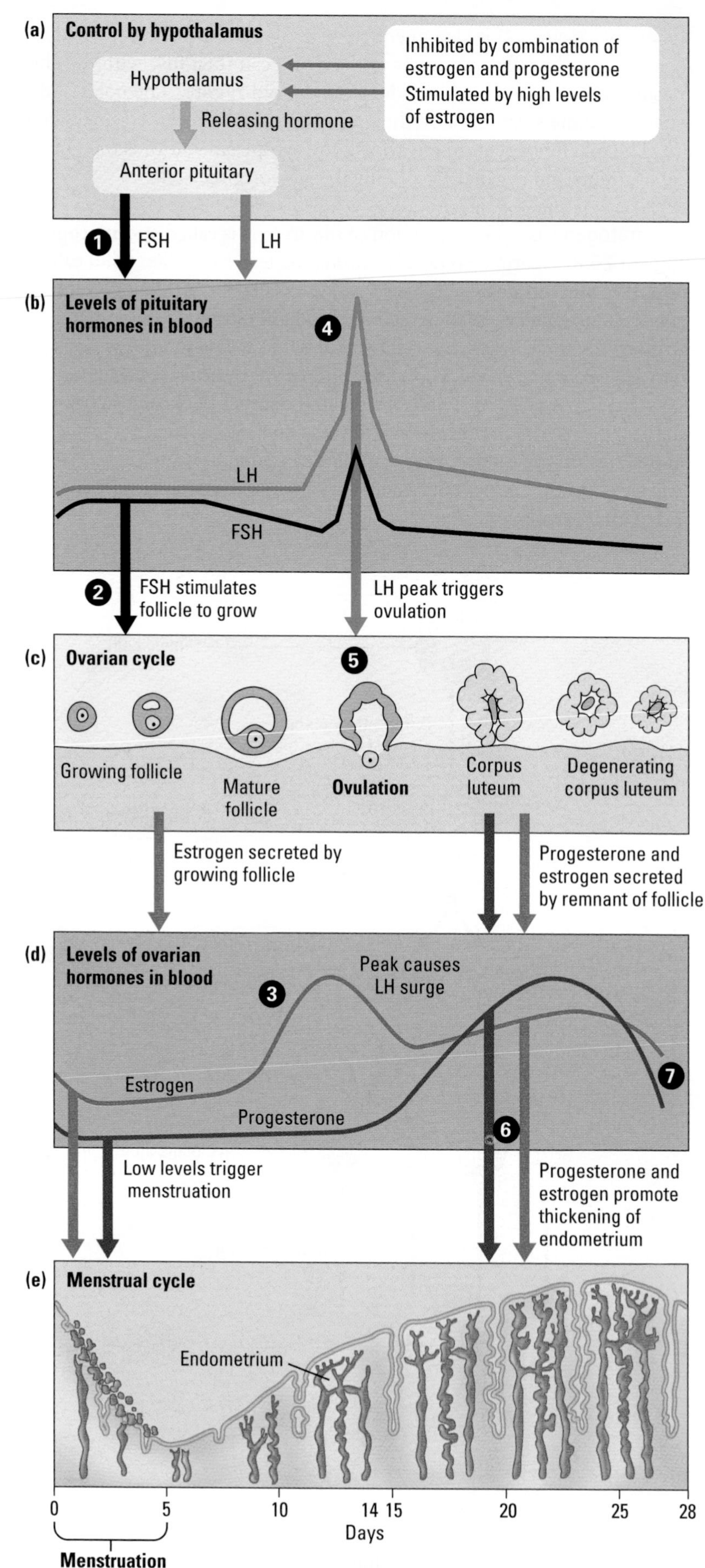

Figure 26.9 The reproductive cycle of the human female. This figure shows how (c) the ovarian cycle and (e) the menstrual cycle are regulated by changing hormonal levels, represented in parts (a), (b), and (d). The time scale at the bottom of the figure applies to parts (b)–(e).

the maturing follicle grows, it secretes estrogen in increasing amounts. After about 12 days, ❸ estrogen levels peak, which causes ❹ a sudden surge of FSH and LH. This stimulates ovulation, and ❺ the developing follicle within the ovary bursts and releases its egg. Ovulation takes place 14 days before the end of the cycle (day 14 of the typical 28-day cycle).

Besides promoting rupture of the follicle, the sudden surge of LH has several other effects. It stimulates the completion of meiosis I, transforming the primary oocyte in the follicle into a secondary oocyte. LH also promotes the secretion of estrogen and progesterone by the corpus luteum. Estrogen and progesterone regulate the menstrual cycle. Rising levels of these two hormones ❻ promote thickening of the endometrium. The combination of estrogen and progesterone also inhibits further secretion of FSH and LH, ensuring that a second follicle does not mature during this cycle. Further degeneration of the corpus luteum causes the levels of estrogen and progesterone to fall off. Once these hormones ❼ fall below a critical level, the endometrium begins to shed, starting menstruation on day 1 of the next cycle. Now that estrogen and progesterone are no longer there to inhibit it, the pituitary secretes FSH and LH, and a new cycle begins.

The description of the reproductive cycle in this section assumes that fertilization has not occurred. If it does, the embryo implants into the endometrium and secretes a hormone called HCG (human chorionic gonadotropin), which can be detected by pregnancy tests (see Chapter 24). HCG maintains the corpus luteum, which continues to secrete progesterone and estrogen, keeping the endometrium intact. We'll return to what happens next—the events of pregnancy—later in the chapter.

CHECKPOINT

1. In which organ of the human female reproductive system does the fetus develop?
2. Arrange the following organs of the male reproductive system in the correct sequence for the travel of sperm: epididymis, testis, urethra, vas deferens.
3. During gametogenesis, a primary oocyte eventually gives rise to __________ ovum/ova, while a primary spermatocyte eventually gives rise to __________ sperm. (Provide two numbers.)
4. What hormonal changes trigger the start of menstruation?

Answers: 1. The uterus **2.** Testis, epididymis, vas deferens, urethra **3.** one; four **4.** Decreasing levels of estrogen and progesterone.

Reproductive Health

Now that you've read about the anatomy and physiology of the human reproductive system, you are in a good position to understand how this knowledge applies to two important issues of reproductive health: contraception and the transmission of disease.

Contraception

Contraception is the deliberate prevention of pregnancy. There are many forms of contraception that interfere with different steps in the process of becoming pregnant. **Table 26.1** lists the most common methods of birth control and their failure rates when used correctly and when used typically. Note that these two rates are often quite different, emphasizing the importance of learning to use contraception correctly. This section presents a

Table 26.1 Contraceptive Methods

	Pregnancies/100 Women/Year	
Method	**Used Correctly**	**Typically**
None		85
Birth control pill (combination)	0.1	5
Vasectomy	0.1	0.15
Tubal ligation	0.2	0.5
Progestin minipill	0.5	5
Rhythm	1–9	20
Withdrawal	4	19
Condom (male)	3	14
Diaphragm and spermicide	6	20
Cervical cap and spermicide	9	20
Spermicide alone	6	26

brief overview of the most common methods of contraception, but it is by no means complete. For more information, consult a health-care professional.

Complete abstinence (avoiding intercourse) is the only totally effective method of contraception, but other methods are effective to varying degrees. Sterilization, an operation that prevents sperm from reaching an egg, is very reliable. A woman may have a **tubal ligation** (having her "tubes tied"). In this procedure, a doctor removes a short section from each oviduct, often tying, or ligating, the remaining ends and thereby blocking the route of sperm to egg. A man may have a **vasectomy,** in which a doctor cuts a section out of each vas deferens to prevent sperm from reaching the urethra. Both forms of sterilization are relatively safe, free from side effects, and permanent. However, if changing life circumstances cause a person to seek a reversal, microsurgical techniques can sometimes (but not always) successfully restore fertility.

The effectiveness of other methods of contraception depends on how they are used. Temporary abstinence, also called the **rhythm method** or **natural family planning,** depends on refraining from intercourse during the days around ovulation, when fertilization is most likely. In theory, the time of ovulation can be determined by monitoring changes in body temperature and the composition of cervical mucus. This requires careful monitoring and record keeping. Additionally, the length of the reproductive cycle can vary from month to month, and sperm can survive for 3–5 days within the female reproductive tract, making natural family planning among the most unreliable methods of contraception in actual practice. **Withdrawal** of the penis from the vagina before ejaculation is also ineffective, because sperm may exit the penis before climax.

If used correctly, **barrier methods** can be quite effective at physically preventing the union of sperm and egg **(Figure 26.10)**. **Condoms** are sheaths, usually made of latex, that fit over the penis or within the vagina. A **diaphragm** is a dome-shaped rubber cap that covers the cervix; a cervical cap is similar but thimble-shaped and smaller. Both require a doctor's visit for proper fitting. To be effective, barrier devices (including condoms) should be used in combination with **spermicides,** sperm-killing chemicals in the form of a jelly, cream, or foam; spermicides used alone are not reliable.

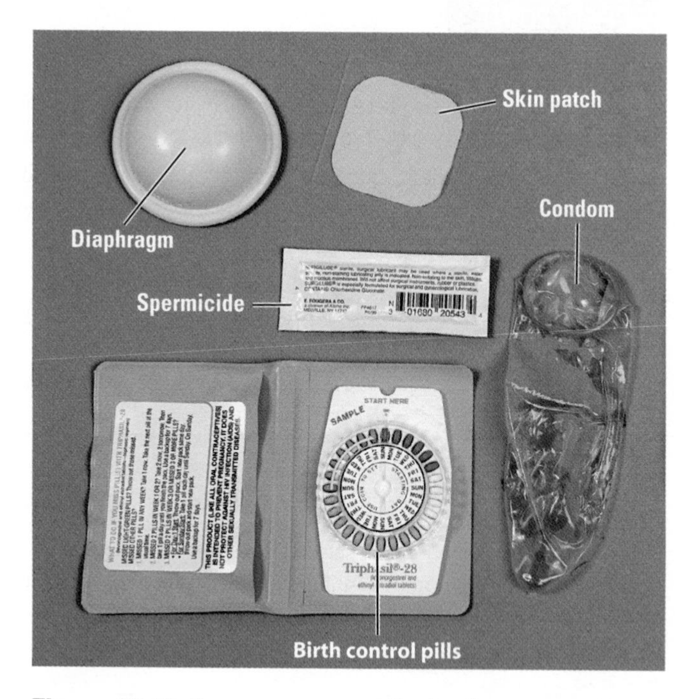

Figure 26.10 Some common methods of contraception.

Some of the most effective methods of contraception prevent the release of gametes. Oral contraceptives, or **birth control pills,** used by millions of women since the 1960s, come in several different forms that contain synthetic estrogen and/or progesterone. In addition to pills, various combinations of these hormones are also available as a shot, a ring inserted into the vagina, or a skin patch. Steady intake of these hormones simulates their constant levels during pregnancy. In response, the hypothalamus fails to send the signals that start development of an ovarian follicle. Ovulation ceases, which prevents pregnancy. (Although an active area of research, there currently is no chemical contraceptive available that prevents the production or release of sperm.)

There are also drugs available that can prevent fertilization or implantation after unprotected intercourse has already occurred. High doses of synthetic estrogen and progesterone can be taken as emergency contraception, often called **morning after pills (MAPs)**. If taken within 3 days of unprotected intercourse, MAPs may disrupt normal hormone signaling enough to prevent fertilization or to prevent a fertilized egg from implanting in the uterus. Such treatments should only be used in emergency situations because they have significant side effects. In 2006, the Food and Drug Administration approved the sale of "Plan B" brand MAPs without a prescription (over-the-counter). If pregnancy has already occurred, the drug RU-486 (mifepristone) can induce an abortion during the first 7 weeks of pregnancy. RU-486 requires a doctor's prescription and several visits to a medical facility.

"Safe sex" involves more than preventing unwanted pregnancies. It also involves preventing the spread of sexually transmitted diseases, the focus of the next section.

Sexually Transmitted Diseases

Sexually transmitted diseases (STDs) are contagious diseases spread by sexual contact. AIDS is caused by HIV (discussed in Chapters 10 and 24); genital herpes and genital warts are caused by other viruses. Viral STDs are generally not curable. They can be controlled by medications, but symptoms and the ability to infect others remain a possibility throughout a person's lifetime. Other STDs are caused by bacteria (chlamydial infections, gonorrhea, and syphilis), protozoans (trichomoniasis), and fungi (candidiasis, or yeast infection); these are all generally curable with drugs, especially if diagnosed early.

Sexually transmitted diseases often cause no apparent symptoms. But an infected person who feels fine may still be capable of infecting partners. Many STDs cause long-term problems if left untreated, and some can be fatal. Anyone who is sexually active should therefore have regular checkups for STDs and seek immediate medical help if any suspicious symptoms appear, even if they are mild. **Table 26.2** lists the STDs most common in the United States, along with their symptoms.

Table 26.2 STDs Common in the United States

Disease	Microbial Agent	Major Symptoms and Effects	Treatment
Bacterial			
Chlamydial infections	*Chlamydia trachomatis*	Genital discharge, itching, and/or painful urination; often no symptoms in women; pelvic inflammatory disease (PID)	Antibiotics
Gonorrhea	*Neisseria gonorrhoeae*	Genital discharge; painful urination; sometimes no symptoms in women; PID	Antibiotics
Syphilis	*Treponema pallidum*	Ulcer (chancre) on genitalia in early stages; spreads throughout body and can be fatal if not treated	Antibiotics can cure in early stages
Viral			
Genital herpes	Herpes simplex virus type 2, occasionally type 1	Recurring symptoms: small blisters on genitalia, painful urination, skin inflammation; linked to cervical cancer, miscarriage, birth defects	Valacyclovir can prevent recurrences
Genital warts	Papillomaviruses	Painless growths on genitalia; some of the viruses linked to cancer	Removal by freezing
AIDS and HIV infection	HIV	See Chapter 24	Combination of drugs
Protozoan			
Trichomoniasis	*Trichomonas vaginalis*	Vaginal irritation, itching, and discharge; usually no symptoms in men	Antiprotozoan drugs
Fungal			
Candidiasis (yeast infections)	*Candida albicans*	Similar to symptoms of trichomoniasis; frequently acquired nonsexually	Antifungal drugs

STDs are most prevalent among teenagers and young adults; nearly two-thirds of infections occur among people under 25. The best way to avoid both unwanted pregnancy and the spread of STDs is, of course, abstinence. For sexually active people, latex condoms provide the best dual protection for "safe sex."

CHECKPOINT

1. __________ is to males as tubal ligation is to females.
2. STDs caused by __________ are generally incurable.
3. Besides limiting sexual contact, what is the best method of preventing the spread of STDs?

Answers: 1. Vasectomy **2.** viruses **3.** Latex condoms

Human Development

Embryonic development begins with **fertilization,** or conception, the union of a sperm and egg to form a zygote. In this section, we will examine the process of fertilization and the subsequent development of a human organism.

Fertilization

Copulation releases hundreds of millions of sperm into the vagina, but only a few hundred survive the several-hour trip to the egg in the oviduct **(Figure 26.11)**. Of these sperm, only a single one can enter and fertilize the egg. All the other millions of sperm die.

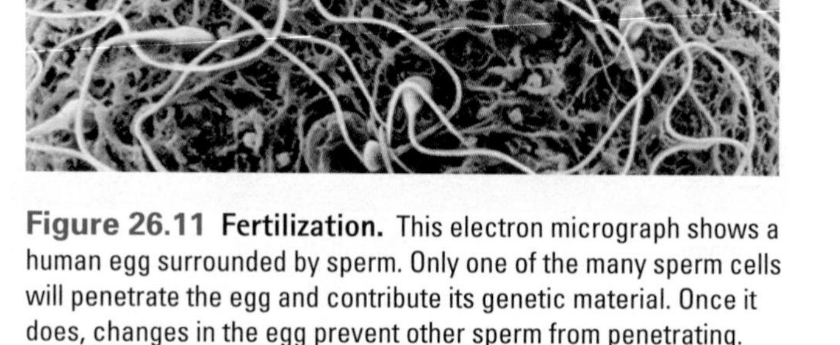

Figure 26.11 Fertilization. This electron micrograph shows a human egg surrounded by sperm. Only one of the many sperm cells will penetrate the egg and contribute its genetic material. Once it does, changes in the egg prevent other sperm from penetrating.

Repeating a theme we've seen again and again during our exploration of anatomy, the shape of sperm is related to what they do (form fits function). A mature human sperm has a streamlined shape that enables it to swim through fluids in the vagina, uterus, and oviduct **(Figure 26.12)**. The sperm's thick head contains a haploid nucleus and is tipped with a membrane-enclosed sac called the **acrosome.** The middle of the sperm contains mitochondria that use high-energy nutrients, especially the sugar fructose, from the semen to fuel movement of the flagellum (tail).

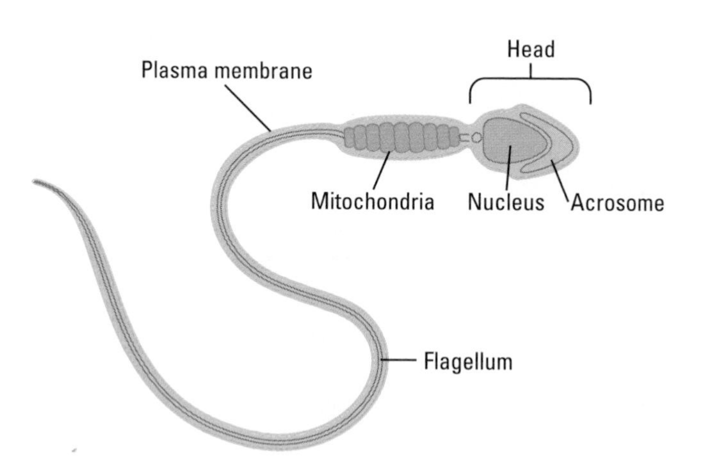

Figure 26.12 A human sperm cell. A human sperm consists of a head, which carries the genetic material in the nucleus, a middle piece with energy-producing mitochondria, and a powerful tail used for swimming.

Figure 26.13 traces one sperm through the events of fertilization. The sperm ❶ approaches and then ❷ contacts the jelly coat (red) that surrounds the egg. The acrosome in the sperm's head releases a cloud of enzymes that digest a hole in the jelly. This allows the sperm head to ❸ fuse its plasma membrane with that of the egg. Fusion of the two membranes makes it possible for the sperm nucleus to ❹ enter the cytoplasm of the egg. Fusion also triggers completion of meiosis II in the egg. Furthermore, contact of sperm with egg triggers a change in the egg's plasma membrane that makes it impenetrable to other sperm cells. This ensures that the zygote that is forming contains only the diploid number of chromosomes. The chromosomes of the egg and sperm nuclei ❺ are eventually

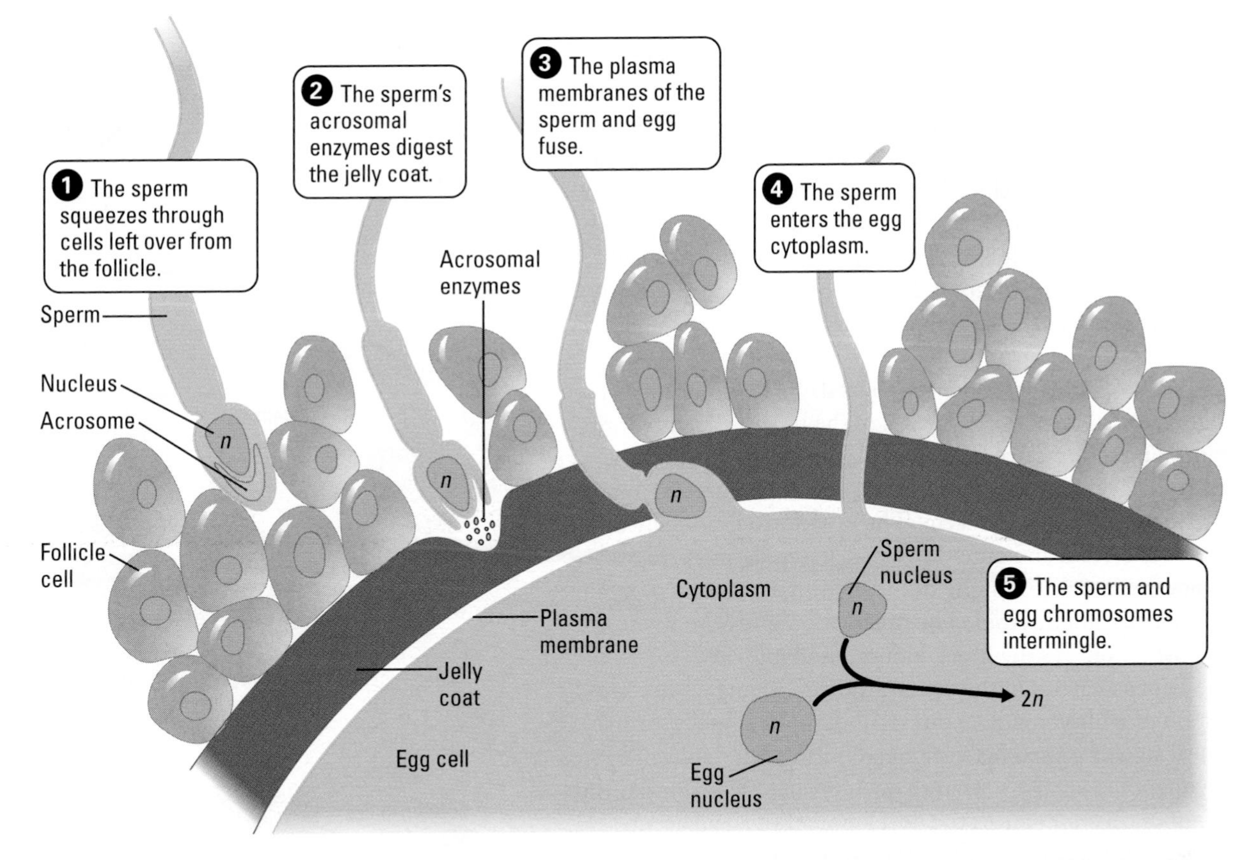
1 The sperm squeezes through cells left over from the follicle.
2 The sperm's acrosomal enzymes digest the jelly coat.
3 The plasma membranes of the sperm and egg fuse.
4 The sperm enters the egg cytoplasm.
5 The sperm and egg chromosomes intermingle.
Sperm
Nucleus
Acrosome
Follicle cell
Acrosomal enzymes
n
Plasma membrane
Jelly coat
Egg cell
Cytoplasm
Sperm nucleus
Egg nucleus
2n

Figure 26.13 The process of fertilization.

enclosed in a single diploid nucleus. In the diploid zygote, the egg's metabolic machinery awakens from dormancy and gears up in preparation for the enormous growth and development that will soon follow.

Basic Concepts of Embryonic Development

A single-celled human zygote formed by fertilization is smaller than the period at the end of this sentence. From this humble start, the zygote develops into a full-blown organism with trillions of cells organized into complex tissues and organs. Clearly, this process requires an astonishing amount of cell division and differentiation to produce all the specialized cells of the adult. The key to development in all organisms is that each stage of development takes place in a highly organized fashion.

Development begins with **cleavage,** a series of rapid cell divisions that results in a multicellular ball. After the zygote divides for the first time, about 36 hours after fertilization, it is called an embryo. Rarely, and apparently at random, the two cells of the early embryo separate from each other. When this happens, each cell may "reset" and act as a zygote; the result is the development of identical (monozygotic) twins. (Nonidentical, or dizygotic, twins result from a completely different mechanism: Two separate eggs fuse with two separate sperm to produce two genetically unique zygotes that develop in parallel.)

During cleavage, DNA replication, mitosis, and cytokinesis occur rapidly, but the total amount of cytoplasm remains unchanged. As a result, the overall size of the embryo does not grow; instead, each cell division partitions the embryo into twice as many smaller cells (**Figure 26.14**, top). Cleavage continues as the embryo moves down the oviduct toward the uterus. A central cavity begins to form in the embryo. About 6–7 days after fertilization, the embryo has reached the uterus as a fluid-filled hollow ball of about 100 cells called a **blastocyst.** Protruding into the central cavity on one side of the human blastocyst is a small clump of cells called the **inner cell mass,** which will eventually form the baby. Occasionally, an embryo does not travel down the oviduct to the uterus. The result is an ectopic pregnancy, one in which the embryo develops in the wrong location. Ectopic pregnancies are invariably fatal to the embryo and can be dangerous to the mother, requiring immediate medical attention.

The next stage of development, under way by 9 days after conception, is **gastrulation** (Figure 26.14, bottom). The cells of the inner cell mass begin an organized migration that produces the **gastrula,** an embryo with three main layers. The three layers produced in gastrulation are embryonic tissues called ectoderm, endoderm, and mesoderm. The **ectoderm** eventually develops into the nervous system and outer layer of skin (epidermis). The **endoderm** becomes the innermost lining of the digestive system and organs such as the liver, pancreas, and thyroid. The **mesoderm** gives rise to most other organs and tissues, such as the heart, kidneys, and muscles.

At this stage, various changes in the cells of the developing embryo result in the formation of embryonic structures. All developmental processes depend on chemical signals passed between neighboring cells and cell layers, telling embryonic cells precisely what to do and when. The mechanism by which one group of cells influences the development of an adjacent group of cells is called **induction.** Its effect is to switch on a set of genes whose expression makes the receiving cells differentiate into a specific cell or tissue type. For example, inductive signals can cause cells to change shape or to migrate from one location to another within the developing embryo. Another key developmental process is **programmed cell death,** a process that kills selected cells. For example, "suicide" genes encode proteins that kill certain cells in developing human hands and feet, separating the fingers and toes **(Figure 26.15)**.

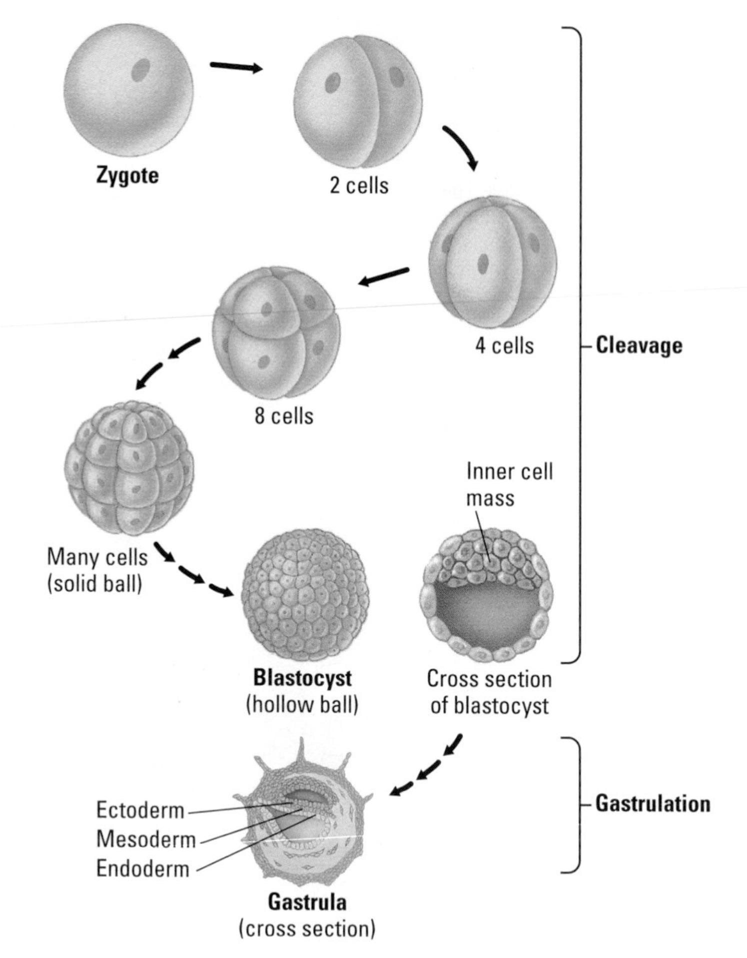

Figure 26.14 Early development of an embryo. A human zygote undergoes a series of rapid cell divisions to form a blastocyst. Notice that each round of division does not change the total size of the embryo. Instead, more and more smaller cells are formed. After the blastocyst is formed, further divisions and specializations of cells produce the three-layered gastrula.

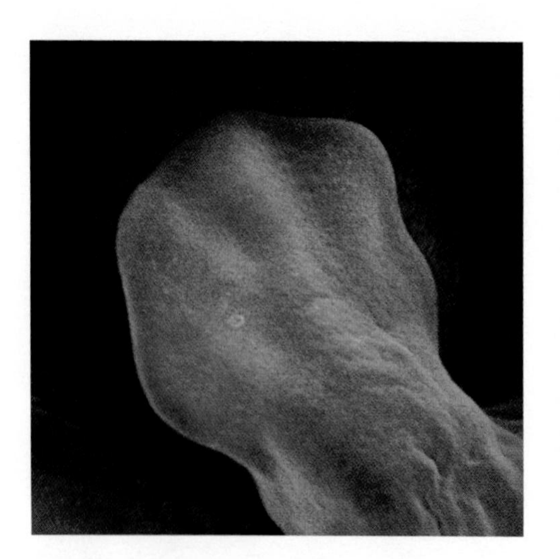

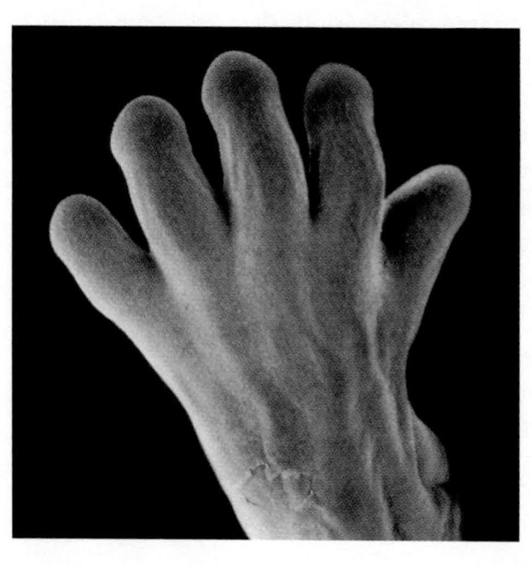

Figure 26.15 Programmed cell death in a developing human hand. Certain proteins produced during human embryonic development destroy cells in the early hand (left), creating the spaces between fingers in more mature hands (right).

During the course of embryonic development, a sequence of inductive signals leads to increasingly greater specialization of cells as organs begin to take shape. The importance of these processes is underscored by birth defects that result from improper signaling. Spina bifida is a condition that results from the failure of a tube of ectoderm cells to properly close and form the spine during the first month of fetal development. Infants born with spina bifida often have permanent nerve damage that results in paralysis of the lower limbs.

Pregnancy and Early Development

Pregnancy, or **gestation,** is the carrying of developing young within the female reproductive tract. It begins at fertilization and continues until the birth of the baby. In humans, gestation lasts about 266 days (38 weeks), but is usually measured as 40 weeks (9 months) from the start of the last menstrual cycle. This is a long time compared with some other mammals; gestation in mice, for instance, lasts only about a month. At the other extreme, elephants have a 22-month gestation period.

The early stages in human development are summarized in **Figure 26.16.** Fertilization and cleavage take place in the oviduct. About a week after conception, the embryo reaches the uterus and implants itself into the endometrium. By this time, the embryo has become a blastocyst, with a central fluid-filled cavity and an inner cell mass **(Figure 26.17a)**. These early embryonic cells are **stem cells,** with the potential to give rise to every type of cell in the body. (As discussed in Chapter 11, embryonic stem cells have great potential as medical tools.) The outer cell layer, called the **trophoblast (Figure 26.17b)**, becomes part of the **placenta,** the organ that provides nourishment and oxygen to the embryo and helps dispose of its metabolic wastes.

Figure 26.18 shows the embryo about a month after conception. Besides the growing embryo, there are now four pieces of life-support equipment: the amnion, the yolk sac, the allantois, and the chorion. The **amnion** is a fluid-filled sac that encloses and protects the embryo. The amnion usually breaks just before childbirth, and the amniotic fluid ("water") leaves the mother's body through her vagina. The **yolk sac** produces the embryo's first blood cells and its first gamete-forming cells in the gonads. The **allantois** forms part of the **umbilical cord**—the lifeline between the embryo and the placenta. The outermost membrane, the **chorion,** becomes part of the placenta.

The placenta contains **chorionic villi,** fingerlike outgrowths with embryonic blood vessels, closely associated with the blood vessels of the mother's endometrium. The chorionic villi absorb nutrients and oxygen from the mother's blood and pass these substances to the embryo. The villi also carry wastes from the embryo to the mother's bloodstream. (As discussed at the beginning of Chapter 9, a small sample of chorionic villus tissue can be removed for prenatal genetic testing.)

The placenta provides for other needs of the embryo as well. For example, it allows protective antibodies to pass from the mother to the fetus. Depending on what is circulating in the mother's blood, however, the placenta can also be a source of trouble. A number of viruses—the German measles virus, for example—can cross the placenta and cause disease. Most

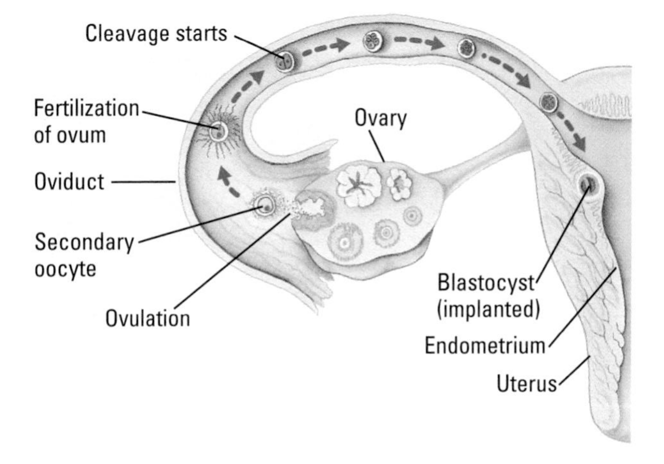

Figure 26.16 Early stages of human development. Conception, or fertilization, takes place in the oviduct. As the zygote travels down the oviduct, cleavage starts. By the time the embryo reaches the uterus, it has become a blastocyst, and it implants into the endometrial lining.

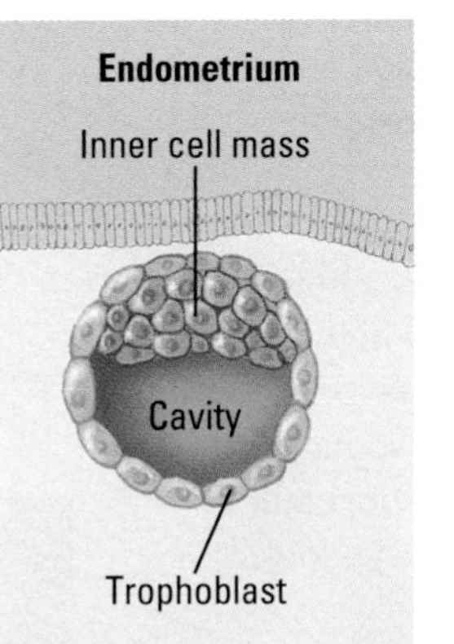

(a) Before implantation. By 6 days after conception, the embryo has developed into a blastocyst.

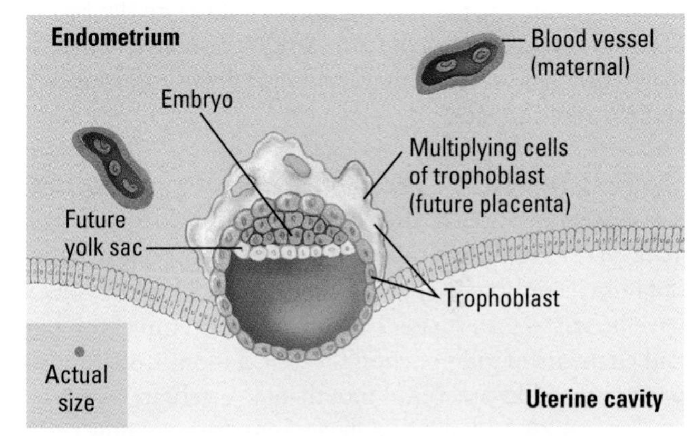

(b) Implantation under way. After implantation, the inner cell mass of embryonic stem cells will develop into the fetus. The outer layer of trophoblast cells will become the embryo's contribution to the placenta.

Figure 26.17 Blastocyst.

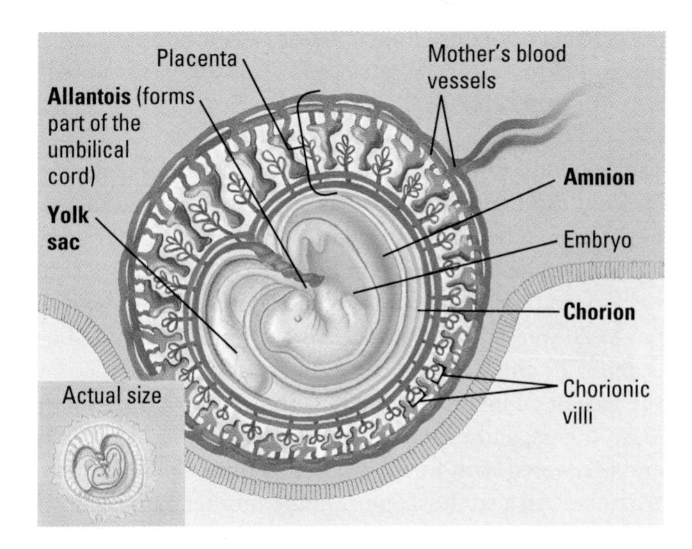

Figure 26.18 A 31-day-old embryo and its life-support equipment.

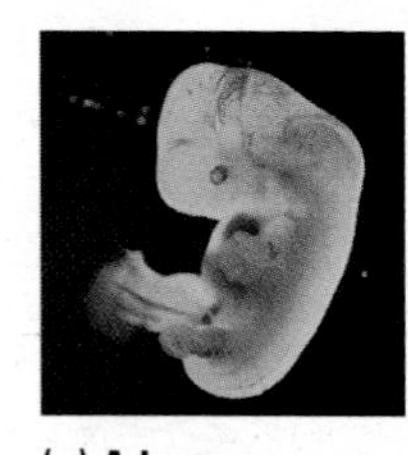

(a) A human embryo at 5 weeks

(b) At 9 weeks

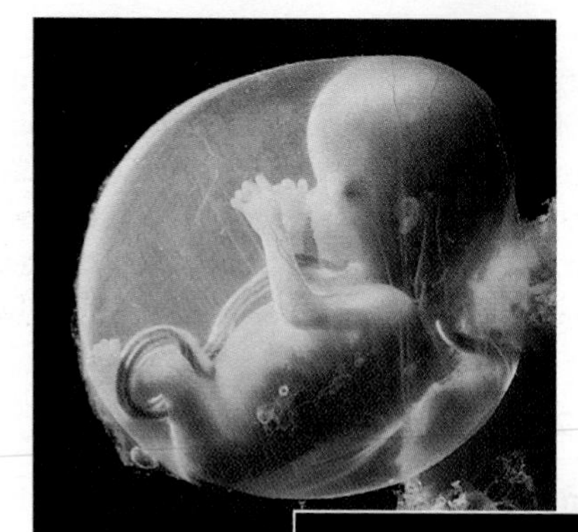

(c) At 14 weeks

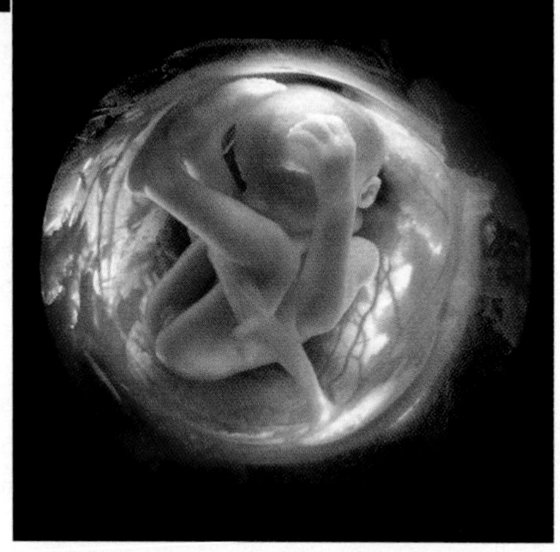

(d) At 20 weeks

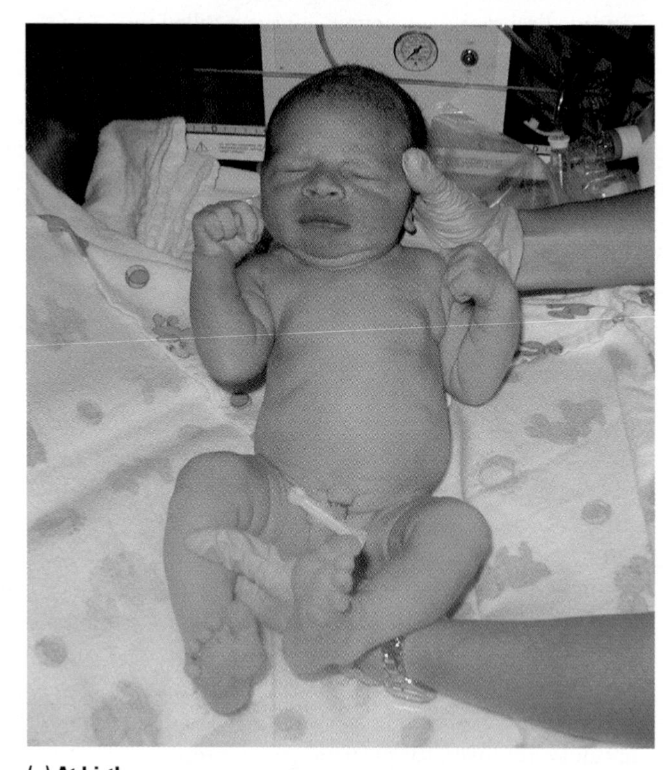

(e) At birth

Figure 26.19 A developing human from 5 weeks until birth.

drugs, both prescription and not, also cross the placenta, and many can harm the embryo. For example, tobacco smoke and alcohol can cause developmental abnormalities and raise the chances of miscarriage.

The Stages of Pregnancy

In this section, we use photographs to illustrate the rest of human development as it takes place in the uterus. For convenience, we divide pregnancy (the period from conception to birth) into three **trimesters** of about 3 months each.

The First Trimester The photograph in **Figure 26.19a** shows a human embryo about 5 weeks after fertilization. In that brief time, this highly organized multicellular embryo has developed from a single cell. A month-old human embryo is about 7 mm (0.28 in.) long. Its brain and spinal cord have begun to take shape. It also has four stumpy limb buds, a short tail, and elements of gill pouches. Overall, a month-old human embryo looks pretty much like any other month-old vertebrate embryo.

Figure 26.19b shows a developing human, now called a fetus, about 9 weeks after fertilization. The large pinkish structure on the left is the placenta, attached to the fetus by the umbilical cord. The clear sac around the fetus is the amnion. The fetus is about 5.5 cm (2.2 in.) long and has all of its organs and major body parts, including muscles and the bones of the back and ribs. The limb buds have become tiny arms and legs with fingers and toes. By the end of the first trimester, the fetus looks like a miniature human being, albeit one with an oversized head. By this time, the sex of the fetus can be reliably determined by an ultrasound exam **(Figure 26.20)**.

The Second Trimester The main developmental changes during the second and third trimesters involve an increase in size and general refinement of the human features—nothing as dramatic as the changes of the first trimester. The photograph in **Figure 26.19c** shows a fetus at 14 weeks, 2 weeks into the second trimester. The fetus is now about 6 cm (2.4 in.) long.

At 20 weeks **(Figure 26.19d)**, well into the second trimester, the fetus is about 19 cm (7.6 in.) long, weighs about half a kilogram (1 lb), and has the face of an infant, complete with eyebrows and eyelashes. Its arms, legs, fingers, and toes have lengthened. It also has fingernails and toenails and is covered with fine hair. By this time, the fetal heartbeat is detectable with a stethoscope, and the mother can usually feel the fetus moving. Because of the limited space in the uterus, the fetus flexes forward into the so-called fetal position, with its head tucked against its knees. By the end of the second trimester, the fetus's eyes are open and its teeth are forming.

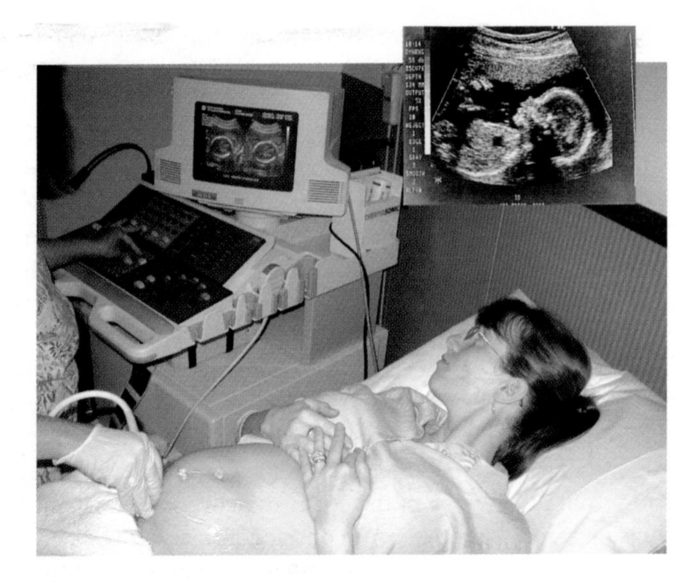

Figure 26.20 Ultrasound imaging. An ultrasound image is produced when high-frequency sounds from an ultrasound scanner held against a pregnant woman's abdomen bounce off the fetus. The inset image shows a fetus in the uterus at about 18 weeks.

The Third Trimester The third trimester is a time of rapid growth as the fetus gains the strength it will need to survive outside the protective environment of the uterus. Babies born prematurely—as early as 24 weeks—may survive, but they require special medical care after birth. During the third trimester, the fetus's circulatory system and respiratory system undergo changes that will allow the switch to air breathing. The fetus gains the ability to maintain its own temperature, and its bones begin to harden and its muscles thicken. It also loses much of its fine body hair, except on its head. The fetus usually rotates so that its head points down toward the cervix. The fetus becomes less active as it fills the space in the uterus. As the fetus grows and the uterus expands around it, the mother's abdominal organs may be squashed, causing frequent urination, digestive troubles, and backaches. At birth **(Figure 26.19e)**, babies average about 50 cm (20 in.) in length and 2.7–4.5 kg (6–10 lb) in weight.

Childbirth

The birth of a child is brought about by a series of strong, rhythmic contractions of the uterus called **labor.** As illustrated in **Figure 26.21**, hormones play a key role in inducing labor. One hormone, estrogen, reaches a high level in the mother's blood during the last weeks of pregnancy. An important effect of this estrogen is to trigger the formation of numerous oxytocin receptors on the uterus. Cells of the fetus produce the hormone oxytocin, and late in pregnancy, the mother's pituitary gland secretes it in increasing amounts. **Oxytocin** is a powerful stimulant for the smooth muscles in the wall of the uterus, causing them to

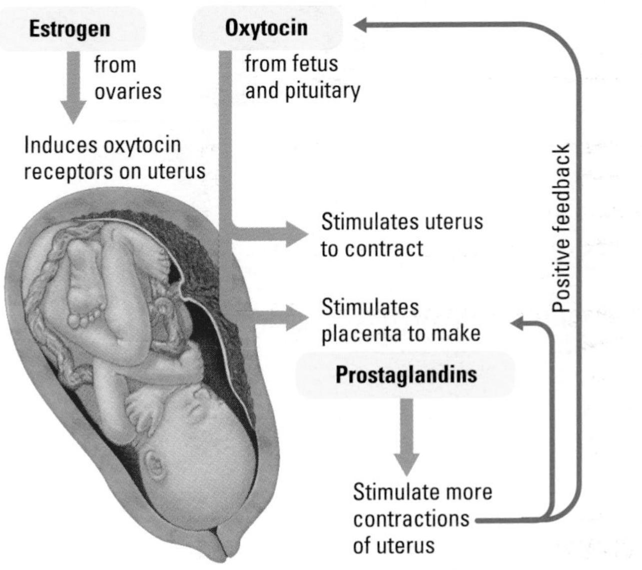

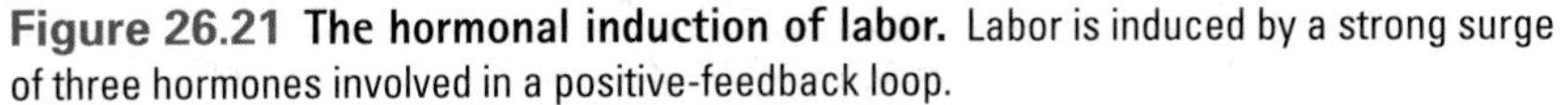

Figure 26.21 The hormonal induction of labor. Labor is induced by a strong surge of three hormones involved in a positive-feedback loop.

contract. It also stimulates the placenta to make prostaglandins, local tissue regulators that also stimulate uterine muscle cells, making the muscles contract even more.

The hormonal induction of labor involves positive-feedback control. Oxytocin and prostaglandins cause uterine contractions that in turn stimulate the release of more and more oxytocin and prostaglandins. The result is climactic—the intense muscle contractions that propel a baby from the womb. If a mother is overdue, or if labor has continued for a long time, a doctor may inject synthetic oxytocin to promote a more rapid birth.

Figure 26.22 shows the three stages of labor. ❶ The first stage, dilation, is the time from the onset of labor until the cervix (neck of the uterus) dilates, or opens, to a width of about 10 cm. Dilation is the longest stage of labor, typically lasting 6–12 hours or even considerably longer. ❷ The period from full dilation of the cervix to delivery of the infant is called the expulsion stage. Strong uterine contractions, lasting about 1 minute each, occur every 2–3 minutes, and the mother feels an increasing urge to push with her abdominal muscles. Within a period of 20 minutes to an hour or so, the infant is forced down and out of the uterus and vagina. After the baby is expelled, the umbilical cord is clamped and cut. The stump of the cord remains for several weeks, then shrivels and falls off, leaving the belly button. ❸ The final stage is the delivery of the placenta ("afterbirth"), usually within 15 minutes after the birth of the baby.

Hormones continue to be important after birth. Decreasing levels of progesterone and estrogen allow the uterus to start returning to its prepregnancy state. The pituitary hormone prolactin promotes milk production by the mammary glands. About 2–3 days after birth, the mother begins to secrete milk. During this delay, a yellowish protein-rich fluid called colostrum is secreted. Besides nutrients, colostrum contains antibodies that help protect the infant from infection.

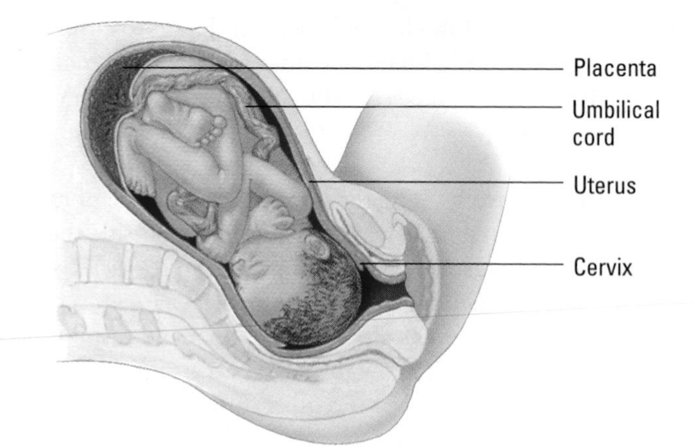

❶ Dilation of the cervix

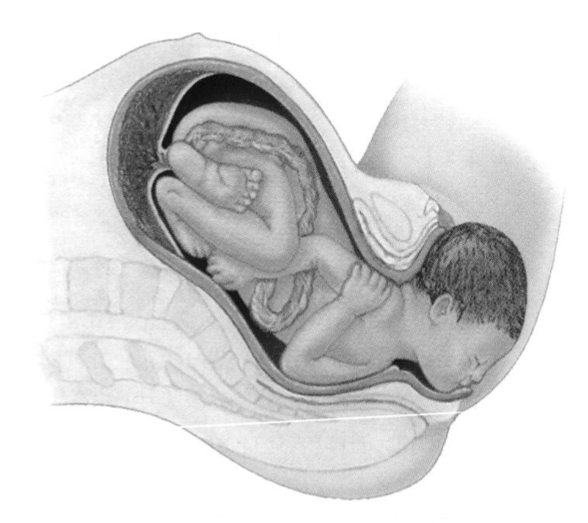

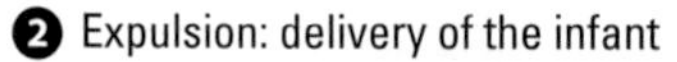

❷ Expulsion: delivery of the infant

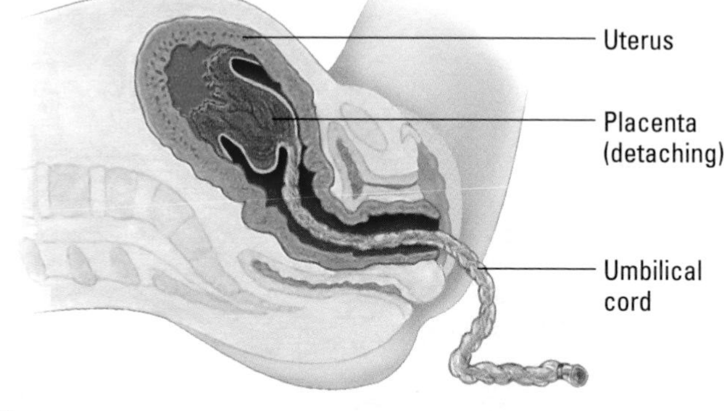

❸ Delivery of the placenta

Figure 26.22 The three stages of labor.

CHECKPOINT

1. In the process of fertilization, two haploid __________ fuse to form a diploid __________.
2. Place these steps of development in order: organ formation, fertilization, gastrulation, cleavage.
3. Would you expect cells of the blastocyst to get a greater or lesser supply of oxygen than the zygote? Why?
4. Why should a woman who is trying to get pregnant avoid drugs such as alcohol and nicotine?
5. Why is it important for the embryo that high levels of estrogen and progesterone be maintained in the mother's bloodstream?
6. Why isn't a newborn immediately susceptible to numerous infections?

Answers: 1. gametes (sperm and egg); zygote **2.** Fertilization, cleavage, gastrulation, organ formation **3.** Cells of the blastocyst will get a greater supply of oxygen because they are smaller than cells of the zygote. Smaller cells have a greater plasma membrane surface area relative to cell volume, facilitating diffusion of oxygen from the environment to the cell's cytoplasm. **4.** Drugs can pass to the developing embryo from the mother's bloodstream and affect it before the mother even knows she is pregnant. **5.** Falling levels of estrogen and progesterone would cause the endometrium to deteriorate, causing a spontaneous abortion of the embryo. **6.** Antibody proteins can pass from the maternal bloodstream to the baby, providing immune protection that lasts several months. Additional antibodies are obtained in colostrum.

Reproductive Technologies

In previous sections, we examined the normal means of human sexual reproduction. Many couples, however, are unable to conceive children because of one or more physical abnormalities. Today, there are reproductive technologies that can solve many of these problems.

Infertility

The first row of Table 26.1 on page 569 indicates that 85% of women become pregnant in one year without contraception. Put another way, about 15% of couples who want children are unable to conceive, even after a year of regular, unprotected sex. Such a condition is called **infertility,** and it can have several different causes.

In most cases, infertility can be traced to problems with the man. His testes may not produce enough sperm (a "low sperm count"). Even if a man produces enough sperm, they may not be vigorous enough to reach an egg, or they may be defective in some other way. In other cases, infertility is caused by **impotence,** also called erectile dysfunction, the inability to maintain an erection. Temporary impotence can result from alcohol or drug use or from psychological problems. Permanent impotence can result from nervous system or circulatory problems.

Female infertility can result from a lack of eggs, a failure to ovulate, or blocked oviducts (often caused by scars after contracting a sexually transmitted disease). Other women are able to conceive but cannot support a growing embryo in the uterus. The resulting multiple miscarriages can take a heavy emotional toll.

Reproductive technologies can help many cases of infertility. For a low sperm count, there is often a simple solution. Underproduction of sperm can frequently be traced to the man's scrotum being too warm, so a switch of underwear from briefs (which hold the scrotum close to the body) to boxers is often a first treatment. If that doesn't work, sperm can be collected, concentrated, and then injected into the woman's uterus via the vagina. Drug therapies (including Viagra) and penile implants can be used to treat impotence. If a man produces no functioning sperm, the couple may elect to use another man's sperm that has been donated to a sperm bank.

If a woman has normal eggs that are not being released properly, hormone injections can induce ovulation. As discussed in the opening essay, such treatments frequently result in multiple pregnancies. If a woman has no eggs of her own, they, too, can be obtained from a donor for fertilization and injection into the uterus. While sperm can be collected without any danger to the donor, the collection of eggs is a surgical procedure that involves pain and risk for the donating woman (see the opening essay in Chapter 8).

If a woman is able to become pregnant but cannot support a growing fetus, she and her partner may hire a surrogate mother. In such cases, the couple enters into a legal contract with another woman who agrees to carry the couple's child to birth. This method has worked for many couples, but serious ethical and legal problems can arise if a surrogate mother changes her mind and wants to keep the baby she has carried for nine months. A number of states have laws restricting surrogate motherhood.

In Vitro Fertilization

A procedure performed in vitro (literally, "in glass") happens under artificial laboratory conditions rather than within the living body (in vivo). Sometimes called creating a "test-tube baby," **in vitro fertilization (IVF)** begins with the surgical removal of eggs and the collection of sperm. The eggs are fertilized in a petri dish **(Figure 26.23)**, allowed to grow for several days into eight-cell embryos, and then carefully injected into a woman's uterus, where one or more of them may successfully implant and develop. The sperm and eggs used for this procedure, as well as the embryos created, can all be used immediately or frozen for later use.

Recent research has shown increased risks in the form of lower birth weights and higher rates of birth defects for babies born from IVF compared with those born from natural conception. Despite such risks and the high cost (around $10,000 for each attempt, whether it succeeds or not), IVF procedures are now routinely performed at medical centers throughout the world and result in the birth of thousands of babies each year.

One major difference between in vivo and in vitro fertilization is that technology offers choices that nature does not. For example, sperm can be sorted based on whether they contain an X or Y chromosome, increasing the likelihood of creating a zygote of a particular sex. Furthermore, a cell can be harmlessly removed from an eight-cell embryo and genetically tested for the presence of disease-causing (or other) genes. Parents can thus choose which of many embryos to implant. As analysis of the human genome progresses, the potential to screen for diseases and physical traits increases.

The concept of parenthood has been greatly complicated by modern reproductive technologies. Mix-ups in fertility clinics have resulted in the implantation of embryos into the wrong women. When this happens, who has what rights? Divorcing couples have sued each other for custody of frozen embryos. Sperm have been harvested from recently deceased men and used to create babies with the widow's eggs. Are such babies entitled to benefits as the dead man's children? An infertile couple could purchase eggs from a woman, sperm from a man, and then hire a surrogate to carry the baby. Who, then, are the baby's true parents? Each new reproductive technology raises moral and legal questions. Many of these have not yet been addressed, much less resolved, by our society. All of us need to understand the science behind these complex issues so that we can make informed decisions as citizen scientists and potential parents.

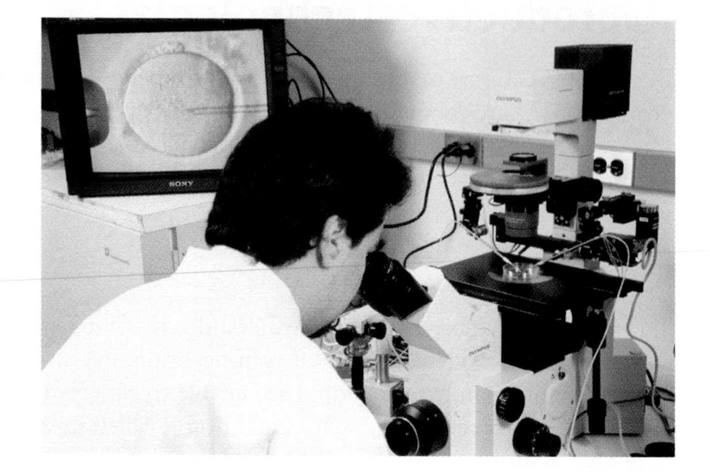

Figure 26.23 In vitro fertilization. In vitro fertilization involves the surgical removal of eggs and fertilization with sperm in a culture dish. In this photo, a technician is piercing an egg with a fine needle (visible on the monitor) to inject sperm.

CHECKPOINT

1. The procedure that creates a "test-tube baby" is ___________.
2. What is the difference between infertility and impotence?
3. Explain how up to five people can be involved in creating one baby.

Answers: 1. in vitro fertilization **2.** Infertility is the inability to produce offspring; impotence is the inability to perform sexual intercourse. **3.** A couple can hire a surrogate mother to carry an embryo created with gametes from a sperm donor and an egg donor.

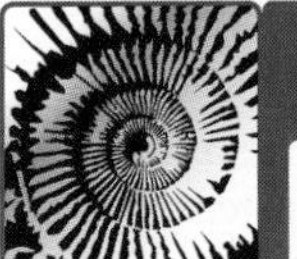

EVOLUTION CONNECTION

Menopause and the Grandmother Hypothesis

Around age 50, changes in hormone levels cause human females to undergo menopause—the cessation of ovulation and menstruation, marking the end of fertility. Even though menopause happens to all human females, it is an unusual feature in the animal kingdom; most species retain their reproductive capacity throughout life. After all, at 50, most of a woman's other body systems are working just fine. Why does the reproductive system shut down?

This question becomes even more intriguing when you consider it from an evolutionary standpoint. Evolutionary fitness is measured by the number of surviving, fertile offspring. Why might natural selection favor females who stopped reproducing?

One group of researchers has offered an intriguing yet controversial hypothesis. University of Utah anthropologist Kristen Hawkes and her colleagues propose that menopause actually increases a woman's evolutionary fitness in the long run. Humans, unlike almost every other species, continue to depend on their mothers to provide food and care for many years after weaning. Perhaps losing the ability to become pregnant allows a woman to focus her energy on caring for the children she has, rather than producing more that might not survive. Furthermore, menopause creates a class of postreproductive women who can contribute to raising grandchildren.

Hawkes and colleagues studied the Hadza, a tribe of hunter-gatherers in northern Tanzania. Among the Hadza, Hawkes noticed, elder women provide a significant amount of the food. Hawkes found that children with caring grandmothers were healthier, gained more weight, and grew faster. Because a woman shares one-half of her genes with each of her children and one-quarter with each grandchild, helping to raise two grandchildren will pass down as many genes to future generations as bearing one more child. Thus, even though menopause may cause a woman to have fewer children, Hawkes's "grandmother hypothesis" suggests that it may actually increase the number of closely related children who will themselves reach maturity, ensuring the continuation of the postmenopausal grandmother's genes. ■

Chapter Review

SUMMARY OF KEY CONCEPTS

For study help and activities, go to campbellbiology.com or the student CD-ROM.

Unifying Concepts of Animal Reproduction

- **Asexual Reproduction** In asexual reproduction, one parent produces genetically identical offspring by binary fission, budding, fission, or fragmentation. Asexual reproduction enables a single individual to produce many offspring rapidly, but the resulting genetically homogeneous population may be less able to survive environmental changes.

- **Sexual Reproduction** Sexual reproduction involves the fusion of gametes (sperm and ova) from two parents to form a zygote. Sexual reproduction increases the variation among offspring, which may enhance reproductive success in changing environments.

Human Reproduction

- **Female Reproductive Anatomy** The human reproductive system consists of a pair of gonads, ducts that carry gametes, and structures for copulation. A woman's ovaries contain follicles that nurture eggs and produce sex hormones. Oviducts convey eggs to the uterus, where a fertilized egg develops into a fetus. The uterus opens into the vagina, which receives the penis during intercourse and forms the birth canal.

Activity *Reproductive System of the Human Female*

- **Male Reproductive Anatomy** Located in an external sac called the scrotum, a man's gonads (the testes) produce sperm, which are expelled through ducts during ejaculation. Several glands contribute to the formation of fluid that carries, nourishes, and protects sperm. This fluid and the sperm constitute semen.

Activity *Reproductive System of the Human Male*

Case Study in the Process of Science *What Might Obstruct the Male Urethra?*

- **Gametogenesis**

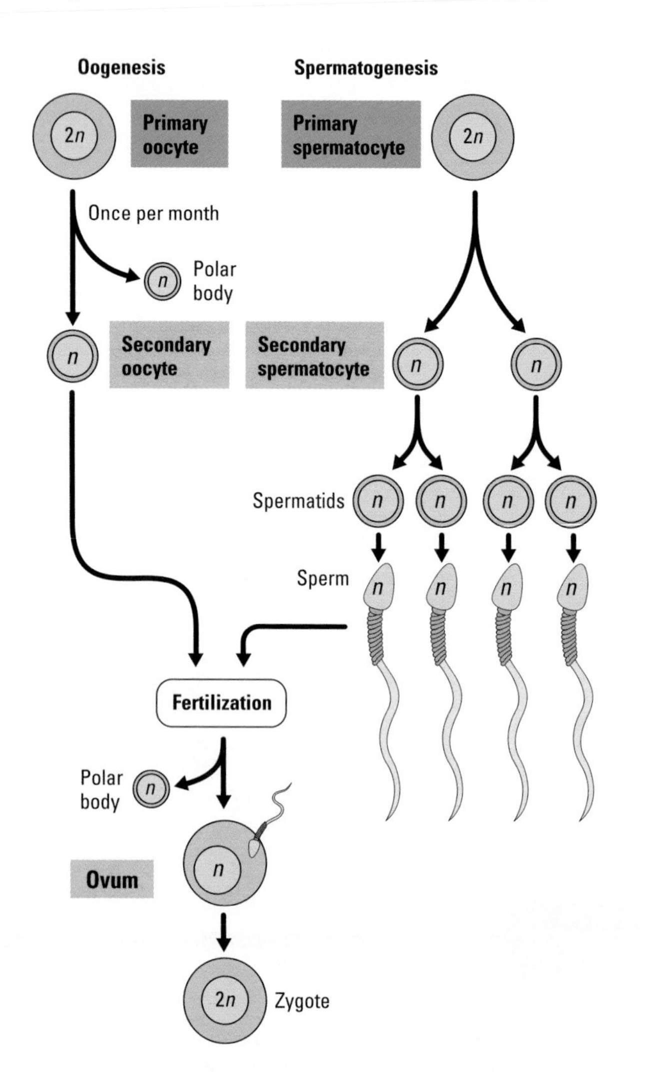

- **The Female Reproductive Cycle** Hormones synchronize cyclic changes in the ovaries and uterus. Approximately every 28 days, the hypothalamus signals the anterior pituitary to secrete FSH and LH, which trigger the growth of a follicle and ovulation, the release of an egg. The follicle secretes estrogen; after ovulation, the corpus luteum secretes both estrogen and progesterone. These two hormones stimulate the endometrium (uterine lining) to thicken, preparing the uterus for implantation. They also inhibit the hypothalamus, reducing FSH and LH secretion. If the egg is not fertilized, the drop in LH triggers menstruation, the breakdown of the endometrium. The hypothalamus and pituitary then stimulate another follicle, starting a new cycle. If fertilization occurs, a hormone from the embryo maintains the uterine lining and prevents menstruation.

MP3 Tutor *Female Reproductive Cycle*

Reproductive Health

- **Contraception** Contraception is the deliberate prevention of pregnancy. Different forms of contraception block different steps in the process of becoming pregnant and have different levels of reliability. Particularly effective methods of contraception include sterilization (tubal ligation for women and vasectomy for men), barrier methods (condom, diaphragm, or cervical cap in conjunction with a spermicide), and oral contraceptives.

- **Sexually Transmitted Diseases** STDs are contagious diseases that can be spread by sexual contact. Viral STDs (AIDS, genital herpes, genital warts) cannot be cured, whereas STDs caused by bacteria, protozoans, and fungi are generally curable with drugs. The use of latex condoms can prevent STDs.

Human Development

- **Fertilization** During fertilization, the union of sperm and egg to form a zygote, a sperm releases enzymes that pierce the egg's membrane; then the sperm and egg plasma membranes fuse, and the two cells unite. Changes in the egg membrane prevent entry of additional sperm, and the fertilized egg is stimulated to develop into an embryo.

- **Basic Concepts of Embryonic Development**

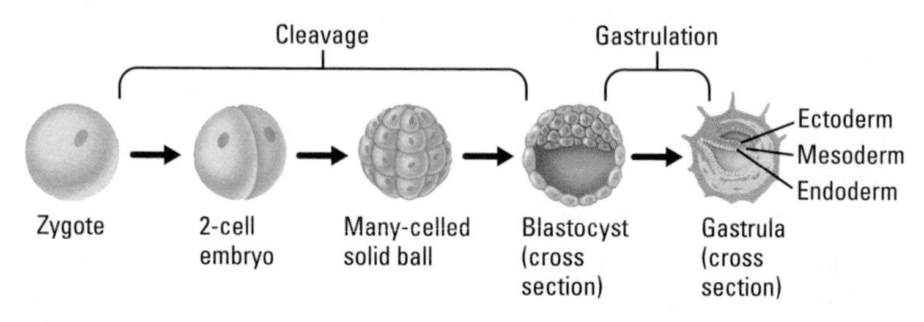

After gastrulation, the three embryonic tissue layers give rise to specific organ systems. In a process called induction, adjacent cells and cell layers influence each other's differentiation via chemical signals. Tissues and organs take shape in a developing embryo as a result of cell shape changes, cell migration, and programmed cell death.

Activity *Sea Urchin Embryonic Development*

Case Study in the Process of Science *What Determines Cell Differentiation in the Sea Urchin?*

Activity *Frog Embryonic Development*

- **Pregnancy and Early Development** Pregnancy, measured as 40 weeks from the start of the last menstrual cycle, is the carrying of the developing young in the uterus. Human development begins with fertilization and cleavage in the oviduct. After about 1 week, the embryo implants itself in the uterine wall. Four structures develop that assist the growing embryo: the amnion, the chorion, the yolk sac, and the allantois.

- **The Stages of Pregnancy** Human embryonic development is divided into three trimesters of about 3 months each. The most rapid changes occur during the first trimester. By 9 weeks, all organs are formed, and the embryo is called a fetus. The second and third trimesters are times of growth and preparation for birth.

- **Childbirth** Hormonal changes induce birth. Estrogen makes the uterus more sensitive to oxytocin, which acts with prostaglandins to initiate labor. The cervix dilates, the baby is expelled by strong muscular contractions, and the placenta follows. Prolactin then stimulates milk secretion.

Reproductive Technologies

- **Infertility** Infertility, the inability to have children after one year of trying, is most often due to problems in the man, such as underproduction of sperm. Female infertility can arise from a lack of eggs or a failure to ovulate. Technologies can help treat many forms of infertility.

- **In Vitro Fertilization**

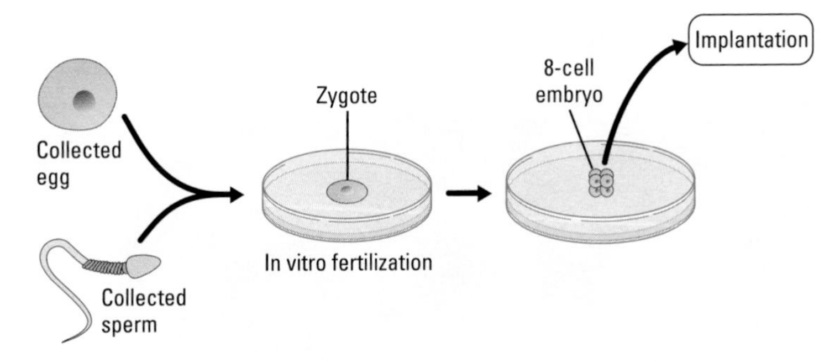

IVF offers choices that the natural pathway does not, raising moral and legal issues.

SELF-QUIZ

1. Some animals, such as rotifers and aphids, are able to alternate between sexual and asexual reproduction. Under what conditions might it be advantageous to reproduce asexually? Sexually?

2. A man is concerned about his fertility. He seeks the help of a fertility specialist, who tells him that he is producing a normal amount of sperm, but his semen does not seem to contain enough of the fluids needed to nourish the sperm. Which structures are most likely responsible for this problem?

3. Match each reproductive structure with its description:

a. uterus	1. female gonad
b. vas deferens	2. sites of spermatogenesis
c. oviduct	3. site of fertilization
d. ovary	4. site of gestation
e. endometrium	5. lining of uterus
f. testes	6. sperm duct

4. A woman has had several miscarriages. Her doctor suspects that a hormonal insufficiency has been causing the lining of the uterus to break down as it does during menstruation, terminating her pregnancies. Treatment with which of the following might help her remain pregnant?
   a. oxytocin
   b. HCG
   c. follicle-stimulating hormone
   d. prolactin

5. Why is it important that the hormones FSH and LH are inhibited after ovulation has occurred in the female cycle? Which female hormones inhibit FSH and LH?

6. What advantage do abstinence and condoms have over other forms of contraception?

7. How does a zygote differ from an ovum?
   a. A zygote has more chromosomes.
   b. A zygote is smaller.
   c. A zygote consists of more than one cell.
   d. A zygote is much larger.

8. If a baby is born missing the outer layer of skin, which of the following germ layers would have been the most likely site of damage in the embryo?
   a. endoderm   c. ectoderm
   b. mesoderm   d. stem cells

9. In an embryo, nerve cells grow out from the spinal cord and form connections with the muscles they will eventually control. What mechanisms described in this chapter might explain how these cells "know" where to go and which cells to connect with?

10. A pregnant female is two weeks past her due date, and the doctor decides to induce her labor. The natural hormone that would be needed is __________. The receptors for this hormone develop in the presence of __________, another hormone.

Answers to the Self-Quiz questions can be found in Appendix D.

Go to the website or CD-ROM for more Self-Quiz questions.

THE PROCESS OF SCIENCE

11. A typical egg cell survives about 24 hours, but some may survive up to 48 hours. A typical sperm survives 48 hours, but some may live up to 5 days. Assume that a female ovulates on day 14 of her cycle. How many days prior to ovulation and after ovulation is there a chance for pregnancy to occur? What is the total fertility window for this female?

12. There is evidence that a male is infertile. His physician is trying to determine the cause of his infertility. Based on what you know about the male reproductive system, what would be the most logical things to test for?

BIOLOGY AND SOCIETY

13. New technology has made it possible for doctors to save a small percentage of babies born 16 weeks prematurely. A baby born this early weighs just over 1 pound and faces months in an intensive-care nursery. The cost for care may be hundreds of thousands of dollars per infant. Some people wonder whether such a huge technological and personnel investment should be devoted to such a small number of babies. They feel that limited resources might be better directed at providing prenatal care that could prevent many premature births. What do you think? In what ways should financial considerations have an impact on prenatal medical decisions?

14. Some infertile couples hire a surrogate mother to have their biological child. When couples enter into an agreement with a surrogate mother, all parties sign a contract. But there have been cases where a surrogate mother signed a contract but then decided not to give up the baby to the couple who hired her. Do you think a surrogate mother should have the right to change her mind and keep the baby, even though it is not genetically hers? In some cases, the surrogate will also donate her eggs to be fertilized by the male's sperm, making the baby genetically hers. Should she have a right to change her mind after she has signed a contract and the baby is born? How should judges decide these cases? Can you suggest any laws or regulations for such situations?

15. When a couple uses in vitro fertilization to produce a baby, they are faced with some novel choices. Typically, more embryos are produced than will be used during any one procedure. Thus, a subset of prepared embryos has to be chosen for implantation. How should the decisions about which embryos to use be made? Should parents have the right to choose embryos based on the presence or absence of disease-causing genes? What about the sex of the embryo? Should parents be able to choose embryos for implantation based on any criteria they like? How would you distinguish acceptable criteria from unacceptable ones? Do you think such options should be legislated?

27 Nervous, Sensory, and Motor Systems

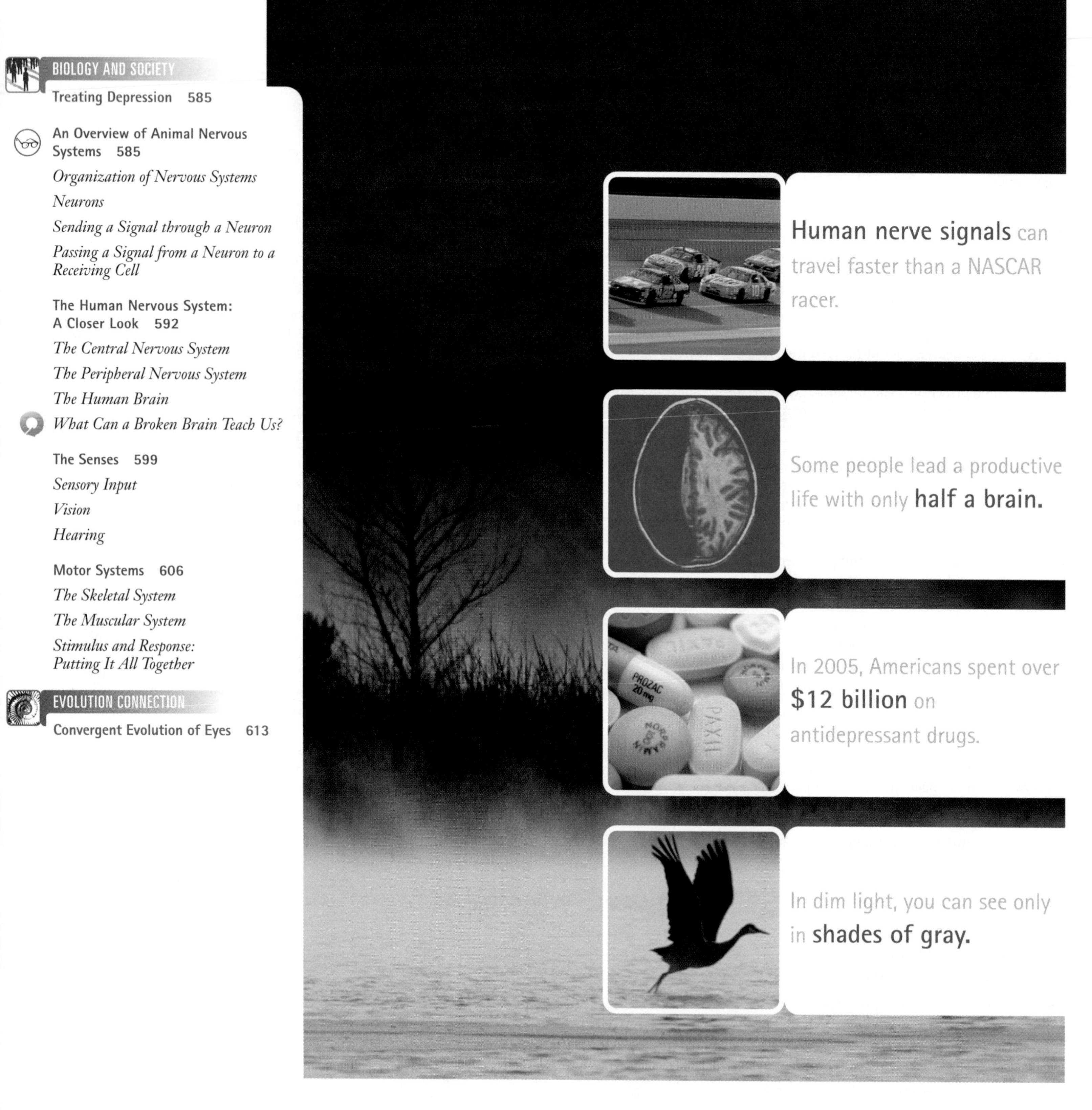
Human nerve signals can travel faster than a NASCAR racer.
Some people lead a productive life with only half a brain.
PROZAC 20 mg
PAXIL
In 2005, Americans spent over $12 billion on antidepressant drugs.
In dim light, you can see only in shades of gray.

BIOLOGY AND SOCIETY

Treating Depression

Nearly 20 million American adults are affected by depression, about two-thirds of them women. Two broad forms of depressive illness have been identified: major depression and bipolar disorder. People with **major depression** may experience persistent sadness, loss of interest in pleasurable activities, changes in body weight and sleep patterns, loss of energy, and suicidal thoughts. While all of us feel sad from time to time, major depression is extreme and more persistent, leaving the sufferer unable to live a normal life. If left untreated, symptoms may become more frequent and severe over time.

Bipolar disorder, or manic-depressive disorder, involves extreme mood swings. The manic phase is characterized by high self-esteem, increased energy, a flow of ideas, and extreme talkativeness, as well as behaviors that often bring disaster, such as increased risk taking, promiscuity, and reckless spending. In its milder forms, this phase is sometimes associated with great creativity, and some well-known artists, musicians, and literary figures (including Keats, Tolstoy, and Hemingway) have had periods of intense output during their manic phases. The depressive phase is marked by sleep disturbances, feelings of worthlessness, and decreased ability to experience interest and pleasure.

In recent years, researchers have begun to learn how brain physiology is involved in depression (**Figure 27.1**). Many depressed people have an imbalance of neurotransmitters, chemicals that help brain cells communicate with each other. A low level of one particular neurotransmitter, serotonin, is frequently associated with depression. For this reason, serotonin is the target of the most commonly prescribed class of antidepressant medications, called selective serotonin reuptake inhibitors (SSRIs). As the name implies, SSRIs increase the amount of time that serotonin is available to stimulate neurons in the brain. Such extra stimulation appears to relieve some symptoms of depression. In the United States, the number of prescriptions for SSRIs—such as fluoxetine (Prozac), paroxetine (Paxil), and sertraline (Zoloft)—has increased dramatically over the last 15 years.

Antidepressants like SSRIs have helped millions of people live healthier, happier lives by altering the level of a chemical in the brain. As you will see in this chapter, the brain is the hub of an intricate network of structures that detect, integrate, and respond to stimuli from the environment. By examining the structure and function of the nervous, sensory, and motor systems, we will explore how stimuli are translated into responses within the body. We'll begin by focusing on the processing center, the nervous system. Next we'll look at the sensory structures that supply input. Finally, we'll see how the motor systems respond. Along the way, we'll consider the consequences to human health when these systems fail. ■

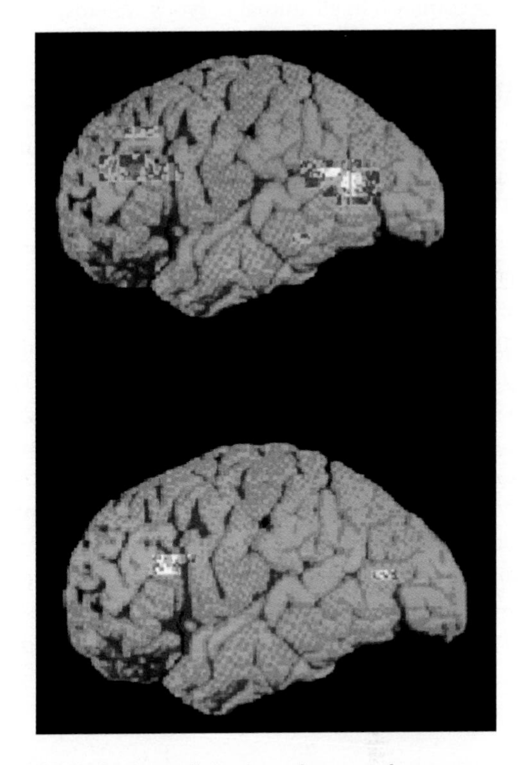

Figure 27.1 Brain activity in a depressed person versus a healthy person. These colored positron emission tomography (PET) scans compare the brains of a depressed patient (top) and healthy patient (bottom). The red and yellow colors indicate areas of low brain activity. Note that the PET scan from the depressed patient shows decreased activity in certain areas of the brain.

An Overview of Animal Nervous Systems

eTutor
How Neurons Work

The **nervous system** forms a communication and coordination network throughout an animal's body. Nervous systems are the most intricately organized data processing systems on Earth. A cubic centimeter of your brain, for instance, may contain well over 50 million **neurons,** nerve cells specialized for carrying electrical signals from one part of the body to another. Each neuron may communicate with thousands of other nerve cells, forming networks that enable us to learn, remember, perceive our surroundings, and move. In this section, we'll examine the organization and cellular basis common to all animal nervous systems.

Organization of Nervous Systems

The nervous systems of most animals have two main divisions. The **central nervous system (CNS)** consists of the brain and, in vertebrates, the spinal cord. The **peripheral nervous system (PNS)** is made up mostly of nerves that carry signals into and out of the CNS. A **nerve** is a communication line made from cable-like bundles of neuron fibers tightly wrapped in connective tissue.

Figure 27.2 highlights the three interconnected functions of the nervous system and the three types of neurons that carry out these functions. Sensory input is the conveyance of signals to the CNS from sensory receptors, such as light-detecting receptors in the eyes. **Sensory neurons** (shown in blue) convey this information. Integration is the interpretation of the sensory signals and the formulation of responses. **Interneurons** (green), located entirely within the CNS, perform this integration. Motor output is the conduction of signals from the integration centers to effectors, such as muscle cells. **Effectors** perform the body's responses (moving the leg in this example). **Motor neurons** (purple) function in motor output.

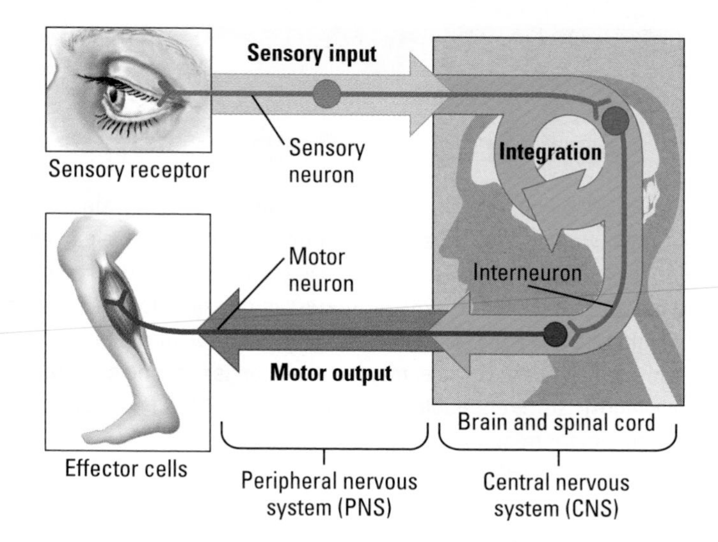

Figure 27.2 Organization of a nervous system. For simplicity, this figure shows only one neuron of each functional type, but body activity actually involves many sensory neurons, interneurons, and motor neurons. A simplified version of this figure will keep us oriented as we explore the nervous, sensory, and motor systems throughout the rest of the chapter.

Neurons

Neurons vary widely in shape, but most of them share some common features. **Figure 27.3** depicts a motor neuron, like those that carry signals

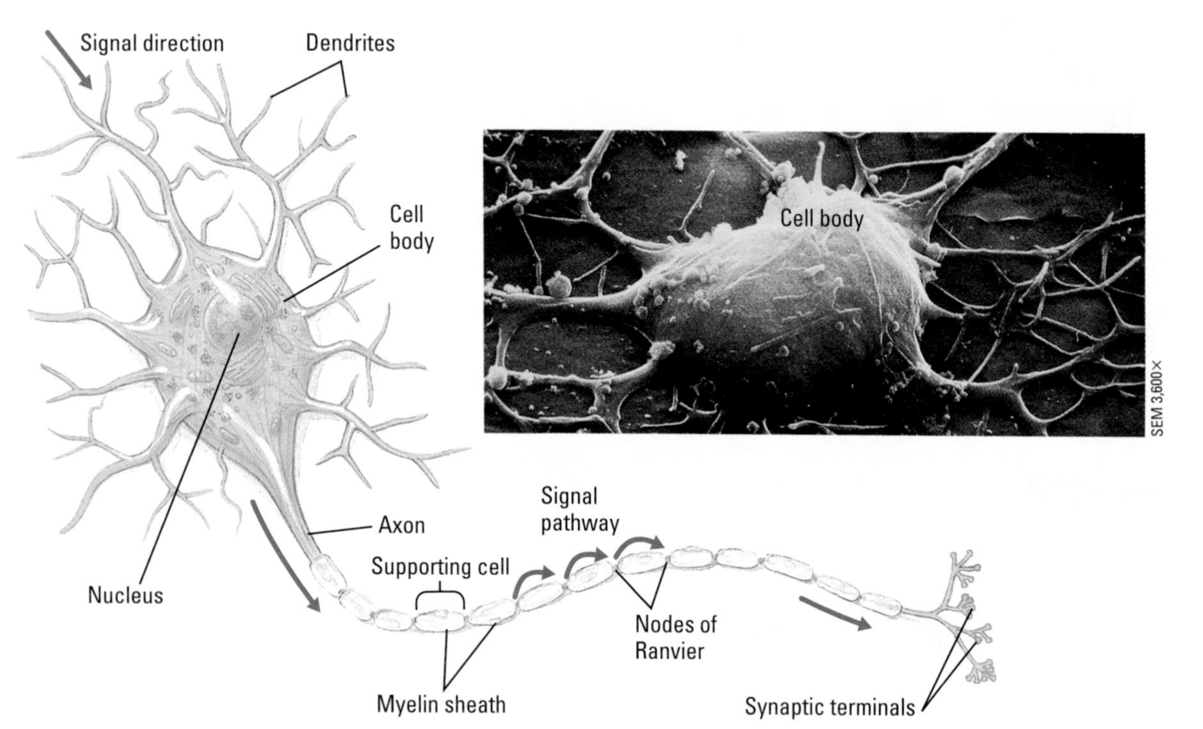

Figure 27.3 Structure of a motor neuron. This diagram and electron micrograph show a motor neuron. The large cell body holds the nucleus and organelles, while branched dendrites and a single axon project outward. The flow of the electrical signal through a neuron follows the path dendrite → cell body → axon. Supporting cells that surround axon fibers contain myelin, an insulating material that helps speed transmission by causing the signal to hop from node to node.

from your spinal cord to your skeletal muscles. A motor neuron has a large **cell body** housing the nucleus and other organelles. Two types of extensions project from the cell body. **Dendrites** are often short, numerous, and highly branched. Dendrites receive incoming messages from other cells or the environment and convey this information toward the cell body. The other type of extension, called an **axon,** conducts signals toward another neuron or toward an effector. Most neurons have a single axon, which can be quite long. Certain axons in your leg, for instance, stretch from the lower part of your spinal cord all the way to the muscles controlling your toes.

Neurons make up only part of a nervous system. Outnumbering neurons by as many as 50 to 1 are **supporting cells** that protect, insulate, and reinforce the neurons. In vertebrates, axons that convey signals very rapidly are enclosed along most of their length by an insulating material called the **myelin sheath.** The myelin sheath is a chain of bead-like supporting cells. The spaces between these cells are called **nodes of Ranvier.** The myelin sheath helps speed electrical transmission along an axon. In the human nervous system, signals can travel along a myelinated axon about 150 m/sec (over 330 mph), which means that a command from your brain can make your fingers move in just a few milliseconds.

The debilitating autoimmune disease multiple sclerosis (MS) leads to a gradual destruction of the myelin sheaths by the immune system. The result is a progressive loss of signal conduction and muscle control. MS has no cure, but drugs that suppress the immune system can relieve symptoms and slow its progress.

Returning to Figure 27.3, notice that the axon ends in a cluster of branches. A typical axon has hundreds or thousands of these branches, each with a bulb-like synaptic terminal at the very end. As we will see later, each **synaptic terminal** relays signals to another neuron or to an effector such as a muscle cell. With the basic structure of a neuron in mind, let's take a closer look at how neurons convey signals.

Sending a Signal through a Neuron

To understand nerve signals, we must first study a resting neuron, one that is not transmitting a signal. A resting neuron contains potential energy that can be put to work to send signals from one part of the body to another. This potential energy resides in an electrical charge difference across the neuron's plasma membrane. The cytoplasm just inside the membrane is negative in charge, and the fluid just outside the membrane is positive. Because opposite charges tend to move toward each other, the membrane stores energy by holding opposite charges apart, much like a battery does. The voltage (potential difference) across the plasma membrane of a resting neuron is called the **resting potential.**

What causes the resting potential? The answer lies with the membrane itself. The membrane keeps dissolved proteins and other large organic molecules inside the cell. Most of these molecules are negatively charged. In addition, the cell membrane has protein channels and pumps embedded within it that regulate the passage of positive ions. Together, the negative molecules trapped inside the cell and the regulation of ion movement create most of the resting potential.

The Action Potential Turning on a flashlight uses the energy stored in a battery to create light. In a similar way, stimulating a neuron's plasma membrane can trigger the release and use of the membrane's potential energy to generate an electrical nerve signal. A **stimulus** is any factor that causes a nerve signal to be generated. Examples of stimuli include light, sound, touching a hot surface, or a chemical signal from another neuron. A stimulus of sufficient strength can trigger an **action potential,** a self-propagating change in the voltage across the plasma membrane of a neuron. It is an action potential that constitutes the signal carried along a nerve cell.

Figure 27.4 shows the series of events involved in generating an action potential. ❶ At first, the membrane is at its resting potential, positively charged outside and negatively charged inside. ❷ A stimulus triggers the opening of a few ion channels in the membrane (represented by the blue arrows), allowing a few positive ions to enter the neuron. This tiny change makes the inside surface of the membrane slightly less negative than before. If the stimulus is strong enough, a sufficient number of protein channels open to reach the threshold. The **threshold** is the minimum change in a membrane's voltage that must occur to trigger the action potential. ❸ Once threshold is reached, additional channels open and more positive ions rapidly move in. As a result, the interior of the cell becomes positively charged with respect to the outside. ❹ This triggers the closing and inactivation of the channels. Meanwhile, another set of channels open (green arrows), allowing other positive ions to diffuse rapidly out, returning the membrane to its resting potential. Within a living neuron, this whole process takes just a few milliseconds.

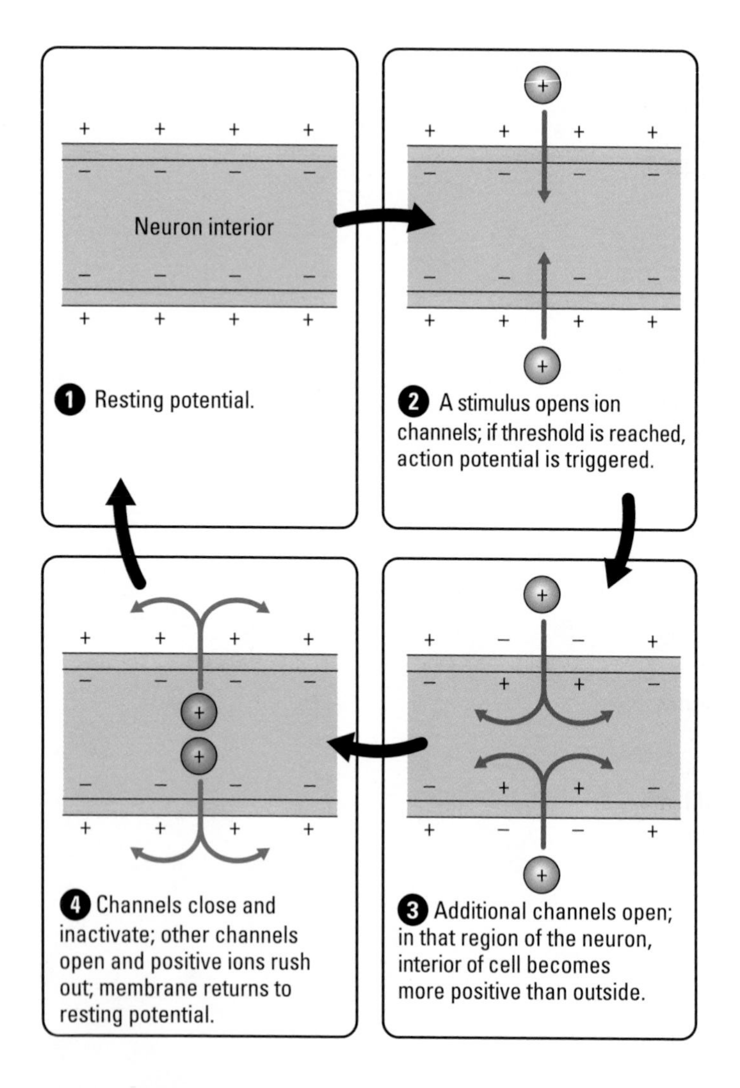

Neuron interior
❶ Resting potential.
❷ A stimulus opens ion channels; if threshold is reached, action potential is triggered.
❸ Additional channels open; in that region of the neuron, interior of cell becomes more positive than outside.
❹ Channels close and inactivate; other channels open and positive ions rush out; membrane returns to resting potential.

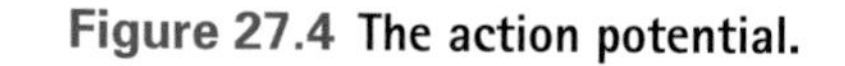
Figure 27.4 The action potential.

Propagation An action potential is a localized electrical event—a rapid change from the resting potential at a specific place along the neuron. A nerve signal starts out as an action potential generated near the cell body of the neuron. To function as a nerve signal, this local event must be passed along the neuron. This is like tipping the first of a row of standing dominoes: The first domino does not travel along the row, but its fall is relayed to the end of the row, one domino at a time.

Figure 27.5 shows the changes that occur in part of an axon at three successive times as a nerve signal passes from left to right. Let's start by focusing only on the axon region at the far left. ❶ When this region of the axon (blue) has its channels open, positive ions diffuse inward (blue arrows), and an action potential is generated. ❷ That same region then opens channels that allow other positive ions to diffuse out of the axon (green arrows) while closing the channels that allow ions in. ❸ A short time later, we would see no signs of an action potential at this (far left) spot because the axon membrane here has returned to its resting potential.

Now let's see how these events lead to the "domino effect" of a nerve signal. In step 1 of the figure, the blue arrows pointing sideways within the axon indicate local spreading of the electrical changes caused by the inflowing positive ions. These changes trigger the opening of channels in the membrane just to the right of the action potential. As a result, a second action potential is generated, as indicated by the blue region in step 2. In the same way, a third action potential is generated in step 3. At each point along the neuron, the action potential triggers changes in the adjacent region that result in propagation of the action potential along the neuron.

So why are action potentials propagated in only one direction along the axon (left to right in the figure)? As the blue arrows indicate, local electrical changes do spread in both directions in the axon. However, these changes cannot open channels and generate an action potential when the channels are inactivated (step 4 in Figure 27.4). Thus, an action potential cannot be generated in the regions where positive charge is leaving the axon (green in Figure 27.5).

Action potentials are all-or-none events; that is, they are the same no matter how strong or weak the stimulus that triggers them (as long as threshold is reached). How, then, do action potentials relay different intensities of information to the central nervous system? It is the frequency of action potentials that changes with the intensity of stimuli. If you rap your finger hard against the desk, for example, your CNS receives many more action potentials per millisecond than after a soft tap. Once your central nervous system receives information in the form of action potentials, it can process the information and formulate a response to it.

Passing a Signal from a Neuron to a Receiving Cell

If an action potential travels in one direction along a neuron, what happens when the signal arrives at the end of the neuron? To continue conveying information, the signal must be passed to another cell. This occurs at a **synapse**, or relay point, between a neuron and a receiving cell. The receiving cell can be another neuron or an effector cell such as a muscle cell or endocrine cell.

Chemical Synapses Synapses come in two varieties: electrical and chemical. In an electrical synapse, action potentials jump directly from one cell to the next. In the human body, electrical synapses are common in the heart and digestive tract, where steady, rhythmic muscle contractions are maintained.

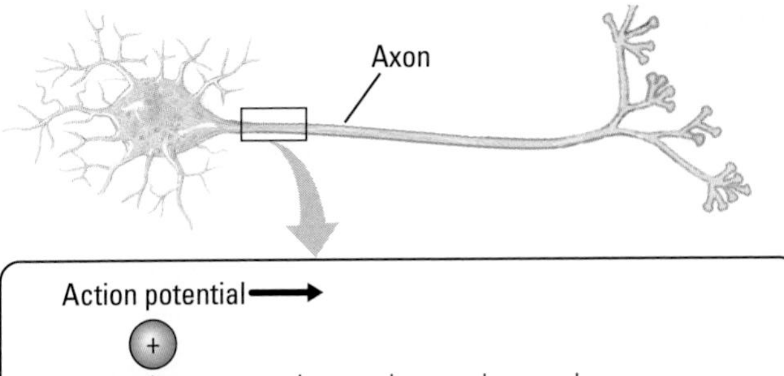

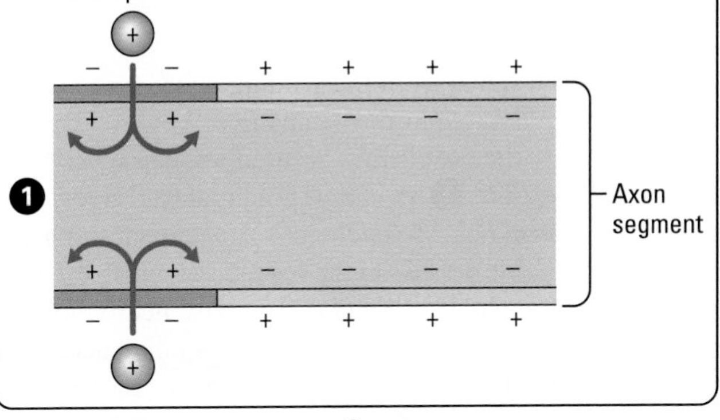

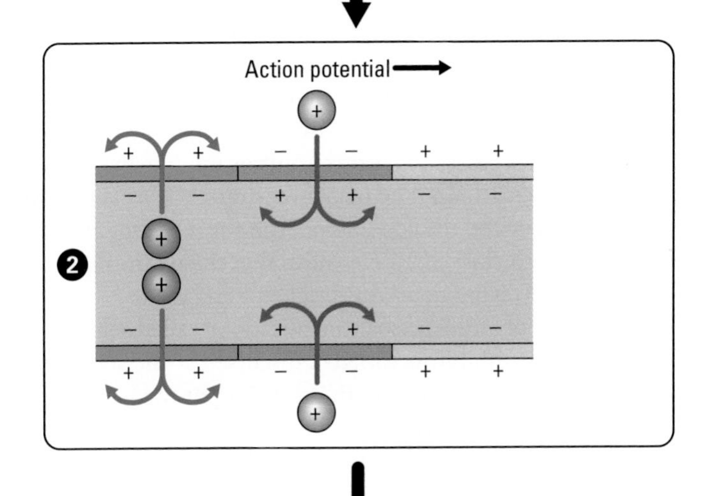

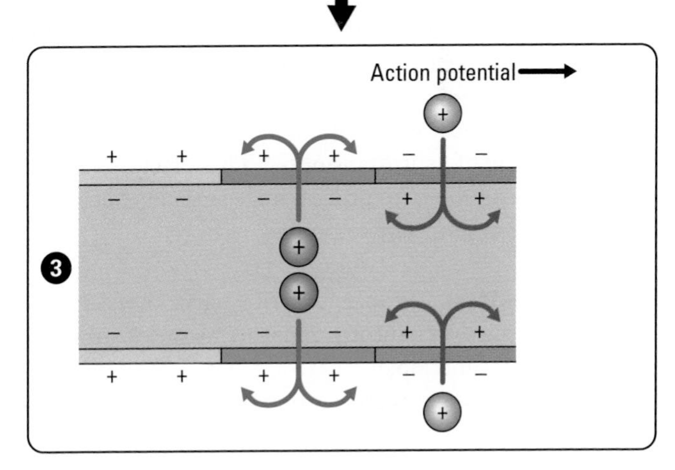

Figure 27.5 Propagation of the action potential along an axon.

Chemical synapses are prevalent in most other organs, including skeletal muscles, and in the central nervous system.

Chemical synapses have a narrow gap, called the **synaptic cleft,** separating the synaptic terminal of the sending neuron from the receiving cell **(Figure 27.6)**. When the action potential (an electrical signal) reaches the end of the sending neuron, it is converted to a chemical signal consisting of molecules of neurotransmitter. A **neurotransmitter** is a chemical that carries information from a nerve cell to another kind of cell that will react, such as another nerve cell or an effector cell. Once a neurotransmitter conveys a chemical signal from the sending neuron, an action potential may then be generated in the receiving cell.

Let's follow the events that occur at a synapse between two neurons in Figure 27.6. ❶ An action potential (red arrow) arrives at the synaptic terminal. Molecules of neurotransmitter are contained within cytoplasmic vesicles in the synaptic terminal. ❷ The action potential triggers chemical changes that make neurotransmitter vesicles fuse with the plasma membrane of the sending neuron. ❸ The fused vesicles release their neurotransmitter molecules (⚬⚬⚬) into the synaptic cleft. ❹ The released neurotransmitter molecules diffuse across the cleft and bind to complementary receptors on ion channel proteins in the receiving neuron's plasma membrane. ❺ The binding of neurotransmitter to receptor opens the ion channels. With the channels open, ions can diffuse into or out of the receiving neuron. In many cases, this triggers a new action potential; in other cases, a neurotransmitter inhibits the generation of an action potential. ❻ The neurotransmitter is broken down or transported back to the sending neuron for recycling, causing the ion channels to close. Step 6 ensures that the neurotransmitter's effect on the receiving neuron is brief and precise.

The class of antidepressant medications called selective serotonin reuptake inhibitors (SSRIs), discussed at the start of the chapter, work by suppressing the recycling of the neurotransmitter serotonin. As a result, there is an increased amount of this mood-altering neurotransmitter available, making it more likely that action potentials will be triggered in receiving neurons.

Chemical synapses can process extremely complex information. A neuron may receive input from hundreds of other neurons via thousands of synaptic terminals **(Figure 27.7)**. The inputs can be highly varied because each sending neuron may secrete a different quantity or kind of neurotransmitter. These factors account for the nervous system's ability to process huge amounts of complex stimuli and formulate appropriate responses.

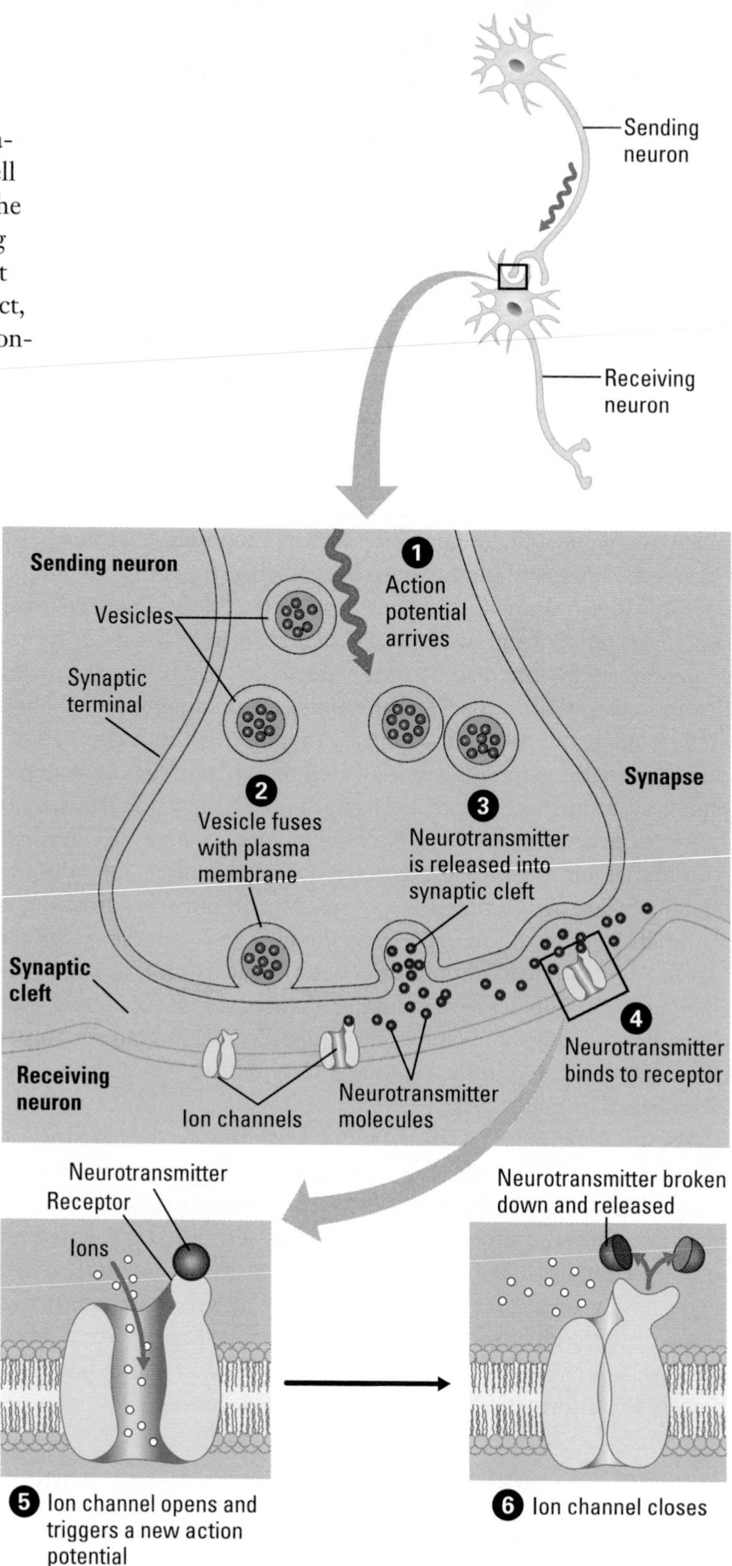

Figure 27.6 Neuron communication at a synaptic cleft.

Neurotransmitters Propagation of a nerve signal across a chemical synapse relies on neurotransmitters. A variety of small molecules serve this function.

The biogenic amines are a group of neurotransmitters derived from amino acids. The biogenic amines include epinephrine, norepinephrine, serotonin, and dopamine, all of which may also function as hormones (see Chapter 25). Serotonin and dopamine affect sleep, mood, attention, and learning. Imbalances of biogenic amines are associated with various disorders. For example, the degenerative illness Parkinson's disease is associated with a lack of dopamine in the brain, and an excess of dopamine is linked to schizophrenia. Reduced levels of norepinephrine and serotonin have been linked with some types of depression.

Many peptides (short chains of amino acids) also serve as neurotransmitters. The endorphins, peptides that function as both neurotransmitters and hormones, decrease our perception of pain. Endorphins may be released in response to a wide variety of stimuli, including traumatic injury, muscle fatigue, and even eating very spicy foods.

Drugs and the Brain Many drugs, even common ones such as caffeine, nicotine, and alcohol, affect the actions of neurotransmitters in the brain's trillions of synapses. Caffeine, found in coffee, tea, chocolate, and many soft drinks, counters the effects of inhibitory neurotransmitters, ones that normally suppress nerve signals. This is why caffeine tends to stimulate you and keep you awake. Nicotine acts as a stimulant by binding to and activating receptors for a neurotransmitter called acetylcholine. Alcohol is a strong depressant; its precise effect on the nervous system is not yet known, but it seems to increase the effects of an inhibitory neurotransmitter called GABA (gamma aminobutyric acid).

Besides SSRIs (discussed in the Biology and Society essay), many other prescription drugs used to treat psychological disorders alter the effects of neurotransmitters. Tranquilizers such as diazepam (Valium) and alprazolam (Xanax) activate the receptors for GABA, increasing the effects of this inhibitory neurotransmitter. In other cases, a drug may physically block a receptor, preventing a neurotransmitter from binding, thereby reducing its effect. For instance, some antipsychotic drugs used to treat schizophrenia block dopamine receptors. Some drugs used to treat attention deficit hyperactivity disorder (ADHD), such as methylphenidate (Ritalin), are chemically similar to the neurotransmitters dopamine and norepinephrine. ADHD medications are believed to block the reuptake of these neurotransmitters, but their precise actions in the brain are poorly understood.

What about illegal drugs? Stimulants such as amphetamines and cocaine increase the release and availability of norepinephrine and dopamine at synapses. Abuse of these drugs can therefore produce symptoms resembling schizophrenia. LSD and mescaline may produce their hallucinatory effects by activating serotonin and dopamine receptors. The active ingredient in marijuana (tetrahydrocannabinol, or THC) binds to brain receptors normally used by another neurotransmitter that seems to play a role in pain, depression, appetite, memory, and fertility. Opiates—morphine, codeine, and heroin—bind to endorphin receptors, reducing pain and producing euphoria. Not surprisingly, opiates are commonly used medicinally to relieve pain. However, abuse of opiates may permanently change the brain's chemical synapses and reduce the normal synthesis of neurotransmitters. This means that a short-term high is often followed by depression. Some researchers believe that this effect contributes to the highly addictive nature of the opiates.

The drugs discussed here are used for a variety of purposes, both medical and recreational. While they have the ability to increase alertness and sense of well-being or to reduce physical and emotional pain, they also have the potential to disrupt the brain's finely tuned neural pathways, altering the chemical balances that are the product of millions of years of evolution.

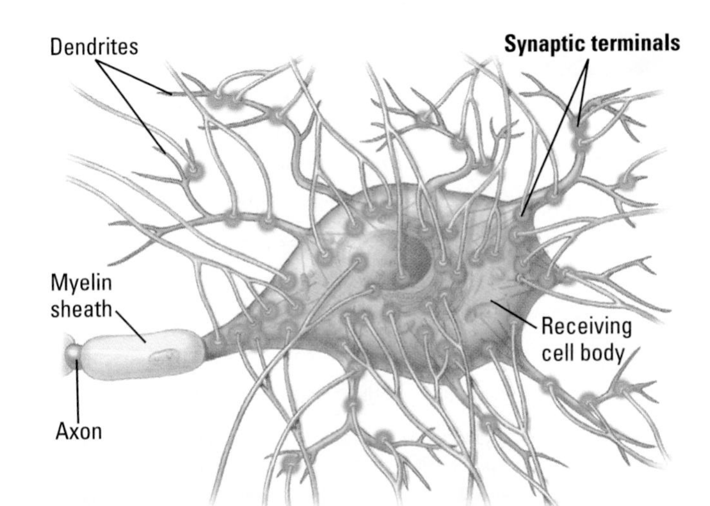

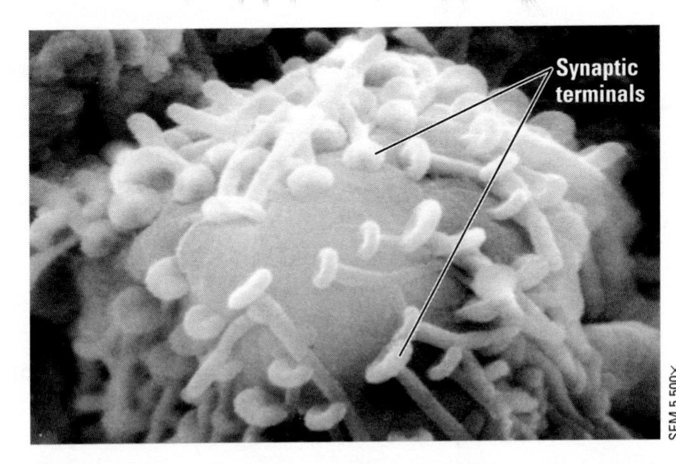

Figure 27.7 A neuron's multiple synaptic inputs. As shown in this drawing and micrograph, a single neuron may receive signals from hundreds of other neurons via thousands of synaptic terminals (shown in orange).

CHECKPOINT

1. Arrange the following neurons into the correct sequence for information flow from stimulus to response: interneuron, sensory neuron, motor neuron. Which of these neuron types is located entirely within the central nervous system?
2. What is the function of the myelin sheath?
3. In what way is an action potential an example of positive feedback?
4. How does the structure of a synapse ensure that signals pass only in one direction, from a sending neuron to a receiving cell?
5. What determines whether a neuron is affected by a specific neurotransmitter?
6. Why are endorphins sometimes called "natural opiates"?

Answers: 1. Sensory neuron, interneuron, motor neuron; interneuron **2.** It speeds up conduction of signals along axons **3.** The opening of channels caused by stimulation of the neuron changes the membrane potential, and this change causes more channels to open **4.** At the synapse, only the sending neuron releases neurotransmitter, and only the receiving cell has receptors to receive the neurotransmitter **5.** To be affected by a particular neurotransmitter, a neuron must have specific receptors for that neurotransmitter built into its membrane **6.** Opiates bind to endorphin receptors. (It would probably be more accurate to call opiates "artificial endorphins.")

The Human Nervous System: A Closer Look

To this point, we have concentrated on cellular mechanisms that are fundamental to nearly all animal nervous systems. Although there is remarkable uniformity throughout the animal kingdom in the way nerve cells function, there is great variety in how nervous systems as a whole are organized. For the rest of this chapter, we'll focus on vertebrates, with particular emphasis on human systems.

The Central Nervous System

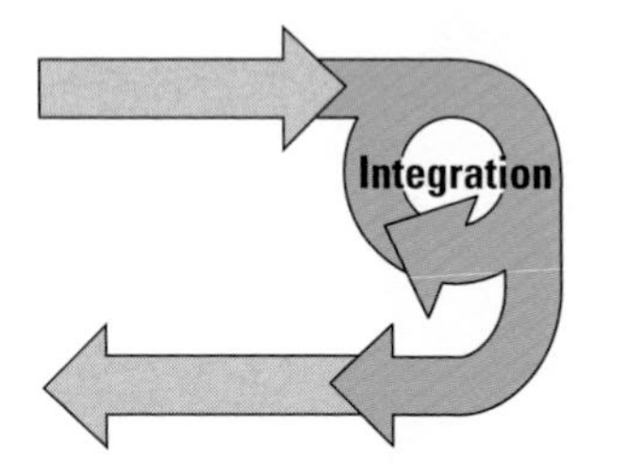

Vertebrate nervous systems are diverse in both structure and level of sophistication. For instance, the brains of dolphins and humans are much more complex than the brains of frogs and fishes. However, some features are common to the nervous systems of all vertebrates and most other animals. One is **cephalization,** the concentration of the nervous system at the head end. Another is **centralization,** the presence of a central nervous system (CNS) distinct from a peripheral nervous system (PNS) **(Figure 27.8)**.

In all vertebrates, the brain and spinal cord make up the CNS. The **spinal cord,** a gelatinous bundle of nerve fibers, lies inside the spinal column. The spinal cord acts as the central communication conduit between the brain and the rest of the body. Millions of nerve fibers within the cord carry motor information from the brain and from the cord itself for delivery to the muscles, while other fibers convey sensory information (such as touch, pain, and body position) from the body to the brain. The spinal cord acts like a telephone cable jam-packed with wires, each of which carries messages between the central hub and relay centers for outlying areas. In humans, the spinal cord is rarely severed because it is protected by the bony spinal column. However, a traumatic blow to the spinal column and subsequent bleeding, swelling, and

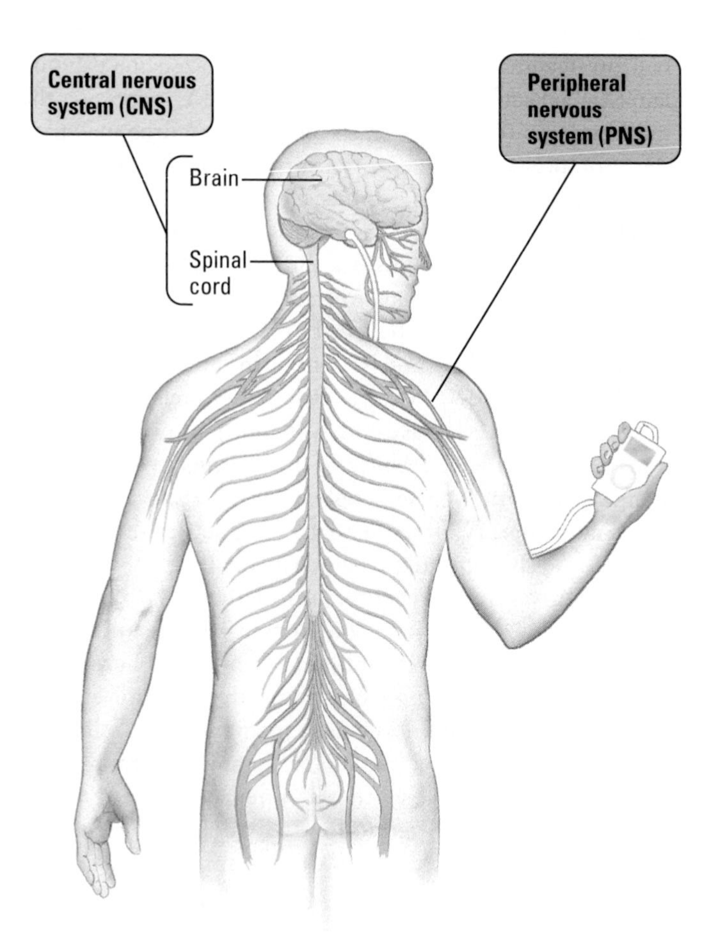

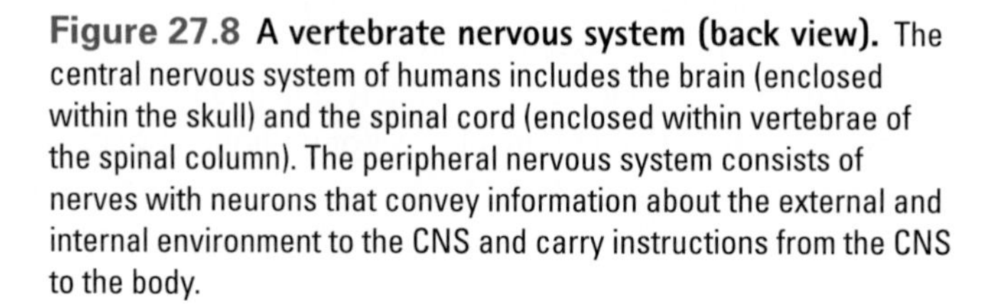
Figure 27.8 A vertebrate nervous system (back view). The central nervous system of humans includes the brain (enclosed within the skull) and the spinal cord (enclosed within vertebrae of the spinal column). The peripheral nervous system consists of nerves with neurons that convey information about the external and internal environment to the CNS and carry instructions from the CNS to the body.

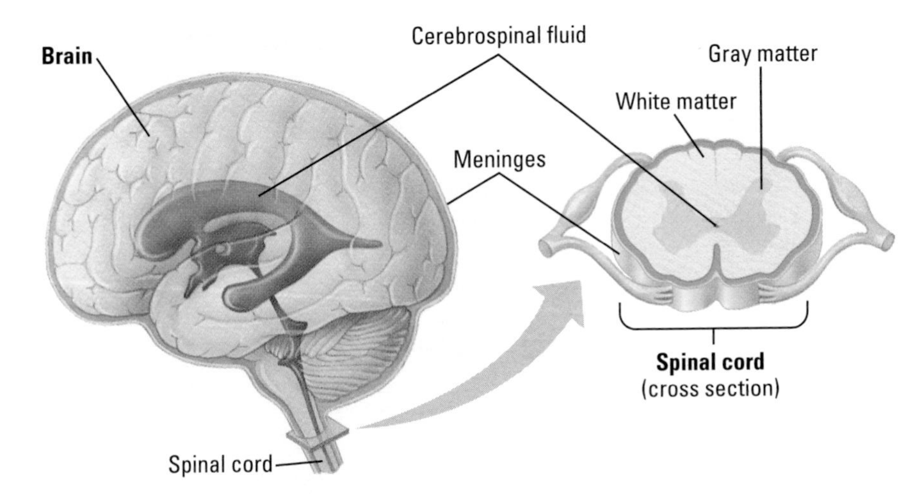

Figure 27.9 Fluid-filled spaces of the vertebrate CNS. Spaces in the brain and spinal cord are filled with cerebrospinal fluid, which helps to cushion the CNS and supply it with needed nutrients, hormones, and white blood cells.

scarring can crush the delicate nerve bundles and prevent signals from passing. The result may be a debilitating injury. Such trauma along the back can cause paraplegia—paralysis of the lower half of the body. Trauma higher up on the spinal column can cause quadriplegia—paralysis from the neck down. Such injuries may be permanent because the nerves of the spinal cord, unlike many other body tissues, cannot be repaired.

The master control center of the nervous system, the **brain,** includes homeostatic centers that keep the body functioning smoothly, sensory centers that integrate data from the sense organs, and centers of emotions and intellect. The brain also sends out motor commands to muscles.

Both the brain and spinal cord contain spaces filled with a liquid called **cerebrospinal fluid** that cushions the CNS and helps supply it with nutrients, hormones, and white blood cells **(Figure 27.9)**. Also protecting the brain and spinal cord are layers of connective tissue, called **meninges.** If the cerebrospinal fluid becomes infected by either bacteria or viruses, the meninges may become inflamed, a condition called meningitis. Viral meningitis is generally not harmful, while bacterial meningitis can have serious consequences if not treated with antibiotics. Infection of the cerebrospinal fluid can be detected by a spinal tap, a procedure that involves inserting a narrow needle through the spinal column to collect a sample of cerebrospinal fluid.

The CNS has two distinct areas, as shown in the cross section of the spinal cord in Figure 27.9. **White matter** is mainly axons (with their whitish myelin sheaths); **gray matter** is mainly neuron cell bodies and dendrites. Within the brain, most of the gray matter is in the cerebral cortex, an outer layer that is the center for higher brain functions.

The Peripheral Nervous System

In addition to the brain and spinal cord of the central nervous system, all vertebrates have a peripheral nervous system. The vertebrate peripheral nervous system can be divided into two components based on the functions of the neurons: the somatic nervous system and the autonomic nervous system **(Figure 27.10)**. Neurons of the **somatic nervous system** carry signals to and from skeletal muscles, mainly in response to external stimuli. When you walk, for instance, these neurons carry commands that make your legs

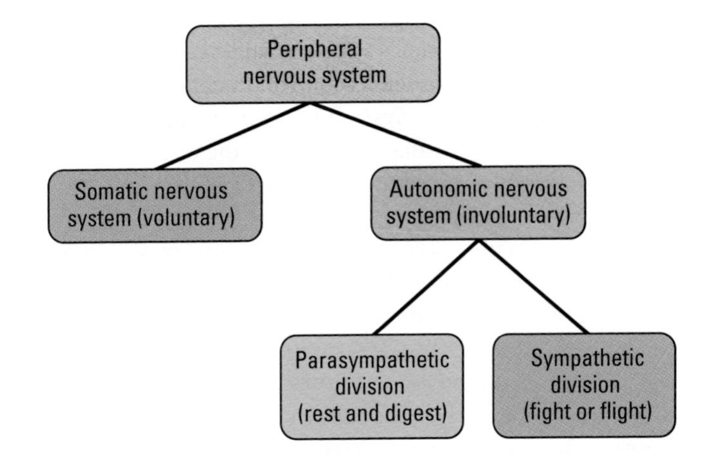

Figure 27.10 Functional divisions of the vertebrate PNS.

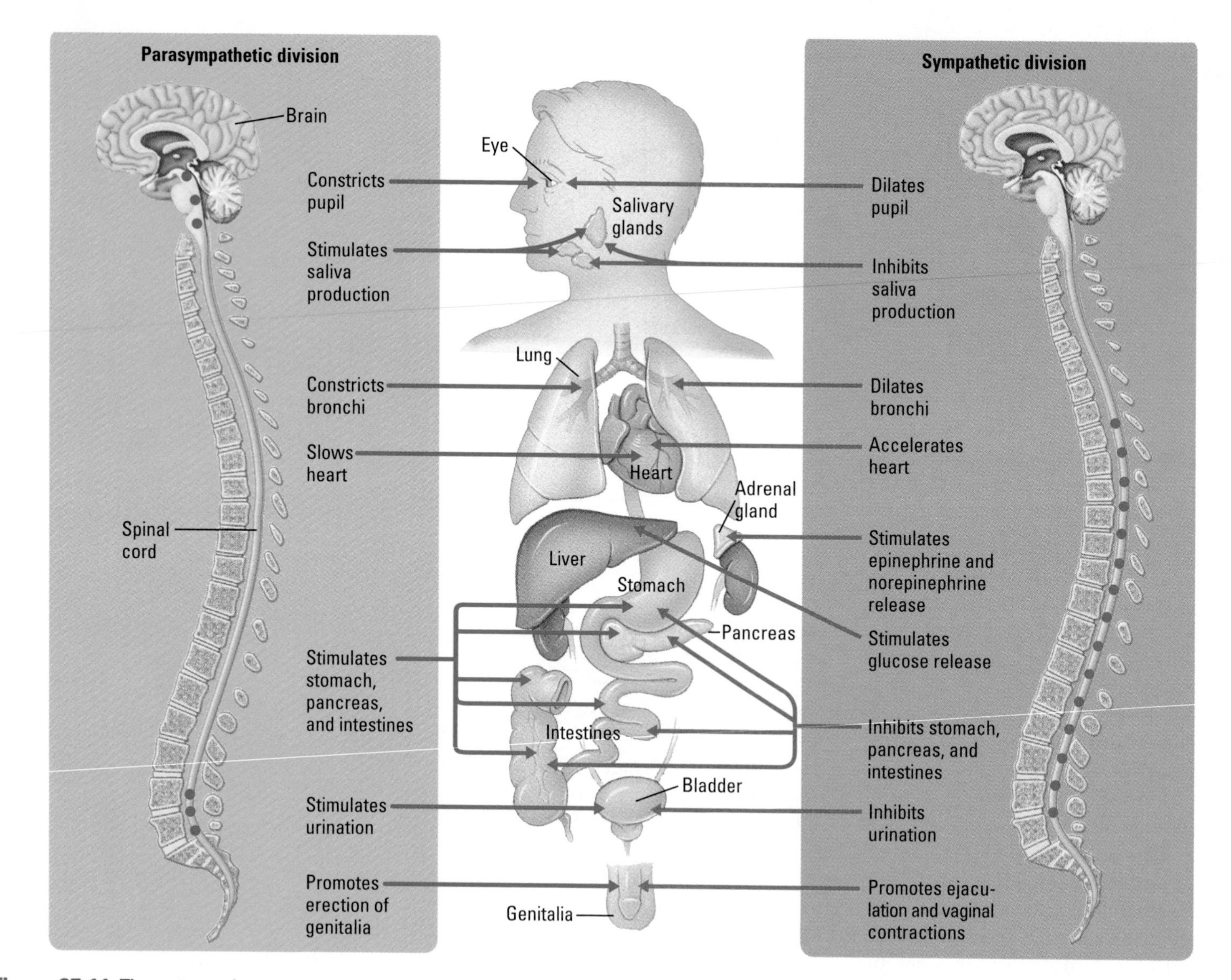

Figure 27.11 The autonomic nervous system. The autonomic nervous system contains two sets of neurons with opposite effects. The parasympathetic division (green) promotes energy-saving activities, such as resting muscles and digesting food. The sympathetic division (red) promotes energy-consuming activities, such as fighting and running. Sympathetic and parasympathetic neurons emerge from different regions of the CNS. Neurons of the parasympathetic system (green dots) emerge from the brain and the lower part of the spinal cord. In contrast, neurons of the sympathetic system (red dots) emerge from the middle regions of the spinal cord.

move. The somatic nervous system is said to be voluntary because many of its actions are under conscious control. In contrast, the motor neurons of the **autonomic nervous system** regulate the internal environment by controlling smooth and cardiac muscles and the organs and glands of the digestive, cardiovascular, excretory, and endocrine systems. This control is generally involuntary.

The autonomic nervous system contains two sets of neurons with opposing effects on most body organs. One set, called the **parasympathetic division,** primes the body for activities that gain and conserve energy for the body ("rest and digest"). A sample of the effects of parasympathetic signals appears on the left in **Figure 27.11**. These include stimulating the digestive organs, such as the salivary glands, stomach, and pancreas; decreasing the heart rate; and narrowing the bronchi, which correlates with a decreased breathing rate. The other set of neurons in the autonomic nervous system, the **sympathetic division,** tends to have the opposite effect, preparing the body for intense, energy-consuming activities, such as fighting, fleeing, or competing in a strenuous game ("fight or flight"). You see some of the effects of sympathetic signals on the right side of the figure. Fight-or-flight and relaxation are opposite extremes. Your body usually operates at intermediate levels, with most of your organs receiving both sympathetic and parasympathetic signals. The opposing signals adjust an organ's activity to a suitable level.

As carriers of command signals, the motor neurons of the sympathetic and parasympathetic systems constitute lower levels of the nervous system's hierarchy. In the next section, we take a closer look at the highest level of the hierarchy, the brain.

The Human Brain

Composed of up to 100 billion intricately organized neurons, with a much larger number of supporting cells, the human brain is more powerful than the most sophisticated computer. The brain is divided structurally into three regions: the hindbrain, the midbrain, and the forebrain (**Figure 27.12** and **Table 27.1**).

Two sections of the hindbrain (blue in Figure 27.12), the **medulla oblongata** and the **pons,** and the midbrain (purple) make up a functional unit called the **brainstem.** All of the sensory and motor neurons carrying information to and from other brain regions pass through the brainstem. The brainstem serves as a sensory filter, selecting which information to pass on. It also regulates sleep and arousal and helps coordinate body movements, such as walking. Table 27.1 lists some of the other functions of the medulla oblongata, pons, and midbrain.

Another part of the hindbrain, the **cerebellum** (light blue in the figure), is a planning center for body movements. There is evidence that it also plays a role in learning and remembering motor responses. The cerebellum receives sensory information about the position of limbs and the

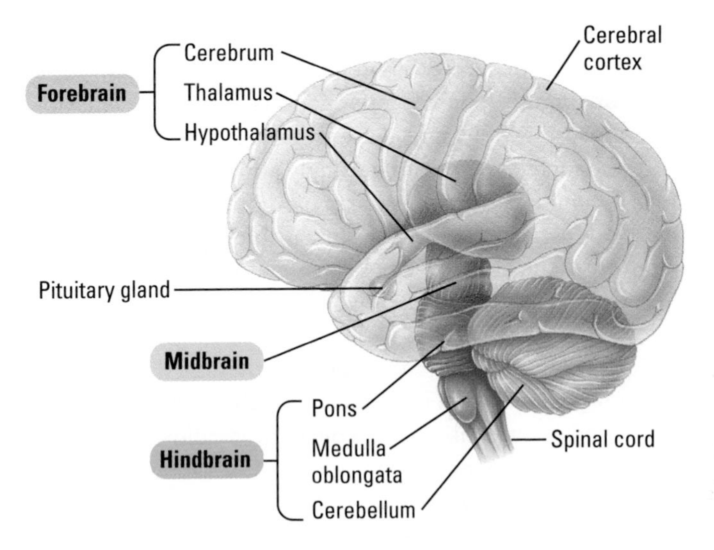

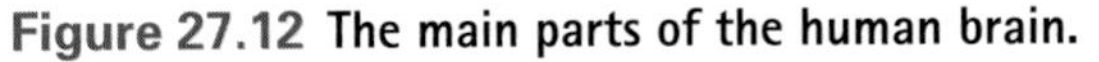

Figure 27.12 The main parts of the human brain.

Table 27.1	Structure of the Brain
Brain Structure	**Major Function**
Brainstem	Conducts data to and from other brain centers; homeostatic control; coordinates body movement
Medulla oblongata	Controls breathing, circulation, swallowing, digestion
Pons	Control breathing
Midbrain	Receives and integrates auditory data; major visual center in nonmammalian vertebrates; coordinates visual reflexes in mammals; sends sensory data to higher brain centers
Cerebellum	Coordinates body movement; learns and remembers motor responses
Thalamus	Input center for sensory data going to the cerebrum; output center for motor responses leaving the cerebrum; data sorting
Hypothalamus	Homeostatic control center; controls pituitary gland; biological clock
Cerebrum	Sophisticated integration; memory, learning, speech; emotions; formulates complex behavioral responses

length of muscles, as well as information from the auditory and visual systems. It also receives input from the motor pathways, telling it which motor actions are being carried out. The cerebellum uses this information to provide coordination of movement and balance. When you step off a curb, for instance, your cerebellum evaluates your body position and relays a plan for your next movements.

The most sophisticated integrating centers are found in the forebrain (orange and tan in Figure 27.12); they are the thalamus, the hypothalamus, and the cerebrum. The **thalamus** contains most of the cell bodies of neurons that relay information to the cerebral cortex, the most extensive portion of the cerebrum. The thalamus first sorts data into categories (all the touch signals from a hand, for instance). It also suppresses some signals and enhances others. The thalamus then sends the sorted information on to the appropriate brain centers for further interpretation and integration.

In Chapter 25, we saw that the hypothalamus controls the pituitary gland and the secretion of many hormones. The **hypothalamus** also regulates body temperature, blood pressure, hunger, thirst, the sex drive, and fight-or-flight responses, and it helps us experience emotions, such as rage and pleasure. A "pleasure center" in the hypothalamus could also be called an addiction center, for it is strongly affected by certain addictive drugs, such as amphetamines and cocaine. Another part of the hypothalamus functions as a timing mechanism, our **biological clock.** Receiving visual input from the eyes, the clock maintains our daily biological rhythms, such as cycles of sleepiness and hunger.

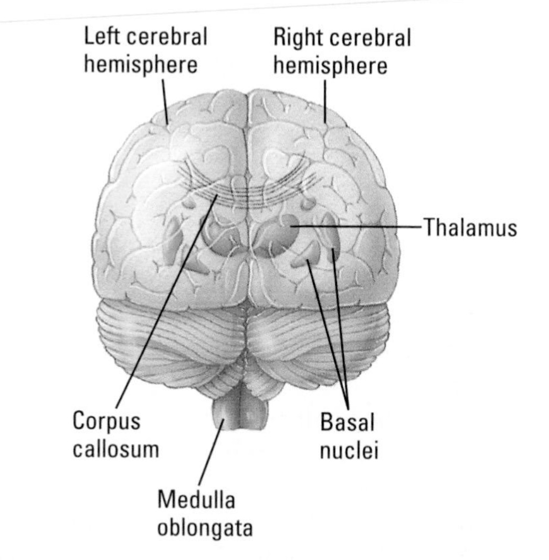

Figure 27.13 A rear view of the brain. The large cerebrum (yellow) consists of left and right cerebral hemispheres connected by a thick band of nerves called the corpus callosum.

The Cerebral Cortex The **cerebrum,** the largest and most sophisticated part of our brain, consists of right and left cerebral hemispheres **(Figure 27.13)**. A thick band of nerve fibers called the **corpus callosum** connects the cerebral hemispheres, enabling them to process information together.

The **cerebral cortex** is a highly folded layer of gray matter forming the surface of the cerebrum (see Figure 27.12). Although less than 5 mm thick, the human cerebral cortex accounts for over 80% of the total brain mass. It contains some 10 billion neurons and hundreds of billions of synapses. Its intricate neural circuitry helps produce our most distinctive human traits: reasoning and mathematical abilities, language skills, imagination, artistic talent, and personality traits. Assembling information it receives from our eyes, ears, nose, taste buds, and touch sensors, the cortex creates our sensory perceptions—what we are actually aware of when we see, hear, smell, taste, or touch. The cerebral cortex also regulates our voluntary movements.

Like the rest of the cerebrum, the cerebral cortex is divided into right and left sides. Because the nerve fibers from the cerebral cortex cross in the medulla oblongata, each hemisphere (right and left) receives information from and controls the movement of the opposite side of the body. The corpus callosum allows communication between the two hemispheres. Under the corpus callosum, small clusters of neuron cell bodies, the basal nuclei, are important in motor coordination.

Each side of the cerebral cortex has four lobes (represented by different colors in **Figure 27.14**). Researchers have identified a number of functional areas within each lobe. One, called the motor cortex, functions mainly in sending commands to

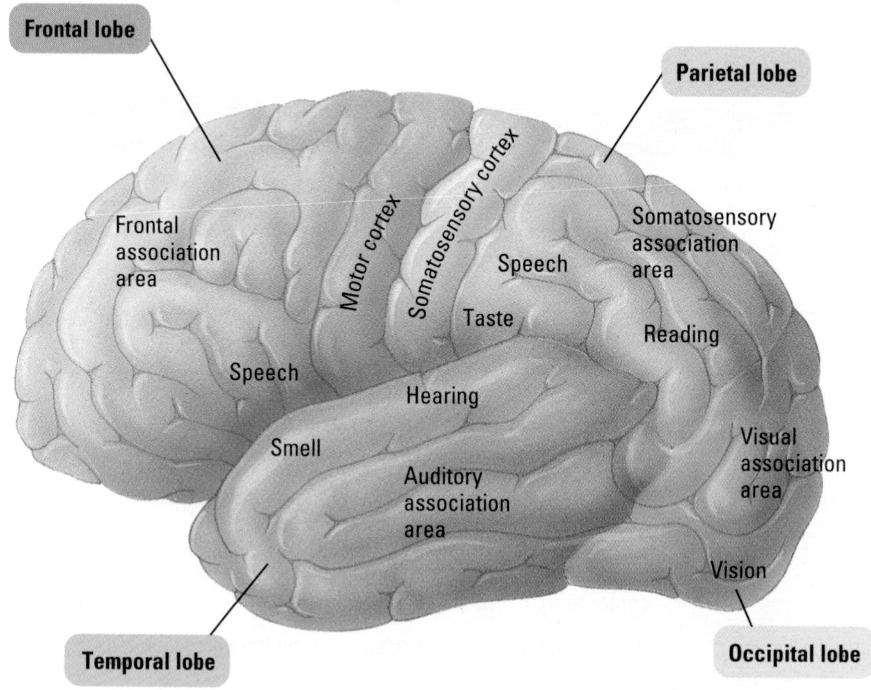

Figure 27.14 Functional areas of the cerebrum's left hemisphere. This figure identifies the main functional areas in the brain's left cerebral hemisphere, which is divided into four lobes.

skeletal muscles. Next to the motor cortex, the somatosensory cortex receives and partially integrates signals from touch, pain, pressure, and temperature receptors throughout the body. The cerebral cortex also has centers that receive and begin processing sensory information concerned with vision, hearing, taste, and smell. Each of these centers cooperates with an adjacent area, called an association area.

Making up most of our cerebral cortex, the **association areas** are the sites of higher mental activities—roughly, what we call thinking. A large association area in the frontal lobe uses varied inputs from many other areas of the brain to evaluate consequences, make considered judgments, and plan for the future. Language results from some extremely complex interactions among several association areas. For instance, the parietal lobe of the cortex has association areas for reading and speech. These areas receive visual information (the appearance of words on a page) from the vision centers in the occipital lobe. Then, if the words are to be spoken aloud, the information is arranged into speech patterns and passed to a speech center in the frontal lobe, which signals the motor cortex to move the tongue, lips, and other muscles to form words.

You may have heard people say that they are "left-brained" or "right-brained." In a phenomenon known as **lateralization,** areas in the two hemispheres become specialized for different functions. In most people, the left hemisphere becomes most adept at language, logic, and mathematical operations. The right hemisphere is stronger at spatial relations, pattern and face recognition, musical ability, and emotional processing in general. In about 10% of us, however, these roles of the left and right hemispheres are reversed or the roles are more alike.

Interestingly, evidence from brain surgery patients indicates that patterns of lateralization are not fixed. One of the most radical surgical alterations of the brain is hemispherectomy—the removal of almost one-half of the brain **(Figure 27.15)**. This procedure is performed to alleviate severe seizure disorders that originate from one of the hemispheres as a result of illness, abnormal development, or stroke. Incredibly, with just half a brain, hemispherectomy patients recover quickly, often leaving the hospital within a few weeks. Although the side of the body opposite the surgery always has permanent partial paralysis, hemispherectomy patients have undiminished intellectual capacities. Higher brain functions that previously originated from the missing half of the brain begin to be controlled by the opposite side. Recovery after hemispherectomy is a striking example of the brain's remarkable plasticity. It also shows that those rare individuals whose brains have been damaged through illness, injury, or surgery can provide significant insight about how healthy human brains operate. One of the most remarkable "broken brains" known to medical science is the topic of the next section.

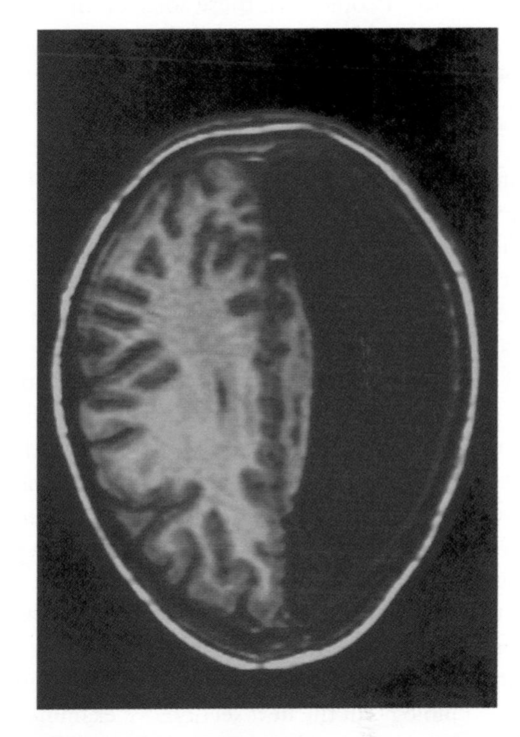

Figure 27.15 X-ray of a hemispherectomy patient after surgery. In a hemispherectomy, nearly half of the patient's brain is removed in order to stop severe seizure disorders. Incredibly, with just half a brain, hemispherectomy patients recover quickly and can lead productive lives.

THE PROCESS OF SCIENCE

What Can a Broken Brain Teach Us? In 1848, a Vermont railroad construction foreman named Phineas Gage accidentally set off an explosion of dynamite that propelled a 3-foot-long, inch-thick spike through his head. The 13-pound steel rod entered his left cheek and traveled upward behind his left eye and out the top of his skull, landing several yards away. Incredibly, Gage walked away from the accident and his intellect appeared to be intact. However, **observations** by his family and associates revealed drastic changes in his personality, with a new propensity toward meanness, vulgarity, irresponsibility, and an inability to control his behavior. One friend commented that "Gage was no longer Gage."

Gage's doctor, John Harlow, raised the **question** of whether these personality changes were due to physical changes in the brain. In 1868, Harlow published the **hypothesis** that damage to the "frontal region" of Gage's brain was responsible for his behavior changes. However, Harlow was never able to perform an autopsy to determine the precise location of the injury, and the 19th-century understanding of the brain was insufficient to accurately predict the effect of a brain injury on personality.

Luckily, after Gage's death, his skull and the offending spike were stored in a medical museum at Harvard University. In 1994, researchers from the University of Iowa made the **prediction** that a virtual re-creation of the injury would show that the spike did indeed pass through Gage's frontal lobes. Their **experiment** involved obtaining detailed X-ray images of Gage's preserved skull and creating a computer simulation of the injury. The **results** supported the original hypothesis, suggesting that the rod had pierced both frontal lobes of Gage's brain **(Figure 27.16)**. People with these sorts of injuries often exhibit irrational decision making and have difficulty processing emotions. The link between the frontal lobes and personality continues to be an active area of research. ■

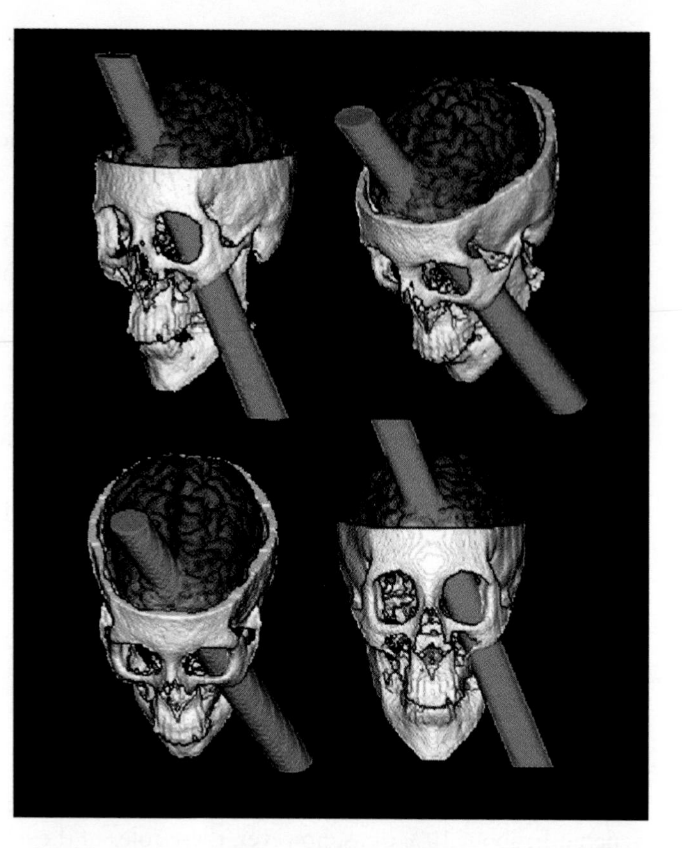

Figure 27.16 Phineas Gage. A 1994 computer simulation of an 1848 injury to railroad worker Phineas Gage suggested that the spike that passed through his skull (shown as a red cylinder in the figure) might have damaged both frontal lobes of his brain. Such an injury could account for Gage's personality change after the accident.

The Limbic System Some regions of the frontal lobes are part of the **limbic system,** a functional group of integrating centers in the thalamus, hypothalamus, and cerebral cortex. Much of human emotion, learning, and memory depends on our limbic system. The limbic system is central to such behaviors as nurturing of infants and emotional bonding to other individuals. Primary emotions that produce laughing and crying are mediated by the limbic system, and it also attaches emotional "feelings" to basic survival mechanisms of the brainstem, such as feeding and aggression. Have you ever had a "scent memory," where a particular smell suddenly triggers a strong memory? This happens because signals from your nose enter your brain through the olfactory bulb, a region of the brain that connects with the limbic system. Thus, a specific scent can immediately trigger emotional reactions and memories.

Unraveling how networks of neurons in the brain store, retrieve, and use memories, control the body's internal environment, and construct our thoughts and feelings is one of the most challenging and engaging aspects of modern biology. In the next section, we examine how the brain receives information from the environment via the sense organs.

CHECKPOINT

1. Name the two structures that make up the vertebrate central nervous system.
2. The central nervous system is protected by ___________, a liquid that cushions it and supplies it with nutrients, and layers of protective tissue called the ___________.
3. How would a drug that inhibits the parasympathetic nervous system affect a person's pulse?
4. The largest and most sophistical part of your brain is the ___________. It is divided into two hemispheres connected by the ___________. Its surface, called the ___________, accounts for most of your brain's mass and is responsible for your higher reasoning abilities.

Answers: 1. Brain and spinal cord **2.** cerebrospinal fluid; meninges **3.** The pulse rate would probably increase **4.** cerebrum; corpus callosum, cerebral cortex

The Senses

In this section, we'll focus on the human body's sensory structures. To begin, we'll examine how these structures gather information and pass it on to the central nervous system (CNS). After that, we'll take a closer look at two of the human senses, vision and hearing.

Sensory Input

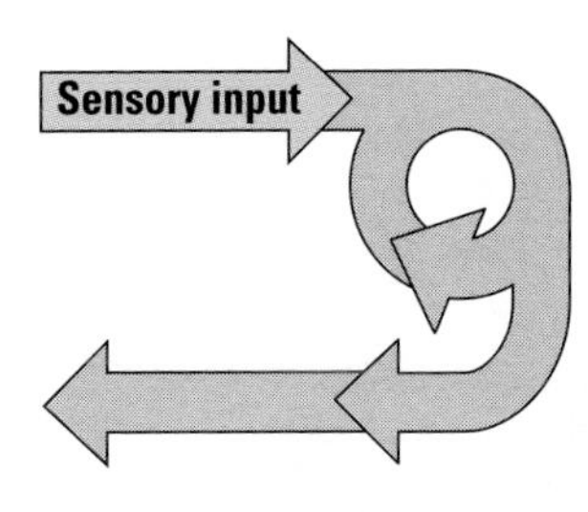

Sensory receptors, such as the sensory cells in your eyes or the taste buds of your tongue, are tuned to the condition of the external world and the internal organs. Sensory receptors detect stimuli such as chemicals, light, tension in a muscle, sounds, electricity, cold, heat, and touch. The sensory receptors in your eyes, for instance, detect light energy; those in your taste buds detect chemicals dissolved in saliva. The sensory receptor's job is completed when it sends information to the CNS by triggering action potentials.

What exactly do we mean when we say that a sensory receptor detects a stimulus? In stimulus detection, the cell converts one type of signal (the stimulus) to an electrical signal. This conversion, called **sensory transduction,** occurs as a change in the membrane potential of the receptor cell.

Sensory Transduction As an example of sensory transduction, **Figure 27.17** shows sensory receptors in a taste bud detecting sugar molecules. (1) When the molecules first come into contact with the taste bud, they bind to specific protein molecules on a taste receptor cell's plasma membrane. (2) The binding causes some ion channels in the membrane to close and others to open. Changes in the flow of ions alter the membrane potential. This change in membrane potential is called the **receptor potential**. In contrast to action potentials, which are all-or-none phenomena, receptor potentials vary in intensity; the stronger the stimulus, the stronger the receptor potential.

Once a receptor cell converts a stimulus to a receptor potential, the signal usually enters the CNS. In Figure 27.17, (3) each receptor cell forms a synapse with a sensory neuron. When there are enough sugar molecules, a strong receptor potential is triggered. This receptor potential makes the receptor cell release enough neurotransmitter to increase the rate of action

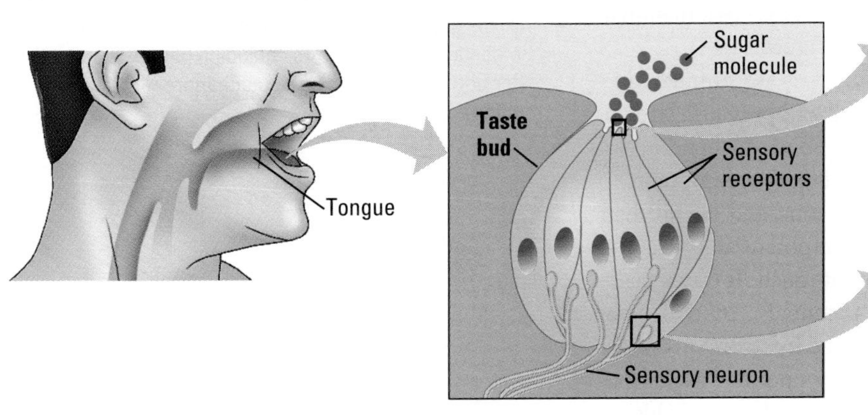

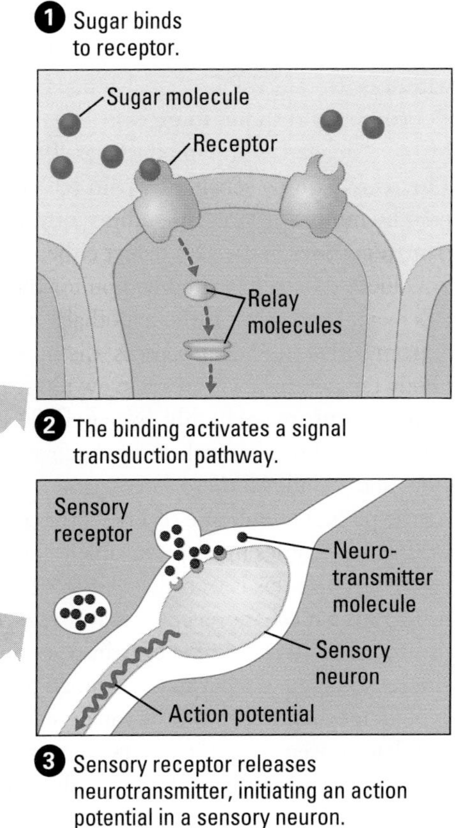

Figure 27.17 Sensory transduction at a taste bud.

potential generation in the sensory neuron. The brain interprets the intensity of the stimulus from the rate at which it receives action potentials. It gains additional information about stimulus intensity by keeping track of how many sensory neurons it receives signals from.

There is an important qualification to what we have just said about stimulus intensity. Have you ever noticed how an odor that is strong at first seems to fade with time? The same effect helps you adjust to a hot or cold shower and enables you to wear clothes without being constantly aware of them. This effect is called **sensory adaptation,** the tendency of some sensory receptors to become less sensitive when they are stimulated repeatedly. When receptors become less sensitive, they trigger fewer action potentials, causing the brain to receive fewer stimuli. Sensory adaptation keeps the body from continuously reacting to normal background stimuli. Without it, our nervous system would become overloaded with useless information.

Types of Sensory Receptors Based on the type of signals to which they respond, we can group sensory receptors into five general categories: pain receptors, thermoreceptors (sensors for heat and cold), mechanoreceptors (sensors for touch, pressure, motion, and sound), chemoreceptors (sensors for chemicals), and electromagnetic receptors (sensors for energy such as light and electricity). These five types of receptors work in various combinations to produce the five human senses.

Figure 27.18, showing a section of human skin, reveals why the surface of our body is sensitive to such a variety of stimuli. Our skin contains pain receptors (labeled in red in the figure), thermoreceptors (blue), and mechanoreceptors (green). Each of these receptors is a modified dendrite of a sensory neuron. The neuron both transduces stimuli and sends action potentials to the CNS. In other words, each receptor serves as both a receptor cell and a sensory neuron.

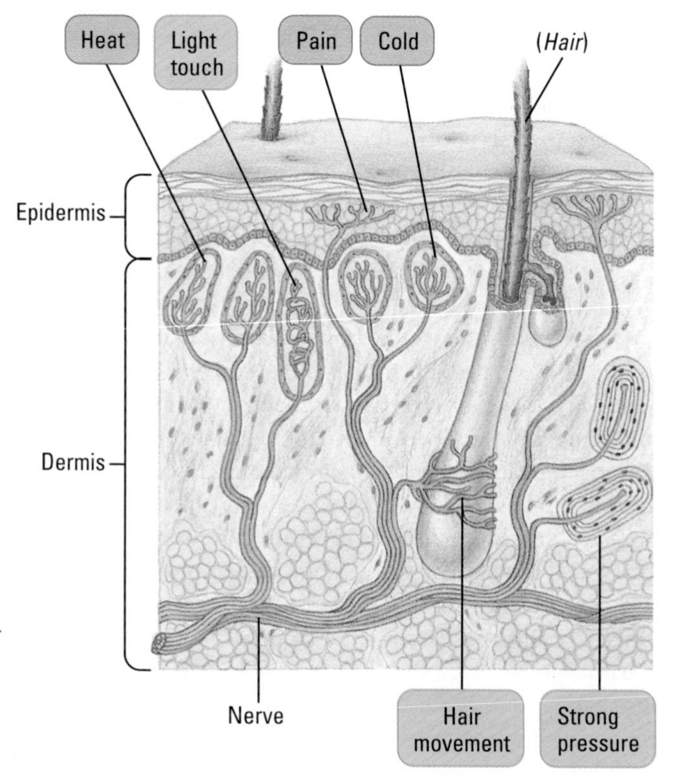

Figure 27.18 Sensory receptors in the human skin. Our skin is sensitive to a wide variety of stimuli because it contains a wide variety of receptors, including pain receptors, thermoreceptors (for both hot and cold), and mechanoreceptors (for different kinds of touch).

All parts of the human body except the brain have pain receptors. Pain is important because it often indicates danger and usually makes an animal withdraw to safety. Pain can also make us aware of injury or disease. **Pain receptors** may respond to excessive heat or pressure or to chemicals released from damaged or inflamed tissues. Prostaglandins are local regulators that increase pain by sensitizing pain receptors. Aspirin and ibuprofen reduce pain by inhibiting prostaglandin synthesis.

Thermoreceptors in the skin detect either heat or cold. Other temperature sensors located deep in the body monitor the temperature of the blood. The body's major thermostat is the hypothalamus. Receiving action potentials both from surface and deep sensors, the hypothalamus keeps a mammal's or bird's body temperature within a narrow range (see Figure 21.14).

Mechanoreceptors are highly diverse. Different types are stimulated by different forms of mechanical energy, such as touch and pressure, stretching, motion, and sound. All these forces produce their effects by bending or stretching the plasma membrane of a receptor cell. When the membrane changes shape, it becomes more permeable to positive ions, and the mechanical energy of the stimulus is transduced into a receptor potential. At the top of Figure 27.18 is a mechanoreceptor that detects light touch. It transduces very slight inputs of mechanical energy into action potentials. Another type of pressure sensor, lying deeper in the skin, is stimulated by strong pressure. A third type of mechanoreceptor, the touch receptor around the base of the hair, detects hair movements. Touch receptors at the base of the stout whiskers on a cat are extremely sensitive and enable the animal to detect close objects by touch in the dark.

Chemoreceptors include the sensory cells in our nose and taste buds, which are attuned to chemicals in the external environment, as well as some internal receptors that detect chemicals in the body's internal environment. Internal chemoreceptors include sensors in some of our arteries that can detect changes in pH and others that can detect changes in concentrations of oxygen in the blood. In all types of chemoreceptors, a receptor cell develops receptor potentials in response to chemicals dissolved in fluid (for instance, blood, saliva, or the fluid coating the inside surface of the nose).

Electromagnetic receptors are sensitive to energy of various wavelengths, which takes such forms as electricity, magnetism, and light. **Photoreceptors,** such as those found in your eyes, are the most common type of electromagnetic receptor. Photoreceptors detect the electromagnetic energy of light. In the next section, we'll focus on one specific organ that uses photoreceptors: the human eye.

Vision

Our eyes are remarkable sense organs, able to detect a multitude of colors, form images of faraway objects, and respond to minute amounts of light energy. In this section, you'll learn about the structure of the human eye and how it processes images. You'll also learn why vision problems occur and how they can be corrected.

Structure of the Human Eye The outer surface of the human eyeball is a tough, whitish layer of connective tissue called the **sclera** **(Figure 27.19)**. At the front of the eye, the sclera becomes the transparent **cornea,** which lets light into the eye and also helps focus light. The sclera surrounds a pigmented layer called the **choroid.** At the front of the eye, the choroid forms the **iris,** which gives the eye its color. The muscles of the iris regulate the size of the **pupil,** the opening in the center of the iris that lets light into the interior of the eye. After going through the pupil, light passes through the disklike **lens,** which is held in position by ligaments. The lens focuses images onto the **retina,** a layer just inside the choroid. Photoreceptor cells of the retina transduce light energy, and action potentials pass via sensory neurons in the **optic nerve** to the visual centers of the brain. Photoreceptor cells are highly concentrated at the retina's center of focus, called the **fovea.** There are no photoreceptor cells in the part of the retina where the optic nerve passes through the back of the eye. We cannot detect light that is focused on this "blind spot," but having two eyes with overlapping fields of view enables us to perceive uninterrupted images.

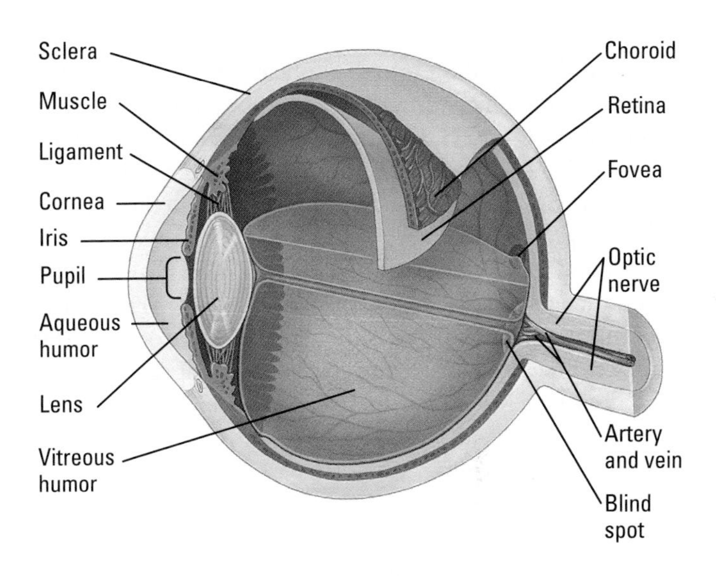

Figure 27.19 The human eye. Light enters the human eye through the cornea, then passes through the pupil and lens. Muscles that surround the lens help focus the image onto the retina, where photoreceptor cells convert the light energy to action potentials.

Two fluid-filled chambers make up the bulk of the eye. The large chamber behind the lens is filled with jellylike **vitreous humor.** The much smaller chamber in front of the lens contains a thinner fluid, **aqueous humor.** The humors help maintain the shape of the eyeball. In addition, the aqueous humor circulates through its chamber. Secreted by capillaries, this fluid supplies nutrients and oxygen to the lens, iris, and cornea and carries off wastes. Blockage of the ducts that drain this fluid can cause glaucoma, increased pressure inside the eye that may lead to blindness. If diagnosed early, glaucoma can be treated with medications that increase the circulation of the aqueous humor.

A thin mucous membrane called the **conjunctiva** helps keep the outside of the eye moist. The conjunctiva lines the inner surface of the eyelids and folds back over the white of the eye (but not the cornea). An infection or allergic reaction may cause inflammation of the conjuctiva, a condition called conjunctivitis or "pink eye." Bacterial conjuctivitis usually clears up after

Figure 27.20 How lenses focus light. When viewing a nearby object, the lens becomes thicker and rounder and focuses the image on the retina. When viewing a distant object, the lens becomes flattened and the image is again focused on the retina.

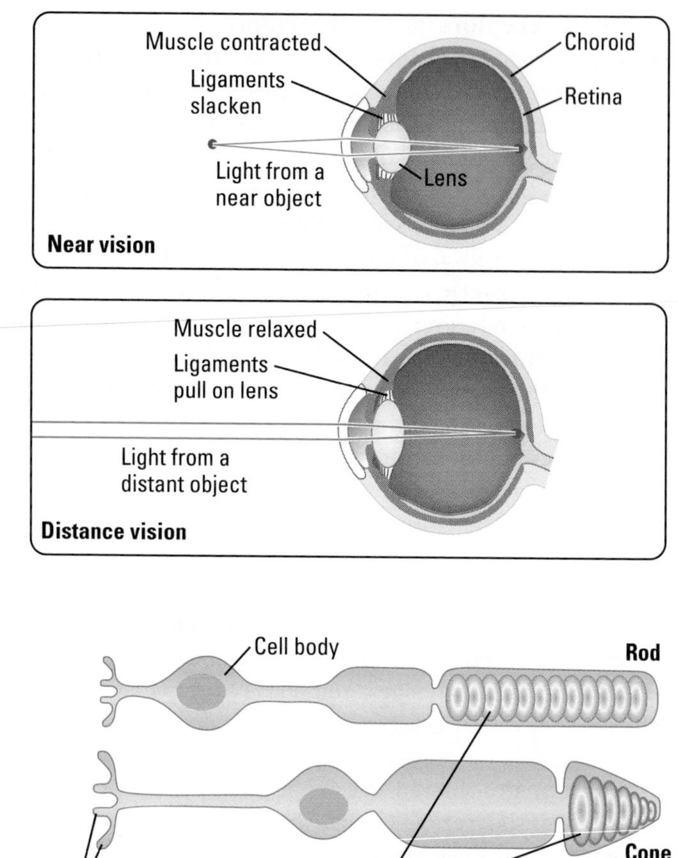

application of antibiotic eyedrops. Viral conjunctivitis usually clears up on its own, although it is very contagious, especially among young children.

The **lacrimal gland** above the eye secretes tears, a dilute salt solution that is spread across the eyeball by blinking and that drains into ducts that lead into the nasal cavities. This fluid cleanses and moistens the eye surface. Excess secretion, in response to eye irritation or strong emotions, causes tears to spill over the eyelid and fill the nasal cavities, producing sniffles.

The lens of the eye focuses light onto the retina by bending (refracting) light rays. Focusing is accomplished through a change in the shape of the lens **(Figure 27.20)**. The thicker the lens, the more sharply it bends light. The shape of the lens is controlled by the muscles attached to the choroid. When the eye focuses on a nearby object, these muscles contract, pulling the choroid layer of the eye inward toward the pupil. This makes the ligaments that suspend the lens slacken. With this reduced tension, the elastic lens becomes thicker and rounder. When the eye focuses on a distant object, the muscles controlling the lens relax, putting tension on the ligaments and flattening the lens.

Seeing in Color Built into the human retina are about 125 million rod cells and 6 million cone cells, two types of photoreceptors named for their shapes **(Figure 27.21)**. **Cones** are stimulated by bright light and can distinguish color, but they do not function in night vision. **Rods** are extremely sensitive to light and enable us to see in dim light, though only in shades of gray.

In humans, rods are found in greatest density at the outer edges of the retina and are completely absent from the fovea. The fovea contains a high concentration of cones and is the retina's center of focus **(Figure 27.22)**. If you face directly toward a dim star in the night sky, the star is hard to see. Looking to the side, however, makes your lens focus the starlight onto the

Figure 27.21 Photoreceptor cells. Within your eyes, there are two types of photoreceptor cells: cones, which are stimulated by bright light and can distinguish color, and rods, which are stimulated by dim light and can only distinguish shades of gray. You can thus see in dim light (at dusk, for example), but only in shades of gray. Your eyes contain many more rods than cones.

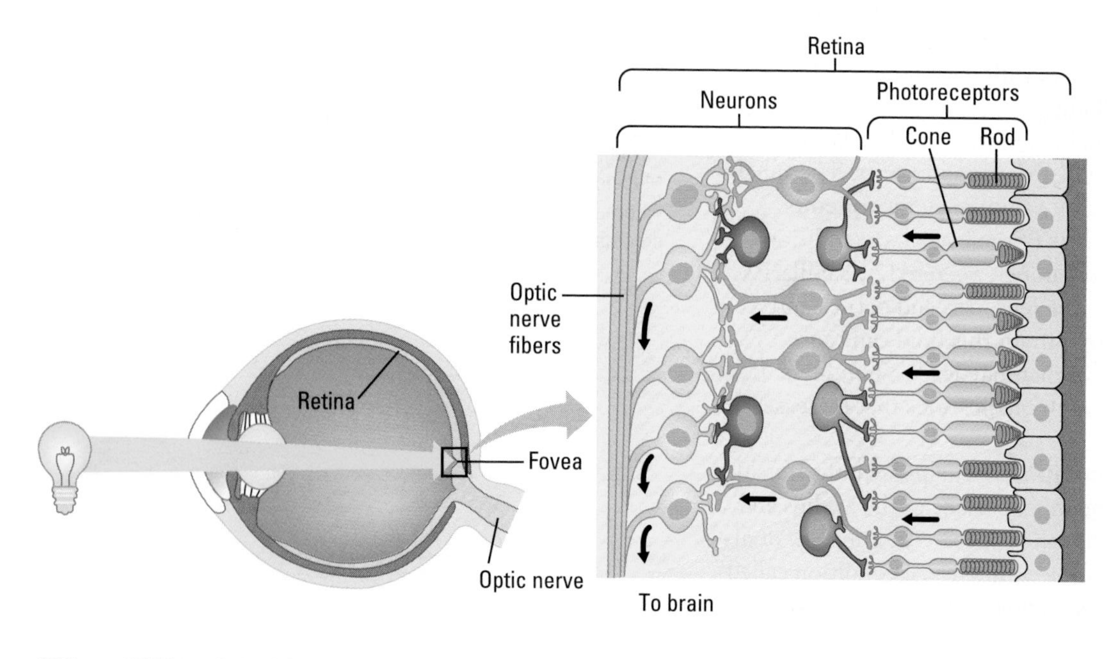

Figure 27.22 The vision pathway from light source to optic nerve. Light strikes photoreceptor cells embedded in the back of the retina. Once stimulated, the rods and cones convert the light energy to receptor potentials that are integrated by other neurons and communicated through the optic nerve as action potentials to the brain.

parts of the retina with the most rods, and you can see the star. By contrast, you achieve your sharpest day vision by looking straight at the object of interest. This is because cones are densest in the fovea. Some birds, such as hawks, have ten times more cones in their fovea than we do, enabling them to spot small prey from high altitudes.

How do rods and cones detect light? As Figure 27.21 shows, each rod and cone includes an array of membranous disks containing light-absorbing visual pigments. Rods contain a visual pigment called rhodopsin, which can absorb dim light. Cones contain visual pigments called photopsins, which absorb bright light. We have three types of cones, each containing a different type of photopsin. These cells are called blue cones, green cones, and red cones, referring to the colors absorbed best by their photopsin. We can perceive a great number of colors because the light from each particular color triggers a unique pattern of stimulation among the three types of cones. Color blindness results from a deficiency in one or more types of cones.

Like all receptor cells, rods and cones are stimulus transducers. When rhodopsin and photopsin absorb light, they change chemically, and the change alters the permeability of the cell's membrane. The resulting receptor potentials trigger a complex integration process that begins in the retina. Action potentials carry the partly integrated information into the brain via the optic nerve. Three-dimensional perceptions (what we actually see) result from further integration in several processing centers of the cerebral cortex.

Vision Problems and Corrections Three of the most common visual problems are nearsightedness, farsightedness, and astigmatism. All three are focusing problems, easily corrected with artificial lenses. Nearsighted people cannot focus well on distant objects, although they can see well at short distances (the condition is named for the type of vision that is unimpaired). A nearsighted eyeball (**Figure 27.23a**, left) is longer than normal. The lens cannot flatten enough to compensate, and it focuses distant objects in front of the retina instead of on it. **Nearsightedness** (also known as myopia) is corrected by glasses or contact lenses that are thinner in the middle than at the outside edge (Figure 27.23a, middle). The corrective lenses make the

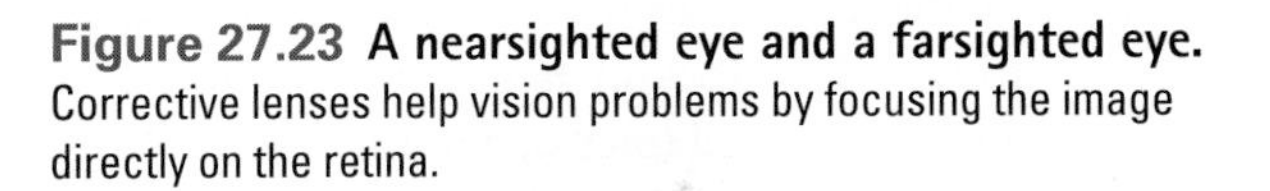
Figure 27.23 A nearsighted eye and a farsighted eye. Corrective lenses help vision problems by focusing the image directly on the retina.

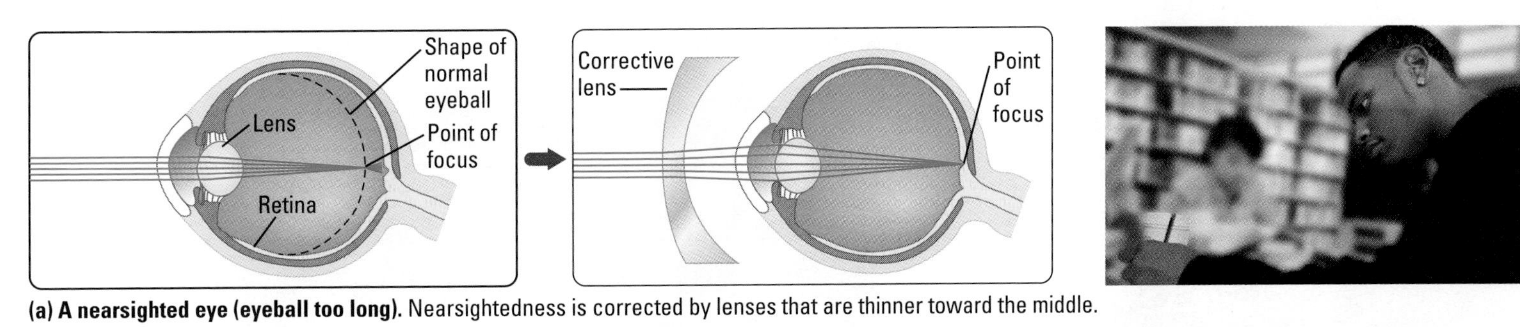

(a) A nearsighted eye (eyeball too long). Nearsightedness is corrected by lenses that are thinner toward the middle.

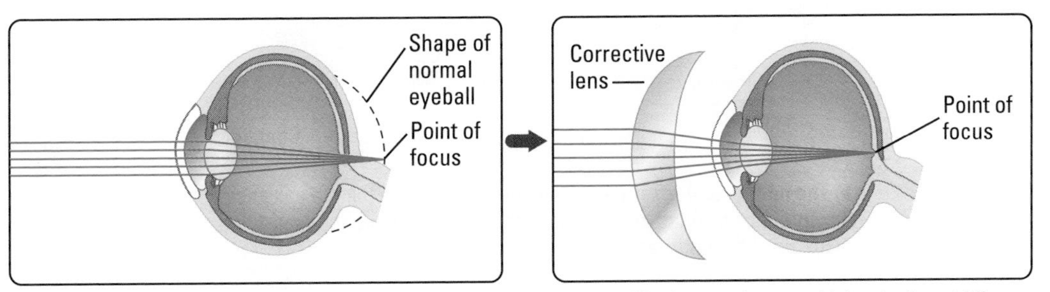

(b) A farsighted eye (eyeball too short). Farsightedness is corrected by lenses that are thicker in the middle.

light rays from distant objects diverge slightly as they enter the eye. The image formed by the lens in the eye is then focused at the correct point.

Farsightedness (also known as hyperopia) is the opposite of nearsightedness. It occurs when the eyeball is shorter than normal, causing the lens to focus images behind the retina **(Figure 27.23b)**. Farsighted people see distant objects normally, but they can't focus at short distances. Corrective lenses that are thicker in the middle than at the outside edge compensate for farsightedness by making light rays from nearby objects converge slightly before they enter the eye. Another type of farsightedness, called presbyopia, develops with age. Beginning around the mid-40s, the lens of the eye becomes less elastic. As a result, the lens gradually loses its ability to change shape and help us focus on nearby objects, and reading without glasses becomes difficult.

Astigmatism is blurred vision caused by a misshapen lens or cornea. Any such distortion makes light rays converge unevenly and not focus at any one point on the retina. Lenses that correct astigmatism are asymmetrical in a way that compensates for the asymmetry in the eye.

In recent years, surgical procedures have become an option for treating vision disorders. In radial keratotomy (RK), a knife is used to cut slits in the cornea to change its shape. In photorefractive keratectomy (PRK) and laser-assisted in situ keratomileusis (LASIK), a laser is used to reshape the cornea and change its focusing power. More than 1 million LASIK procedures are performed each year to treat nearsightedness, farsightedness, and astigmatism.

Hearing

The ear is composed of three regions: the outer ear, the middle ear, and the inner ear **(Figure 27.24a)**. The **outer ear** consists of the flaplike **pinna**—the bendable structure we commonly refer to as our "ear"—and the **auditory canal.** The pinna and the auditory canal collect sound waves and channel them to the **eardrum,** a sheet of tissue that separates the outer ear from the **middle ear.** When sound waves strike the eardrum, it vibrates and passes the sound waves to three small bones **(Figure 27.24b)**: the hammer (more formally, the malleus), the anvil (incus), and the stirrup

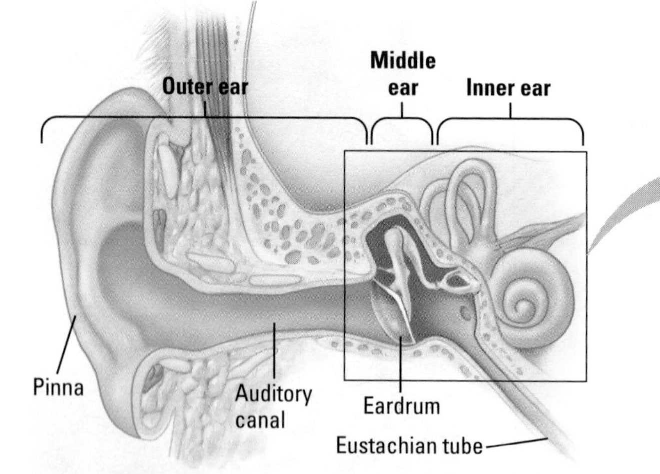

(a) Overview. The human ear is divided into three regions. The outer ear conveys sounds to the eardrum, which vibrates and passes the sounds into the middle ear.

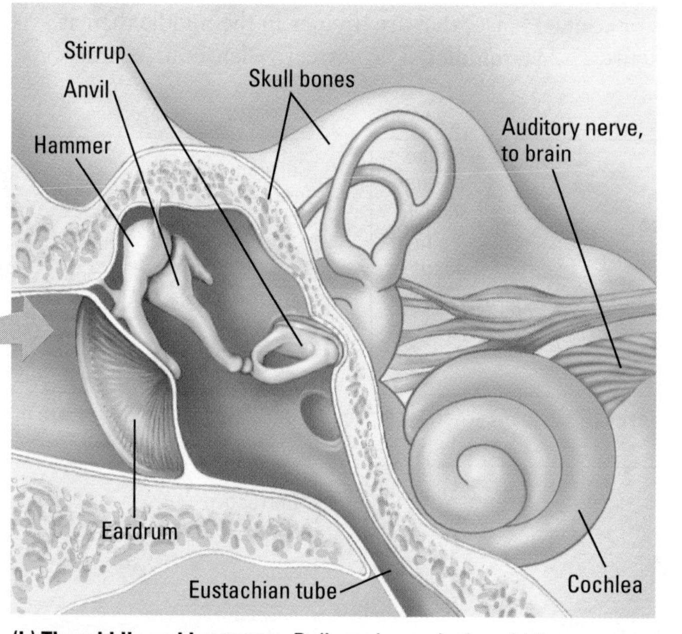

(b) The middle and inner ears. Delicate bones in the middle ear pass the vibrations through an opening in the skull bone into the fluid-filled channels of the inner ear.

Figure 27.24 An overview of the human ear.

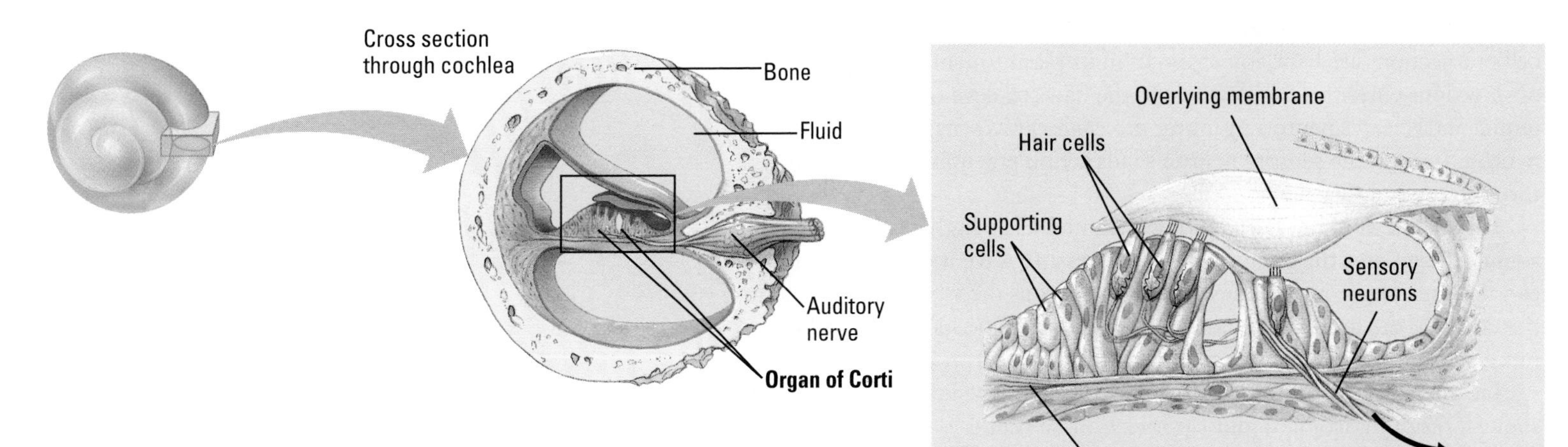

(stapes). The stirrup passes vibrations into the inner ear through an opening in the skull bone. The **Eustachian tube** conducts air between the middle ear and the back of the throat, ensuring that air pressure is kept equal on either side of the eardrum. This tube is what enables you to move air in or out to equalize pressure ("pop" your ears) when changing altitude rapidly in an airplane or diving underwater.

The **inner ear** consists of fluid-filled channels in the bones of the skull. Sound waves set the fluid in motion. One of the channels, the **cochlea,** is a long, coiled tube. Our actual hearing organ, the **organ of Corti,** is located within a fluid-filled canal inside the cochlea **(Figure 27.25)**. The organ of Corti consists of an array of hair cells embedded in a **basilar membrane.** The hair cells are the receptor cells of the ear. Notice that they project into the fluid and that the tips of most are in contact with the overlying gel-like membrane. Sensory neurons at the base of the hair cells carry action potentials from the organ of Corti into the brain via the auditory nerve.

Now let's see how the parts of the ear function in hearing. As indicated in **Figure 27.26**, a vibrating object, such as a plucked guitar string, creates pressure waves in the surrounding air, represented by the up-and-down waves in the figure. Collected by the pinna and auditory canal of the outer ear, these waves make your eardrum vibrate with the same frequency as the sound. (Frequency is the number of vibrations per second.) From the eardrum, the vibrations pass through the hammer, anvil, and stirrup in the middle ear. The stirrup transmits the vibrations to the inner ear, producing pressure waves in the fluid within the cochlea. As a pressure wave passes through the cochlea, it makes the basilar membrane vibrate. Vibration of the basilar membrane makes the hairlike projections on the hair cells alternately

Figure 27.25 The organ of Corti, within the cochlea. The organ of Corti, the actual hearing organ of the ear, is located within a fluid-filled canal inside the spiral cochlea. Receptors cells (hair cells) communicate with sensory neurons. From there, action potentials pass to the auditory nerve and to the brain.

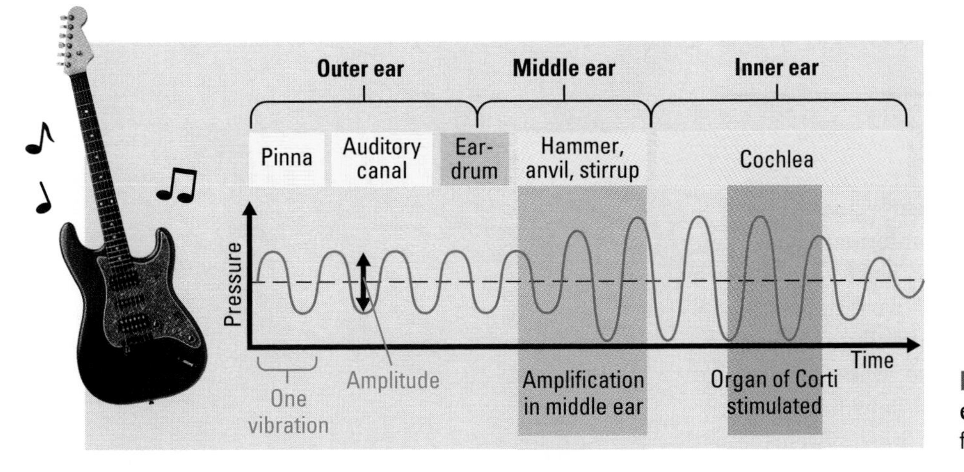

Figure 27.26 The route of sound waves through the ear. This figure traces the path of sound stimulus as it passes from the environment to the brain.

brush against and draw away from the overlying membrane. When a hair cell's projections are bent, ion channels in its plasma membrane open, and positive ions enter the cell. As a result, the hair cell develops a receptor potential and releases neurotransmitter molecules at its synapse with a sensory neuron. In turn, the sensory neuron sends action potentials to the brain through the auditory nerve.

How is the volume of a sound determined? The higher the volume of sound, the greater the amplitude of the pressure wave it generates. In the ear, the greater the amplitude, the more vigorous the vibrations of fluid in the cochlea, the more pronounced the bending of the hair cells, and the more action potentials generated in the sensory neurons.

Deafness, the loss of hearing, can be caused by the inability to conduct sounds, resulting from middle-ear infections, a ruptured eardrum, or stiffening of the middle-ear bones (a common age-related problem). Deafness can also result from damage to receptor cells or neurons. Few parts of our anatomy are more delicate than the organ of Corti. Frequent or prolonged exposure to very loud sounds can damage or destroy hair cells. For example, you may have noticed that a few hours in a very loud environment leaves you with ringing or buzzing sounds in your ears, a condition called tinnitus.

In this section, you learned how the body perceives stimuli via sensory receptors. The various forms of sensory signals are passed to the central nervous system, which couples them to body responses. In the next section, we'll see how the body's motor systems complete the neural circuit.

CHECKPOINT

1. What is meant by sensory transduction?
2. For each of the following senses in humans, identify the type of receptor: seeing, tasting, hearing, smelling.
3. Arrange the following eye parts into the correct sequence encountered by photons of light traveling into the eye: pupil, retina, cornea, lens, vitreous humor, aqueous humor.
4. As you read this text, the lenses of your eyes are relatively thick/thin (choose one).
5. Explain why our night vision is mostly in shades of gray rather than in color.
6. How does the ear convert sound waves in the air to pressure waves in the fluid in the cochlea?

Answers: 1. The conversion of a stimulus detected by a sensory receptor to an electrical signal (a receptor potential) **2.** Photoreceptor; chemoreceptor; mechanoreceptor; chemoreceptor **3.** Cornea → aqueous humor → pupil → lens → vitreous humor → retina **4.** Thick **5.** Rods are more sensitive than cones to dim light, and thus the low light intensity at night stimulates far more rods than cones. **6.** Sound waves in the air cause the eardrum to vibrate. The small bones attached to the inside of the eardrum transmit vibrations to the fluid in the inner ear, which includes the cochlea.

Motor Systems

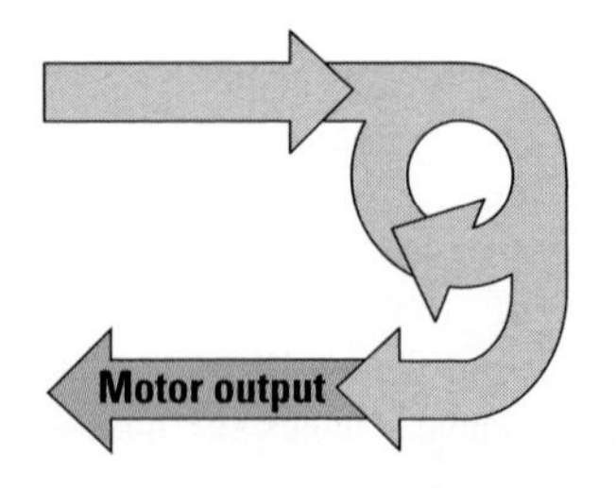

So far, we've seen how humans sense the environment and process this information. Movement in response to such external stimuli is one of the most distinctive features of animals. Whether an animal walks or runs on two, four, or six legs or more, swims, crawls, flies, or sits in one place and only moves its mouthparts, an interplay of organ systems provides its movement. The nervous system plays a key role in issuing commands to the muscular system. The muscular system exerts the force that actually makes an animal or its parts move. The skeletal system is essential in many animals because the force exerted by muscles produces movement only when the force is applied against a firm structure, the skeleton. In this section, we focus on the human skeleton and muscles and the movement their interactions produce.

The Skeletal System

A skeleton has many functions. Most land animals would sag from their own weight if they had no skeleton to support them. Skeletons also may protect an animal's soft organs. For example, the vertebrate skull protects the brain, and the ribs form a cage around the heart and lungs.

Organization of the Human Skeleton Humans, like all vertebrates, have an **endoskeleton**—hard supporting elements situated among soft tissues. The human endoskeleton is a combination of cartilage and the 206 bones that make up the skeletal system **(Figure 27.27)**; the cartilage provides flexibility in certain areas.

The human skeleton is organized into two basic units: the axial skeleton and the appendicular skeleton. The **axial skeleton** supports the axis, or trunk, of the body and includes the skull, enclosing and protecting the brain; the vertebrae of the spinal column, enclosing the spinal cord; and a rib cage around the lungs and heart. The **appendicular skeleton** is made up of the bones of the limbs, shoulders, and pelvis.

The bones of the skeleton are held together at movable joints by strong fibrous tissue called **ligaments.** Much of the versatility of our skeleton comes from these movable joints (**Figure 27.28** and the numbered locations in Figure 27.27). Humans have ❶ **ball-and-socket joints** in the shoulder where the humerus joins to the scapula and in the hip where the femur joins to the pelvis. These joints enable us to rotate our arms and legs and move them in several planes. Two other kinds of joints provide flexibility at the elbow and knee. Shown in the arm, ❷ a **hinge joint** between the humerus and the head of the ulna permits movement in a single plane. ❸ A **pivot joint** enables us to rotate the forearm at the elbow. Hinge and pivot joints between the bones in our wrists and hands enable us to make precise manipulations.

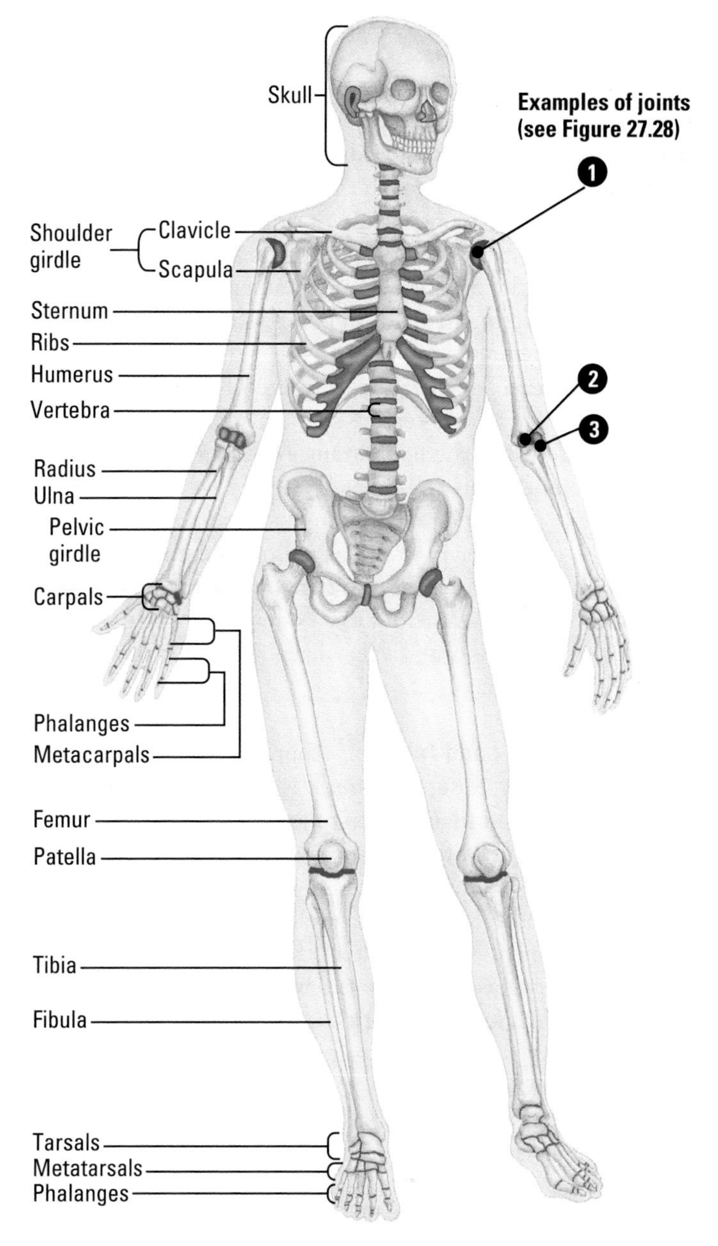

Figure 27.27 The human endoskeleton. The human skeleton, like most vertebrate skeletons, can be divided into the axial skeleton (supporting the trunk, green) and the appendicular skeleton (supporting the limbs, gold). Cartilage provides flexibility at various points (blue).

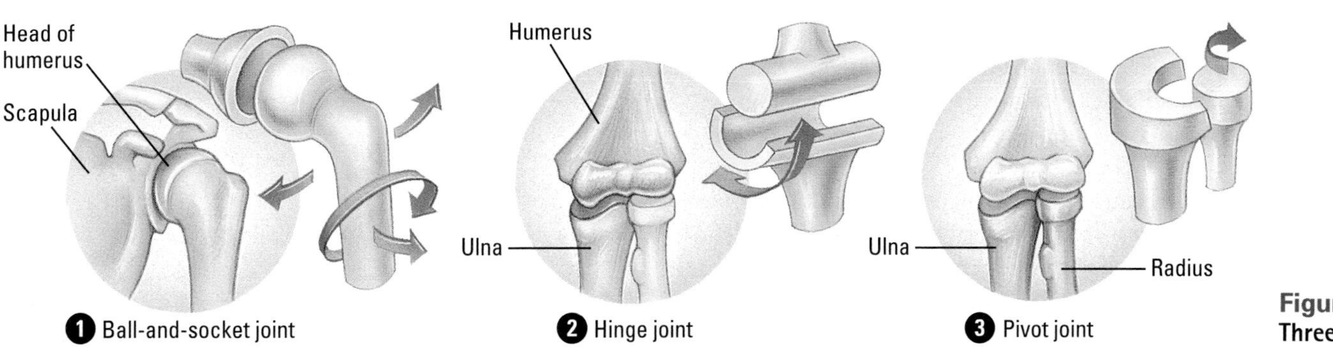

Figure 27.28 Three kinds of joints.

The Structure of Bones Bones are complex organs consisting of several kinds of tissues. You can get a sense of some of a bone's complexity from the human humerus (the upper arm bone, shown in **Figure 27.29**). A sheet of fibrous connective tissue, shown in pink (most visible in the enlargement on the lower right), covers most of the outside surface. The cells of this tissue can form new bone during normal growth or in the event of a fracture. At either end of the bone is a thin sheet of cartilage (blue), also living tissue, that forms a cushion-like surface for joints, protecting the ends of bones as they move against each other. The bone itself contains cells that secrete a surrounding material, or matrix (see Figure 21.5). Like all living tissues, tissues in a bone require servicing. Blood vessels course through channels in the bone, transporting nutrients and regulatory hormones to its cells.

Notice that the shaft of this long bone surrounds a central cavity. The central cavity contains **yellow bone marrow,** which is mostly stored fat brought into the bone by the blood. The ends of the bone have cavities that contain **red bone marrow** (not shown in the figure), a specialized tissue that produces blood cells (see Chapter 23).

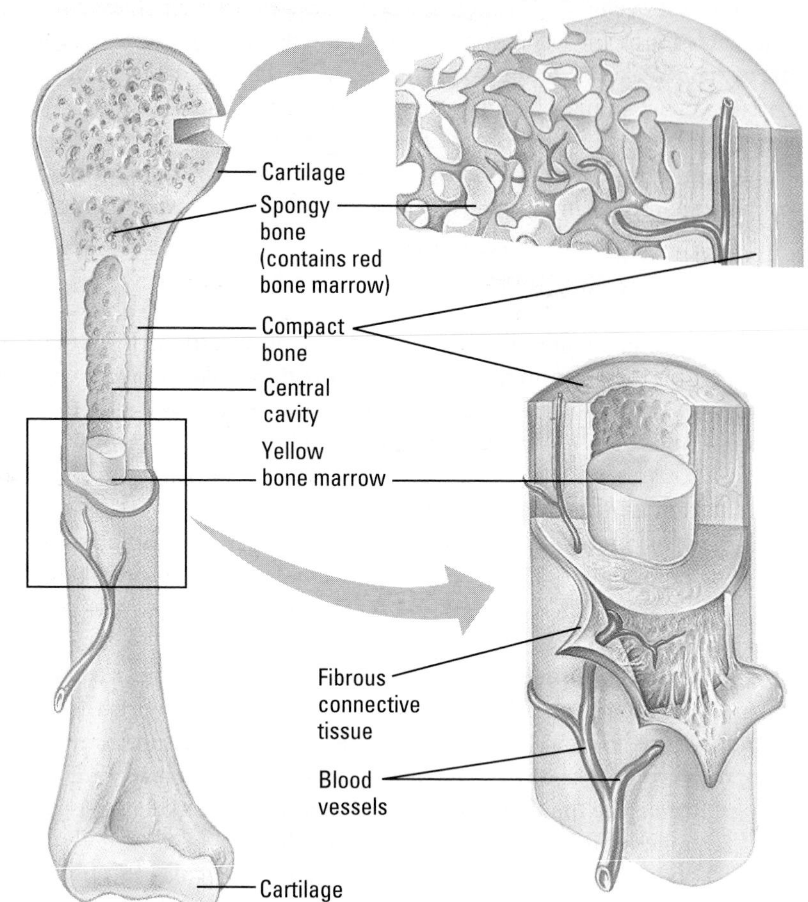

Figure 27.29 The structure of an arm bone.

Skeletal Diseases and Injuries Your skeleton is quite strong and provides reliable support, but it is susceptible to disease and injury. **Arthritis**—inflammation of joints—affects one out of every seven people in the United States. The most common form of arthritis seems to occur as a result of aging. The joints become stiff and sore and often swell as the cartilage between the bones wears down. Sometimes the bones thicken at the joints, producing crunching noises when they rub together and restricting movement. This form of arthritis is irreversible but not crippling in most cases, and moderate exercise, rest, and over-the-counter pain medications usually relieve most symptoms.

A much more debilitating form of arthritis, **rheumatoid arthritis,** is an autoimmune disease. The joints become highly inflamed, and their tissues may be destroyed by the body's immune system (see Chapter 24). Rheumatoid arthritis usually begins between ages 40 and 50 and affects more women than men. Anti-inflammatory drugs relieve symptoms, but there is no cure.

Osteoporosis is another serious bone disorder. It is most common in women after menopause: Estrogen contributes to normal bone maintenance, and with lowered production of the hormone, bones may become thinner, more porous, and easily broken. Insufficient exercise, smoking, diabetes mellitus, and an inadequate intake of protein and calcium may also contribute to the disease. Treatments include calcium and vitamin supplements, hormone replacement therapy, and drugs that slow bone loss or increase bone formation. Prevention of osteoporosis begins with sufficient calcium intake while bones are still increasing in density (up until about age 35). Weight-bearing exercise (walking, jogging, lifting weights) builds bone mass and is beneficial throughout life.

Bones are rigid but not inflexible; they will bend in response to external forces. However, as many of us know from personal experience, the skeletal system has its limits. If a force is applied that exceeds a bone's capacity to bend, the result is a broken bone, or fracture. The average American will

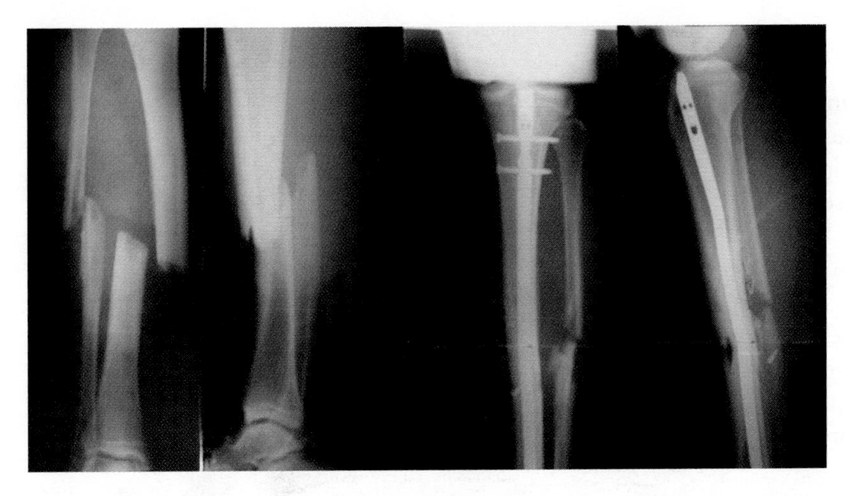

Figure 27.30 Broken bones. These X-rays show the same set of broken leg bones fractured (left) and after they have been held together with screws and a plate (right).

break two bones during his or her lifetime, most commonly the forearm or, for people over 75, the hip.

Treatment of a fracture involves two steps: putting the bone back into its natural shape and then immobilizing it until the body's natural bone-building cells can repair the fracture. A splint or a cast is usually sufficient to protect the area, prevent movement, and promote healing. Sometimes, the application of external pressure (traction) is used to align the broken parts. In more severe cases, a fracture can only be repaired surgically by inserting plates, rods, and/or screws that hold the broken pieces together **(Figure 27.30)**.

The Muscular System

Now that you've learned about the skeletal system, we can focus on **skeletal muscles,** muscles that are attached to the skeleton and produce body movement by interacting with the skeleton. **Figure 27.31** illustrates how muscles interact with bones to raise and lower the human forearm. As shown in the figure, **tendons** connect muscles to bones. For instance, one end of the biceps muscle is attached by tendons to bones of the shoulder. The other end is attached by a tendon to one of the bones in the forearm.

The ability to move the forearm in opposite directions requires that two muscles work as an antagonistic pair—that is, the two muscles must perform opposite tasks. For example, contraction of the biceps muscle shortens the muscle and pulls up the forearm. The triceps muscle is the biceps's antagonist; contraction of the triceps pulls the forearm down. Notice in Figure 27.31 that when one muscle is contracted, the antagonistic (opposite) muscle is relaxed.

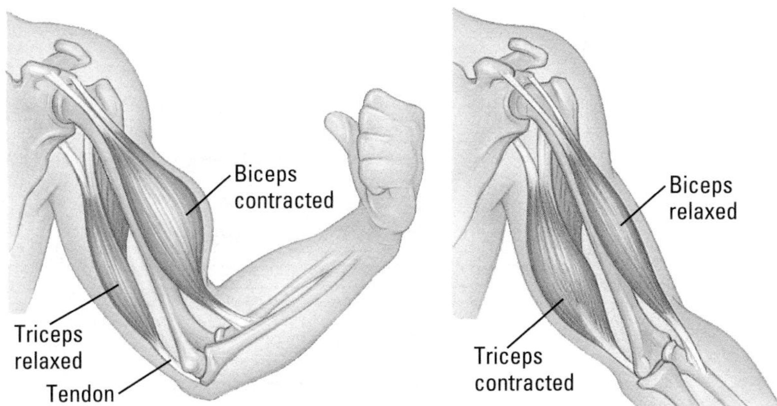

Figure 27.31 Antagonistic action of muscles in the human arm. Many movements are produced through the action of two opposing muscles. In this example, contraction of the biceps muscle raises the forearm, while contraction of the triceps muscle lowers the forearm. Tendons attach muscles to bones.

The Cellular Basis of Muscle Contraction Skeletal muscle is made up of a hierarchy of smaller and smaller parallel strands. As shown at the top of **Figure 27.32**, a muscle consists of bundles of parallel muscle fibers. Each muscle fiber is a single cell with many nuclei. Farther down in the figure, notice that each muscle fiber is itself a bundle of smaller **myofibrils**. Skeletal muscle is also called striated (striped) muscle because the myofibrils exhibit alternating light and dark bands when viewed with a light microscope. A myofibril consists of repeating units called **sarcomeres.** Structurally, a sarcomere is the region between two dark, narrow lines called **Z lines** in the myofibril. Functionally, the sarcomere is the contractile apparatus in a myofibril—the muscle fiber's fundamental unit of action.

The micrograph and the diagram below it reveal the structure of a sarcomere in more detail. A myofibril is composed of regular arrangements of two kinds of filaments: **thin filaments** made primarily from the protein actin (blue in the diagram) and **thick filaments** made from the protein myosin (red).

A sarcomere contracts (shortens) when its thin filaments slide past its thick filaments. **Figure 27.33**, a simplified diagram of the sliding-filament model, shows a sarcomere in a relaxed muscle, a contracting muscle, and a fully contracted muscle. Notice in the contracting sarcomere that the thin filaments (blue) have moved toward the middle of the sarcomere. When the muscle is fully contracted, the thin filaments overlap in the middle of the sarcomere. Contraction only shortens the sarcomere; it does not change the lengths of the thick and thin filaments. A muscle can shorten to about 35% of its resting length when all its sarcomeres contract.

What makes the thin filaments slide? The key event is the binding between parts (called heads) of the myosin molecules in the thick filaments and specific sites on actin molecules in the thin filaments. The process of contraction requires energy, provided by ATP.

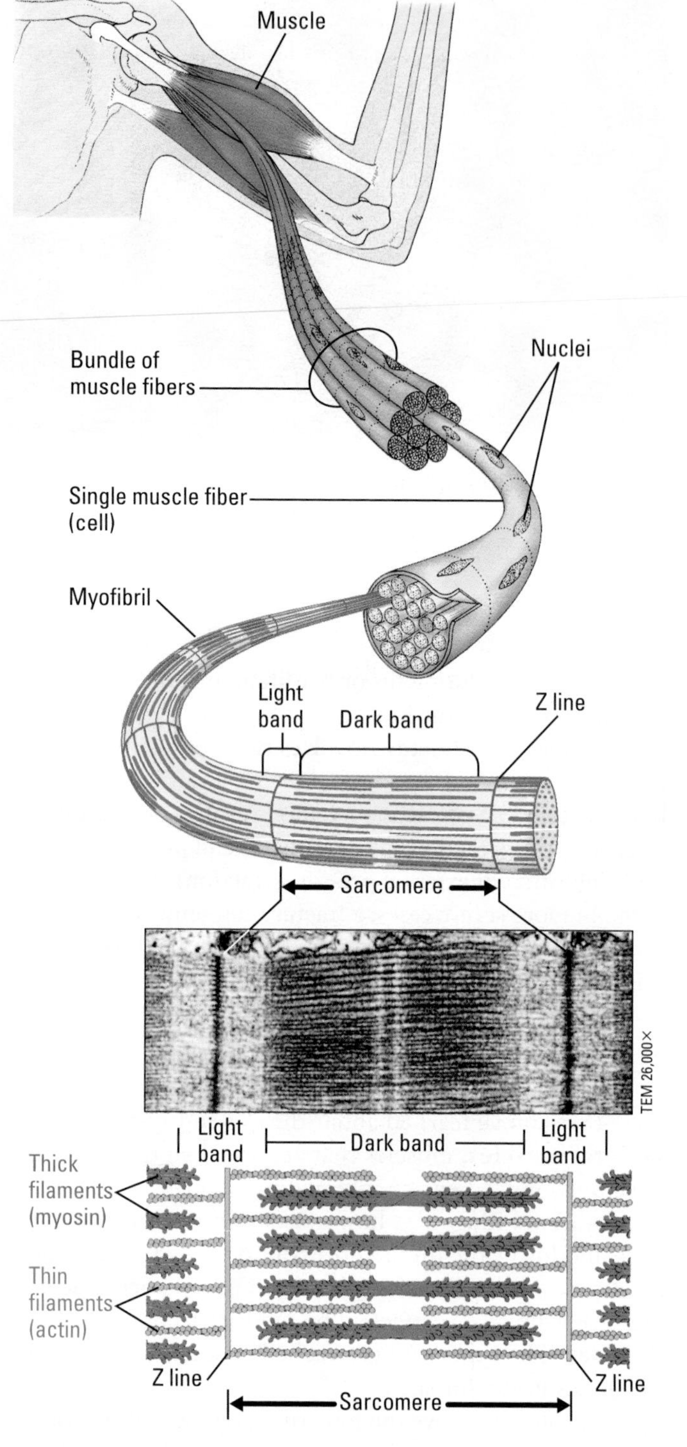

Figure 27.32 The contractile apparatus of skeletal muscle.

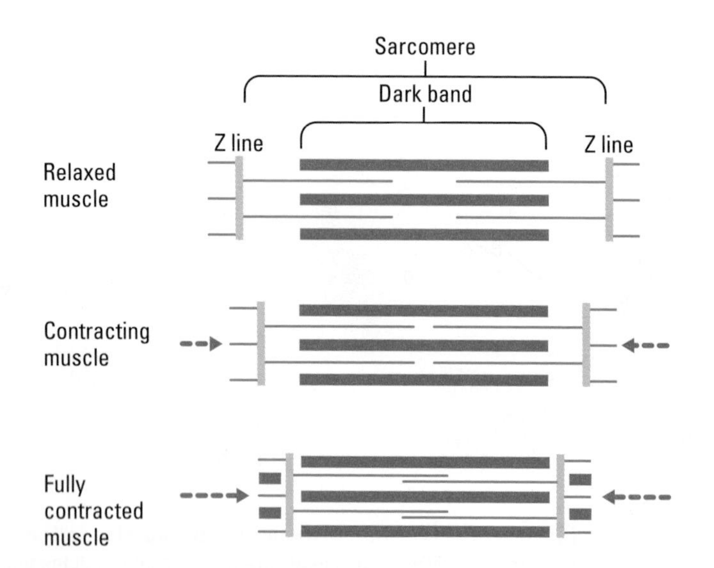

Figure 27.33 The sliding-filament model of muscle contraction.

Figure 27.34 indicates how sliding is thought to work. ❶ ATP binds to a myosin head, causing the head to detach from a binding site on actin (●). ❷ A myosin head gains energy from the breakdown of ATP and changes shape. In this high-energy position, the myosin head is cocked like a pistol ready to fire (actually, ready to bind with another site on the actin molecule). ❸ The energized myosin head binds to an exposed binding site on actin. ❹ The molecular event that actually causes sliding is called the power stroke. The myosin head bends back to its low-energy position, pulling the thin filament toward the center of the sarcomere. After the power stroke, more ATP binds with the myosin head, and the whole process repeats. On the next power stroke, the myosin head attaches to another binding site ahead of the previous one on the thin filament.

This sequence—detach, cock, attach, bend—occurs again and again in a contracting muscle. Though we show only one myosin head in the figure, a typical thick filament has about 350 heads, each of which can bind and unbind to a thin filament about five times per second. As long as sufficient ATP is present, the process continues until the muscle is fully contracted or until the signal to contract stops.

Motor Neurons: Control of Muscle Contraction The sarcomeres of a muscle fiber do not contract on their own. They must be stimulated to contract by motor neurons. A typical motor neuron can stimulate more than one muscle fiber because each neuron has many branches. In the example shown in **Figure 27.35**, you see two so-called **motor units,** each consisting of a neuron and all the muscle fibers it controls (two or three, in this case). A motor neuron has its dendrites and cell bodies in the central nervous system (here, the spinal cord). Its axon forms synapses, called neuromuscular junctions, with the muscle fibers. When

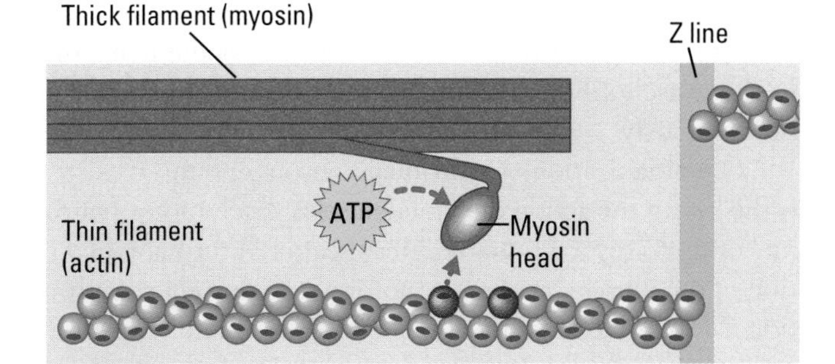

❶ ATP binds to a myosin head, which is released from an actin filament.

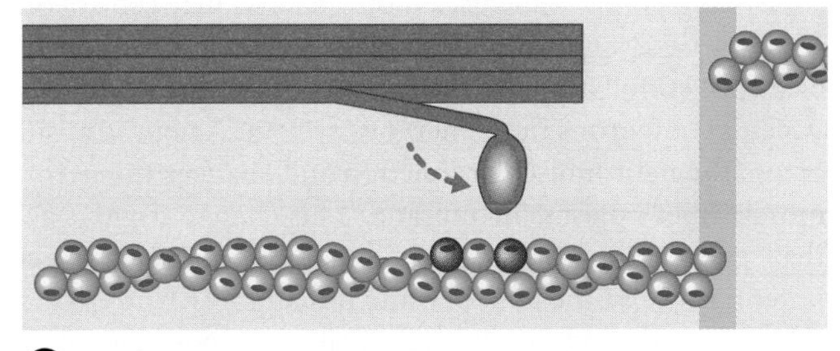

❷ The breakdown of ATP cocks the myosin head.

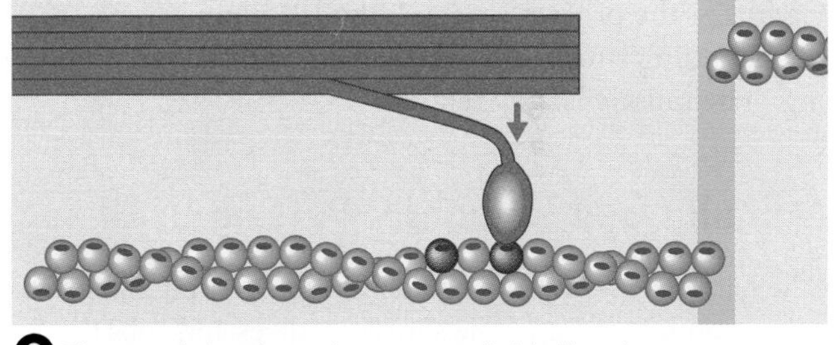

❸ The myosin head attaches to an actin binding site.

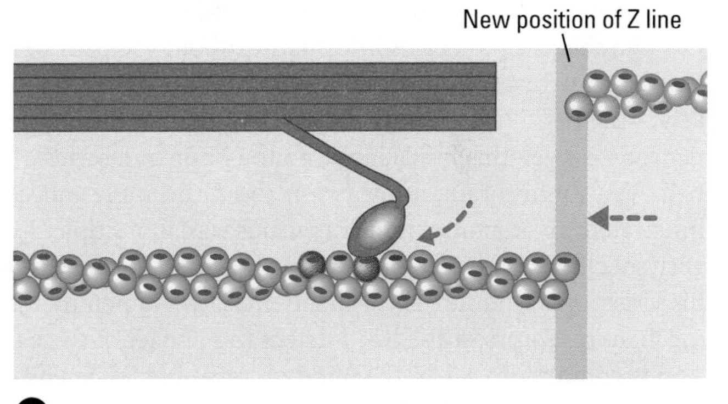

❹ The power stroke slides the actin (thin) filament toward the center of the sarcomere.

Figure 27.34 The mechanism of filament sliding.

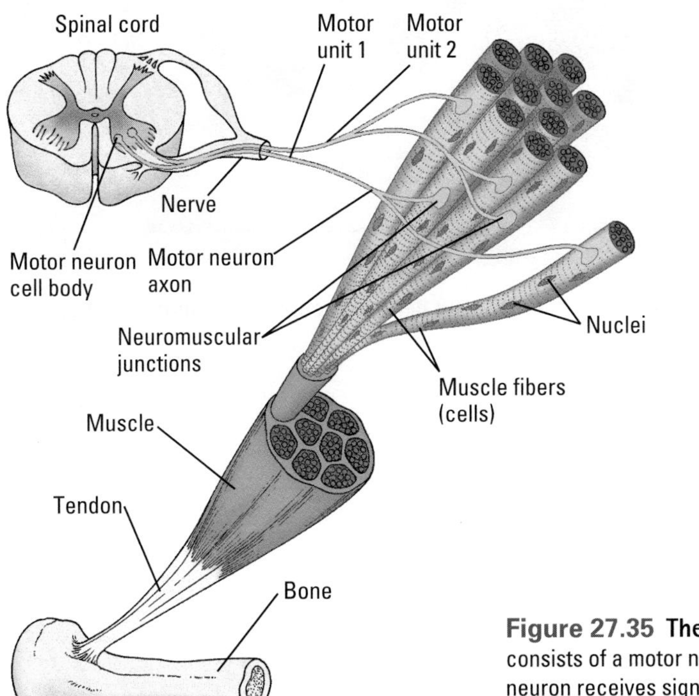

Figure 27.35 The relation between motor neurons and muscle fibers. A motor unit consists of a motor neuron and the one or more muscle fibers that it stimulates. When a motor neuron receives signals from the CNS, it passes them via synapses called neuromuscular junctions to the muscle fibers, causing them to contract.

a motor neuron sends out an action potential, its synaptic terminals release a neurotransmitter that diffuses across the neuromuscular junctions to the muscle fibers, making all the fibers of the motor unit contract simultaneously.

The organization of individual neurons and muscle cells into motor units is the key to the action of whole muscles. Each motor neuron may serve just one or up to several hundred fibers scattered throughout a muscle. Stimulation of the muscle by a single motor neuron produces only a weak contraction. More forceful contractions result when additional motor units are activated. Thus, depending on how many motor units your brain commands to contract, you can apply a small amount of force to lift a fork or considerably more to lift a barbell. In muscles requiring precise control, such as those controlling eye movements, a motor neuron may control only one fiber.

How does a motor neuron make a muscle fiber contract? The initial events of stimulation are the same as those that occur at a synapse between two neurons in the nervous system: A neurotransmitter diffuses across the neuromuscular junction and changes the permeability of the muscle fiber's plasma membrane. This change triggers action potentials that sweep across the muscle cell membrane. Inside the cell, the action potentials cause the endoplasmic reticulum (ER) to release calcium (Ca^{2+}) into the cytoplasm. Ca^{2+} helps to expose the binding sites on the thin filaments so that myosin heads can attach to actin, initiating filament sliding. When motor neurons stop sending action potentials to the muscle fibers, the process reverses: the ER pumps Ca^{2+} out of the cytoplasm, actin binding sites are blocked again, and the sarcomeres stop contracting. The muscle is now relaxed.

Stimulus and Response: Putting It All Together

The batter steps up to the plate. On the first pitch, he drives the ball hard toward an infield gap. The shortstop dives and robs the batter of a base hit **(Figure 27.36)**. We can now appreciate these events in terms of sensory input, nervous system communication, and motor response.

Figure 27.36 The nervous, sensory, and motor systems in action.

The shortstop receives information about the path of the ball from his sensory structures. Photoreceptors in his eyes detect the ball as soon as it leaves the bat. His ears hear the crack of the bat hitting the ball. Sensory neurons convey this information to the brain as a series of action potentials. Within his brain, information about the angle and speed of the ball is integrated. In response, motor neurons signal multiple muscles to perform very specific actions. He dives toward the ball, extending his arm, turning his wrist joint, and flexing his hand muscles to open the glove. When mechanoreceptors in the hand detect the impact of the ball, this information is communicated and integrated, resulting in signals to hand muscles that move against finger bones, squeezing the glove and capturing the ball. The fans see a spectacular play by a talented athlete. But underlying it all is the nervous system directing muscles to respond to information that sensory receptors have gathered about the environment.

We have now completed the unit on animal structure and function. Our overriding theme has been that structure underlies function—that the structural adaptations of a cell, tissue, organ, or organ system determine the job it can perform. We'll see the structure-function theme emerge again in the context of adaptations to the environment when we take up the study of plants in the next unit.

CHECKPOINT

1. Of the following, which is not part of the axial skeleton: skull, ribs, pelvis, vertebrae?
2. What two things can a young woman do to help prevent osteoporosis?
3. What two steps are necessary to heal a broken bone?
4. When exercising to strengthen muscles, why is it important to impose resistance while both flexing and extending the limbs?
5. The most abundant protein in a thin filament of a myofibril is ___________, and the most abundant protein in a thick filament is ___________.
6. How does the endoplasmic reticulum help regulate muscle contraction?

Answers: 1. Pelvis **2.** Ingest plenty of calcium and exercise regularly **3.** The broken bone must be returned to its natural position and then immobilized. **4.** This exercises both muscles of antagonistic pairs. **5.** actin; myosin **6.** By reversibly storing and releasing Ca^{2+}, the ER regulates the cytoplasmic concentration of this ion, which is required in the cytoplasm for the binding of myosin to actin.

EVOLUTION CONNECTION

Convergent Evolution of Eyes

The human eye is a refined sensory organ containing multiple parts that work together to form an image and send that visual information to the brain. Many people find it hard to believe that an organ as sophisticated as the eye could be the product of gradual evolution—"descent with modification"—rather than a complex structure that appeared all at once. Some people wonder how a partial or unsophisticated eye could be of any use as an evolutionary stage.

The answer lies in the observation that even very simple eyes can confer a survival advantage to an organism. Consider eye cups, the visual sensory organs of flatworms (phylum Platyhelminthes), specifically within the genus *Dugesia,* free-living nonparasitic flatworms known as planarians **(Figure 27.37)**. Planarian eye cups provide information about light intensity and direction but not enough data for the brain to form an image. Nevertheless, this information is enough to tell planarians to move directly away from a light source. Thus eye cups provide planarians with a clear survival advantage: the ability to find the undersides of rocks and other dark hiding places.

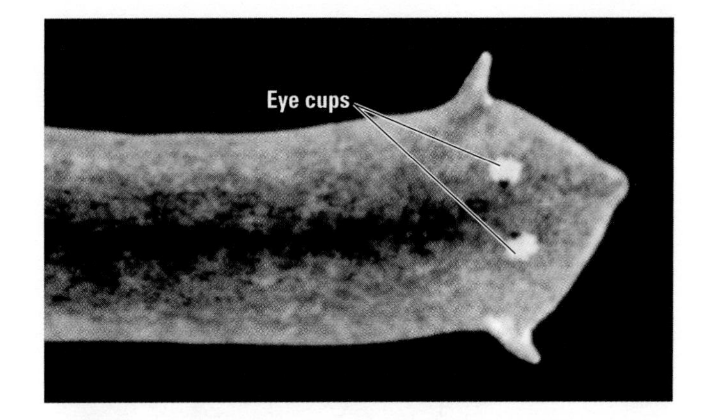

Figure 27.37 A planarian. You can see the eye cups on this planarian, a member of the phylum Platyhelminthes (flatworms). Planarian eye cups contain the simplest photoreceptors in the animal kingdom; yet even these simple structures provide a survival advantage.

Human eyes did not evolve from planarian eye cups. Rather, human and planarian visual organs are examples of analogous structures that arose through convergent evolution, the independent development of similar structures (see Chapter 14). An even better example of convergent evolution is the striking similarity between human eyes and the eyes of squids and octopuses. With more complicated lifestyles than planarians, vertebrates and cephalopods independently evolved similar, more sophisticated solutions to the problem of receiving useful visual information from the environment. But the relatively simple eye cups of planarians make an important point about evolution: Even very simple structures can confer a survival advantage to those organisms that have them. ■

Chapter Review

SUMMARY OF KEY CONCEPTS

For study help and activities, go to campbellbiology.com or the student CD-ROM.

An Overview of Animal Nervous Systems

eTutor *How Neurons Work*

• The nervous system is a communication and coordination network composed of neurons, specialized nerve cells capable of carrying signals throughout the body.

• **Organization of Nervous Systems**

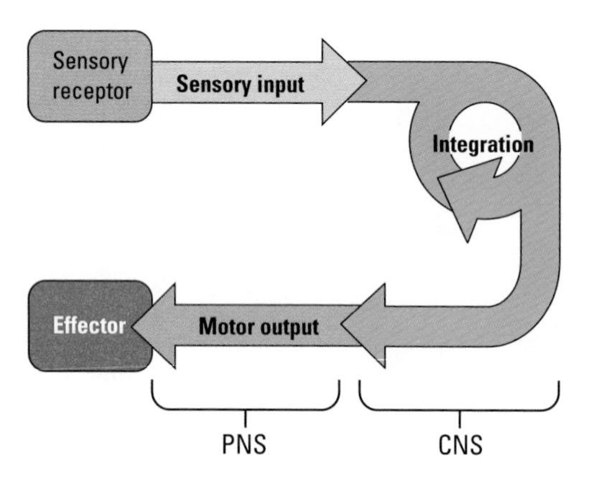

• **Neurons** Neurons are the functional units of the nervous system.

Action potential signal
Dendrites
Cell body
Axon
Myelin (speeds signal transmission)
Synaptic terminals

Activity *Neuron Structure*

• **Sending a Signal through a Neuron** At rest, a neuron's plasma membrane has a resting potential caused by the membrane's ability to maintain a positive charge on its outer surface opposing a negative charge on its inner surface. A stimulus alters a portion of the membrane, allowing ions to enter and exit the neuron and creating an action potential. Action potentials are self-propagated in a one-way chain reaction along a neuron. An action potential is an all-or-none event; its size is not affected by differences in stimulus strength. The frequency of action potentials does change with the strength of the stimulus.

Activity *Nerve Signals: Action Potentials*

Case Study in the Process of Science *What Triggers Nerve Impulses?*

• **Passing a Signal from a Neuron to a Receiving Cell** The transmission of signals beyond an individual neuron occurs at relay points called synapses. At chemical synapses, the sending cell secretes a chemical signal, a neurotransmitter, which crosses the synaptic cleft (a gap between the cells) and binds to a specific receptor on the surface of the receiving cell. A cell may receive different signals from many neurons. A wide variety of small molecules can act as neurotransmitters. Many drugs act at synapses by increasing or decreasing the normal effect of neurotransmitters.

Activity *Neuron Communication*

The Human Nervous System: A Closer Look

• **The Central Nervous System** and **The Peripheral Nervous System**

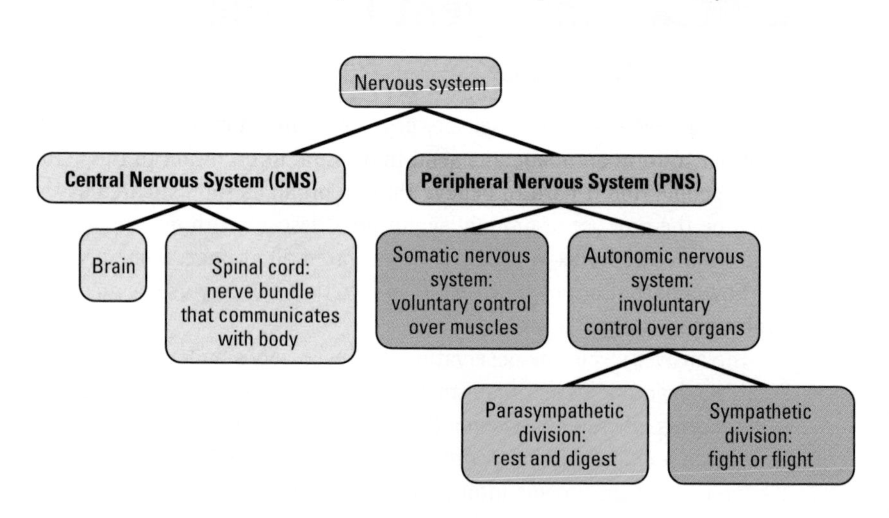

• **The Human Brain**

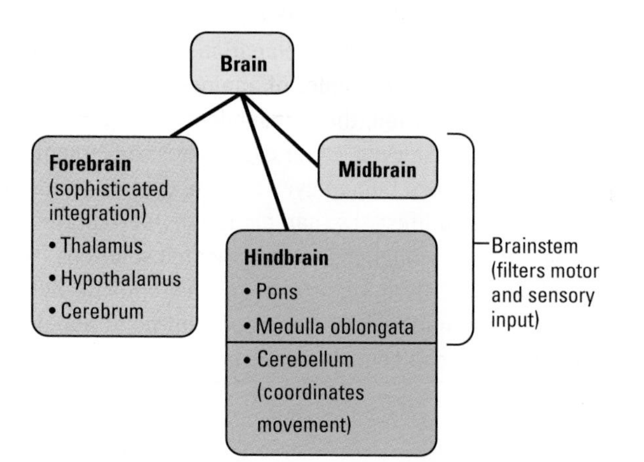

• The forebrain's cerebrum, divided into right and left hemispheres, is the largest and most complex part of the brain. In particular, the surface of the cerebrum, the cerebral cortex, is responsible for many of the most distinctive human traits. The right and left cerebral hemispheres specialize in different mental tasks. The limbic system, a functional group of integrating centers in the cerebral cortex, thalamus, and hypothalamus, is involved in emotions, memory, and learning.

The Senses

• **Sensory Input** The process of sensory transduction is the conversion by sensory receptors of stimuli to electrical signals called receptor potentials. Action potentials representing the stimuli are transmitted via sensory neurons to the central nervous system for processing. The strength of the stimulus alters the rate of action potential transmission. Some sensory neurons tend to become less sensitive when stimulated repeatedly, a phenomenon known as sensory adaptation. There are five categories of sensory receptors. Pain receptors sense stimuli that may indicate tissue damage; thermoreceptors detect heat or cold; mechanoreceptors respond to mechanical energy (such as touch, pressure, and sound); chemoreceptors respond to chemicals in the external environment or body fluids; and electromagnetic receptors respond to electricity, magnetism, and light. Photoreceptors, which sense light, are the most common type of electromagnetic receptor.

• **Vision** In the human eye, the cornea and lens focus light on photoreceptor cells in the retina. The human lens changes shape to bring objects at different distances into sharp focus. Photoreceptor cells called rods contain the visual pigment rhodopsin and function in dim light. Cones are photoreceptor cells that contain photopsin, which enables us to see color in full light. Nearsightedness, farsightedness, and astigmatism result from focusing problems in the lens. Corrective lenses bend the light rays to compensate, and surgical procedures can change the shape of the cornea.

Activity *Structure and Function of the Eye*

• **Hearing** The waves generated in the cochlear fluid move hair cells (mechanoreceptors) of the organ of Corti against an overlying membrane. Bending of the hair cells triggers nerve signals to the brain. Louder sounds cause greater movement and more action potentials; sounds of different pitches stimulate hair cells in different parts of the organ of Corti. Deafness can be caused by infections, injury, or overexposure to loud noises.

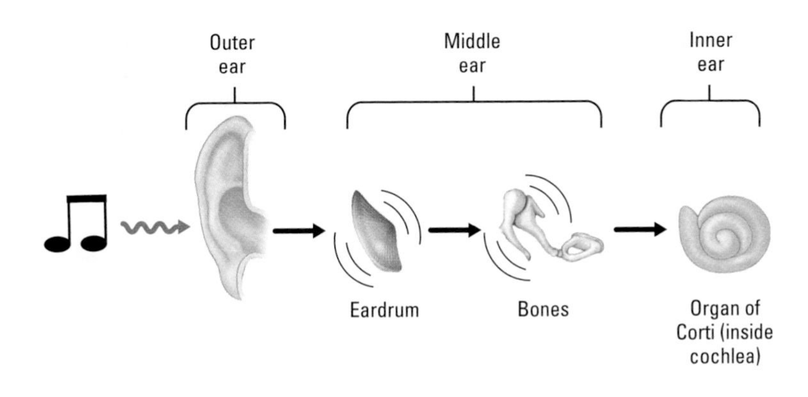

Motor Systems

• Movement is one of the most distinctive features of animals. The nervous system issues commands to the muscular system, and the muscular system exerts propulsive force against the skeleton.

• **The Skeletal System** The skeletal system functions in support, movement, and the protection of internal organs. The human endoskeleton is composed of cartilage and bone. The axial skeleton (skull, vertebrae, and ribs) supports the body's trunk. The appendicular skeleton consists of bones of the shoulder, limbs, and pelvis. Movable joints provide flexibility. A bone is a living organ containing several kinds of tissues. It is covered with a connective tissue membrane. Cartilage at the ends of the bone cushions the joints. Long bones have a central cavity that stores fat; they also have cavities at their ends that contain red marrow, where blood cells form. The human skeleton is versatile, but it is also subject to problems, such as arthritis and osteoporosis. Broken bones are realigned and immobilized by splints or casts; bone cells then build new bone and repair the break.

Activity *The Human Skeleton*

• **Muscular System** Skeletal muscles pull on bones to produce movements. Antagonistic pairs of muscles produce opposite movements. Each skeletal muscle cell, or fiber, contains bundles of myofibrils, which in striated muscle exhibit alternating light and dark bands. Each myofibril contains bundles of overlapping thick (myosin) and thin (actin) protein filaments. Repeating units of thick and thin filaments, called sarcomeres, are the muscle fiber's contractile units. The sliding-filament model explains the molecular process of muscle contraction. Using ATP, the myosin heads of the thick filaments attach to binding sites on the actin molecules and pull the thin filaments toward the center of the sarcomere. Motor neurons carry action potentials that initiate muscle contraction. A neuron can branch to a number of muscle fibers; the neuron and the muscle fibers it controls constitute a motor unit. The strength of a muscle contraction depends on the number of motor units activated.

Activity *Skeletal Muscle Structure*

Activity *Muscle Contraction*

Case Study in the Process of Science *How Do Electrical Stimuli Affect Muscle Contraction?*

• **Stimulus and Response: Putting It All Together** An animal's nervous system connects sensations derived from environmental stimuli to responses carried out by its muscles.

SELF-QUIZ

1. The large, central hub of a neuron, called the __________, contains the nucleus and organelles. Short, numerous fibers called __________ receive incoming messages, while a single, long fiber called a(an) __________ conducts signals toward other cells. A neuron ends in many branches, each with a bulb-like __________ that contains neurotransmitters, which can be used to communicate signals to other neurons.

2. Your nervous system can be divided into two broad subsystems, the __________ and the __________. The latter of these is made up of nerves that conduct signals to and from the other structures. The former consists of the __________ and __________.

3. When people say that alcohol lowers a person's inhibitions, it is a behavioral description. At the neurological level, it is probably more accurate to say that alcohol raises inhibitions. Why?

4. While driving down the interstate at rush hour, you get cut off by an 18-wheeler and have to slam on the brakes. In addition to a major case of road rage, you also develop a rapid heart rate and a rapid breathing rate. What caused this increase? Your answer should indicate the branch of the nervous system involved.

5. A victim of a severe head injury can live for years in a nonresponsive state in which the cerebral cortex is not functioning but the person is still alive and performing metabolic functions. Based on your knowledge of brain structure and function, how can this be possible? Make sure your answer indicates which brain structures might be keeping the person alive.

6. How is an action potential different from a receptor potential? Your answer should include an explanation of why receptor potentials are needed.

7. Mr. Johnson was becoming slightly deaf. To test his hearing, his doctor held a vibrating tuning fork tightly against the back of Mr. Johnson's skull. This sent vibrations through the bones of the skull, setting the fluid in the cochlea in motion. Mr. Johnson could hear the tuning fork this way, but not when it was held away from the skull a few inches from his ear. Where was Mr. Johnson's hearing problem located? (Explain your answer.)

a. In the auditory nerve leading to the brain
b. In the hair cells in the cochlea
c. In the bones of the middle ear
d. In the fluid of the cochlea

8. A human's soft internal organs are protected by the __________, while the bones of the limbs make up the __________.

9. Arm and leg muscles are arranged in antagonistic pairs. How does this affect their functioning?

a. It provides a backup if one of the muscles is injured.
b. One muscle of the pair pushes while the other pulls.
c. A single neuron controls both of them.
d. It allows the muscles to produce opposing movements.

10. Muscle A and muscle B are the same size, but muscle A is capable of much finer control than muscle B. Which of the following is likely to be true of muscle A? (Explain your answer.)

a. It is controlled by more neurons than muscle B.
b. It contains fewer motor units than muscle B.
c. It is controlled by fewer neurons than muscle B.
d. Each of its motor units consists of more cells than the motor units of muscle B.

Answers to the Self-Quiz questions can be found in Appendix D.

Go to the website or CD-ROM for more Self-Quiz questions.

THE PROCESS OF SCIENCE

11. Much of what we know about brain function has come from rare individuals whose brains were altered as the result of surgery, accidents, or disease. Suppose a patient has brain surgery on the cerebral cortex, and immediately following the surgery, she appears normal. As she recovers over the next few days, some unusual things happen, such as being unable to name objects while looking at them with one eye covered. Her sensory perception seems normal except for when she tries to integrate information. What brain structure might have been damaged during the surgery that would account for these symptoms? Justify your answer.

12. Sensory organs tend to come in pairs. We have two eyes and two ears. Similarly, a planarian worm has two eye cups, a rattlesnake has two infrared receptors, and a butterfly has two antennae. Propose a testable hypothesis that could explain the advantage of having two ears or eyes instead of one.

13. When a person dies, muscles become rigid and fixed in position—a condition known as rigor mortis, which often figures importantly in mystery novels. Rigor mortis occurs because muscle cells are no longer supplied with ATP (when breathing stops, ATP synthesis ceases). Also, calcium flows freely into dying cells. Explain, in terms of the mechanism of muscle contraction described in this chapter, why the presence of calcium and the lack of ATP would cause muscles to become rigid, rather than limp, after death.

BIOLOGY AND SOCIETY

14. Brain injuries tend to be so severe because most neurons will not be regenerated once they have been damaged. The use of embryonic stem cells (see Chapter 11) has been proposed as a potential solution to many neurological diseases. Ideally, stem cells could be placed into the brain of a person with Alzheimer's or Parkinson's disease. Theoretically, those cells could differentiate into neurons, replacing those that had been killed by the disease. Do you favor or oppose research along these lines? Why? Would your opinion change if the person to be treated with stem cells were a close family member who would die if not treated?

15. Alcohol's depressant effects on the nervous system cloud judgment and slow reflexes. Alcohol consumption is a factor in many fatal traffic accidents in the United States. What are some other impacts of alcohol abuse on society? What are some of the responses of people and society to alcohol abuse? Do you think that alcohol abuse is primarily an individual or societal problem? Do you think our responses to alcohol abuse are appropriate and proportional to the seriousness of the problem?

16. Have you ever felt your ears ringing after listening to loud music from a stereo or at a concert? Are you worried that this music may be loud enough to permanently impair your hearing? Do you think that people are aware of the possible danger of prolonged exposure to loud music? Should anything be done to warn or protect them? If so, what action would you suggest? What effect might warnings have?

UNIT 6

Plant Structure and Function

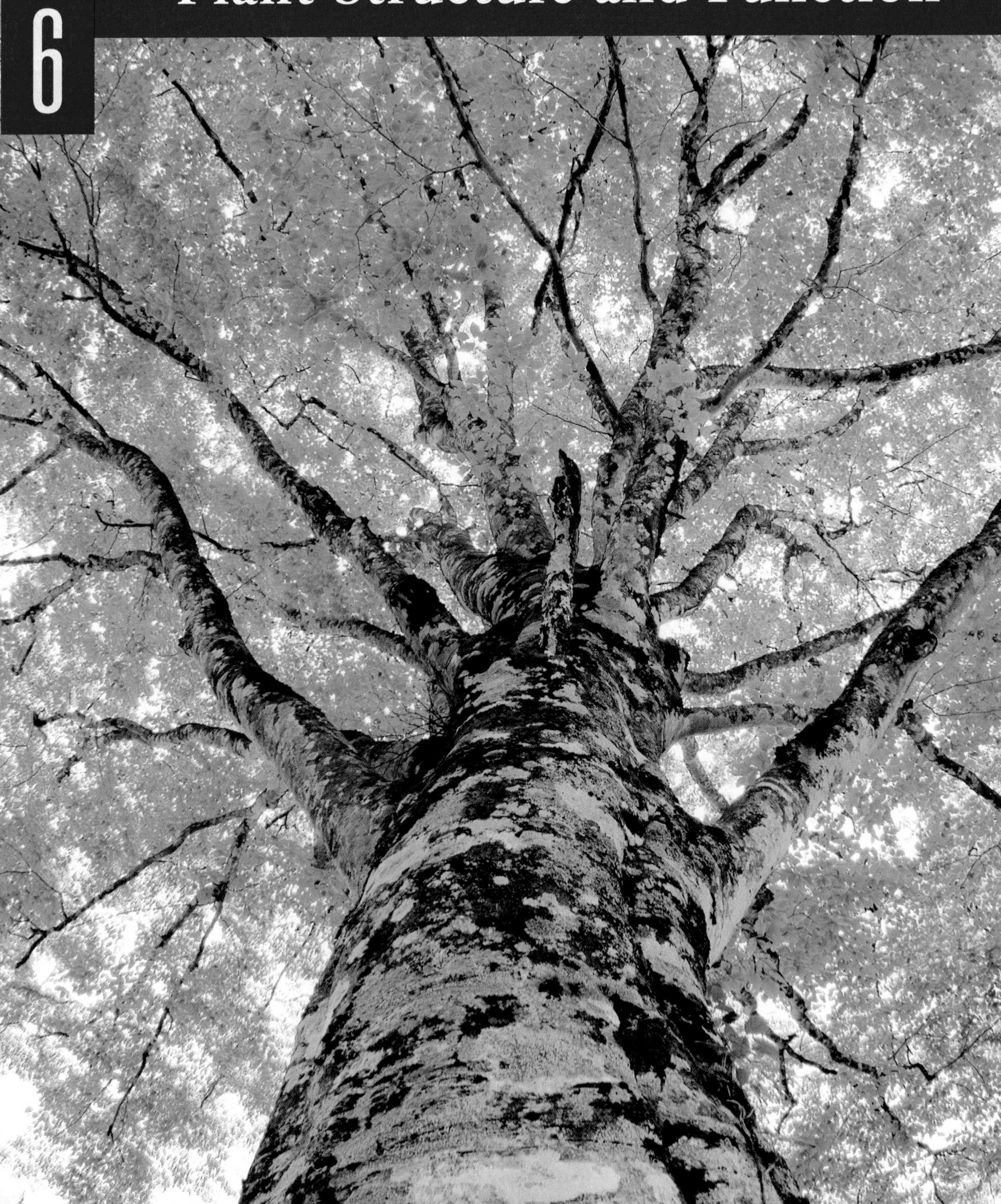

28 The Life of a Flowering Plant

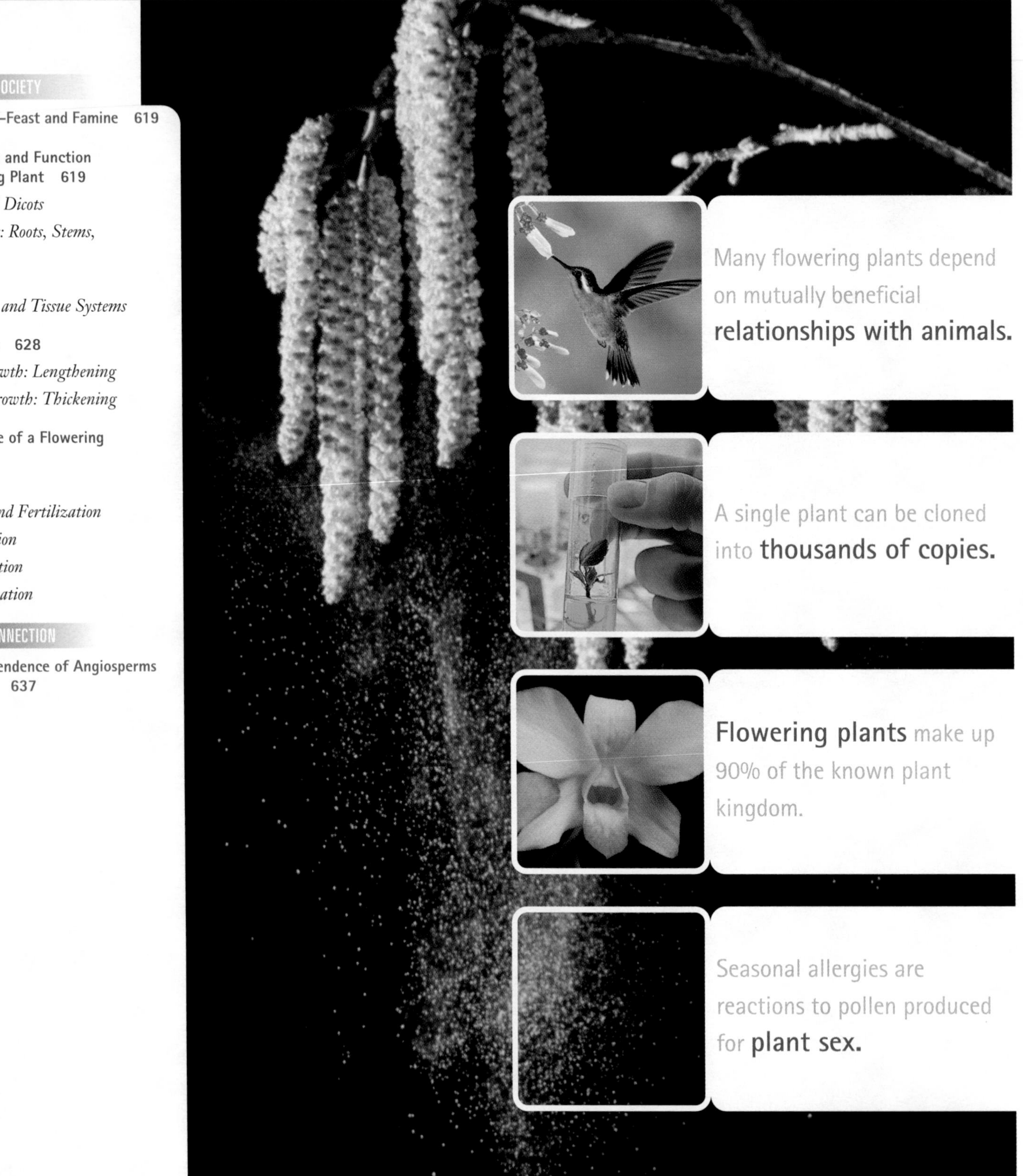

Many flowering plants depend on mutually beneficial **relationships with animals.**
A single plant can be cloned into **thousands of copies.**
Flowering plants make up 90% of the known plant kingdom.
Seasonal allergies are reactions to pollen produced for **plant sex.**

BIOLOGY AND SOCIETY

Plant Cloning—Feast and Famine

Humans depend on plants. Nearly every comfort of modern civilization has roots in the soil. Plants provide us with lumber, fabric, paper, industrial chemicals, and medicines. In particular, our diet—like the diets of virtually all land animals—consists almost entirely of plants and of animals that ate plants.

For the past 10,000 years, humankind has cultivated plants to ensure an adequate food supply, often with great success. Consider the potato. Grown in South America for thousands of years, potatoes were introduced to Europe in the 16th century. The transplanted crop was particularly well suited to grow in the rocky soil of Ireland, and there it thrived. The potato soon dominated Irish agriculture because of its high yield per acre and because it could be easily cloned: When a mature potato is split and the parts planted, each "eye" can develop into a new potato plant that is genetically identical to the parent plant. Eventually, the potato accounted for 80% of the calories in a typical Irish diet.

Cloned potatoes fed enormous numbers of people, allowing Ireland's population to grow significantly. However, a problem lurked underground. While cloning allows for rapid propagation, it creates a genetically homogeneous population. When the environment changes, such a crop is more likely to be wiped out than a crop of genetically varied individuals.

In the 1840s, late blight (*Phytophthora infestans*), a fungus-like organism that causes the disease potato rot, began to infect Irish potato plants (**Figure 28.1**). Because of their genetic similarity, virtually all the plants were equally susceptible to infection. The disease spread throughout Ireland within a few years, devastating the potato crop. The resulting famine caused Ireland's population to drop by about 25% in just a few years. Over 1 million people died of starvation and even more fled, primarily to the United States and Canada. In this way, world history was profoundly altered by a plant. Despite the development of fungicides against *P. infestans*, late blight continues to be a problem today: New fungicide-resistant varieties of *P. infestans* have recently appeared in Russia, where potatoes remain a staple crop.

In addition to their importance to humans, plants are vital to the well-being of Earth's biosphere. Above and below the ground, plants provide shelter, food, and breeding areas for animals, fungi, and microorganisms. Plant roots prevent soil erosion, and photosynthesis in plant leaves reduces carbon dioxide levels in the atmosphere and adds oxygen to the air.

In this unit, you will learn how the structure-fits-function theme applies to plants. Because **angiosperms** (flowering plants) make up about 90% of the plant kingdom, they are the focus of this unit. In this chapter, we begin by examining angiosperm structure, both at the level of the whole plant and at the microscopic level of cells and tissues. We'll then see how plant structures function in growth and reproduction. ■

Figure 28.1 Potato rot. The disease potato rot, which devastated Ireland's food supply in the 1840s, is caused by a fungus-like organism called late blight (*Phytophthora infestans*, inset).

The Structure and Function of a Flowering Plant

Angiosperms have dominated the land for over 100 million years, and about 250,000 species of flowering plants are known to inhabit Earth today (see Chapter 16). Most of our foods come from a few hundred domesticated species of flowering plants. Among these foods are roots, such as beets and carrots; the fruits of trees and vines, such as apples, nuts, berries, and squashes; the fruits and seeds of legumes, such as peas, peanuts, and beans; and grains, the fruits of grasses such as rice, wheat, and corn.

Monocots and Dicots

Botanists have traditionally classified angiosperms into two groups, called monocots and dicots, on the basis of several structural differences **(Figure 28.2)**. The names of the groups refer to embryonic structures called **cotyledons** (also called seed leaves), the first leaves to emerge from a growing seedling. A **monocot** embryo has one seed leaf; a **dicot** embryo has two seed leaves.

Monocots (about 70,000 species) include the orchids, palms, and lilies, as well as the grains and other grasses. Most monocots have leaves with parallel veins. Monocot stems have vascular tissues (tissues that transport water and nutrients) organized into bundles that are arranged in a scattered pattern. The flowers of most monocots have petals and floral parts in multiples of three. Monocot roots form a fibrous system—a mat of threads—that spreads out below the soil surface. With most of their roots in the top few centimeters of soil, monocots, especially grasses, make excellent ground covers and can help reduce soil erosion.

Most angiosperms are dicots (about 170,000 species). The dicots include most shrubs and trees (except for the conifers, which are cone-bearing gymnosperms), the majority of ornamental plants, and many food crops. Dicot leaves have a multibranched network of veins, and dicot stems have vascular bundles arranged in a ring. Dicot flowers usually have petals and floral parts in multiples of four or five. The large, vertical root of a

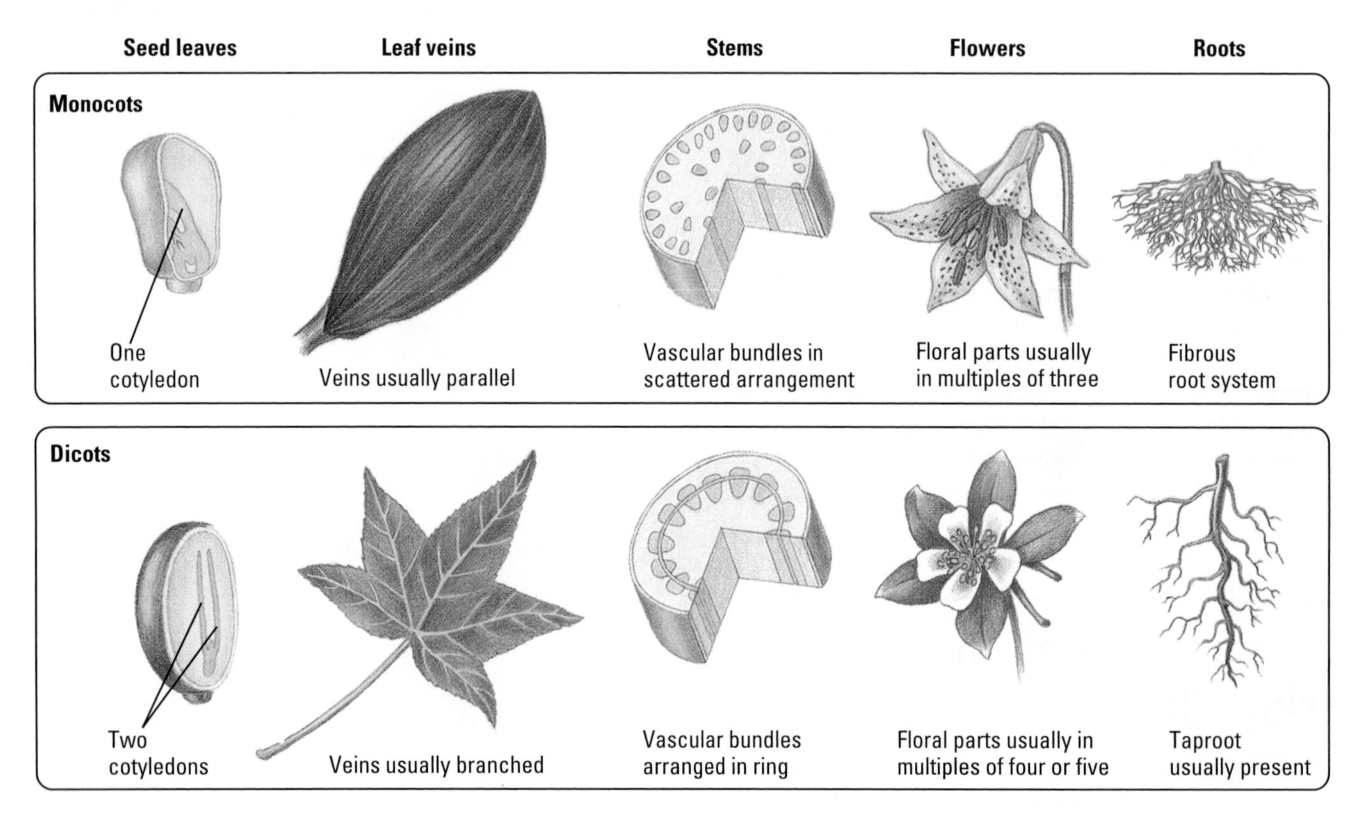

Figure 28.2 Comparing monocots and dicots.

dicot, known as a taproot, goes deep into the soil, as you know if you've ever tried to pull up a dandelion.

Plant Organs: Roots, Stems, and Leaves

Among the evolutionary adaptations that made it possible for plants to thrive on land were structures for absorbing water and minerals from the soil; a large light-collecting surface; the ability to take in carbon dioxide from the air for photosynthesis; and adaptations for surviving dry conditions. In a land plant such as the generalized angiosperm shown in **Figure 28.3**, the roots (the below-ground structures) and the shoots (the above-ground structures, including stems, leaves, and flowers) perform all these vital functions. Neither roots nor shoots can survive without the other. Lacking chloroplasts and living in the dark, the roots would starve without sugar and other organic nutrients transported from photosynthetic stems and leaves. Conversely, stems and leaves depend on the water and minerals absorbed by roots.

Roots A plant's **root system** anchors it in the soil, absorbs and transports minerals and water, and stores food. The fibrous root system of a monocot provides broad exposure to soil water and minerals as well as firm anchorage. The root system of a dicot, with many small secondary roots growing out from one large taproot, absorbs water and minerals from deeper in the soil. Near the root tips in both dicots and monocots, tiny projections called **root hairs** increase the surface area of the root (shown in the insets in Figure 28.3), providing an extensive outer layer for absorption. Each root hair is an outgrowth of a cell in the epidermal tissue of the root. It is difficult to move an established plant without injuring it because transplantation often damages the plant's delicate root hairs.

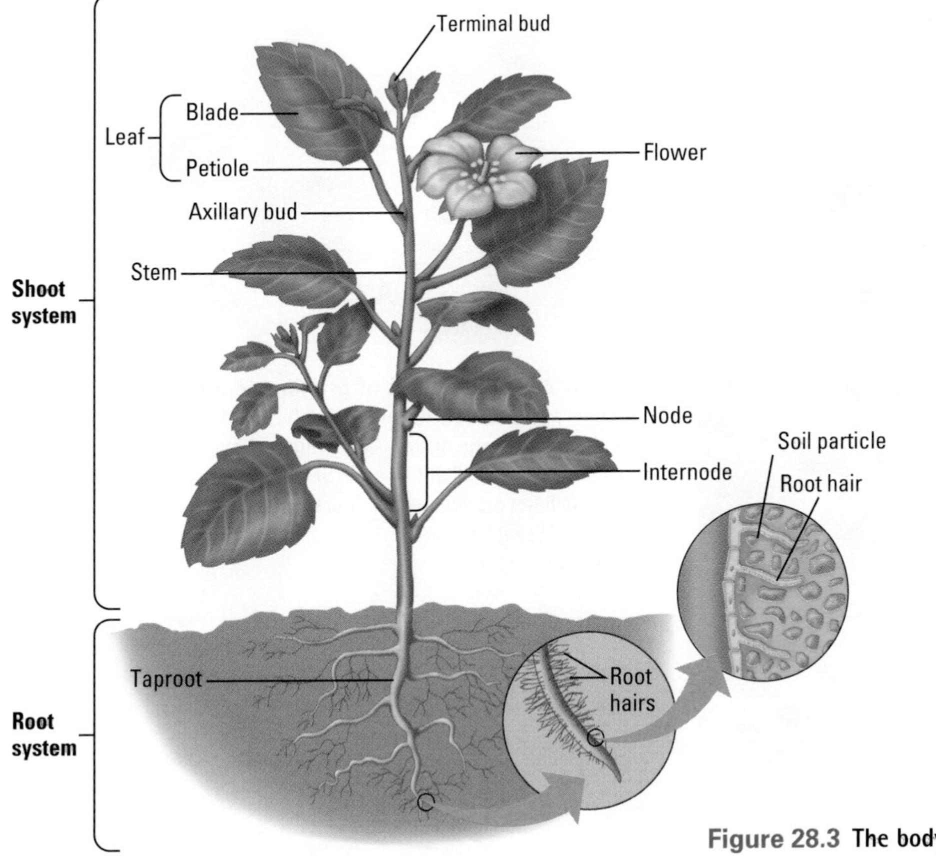

Figure 28.3 The body plan of a flowering plant.

Large taproots, such as those found in carrots, turnips, sugar beets, and sweet potatoes, store food in the form of carbohydrates such as starch **(Figure 28.4)**. The plants use these stored sugars during periods of active growth and when they are producing flowers and fruit.

Stems The **shoot system** of an angiosperm is made up of stems, leaves, and structures for reproduction (flowers). As indicated in Figure 28.3, the **stems** generally grow above-ground and support the leaves and flowers. In the case of a tree, the stems are the trunk and all the branches. A stem has **nodes,** the points at which leaves are attached, and **internodes,** the portions of the stem between nodes.

The two types of buds you see in Figure 28.3 are undeveloped shoots. When a plant stem is growing in length, the **terminal bud,** at the apex (tip) of the stem, has developing leaves and a compact series of nodes and internodes. The **axillary buds,** one in each of the angles formed by a leaf and the stem, usually remain dormant. In many plants, the terminal bud produces hormones that inhibit growth of the axillary buds, a phenomenon called **apical dominance.** By concentrating the plant's resources on growing taller, apical dominance is an evolutionary adaptation that increases the plant's exposure to light. Branching is also important for increasing the exposure of the shoot system to the environment, and under certain conditions, the axillary buds begin growing and developing into branches. In some plants, such as fruit trees and many houseplants, removing the terminal bud through pruning or "pinching back" stimulates the growth of axillary buds **(Figure 28.5)**.

As Figure 28.6 shows, stems can take many forms. Strawberry plants have horizontal stems, or runners, that grow along the ground surface. A

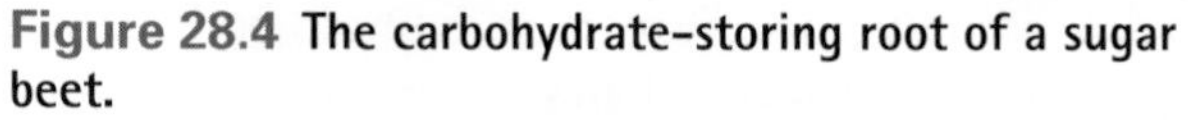

Figure 28.4 The carbohydrate-storing root of a sugar beet.

Figure 28.5 The effect of pruning on a basil plant. The terminal bud of the basil plant on the left produces hormones that inhibit growth of the axillary buds. Such apical dominance causes the plant to grow tall and sparse, or "leggy." The basil plant on the right has been pruned, which decreases apical dominance and increases branching and leaf production. The result is an increase in the useful yield.

Figure 28.6 Modified stems.

runner is a means of asexual reproduction; as shown in **Figure 28.6a**, a new plant can emerge from it. This is why strawberries, if left unchecked, can rapidly fill your garden. If you dig up an iris or ginger plant, you'll see a different stem modification; the large, brownish, rootlike structures near the soil surface are actually horizontal underground stems called rhizomes **(Figure 28.6b)**. Rhizomes store food, and because they have buds, they can also form new plants. About every three years, gardeners can dig up iris rhizomes, split them, and plant the partial rhizomes to get multiple identical plants. A white potato plant has rhizomes ending in enlarged structures called **tubers** (the potatoes we eat), where food is stored in the form of starch **(Figure 28.6c)**. The "eyes" of a potato are axial buds, which can grow when planted. As mentioned in the chapter opening essay, the axial buds allow potatoes to be easily propagated.

Leaves **Leaves** are the primary sites of photosynthesis in most plants, although many plants also have green, photosynthetic stems. A leaf consists of a flattened **blade** and a stalk, or **petiole,** which joins the leaf to the stem (see Figure 28.3).

Plant leaves are highly varied in their arrangements and in their shapes **(Figure 28.7)**. Grasses and most other monocots, for instance, have long leaves without petioles. Some dicots have enormous petioles that contain a lot of water and stored food; the edible stalks of celery are one example. A modified leaf called a tendril can help plants such as sweet peas or grapes climb up their supports **(Figure 28.8a)**. And the spines of the barrel cactus **(Figure 28.8b)** are modified leaf parts that may protect the plant from predatory animals. In this and many other cactus species, the main photosynthetic organ is the large green stem, which also stores water.

So far we have examined plants as we see them with the unaided eye. In the next section, we begin to dissect a plant and explore its microscopic organization.

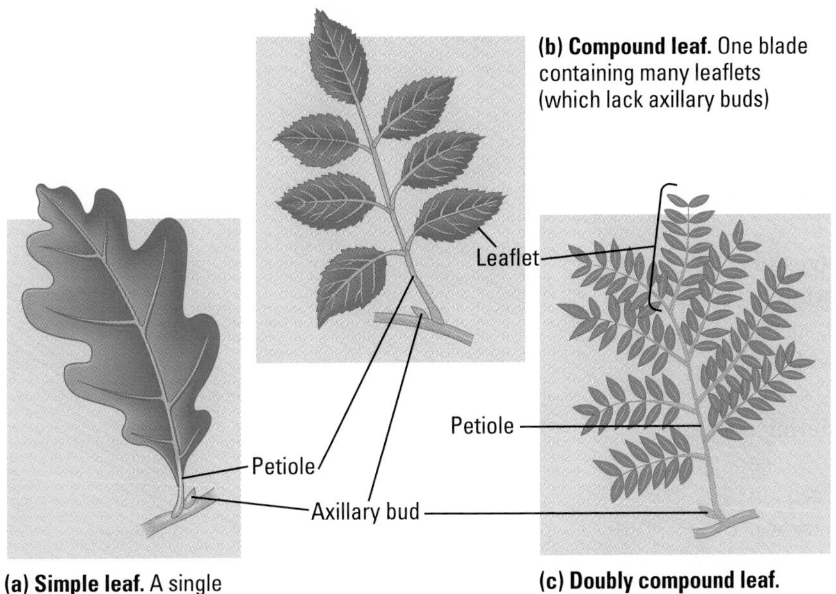

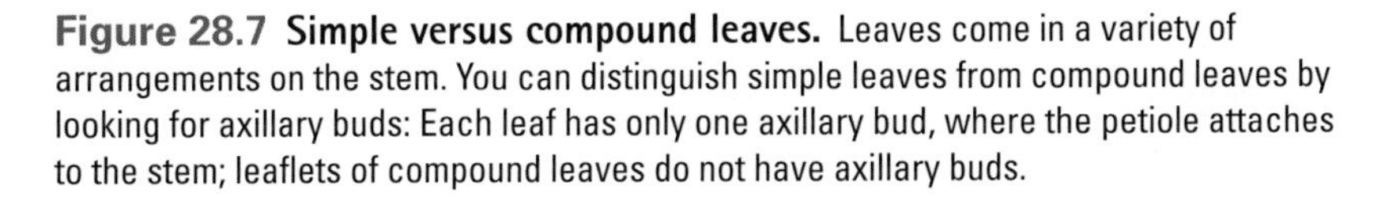

Figure 28.7 Simple versus compound leaves. Leaves come in a variety of arrangements on the stem. You can distinguish simple leaves from compound leaves by looking for axillary buds: Each leaf has only one axillary bud, where the petiole attaches to the stem; leaflets of compound leaves do not have axillary buds.

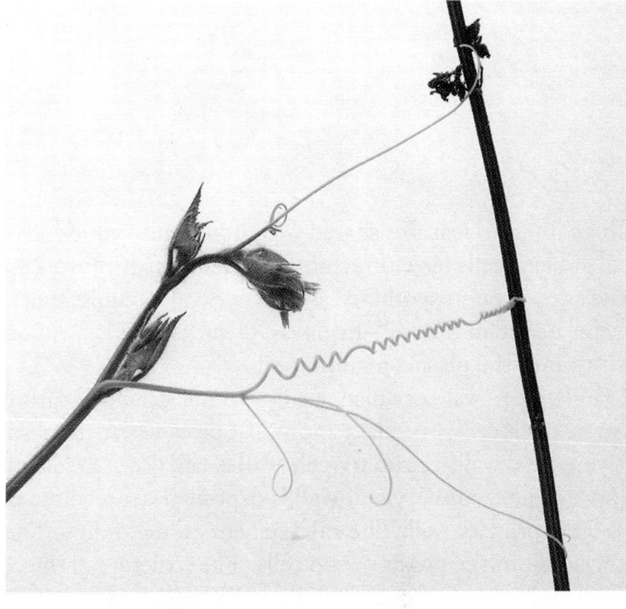

(a) Sweet pea tendrils

(b) Barrel cactus spines

Figure 28.8 Modified leaves.

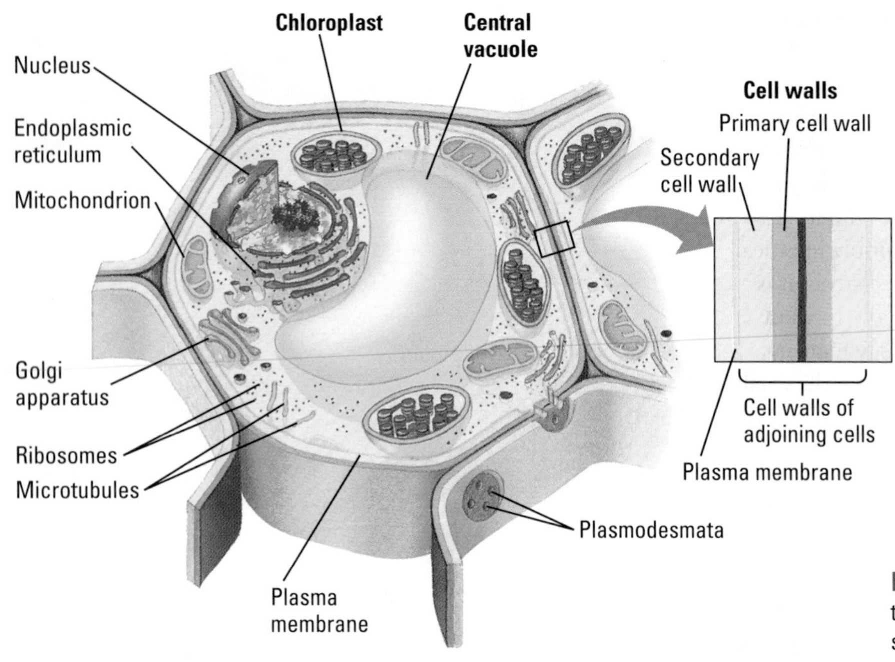

Figure 28.9 The structure of a plant cell. The boldface terms highlight the unique features of plant cells. The enlargement shows the interface of two cell walls from adjoining cells.

Plant Cells

In addition to features shared with other eukaryotic cells (see Figure 4.6), most plant cells have three unique structures **(Figure 28.9)**: **chloroplasts,** the sites of photosynthesis; a large **central vacuole** containing fluid that helps maintain the cell's firmness, or turgor; and a supportive **cell wall** that surrounds the plasma membrane.

Plant cell walls are made mainly of the structural carbohydrate cellulose. Some plant cells, especially those that provide structural support, have a two-layered cell wall; a **primary cell wall** is laid down first, and then a thicker, more rigid **secondary cell wall** is deposited between the plasma membrane and the primary wall. The enlargement on the right in Figure 28.9 shows the adjoining cell walls of two cells. The structure of the cell and the nature of its wall often correlate with the cell's main functions. This is evident in the various types of plant cells shown in Figure 28.10.

Parenchyma cells (Figure 28.10a) are the most abundant type of cell in most plants. They remain alive when mature and have only primary (often thin) walls. Parenchyma cells perform a variety of functions, such as food storage and photosynthesis. Most parenchyma cells can divide and differentiate (become specialized in their structure and function; see Chapter 11) into other types of plant cells, which they may do during repair of an injury.

Collenchyma cells (Figure 28.10b) resemble parenchyma cells in lacking secondary walls, but they have unevenly thickened primary walls. Their main function is to provide support in parts of the plant that are actively growing. Young stems and petioles, for example, often have collenchyma cells that elongate with the growing stem.

Sclerenchyma cells have thick secondary cell walls (colored yellow in **Figure 28.10c**) that are sometimes hardened with **lignin,** a chemical component of wood. Mature sclerenchyma cells cannot elongate, and they develop

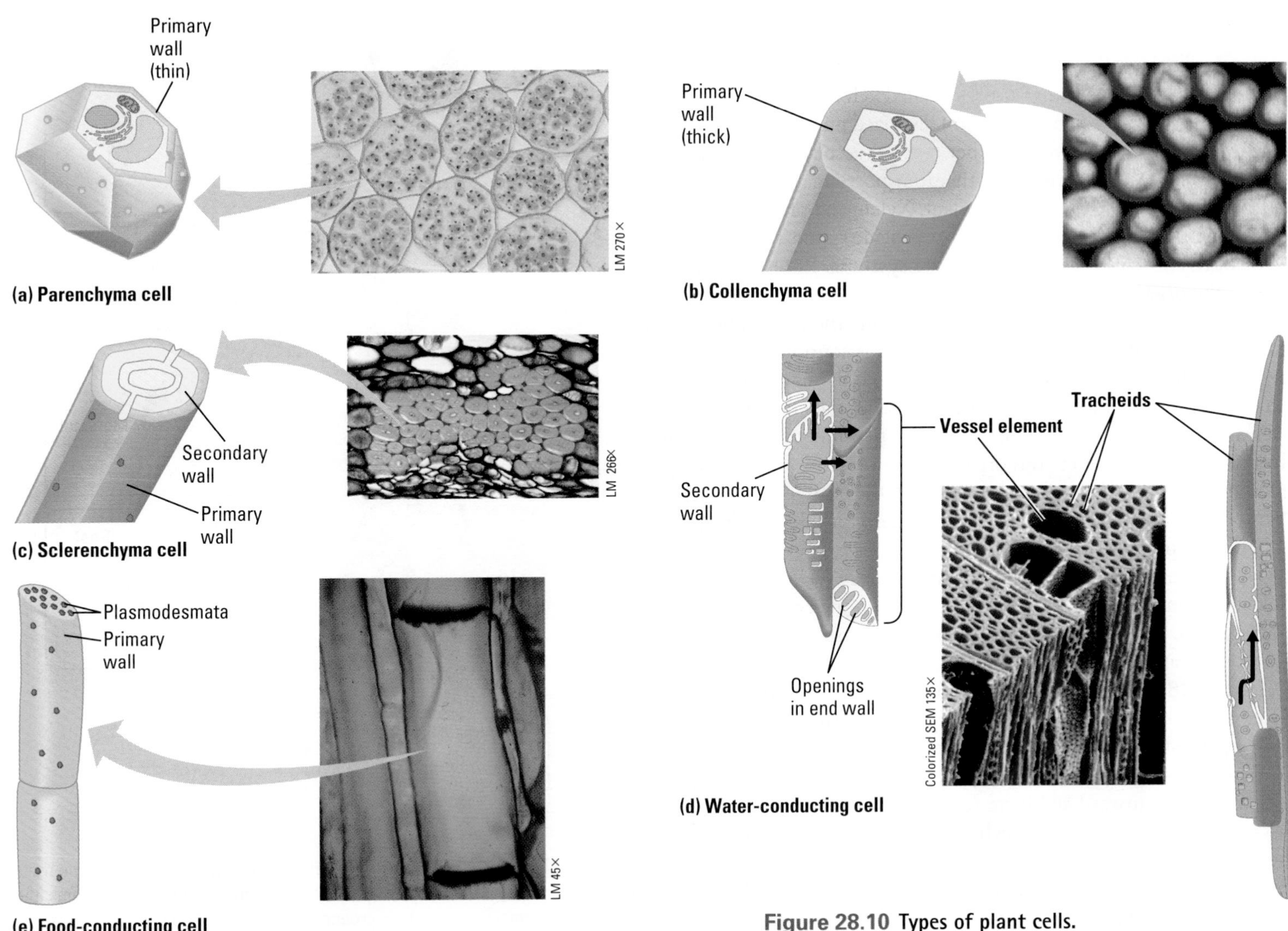

Figure 28.10 Types of plant cells.

only in regions that have stopped growing in length. When mature, most sclerenchyma cells are dead, and their rigid cell walls support the plant much like steel beams in a building. Sclerenchyma cells make up some commercially important plant products, such as the highly versatile hemp fiber used to make rope and clothing.

Water-conducting cells are of two types—tracheids and vessel elements—both of which have rigid, lignin-containing secondary cell walls **(Figure 28.10d)**. **Tracheids** are long, narrow cells with tapered ends; **vessel elements** are wider, shorter, and less tapered. Water-conducting cells are arranged in chains with overlapping ends to form a system of water-carrying tubes. The tubes are hollow because water-conducting cells are dead when mature; only their cell walls remain.

Food-conducting cells are also arranged end to end, forming tubes **(Figure 28.10e)**. Unlike water-conducting cells, however, these cells remain alive at maturity. Their end walls, which are perforated with large plasmodesmata (open channels through which cytoplasm can flow from cell to cell), allow sugars, other compounds, and some minerals to move between adjacent food-conducting cells.

Plant Tissues and Tissue Systems

As in animals, the cells of plants are grouped into tissues with characteristic functions. For example, vascular tissue called **xylem** contains the water-conducting cells that convey water and dissolved minerals upward from the roots to the stems and leaves. Another type of vascular tissue called **phloem** contains the food-conducting cells that transport sugars from leaves or storage tissues to other parts of the plant. Xylem and phloem also contain sclerenchyma and parenchyma cells that provide support and storage.

Plant tissues are organized into **tissue systems.** Roots, stems, and leaves—the organs of plants—are made up of three tissue systems: the dermal tissue system, the vascular tissue system, and the ground tissue system. As shown in **Figure 28.11**, each tissue system is continuous throughout the plant. The **dermal tissue system** (brown in the figure) covers and protects leaves, stems, and roots. Like our own skin, the epidermis (the outer layer of the dermal tissue system) is a plant's first line of defense against physical damage and infectious organisms. On leaves and on most stems, epidermal cells secrete a waxy coating called the **cuticle,** which helps the plant retain water. The **vascular tissue system** (purple), made up of xylem and phloem, provides support and transports water, nutrients, and food throughout the plant. The **ground tissue system** (yellow) makes up the bulk of a young plant, filling the spaces between the epidermis and vascular tissue system. The ground tissue system has diverse functions, including photosynthesis, storage, and support.

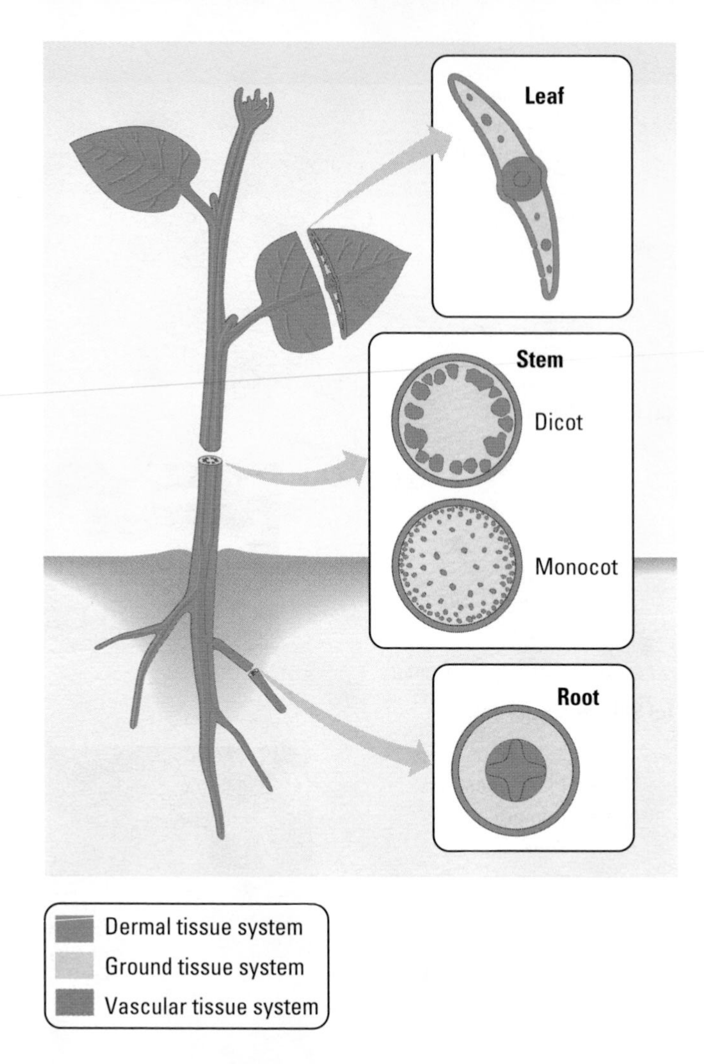

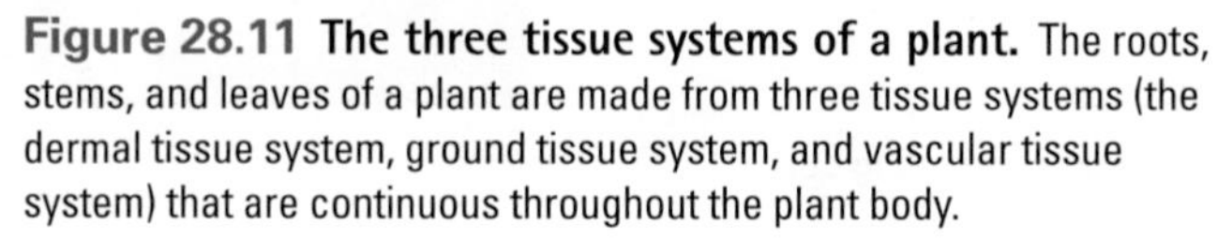

Figure 28.11 The three tissue systems of a plant. The roots, stems, and leaves of a plant are made from three tissue systems (the dermal tissue system, ground tissue system, and vascular tissue system) that are continuous throughout the plant body.

Figure 28.12, a cross section of a root, shows what the three tissue systems look like under a microscope. The epidermis is a single layer of tightly packed cells covering the entire root. Water and minerals enter the plant from the soil through these cells. Some of the young epidermal cells will grow outward and form root hairs. In the center of the root, the vascular tissue system forms a cylinder, with xylem cells radiating from the center like the spokes of a wheel and phloem cells filling in the wedges between the spokes. The ground tissue system of the root forms the **cortex,** where cells store food and take up water and minerals. The innermost layer of cortex is the **endodermis,** a thin cylinder one cell thick. The endodermis is a selective barrier that regulates the passage of substances between the cortex and the vascular tissue.

All plant stems have vascular tissue systems arranged in numerous vascular bundles; each bundle contains both xylem and phloem. The location and arrangement of these bundles differ between monocots and dicots (see Figure 28.2). In dicots, another part of the ground tissue system, known as the **pith,** fills the center of the stem and is often important in food storage.

Figure 28.13 illustrates the arrangement of the three tissue systems found in many leaves. In the epidermis of leaves (and some stems) are **stomata** (singular, *stoma*), tiny pores between two specialized epidermal cells called guard cells (see Figure 28.13, right). Stomata are more numerous in the leaf's lower epidermis, an adaptation that minimizes water loss, since water evaporates more rapidly from the sunlit upper surface of the leaf. **Guard cells** regulate the size of the stomata (see Figure 29.11), allowing gas exchange between the surrounding air and the photosynthetic cells inside the leaf.

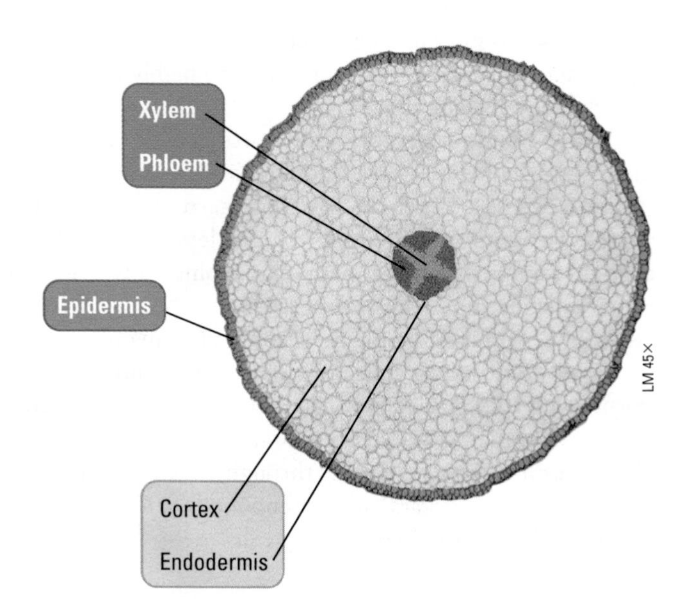

Figure 28.12 Tissue systems in a dicot root. This micrograph cross section of a young buttercup root provides a view of the single-layered epidermis, the cylindrical vascular tissue system with xylem and phloem, and the ground tissue system that makes up the bulk of the root.

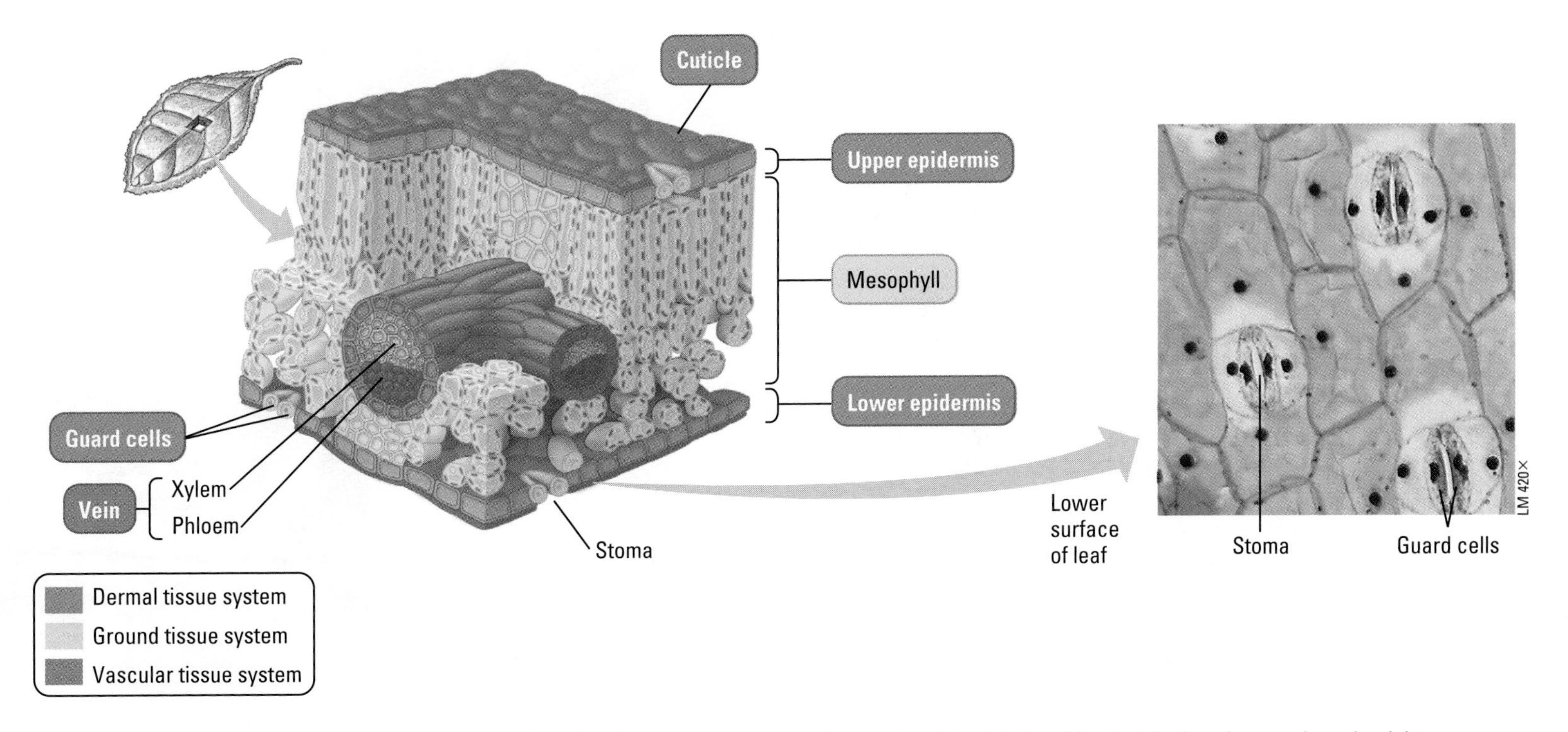

Figure 28.13 Tissue systems in a generalized dicot leaf. Notice the stoma, a pore on the bottom surface that allows gas exchange between the environment and the interior of the leaf. In the micrograph on the right, you can clearly see the stomata and a surface view of the epidermis.

The main site of photosynthesis is the **mesophyll,** which is the ground tissue system of a leaf. Mesophyll consists mainly of parenchyma cells, each equipped with many chloroplasts (green in Figure 28.13).

The leaf's vascular tissue system is made up of a network of veins. As you can see in the figure, each vein is a vascular bundle composed of xylem and phloem surrounded by parenchyma cells. The xylem and phloem, continuous with the vascular bundles of the stem, are in close contact with the leaf's photosynthetic tissues. This ensures that those tissues are supplied with water and mineral nutrients from the soil and that sugars made in the leaves are transported throughout the plant. These nutrients can then be used by the plant to perform all its life functions, including growth, our next topic.

CHECKPOINT

1. How are monocots and dicots different in terms of their seed leaves? Roots? Flower petals?
2. Explain why pruning certain types of fruit trees increases future fruit harvest.
3. Which of the following cell types has the potential to give rise to all others in the list: collenchyma, sclerenchyma, parenchyma, water-conducting cells?
4. The tissue system that covers and protects the outside of the plant is called the __________ tissue system, while the __________ tissue system is the main site of photosynthesis.

Answers: 1. A monocot embryo has a single seed leaf (cotyledon), while a dicot embryo has two. Monocots have fibrous roots that form a wide mat, while dicots have a vertical taproot. The flower petals of monocots usually come in multiples of three, while in dicots they come in multiples of four or five. **2.** Removal of terminal buds from major branches results in more branching by reducing inhibition of axillary buds. More branches produce more flowers and hence more fruit. **3.** Parenchyma **4.** dermal; ground

Plant Growth

The growth of a plant differs from that of an animal in a fundamental way. Most animals are characterized by **determinate growth;** that is, they cease growing after reaching a certain size. Most species of plants, in contrast, continue to grow as long as they live, a condition known as **indeterminate growth.**

Indeterminate growth does not mean that plants are immortal. In fact, most plants have a finite life span. Plants called **annuals** emerge from seed, mature, reproduce, and die in a single year or growing season. Our most important food crops—wheat, corn, and rice, for example—are annuals **(Figure 28.14a)**. **Biennials** live for two years; flowering and seed production usually occur during the second year. Carrots are biennials, but we usually harvest them in their first year and so miss seeing their flowers **(Figure 28.14b)**. Plants that live and reproduce for many years, including trees, shrubs, and some grasses, are known as **perennials** **(Figure 28.14c)**.

(a) Wheat, an annual

(b) Carrot, a biennial, flowering in its second year

(c) Grass, a perennial

Figure 28.14 Plants exhibit varying life spans.

Primary Growth: Lengthening

Growth in all plants is made possible by tissues called meristems. A **meristem** consists of unspecialized cells that divide frequently, generating new cells. Meristems at the tips of roots and in the terminal and axillary buds of shoots are called **apical meristems.** Cell division in apical meristems of both roots and shoots produces new cells that enable a plant to grow in length, a process called **primary growth (Figure 28.15)**. Tissues produced by primary growth are called primary tissues.

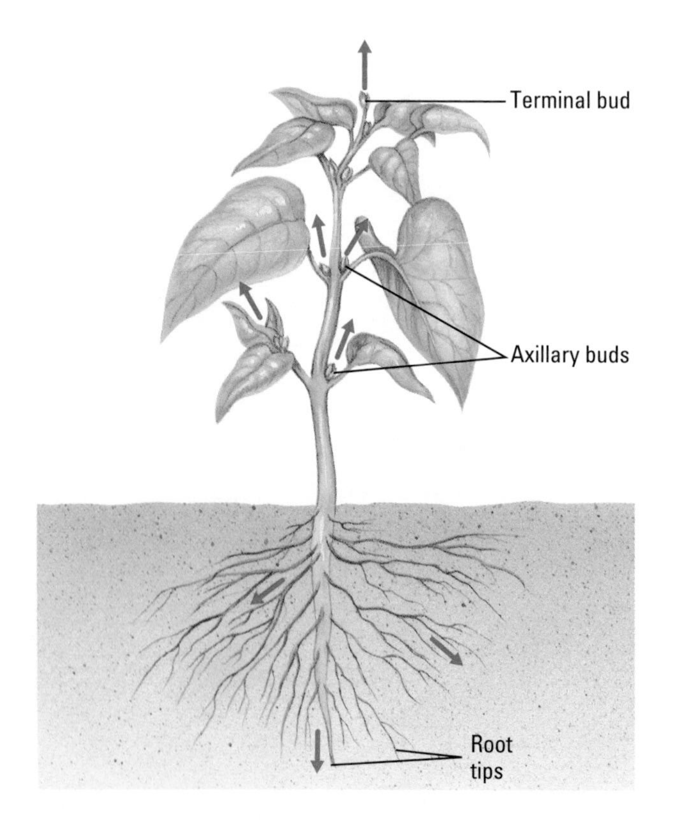

Figure 28.15 Primary growth in apical meristems. Meristems are tissues responsible for growth in all plants. The apical meristems are located at the root tips and in the terminal and axillary buds of shoots. The red arrows indicate the directions of primary growth.

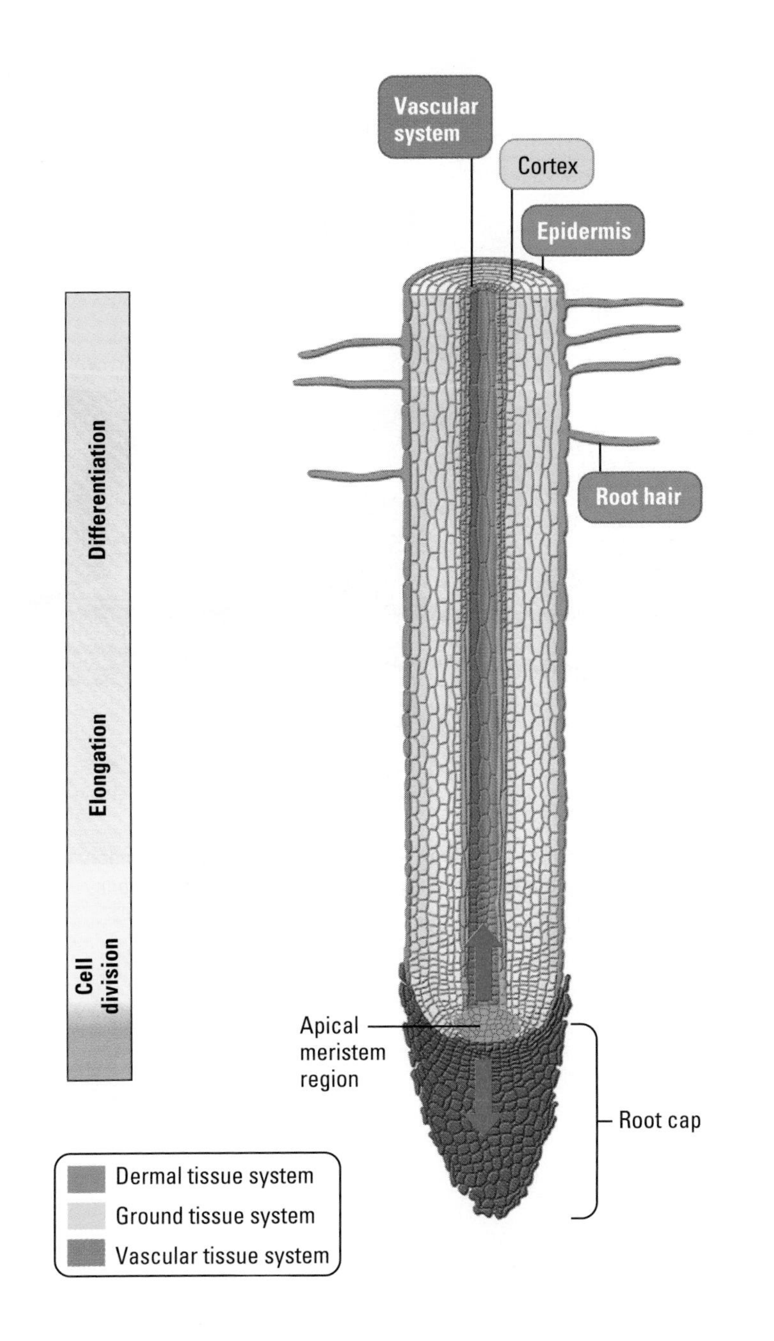

Figure 28.16 Primary growth in a root. Primary growth is achieved through two mechanisms. Cells within the apical meristem actively divide, replacing cells of the root cap (downward red arrow) and producing new cells for growth (upward red arrow). Additionally, cells behind the apical meristem elongate up to tenfold, mainly by taking in water and expanding.

Figure 28.16 shows a longitudinal section through a growing onion root. Primary growth enables roots to push through the soil. (A very similar process results in the upward growth of shoots.) At the very tip of the root is the **root cap,** a thimble-like cone of cells that protects the delicate, actively dividing cells of the apical meristem. The root's apical meristem (orange oval in the figure) has two roles: It replaces cells of the root cap that are scraped away by the soil (downward arrow), and it produces cells for primary growth (upward arrow). Primary growth is achieved not only by cell division but also by the lengthening of cells just above the apical meristem (see Figure 28.16, center). These cells can undergo a tenfold increase in length, mainly by taking up water. Elongation of these cells is what actually forces a root down through the soil. The elongating cells begin to differentiate, forming primary tissues that develop into the epidermis, cortex, and vascular tissue (see Figure 28.16, top). Cells of this last type eventually differentiate into vascular tissues called primary xylem and primary phloem.

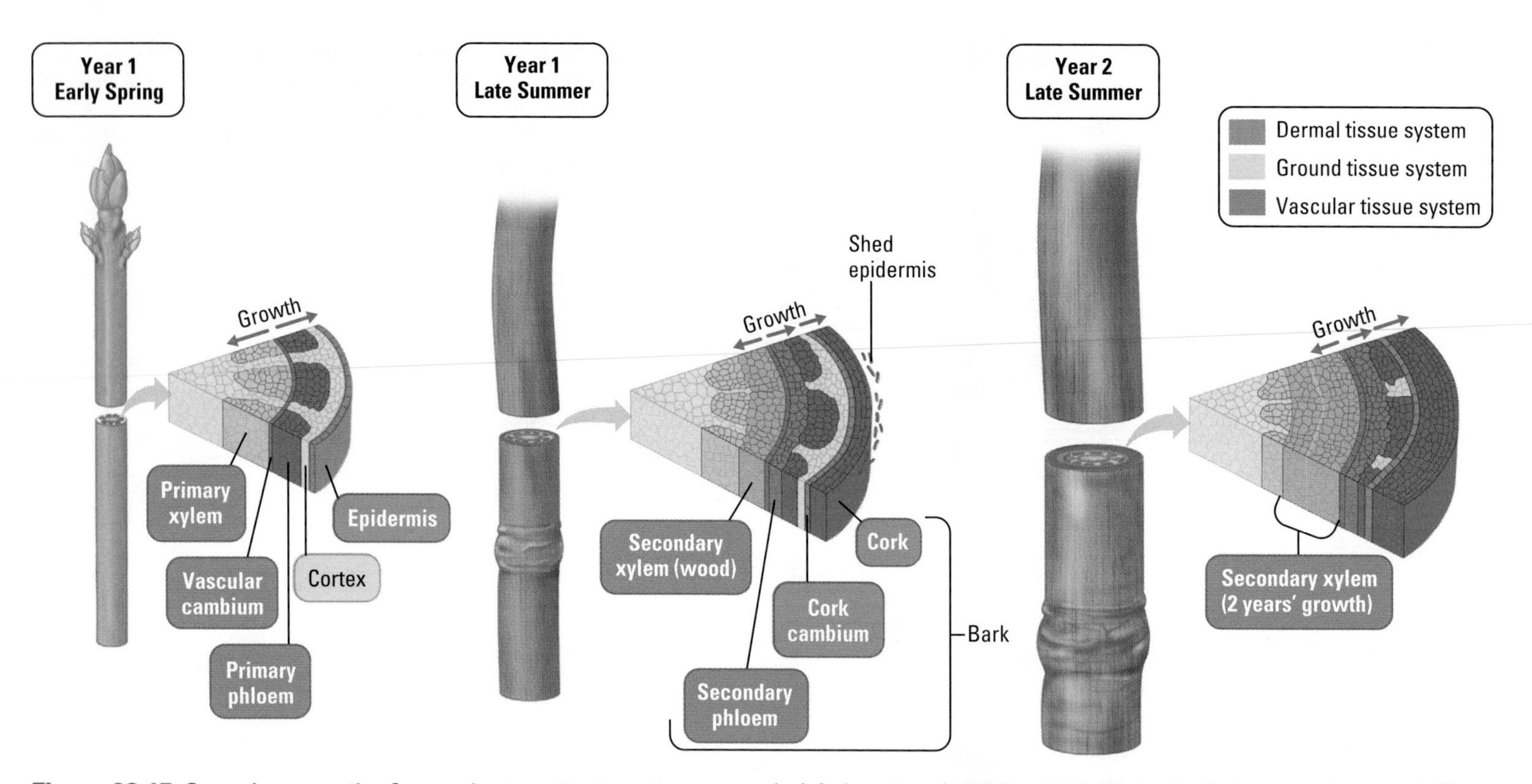

Figure 28.17 Secondary growth of a woody stem. The branches of most woody plants are made up of tissues of varying ages, with the youngest regions near the tip and the older regions nearer the trunk. The cross section at the left shows a region of the stem that is just beginning secondary growth. The middle and right cross sections show progressively older regions.

Secondary Growth: Thickening

In addition to lengthwise primary growth, the stems and roots of many plant species also thicken by a process called **secondary growth.** Such thickening is most evident in the woody plants—trees, shrubs, and vines—whose stems last from year to year and consist mainly of thick layers of mature, mostly dead xylem tissue, called **wood.** Tissues produced by secondary growth are called secondary tissues.

Secondary growth involves cell division in two meristems we have not yet discussed: the vascular cambium and the cork cambium. The **vascular cambium** (blue in **Figure 28.17**) first appears as a cylinder of actively dividing cells between the primary xylem and primary phloem, as you can see in the pie-shaped section at the left of the figure. Secondary growth adds cells on either side of the vascular cambium, as indicated by the red arrows.

The center and right drawings in Figure 28.17 show the results of secondary growth. In the center drawing, the vascular cambium has given rise to two new tissues: secondary phloem to its exterior and secondary xylem to its interior. Yearly production of a new layer of secondary xylem accounts for most of the growth in thickness of a perennial plant.

Consisting of xylem cells and fibers that have thick walls rich in lignin, secondary xylem makes up the wood of a tree, shrub, or vine. Over the years, a woody stem gets thicker and thicker as its vascular cambium produces layer upon layer of secondary xylem. This process results in annual

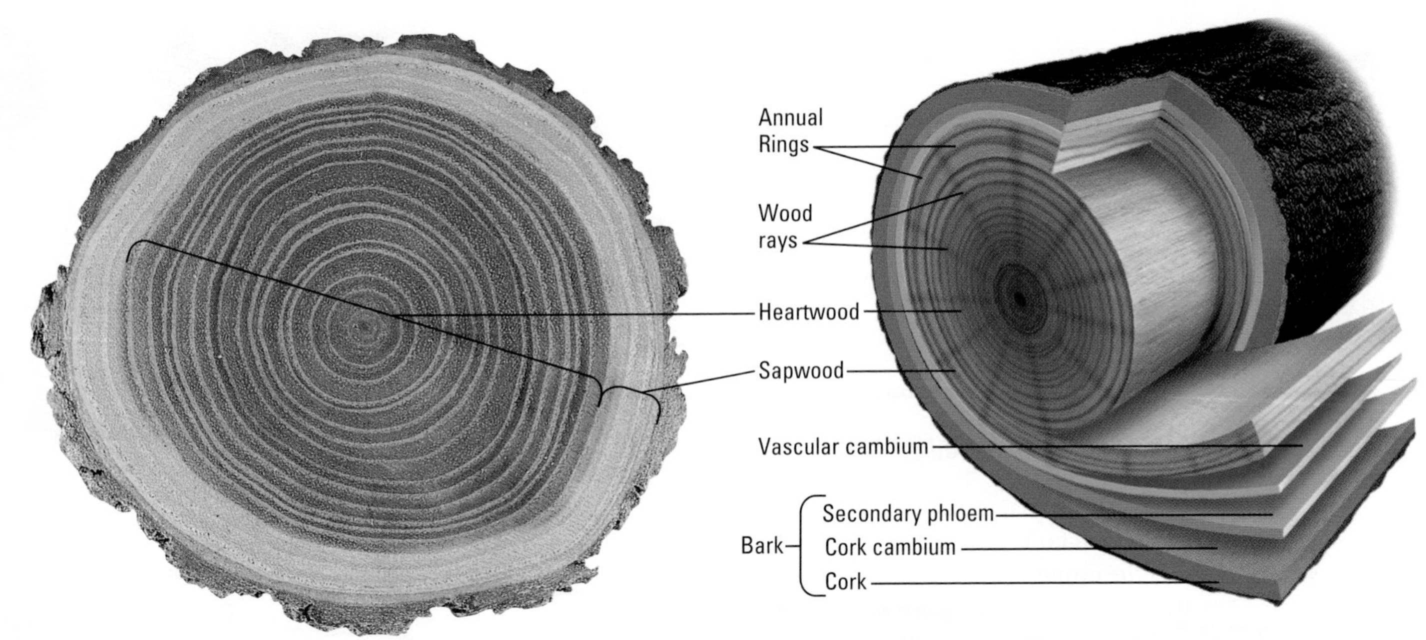

Figure 28.18 Anatomy of a log. On the left, you see a cross section of a locust log, with several decades of growth rings visible. On the right, the various layers of a mature log are separated for easier viewing.

growth rings **(Figure 28.18)**. The visibility of the rings is due to uneven activity of the vascular cambium during each year. In temperate regions, such as most of the United States, the vascular cambium of woody plants becomes dormant during winter, and secondary growth is interrupted. When secondary growth resumes in the spring, a cylinder of early wood forms. Made up of the first new xylem cells to develop, early wood cells are usually larger in diameter and thinner-walled than those produced later in the summer. The boundary between the large cells of early wood and the smaller cells of the late wood from the year before is usually visible as a distinct ring. You can estimate a tree's age by counting its annual rings.

Going back to the left part of Figure 28.17, notice that the epidermis and cortex, both the result of primary growth, make up the young stem's external covering. When secondary growth begins, the epidermis is sloughed off and replaced with a new outer layer called **cork.** Mature cork cells are dead and have thick, waxy walls, which protect the underlying tissues of the stem. Cork is produced by a meristem tissue called the **cork cambium.** Everything external to the vascular cambium (the secondary phloem, cork cambium, and cork) is called **bark.**

The log on the left in Figure 28.18 shows the results of several decades of secondary growth. The bulk of a trunk like this is dead tissue. The exceptions are wood rays (radiating from the center of the log in the drawing on the right), which consist of living parenchyma cells that function in nutrient storage and radial transport of water and nutrients within the wood. The heartwood, in the center of the trunk, consists of older layers of secondary xylem. These cells no longer transport water; they are clogged with resins and other metabolic by-products that make the heartwood resistant to rotting. The lighter-colored sapwood consists of younger secondary xylem that does conduct water.

Thousands of useful products are made from wood—from construction lumber to fine furniture, musical instruments, paper, and a long list of chemicals, including turpentine, alcohols, and artificial vanilla flavoring. Among the qualities that make wood so useful are a unique combination of strength, hardness, lightness, durability, and workability. In many cases, there is simply no good substitute for wood. A wooden oboe, for instance, produces far richer sounds than a plastic one, and fence posts made of heartwood last much longer in the ground than metal ones.

CHECKPOINT

1. What term describes a plant that produces flowers only at the end of its second growing season?
2. What two basic cellular mechanisms account for primary growth?
3. What type of plant tissue makes up wood? What is bark?

Answers: 1. A biennial **2.** Cell division and cell lengthening **3.** Secondary xylem; all tissues exterior to the vascular cambium: secondary phloem, cork cambium, and cork

The Life Cycle of a Flowering Plant

MP3 Tutor
From Flower to Fruit

Many flowering plants can reproduce both sexually and asexually, and both modes have played an important role in the evolutionary adaptation of plant populations to their environments. **Figure 28.19** shows three examples of **asexual reproduction,** the creation of offspring derived from a single parent without fertilization. Through asexual reproduction, a single plant can produce many offspring quickly and efficiently. This ability of plants is useful in agriculture because it allows farmers to grow a large crop of identical plants from a parent with desirable traits. But as described in the chapter opening, a genetically homogeneous population can be at a disadvantage when the environment changes quickly.

The Flower

As with animals, sexual reproduction in plants produces genetically distinct offspring. In angiosperms, the structure specific to sexual reproduction is the flower (**Figure 28.20**; also see Figure 28.3). The main

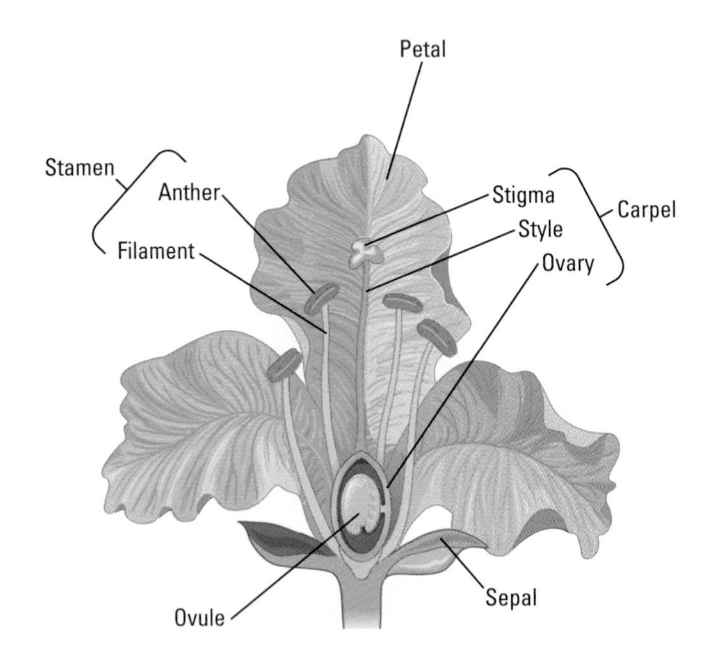

Figure 28.20 The structure of a flower. The flower is the structure used by angiosperms for sexual reproduction. The stamens are the sperm-producing reproductive organs, while the carpel is the egg-producing organ.

(a) Garlic. This garlic bulb is actually an underground stem that functions in storage. A single large bulb fragments into several parts, called cloves. Each clove can give rise to a separate plant, as indicated by the green shoots emerging from some of them.

(b) Holly trees. Each of the small trees is a sprout from the roots of a single holly tree. Eventually, one or more of these root sprouts may take the place of its parent.

(c) Creosote bushes. This ring of plants is a clone of creosote bushes growing in the Mojave Desert in southern California. All these bushes came from generations of vegetative reproduction by roots. This clone apparently began with a single plant that germinated from a seed about 12,000 years ago. The original plant probably occupied the center of the ring.

Figure 28.19 Asexual reproduction in plants.

parts of a flower—the sepals, petals, stamens, and carpels—are modified leaves. The **sepals** enclose and protect the flower bud. The **petals** are often bright and colorful and advertise the flower to insects and other pollinators.

The flower's reproductive organs are the stamens and carpel. A **stamen** consists of a stalk **(filament)** tipped by an **anther.** Within the anther are sacs where meiosis occurs and pollen grains develop. Pollen grains house the cells that develop into sperm. A **carpel** (more than one in some plants) has a long slender **style** (neck) with a sticky stigma at its tip. The **stigma** is the landing platform for pollen grains from the same flower or from other flowers (carried by wind or animals). The base of the stigma is the **ovary.** Within the ovary are reproductive structures called **ovules,** each containing one developing egg and cells that support it.

Pollination and Fertilization

The life cycle (the sequence of stages from the adults of one generation to the adults of the next) of a sexually reproducing angiosperm is shown in **Figure 28.21**. Fertilization occurs in the ovule of a flower, which then matures into a seed containing the embryo. Meanwhile, the ovary develops into a fruit, which protects the seed and aids in dispersing it. Completing the life cycle, the seed **germinates** (begins to grow) in a suitable habitat, the embryo develops into a seedling, and the seedling grows into a mature plant. In the rest of this section, we examine key stages in the angiosperm sexual life cycle in more detail.

As discussed in Chapter 16, the life cycles of all plants include alternation of haploid (n) and diploid ($2n$) generations (see Figure 16.10). The roots, stems, leaves, and most of the reproductive structures of all seed plants, including angiosperms, are diploid. The diploid plant body is called the **sporophyte.** A sporophyte produces special structures (in angiosperms these are the anthers and ovules) in which cells undergo meiosis to produce haploid spores. Each spore then divides mitotically and becomes a multicellular **gametophyte,** the plant's haploid generation. The gametophyte produces gametes through the process of mitosis. **Fertilization** occurs when the male and female gametes (sperm and egg, respectively) unite, producing a diploid zygote. The life cycle is completed when the zygote divides by mitosis and develops into a new sporophyte.

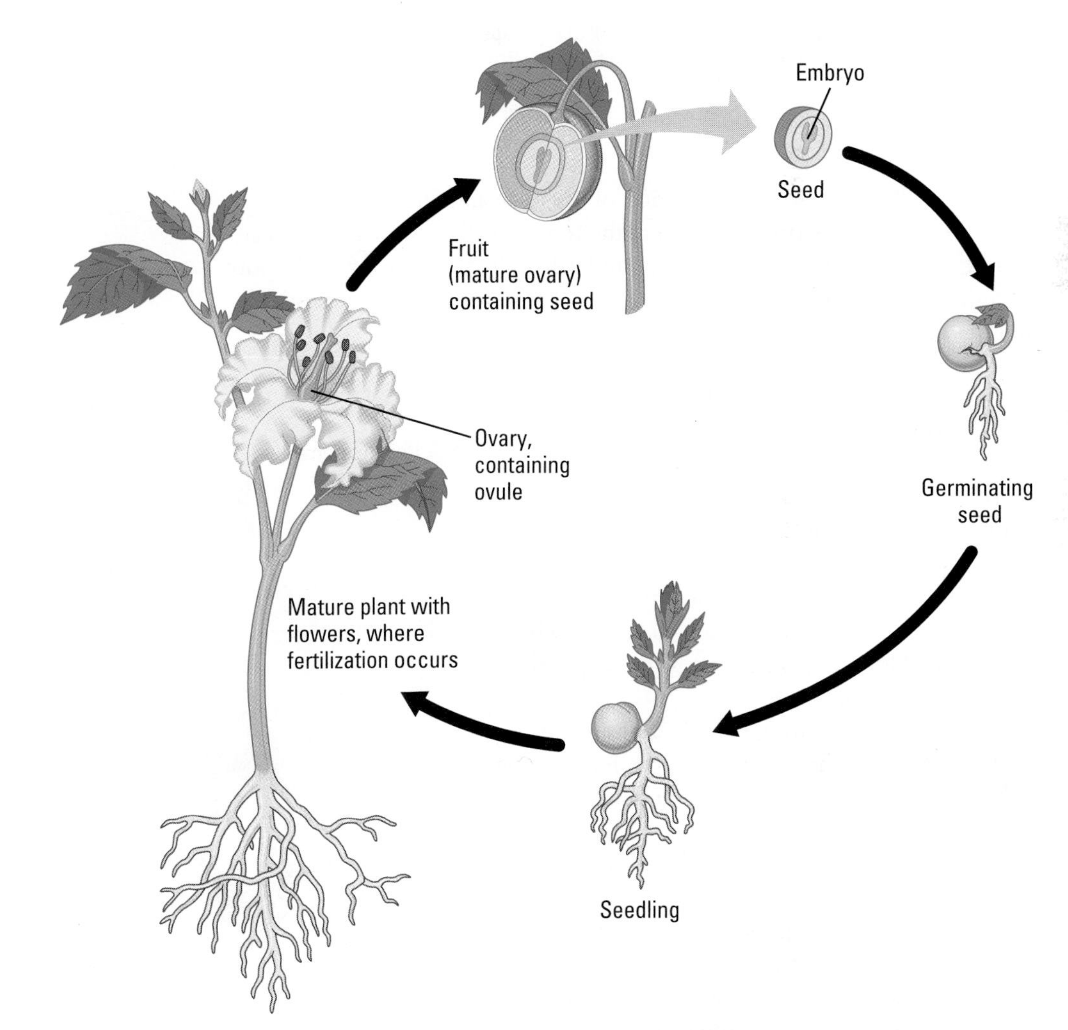

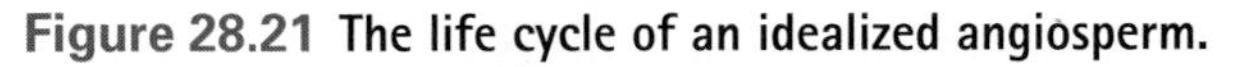
Figure 28.21 **The life cycle of an idealized angiosperm.**

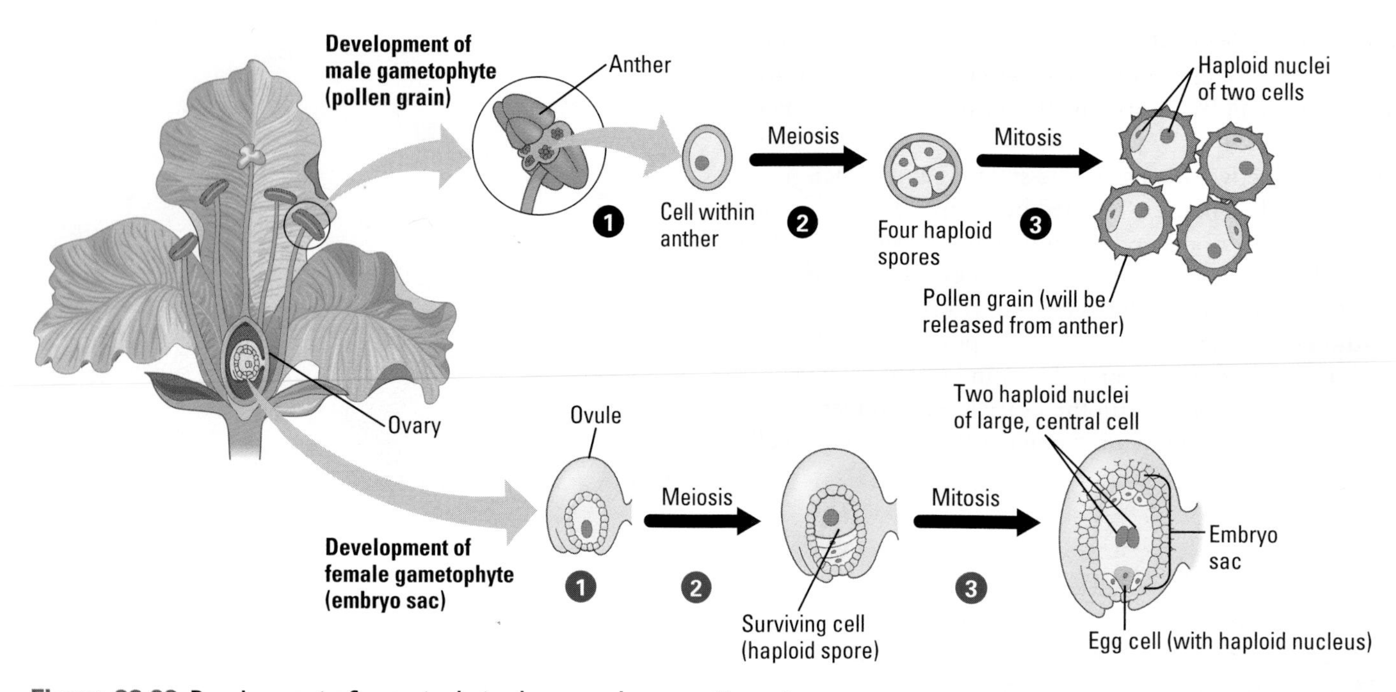

Figure 28.22 Development of gametophytes in an angiosperm. The male gametotype (pollen grain) develops within the anther (top). The female gametophyte (embryo sac) develops within the ovule (bottom). In most species, the ovary of a flower contains several ovules, but only one is shown here.

Figure 28.22 shows the formation of the male and female gametophytes of an angiosperm. As shown at the top, ❶ cells within a flower's anthers ❷ undergo meiosis to form haploid cells called spores. ❸ Each spore then divides by mitosis into two cells. A thick wall forms around these cells, and the resulting male gametophyte—the **pollen grain**—is ready for release from the anther.

Moving to the bottom of the figure, we see that ❶ within an ovule, a central cell enlarges and ❷ undergoes meiosis, producing four haploid spores. Three of the spores usually degenerate, but the surviving one enlarges and ❸ divides by mitosis, producing the female gametophyte, a multicellular structure called the **embryo sac.** The sac contains a large central cell with two haploid nuclei. One of its other cells is the haploid egg, ready to be fertilized.

How does fertilization occur? As shown in **Figure 28.23**, the first step leading to fertilization is ❶ **pollination,** the delivery of pollen to the stigma of a carpel. Many angiosperms are dependent on animals to transfer their pollen, but the pollen of some plants, such as grasses, is windborne (as anyone with pollen allergies knows!). ❷ After pollination, the pollen grain germinates on the stigma. It divides by mitosis, forming ❸ two sperm that travel to the ovule through a pollen tube. ❹ One sperm fertilizes the egg,

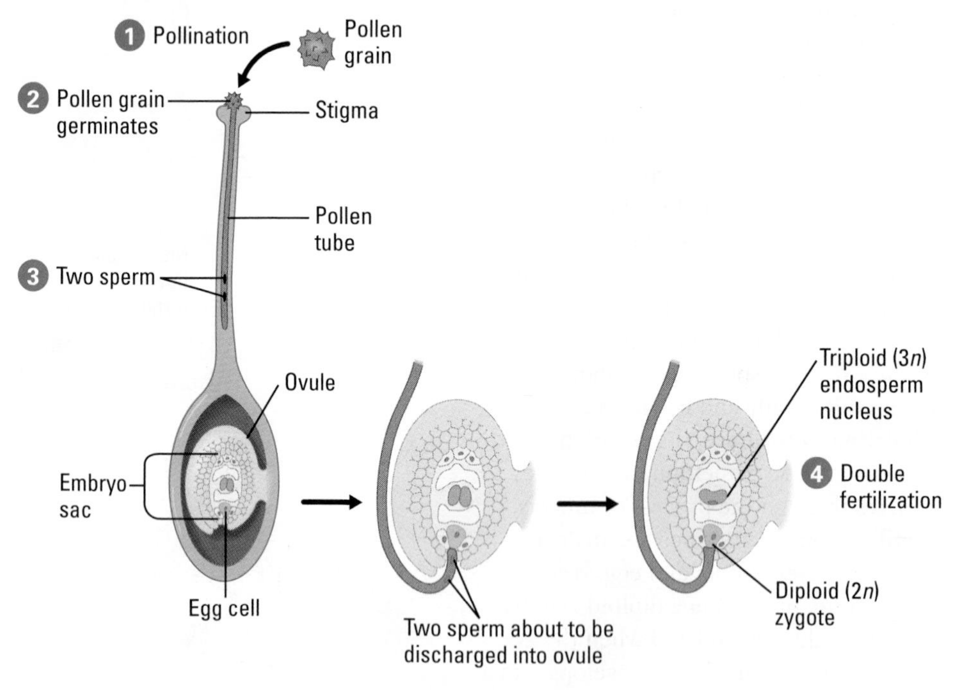

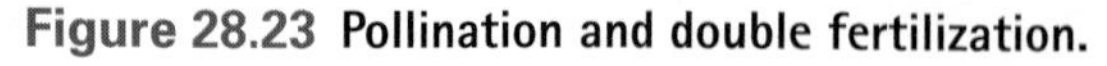
Figure 28.23 Pollination and double fertilization.

forming the diploid zygote. The other sperm contributes its haploid nucleus to the large central cell of the embryo sac, resulting in a triploid ($3n$) nucleus. The formation of both a zygote and a fertilized central cell with a triploid nucleus is called **double fertilization.** This occurs only in plants, mainly in angiosperms.

Seed Formation

After fertilization, the ovule, containing the zygote and the triploid central cell, begins developing into a seed **(Figure 28.24)**. The zygote divides via mitosis into a ball of cells that becomes the embryo. Meanwhile, the triploid cell divides and develops into an **endosperm,** a multicellular mass that nourishes the embryo after germination. As cotyledons develop, they absorb nutrients from the endosperm. In some plants, such as beans, the cotyledons become very fleshy.

The result of embryonic development in the ovule is a mature seed (see Figure 28.24, bottom) with a tough protective **seed coat** that encloses the embryo and the endosperm. At this point, the embryo stops developing and the seed becomes dormant; growth and development are suspended until the seed germinates. Seed dormancy is an important evolutionary adaptation. It allows time for seed dispersal and increases the chance that a new generation of plants will begin growing only when environmental conditions, such as temperature and moisture, favor survival.

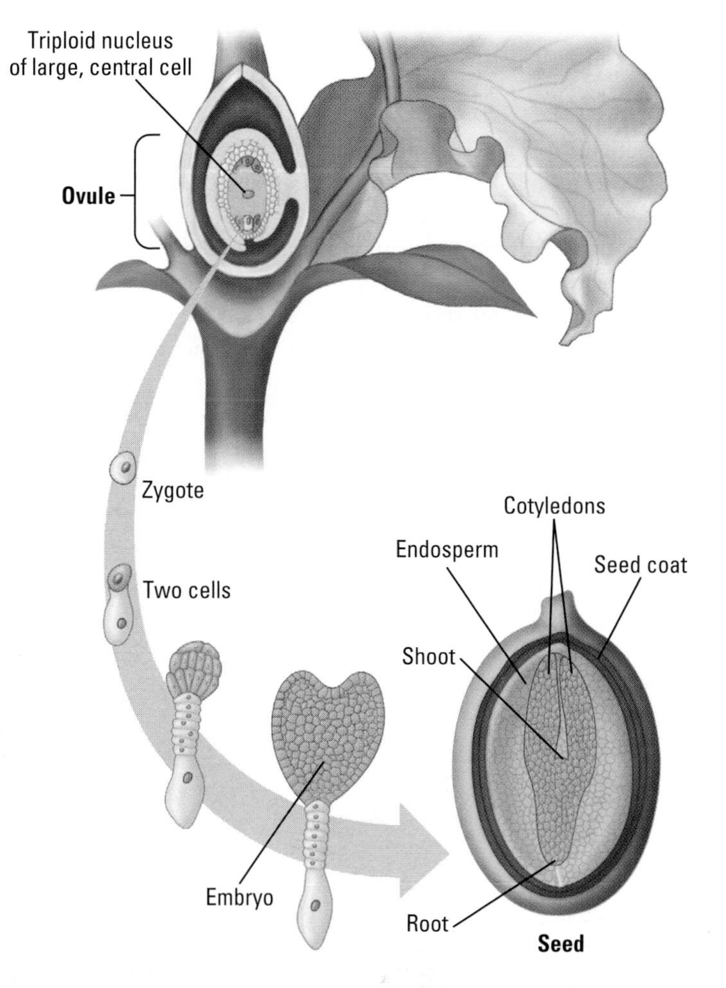

Figure 28.24 Development of a seed. The plant in this drawing is a dicot; a monocot would have only one cotyledon. The bulges you see on the embryo are the cotyledons starting to form.

Fruit Formation

Fruits, one of the most distinctive features of angiosperms, develop at the same time seeds do. A **fruit** is a mature ovary that acts as a vessel, housing and protecting seeds and helping disperse them from the parent plant. A corn kernel is a fruit, as is a peach, tomato, or pea pod.

The photographs in **Figure 28.25** illustrate the changes in a pea plant that lead to pod formation. ❶ Soon after pollination, ❷ the flower drops its petals and the ovary starts growing. The ovary expands tremendously, and its wall thickens, ❸ forming the pod, or fruit, which holds the seeds. Peas are usually harvested at this stage of fruit development. If the pods are allowed to develop further, they become dry and brownish and will split open, releasing the dried seeds.

❶

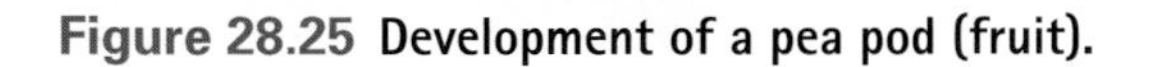

Figure 28.25 Development of a pea pod (fruit).

Fruits are highly varied, as **Figure 28.26** illustrates. In some cases, what we commonly call fruits contain edible tissues in addition to the mature ovaries. In an apple, for instance, the part we discard, the core, is the thickened ovary and therefore the true fruit. Fruits with fleshy, edible tissues, such as apples, serve as enticements to animals that help spread seeds (see Figure 16.19). Some of the structures we call fruits, such as pineapples, actually consist of many individual fruits fused together.

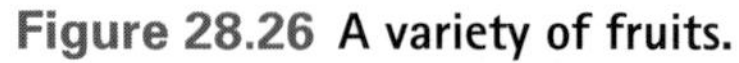

Figure 28.26 A variety of fruits.

Seed Germination

Germination of a seed usually begins when the seed takes up water. The hydrated seed expands, rupturing its coat, and the embryo resumes the growth and development that was temporarily suspended during seed dormancy.

Figure 28.27 traces germination of a garden bean. The embryonic root of the bean emerges first and grows downward from the germinating seed. Next, the embryonic shoot emerges, and a hook forms near its tip. The hook protects the delicate shoot tip by holding it downward as it pushes through the abrasive soil. As the shoot breaks through the soil surface, exposure to light stimulates the hook to straighten, and the tip is lifted. The first foliage leaves ("true leaves") then expand from the shoot tip and begin making food by photosynthesis. In a bean plant, the cotyledons emerge from the soil and become leaflike photosynthetic structures. In many other plants, such as peas, the cotyledons remain behind in the soil and decompose.

A germinating seed is fragile. In the wild, only a small fraction of seedlings endure long enough to reproduce. The great numbers of seeds produced by most plants compensate for the odds against each seedling, increasing the chances of reproductive success.

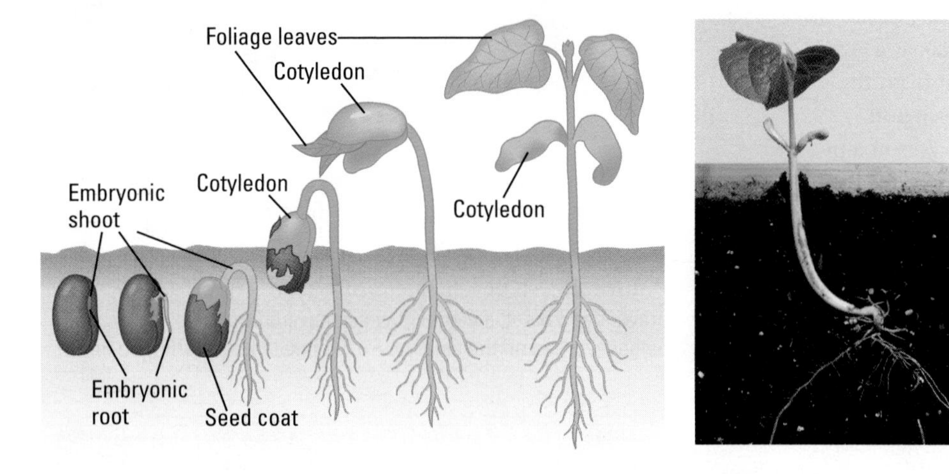

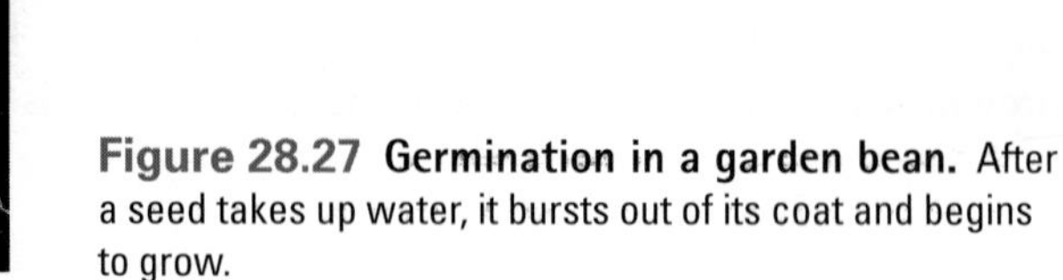

Figure 28.27 Germination in a garden bean. After a seed takes up water, it bursts out of its coat and begins to grow.

CHECKPOINT

1. Many plant species can reproduce both sexually and asexually. Which mode of reproduction would generally be more advantageous in a location where the composition of the soil is changing rapidly? Why?
2. Pollen develops within the __________ of reproductive organs called __________. Ovules develop within the __________ of egg-producing organs called __________.
3. The two products of double fertilization are the __________, which gives rise to the embryo, and the __________, which stores nutrients.

Answers: 1. Sexual, because it generates genetic variation among the offspring, enhancing the potential for adaptation to a changing environment **2.** anthers; stamens; ovaries; carpels **3.** zygote; endosperm

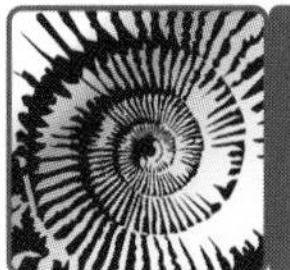

EVOLUTION CONNECTION

The Interdependence of Angiosperms and Animals

Flowering plants and land animals have had mutually beneficial relationships throughout their evolutionary history. Most angiosperms depend on insects, birds, or mammals for pollination and seed dispersal (see Figure 16.19). And most land animals depend on angiosperms for food and shelter. Such mutual dependencies tend to improve the reproductive success of both the plants and the animals and are thus favored by natural selection.

The flowers of many angiosperms attract pollinators that rely entirely on the flowers' nectar and pollen for food **(Figure 28.28)**. Nectar is a high-energy fluid that apparently functions only to attract pollinators. Evolutionary adaptations for advertising the presence of nectar include flower color and fragrance that are keyed to pollinators' senses of sight and smell. For example, many flowers pollinated by birds are red or pink, colors to which bird eyes are especially sensitive. Flowers may also have markings that attract pollinators, leading them past pollen-bearing organs on their way to gathering nectar. For example, flowers pollinated by bees often have markings that reflect ultraviolet light. Such markings are invisible to us, but vivid to bees. Many other animals—including hummingbirds, butterflies, and fruit bats—have similar relationships with other flower species. To a large extent, flowering plants are as diverse and successful as they are today because of their close connections with animals. ■

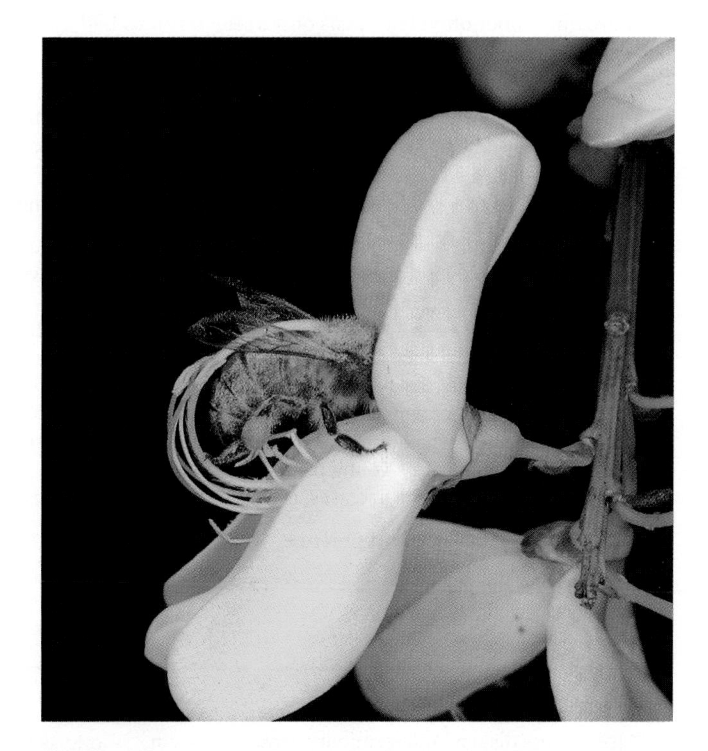

Figure 28.28 A busy bee. The flower known as the Scottish broom has a trapping mechanism that dusts pollen onto the back of a visiting bee that is feeding on nectar. Some of the pollen the bee picks up here will rub off onto the egg-carrying organs of the next flower it visits.

Chapter Review

SUMMARY OF KEY CONCEPTS

For study help and activities, go to campbellbiology.com or the student CD-ROM.

The Structure and Function of a Flowering Plant

- Flowering plants, or angiosperms, account for nearly 90% of the plant kingdom.
- **Monocots and Dicots** Angiosperms can be grouped into two categories based on the number of cotyledons (seed leaves) found in the embryo and other structural differences.

	Seed leaves	Leaf veins	Vascular bundles	Floral parts	Roots
Monocots	One cotyledon	Parallel	Scattered	Multiples of 3	Fibrous
Dicots	Two cotyledons	Branched	Ring	Multiples of 4 or 5	Taproot

- **Plant Organs: Roots, Stems, and Leaves** A plant body consists of a root system and a shoot system, each depending on the other.

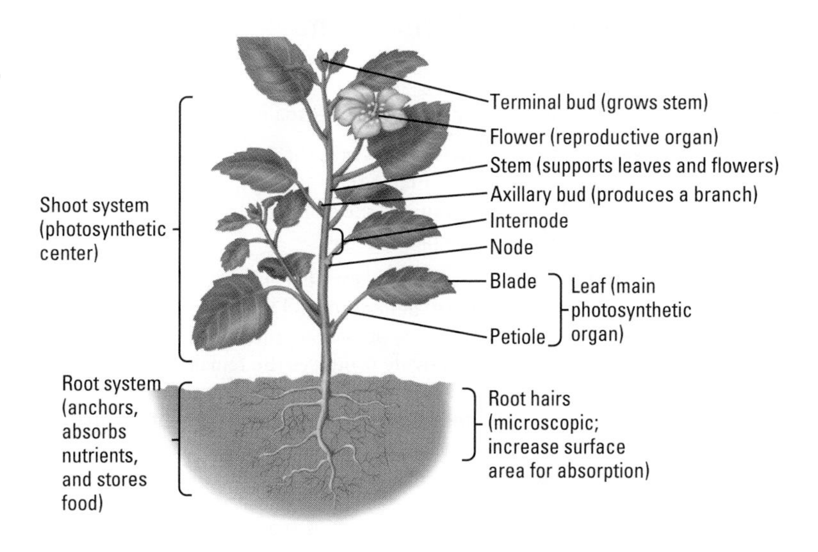

• A growing shoot often lengthens only at the terminal bud; its axillary buds are dormant. This condition is called apical dominance and allows the plant to grow tall quickly.

• **Plant Cells** Plant cells are distinguished by the presence of chloroplasts, a large central vacuole, and cell walls. There are several types of plant cells: parenchyma cells, collenchyma cells, sclerenchyma cells, water-conducting cells (tracheids and vessel elements), and food-conducting cells.

• **Plant Tissues and Tissue Systems** Roots, stems, and leaves are made up of tissues organized into three tissue systems: the dermal tissue system, the vascular tissue system, and the ground tissue system. The epidermis, the outer layer of the dermal tissue system, covers and protects the plant. Within the epidermis are pores called stomata, each surrounded by two guard cells, that regulate gas exchange between the environment and leaf cells. The vascular tissue system contains xylem, which conveys water and dissolved minerals, and phloem, which transports sugars. The ground tissue system, which makes up the bulk of a young plant, functions mainly in storage. Mesophyll, the ground tissue of leaves, is where most photosynthesis occurs.

Case Study in the Process of Science *How Do the Tissues in Monocot and Dicot Stems Compare?*

Activity *Roots, Stems, and Leaves*

Plant Growth

• Most plants display indeterminate growth, growing as long as they are alive. Some plants grow, reproduce, and die in one year (annuals); some live for two years (biennials); and some live for many years (perennials).

• **Primary Growth: Lengthening** Plant growth originates in meristems, areas of unspecialized, dividing cells. Apical meristems are located at the tips of roots and in the terminal and axillary buds of shoots. These meristems initiate lengthwise growth by producing new cells. A root or shoot lengthens further as the new cells elongate. This cell division and elongation is called primary growth. The new cells eventually differentiate into specialized tissues.

• **Secondary Growth: Thickening** An increase in a plant's girth, called secondary growth, arises from cell division in a cylindrical meristem called the vascular cambium. The vascular cambium produces layers of secondary xylem, or wood, next to its inner surface. Outside the vascular cambium is bark, which consists of secondary phloem, a meristem called the cork cambium, and cork cells produced by the cork cambium. The outer layers of bark are sloughed off as the plant thickens.

Activity *Primary and Secondary Growth*

The Life Cycle of a Flowering Plant

MP3 Tutor *From Flower to Fruit*

• Many flowering plants can reproduce both sexually and asexually.

• **The Flower** In angiosperms, the flower is the structure specific to sexual reproduction. The stamen contains the anther, in which pollen grains are produced. The carpel contains the stigma, which receives pollen, and the ovary, which contains the ovule.

• **Pollination and Fertilization** The plant life cycle alternates between diploid ($2n$) and haploid (n) generations. The spores in the anthers give rise to male gametophytes, or pollen grains, each of which produces two sperm. A spore in an ovule produces the female gametophyte, called an embryo sac. Each embryo sac contains an egg cell. After pollen lands on the stigma (pollination), the sperm pass down the stigma into the ovule. One sperm combines with the egg and the other combines with a second, diploid, cell. This process, called double fertilization, is unique to plants.

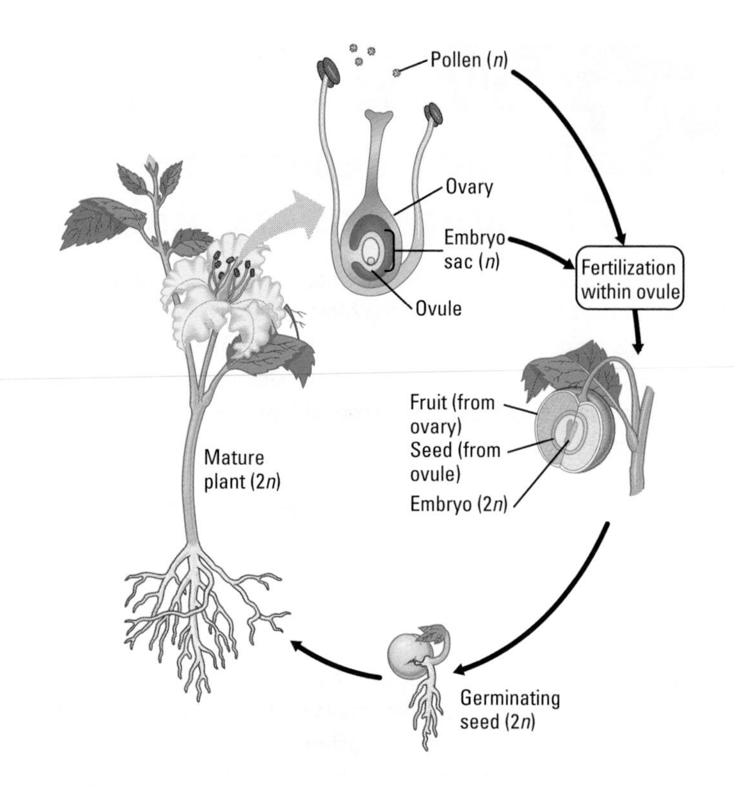

• **Seed Formation** After fertilization, the ovule becomes a seed, and the fertilized egg within it divides to become an embryo. The other fertilized cell (now triploid) develops into the endosperm, which stores food for the embryo. A tough seed coat protects the embryo and endosperm. The embryo develops cotyledons (seed leaves), which absorb food from the endosperm.

• **Fruit Formation** While the ovule becomes a seed, the ovary develops into a fruit, which helps protect and disperse the seeds.

Activity *The Life Cycle of a Flowering Plant*

• **Seed Germination** A seed starts to germinate when it takes up water, expands, and bursts its seed coat. The embryo resumes growth, an embryonic root emerges, and a shoot pushes upward and expands its leaves.

Activity *Seed and Fruit Formation and Germination*

Case Study in the Process of Science *What Tells Desert Seeds When to Germinate?*

SELF-QUIZ

1. While walking in the woods, you encounter an unfamiliar flowering plant. Which of the following plant features would help you determine whether that plant is a monocot or a dicot? (Choose all that are appropriate.)
   a. size of the plant
   b. number of seed leaves
   c. shape of its root system
   d. number of petals in its flowers
   e. arrangement of vascular bundles in its stem
   f. whether or not it produces seed-bearing cones

2. Your friend is planting a new herb garden. During the first year, she allows the plants to grow naturally. Her plants grow tall and relatively spindly during the growing season, and when it comes time to harvest, she doesn't have a large yield. What phenomenon is responsible for her poor yield? What could your friend do to increase her yield next year?

3. A pea pod is formed from a(n) __________. A pea inside the pod is formed from a(n) __________.

4. Match each flower structure with its function.

a. pollen grain	1. Attracts pollinators
b. ovule	2. Develops into seed
c. anther	3. Protects an unopened flower
d. ovary	4. Produces sperm
e. sepal	5. Produces pollen
f. petal	6. Contains ovules

5. What part of a plant are you eating when you consume each of the following?
   a. tomato
   b. celery stalk
   c. green bean
   d. lettuce
   e. beet

6. Name three kinds of asexual reproduction found in plants. What advantage for the plant might asexual reproduction have over sexual reproduction? What disadvantage might it have?

7. In angiosperms, each pollen grain produces two sperm. What do these sperm do?
   a. Each one fertilizes a separate egg cell to produce two zygotes.
   b. One fertilizes an egg to form a zygote, and the other fertilizes the fruit.
   c. One fertilizes an egg to form a zygote, and the other is kept in reserve.
   d. One fertilizes an egg, and the other fertilizes a cell that develops into stored food.

8. In the angiosperm life cycle, which of the following processes is directly dependent on meiosis?
   a. production of gametophytes
   b. production of gametes
   c. production of spores
   d. all of the above

9. Which of the following is closest to the center of a woody stem? Which is farthest from the center?
   a. cork cambium
   b. primary xylem
   c. primary phloem
   d. secondary xylem
   e. secondary phloem

10. A gardener is concerned that his perennial plants are not growing well. He notes that the stems on his plants are not thickening adequately over the years. This suggests reduced activity of which plant tissue?

Answers to the Self-Quiz questions can be found in Appendix D.

Go to the website or CD-ROM for more Self-Quiz questions.

THE PROCESS OF SCIENCE

11. During a recent ice storm, many old trees were snapped by the weight of the ice. A botanist in the area would like to analyze the remains of the trees and determine how old they really were. Describe a procedure that could be used to determine the approximate age of the trees. Explain the basis of this method in terms of plant tissue growth.

12. Plants can alter the shape of their guard cells, thereby regulating the size of the stomata on their leaves. Stomata must be open to allow entrance of carbon dioxide, which is necessary for photosynthesis. During the summer, the stomata are often closed around noon, when the most sunlight is likely to reach the plant. Propose an explanation for why it may be advantageous for stomata to be closed when conditions seem optimal for photosynthesis. How could you test your hypothesis?

BIOLOGY AND SOCIETY

13. Tropical forests contain a wealth of plants that are potential new sources of food, medicine, and other useful products. As a result of growing populations and debt, however, many tropical forests in developing countries are being cut for lumber and farmland, and many species are disappearing. Developed countries are pressuring the tropical countries to protect the forests before even more species are lost. Many people in the developing nations see little incentive to preserve the forests only to have corporations from industrialized countries profit from their products. Is there a way to preserve the tropical forests so that both the developed and developing nations will benefit from their abundance?

14. Cloning is a fiercely debated topic in our country. Most of the ethical debates concerning cloning center on the cloning of animals. There is not much talk about the cloning of plants. Do you have the same opinion about cloning plants and animals? Why or why not? Do you think that they should be similarly regulated? What unique problems might each present?

29 The Working Plant

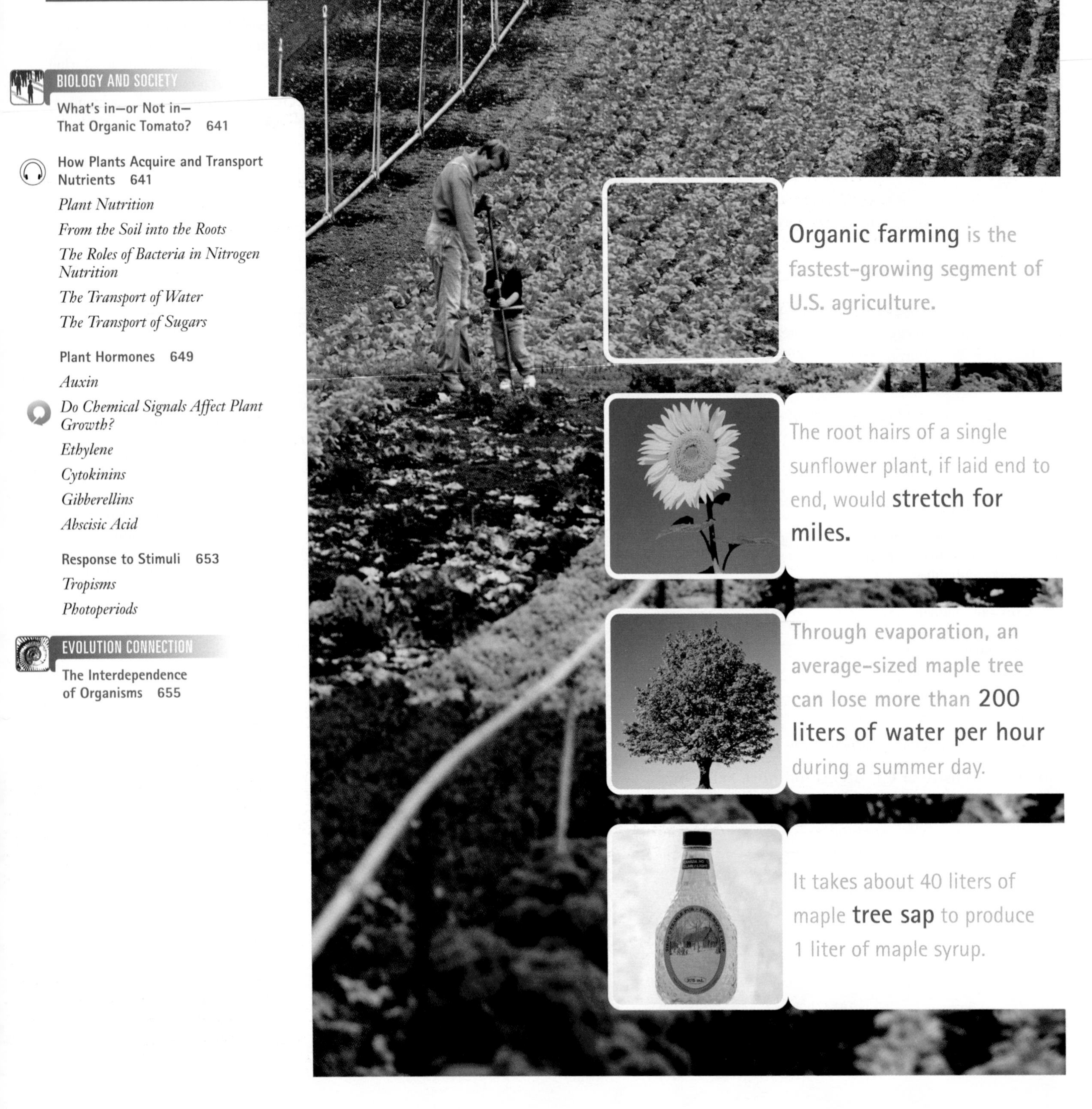

Organic farming is the fastest-growing segment of U.S. agriculture.

The root hairs of a single sunflower plant, if laid end to end, would **stretch for miles.**

Through evaporation, an average-sized maple tree can lose more than **200 liters of water per hour** during a summer day.

It takes about 40 liters of maple **tree sap** to produce 1 liter of maple syrup.

BIOLOGY AND SOCIETY

What's in—or Not in—That Organic Tomato?

If you buy tomatoes labeled "organic" at a grocery store, what does that label actually mean? To use the term *organic* or to bear the "USDA Organic" seal on a label, food producers must grow and process their products according to strict guidelines established and regulated by the U.S. Department of Agriculture (USDA). These guidelines are intended to build a sustainable agricultural system that protects biological diversity, maintains and replenishes soil quality (as by crop rotation), manages pests with no or few synthetic pesticides (by, for example, providing habitat for predators and parasites of crop pests), avoids genetically modified organisms, conserves water, and uses no or few synthetic fertilizers. Yearly inspections ensure proper organic farming practices, accurate record keeping, and a buffer of land between organic farms and neighboring conventional farms.

In the United States, over 2 million acres are dedicated to organic farming, some managed by large businesses, others by small family operations **(Figure 29.1)**. The U.S. organic farming industry is growing at a rate of 20% per year, making it one of the fastest-growing segments of agriculture. But while organic farming practices are spreading, they currently occupy only about 0.3% of U.S. cropland, and only about 2% of the U.S. food supply is grown using organic methods.

Figure 29.1 Organic farming. An organic farmer harvests her crop of peppers. Only products that conform to a strict set of guidelines can display the USDA organic seal.

The benefits of organic farming are clear: fewer synthetic chemicals in the environment and less risk of exposing farm workers and wildlife to potential toxins. But does an organic designation mean that a food is better for you? In fact, an organic label is no guarantee of extra health benefits. Scientists disagree, for example, about the nutritional differences, if any, between organic and conventional produce. And organic snacks (cookies, say, or potato chips) can be just as full of sugar, salt, and unhealthy fat as their nonorganic counterparts.

Nevertheless, a guiding principle of organic farming is that the nutritional health of plants affects the nutritional health of those who eat the plants. Keep this in mind as we begin this chapter with a discussion of plant nutrition. Then we'll examine other aspects of the life of a working plant, including the crucial roles played by hormones and how certain plant activities are affected by stimuli from the environment. ■

How Plants Acquire and Transport Nutrients

MP3 Tutor Transpiration

Watch a plant grow from a tiny seed, and you can't help wondering where all the mass comes from. About 96% of a plant's dry weight is organic material synthesized from inorganic nutrients extracted from the surroundings. (In this context, "organic" means a molecule containing carbon; this use of the term is unrelated to "organic" farming.) Plants obtain carbon dioxide (CO_2) from the air and water (H_2O) and **minerals** (inorganic ions) from the soil **(Figure 29.2)**. From the CO_2 and H_2O, plants

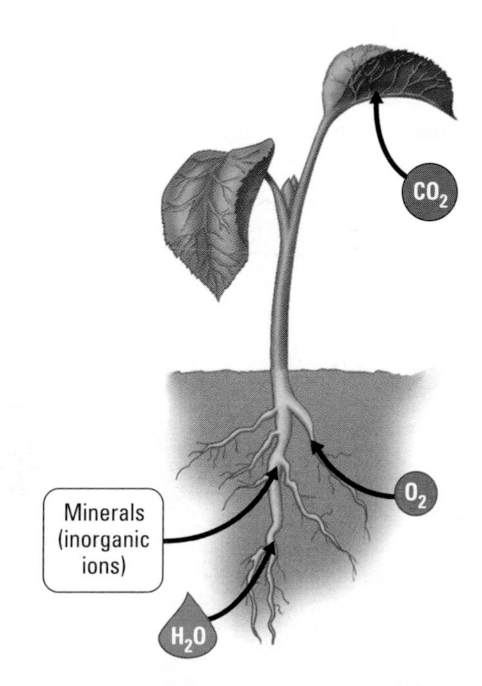

Figure 29.2 The uptake of nutrients by a plant. A plant absorbs the CO_2 it needs from the air and minerals and H_2O from the soil. A plant's roots also require O_2 from the soil for cellular respiration, but a plant generally releases more O_2 from photosynthesis than it consumes by respiration.

produce sugars via photosynthesis. These sugars, combined with minerals, are used to construct all the other organic materials a plant needs **(Figure 29.3)**.

Figure 29.3 A Bavaria beech tree, a giant product of photosynthesis. Plants use sugars created by photosynthesis to produce all the organic materials they need. The giant trunk of this tree consists mainly of sugar derivatives.

Plant Nutrition

A chemical element is considered an **essential element** if a plant must obtain it from its environment to complete its life cycle. Seventeen elements are essential to all plants, and a few others are essential to certain types of plants. Of the 17 essential elements, 9 are called **macronutrients** because plants require relatively large amounts of them. Elements that plants need in extremely small amounts are called **micronutrients.**

Macronutrients Six of the nine macronutrients—carbon, oxygen, hydrogen, nitrogen, sulfur, and phosphorus—make up almost 98% of a plant's dry weight. The other three macronutrients—calcium, potassium, and magnesium—make up another 1.7%.

What does a plant do with macronutrients? Carbon, oxygen, and hydrogen are the basic ingredients of a plant's organic compounds. Nitrogen is a component of all nucleic acids and proteins, as well as ATP, chlorophyll (the key light-absorbing molecule in plants), and many plant hormones. Sulfur is a component of most proteins. Phosphorus is a major component of nucleic acids, phospholipids, and ATP. The other macronutrients, though present in smaller amounts, play similarly important roles. For example, magnesium is an essential component of chlorophyll.

Micronutrients Eight micronutrients make up the remaining 0.3% of a plant's dry weight and are required by all plants. The eight micronutrients are iron, chlorine, copper, manganese, zinc, molybdenum, boron, and nickel. A plant recycles the atoms of micronutrients over and over, so it needs them in only minute quantities. Yet a deficiency of any micronutrient can kill a plant.

Nutrient Deficiencies The quality of soil, especially the availability of nutrients, affects the quality of our own nutrition. Nitrogen-deficient plants **(Figure 29.4)** may produce grain, but the crop will have a lower nutritional value, and its nutrient deficiencies will be passed on to livestock or human consumers.

Nitrogen deficiency is the most common nutrient deficiency in plants, followed by phosphorus and potassium deficiencies. Many growers diagnose their ailing plants visually, then check their conclusions by having soil samples

Figure 29.4 The effect of nitrogen availability on corn growth. This photo shows two experimental corn crops. The plants on the left are growing in soil rich in the nitrogen-containing compounds they use to build proteins. The smaller, lighter-colored plants on the right are growing in nitrogen-deficient soil.

Figure 29.5 **Nutrient deficiencies.** The photographs compare a healthy tomato plant with genetically identical plants suffering from three different macronutrient deficiencies.

(a) A healthy tomato plant.

(b) Nitrogen deficiency. Stunted growth and yellow-green leaves are signs of nitrogen deficiency. The older leaves usually show the signs first.

(c) Phosphorus deficiency. A phosphorus-deficient plant may have green leaves, but its growth rate is markedly reduced, and its new growth is often spindly and brittle. Also, in some plants, phosphorus deficiency produces a purplish color on the undersides of the leaves.

(d) Potassium deficiency. The signs of a potassium shortage are generally more localized than those of nitrogen and phosphorus deficiencies. The older leaves usually show the most obvious signs; they often turn yellow and develop dead, brownish tissue at the edges or in spots. Stems and roots are also weakened, leading to stunting.

chemically analyzed at a state or local laboratory. As shown in **Figure 29.5**, you can diagnose nutrient deficiencies in your own plants. Once you've made a diagnosis, treating the problem is usually simple: You can choose from a number of fertilizers, natural or manufactured, to enrich the soil. If you are unsure of how and when to fertilize your plants, your local agricultural extension office (often run by the state university) can provide helpful advice.

From the Soil into the Roots

A plant absorbs water and essential nutrients from the soil through its roots. Plant roots have a remarkable capacity for extracting nutrients from soil because of their root hairs, extensions of epidermal cells that dramatically increase the surface area available for absorption **(Figure 29.6)**. The root hairs of a single sunflower plant, for example, if laid end to end, could stretch many miles, providing a huge surface area in contact with nutrient-containing soil.

All mineral nutrients that enter a plant root are dissolved in water. To be transported throughout the plant, water and solutes must

Figure 29.6 **Root hairs of a radish seedling.** The root hairs of a plant, clearly visible in this photograph, are extensions of epidermal cells on the outer surface of the root.

move from the soil through the epidermis and cortex of the root, then into the water-conducting xylem tissue in the root's central cylinder (see Figure 28.12). To reach the xylem, the solution must pass through the plasma membranes of root cells. Because these membranes are selectively permeable, only certain solutes reach the xylem. This selectivity helps regulate the mineral composition of a plant's vascular system.

Many plants gain even more absorptive surface than their root hairs provide through symbiotic associations with fungi **(Figure 29.7)**. Together, the plant's roots and the fungus comprise a mutually beneficial association called a **mycorrhiza** (plural, *mycorrhizae*). Fungal filaments around the roots absorb water and minerals much more rapidly than the roots alone can. Some of the water and minerals taken up by the fungus are transferred to the plant, while the plant's photosynthetic products nourish the fungus.

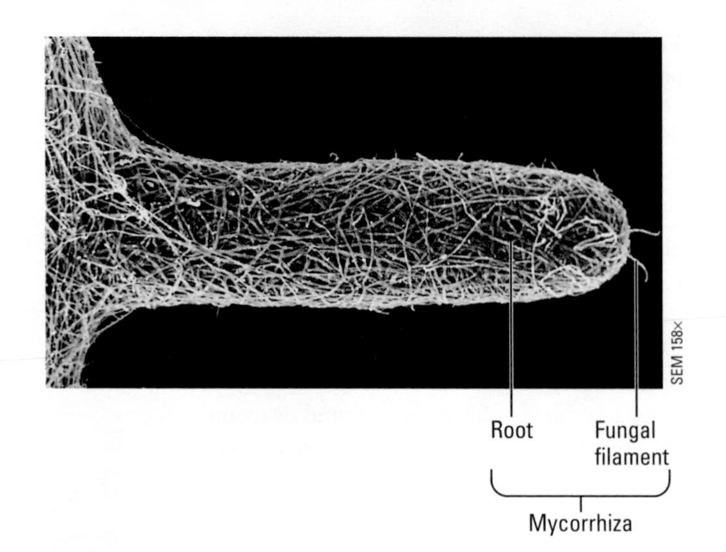

Figure 29.7 A eucalyptus mycorrhiza. This micrograph shows a small root of a eucalyptus tree. The root is covered with a twisted mat of fungal filaments. Together, the plant and fungus form a mutually beneficial association called a mycorrhiza.

The Roles of Bacteria in Nitrogen Nutrition

Air penetrates the soil around roots, but plants cannot use the form of nitrogen found in air (gaseous N_2). For plants to absorb nitrogen from the soil, N_2 must first be converted to ammonium ions (NH_4^+) or nitrate ions (NO_3^-). Most plants rely on bacteria to supply them with usable nitrogen.

Soil Bacteria and Nitrogen **Figure 29.8** shows three types of soil bacteria that play an essential role in supplying plants with nitrogen. One type, called nitrogen-fixing bacteria (⬬) converts atmospheric N_2 to ammonium,

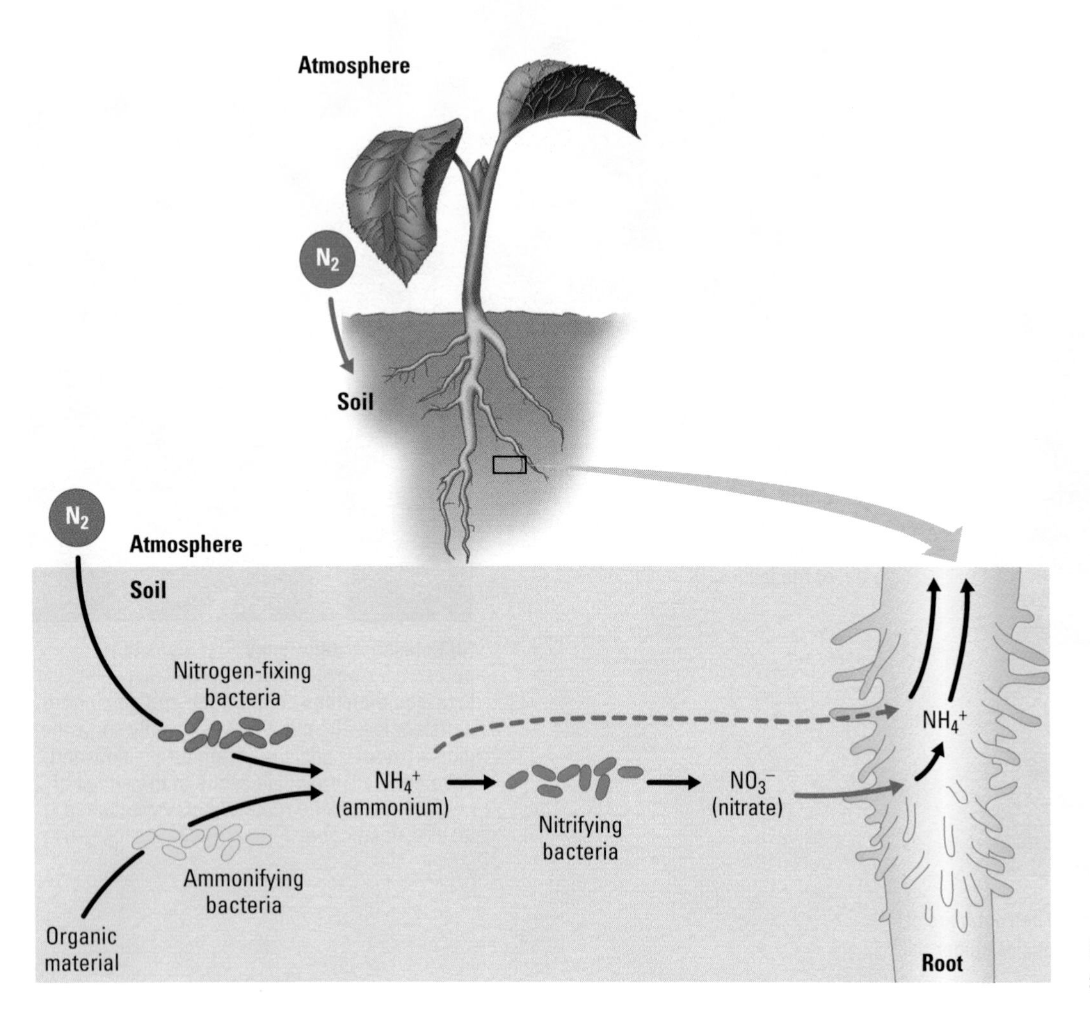

Figure 29.8 The roles of bacteria in supplying nitrogen to plants.

a process called **nitrogen fixation.** A second type, called ammonifying bacteria ( ), adds to the soil's supply of ammonium by decomposing organic matter. However, plant roots can absorb only a small amount of ammonium from the soil. A third type of soil bacteria, called nitrifying bacteria ( ), converts soil ammonium to nitrate, and plants take up most of their nitrogen in this form. Plants then convert nitrate back to ammonium, which they incorporate into proteins and other nitrogen-containing organic molecules.

Root Nodule Bacteria and Nitrogen Some plant families, including legumes—peas, beans, peanuts, and many other plants that produce their seeds in pods—have their own built-in source of ammonium: nitrogen-fixing bacteria that live in swellings called **root nodules** **(Figure 29.9a)**. Within these nodules, plant cells have been "infected" by nitrogen-fixing bacteria that reside inside cytoplasmic vesicles **(Figure 29.9b)**. The symbiotic relationship between a plant and its nitrogen-fixing bacteria is mutually beneficial. The bacteria have enzymes that catalyze the conversion of atmospheric N_2 to ammonium ions (NH_4^+); the plant, in turn, provides the bacteria with carbohydrates and other organic compounds. When conditions are favorable, root nodule bacteria fix so much nitrogen that the nodules secrete excess ammonia, which increases the fertility of the soil. This is one reason farmers practice crop rotation; legumes planted in alternate years can be plowed under to decompose into "green manure," reducing the need for manufactured fertilizer.

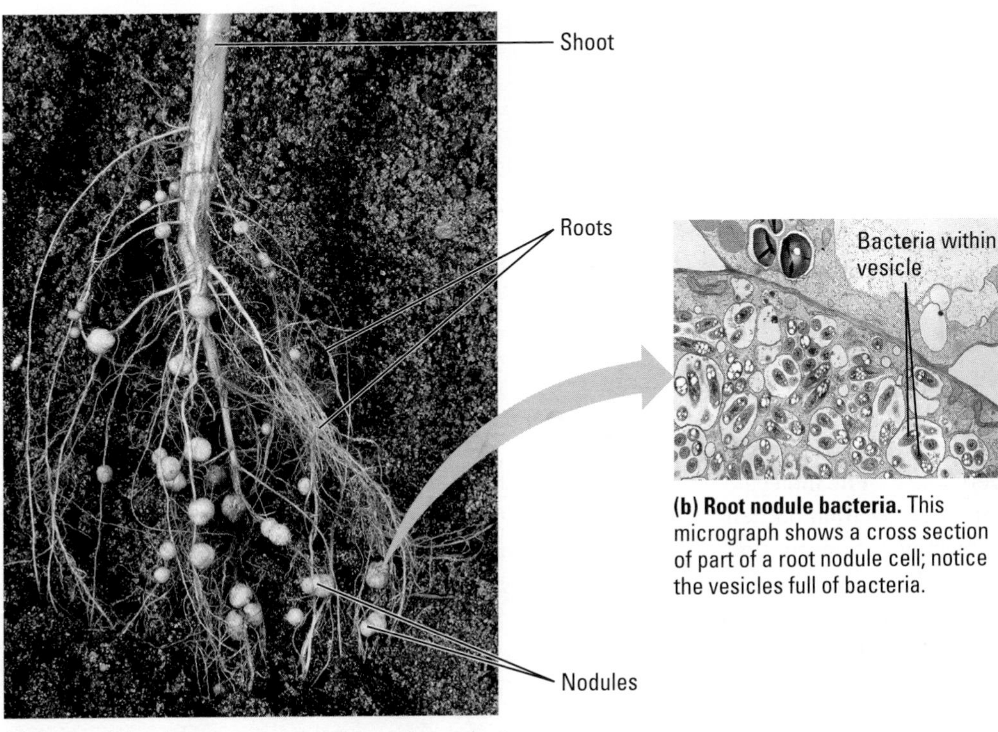

(b) Root nodule bacteria. This micrograph shows a cross section of part of a root nodule cell; notice the vesicles full of bacteria.

(a) Root nodules. Legumes, such as this soybean plant, have root nodules that contain nitrogen-fixing bacteria.

Figure 29.9 Root nodules on a pea plant.

The Transport of Water

As a plant grows upward toward sunlight, it needs an increasing supply of resources from the soil. To thrive, a plant must be able to transport water and dissolved ions from its roots to the rest of the plant.

We saw in Chapter 28 that mature water-conducting cells of the xylem consist only of cell walls with openings in their ends, arranged end to end to form very thin vertical tubes (see Figure 28.10d). A solution of water and nutrients, called **xylem sap,** flows through these tubes all the way from a plant's roots to the tips of its leaves.

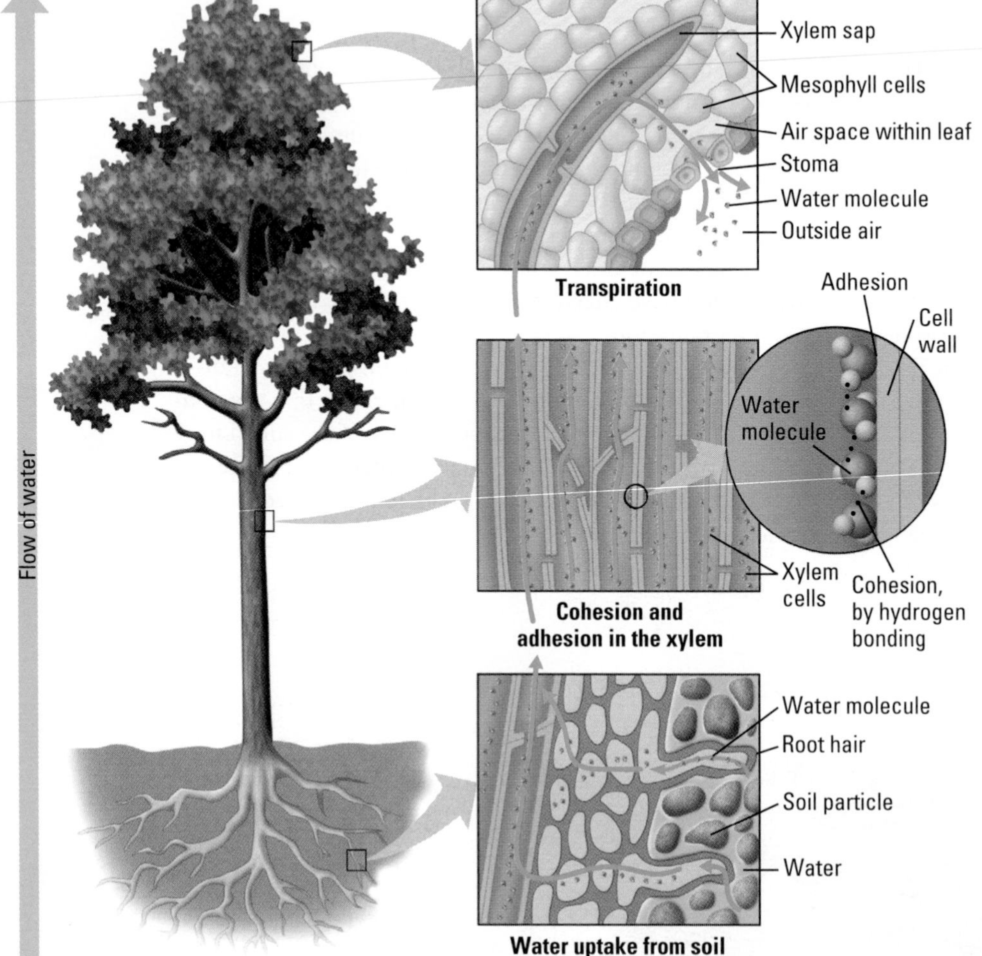

Figure 29.10 The flow of water up a tree.

The Ascent of Xylem Sap What force moves xylem sap up against the downward pull of gravity? For the most part, xylem sap is pulled upward by **transpiration,** the loss of water vapor from a plant **(Figure 29.10)**. Most transpiration occurs through the stomata of leaves. The stomata open into air spaces, which are filled with water molecules that have evaporated from the surrounding mesophyll cells. Water vapor diffuses out of the stomata because its concentration is higher inside the leaf's air spaces than in the outside air.

Transpiration can pull xylem sap up a tree because of two special properties of water: cohesion and adhesion. **Cohesion** is the sticking together of molecules of the same kind. In the case of water, hydrogen bonds make the H_2O molecules stick to one another (as described in Chapter 2). **Adhesion** is the sticking together of molecules of different kinds. Water molecules tend to adhere to cellulose molecules in the walls of xylem cells. Together, adhesion and cohesion create a continuous string of water molecules that stick to each other and to the inside walls of the xylem tubes (see Figure 29.10).

Before a water molecule can exit the leaf, it must break off from the top of the string. In effect, the water molecule is pulled off by a concentration gradient: The air outside the leaf is much drier than the moist interior of the leaf, causing the water molecule to diffuse outward. Cohesion resists this pulling force but is not strong enough to overcome it. The molecule breaks off, and the opposing forces of cohesion and transpiration put tension on the rest of the string of water molecules. As long as transpiration continues, the string is kept tense and is pulled upward as one molecule exits the leaf and the one behind it is tugged up into its place. Plant biologists call this explanation for the ascent of xylem sap the **transpiration-cohesion-tension mechanism.** Transpiration exerts a pull on a tense string of water molecules that is held together by cohesion and held up by adhesion.

Note that the transport of xylem sap requires no energy expenditure by the plant. Physical properties—cohesion, adhesion, and the evaporating effect of the sun—move water and dissolved minerals from a plant's roots to

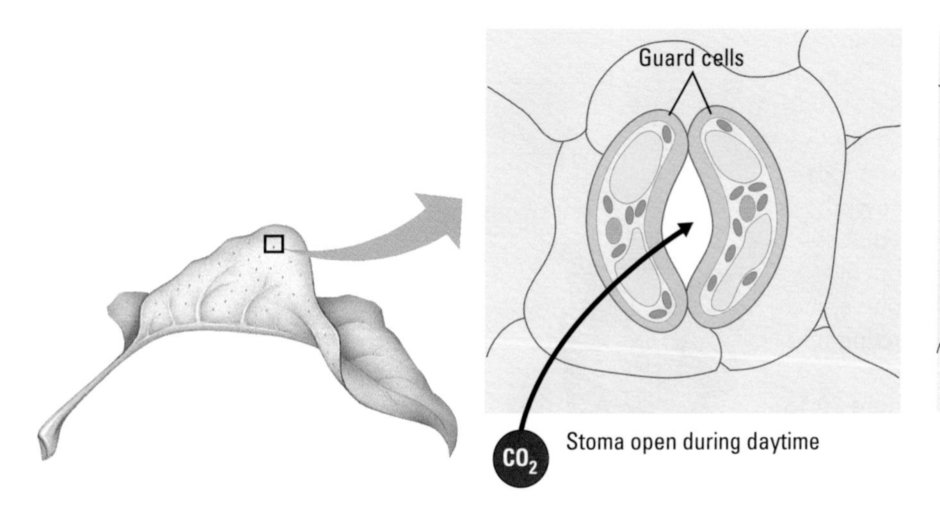

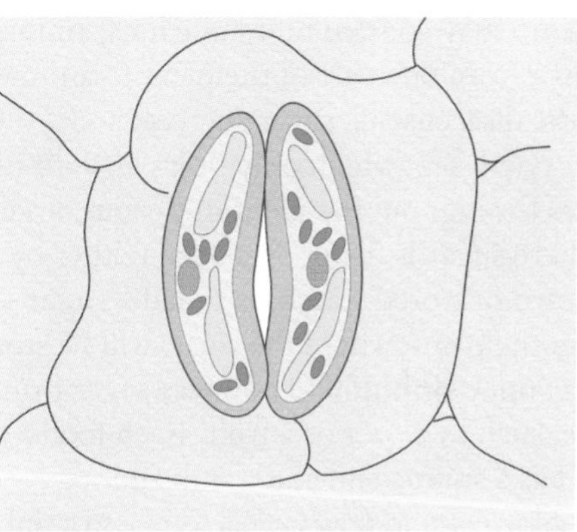

Figure 29.11 How guard cells control stomata. A pair of guard cells surrounding each stoma can change shape in response to environmental signals, regulating the rate of transpiration and CO_2 uptake.

its shoots. Transpiration is thus a highly efficient means of moving a lot of water upward through the body of a plant.

You have probably tasted xylem sap moved via transpiration yourself. When deciduous trees, such as maples, resume growth in spring after a leafless winter, starch that was produced the previous summer and stored in the roots is converted to sugar, which is transported upward in the xylem sap via transpiration to nourish the developing leaf buds. This sap can be harvested in buckets and boiled to concentrate the sugar; it takes about 30–50 L of xylem sap to produce 1 L of maple syrup.

The Regulation of Transpiration by Stomata Transpiration works both for and against plants. Although it helps to transport materials from the soil to the rest of the plant, it can result in the plant's losing an astonishing amount of water. Transpiration is usually greatest on days that are sunny, warm, dry, and windy, because these climatic conditions increase evaporation. An average-sized maple tree (about 20 m high) can lose more than 200 L of water per hour during a summer day. As long as water moves up from the soil fast enough to replace the lost water, this presents no problem. But if the soil dries out and transpiration exceeds the delivery of water, the leaves will wilt. Unless the soil and leaves are rehydrated, the plant will eventually die.

The leaf stomata, which can open and close, are evolutionary adaptations that help plants adjust their transpiration rates to changing environmental conditions. Stomata are usually open during the day, which allows CO_2 to enter the leaf from the atmosphere and thus keep photosynthesis going when sunlight is available. However, the plant also loses water through transpiration. At night, when there is no light for photosynthesis and therefore no need for CO_2, the stomata are closed, saving water. Stomata may also close during the day if a plant is losing water too fast. Stomatal opening and closing is controlled by the changing shape of the two guard cells flanking each stoma **(Figure 29.11)**.

The Transport of Sugars

In addition to transporting materials absorbed from the soil, a plant needs to transport the sugars it makes by photosynthesis. This is the main function of phloem.

Phloem consists of living food-conducting cells arranged end to end into tubes **(Figure 29.12)**. Through perforations in the end walls of these cells, a sugary solution called **phloem sap** moves freely from one cell to

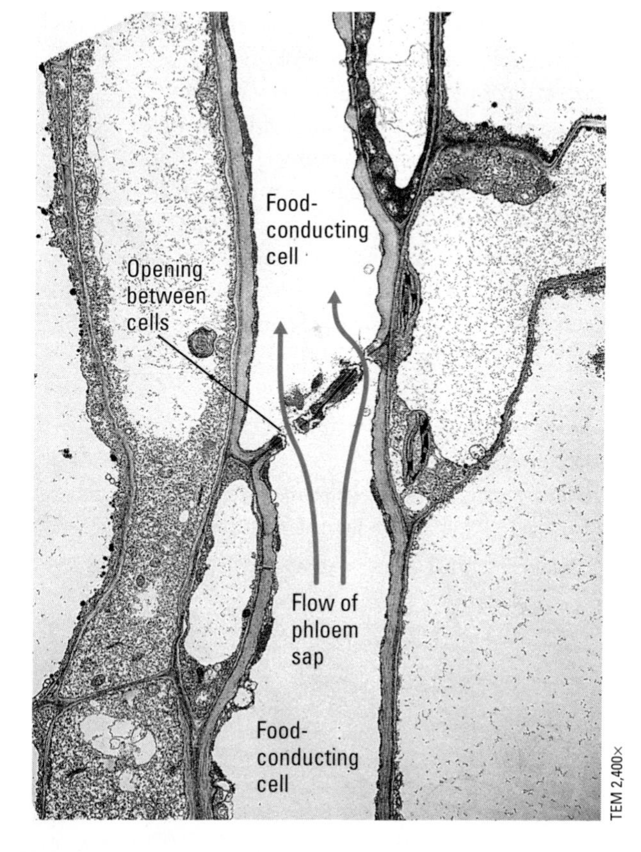

Figure 29.12 Food-conducting cells of phloem. The main function of phloem is to transport sugars through live food-conducting cells arranged into tubes. Phloem sap moves through openings between cells.

the next. Phloem sap may contain inorganic ions, amino acids, and hormones in transit from one part of the plant to another, but its main solute is usually the disaccharide sugar sucrose.

In contrast to xylem sap, which only flows upward from the roots, phloem sap moves throughout the plant in various directions. A location in a plant where sugar is being produced (either by photosynthesis or by the breakdown of stored starch) is called a **sugar source.** A recipient location in the plant, where the sugar will be stored or consumed, is called a **sugar sink.** Phloem moves sugar from a source, such as a leaf, to a sink, such as a root or a fruit. Each food-conducting tube in phloem tissue has a source end and a sink end.

What causes phloem sap to flow from a source to a sink? **Figure 29.13** uses a beet plant to illustrate a widely accepted model called the **pressure-flow mechanism.** At the sugar source (the beet leaves), ❶ sugar is loaded from a photosynthetic cell into a phloem tube via active transport. Sugar loading at the source end raises the solute concentration inside the phloem tube. ❷ The high solute concentration draws water into the tube by osmosis, usually from the xylem. The inward flow of water from the xylem into the phloem raises the water pressure at the source end of the tube.

At the sugar sink (the beet root), both sugar and water leave the phloem tube. ❸ As sugar leaves the phloem, lowering the sugar concentration at the sink end, ❹ water follows by osmosis back into the xylem. The exit of water lowers the water pressure in the tube.

The building of water pressure at the source end of the phloem tube and the reduction of water pressure at the sink end cause phloem sap to flow from source to sink. This pressure-flow mechanism explains why phloem sap always flows from a sugar source to a sugar sink, regardless of their locations in the plant.

We now have a broad picture of how a plant absorbs substances from the soil and transports materials from one part of its body to another: Water and inorganic ions enter from the soil and are distributed by xylem. The xylem sap is pulled upward by transpiration. Carbon dioxide enters leaves through the stomata and is incorporated into sugars, which are distributed by phloem. Pressure flow drives the phloem sap from leaves and storage sites to other parts of the plant, where the sugars are used or stored.

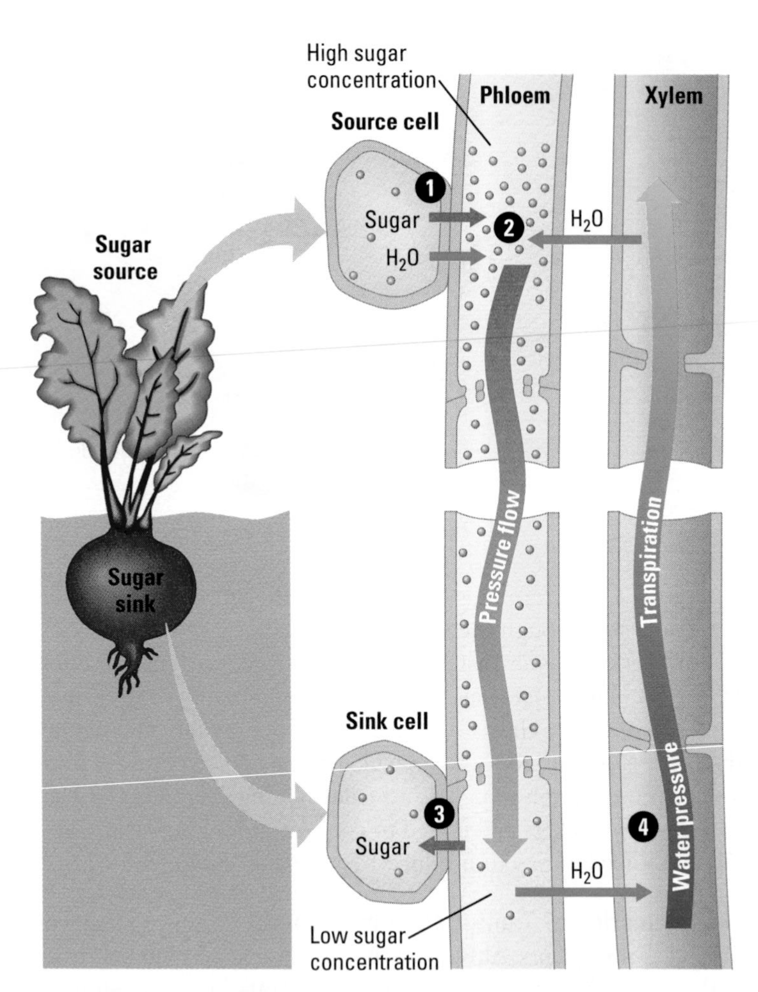

Figure 29.13 Pressure flow in plant phloem. The red dots in the phloem tube represent sugar molecules; notice that a concentration gradient decreasing from top to bottom creates a pressure flow (red arrow). The blue color represents a parallel gradient of water pressure due to transpiration in the phloem sap; water pressure decreases from bottom to top. As indicated on the right side, xylem tubes transport the water back from sink to source.

CHECKPOINT

1. Plants require nutrients, which they acquire from the atmosphere in the form of __________ and from the soil in the form of __________ and __________.
2. What is the most common nutrient deficiency in plants?
3. What are mycorrhizae?
4. Why might a pollutant that kills soil bacteria result in nitrogen deficiency in plants?
5. Contrast cohesion and adhesion and describe the role of each in the ascent of xylem sap.

Answers: 1. carbon dioxide; water; minerals (inorganic ions) **2.** Nitrogen deficiency **3.** Symbiotic associations of roots and fungi **4.** Because certain soil bacteria make nitrogen available to plants in forms they can use **5.** Cohesion is the sticking together of identical molecules—water molecules in the case of xylem sap. Adhesion is the sticking together of different kinds of molecules, as in the adhesion of water to the cellulose of xylem walls. Cohesion enables transpiration to pull xylem sap up without the water in the vessels separating; adhesion helps to support xylem sap against the downward pull of gravity.

Plant Hormones

You learned in Chapter 25 that animal **hormones** are regulatory chemicals that travel from their production sites to affect other parts of an animal's body. Plants, like animals, use hormones to regulate their internal activities. Plants produce hormones in very small amounts, but a minute amount of any of these chemicals can have profound effects on target cells. Hormones control plant growth and development by affecting the division, elongation, and differentiation of cells.

Plant biologists have identified five major types of plant hormones, which are listed in **Table 29.1**. Each type of hormone can produce a variety of effects, depending on the species and developmental stage of the plant as well as the hormone's concentration and site of action. In most situations, no single hormone acts alone. Instead, it is usually the relative concentration of several plant hormones that controls a plant's growth and development.

Auxin

Phototropism is the directional growth of a plant shoot in response to light. For example, a houseplant on a windowsill grows toward light **(Figure 29.14)**. If you rotate the plant, it will soon reorient its growth until its leaves again face the window. Phototropism directs both growing seedlings and the shoots of mature plants toward the sunlight they use for photosynthesis.

Figure 29.14 A houseplant growing toward light. Phototropism guides growing plants toward the sunlight needed for photosynthesis.

Table 29.1 Major Types of Plant Hormones

Hormone	Major Functions	Where Produced or Found in Plant
Auxin	Stimulates stem elongation; affects root growth, differentiation, and branching; stimulates development of fruit, apical dominance, phototropism, and gravitropism	Meristems of apical buds; young leaves; embroyos within seeds
Ethylene	Promotes fruit ripening; opposes some auxin effects; promotes or inhibits growth and development of roots, leaves, and flowers, depending on species	Ripening fruit, nodes of stems, aging leaves and flowers
Cytokinins	Affect root growth and differentiation; stimulate cell division and growth; stimulate germination; delay aging	Made in roots and transported to other organs
Gibberellins	Promote seed germination, bud development, stem elongation, and leaf growth; stimulate flowering and fruit development; affect root growth and differentiation	Meristems of apical buds and roots; young leaves; embryos
Abscisic acid (ABA)	Inhibits growth; closes stomata during water stress; helps maintain dormancy	Leaves, stems, roots, green fruit

Cell Elongation How does a plant grow in a particular direction? Microscopic observations of growing plants reveal the cellular mechanism that underlies phototropism **(Figure 29.15)**. Cells on the darker side of the stem are larger—actually, they have elongated faster—than those on the brighter side, causing the shoot to bend toward the light. If a seedling is illuminated uniformly from all sides or if it is kept in the dark, the cells all elongate at a similar rate and the seedling grows straight upward.

What causes plant cells on the dark side of a shoot to grow faster than those on the bright side? Our present understanding of this phenomenon emerged from a series of classic experiments conducted by two scientists with a very familiar name.

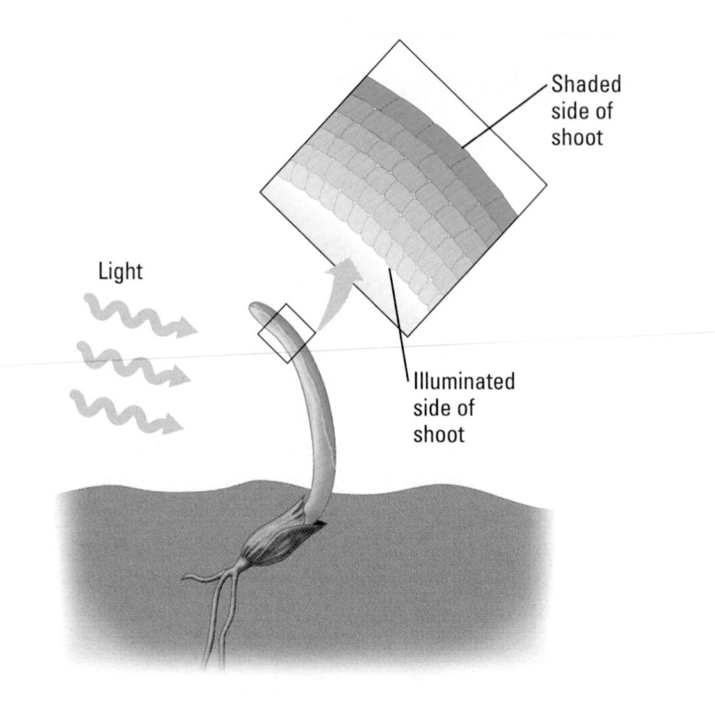

Figure 29.15 Phototropism in a grass seedling. In this figure, a grass seedling curves toward light coming from one side because of the faster elongation of cells on the shaded side.

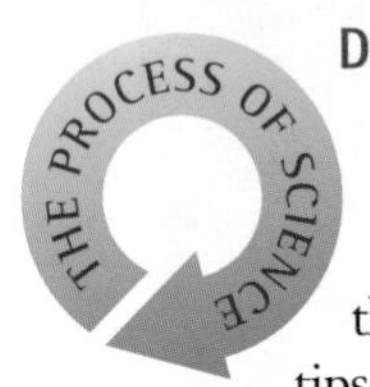

Do Chemical Signals Affect Plant Growth? In the late 19th century, Charles Darwin and his son Francis performed some of the earliest experiments on phototropism **(Figure 29.16a)**. ⓐ They began with the **observation** that grass seedlings seemed to bend toward light only if the tips of their shoots were present. This led them to **question** whether the tips of the seedlings produced some kind of growth signal. Their **hypothesis** was that the plant tips sensed light and produced a growth signal in response. They made a **prediction**: Removing a shoot tip or blocking access to light would prevent phototropism. This was the first of several **experiments**, which are summarized in Figure 29.16.

The Darwins' **results** showed that ⓑ removing the tip of a grass shoot did prevent growth toward light. The shoot also remained straight when they placed ⓒ an opaque cap on its tip. However, the shoot curved normally when they placed ⓓ a transparent cap on its tip or ⓔ an opaque shield around its base. The Darwins concluded that the tip of the shoot was responsible for sensing light. They also recognized that the growth response, the bending of the shoot, occurs below the tip. Therefore, they speculated that some growth signal was transmitted downward from the tip to the growing region of the shoot. Later experiments by other researchers demonstrated that a mobile chemical produced in the shoot tip acts as the signal for phototropism. ■

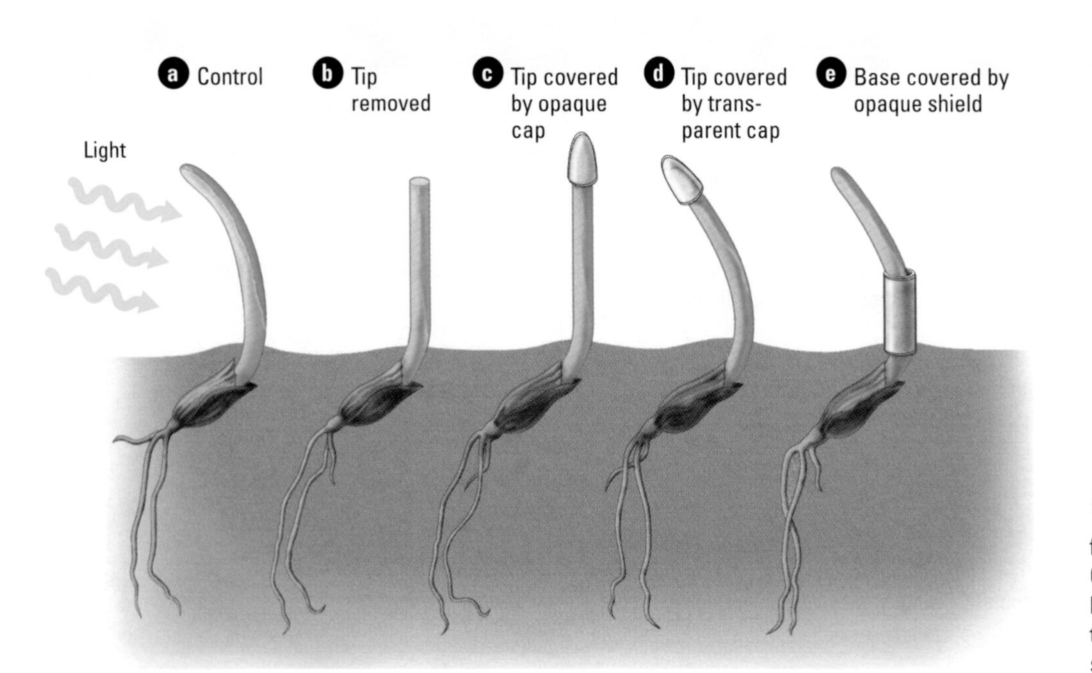

Figure 29.16 The Darwins' classic experiments on phototropism. This figure summarizes experiments conducted by Charles Darwin and his son Francis during the late 19th century. Together, these results suggest that a substance produced by the tip of a seedling controls phototropism.

The Action of Auxin The chemical responsible for phototropism is a hormone called **auxin.** Auxin is produced by the apical meristem at the tip of a shoot. Illuminating one side of a shoot results in a higher auxin concentration on the dark side compared to the light side. This uneven distribution of auxin makes cells on the dark side elongate, causing the shoot to bend.

Auxin promotes cell elongation in stems only when it is present within a certain concentration range. Interestingly, this same concentration range inhibits cell elongation in roots. Above this range, auxin inhibits cell elongation in stems; and below this range, it causes root cells to elongate. This complex set of effects of auxin on cell elongation reinforces two points: (1) The same hormone may have different effects at different concentrations in the same target cell, and (2) a particular concentration of a hormone may have different effects on different target cells.

In addition to causing stems and roots to lengthen, auxin induces cell division in the vascular cambium, thus promoting growth in stem diameter. Furthermore, auxin is produced by developing seeds and promotes the growth of fruit. Farmers sometimes produce seedless tomatoes, cucumbers, and eggplants by spraying unpollinated plants with synthetic auxin.

Ethylene

Ethylene is a hormone, released as a gas, that triggers a variety of aging responses in plants, including fruit ripening and dropping of leaves.

Fruit Ripening A burst of ethylene production in a fruit triggers its ripening. Because ethylene is a gas, the signal to ripen spreads from fruit to fruit: One bad apple does indeed spoil the lot. You can make some fruits ripen faster if you store them in a plastic bag so that the ethylene gas accumulates **(Figure 29.17)**. On a commercial scale, many kinds of fruit—tomatoes, for instance—are often picked green and then partially ripened in huge storage bins into which ethylene gas is piped. In other cases, growers take measures to retard the ripening action of natural ethylene. Stored apples are often flushed with CO_2, which inhibits the action of ethylene, so that apples picked in autumn can be stored for sale the following summer.

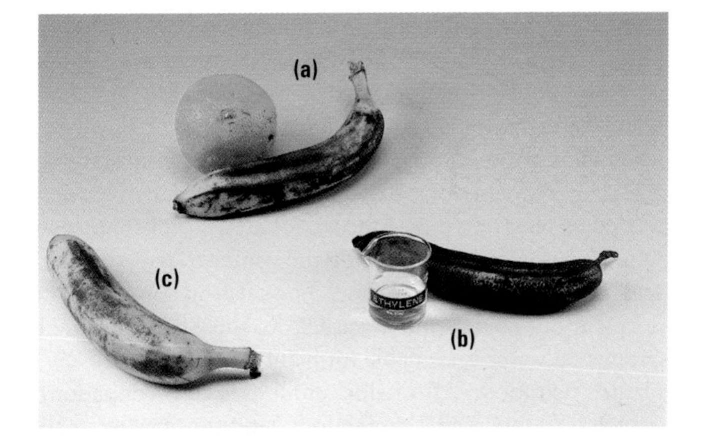

Figure 29.17 The effect of ethylene on the ripening of bananas. These three photos show the results of a fruit-ripening demonstration. Three unripe bananas were stored for the same time period in plastic bags: **(a)** with an ethylene-releasing orange, **(b)** with a beaker of an ethylene-releasing chemical, and **(c)** alone. As you can see, the more ethylene present, the riper (darker) the banana.

Leaf Drop The loss of leaves in autumn is affected by ethylene. Leaf drop is triggered by environmental stimuli, including the shortening days of autumn and cooler temperatures. These stimuli cause a change in the balance of ethylene and auxin that weakens cell walls in a layer of cells at the base of the leaf stalk. The weight of the leaf, often assisted by wind, causes this layer to split, releasing the leaf.

Cytokinins

Cytokinins are growth regulators that promote cell division, or cytokinesis. Cytokinins are produced in actively growing tissues, particularly in roots, embryos, and fruits.

Cytokinins stimulate the growth of axillary buds, making a plant grow more branches and become bushy. Cytokinins entering the shoot system from the roots counter the inhibitory effects of auxin coming down from the terminal buds. The complex growth patterns of most plants probably result from the relative concentrations of auxin and cytokinins. Christmas tree growers sometimes use cytokinins to produce attractive branching.

Gibberellins

Gibberellins are another group of growth-regulators. Researchers have identified more than 100 gibberellins in various plant species. Roots and young leaves are major sites of gibberellin production. One of the main effects of gibberellins is to stimulate cell elongation and cell division in stems. This action generally enhances that of auxin. Also in combination with auxin, gibberellins can influence fruit development, and gibberellin-auxin sprays can make apples, currants, and eggplants develop without pollination and seed production. Gibberellins are used commercially to make seedless grapes grow larger and farther apart in a cluster **(Figure 29.18)**.

Gibberellins are also important in seed germination in many plants. Many seeds that require special environmental conditions to germinate, such as exposure to light or cold, will germinate on demand when sprayed with gibberellins. In nature, gibberellins in seeds are probably the link between environmental cues and the metabolic processes that reactivate growth of the embryo after a period of dormancy. For example, when water becomes available to a grass seed, it causes the embryo in the seed to release gibberellins, which promote germination by mobilizing nutrients stored in the seed.

Figure 29.18 The effect of gibberellins on grapes. The left cluster of grapes is untreated; the right cluster shows the effect of gibberellin treatment: larger grapes farther apart in the cluster.

Abscisic Acid

Unlike the growth-stimulating hormones we have studied so far, **abscisic acid (ABA)** generally slows growth. One of the times in a plant's life when it is advantageous to suspend growth is the onset of seed dormancy. Seed dormancy is an evolutionary adaptation that ensures that a seed will germinate only when there are appropriate conditions of light, temperature, and moisture.

Many types of dormant seeds will germinate only when ABA is removed or inactivated in some way. The seeds of some plants require prolonged exposure to cold during the winter to trigger ABA inactivation and seed germination in the spring. The seeds of some desert plants remain dormant until a downpour washes out the ABA, allowing them to germinate when water is available for plant growth **(Figure 29.19)**. For many plants, the ratio of ABA (which inhibits seed germination) to gibberellins (which promote seed germination) determines whether the seed will remain dormant or germinate.

Figure 29.19 The effect of ABA removal on seed dormancy. These evening primroses and purple sand verbena, photographed in the Mojave Desert in California, grew just after a hard rain washed ABA out of the seeds, allowing them to germinate.

CHECKPOINT

1. The status of axillary buds—dormant or growing—depends on the relative concentrations of ___________ moving down from the shoot tip and ___________ moving up from the roots.
2. When you buy a bunch of bananas, you might bring them home in a plastic bag. If the bananas are too green to eat, would it be better to leave the bananas in the bag or not? Why?
3. Researchers working with pea plants have connected a mutation that blocks gibberellin synthesis to one of the recessive traits Gregor Mendel studied in his famous experiments. Return to Figure 9.5 and identify the gibberellin-deficient mutant among Mendel's pea varieties.
4. Which two hormones regulate seed dormancy and germination? What are their opposing effects?

Answers: 1. auxin; cytokinins **2.** It would be better to leave the bananas in the bag because this will cause ethylene gas to accumulate, which promotes ripening. **3.** Dwarf variety **4.** Abscisic acid maintains seed dormancy; gibberellins promote germination.

Response to Stimuli

In this section, we consider how a plant responds to physical stimuli from the environment. In particular, we will focus on the most important stimulus in the life of a plant: light.

Tropisms

Tropisms are directed growth responses that cause parts of a plant to grow toward or away from a stimulus. Phototropism, as you saw in Figure 29.14, is a particularly important example, but there are other directed growth responses. For example, **thigmotropism** is a response to touch, as when a pea tendril (actually a modified leaf) contacts a string or wire and coils around it for support. The tendril in **Figure 29.20a** grew straight until it touched the support. Contact then stimulated the cells to grow at different rates on opposite sides of the tendril (slower in the contact area), making the tendril coil around the wire.

Gravitropism is the directional growth of a plant organ in response to gravity: Shoots grow upward and roots grow downward, regardless of the orientation of the seed **(Figure 29.20b)**. Auxin plays an important role in gravitropism. The seedling shown on the right in Figure 29.20b responded to gravity by redistributing auxin to the lower side of its horizontally growing root. A high concentration of auxin inhibits the growth of root cells. As growth of the lower side of the root was slowed, cells on the upper side continued to elongate normally, and the root curved downward.

(a) Thigmotropism. Thigmotropism, growth in response to touch, allows vining plants to wrap around support structures. In this photo, the tendril of a pea plant coils around a wire fence.

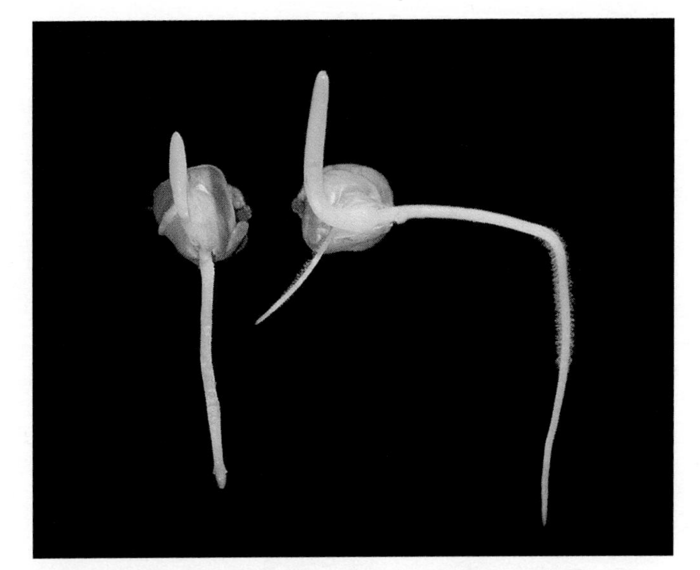

(b) Gravitropism. These seedlings were both germinated in the dark. The one on the left was left alone; notice that its shoot grew straight up and its root straight down. The seedling on the right was germinated in the same way, but two days later it was turned on its side so that the shoot and root were horizontal. By the time the photo was taken, the shoot had turned back upward and the root had turned down.

Figure 29.20 Tropisms.

Photoperiods

In addition to providing energy for photosynthesis and directing growth, light helps regulate a plant's life cycle. Flowering, seed germination, and the onset and ending of dormancy are stages in plant development that usually occur at specific times of the year. The environmental stimulus that plants most often use to detect the time of year is called **photoperiod,** the relative lengths of day and night.

Plants whose flowering is triggered by photoperiod fall into two groups: long-night plants and short-night plants **(Figure 29.21)**. Long-night plants, such as chrysanthemum and poinsettia, generally flower in late summer, fall, or winter, when night lengthens. Short-night plants, such as lettuce, iris, and many cereal grains, usually flower in late spring or early summer, when nights are briefer. Some plants, such as dandelions, are night-neutral; their flowering is unaffected by photoperiod. Florists apply this knowledge to bring us flowers out of season. The blooming of chrysanthemums, for instance, can be stalled until spring by interrupting each long night with a flash of light, thus turning one long night into two short nights. (Long-night and short-night plants were historically called short-day and long-day plants, respectively; the newer terminology that we use here reflects the present understanding of the mechanism behind this effect.)

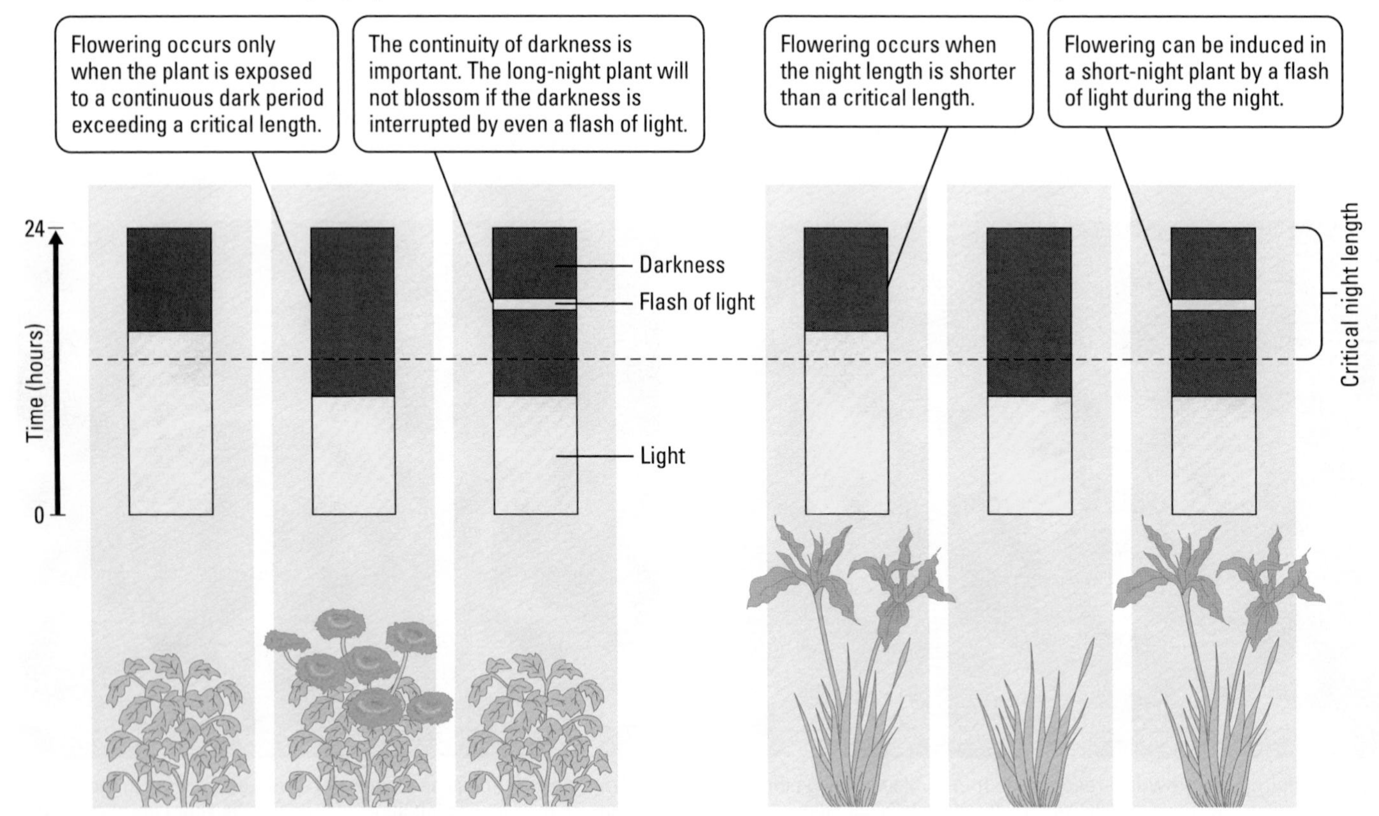
Long-night plants
Flowering occurs only when the plant is exposed to a continuous dark period exceeding a critical length.
The continuity of darkness is important. The long-night plant will not blossom if the darkness is interrupted by even a flash of light.
Short-night plants
Flowering occurs when the night length is shorter than a critical length.
Flowering can be induced in a short-night plant by a flash of light during the night.
24
0
Time (hours)
Darkness
Flash of light
Light
Critical night length

Figure 29.21 Photoperiod and flowering. Photoperiod, the relative lengths of day and night, can trigger flowering.

CHECKPOINT

1. Why are tropisms called growth responses?
2. A particular long-night plant won't flower in the spring. You try to induce flowering by using a short, dark interruption to split a long-day period of spring into two short-day periods. What result do you predict?

Answers: 1. Because the movement of a plant organ toward or away from an environmental stimulus takes place by growing. An organ bends when cells on one side grow faster than cells on the other side, and it extends in one direction when cells grow evenly. **2.** The plants still won't flower, because it is actually night length, not day length, that controls flowering.

EVOLUTION CONNECTION

The Interdependence of Organisms

Although plants make their own food molecules, they are not independent of other organisms. As you learned in the last chapter, most flowering plants depend on insects or other animals for pollination and seed dispersal. In this chapter, you saw that plants rely on organisms from two other kingdoms to help acquire nutrients: soil bacteria and the fungi of mycorrhizae. The fossil record shows that mycorrhizae have existed since plants first evolved. Indeed, the mycorrhizal connection may have altered the entire course of evolution by helping make possible the colonization of land.

Nearly all animals depend on plants or other photosynthetic organisms for food. As animals have evolved in tandem with plants, they have developed many specialized feeding mechanisms. For example, small insects called aphids have adaptations that allow them to tap into sugar-containing phloem sap (**Figure 29.22**, top). And the aphids themselves are food to predators—among them the devil's flower mantis (*Blepharopsis mendica*), pictured on this book's cover (Figure 29.22, bottom). Using long forearms and busy mouthparts, mantids are voracious consumers of aphids.

Thus, we close this chapter—and this book—illustrating one of biology's overarching themes: the interrelatedness of organisms. Animals depend on other animals and plants, which in turn depend on animals, fungi, and prokaryotes. This reminds us once again that it is impossible to separate ourselves from all of the living creatures that share the biosphere. We hope that *Essential Biology with Physiology* has given you a new appreciation of your place in the living world. ■

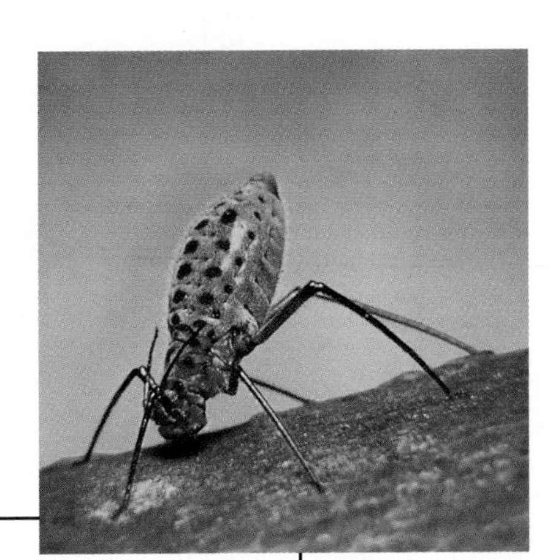

Figure 29.22 The devil's flower mantis. Like most mantids, the devil's flower mantis (*Blepharopsis mendica*) feeds on aphids (top), which feed on plant sap, which in turn is created using nutrients obtained with the help of bacteria and fungi. These relationships emphasize the interrelatedness of all life on Earth.

Chapter Review

SUMMARY OF KEY CONCEPTS

For study help and activities, go to campbellbiology.com or the student CD-ROM.

How Plants Acquire and Transport Nutrients

MP3 Tutor *Transpiration*

- As a plant grows, its roots absorb water, minerals (inorganic ions), and some O_2 from the soil. Its leaves absorb CO_2 from the air.

- **Plant Nutrition** A plant must obtain from its surroundings usable sources of the chemical elements—nutrients—it requires.

Macronutrients	Micronutrients
• Needed in large amounts • Used to build organic molecules	• Reusable, so needed in much smaller amounts • Used as enzyme cofactors/components
Carbon, oxygen, hydrogen, nitrogen, sulfur, phosphorus, calcium, potassium, magnesium	Iron, chlorine, copper, manganese, zinc, molybdenum, boron, nickel

Nutrient deficiencies affect the health of a plant and its nutritional value.

Case Study in the Process of Science *How Does Acid Precipitation Affect Mineral Deficiency?*

- **From the Soil into the Roots** Root hairs greatly increase a root's absorptive surface. Water and solutes move through the root's epidermis and cortex, then into the xylem (water-conducting tissue) for transport upward. Relationships with other organisms help plants obtain nutrients. Many plants form mycorrhizae, mutually beneficial associations with fungi. A network of fungal threads increases a plant's absorption of nutrients, and the fungus receives some nutrients from the plant.

Activity *Absorption of Nutrients from Soil*

Case Study in the Process of Science *What Determines If Water Moves Into or Out of a Plant Cell?*

- **The Roles of Bacteria in Nitrogen Nutrition** Most plants depend on bacteria to convert nitrogen to a form usable by the plant.

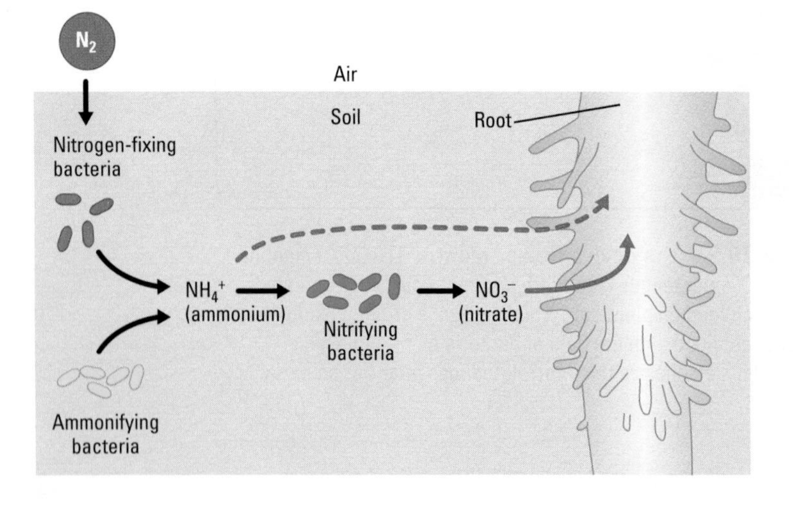

Legumes and certain other plants have nodules in their roots that house vesicles filled with nitrogen-fixing bacteria.

- **The Transport of Water** Most of the force that moves water and solutes upward in the xylem comes from transpiration.

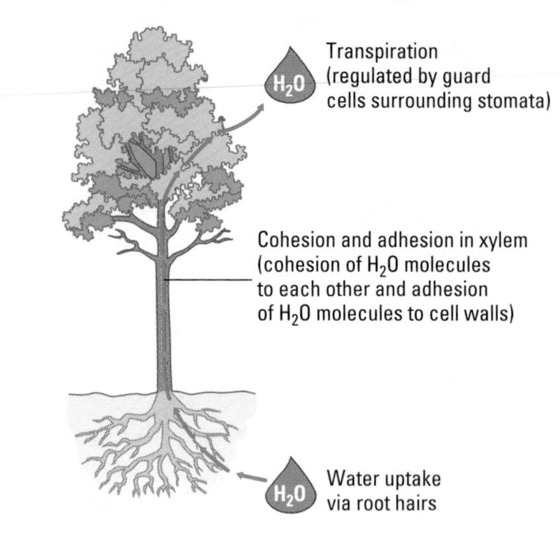

Activity *Transpiration*

Case Study in the Process of Science *How Is the Rate of Transpiration Calculated?*

- **The Transport of Sugars** Phloem uses a pressure-flow mechanism to transport food molecules made by photosynthesis. At a sugar source, such as a leaf, sugar is loaded into a phloem tube. The sugar raises the solute concentration in the tube, and water follows by osmosis, raising the pressure in the tube. As sugar is removed and stored or used in a sugar sink, such as the root, water follows. The increase in pressure at the sugar source and decrease at the sugar sink causes phloem sap to flow from source to sink.

Activity *Transport in Phloem*

Plant Hormones

- Hormones coordinate the internal activities of plants.

- **Auxin** Bending toward light results from faster cell growth on the shaded side of the shoot than on the lighted side. This effect is regulated by the hormone auxin. Plants produce auxin in the apical meristems at the tips of shoots. At different concentrations, auxin stimulates or inhibits the elongation of shoots and roots.

- **Ethylene** As a fruit ages, it gives off ethylene gas, which hastens ripening. Fruit growers use ethylene to control ripening. A changing ratio of auxin to ethylene, triggered mainly by longer nights, probably causes the loss of leaves from deciduous trees.

Activity *Leaf Drop*

- **Cytokinins** Cytokinins, produced by growing roots, embryos, and fruits, are hormones that promote cell division. The ratio of auxin to cytokinins helps coordinate the growth of roots and shoots.

- **Gibberellins** Gibberellins stimulate the elongation of stems and leaves and the development of fruits. Gibberellins released from embryos function in some of the early events of seed germination. Auxin and gibberellins are used to produce seedless fruits.

• **Abscisic Acid** Abscisic acid (ABA) inhibits the germination of seeds. The ratio of ABA to gibberellins often determines whether a seed will remain dormant or germinate. Seeds of many plants remain dormant until their ABA is inactivated or washed away.

Case Study in the Process of Science *What Plant Hormones Affect Organ Formation?*

Response to Stimuli

• **Tropisms** Tropisms are growth responses that make a plant grow toward or away from a stimulus. Important tropisms include phototropism, directional growth in response to light, gravitropism, a response to gravity, and thigmotropism, a response to touch.

• **Photoperiods** Plants mark the seasons by measuring photoperiod, the relative lengths of night and day. The timing of flowering is one of the seasonal responses to photoperiod. Long-night plants flower when nights exceed a certain critical length; short-night plants flower when nights are shorter than a critical length.

Activity *Flowering Lab*

SELF-QUIZ

1. Certain fungi cause diseases in plants. There are a variety of antifungal sprays that can be used to control this problem. Some gardeners constantly spray their plants with fungicides, even when no signs of disease are evident. How might this be disadvantageous to the plant?

2. Which of the following activities of soil bacteria does not contribute to creating usable nitrogen supplies for plant use?
   a. the fixation of atmospheric nitrogen
   b. the conversion of ammonium ions to nitrate ions
   c. the assembly of amino acids into proteins
   d. the generation of ammonium from proteins in dead leaves

3. Plants transport two types of sap. __________ sap, a solution of mostly water and sucrose, is transported from sites of sugar production to other parts of the plant. __________ sap, a solution of mostly water and inorganic ions, is transported from the roots to the rest of the plant body.

4. Auxin causes a shoot to bend toward light by
   a. causing cells to shrink on the dark side of the shoot.
   b. stimulating growth on the dark side of the shoot.
   c. causing cells to shrink on the lighted side of the shoot.
   d. stimulating growth on the lighted side of the shoot.

5. Match each of the following hormones to its primary role in a plant.

a. auxin	1. At different concentrations, stimulates or inhibits the elongation of shoots and roots
b. ethylene	2. Produced by roots, promotes cell division
c. cytokinins	3. Inhibits seed germination
d. gibberellins	4. Hastens fruit ripening
e. abscisic acid (ABA)	5. Promotes fruit development

6. The hormone that promotes leaf dropping in fall and fruit ripening is __________.

7. Buds and sprouts often form on tree stumps as a result of the action of the hormone(s) __________.

8. Why does a houseplant become bushier if you pinch off its terminal buds? Explain this effect in terms of plant hormones.

9. Match the following terms to their meanings.

a. gravitropism	1. Growth response to light
b. photoperiod	2. Growth response to touch
c. phototropism	3. Relative lengths of night and day
d. thigmotropism	4. Growth response to gravity

10. Jon just started a new job as night watchman at a plant nursery. His boss told him to stay out of a room where chrysanthemums (which are long-night plants) were about to flower. Around midnight, looking for the rest room, Jon accidentally opened the door to the chrysanthemum room and turned on the lights for a moment. How might this affect the chrysanthemums?

Answers to the Self-Quiz questions can be found in Appendix D.

Go to the website or CD-ROM for more Self-Quiz questions.

THE PROCESS OF SCIENCE

11. In some situations, the application of nitrogen fertilizer to crops has to be increased each year because the fertilizer decreases the rate of nitrogen fixation in the soil. Propose a hypothesis to explain this phenomenon. Describe a test for your hypothesis. What results would you expect from your test?

BIOLOGY AND SOCIETY

12. Agriculture is by far the biggest user of water in arid western states, including Colorado, Arizona, and California. The populations of these states are growing, and there is an ongoing conflict between cities and farm regions over water. To ensure water supplies for urban growth, cities are purchasing water rights from farmers. This is often the least expensive way for a city to obtain more water, and some farmers can make more money selling water than growing crops. Discuss the possible consequences of this trend. Is this the best way to allocate water for all concerned? Why or why not?

13. Imagine the following scenario: A plant scientist has developed a synthetic chemical that mimics the effects of a plant hormone. The chemical can be sprayed on apples before harvest to prevent flaking of the natural wax that is formed on the skin. This makes the apples shinier and gives them a deeper red color. What kinds of questions do you think should be answered before farmers start using this chemical on apples? How might the scientist go about finding answers to these questions?

A Appendix A: Metric conversion table

measurement	Unit and Abbreviation	Metric Equivalent	Approximate Metric-to-English Conversion Factor	Approximate English-to-Metric Conversion Factor
Length	1 kilometer (km)	= 1,000 (10^3) meters	1 km = 0.6 mile	1 mile = 1.6 km
	1 meter (m)	= 100 (10^2) centimeters = 1,000 millimeters	1 m = 1.1 yards 1 m = 3.3 feet 1 m = 39.4 inches	1 yard = 0.9 m 1 foot = 0.3 m
	1 centimeter (cm)	= 0.01 (10^{-2}) meter	1 cm = 0.4 inch	1 foot = 30.5 cm 1 inch = 2.5 cm
	1 millimeter (mm)	= 0.001 (10^{-3}) meter	1 mm = 0.04 inch	
	1 micrometer (µm)	= 10^{-6} meter (10^{-3} mm)		
	1 nanometer (nm)	= 10^{-9} meter (10^{-3} µm)		
	1 angstrom (Å)	= 10^{-10} meter (10^{-4} µm)		
Area	1 hectare (ha)	= 10,000 square meters	1 ha = 2.5 acres	1 acre = 0.4 ha
	1 square meter (m^2)	= 10,000 square centimeters	1 m^2 = 1.2 square yards 1 m^2 = 10.8 square feet	1 square yard = 0.8 m^2 1 square foot = 0.09 m^2
	1 square centimeter (cm^2)	= 100 square millimeters	1 cm^2 = 0.16 square inch	1 square inch = 6.5 cm^2
Mass	1 metric ton (t)	= 1,000 kilograms	1 t = 1.1 tons	1 ton = 0.91 t
	1 kilogram (kg)	= 1,000 grams	1 kg = 2.2 pounds	1 pound = 0.45 kg
	1 gram (g)	= 1,000 milligrams	1 g = 0.04 ounce 1 g = 15.4 grains	1 ounce = 28.35 g
	1 milligram (mg)	= 10^{-3} gram	1 mg = 0.02 grain	
	1 microgram (µg)	= 10^{-6} gram		
Volume (Solids)	1 cubic meter (m^3)	= 1,000,000 cubic centimeters	1 m^3 = 1.3 cubic yards	1 cubic yard = 0.8 m^3
	1 cubic centimeter (cm^3 or cc)	= 10^{-6} cubic meter	1 m^3 = 35.3 cubic feet	1 cubic foot = 0.03 m^3
	1 cubic millimeter (mm^3)	= 10^{-9} cubic meter (10^{-3} cubic centimeter)	1 cm^3 = 0.06 cubic inch	1 cubic inch = 16.4 cm^3
Volume (Liquids and Gases)	1 kiloliter (kL or kl)	= 1,000 liters	1 kL = 264.2 gallons	1 gallon = 3.79 L
	1 liter (L)	= 1,000 millimeters	1 L = 0.26 gallon 1 L = 1.06 quarts	1 quart = 0.95 L
	1 milliliter (mL or ml)	= 10^{-3} liter = 1 cubic centimeter	1 mL = 0.03 fluid ounce 1 mL = approx. $\frac{1}{4}$ teaspoon 1 mL = approx. 15–16 drops	1 quart = 946 mL 1 pint = 473 mL 1 fluid ounce = 29.6 mL 1 teaspoon = approx. 5 mL
Volume (Liquids and Gases)	1 microliter (µl or µL)	= 10^{-6} liter (10^{-3} milliliters)		
Time	1 second (s)	= $\frac{1}{60}$ minute		
	1 millisecond (ms)	= 10^{-3} second		
Temperature	Degrees Celsius (°C)		$°F = \frac{9}{5}°C + 32$	$°C = \frac{5}{9}(°F - 32)$

B Appendix B: Periodic table of elements

Representative elements

Alkali metals · Alkaline earth metals · Halogens · Noble gases · Transition elements

Period number	1 Group 1A	2 Group 2A	3 3B	4 4B	5 5B	6 6B	7 7B	8 8B	9 8B	10 8B	11 1B	12 2B	13 Group 3A	14 Group 4A	15 Group 5A	16 Group 6A	17 Group 7A	18 Group 8A
1	1 H 1.008																	2 He 4.003
2	3 Li 6.941	4 Be 9.012											5 B 10.81	6 C 12.01	7 N 14.01	8 O 16.00	9 F 19.00	10 Ne 20.18
3	11 Na 22.99	12 Mg 24.31											13 Al 26.98	14 Si 28.09	15 P 30.97	16 S 32.07	17 Cl 35.45	18 Ar 39.95
4	19 K 39.10	20 Ca 40.08	21 Sc 44.96	22 Ti 47.87	23 V 50.94	24 Cr 52.00	25 Mn 54.94	26 Fe 55.85	27 Co 58.93	28 Ni 58.69	29 Cu 63.55	30 Zn 65.41	31 Ga 69.72	32 Ge 72.64	33 As 74.92	34 Se 78.96	35 Br 79.90	36 Kr 83.80
5	37 Rb 85.47	38 Sr 87.62	39 Y 88.91	40 Zr 91.22	41 Nb 92.91	42 Mo 95.94	43 Tc (98)	44 Ru 101.1	45 Rh 102.9	46 Pd 106.4	47 Ag 107.9	48 Cd 112.4	49 In 114.8	50 Sn 118.7	51 Sb 121.8	52 Te 127.6	53 I 126.9	54 Xe 131.3
6	55 Cs 132.9	56 Ba 137.3	57* La 138.9	72 Hf 178.5	73 Ta 180.9	74 W 183.8	75 Re 186.2	76 Os 190.2	77 Ir 192.2	78 Pt 195.1	79 Au 197.0	80 Hg 200.6	81 Tl 204.4	82 Pb 207.2	83 Bi 209.0	84 Po (209)	85 At (210)	86 Rn (222)
7	87 Fr (223)	88 Ra (226)	89† Ac (227)	104 Rf (261)	105 Db (262)	106 Sg (266)	107 Bh (264)	108 Hs (269)	109 Mt (268)	110 Ds (271)	111 — (272)	112 — (285)	113 — (284)	114 — (289)	115 — (288)			

*Lanthanides	58 Ce 140.1	59 Pr 140.9	60 Nd 144.2	61 Pm (145)	62 Sm 150.4	63 Eu 152.0	64 Gd 157.3	65 Tb 158.9	66 Dy 162.5	67 Ho 164.9	68 Er 167.3	69 Tm 168.9	70 Yb 173.0	71 Lu 175.0
†Actinides	90 Th 232.0	91 Pa 231.0	92 U 238.0	93 Np (237)	94 Pu (244)	95 Am (243)	96 Cm (247)	97 Bk (247)	98 Cf (251)	99 Es 252	100 Fm 257	101 Md 258	102 No 259	103 Lr 260

Metals · Metalloids · Nonmetals

Name	Symbol
Actinium	Ac
Aluminum	Al
Americium	Am
Antimony	Sb
Argon	Ar
Arsenic	As
Astatine	At
Barium	Ba
Berkelium	Bk
Beryllium	Be
Bismuth	Bi
Bohrium	Bh
Boron	B
Bromine	Br
Cadmium	Cd
Calcium	Ca
Californium	Cf
Carbon	C
Cerium	Ce
Cesium	Cs
Chlorine	Cl
Chromium	Cr
Cobalt	Co
Copper	Cu
Curium	Cm
Darmstadtium	Ds
Dubnium	Db
Dysprosium	Dy
Einsteinium	Es
Erbium	Er
Europium	Eu
Fermium	Fm
Fluorine	F
Francium	Fr
Gadolinium	Gd
Gallium	Ga
Germanium	Ge
Gold	Au
Hafnium	Hf
Hassium	Hs
Helium	He
Holmium	Ho
Hydrogen	H
Indium	In
Iodine	I
Iridium	Ir
Iron	Fe
Krypton	Kr
Lanthanum	La
Lawrencium	Lr
Lead	Pb
Lithium	Li
Lutetium	Lu
Magnesium	Mg
Manganese	Mn
Meitnerium	Mt
Mendelevium	Md
Mercury	Hg
Molybdenum	Mo
Neodymium	Nd
Neon	Ne
Neptunium	Np
Nickel	Ni
Niobium	Nb
Nitrogen	N
Nobelium	No
Osmium	Os
Oxygen	O
Palladium	Pd
Phosphorus	P
Platinum	Pt
Plutonium	Pu
Polonium	Po
Potassium	K
Praseodymium	Pr
Promethium	Pm
Protactinium	Pa
Radium	Ra
Radon	Rn
Rhenium	Re
Rhodium	Rh
Rubidium	Rb
Ruthenium	Ru
Rutherfordium	Rf
Samarium	Sm
Scandium	Sc
Seaborgium	Sg
Selenium	Se
Silicon	Si
Silver	Ag
Sodium	Na
Strontium	Sr
Sulfur	S
Tantalum	Ta
Technetium	Tc
Tellurium	Te
Terbium	Tb
Thallium	Tl
Thorium	Th
Thulium	Tm
Tin	Sn
Titanium	Ti
Tungsten	W
Uranium	U
Vanadium	V
Xenon	Xe
Ytterbium	Yb
Yttrium	Y
Zinc	Zn
Zirconium	Zr

C Appendix C: Credits

Photo Credits

Unit Openers: **Unit I** Jan Hinsch/SPL/Photo Researchers **Unit II** Barry Runk/Grant Heilman Photography **Unit III** Mark Blum/Image Quest **Unit IV** Gary Braasch/Corbis **Unit V** Jonathan Ferrey/Getty Images **Unit VI** Kazuyo Shiota/Jupiterimages

Chapter 1: **Chapter opening photos** Brian Lightfoot/Nature Picture Library; Dave King/Dorling Kindersley; Dr. Gopal Murti/Visuals Unlimited; N.L. Max/Biological Photo Service **1.1** Pearson Education/Benjamin Cummings Publishing Company **1.2a** ImageState/International Stock Photography Ltd. **1.2b** Joe McDonald/CORBIS **1.2c** Frans Lanting/Minden Pictures **1.2d** Michael & Patricia Fogden/CORBIS **1.2e** Kim Taylor and Jane Burton/Dorling Kindersley **1.2f** Gerry Ellis/Minden Pictures **1.2g** Fred Bavendam/Minden Pictures **1.3** Photodisc/Getty Images; James Gritz/Photodisc/Getty Images **1.7** Volker Steger/SPL/Photo Researchers **1.8** USDA/APHIS/Animal and Plant Health Inspection Service **1.9 top left** Olive Meckes/Nicole Ottawa/Photo Researchers **1.9 bottom left** Ralph Robinson/Visuals Unlimited **1.9 top center** D. P. Wilson/Photo Researchers **1.9 bottom center** Corbis **1.9 top right** Corbis **1.9 bottom right** Digital Vision/Getty Images **1.10** Mike Hettwer/Project Exploration **1.12 left** Science Source/Photo Researchers **1.12 right** American Museum of Natural History **01.15a** Anne Dowie, Pearson Education/Benjamin Cummings Publishing Company **inset** Inga Spence/Tom Stack & Assoc. **1.15b** Chris Colling/Corbis **inset** Theo Allofs/zefa/Corbis **1.16 left** S. Lowry/University of Ulster/Stone/Getty Images **1.16 right** Simon Fraser/Photo Researchers **1.17** Karl Ammann/CORBIS **inset** Tim Ridley/Dorling Kindersley, Courtesy of the Jane Goodall Institute, Clarendon Park, Hampshire **1.20a** E R Degginger/Photo Researchers **1.20b** Breck P. Kent **1.22** Jeanne Strongin, Courtesy of NYU **1.23** James Holmes/Cellmark Diagnostics/SPL/Photo Researchers

Chapter 2: **Chapter opening photos** Visuals Unlimited/Corbis; Michael Newman/PhotoEdit; Canstock Images Inc/Jupiter Images; Bernard Photo Productions/Animals Animals **2.1 top** NIBSC/Science Photo Library/Photo Researchers **bottom** Kwong PD, Wyatt R, Robinson J, Sweet RW, Sodroski J, Hendrickson WA. Structure of an HIV gp120 envelope glycoprotein in complex with the CD4 receptor and a neutralizing human antibody. *Nature.* 1998 Jun 18;393(6686):648-59. **2.4a** Alison Wright/CORBIS **2.4b** Anne Dowie, Pearson Education/Benjamin Cummings Publishing Company **2.6** W. E. Klunk and C. A. Mathis, University of Pittsburgh **2.11** NASA **2.12 top** Photodisc/Getty Images **bottom** R. Kessel-Shih/Visuals Unlimited **2.13** Bernard Photo Productions/Animals Animals **2.14** Michael Cole/CORBIS **2.18** Oliver Strewe/Stone/Getty Images **2.19** Peter Sawyer/NMNH Smithsonian Inst.

Chapter 3: **Chapter opening photos** Mark Harwood/Stone/Getty Images; Maximilian Weinzierl/Alamy; Donna Day/Stone/Getty Images; Juhn Giustina/PhotoDisc/Getty Images; **3.1** Ryan McVay/Photodisc/Getty Images **3.4 left** Anne Dowie, Pearson Education/Benjamin Cummings Publishing Company **right** Photodisc/Getty Images **3.7** Scott Camazine/Photo Researchers **03.11** Donna Day/Stone/Getty Images **inset** PLG Studios, Pearson Education/Benjamin Cummings Publishing Company **3.12** Anne Dowie, Pearson Education/Benjamin Cummings Publishing Company **3.13a** Biophoto Associates/Photo Researchers **3.13b** L. M. Beidler, Florida State University **3.13c** Biophoto Associates/Photo Researchers **3.14** Photodisc/Getty Images **inset** T. J. Beveridge/Visuals Unlimited **3.16** Richard Megna, Fundamental Photographs, NYC **3.18** Pool Getty Images/Associated Press **3.19a** Photodisc/Getty Images, Inc. **3.19b** Ian O'Leary/Dorling Kindersley **03.19c** William Sallaz/Duomo/Corbis **03.19d** Motta & S. Correr/Science Photo Library/Photo Researchers **3.23** Stanley Flegler/Visuals Unlimited

Chapter 4: **Chapter opening photos** Tony Brain & David Parker/Science Photo Library/Photo Researchers; Anne Dowie, Pearson Education/Benjamin Cummings; Fred Felleman/Stone/Getty Images; R. M. Motta & S. Correr/Science Source/Photo Researchers **4.1 top** Dept. of Biological Science, Ajou University, Korea, http://bio.ajou.ac.kr **bottom** Ohio State University Medical Center **4.2a** Michael Abbey/Visuals Unlimited **4.2b** Andrew Syred/Photo Researchers **4.2c** Dennis Kunkel/Visuals Unlimited **4.4 left** S. C. Holt/Biological Photo Service **right** Dr. Gopal Murti/Visuals Unlimited **4.10** Barry King/Biological Photo Service **4.12** Garry Cole/Biological Photo Service **4.14a** Roland Birke/Peter Arnold, Inc. **4.14b** Dr. Henry C. Aldrich/Visuals Unlimited **4.16** E.H. Newcomb & W. P. Wergin/Biological Photo Service **4.17** Daniel S. Friend **4.18a** Manfred Schliwa/Visuals Unlimited **4.18b** Dr. D. J. Patterson/SPL/Photo Researchers **4.19a** Dr. Dennis Kunkel/Phototake NYC **4.19b** Karl Aufderheide/Visuals Unlimited **4.19c** Science Photo Library/Photo Researchers **4.20** Peter B. Armstrong

Chapter 5: **Chapter opening photos** Pete Saloutos/Corbis; C Squared Studios/Photodisc/Getty Images; Susanna Price/Dorling Kindersley; Paul Barton/Corbis **5.1** Greg Kuchik/Photodisc/Getty Images **5.2** David W. Hamilton/Getty Images **5.15** Nigel Cattlin/Holt Studios International/Photo Researchers **5.18** M. Abbey/Visuals Unlimited **5.20** Eyewire/Photodisc/Getty Images **5.21** The Protein Data Bank/RCSB

Chapter 6: **Chapter opening photos** Eyewire/Getty Images; Ian O'Leary/Dorling Kindersley; Larry Lee Photography/Corbis; Clive Brunskill/Liaison/Getty Images **6.1** Ted Spiegel/Corbis **6.2** Gail Shumway/Taxi/Getty Images **6.4** Corbis **6.16a** David L. Moore/Alamy Images **6.16b** Ian O'Leary/Dorling Kindersley

Chapter 7: **Chapter opening photos** Craig Tuttle/Corbis; Digital Vision/PictureQuest; Anne Rippy/Iconica/Getty Images; David Middleton/Bill Coster/NHPA **7.1** Tim Volk **7.2a** Renee Lynn/Photo Researchers **7.2b** Bob Evans/Peter Arnold, Inc. **7.2c** Dwight Kuhn/Dwight R. Kuhn Photography **7.2d** Sinclair Stammers/Photo Researchers **7.3 left** M. Eichelberger/Visuals Unlimited **right** E.H. Newcomb & W. P. Wergin/Biological Photo Service **7.6** Eric V. Grave/Photo Researchers **7.7** Adam Smith/Taxi/Getty Images **7.8** Siegfried Layda/Stone/Getty Images **7.9b** Tony Freeman/PhotoEdit, Inc. **7.15a** C. F. Miescke/Biological Photo Service **7.15b** Stone/Getty Images **7.17** Raymond Gehman/Corbis **7.19** Tom Adams/Visuals Unlimited

Chapter 8: **Chapter opening photos** Luke Dodd/Science Photo Library/Photo Researchers; Adrian T. Sumner/Stone/Getty Images; David M. Phillips/Photo Researchers; Bill Greenblatt/UPI/Landov **8.1** Phototake NYC **8.2a** Biophoto Associates/Photo Researchers **8.2b** Brian Parker/Tom Stack & Associates, Inc. **8.2c** Eric Simon **8.4** Andrew S. Bajer **8.5 top** A. L. Olins/Biological Photo Service **bottom** Biophoto Associates/Photo Researchers **8.8** Conly L. Rieder, Ph.D. **8.9a** David M. Phillips/Visuals Unlimited **8.9b** Carolina Biological Supply/Phototake NYC **8.11** James King-Holmes/Science Photo Library/Photo Researchers **8.12** Bob Thomas/Photographers Choice/Getty Images **8.13** CNRI/SPL/Photo Researchers **8.22 left** Lauren Shear/Photo Researchers **right** CNRI/SPL/Photo Researchers **8.24** Milton H. Gallardo **page 142** Carolina Biological Supply/Phototake

Chapter 9: **Chapter opening photos** Jacqui Hurst/Dorling Kindersley; Photodisc; AKG-Images; Sinclair Stammers/Science Photo Library/Photo Researchers **9.1** Yoav Levy/Phototake NYC **9.2** Hulton Archive/Getty Images **9.12 top left** Corbis **9.12 top right** Miep van Damm/Masterfile **9.12 center both** PhotoDisc/Getty Images **9.12 bottom both** Anthony Loveday, Pearson Education/Benjamin Cummings Publishing Company **9.15** Michael Ciesielski **9.16** Andy Jackson/Alamy **9.20** Lawrence Berkeley National Laboratory **9.22** Eric Simon **9.29a** Jean Claude Revy/Phototake NYC **9.29b** Carolina Biological Supply Company/Phototake NYC **9.32** FPG International/Taxi/Getty Images **9.33** Tudor Parfitt **page 170** USDA/APHIS/Animal and Plant Health Inspection Service

Chapter 10: **Chapter opening photos** Chris Bjornbert/Photo Researchers; Juda Ngwenya/Reuters/Getty Images; Richard Wagner; Keith V. Wood **10.1b** Impact Visuals/Phototake NYC **10.3a** Barrington Brown/Photo Researchers **10.3b left** Library of Congress **right** Cold Spring Harbor Laboratory Archives **10.5c** Richard Wagner **10.7** Ingram Publishing/Alamy **10.12** Keith V. Wood **10.23** Jeff Hunter/Image Bank/Getty Images **10.24** Francis Leroy, Biocosmos/Science Photo Library/Photo Researchers **10.25** Oliver Meckes/Eye of Science/Photo Researchers **10.27** N. Thomas/Photo Researchers **10.30c** NIBSC/Science Photo Library/Photo Researchers **10.31a** CDC/Phototake, NYC **10.31b** Chris Bjornbert/Photo Researchers **10.32** Christian

Keenan/Liaison/Getty Images **inset** Dr. Linda Stannard/SPL/Photo Researchers **10.33** KHAM/Reuters/Corbis **inset** NIBSC/Science Photo Library/Photo Researchers

Chapter 11: **Chapter opening photos** John Burwell/Foodpix/Jupiterimages; Photodisc/Getty Images; TK; Lee Snider/The Image Works **11.1 top** Advanced Cell Technology **11.1 center** Lino Loi, Dipartimento di Scienze Biomediche Comparate, Facoltà di Medicina Veterinaria, Teramo **11.1 bottom** Advanced Cell Technology **11.2 all** Ed Reschke **11.6** Dave King/Dorling Kindersley **11.11** American Association for the Advancement of Science **11.13** Roslin Institute **11.14** University of Missouri **11.16** Craig Hammell/Stock Market/Corbis **inset** Cord Blood Registry **11.17** Jose Cibelli **11.21 left top and bottom** Edward B. Lewis **right top and bottom** F. Rudolf Turner

Chapter 12: **Chapter opening photos** Don Tremain/Alamy; Andrew Brookes/Corbis; Photodisc/Getty Images; T. J. Berveridge and S. Schultze/Biological Photo Service **12.1** Peter Marlow/Magnum Photos, Inc. **inset** Neville Chadwick/Photo Researchers, Inc. **page 222** SIU/Visuals Unlimited **12.3** Novo Nordisk **12.4** Corbis **inset** AFP/Agence France Presse/Getty Images **12.5** Peter Berger **12.6** PPL Therapeutics **12.7 left** Prof. S. Cohen/Science Photo Library/Photo Researchers **right** Huntington Potter, University of South Florida College of Medicine **12.13** South West News Service, Bristol, England **12.14** Applied Biosystems **12.19** AP Wide World Photos **12.20** Douglas Graham/Roll Call/Sygma/Corbis **12.21b** David Parker/Science Photo Library/Photo Researchers **12.22** Paez JG, Janne PA, Lee JC, Tracy S, Greulich H, Gabriel S, Herman P, Kaye FJ, Lindeman N, Boggon TJ, Naoki K, Sasaki H, Fujii Y, Eck MJ, Sellers WR, Johnson BE, Meyerson M. EGFR mutations in lung cancer: correlation with clinical response to gefitinib **12.24** Philippe Plailly/Eurelios/Science Photo Library/Photo Researchers **12.25** AFP/Getty Images **12.26** Selna Kaplan

Chapter 13: **Chapter opening photos** Bruce Davidson/Nature Picture Library; Lara Jo Regan/Liaison/Getty Images; John Edwards/Stone/Getty Images; Science Source/Photo Researchers **13.1** Corbis **13.2a** Ken G. Preston-Mafham/Animals Animals **13.2b** P. Ward/Animals Animals **13.2c** Edward S. Ross, California Academy of Sciences **13.4 left** Science Source/Photo Researchers **right** National Maritime Museum Picture Library, London, England **13.5** Tui de Roy/Minden Pictures **inset** Claus Meyer/Minden Pictures **13.8** Corbis **13.9** Discover Magazine **13.10 left** Bill Ling/Dorling Kindersley **right** Ken Findlay/Dorling Kindersley **13.12a** Dwight R. Kuhn Photography **13.12b** Lennart Nilsson/Albert Bonniers Forlag AB **13.14a** Tui DeRoy/Bruce Coleman Inc. **13.14b** Mark Moffett/Minden Pictures **13.14c** Tui DeRoy/Bruce Coleman Inc. **13.15** Michael Fogden/DRK Photo **13.16** Laura Jesse, Extension Entomologist, Iowa State University **13.17 both** Young KV, Brodie ED Jr, Brodie ED 3rd. How the horned lizard got its horns. *Science.* 2004 Apr 2;304(5667):65. Brevia, Fig. 1. **13.18a** PhotoAlto/Getty Images **13.18b** Earth Imaging/Getty Images **13.19** Edmund D. Brodie III **13.21** Anne Dowie, Pearson Education/Benjamin Cummings Publishing Company **13.24** EyeWire/Photodisc/Getty Images **13.25** Hulton-Deutsch Collection/Corbis **13.26** Time Inc. Magazines/Sports Illustrated **13.27** Michael Fogden/DRK Photo **13.29** Bill Longcore/Photo Researchers

Chapter 14: **Chapter opening photos** Gerry Ellis/Minden Pictures; Rodin Auguste (1840–1917). The Thinker (Le Penseur) (bronze). ©The Bridgeman Art Library International Ltd.; Andreas Lykke-Olesen **14.1a** Richard D. Nowitz/Corbis **14.3a top** USDA/APHIS/Animal and Plant Health Inspection Service **bottom** W. J. Weber/Visuals Unlimited **14.3b all** Photodisc/Getty Images **14.05** Wolfgang Kaehler/Corbis **14.6 top left** Ralph A. Reinhold/Animals Animals/Earth Scenes **top right** EyeWire Collection/Photodisc/Getty Images **bottom** Grant Heilman/Grant Heilman Photography, Inc; **14.8** Corbis **left inset** John Shaw/Bruce Coleman, Inc. **right inset** Michael Fogden/Bruce Coleman, Inc. **14.10 all** University of Amsterdam **14.13** Chip Clark **14.14** Stephen Dalton/Photo Researchers **14.15 both** Photodisc/Getty Images **14.16a** George Gerster/Photo Researchers **14.16b** F. Latreille/Cerpolex/Cercles Polaires Expeditions **14.16c** Tom Bean/Corbis **14.16d** Jeff Daly/Visuals Unlimited **14.16e** Martin Lockley **14.18** Will & Deni McIntyre/Photo Researchers **14.20** Mark Pilkington/Geological Survey of Canada/SPL/Photo Researchers **14.21** Photodisc/Getty Images **14.23** Hanny Paul/Gamma Press USA, Inc.

Chapter 15: **Chapter opening photos** Michael Rosenfeld/Stone/Getty Images, Inc; Frederick P. Mertz/Visuals Unlimited; joeysworld.com/Alamy; P. Motta/Dept. of Anatomy/University "La Sapienza" Rome/Photo Researchers **15.1** Alex Wong/Liaison/Getty Images **15.3** Artist: Peter Sawyer/NMNH Smithsonian Inst. **15.4a** Roger Ressmeyer/Corbis **15.7a** Sidney Fox, University of Miami/Visuals Unlimited **15.7b** F. M. Menger and Kurt Gabrielson **15.8a** Stanley Awramik/Biological Photo Service **15.8b** Dr. Tony Brain and David Parker/Science Photo Library/Photo Researchers **15.9** Helen E. Carr/Biological Photo Service **15.10a, b** David M. Phillips/Visuals Unlimited **15.10c** CNRI/SPL/Photo Researchers **15.11a** David M. Phillips/Science Source/Photo Researchers **15.11b** Susan M. Barns, Ph.D. **15.11c** Heide Schulz/Max-Planck-Institut fur Marine Mikrobiologie **15.12** Lee D. Simon/Science Source/Photo Researchers **15.13** H. S. Pankratz, T. C. Beaman/Biological Photo Service **15.14** Dr. Tony Brain/Science Photo Library/Photo Researchers **15.15a** Scott Camazine/Photo Researchers; R. Calentine/Visuals Unlimited **15.15b** Centers for Disease Control and Prevention (CDC) **15.16** Martin Bond/Science Photo Library/Photo Researchers **15.17** ExxonMobil Corporation **15.19a** Oliver Meckes/Science Photo Library/Photo Researchers **15.19b** M. Abbey/Visuals Unlimited **15.19c** Manfred Kage/Peter Arnold, Inc. **15.19d** Dr. Masamichi Aikawa **15.19e** Michael Abbey/Science Photo Library/Photo Researchers **15.20** The Hidden Forest, www.hiddenforest.co.nz **15.21 top** Robert Kay, MRC Cambridge **center** Matt Springer, Stanford University **bottom** Robert Kay, MRC Cambridge **15.22a** Biophoto Associates/Photo Researchers **15.22b** Kent Wood/Photo Researchers **15.22c** Herb Charles Ohlmeyer, Fran Heyl Associates **15.22d** Manfred Kage/Peter Arnold, Inc. **15.23a** Laurie Campbell/NHPA **15.23b** Gary Robinson/Visuals Unlimited **15.23c** David Hall/Photo Researchers

Chapter 16: **Chapter opening photos** PHONE Labat Jean-Michel/Peter Arnold, Inc.; Corbis; G. Prance/Visuals Unlimited; Sequoia National Park Service **16.1** Great Smoky Mountains National Park Library, Gatlinburg, Tennessee **16.3** Dana Richter/Visuals Unlimited **16.4** David Middleton/Bill Coster/NHPA **16.5** Graham Kent **16.6a** E. R. Degginger/Photo Researchers **16.6b** Linda E. Graham **16.8** John Shaw **inset** Eric Simon **16.9** Dwight Kuhn **16.11** John Shaw/Tom Stack & Associates **left inset** Milton Rand/Tom Stack & Associates **right inset** Glenn Oliver/Visuals Unlimited **16.12** The Field Museum, Chicago **16.13** Andrew Gunners/Digital Vision/Getty Images **16.15** Doug Sokell/Visuals Unlimited **left inset** Derrick Ditchburn/Visuals Unlimited **right inset** Gerald & Buff Corsi/Visuals Unlimited **16.19a** Scott Camazine/Photo Researchers **16.19b** Taxi/Getty Images **16.19c** Dwight R. Kuhn **16.20a** Tierbild Okapia/Photo Researchers **16.20b** Rob Simpson/Visuals Unlimited **16.20c** G. L. Barron/Biological Photo Service **16.20d left** M. F. Brown/Visuals Unlimited **right** Jack Bostrack/Visuals Unlimited **16.20e** N. Allin & G. L. Barron, University of Guelph/BPS **16.20f** J. Forsdyke/Gene Cox/Science Photo Library/Photo Researchers **16.21** Fred Rhoades/Mycena Consulting **16.22a** Stuart Bebb/Oxford Scientific Films/Animals Animals/Earth Scenes **16.22b** D. Cavagnaro/Visuals Unlimited **16.23a** Robb Walsh/Austin Chronicle **16.23b** Photodisc/Getty Images **16.24** Christine Case **16.25 top** Jeremy Burgess/Science Photo Library/Photo Researchers **bottom** V. Ahmadijian/Visuals Unlimited

Chapter 17: **Chapter opening photos** Craig K. Lorenz/Photo Researchers; Digital Vision/age fotostock; Jeremy Woodhouse/Getty Images; R. Calentine/Visuals Unlimited; **17.1a** Gareth Morgan/AP Photo **17.1b** Astrid & Hanns-Frieder Michler/Photo Researchers **17.2** Gunter Ziesler/Peter Arnold, Inc. **17.5** J. Sibbick/The Natural history Museum, London **inset** Chip Clark **17.9** Charles R. Wyttenbach/Biological Photo Service **17.11a** Corbis **17.11b** Claudia E. Mills **17.12** Mike Bacon/Tom Stack & Associates, Inc. **17.15** Stanley Fleger/Visuals Unlimited **17.16a** Reprinted with permission from A. Eizinger and R. Sommer, Max-Planck-Institut fur Entwicklungsbiologie, Tubingen. Copyright 2000 American Association for the Advancement of Science **17.16b** Science Photo Library/Photo Researchers **17.18a** A.N.T./NHPA **17.18b left** R. DeGoursey/Visuals Unlimited **right** Corbis **17.20a** William Dow/Corbis **17.20b** D. Suzio/Photo Reseachers **17.20c** Oliver Meckes/Ottawa/Photo Reseachers **17.21a** Corbis **17.21b** A. Kerstitch/Visuals Unlimited **17.22** Carolina Biological Supply Company/Phototake NYC **17.23a** Gerry Ellis/Minden Pictures **17.23b** Hans Christoph Kappel/naturepl.com **17.23c** Mitsuhiko Imamori/Minden Pictures **17.25 all** John Shaw/Tom Stack & Associates **17.27a** Tony Craddock/Photo Researchers **17.27b** Harold W. Pratt/BPS **17.27c** Norbert Wu/Minden Pictures **17.28a left** Gerald Corsi/Visuals Unlimited **right** Gary Milburn/Tom Stack & Associates **17.28b** David Wrobel **17.28c** Fred Bavendam/Peter Arnold, Inc. **17.29** Biophoto Associates/Photo Researchers **17.31a** Runk/Schoenberger/Grant Heilman Photography, Inc. **17.31b** Robert Brons/BPS **17.33a** Breck P. Kent/Animals Animals **inset** Gary Meszaros/Photo Reseachers **17.33b** George Grall/National Geographic Image Collection **17.33c** J. M. Labat/Jacana/Photo Researchers **17.34a** Geoff Brightling/Dorling Kindersley **17.34b** Hans Pfletschinger/Peter Arnold, Inc. **17.34c** Dr. Eckart Pott/Bruce Coleman Inc. **17.36** Robert and Linda Mitchell **17.37** The Natural History Museum, London **17.38** Stephen J. Kraseman/DRK Photo **17.39a** Mervyn Griffiths/Commonwealth Scientific and Industrial Research Organization **inset** D. Parer and E. Parer Cook/Auscape **17.39b** Dan Hadden/Ardea Ltd **17.39c** Mitch Reardon/Photo Researchers **17.40** Digital Vision/Getty Images **17.42a** E. H. Rao/Photo Researchers **17.42b**

Reinhard Dirscherl/Alamy **17.42c left** Kevin Schafer/Photo Researchers **right** Digital Vision/Getty Images **17.42d clockwise** Digital Vision/Getty Images; Digital Vision/Getty Images; Nancy Adams/Tom Stack & Associates, Inc.; Digital Vision/Getty Images **17.42e** Karl Weatherly/Photodisc/Getty Images **17.44a** The Cleveland Museum of Natural History **17.44b** John Reader/SPL/Photo Researchers **17.44c** Donald Johanson/Institute of Human Origins **17.45 top to bottom** Jean Clottes/Sygma/Corbis; Jean Clottes/Sygma/Corbis; Jean-Marie Chauvet/Sygma/Corbis; Jean Clottes/Sygma/Corbis; Chauvet/Le Seuil/Sygma/Corbis **17.46** Lara Hartley/Lara Hartley Photography

Chapter 18: **Chapter opening photos** Roy Rainford/Robert Harding World Imagery/Getty Images; Mark Edwards/Still Pictures/Peter Arnold, Inc.; Raymond Gehman/Corbis; John D. Cunningham/Visuals Unlimited **18.1** Corbis **18.2 both** Raphael Gaillarde/Liaison/Getty Images **18.3a** Francois Gohier/Photo Researchers **18.3b** Ingrid Van Den Berg/Animals Animals **18.3c** David Lazenby/Planet Earth Pictures/Getty Images **18.3d** Jeremy Woodhouse/Photodisc/Getty Images **18.4** Erich Hartmann/Magnum Photos, Inc. **18.5** Joel W. Rogers/Corbis **18.6** NASA **18.7** Stephen Krasemann/Photo Researchers **18.8** Corbis **18.9** Brian Parker/Tom Stack & Associates, Inc. **18.10 left** Raymond Gehman/Corbis **right** Scott T. Smith/Corbis **18.12** Robert Brenner/PhotoEdit Inc. **18.13** Jim Brandenburg/Minden Pictures **18.14 both** Frans Lanting/Minden Pictures **18.15a** Sophie de Wilde/Jacana/Photo Researchers **18.15b** Art Wolfe/Image Bank/Getty Images **18.15c** Will & Deni McIntryre/Photo Researchers **18.16 left** Tim Thompson/Stone/Getty Images **center** Earth Satellite Corporation/SPL/Photo Researchers **right** Randy Wells/Photo Researchers **18.18** Roy Corral/Corbis **18.21** Runk/Schoenberg/Grant Heilman Photography, Inc. **18.22** Holt Studios International/Photo Researchers **18.23** Alan Carey/Photo Researchers **18.25** Mahaux Photography/Getty Images **18.27a** Alain Evrard/Photo Researchers **18.27b** Pete Seaward/Tony Stone Images/Getty Images **18.29** Tom Bean/Corbis **Table 18.2 page 400 left** Corbis **right** Ernest Manewal/Visuals Unlimited **18.30** David Reznick **18.32** Corbis

Chapter 19: **Chapter opening photos** Chris Newbert/Minden Pictures; Lester Lefkowitz/Corbis; Jeffrey L. Rotman/Corbis; USDA/Science Photo Library/Photo Researchers **19.1** A. Witte/C. Mahaney/Tony Stone Images/Getty Images **19.2** Richard D. Estes/Photo Researchers **19.4** Galen Rowell/Mountain Light/Explorer/Photo Researchers **19.6 both** Heather Angel/Biofotos/Natural Visions **19.07 top** Joseph T. Collins/Photo Researchers **bottom** Kevin de Queiroz **19.8** Arthur Morris/VIREO/The Academy of Natural Sciences **19.9** Jeff Lepore/Photo Researchers **19.10** Jerry Young/Dorling Kindersley **19.11** Gail Shumway/Taxi/Getty Images **19.12a** Lincoln P. Brower **19.12b** Peter J. Mayne **19.13a** Edward S. Ross **19.13b** Runk/Schoenberger/Grant Heilman Photography, Inc. **19.14** William E. Townsend/Photo Researchers **19.15** Embassy of Australia **19.16** Michael Fogden/DRK Photo **19.18.1** David Muench/Corbis **19.18.2** Tom Bean **19.18.3** Andrew Brown/Ecoscene/Corbis **19.18.4** Tom Bean **19.18.5** Tom Bean **19.18.6** Tom Bean **19.19** John Sohlden/Visuals Unlimited **19.22** Gregory G. Dimijian/Photo Researchers **19.34a** Davis Samuel Robbins/Corbis **19.34b** Joe McDonald/Corbis **19.34c** Joe McDonald/Corbis **19.34d** John D. Cunningham/Visuals Unlimited **19.34e** Jake Rajs/Stone/Getty Images **19.34f** Kennan Ward/Corbis **19.34g** Richard Hamilton Smith/Corbis **19.34h** Darrell Gulin/Corbis **19.35a** WorldSat International/Science Source/Photo Researchers **19.35b** Jay Dickman/CORBIS **19.35c** David Muench/CORBIS **19.36** Ilene MacDonald/Alamy **19.37** M. E. Warren/Photo Researchers **19.39** Frans Lanting/Photo Researchers **19.40 left** Emory Kristof/National Geographic Image Collection **right** D. Foster, Woods Hole Oceanographic Institution/Science/Visuals Unlimited **19.41 all** Lawrence E. Gilbert/Biological Photo Science

Chapter 20: **Chapter opening photos** Michael & Patricia Fogden/Minden Pictures; AFP/Agence France Presse/Getty Images; J. Lubner/Wisconsin Sea Grant Institute; Pete Oxford/Minden Pictures **20.1a** Ed Wray/AP Photo **20.1b** Rachel Woodfield, Merkel Associates **20.2** Susan Barnett **20.3a** Buddy Mays/Corbis **20.3b** Lynda Richardson/Corbis **20.3c** Marc Dantzker **20.3d** J. Lubner/Wisconsin Sea Grant Institute **inset** Scott Camazine/Photo Researchers **20.4** Corbis **20.5** D. W. Schindler **20.6** Nigel Tucker **20.7a** John D. Cunningham/Visuals Unlimited **20.7b** USDA Forest Service **20.9** NASA Earth Observing System **20.10** Corbis **20.12b both** NASA/Goddard Space Flight Center **20.13** SA Team/FotoNatura/Minden Pictures **20.14a** AFP/Getty Images **20.14b** Mark Carwardine/Still Pictures/Peter Arnold, Inc. **20.14c** MCMXCI/Dieter & Mary Plage/Bruce Coleman Inc. **20.15a** Phil Schermeister/Stone/Getty Images **20.15b** Gary Kramer **20.15c** Richard Vogel/Liaison/Getty Images **20.16** Scott Camazine/Photo Researchers **20.18 top** Gary Braasch/Woodfin Camp & Associates **bottom** Greg Vaughn/Tom Stack & Associates **20.19a** Rob Curtis/The Early Birder **20.19b** David Sieren/Visuals Unlimited **20.19c** USDA Forest Service **20.20a** David Hosking/Photo Researchers **20.20b** James P. Blair/National Geographic Society **20.21** Florida Department of Transportation **20.22b** Frans Lanting/Minden Pictures **20.23** Frans Lanting/Minden Pictures **20.24** Frans Lanting/Minden Pictures

Chapter 21: **Chapter opening photos** Steve Skjold/Alamy; Dr. Rodolfo Llinas/Peter Arnold, Inc; K. M. Highfill/Photo Researchers; Rodney Hyett/Elizabeth Whiting & Associates/Corbis **21.1** Eddie Keogh/Reuters/Corbis **21.3a** Photodisc/Getty Images **21.3b** Janice Sheldon **21.3c** T.D. Parsons, D. Kleinfeld, F. Raccuia-Behling and B. Salzberg. Biophysical Journal, July 1989. Photo courtesy of Brian Salzberg **21.5a-b** Nina Zanetti/Pearson Education/Benjamin Cummings Publishing Company **21.5c** Dr. Gopal Murti/SPL/Photo Researchers **21.5d** Science VU/Visuals Unlimited **21. 5e** Chuck Brown/Photo Researchers **21.5f** Nina Zanetti/Pearson Education/Benjamin Cummings Publishing Company **21.6a** Nina Zanetti/Pearson Education/Benjamin Cummings Publishing Company **21.6b** Manfred Kage/Peter Arnold, Inc. **21.6c** Gladden Willis, M. D./Visuals Unlimited **21.10** Science Photo Library/Photo Researchers **21.15a** Art Wolfe/Image Bank/Getty Images **21.15b** Duomo/Corbis **21.18** BSIP/Phototake **21.19a** Photodisc/Getty Images **21.19b** Masa Ushioda/Stephen Frink Collection/Alamy **21.19c** Corbis **21.19d** Amos Nachoum/Corbis **21.19e** Tui de Roy/Minden Pictures **21.19f** George Hall/Corbis

Chapter 22: **Chapter opening photos** Garry Gay/Alamy; Innerspace Imaging/SPL/Photo Researchers; Oliver Meckes/Ottawa/Photo Researchers; Pearson Education, Benjamin Cummings **22.1** Frank Micelotta/Getty Images **22.2a** Thomas Eisner, Cornell University **22.2b** Alissa Crandall/Corbis **22.2c** Burke/Triolo Productions/FoodPix/jupiterimages.com **22.3** Eric Simon **22.11** Painting by Dean Cornwell, by permission of Wyeth **22.12** Oliver Meckes/Ottawa/Photo Researchers **22.19a** Photodisc/Getty Images **22.19b left to right** Nick Rowe/Photodisc/Getty Images; Dorling Kindersley; Dorling Kindersley **22.21** Dagmar Fabricius/Stock Boston **22.23** Jackson Laboratory **22.24** AP Wide World Photos

Chapter 23: **Chapter opening photos** Eye of Science/Photo Researchers; NASA/Goddard Space Flight Center; Laurent Delhourme/Taxi/Getty Images; Science Photo Library/Photo Researchers **23.1** AP Wide World Photos **23.7b** Photodisc/Getty Images **23.09a** Lennart Nilsson/Boehringer Ingelheim International **23.12a** Andrew Syred/Stone/Getty Images **23.12b** R. Kessel & R. Kardon/ Visuals Unlimited **23.12c** Courtesy, The Gillette Company **23.14a** Ed Reschke **23.14b** W. Ober/Visuals Unlimited **23.23 both** Martin M. Rotker **23.24** Zig Leszczynski/Animals Animals

Chapter 24: **Chapter opening photos** Steve Gschmeissner/Photo Researchers; Aaron Haupt/Photo Researchers; Steve Prezant/Corbis; Henri Tullio/Corbis **24.1a** John Cardamone Jr., University of Pittsburgh/Biological Photo Service **24.01b** National Archives and Records Administration, Still Picture Branch **24.3** Photo Researchers **24.6** Photodisc/Getty Images **24.8** Aaron Haupt/Photo Researchers **24.16** Dey, L.P. **24.17** M. English, MD/Custom Medical Stock Photo **24.18** Laborers' Health and Safety Fund of North America **24.19** NIBSC/Science Photo Library/Photo Researchers

Chapter 25: **Chapter opening photos** Charlie Newham/Alamy; National Institutes of Health/Science Source/Photo Researchers; Lennart Nilsson/Boehringer Ingelheim International; Michael Abramson/Liaison/Getty Images **25.1** Masaji Ishiguro, Suntory Institute for Bioorganic Research, Osaka, Japan **25.8a** Michael Abramson/Liaison **25.8b** American Journal of Medicine 20:133, 1956. **25.8c** Lee Celano/ AFP/Getty Images **25.9** NMSB/Custom Medical Stock Photo **25.13** Corbis

Chapter 26: **Chapter opening photos** Gerry Ellis/Minden Pictures; Yorgos Nikas/Stone/Getty Images; Yorgos Nikas/Stone/Getty Images; C. Edelman/La Vilette/Photo Researchers **26.1** Brooks Kraft/Sygma/Corbis **26.2a** David Wrobel **26.2b** Tierbild Okapia/Photo Researchers **26.3a** Jim Solliday/Biological Photo Service **26.3b** Robin Chittenden/Corbis **26.3c** Dwight Kuhn/Dwight R. Kuhn Photography **26.5** C. Edelman/La Vilette/Photo Researchers **26.10** Al Dodge/Pearson Education/PH College **26.11** Yorgos Nikas/Stone/Getty Images **26.15 left** David Barlow **right** David Barlow/BBC **26.19a-d** Lennart Nilsson/Boehringer Ingelheim International **26.19e** Eric Simon **26.20** Eric Simon **inset** Stone/Getty Images, Inc. **26.23** Michael Tamborrino/Taxi/Getty Images

Chapter 27: **Chapter opening photos** Arthur Morris/Corbis; Kevin Fleming/Corbis; Johns Hopkins University; Jonathan Nourok/PhotoEdit Inc. **27.1** WDCN/Univ. College London/Photo Researchers **27.3** Manfred Kage/Peter Arnold, Inc. **27.7** Edwin R. Lewis, Professor Emeritus **27.15** Johns Hopkins University **27.16** Hanna

Damasio, M.D. **27.23** Corbis **27.26** C Squared Studios/Photodisc/Getty Images **27.30** Reuters NewMedia Inc./Corbis **27.32** Professor Clara Franzini-Armstrong **27.36** Donald Miralle/Getty Images **27.37** M. Peres/Custom Medical Stock Photo, Inc.

Chapter 28: **Chapter opening photos** Dr. Jeremy Burgess/SPL/Photo Researchers; Michael Fogden/DRK Photo; Eric Simon; Runk/Schoenberg/Grant Heilman, Inc. **28.1** N. Cattlin/Photo Researchers **inset** Andrew Syred/SPL/Photo Researchers **28.4** Eric J. Simon **28.5 both** Chandoha Photography **28.6a** Dorling Kindersley **26.6b** Seelevel.com **28.6c** Grant Heilman Photography, Inc. **28.8a** Kim Taylor and Jane Burton/Dorling Kindersley **28.8b** Corbis **28.10a** Dwight Kuhn/Dwight R. Kuhn Photography **28.10b** Graham Kent **28.10c** Graham Kent **28.10d** R. Kessel-Shih/ Visuals Unlimited **28.10e** Visuals Unlimited **28.12** Ed Reschke **28.13** Ed Reschke **28.14a** Luis Castaneda/Qwentes Italia srl/Tips Images **28.14b** Fred Crowe, Dept. of Botany & Plant Pathology, Oregon State University, Central Oregon Agricultural Research Center **28.14c** AGE fotostock/SuperStock **28.18** Runk/ Schoenberger/ Grant Heilman Photography **28.19a** Kevin Schafer Photography **28.19b** Duncan Smith/SPL/Photo Researchers **28.19c** Galen Rowell/Mountain Light Photography, Inc. **28.25** W. H. Hodge/Peter Arnold, Inc. **28.26** Photodisc/ Getty Images **28.27** Stephen J. Krasemann/Photo Researchers **28.28** D. Wilder/ Wilder Nature Photography

Chapter 29: **Chapter opening photos** Steve Satushek/The Image Bank/Getty Images; Oliver Benn/Stone/Getty Images, Inc.; Rich Iwasaki/Stone/Getty Images, Inc.; Alan Keohane/Dorling Kindersley **29.1** Jacques Pavlovsky/Sygma/Corbis **29.3** Herbert Spichtinger/zefa/Corbis **29.4** Grant Heilman Photography, Inc. **29.05a** Nigel Cattlin/Holt Studios/Photo Researchers **29.05b** Holt Studio/Animals Animals **29.5c** James Pushnik **29.5d** James Pushnik **29.6** Brian Capon **29.7** R. L. Peterson/Biological Photo Service **29.9a** Wally Eberhart/Visuals Unlimited **29.9b** E. H. Newcomb and S. R. Tandon/Biological Photo Service **29.12** Professor Ray F. Evert **29.14** M.T. Frazier/Photo Researchers **29.17** Runk/Schoenberg/Grant Heilman, Inc **29.18** Fred Jensen **29.19** Richard Cummins/Corbis **29.20a** Scott Camazine/ Photo Researchers **29.20b** Michael Evans **29.22 top** P. B. Tomlinson **bottom** Mike Hill/Image Bank/Getty Images

Illustration and Text Credits

This page constitutes an extension of the copyright page.

The following figures are adapted from C. K. Mathews and K. E. van Holde, *Biochemistry*, 2nd ed. (Menlo Park, CA: Benjamin/Cummings, 1996), © 1996 The Benjamin/Cummings Publishing Company: **6.12** and **8.5.**

Figure 1.21: Data in graph based on D. W. Pfennig et al., 2001. Frequency-dependent Batesian mimicry. *Nature* 410: 323. Figure adapted from Neil A. Campbell, Brad Williamson, and Robin J. Heyden, *Biology: Exploring Life* (Upper Saddle River, NJ: Prentice Hall, 2004) © 2004 Pearson Education, Inc.

Figure 4.17: Adapted from W. M. Becker, L. J. Kleinsmith, and J. Hardin, *The World of the Cell* (San Francisco, CA: Benjamin Cummings, 2003) © 2003 Pearson, Inc., publishing as Benjamin Cummings.

Figure 5.4: Data from Susan E. Gebhardt and Robin G. Thomas, *Nutritive Values of Foods* (USDA, 2002); Sharon A. Plowman and Denise L. Smith, *Exercise Physiology for Health, Fitness, and Performance*, 2nd ed. (San Francisco: Benjamin Cummings, 2003) © 2003 Pearson, Inc., publishing as Benjamin Cummings.

Figure 6.6: Copyright © 2002 from *Molecular Biology of the Cell*, 4th ed. by Bruce Alberts et al., fig. 2.69, p. 92. Garland Science/Taylor & Francis Books, Inc.

Figure 7.12: Adapted from Richard and David Walker, *Energy, Plants and Man*, fig. 4.1, p. 69. Oxygraphics. Copyright © Richard Walker. Used courtesy of Richard Walker, http://www.oxygraphics.co.uk.

Figure 9.13: Adapted from *Everyone Here Spoke Sign Language* by Nora Ellen Groce. Copyright 1985 by Nora Ellen Groce. Harvard University Press.

Figure 9.20: Highly adapted from *Introduction to Genetic Analysis*, 4th ed. by Suzuki, Griffiths, Miller, and Lewontin. Copyright 1976, 1981, 1986, 1989, 1993, 1996 by W.H. Freeman and Company.

Chapter 10, page 194: Text quotation by Joshua Lederberg, from Barbara J. Culliton, "Emerging Viruses, Emerging Threat," *Science*, vol. 247, p. 279 (19 January 1990). Copyright © 1990 American Association for the Advancement of Science.

Table 11.1: Data from the American Cancer Society website, "Cancer Facts & Figures 2005,": http://www.cancer.org/docroot/STT/stt_0.asp.

Figure 11.22: Adapted with permission from an illustration by William McGinnis, UCSD.

Figure 13.7c: Data for graph from K. V. Young et al., "How the horned lizards got its horns," in *Science, 304*, 4/2/2004, p. 65. American Association for the Advancement of Science.

Figure 13.29: Adapted from A. C. Allison, "Abnormal Hemoglobins and Erythrovute Enzyme-Deficiency Traits," in *Genetic Variation in Human Populations*, by G. A. Harrison, ed. (Oxford: Elsevier Science, 1961).

Figure 14.1b: Data for the distribution of West Nile virus from Centers for Disease Control and Prevention at www.cdc.gov.

Chapter 16 opening: "Evil Tendencies Cancel" from *The Poetry of Robert Frost* edited by Edward Connery Lathem. Copyright 1969 by Henry Holt and Company. Copyright 1938 by Robert Frost, copyright 1964 by Lesley Frost Ballantine.

Table 16.1: Adapted from Randy Moore et al., *Botany*, 2nd ed. Dubuque, IA: Brown, 1998, Table 2.2, p. 37.

Figure 17.43: *H. neanderthalensis* adapted from *The Human Evolution Coloring Book. P. boisei* drawn from a photo by David Bill.

Figure 18.24: Adapted from data from J. Wackernagel et al., "National Natural Capital Accounting with the Ecological Footprint Concept," *Ecological Economics*, vol. 29, pp. 375–390 (1999).

Figure 18.25: Data from Population Division of the Department of Economic and Social Affairs of the United Nations Secretariat, *World Population Prospects: The 2004 Revision* and *World Urbanization Prospects: The 2003 Revision.*

Figure 18.26: Data from U.S. Census Bureau International Data Base 2003.

Table 18.1: Data from Centers for Disease Control and Prevention website, "United States Life Tables, 2003," *National Vital Statistics Report*, vol. 50, no. 14, April 19, 2006. www.cdc.gov.

Table 18.2: Adapted from E. R. Pianka, *Evolutionary Ecology*, 6th ed. (San Francisco, CA: Benjamin/Cummings, 2000), p. 186. © 2000 The Benjamin Cummings Publishing Company.

Figure 19.7: A. S. Rand and E. E. Williams, 1969. The anoles of La Palma: Aspects of their ecological relationships. Breviora 327. Museum of Comparative Zoology, Harvard University. © Presidents and Fellows of Harvard College.

Figure 19.17: Adapted with permission from Edward Farmer, *Science*, vol. 276 (1997): 912. Copyright © 1997 American Association for the Advancement of Science.

Figure 19.23: From Robert Leo Smith, *Ecology and Field Biology*, 4th ed. Copyright © 1990 by Robert Leo Smith. Reprinted by permission of Addison Wesley Longman, Inc.

Figure 19.29d: Adapted from R. E. Ricklefs. 1997. *The Economy of Nature*, 4th ed. © 1997 by W.H. Freeman and Company. Used with permission.

Figure 20.7c: From C. E. Likens, et al., "Effects of Forest Cutting and Herbicide Treatment on Nutrient Budgets in the Hubbard Book Watershed Ecosystem," *Ecological Monographs*, 1966, 50: 22–32. Copyright 1966 The Ecological Society of America. Reprinted with permission.

Figure 20.10: Temperature data from U.S. National Climate Data Center, NOAA. CO_2 data from C. D. Keeling and T. P. Whorf, Scripps Institute of Oceanography.

Figure 20.12a: Data from British Antarctic Survey.

Figure 20.17: From N. Myers et al., "Biodiversity hotspots for conservation priorities," *Nature*, vol. 403, p. 853, 2/24/2000. Reprinted by permission from Macmillan Publishers Ltd. Copyright © 2000 Nature Publishing, Inc.

Figure 20.22a: Adapted from W. Purves and G. Orians, *Life: The Science of Biology*, 5th ed., fig. 55.23, p. 1239. Copyright © 1998 by Sinauer and W.H. Freeman and Company. Reprinted by permission.

Figure 22.17: U.S. Department of Agriculture, Center for Nutrition Policy and Promotion, April 2005.

Figure 22.22: Data adapted from Clinical Guidelines on the Identification, Evaluation, and Treatment of Overweight and Obesity in Adults: The Evidence Report. Data found at the National Heart, Lung, and Blood Institute: http://www.nhlbi.nih.gov/guidelines/obesity/bmi_tbl.htm.

Table 22.1: Data from Susan E. Gebhardt and Robin G. Thomas, *Nutritive Values of Foods* (USDA, 2002); Sharon A. Plowman and Denise L. Smith, *Exercise Physiology for Health, Fitness, and Performance*, 2nd ed. (San Francisco: Benjamin Cummings, 2003) © 2003 Pearson, Inc. publishing as Benjamin Cummings.

Figure 23.22: Illustration from Irving Geis. Rights owned by Howard Hughes Medical Institute. Not to be reproduced without permission.

Figure 25.1: Graph republished with permission of American Diabetes Association, from D.E. Cummings et al., "A preprandial rise in plasma ghrelin levels suggests a role in meal initiation in humans." Diabetes, Vol. 50, August, 2001, p. 1715. Copyright © 2001. Permission conveyed through Copyright Clearance Center, Inc.

Table 26.1: Data from R. Hatcher et al., *Contraceptive Technology: 1990–1992*, p. 134 (New York: Irvington, 1990).

Figure 27.1: Data from National Center for Health Statistics.

Figure 27.6: Adapted from W. M. Becker and D. W. Deamer, *The World of the Cell*, 2nd ed. Figure 20.16 (Redwood City, CA: Benjamin/Cummings, 1991) © 1991 The Benjamin Cummings Publishing Company.

Figure 29.1: Organic seal from U.S. Department of Agriculture.

D Appendix D: Self-quiz answers

Chapter 2

1. d
2. $MgCL_2$ and C_2H_6
3. element
4. electrons; neutrons; protons
5. two
6. An ionic bond involves the transfer of electrons from one atom to another, while a covalent bond involves sharing electrons.
7. d
8. a
9. The positive and negative poles cause adjacent water molecules to become attracted to each other, forming hydrogen bonds. The properties of water all arise from this atomic stickiness.
10. The cola is an aqueous solution, with water as the solvent, sugar as the solute, and the CO_2 making the solution acidic.

Chapter 3

1. dehydration; hydrolysis
2. b
3. fatty acid; glycerol
4. b
5. c
6. d
7. Hydrophobic amino acids are most likely to be found within the interior of a protein, far from the watery environment.
8. a
9. starch (or glycogen or cellulose); nucleotide
10. Both DNA and RNA are polynucleotides; both have the same phosphate group along the backbone; and both use A, C, and G bases. But DNA uses T while RNA uses U as a base; the sugar differs between them; and DNA is usually double-stranded while RNA is usually single-stranded.

Chapter 4

1. d
2. b
3. Bacteria; Archaea
4. b
5. smooth ER; rough ER
6. rough ER; Golgi; plasma membrane
7. Both organelles use membranes to organize enzymes, and both provide energy to the cell. But chloroplasts use pigments to capture energy from sunlight in photosynthesis, whereas mitochondria release energy from glucose using oxygen in cellular respiration. Chloroplasts are only in photosynthetic plants and protists, whereas mitochondria are in almost all eukaryotic cells.
8. a3, b1, c5, d2, e4
9. nucleus, nuclear pores, ribosomes, rough ER, Golgi

Chapter 5

1. It is converted to potential energy and heat.
2. Energy; entropy
3. 10,000 g (or 10 kg); remember that 1 Calorie on a food label equals 1,000 calories of heat energy.
4. Negatively charged phosphate groups crowded together in the triphosphate tail repel each other. The release of a phosphate group makes some of this potential energy available to cells to perform work.
5. Hydrolases are enzymes that participate in hydrolysis reactions, breaking down polymers into the monomers that make them up.
6. An inhibitor's binding to another site on the enzyme can cause the enzyme's active site to change shape.
7. b
8. a
9. b
10. signal transduction pathway

Chapter 6

1. d
2. The electron transport chain
3. glucose; NAD^+
4. O_2
5. The majority of the energy provided by cellular respiration is generated during the electron transport chain. Shutting down that pathway will deprive cells of energy very quickly.
6. b
7. d
8. Glycolysis
9. b
10. Because fermentation supplies only 2 ATP per glucose molecule compared with 38 from cellular respiration, the yeast will have to consume 19 times as much glucose to produce the same amount of ATP.

Chapter 7

1. thylakoids; stroma
2. inputs: a, d, e; outputs: b, c
3. Green light is reflected by leaves, not absorbed, and therefore cannot drive photosynthesis.
4. H_2O
5. c
6. The reactions of the Calvin cycle require the outputs of the light reactions (ATP and NADPH) to proceed.
7. In hot, dry environments, most plants close their stomata, which saves water but decreases the amount of available CO_2.
8. C_4 and CAM plants can close their stomata and save water without shutting down photosynthesis.
9. c
10. CO_2

Chapter 8

1. c
2. They are in the form of very long, thin strands.
3. b
4. Prophase and telophase
5. b
6. 39
7. Prophase II or metaphase II. It cannot be during meiosis I because then you would see an even number of chromosomes. It cannot be during a later stage in meiosis II because then you would see the sister chromatids separated.
8. benign; malignant
9. 16 ($2n = 8$, so $n = 4$, and $2^n = 2^4 = 16$)
10. Nondisjunction would create just as many gametes with an extra copy of chromosome 3 or 16, but extra copies of chromosome 3 or 16 are probably fatal.

Chapter 9

1. alleles; homozygous; heterozygous
2. genotype; phenotype
3. c
4. c
5. d
6. d
7. d

More Genetics Problems

8. The parental-type gametes are *WS* and *ws*. Recombinant gametes are *Ws* and *wS*, produced by crossing over.
9. Height appears to result from polygenic inheritance, like human skin color. See Figure 9.21.
10. The brown allele appears to be dominant, the white allele recessive. The brown parent appears to be homozygous dominant, *BB*, and the white mouse is homozygous recessive, *bb*. The F_1 mice are all heterozygous, *Bb*. If two of the F_1 mice are mated, $\frac{3}{4}$ of the F_2 mice will be brown.
11. The best way to find out whether a brown F_2 mouse is homozygous dominant or heterozygous is to do a testcross: Mate the brown mouse with a white mouse. If the brown mouse is homozygous, all the offspring will be brown. If the brown mouse is heterozygous, you would expect half the offspring to be brown and half to be white.
12. Freckles is dominant, so Tim and Jan must both be heterozygous. There is a $\frac{3}{4}$ chance that they will produce a child with freckles, a $\frac{1}{4}$ chance that they will produce a child without freckles. The probability that the next two children will have freckles is $\frac{3}{4} \times \frac{3}{4} = \frac{9}{16}$.
13. Half their children will be heterozygous and have elevated cholesterol levels. There is a $\frac{1}{4}$ chance that their next child will be homozygous,

bb, and have an extremely high cholesterol level, like Katerina.

14. The bristle-shape alleles are sex-linked, carried on the X chromosome. Normal bristles is dominant *(F)* and forked is recessive *(f)*. The genotype of the female parent is X^fX^f. The genotype of the male parent is X^FY. Their female offspring are X^FX^f; their male offspring are X^fY.
15. The mother is a heterozygous carrier, and the father is normal. See Figure 9.32 for a pedigree. $\frac{1}{4}$ of their children will be boys suffering from hemophilia; $\frac{1}{4}$ will be female carriers.
16. For a woman to be color-blind, she must inherit X chromosomes bearing the color-blindness allele from both parents. Her father has only one X chromosome, which he passes on to all his daughters, so he must be color-blind. A male only needs to inherit the color-blindness allele from a carrier mother; both his parents are usually phenotypically normal.

Chapter 10

1. polynucleotides; nucleotides; sugar (deoxyribose); phosphate; nitrogenous base
2. b
3. Each daughter DNA molecule will have half the radioactivity of the parent molecule, since one polynucleotide from the original parental DNA molecule winds up in each daughter DNA molecule.
4. CAU; GUA; histidine (His)
5. A gene is the polynucleotide sequence with information for making one polypeptide. Each codon—a triplet of bases in DNA or RNA—codes for one amino acid. Transcription occurs when RNA polymerase produces mRNA using one strand of DNA as a template. A ribosome is the site of translation, or polypeptide synthesis, and tRNA molecules serve as interpreters of the genetic code. Each tRNA molecule has an amino acid attached at one end and a three-base anticodon at the other end. Beginning at the start codon, mRNA moves relative to the ribosome a codon at a time. A tRNA with a complementary anticodon pairs with each codon, adding its amino acid to the polypeptide chain. The amino acids are linked by peptide bonds. Translation stops at a stop codon, and the finished polypeptide is released. The polypeptide folds to form a functional protein, sometimes in combination with other polypeptides.
6. a3, b3, c1, d2, e2 and 3
7. d
8. d
9. reverse transcriptase
10. The process of reverse transcription occurs only in RNA-containing retroviruses like HIV. Human cells never undergo reverse transcription, so reverse transcriptase can be knocked out without harming the human host.

Chapter 11

1. c
2. The ability of these cells to produce entire organisms through cloning
3. nuclear transplantation
4. Which genes are active in a particular sample of cells
5. b
6. operon
7. b
8. a
9. Proto-oncogenes are normal genes involved in the control of the cell cycle. Mutation or viruses can cause them to be converted to oncogenes, or cancer-causing genes. Proto-oncogenes are necessary for normal control of cell division.
10. Homeotic genes are master control genes that regulate many other genes during development.

Chapter 12

1. c, d, b, a
2. Because it does not contain introns
3. vector
4. Such an enzyme creates DNA fragments with "sticky ends," single-stranded regions whose unpaired bases can hydrogen-bond to the complementary sticky ends of other fragments created by the same enzyme.
5. PCR
6. Because different people tend to have different numbers of repeats within the STR site. DNA fragments prepared from the STR sites will thus have different lengths, causing them to migrate to different locations on a gel.
7. a
8. b
9. the whole-genome shotgun method
10. c, b, a, d

Chapter 13

1. b
2. c
3. *Bb:* 0.42; *BB:* 0.49; *bb:* 0.09
4. The fitness of an individual (or of a particular genotype) is measured by the relative number of genes that it contributes to the gene pool of the next generation compared with the contribution of others. Thus the number of fertile offspring produced determines an individual's fitness.
5. e
6. c
7. A small founding population is subject to extensive sampling error in the composition of its gene pool.
8. stabilizing selection
9. b, c, d
10. The prevalence of malaria

Chapter 14

1. b
2. prezygotic: a, b, c, e; postzygotic: d
3. Because a small gene pool is more likely to be changed substantially by genetic drift and natural selection
4. exaptation
5. d
6. d
7. species, genus, family, order, class, phylum, kingdom, domain
8. 2.6
9. Archaea and Bacteria

Chapter 15

1. d, c, a, b, e
2. d, c, b, a, e
3. DNA polymerase is a protein, which must be transcribed from a gene. But a DNA gene requires DNA polymerase to be replicated. This creates a paradox about which came first—DNA or protein. But RNA can act as both an information storage molecule and an enzyme, suggesting that dual-role RNA may have preceded both DNA and proteins.
4. bacteria; archaea; bacteria
5. By preventing the bacteria from making cell walls
6. Exotoxins are poisons secreted by pathogenic bacteria; endotoxins are components of the cell walls of pathogenic bacteria.
7. Autotrophs make their own organic compounds from CO_2, while heterotrophs must obtain at least one type of organic compound from another organism.
8. They can form endospores.
9. Chemoheterotroph
10. c
11. b

Chapter 16

1. flowers
2. fruit
3. b
4. b
5. a fern
6. vascular plant
7. a
8. Answers might include nearly any food, such as grains (wheat, corn, oats, barley, and so on), fruits from trees, or garden vegetables; textile fibers, such as cotton; or products made from hardwoods such as oak, cherry, or walnut.
9. algae; fungi
10. absorption

Chapter 17

1. c
2. flatworms (Platyhelminthes) and roundworms (Nematoda)
3. arthropod
4. b
5. Chordata; notochord; cartilage disks between your vertebrae
6. bipedalism
7. a
8. *Australopithecus* species, *Homo habilis, Homo erectus, Homo sapiens*
9. a4, b5, c1, d2, e3

Chapter 18

1. Organismal ecology, population ecology, community ecology, ecosystem ecology
2. Light, water, temperature, chemicals added
3. physiological; behavioral
4. The number of people and the land area in which they live
5. Uniform
6. d
7. intermediate (b)
8. d
9. c
10. III; I

Chapter 19

1. c
2. a2, b5, c5, d3 or 4, e1
3. d

4. By preying on the dominant competitor
5. a3, b2, c1, d4
6. succession
7. Only 10% of the energy trapped by photosynthesis is turned into biomass by the plant, and only 10% of that energy is turned into meat of a grazing animal. Therefore, eating grain-fed beef obtains only about 1% of the energy of photosynthesis.
8. Many nutrients come from the soil, but carbon comes from the air.
9. chaparral
10. d

Chapter 20

1. An introduced species is a species that has been accidentally or intentionally transferred by humans from one location to another location where it does not occur naturally. Introduced species usually have relatively few pathogens, parasites, and predators to slow population growth and so may outcompete native organisms for resources.
2. c
3. d
4. Because the pesticides become concentrated in their prey.
5. Carbon dioxide and other gases in the atmosphere absorb infrared radiation and thus slow the escape of heat from Earth. This is called the greenhouse effect. As the carbon dioxide concentration in the atmosphere increases, more heat is retained, which could account for increased global temperatures (global warming).
6. a
7. c
8. a
9. habitat destruction
10. landscape

Chapter 21

1. Organism, organ system, organ, tissue, cell
2. A roundworm is more likely to have a circulatory system because there is not enough body surface area for exchange with the external environment to provide for all its internal cells.
3. c
4. interstitial fluid
5. positive; negative
6. Endotherms get most of their body heat from metabolic reactions, while ectotherms get most of their body heat from the environment.
7. Heat stroke
8. d
9. b
10. Most of it is reabsorbed into the blood; the rest is excreted in the urine.

Chapter 22

1. Large food items could not be digested; it would be difficult to eliminate wastes, and it would be difficult to nourish the many cells that compose a large animal.
2. It is. Absorption is the uptake of nutrients into body cells. Without absorption, you would eat forever but never gain any calories (or nutritional value) from your food.
3. Digestion; absorption
4. Bacteria (specificially, *Helicobacter pylori*); by using antibiotics and bismuth
5. d
6. b
7. Because cellular respiration requires oxygen to break down food
8. Because vitamins generally have catalytic functions as enzyme helpers, and thus each vitamin molecule can repeat its function many times
9. Many plant proteins are incomplete, meaning that they lack certain essential amino acids. Animal products tend to have complete proteins that contain all the essential amino acids. To deal with this problem, vegetarians must eat a variety of plant products to receive all the essential amino acids.
10. c

Chapter 23

1. In an open circulatory system, vessels have open ends and the circulating fluid flows directly around the body cells. In a closed circulatory system, the fluid stays within a closed system of vessels.
2. Arteries carry oxygen-rich blood to the body, but they carry oxygen-poor blood to the lungs. Veins carry oxygen-poor blood from the body to the heart, but they carry oxygen-rich blood from the lungs to the heart. Arteries are defined as vessels that carry blood away from the heart; veins carry blood to the heart.
3. a3, b1, c4, d2
4. anemia
5. a3, b4, c1, d2
6. d
7. Tracheae deliver O_2 directly to body cells; lungs deliver O_2 to the circulatory system, which then transports O_2 to body cells.
8. alveoli
9. b
10. b

Chapter 24

1. antigens
2. Nonspecific: a, b, d, e Specific: c, f, g
3. b
4. d
5. a4, b1, c6, d3, e2, f5
6. Swelling dilutes toxins, delivers more oxygen, promotes scabbing; pain changes behavior to discourage further damage; fever may inhibit bacterial growth.
7. HIV destroys helper T cells, thereby impairing both the cell-mediated and the humoral immune response.
8. B cells
9. It has made antibodies against the person's own (self) molecules.
10. memory cells; clonal selection

Chapter 25

1. d
2. b
3. The first two cells could have membrane receptors for that hormone, but they might trigger different signal transduction pathways. The third cell could lack membrane receptors for that particular hormone.
4. b
5. hypothyroidism; thyroid; the thyroid cannot produce hormones without iodine
6. d
7. c
8. At high levels, glucocorticoids suppress the immune system. Although this is good for relieving pain and inflammation, the immune system can no longer respond appropriately to threats such as infection. Additionally, glucocorticoids may mask pain, leading to behaviors that aggravate the injury.
9. androgens
10. a. endorphins b. GH c. estrogens (or progestins) d. ADH

Chapter 26

1. Asexual reproduction is favorable in a stable environment with many resources. Sexual reproduction may enhance success in a changing environment.
2. The seminal vesicles and/or the prostate
3. a4, b6, c3, d1, e5, f2
4. b
5. Inhibiting FSH and LH after ovulation ensures that another egg will not be released during this cycle. Estrogen and progesterone inhibit FSH and LH.
6. They also help prevent the spread of sexually transmitted diseases.
7. a
8. c
9. The nerve cells may follow chemical trails to the muscle cells (induction).
10. oxytocin; estrogen

Chapter 27

1. cell body; dendrites; axon; synaptic terminal
2. central nervous system; peripheral nervous system; brain; spinal cord.
3. Alcohol probably depresses the brain by enhancing the inhibitory effects of GABA.
4. This is caused by activation of the sympathetic branch of the peripheral nervous system.
5. The cerebral cortex deals with sensory information but not vital functions. It can be damaged and the person may still be alive. The structures in the midbrain and brainstem are responsible for maintaining the vital functions that keep this person alive.
6. A receptor potential is different from an action potential (used by neurons) in that an action potential is an all-or-none situation. Action potentials do not vary in the intensity of their strength, whereas sensory receptor cells convey information about the strength of the stimulus.
7. c (He could hear the tuning fork against his skull, so the cochlea, nerve, and brain are all functioning properly. Apparently, sounds are not being transmitted to the cochlea; therefore, the bones of the middle ear are the problem.)
8. axial skeleton; appendicular skeleton
9. d
10. a (Each neuron controls a smaller number of muscle fibers.)

Chapter 28

1. b, c, d, e
2. When most plants grow naturally, they exhibit apical dominance, which inhibits the outgrowth

of the axillary buds. If your friend were to pinch back the terminal buds, there would be increased growth from the axillary buds. This would result in more branches and leaves, which would increase next year's yield.
3. ovary; ovule
4. a4, b2, c5, d6, e3, f1
5. a. fruit (ripened ovary); b. leaf stalk (petiole); c. fruit; d. leaf; e. root
6. Bulbs, root sprouts, and runners are all examples of asexual reproduction. Asexual reproduction has the advantage of propagating exact copies of a well-adapted plant, but has the disadvantage of being less able to tolerate environmental changes.
7. d
8. c
9. b; a
10. vascular cambium

Chapter 29

1. Excessive amounts of fungicides could destroy mycorrhizae, symbiotic associations of fungi and plant root hairs. The fungal filaments provide lots of surface area for absorption of water and nutrients. Destroying the mycorrhizae could cause a water or nutrient deficiency in the plant.
2. c
3. Phloem; Xylem
4. b
5. a1, b4, c2, d5, e3
6. ethylene
7. cytokinins
8. The terminal bud produces auxins, which counters the effects of cytokinins from the roots and inhibits the growth of axillary buds. If the terminal bud is removed, the cytokinins predominate, and branching occurs at the axillary buds.
9. a3, b4, c2, d1
10. The room lights will inhibit flowering in the chrysanthemums.

Glossary

A

abiotic component (ā′-bī-ot′-ik) The nonliving factors of an ecosystem, such as air, water, and temperature.

abiotic reservoir The part of an ecosystem where a chemical, such as carbon or nitrogen, accumulates or is stockpiled outside of living organisms.

ABO blood groups Genetically determined classes of human blood that are based on the presence or absence of carbohydrates A and B on the surface of red blood cells. The ABO blood group phenotypes, also called blood types, are A, B, AB, and O.

abscisic acid (ABA) (ab-sis′-ik) A plant hormone that inhibits cell division and promotes dormancy. ABA interacts with gibberellins in regulating seed germination.

absorption The uptake of small nutrient molecules by an organism's own body; the third main stage of food processing, following digestion.

acclimation (ak-li-mā-shun) A long-term but reversible physiological response to an environmental change.

acetyl CoA (acetyl coenzyme A) The entry compound for the Krebs cycle in cellular respiration; formed from a fragment of pyruvate attached to a coenzyme.

acetylcholine (as′-uh-til-kō′-lēn) A nitrogen-containing neurotransmitter; among other effects, it slows the heart rate and makes skeletal muscles contract.

achondroplasia (uh-kon′-druh-plā′-zhuh) A form of human dwarfism caused by a single dominant allele. The homozygous condition is lethal.

acid A substance that increases the hydrogen ion (H^+) concentration in a solution.

acid chyme (kīm) A mixture of recently swallowed food and gastric juice.

acid precipitation Rain, snow, sleet, hail, drizzle, etc., with a pH below 5.6; can damage or destroy organisms by acidifying lakes, streams, and possibly land habitats.

acrosome (ak′-ruh-sōm) A membrane-enclosed sac at the tip of a sperm; contains enzymes that help the sperm penetrate an egg.

actinomycete (ak-tin′-ō-mī′-sēt) One of a group of bacteria characterized by a mass of branching cell chains called filaments.

action potential A self-propagating change in the voltage across the plasma membrane of a neuron; a nerve signal.

activation energy The amount of energy that reactants must absorb before a chemical reaction will start.

activator A protein that switches on a gene or group of genes.

active immunity Immunity that involves the production of antibodies by the body, conferred by recovering from an infectious disease.

active site The part of an enzyme molecule where a substrate molecule attaches (by means of weak chemical bonds); typically, a pocket or groove on the enzyme's surface.

active transport The movement of a substance across a biological membrane against its concentration gradient, aided by specific transport proteins and requiring input of energy (often as ATP).

adaptation An inherited characteristic that enhances an organism's ability to survive and reproduce in a particular environment.

adaptive radiation The emergence of numerous species from a common ancestor introduced to new and diverse environments.

adenine (A) (ad′-uh-nēn) A double-ring nitrogenous base found in DNA and RNA.

adhesion The attraction between different kinds of molecules.

adipose tissue A type of connective tissue whose cells contain fat.

ADP Adenosine diphosphate (a-den′-ō-sēn dī-fos′-fāt). A molecule composed of an adenine (a ribose sugar) and two phosphate groups. The high-energy molecule ATP is made by combining a molecule of ADP with a third phosphate in an energy-consuming reaction.

adrenal cortex (uh-drē′-nul) The outer portion of an adrenal gland, controlled by ACTH from the anterior pituitary; secretes hormones called glucocorticoids and mineralocorticoids.

adrenal gland One of a pair of endocrine glands, located atop each kidney in mammals, composed of an outer cortex and a central medulla.

adrenal medulla (uh-drē′-nul muh-dul′-uh) The central portion of an adrenal gland, controlled by nerve signals; secretes the fight-or-flight hormones epinephrine and norepinephrine.

adrenocorticotropic hormone (ACTH) (uh-drē′-nō-cōr′-ti-kō-trop′-ik) A protein hormone secreted by the anterior pituitary that stimulates the adrenal cortex to secrete corticosteroids.

adult stem cell A cell present in adult tissues that generates replacements for nondividing differentiated cells.

aerobic (ār-ō′-bik) Containing or requiring molecular oxygen (O_2).

aerobic capacity The maximum rate at which oxygen (O_2) can be taken in and used by muscle cells.

age structure The relative number of individuals of each age in a population.

aggregate fruit A fruit such as a blackberry that develops from a single flower with many carpels.

AIDS Acquired immunodeficiency syndrome; the late stages of HIV infection, characterized by a reduced number of T cells; usually results in death caused by opportunistic infections.

alcohol fermentation The conversion of the acid produced by glycolysis to carbon dioxide and ethyl alcohol.

alga (al′-guh) (plural, **algae**) Any of a great variety of protists, most of which are unicellular or colonial photosynthetic autotrophs with chloroplasts containing the pigment chlorophyll *a*. Heterotrophic and multicellular protists closely related to unicellular autotrophs are also regarded as algae.

alimentary canal (al′-uh-men′-tuh-rē) A digestive tract consisting of a tube running between a mouth and an anus.

allantois (al′-an-tō′-is) In animals, an extraembryonic membrane that develops from the yolk sac; helps dispose of the embryo's nitrogenous wastes and forms part of the umbilical cord in mammals.

allele (uh-lē′-ul) An alternative form of a gene.

allergen (al′-er-jen) An antigen that causes an allergic reaction.

allergy An abnormal sensitivity to an antigen. Symptoms are triggered by histamines released from mast cells.

allopatric speciation The formation of a new species as a result of an ancestral population's becoming isolated by a geographic barrier. *See also* sympatric speciation.

alpha helix (al′-fuh hē′-liks) The spiral shape resulting from the coiling of a polypeptide in a protein's secondary structure.

alternation of generations A life cycle in which there is both a multicellular diploid form, the sporophyte, and a multicellular haploid form, the gametophyte; a characteristic of plants and multicellular green algae.

alternative RNA splicing A type of regulation at the RNA-processing level in which different mRNA molecules are produced from the same primary transcript, depending on which RNA segments are treated as exons and which as introns.

alveolus (al-vē′-oh-lus) (plural, **alveoli**) One of millions of tiny sacs within the vertebrate lungs where gas exchange occurs.

amine (uh-mēn′) An organic compound with one or more amino groups.

amino acid (uh-mēn′-ō) An organic molecule containing a carboxyl group and an amino group; serves as the monomer of proteins.

amino acid sequencing Determining the sequence of amino acids in a polypeptide.

amino group In an organic molecule, a functional group consisting of a nitrogen atom bonded to two hydrogen atoms.

ammonia A small and very toxic nitrogenous waste produced by metabolism.

amniocentesis (am′-nē-ō-sen-tē′-sis) A technique for diagnosing genetic defects while a fetus is in the uterus; a sample of amiotic fluid, obtained via a needle inserted into the amnion, is analyzed for telltale chemicals and defective fetal cells.

amnion (am′-nē-on) In vertebrate animals, the extraembryonic membrane that encloses the fluid-filled amniotic sac containing the embryo.

amniote Member of a clade of tetrapods that have an amniotic egg containing specialized membranes that protect the embryo. Amniotes include mammals and birds and other reptiles.

amniotic egg (am′-nē-ot′-ik) A shelled egg in which an embryo develops within a fluid-filled amniotic sac and is nourished by yolk. Produced by reptiles (including birds) and egg-laying mammals, it enables them to complete their life cycles on dry land.

amoeba (uh-mē′-buh) A type of protist characterized by great flexibility and the presence of pseudopodia.

amoebocyte (uh-mē′-buh-sīt) An amoeba-like cell that moves by pseudopodia, found in most animals; depending on the species, may digest and distribute food, dispose of wastes, form skeletal fibers, fight infections, and change into other cell types.

amphibian Member of a class of vertebrate animals that include frogs, toads, and salamanders.

anabolic steroid (an′-uh-bol′-ik stār′-oyd) A synthetic variant of the male hormone testosterone that mimics some of its effects.

anaerobic (an′-ār-ō′-bik) Lacking or not requiring molecular oxygen (O_2).

analogy The similarity of structure between two species that are not closely related, attributable to convergent evolution.

anaphase The third stage of mitosis, beginning when sister chromatids separate from each other and ending when a complete set of daughter chromosomes has arrived at each of the two poles of the cell.

anaphylactic shock (an′-uh-fi-lak′-tik) A potentially fatal allergic reaction caused by extreme sensitivity to an allergen; involves an abrupt dilation of blood vessels and a sharp drop in blood pressure.

anatomy The study of the structure of an organism and its parts.

androgen (an′-drō-jen) A steroid sex hormone secreted by the gonads that promotes the development and maintenance of the male reproductive system and male body features.

anemia (uh-nē′-me-ah) A condition in which an abnormally low amount of hemoglobin or a low number of red blood cells results in the body cells not receiving enough oxygen.

angiosperm (an′-jē-ō-sperm) A flowering plant, which forms seeds inside a protective chamber called an ovary.

animal A eukaryotic, multicellular, heterotrophic organism that obtains nutrients by ingestion.

Animalia (an-eh-mal′-ē-uh) The kingdom that contains the animals.

annelid (an′-uh-lid) A segmented worm. Annelids include earthworms, polychaetes, and leeches.

annual A plant that completes its life cycle in a single year or growing season.

anorexia An eating disorder that results in self-starvation due to an intense fear of gaining weight, even when the person is underweight.

antagonistic hormones Two hormones that have opposite effects.

anterior Pertaining to the front, or head, of a bilaterally symmetrical animal.

anterior pituitary (puh-tū′-uh-tār-ē) An endocrine gland, adjacent to the hypothalamus and the posterior pituitary, that synthesizes several hormones, including some that control the activity of other endocrine glands.

anther A sac in which pollen grains develop, located at the tip of a flower's stamen.

anthropoid (an′-thruh-poyd) A member of a primate group made up of the apes (gibbons, orangutans, gorillas, chimpanzees, and bonobos), monkeys, and humans.

antibody (an′-tih-bod′-ē) A protein dissolved in blood plasma that attaches to a specific kind of antigen and helps counter its effects.

anticodon (an′-tī-kō′-don) On a tRNA molecule, a specific sequence of three nucleotides that is complementary to a codon triplet on mRNA.

antidiuretic hormone (ADH) (an′-tē-dī′-yū-ret′-ik) A hormone made by the hypothalamus and secreted by the posterior pituitary that promotes water retention by the kidneys.

antigen (an′-tuh-jen) A foreign (nonself) molecule that elicits an immune response.

antigen receptor Transmembrane versions of antibody molecules that B cells and T cells use to recognize specific antigens. Also called membrane antibodies.

antigen-binding site A region of the antibody molecule responsible for its recognition and binding function.

antihistamine (an′-tē-his′-tuh-mēn) A drug that interferes with the action of histamine, providing temporary relief from an allergic reaction.

anus The digestive system opening through which undigested materials are expelled.

aorta (ā-or′-tuh) An artery that conveys blood directly from the heart to other arteries.

aphotic zone (ā-fō′-tik) The region of an aquatic ecosystem beneath the photic zone, where light does not penetrate enough for photosynthesis to take place.

apical dominance (ā′-pik-ul) In a plant, the hormonal inhibition of axillary buds by a terminal bud.

apical meristem (ā′-pik-ul mār′-uh-stem) A meristem at the tip of a plant root or in the terminal or axillary bud of a shoot.

apicomplexan (ap′-ē-kom-pleks′-un) A type of parasitic protozoan. Some apicomplexans cause serious human disease.

appendicular skeleton (ap′-en-dik′-yū-ler) Components of the skeletal system that support the arms and legs of a land vertebrate; cartilage and bones of the shoulder girdle, pelvic girdle, and the forelimbs and hind limbs. *See also* axial skeleton.

appendix (uh-pen′-dix) A small, fingerlike extension near the union of the small and large intestines.

aqueous humor (ā′-kwē-us hyū′-mer) A plasma-like liquid in the space between the lens and the cornea in the vertebrate eye; helps maintain the shape of the eye, supplies nutrients and oxygen to its tissues, and disposes of its wastes.

aqueous solution (ā′-kwē-us) A solution in which water is the solvent.

arachnid A member of a major arthropod group that includes spiders, scorpions, ticks, and mites.

Archaea (ar′-kē-uh) One of two prokaryotic domains of life, the other being Bacteria.

archaean (plural, **archaea**) An organism that is a member of the domain Archaea.

arteriole (ar-tār′-ē-ōl) A vessel that conveys blood between an artery and a capillary bed.

artery A vessel that carries blood away from the heart to other parts of the body.

arthritis (ar-thrī′-tis) A skeletal disorder characterized by inflamed joints and deterioration of the cartilage between bones.

arthropod (ar′-thruh-pod) A member of the most diverse phylum in the animal kingdom; includes the horseshoe crab, arachnids (e.g., spiders, ticks, scorpions, and mites), crustaceans (e.g., crayfish, lobsters, crabs, and barnacles), millipedes, centipedes, and insects. Arthropods are characterized by a chitinous exoskeleton, molting, jointed appendages, and a body formed of distinct groups of segments.

artifical pacemaker A tiny electronic device surgically implanted near the AV node that emits electronic signals that trigger normal heartbeats.

artificial selection Selective breeding of domesticated plants and animals to promote the occurrence of desirable inherited traits in offspring.

asexual reproduction The creation of offspring by a single parent, without the participation of sperm and egg.

association area Area of the cerebral cortex where different types of incoming sensory information are integrated.

astigmatism (uh-stig′-muh-tizm) Blurred vision caused by a misshapen lens or cornea.

atherosclerosis (ath′-uh-rō′-skluh-rō′-sis) A cardiovascular disease in which growths called plaques develop on the inner walls of the arteries, narrowing their inner diameters.

atom The smallest unit of matter that retains the properties of an element.

atomic number The number of protons in each atom of a particular element.

ATP Adenosine triphosphate (a-den′-ō-sēn trī-fos′-fāt). The main energy source for cells.

ATP synthase A complex (cluster) of several proteins found in a cellular membrane (including the inner membrane of mitochondria, the thylakoid membrane of chloroplasts, and the plasma membrane of prokaryotes) that functions in chemiosmosis with adjacent electron transport chains, using the energy of a hydrogen ion concentration gradient to make ATP. An ATP synthase provides a port through which hydrogen ions (H^+) diffuse.

atrium (ā′-trē-um) (plural, **atria**) A heart chamber that receives blood from the veins.

auditory canal Part of the vertebrate outer ear that channels sound waves from the pinna or outer body surface to the eardrum.

autoimmune disease An immunological disorder in which the immune system attacks the body's own molecules.

autonomic nervous system (ot′-ō-nom′-ik) A component of the peripheral nervous system of vertebrates that regulates the internal environment. The autonomic nervous system is in turn made up of the sympathetic and parasympathetic subdivisions.

autosome A chromosome not directly involved in determining the sex of an organism; in mammals, for example, any chromosome other than X or Y.

autotroph (ot′-ō-trōf) An organism that makes its own food, thereby sustaining itself without eating other organisms or their molecules. Plants, algae, and photosynthetic bacteria are autotrophs.

auxin (ok′-sin) A plant hormone whose chief effect is to promote the growth and development of shoots.

AV (atrioventricular) node A region of specialized muscle tissue between the heart's right atrium and right ventricle. It generates electrical impulses that primarily cause the ventricles to contract.

axial skeleton (ak′-sē-ul) Components of the skeletal system that support the central trunk of the body; the skull, backbone, and rib cage in a vertebrate. *See also* appendicular skeleton.

axillary bud (ak′-sil-ār-ē) An embryonic shoot present in the angle formed by a leaf and stem.

axon (ak′-son) A neuron fiber that conducts signals to another neuron or to an effector cell.

B

B cell A type of lymphocyte that matures in the bone marrow and later produces antibodies; responsible for the humoral immune response. *See also* T cell.

bacillus (buh-sil′-us) (plural, **bacilli**) A rod-shaped prokaryotic cell.

backbone A series of segmental units called vertebrae, present in all vertebrates.

Bacteria One of two prokaryotic domains of life, the other being Archaea.

bacterial chromosome The single, circular DNA molecule found in bacteria.

bacteriophage (bak-tēr′-ē-ō-fāj) A virus that infects bacteria; also called a phage.

bacterium (plural, **bacteria**) An organism that is a member of the domain Bacteria.

ball-and-socket joint A joint that allows rotation and movement in several planes. Examples in humans are the hip and shoulder joints.

bark All the tissues external to the vascular cambium in a plant that is growing in thickness. Bark is made up of secondary phloem, cork cambium, and cork.

barrier method Contraception that relies on a physical barrier to block the passage of sperm. Examples include condoms and diaphragms.

basal ganglia (gang′-lē-uh) Clusters of nerve cell bodies located deep within the cerebrum that are important in motor coordination.

basal metabolic rate (BMR) The number of kilocalories a resting animal requires to fuel its essential body processes for a given time.

base A substance that decreases the hydrogen ion (H^+) concentration in a solution.

basilar membrane The floor of the middle canal of the inner ear.

Batesian mimicry (bāt′-zē-un mim′-uh-krē) A type of mimicry in which a species that a predator can eat looks like a different species that is poisonous or otherwise harmful to the predator.

behavioral isolation A type of prezygotic barrier between species; two species remain isolated because individuals of neither species are sexually attracted to individuals of the other species.

benign tumor An abnormal mass of cells that remains at its original site in the body.

benthic zone A seafloor or the bottom of a freshwater lake, pond, river, or stream.

biennial A plant that completes its life cycle in two years.

bilateral symmetry An arrangement of body parts such that an organism can be divided equally by a single cut passing longitudinally through it. A bilaterally symmetrical organism has mirror-image right and left sides.

bile A solution of salts secreted by the liver that emulsifies fats and aids in their digestion.

binary fission A means of asexual reproduction in which a parent organism, often a single cell, divides into two individuals of about equal size.

binomial A two-part latinized name of a species; for example, *Homo sapiens*.

biodiversity All of the variety of life; usually refers to the variety of species that make up a community; concerns both species richness (the total number of different species) and the relative abundance of the different species.

biodiversity crisis The current rapid decline in the variety of life on Earth, largely due to the effects of human culture.

biodiversity hot spot A small geographic area with an exceptional concentration of species, especially endemic species (those found nowhere else).

biogenesis The principle that all life arises by the reproduction of preexisting life.

biogenic amines Neurotransmitters derived from amino acids.

biogeochemical cycle Any of the various chemical circuits occurring in an ecosystem, involving both biotic and abiotic components of the ecosystem.

biogeography The geographic distribution of species.

biological clock An internal timekeeper that controls an organism's biological rhythms; marks time with or without environmental cues but often requires signals from the environment to remain tuned to an appropriate period.

biological community *See* community.

biological magnification The accumulation of persistent chemicals in the living tissues of consumers in food chains.

biological species concept The definition of a species as a population or group of populations whose members have the potential in nature to interbreed and produce fertile offspring.

biology The scientific study of life.

biomass The amount, or mass, of organic material in an ecosystem.

biome (bī′-ōm) A major type of ecosystem that covers a large geographic region and that is largely determined by climate, usually classified according to predominant vegetation, and characterized by organisms adapted to the particular environments.

biophilia The human desire to affiliate with other life in its many forms.

bioremediation The use of living organisms to detoxify and restore polluted and degraded ecosystems.

biosphere The global ecosystem; that portion of Earth that is alive; all of life and where it lives.

biotechnology The use of living organisms (often microbes) to perform useful tasks; today, usually involves DNA technology.

biotic component (bī-ot′-ik) The living factors of a biological community; all the organisms that are part of an individual's environment.

bipolar disorder Depressive mental illness characterized by swings of mood from high to low; also called manic-depressive disorder.

bird Member of a group of reptiles with feathers and adaptations for flight.

birth control pill A chemical contraceptive that inhibits ovulation, retards follicular development, or alters a woman's cervical mucus to prevent sperm from entering the uterus.

bivalve A member of a group of molluscs that includes clams, mussels, scallops, and oysters.

blade The flattened portion of a typical leaf.

blastocyst (blas′-tō-sist) A mammalian embryo made up of a hollow ball of cells that results from cleavage and that implants in the mother's endometrium.

blastula (blas′-tyū-luh) An embryonic stage that marks the end of cleavage during animal development; a hollow ball of cells in many species.

blind spot The place on the retina of the vertebrate eye where the optic nerve passes through the eyeball and where there are no photoreceptor cells.

blood A type of connective tissue with a fluid matrix called plasma in which blood cells are suspended.

blood pressure The force that blood exerts against the walls of the blood vessels.

body cavity A fluid-containing space between the digestive tract and the body wall.

body mass index (BMI) A ratio of weight to height.

bone A type of connective tissue consisting of living cells held in a rigid matrix of collagen fibers embedded in calcium salts.

bone marrow Blood-cell forming tissue (red bone marrow) or stored fat (yellow bone marrow) found in cavities within bones.

bony fish A fish that has a stiff skeleton reinforced by calcium salts.

bottleneck effect Genetic drift resulting from a drastic reduction in population size.

brain The part of the central nervous system involved in regulating and controlling body activity and interpreting information from the senses transmitted through the nervous system.

brainstem A functional unit of the vertebrate brain, composed of the midbrain, medulla oblongata, and pons; serves mainly as a sensory filter, selecting which information reaches higher brain centers.

breathing The alternation of inhalation and exhalation, supplying a lung or gill with O_2-rich air or water and expelling CO_2-rich air or water.

breathing control center A brain center that directs the activity of organs involved in breathing.

bronchiole (bron′-kē-ōl) A thin breathing tube that branches from a bronchus within a lung.

bronchus (bron′-kus) (plural, **bronchi**) One of a pair of breathing tubes that branch from the trachea into the lungs.

brown alga One of a group of marine, multicellular, autotrophic protists, the most common and largest type of seaweed. Brown algae include the kelps.

bryophyte (brī′-uh-fīt) A type of plant that lacks xylem and phloem; a nonvascular plant. Bryophytes include mosses and their close relatives.

budding A means of asexual reproduction in which a new individual develops from an outgrowth of a parent.

buffer A chemical substance that resists changes in pH by accepting hydrogen ions from or donating hydrogen ions to solutions.

bulimia An eating disorder characterized by episodic binge eating followed by purging through induced vomiting, abuse of laxatives, or excessive exercise.

C

C_3 plant A plant that uses the Calvin cycle for the initial steps that incorporate CO_2 into organic material, forming a three-carbon compound as the first stable intermediate.

C_4 plant A plant that prefaces the Calvin cycle with reactions that incorporate CO_2 into four-carbon compounds, the end product of which supplies CO_2 for the Calvin cycle.

calcitonin (kal′-sih-tōn′-in) A peptide hormone secreted by the thyroid gland that lowers the blood calcium level.

calorie The amount of energy that raises the temperature of 1 g of water by 1°C.

Calvin cycle The second of two stages of photosynthesis; a cyclic series of chemical reactions that occur in the stroma of a chloroplast, using the carbon in CO_2 and the ATP and NADPH produced by the light reactions to make the energy-rich sugar molecule G3P.

CAM plant A plant that uses crassulacean acid metabolism, an adaptation for photosynthesis in arid conditions. Carbon dioxide entering open stomata during the night is converted to organic acids, which release CO_2 for the Calvin cycle during the day, when stomata are closed.

cancer A malignant growth or tumor caused by abnormal and uncontrolled cell division.

cancer cell A cell that is not subject to normal cell cycle control mechanisms and that will therefore divide continuously.

cap Extra nucleotides added to the beginning of an RNA transcript in the nucleus of a eukaryotic cell.

capillary (kap′-il-ār-ē) A microscopic blood vessel that conveys blood between an artery and a vein or between an arteriole and a venule; enables the exchange of nutrients and dissolved gases between the blood and interstitial fluid.

capillary bed A network of capillaries that infiltrate every organ and tissue in the body.

carbohydrate (kar′-bō-hī′-drāt) A biological molecule consisting of simple single-monomer sugars (monosaccharides), two-monomer sugars (disaccharides), and other multi-unit sugars (polysaccharides).

carbon skeleton The chain of carbon atoms that forms the structural backbone of an organic molecule.

carbonyl group (kar′-buh-nēl′) In an organic molecule, a functional group consisting of a carbon atom linked by a double bond to an oxygen atom.

carboxyl group (kar-bok′-sil) In an organic molecule, a functional group consisting of an oxygen atom double bonded to a carbon atom that is also bonded to a hydroxyl group.

carcinogen (kar-sin′-uh-jin) A cancer-causing agent, either high-energy radiation (such as X-rays or UV light) or a chemical.

carcinoma (kar′-sih-nō′-muh) Cancer that originates in the coverings of the body, such as the skin or the lining of the intestinal tract.

cardiac cycle (kar′-dē-ak) The alternating contractions and relaxations of the heart.

cardiac muscle Striated muscle that forms the contractile tissue of the heart.

cardiovascular disease (kar′-dē-ō-vas′-kyū-ler) A set of diseases of the heart and blood vessels.

cardiovascular system A closed circulatory system with a heart and a branching network of arteries, capillaries, and veins.

carnivore An animal that eats other animals. *See also* herbivore; omnivore.

carpel (kar′-pul) The female part of a flower, consisting of a stalk with an ovary at the base and a stigma, which traps pollen, at the tip.

carrier An individual who is heterozygous for a recessively inherited disorder and who therefore does not show symptoms of that disorder.

carrying capacity In a population, the number of individuals that an environment can sustain.

cartilage (kar′-ti-lij) A type of connective tissue consisting of living cells embedded in a rubbery matrix with collagenous fibers.

cartilaginous fish (kar-ti-laj′-uh-nus) A fish that has a flexible skeleton made of cartilage.

case study An in-depth examination of an actual investigation.

cell A basic unit of living matter separated from its environment by a plasma membrane; the fundamental structural unit of life.

cell body The part of a cell, such as a neuron, that houses the nucleus.

cell cycle An orderly sequence of events (including interphase and the mitotic phase) that extends from the time a eukaryotic cell divides to form two daughter cells to the time those daughter cells divide again.

cell cycle control system A cyclically operating set of proteins that triggers and coordinates events in the eukaryotic cell cycle.

cell division The reproduction of a cell.

cell junction A structure that connects tissue cells to one another.

cell plate A double membrane across the midline of a dividing plant cell, between which the new cell wall forms during cytokinesis.

cell theory The theory that all living things are composed of cells and that all cells come from other cells.

cell wall A protective layer external to the plasma membrane in plant cells, bacteria, fungi, and some protists; protects the cell and helps maintain its shape.

cell-mediated immune response The type of specific immune response brought about by T cells; fights body cells infected with pathogens. *See also* humoral immune response.

cellular differentiation The specialization in the structure and function of cells that occurs during the development of an organism; results from selective activation and deactivation of the cells' genes.

cellular respiration The aerobic harvesting of energy from food molecules; the energy-releasing chemical breakdown of food molecules, such as glucose, and the storage of potential energy in a form that cells can use to perform work; involves glycolysis, the citric acid cycle, the electron transport chain, and chemiosmosis.

cellular slime mold A type of protist that has unicellular amoeboid cells and a multicellular reproductive body in its life cycle.

cellulose (sel′-yū-lōs) A large polysaccharide composed of many glucose monomers linked into cable-like fibrils that provide structural support in plant cell walls.

centipede A carnivorous terrestrial arthropod that has one pair of long legs for each of its numerous body segments, with the front pair modified as poison claws.

central nervous system (CNS) The integration and command center of the nervous system; the brain and, in vertebrates, the spinal cord.

central vacuole (vak′-yū-ōl) A membrane-enclosed sac occupying most of the interior of a mature plant cell, having diverse roles in reproduction, growth, and development.

centralization The presence of a central nervous system (CNS) distinct from a peripheral nervous system.

centriole (sen′-trē-ōl) A structure in an animal cell, composed of microtubule triplets arranged in a 9 + 0 pattern. An animal cell usually has a pair of centrioles within each of its centrosomes.

centromere (sen′-trō-mēr) The region of a chromosome where two sister chromatids are joined and where spindle microtubules attach during mitosis and meiosis. The centromere divides at the onset of anaphase during mitosis and anaphase II of meiosis.

centrosome (sen′-trō-sōm) Material in the cytoplasm of a eukaryotic cell that gives rise to microtubules; important in mitosis and meiosis; also called microtubule-organizing center.

cephalization (sef′-uh-luh-zā′-shun) The concentration of the nervous system at the anterior end.

cephalopod A member of a group of molluscs that includes squids and octopuses.

cerebellum (sār′-ruh-bel′-um) Part of the vertebrate hindbrain; mainly a planning center that interacts closely with the cerebrum in coordinating body movement.

cerebral cortex (suh-rē′-brul kor′-teks) A folded sheet of gray matter forming the surface of the cerebrum. In humans, it contains integrating centers for higher brain functions, such as reasoning, speech, language, and imagination.

cerebral hemisphere The right or left half of the vertebrate cerebrum.

cerebrospinal fluid (suh-rē′-brō-spī′-nul) Blood-derived fluid that surrounds, protects against infection, nourishes, and cushions the brain and spinal cord.

cerebrum (suh-rē′-brum) The largest, most sophisticated, and most dominant part of the vertebrate forebrain, made up of right and left cerebral hemispheres.

cervix (ser′-viks) The neck of the uterus, which opens into the vagina.

chaparral (shap′-uh-ral′) A biome dominated by spiny evergreen shrubs adapted to periodic drought and fires; found where cold ocean currents circulate offshore, creating mild, rainy winters and long, hot, dry summers.

charophycean (kār′-uh-fī′-sē-un) A member of the green algal group that shares features with land plants. Charophyceans are considered the closest relatives of land plants.

chemical bond An attraction between two atoms resulting from a sharing of outer-shell electrons or the presence of opposite charges on the atoms. The bonded atoms gain complete outer electron shells.

chemical cycling The use and reuse of chemical elements such as carbon within an ecosystem.

chemical digestion The breakdown of food molecules through the action of enzymes into small molecules that can be absorbed by the body.

chemical energy Energy stored in the chemical bonds of molecules; a form of potential energy.

chemical reaction A process leading to chemical changes in matter, involving the making and/or breaking of chemical bonds.

chemoautotroph An organism that obtains both energy and carbon from inorganic chemicals, making its own organic compounds from CO_2 without using light energy.

chemoheterotroph (kē′-mō-het′-er-ō-trōf) An organism that obtains energy and carbon from organic molecules.

chemoreceptor (kē′-mō-rē-sep′-ter) A sensory receptor that detects chemical changes within the body or a specific kind of molecule in the external environment.

chemotherapy (kē′-mo-thār′-uh-pē) Treatment for cancer in which drugs are administered to disrupt cell division of the cancer cells.

chiasma (kī-az′-muh) (plural, **chiasmata**) The microscopically visible site where crossing over has occurred between chromatids of homologous chromosomes during prophase I of meiosis.

chlorophyll *a* (klor′-ō-fil ā) A green pigment in chloroplasts that participates directly in the light reactions.

chloroplast (klō′-rō-plast) An organelle found in plants and photosynthetic protists. Enclosed by two concentric membranes, a chloroplast absorbs sunlight and uses it to power the synthesis of organic food molecules (sugars).

choanocyte (kō-an′-uh-sīt) A flagellated feeding cell found in sponges. Also called a collar cell, it has a collarlike ring that traps food particles around the base of its flagellum.

choanoflagellate An ancestral colonial protist from which sponges, and possibly all animals, probably arose.

chondrichthyan (kon-drik′-thē-an) Member of a class of cartilaginous fishes that includes sharks and rays.

chordate (kōr′-dāt) An animal that at some point during its development has a dorsal, hollow nerve cord, a notochord, gill structures, and a post-anal tail. Chordates include lancelets, tunicates, and vertebrates.

chorion (kō′r-ē-on) In animals, the outermost extraembryonic membrane, which becomes the mammalian embryo's part of the placenta.

chorionic villus (kōr′-ē-on′-ik vil′-us) (plural, **villi**) An outgrowth of the chorion, containing embryonic blood vessels. As part of the placenta, chorionic villi absorb nutrients and oxygen from, and pass wastes into, the mother's bloodstream.

chorionic villus sampling (CVS) A technique for diagnosing genetic defects while the fetus is in the uterus. A small sample of the fetal portion of the placenta is removed and analyzed.

choroid (kōr′-oyd) A thin, pigmented layer in the vertebrate eye, surrounded by the sclera. The iris is part of the choroid.

chromatin (krō′-muh-tin) The combination of DNA and proteins that constitute chromosomes; often used to refer to the diffuse, very extended form taken by the chromosomes when a eukaryotic cell is not dividing.

chromosome (krō′-muh-sōm) A threadlike, gene-carrying structure found in the nucleus of a eukaryotic cell and most visible during mitosis and meiosis; also, the main gene-carrying structure of a prokaryotic cell. Chromosomes consist of chromatin.

chromosome theory of inheritance A basic principle in biology stating that genes are located on chromosomes and that the behavior of chromosomes during meiosis accounts for inheritance patterns.

GLOSSARY

ciliate (sil′-ē-it) A type of protozoan that moves by means of cilia.

cilium (sil′-ē-um) (plural, **cilia**) A short appendage that propels some protists through the water and moves fluids across the surface of many tissue cells in animals. In common with eukaryotic flagella, cilia have a 9 + 2 arrangement of microtubules covered by the cell's plasma membrane.

circulatory system The organ system that transports materials such as nutrients, O_2, and hormones to body cells and transports CO_2 and other wastes from body cells.

citric acid cycle The metabolic cycle that is fueled by acetyl CoA formed after glycolysis in cellular respiration. Chemical reactions in the cycle complete the metabolic breakdown of glucose molecules to carbon dioxide. The cycle occurs in the matrix of mitochondria and supplies most of the NADH molecules that carry energy to the electron transport chains. Also referred to as the Krebs cycle.

clade An evolutionary branch that consists of an ancestor and all its descendants.

cladistics (kluh-dis′-tiks) The study of evolutionary history; specifically, the scientific search for monophyletic taxa (clades), taxonomic groups composed of an ancestor and all its descendants.

cladogram (klad′-uh-gram) A dichotomous phylogenetic tree that branches repeatedly, suggesting a classification of organisms based on the time sequence in which evolutionary branches arise.

class In classification, the taxonomic category above order.

cleavage (klē-vij) (1) Cytokinesis in animal cells and in some protists, characterized by pinching in of the plasma membrane. (2) In animal development, the succession of rapid cell divisions without cell growth, converting the animal zygote to a ball of cells.

cleavage furrow The first sign of cytokinesis during cell division in an animal cell; a shallow groove in the cell surface near the old metaphase plate.

clitoris (klit′-uh-ris) An organ in the female that engorges with blood and becomes erect during sexual arousal.

clonal selection (klōn′-ul) The production of a lineage of genetically identical cells that recognize and attack the specific antigen that stimulated their proliferation. Clonal selection is the mechanism that underlies the immune system's specificity and memory of antigens.

clone As a verb, to produce genetically identical copies of a cell, organism, or DNA molecule. As a noun, the collection of cells, organisms, or molecules resulting from cloning; also (colloquially), a single organism that is genetically identical to another because it arose from the cloning of a somatic cell.

closed circulatory system A circulatory system in which blood is confined to vessels and is kept separate from the interstitial fluid.

cnidarian (nī-dār′-ē-an) An animal characterized by cnidocytes, radial symmetry, a gastrovascular cavity, and a polyp and medusa body form. Cnidarians includes hydras, jellies, sea anemones, corals, and related animals.

cnidocyte (nī′-duh-sīt) A specialized cell for which the phylum Cnidaria is named; consists of a capsule containing a fine coiled thread, which, when discharged, functions in defense and prey capture.

coccus (kok′-us) (plural, **cocci**) A spherical prokaryotic cell.

cochlea (kok′-lē-uh) A coiled tube in the inner ear of birds and mammals that contains the hearing organ, the organ of Corti.

codominance The expression of two different alleles of a gene in a heterozygote.

codon (kō′-don) A three-nucleotide sequence in mRNA that specifies a particular amino acid or polypeptide termination signal; the basic unit of the genetic code.

coelom (sē′-lōm) A body cavity completely lined with mesoderm.

coenzyme (kō-en′-zīm) An organic molecule (usually a vitamin or a compound synthesized from a vitamin) that acts as a cofactor, helping an enzyme catalyze a metabolic reaction.

coevolution Evolutionary change in which adaptations in one species act as a selective force on a second species, inducing adaptations that in turn act as a selective force on the first species; mutual influence on the evolution of two different interacting species.

cofactor A nonprotein substance (such as a copper, iron, or zinc atom, or an organic molecule) that helps an enzyme catalyze a metabolic reaction. *See also* coenzyme.

cohesion (kō-hē′-zhun) The attraction between molecules of the same kind.

collecting duct A tube in the vertebrate kidney that concentrates urine while conveying it to the renal pelvis.

collenchyma cell (kuh-leng′-kuh-muh) In plants, a cell with a thick primary wall and no secondary wall, functioning mainly in supporting growing parts.

colon (kō′-lun) Most of the length of the large intestine, the tubular portion of the vertebrate alimentary canal between the small intestine and the rectum; functions mainly in water absorption and the formation of feces.

community An assemblage of all the organisms living together and potentially interacting in a particular area.

community ecology The study of how interactions between species affect community structure and organization.

comparative anatomy The study of body structures in different organisms.

comparative embryology The study of the formation, early growth, and development of different organisms.

competitive exclusion principle The concept that populations of two species cannot coexist in a community if their niches are nearly identical. Using resources more efficiently and having a reproductive advantage, one of the populations will eventually outcompete and eliminate the other.

complement protein A nonspecific defensive blood protein that cooperates with other components of the vertebrate defense system to protect against microbes. Complement proteins can enhance phagocytosis, directly lyse pathogens, and amplify the inflammatory response.

complementary DNA (cDNA) A DNA molecule made in vitro using mRNA as a template and the enzyme reverse transcriptase. A cDNA molecule therefore corresponds to a gene but lacks the introns present in the DNA of the genome.

complete digestive tract A digestive tube with two openings, a mouth and an anus.

compound A substance containing two or more elements in a fixed ratio; for example, table salt (NaCl) consists of one atom of the element sodium (Na) for every atom of chlorine (Cl).

compound eye The photoreceptor in many invertebrates; made up of many tiny light detectors, each of which detects light from a tiny portion of the field of view.

computed tomography (CT) A technology that uses a computer to create X-ray images of a series of sections through the body.

concentration gradient An increase or decrease in the density of a chemical substance in an area. Cells often maintain concentration gradients of H ions across their membranes. When a gradient exists, the ions or other chemical substances involved tend to move from where they are more concentrated to where they are less concentrated.

condom A flexible sheath, usually made of thin rubber or latex, designed to cover the penis during sexual intercourse for contraceptive purposes or as a means of preventing sexually transmitted diseases.

cone (1) In vertebrates, a photoreceptor cell in the retina, stimulated by bright light and enabling color vision. (2) In conifers, a reproductive structure bearing pollen or ovules.

conifer A gymnosperm, or naked-seed plant, that produces cones.

coniferous forest A biome characterized by conifers, cone-bearing evergreen trees.

conjunctiva A mucous membrane that helps keep the eye moist; lines the inner surface of the eyelids and covers the front of the eyeball, except the cornea.

connective tissue Tissue consisting of cells held in an abundant extracellular matrix, which they produce.

conservation biology The science of species preservation; the scientific study of ways to slow the current high rate of species loss.

conservation of energy The principle that energy can neither be created nor destroyed.

consumer An organism that obtains its food by eating plants or by eating animals that have eaten plants.

continental drift A change in the position of continents resulting from the incessant slow movement (floating) of the plates of Earth's crust on the underlying molten mantle. It has caused continents to fuse and break apart periodically throughout geologic history.

continental shelves The submerged parts of continents.

contraception The deliberate prevention of pregnancy.

control center A mechanism within a homeostatic feedback system that causes changes to that system in response to stimuli. Examples include a home thermostat and the thermoregulatory control center of the brain.

controlled experiment A component of the process of science whereby a scientist carries out two parallel tests, an experimental test and a control test. The experimental test differs from the control by one factor, the variable.

convergent evolution Adaptive change resulting in nonhomologous (analogous) similarities among organisms. Species from different evolutionary lineages come to resemble each other (evolve analogous structures) as a result of living in very similar environments.

copulation Sexual intercourse, necessary for internal fertilization to occur.

coral reefs Warm water, tropical, ecosystems dominated by the hard skeletal structures secreted primarily by the resident cnidarians.

cork The outermost protective layer of a plant's bark, produced by the cork cambium.

cork cambium Meristematic tissue that produces cork cells during secondary growth of a plant.

cornea (kor′-nē-uh) A transparent frontal portion of the sclera that admits light into the vertebrate eye.

coronary artery (kōr′-uh-nār-ē) A large blood vessel that conveys blood from the aorta to the tissues of the heart.

corpus callosum (kor′-pus kuh-lō′-sum) The thick band of nerve fibers that connects the right and left cerebral hemispheres in placental mammals, enabling the hemispheres to process information together.

corpus luteum (kor′-pus lū′-tē-um) A small body of endocrine tissue that develops from an ovarian follicle after ovulation; secretes progesterone and estrogen during pregnancy.

cortex In plants, the ground tissue system of a root, made up mostly of parenchyma cells, which store food and absorb minerals that have passed through the epidermis. *See also* adrenal cortex; cerebral cortex.

corticosteroid One of a family of hormones synthesized and secreted by the adrenal cortex, consisting of the mineralocorticoids and the glucocorticoids.

cotyledon (kot′-uh-lē′-don) The first leaf that appears on an embryo of a flowering plant; a seed leaf. Monocot embryos have one cotyledon; dicot embryos have two.

covalent bond (kō-vā′-lent) An attraction between atoms that share one or more pairs of outer-shell electrons; symbolized by a single line between the atoms.

crista (kris′-tuh) (plural, **cristae**) A fold of the inner membrane of a mitochondrion. Enzyme molecules embedded in cristae make ATP.

cross The cross-fertilization of two different varieties of an organism or of two different species; also called hybridization.

cross-fertilization The fusion of sperm and egg derived from two different individuals.

crossing over The exchange of segments between chromatids of homologous chromosomes during synapsis in prophase I of meiosis; also, the exchange of segments between DNA molecules in prokaryotes.

crustacean A member of a major arthropod group that includes lobsters, crayfish, crabs, shrimps, and barnacles.

cryptic coloration A type of camouflage that makes potential prey difficult to spot against its background.

CT *See* computed tomography.

culture The accumulated knowledge, customs, beliefs, arts, and other human products that are socially transmitted over the generations.

cuticle (kyū′-tuh-kul) (1) In animals, a tough, nonliving outer layer of the skin. (2) In plants, a waxy coating on the surface of stems and leaves that helps retain water.

cyanobacteria (sī-an′-ō-bak-tēr′-ē-uh) Photosynthetic, oxygen-producing bacteria, formerly called blue-green algae.

cystic fibrosis (sis′-tik fī-brō′-sis) A genetic disease that occurs in people with two copies of a certain recessive allele; characterized by an excessive secretion of mucus and consequent vulnerability to infection; fatal if untreated.

cytokinesis (sī′-tō-kuh-nē′-sis) The division of the cytoplasm to form two separate daughter cells. Cytokinesis usually occurs during telophase of mitosis, and the two processes (metosis and cytokinesis) make up the mitotic (M) phase of the cell cycle.

cytokinin (sī′-tō-kī′-nin) One of a family of plant hormones that promote cell division, retard aging in flowers and fruits, and may interact antagonistically with auxins in regulating plant growth and development.

cytoplasm (sī′-tō-plaz′-um) Everything inside a cell between the plasma membrane and the nucleus; consists of a semifluid medium and organelles.

cytosine (C) (sī′-tuh-sin) A single-ring nitrogenous base found in DNA and RNA.

cytoskeleton A meshwork of fine fibers in the cytoplasm of a eukaryotic cell; includes microfilaments, intermediate filaments, and microtubules.

cytosol (sī-tuh-sol) The semifluid medium of a cell's cytoplasm.

cytotoxic T cell (sī′-tō-tok′-sik) A type of lymphocyte that attacks body cells infected with pathogens.

D

decomposer An organism that derives its energy from organic wastes and dead organisms; also called a detritivore.

decomposition The breakdown of organic materials into inorganic ones.

dehydration reaction (dē-hī-drā′-shun) A chemical process in which a polymer forms as monomers are linked by the removal of water molecules. One molecule of water is removed for each pair of monomers linked. Also called condensation.

denaturation (dē-nā′-chur-ā′-shun) A process in which a protein unravels, losing its specific conformation and hence function; can be caused by changes in pH or salt concentration or by high temperature; also refers to the separation of the two strands of the DNA double helix, caused by similar factors.

dendrite (den′-drīt) A neuron fiber that conveys signals from its tip inward, toward the rest of the neuron; in a motor neuron, one of several short, branched extensions that convey nerve signals toward the cell body.

density-dependent factor A population-limiting factor whose effects depend on population density.

density-independent factor A population-limiting factor whose occurrence and effects are not affected by population density.

dermal tissue system In plants, the tissue system forming the protective outer covering of leaves, young stems, and young roots.

descent with modification Darwin's initial phrase for the general process of evolution.

desert A biome characterized by organisms adapted to sparse rainfall (less than 30 cm per year) and rapid evaporation.

desertification The conversion of semi-arid regions to desert.

GLOSSARY

determinate growth Termination of growth after reaching a certain size, as in most animals. *See also* indeterminate growth.

detritivore (duh-trī′-tuh-vor) An organism that derives its energy from organic wastes and dead organisms; also called a decomposer.

detritus (duh-trī′-tus) Dead organic matter.

deuterostome (dū-ter′-ō-stōm) An animal with a coelom that forms from hollow outgrowths of the digestive tube of the early embryo. The deuterostomes include the echinoderms and the chordates.

diabetes mellitus (dī′-uh-bē′-tis me-lī′-tis) A human hormonal disease in which body cells cannot absorb enough glucose from the blood and become energy starved. Body fats and proteins are then consumed for their energy. Insulin-dependent diabetes results when the pancreas does not produce insulin; non-insulin-dependent diabetes results when body cells fail to respond to insulin.

dialysis (dī-al′-uh-sis) Separation and disposal of metabolic wastes from the blood by mechanical means; an artificial method of performing the functions of the kidneys.

diaphragm (dī′-uh-fram) (1) The sheet of muscle separating the chest cavity from the abdominal cavity in mammals. Its contraction expands the chest cavity, and its relaxation reduces it. (2) A dome-shaped rubber cap that covers a woman's cervix, serving as a barrier method of contraception.

diastole (dī-as′-tō-lē) The stage of the heart cycle in which the heart muscle is relaxed, allowing the chambers to fill with blood. *See also* systole.

diatom (dī′-uh-tom) A unicellular photosynthetic alga with a unique glassy cell wall containing silica.

dicot (dī′-kot) A flowering plant whose embryo has two seed leaves, or cotyledons.

differentiation *See* cellular differentiation.

diffusion The spontaneous movement of particles of any kind from where they are more concentrated to where they are less concentrated.

digestion The mechanical and chemical breakdown of food into molecules small enough for the body to absorb; the second main stage of food processing, following ingestion.

digestive system The organ system that ingests food, breaks it down into smaller chemical units, and absorbs the nutrient molecules.

digestive tube A digestive compartment with two openings: a mouth for the entrance of food and an anus for the exit of undigested wastes. Most animals, including humans, have a digestive tube.

dihybrid cross (dī′-hī′-brid) An experimental mating of individuals differing at two genetic loci.

dikaryotic phase (dī-kār′-ē-ot′-ik) A series of stages in the life cycle of many fungi in which cells contain two nuclei.

dinoflagellate (dī′-nō-flaj′-uh-let) A unicellular photosynthetic alga with two flagella situated in perpendicular grooves in cellulose plates covering the cell.

diploid (dip′-loid) Containing two sets of chromosomes (homologous pairs) in each cell, one set inherited from each parent; referring to a $2n$ cell.

directional selection Natural selection that acts in favor of the individuals at one end of a phenotypic range.

disaccharide (dī-sak′-uh-rīd) A sugar molecule consisting of two monosaccharides linked by a dehydration reaction.

discovery science The process of scientific inquiry that focuses on describing nature. *See also* hypothesis-driven science.

dispersion pattern The manner in which individuals in a population are spaced within their area. Three common dispersion patterns are clumped (individuals are aggregated in patches), uniform (individuals are evenly distributed), and random (unpredictable distribution).

disruptive selection Natural selection that favors extreme over intermediate phenotypes.

disturbance In an ecological sense, a force that changes a biological community and usually removes organisms from it. Disturbances, such as fires and storms, play a pivotal role in structuring many biological communities.

DNA Deoxyribonucleic acid (dē-ok′-sē-rī′-bō-nū-klā′-ik). The genetic material that organisms inherit from their parents; a double-stranded helical macromolecule consisting of nucleotide monomers with deoxyribose sugar and the nitrogenous bases adenine (A), cytosine (C), guanine (G), and thymine (T). *See also* gene.

DNA fingerprinting A procedure that analyzes an individual's unique collection of DNA restriction fragments, detected by electrophoresis and nucleic acid probes. DNA fingerprinting can be used to determine whether two samples of genetic material are from the same individual.

DNA ligase (lī′-gās) An enzyme, essential for DNA replication, that catalyzes the covalent bonding of adjacent DNA nucleotides; used in genetic engineering to paste a specific piece of DNA containing a gene of interest into a bacterial plasmid or other vector.

DNA microarray A glass slide containing thousands of different kinds of single-stranded DNA fragments arranged in an array. Tiny amounts of DNA fragments, representing different genes, are fixed to the glass slide. These fragments are tested for hybridization with various samples of cDNA molecules, thereby measuring the expression of thousands of genes at one time.

DNA polymerase (puh-lim′-er-ās) An enzyme that assembles DNA nucleotides into polynucleotides using a preexisting strand of DNA as a template.

DNA technology Methods used to study and/or manipulate DNA, including recombinant DNA technology.

domain A taxonomic category above the kingdom level; the three domains of life are Archaea, Bacteria, and Eukarya.

dominant allele In a heterozygote, the allele that determines the phenotype with respect to a particular gene.

dorsal Pertaining to the back of a bilaterally symmetrical animal.

dorsal, hollow nerve cord One of the four hallmarks of chordates; the chordate brain and spinal cord.

double bond A type of covalent bond in which two atoms share two pairs of electrons; symbolized by a pair of lines between the bonded atoms.

double circulation system A circulation scheme with separate pulmonary and systemic circuits, which ensures vigorous blood flow to all organs.

double fertilization In flowering plants, the formation of both a zygote and a cell with a triploid nucleus, which develops into the endosperm.

double helix The form of native DNA, referring to its two adjacent polynucleotide strands wound into a spiral shape.

Down syndrome A human genetic disorder resulting from the presence of an extra chromosome 21; characterized by heart and respiratory defects and varying degrees of mental retardation.

Duchenne muscular dystrophy (duh-shen′ dis′-truh-fē) A human genetic disease caused by a sex-linked recessive allele and characterized by progressive weakening and a loss of muscle tissue.

duodenum (dū-ō-dē′-num) The first portion of the vertebrate small intestine after the stomach, where acid chyme from the stomach is mixed with bile and digestive enzymes.

duplication Repetition of part of a chromosome resulting from fusion with a fragment from a homologous chromosome; can result from an error in meiosis or from mutagenesis.

E

eardrum A sheet of connective tissue separating the outer ear from the middle ear that vibrates when stimulated by sound waves and passes the waves to the middle ear.

echinoderm (uh-kī′-nō-derm) Member of a group of slow-moving or sessile marine animals characterized by a rough or spiny skin, a water vascular system, an endoskeleton, and radial symmetry in adults. Echinoderms include sea stars, sea urchins, and sand dollars.

ecological niche A population's role in its community; the sum total of a species' use of the biotic and abiotic resources of its habitat.

ecological species concept The idea that ecological roles (niches) define species.

ecological succession The process of biological community change resulting from disturbance; transition in the species composition of a biological community, often following a flood, fire, or volcanic eruption. *See also* primary succession; secondary succession.

ecology The scientific study of how organisms interact with their environments.

ecosystem (ē′-kō-sis-tem) All the organisms in a given area, along with the nonliving (abiotic) factors with which they interact; a biological community and its physical environment.

ecosystem ecology The study of energy flow and the cycling of chemicals among the various biotic and abiotic factors in an ecosystem.

ectoderm (ek′-tō-derm) The outer layer of three embryonic cell layers in a gastrula; forms the skin of the gastrula and gives rise to the epidermis and nervous system in the adult.

ectopic pregnancy (ek-top′-ik) The implantation and development of an embryo outside the uterus.

ectotherm (ek′-tō-therm) An animal that warms itself mainly by absorbing heat from its surroundings.

effector A cell, tissue, or organ capable of carrying out some action in response to a command from the nervous system.

effector cell A muscle cell or gland cell that performs the body's responses to stimuli; responds to signals from the brain or other processing center of the nervous system.

ejaculation (ih-jak′-yū-lā′-shun) Discharge of semen from the penis.

ejaculatory duct The short section of the ejaculatory route in mammals formed by the convergence of the vas deferens and a duct from the seminal vesicle. The ejaculatory duct transports sperm from the vas deferens to the urethra.

electrocardiogram (ECG, EKG) A record of the electrical impulses that travel through cardiac muscle during the heart cycle.

electromagnetic energy Solar energy, or radiation, which travels in space as rhythmic waves and can be measured in photons.

electromagnetic receptor A sensory receptor that detects energy of different wavelengths, such as electricity, magnetism, and light.

electromagnetic spectrum The full range of radiation, from the very short wavelengths of gamma rays to the very long wavelengths of radio signals.

electron A subatomic particle with a single unit of negative electrical charge. One or more electrons move around the nucleus of an atom.

electron carrier A molecule that conveys electrons within a cell; one of several membrane molecules that make up electron transport chains. Electron carriers shuttle electrons during the redox reactions that release energy ultimately used for ATP synthesis.

electron microscope (EM) An instrument that focuses an electron beam through or onto the surface of a specimen. An electron microscope achieves a thousandfold greater resolving power than a light microscope; the most powerful EM can distinguish objects as small as 0.2 nm (2×10^{-10} m).

electron shell An energy level representing the distance of an electron from the nucleus of an atom.

electron transport A redox (oxidation-reduction) reaction in which one or more electrons are transferred to carrier molecules. A series of such reactions, called an electron transport chain, can release the energy stored in high-energy molecules, such as glucose.

electron transport chain A series of electron carrier molecules that shuttle electrons during the redox reactions that release energy used to make ATP; located in the inner membrane of mitochondria, the thylakoid membrane of chloroplasts, and the plasma membrane of prokaryotes.

electronegativity The tendency for an atom to pull electrons toward itself.

electrophoresis *See* gel electrophoresis.

element A substance that cannot be broken down to other substances by chemical means. Scientists recognize 92 chemical elements occurring in nature.

elimination The passing of undigested material out of the digestive compartment; the fourth main stage of food processing, following absorption.

embolus A thrombus that dislodges from its point of origin and travels in the blood.

embryo (em′-brē-ō) A developing stage of a multicellular organism. In humans, the stage in the development of offspring from the first division of the zygote until body structures begin to appear, about the 9th week of gestation.

embryo sac The female gametophyte contained in the ovule of a flowering plant.

embryonic stem cell (ES cell) Any of the cells in the early animal embryo that differentiate during development to give rise to all the different kinds of specialized cells in the body.

emerging virus A virus that has appeared suddenly or has recently come to the attention of medical scientists.

emphysema (em′-fuh-sē′-muh) A respiratory disease in which the alveoli become brittle and rupture, reducing the lungs' capacity for gas exchange.

endangered species As defined in the U.S. Endangered Species Act, a species that is in danger of extinction throughout all or a significant portion of its range.

endemic species A species whose distribution is limited to a specific geographic area.

endocrine gland (en′-dō-krin) A ductless gland that synthesizes hormone molecules and secretes them directly into the bloodstream.

endocrine system The organ system consisting of ductless glands that secrete hormones and the molecular receptors on or in target cells that respond to the hormones; cooperates with the nervous system in regulating body functions and maintaining homeostasis.

endocytosis (en′-dō-sī-tō′-sis) The movement of materials into the cytoplasm of a cell via membranous vesicles or vacuoles.

endoderm (en′-dō-derm) The innermost of three embryonic cell layers in a gastrula; forms the archenteron in the gastrula and gives rise to the innermost linings of the digestive tract and other hollow organs in the adult.

endodermis The innermost layer (a one-cell-thick cylinder) of the cortex of a plant root; forms a selective barrier, determining which substances pass from the cortex into the vascular tissue.

endomembrane system A network of membranous organelles that partition the cytoplasm of eukaryotic cells into functional compartments. Some of the organelles are structurally connected to each other, whereas others are structurally separate but functionally connected by the traffic of membranous vesicles between them.

endometrium (en′-dō-mē′-trē-um) The inner lining of the uterus in mammals, richly supplied with blood vessels that provide the maternal part of the placenta and nourish the developing embryo.

endoplasmic reticulum (ER) An extensive membranous network in a eukaryotic cell, continuous with the outer nuclear membrane and composed of ribosome-studded (rough) and ribosome-free (smooth) regions. *See also* rough ER; smooth ER.

endorphin (en-dōr′-fin) A pain-inhibiting hormone produced by the brain and anterior pituitary; also serves as a neurotransmitter.

endoskeleton A hard skeleton located within the soft tissues of an animal; includes spicules of sponges, the hard plates of echinoderms, and the cartilage and bony skeletons of many vertebrates.

endosperm In flowering plants, a nutrient-rich mass formed by the union of a sperm cell with two polar nuclei during double fertilization; provides nourishment to the developing embryo in the seed.

endospore A thick-coated, protective cell produced within a bacterial cell exposed to harsh conditions.

endosymbiosis (en′-dō-sim′-bē-ō-sis) Symbiotic relationship in which one species resides within another species. The mitochondria and chloroplasts of eukaryotic cells probably evolved from symbiotic associations between small prokaryotic cells living inside larger ones.

endotherm An animal that derives most of its body heat from its own metabolism.

endotoxin A poisonous component of the cell walls of certain bacteria.

energy The capacity to perform work, or to move matter in a direction it would not move if left alone.

energy coupling In cellular metabolism, the use of energy released from an exergonic reaction to drive an endergonic reaction.

energy flow The passage of energy through the components of an ecosystem.

energy pyramid A diagram depicting the cumulative loss of energy from a food chain.

enhancer A eukaryotic DNA sequence that helps stimulate the transcription of a gene at some distance from it. An enhancer functions by means of a transcription factor called an activator, which binds to it and then to the rest of the transcription apparatus. *See* silencer.

entomology The branch of biology that specializes in the study of insects.

entropy (en′-truh-pē) A measure of disorder. One form of disorder is heat, which is random molecular motion.

enzyme (en′-zīm) A protein that serves as a biological catalyst, changing the rate of a chemical reaction without itself being changed in the process.

enzyme inhibitor A chemical that interferes with an enzyme's activity.

epidermis (ep′-uh-der′-mis) (1) In animals, the living layer or layers of cells forming the protective covering, or outer skin. (2) In plants, the tissue system forming the protective outer covering of leaves, young stems, and young roots.

epididymis (ep′-uh-did′-uh-mus) A long coiled tube into which sperm pass from the testis and are stored until mature and ejaculated.

epinephrine (ep′-uh-nef′-rin) An amine hormone (also called adrenaline) secreted by the adrenal medulla that prepares body organs for "fight or flight"; also serves as a neurotransmitter.

epithelial tissue (ep′-uh-thē′-lē-ul) A sheet of tightly packed cells lining organs and cavities; also called epithelium.

epithelium (plural, **epithelia**) *See* epithelial tissue.

equilibrial life history (ē-kwi-lib-rē-ul) Often seen in larger-bodied species, the pattern of reproducing when mature and producing few offspring but caring for the young.

erythrocyte (eh-rith′-rō-sīt) *See* red blood cell.

esophagus (eh-sof′-uh-gus) The channel through which food passes in a digestive tract, connecting the pharynx to the stomach.

essential amino acid Any amino acid that an animal cannot synthesize itself and must obtain from food. Eight amino acids are essential for the human adult.

essential element In plants, a chemical element required for the plant to complete its life cycle (to grow from a seed and produce another generation of seeds).

essential fatty acid An unsaturated fatty acid that animals need but cannot make.

essential nutrient A substance that an organism must absorb in preassembled form because it cannot synthesize the nutrient from any other material. In humans, there are essential vitamins, minerals, amino acids, and fatty acids.

estrogen (es′-trō-jen) One of several chemically similar steroid hormones secreted by the gonads; maintains the female reproductive system and promotes the development of female body features.

estuary (es′-chū-ār-ē) An area where fresh water merges with seawater.

ethylene A gas that functions as a hormone in plants, triggering aging responses such as fruit ripening and leaf drop.

Eukarya (yū-kār-ē-uh) The domain of eukaryotes, organisms made of eukaryotic cells; includes all of the protists, plants, fungi, and animals.

eukaryote (yū-kār′-ē-ōt) An organism characterized by eukaryotic cells. *See* eukaryotic cell.

eukaryotic cell (yū-kār-ē-ot′-ik) A type of cell that has a membrane-enclosed nucleus and other membrane-enclosed organelles. All organisms except bacteria and archaea are composed of eukaryotic cells.

Eustachian tube (yū-stā′-shun) An air passage between the middle ear and throat of vertebrates that equalizes air pressure on either side of the eardrum.

eutherian (yū-thēr′-ē-un) A placental mammal; a mammal whose young complete their embryonic development within the uterus, joined to the mother by the placenta.

eutrophication (yū-trō-fuh-kā′-shun) Overfertilization of an aquatic ecosystem, leading to an increase in productivity.

evaporative cooling The property of a liquid whereby the surface becomes cooler during evaporation, owing to a loss of highly kinetic molecules to the gaseous state.

evo-devo The research field that combines evolutionary biology with developmental biology.

evolution Genetic change in a population or species over generations; all the changes that transform life on Earth; the heritable changes that have produced Earth's diversity of organisms.

exaptation (ek′-sap-tā′-shun) A structure that has evolved in one environmental context and later becomes adapted for a different function in a different environmental context.

excretion (ek-skrē′-shun) The disposal of nitrogen-containing metabolic wastes.

exocytosis (ek′-sō-sī-tō′-sis) The movement of materials out of the cytoplasm of a cell via membranous vesicles or vacuoles.

exon (ek′-son) In eukaryotes, a coding portion of a gene. *See also* intron.

exoskeleton A hard, external skeleton that protects an animal and provides points of attachment for muscles.

exotoxin A poisonous protein secreted by certain bacteria.

exponential growth model A mathematical description of idealized, unregulated population growth.

external fertilization The fusion of gametes that parents have discharged into the environment.

extracellular matrix A substance in which the cells of an animal tissue are embedded; consists of protein and polysaccharides.

extreme halophiles Microorganisms that live in unusually highly saline environments such as the Great Salt Lake or the Dead Sea.

extreme thermophiles Microorganisms that thrive in hot environments (often 60–80°C).

eye cup The simplest type of photoreceptor, a cluster of photoreceptor cells shaded by a cuplike cluster of pigmented cells; detects light intensity and direction.

F

F_1 generation The offspring of two parental (P generation) individuals. F_1 stands for first filial.

F_2 generation The offspring of the F_1 generation. F_2 stands for second filial.

facilitated diffusion The passage of a substance across a biological membrane down its concentration gradient, aided by specific transport proteins.

facultative anaerobe (fak′-ul-tā′-tiv an′-uh-rōb) A microorganism that makes ATP by aerobic respiration if oxygen is present, but that switches to fermentation when oxygen is absent.

family In classification, the taxonomic category above genus.

farsightedness An inability to focus on close objects; occurs when the eyeball is shorter than normal and the focal point of the lens is behind the retina; also called hyperopia.

fat A large lipid molecule made from an alcohol called glycerol and three fatty acids; a triglyceride. Most fats function as energy-storage molecules.

feces The wastes of the digestive tract.

feedback regulation A method of metabolic control in which the end product of a metabolic pathway acts as an inhibitor of an enzyme within that pathway.

fermentation The anaerobic harvest of food by some cells.

fern Any of a group of seedless vascular plants.

fertilization The union of the nucleus of a sperm cell with the nucleus of an egg cell, producing a zygote.

fetus (fē′-tus) A developing human from the 9th week of gestation until birth; has all the major structures of an adult.

fever An abnormally high internal body temperature, usually the result of an infection.

fiber (1) In animals, an elongate, supportive thread in the matrix of connective tissue; an extension of a neuron; a muscle cell. (2) In plants, a long, slender sclerenchyma cell that usually occurs in a bundle.

fibrin (fī′-brin) The activated form of the blood-clotting protein fibrinogen, which aggregates into threads that form the fabric of a blood clot.

fibrinogen (fī′-brin′-uh-jen) The plasma protein that is activated to form a clot when a blood vessel is injured.

fibrous connective tissue A dense tissue with large numbers of collagenous fibers organized into parallel bundles. This is the dominant tissue in tendons and ligaments.

filament In a flowering plant, the stalk of a stamen.

filtrate Fluid extracted by the excretory system from the blood or body cavity. The excretory system produces urine from the filtrate after extracting valuable solutes from it and concentrating it.

filtration In the vertebrate kidney, the extraction of water and small solutes, including metabolic wastes, from the blood by the nephrons.

fission A means of asexual reproduction whereby a parent separates into two or more genetically identical individuals of about equal size.

fitness The contribution an individual makes to the gene pool of the next generation relative to the contribution of other individuals in the population.

flagellate (flaj′-uh-lit) A protist (protozoan) that moves by means of one or more flagella.

flagellum (fluh-jel′-um) (plural, **flagella**) A long appendage that propels protists through the water and moves fluids across the surface of many tissue cells in animals. A cell may have one or more flagella. Like cilia, flagella have a 9 + 2 arrangement of microtubules covered by the cell's plasma membrane.

flatworm A bilateral animal with a thin, flat body form, gastrovascular cavity with a single opening, and no body cavity. Flatworms include planarians, flukes, and tapeworms.

flower In an angiosperm, a short stem with four sets of modified leaves, bearing structures that function in sexual reproduction.

fluid feeder An animal that lives by sucking nutrientrich fluids from another living organism.

fluid mosaic A description of membrane structure, depicting a cellular membrane as a mosaic of diverse protein molecules embedded in a fluid bilayer made of phospholipid molecules.

follicle (fol′-uh-kul) A cluster of cells surrounding, protecting, and nourishing a developing egg cell in the ovary; also secretes estrogen.

follicle-stimulating hormone (FSH) A protein hormone secreted by the anterior pituitary that stimulates the production of eggs by the ovaries and sperm by the testes.

food chain A sequence of food transfers from producers through several levels of consumers in an ecosystem.

food vacuole (vak-ū-ōl) The simplest type of digestive cavity, found in single-celled organisms.

food web A network of interconnecting food chains.

food-conducting cell A specialized living plant cell with a thin primary wall. Arranged end to end, such cells collectively form phloem tissue.

foram A marine protozoan that secretes a shell and extends pseudopodia through pores in its shell.

forebrain One of three ancestral and embryonic regions of the vertebrate brain; develops into the thalamus, hypothalamus, and cerebrum.

forensics The scientific analysis of evidence for crime scene investigations and other legal proceedings.

fossil A preserved remnant or impression of an organism that lived in the past.

fossil fuel An energy deposit formed from the remains of extinct organisms.

fossil record The chronicle of evolution over millions of years of geologic time engraved in the order in which fossils appear in rock strata.

founder effect Random change in the gene pool that occurs in a small colony of a population.

fovea (fō′-vē-uh) An eye's center of focus and the place on the retina where photoreceptors are highly concentrated.

fragmentation A means of asexual reproduction whereby a single parent breaks into parts that regenerate into whole new individuals.

fruit A ripened, thickened ovary of a flower, which protects dormant seeds and aids in their dispersal.

fruiting body A stage in an organism's life cycle that functions only in reproduction; for example, a mushroom is a fruiting body of many fungi.

functional group The atoms that form the chemically reactive part of an organic molecule.

Fungi (fun′-jē) The kingdom that contains the fungi.

fungus (plural, **fungi**) A heterotrophic eukaryote that digests its food externally and absorbs the resulting small nutrient molecules. Most fungi consist of a netlike mass of filaments called hyphae. Molds, mushrooms, and yeasts are examples of fungi.

G

gallbladder An organ that stores bile and releases it as needed into the small intestine.

gametangium (gam′-uh-tan′-jē-um) (plural, **gametangia**) A reproductive organ that houses and protects the gametes of a plant.

gamete (gam′-ēt) A sex cell; a haploid egg or sperm. The union of two gametes of opposite sex (fertilization) produces a zygote.

gametic isolation (guh-mē′-tik) A type of prezygotic barrier between species; the species remain isolated because male and female gametes of the different species cannot fuse, or they die before they unite.

gametogenesis The creation of gametes within the gonads.

gametophyte (guh-mē′-tō-fīt) The multicellular haploid form in the life cycle of organisms undergoing alternation of generations; mitotically produces haploid gametes that unite and grow into the sporophyte generation.

ganglion (gang′-glē-un) (plural, **ganglia**) A cluster (functional group) of nerve cell bodies in a centralized nervous system.

gas exchange *See* respiration.

gastric juice The collection of fluids secreted by the epithelium lining the stomach.

gastric ulcer An open sore in the lining of the stomach, resulting when pepsin and hydrochloric acid destroy the lining tissues faster than they can regenerate.

gastropod A member of the largest group of molluscs, including snails and slugs.

gastrovascular cavity A digestive compartment with a single opening that serves as both the entrance for food and the exit for undigested wastes; mouth; may also function in circulation, body support, and gas exchange. Jellies and hydras are examples of animals with a gastrovascular cavity.

gastrula (gas′-trū-luh) The embryonic stage resulting from gastrulation in animal development. Most animals have a gastrula made up of three layers of cells: ectoderm, endoderm, and mesoderm.

gastrulation (gas′-trū-lā′-shun) The phase of embryonic development that transforms the blastula (blastocyst in mammals) into a gastrula. Gastrulation adds more cells to the embryo and sorts the cells into distinct cell layers.

gel electrophoresis (jel′ ē-lek′-trō-fōr-ē′-sis) A technique for separating and purifying macromolecules. A mixture of molecules is placed on a gel between a positively charged electrode and a negatively charged one; negative charges on the molecules are attracted to the positive electrode, and the molecules migrate toward that electrode; the molecules separate in the gel according to their rates of migration.

gene A discrete unit of hereditary information consisting of a specific nucleotide sequence in DNA (or RNA, in some viruses). Most of the genes of a eukaryote are located in its chromosomal DNA; a few are carried by the DNA of mitochondria and chloroplasts.

gene cloning The production of multiple copies of a gene.

gene expression The process whereby genetic information flows from genes to proteins; the flow of genetic information from the genotype to the phenotype.

gene flow The gain or loss of alleles from a population by the movement of individuals or gametes into or out of the population.

gene pool All the genes in a population at any one time.

gene regulation The turning on and off of specific genes within a living organism.

genetic code The set of rules giving the correspondence between nucleotide triplets (codons) in mRNA and amino acids in protein.

genetic drift A change in the gene pool of a population due to chance.

genetic marker (1) An allele tracked in a genetic study. (2) A specific section of DNA that earmarks a particular allele; may contain specific restriction sites (points where restriction enzymes cut the DNA) that occur only in DNA that contains the allele.

genetic recombination The production, by crossing over and/or independent assortment of chromosomes during meiosis, of offspring with allele combinations different from those in the parents. The term may also be used more specifically to mean the production by crossing over of eukaryotic or prokaryotic chromosomes with gene combinations different from those in the original chromosomes.

genetically modified (GM) organism An organism that has acquired one or more genes by artificial means. If the gene is from another species, the organism is also known as a transgenic organism.

genetics The scientific study of heredity and hereditary variations.

genome (jē′-nōm) A complete (haploid) set of an organism's genes; an organism's genetic material.

genomic library (juh-nō′-mik) A set of DNA segments from an organism's genome. Each segment is usually carried by a plasmid or phage.

genomics The study of whole sets of genes and their interactions.

genotype (jē′-nō-tīp) The genetic makeup of an organism.

genus (jē′-nus) (plural, **genera**) In classification, the taxonomic category above species; the first part of a species' binomial; for example, *Homo*.

geologic time scale A time scale established by geologists that reflects a consistent sequence of historical periods, grouped into four eras: Precambrian, Paleozoic, Mesozoic, and Cenozoic.

germinate To initiate growth, as in a plant seed.

gestation (jes-tā′-shun) Pregnancy; the state of carrying developing young within the female reproductive tract.

gibberellin (jib′-uh-rel′-in) One of a family of plant hormones that trigger the germination of seeds and interact with auxins in regulating growth and fruit development.

gill An extension of the body surface of an aquatic animal, specialized for gas exchange and/or suspension feeding.

glans The rounded, highly sensitive head of the clitoris in females and penis in males.

glaucoma A disorder of the vertebrate eye in which increased pressure may lead to blindness.

global warming A slow but steady rise in Earth's surface temperature, caused by increasing concentrations of greenhouse gases (such as CO_2 and CH_4) in the atmosphere.

glucagon (glū′-kuh-gon) A peptide hormone secreted by islet cells in the pancreas that raises the level of glucose in the blood.

glucocorticoid (glū′-kuh-kor′-tih-koyd) A corticosteroid hormone secreted by the adrenal cortex that increases the blood glucose level and helps maintain the body's response to long-term stress.

glycogen (glī′-kō-jen) A complex, extensively branched polysaccharide made up of many glucose monomers; serves as an energy-storage molecule in liver and muscle cells.

glycolysis (glī-kol′-uh-sis) The multistep chemical breakdown of a molecule of glucose into two molecules of pyruvic acid; the first stage of cellular respiration in all organisms; occurs in the cytoplasmic fluid.

goiter An enlargement of the thyroid gland resulting from a dietary iodine deficiency.

Golgi apparatus (gol′-jē) An organelle in eukaryotic cells consisting of stacks of membranous sacs that modify, store, and ship products of the endoplasmic reticulum.

gonad A sex organ in an animal; an ovary or a testis.

Gondwana (gon-dwa′-na) The southern landmass formed during the Mesozoic era when continental drift split Pangaea (all land masses fused). *See also* Laurasia.

granum (gran′-um) (plural, **grana**) A stack of hollow disks formed of thylakoid membrane in a chloroplast. Grana are the sites where light energy is trapped by chlorophyll and converted to chemical energy during the light reactions of photosynthesis.

gravitropism (grav′-uh-trō′-pizm) A plant's growth response to gravity.

gray matter Regions of dendrites and clusters of nerve cell bodies within the CNS.

green alga One of a group of photosynthetic protists that includes unicellular, colonial, and multicellular species. Green algae are plantlike in having biflagellated cells (gametes in colonial and multicellular species), chloroplasts with chlorophyll *a*, cellulose cell walls, and starch.

greenhouse effect The warming of the atmosphere caused by CO_2, CH_4, and other gases that absorb infrared radiation and slow its escape from Earth's surface.

greenhouse gas Any of the gases in the atmosphere that absorb heat radiation, contributing to the greenhouse effect.

ground tissue system A tissue of mostly parenchyma cells that makes up the bulk of a young plant and is continuous throughout its body. The ground tissue system fills the space between the epidermis and the vascular tissue system.

growth factor A protein secreted by certain body cells that stimulates other cells to divide.

growth hormone (GH) A protein hormone secreted by the anterior pituitary that promotes development and growth and stimulates metabolism.

guanine (G) (gwa′-nēn) A double-ring nitrogenous base found in DNA and RNA.

guard cell A specialized epidermal cell in plants that regulates the size of a stoma, allowing gas exchange between the surrounding air and the photosynthetic cells in the leaf.

gymnosperm (jim′-nō-sperm) A naked-seed plant. Its seed is said to be naked because it is not enclosed in a fruit.

H

habitat A place where an organism lives; a specific environment in which an organism lives.

habitat isolation A type of prezygotic barrier between species; the species remain isolated because they breed in different habitats.

haploid Containing a single set of chromosomes; referring to an *n* cell.

Hardy-Weinberg equilibrium The principle that the shuffling of genes that occurs during sexual reproduction, by itself, cannot change the overall genetic makeup of a population.

Hardy-Weinberg formula A formula for calculating the frequencies of genotypes in a gene pool from the frequencies of alleles, and vice versa.

HDL *See* high-density lipoprotein.

heart (1) The chambered muscular organ in vertebrates that pumps blood received from the veins into the arteries, thereby maintaining the flow of blood through the entire circulatory system. (2) A similarly functioning structure in invertebrates.

heart attack Death of cardiac muscle cells and the resulting failure of the heart to deliver enough blood to the body.

heart murmur A hissing sound often emitted by an abnormal heart, usually caused by a defective heart valve.

heart rate The number of heartbeats per minute.

heartwood In the center of trees, the darkened, older layers of secondary xylem made up of cells that no longer transport water and are clogged with resins. *See also* sapwood.

heat The amount of energy associated with the movement of the atoms and molecules in a body of matter. Heat is energy in its most random form.

heat exhaustion A condition caused by exposure to heat, resulting in the depletion of body fluids and causing weakness, dizziness, nausea, and often collapse.

heat stroke A severe condition caused by impairment of the body's temperature-regulating abilities, resulting from prolonged exposure to excessive heat and characterized by cessation of sweating, severe headache, high fever, hot dry skin, and in serious cases collapse and coma.

helper T cell A type of lymphocyte that helps activate other types of T cells and may help stimulate B cells to produce antibodies.

hemoglobin (hē′-mō-glō-bin) An iron-containing protein in red blood cells that reversibly binds O_2 and transports it to body tissues.

hemophilia (hē′-mō-fil′-ē-uh) A human genetic disease caused by a sex-linked recessive allele and characterized by excessive bleeding following injury.

herbivore An animal that eats mainly plants or algae. *See also* carnivore; omnivore.

hermaphrodite (her-maf′-rō-dīt) An individual that has both female and male gonads and that functions as both a male and female in sexual reproduction by producing both sperm and eggs.

heterotroph (het′-er-ō-trōf) An organism that cannot make its own organic food molecules and must obtain them by consuming other organisms or their organic products; a consumer or a decomposer in a food chain.

heterozygous (het′-er-ō-zī′-gus) Having two different alleles for a given gene.

high-density lipoprotein (HDL) A cholesterol-carrying particle in the blood, made up of cholesterol and other lipids surrounded by a single layer of phospholipids in which proteins are embedded. An HDL particle carries less cholesterol than a related lipoprotein, LDL, and may be correlated with a decreased risk of blood vessel blockage.

hindbrain One of three ancestral and embryonic regions of the vertebrate brain; develops into the medulla oblongata, pons, and cerebellum.

hinge joint A joint that allows movement in only one plane. In humans, examples include the elbow and knee.

histamine (his′-tuh-mēn) A chemical alarm signal released by injured cells that causes blood vessels to dilate during an inflammatory response.

histone (his′-tōn) A small basic protein molecule associated with DNA and important in DNA packing in the eukaryotic chromosome.

HIV Human immunodeficiency virus, the retrovirus that attacks the human immune system and causes AIDS.

homeobox (hō′-mē-ō-boks′) A 180-nucleotide sequence within a homeotic gene encoding the part of the protein that binds to the DNA of the genes regulated by the protein.

homeostasis (hō′-mē-ō-stā′-sis) The steady state of body functioning; a state of equilibrium characterized by a dynamic interplay between outside forces that tend to change an organism's internal environment and the internal control mechanisms that oppose such changes.

homeotic gene (hō′-mē-ot′-ik) A master control gene that determines the identity of a body structure of a developing organism, presumably by controlling the developmental fate of groups of cells. (In plants, such genes are called organ identity genes.)

hominid (hah′-mi-nid) A species on the human branch of the evolutionary tree; a member of the family Hominidae, including *Homo sapiens* and our ancestors.

hominoid A term that refers to great apes and humans.

homologous chromosomes (hō-mol′-uh-gus) The two chromosomes that make up a matched pair in a diploid cell. Homologous chromosomes are of the same length, centromere position, and staining pattern and possess genes for the same characteristics at corresponding loci. One homologous chromosome is inherited from the organism's father, the other from the mother.

homologous structures Structures that are similar in different species of common ancestry.

homology (hō-mol′-uh-jē) Anatomical similarity due to common ancestry.

homozygous (hō′-mō-zī′-gus) Having two identical alleles for a given gene.

hormone A regulatory chemical that travels in the blood from its production site, usually an endocrine gland, to other sites, where target cells respond to the regulatory signal.

host The larger participant in a symbiotic relationship, serving as home and feeding ground to the parasite.

human chorionic gonadotropin (HCG) (kōr′-ē-on′-ik gon′-uh-dō-trō′-pin) A hormone secreted by the chorion that maintains the corpus luteum of the ovary during the first three months of pregnancy.

human gene therapy Treatment for a disease in which the patient is provided with a new gene.

Human Genome Project An international collaborative effort to map and sequence the DNA of the entire human genome.

humoral immune response The type of specific immune response brought about by antibody-producing B cells; fights bacteria and viruses in body fluids. *See also* cell-mediated immune response.

Huntington's disease A human genetic disease caused by a dominant allele; characterized by uncontrollable body movements and degeneration of the nervous system; usually fatal 10 to 20 years after the onset of symptoms.

hybrid The offspring of parents of two different species or of two different varieties of one species; the offspring of two parents that differ in one or more inherited traits; an individual that is heterozygous for one or more pairs of genes.

hybrid inviability A type of postzygotic barrier between species; the species remain isolated because hybrid zygotes do not develop or hybrids do not become sexually mature.

hybrid sterility A type of postzygotic barrier between species; the species remain isolated because hybrids fail to produce functional gametes.

hybridization The cross-fertilization of two different varieties of an organism or of two different species; also called a cross.

hydrocarbon A chemical compound composed only of the elements carbon and hydrogen.

hydrogen bond A type of weak chemical bond formed when the partially positive hydrogen atom participating in a polar covalent bond in one molecule is attracted to the partially negative atom participating in a polar covalent bond in another molecule (or in another part of the same macromolecule).

hydrogenation The process of converting unsaturated fats to saturated fats by the addition of hydrogen.

hydrolase A general term for any enzyme that catalyzes a hydrolysis reaction, the chemical breakdown of polymers into smaller molecules through the addition of water molecules. For example, lactase is a hydrolase that catalyzes the breakdown of the disaccharide lactose into glucose and galactose.

hydrolysis (hī-drol′-uh-sis) A chemical process in which macromolecules are broken down by the chemical addition of water molecules to the bonds linking their monomers; an essential part of digestion.

hydrophilic (hī′-drō-fil′-ik) "Water-loving"; pertaining to polar or charged molecules (or parts of molecules) that are soluble in water.

hydrophobic (hī′-drō-fō′-bik) "Water-fearing"; pertaining to nonpolar molecules (or parts of molecules) that do not dissolve in water.

hydrothermal vent community A seafloor community powered by chemical energy from Earth's interior rather than by sunlight.

hydroxyl group (hī-drok′-sil) In an organic molecule, a functional group consisting of a hydrogen atom bonded to an oxygen atom.

hymen A thin membrane that partly covers the vaginal opening in the human female and is ruptured by sexual intercourse or other vigorous activity.

hypercholesterolemia (hī′-per-kō-les′-tur-ah-lēm′-ē-uh) An inherited human disease characterized by an excessively high level of cholesterol in the blood.

hypertension Abnormally high blood pressure; a persistent blood pressure of 140/90 or higher.

hypertonic In comparing two solutions, referring to the one with the greater concentration of solutes.

hypha (hī′-fuh) (plural, **hyphae**) One of many filaments making up the body of a fungus.

hypoglycemia (hī′-pō-glī-sē′-mē-uh) An abnormally low level of glucose in the blood that results when the pancreas secretes too much insulin into the blood.

hypothalamus (hī′-pō-thal′-uh-mus) The master control center of the endocrine system, located in the ventral portion of the vertebrate forebrain. The hypothalamus functions in maintaining homeostasis, especially in coordinating the endocrine and nervous systems; secretes hormones of the posterior pituitary and releasing hormones that regulate the anterior pituitary.

hypothesis (hī-poth′-uh-sis) (plural, **hypotheses**) A tentative explanation that a scientist proposes for a specific phenomenon that has been observed.

hypothesis-driven science The process of scientific inquiry that uses the steps of the scientific method to answer questions about nature. *See also* discovery science; scientific method.

hypotonic In comparing two solutions, referring to the one with the lower concentration of solutes.

I

immune system The system of cells that protects the body by recognizing and attacking specific kinds of pathogens and cancer cells.

immunity Resistance to specific body invaders.

immunodeficiency disease An immunological disorder in which the immune system lacks one or more components, making the body susceptible to infectious agents that would ordinarily not be pathogenic.

impotence The inability to maintain an erection; also called erectile dysfunction.

in vitro fertilization (IVF) (vē′-tro) Uniting sperm and egg in a laboratory container, followed by the placement of a resulting early embryo in the mother's uterus.

inbreeding The mating of close relatives.

incomplete dominance A type of inheritance in which the phenotype of a heterozygote (*Aa*) is intermediate between the phenotypes of the two types of homozygotes (*AA* and *aa*).

indeterminate growth Growth that continues throughout life, as in most plants. *See also* determinate growth.

induced fit The interaction between a substrate molecule and the active site of an enzyme, which changes shape slightly to embrace the substrate and catalyze the reaction.

induction During embryonic development, the influence of one group of cells on another group of cells.

inferior vena cava (vē′-nuh kā′-vuh) A large vein that returns O_2-poor blood to the heart from the lower, or posterior, part of the body. *See also* superior vena cava.

infertility The inability to conceive after one year of regular, unprotected intercourse.

inflammatory response A nonspecific body defense caused by a release of histamine and other chemical alarm signals, which trigger increased blood flow, a local increase in white blood cells, and fluid leakage from the blood. The results include redness, heat, and swelling in the affected tissues.

ingestion The act of eating; the first main stage of food processing.

inner cell mass A cluster of cells in a mammalian blastocyst that protrudes into one end of the cavity and subsequently develops into the embryo proper and some of the extraembryonic membranes.

inner ear One of three main regions of the vertebrate ear; includes the cochlea, organ of Corti, and semicircular canals.

insect An arthropod that usually has three body segments (head, thorax, and abdomen), three pairs of legs, and one or two pairs of wings.

insulin A protein hormone, secreted by islet cells in the pancreas, that lowers the level of glucose in the blood.

integration The interpretation of sensory signals and the formulation of responses within neural processing centers of the central nervous system.

integumentary system (in-teg′-yū-men′-ter-ē) The organ system consisting of the skin and its derivatives, such as hair and nails in mammals; helps protect the body from drying out, mechanical injury, and infection.

interferon (in′-ter-fēr′-on) A nonspecific defensive protein produced by virus-infected cells and capable of helping other cells resist viruses.

intermembrane space One of the two fluid-filled internal compartments of the mitochondrion, the narrow region between the inner and outer membranes.

internal fertilization Reproductive process in which sperm are typically deposited in or near the female reproductive tract, with fertilization occurring within the tract.

interneuron (in′-ter-nūr′-on) A nerve cell, entirely within the central nervous system, that integrates sensory signals and may relay command signals to motor neurons.

internode The portion of a plant stem between two nodes.

interphase The period in the eukaryotic cell cycle when the cell is not actually dividing. *See also* mitosis.

interspecific competition Competition between populations of two or more species that require similar limited resources. Interspecific competition may inhibit population growth and help structure communities.

interspecific interaction Any interaction between members of different species.

interstitial fluid (in′-ter-stish′-ul) An aqueous solution that surrounds body cells and through which materials pass back and forth between the blood and the body tissues.

intertidal zone (in′-ter-tīd′-ul) A shallow zone where the waters of an estuary or ocean meet land.

intestine The region of a digestive tract between the gizzard or stomach and the anus, where chemical digestion and nutrient absorption usually occur.

intraspecific competition Competition between individuals of the same species for a limited resource.

introduced species A species that humans move from the species' native location to a new geographic region; sometimes called an exotic species.

intron (in′-tron) In eukaryotes, a nonexpressed (noncoding) portion of a gene that is excised from the RNA transcript. *See also* exon.

invertebrate An animal that lacks a backbone.

ion (ī′-on) An atom or molecule that has gained or lost one or more electrons, thus acquiring an electrical charge.

ionic bond (ī-on′-ik) An attraction between two ions with opposite electrical charges. The electrical attraction of the opposite charges holds the ions together.

iris The colored part of the vertebrate eye, formed by the anterior portion of the choroid.

islet cells (ī′-lit) Clusters of endocrine cells in the pancreas that produce insulin or glucagon.

isomers (ī′-sō-mers) Organic compounds with the same molecular formula but different structures and thus different properties.

isotonic (ī-sō-ton′-ik) Having the same solute concentration as another solution.

isotope (ī′-sō-tōp) A variant form of an atom. Isotopes of an element have the same number of protons and electrons but different numbers of neutrons.

K

karyotype (kār′-ē-ō-tīp) A display of micrographs of the metaphase chromosomes of a cell, arranged by size and centromere position.

keystone predator A predator species that reduces the density of the strongest competitors in a community, thereby helping maintain species diversity.

keystone species Species that are not usually abundant in a community yet exert strong control on community structure by the nature of their ecological roles or niches.

kilocalorie A quantity of heat equal to 1,000 calories. Used to measure the energy content of food, it is usually called a "Calorie."

kinetic energy (kuh-net′-ik) Energy that is actually doing work; the energy of a mass of matter that is moving. Moving matter performs work by transferring its motion to other matter, such as leg muscles pushing bicycle pedals.

kingdom In classification, the broad taxonomic category above phylum or division.

Krebs cycle *See* citric acid cycle.

L

labia majora (lā′-bē-uh muh-jor′-uh) A pair of outer thickened folds of skin that protect the female genital region.

labia minora (lā′-bē-uh mi-nor′-uh) A pair of inner folds of skin bordering and protecting the female genital region.

labor A series of strong, rhythmic contractions of the uterus that expel a baby out of the uterus and vagina during childbirth.

lacrimal gland A gland above the eye that secretes tears.

lactic acid fermentation The conversion of pyruvate to lactate with no release of carbon dioxide.

lancelet One of a group of invertebrate chordates.

landscape A regional assemblage of interacting ecosystems.

landscape ecology The application of ecological principles to the study of land-use patterns; the scientific study of the biodiversity of interacting ecosystems.

large intestine The tubular portion of the vertebrate alimentary canal between the small intestine and the anus. *See also* colon.

larva (lar′-vuh) (plural, **larvae**) A free-living, sexually immature form in some animal life cycles that may differ from the adult in morphology, nutrition, and habitat.

larynx (lār′-inks) The voice box, containing the vocal cords.

lateral Pertaining to the side of a bilaterally symmetrical animal.

lateral line system A row of sensory organs along each side of a fish's body. Sensitive to changes in water pressure, it enables a fish to detect minor vibrations in the water.

lateralization The phenomenon in which the two hemispheres of the brain become specialized for different functions.

Laurasia (lah-rā′-zhuh) The northern landmass formed when continental drift split Pangaea (all land masses fused) during the Mesozoic era. *See also* Gondwana.

law of independent assortment A general rule in inheritance that when gametes form during meiosis, each pair of alleles for a particular characteristic segregate independently; also known as Mendel's second law of inheritance.

law of segregation A general rule in inheritance that individuals have two alleles for each gene and that when gametes form by meiosis, the two alleles separate, and each resulting gamete ends up with only one allele of each gene; also known as Mendel's first law of inheritance.

LDL *See* low-density lipoprotein.

leaf The main site of photosynthesis in a plant; consists of a flattened blade and a stalk (petiole) that joins the leaf to the stem.

lens The structure in an eye that focuses light rays onto the retina.

leukemia (lū-kē′-mē-ah) A type of cancer of the blood-forming tissues, characterized by an excessive production of white blood cells and an abnormally high number of them in the blood; cancer of the bone marrow cells that produce leukocytes.

leukocyte (lū′-kō-sīt) *See* white blood cell.

lichen (lī′-ken) A mutualistic association between a fungus and an alga or between a fungus and a cyanobacterium.

life The set of common characteristics that distinguish living organisms, including such properties and processes as order, regulation, growth and development, metabolism, response to the environment, reproduction, and the capacity to evolve over time.

life cycle The entire sequence of stages in the life of an organism, from the adults of one generation to the adults of the next.

life history The series of events from birth through reproduction to death.

life table A listing of survivals and deaths in a population in a particular time period and predictions of how long, on average, an individual of a given age will live.

ligament A type of fibrous connective tissue that joins bones together at joints.

light microscope (LM) An optical instrument with lenses that refract (bend) visible light to magnify images and project them into a viewer's eye or onto photographic film.

light reactions The first of two stages in photosynthesis, the steps in which solar energy is absorbed and converted to chemical energy in the form of ATP and NADPH. The light reactions power the sugar-producing Calvin cycle but produce no sugar themselves.

lignin (lig′-nin) A chemical that hardens the cell walls of plants.

limbic system (lim′-bik) A functional unit of several integrating and relay centers located deep in the human forebrain; interacts with the cerebral cortex in creating emotions and storing memories.

linkage map A map of a chromosome showing the relative positions of genes.

linked genes Genes located close enough together on a chromosome to be usually inherited together.

lipid An organic compound consisting mainly of carbon and hydrogen atoms linked by nonpolar convalent bonds and therefore mostly hydrophobic and insoluble in water. Lipids include fats, waxes, phospholipids, and steroids.

liver The largest organ in the vertebrate body. The liver performs diverse functions such as producing bile, preparing nitrogenous wastes for disposal, and detoxifying poisonous chemicals in the blood.

lobe-finned fish A bony fish with strong, muscular fins supported by bones. Lobe-fins are extinct except for one species, the coelacanth.

locus (plural, **loci**) The particular site where a gene is found on a chromosome. Homologous chromosomes have corresponding gene loci.

logistic growth model A mathematical description of idealized population growth that is restricted by limiting factors.

loose connective tissue The most widespread connective tissue in the vertebrate body. It binds epithelia to underlying tissues and functions as packing material, holding organs in place.

low-density lipoprotein (LDL) A cholesterol-carrying particle in the blood, made up of cholesterol and other lipids surrounded by a single layer of phospholipids in which proteins are embedded. An LDL particle carries more cholesterol than a related lipoprotein, HDL, and high LDL levels in the blood correlate with a tendency to develop blocked blood vessels and heart disease.

lung An internal sac, lined with moist epithelium, where gases are exchanged between inhaled air and the blood.

lungfish A bony fish that generally inhabits stagnant waters and gulps air into lungs connected to a pharynx.

luteinizing hormone (LH) (lū′-tē-uh-nī′-zing) A protein hormone secreted by the anterior pituitary that stimulates ovulation in females and androgen production in males.

Lyme disease A debilitating human disease caused by the bacterium *Borrelia burgdorferi;* characterized at first by a red rash at the site of a tick bite and, if not treated, by heart disease, arthritis, and nervous disorders.

lymph A fluid similar to interstitial fluid that circulates in the lymphatic system.

lymph node A small organ that is located along a lymph vessel and that filters lymph and helps attach viruses and bacteria.

lymphatic system (lim-fat′-ik) The organ system through which lymph circulates; includes lymph vessels, lymph nodes, and the spleen. The lymphatic system helps remove toxins and pathogens from the blood and interstitial fluid and returns fluid and solutes from the interstitial fluid to the circulatory system.

lymphocyte (lim′-fuh-sīt) A type of white blood cell, found mostly in the lymphatic system, that is chiefly responsible for the immune response. *See also* B cell; T cell.

lymphoma (lim-fō′-muh) Cancer of the tissues that form white blood cells.

lysogenic cycle (lī-sō-jen′-ik) A bacteriophage replication cycle in which the viral genome is incorporated into the bacterial host chromosome as a prophage. New phages are not produced, and the host cell is not killed or lysed unless the viral genome leaves the host chromosome.

lysosomal storage disease A hereditary disorder associated with abnormal lysosomes, where the sufferer is missing one of the lysosomal digestive enzymes.

lysosome (lī′-sō-sōm) A digestive organelle in eukaryotic cells; contains hydrolytic enzymes that digest the cell's food and wastes.

lytic cycle (lit′-ik) A viral replication cycle resulting in the release of new viruses by lysis (breaking open) of the host cell.

M

macroevolution Evolutionary change on a grand scale, encompassing the origin of new taxonomic groups, evolutionary trends, adaptive radiation, and mass extinction.

macromolecule A giant molecule in a living organism: a protein, polysaccharide, or nucleic acid.

macronutrient A chemical substance that an organism must obtain in relatively large amounts. *See also* micronutrient.

macrophage (mak′-rō-fāj) A large, amoeboid, phagocytic white blood cell that develops from a monocyte.

magnetic resonance imaging (MRI) Imaging technology that uses magnetism and radio waves to induce hydrogen nuclei in water molecules to emit faint radio signals. A computer creates images of the body from the radio signals.

magnification An increase in the apparent size of an object.

major depression Depressive mental illness characterized by persistant sadness and loss of interest in pleasurable activities.

malignant tumor An abnormal tissue mass that can spread into neighboring tissue and to other parts of the body; a cancerous tumor.

malnutrition The absence of one or more essential nutrients from the diet.

mammal Member of a class of endothermic amniotes that possess mammary glands and hair.

mantle In molluscs, the outgrowth of the body surface that drapes over the animal. The mantle produces the shell and forms the mantle cavity.

mark-recapture method A sampling technique used to estimate wildlife populations.

marsupial (mar-sū′-pē-ul) A pouched mammal, such as a kangaroo, opossum, or koala. Marsupials give birth to embryonic offspring that complete development while housed in a pouch and attached to nipples on the mother's abdomen.

mass A measure of the amount of material in an object.

mass number The sum of the number of protons and neutrons in an atom's nucleus.

mast cell A vertebrate body cell that produces histamine and other molecules that trigger the inflammatory response.

matrix The thick fluid contained within the inner membranes of the mitochondrion.

matter Anything that occupies space and has mass.

mechanical digestion The physical breakdown of food into smaller pieces, as by chewing.

mechanical isolation A type of prezygotic barrier between species; the species remain isolated because structural differences between them prevent fertilization.

mechanoreceptor (mek′-uh-nō-ri-sep′-ter) A sensory receptor that detects physical deformations in the environment, associated with pressure, touch, stretch, motion, and sound.

medulla oblongata (meh-duh′-luh ob′-long-got′-uh) Part of the vertebrate hindbrain continuous with the spinal cord; passes data between the spinal cord and forebrain and controls autonomic, homeostatic functions, including breathing, heart rate, swallowing, and digestion.

medusa (med-ū′-suh) (plural, **medusae**) One of two types of cnidarian body forms; an umbrella-like body form; also called a jelly.

meiosis (mī-ō′-sis) In a sexually reproducing organism, the division of a single diploid nucleus into four haploid daughter nuclei. Meiosis and cytokinesis produce haploid gametes from diploid cells in the reproductive organs of the parents.

membrane infolding A process by which the eukaryotic cell's endo-membrane system evolved from inward folds of the plasma membrane of a prokaryotic cell.

memory cell One of a clone of long-lived lymphocytes formed during the primary immune response; remains in a lymph node until activated by exposure to the same antigen that triggered its formation. When activated, a memory cell forms a large clone that mounts the secondary immune response.

meninges (muh-nin′-jēz) Layers of connective tissue that enwrap and protect the brain and spinal cord.

menstrual cycle (men′-strū-ul) The hormonally synchronized cyclic buildup and breakdown of the endometrium in some primates, including humans.

menstruation (men′-strū-ā′-shun) Uterine bleeding resulting from shedding of the endometrium during a menstrual cycle.

meristem (mār′-eh-stem) Plant tissue consisting of undifferentiated cells that divide and generate new cells and tissues.

mesoderm (mez′-ō-derm) The middle layer of the three embryonic cell layers in a gastrula; gives rise to muscles, bones, the dermis of the skin, and most other organs in the adult.

mesophyll (mes′-ō-fil) The green tissue in the interior of a leaf; a leaf's ground tissue system, the main site of photosynthesis.

messenger RNA (mRNA) The type of ribonucleic acid that encodes genetic information from DNA and conveys it to ribosomes, where the information is translated into amino acid sequences.

metabolic rate Energy expended by the body per unit time.

metabolism (muh-tab′-uh-liz-um) The sum total of all the chemical reactions that occur in organisms.

metamorphosis (met′-uh-mōr′-fuh-sis) The transformation of a larva into an adult.

metaphase (met′-eh-fāz) The second stage of mitosis. During metaphase, all the cell's duplicated chromosomes are lined up at an imaginary plane equidistant between the poles of the mitotic spindle.

metastasis (muh-tas′-tuh-sis) The spread of cancer cells beyond their original site.

microevolution A change in a population's gene pool over a succession of generations;

evolutionary changes in species over relatively brief periods of geologic time.

micrograph A photograph taken through a microscope.

micronutrient An element that an organism needs in very small amounts and that functions as a component or cofactor of enzymes. *See also* macronutrient.

microtubule The thickest of the three main kinds of fibers making up the cytoskeleton of a eukaryotic cell; a straight, hollow tube made of globular proteins called tubulins. Microtubules form the basis of the structure and movement of cilia and flagella.

microvillus (plural, **microvilli**) A microscopic projection on the surface of a cell. Microvilli increase a cell's surface area.

midbrain One of three ancestral and embryonic regions of the vertebrate brain; develops into sensory integrating and relay centers that send sensory information to the cerebrum.

middle ear One of three main regions of the vertebrate ear; a chamber containing three small bones (the hammer, anvil, and stirrup) that convey vibrations from the eardrum to the inner ear.

millipede A terrestrial arthropod that has two pairs of short legs for each of its numerous body segments and that eats decaying plant matter.

mineral In nutrition, a chemical element other than carbon, hydrogen, oxygen, or nitrogen that an organism requires for proper body functioning.

mitochondrion (mī′-tō-kon′-drē-on) (plural, **mitochondria**) An organelle in eukaryotic cells where cellular respiration occurs. Enclosed by two concentric membranes, it is where most of the cell's ATP is made.

mitosis (mī′-tō-sis) The division of a single nucleus into two genetically identical daughter nuclei. Mitosis and cytokinesis make up the mitotic (M) phase of the cell cycle.

mitotic phase The part of the cell cycle when mitosis divides the nucleus and distributes its chromosomes to the daughter nuclei and cytokinesis divides the cytoplasm, producing two daughter cells.

mitotic spindle A spindle-shaped structure formed of microtubules and associated proteins that is involved in the movement of chromosomes during mitosis and meiosis. (A spindle is shaped roughly like a football.)

modern synthesis A comprehensive theory of evolution that incorporates genetics and includes most of Darwin's ideas, focusing on populations as the fundamental units of evolution.

molecular biology The study of the molecular basis of genes and gene expression; molecular genetics.

molecule A group of two or more atoms held together by covalent bonds.

mollusc (mol-lusk′) A soft-bodied animal characterized by a muscular foot, mantle, mantle cavity, and radula. Molluscs include gastropods (snails and slugs), bivalves (clams, oysters, and scallops), and cephalopods (squids and octopuses).

molting In arthropods, the process of shedding an old exoskeleton and secreting a new, larger one.

monoclonal antibody (mon′-ō-klōn′-ul) An antibody secreted by a clone of cells and consequently specific for the one antigen that triggered the development of the clone.

monocot (mon′-ō-kot) A flowering plant whose embryos have a single seed leaf, or cotyledon.

monoculture The cultivation of a single plant variety in a large land area.

monohybrid cross An experimental mating of individuals differing at one genetic locus.

monomer (mon′-uh-mer) A chemical subunit that serves as a building block of a polymer.

monosaccharide (mon′-ō-sak′-uh-rīd) The smallest kind of sugar molecule; a single-unit sugar; also known as a simple sugar. Monosaccharides are the building blocks of more complex sugars and polysaccharides.

monotreme (mon′-uh-trēm) An egg-laying mammal, such as the duck-billed platypus.

morning after pill (MAP) A birth control pill taken within three days of unprotected intercourse to prevent fertilization or implantation.

moss Any of a group of seedless nonvascular plants.

motor neuron A nerve cell that conveys command signals from the central nervous system to effector cells, such as muscle cells or gland cells.

motor output The conduction of signals from the central nervous system to effector cells.

motor unit A motor neuron and all the muscle fibers it controls.

mouth An opening through which food is taken into an animal's body.

movement corridor A series of small clumps or a narrow strip of quality habitat (usable by organisms) that connects otherwise isolated patches of quality habitat.

mRNA *See* messenger RNA.

mucous membrane (myū′-kus) Smooth, moist epithelium that lines the digestive tract and air tubes leading to the lungs.

Müllerian mimicry (myū-lār′-ē-un mim′-uh-krē) A mutual mimicry by two species, both of which are poisonous or otherwise harmful to a predator.

muscle tissue Tissue consisting of long muscle cells that are capable of contracting when stimulated by nerve impulses; the most abundant tissue in a typical animal. *See also* skeletal muscle; cardiac muscle; smooth muscle.

muscular system All the skeletal muscles in the body. (Cardiac muscle and smooth muscle are components of other organ systems.)

mutagen (myū′-tuh-jen) A chemical or physical agent that interacts with DNA and causes a mutation.

mutation A change in the nucleotide sequence of DNA; the ultimate source of genetic diversity.

mutualism A symbiotic relationship in which both partners benefit.

mycelium (mī-sē′-lē-um) (plural, **mycelia**) The densely branched network of hyphae in a fungus.

mycorrhiza (mī′-kō-rī′-zuh) (plural, **mycorrhizae**) A mutualistic association of plant roots and fungi.

myelin sheath (mī′-uh-lin) A series of cells, each wound around, and thus insulating, the axon of a nerve cell in vertebrates. Each pair of cells in the sheath is separated by a space called a node of Ranvier.

myofibril (mī′-ō-f ī′-bril) A contractile thread in a muscle cell (fiber) made up of many sarcomeres. Longitudinal bundles of myofibrils make up a muscle fiber.

N

NAD^+ Nicotinamide adenine dinucleotide; a coenzyme that assists enzymes by conveying electrons (from hydrogen atoms) during the redox reactions of cellular metabolism. The plus sign indicates that the molecule is oxidized and ready to pick up hydrogens; the reduced, hydrogen (electron)-carrying form is NADH.

NADH A molecule that carries electrons from glucose and other fuel molecules and deposits them at the top of an electron transport chain. NADH is generated during glycolysis and the citric acid cycle.

NADPH An electron carrier involved in photosynthesis. Light drives electrons from chlorophyll to $NADP^+$, forming NADPH, which provides the high-energy electrons for the reduction of carbon dioxide to sugar in the Calvin cycle.

natural family planning A form of contraception that relies on refraining from sexual intercourse when conception is most likely to occur; also called the rhythm method.

natural killer cell (NK cell) A nonspecific defensive cell that attacks cancer cells and infected body cells, especially those harboring viruses.

natural selection Differential success in reproduction by different phenotypes resulting from interactions with the environment. Evolution occurs when natural selection produces changes in the relative frequencies of alleles in a population's gene pool.

nearsightedness An inability to focus on distant objects; occurs when the eyeball is longer than normal and the lens focuses distant objects in front of the retina; also called myopia.

negative feedback A control mechanism in which a chemical reaction, metabolic pathway, or hormone-secreting gland is inhibited by the products of the reaction, pathway, or gland. As the concentration of the products builds up, the product molecules themselves inhibit the process that produced them.

negative pressure breathing A breathing system in which air is pulled into the lungs.

nematode (nem′-uh-tōd) An animal characterized by a pseudocoelom, a cylindrical, wormlike body form, and a complete digestive tract; also called a roundworm.

nephron The tubular excretory unit and associated blood vessels of the vertebrate kidney. The nephron extracts filtrate from the blood and refines it into urine.

nerve A cable-like bundle of neuron fibers (axons and dendrites) tightly wrapped in connective tissue.

nervous system The organ system that forms a communication and coordination network throughout an animal's body.

nervous tissue Tissue made up of neurons and supportive cells.

neuron (nyūr′-on) A nerve cell; the fundamental structural and functional unit of the nervous system, specialized for carrying signals from one location in the body to another.

neurosecretory cell A nerve cell that synthesizes hormones and secretes them into the blood, as well as conducting nerve signals.

neurotransmitter A chemical messenger that carries information from a transmitting neuron to a receiving cell, either another neuron or an effector cell.

neutron An electrically neutral particle (a particle having no electrical charge), found in the nucleus of an atom.

neutrophil (nyū′-truh-fil) Phagocytic white blood cell that can engulf bacteria and viruses in infected tissue; part of the body's nonspecific defense system.

niche (nich) A population's role in its community; the sum total of a population's use of the biotic and abiotic resources of its habitat.

nitrogen fixation The conversion of atmospheric nitrogen (N_2) to nitrogen compounds (NH_4, NO_3) that plants can absorb and use.

nitrogenous base (nī-troj′-en-us) An organic molecule that is a base and that contains the element nitrogen.

node The point of attachment of a leaf on a stem.

node of Ranvier (ron′-vē-ā) An unmyelinated region on a myelinated axon of a nerve cell, where signal transmission occurs.

nondisjunction An accident of meiosis or mitosis in which a pair of homologous chromosomes or a pair of sister chromatids fail to separate at anaphase.

nonpolar covalent bond An attraction between atoms that share one or more pairs of electrons equally because the atoms have similar electronegativity.

nonself molecule A foreign antigen; a protein or other macromolecule that is not part of an organism's body. *See also* self protein.

norepinephrine (nor′-ep-uh-nef′-rin) An amine hormone (also called noradrenaline) secreted by the adrenal medulla that prepares body organs for fight or flight; also serves as a neurotransmitter.

notochord (nō′-tuh-kord) A flexible, cartilage-like, longitudinal rod located between the digestive tract and nerve cord in chordate animals, present only in embryos in many species.

nuclear envelope A double membrane, perforated with pores, that encloses the nucleus and separates it from the rest of the eukaryotic cell.

nuclear transplantation A technique in which the nucleus of one cell is placed into another cell that already has a nucleus or in which the nucleus has been previously destroyed.

nucleic acid (nū-klā′-ik) A polymer consisting of many nucleotide monomers; serves as a blueprint for proteins and, through the actions of proteins, for all cellular structures and activities. The two types of nucleic acids are DNA and RNA.

nucleic acid probe (nū-klā′-ik) In DNA technology, a labeled single-stranded nucleic acid molecule used to find a specific gene or other nucleotide sequence within a mass of DNA. The probe hydrogen-bonds to the complementary sequence in the targeted DNA.

nucleoid region (nū′-klē-oyd) The region in a prokaryotic cell consisting of a concentrated mass of DNA.

nucleolus (nū-klē′-ō-lus) A structure within the nucleus of a eukaryotic cell where ribosomal RNA is made and assembled with proteins to make ribosomal subunits; consists of parts of the chromatin DNA, RNA transcribed from the DNA, and proteins imported from the cytoplasm.

nucleosome (nū′-klē-ō-sōm) The bead-like unit of DNA packing in a eukaryotic cell; consists of DNA wound around a protein core made up of eight histone molecules.

nucleotide (nū′-klē-ō-tīd) An organic monomer consisting of a five-carbon sugar covalently bonded to a nitrogenous base and a phosphate group. Nucleotides are the building blocks of nucleic acids.

nucleus (plural, **nuclei**) (1) An atom's central core, containing protons and neutrons. (2) The genetic control center of a eukaryotic cell.

O

obesity An excessively high body mass index, a ratio of weight to height.

obligate aerobe (ob′-li-get ār′-ōb) An organism that cannot survive without oxygen (O_2).

obligate anaerobe (ob′-li-get an′-uh-rōb) An organism that cannot survive in the presence of oxygen (O_2).

omnivore An animal that eats both plants and animals. *See also* carnivore; herbivore.

oncogene (on′-kō-jēn) A cancer-causing gene; usually contributes to malignancy by abnormally enhancing the amount or activity of a growth factor made by the cell.

oogenesis (ō′-uh-jen′-uh-sis) The formation of ova (egg cells) within the ovaries.

open circulatory system A circulatory system in which blood is pumped through open-ended vessels and out among the body cells. In an animal with an open circulatory system, blood and interstitial fluid are one and the same.

open system Any system that exchanges chemicals and energy with its surroundings. All organisms are open systems.

operator In prokaryotic DNA, a sequence of nucleotides near the start of an operon to which an active repressor can attach. The binding of repressor prevents RNA polymerase from attaching to the promoter and transcribing the genes of the operon.

operculum (ō-per′-kyū-lum) (plural, **opercula**) A protective flap on each side of a bony fish's head that covers a chamber housing the gills.

operon (op′-er-on) A unit of genetic regulation common in prokaryotes; a cluster of genes with related functions, along with the promoter and operator that control their transcription.

opportunistic life history Often seen in small-bodied species, the pattern of reproducing when young and producing many offspring.

optic nerve Either of the second pair of cranial nerves that arise from the retina and carry visual information to the thalamus and other parts of the brain.

oral cavity *See* mouth.

order In classification, the taxonomic category above family.

organ A structure consisting of several tissues adapted as a group to perform specific functions.

organ of Corti (kor′-tē) The hearing organ in birds and mammals, located within the cochlea.

organ system A group of organs that work together in performing vital body functions.

organelle (ōr-guh-nel′) A structure with a specialized function within a cell.

organic chemistry The study of carbon compounds.

organic compound A chemical compound containing the element carbon and usually synthesized by cells.

organism An individual living thing, such as a bacterium, fungus, protist, plant, or animal.

organismal ecology The study of the evolutionary adaptations that enable individual organisms to meet the challenges posed by their abiotic environments.

orgasm Rhythmic contractions of the reproductive structures, accompanied by extreme pleasure, at the peak of sexual excitement in both sexes; includes ejaculation by the male.

osmoconformer (oz′-mō-con-form′-er) An organism whose body fluids have a solute concentration equal to that of its surroundings. Osmoconformers do not have a net gain or loss of water by osmosis.

osmoregulation The control of the gain or loss of water and dissolved solutes in an organism.

osmoregulator An organism whose body fluids have a solute concentration different from that of its environment and that must use energy in controlling water loss or gain.

osmosis (oz-mō′-sis) The passive transport of water across a selectively permeable membrane.

osteichthyan (os-tē-ik′-thē-un) Member of the vertebrate class of bony fishes that includes trout and goldfish.

osteoporosis (os′-tē-ō-puh-rō′-sis) A skeletal disorder characterized by thinning, porous, and easily broken bones; most common among women after menopause and often related to low estrogen levels.

outer ear One of three main regions of the ear in reptiles (including birds) and mammals; made up of the auditory canal and, in many birds and mammals, the pinna.

ovarian cycle (ō-vār′-ē-un) Hormonally synchronized cyclic events in the mammalian ovary, culminating in ovulation.

ovary (1) In animals, the female gonad, which produces egg cells and reproductive hormones. (2) In flowering plants, the basal portion of a carpel in which the egg-containing ovules develop.

oviduct (ō′-vuh-dukt) The tube that conveys egg cells away from an ovary; also called a fallopian tube.

ovulation (ah′-vyū-lā′-shun) The release of an egg cell from an ovarian follicle.

ovule (ō′-vyūl) A reproductive structure in a seed plant, containing the female gametophyte and the developing egg. An ovule develops into a seed.

ovum (ō′-vum) (plural, **ova**) An unfertilized egg, or female gamete.

oxidation The loss of electrons from a substance involved in a redox reaction; always accompanies reduction.

oxytocin (ok′-si-tō′-sin) A peptide hormone, made by the hypothalamus and secreted by the posterior pituitary, that stimulates contraction of the uterus and mammary gland cells.

ozone layer The layer of O_3 in the upper atmosphere that protects life on Earth from the harmful ultraviolet rays in sunlight.

P

P generation The parent individuals from which offspring are derived in studies of inheritance. P stands for parental.

pacemaker The SA (sinoatrial) node; a specialized region of cardiac muscle that maintains the heart's pumping rhythm (heartbeat) by setting the rate at which the heart contracts.

paedomorphosis (pē′-duh-mōr′-fuh-sis) The retention in the adult of features that were juvenile in its ancestors.

pain receptor A sensory receptor that detects painful stimulus.

paleoanthropology (pā′-lē-ō-an′-thruh-pol′-uh-jē) The study of human origins and evolution.

pancreas (pan′-krē-us) A gland with dual functions: The nonendocrine portion secretes digestive enzymes and an alkaline solution into the small intestine via a duct; the endocrine portion secretes the hormones insulin and glucagon into the blood.

pandemic A worldwide outbreak of a disease.

Pangaea (pan-jē′-uh) The supercontinent consisting of all the major landmasses of Earth fused together. Continental drift formed Pangaea near the end of the Paleozoic era.

parasite An organism that benefits at the expense of another organism, the host, which is harmed in the process.

parasitism (pār′-uh-sit-izm) A symbiotic relationship in which the parasite, a type of predator, lives within or on the surface of a host, from which it derives its food.

parasympathetic division One of two sets of neurons in the autonomic nervous system; generally promotes body activities that gain and conserve energy, such as digestion and reduced heart rate. *See also* sympathetic division.

parathyroid gland (pār′-uh-thī′-royd) One of four endocrine glands embedded in the surface of the thyroid gland that secrete parathyroid hormone.

parathyroid hormone (PTH) A peptide hormone secreted by the parathyroid glands that raises blood calcium level.

parenchyma cell (puh-reng′-kuh-muh) In plants, a relatively unspecialized cell with a thin primary wall and no secondary wall; functions in photosynthesis, food storage, and aerobic respiration and may differentiate into other cell types.

passive immunity Temporary immunity obtained by acquiring ready-made antibodies or immune cells; lasts only a few weeks or months because the immune system has not been stimulated by antigens.

passive transport The diffusion of a substance across a biological membrane, without any input of energy.

pathogen A disease-causing organism.

pedigree A family tree representing the occurrence of heritable traits in parents and offspring across a number of generations.

pelagic zone (puh-laj′-ik) The region of an ocean occupied by seawater; the open ocean.

penis The copulatory structure of male mammals.

pepsin An enzyme present in gastric juice that begins the hydrolysis of proteins.

peptide bond The covalent linkage between two amino acid units in a polypeptide; formed by a dehydration reaction.

peptidoglycan (pep′-tid-ō-glī′-kan) A polymer of complex sugars cross-linked by short polypeptides; a material unique to eubacterial cell walls.

perennial (puh-ren′-ē-ul) A plant that lives for many years.

perforin (per′-fuh-rin) A protein secreted by a cytotoxic T cell that lyses (ruptures) an infected cell by perforating its membrane.

peripheral nervous system (PNS) The network of nerves and ganglia carrying signals into and out of the central nervous system.

peristalsis (pār′-uh-stal′-sis) Rhythmic waves of contraction of smooth muscles. Peristalsis propels food through a digestive tract and also enables many animals, such as earthworms, to crawl.

permafrost Continuously frozen ground found in the tundra.

PET *See* positron-emission tomography.

petal A modified leaf of a flowering plant. Petals are the often colorful parts of a flower that advertise it to insects and other pollinators.

petiole (pet′-ē-ōl) The stalk of a leaf, which joins the leaf to a node of the stem.

pH scale A measure of the relative acidity of a solution, ranging in value from 0 (most acidic) to 14 (most basic). pH stands for potential hydrogen and refers to the concentration of hydrogen ions (H^+).

phage (fāj) *See* bacteriophage.

phagocyte (fag′-ō-sīt) A white blood cell (e.g., a neutrophil or a monocyte) that engulfs bacteria, foreign proteins, and the remains of dead body cells.

phagocytosis (fag′-ō-sī-tō′-sis) Cellular "eating"; a type of endocytosis whereby a cell engulfs macromolecules, other cells, or particles into its cytoplasm.

pharyngeal slit (fā-rin′-jē-ul) A gill structure in the pharynx, found in chordate embryos and some adult chordates.

pharynx (fār′-inks) The organ in a digestive tract that receives food from the oral cavity; in terrestrial vertebrates, the throat region where the air and food passages cross.

phenotype (fē′-nō-tīp) The expressed traits of an organism.

phenylketonuria (PKU) (fen′-ul-kē′-tuh-nūr′-ē-uh) A recessive genetic disorder characterized by an inability to properly break down the amino acid phenylalanine; if untreated results in mental retardation.

phloem (flō′-um) The portion of a plant's vascular system that conveys sugars, nutrients, and hormones throughout a plant; made up of food-conducting cells.

phloem sap The solution of sugars, other nutrients, and hormones conveyed throughout a plant via phloem tissue.

phosphate group (fos′-fāt) A functional group consisting of a phosphorus atom covalently bonded to four oxygen atoms.

phospholipid (fos′-fō-lip′-id) A molecule that is a constituent of the inner bilayer of biological membranes, having a polar, hydrophilic head and a nonpolar, hydrophobic tail.

phospholipid bilayer A double layer of phospholipid molecules (each molecule consisting of a phosphate group bonded to two fatty acids) that is the primary component of all cellular membranes.

phosphorylation (fos′-fōr-uh-lā′-shun) The transfer of a phosphate group, usually from ATP, to a molecule. Nearly all cellular work depends on ATP energizing other molecules by phosphorylation.

photic zone (fō′-tik) The region of an aquatic ecosystem into which light penetrates and where photosynthesis occurs.

photoautotroph An organism that obtains energy from sunlight and carbon from CO_2 by photosynthesis.

photoheterotroph An organism that obtains energy from sunlight and carbon from organic sources.

photon (fō′-ton) A fixed quantity of light energy. The shorter the wavelength of light, the greater the energy of a photon.

photoperiod The length of the day relative to the length of the night; an environmental stimulus that plants use to detect the time of year.

photopsin (fō-top′-sin) One of a family of visual pigments in the cones of the vertebrate eye that absorb bright, colored light.

photoreceptor A type of electromagnetic receptor that detects light.

photosynthesis (fō′-tō-sin′-thuh-sis) The process by which plants, autotrophic protists, and some bacteria use light energy to make sugars and other organic food molecules from carbon dioxide and water.

photosystem A light-harvesting unit of a chloroplast's thylakoid membrane; consists of several hundred antenna molecules, a reaction-center chlorophyll, and a primary electron acceptor.

phototropism (fō′-tō-trō′-pizm) The growth of a plant shoot in response to light.

phylogenetic tree (fī′-lō-juh-net′-ik) A branching diagram that represents a hypothesis about evolutionary relationships among organisms.

phylogeny (fī-loj′-uh-nē) The evolutionary history of a species or group of related species.

phylum (fī′-lum) (plural, **phyla**) In classification, the taxonomic category above class and below kingdom. Members of a phylum all have a similar general body plan.

physiology (fi′-zi-ol′-uh-ji) The study of the function of an organism's structural equipment.

phytoplankton (fī′-tō-plank′-ton) Algae and photosynthetic bacteria that drift passively in aquatic environments.

pili (pī′-lī) (singular, **pilus**) Short projections on the surface of prokaryotic cells that help prokaryotes attach to other surfaces; specialized sex pili are used in conjugation to hold the mating cells together.

pinna (pin′-uh) The flap-like part of the outer ear, projecting from the body surface of many birds and mammals. The pinna collects sound waves and channels them to the auditory canal.

pinocytosis (pī′-nō-sī-tō′-sis) Cellular "drinking"; a type of endocytosis in which the cell takes fluid and dissolved solutes into small membranous vesicles.

pith Part of the ground tissue system of a dicot plant. Pith fills the center of a stem and may store food.

pituitary gland (puh-tū′-uh-tār′-ē) An endocrine gland at the base of the hypothalamus; consists of a posterior lobe, which stores and releases two hormones produced by the hypothalamus, and an anterior lobe, which produces and secretes many hormones that regulate diverse body functions.

pivot joint A joint that allows precise rotations in multiple planes. An example in humans is the wrist.

placenta (pluh-sen′-tuh) In most mammals, the organ that provides nutrients and oxygen to the embryo and helps dispose of its metabolic wastes; formed of the embryo's chorion and the mother's endometrial blood vessels.

placental mammal (pluh-sen′-tul) Mammal whose young complete their embryonic development in the uterus, nourished via the mother's blood vessels in the placenta; also called eutherian.

plankton Algae and other organisms, mostly microscopic, that drift passively in ponds, lakes, and oceans.

plant A multicellular eukaryote that carries out photosynthesis.

Plantae (plan′-tā) The kingdom that contains the plants.

plasma The liquid matrix of the blood in which the blood cells are suspended.

plasma cell An antibody-secreting B cell.

plasma membrane The thin layer of lipids and proteins that sets a cell off from its surroundings and acts as a selective barrier to the passage of ions and molecules into and out of the cell; consists of a phospholipid bilayer in which are embedded molecules of protein and cholesterol.

plasmid A small ring of DNA separate from the chromosome(s). Plasmids are found in prokaryotes and yeasts.

plasmodesma (plaz′-mō-dez′-muh) (plural, **plasmodesmata**) An open channel in a plant cell wall, through which strands of cytoplasm connect from adjacent walls.

plasmodial slime mold (plaz-mō′-dē-ul) A type of protist that has amoeboid cells, flagellated cells, and an amoeboid plasmodial feeding stage in its life cycle.

plasmodium (1) A single mass of cytoplasm containing many nuclei. (2) The amoeboid feeding stage in the life cycle of a plasmodial slime mold.

plasmolysis (plaz-mol′-uh-sis) A phenomenon that occurs in plant cells in a hypertonic environment. The cell loses water and shrivels, and its plasma membrane pulls away from the cell wall, usually killing the cell.

plate tectonics Forces within planet Earth that cause movements of the crust, resulting in continental drift, volcanoes, and earthquakes.

platelet A piece of membrane-enclosed cytoplasm from a large cell in the bone marrow of a mammal; a blood-clotting element, also called a thrombocyte.

Platyhelminthes (plat′-ē-hel-min′-thēz) The phylum that contains the flatworms, the bilateral animals with a thin, flat body form, gastrovascular cavity or no digestive system, and no body cavity; the free-living flatworms, flukes, and tapeworms.

pleated sheet The folded arrangement of a polypeptide in a protein's secondary structure.

pleiotropy (plī′-uh-trō-pē) The control of more than one phenotypic characteristic by a single gene.

polar body The smaller of two daughter cells produced during meiosis of oogenesis.

polar covalent bond An attraction between atoms that share electrons unequally. The shared electrons are pulled closer to one atom, making it partially negative and the other atom partially positive.

polar molecule A molecule containing polar covalent bonds, (having opposite charges on opposite ends).

pollen *See* pollen grain.

pollen grain In a seed plant, the male gametophyte that develops within the anther of a stamen.

pollination In seed plants, the delivery, by wind or animals, of pollen from the male parts of a plant to the stigma of a carpel on the female.

polygenic inheritance (pol′-ē-jen′-ik) The additive effect of two or more gene loci on a single phenotypic characteristic.

polymer (pol′-uh-mer) A large molecule consisting of many identical or similar molecular units, called monomers, covalently joined together in a chain.

polymerase chain reaction (PCR) (puh-lim′-uh-rās) A technique used to obtain many copies of a DNA molecule or many copies of part of a DNA molecule. A small amount of DNA mixed with the enzyme DNA polymerase, DNA nucleotides, and a few other ingredients replicates repeatedly in a test tube.

polymorphic (pol′-ē-mōr′-fik) Referring to a population in which two or more physical forms are present in readily noticeable frequencies.

polymorphism (pol′-ē-mōr′-fizm) The coexistence of two or more distinct forms of individuals (polymorphic characters) in the same population.

polynucleotide (pol′-ē-nū′-klē-ō-tīd) A polymer made up of many nucleotides covalently bonded together.

polyp (pol′-ip) One of two types of cnidarian body forms; a columnar, hydra-like body.

polypeptide A chain of amino acids linked by peptide bonds.

polyploid (pol′-ē-ploid) Containing more than two complete sets of chromosomes in each somatic cell.

polysaccharide (pol′-ē-sak′-uh-rīd) A carbohydrate polymer consisting of hundreds to thousands of monosaccharides (sugars) linked by covalent bonds.

pons (pahnz) Part of the vertebrate hindbrain that functions with the medulla oblongata in passing data between the spinal cord and forebrain and in controlling autonomic, homeostatic functions.

population A group of interacting individuals belonging to one species and living in the same geographic area.

population density The number of individuals of a species per unit area or volume.

population ecology The study of how members of a population interact with their environment, focusing on factors that influence population density and growth.

population fragmentation The splitting and consequent isolation of portions of a biological population, usually by human-caused habitat degradation.

population genetics The study of genetic changes in populations; the science of microevolutionary changes in populations.

population-limiting factor An environmental factor that restricts population growth.

positive feedback A control mechanism in which the products of a process stimulate the process that produced them.

positron-emission tomography (PET) Imaging technology that uses radioactively labeled biological molecules, such as glucose, to obtain information about metabolic processes at specific locations in the body. The labeled molecules are injected into the bloodstream, and a PET scan for radioactive emissions determines which tissues have taken up the molecules.

post-anal tail A tail posterior to the anus, found in chordate embryos and most adult chordates.

posterior Pertaining to the rear, or tail, of a bilaterally symmetrical animal.

posterior pituitary An extension of the hypothalamus composed of nervous tissue that secretes hormones made in the hypothalamus; a temporary storage site for hypothalamic hormones.

postzygotic barrier (pōst′-zī-got′-ik) A reproductive barrier that operates should interspecies mating occur and form hybrid zygotes.

potential energy Stored energy; the capacity to perform work that matter possesses because of its location or arrangement. Water behind a dam and chemical bonds both possess potential energy.

predation An interaction between species in which one species, the predator, eats the other, the prey.

predator A consumer in a biological community.

prepuce (prē′-pyūs) A fold of skin covering the head of the clitoris or penis.

pressure-flow mechanism The method by which phloem sap is transported through a plant from a sugar source, where sugars are produced, to a sugar sink, where sugars are used.

prey An organism eaten by a predator.

prezygotic barrier (prē′-zī-got′-ik) A reproductive barrier that impedes mating between species or hinders fertilization of eggs if members of different species should attempt to mate.

primary cell wall A relatively thin and flexible layer first secreted by a young plant cell.

primary consumer An organism that eats only autotrophs; an herbivore.

primary growth Growth in the length of a plant root or shoot produced by an apical meristem.

primary immune response The initial immune response to an antigen, which appears after a lag of several days.

primary oocyte (ō′-uh-sīt) A diploid cell, in prophase I of meiosis, that can be hormonally triggered to develop into an ovum.

primary production The amount of solar energy converted to chemical energy (organic compounds) by autotrophs in an ecosystem during a given time period.

primary productivity The rate at which an ecosystem's plants and other producers build biomass, or organic matter.

primary spermatocyte (sper-mat′-eh-sīt′) A diploid cell in the testis that undergoes meiosis I.

primary structure The first level of protein structure; the specific sequence of amino acids making up a polypeptide chain.

primary succession A type of ecological succession in which a biological community arises in an area without soil. *See also* secondary succession.

primate Member of the mammalian group that includes lorises, pottos, lemurs, tarsiers, monkeys, apes, and humans.

probe In DNA technology, a labeled single-stranded nucleic acid molecule used to find a specific gene, or other nucleotide sequence, within a mass of DNA. The probe hydrogen-bonds to the complementary sequence in the targeted DNA.

producer An organism that makes organic food molecules from CO_2, H_2O, and other inorganic raw materials: a plant, alga, or autotrophic bacterium.

product An ending material in a chemical reaction.

progesterone (prō-jes′-teh-rōn) A steroid hormone secreted by the corpus luteum of the ovary; maintains the uterine lining during pregnancy.

progestin (prō-jes′-tin) One of a family of steroid hormones, including progesterone, produced by the mammalian ovary. Progestins prepare the uterus for pregnancy.

programmed cell death The timely death (and disposal of the remains) of certain cells, triggered by certain genes; an essential process in normal development; also called apoptosis.

prokaryote (prō-kār′-ē-ōt) An organism characterized by prokaryotic cells. *See also* prokaryotic cell.

prokaryotic cell (prō-kār′-ē-ot′-ik) A type of cell lacking a membrane-enclosed nucleus and other membrane-enclosed organelles; found only in the domains Bacteria and Archaea.

prokaryotic cell wall A fairly rigid, chemically complex wall that protects the prokaryotic cell and helps maintain its shape.

prokaryotic flagellum (plural, **flagella**) A long surface projection that propels a prokaryotic cell through its liquid environment; totally different from the flagellum of a eukaryotic cell.

prolactin (PRL) (pro-lak′-tin) A protein hormone secreted by the anterior pituitary that stimulates milk production in mammals.

promoter A specific nucleotide sequence in DNA, located at the start of a gene, that is the binding site for RNA polymerase and the place where transcription begins.

prophage (prō′-fāj) Phage DNA that has inserted by genetic recombination into the DNA of a prokaryotic chromosome.

prophase The first stage of mitosis, during which duplicated chromosomes condense to form structures visible with a light microscope and the mitotic spindle forms and begins moving the chromosomes toward the center of the cell.

prostaglandin (pros′-tuh-glan′-din) One of a large family of local regulators secreted by virtually all tissues and performing a wide variety of regulatory functions.

prostate (pros′-tāt) A gland in human males that secretes an acid-neutralizing component of semen.

protein A biological polymer constructed from amino acid monomers.

proteomics The systematic study of the full protein sets (proteomes) encoded by genomes.

protist (prō′-tist) Any eukaryote that is not a plant, animal, or fungus.

proton A subatomic particle with a single unit of positive electrical charge, found in the nucleus of an atom.

proto-oncogene (prō′-tō-on′-kō-jēn) A normal gene that can be converted to a cancer-causing gene.

protozoan (prō′-tō-zō′-un) A protist that lives primarily by ingesting food; a heterotrophic, animal-like protist.

provirus Viral DNA that inserts into a host genome.

pseudocoelom (sū′-dō-sē′-lōm) A body cavity that is in direct contact with the wall of the digestive tract.

pseudopodium (sū′-dō-pō′-dē-um) (plural, **pseudopodia**) A temporary extension of an amoeboid cell. Pseudopodia function in moving cells and engulfing food.

pulmonary artery A large blood vessel that conveys blood from the heart to a lung.

pulmonary circuit One of two main blood circuits in terrestrial vertebrates; conveys blood between the heart and the lungs. *See also* systemic circuit.

pulmonary vein A blood vessel that conveys blood from a lung to the heart.

pulse The rhythmic stretching of the arteries caused by the pressure of blood forced through the arteries by contractions of the ventricles during systole.

punctuated equilibrium The idea that speciation occurs in spurts followed by long periods of little change.

Punnett square A diagram used in the study of inheritance to show the results of random fertilization.

pupil The opening in the iris that admits light into the interior of the vertebrate eye. Muscles in the iris regulate its size.

pyrogen A chemical, released by certain leukocytes, that sets the body's thermostat to a higher temperature.

Q

quaternary consumer (kwot′-er-nār-ē) An organism that eats tertiary consumers.

quaternary structure The fourth level of protein structure; the shape resulting from the association of two or more polypeptide subunits.

R

radial symmetry An arrangement of the body parts of an organism like pieces of a pie around an imaginary central axis. Any slice passing longitudinally through a radially symmetrical organism's central axis divides it into mirror-image halves.

radiation therapy Treatment for cancer in which parts of the body that have cancerous tumors are exposed to high-energy radiation to disrupt cell division of the cancer cells.

radioactive isotope An isotope whose nucleus decays spontaneously, giving off particles and energy.

radiometric dating A method for determining the age of fossils and rocks from the ratio of a radioactive isotope to the nonradioactive istope(s) of the same element in the sample.

radula (rad′-yū-luh) A toothed, rasping organ found in many molluscs, used to scrape up or shred food.

ray-finned fish A bony fish having fins supported by thin, flexible skeletal rays. All but one living species of bony fishes are ray-fins. *See* lobe-finned fish.

reabsorption In the vertebrate kidney, the reclaiming of water and valuable solutes from the filtrate.

reactant A starting material in a chemical reaction.

reaction center In a photosystem in a chloroplast, the chlorophyll *a* molecule and the primary electron acceptor that trigger the light reactions of photosynthesis. The chlorophyll donates an electron excited by light energy to the primary electron acceptor, which passes an electron to an electron transport chain.

reading frame The way a cell's mRNA-translating machinery groups the mRNA nucleotides into codons.

receptor On or in a cell, a specific protein molecule whose shape fits that of a specific molecular messenger, such as a hormone.

receptor potential The change in membrane potential that results from sensory transduction.

receptor-mediated endocytosis (en′-dō-sī-tō′-sis) The movement of specific molecules into a cell by the inward budding of membranous vesicles. The vesicles contain proteins with receptor sites specific to the molecules being taken in.

recessive allele In a heterozygous individual, the allele that has no noticeable effect on the phenotype.

recombinant DNA A DNA molecule carrying genes derived from two or more sources.

recombinant DNA technology A set of techniques for synthesizing recombinant DNA in vitro and transferring it into cells, where it can be replicated and may be expressed; also known as genetic engineering.

recombination frequency With respect to two given genes, the number of recombinant progeny from a mating divided by the total number of progeny. Recombinant progeny carry combinations of alleles different from those in either of the parents as a result of independent assortment of chromosomes and crossing over.

recommended daily allowance (RDA) A recommendation for daily nutrient intake established by nutritionists.

rectum The terminal portion of the large intestine, where the feces are stored until they are eliminated.

red blood cell A blood cell containing hemoglobin, which transports O_2; also called an erythrocyte.

red bone marrow A specialized tissue found in cavities in the ends of bones that produces blood cells.

red-green color blindness A category of common sex-linked human disorders involving several genes on the X chromosome and characterized by a malfunction of light-sensitive cells in the eyes; affects mostly males but also homozygous females.

redox reaction Short for oxidation-reduction reaction; a chemical reaction in which electrons are lost from one substance (oxidation) and added to another (reduction). Oxidation and reduction always occur together.

reduction The gain of electrons by a substance involved in a redox reaction; always accompanies oxidation.

regeneration The regrowth of body parts from pieces of an organism.

regulatory gene A gene that codes for a protein, such as a repressor, that controls the transcription of another gene or group of genes.

releasing hormone A hormone, secreted by the hypothalamus, that makes the anterior pituitary secrete hormones.

repetitive DNA Nucleotide sequences that are present in many copies in the DNA of a genome. The repeated sequences may be long or short and may be located next to each other or dispersed in the DNA.

repressor A protein that blocks the transcription of a gene or operon.

reproduction The creation of new individuals from existing ones.

reproductive barrier A biological feature of a species that prevents it from interbreeding with other species even when populations of the two species live together.

reproductive cloning Using a somatic cell from a multicellular organism to make one or more genetically identical individuals.

reproductive cycle In females, a recurring series of events that produces gametes, makes them available for fertilization, and prepares the body for pregnancy.

reproductive system The body organ system responsible for reproduction.

reptile Member of the clade of amniotes that includes snakes, lizards, turtles, crocodiles, alligators, birds, and a number of extinct groups (most of the dinosaurs).

resolving power A measure of the clarity of an image; the ability of an optical instrument to show two objects as separate.

resource partitioning The division of environmental resources by coexisting species such that the niche of each species differs by one or more significant factors from the niches of all coexisting species.

respiration (1) Gas exchange, or breathing; the exchange of O_2 and CO_2 between an organism and its environment. An aerobic organism takes up O_2 and gives off CO_2. (2) Cellular respiration; the aerobic harvest of energy from food molecules by cells.

respiratory surface The part of an animal where gases are exchanged with the environment.

respiratory system The organ system that functions in exchanging gases with the environment; it supplies the blood with O_2 and disposes of CO_2.

resting potential The voltage across the plasma membrane of a resting neuron.

restriction enzyme A bacterial enzyme that cuts up foreign DNA, thus protecting bacteria against intruding DNA from phages and other organisms. Restriction enzymes are used in DNA technology to cut DNA molecules in reproducible ways.

restriction fragments Molecules of DNA produced from a longer DNA molecule cut up by a restriction enzyme; used in genome mapping and other applications.

restriction site A specific sequence on a DNA strand that is recognized as a "cut site" by a restriction enzyme.

retina (ret′-uh-nuh) The light-sensitive layer in an eye, made up of photoreceptor cells and sensory neurons.

retrovirus An RNA virus that reproduces by means of a DNA molecule. It reverse-transcribes its RNA into DNA, inserts the DNA into a cellular chromosome, and then transcribes more copies of the RNA from the viral DNA. HIV and a number of cancer-causing viruses are retroviruses.

reverse transcriptase (tran-skrip′-tās) An enzyme that catalyzes the synthesis of DNA on an RNA template.

RFLP analysis A common method of DNA fingerprinting, the comparison of the set of restriction fragments produced by DNA from different individuals.

RFLPs (rif′-lips) Restriction fragment length polymorphisms; the differences in homologous DNA sequences that are reflected in different lengths of restriction fragments produced when the DNA is cut up with restriction enzymes.

rheumatoid arthritis An autoimmune disease in which the joints become highly inflamed.

rhizome (rī′-zōm) A horizontal stem that grows below the ground.

rhodopsin (rō-dop′-sin) A visual pigment in the rods of the vertebrate eye that absorbs dim light.

rhythm method A form of contraception that relies on refraining from sexual intercourse when conception is most likely to occur; also called natural family planning.

ribosomal RNA (rRNA) (rī′-buh-sōm′-ul) The type of ribonucleic acid that, together with proteins, makes up ribosomes; the most abundant type of RNA.

ribosome (rī′-buh-sōm) A cell organelle consisting of RNA and protein organized into two subunits and functioning as the site of protein synthesis in the cytoplasm. The ribosomal subunits are constructed in the nucleolus.

ribozyme (rī′-bō-zīm) An enzymatic RNA molecule that catalyzes chemical reactions.

RNA Ribonucleic acid (rī′-bō-nū-klā′-ik). A type of nucleic acid consisting of nucleotide monomers with a ribose sugar and the nitrogenous bases adenine (A), cytosine (C), guanine (G), and uracil (U); usually single-stranded; functions in protein synthesis and as the genome of some viruses.

RNA polymerase (puh-lim′-uh-rās) An enzyme that links together the growing chain of RNA nucleotides during transcription, using a DNA strand as a template.

RNA splicing The removal of introns and joining of exons in eukaryotic RNA, forming an mRNA molecule with a continuous coding sequence; occurs before mRNA leaves the nucleus.

RNA world A hypothetical period in the evolution of life when RNA served as rudimentary genes and the sole catalytic molecules.

rod A photoreceptor cell in the vertebrate retina, enabling vision in dim light.

root A plant structure that anchors the plant in the soil, absorbs and transports minerals and water, and stores food.

root cap A cone of cells at the tip of a plant root that protects the root's apical meristem.

root hair An outgrowth of an epidermal cell on a root, which increases the root's absorptive surface area.

root nodule A swelling on a plant root consisting of plant cells that contain nitrogen-fixing bacteria.

root system All of a plant's roots, which anchor it in the soil, absorb and transport minerals and water, and store food.

rough ER (rough endoplasmic reticulum) (reh-tik′-yuh-lum) A network of interconnected membranous sacs in a eukaryotic cell's cytoplasm. Rough ER membranes are studded with ribosomes that make membrane proteins and secretory proteins. The rough ER constructs membrane from phospholipids and proteins.

roundworm A nematode. *See* nematode.

rRNA *See* ribosomal RNA.

rule of multiplication A rule stating that the probability of a compound event is the product of the separate probabilities of the independent events.

S

SA (sinoatrial) node (sī′-nō-ā′-trē-ul) The pacemaker of the heart, located in the wall of the right atrium. At the base of the wall separating the two atria is another patch of nodal tissue called the atrioventricular (AV) node. *See also* pacemaker.

salivary amylase A salivary gland enzyme that hydrolyzes starch.

salt A compound resulting from the formation of ionic bonds, also called an ionic compound.

sapwood Light-colored, water-conducting secondary xylem in a tree. *See also* heartwood.

sarcoma (sar-kō′-muh) Cancer of the supportive tissues, such as bone, cartilage, and muscle.

sarcomere (sar′-kō-mēr) The fundamental unit of muscle contraction, composed of thin filaments of actin and thick filaments of myosin; the region between two narrow, dark lines, called Z lines, in the myofibril.

saturated Pertaining to fats and fatty acids whose hydrocarbon chains contain the maximum number of hydrogens and therefore have no double covalent bonds. Saturated fats and fatty acids solidify at room temperature.

savanna A biome dominated by grasses and scattered trees.

scanning electron microscope (SEM) A microscope that uses an electron beam to study the surface architecture of a cell or other specimen.

science Any method of learning about the natural world that follows the scientific method. *See* discovery science and hypothesis-driven science.

scientific method Scientific investigation involving the observation of phenomena, the formulation of a hypothesis concerning the phenomena, experimentation to demonstrate the truth or falseness of the hypothesis, and results that validate or modify the hypothesis.

sclera (sklār′-uh) A layer of connective tissue forming the outer surface of the vertebrate eye. The cornea is the frontal part of the sclera.

sclerenchyma cell (skli-reng′-kuh-muh) In plants, a supportive cell with a rigid secondary wall hardened with lignin.

scrotum A pouch of skin outside the abdomen that houses a testis; functions in cooling sperm, thereby keeping them viable.

seaweed A large, multicellular marine alga.

secondary cell wall A strong and durable matrix often deposited in several laminated layers for plant cell protection and support.

secondary consumer An organism that eats primary consumers.

secondary growth An increase in a plant's girth, involving cell division in the vascular cambium and cork cambium.

secondary immune response The immune response elicited when an animal encounters the same antigen at some later time. The secondary immune response is more rapid, of greater magnitude, and of longer duration than the primary immune response.

secondary oocyte (ō′-uh-sīt′) A haploid cell that results from meiosis I in oogenesis and that will become an ovum after meiosis II.

secondary spermatocyte (sper-mat′-uh-sīt′) A haploid cell that results from meiosis I in spermatogenesis and that will become a sperm cell after meiosis II.

secondary structure The second level of protein structure; the regular patterns of coils or folds of a polypeptide chain.

secondary succession A type of ecological succession that occurs where a disturbance has destroyed an existing biological community but left the soil intact. *See also* primary succession.

secretion In the vertebrate kidney, the discharge of wastes from the blood into the filtrate from the nephron tubules.

secretory protein A protein that is secreted by a cell, such as an antibody.

seed A plant embryo packaged with a food supply within a protective covering.

seed coat A tough outer covering of a seed, formed from the outer coat of an ovule. In a flowering plant, it encloses and protects the embryo and endosperm.

seed dormancy The temporary suspension of growth and development of a seed.

segmentation Subdivision along the length of an animal body into a series of repeated parts called segments.

selectively permeable (per′-mē-uh-bul) Allowing some substances to cross a biological membrane more easily than others and blocking the passage of other substances altogether.

self protein A protein on the surface of an antigenpresenting cell that can hold a foreign antigen and display it to helper T cells. Each individual has a unique set of self proteins that serve as molecular markers for the body. Lymphocytes do not attack self proteins unless the proteins are displaying foreign antigens; therefore, self proteins mark normal body cells as off-limits to the immune system. *See also* nonself molecule.

self-fertilization The fusion of sperm and egg that are produced by the same individual organism.

self-fertilize To form a zygote through the fusion of sperm and egg produced by the same individual organism.

semen (sē′-mun) The sperm-containing fluid that is ejaculated by the male during orgasm.

seminal vesicle (sem′-uh-nul ves′-uh-kul) A gland in males that secretes a fluid component of semen that lubricates and nourishes sperm.

seminiferous tubule (sem′-uh-nif′-uh-rus) A coiled sperm-producing tube in a testis.

sensory adaptation The tendency of sensory neurons to become less sensitive when they are stimulated repeatedly.

sensory division The afferent neurons of the PNS that convey information to the CNS from the sensory receptors that monitor the external and internal environments.

sensory input The conduction of signals from sensory receptors to processing centers in the central nervous system.

sensory neuron A nerve cell that receives information from sensory receptors and conveys signals into the central nervous system.

sensory transduction The conversion of a stimulus signal to an electrical signal by a sensory receptor cell.

sepal (sē′-pul) A modified leaf of a flowering plant. A whorl of sepals encloses and protects the flower bud before it opens.

sex chromosome A chromosome that determines whether an individual is male or female.

sex-linked gene A gene located on a sex chromosome.

sexual reproduction The creation of offspring by the fusion of two haploid sex cells (gametes), forming a diploid zygote.

sexually transmitted disease (STD) A contagious disease spread by sexual contact.

shoot The stem and leaves of a plant.

shoot system All of a plant's stems, leaves, and reproductive structures.

short tandem repeats (STRs) Short sequences of DNA repeated many times in a row (tandomly).

sickle-cell disease A genetic disorder in which the red blood cells have abnormal hemoglobin molecules and take on an abnormal shape.

signal transduction pathway A series of molecular changes that converts a signal on a target cell's surface to a specific response inside the cell.

silencer A eukaryotic DNA sequence that inhibits the start of gene transcription; may act analogously to an enhancer, binding a repressor.

single-lens eye The cameralike eye found in some jellies, polychaetes, spiders, and many molluscs.

sink habitat An area of habitat where a species' death rate exceeds its reproductive success.

sister chromatid (krō′-muh-tid) One of the two identical parts of a duplicated chromosome in a eukaryotic cell.

skeletal muscle Striated muscle attached to the skeleton. The contraction of striated muscles produces voluntary movements of the body.

skeletal system The organ system that provides body support and protects body organs such as the brain, heart, and lungs.

skull The bony framework of the head.

sliding-filament model The theory explaining how muscle contracts, based on change within a sarcomere, the basic unit of muscle organization, stating that thin (actin) filaments slide across thick (myosin) filaments, shortening the sarcomere; the shortening of all sarcomeres in a myofibril shortens the entire myofibril.

slime mold *See* cellular slime mold; plasmodial slime mold.

small intestine The longest section of the alimentary canal. It is the principal site of the enzymatic hydrolysis of food macromolecules and absorption of nutrients.

smooth ER (smooth endoplasmic reticulum) A network of interconnected membranous tubules in a eukaryotic cell's cytoplasm. Smooth ER lacks ribosomes. Enzymes embedded in the smooth ER membrane function in the synthesis of certain kinds of molecules, such as lipids.

smooth muscle Muscle made up of cells without striations, found in the walls of organs such as the digestive tract, urinary bladder, and arteries.

solar energy Energy obtained from the sun.

solute (sol′-yūt) A substance that is dissolved in a solution.

solution A liquid consisting of a homogeneous mixture of two or more substances: a dissolving agent, the solvent, and a substance that is dissolved, the solute.

solvent The dissolving agent in a solution. Water is the most versatile known solvent.

somatic cell (sō-mat′-ik) Any cell in a multicellular organism except a sperm or egg cell or a cell that develops into a sperm or egg.

somatic nervous system A component of the peripheral nervous system of vertebrates composed of neurons that carry signals to and from skeletal muscles.

source habitat An area of habitat where a species' reproductive success exceeds its death rate and from which new individuals often disperse to other areas.

speciation (spē′-sē-ā′-shun) The formation of new species.

species A group whose members possess similar anatomical characteristics and have the ability to interbreed. *See* biological species concept.

species diversity The number and relative abundance of species in a biological community.

species richness The total number of different species in a community.

sperm A male gamete.

spermatid A haploid cell formed via meiosis II of spermatogenesis. Spermatids develop into sperm cells, the male gametes.

spermatogenesis (sper-mat′-ō-jen′-uh-sis) The formation of sperm cells.

spermicide A sperm-killing chemical, in the form of a cream, jelly, or foam, that works with a barrier device as a method of contraception.

sphincter (sfink′-ter) A ringlike valve, consisting of modified muscles in a muscular tube, such as a digestive tract; closes off the tube like a drawstring.

spinal cord In vertebrates, a gelatinous bundle of nerve fibers located within the vertebral column; with the brain, makes up the central nervous system.

spirochete (spī′-rō-kēt) A large spiral-shaped (curved) prokaryotic cell.

sponge An aquatic animal characterized by a highly porous body, choanocytes, and no true tissues.

spontaneous generation The incorrect notion that life can emerge from inanimate material.

spore (1) In plants and algae, a haploid cell that can develop into a multicellular individual without fusing with another cell. (2) In prokaryotes, protists, and fungi, any of a variety of thick-walled life cycle stages capable of surviving unfavorable environmental conditions.

sporophyte (spōr′-uh-fit) The multicellular diploid form in the life cycle of organisms undergoing alternation of generations; results from a union of gametes and meiotically produces haploid spores that grow into the gametophyte generation.

stability In an ecological sense, the tendency of a biological community to remain in more or less constant balance due largely to interactions among organisms.

stabilizing selection Natural selection that favors intermediate variants by acting against extreme phenotypes.

stamen (stā′-men) A pollen-producing (male) reproductive part of a flower, consisting of a stalk and an anther.

starch A storage polysaccharide found in the roots of plants and certain other cells; a polymer of glucose.

start codon (kō′-don) On mRNA, the specific three nucleotide sequence (AUG) to which an initiator tRNA molecule binds, starting translation of genetic information.

stem That part of a plant's shoot system that supports the leaves and reproductive structures.

stem cell A relatively unspecialized cell that can give rise to one or more types of specialized cells. *See* embryonic stem cell (ES cell); adult stem cell.

steroid (stār′-oyd) A type of lipid whose carbon skeleton is in the form of four fused rings: three 6-sided rings and one 5-sided ring; examples are cholesterol, testosterone, and estrogen.

steroid hormone A lipid made from cholesterol that activates the transcription of specific genes in target cells.

stigma (stig′-muh) (plural, **stigmata**) The sticky tip of a flower's carpel, which traps pollen grains.

stimulus (plural, **stimuli**) In a nervous system, a factor that triggers sensory transduction.

stoma (stō′-muh) (plural, **stomata**) A pore surrounded by guard cells in the epidermis of a leaf. When stomata are open, CO_2 enters a leaf, and water and O_2 exit. A plant conserves water when its stomata are closed.

stomach A pouch-like organ in a digestive tract that grinds and churns food and may store it temporarily.

stop codon In mRNA, one of three triplets (UAG, UAA, UGA) that signal gene translation to stop.

STR analysis A method for producing and comparing DNA fingerprints that reflect the lengths of STR sequences at specific sites in the genome.

stroma (strō′-muh) A thick fluid enclosed by the inner membrane of a chloroplast. Sugars are made in the stroma by the enzymes of the Calvin cycle.

style The stalk of a flower's carpel, with the ovary at the base and the stigma at the top.

substrate (1) A specific substance (reactant) on which an enzyme acts. Each enzyme recognizes only the specific substrate of the reaction it catalyzes. (2) A surface in or on which an organism lives.

succession *See* ecological succession; primary succession; secondary succession.

sugar sink A plant organ that is a net consumer or storer of sugar. Growing roots, shoot tips, stems, and fruit are sugar sinks supplied by phloem.

sugar source A plant organ in which sugar is being produced by either photosynthesis or the breakdown of starch. Mature leaves are the primary sugar sources of plants.

sugar-phosphate backbone The alternating chain of sugar and phosphate to which DNA and RNA nitrogenous bases are attached.

superior vena cava (vē′-nuh kā′-vuh) A large vein that returns O_2-poor blood to the heart from the upper body and head. *See also* inferior vena cava.

supporting cell In the nervous system, a cell that protects, insulates, and reinforces a neuron.

surface tension A measure of how difficult it is to stretch or break the surface of a liquid.

survivorship curve A plot of the number of members of a cohort that are still alive at each age; one way to represent age-specific mortality.

suspension feeder An animal that extracts food particles suspended in the surrounding water.

sustainable development The long-term prosperity of human societies and the ecosystems that support them.

swim bladder A gas-filled internal sac that helps bony fishes maintain buoyancy.

symbiont (sim′-bē-unt) The smaller participant in a symbiotic relationship, living in or on the host.

symbiosis (sim′-bē-ō′-sis) An interspecific interaction in which one species, the symbiont, lives in or on another species, the host.

symbiotic relationship An interaction between two or more species that live together in direct contact.

sympathetic division One of two sets of neurons in the autonomic nervous system; generally prepares the body for energy-consuming activities, such as fleeing or fighting. *See also* parasympathetic division.

sympatric speciation The formation of a new species as a result of a genetic change that produces a reproductive barrier between the changed population (mutants) and the parent population. Sympatric speciation occurs without a geographic barrier. *See also* allopatric speciation.

synapse (sin′-aps) A junction, or relay point, between two neurons or between a neuron and an effector cell. Electrical and chemical signals are relayed from one cell to another at a synapse.

synaptic cleft (sin-ap′-tik) A narrow gap separating the synaptic terminal of a transmitting neuron from a receiving neuron or an effector cell.

synaptic terminal The relay point at the tip of a transmitting neuron's axon, where signals are sent to another neuron or to an effector.

systematics The scientific study of biological diversity and its classification.

systemic circuit One of two main blood circuits in terrestrial vertebrates; conveys blood between the heart and the rest of the body. *See also* pulmonary circuit.

systole (sis′-tō-lē) The contraction stage of the heart cycle, when the heart chambers actively pump blood. *See also* diastole.

T cell A type of lymphocyte that matures in the thymus and is responsible for the cell-mediated immune response; also involved in the humoral immune response. *See also* B cell.

taiga (tī′-guh) The northern (boreal) coniferous forest, which extends across North America and Eurasia, to the southern border of the arctic tundra; also found just below alpine tundra on mountainsides in temperate zones.

tail Extra nucleotides added at the end of an RNA transcript in the nucleus of a eukaryotic cell.

taproot A root system common to dicots consisting of one large, vertical root that produces many smaller lateral, or branch, roots.

target cell A cell that responds to a regulatory signal, such as a hormone.

taxonomy The branch of biology concerned with identifiying, naming, and classifying species.

technology The practical application of scientific knowledge.

telophase The fourth and final stage of mitosis, during which daughter nuclei form at the two poles of a cell. Telophase usually occurs together with cytokinesis.

temperate deciduous forest A biome located throughout midlatitude regions where there is sufficient moisture to support the growth of large, broadleaf deciduous trees.

temperate grassland Grassland region maintained by seasonal drought, occasional fires, and grazing by large mammals.

temperate zones Latitudes between the tropics and the Arctic Circle in the north and the Antarctic Circle in the south; regions with milder climates than the tropics or polar regions.

temperature A measure of the intensity of heat, reflecting the average kinetic energy or speed of molecules.

temporal isolation A type of prezygotic barrier between species; the species remain isolated because they breed at different times.

tendon Fibrous connective tissue connecting a muscle to a bone.

terminal bud Embryonic tissue at the tip of a shoot, made up of developing leaves and a compact series of nodes and internodes.

terminator A special sequence of nucleotides in DNA that marks the end of a gene. It signals RNA polymerase to release the newly made RNA molecule, which then departs from the gene.

territory An area that an individual or individuals defend and from which other members of the same species are usually excluded.

tertiary consumer (ter′-shē-ār-ē) An organism that eats secondary consumers.

tertiary structure The third level of protein structure; the overall, three-dimensional shape of a polypeptide in a protein.

testcross The mating between an individual of unknown genotype for a particular characteristic and an individual that is homozygous recessive for that same characteristic.

testis (plural, **testes**) The male gonad in an animal; produces sperm and, in many species, reproductive hormones.

testosterone (tes-tos′-tuh-rōn) An androgen hormone that stimulates an embryo to develop into a male and promotes male body features.

tetrad A paired set of homologous chromosomes, each composed of two sister chromatids. Tetrads form during prophase I of meiosis.

tetrapod A vertebrate with two pairs of limbs. Tetrapods include mammals, amphibians, and birds and other reptiles.

thalamus (thal′-uh-mus) An integrating and relay center of the vertebrate forebrain; sorts and relays selected information to specific areas in the cerebral cortex.

theory A widely accepted explanatory idea that is broad in scope and supported by a large body of evidence.

therapeutic cloning The cloning of human cells by nuclear transplantation for therapeutic purposes, such as the replacement of body cells that have been irreversibly damaged by disease or injury. *See also* nuclear transplantation; reproductive cloning.

thermoreceptor A sensor (sensory receptor) that detects heat or cold.

thermoregulation The maintenance of internal temperature within a range that allows cells to function efficiently.

thick filament A filament composed of staggered arrays of myosin molecules; a component of myofibrils in muscle fibers.

thigmotropism (thig′-mō-trō′-pizm) Growth movement of a plant in response to touch.

thin filament The smaller of the two myofilaments consisting of two strands of actin and two strands of regulatory protein coiled around one another.

threatened species As defined in the U.S. Endangered Species Act, a species that is likely to become endangered in the foreseeable future throughout all or a significant portion of its range.

three-domain system A system of taxonomic classification based on three basic groups: Bacteria, Archaea, and Eukarya.

threshold The minimum change in a membrane's voltage that must occur to generate a nerve signal (action potential).

thrombocyte *See* platelet.

thrombus A blood clot that forms in the absence of injury and blocks the flow of blood through a blood vessel.

thylakoid (thī′-luh-koyd) One of a number of disk-shaped membranous sacs inside a chloroplast. Thylakoid membranes contain chlorophyll and the enzymes of the light reactions of photosynthesis. A stack of thylakoids is called a granum.

thymine (T) (thī′-min) A single-ring nitrogenous base found in DNA.

thymus gland (thī′-mus) An endocrine gland in the neck region of mammals that is active in establishing the immune system; secretes several hormones that promote the development and differentiation of T cells.

thyroid gland (thī′-royd) An endocrine gland located in the neck that secretes thyroxine (T_4), triiodothyronine (T_3), and calcitonin.

thyroid-stimulating hormone (TSH) A protein hormone secreted by the anterior pituitary that stimulates the thyroid gland to secrete its hormones.

tissue An integrated group of similar cells that perform a specific function within a multicellular organism.

tissue system Organized collection of plant tissues. The organs of plants (such as roots, stems, and leaves) are formed from the dermal, vascular, and ground tissue systems.

tongue A muscular organ of the mouth that helps to taste and swallow food.

topsoil A mixture of particles derived from rock, living organisms, and humus.

trace element An element that is essential for the survival of an organism but is needed in only minute quantities.

trachea (trā′-kē-uh) (plural, **tracheae**) (1) The windpipe; the portion of the respiratory tube between the larynx and the bronchi. (2) One of many tiny tubes that branch throughout an insect's body, enabling gas exchange between outside air and body cells.

tracheid (trā′-kē-id) A tapered, porous, water-conducting and supportive cell in plants. Chains of tracheids or vessel elements make up the water-conducting, supportive tubes in xylem.

trans fat An unsaturated fatty acid produced by the partial hydrogenation of vegetable oils and present in hardened vegetable oils, most margarines, commercial baked foods, and many fried foods.

transcription The synthesis of RNA on a DNA template.

transcription factor In the eukaryotic cell, a protein that functions in initiating or regulating transcription. Transcription factors bind to DNA or to other proteins that bind to DNA.

transfer RNA (tRNA) A type of ribonucleic acid that functions as an interpreter in translation. Each tRNA molecule has a specific anticodon, picks up a specific amino acid, and conveys the amino acid to the appropriate codon on mRNA.

transgenic organism An organism that contains genes from another species.

translation The synthesis of a polypeptide using the genetic information encoded in an mRNA molecule. There is a change of "language" from nucleotides to amino acids.

translocation (1) During protein synthesis, the movement of a tRNA molecule carrying a growing polypeptide chain from the A site to the P site on a ribosome. (The mRNA travels with it.) (2) A change in a chromosome resulting from a chromosomal fragment attaching to a nonhomologous chromosome; can occur as a result of an error in meiosis or from mutagenesis.

transmission electron microscope (TEM) A microscope that uses an electron beam to study the internal structure of thinly sectioned specimens.

transpiration The evaporative loss of water from a plant.

transpiration-cohesion-tension mechanism The transport mechanism whereby transpiration exerts a pull that is relayed downward along a string of water molecules held together by cohesion and helped upward by adhesion.

transport protein A membrane protein that helps move substances across a cell membrane.

transport vesicle A tiny membranous sac in a cell's cytoplasm carrying molecules produced by the cell. The vesicle buds from the endoplasmic reticulum or Golgi and eventually fuses with another membranous organelle or the plasma membrane, releasing its contents.

triglyceride (trī-glis′-uh-rīd) A fat, which consists of a molecule of glycerol linked to three molecules of fatty acid.

trimester In human development, one of three 3-month-long periods of pregnancy.

triplet code A set of three-nucleotide-long words that specify the amino acids for polypeptide chains. *See* genetic code.

trisomy 21 *See* Down syndrome.

tRNA *See* transfer RNA.

trophic level (trō-fik) A level in a food chain.

trophic structure (trō′-fik) The feeding relationships in an ecosystem. Trophic structure determines the route of energy flow and the pattern of chemical cycling in an ecosystem.

trophoblast (trōf′-ō-blast) In mammalian development, the outer portion of a blastocyst. Cells of the trophoblast secrete enzymes that enable the blastocyst to implant in the endometrium of the mother's uterus.

tropical forest A terrestrial biome characterized by warm temperatures year-round.

tropics Latitudes between 23.5° north and south (from the Tropic of Cancer to the Tropic of Capricorn).

tropism (trō′-pizm) A growth response that makes a plant grow toward or away from a stimulus.

true-breeding Referring to organisms for which sexual reproduction produces offspring with inherited traits identical to those of the parents. The organisms are homozygous for the characteristics under consideration.

tubal ligation A means of sterilization in which a woman's two oviducts (fallopian tubes) are tied closed to prevent eggs from reaching the uterus. A segment of each oviduct is removed.

tuber An enlargement at the end of a rhizome, in which food is stored.

tubule A fine tube within the internal structure of the human kidney.

tumor An abnormal mass of cells that forms within otherwise normal tissue.

tumor-suppressor gene A gene whose product inhibits cell division, thereby preventing uncontrolled cell growth.

tundra A biome at the northernmost limits of plant growth and at high altitudes, characterized by dwarf woody shrubs, grasses, mosses, and lichens.

tunicate One of a group of invertebrate chordates.

U

ultrasound imaging A technique for examining a fetus in the uterus. High-frequency sound waves echoing off the fetus are used to produce an image of the fetus.

umbilical cord A structure containing arteries and veins that connects a developing embryo to the placenta of the mother.

undernutrition Inadequate nutrition resulting from lack of food or failure of the body to absorb or assimilate nutrients properly.

unsaturated Pertaining to fats and fatty acids whose hydrocarbon chains lack the maximum number of hydrogen atoms and therefore have one or more double covalent bonds. Unsaturated fats and fatty acids do not solidify at room temperature.

uracil (U) (yū′-ruh-sil) A single-ring nitrogenous base found in RNA.

urea (yū-rē′-ah) A soluble form of nitrogenous waste excreted by mammals and most adult amphibians.

ureter (yū-rē′-ter or yū′-reh-ter) A duct that conveys urine from the kidney to the urinary bladder.

urethra (yū-rē′-thruh) A duct that conveys urine from the urinary bladder to the outside. In the male, the urethra also conveys semen out of the body during ejaculation.

uric acid (yū′-rik) An insoluble precipitate of nitrogenous waste excreted by land snails, insects, birds, and some reptiles.

urinary bladder The pouch where urine is stored prior to elimination.

urine Concentrated filtrate produced by the kidneys and excreted via the bladder.

uterus (yū′-ter-us) In the reproductive system of a mammalian female, the organ where the development of young occurs; the womb.

V

vaccination (vak′-suh-nā′-shun) A procedure that presents the immune system with a harmless variant or derivative of a pathogen, thereby stimulating the immune system to mount a long-term defense against the pathogen.

vaccine (vak-sēn′) A harmless variant or derivative of a pathogen used to stimulate a host organism's immune system to mount a long-term defense against the pathogen.

vacuole (vak′-ū-ōl) A membrane-enclosed sac, part of the endomembrane system of a eukaryotic cell, having diverse functions.

vagina (vuh-jī′-nuh) Part of the female reproductive system between the uterus and the outside opening; the birth canal in mammals. The vagina accommodates the male's penis and receives sperm during copulation.

vas deferens (vas def′-er-enz) (plural, **vasa deferentia**) Part of the male reproductive system that conveys sperm away from the testis; the sperm duct; in humans, the tube that conveys sperm between the epididymis and the common duct that leads to the urethra.

vascular bundle (vas′-kyū-ler) A strand of vascular tissues (both xylem and phloem) in a plant stem.

vascular cambium (vas′-kyū-ler kam′-bē-um) During secondary growth of a plant, the cylinder of meristematic cells, surrounding the xylem and pith, that produces secondary xylem and phloem.

vascular plant A plant with xylem and phloem.

vascular tissue Plant tissue consisting of cells joined into tubes that transport water and nutrients throughout the plant body.

vascular tissue system A system formed by xylem and phloem throughout a plant, serving as a transport system for water and nutrients, respectively.

vasectomy (vuh-sek′-tuh-mē) Surgical removal of a section of the two sperm ducts (vasa deferentia) to prevent sperm from reaching the urethra; a means of sterilization in the male.

vector In molecular biology, a piece of DNA, usually a plasmid or a viral genome, that is used to move genes from one cell to another.

vegetative reproduction Asexual reproduction by a plant.

vein (1) In animals, a vessel that returns blood to the heart. (2) In plants, a vascular bundle in a leaf, composed of xylem and phloem.

ventral Pertaining to the underside, or bottom, of a bilaterally symmetrical animal.

ventricle (ven′-truh-kul) A heart chamber that pumps blood out of the heart.

venule (ven′-yūl) A vessel that conveys blood between a capillary bed and a vein.

vertebra (ver′-tuh-bruh) (plural, **vertebrae**) One of a series of segmented units making up the backbone of a vertebrate animal.

vertebrate (ver′-tuh-brāt) A chordate animal with a backbone. Vertebrates include agnathans, cartilaginous fishes, bony fishes, amphibians, reptiles, birds, and mammals.

vessel element A short, open-ended, water-conducting and supportive cell in plants. Chains of vessel elements or tracheids make up the water-conducting, supportive tubes in xylem.

vestigial organ A structure of marginal, if any, importance to an organism. Vestigial organs are historical remnants of structures that had important functions in ancestors.

villus (vil′-us) (plural, **villi**) (1) A fingerlike projection of the inner surface of the small intestine. (2) A fingerlike projection of the chorion of the mammalian placenta. Large numbers of villi increase the surface areas of these organs.

virus A microscopic particle capable of infecting cells of living organisms and inserting its genetic material. Viruses are generally not considered to be alive because they do not display all of the characteristics associated with life.

vitamin An organic nutrient that an organism requires in very small quantities. Vitamins generally function as coenzymes.

vitreous humor (vit′-rē-us hyū′-mer) A jellylike substance filling the space behind the lens in the vertebrate eye; helps maintain the shape of the eye.

vocal cord One of a pair of string-like tissues in the larynx. Air rushing past the tensed vocal cords makes them vibrate, producing sounds.

vulva The outer features of the female reproductive anatomy.

warning coloration The often brightly colored markings of animals possessing chemical defenses. The coloration provides a caution to their predators.

water vascular system In echinoderms, a radially arranged system of water-filled canals that branch into extensions called tube feet. The system provides movement and circulates water, facilitating gas exchange and waste disposal.

water-conducting cell A specialized dead plant cell with a lignin-containing secondary wall. These cells are arranged end to end, forming xylem tissue. *See also* tracheid; vessel element.

wavelength The distance between crests of adjacent waves, such as those of the electromagnetic spectrum.

wetland An ecosystem intermediate between an aquatic one and a terrestrial one. Wetland soil is saturated with water permanently or periodically.

white blood cell A blood cell that functions in defending the body against infections and cancer cells; also called a leukocyte.

white matter Tracts of axons within the CNS.

wild-type traits Traits most commonly found in nature.

withdrawal The withdrawal of the penis from the vagina before ejaculation, an unreliable method of contraception.

wood Secondary xylem of a plant. *See also* heartwood; sapwood.

wood ray A column of parenchyma cells that radiates from the center of a log and transports water to its outer living tissues.

X chromosome inactivation In female mammals, the inactivation of one X chromosome in each somatic cell.

X-Ray Diagnostic test in which an image is created using low doses of radiation.

xylem (zī′-lum) The nonliving portion of a plant's vascular system that provides support and conveys water and inorganic nutrients from the roots to the rest of the plant. Xylem is made up of vessel elements and/or tracheids, water-conducting cells.

xylem sap The solution of inorganic nutrients conveyed in xylem tissue from a plant's roots to its shoots.

yellow bone marrow A tissue found within the central cavities of long bones, consisting mostly of stored fat.

yolk sac An extraembryonic membrane that develops from endoderm; produces the embryo's first blood cells and germ cells and gives rise to the allantois.

Z line A border of a sarcomere.

zoned reserve An extensive region of land that includes one or more areas that are undisturbed by humans. The undisturbed areas are surrounded by lands that have been altered by human activity.

zooplankton (zō′-ō-plank′-tun) Animals that drift in aquatic environments.

zygote (zī′-gōt) The fertilized egg, which is diploid, that results from the union of a sperm cell nucleus and an egg cell nucleus.

Index

Page numbers with *f* indicate a figure, and those with *t* indicate a table

A

B

INDEX

C

D

E

F

G

INDEX

H

I

J

K

L

M

N

O

P

Q

R

S

T

U

V

W

X

Y

Z

Student cd-rom and website activities

	eTutors	MP3 Tutors	Discovery Channel Video Clips	Activities	You Decide	Graph It!
Chapter 1			Antibiotic Resistance	The Levels of Life Card Game Energy Flow and Chemical Cycling Classification Schemes Darwin and the Galápagos Islands Science and Technology: DDT	What Can We Do About Antibiotic-Resistant Bacteria?	An Introduction to Graphing
Chapter 2		The Properties of Water	Early Life	The Structure of Atoms Electron Arrangement Build an Atom Ionic Bonds Covalent Bonds The Structure of Water The Cohesion of Water in Trees Acids, Bases, and pH		
Chapter 3		Protein Structure and Function		Diversity of Carbon-Based Molecules Functional Groups Making and Breaking Polymers Models of Glucose Carbohydrates Lipids Protein Functions Protein Structure Nucleic Acid Functions Nucleic Acid Structure	Low-Fat or Low-Carb Diets: Which is Healthier?	
Chapter 4	A Tour of the Cell		Cells	Metric System Review Prokaryotic Cell Structure and Function Comparing Cells Build an Animal Cell and a Plant Cell Membrane Structure Overview of Protein Synthesis The Endomembrane System Build a Chloroplast and a Mitochondrion Cilia and Flagella Review: Animal Cell Structure and Function Review: Plant Cell Structure and Function		

	eTutors	MP3 Tutors	Discovery Channel Video Clips	Activities	You Decide	Graph It!
Chapter 5		Basic Energy Concepts		Energy Concepts The Structure of ATP How Enzymes Work Membrane Structure Diffusion Facilitated Diffusion Osmosis and Water Balance in Cells Active Transport Exocytosis and Endocytosis Cell Signaling		
Chapter 6	Cellular Respiration	Cellular Respiration Part 1: Glycolysis Cellular Respiration Part 2: Citric Acid Cycle and Electron Transport Chain		Build a Chemical Cycling System Overview of Cellular Respiration Glycolysis The Citric Acid Cycle Electron Transport Fermentation		
Chapter 7	Photosynthesis	Photosynthesis	Space Plants	The Plants in Our Lives The Sites of Photosynthesis Overview of Photosynthesis Light Energy and Pigments The Light Reactions The Calvin Cycle Photosynthesis in Dry Climates The Greenhouse Effect		
Chapter 8	Mitosis Meiosis	Mitosis Meiosis Mitosis-Meiosis Comparison		Asexual and Sexual Reproduction The Cell Cycle Mitosis and Cytokinesis Animation Mitosis and Cytokinesis Video Causes of Cancer Human Life Cycle Meiosis Animation Origins of Genetic Variation Polyploid Plants		
Chapter 9		Chromosomal Basis of Inheritance	Colored Cotton Novelty Gene	Monohybrid Cross Dihybrid Cross Gregor's Garden Incomplete Dominance Linked Genes and Crossing Over Sex-Linked Genes		

	eTutors	MP3 Tutors	Discovery Channel Video Clips	Activities	You Decide	Graph It!
Chapter 10	Protein Synthesis	DNA to RNA to Protein	Emerging Diseases	The Hershey-Chase Experiment DNA and RNA Structure DNA Double Helix DNA Replication Overview of Protein Synthesis Transcription and RNA Processing Translation Simplified Reproductive Cycle of a Virus Phage Lytic Cycle Phage Lysogenic and Lytic Cycles HIV Reproductive Cycle		
Chapter 11		Control of Gene Expression	Cloning Fighting Cancer	The *lac* Operon in *E. coli* Gene Regulation in Eukaryotes Review: Gene Regulation in Eukaryotes Signal-Transduction Pathway Causes of Cancer	Do Cell Phones Cause Brain Cancer? Is Second-hand Smoke Dangerous?	
Chapter 12			Transgenics DNA Forensics Colored Cotton	Applications of DNA Technology DNA Technology and Golden Rice Cloning a Gene in Bacteria Restriction Enzymes DNA Fingerprinting Gel Electrophoresis of DNA Analyzing DNA Fragments Using Gel Electrophoresis The Human Genome Project: Human Chromosome 17		
Chapter 13		Natural Selection	Charles Darwin	The Voyage of the *Beagle:* Darwin's Trip Around the World Darwin and the Galápagos Islands Reconstructing Forelimbs Genetic Variation from Sexual Recombination Causes of Microevolution	Is Ephedra Safe and Effective?	
Chapter 14		Speciation	Mass Extinction	Mechanisms of Macroevolution Exploring Speciation on Islands Polyploid Plants Paedomorphosis: Morphing Chimps and Humans The Geologic Time Scale Classification Schemes	Can We Prevent Species Extinction?	

	eTutors	MP3 Tutors	Discovery Channel Video Clips	Activities	You Decide	Graph It!
Chapter 15			Bacteria Early Life Tasty Bacteria	The History of Life Prokaryotic Cell Structure and Function Diversity of Prokaryotes		
Chapter 16			Plant Pollination Fungi	Terrestrial Adaptations of Plants Highlights of Plant Evolution Moss Life Cycle Fern Life Cycle Pine Life Cycle Angiosperm Life Cycle Madagascar and the Biodiversity Crisis Fungal Reproduction and Nutrition		
Chapter 17		Human Evolution	Invertebrates	Overview of Animal Phylogeny Characteristics of Invertebrates Characteristics of Chordates Primate Diversity Human Evolution		
Chapter 18				DDT and the Environment Evolutionary Adaptations Techniques for Estimating Population Density and Size Human Population Growth Analyzing Age-Structure Diagrams Investigating Survivorship Curves		Age Pyramids and Population Growth
Chapter 19		Energy Flow in Ecosystems	Leafcutter Ants Rain Forests Space Plants Trees	Interspecific Interactions Primary Succession Energy Flow and Chemical Cycling Food Webs Energy Pyramids The Carbon Cycle The Nitrogen Cycle Terrestrial Biomes Aquatic Biomes		Species Area Effect and Island Biogeography Animal Food Production Efficiency and Food Policy Atmospheric CO_2 and Temperature Changes
Chapter 20		Global Warming	Introduced Species Trees	Fire Ants: An Introduced Species Water Pollution from Nitrates DDT and the Environment The Greenhouse Effect Madagascar and the Biodiversity Crisis Conservation Biology Review	Does Human Activity Cause Global Warming? Can We Prevent Species Extinction?	Forestation Change Global Fisheries and Overfishing Municipal Solid Waste Trends in the U.S. Global Fresh Water Resources Prospects for Renewable Energy

	eTutors	MP3 Tutors	Discovery Channel Video Clips	Activities	You Decide	Graph It!
Chapter 21			An Introduction to the Human Body	Correlating Structure and Function of Cells The Levels of Life Card Game Overview of Animal Tissues Epithelial Tissue Connective Tissue Muscle Tissue Nervous Tissue Regulation: Negative and Positive Feedback Human Urinary System Nephron Function Control of Water Reabsorption		
Chapter 22		The Human Digestive System	Nutrition	How Animals Eat Food Human Digestive System Analyzing Food Labels Case Studies of Nutritional Disorders	Low-Fat or Low-Carb Diets: Which is Healthier?	
Chapter 23			Blood	Path of Blood Flow Cardiovascular System Structure Cardiovascular System Function The Human Respiratory System Transport of Respiratory Gases	Is Second-Hand Smoke Dangerous?	
Chapter 24		The Human Immune System	Vaccines	Immune System Responses HIV Reproductive Cycle		
Chapter 25			The Endocrine System	Overview of Cell Signaling Action of Amino-Acid-Based Hormones Action of Steroid Hormones Human Endocrine Glands and Hormones		
Chapter 26		The Female Reproductive Cycle		Reproductive System of the Human Female Reproductive System of the Human Male Sea Urchin Embryonic Development Frog Embryonic Development		

	eTutors	MP3 Tutors	Discovery Channel Video Clips	Activities	You Decide	Graph It!
Chapter 27	How Neurons Work		Teen Brains Muscles and Bones	Neuron Structure Nerve Signals: Action Potentials Neuron Communication Structure and Function of the Eye The Human Skeleton Skeletal Muscle Structure Muscle Contraction		
Chapter 28		From Flower to Fruit	Plant Pollination	Roots, Stems, and Leaves The Life Cycle of a Flowering Plant Seed and Fruit Formation and Germination		
Chapter 29		Transpiration		Absorption of Nutrients from Soil Transpiration Transport in Phloem Leaf Drop Flowering Lab	Is Ephedra Safe and Effective?	Global Soil Degradation